32. $\int \sqrt{p^2 - x^2}\, dx = \frac{1}{2}\left(x\sqrt{p^2 - x^2} + p^2 \sin^{-1}\frac{x}{p}\right) \qquad p > 0$

33. $\int \frac{dx}{\sqrt{x^2 \pm p^2}} = \ln |x + \sqrt{x^2 \pm p^2}|$

34. $\int \sqrt{(p^2 - x^2)^3}\, dx = \frac{1}{4}\left[x\sqrt{(p^2 - x^2)^3} + \frac{3p^2x}{2}\sqrt{p^2 - x^2} + \frac{3p^4}{2}\sin^{-1}\frac{x}{p}\right] \qquad p > 0$

Expressions Containing $ax^2 + bx + c$

35. $\int \frac{dx}{ax^2 + bx + c} = \frac{1}{\sqrt{b^2 - 4ac}} \ln \left|\frac{2ax + b - \sqrt{b^2 - 4ac}}{2ax + b + \sqrt{b^2 - 4ac}}\right| \qquad b^2 > 4ac$

36. $\int \frac{dx}{ax^2 + bx + c} = \frac{2}{\sqrt{4ac - b^2}} \tan^{-1} \frac{2ax + b}{\sqrt{4ac - b^2}} \qquad b^2 < 4ac$

37. $\int \frac{dx}{ax^2 + bx + c} = -\frac{2}{2ax + b} \qquad b^2 = 4ac$

38. $\int \frac{dx}{(ax^2 + bx + c)^{n+1}} = \frac{2ax + b}{n(4ac - b^2)(ax^2 + bx + c)^n} + \frac{2(2n - 1)a}{n(4ac - b^2)} \int \frac{dx}{(ax^2 + bx + c)^n}$

39. $\int \frac{x\, dx}{ax^2 + bx + c} = \frac{1}{2a} \ln |ax^2 + bx + c| - \frac{b}{2a} \int \frac{dx}{ax^2 + bx + c}$

40. $\int \frac{dx}{\sqrt{ax^2 + bx + c}} = \frac{1}{\sqrt{a}} \ln |2ax + b + 2\sqrt{a}\sqrt{ax^2 + bx + c}| \qquad a > 0$

41. $\int \frac{dx}{\sqrt{ax^2 + bx + c}} = \frac{1}{\sqrt{-a}} \sin^{-1} \frac{-2ax - b}{\sqrt{b^2 - 4ac}} \qquad a < 0$

42. $\int \frac{x\, dx}{\sqrt{ax^2 + bx + c}} = \frac{\sqrt{ax^2 + bx + c}}{a} - \frac{b}{2a} \int \frac{dx}{\sqrt{ax^2 + bx + c}}$

43. $\int \sqrt{ax^2 + bx + c}\, dx = \frac{2ax + b}{4a}\sqrt{ax^2 + bx + c} + \frac{4ac - b^2}{8a} \int \frac{dx}{\sqrt{ax^2 + bx + c}}$

Expressions Containing $\sin ax$

44. $\int \sin^2 ax\, dx = \frac{x}{2} - \frac{\sin 2ax}{4a}$

45. $\int \sin^3 ax\, dx = -\frac{1}{a} \cos ax + \frac{1}{3a} \cos^3 ax$

46. $\int \sin^n ax\, dx = -\frac{\sin^{n-1} ax \cos ax}{na} + \frac{n - 1}{n} \int \sin^{n-2} ax\, dx \qquad n$ positive integer

47. $\int \frac{dx}{1 \pm \sin ax} = \mp \frac{1}{a} \tan \left(\frac{\pi}{4} \mp \frac{ax}{2}\right)$

Expressions Containing $\cos ax$

48. $\int \cos^2 ax\, dx = \frac{x}{2} + \frac{\sin 2ax}{4a}$

49. $\int \cos^3 ax\, dx = \frac{1}{a} \sin ax - \frac{1}{3a} \sin^3 ax$

50. $\int \cos^n ax\, dx = \frac{\cos^{n-1} ax \sin ax}{na} + \frac{n - 1}{n} \int \cos^{n-2} ax\, dx$

Recursions for $\int \tan^n ax\, dx$ *and* $\int \sec^n ax\, dx$

51. $\int \tan^n ax\, dx = \frac{\tan^{n-1} ax}{a(n - 1)} - \int \tan^{n-2} ax\, dx \qquad n \neq 1$

52. $\int \sec^n ax\, dx = \frac{\sec^{n-2} ax \tan ax}{a(n - 1)} + \frac{n - 2}{n - 1} \int \sec^{n-2} ax\, dx \qquad n \neq 1$

$\int \sec^3 x\, dx = \frac{1}{2} \sec x \tan x + \frac{1}{2} \ln|\sec x + \tan x|$

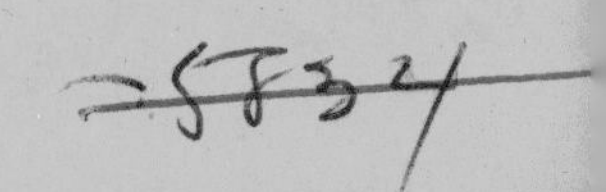

Expressions Containing Algebraic and Trigonometric Functions

53. $\int x \sin ax\, dx = \frac{1}{a^2} \sin ax - \frac{1}{a} x \cos ax$

54. $\int x \cos ax\, dx = \frac{1}{a^2} \cos ax + \frac{1}{a} x \sin ax$

55. $\int x^n \sin ax\, dx = -\frac{1}{a} x^n \cos ax + \frac{n}{a} \int x^{n-1} \cos ax\, dx$ $\quad n$ positive

56. $\int x^n \cos ax\, dx = \frac{1}{a} x^n \sin ax - \frac{n}{a} \int x^{n-1} \sin ax\, dx$ $\quad n$ positive

57. $\int \sin ax \cos bx\, dx = -\frac{\cos (a-b)x}{2(a-b)} - \frac{\cos (a+b)x}{2(a+b)}$ $\quad a^2 \neq b^2$

Expressions Containing Exponential and Logarithmic Functions

58. $\int xe^{ax}\, dx = \frac{e^{ax}}{a^2}(ax - 1),$

$\int xb^{ax}\, dx = \frac{xb^{ax}}{a \ln b} - \frac{b^{ax}}{a^2(\ln b)^2} \quad b > 0$

59. $\int x^n e^{ax}\, dx = \frac{1}{a} x^n e^{ax} - \frac{n}{a} \int x^{n-1} e^{ax}\, dx \quad n$ positive

60. $\int e^{ax} \sin bx\, dx = \frac{e^{ax}}{a^2 + b^2}(a \sin bx - b \cos bx)$

61. $\int e^{ax} \cos bx\, dx = \frac{e^{ax}}{a^2 + b^2}(a \cos bx + b \sin bx)$

62. $\int x^n \ln ax\, dx = x^{n+1}\left[\frac{\ln ax}{n+1} - \frac{1}{(n+1)^2}\right] \quad n \neq -1$

Expressions Containing Inverse Trigonometric Functions

63. $\int \sin^{-1} ax\, dx = x \sin^{-1} ax + \frac{1}{a}\sqrt{1 - a^2x^2}$

64. $\int \cos^{-1} ax\, dx = x \cos^{-1} ax - \frac{1}{a}\sqrt{1 - a^2x^2}$

65. $\int \csc^{-1} ax\, dx = x \csc^{-1} ax + \frac{1}{a} \ln |ax + \sqrt{a^2x^2 - 1}|$

66. $\int \sec^{-1} ax\, dx = x \sec^{-1} ax - \frac{1}{a} \ln |ax + \sqrt{a^2x^2 - 1}|$

67. $\int \tan^{-1} ax\, dx = x \tan^{-1} ax - \frac{1}{2a} \ln (1 + a^2x^2)$

68. $\int \cot^{-1} ax\, dx = x \cot^{-1} ax + \frac{1}{2a} \ln (1 + a^2x^2)$

ALGEBRA

Roots of $ax^2 + bx + c = 0$: $x = \frac{-b \pm \sqrt{b^2 - 4ac}}{2a}$ (quadratic formula)

$d^n - c^n = (d - c)(d^{n-1} + d^{n-2}c + d^{n-3}c^2 + \cdots + c^{n-1})$

$a + ar + ar^2 + \cdots + ar^{n-1} = \frac{a(1 - r^n)}{1 - r}, r \neq 1$ (sum of finite geometric series)

$k! = 1 \cdot 2 \cdot 3 \cdots k$

$(1 + x)^n = 1 + nx + \binom{n}{2} x^2 + \cdots + \binom{n}{k} x^k + \cdots + x^n$ (binomial theorem)

$\binom{n}{k} = \frac{n!}{k!(n-k)!} = \frac{n}{1}\frac{(n-1)}{2}\frac{(n-2)}{3}\cdots\frac{(n-k+1)}{k}$

BRIEF EDITION

CALCULUS

AND ANALYTIC GEOMETRY

BRIEF EDITION

CALCULUS AND ANALYTIC GEOMETRY

SHERMAN K. STEIN
Professor of Mathematics
University of California, Davis

FOURTH EDITION

McGRAW-HILL BOOK COMPANY

New York St. Louis San Francisco Auckland Bogotá Hamburg London Madrid Mexico Milan Montreal New Delhi Panama Paris São Paulo Singapore Sydney Tokyo Toronto

CALCULUS

AND
ANALYTIC
GEOMETRY

34567890VNHVNH8921098

ISBN 0-07-061162-9

This book was set in Times Roman by General Graphic Services, Inc.
The editors were Robert A. Weinstein and James W. Bradley;
the designer was Joan E. O'Connor;
the production supervisor was Diane Renda.
The drawings were done by Oxford Illustrators Limited.
Von Hoffmann Press, Inc., was printer and binder.

Library of Congress Catalog Card Number: 87-61754

To
Joshua,
Rebecca,
and
Susanna

The great body of physical, a great deal of the essential fact of financial science, and endless social and political problems are only accessible and only thinkable to those who have had a sound training in mathematical analysis, and the time may not be very remote when it will be understood that for complete initiation as an efficient citizen of one of the new great complex world-wide states that are now developing, it is as necessary to be able to compute, to think in averages and maxima and minima, as it is now to be able to read and write.

H.G. Wells, Mankind in the Making, *p. 192, Scribner's, New York, 1904.*

. . . at about the age of sixteen, I was offered a choice which, in retrospect, I can see that I was not mature enough, at the time, to make wisely. The choice was between starting on the calculus and, alternatively, giving up mathematics altogether and spending the time saved from it on reading Latin and Greek literature more widely. I chose to give up mathematics, and I have lived to regret this keenly after it has become too late to repair my mistake. The calculus, even a taste of it, would have given me an important and illuminating additional outlook on the Universe, whereas, by the time at which the choice was presented to me, I had already got far enough in Latin and Greek to have been able to go farther with them unaided. So the choice that I made was the wrong one, yet it was natural that I should choose as I did. I was not good at mathematics; I did not like the stuff. . . . Looking back, I feel sure that I ought not to have been offered the choice; the rudiments, at least, of the calculus ought to have been compulsory for me. One ought, after all, to be initiated into the life of the world in which one is going to have to live. I was going to have to live in the Western World . . . and the calculus, like the full-rigged sailing ship, is . . . one of the characteristic expressions of the modern Western genius.

Arnold Toynbee, Experiences, *pp. 12–13, Oxford University Press, London, 1969.*

Contents

APPENDICES S-1

To the Instructor

"EDITOR'S NOTE:

This brief version of CALCULUS AND ANALYTIC GEOMETRY is designed for a one-year course in the calculus of a single variable. The following preface is from the three-semester version, which includes five chapters covering algebraic operations on vectors, partial derivatives, definite integrals over plane and solid regions, the derivative of a vector function, Green's theorem, the divergence theorem, and Stokes' theorem. All of the appendices from the full-length edition have been included in this brief version."

The goal of this edition remains the same as that of the first three editions: To give both student and instructor a readable, flexible text that covers the main topics in a three-semester calculus course as clearly as possible.

At the urging of colleagues and students I have made this edition shorter than its predecessor, even though this violates the old axiom, "A calculus text always gets longer on revision." However, I have kept the expository style that distinguished the earlier editions ("leisurely" in the view of instructors, "understandable" in the view of students).

While a quick riffling through the pages will show that the text has an abundance of examples and exercises, there are some unique features that might be overlooked in a casual glance, to which I would like to call the reader's attention.

EXERCISES

In addition to the routine exercises that come in pairs (before the box, ■), there are an unusual number of exercises aimed at developing understanding and self-reliance, as contrasted to skill in rote applications of algorithms. Exercise 20 in Sec. 2.4, which tests the limits of intuition, is one example. Exercises 49 and 50 in Sec. 6.9 take it further. Finally, Exercise 183 in Sec. 6.S completes the development. The Instructor's Manual comments on such exercises in detail.

SUMMARIES

The chapter summaries are the most extensive of any calculus text. Although instructors may use them only as another source of exercises, students find their overviews very helpful. Taken together, they comprise a built-in study guide.

EPSILON, DELTA

The precise definitions of limits are found in Secs. 2.7 and 2.8. Section 2.7 defines $\lim_{x\to\infty} f(x) = \infty$ and $\lim_{x\to\infty} f(x) = L$, which are easier for the students to understand, since the diagrams and the algebra are simpler than for $\lim_{x\to a} f(x) = L$, which is treated in Sec. 2.8. These sections can be covered in two or three lectures or by weaving them more slowly into the rest of the course in smaller portions. Appendix G provides formal proofs of some important limit theorems.

LOGARITHMS

Since many students are not comfortable with logarithms, I include a thorough review in Sec. 6.1. Moreover, since students have trouble grasping that a logarithm is an exponent, I do not feel that telling them that it is an area will help them. The suggestion that logarithms be presented as integrals goes back to J. W. Bradshaw, The Logarithm as a Direct Function, *Annals of Mathematics,* vol. 4, pp. 51–62, 1902. However, we should keep in mind Osgood's introduction to this article, in which he remarks, "How simple the analysis is appears from a casual glance at the following pages, in which Mr. Bradshaw has carried through all the details in a rigourous development of the Logarithm. . . . It is hoped that this presentation may prove attractive to students who have *finished a thorough course in elementary calculus*" [emphasis added]. The integral approach is now in Appendix H, for the convenience of instructors who wish to present it to students who are prepared to grasp it.

COMPLEX NUMBERS

It is often assumed that students have already mastered the algebra and geometry of complex numbers "somewhere else." I do not feel that this assumption is justified. For that reason, and because the complex numbers are useful in a variety of courses, Sec. 10.9 is devoted only to them. The important relation $e^{i\theta} = \cos\theta + i\sin\theta$ merits all of Sec. 10.10, which provides another application of power series, as well as a tool needed in Sec. 10.11 for dealing with certain differential equations.

TECHNIQUES OF INTEGRATION

Integration by substitution and by parts are the most important techniques in Chap. 7. Since integration by parts will be used in subsequent courses, e.g., in differential equations to treat the Laplace transform, I give it more than the usual emphasis. In particular, I use it in Sec. 10.8 to obtain an exact formula for the error in Taylor's series. I feel that this argument, as presented here, is more instructive and understandable than the tricks by which the derivative form of the remainder is customarily obtained. [It is shown in Sec. 10.8 that the latter form is not even strong enough to show that $1 - x + x^2 - \cdots$ converges to $1/(1 + x)$ when $-1 < x \leq -\frac{1}{2}$.]

How thoroughly the more specialized techniques of integration in Chap. 8 are covered is a matter of priorities and time constraints. On the one hand, computer programs such as MACSYMA can carry out formal integration; on the other, students should have some feeling for what the integrals of various types of elementary functions look like.

ART

The introduction of a second color has made many of the illustrations much clearer. In addition, Howard Dwyer, trained as an engineer and applied mathematician, has used his drafting skills to redraw several hundred of the diagrams. The techniques employed have been kept simple, so that students can use the illustrations as models for their own work.

HISTORY

I have added The Long Road to Calculus, which describes the development of calculus from Archimedes through Newton and Leibniz. This comes in Sec. 5.4, right after the fundamental theorem of calculus. The events that stretch from Kepler's announcement of his laws to the appearance of Newton's *Principia* are described in detail in Sec. 14.8. Other historical remarks are included where they add interest and illuminate the mathematics. (See, for example, Cicero's words about Archimedes, p. 472; Steinmetz's comments on the complex numbers, p. 554; and Maxwell's letter introducing the term *curl,* p. 864.)

COMPUTERS

I have added descriptions of specific software for computing derivatives (in Sec. 3.6), integrals, and definite integrals (in Sec. 7.S). Instructors who wish to incorporate more computing may consider using Rothenberg's *Basic Computing for Calculus* (McGraw-Hill), Oberlie's *Calculus and the Computer* (Addison-Wesley), or Stroyan's *Computer Explorations in Calculus* (Harcourt), which are designed to supplement any calculus text.

APPENDICES

Appendices A through E cover precalculus material, including the definition of closed and open intervals and the slope of a line. You may simply call the students' attention to them, assign them as reading, draw on their exercises for review problems, or incorporate them in lectures as needed. Appendix F is devoted to conic sections, Appendix G to the theory of limits, Appendix H (as already noted) to the definite integral definition of the natural logarithm, and Appendix I to the Taylor series for $f(x, y)$.

Appendix J serves as a reference for some postcalculus material, providing elementary proofs concerning the equality of mixed partial derivatives and differentiation under the integral sign. Because students get into trouble by not distinguishing between a statement and its converse, I have added Appendix K, The Converse of a Statement, to which you may direct your students on appropriate occasions.

SUPPLEMENTS

The Student's Solutions Manual contains solutions to the odd-numbered exercises, except for a few that would lose their interest if their solutions

	Maximum		*Minimum*	
	Comment	*Lectures*	*Comment*	*Lectures*
Chapter				
1		3	Precalculus	0
2		8	Omit Secs. 2.7 and 2.8	6
3		6	Secs. 3.1 to 3.3 in two lectures	5
4	Two lectures on Sec. 4.6	9		8
5		6	Omit Sec. 5.6	5
6	Two lectures on Sec. 6.9	12	Omit Secs. 6.6 and 6.11	9
7		9	Treat special techniques lightly	6
8		8	Omit Sec. 8.8	7
9		7	Omit Secs. 9.5 and 9.6	5
10	Two lectures on Secs. 10.4 and 10.8	14	Omit Secs. 10.9 to 10.12	8
11		6		6
12		8	Omit Sec. 12.8	7
13	Two lectures on Sec. 13.3	10	Omit Secs. 13.8 and 13.9	7
14		8	Cover only Secs. 14.1 to 14.3 and 14.6, which are used in Chap. 15	3
15	Two lectures on each of Secs. 15.3 to 15.7	12	Omit Sec. 15.8	7
	Total:	126	Total:	89
Appendices				
A	Precalculus (A–F)	1	Assign appendices to be read as needed	
B		2		
C		1		
D		1		
E		1		
F		4		
G		2		
H		2		
I		1		
J		3		
K		1		
	Total:	19		

were too easily available. Solutions to the even-numbered exercises and a few of the odd are published separately in an Instructor's Manual, which also contains advice on how to treat the sections and chapters. This advice should be especially helpful to teaching assistants.

A computerized test generator, SteinTest, accompanies this text. SteinTest will produce many variations on problems related to the learning objectives in the book. This program prints final exams and quizzes, or can be used to create practice material for students.

FLEXIBILITY

Chapters, sections, and appendices are designed to permit maximum flexibility in their use. For instance, if one wants to complete the elementary derivatives before beginning integration, one can easily switch Chaps. 5 and 6, except Secs. 6.7 and 6.8. If one prefers to introduce the derivative on "day 1," Sec. 3.1 can be covered first. The Instructor's Manual describes such options in detail. The table on the preceding page also indicates possible options by contrasting two hypothetical courses: maximum (the most thorough) and minimum (just the essentials).

ACKNOWLEDGMENTS

I wish to acknowledge the advice of reviewers Vladimir Akis, San Jose State University; Thomas W. Cusick, SUNY Buffalo; Frank Deane, Berkshire Community College; Dan Drucker, Wayne State University; Larry Gerstein, University of California at Santa Barbara; Murray B. Peterson, College of Marin; Ken Seydel, Skyline College, San Bruno; Bruce Edwards, University of Florida; William R. Fuller, Purdue University; Richard Bagby, New Mexico State University; James T. Vance, Jr., Wright State University; Stuart Goldenberg, California Polytechnic State University; and John Wenger, Loop College, Chicago.

I also received advice from fellow instructors on my campus: Glen Ericson and Ken Greider of the physics department, Mott Hubbard and Gary Ford of the college of engineering, Henry Adler, Carlos Borges, Howard Dwyer, Shirley Goldman, Dorothy Hawkes, Lawrence Marx, and David G. Mead of the mathematics department.

At each stage of this revision two former graduate students at Davis, Anthony Barcellos and Dean Hickerson, scrutinized every sentence, exercise, diagram, and marginal note. They did much to improve the book.

Peter Devine, former senior editor of mathematics at McGraw-Hill, offered advice essential to the shaping of this edition.

Sherman K. Stein

To the Student

When you do your homework, follow these steps in the order shown:

1 Read the text.
2 Study the examples, first trying to work them without looking at their solutions.
3 Begin the homework problems.

I emphasize this order because it is the very reverse of the order that many students follow. The time spent on steps 1 and 2 will make step 3 more efficient and instructive.

In high school, the homework exercises were almost always identical with the examples done in the text or by the teacher. It was seldom necessary to read the text. A student would start with step 3; then, if stuck, the student would go to step 2; and if that did not do the trick, the student would reluctantly go to step 1.

If you highlight with a felt pen, do not wait until the night before a test to think about what you marked. Mastering the material as you go along will deepen your understanding and reduce stress. After all, doing well in mathematics depends on understanding, not memorizing. And understanding cannot be achieved through cramming. The chapter summaries will help you develop a larger view than the individual sections can offer.

Answers to almost all the odd-numbered exercises are in the back of the book; complete solutions are in the Student's Solutions Manual by Anthony Barcellos and Dean Hickerson, published by McGraw-Hill for students who may find it useful.

Before you begin your study, read the remarks of Arnold Toynbee, the historian, and of H. G. Wells, the social philosopher, quoted in the front of the text. The Long Road to Calculus, pp. 218–219, will repay reading now and also when you complete Chap. 5.

I believe that this reduction violates the old axiom, "A text always expands when revised." We have also made a special effort to root out mistakes, which are a nuisance to instructor and student alike and an embarrassment to the author. Nevertheless, some mistakes may have slipped by, and I encourage you to inform me of any you may find.

POSTSCRIPT

Tony Wexler, a mechanical engineer, suggested that I pass on this advice to engineering students:

> Learn the theory. Don't keep asking, "What's the use of this to me?" You can't predict what math will be of practical use after you graduate. In my eight years in the "real world" I've worked on the flow of jet fuel, solar energy, computers, statistical software, and modeling the physiological functions of the kidney. When I was a student I could never have guessed that some of the things I thought were far out would someday be of use to me.

BRIEF EDITION

CALCULUS

AND ANALYTIC GEOMETRY

1 Functions

This chapter develops the notion of a function and some important ideas connected with it.

1.1 FUNCTIONS

The area A of a square depends on the length of its side x and is given by the formula

$$A = x^2.$$

Similarly, the distance s (in feet) that a freely falling object drops in the first t seconds is described by the formula

$$s = 16t^2.$$

Each choice of t determines a unique value for s. For instance, when $t = 3$, $s = 16 \cdot 3^2 = 144$.

Both these formulas illustrate the mathematical notion of a function.

Definition Let X and Y be sets. A **function** from X to Y is a rule or method for assigning to each element in X a unique element in Y.

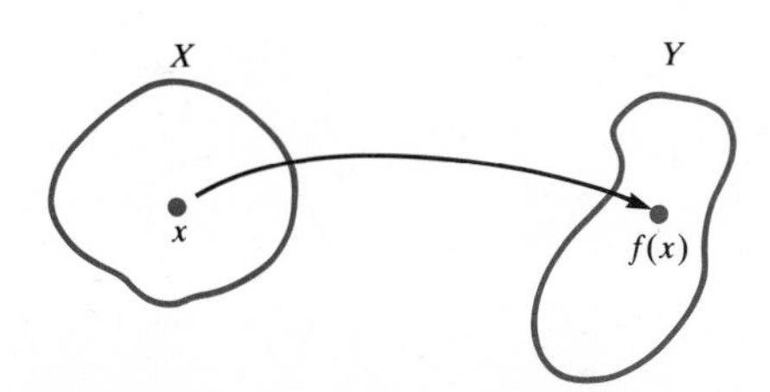

Figure 1.1

The notion of a function is schematized in Fig. 1.1.

A function may be given by a formula, as are the functions $A = x^2$ and $s = 16t^2$. In daily life a function is often indicated by a table. For instance, a table of populations of the cities in the United States can be thought of as a function that assigns to each city the size of its population. In this case X is the set of cities and Y is the set of positive integers.

A function is often denoted by the symbol f. The element that the

Calling "f(x)" a function is known as an "abuse of notation."

function assigns to the element x is denoted $f(x)$ (read "f of x"). In practice, though, almost everyone speaks interchangeably of the function f or the function $f(x)$.

EXAMPLE 1 Let $f(x) = x^2$ for each real number x. Compute (*a*) $f(3)$, (*b*) $f(2)$, and (*c*) $f(-2)$.

SOLUTION
(*a*) $f(3) = 3^2 = 9$.
(*b*) $f(2) = 2^2 = 4$.
(*c*) $f(-2) = (-2)^2 = 4$. ■

Domain and range

Definitions Let X and Y be sets and let f be a function from X to Y. The set X is called the **domain** of the function. If $f(x) = y$, y is called the **value** of f at x. The set of all values of the function is called the **range** of the function.

When the function is given by a formula, the domain is usually understood to consist of all the numbers for which the formula is defined.

EXAMPLE 2 Find the domain and range of the function given by the formula

$$f(x) = x^2.$$

SOLUTION Since x^2 is defined for every real number x, the domain of the squaring function x^2 consists of the entire x axis.

The range of f consists of all real numbers that are of the form x^2 for some real number x. The number 9 is in the range since $9 = 3^2$; the number 5 is in the range since $5 = (\sqrt{5})^2$. Zero is also in the range since $0 = 0^2$. In fact, every nonnegative real number r is in the range since r can be expressed as the square of a real number: $r = (\sqrt{r})^2$.

However, no negative number is in the range, since no negative number is the square of a real number.

Thus the range consists precisely of the nonnegative real numbers. ■

EXAMPLE 3 Let $f(x) = 1/x$. Find the domain and range of f.

SOLUTION Since division by 0 is meaningless, the domain consists of all real numbers except 0.

What is the range of $1/x$? For instance, $\frac{1}{2}$ is in the range since

$$f(2) = \tfrac{1}{2}.$$

Also, -3 is in the range since

$$f(-\tfrac{1}{3}) = \frac{1}{-\frac{1}{3}} = -3.$$

More generally, any real number other than 0 is in the range. To show this, consider a real number b, $b \neq 0$. We must show that there is a real number x such that

$$f(x) = b,$$

that is,

$$\frac{1}{x} = b.$$

This last equation is equivalent to $1 = xb$ or

$$x = \frac{1}{b}.$$

So, if $x = 1/b$, $f(x) = b$. [To check this, compute $f(1/b)$: $f(1/b) = 1/(1/b) = b$.]

The number 0 is not in the range, since there is no number x such that $1/x = 0$.

By coincidence the range and domain of this function $1/x$ coincide; both consist of all real numbers except 0. ■

EXAMPLE 4 Find the domain and range of $f(x) = 2 + \sqrt{x - 1}$.

SOLUTION For $2 + \sqrt{x - 1}$ to be meaningful, the square root of $x - 1$ must make sense; that is, $x - 1$ must not be negative. Thus the domain consists of all numbers x such that

$$x - 1 \geq 0,$$

or, equivalently,

$$x \geq 1.$$

That is, the domain is the interval $[1, \infty)$. (See Appendix A for this notation.)

As x varies from 1 to larger numbers, $f(x)$ increases from $f(1) = 2 + \sqrt{1 - 1}$ to arbitrarily large values. Thus the range of f consists of all numbers greater than or equal to 2, that is, the interval $[2, \infty)$. ■

Input and Output

Independent and dependent variables

The value $f(x)$ of a function f at x is also called the **output**; x is called the **input** or **argument**. If $y = f(x)$, the symbol x is called the **independent variable** and the symbol y is called the **dependent variable**.

If both the inputs and outputs of a function are numbers, we shall call the function **numerical**. In some more advanced courses such a function is also called a real function of a real variable.

Several of the keys on a hand-held calculator correspond to functions. One such function is the $\sqrt{x}$-key. The input can be any nonnegative number. If you punch in a negative number, the calculator may flash a warning that you have given it a number not in the domain of the function. The reciprocal or $1/x$-key represents another function. The sum or +-key represents a function of a different type. In this case the input is a pair of numbers x and y, and the output is their sum, a single number $x + y$.

If both the domain and range of a function consist of real numbers, it is possible to draw a picture that displays the behavior of the function.

Definition *Graph of a numerical function.* Let f be a numerical function. The **graph** of f consists of those points (x, y) such that $y = f(x)$.

For instance, the graph of the squaring function $f(x) = x^2$ consists of the points (x, y) such that $y = x^2$. It is the parabola shown in Fig. B.7 in Sec. B.1. The graph of the cubing function $y = x^3$ is also shown in that section. The next example shows how to use a table to graph a function.

EXAMPLE 5 Graph the function $f(x) = 1/(1 + x^2)$.

SOLUTION Since $1 + x^2$ is never 0, the domain of the function consists of the entire x axis. Pick a few convenient inputs x and calculate the corresponding outputs, as shown in this table:

x	0	1	2	3
$f(x) = \dfrac{1}{1 + x^2}$	1	$\frac{1}{2}$	$\frac{1}{5}$	$\frac{1}{10}$

For any x, $x^2 \geq 0$, so $1 + x^2 \geq 1$ and $\dfrac{1}{1 + x^2} \leq 1$.

See Appendix A for the definition of $|x|$.

Since x appears only to an even power, the graph is symmetric with respect to the y axis; there is no need to evaluate the function for negative x. Plotting the four calculated points suggests the general shape of the graph. (See Fig. 1.2.) When $|x|$ is large, $y = 1/(1 + x^2)$ is small. This means that for large $|x|$ the graph approaches the x axis.

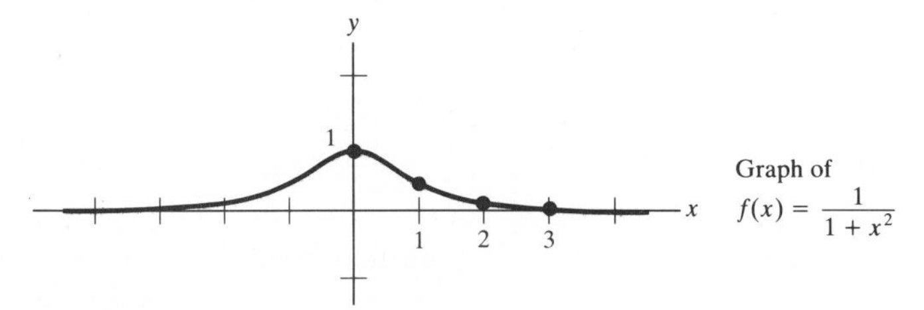

Figure 1.2

Note that the range of the function consists of all positive numbers less than or equal to 1. ■

How to tell whether a curve is the graph of a function

Not every curve is the graph of a function. For instance, the curve in Fig. 1.3 is not the graph of a function. The reason is that a function assigns to a given input a *single* number as the output. A line parallel to the y axis therefore meets the graph of a function in at most one point. This observation provides a visual test for deciding whether a curve in a plane is the graph of a function $y = f(x)$. If some line parallel to the y axis meets the curve more than once, then the curve is *not* the graph of a function. Otherwise it is the graph of a function. The curve in Fig. 1.4 is the graph of a function.

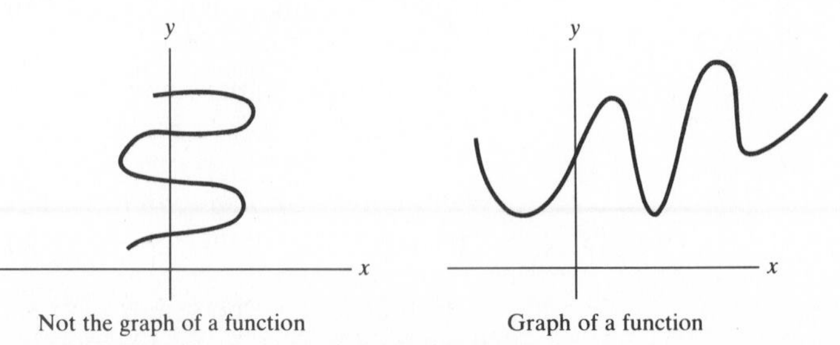

Not the graph of a function
Figure 1.3

Graph of a function
Figure 1.4

A function as a table with two rows

A function can also be thought of as a table consisting of two rows, one row for inputs and one row for outputs. There may be repetitions in the outputs but not in the inputs. Many functions are given this way. For instance, the following table records the rate in millions of barrels per day at which the United States imported crude petroleum from 1948 to 1984. The corresponding graph is shown in Fig. 1.5.

Year	1948	1949	1950	1951	1952	1953	1954	1955	1956	1957	1958	1959	1960
Imports	0.51	0.65	0.85	0.84	0.95	1.03	1.05	1.25	1.44	1.57	1.70	1.78	1.82

Year	1961	1962	1963	1964	1965	1966	1967	1968	1969	1970	1971	1972
Imports	1.92	2.08	2.12	2.26	2.47	2.57	2.54	2.84	3.17	3.42	3.93	4.74

Year	1973	1974	1975	1976	1977	1978	1979	1980	1981	1982	1983	1984
Imports	6.26	6.11	6.06	7.31	8.81	8.36	8.46	6.91	6.00	5.11	5.05	5.46

Source: Statistical Abstract of the United States.

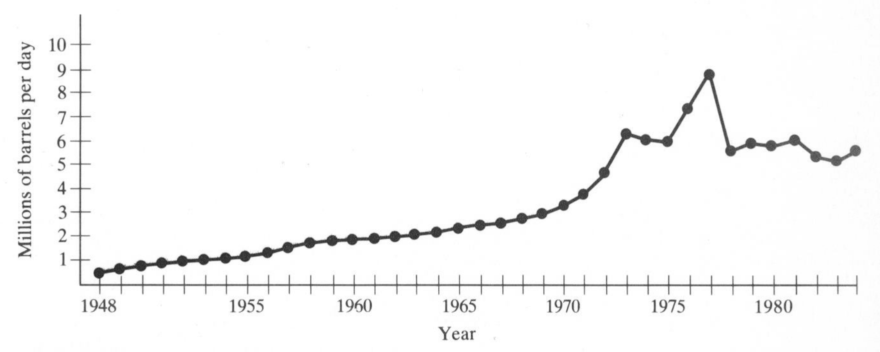

Figure 1.5

A function as an input-output machine

It is sometimes helpful to think of a function as a machine. When you insert the number x into the machine, the number $f(x)$ falls out, as illustrated in Fig. 1.6.

A function as a projector from line to line

On several occasions it will be illuminating to picture a function f as a projection from a slide to a screen. Both the slide and screen are lines. The function, through some ingenious lens, projects the point on the slide given by the number x to the point on the screen given by the number $f(x)$, as shown in Fig. 1.7.

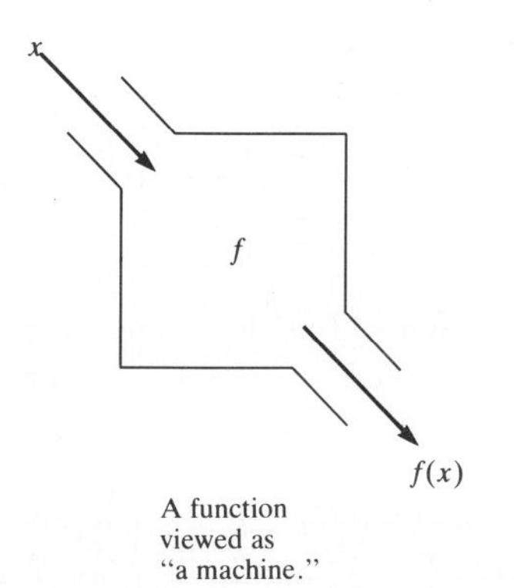

A function viewed as "a machine."

Figure 1.6

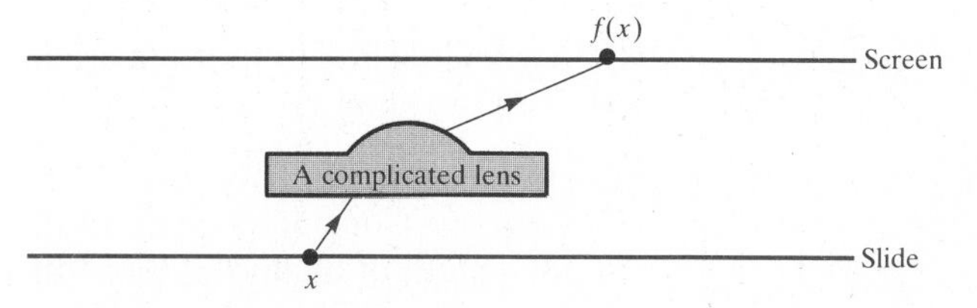

A function viewed as a "projector."

Figure 1.7

Beginning in Chap. 2 we shall examine how the output of a function changes as we change the input. To prepare for the algebraic manipulations that will be needed, we pause now to illustrate some of the techniques.

EXAMPLE 6 Let f be the squaring function $f(x) = x^2$. Compute (*a*) $f(2 + 3)$ and (*b*) $f(2 + h)$.

SOLUTION
(*a*) For any number x, $f(x)$ is the square of that number. Thus

$$f(2 + 3) = (2 + 3)^2 = 5^2 = 25.$$

(*b*) Similarly,

$$f(2 + h) = (2 + h)^2 = 4 + 4h + h^2. \quad \blacksquare$$

Warning about the value of a function when the input is a sum

Warning: A common error is to assume that $f(2 + 3)$ is somehow related to $f(2) + f(3)$. For most functions there is no relation between the two numbers. In the case of the function x^2, $f(2) + f(3) = 2^2 + 3^2 = 4 + 9 = 13$, but, as was shown in Example 6, $f(2 + 3) = 25$.

EXAMPLE 7 Let f be the cubing function $f(x) = x^3$. Evaluate the difference $f(2 + 0.1) - f(2)$.

SOLUTION

$$\begin{aligned} f(2 + 0.1) - f(2) &= f(2.1) - f(2) \\ &= (2.1)^3 - 2^3 \\ &= 9.261 - 8 \\ &= 1.261. \quad \blacksquare \end{aligned}$$

In Examples 6 and 7 the inputs are specific numbers. Often the inputs will be indicated by algebraic expressions instead. The next example shows how to deal with such inputs.

EXAMPLE 8 Let f be the squaring function $f(x) = x^2$. Simplify the expression $f(a + b) - f(a) - f(b)$ as far as possible.

SOLUTION

$$\begin{aligned} f(a + b) - f(a) - f(b) &= (a + b)^2 - a^2 - b^2 \\ &= a^2 + 2ab + b^2 - a^2 - b^2 \\ &= 2ab. \quad \blacksquare \end{aligned}$$

In calculus, functions often originate in geometric or physical problems. The next example is typical.

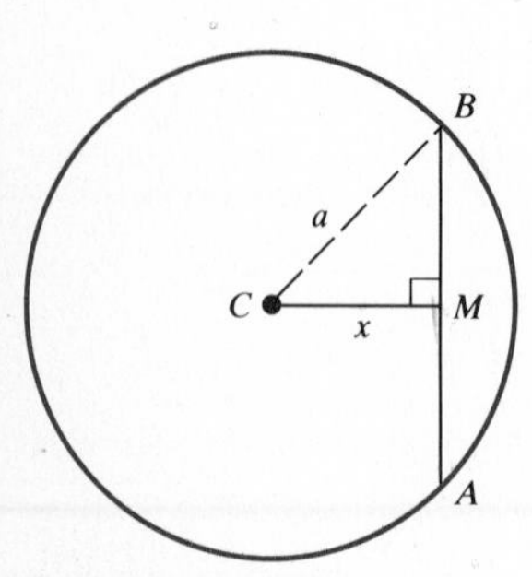

$f(x)$ = length of AB

Figure 1.8

EXAMPLE 9 Consider a circle of radius a, as shown in Fig. 1.8. Let $f(x)$ be the length of a chord AB of this circle at a distance x from the center of the circle. Find a formula for $f(x)$.

SOLUTION Let M be the midpoint of the chord AB and let C be the center of the circle. Observe that $\overline{CM} = x$ and $\overline{CB} = a$. By the pythagorean theorem, $\overline{BM} = \sqrt{a^2 - x^2}$. Hence $\overline{AB} = 2\sqrt{a^2 - x^2}$. Thus

$$f(x) = 2\sqrt{a^2 - x^2}$$

describes the function f in algebraic terms.

The notation $[0, a]$ *is defined in Appendix A.*

As a check, note that $f(0) = 2\sqrt{a^2 - 0^2} = 2a$, which is correct, as a glance at Fig. 1.8 shows. Similarly, $f(a) = 2\sqrt{a^2 - a^2} = 0$, which is also correct. The domain of the function f is the interval $[0, a]$. ■

EXERCISES FOR SEC. 1.1: FUNCTIONS

In Exercises 1 to 10 graph the functions.

1 $f(x) = 3x$
2 $f(x) = -2x$
3 $f(x) = 3x^2$
4 $f(x) = 1 + x^2$
5 $f(x) = -x^2 + 1$
6 $f(x) = -3x^2 + 2$
7 $f(x) = x^2 - x$
8 $f(x) = x^2 + 2x + 1$
9 $f(x) = 2/(1 + x^2)$
10 $f(x) = 1/(1 + 2x^2)$

In Exercises 11 to 20 describe the domain and range of each function.

11 $f(x) = \sqrt{x}$
12 $f(x) = \sqrt{x + 1}$
13 $f(x) = \sqrt{4 - x^2}$
14 $f(x) = \sqrt{4 + x^2}$
15 $f(x) = 3/x^2$
16 $f(x) = 1/(x + 1)$
17 $f(x) = 1/x^3$
18 $f(x) = 1/x^4$
19 $f(x) = 1/\sqrt{x}$
20 $f(x) = 1/(1 - x^2)$

In Exercises 21 and 22 which curves are the graphs of functions and which are not?

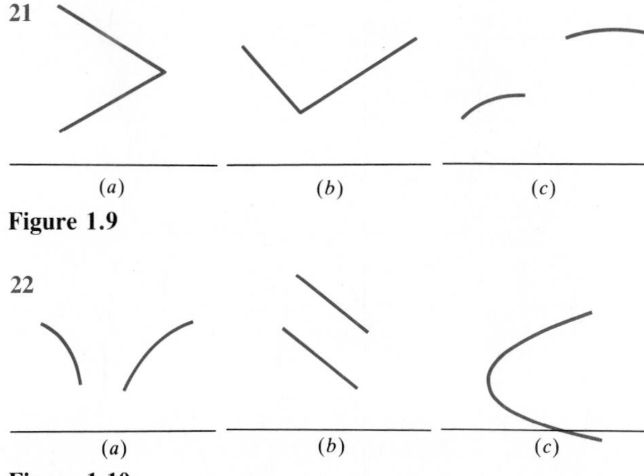

Figure 1.9

Figure 1.10

In each of Exercises 23 to 26 compute as decimals the outputs of the given function for the given inputs.

23 $f(x) = x + 1$ (*a*) -1 (*b*) 3 (*c*) 1.25 (*d*) 0
24 $f(x) = 1/(1 + x)$ (*a*) -3 (*b*) 3 (*c*) 9 (*d*) 99
25 $f(x) = x^3$ (*a*) $1 + 2$ (*b*) $4 - 1$
26 $f(x) = 1/x^2$ (*a*) $5 - 3$ (*b*) $4 - 6$

In Exercises 27 to 32 for the given functions evaluate and simplify the given expressions. (Assume that no denominator is 0.)

27 $f(x) = x^3$; $f(a + 1) - f(a)$
28 $f(x) = 1/x$; $f(a + h) - f(a)$
29 $f(x) = \dfrac{1}{x^2}$; $\dfrac{f(d) - f(c)}{d - c}$
30 $f(x) = \dfrac{1}{2x + 1}$; $\dfrac{f(x + h) - f(x)}{h}$
31 $f(x) = x + \dfrac{1}{x}$; $\dfrac{f(u) - f(v)}{u - v}$
32 $f(x) = 3 - \dfrac{1}{x}$; $\dfrac{f(x + h) - f(x)}{h}$

■

In each of Exercises 33 to 35 give an algebraic formula for the function.

33 For $0 \le x \le 4$, $f(x)$ is the length of the path from A to B to C in Fig. 1.11.

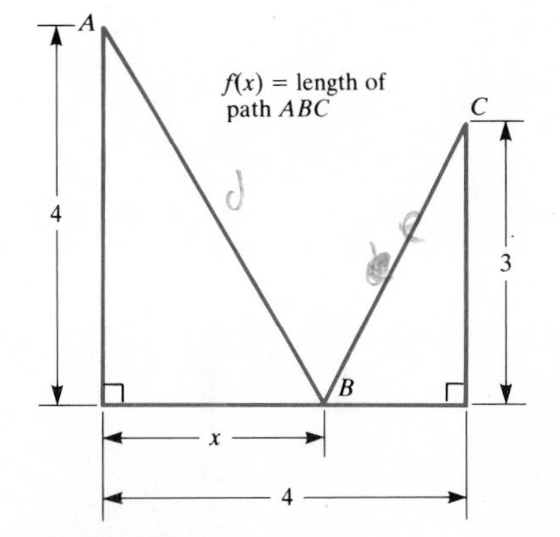

Figure 1.11

34 For $0 \le x \le 10$, $f(x)$ is the perimeter of the rectangle $ABCD$, one side of which has length x, inscribed in the circle of radius 5 shown in Fig. 1.12.

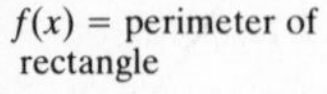

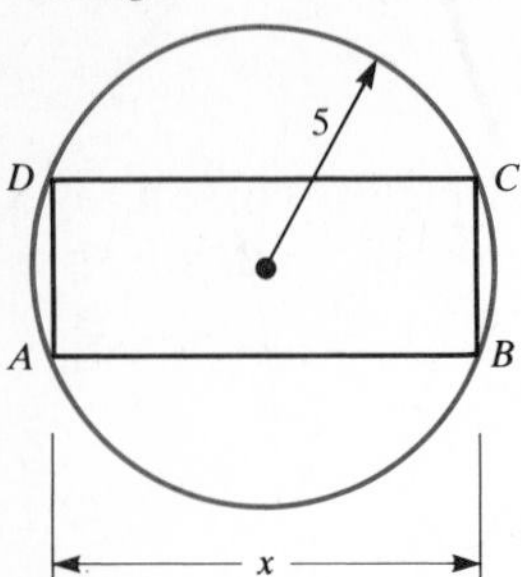

Figure 1.12

35 For positive numbers x and y, $f(x, y)$ is the total surface area of a cylindrical tin can of radius x and height y (top and bottom included).

36 (See Fig. 1.13.) A person at point A in a lake is going to swim to the shore ST and then walk to point B. She swims at 1.5 miles per hour and walks at 4 miles per hour. If she reaches the shore at a point P, x miles from S, let $f(x)$ denote the time for her combined swim and walk. Obtain an algebraic formula for $f(x)$.

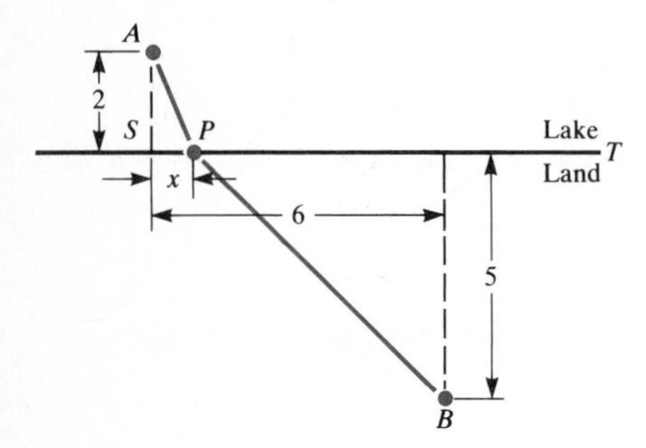

Figure 1.13

37 Graph $y = x^2 + x$ after filling in this table:

x	-2	-1	0	$\frac{1}{2}$	1	2	3
$x^2 + x$							

38 Graph $f(x) = x(x + 1)(x - 1)$.
(*a*) For which values of x is $f(x) = 0$?
(*b*) Where does the graph cross the x axis?
(*c*) Where does the graph cross the y axis?

39 A complicated lens projects a linear slide to a linear screen as shown in Fig. 1.14, which indicates the paths of four of the light rays. Let $f(x)$ be the image on the screen of x on the slide.
(*a*) What are $f(0)$, $f(1)$, $f(2)$, $f(3)$?
(*b*) Fill in this table:

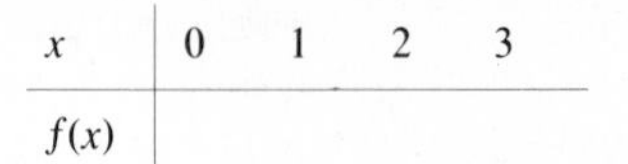

x	0	1	2	3
$f(x)$				

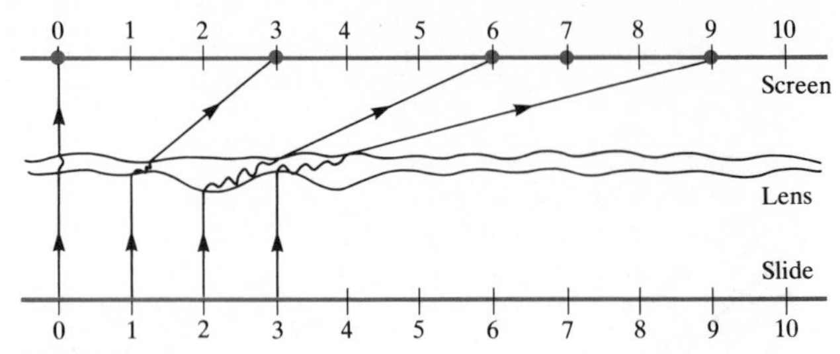

Figure 1.14

(*c*) Plot the four points in (*b*).
(*d*) Which is larger, $f(3) - f(2)$ or $f(2) - f(1)$?

40 The amount of automobile traffic and the rate of fatal automobile accidents vary with the day of the week and the time of the day. Let $f(x)$ be the number of automobiles on the roads at time x. Let $g(x)$ be the number of fatal accidents per hour at time x.
(*a*) What function defined in terms of f and g would best represent the danger or risk of driving at time x?
(*b*) Figure 1.15 is the graph of f and g in a typical 48-hour period, Friday and Saturday. (The vertical scale is proportional to the numbers of cars and fatal accidents.)

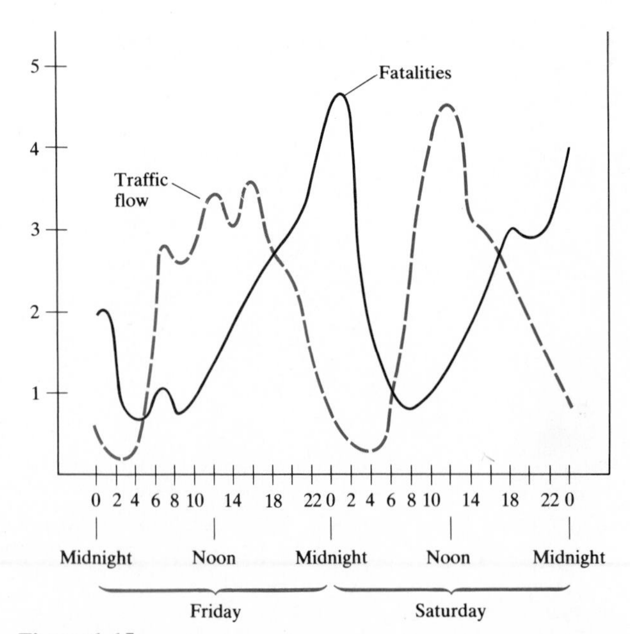

Figure 1.15

On the basis of Fig. 1.15, estimate what is the most dangerous time to drive and what is the safest. How many times as risky is the first in comparison to the second?

41 (Calculator) Let $f(x) = x^2$. Evaluate

$$\frac{f(3 + h) - f(3)}{h}$$

to four decimal places for h equal to (*a*) 1, (*b*) 0.01, (*c*) -0.01, and (*d*) 0.0001. (*e*) What do you think happens to $[f(3 + h) - f(3)]/h$ as h gets smaller and smaller?

42 (Calculator) Let $f(x) = x^3$. Evaluate

$$\frac{f(2 + h) - f(2)}{h}$$

to four decimal places for h equal to (*a*) 1, (*b*) 0.05, (*c*) 0.0001, and (*d*) -0.001. (*e*) What do you think happens to $[f(2 + h) - f(2)]/h$ as h gets smaller and smaller?

43 For which of the following functions is $f(a + b)$ equal to $f(a) + f(b)$ for all positive numbers a and b?
(*a*) $f(x) = x^2$ (*b*) $f(x) = 3x$ (*c*) $f(x) = -4x$
(*d*) $f(x) = \sqrt{x}$ (*e*) $f(x) = 2x + 1$

44 For which of the following functions is $f(ab)$ equal to $f(a)f(b)$ for all positive numbers a and b?
(*a*) $f(x) = 3x$ (*b*) $f(x) = x^3$ (*c*) $f(x) = 1/x$
(*d*) $f(x) = \sqrt{x}$ (*e*) $f(x) = x + 1$

45 Let f have as its domain the set of all integers . . . , $-3, -2, -1, 0, 1, 2, 3, \ldots$. For each integer x, $f(x)$ is a real number. Assume that $f(x + y) = f(x) + f(y)$ for all integers x and y. Assume also that $f(1) = 3$.
(*a*) Show that $f(2) = 6$.
(*b*) Show that $f(0) = 0$.
(*c*) What can you say about $f(-1)$?
(*d*) What can you say about $f(3)$? $f(-3)$?
(*e*) What possibilities are there for f?

In Exercises 46 to 49 give three examples of numerical functions f that meet the given condition.

46 $f(-x) = 1/f(x)$ for all numbers x

47 $f(x + 1) = 2f(x)$ for all numbers x

48 $f(xy) = f(x)f(y)$ for all numbers x and y

49 $f(xy) = f(x) + f(y)$ for all positive numbers x and y

50 (See Fig. 1.13.) Let $g(x)$ be the total distance traversed in the combined swim and walk in Exercise 36. Find a formula for $g(x)$.

1.2 COMPOSITE FUNCTIONS

This section describes a way of building up functions by applying one function to the output of another. For instance, the function

$$y = (1 + x^2)^{100}$$

is built up by raising $1 + x^2$ to the one-hundredth power. That is,

$$y = u^{100}, \qquad \text{where } u = 1 + x^2.$$

Similarly, the function

$$y = \sqrt{1 + 2x^2}$$

is built up by taking the square root of $1 + 2x^2$. That is,

$$y = \sqrt{u}, \qquad \text{where } u = 1 + 2x^2.$$

The theme common to these two examples is spelled out in the following definition.

Definition *Composition of functions.* Let f and g be functions. Suppose that x is such that $g(x)$ is in the domain of f. Then the function that assigns to x the value

$$f(g(x))$$

is called the **composition of f and g**. It is denoted $f \circ g$.

Thus if $g(x) = u$ and $f(u) = y$, then $(f \circ g)(x) = y$. ($f \circ g$ is read as

"f circle g" or as "f composed with g.") In practical terms, the definition says: "To compute $f \circ g$, first apply g and then apply f to the result."

Thinking of functions as input-output machines, we may consider $f \circ g$ as the machine built by hooking the machine for f onto the machine for g, as shown in Fig. 1.16.

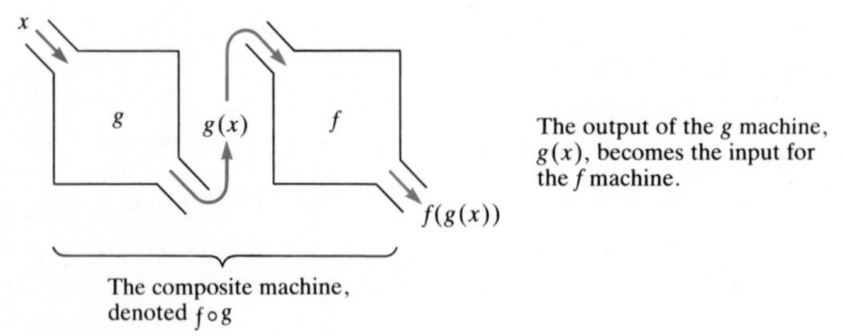

Figure 1.16

For our purposes, it is also instructive to think of composite functions in terms of projections from slides to screens. If g is interpreted as some complicated projection from a slide to a screen (see Sec. 1.1) and f as a projection from that screen to a second screen, then $f \circ g$ is the projection from the slide to the second screen, as in Fig. 1.17.

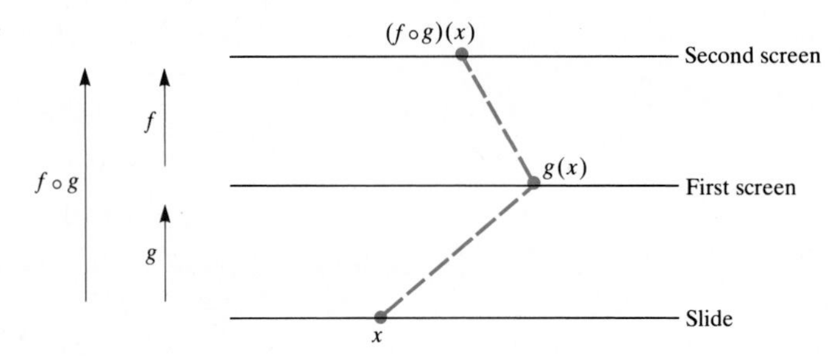

Figure 1.17

A function can be the composition of more than two functions. For example, $\sqrt{(1 + x^2)^5}$ is the composition of three functions. First $1 + x^2$ is formed; then the fifth power; then a square root. More formally, the assertion that $y = \sqrt{(1 + x^2)^5}$ is the same as saying that

$$y = \sqrt{u}, \qquad u = v^5, \qquad \text{and} \qquad v = 1 + x^2.$$

EXAMPLE 1 Write $y = 2^{(x^2)}$ as a composition of functions.

SOLUTION $y = 2^u$, where $u = x^2$. ■

EXAMPLE 2 Let $f(x) = 1 + 2x$ and $g(x) = x^2$. Compute $(f \circ g)(x)$ and $(g \circ f)(x)$. Are they equal?

SOLUTION

$$
\begin{aligned}
(f \circ g)(x) &= f(g(x)) \\
&= f(x^2) \\
&= 1 + 2x^2. \qquad (f \text{ of } \textit{any} \text{ input is } 1 + \text{twice that input.}) \\
(g \circ f)(x) &= g(f(x)) \\
&= g(1 + 2x) \\
&= (1 + 2x)^2. \qquad (g \text{ of } \textit{any} \text{ input is the square of that input.})
\end{aligned}
$$

Warning: $f \circ g$ is usually not equal to $g \circ f$.

Since the function $(1 + 2x)^2$ is not equal to $1 + 2x^2$, $f \circ g$ is not equal to $g \circ f$. This shows that $f \circ g$ is not necessarily equal to $g \circ f$. ■

EXAMPLE 3 Let $f(x) = -x$. Compute $(f \circ f)(x)$.

SOLUTION $(f \circ f)(x) = f(f(x)) = f(-x) = -(-x) = x.$

Thus $(f \circ f)(x) = x$. ■

EXAMPLE 4 Let f be the cubing function, $f(x) = x^3$, and g the cube root function, $g(x) = \sqrt[3]{x}$. Compute $(f \circ g)(x)$ and $(g \circ f)(x)$.

SOLUTION

$$(f \circ g)(x) = f(g(x)) = f(\sqrt[3]{x}) = (\sqrt[3]{x})^3 = x.$$

$$(g \circ f)(x) = g(f(x)) = g(x^3) = \sqrt[3]{x^3} = x. \quad ■$$

In Example 4, f and g "reverse" the effect of each other. For instance, $f(2) = 2^3 = 8$ and $g(8) = \sqrt[3]{8} = 2$. Two functions related in this way are said to be "inverses" of each other. The concept of inverse functions will be developed more fully in the next section.

Even and odd functions

Certain functions behave nicely when composed with the function $-x$. That is, their values at $-x$ are closely related to their values at x. The following definitions make this precise.

Definition *Even function.* A function f such that $f(-x) = f(x)$ is called **an even function**.

Consider, for instance, $f(x) = x^4$. We have

Recall that $(-1)^4 = 1$.

$$f(-x) = (-x)^4 = x^4 = f(x).$$

Thus $f(x) = x^4$ is an even function. In fact, for any *even* integer, n, $f(x) = x^n$ is an even function (hence the name).

Definition *Odd function.* A function f such that $f(-x) = -f(x)$ is called an **odd function**

The function $f(x) = x^3$ is odd since

$$f(-x) = (-x)^3 = -(x^3) = -f(x).$$

For any odd integer n, $f(x) = x^n$ is an odd function.

Most functions are neither even nor odd. For instance, $x^3 + x^4$ is neither even nor odd since $(-x)^3 + (-x)^4 = -x^3 + x^4$, which is neither $x^3 + x^4$ nor $-(x^3 + x^4)$. However, many functions used in calculus happen to be even or odd. The graph of such a function is symmetric with respect to the y axis or with respect to the origin, as will now be shown.

Symmetry of an even function

Consider an even function f. Assume that the point (a, b) is on the graph of f. That means that $f(a) = b$. Since f is even, $f(-a) = b$. Consequently, the point $(-a, b)$ is also on the graph of f. In other words, the graph of an even function is symmetric with respect to the y axis. (See Fig. 1.18.)

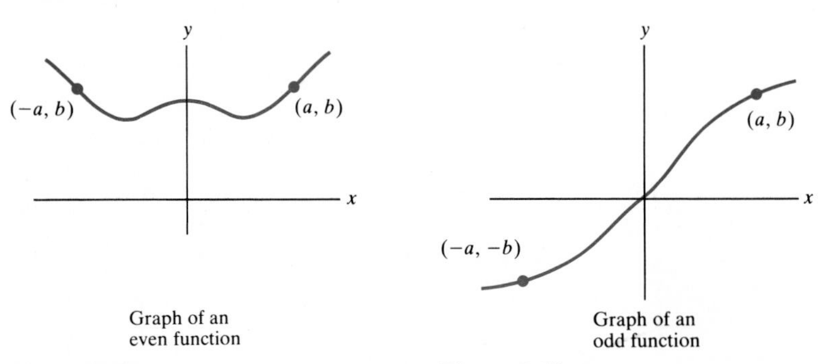

Figure 1.18 **Figure 1.19**

Symmetry of an odd function

Let (a, b) be on the graph of an odd function f. That is, $f(a) = b$. Since f is odd, $f(-a) = -b$, which tells us that the point $(-a, -b)$ is also on the graph of f. Consequently, the graph of an odd function is symmetric with respect to the origin. (See Fig. 1.19.)

EXERCISES FOR SEC. 1.2: COMPOSITE FUNCTIONS

In each of Exercises 1 to 4 find the function $y = f(x)$ defined by the composition of the given functions.

1 $y = u^3$, $u = x^2$

2 $y = 1 + u^2$, $u = 1 + x$

3 $y = \sqrt{u}$, $u = 1 + 2v$, $v = x^3$

4 $y = 1/u$, $u = 3 + v$, $v = x^2$

In Exercises 5 to 8 express the given functions as compositions of two or more simpler functions.

5 $y = (1 + x^3)^2$

6 $y = \dfrac{1}{1 + x^2}$

7 $y = \sqrt[3]{\dfrac{1}{1 + x^2}}$

8 $y = [(1 + x^2)^5]^3$

In Exercises 9 to 12 compute the compositions of the given functions.

9 $w = y^3$, $y = 3t$

10 $w = v^{10}$, $v = u^4$, $u = 1 + x$

11 $w = \sqrt[5]{y}$, $y = 1 + u^2$, $u = 1/x$

12 $y = w^3$, $w = 1/u$, $u = 1 + 3x$

■

13 Let $f(x) = 1 + 2x$. For which constants a and b does the function $g(x) = a + bx$ commute with f, that is, $(g \circ f)(x) = (f \circ g)(x)$?

14 Which of the following functions are even? odd? neither?
(*a*) $3x^2 + 5x^4$ (*b*) $|x|$ (*c*) $|x + 1|$
(*d*) $x^3 - 4x^5$ (*e*) $x^2/(1 + x)$ (*f*) $x^3/(1 + x^5)$
(*g*) $1 + x^2 + x^5$ (*h*) $4 + x^2$ (*i*) $3 + x^5$
(*j*) $x^3/(1 + x^4)$ (*k*) $x^2/(1 + x^4)$

15 (*a*) Show that the constant function $f(x) = 0$ is both odd and even.
(*b*) Are there other functions with that property?

16 Let $f(x) = 2x^2 - 1$ and $g(x) = 4x^3 - 3x$. Show that $(f \circ g)(x) = (g \circ f)(x)$. [Rare indeed are pairs of polynomials that commute with each other, as you may convince yourself by trying to find more. Of course, any two powers, such as x^3 and x^4, commute. (The composition of x^3 and x^4 in either order is x^{12}, as may be checked.)]

17 Assuming that 0 is in the domain of the odd function f, find $f(0)$.

18 Let $f(x) = 1/(1 - x)$. What is the domain of f? of $f \circ f$? of $f \circ f \circ f$? Show that $(f \circ f \circ f)(x) = x$ for all x in the domain of $f \circ f \circ f$.

19 Let $f(x) = x^5$. Is there a function $g(x)$ such that $(f \circ g)(x) = x$ for all numbers x? If so, how many such functions are there?

20 Let $f(x) = x^4$. Is there a function $g(x)$ such that $(f \circ g)(x) = x$ for all numbers x? If so, how many such functions are there?

21 Let $f(x) = 2x + 3$. How many functions are there of the form $g(x) = ax + b$, a and b constants, such that $f \circ g = g \circ f$?

22 Let $f(x) = 2x + 3$. How many functions are there of the form $g(x) = ax^2 + bx + c$, a, b, and c constants, a not 0, such that $f \circ g = g \circ f$?

23 Which of these functions are even? odd? neither?
(*a*) $\sin x$ (*b*) $\cos x$ (*c*) $\sin x^2$
(*d*) $\cos x + \sin x$ (*e*) $\sqrt[3]{1 + x^2}$
(*f*) $\sqrt{1 + x^3}$ (*g*) 2^x (*h*) $\tan 3x$

1.3 ONE-TO-ONE FUNCTIONS AND THEIR INVERSE FUNCTIONS

With some functions, "the output determines the input." For instance, the cubing function, $f(x) = x^3$, has this property. If we are told that the output of this function is, say, 64, then we know that the input must have been 4. However, the squaring function, $f(x) = x^2$, does not have this property. If we are told that the output of this function is, say, 25, then we do not know what the input is. It could be 5 or -5, since $5^2 = 25$ and $(-5)^2 = 25$.

Definition A function f that assigns distinct outputs to distinct inputs is called a **one-to-one function**.

For instance, x^3 is a one-to-one function, but x^2 (with domain taken to be the entire x axis) is *not* one-to-one.

The function that assigns to each U.S. citizen a Social Security number is supposed to be one-to-one. If it were not, different people would have the same Social Security number, a circumstance that could have awkward consequences.

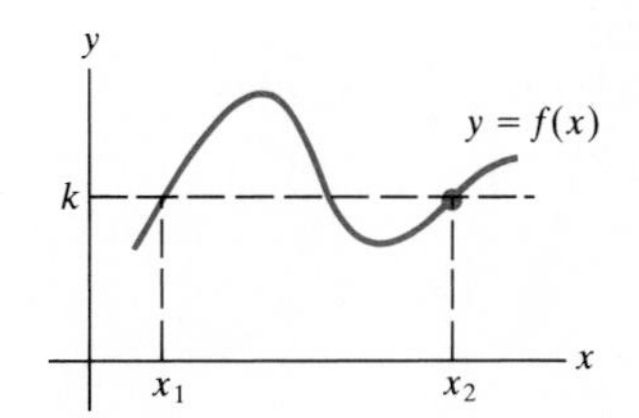

Not the graph of a one-to-one function

Figure 1.20

The graph of a one-to-one numerical function has the property that *every horizontal line meets it in at most one point*. To see why, consider the line $y = k$ in Fig. 1.20. If it meets the graph of a function f in at least two distinct points, say (x_1, k) and (x_2, k), then $f(x_1) = k$ and $f(x_2) = k$. This means that f is not a one-to-one function, since the outputs corresponding to the inputs x_1 and x_2 are equal, namely, k.

On the other hand, if *each horizontal line meets the graph of a function f in at most one point, then f is one-to-one.*

The most important one-to-one functions are described in this definition.

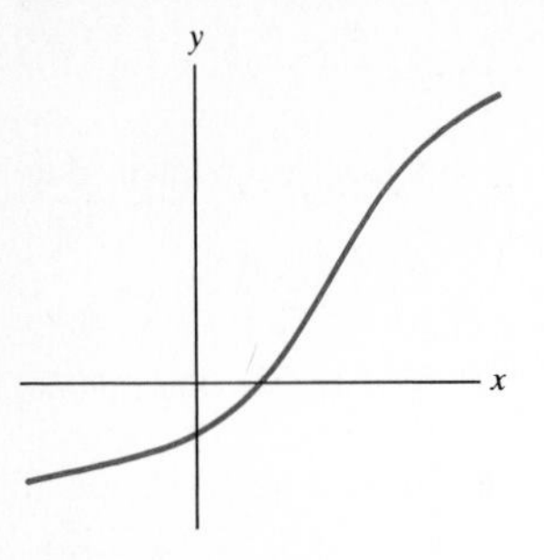

An increasing function (necessarily one-to-one)

Figure 1.21

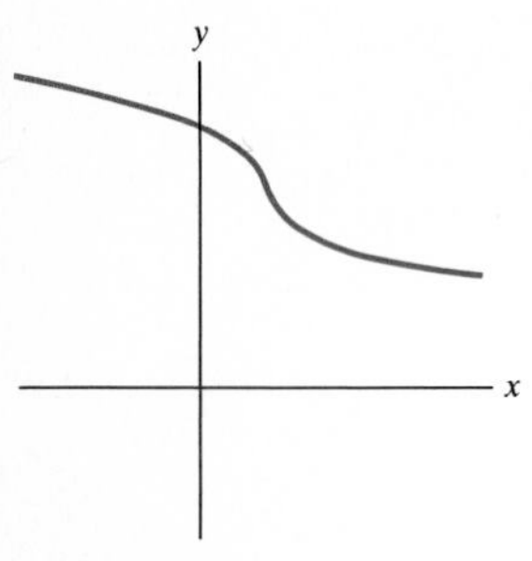

A decreasing function (necessarily one-to-one)

Figure 1.22

Definition If $f(x_1) < f(x_2)$ whenever $x_1 < x_2$, then f is an **increasing function**. If $f(x_1) > f(x_2)$ whenever $x_1 < x_2$, then f is a **decreasing function**.

These are illustrated in Figs. 1.21 and 1.22. (These two types of functions are also called **monotonic**.)

The function $y = f(x) = x^2$ is not increasing if its domain is taken to be the entire x axis. However, it is an increasing function if it is considered only for $x \geq 0$.

It is generally possible to restrict the domain of a function met in calculus to some interval so that the function, considered only on that interval, is one-to-one. (See Fig. 1.23.)

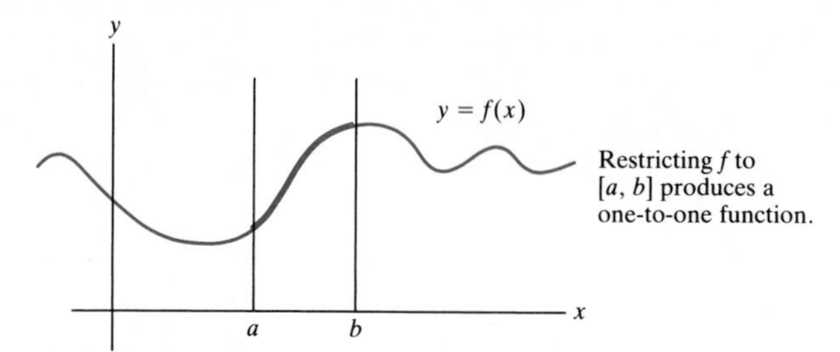

Restricting f to $[a, b]$ produces a one-to-one function.

Figure 1.23

Associated with a one-to-one function f is a second function g that records the unique input that yields a given output for the function f.

Definition Let $y = f(x)$ be a one-to-one function. The function g that assigns to each output of f the corresponding unique input is called the **inverse** of f. That is, if $y = f(x)$, then $x = g(y)$.

For example, $y = x^3$ is a one-to-one function. Its inverse is found by solving for x in terms of y; that is, $x = \sqrt[3]{y}$.

EXAMPLE 1 Determine the inverse of the "doubling" function f defined by $f(x) = 2x$ and then graph it.

SOLUTION If $y = 2x$, there is only one value of x for each value of y, and it is obtained by solving the equation $y = 2x$ for x: $x = y/2$. Thus f is one-to-one and its inverse function g is the "halving" function: If y is the input in the function g, then the output is $y/2$.

For instance, $f(3) = 6$ and $g(6) = 3$. Thus (3, 6) is on the graph of f, and (6, 3) is on the graph of g. Since it is customary to reserve the x axis for inputs, we should write the formula for g, the "halving" function, as

$$g(x) = \frac{x}{2}.$$

Thus f has the formula $y = 2x$, doubling, and g has the formula $y = x/2$, halving.

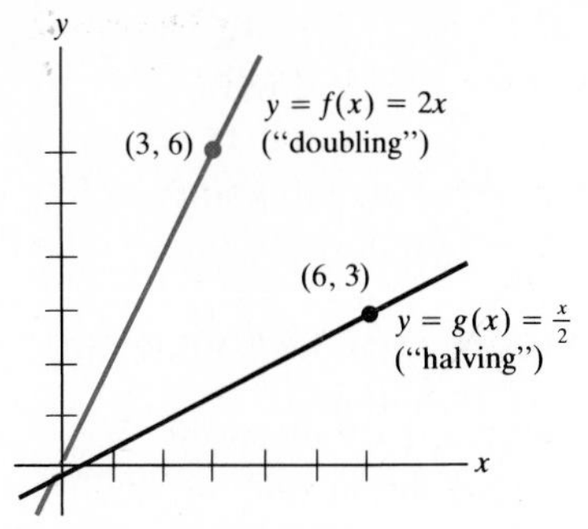

Figure 1.24

The graphs of f and g are lines (see Fig. 1.24). ■

EXAMPLE 2 Graph the cubing function, $y = x^3$, and its inverse (the cube root function, $y = \sqrt[3]{x}$) on the same axes.

SOLUTION We first prepare brief tables:

x	-2	-1	0	1	2
x^3	-8	-1	0	1	8

x	-8	-1	0	1	8
$\sqrt[3]{x}$	-2	-1	0	1	2

These data suggest the general shape of the graphs, as sketched in Fig. 1.25.

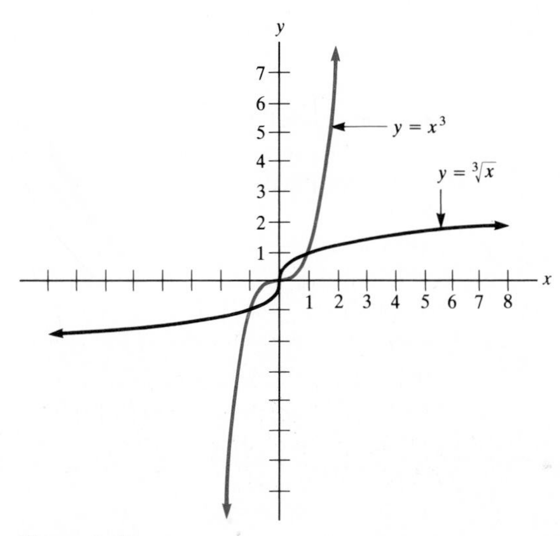

Figure 1.25 ■

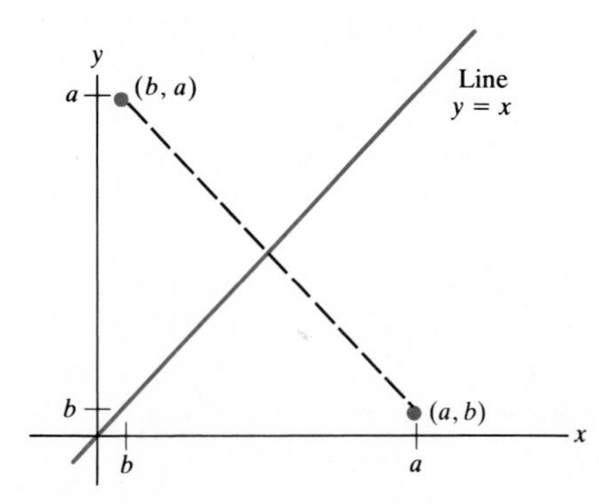

Figure 1.26

Note the relation between the two tables in Example 2. One is obtained from the other by switching inputs and outputs. This is the case for any one-to-one function and its inverse. If a and b are the numerical entries in a column for one function, then b and a are the entries in a column for the inverse function. Also note the relation between the two graphs in Examples 1 and 2. One graph is obtained from the other by reflecting it across the line $y = x$. This can be done because, if (a, b) is on the graph of one function, then (b, a) is on the graph of the other. If you fold the paper along the line $y = x$, the point (b, a) comes together with the point (a, b), as you will note in Fig. 1.26. This relation between the graphs holds for any one-to-one function and its inverse.

These examples are typical of the correspondence between a one-to-one function and its inverse. Perhaps the word "reverse" might be more descriptive than "inverse." One final matter of notation: We have used the letter g to denote the inverse of f. It is common to use the symbol

The symbol f^{-1} denotes the inverse function.

f^{-1} (read as "f inverse") to denote the inverse function. We preferred to delay its use because its resemblance to the reciprocal notation might cause confusion. It should be clear from the examples that f^{-1} does *not* mean to divide 1 by f. The symbol inv f would be unambiguous. However, it is longer than the symbol f^{-1} and the weight of tradition is behind f^{-1}.

Inverse functions come in pairs, each reversing the effect of the other. This table lists some pairs of inversely related functions:

Function f	*Inverse Function g*
Cubing, $y = x^3$	Cube root, $x = \sqrt[3]{y}$
Cube root, $y = \sqrt[3]{x}$	Cubing, $x = y^3$
Squaring, $y = x^2, x \geq 0$	Square root, $x = \sqrt{y}, y \geq 0$
Square root, $y = \sqrt{x}, x \geq 0$	Squaring, $x = y^2, y \geq 0$

Inverse Functions on a Calculator Most scientific calculators have a $\sqrt{x}$-key and an x^2-key. Each "undoes" the effect of the other. For instance, if you press the

1 5-key
2 then the $\sqrt{x}$-key
3 then the x^2-key

the calculator will end up displaying 5.

If you reverse the order of the square root and squaring keys, a similar effect results. If you press

1 the 5-key
2 then the x^2-key
3 then the $\sqrt{x}$-key

Why doesn't this work for -5?

the final output, the composition of the two functions, should be 5.

The inverse key might be labeled "2nd F" or have a special color and no label.

To minimize the number of keys, many calculators have a special "inverse" key, sometimes labeled "inv." The inv-key, in combination with some other function key, produces the inverse of that function (if the domain of the function is restricted to make the function one-to-one).

Warning: When you enter a number x in a calculator, then press the squaring key and then the square root key, you do not necessarily get exactly what you started with, x. This discrepancy is due to round off errors in the calculations performed. Since calculators usually carry out computations to one or two more digits than they display, such a discrepancy is rare.

The notion of an inverse function also can be expressed in terms of composition of functions. Let f be a one-to-one function from A to B, with the range of f being all of B. Then the inverse function g from B to A has the property that $(g \circ f)(a) = a$ for all a in A and $(f \circ g)(b) = b$ for all b in B. (See Fig. 1.27.)

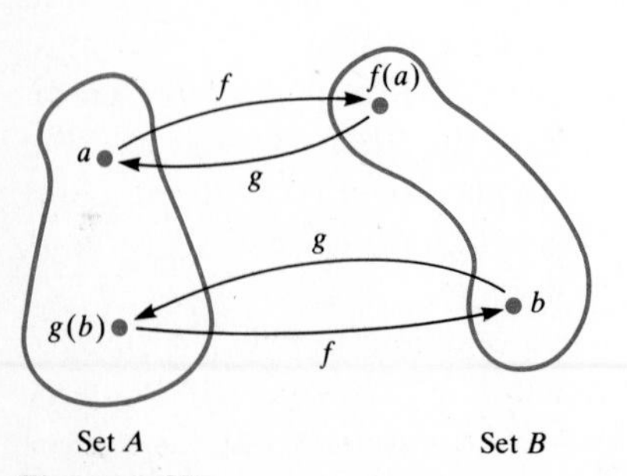

Figure 1.27

EXERCISES FOR SEC. 1.3: ONE-TO-ONE FUNCTIONS AND THEIR INVERSES

In Exercises 1 to 4 determine whether the given function is one-to-one on the given domain. If it is, obtain a formula for its inverse.

1 $y = x^4$ (*a*) $[-1, 1]$ (*b*) $[0, 2]$
2 $y = (x - 1)^2$ (*a*) $[0, 2]$ (*b*) $[1, 3]$
3 $y = 1 + x^5$ (*a*) $[0, 1]$ (*b*) $[-100, 100]$
4 $y = (x + 1)^5$ (*a*) $[0, 1]$ (*b*) $[-6, 6]$

In Exercises 5 and 6 decide whether the given graph is the graph of a one-to-one function. If it is, sketch the graph of the inverse function.

5

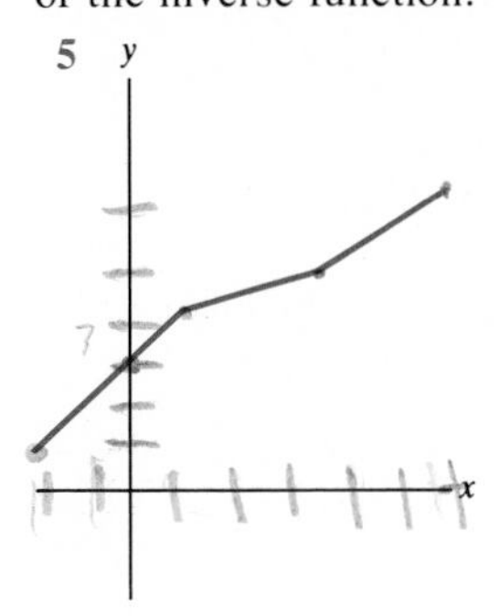

Figure 1.28

6

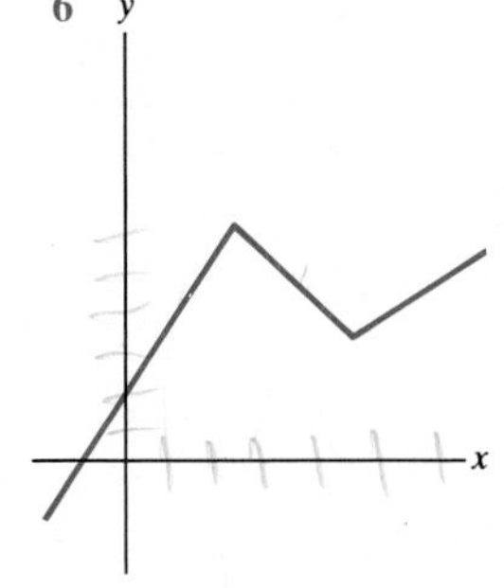

Figure 1.29

7 This table records some of the data about a one-to-one function f:

x	-1	0	1	2
$f(x)$	3	5	6	7

Use the data to plot four points on the graph of (*a*) f and (*b*) g, the inverse of f.

8 This table records some of the data about a one-to-one function f:

x	-5	2	π
$f(x)$	2	1	0

Use the data to plot three points on the graph of (*a*) f and (*b*) g, the inverse of f.

■

9 Is the function $f(x) = x^3 + 2x$ one-to-one?

10 Let $A = \{1, 2\}$ and $B = \{a, b, c\}$.
(*a*) How many functions are there from A to B?
(*b*) How many of them are one-to-one?

11 Let $A = \{1, 2, 3\}$ and $B = \{a, b, c\}$.
(*a*) List the tables for all the one-to-one functions from A to B. (There are six such functions.)
(*b*) List the table for the inverse of one of the functions in (*a*).

12 (*a*) For which choices of the constants a and b is the function $f(x) = ax + b$ one-to-one?
(*b*) In case a function in (*a*) is one-to-one, give the formula for its inverse.

In Exercises 13 to 16 the domains of the functions are the entire x axis. The values of the functions are real numbers. If you believe the statement is true, explain why. If you feel the statement is false, then give a specific "counter-example" to show why.

13 If f and g are one-to-one, then so is $f \circ g$.

14 If $f \circ g$ is one-to-one, then g is one-to-one.

15 If $f \circ g$ is one-to-one, then f is one-to-one.

16 If $(f \circ g)(x) = x$ and $(g \circ f)(x) = x$ for all x, then f is one-to-one and g is its inverse.

In Exercises 17 to 19 f and g are numerical functions and the range of g is included in the domain of f. What can be said about $f \circ g$ if:

17 f and g are increasing functions?

18 f and g are decreasing functions?

19 f is increasing and g is decreasing?

20 Is there a function g such that $(g(x))^2 = x$ for all real numbers x?

21 Is there a function g such that $(g(x))^5 = x$ for all real numbers x?

In Exercises 22 to 24 determine whether the function is one-to-one.

22 $x^6 + 3x^2 + 2$; domain $x \geq 0$

23 $x/(x + 1)$; domain x axis except -1

24 $x^2 + x + 1$; domain x axis

25 Let $f(x) = \text{Sin } x$ (angle in degrees). Is f one-to-one if the domain is taken to be
(*a*) the entire x axis?
(*b*) the interval $[0, 360]$?
(*c*) the interval $[0, 180]$?
(*d*) the interval $[-90, 90]$?

26 Let $f(x) = \text{Cos } x$ (angle in degrees). Answer the same questions as in the preceding exercise.

27 Let A be the set of positive integers.
(*a*) Explain why each integer in A can be written uniquely in the form $2^{a-1}(2b - 1)$, where a and b are positive integers.
(*b*) Let B be the set of fractions of the form u/v, where u and v are positive integers. Show that there is a one-to-one function from A to B with range B. (This means that there are just as many positive integers as there are fractions.) Incidentally, each positive rational number is represented by an infinite number of such fractions, for example, $\frac{1}{2} = \frac{2}{4} = \frac{3}{6} = \cdots$.

1.S SUMMARY

This chapter discussed functions. A function is any method for assigning to each element in some set X a unique element in some set Y. If both the inputs and the outputs are real numbers, the function is called numerical or a real function of a real variable.

The graph of a numerical function f consists of all points of the form $(x, f(x))$.

Calculus is concerned mainly with numerical functions that describe how one quantity depends on another.

Vocabulary

function
domain
range
input (argument)
output (value)
graph
composite function
even function
odd function
one-to-one function
increasing function
decreasing function
inverse of a one-to-one function
symmetry

Key Facts

A curve in the xy plane which meets each vertical line at most once is the graph of a function. If it is also true that each horizontal line meets the graph at most once, the function is one-to-one.

A function can be viewed many ways: as a table of values, as a graph, as an input-output machine, as a projection from one line to another.

Many functions are built up from other functions by composition. If $u = g(x)$ and $y = f(u)$, then $y = (f \circ g)(x)$.

The graph of an even function is symmetric with respect to the y axis; the graph of an odd function is symmetric with respect to the origin.

With a one-to-one function, f, is associated an inverse function, denoted g, or f^{-1}, or inv f. A numerical function that is increasing (or decreasing) throughout some interval is a one-to-one function. Functions considered in calculus can often be made one-to-one if their domains are restricted.

In case f is a numerical one-to-one function, the graph of its inverse function is obtained by reflecting the graph of f about the line $y = x$.

GUIDE QUIZ ON CHAP. 1: FUNCTIONS

1 This table records some of the values of a function f:

x	1	2	3	4	5
$f(x)$	3	1	4	3	2

(*a*) Is f one-to-one?
(*b*) Is f one-to-one when the domain is restricted to the set $\{2, 3, 4, 5\}$?

2 (*a*) Graph $y = x^3 + 1$.
(*b*) Why is the function in (*a*) one-to-one?
(*c*) Use the graph in (*a*) to graph the inverse of the function $y = x^3 + 1$.
(*d*) Give the formula for the inverse function.

3 Simplify the expression

$$\frac{f(x + h) - f(x)}{h}$$

for $h \neq 0$ as far as possible if (*a*) $f(x) = x^3$ for all x, (*b*) $f(x) = 1/x$ for all $x \neq 0$. Assume that $x + h \neq 0$.

4 Let $f(x) = \sqrt{x^2 + 10} - x$. Using a calculator, examine what happens to $f(x)$ as x gets very large.

5 Express the function $y = \sqrt{(1 + 2x)^3} + \sqrt{1 + 2x}$ as the composition of three functions.

REVIEW EXERCISES FOR CHAP. 1: FUNCTIONS

1 Let $f(x)$ be the volume of a right circular cylinder of radius x inscribed in a sphere of radius a, as shown in Fig. 1.30. Find a formula for $f(x)$.

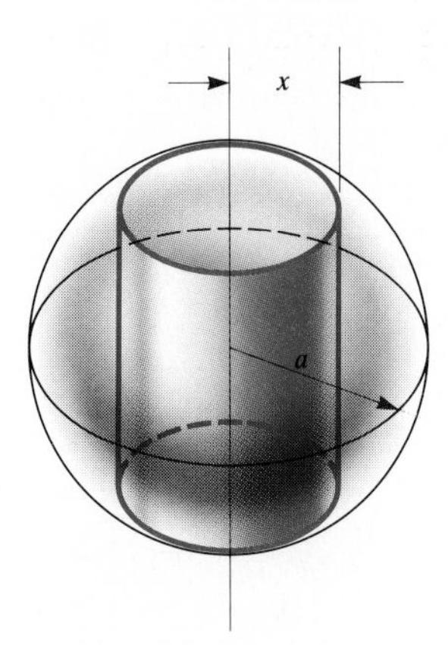

Figure 1.30

2 Give examples of three numerical functions f such that $f(x + 2) = f(x) + 2$ for all x.

3 A function from the xy plane to the xy plane is given by the formula $f(x, y) = (ax + by, cx + dy)$, where a, b, c, and d are constants. For which choices of a, b, c, d is f one-to-one?

4 Let $f(x) = 2x^2 - 1/x$. Express the fraction

$$\frac{f(x + h) - f(x)}{h}$$

as simply as possible. (Assume that none of h, x, and $x + h$ are 0.)

In each of Exercises 5 to 10 give the domain and range of the function.

5 $x^{1/5}$ **6** $x^{1/6}$
7 $x^{2/5}$ **8** x^2
9 $1/\sqrt{x + 1}$ **10** $\sqrt{4 - x^2}$

In Exercises 11 and 12 evaluate and express as decimals.

11 $f(2 + 0.1) - f(2)$ if $f(x) = x^2$
12 $f(2 - 0.1) - f(2)$ if $f(x) = x^2 + 3$

In Exercises 13 to 16 evaluate and simplify using algebra.

13 $\dfrac{f(a + h) - f(a)}{h}$ if $f(x) = 1/(x + 1)$

14 $\dfrac{f(2 + h) - f(2)}{h}$ if $f(x) = 3 + 2x + x^2$

15 $\dfrac{f(u) - f(v)}{u - v}$ if $f(x) = x^3 - 3x - 2$

16 $f(x) + f(-x)$ if $f(x) = x$

In Exercises 17 to 26 graph the functions.

17 $\sqrt{4 - x^2}$ **18** $-\sqrt{9 - x^2}$
19 $\sqrt{x^2 - 1}$ **20** $-\sqrt{x^2 - 5}$
21 $1/(x - 1)^2$ **22** $1/2^x$
23 3^{-x} **24** 1^x
25 $(x - 1)^2$ **26** $-x^2 + x + 4$

27 For which of the following functions is $f(b + 1)$ equal to $3f(b)$?
(*a*) $f(x) = 2^x$ (*b*) $f(x) = 3^x$
(*c*) $f(x) = 3^x/2$ (*d*) $f(x) = x/3$

28 For which of the following functions is $f(a + b) = f(a)f(b)$?
(*a*) $f(x) = -1$ for all x (*b*) $f(x) = 2^x$
(*c*) $f(x) = 5x$ (*d*) $f(x) = 3^{-x}$
(*e*) $f(x) = 1$ for all x (*f*) $f(x) = 3^x/7$

29 (*a*) Graph the function $f(x) = 2^x$.
(*b*) By reflecting the graph in (*a*) across the line $y = x$, obtain the graph of the inverse function g.
(*c*) What well-known function is g?

30 (Calculator) Graph $f(x) = x/2^x$.

31 (Calculator) Let $f(x) = (1 + x)^{1/x}$ for $x > 0$.
(*a*) Fill in this table:

x	0.001	0.01	0.1	1	2	10	100
$f(x)$							

(*b*) With the aid of (*a*) graph f.

32 (Calculator) Let $f(x) = x^x$ for $x > 0$.
(*a*) Evaluate the function for the inputs 1, 0.5, 0.4, 0.3, 0.1, and 0.001.
(*b*) With the aid of (*a*) graph f.

33 It is known that the perimeter of a rectangle is 100 inches. If one side has length x inches, denote the area of that rectangle as $f(x)$ square inches.
(*a*) Find a formula for $f(x)$.
(*b*) What is the domain of the function in the given context?

34 A rectangular box has sides of lengths x, y, and z inches. Obtain a formula for the function that describes (*a*) the total surface area of the box, (*b*) the total volume of the box, and (*c*) the total length of the edges of the box.

35 Define a numerical function f as follows: If x is an integer, let $f(x) = -x$; if x is not an integer, let $f(x) = x$.
(*a*) Show that $(f \circ f)(x) = x$ for all numbers x.
(*b*) How many functions g are there from the x axis to the x axis such that $(g \circ g)(x) = x$ for all numbers x?

36 Let f be a function from the xy plane to the x axis. For each ordered pair of numbers x and y, $f(x, y)$ is a number. The addition function $f(x, y) = x + y$ is a famous example. In this case, the function symbol + is placed between x and y rather than in front of them. Now consider the function $\sqrt{x^2 + y^2}$. For convenience denote it $x * y$. (That is, $x * y = \sqrt{x^2 + y^2}$.) Does $x * (y * z)$ equal $(x * y) * z$?

Two functions f and g are said to commute if $f \circ g = g \circ f$. For instance, x^2 and x^3 commute since $(x^3)^2 = (x^2)^3$. However, x^2 and $x^2 + 1$ do not, since $(x^2 + 1)^2$ is not the same as $(x^2)^2 + 1$. This definition is used in Exercises 37 through 40.

37 The polynomial x^3 commutes with x^2. Are there any other polynomials of degree 3 that commute with x^2?

38 Let $f(x) = 2x^2 - 1$.
(*a*) How many polynomials of degree 2 commute with f?
(*b*) Are there any of degree 3? If so, how many?

39 Like Exercise 38, but with $f(x) = x^2 + 1$.

40 Let $f(x) = 2x^2 - 1$.
(*a*) Show that $f \circ f$ commutes with f.
(*b*) Find all polynomials of degree 4 that commute with f.

41 Intuition tells us that there is only one size of "infinity." In other words, given infinite sets X and Y, there should be a one-to-one function f from X to Y whose range is all of Y. Such a function "pairs off" the elements of X with the elements of Y. Georg Cantor, in 1873, proved that intuition is wrong. Here is his example. Let X be the set of positive integers. Let Y be the set of real numbers in the interval $[0, 1]$, each expressed in decimal form. Let f be *any function* from X to Y. Cantor used f itself to construct a number in Y that is *not* in the range of f. Here is how he did it.

For the positive integer n, let $f(n) = 0.a_{n1}a_{n2}a_{n3} \cdots$, where a_{ni} is the ith digit in the decimal representation of $f(n)$, and so is one of the digits $0, 1, \ldots, 9$. Imagine the table for f:

1	$0.a_{11}a_{12}a_{13}a_{14} \cdots$
2	$0.a_{21}a_{22}a_{23}a_{24} \cdots$
3	$0.a_{31}a_{32}a_{33}a_{34} \cdots$
$\vdots$	$\vdots$

Construct a real number $b = 0.b_1b_2b_3b_4 \cdots$ as follows. The nth digit of b is 5 if a_{nn} is not 5; the nth digit of b is 4 if a_{nn} is 5. Explain why b is not in the range of f.

This shows that there is a scale of infinite numbers, just as there is a scale of finite numbers, $0 < 1 < 2 < 3 < \cdots$. In particular, the infinitude of the set of real numbers is greater than the infinitude of the set of positive integers.

The idea behind this proof, known as **Cantor's diagonal argument**, appears in several branches of mathematics and in the general theory of computability.

2 Limits and Continuous Functions

This chapter develops the concept of a limit, which lies at the heart of calculus and provides the foundation for both the derivative and the integral.

2.1 THE LIMIT OF A FUNCTION

Three examples will introduce the notion of the limit of a numerical function. After them, the concept of a limit will be defined.

EXAMPLE 1 Let $f(x) = 2x^2 + 1$. What happens to $f(x)$ as x is chosen closer and closer to 3?

SOLUTION Let us make a table of the values of $f(x)$ for some choices of x near 3.

x	3.1	3.01	3.001	2.999	2.99	2.9
$f(x)$	20.22	19.1202	19.012002	18.988002	18.8802	17.82

When x is close to 3, $2x^2 + 1$ is close to $2 \cdot 3^2 + 1 = 19$. We say that "the limit of $2x^2 + 1$ as x approaches 3 is 19" and write

The "limit" notation

$$\lim_{x \to 3} (2x^2 + 1) = 19. \quad \blacksquare$$

Example 1 presented no obstacle. The next example offers a slight challenge.

EXAMPLE 2 Let $f(x) = (x^3 - 1)/(x^2 - 1)$. Note that this function is not defined when $x = 1$, for when x is 1, both numerator and denominator are 0. But we have every right to ask: How does $f(x)$ behave when x is *near* 1 but is *not* 1 itself?

SOLUTION First make a brief table of values of $f(x)$, to four decimal places, for x near 1. Choose some x larger than 1 and some x smaller than 1. For instance,

$$f(1.01) = \frac{1.01^3 - 1}{1.01^2 - 1} = \frac{1.030301 - 1}{1.0201 - 1} = \frac{0.030301}{0.0201} \approx 1.5075.$$

x	1.1	1.01	0.9	0.99
$\dfrac{x^3 - 1}{x^2 - 1}$	1.5762	1.5075	1.4263	1.4925

[If you have a calculator handy, evaluate $(x^3 - 1)/(x^2 - 1)$ at 1.001 and 0.999 as well.]

Two influences operate on $\dfrac{x^3 - 1}{x^2 - 1}$.

There are two influences acting on the fraction $(x^3 - 1)/(x^2 - 1)$ when x is near 1. *On the one hand, the numerator* $x^3 - 1$ *approaches* 0; *thus there is an influence pushing the fraction toward* 0. *On the other hand, the denominator* $x^2 - 1$ *also approaches* 0; *division by a small number tends to make a fraction large.* How do these two opposing influences balance out?

The algebraic identities

$$x^3 - 1 = (x^2 + x + 1)(x - 1)$$

and

$$x^2 - 1 = (x + 1)(x - 1)$$

enable us to answer the question.

Rewrite the quotient $(x^3 - 1)/(x^2 - 1)$ as follows: When $x \neq 1$, we have

$$\frac{x^3 - 1}{x^2 - 1} = \frac{(x^2 + x + 1)(x - 1)}{(x + 1)(x - 1)} = \frac{x^2 + x + 1}{x + 1},$$

so the behavior of $(x^3 - 1)/(x^2 - 1)$ for x near 1, but not equal to 1, is the same as the behavior of $(x^2 + x + 1)/(x + 1)$ for x near 1, but not equal to 1. Thus

$$\lim_{x \to 1} \frac{x^3 - 1}{x^2 - 1} = \lim_{x \to 1} \frac{x^2 + x + 1}{x + 1}.$$

Now, as x approaches 1, $x^2 + x + 1$ approaches 3 and $x + 1$ approaches 2. Thus

$$\lim_{x \to 1} \frac{x^2 + x + 1}{x + 1} = \frac{3}{2},$$

from which it follows that

$$\lim_{x\to 1} \frac{x^3 - 1}{x^2 - 1} = \frac{3}{2}.$$

Note that $\frac{3}{2} = 1.5$, which is closely approximated by $f(1.01)$ and $f(0.99)$. ■

The notation → for "approaches"

The arrow → will stand for "approaches." According to Example 2,

$$\text{as } x \to 1, \qquad \frac{x^3 - 1}{x^2 - 1} \to \frac{3}{2}.$$

This notation will be used in the next example and often later.

EXAMPLE 3 Consider the function f defined by $f(x) = x/|x|$. The domain of this function consists of every number except 0. For instance,

$$f(3) = \frac{3}{|3|} = \frac{3}{3} = 1$$

and

$$f(-2) = \frac{-2}{|-2|} = \frac{-2}{2} = -1.$$

The "hollow dot" notation for a missing point

When x is positive, $f(x) = 1$. When x is negative, $f(x) = -1$. This is shown in Fig. 2.1. The graph does not intersect the y axis, since f is not defined for $x = 0$. The hollow circles at (0, 1) and (0, −1) indicate that those points are not on the graph. What happens to $f(x)$ as $x \to 0$?

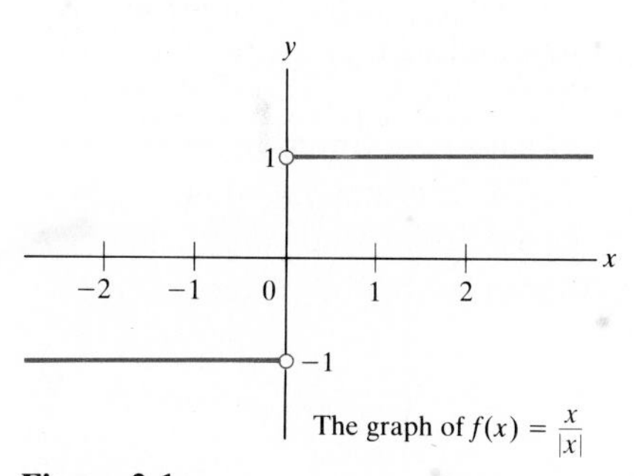

The graph of $f(x) = \frac{x}{|x|}$

Figure 2.1

SOLUTION As $x \to 0$ through positive numbers, $f(x) \to 1$, since $f(x) = 1$ for any positive number. As $x \to 0$ through negative numbers, $f(x) \to -1$, since $f(x) = -1$ for any negative number.

When x is near 0, it is *not* the case that $f(x)$ is near one specific number. Thus

$$\lim_{x\to 0} f(x)$$

does *not* exist, that is,

$$\lim_{x\to 0} \frac{x}{|x|}$$

does *not* exist. However, if $a \neq 0$,

$$\lim_{x\to a} f(x)$$

does exist, being 1 when a is positive and −1 when a is negative. Thus $\lim_{x\to a} f(x)$ exists for all a other than 0. ■

Whether a function f has a limit at a has nothing to do with $f(a)$ itself. In fact, a might not even be in the domain of f. See, for instance, Examples 2 and 3. In Example 1, a (=3) happened to be in the domain of f, but that fact did not influence the reasoning. It is only the behavior of $f(x)$ for x *near* a that concerns us.

A precise definition of limit is to be found in Sec. 2.8.

These three examples provide a background for describing the limit concept which will be used throughout the text.

Consider a function f and a number a which may or may not be in the domain of f. In order to discuss the behavior of $f(x)$ for x near a, we must know that the domain of f contains numbers arbitrarily close to a. Note how this assumption is built into each of the following definitions.

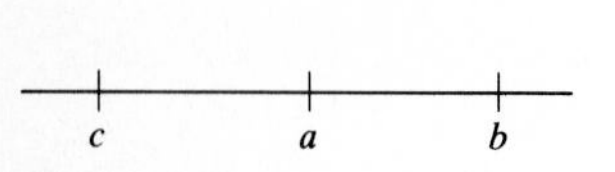

Figure 2.2

Definition *Limit of $f(x)$ at a.* Let f be a function and a some fixed number. Assume that the domain of f contains open intervals (c, a) and (a, b), as shown in Fig. 2.2. If there is a number L such that as x approaches a, either from the right or from the left, $f(x)$ approaches L, then L is called **the limit of $f(x)$ as x approaches a**. This is written

$$\lim_{x \to a} f(x) = L$$

or

$$f(x) \to L \qquad \text{as } x \to a.$$

"$\lim_{x\to 3}$" *is text version of* "$\lim\limits_{x\to 3}$"

In Example 1 we found $\lim_{x\to 3} (2x^2 + 1) = 19$, which illustrates this definition for $a = 3$ and $f(x) = 2x^2 + 1$. The fact that 3 happens to be in the domain of f is irrelevant. Example 2 showed that

$$\lim_{x\to 1} \frac{x^3 - 1}{x^2 - 1} = \frac{3}{2},$$

which illustrates this definition for $a = 1$ and $f(x) = (x^3 - 1)/(x^2 - 1)$. The fact that $f(x)$ is not defined for $x = 1$ did not affect the reasoning.

Example 3 concerns the behavior of $f(x) = x/|x|$ when x is near 0. As $x \to 0$, $f(x)$ does not approach a specific number. However, as x approaches 0 through positive numbers, $f(x) \to 1$. Also, as x approaches 0 through negative numbers, $f(x) \to -1$. This behavior illustrates the idea of a one-sided limit, which will now be defined.

Definition *Right-hand limit of $f(x)$ at a.* Let f be a function and a some fixed number. Assume that the domain of f contains an open interval (a, b). If, as x approaches a from the right, $f(x)$ approaches a specific number L, then L is called the **right-hand limit of $f(x)$ as x approaches a**. This is written

Right-hand limit

$$\lim_{x\to a^+} f(x) = L$$

or

$$\text{as } x \to a^+, \qquad f(x) \to L.$$

The assertion that

$$\lim_{x\to a^+} f(x) = L$$

is read "the limit of f of x as x approaches a from the right is L" or "as x approaches a from the right, $f(x)$ approaches L."

Left-hand limit

The left-hand limit is defined similarly. The only differences are that the domain of f must contain an open interval of the form (c, a) and $f(x)$ is examined as x approaches a from the left. The notations for the left-hand limit are

$$\lim_{x\to a^-} f(x) = L$$

or

$$\text{as } x \to a^-, \qquad f(x) \to L.$$

As Example 3 showed,

$$\lim_{x \to 0^+} \frac{x}{|x|} = 1 \qquad \text{and} \qquad \lim_{x \to 0^-} \frac{x}{|x|} = -1.$$

We could also write, for instance,

$$\text{as } x \to 0^+, \qquad \frac{x}{|x|} \to 1.$$

Note that if both the right-hand and the left-hand limits of f exist at a and are equal, then $\lim_{x \to a} f(x)$ exists. But if the right-hand and left-hand limits are not equal, then $\lim_{x \to a} f(x)$ does not exist. For instance, $\lim_{x \to 0} x/|x|$ does not exist.

The next example reviews the three limit concepts.

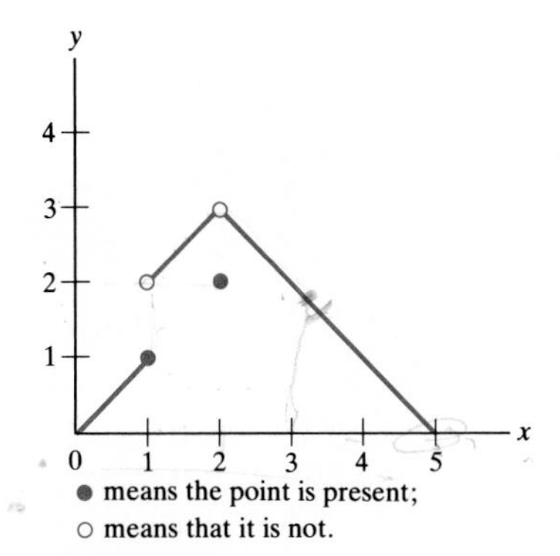

Figure 2.3

EXAMPLE 4 Figure 2.3 shows the graph of a function f whose domain is the closed interval [0, 5].

(*a*) Does $\lim_{x \to 1} f(x)$ exist?
(*b*) Does $\lim_{x \to 2} f(x)$ exist?
(*c*) Does $\lim_{x \to 3} f(x)$ exist?

SOLUTION
(*a*) Inspection of the graph shows that

$$\lim_{x \to 1^-} f(x) = 1 \qquad \text{and} \qquad \lim_{x \to 1^+} f(x) = 2.$$

Although the two one-sided limits exist, they are not equal. Thus $\lim_{x \to 1} f(x)$ does not exist. In short, "f does not have a limit as $x \to 1$."

(*b*) Inspection of the graph shows that

$$\lim_{x \to 2^-} f(x) = 3 \qquad \text{and} \qquad \lim_{x \to 2^+} f(x) = 3.$$

Thus $\lim_{x \to 2} f(x)$ exists and is 3. Incidentally, the solid dot at (2, 2) shows that $f(2) = 2$. This information, however, plays no role in our examination of the limit of $f(x)$ as $x \to 2$.

(*c*) Inspection shows that

$$\lim_{x \to 3^-} f(x) = 2 \qquad \text{and} \qquad \lim_{x \to 3^+} f(x) = 2.$$

Thus $\lim_{x \to 3} f(x)$ exists and is 2. Incidentally, the fact that $f(3)$ is equal to 2 is irrelevant in determining $\lim f(x)$. ■

EXAMPLE 5 Figure 2.4 shows the graph of a function whose domain is the closed interval [0, 6].

(*a*) Does $\lim_{x \to 3^-} f(x)$ exist?
(*b*) Does $\lim_{x \to 3^+} f(x)$ exist?

SOLUTION
(*a*) As $x \to 3$ from the left, $f(x) \to 2$. Thus $\lim_{x \to 3^-} f(x)$ exists and is 2.

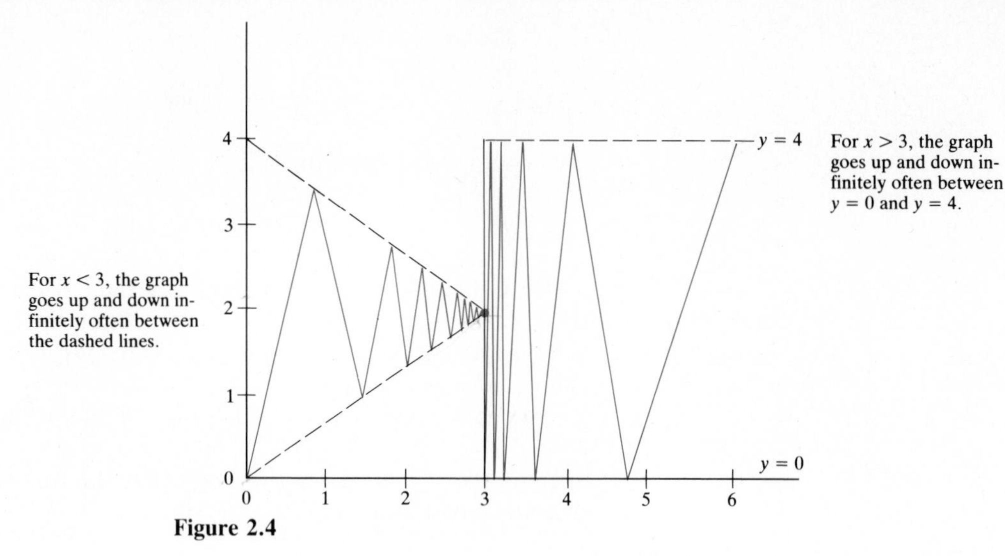

Figure 2.4

(*b*) As $x \to 3$ from the right, $f(x)$ does not approach one single number. Thus $\lim_{x \to 3^+} f(x)$ does not exist. ■

A function as wild as the one described in Example 5 will not be of major concern in calculus. However, it does serve to clarify the definitions of right-hand limit and left-hand limit, just as the notion of sickness illuminates our understanding of health.

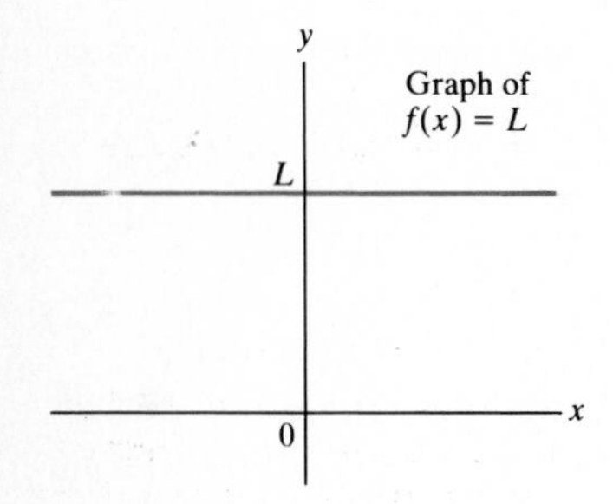

Figure 2.5

By contrast, the tamest functions are the "constant" functions. A **constant function** assigns the same output to all inputs. If that fixed output is, say, L, then $f(x) = L$ for all x. The graph of this function is a line parallel to the x axis, as in Fig. 2.5. We have

$$\lim_{x \to a} f(x) = L.$$

It may seem strange to say that "the limit of L is L," but in practice this offers no difficulty. For instance,

$$\lim_{x \to 5} \frac{1 + x^2}{1 + x^2} = \lim_{x \to 5} 1 = 1$$

and

$$\lim_{x \to 3} 1^x = \lim_{x \to 3} 1 = 1.$$

EXERCISES FOR SEC. 2.1: THE LIMIT OF A FUNCTION

In Exercises 1 to 14 find the limits, all of which exist. Use intuition and, if needed, algebra.

1 $\lim_{x \to 5} (x + 7)$

2 $\lim_{x \to 1} (4x - 2)$

3 $\lim_{x \to 2} \frac{x^2 - 4}{x - 2}$

4 $\lim_{x \to 3} \frac{x^2 - 9}{x - 3}$

5 $\lim_{x \to 1} \frac{x^4 - 1}{x^3 - 1}$

6 $\lim_{x \to 1} \frac{x^6 - 1}{x^3 - 1}$

7 $\lim_{x \to 3} \frac{1}{x + 2}$

8 $\lim_{x \to 5} \frac{3x + 5}{4x}$

9 $\lim_{x\to 3} 25$

10 $\lim_{x\to 3} \pi^2$

11 $\lim_{x\to 0^+} \sqrt{x}$

12 $\lim_{x\to 1^+} \sqrt{4x - 4}$

13 $\lim_{x\to 1^+} \dfrac{x - 1}{|x - 1|}$

14 $\lim_{x\to 1^-} \dfrac{x - 1}{|x - 1|}$

In Exercises 15 to 24 decide whether the limits exist and, if they do, evaluate them.

15 $\lim_{h\to 1} \dfrac{(1 + h)^2 - 1}{h}$

16 $\lim_{h\to 0} \dfrac{(1 + h)^2 - 1}{h}$

17 $\lim_{x\to 2} \dfrac{\dfrac{1}{x} - \dfrac{1}{2}}{x - 2}$

18 $\lim_{x\to 3} \dfrac{\dfrac{1}{x} - \dfrac{1}{2}}{x - 2}$

19 $\lim_{x\to 0} \dfrac{\sqrt{x + 4} - 2}{x}$ (*Hint:* Rationalize the numerator.)

20 $\lim_{x\to 0} \dfrac{\sqrt{x + 4} + 2}{x}$

21 $\lim_{x\to 4^+} (\sqrt{x - 4} + 2)$

22 $\lim_{x\to 9} (\sqrt{x - 4} + \sqrt{x + 4})$

23 $\lim_{x\to 0} 64^x$

24 $\lim_{x\to 1} \dfrac{3^x - 3}{2^x}$

In each of Exercises 25 and 26 there is a graph of a function. Decide which of the given limits exist, and evaluate those which do.

25 (See Fig. 2.6.) (*a*) $\lim_{x\to 0^+} f(x)$ (*b*) $\lim_{x\to 1} f(x)$ (*c*) $\lim_{x\to 2^-} f(x)$ (*d*) $\lim_{x\to 2^+} f(x)$

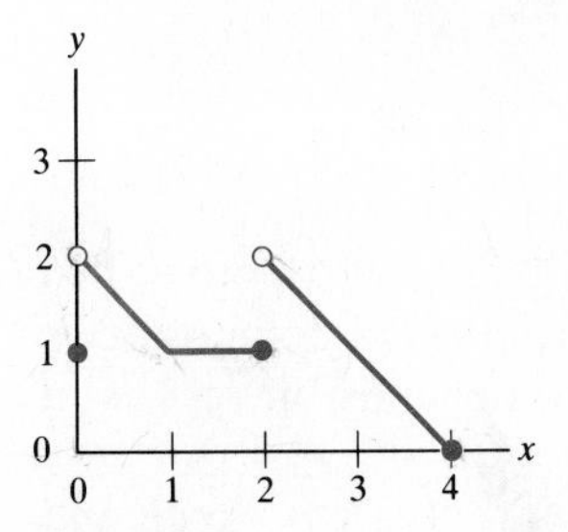

Figure 2.6

26 (See Fig. 2.7.) (*a*) $\lim_{x\to 1} f(x)$ (*b*) $\lim_{x\to 2} f(x)$ (*c*) $\lim_{x\to 3} f(x)$ (*d*) $\lim_{x\to 4^-} f(x)$

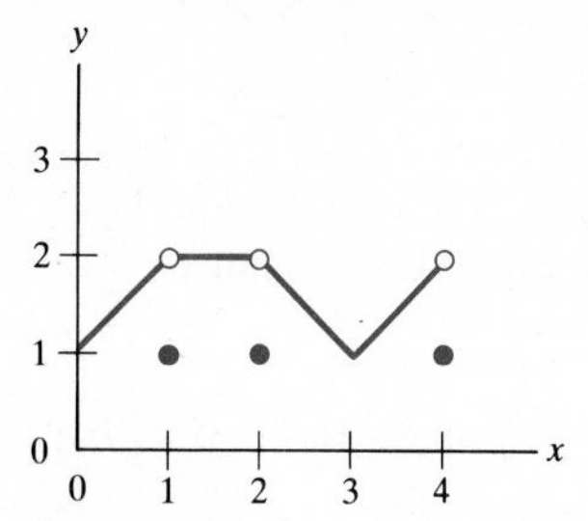

Figure 2.7

27 Find (*a*) $\lim_{x\to 4} (x - 4)$, (*b*) $\lim_{x\to 4} (\sqrt{x} - 2)$, and (*c*) $\lim_{x\to 4} [(x - 4)/(\sqrt{x} - 2)]$.

28 Find (*a*) $\lim_{x\to -2} (x + 2)$, (*b*) $\lim_{x\to -2} (x^3 + 8)$, and (*c*) $\lim_{x\to -2} [(x^3 + 8)/(x + 2)]$.

■

29 (Calculator) Let $f(x) = (3^x - 1)/(2^x - 1)$.
(*a*) Fill in this table:

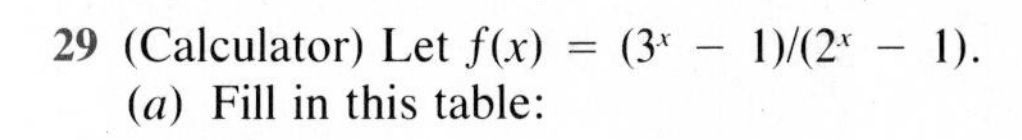

x	1	0.1	0.01	0.001	−1	−0.01	−0.001
$f(x)$							

(*b*) On the basis of (*a*) do you think $\lim_{x\to 0} f(x)$ exists? If so, estimate this limit.

30 (Calculator) Let $f(x) = (1 + x)^{1/x}$ for $x > -1$, $x \neq 0$.
(*a*) Compute $f(x)$ for $x = 1, 0.1, 0.01$, and 0.001.
(*b*) Compute $f(x)$ for $x = -0.1, -0.01, -0.001$.

31 Figure 2.8 shows a graph of a function that goes up and down infinitely often between the dashed lines, both to the right and to the left of 3. Does $\lim_{x\to 3} f(x)$ exist? If so, what is it?

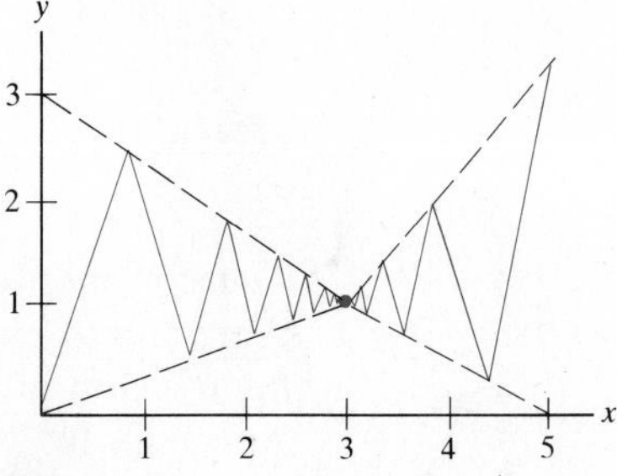

Figure 2.8

32 Define a certain function f as follows:

$$f(x) = \begin{cases} 1 & \text{if } x \text{ is an integer,} \\ 0 & \text{if } x \text{ is not an integer.} \end{cases}$$

(*a*) Graph f.
(*b*) Does $\lim_{x\to 3} f(x)$ exist?
(*c*) Does $\lim_{x\to 3.5} f(x)$ exist?
(*d*) For which numbers a does $\lim_{x\to a} f(x)$ exist?

33 (*a*) Graph the function f given by the formula

$$f(x) = |x| - x.$$

(*b*) For which numbers a does $\lim_{x\to a} f(x)$ exist?

34 Define f as follows:

$$f(x) = \begin{cases} x & \text{if } x \text{ is rational,} \\ -x & \text{if } x \text{ is not rational.} \end{cases}$$

(*a*) What does the graph of f look like? (A dotted curve may be used to indicate that points are missing.)
(*b*) Does $\lim_{x\to 1} f(x)$ exist?
(*c*) Does $\lim_{x\to\sqrt{2}} f(x)$ exist?
(*d*) Does $\lim_{x\to 0} f(x)$ exist?
(*e*) For which numbers a does $\lim_{x\to a} f(x)$ exist?

35 Define $f(x) = \begin{cases} x^2 & \text{if } x \text{ is rational,} \\ x^3 & \text{if } x \text{ is irrational.} \end{cases}$

(*a*) What does the graph of f look like? [See the advice for Exercise 34(*a*).]
(*b*) Does $\lim_{x\to 2} f(x)$ exist?
(*c*) Does $\lim_{x\to 1} f(x)$ exist?
(*d*) Does $\lim_{x\to 0} f(x)$ exist?
(*e*) For which numbers a does $\lim_{x\to a} f(x)$ exist?

36 (Calculator) Let $f(x) = x^x$ for $x > 0$.
(*a*) Fill in this table:

x	1.0	0.5	0.4	0.3	0.2	0.1	0.01
x^x							

(*b*) What do you think is the smallest value of x^x for x in (0, 1)?
(*c*) Do you think that $\lim_{x\to 0^+} x^x$ exists? If so, what do you think it is?

2.2 COMPUTATIONS OF LIMITS

Certain frequently used properties of limits should be put on the record.

Theorem Let f and g be two functions and assume that

$$\lim_{x\to a} f(x) \quad \text{and} \quad \lim_{x\to a} g(x)$$

both exist. Then

Properties of limits

1 $\lim_{x\to a} (f(x) + g(x)) = \lim_{x\to a} f(x) + \lim_{x\to a} g(x).$

2 $\lim_{x\to a} (f(x) - g(x)) = \lim_{x\to a} f(x) - \lim_{x\to a} g(x).$

3 $\lim_{x\to a} kf(x) = k \lim_{x\to a} f(x)$ for any constant k.

4 $\lim_{x\to a} f(x)g(x) = \lim_{x\to a} f(x) \lim_{x\to a} g(x).$

5 $\lim_{x\to a} \dfrac{f(x)}{g(x)} = \dfrac{\lim_{x\to a} f(x)}{\lim_{x\to a} g(x)}$ if $\lim_{x\to a} g(x) \neq 0$.

6 $\lim_{x\to a} f(x)^{g(x)} = \left(\lim_{x\to a} f(x)\right)^{\lim_{x\to a} g(x)}$ if $\lim_{x\to a} f(x) > 0$. ■

These properties have been tacitly assumed in Sec. 2.1. Properties 1 to 5 are treated in Appendix G, which employs the precise definitions of limits given in the Secs. 2.7 and 2.8. Property 6 depends on results in Appendix H.

Property 1, for instance, asserts that $\lim_{x\to a} (f(x) + g(x))$ exists and equals the sum of the two given limits. This property extends to any finite sum of functions. For example, if $\lim_{x\to a} f(x)$, $\lim_{x\to a} g(x)$, and $\lim_{x\to a} h(x)$ exist, then

$$\lim_{x \to a} (f(x) + g(x) + h(x)) \quad \text{exists,}$$

and $$\lim_{x \to a} (f(x) + g(x) + h(x)) = \lim_{x \to a} f(x) + \lim_{x \to a} g(x) + \lim_{x \to a} h(x).$$

Similarly, property 4 extends to the product of any finite number of functions.

EXAMPLE 1 Suppose that $\lim_{x \to 3} f(x) = 4$ and $\lim_{x \to 3} g(x) = 5$; discuss $\lim_{x \to 3} f(x)/g(x)$.

SOLUTION By property 5, $\lim_{x \to 3} f(x)/g(x)$ exists and

$$\lim_{x \to 3} \frac{f(x)}{g(x)} = \frac{4}{5}.$$

No further information about f and g is needed to determine the limit of $f(x)/g(x)$ as $x \to 3$. ■

EXAMPLE 2 Suppose that $\lim_{x \to 3} f(x) = 0$ and $\lim_{x \to 3} g(x) = 0$; discuss $\lim_{x \to 3} f(x)/g(x)$.

SOLUTION In contrast to Example 1, in this case property 5 gives no information, since $\lim_{x \to 3} g(x) = 0$. It is necessary to have more information about f and g.

For instance, if

$$f(x) = x^2 - 9 \quad \text{and} \quad g(x) = x - 3,$$

then $$\lim_{x \to 3} f(x) = 0 \quad \text{and} \quad \lim_{x \to 3} g(x) = 0$$

and the limit of the quotient is

$$\lim_{x \to 3} \frac{x^2 - 9}{x - 3} = \lim_{x \to 3} \frac{(x + 3)(x - 3)}{x - 3}$$
$$= \lim_{x \to 3} (x + 3) = 6.$$

Loosely put, "when x is near 3, $x^2 - 9$ is about 6 times as large as $x - 3$."

A different choice of f and g could produce a different limit for the quotient $f(x)/g(x)$. To be specific, let

$$f(x) = (x - 3)^2 \quad \text{and} \quad g(x) = x - 3.$$

Then $$\lim_{x \to 3} f(x) = 0 \quad \text{and} \quad \lim_{x \to 3} g(x) = 0,$$

and the limit of the quotient is

$$\lim_{x \to 3} \frac{(x - 3)^2}{x - 3} = \lim_{x \to 3} (x - 3)$$
$$= 0.$$

If we know only that $f(x) \to 0$ and $g(x) \to 0$ as $x \to a$, we do not know how $f(x)/g(x)$ behaves as $x \to a$.

In this case we could say "$(x - 3)^2$ approaches 0 much faster than does $x - 3$, when $x \to 3$."

In short, the information that $\lim_{x\to 3} f(x) = 0$ and $\lim_{x\to 3} g(x) = 0$ is not enough to tell us how $f(x)/g(x)$ behaves as $x \to 3$. ■

Sometimes it is useful to know how $f(x)$ behaves when x is a very large positive number (or a negative number of large absolute value). Example 3 serves as an illustration and introduces a variation on the theme of limits.

EXAMPLE 3 Determine how $f(x) = 1/x$ behaves for (*a*) large positive inputs and (*b*) negative inputs of large absolute value.

x	10	100	1000
$1/x$	0.1	0.01	0.001

SOLUTION

(*a*) First make a table of values as shown in the margin. As x gets arbitrarily large, $1/x$ approaches 0.

(*b*) This is similar to (*a*). For instance,

$$f(-1000) = -0.001.$$

As negative numbers x are chosen of arbitrarily large absolute value, $1/x$ approaches 0. ■

The notation $\lim_{x\to\infty} f(x) = L$

Rather than writing "as x gets arbitrarily large through positive values, $f(x)$ approaches the number L," it is customary to use the shorthand

$$\lim_{x\to\infty} f(x) = L.$$

Since ∞ *is not a number, the case* $x \to \infty$ *is distinct from the case* $x \to a$.

This is read "as x approaches infinity, $f(x)$ approaches L," or "the limit of $f(x)$ as x approaches infinity is L." For instance,

$$\lim_{x\to\infty} \frac{1}{x} = 0.$$

More generally, for any fixed positive exponent a,

$$\lim_{x\to\infty} \frac{1}{x^a} = 0.$$

The notation $\lim_{x\to -\infty} f(x) = L$

Similarly, the assertion that "as negative numbers x are chosen of arbitrarily large absolute value, $f(x)$ approaches the number L" is abbreviated to

$$\lim_{x\to -\infty} f(x) = L.$$

For instance,
$$\lim_{x\to -\infty} \frac{1}{x} = 0.$$

The six properties of limits stated at the beginning of the section hold when "$x \to a$" is replaced by "$x \to \infty$" or by "$x \to -\infty$."

It could happen that as $x \to \infty$, a function $f(x)$ becomes and remains arbitrarily large and positive. For instance, as $x \to \infty$, x^3 gets arbitrarily large. The shorthand for this is

The notation $\lim_{x\to\infty} f(x) = \infty$

$$\lim_{x\to\infty} f(x) = \infty.$$

For instance, $$\lim_{x\to\infty} x^3 = \infty.$$

It is important, when reading the shorthand

$$\lim_{x\to\infty} f(x) = \infty,$$

to keep in mind that "∞" is not a number. *The limit does not exist.* Properties 1 to 6 cannot, in general, be applied in such cases.

Other notations, such as $\lim_{x\to\infty} f(x) = -\infty$ or $\lim_{x\to-\infty} f(x) = \infty$ are defined similarly. For instance,

$$\lim_{x\to-\infty} x^3 = -\infty.$$

It can be shown that if, as $x \to \infty$, $f(x) \to \infty$ and $g(x) \to L > 0$, then $\lim_{x\to\infty} f(x)g(x) = \infty$. This fact is used in the next example.

EXAMPLE 4 Discuss the behavior of $2x^3 - 11x^2 + 12x$ when x is large.

SOLUTION First consider x positive and large. The three terms, $2x^3$, $-11x^2$, and $12x$, all become of large absolute value. To see how the function $2x^3 - 11x^2 + 12x$ behaves for large positive x, factor out x^3:

This factoring shows the importance of the highest power.

$$2x^3 - 11x^2 + 12x = x^3\left(2 - \frac{11}{x} + \frac{12}{x^2}\right). \tag{1}$$

Now, since $11/x$ and $12/x^2 \to 0$ as $x \to \infty$,

$$\lim_{x\to\infty}\left(2 - \frac{11}{x} + \frac{12}{x^2}\right) = 2.$$

Moreover, as $x \to \infty$, $x^3 \to \infty$. Thus

$$\lim_{x\to\infty} x^3\left(2 - \frac{11}{x} + \frac{12}{x^2}\right) = \infty;$$

hence $$\lim_{x\to\infty} (2x^3 - 11x^2 + 12x) = \infty.$$

Now consider x negative and of large absolute value. The argument is similar. Use Eq. (1), and notice that $\lim_{x\to-\infty} x^3 = -\infty$ and

$$\lim_{x\to-\infty}\left(2 - \frac{11}{x} + \frac{12}{x^2}\right) = 2.$$

It follows that $$\lim_{x\to-\infty} (2x^3 - 11x^2 + 12x) = -\infty.$$

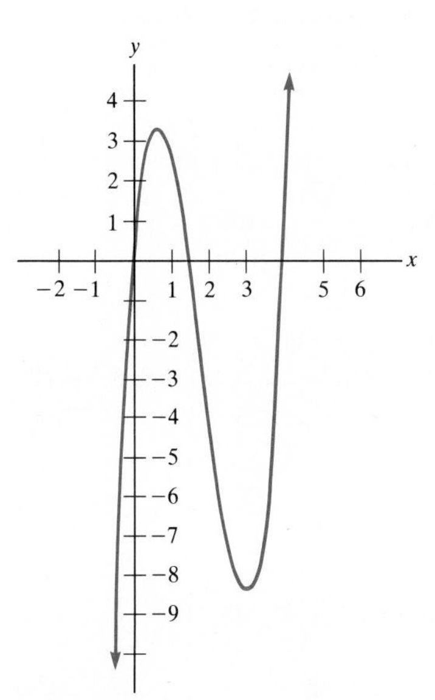

Figure 2.9

This completes the discussion. It is of interest, however, to graph $f(x) = 2x^3 - 11x^2 + 12x$ and see what is happening for $|x|$ large. This is done in Fig. 2.9. Since $\lim_{x\to\infty} (2x^3 - 11x^2 + 12x) = \infty$ the graph rises arbitrarily high as $x \to \infty$. Since $\lim_{x\to-\infty} (2x^3 - 11x^2 + 12x) = -\infty$, the graph goes arbitrarily far down as $x \to -\infty$. ■

General form of a polynomial

Example 4 generalizes to any polynomial. (A **polynomial** is a function of the form $a_nx^n + a_{n-1}x^{n-1} + \cdots + a_0$, where $a_0, a_1, \ldots, a_n$ are fixed real numbers and n is a nonnegative integer. If a_n is not 0, n is the **degree** of the polynomial. The numbers $a_0, a_1, \ldots, a_n$ are the **coefficients**.)

Limits of a polynomial as $x \to \infty$ or as $x \to -\infty$

Let $f(x)$ be a polynomial of degree at least 1 and with the lead coefficient a_n positive. Then

$$\lim_{x \to \infty} f(x) = \infty.$$

If the degree of f is even, then

$$\lim_{x \to -\infty} f(x) = \infty.$$

But if the degree of f is odd, then

$$\lim_{x \to -\infty} f(x) = -\infty.$$

EXAMPLE 5 Determine how $f(x) = (x^3 + 6x^2 + 10x + 2)/(2x^3 + x^2 + 5)$ behaves for arbitrarily large positive numbers x.

A contest between a large numerator and a large denominator

SOLUTION As x gets large, the numerator $x^3 + 6x^2 + 10x + 2$ grows large, influencing the quotient to become large. On the other hand, the denominator also grows large, influencing the quotient to become small. An algebraic device will help reveal what happens to the quotient. We have

$$f(x) = \frac{x^3 + 6x^2 + 10x + 2}{2x^3 + x^2 + 5} = \frac{x^3\left(1 + \dfrac{6}{x} + \dfrac{10}{x^2} + \dfrac{2}{x^3}\right)}{x^3\left(2 + \dfrac{1}{x} + \dfrac{5}{x^3}\right)}$$

$$= \frac{1 + \dfrac{6}{x} + \dfrac{10}{x^2} + \dfrac{2}{x^3}}{2 + \dfrac{1}{x} + \dfrac{5}{x^3}} \qquad \text{for } x \neq 0.$$

Now we can see what happens to $f(x)$ when x is large.

As x increases, $6/x \to 0$, $10/x^2 \to 0$, $2/x^3 \to 0$, $1/x \to 0$, and $5/x^3 \to 0$. Thus

$$f(x) \to \frac{1 + 0 + 0 + 0}{2 + 0 + 0} = \frac{1}{2}.$$

So, as x gets arbitrarily large through positive values, the quotient $(x^3 + 6x^2 + 10x + 2)/(2x^3 + x^2 + 5)$ approaches $\frac{1}{2}$. In short,

$$\lim_{x \to \infty} \frac{x^3 + 6x^2 + 10x + 2}{2x^3 + x^2 + 5} = \frac{1}{2}. \quad \blacksquare$$

The technique used in Example 5 applies to any function that can be written as the quotient of two polynomials. Such a function is called a **rational function**.

How to find the limit of a rational function as $x \to \infty$ or as $x \to -\infty$

Let $f(x)$ be a polynomial and let ax^n be its term of highest degree. Let $g(x)$ be another polynomial and let bx^m be its term of highest degree.

Then

$$\lim_{x\to\infty} \frac{f(x)}{g(x)} = \lim_{x\to\infty} \frac{ax^n}{bx^m} \quad \text{and} \quad \lim_{x\to-\infty} \frac{f(x)}{g(x)} = \lim_{x\to-\infty} \frac{ax^n}{bx^m}.$$

(The proofs of these facts are similar to the argument used in Example 5.) In short, when working with the limit of a quotient of two polynomials as $x \to \infty$ or as $x \to -\infty$, disregard all terms except the one of highest degree in each of the polynomials. The next example illustrates this technique.

EXAMPLE 6 Examine the following limits:

(*a*) $\lim_{x\to\infty} \dfrac{3x^4 + 5x^2}{-x^4 + 10x + 5}$ (*b*) $\lim_{x\to\infty} \dfrac{x^3 - 16x}{5x^4 + x^3 - 5x}$

(*c*) $\lim_{x\to-\infty} \dfrac{x^4 + x}{6x^3 - x^2}$

SOLUTION By the preceding observations,

(*a*) $$\lim_{x\to\infty} \frac{3x^4 + 5x^2}{-x^4 + 10x + 5} = \lim_{x\to\infty} \frac{3x^4}{-x^4} = \lim_{x\to\infty} (-3) = -3.$$

(*b*) $$\lim_{x\to\infty} \frac{x^3 - 16x}{5x^4 + x^3 - 5x} = \lim_{x\to\infty} \frac{x^3}{5x^4} = \lim_{x\to\infty} \frac{1}{5x} = 0.$$

(*c*) $$\lim_{x\to-\infty} \frac{x^4 + x}{6x^3 - x^2} = \lim_{x\to-\infty} \frac{x^4}{6x^3} = \lim_{x\to-\infty} \frac{x}{6} = -\infty. \quad \blacksquare$$

The technique of factoring out a power of x applies more generally than just to polynomials, as the next example illustrates.

EXAMPLE 7 Examine (*a*) $\lim_{x\to\infty} (\sqrt{3x^2 + x}/x)$ and (*b*) $\lim_{x\to-\infty} (\sqrt{3x^2 + x}/x)$.

SOLUTION Before beginning the solution, note that if x is positive, $\sqrt{x^2} = x$, but if x is negative, $\sqrt{x^2} = -x$.

Recall that $\sqrt{a^2} = |a|$.

(*a*) $$\lim_{x\to\infty} \frac{\sqrt{3x^2 + x}}{x} = \lim_{x\to\infty} \frac{\sqrt{x^2(3 + x/x^2)}}{x} = \lim_{x\to\infty} \frac{x\sqrt{3 + 1/x}}{x}$$

$$= \lim_{x\to\infty} \sqrt{3 + 1/x} = \sqrt{3}.$$

(*b*) $$\lim_{x\to-\infty} \frac{\sqrt{3x^2 + x}}{x} = \lim_{x\to-\infty} \frac{\sqrt{x^2(3 + x/x^2)}}{x} = \lim_{x\to-\infty} \frac{-x\sqrt{3 + 1/x}}{x}$$

$$= \lim_{x\to-\infty} -\sqrt{3 + 1/x} = -\sqrt{3}. \quad \blacksquare$$

The final step in Example 7 deserves some comment, for it is a big leap that otherwise may pass unnoticed. It is assumed that

$$\lim_{x\to-\infty} \sqrt{3 + \frac{1}{x}} = \sqrt{3}$$

because

$$\lim_{x\to-\infty} \left(3 + \frac{1}{x}\right) = 3.$$

This conclusion depends on a property of the square root function, which will now be described.

Let $f(x) = \sqrt{x}$ and let $g(x) = 3 + 1/x$. Then in Example 7 it was assumed that

Taking the limit "inside"

$$\lim_{x\to-\infty} f(g(x)) = f\left(\lim_{x\to-\infty} g(x)\right).$$

For the functions f commonly met in calculus this switch of the order of "lim" and "f" is justified. Section 2.5 investigates this type of function f in detail.

EXAMPLE 8 Examine $\lim_{x\to\infty} (\sqrt{x^2 + x} - x)$.

SOLUTION As $x \to \infty$, both $\sqrt{x^2 + x}$ and x approach ∞. It is not immediately clear how their difference $\sqrt{x^2 + x} - x$ behaves. It is necessary to use a little algebra and rationalize the expression:

$$\begin{aligned}\lim_{x\to\infty} (\sqrt{x^2 + x} - x) &= \lim_{x\to\infty} (\sqrt{x^2 + x} - x)\frac{(\sqrt{x^2 + x} + x)}{(\sqrt{x^2 + x} + x)}\\ &= \lim_{x\to\infty} \frac{x^2 + x - x^2}{\sqrt{x^2 + x} + x}\\ &= \lim_{x\to\infty} \frac{x}{\sqrt{x^2(1 + 1/x)} + x}\\ &= \lim_{x\to\infty} \frac{x}{x(\sqrt{1 + 1/x} + 1)}\\ &= \lim_{x\to\infty} \frac{1}{\sqrt{1 + 1/x} + 1} = \frac{1}{2}. \quad \blacksquare\end{aligned}$$

The result in Example 8 may be surprising, but a little calculation would have suggested that the limit is $\frac{1}{2}$. For instance, $\sqrt{100^2 + 100} - 100 \approx 100.499 - 100 = 0.499$.

The next example concerns a case in which $f(x)$ becomes arbitrarily large as x approaches a fixed real number.

EXAMPLE 9 How does $f(x) = 1/x$ behave when x is near 0?

SOLUTION The reciprocal of a small number x has a large absolute value. For instance, when $x = 0.01$, $1/x = 100$; when $x = -0.01$, $1/x = -100$. Thus, as x approaches 0 from the right, $1/x$, which is positive, becomes arbitrarily large. The notation for this is

$$\lim_{x\to0^+} \frac{1}{x} = \infty.$$

As x approaches 0 from the left, $1/x$, which is negative, has arbitrarily large absolute values. The notation for this is

$$\lim_{x\to0^-} \frac{1}{x} = -\infty. \quad \blacksquare$$

The behavior of $1/x$, described in Example 9, is quite different from that of $1/x^2$. Since x^2 is positive whether x is positive or negative, and since $1/x^2$ is large when x is near 0, we have

The notation "$\lim_{x \to 0} \frac{1}{x^2} = \infty$" is useful, though the limit does not exist since ∞ is not a number.

$$\lim_{x \to 0^+} \frac{1}{x^2} = \infty \quad \text{and} \quad \lim_{x \to 0^-} \frac{1}{x^2} = \infty.$$

In this case we may write

$$\lim_{x \to 0} \frac{1}{x^2} = \infty,$$

meaning that "as $x \to 0$, both from the right and from the left, $1/x^2$ becomes arbitrarily large through positive values." We can also write

$$\lim_{x \to 0} \frac{1}{|x|} = \infty,$$

but there is no corresponding statement for $\lim_{x \to 0} 1/x$.

A function may not have a limit at certain numbers even if it does not become arbitrarily large, as the next example illustrates.

EXAMPLE 10 Examine $\lim_{x \to 3} \dfrac{x - 3}{|x - 3|}$.

SOLUTION Let's see how $f(x) = (x - 3)/|x - 3|$ behaves for some choices of x near 3. When $x = 3.01$, we have

$$f(3.01) = \frac{3.01 - 3}{|3.01 - 3|} = \frac{0.01}{|0.01|} = \frac{0.01}{0.01} = 1.$$

When $x = 2.99$, we have

$$f(2.99) = \frac{2.99 - 3}{|2.99 - 3|} = \frac{-0.01}{|-0.01|} = \frac{-0.01}{0.01} = -1.$$

For $x > 3$, $f(x) = 1$; for $x < 3$, $f(x) = -1$. The right-hand limit, $\lim_{x \to 3^+} f(x)$, exists and is 1. The left-hand limit, $\lim_{x \to 3^-} f(x)$, exists and is -1. However, $\lim_{x \to 3} f(x)$ does not exist. The graph of $f(x)$ in Fig. 2.10 displays the misbehavior of $f(x)$ for x near 3.

This example is closely related to Example 3 in Sec. 2.1.

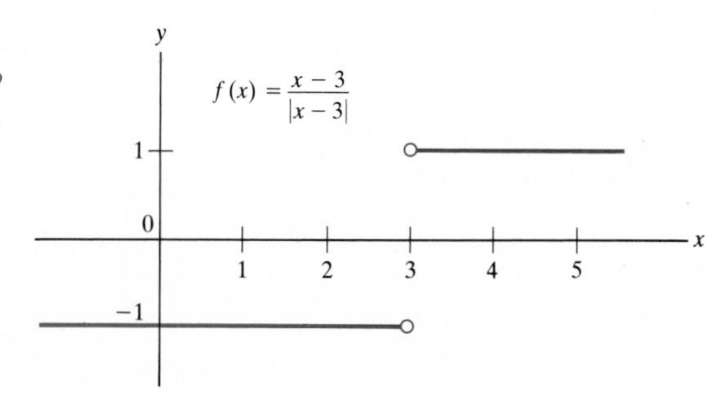

Figure 2.10 ■

The many different types of limits all have the same flavor. Rather than spell each out in detail, we list some typical cases.

Notation	*In Words*	*Concept*	*Example*
$\lim_{x\to a} f(x) = L$	As x approaches a, $f(x)$ approaches L	$f(x)$ is defined in some open intervals (c, a) and (a, b) and, as x approaches a from the right or from the left, $f(x)$ approaches L.	$\lim_{x\to 3} (2x + 1) = 7$
$\lim_{x\to\infty} f(x) = L$	As x approaches (positive) infinity, $f(x)$ approaches L	$f(x)$ is defined for all x beyond some number and, as x gets large through positive values, $f(x)$ approaches L.	$\lim_{x\to\infty} \frac{1}{x} = 0$
$\lim_{x\to-\infty} f(x) = L$	As x approaches negative infinity, $f(x)$ approaches L	$f(x)$ is defined for all x to the left of some number and, as the negative number x takes on large absolute values, $f(x)$ approaches L.	$\lim_{x\to-\infty} \frac{x + 1}{x} = 1$
$\lim_{x\to\infty} f(x) = \infty$	As x approaches infinity, $f(x)$ approaches positive infinity.	$f(x)$ is defined for all x beyond some number and, as x gets large through positive values, $f(x)$ becomes and remains arbitrarily large and positive.	$\lim_{x\to\infty} x^3 = \infty$
$\lim_{x\to a+} f(x) = \infty$	As x approaches a from the right, $f(x)$ approaches (positive) infinity.	$f(x)$ is defined in some open interval (a, b), and, as x approaches a from the right, $f(x)$ becomes and remains arbitrarily large and positive.	$\lim_{x\to 0+} \frac{1}{x} = \infty$
$\lim_{x\to a+} f(x) = -\infty$	As x approaches a from the right, $f(x)$ approaches negative infinity.	$f(x)$ is defined in some open interval (a, b), and, as x approaches a from the right, $f(x)$ becomes negative and $\lvert f(x)\rvert$ becomes and remains arbitrarily large.	$\lim_{x\to 1+} \frac{1}{1 - x} = -\infty$
$\lim_{x\to a} f(x) = \infty$	As x approaches a, $f(x)$ approaches (positive) infinity.	$f(x)$ is defined for some open intervals (c, a) and (a, b), and, as x approaches a from either side, $f(x)$ becomes and remains arbitrarily large and positive.	$\lim_{x\to 0} \frac{1}{x^2} = \infty$

Other notations, such as

$$\lim_{x\to a} f(x) = -\infty, \qquad \lim_{x\to a^-} f(x) = \infty, \qquad \text{and} \qquad \lim_{x\to\infty} f(x) = -\infty$$

are defined similarly.

EXERCISES FOR SEC. 2.2: COMPUTATIONS OF LIMITS

In Exercises 1 to 26 examine the given limits and compute those which exist.

1 $\lim_{x\to\infty} (x^5 - 100x^4)$

2 $\lim_{x\to\infty} (-4x^5 + 35x^2)$

3 $\lim_{x\to-\infty} (6x^5 + 21x^3)$

4 $\lim_{x\to-\infty} (19x^6 + 5x - 300)$

5 $\lim_{x\to-\infty} (-x^3)$

6 $\lim_{x\to-\infty} (-x^4)$

7 $\lim_{x\to\infty} \dfrac{6x^3 - x}{2x^{10} + 5x + 8}$

8 $\lim_{x\to\infty} \dfrac{100x^9 + 22}{x^{10} + 21}$

9 $\lim_{x\to\infty} \dfrac{x^5 + 1066x^2 - 1492x}{2x^4 - 1984}$

10 $\lim_{x\to\infty} \dfrac{6x^3 - x^2 + 5}{3x^3 - 100x + 1}$

11 $\lim_{x\to\infty} \dfrac{x^3 + 1}{x^4 + 2}$

12 $\lim_{x\to-\infty} \dfrac{5x^3 + 2x}{x^{10} + x + 7}$

13 $\lim_{x\to 0^+} \dfrac{1}{x^4}$

14 $\lim_{x\to 0^-} \dfrac{1}{x^4}$

15 $\lim_{x\to 0^+} \dfrac{1}{x^3}$

16 $\lim_{x\to 0^-} \dfrac{1}{x^3}$

17 $\lim_{x\to\infty} (\sqrt{x^2 + 100} - x)$

18 $\lim_{x\to 1} (\sqrt{x^2 + 5} - \sqrt{x^2 + 3})$

19 $\lim_{x\to\infty} (\sqrt{x^2 + 100x} - x)$

20 $\lim_{x\to\infty} (\sqrt{x^2 + 100x} - \sqrt{x^2 + 50x})$

21 $\lim_{x\to\infty} \dfrac{\sqrt{4x^2 + 2x + 1}}{3x}$ **22** $\lim_{x\to-\infty} \dfrac{\sqrt{9x^2 + x + 3}}{6x}$

23 $\lim_{x\to\infty} \dfrac{\sqrt{4x^2 + x}}{\sqrt{9x^2 - 3x}}$ **24** $\lim_{x\to-\infty} \dfrac{\sqrt{x^2 + 3x + 1}}{\sqrt{16x^2 + x + 2}}$

25 (*a*) $\lim_{x\to 1^+} \dfrac{1}{x - 1}$ (*b*) $\lim_{x\to 1^-} \dfrac{1}{x - 1}$

(*c*) $\lim_{x\to 1} \dfrac{1}{x - 1}$

26 (*a*) $\lim_{x\to -1^-} \dfrac{1}{(x + 1)^2}$ (*b*) $\lim_{x\to -1^+} \dfrac{1}{(x + 1)^2}$

(*c*) $\lim_{x\to -1} \dfrac{1}{(x + 1)^2}$

■

27 Two citizens are arguing about

$$\lim_{x\to\infty} \left(\frac{3x^2 + 2x}{x + 5} - 3x\right).$$

The first claims, "For large x, $2x$ is small in comparison to $3x^2$, and 5 is small in comparison to x. So $(3x^2 + 2x)/(x + 5)$ behaves like $3x^2/x = 3x$. Hence the limit in question is 0." Her companion replies, "Nonsense. After all,

$$\frac{3x^2 + 2x}{x + 5} = \frac{3x + 2}{1 + (5/x)},$$

which clearly behaves like $3x + 2$ for large x. Thus the limit in question is 2, not 0."

Settle the argument.

In Exercises 28 to 30 information is given about functions f and g. In each case decide whether the limit asked for can be determined on the basis of that information. If it can, give its value. If it cannot, show by specific choices of f and g that it cannot.

28 Given that $\lim_{x\to\infty} f(x) = 0$ and $\lim_{x\to\infty} g(x) = 1$, discuss

(*a*) $\lim_{x\to\infty} (f(x) + g(x))$ (*b*) $\lim_{x\to\infty} (f(x)/g(x))$

(*c*) $\lim_{x\to\infty} f(x)g(x)$ (*d*) $\lim_{x\to\infty} (g(x)/f(x))$

(*e*) $\lim_{x\to\infty} g(x)/|f(x)|$

29 Given that $\lim_{x\to\infty} f(x) = \infty$ and $\lim_{x\to\infty} g(x) = \infty$, discuss

(*a*) $\lim_{x\to\infty} (f(x) + g(x))$ (*b*) $\lim_{x\to\infty} (f(x) - g(x))$

(*c*) $\lim_{x\to\infty} f(x)g(x)$ (*d*) $\lim_{x\to\infty} (g(x)/f(x))$

30 Given that $\lim_{x\to\infty} f(x) = 1$ and $\lim_{x\to\infty} g(x) = \infty$, discuss

(*a*) $\lim_{x\to\infty} (f(x)/g(x))$ (*b*) $\lim_{x\to\infty} f(x)g(x)$

(*c*) $\lim_{x\to\infty} (f(x) - 1)g(x)$

31 Let $P(x)$ be a polynomial of degree n, with lead term ax^n, $a > 0$, and let $Q(x)$ be a polynomial of degree m, with lead term bx^m, $b > 0$. Examine $\lim_{x\to\infty} P(x)/Q(x)$ if (*a*) $m = n$, (*b*) $m < n$, (*c*) $m > n$.

32 A function f is defined as follows: $f(x) = x$ if x is an integer, and $f(x) = -x$ if x is not an integer. (*a*) Graph f. (*b*) Examine $\lim_{x\to\infty} f(x)$. (*c*) Examine $\lim_{x\to\infty} |f(x)|$.

33 Examine $\lim_{x\to\infty} \cos x$. (*Hint:* Graph $y = \cos x$. See Appendix E if necessary.)

34 A function f is defined as follows: $f(x) = 2$ if x is an integer and $f(x) = 3$ if x is not an integer. (*a*) Graph f. (*b*) Discuss $\lim_{x\to\infty} f(x)$. (*c*) Discuss $\lim_{x\to 2} f(x)$.

35 Examine (*a*) $\lim_{x\to 1} 2^{x/|x|}$ and (*b*) $\lim_{x\to 0} 2^{x/|x|}$.

2.3 ASYMPTOTES AND THEIR USE IN GRAPHING

Horizontal asymptotes

If $\lim_{x\to\infty} f(x) = L$, where L is a real number, the graph of $y = f(x)$ gets arbitrarily close to the horizontal line $y = L$ as x increases. The line $y = L$ is called a **horizontal asymptote** of the graph of f. An asymptote is defined similarly if $f(x) \to L$ as $x \to -\infty$.

Vertical asymptotes

If $\lim_{x\to a^+} f(x) = \infty$ or if $\lim_{x\to a^-} f(x) = \infty$, the graph of $y = f(x)$ resembles the vertical line $x = a$ for x near a. The line $x = a$ is called a **vertical asymptote** of the graph of f. A similar definition holds if $\lim_{x\to a^+} f(x) = -\infty$ or if $\lim_{x\to a^-} f(x) = -\infty$.

Figure 2.11 shows some of these asymptotes.

Horizontal and Vertical Asymptotes in Graphing Some examples of graphing rational functions will show the usefulness of asymptotes.

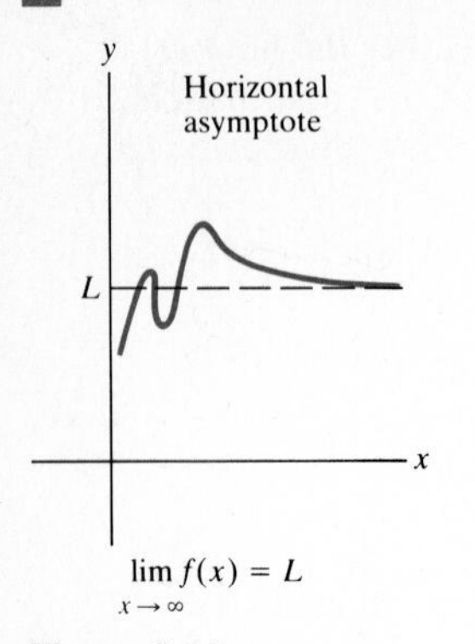

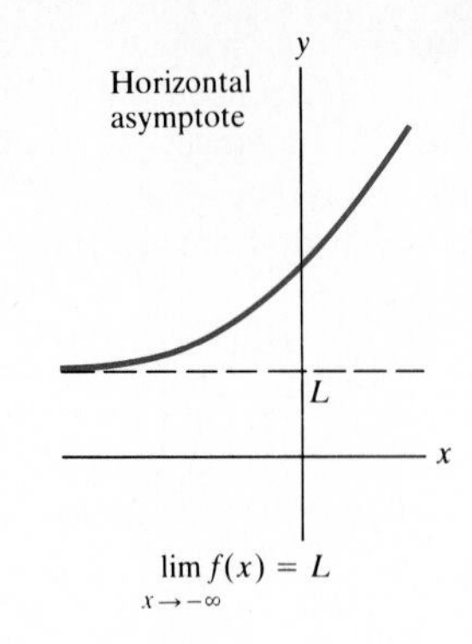

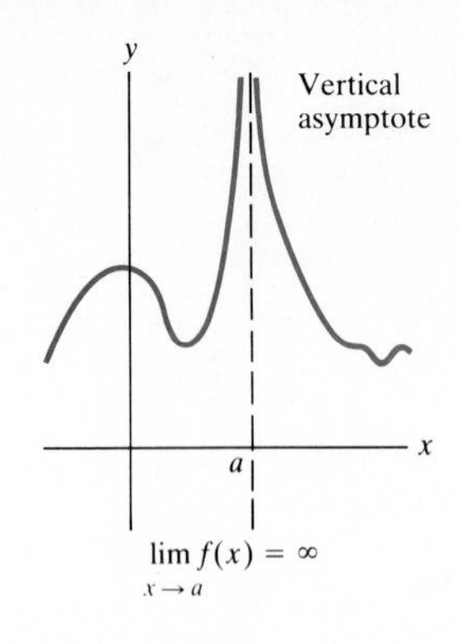

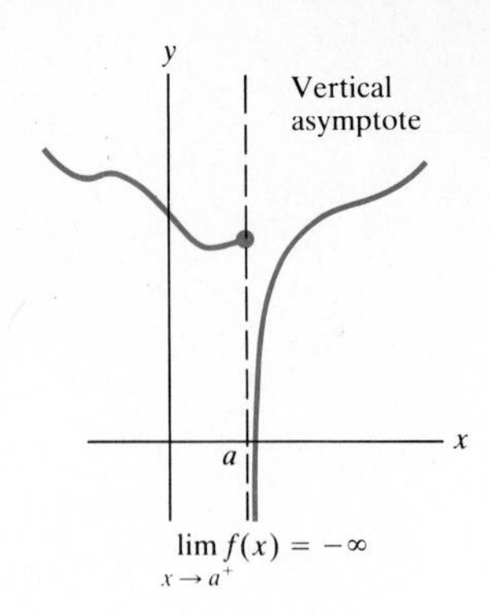

Figure 2.11

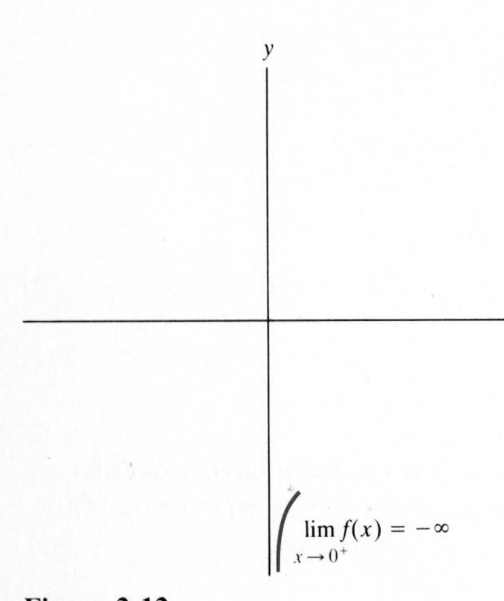

Figure 2.12

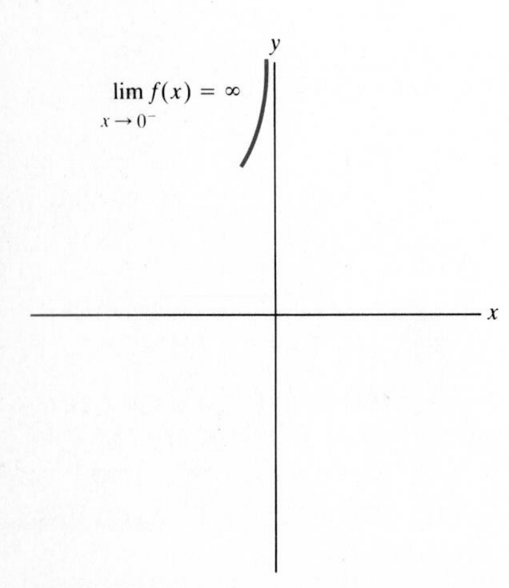

Figure 2.13

EXAMPLE 1 Graph $f(x) = \dfrac{1}{x(x-1)}$.

SOLUTION Note that when $x = 0$ or when $x = 1$ the denominator $x(x-1)$ is 0. Thus 0 and 1 are not in the domain of the function. More important, when x is near 0 or near 1 the quotient $1/[x(x-1)]$ has large absolute value. Thus the lines $x = 0$ and $x = 1$ are vertical asymptotes for the graph. To decide how the graph approaches these asymptotes, it is necessary to examine the sign of $f(x)$ for x near 0 and for x near 1.

Consider x near 0. If x is a very small *positive* number, then

$$f(x) = \frac{1}{x}\frac{1}{x-1}$$

is the product of $1/x$, which is a large positive number, and $1/(x-1)$, which is near $1/(-1) = -1$. So, for x small and positive, $f(x)$ is negative and of large absolute value; that is, $\lim_{x\to 0^+} f(x) = -\infty$. This fact is recorded in Fig. 2.12.

This fact is recorded in Fig. 2.12.

If x is near 0, but negative, then $1/x$ is negative and of large absolute value. Again, $1/(x-1)$ is near -1. Thus

$$\lim_{x\to 0^-} f(x) = \lim_{x\to 0^-} \frac{1}{x}\frac{1}{x-1} = \infty.$$

This fact is recorded in Fig. 2.13.

Next, how does $f(x)$ behave near $x = 1$? Consider first x near 1 but larger than 1. Then

$$f(x) = \frac{1}{x}\frac{1}{x-1}$$

is a large positive number, since $1/x$ is near 1 and $x - 1$ is a small positive number. Thus

$$\lim_{x\to 1^+} f(x) = \infty.$$

Similarly,

$$\lim_{x\to 1^-} f(x) = -\infty.$$

These two facts are recorded in Fig. 2.14. Piecing together these three figures suggests that the graph of f for x in or near the interval $(0, 1)$ looks

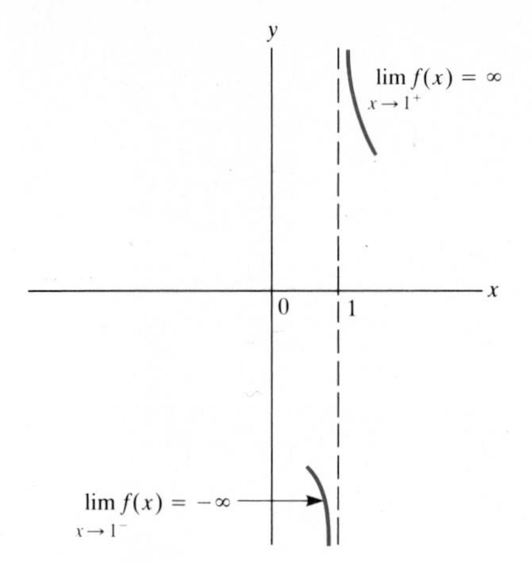

Figure 2.14

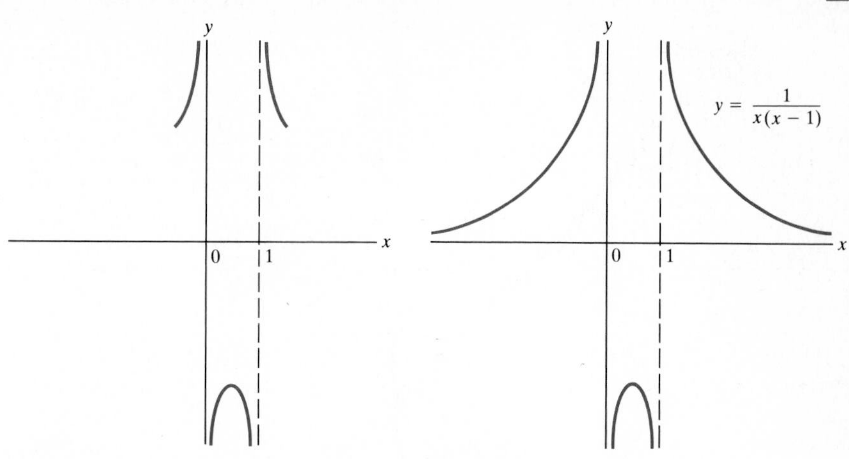

Figure 2.15

Figure 2.16

something like Fig. 2.15. [How high the curve goes for x in (0, 1) is the type of question answered in Chap. 4 with the aid of the derivative. However, Exercise 25 describes a method that works for this particular function.]

How does $f(x)$ behave when $|x|$ is large? Since

$$\lim_{x\to\infty} \frac{1}{x(x-1)} = 0 \qquad \text{and} \qquad \lim_{x\to-\infty} \frac{1}{x(x-1)} = 0,$$

the x axis is a horizontal asymptote (both for x positive and for x negative). In both cases the graph approaches the x axis from above, not from below, since the function is positive when $|x|$ is large. The graph of f must look something like Fig. 2.16. ■

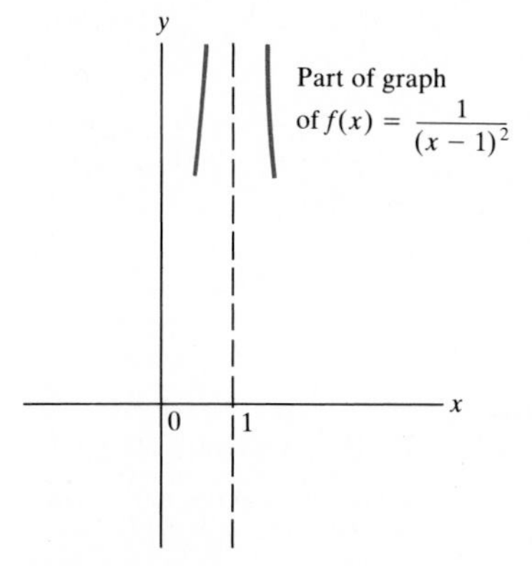

Figure 2.17

EXAMPLE 2 Using asymptotes, graph $f(x) = 1/(x - 1)^2$.

SOLUTION When $x = 1$, the function is undefined. However, when x is near 1, $1/(x - 1)^2$ is a large positive number, since $(x - 1)^2$ is a small positive number. Thus

The square of any nonzero real number is positive.

$$\lim_{x\to1} \frac{1}{(x-1)^2} = \infty.$$

This means that the graph of $f(x) = 1/(x - 1)^2$ approaches the upper part of the vertical asymptote $x = 1$ both from the right and from the left, as shown in Fig. 2.17.

Since

$$\lim_{x\to\infty} \frac{1}{(x-1)^2} = 0 \qquad \text{and} \qquad \lim_{x\to-\infty} \frac{1}{(x-1)^2} = 0,$$

the x axis is a horizontal asymptote.

All this information is incorporated in Fig. 2.18. ■

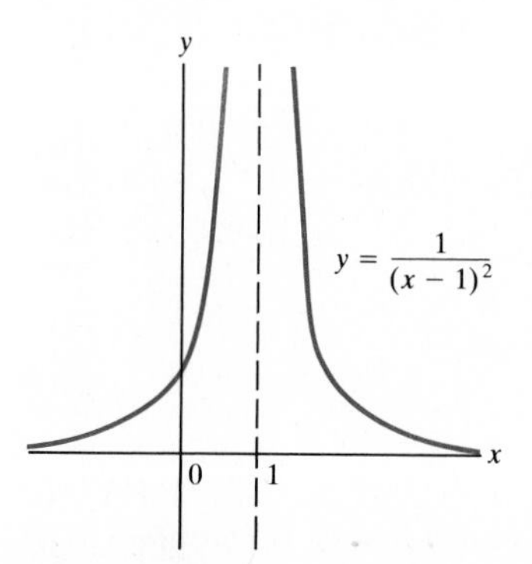

Figure 2.18

Fairly simple functions can have graphs with tilted asymptotes, as Example 3 illustrates.

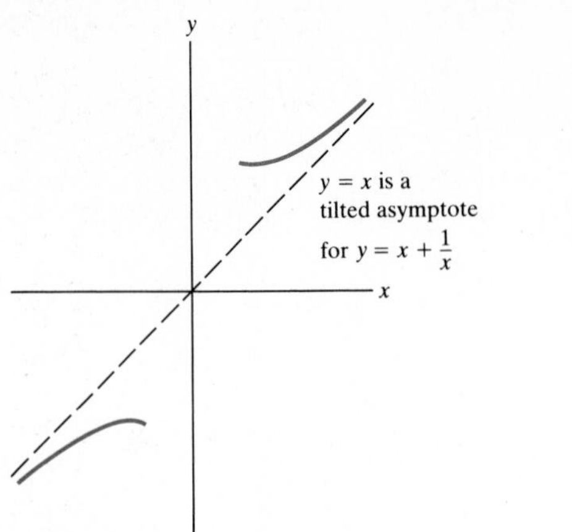

Figure 2.19

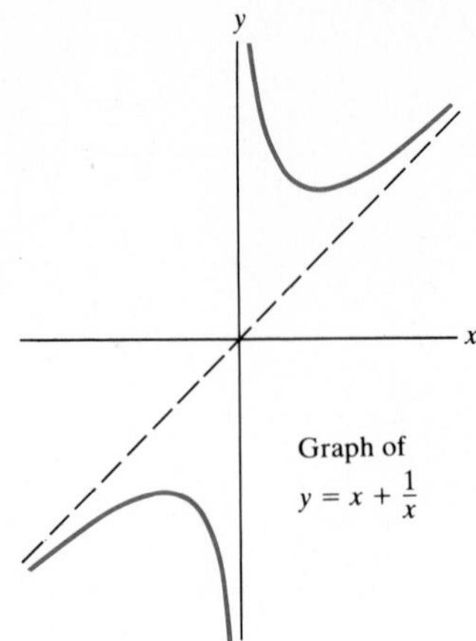

Figure 2.20

EXAMPLE 3 Graph $f(x) = (x^2 + 1)/x$.

SOLUTION After dividing x into $x^2 + 1$, we can write

$$f(x) = x + \frac{1}{x}.$$

A tilted asymptote

When $|x|$ is large, $f(x)$ differs from x by the small quantity $1/x$. So when $|x|$ is large, the graph of f is close to the line $y = x$. When x is negative, $f(x) = x + 1/x$ is smaller than x, since $1/x$ is negative. So for x negative the graph of f lies below the line $y = x$. Similar reasoning shows that for x positive the graph of f lies above the line $y = x$. This information is recorded in Fig. 2.19.

Next search for any vertical asymptotes. Near $x = 0$ the function becomes arbitrarily large. In fact,

$$\lim_{x \to 0^+} \left(x + \frac{1}{x}\right) = \infty \quad \text{and} \quad \lim_{x \to 0^-} \left(x + \frac{1}{x}\right) = -\infty.$$

It is shown in Appendix F.3 that the graph in Fig. 2.20 is a hyperbola.

The y axis is a vertical asymptote. The graph in Fig. 2.20 incorporates the information about the tilted and vertical asymptotes. ■

Example 3 illustrates how to graph $y = A(x)/B(x)$, where $A(x)$ and $B(x)$ are polynomials and the degree of $A(x)$ is at least as large as the degree of $B(x)$. First carry out the long division of $B(x)$ into $A(x)$, obtaining a quotient $Q(x)$ (a polynomial) and a remainder $R(x)$ [a polynomial of degree less than the degree of $B(x)$]. Then

$$A(x) = Q(x)B(x) + R(x)$$

and

$$\frac{A(x)}{B(x)} = Q(x) + \frac{R(x)}{B(x)}.$$

Since $\lim_{x\to\infty} R(x)/B(x) = 0$, the graph of $y = A(x)/B(x)$ is asymptotic to the graph of $y = Q(x)$. If the degree of $Q(x)$ is 1, as in Example 3, the graph of $y = Q(x)$ is a line. If the degree of $Q(x)$ is at least 2, then the graph of $y = Q(x)$ is a curve. Example 4 illustrates this case.

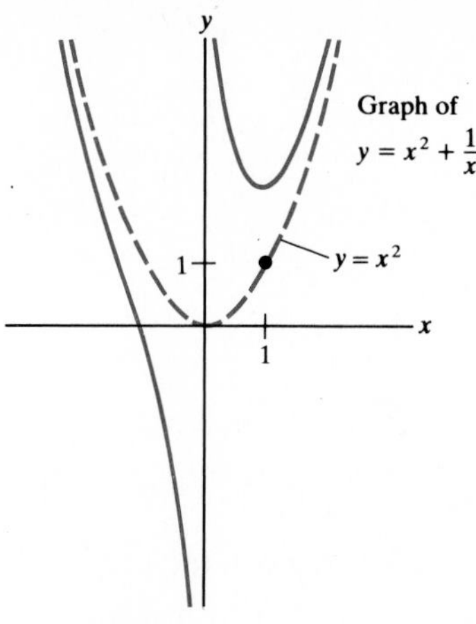

Figure 2.21

EXAMPLE 4 Graph $f(x) = (x^3 + 1)/x$.

SOLUTION First divide x into $x^3 + 1$, obtaining $(x^3 + 1)/x = x^2 + 1/x$. When x is large, $1/x$ is near 0. The graph of $f(x) = x^2 + 1/x$, therefore, resembles the graph of the parabola $y = x^2$ when x is large. Moreover, for positive x, $f(x)$ is larger than x^2, and for negative x, $f(x)$ is smaller than x^2.

At $x = 0$ there is a vertical asymptote, and

$$\lim_{x\to 0^+} f(x) = \infty \qquad \text{and} \qquad \lim_{x\to 0^-} f(x) = -\infty.$$

The graph of f is shown in Fig. 2.21. ■

EXERCISES FOR SEC. 2.3: ASYMPTOTES AND THEIR USE IN GRAPHING

In Exercises 1 to 14 use asymptotes to sketch the graphs of the functions.

1 $f(x) = \dfrac{1}{x - 2}$ **2** $f(x) = \dfrac{1}{x + 3}$

3 $f(x) = \dfrac{1}{(x + 1)^2}$ **4** $f(x) = \dfrac{1}{(x + 1)^3}$

5 $y = \dfrac{1}{x^2 - x}$ (*Hint:* Factor the denominator.)

6 $y = \dfrac{1}{x^3 - x}$ **7** $y = \dfrac{1}{x^4 - x^2}$

8 $y = \dfrac{1}{x^3 + x^2}$

9 $y = \dfrac{x(x - 1)}{x^2 + 1}$ (Also show the x intercepts, that is, where the function has the value 0.)

10 $y = \dfrac{(x - 1)(x - 2)}{x^2(x - 3)}$ (Also show the x intercepts.)

11 $y = \dfrac{x^3 + 2x^2 + x + 4}{x^2}$

12 $y = \dfrac{x^2 - 4}{x + 4}$ (Also show the x intercepts.)

13 $y = \dfrac{x^3}{x^2 + 1}$ **14** $y = \dfrac{x^3}{x^2 - 1}$

15 Graph $y = 2^x + 2^{-x}$. **16** Graph $y = 2^x - 2^{-x}$.

17 Graph $y = \dfrac{x^2}{x^2 + 1}$. **18** Graph $y = x^3 + x^{-1}$.

19 Graph $y = \dfrac{4x^2 + 3x}{2x + 1}$.

20 Graph $y = \dfrac{x^2 + x + 1}{x - 2}$.

21 Graph $y = \dfrac{x^3 + 1}{x^2 + 2}$.

22 Graph $y = \dfrac{x^4 + x^3 - x + 1}{x^2 + x + 2}$.

■

23 Let a, b, and c be constants, $a \neq 0$. Show that $y = (ax^2 + bx + c)/(x + 1)$ has a tilted asymptote and give its equation.

24 Let $P(x)$ be a polynomial of degree m and $Q(x)$ be a polynomial of degree n. For which values of m and n does the graph of $y = P(x)/Q(x)$ have (*a*) a horizontal asymptote? (*b*) a tilted asymptote?

25 In Chap. 4 a general method for finding low and high points on curves will be given. However, elementary algebra can find the high point on the graph of $f(x) = 1/[x(x - 1)]$ in Example 1 for x in (0, 1).

(*a*) Show that $x(x - 1) = (x - \frac{1}{2})^2 - \frac{1}{4}$.

(*b*) Using (*a*), find the minimum value of $x(x - 1)$ for all possible values of x.

(*c*) Using (*b*), find the maximum value of $1/[x(x - 1)]$ for x in (0, 1).

(*d*) What are the coordinates of the highest point P on that part of the graph that lies between $x = 0$ and $x = 1$?

2.4 THE LIMIT OF $(\sin\theta)/\theta$ AS θ APPROACHES 0

The limits evaluated in Secs. 2.1 and 2.2 were found by algebraic means, such as factoring, rationalizing, or canceling. But some of the most important limits in calculus cannot be found so easily. To reinforce the concept of a limit and also to prepare for the calculus of trigonometric functions, we shall determine

$$\lim_{\theta\to 0}\frac{\sin\theta}{\theta}.$$

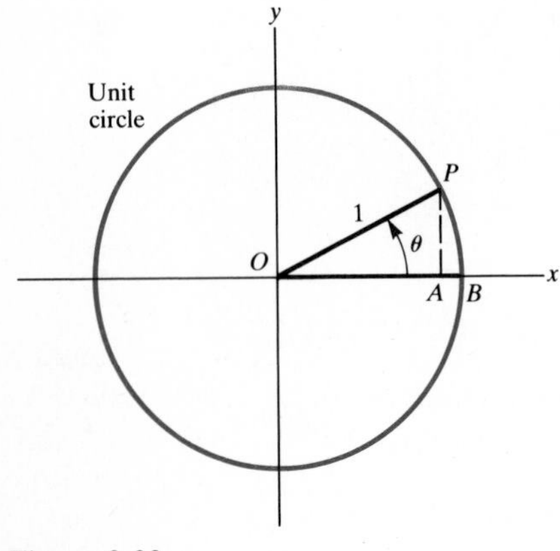

Figure 2.22

Since both the numerator, $\sin\theta$, and the denominator, θ, approach 0, this is a challenging limit.

To begin, a sketch (shown in Fig. 2.22) of a small angle in the unit circle suggests what the limit is. Since an angle is measured in radians, and the circle has radius 1, the length of the arc PB is θ. By the definition of $\sin\theta$, the length of PA is $\sin\theta$. Let $\overline{PA}$ denote the length of the segment PA and let $\overset{\frown}{PB}$ be the length of the arc PB. Thus

$$\frac{\sin\theta}{\theta}=\frac{\overline{PA}}{\overset{\frown}{PB}}.$$

When θ is small, so are $\overline{PA}$ and $\overset{\frown}{PB}$. However, for small θ, PA looks so much like the arc PB that it seems likely that the quotient $\overline{PA}/\overset{\frown}{PB}$ is near 1. This suggests that

$$\lim_{\theta\to 0}\frac{\sin\theta}{\theta}=1.$$

A few computations easily performed on a calculator also suggest that $\lim_{\theta\to 0}(\sin\theta)/\theta = 1$. This table shows $\sin\theta$ and $(\sin\theta)/\theta$ to five significant figures.

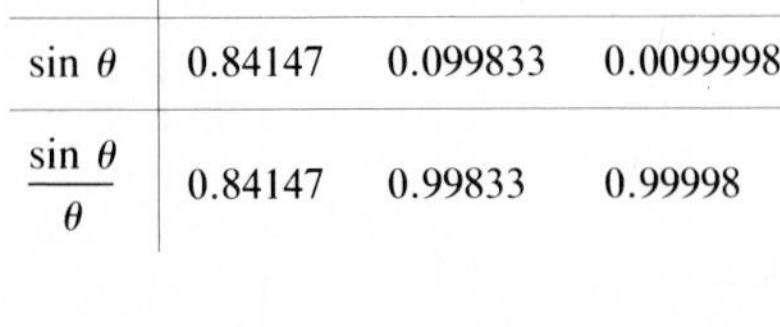

θ	1	0.1	0.01
$\sin\theta$	0.84147	0.099833	0.0099998
$\frac{\sin\theta}{\theta}$	0.84147	0.99833	0.99998

The reader is invited to try $\theta = 0.001$ and 0.0001. Be sure to set the calculator for radians, not degrees.

The geometric argument that $\lim_{\theta\to 0}(\sin\theta)/\theta = 1$ depends on a comparison of three areas. For this reason it will be necessary to develop a formula for the area of a sector of a circle subtended by an angle of θ radians, as in Fig. 2.23. When the angle is 2π, the sector is the entire circle of radius r; hence it has an area of πr^2. Since the area of a sector is proportional to θ, it follows that

$$\frac{\text{Area of sector}}{\pi r^2}=\frac{\theta}{2\pi}.$$

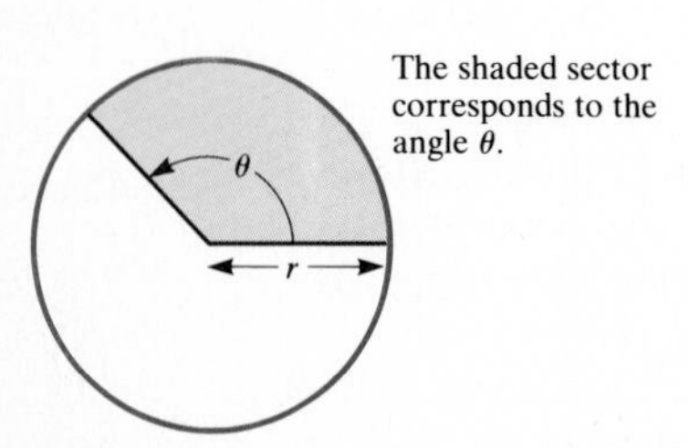

The shaded sector corresponds to the angle θ.

Figure 2.23

From this equation it follows that

Formula for area of a sector

$$\text{Area of sector} = \frac{\theta}{2\pi}\pi r^2 = \frac{\theta r^2}{2}.$$

(This formula will be used in later chapters. It is safer to memorize the proportion that led to it than the formula itself. It is easy to forget the denominator 2 and also that the number π does not appear.)

The next theorem describes the behavior of $(\sin\theta)/\theta$ when θ is near 0.

Theorem 1 Let $\sin\theta$ denote the sine of an angle of θ radians. Then

$$\lim_{\theta\to 0}\frac{\sin\theta}{\theta} = 1.$$

Proof It will be enough to consider only $\theta > 0$, since

$$\frac{\sin(-\theta)}{-\theta} = \frac{-\sin\theta}{-\theta} = \frac{\sin\theta}{\theta}.$$

Moreover, it will be convenient to restrict θ to be less than $\pi/2$.

We shall compare the areas of three regions in Fig. 2.24, two sectors and a triangle. One sector, OAC, has angle θ and radius equal to $\cos\theta$. Sector OBP has angle θ and radius 1. Triangle OAP has base $\cos\theta$ and altitude $\sin\theta$.

As inspection of Fig. 2.24 shows,

$$\text{Area of sector } OAC < \text{Area of triangle } OAP < \text{Area of sector } OBP. \quad (1)$$

Now,

$$\text{Area of sector } OAC = \frac{\theta(\cos\theta)^2}{2},$$

$$\text{Area of triangle } OAP = \tfrac{1}{2}\cdot\text{Base}\cdot\text{Altitude} = \frac{\cos\theta\cdot\sin\theta}{2},$$

and $$\text{Area of sector } OBP = \frac{\theta\cdot 1^2}{2} = \frac{\theta}{2}.$$

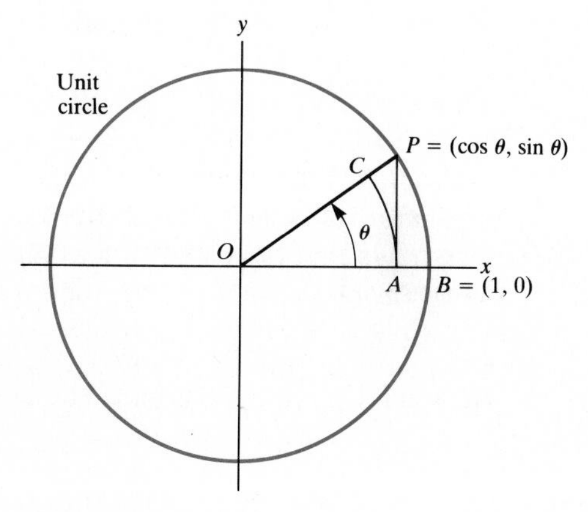

Figure 2.24

So inequalities (1) take the form

$$\frac{\theta \cos^2 \theta}{2} < \frac{\cos \theta \sin \theta}{2} < \frac{\theta}{2}. \tag{2}$$

Multiplying by the positive number 2 and dividing by the positive number $\theta \cos \theta$ yields the inequalities

$$\cos \theta < \frac{\sin \theta}{\theta} < \frac{1}{\cos \theta}. \tag{3}$$

A glance at Fig. 2.24 shows that

$$\lim_{\theta \to 0} \cos \theta = 1.$$

We have

$$\lim_{\theta \to 0} \frac{1}{\cos \theta} = \frac{1}{1} = 1.$$

For a limit that may surprise you, see Exercise 20.

Thus, as $\theta \to 0$, both $\cos \theta$ and $1/\cos \theta$ approach 1. Hence $(\sin \theta)/\theta$, squeezed between $\cos \theta$ and $1/\cos \theta$, must also approach 1. Thus

$$\lim_{\theta \to 0} \frac{\sin \theta}{\theta} = 1,$$

as had been anticipated. ■

In this proof we used the fact that $(\sin \theta)/\theta$, being squeezed between two functions that approach 1, must also approach 1. For later use we state this observation as a general principle:

The Squeeze Principle If $g(x) \leq f(x) \leq h(x)$ and

$$\lim_{x \to a} g(x) = L = \lim_{x \to a} h(x),$$

then

$$\lim_{x \to a} f(x) = L.$$

We shall also need the limit

$$\lim_{\theta \to 0} \frac{1 - \cos \theta}{\theta}$$

in the next chapter. It is not obvious what this limit is, if indeed it exists. As $\theta \to 0$, the numerator $1 - \cos \theta$ approaches 0; so does the denominator. The numerator influences the quotient to become small, while the denominator influences the quotient to become large. The following theorem shows that the numerator is the stronger influence, causing the quotient to approach 0 as θ approaches 0.

Theorem 2 Let $\cos \theta$ denote the cosine of an angle of θ radians. Then

$$\lim_{\theta \to 0} \frac{1 - \cos \theta}{\theta} = 0.$$

Proof

$$\frac{1-\cos\theta}{\theta} = \frac{1-\cos\theta}{\theta}\,\frac{1+\cos\theta}{1+\cos\theta} \qquad (\theta \neq 0,\ \cos\theta \neq -1)$$

$$= \frac{1-\cos^2\theta}{\theta(1+\cos\theta)} = \frac{\sin^2\theta}{\theta(1+\cos\theta)}$$

$$= \frac{\sin\theta}{\theta}\,\frac{\sin\theta}{1+\cos\theta}.$$

Thus

$$\lim_{\theta\to 0}\frac{1-\cos\theta}{\theta} = \lim_{\theta\to 0}\left(\frac{\sin\theta}{\theta}\,\frac{\sin\theta}{1+\cos\theta}\right)$$

$$= 1\,\frac{0}{1+1} = 0.$$

Consequently,

$$\lim_{\theta\to 0}\frac{1-\cos\theta}{\theta} = 0.$$

This implies that when θ is small, $1 - \cos\theta$ is much smaller than θ. ■

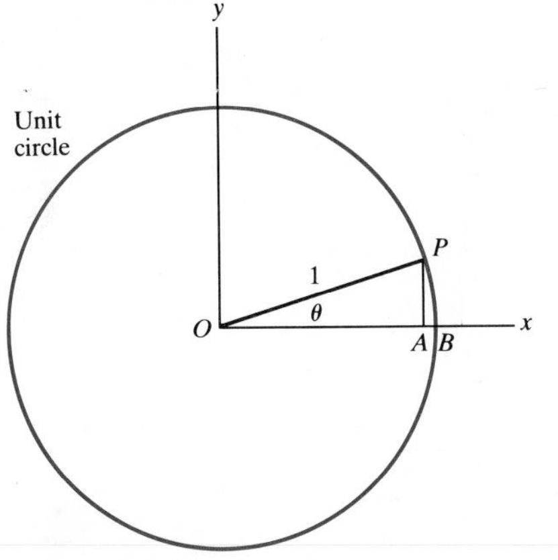

Figure 2.25

A sketch of the unit circle will make Theorem 2 plausible. Figure 2.25 shows a small angle θ in the unit circle. The arc length along the circle from P to B is θ. Since $\overline{OA} = \cos\theta$, $\overline{AB} = 1 - \cos\theta$. It appears from Fig. 2.25 that the ratio $\overline{AB}/\widehat{PB}$ is small when θ is small. This suggests that $\lim_{\theta\to 0}(1 - \cos\theta)/\theta = 0$, as Theorem 2 asserts.

Some other limits involving sine and cosine can be found with the aid of $\lim_{\theta\to 0}(\sin\theta)/\theta$ and $\lim_{\theta\to 0}(1 - \cos\theta)/\theta$, as shown in the following examples.

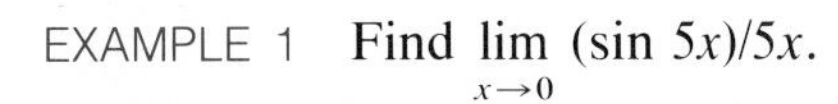

EXAMPLE 1 Find $\lim_{x\to 0}(\sin 5x)/5x$.

SOLUTION Observe that as $x \to 0$, $5x \to 0$. Let $\theta = 5x$. Thus

$$\lim_{x\to 0}\frac{\sin 5x}{5x} = \lim_{\theta\to 0}\frac{\sin\theta}{\theta} = 1. \quad ■$$

EXAMPLE 2 Find $\lim_{x\to 0}(\sin 5x)/2x$.

SOLUTION A little algebra permits one to exploit the result found in Example 1:

$$\lim_{x\to 0}\frac{\sin 5x}{2x} = \lim_{x\to 0}\frac{\sin 5x}{5x}\cdot\frac{5x}{2x}$$

$$= \lim_{x\to 0}\frac{\sin 5x}{5x}\cdot\frac{5}{2}$$

$$= 1\cdot\frac{5}{2} = \frac{5}{2}. \quad ■$$

EXAMPLE 3 Find $\lim_{x\to 0}(\sin 3x)/(\sin 2x)$.

SOLUTION First rewrite the quotient as follows:

$$\frac{\sin 3x}{\sin 2x} = \frac{\sin 3x}{3x} \cdot 3x \cdot \frac{2x}{\sin 2x} \cdot \frac{1}{2x}$$

$$= \frac{3}{2} \cdot \frac{\sin 3x}{3x} \cdot \frac{2x}{\sin 2x}.$$

Now, $\lim_{\theta \to 0} (\sin \theta)/\theta = 1$ and $\lim_{\theta \to 0} \theta/\sin \theta = 1$. Thus

$$\lim_{x \to 0} \frac{3}{2} \cdot \frac{\sin 3x}{3x} \cdot \frac{2x}{\sin 2x} = \frac{3}{2} \cdot \lim_{x \to 0} \frac{\sin 3x}{3x} \cdot \lim_{x \to 0} \frac{2x}{\sin 2x}$$

$$= \frac{3}{2} \cdot 1 \cdot 1 = \frac{3}{2}.$$

Consequently,

$$\lim_{x \to 0} \frac{\sin 3x}{\sin 2x} = \frac{3}{2}. \quad \blacksquare$$

From a practical point of view this section showed that if angles are measured in radians, then the sine of a small angle is "roughly" the angle itself; that is

$$\sin x \approx x.$$

A useful fact: for small x, $\sin x \approx x$, $\tan x \approx x$.

This is another way of saying that when x is small, the quotient $(\sin x)/x$ is close to 1. In engineering and physics $\sin x$ is often replaced by x when x is small. Moreover, $\tan x$ may also be replaced by x for small x. That this is a reasonable estimate is justified by the fact that

$$\lim_{x \to 0} \frac{\tan x}{x} = \lim_{x \to 0} \frac{\frac{\sin x}{\cos x}}{x}$$

$$= \left(\lim_{x \to 0} \frac{\sin x}{x}\right)\left(\lim_{x \to 0} \frac{1}{\cos x}\right) = 1 \cdot 1 = 1.$$

So $\tan x \approx x$ for small x. Similarly, $\tan x \approx \sin x$ for small x.

EXERCISES FOR SEC. 2.4: THE LIMIT OF $(\sin \theta)/\theta$ AS θ APPROACHES 0

1 What is the area of a sector of a circle of (*a*) radius 3 and angle $\pi/2$? (*b*) radius 1 and angle θ? (*c*) radius 2 and angle θ?

2 What is the area of the sector of a circle of radius 6 inches subtended by an angle of (*a*) $\pi/4$ radians? (*b*) 3 radians? (*c*) 45°?

In Exercises 3 to 16 examine the limits.

3 $\lim_{x \to 0} \frac{\sin x}{2x}$

4 $\lim_{x \to 0} \frac{\sin 2x}{x}$

5 $\lim_{x \to 0} \frac{\sin 3x}{5x}$

6 $\lim_{x \to 0} \frac{2x}{\sin 3x}$

7 $\lim_{\theta \to 0} \frac{\sin^2 \theta}{\theta}$

8 $\lim_{h \to 0} \frac{\sin h^2}{h^2}$

9 $\lim_{\theta \to 0} \frac{\tan^2 \theta}{\theta}$

10 $\lim_{\theta \to 0} \theta \cot \theta$

11 $\lim_{\theta \to 0} \frac{1 - \cos \theta}{\theta^2}$

12 $\lim_{\theta \to 0^-} \frac{1 - \cos \theta}{\theta^3}$

13 $\lim_{\theta \to 0^+} \frac{1 - \cos \theta}{\theta^3}$

14 $\lim_{x \to 0} \frac{\sin^2 x}{x^2}$

15 $\lim_{\theta \to 0^+} \frac{1}{\sin \theta}$

16 $\lim_{\theta \to 0^-} \frac{1}{\sin \theta}$

■

17 (*a*) What is the domain of the function $f(x) = (\sin x)/x$?
(*b*) Show that f is an even function, that is, $f(-x) = f(x)$.
(*c*) Find $\lim_{x\to\infty} f(x)$. *Hint:* For all x, $|\sin x| \le 1$.
(*d*) For which x is $f(x) = 0$?
(*e*) Fill in the following table to two decimal places using a calculator.

x	0.1	$\pi/2$	$3\pi/2$	2π	$5\pi/2$	3π	$7\pi/2$
$\sin x$							
$\dfrac{\sin x}{x}$							

(*f*) Graph f for $x > 0$.
(*g*) Graph f for $x < 0$.
(*h*) What is $\lim_{x\to 0} f(x)$?

18 (*a*) What is the domain of $g(x) = (1 - \cos x)/x$?
(*b*) Show that g is an odd function, that is, $g(-x) = -g(x)$.
(*c*) Find $\lim_{x\to\infty} g(x)$. *Hint:* $0 \le 1 - \cos x \le 2$ for all x.
(*d*) For which x is $g(x) = 0$?
(*e*) Fill in the following table to two decimal places, using a calculator.

x	0.1	$\pi/2$	$3\pi/2$	2π	3π
$1 - \cos x$					
$g(x) = \dfrac{1 - \cos x}{x}$					

(*f*) Graph g for $x > 0$.
(*g*) Graph g for $x < 0$.
(*h*) What is $\lim_{x\to 0} g(x)$?

19 Examine $\lim_{x\to\pi/2} \dfrac{1 - \sin x}{x - (\pi/2)}$. $\left(\textit{Hint: } \text{Let } \theta = x - \dfrac{\pi}{2}.\right)$

20 (A test of intuition) An intuitive argument suggested that $\lim_{\theta\to 0} (\sin \theta)/\theta = 1$, which turned out to be correct. Try your intuition on another limit associated with the unit circle shown in Fig. 2.26.
(*a*) What do you think happens to the quotient

$$\frac{\text{Area of triangle } ABC}{\text{Area of shaded region}}$$

as $\theta \to 0$? More precisely what does your intuition suggest is the limit of that quotient as $\theta \to 0$?
(*b*) Estimate the limit in (*a*) using $\theta = 0.01$. *Note:* The limit is determined in Chap. 6. This question arose during some research in geometry. I guessed wrong, as has everyone I have asked.

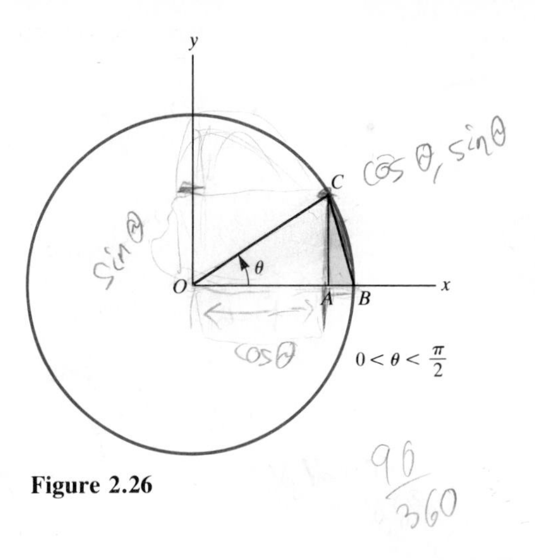

Figure 2.26

Exercises 21 to 25 concern the function $f(x) = \sin(1/x)$ and its graph. Use the same scale on both axes, with the distance from 0 to 1 at least 10 centimeters.

21 (*a*) Show that $f(1/(n\pi)) = 0$ for any nonzero integer n.
(*b*) Plot the points on the graph of f with x coordinate $1/\pi$, $1/(2\pi)$, $1/(3\pi)$, $1/(4\pi)$, and $1/(5\pi)$.

22 (*a*) Show that $f(1/[(2n + \frac{3}{2})\pi]) = -1$ for any integer n.
(*b*) Plot the points on the graph of f with $x = 1/(\frac{3}{2}\pi)$, $1/(\frac{7}{2}\pi)$, and $1/(\frac{11}{2}\pi)$.

23 (*a*) Show that, for any integer n, $f(1/[(2n + \frac{1}{2})\pi]) = 1$.
(*b*) Plot the points on the graph of f with $x = 1/(\pi/2)$, $1/(5\pi/2)$, and $1/(9\pi/2)$.

24 What is $\lim_{x\to\infty} \sin(1/x)$?

25 (*a*) With the aid of Exercises 21 to 24, graph f for $x > 0$.
(*b*) Does $\lim_{x\to 0^+} f(x)$ exist?

26 This exercise concerns the graph of $f(x) = x \sin x$.
(*a*) For which x is $f(x) = 0$?
(*b*) For which x is $f(x) = x$?
(*c*) For which x is $f(x) = -x$?
(*d*) For $x \ge 0$, plot the points given in (*a*), (*b*), and (*c*). (There are an infinite number of them; just sketch a few.)
(*e*) With the aid of (*d*) graph f for $x \ge 0$.
(*f*) Does $\lim_{x\to 0^+} f(x)$ exist?
(*g*) Does $\lim_{x\to\infty} f(x)$ exist?

27 (Calculator) Examine the behavior of $(\theta - \sin\theta)/\theta^3$ for θ near 0.

28 (Calculator) Examine the behavior of the quotient $(\cos\theta - 1 + \theta^2/2)/\theta^4$ for θ near 0.

29 (See Exercise 11.)
(*a*) Show that for small θ, $\cos\theta \approx 1 - \theta^2/2$.
(*b*) Use (*a*) to estimate cos 0.1 and cos 0.01.
(*c*) Compare the values in (*b*) to the values for $\cos\theta$ found by a calculator.

2.5 CONTINUOUS FUNCTIONS

Imagine that you had the information shown in the table about some function f. What would you expect the output $f(1)$ to be?

x	0.9	0.99	0.999
$f(x)$	2.93	2.9954	2.9999997

It would be quite a shock to be told that $f(1)$ is, say, 625. A reasonable function should present no such surprise. The expectation is that $f(1) = 3$. More generally, we expect the output of a function at the input a to be closely connected with the outputs of the function at inputs that are near a. The functions of interest in calculus usually behave in the expected way; they offer no spectacular gaps or jumps. The graphs of these functions consist of curves or lines, not wildly scattered points. The technical term for these functions is "continuous," which will be defined in this section.

The following three definitions express in terms of limits our expectation that $f(a)$ is determined by the values of $f(x)$ for x near a.

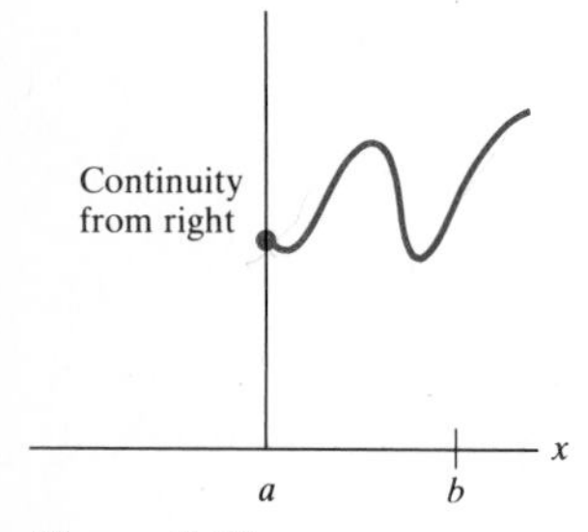

Figure 2.27

Definition *Continuity from the right at a number a.* Assume that $f(x)$ is defined at a and in some open interval (a, b). Then the function f is **continuous at a from the right** if $\lim_{x \to a^+} f(x) = f(a)$. This means that

1 $\lim_{x \to a^+} f(x)$ exists and

2 that limit is $f(a)$.

Figure 2.27 illustrates right-sided continuity at a.

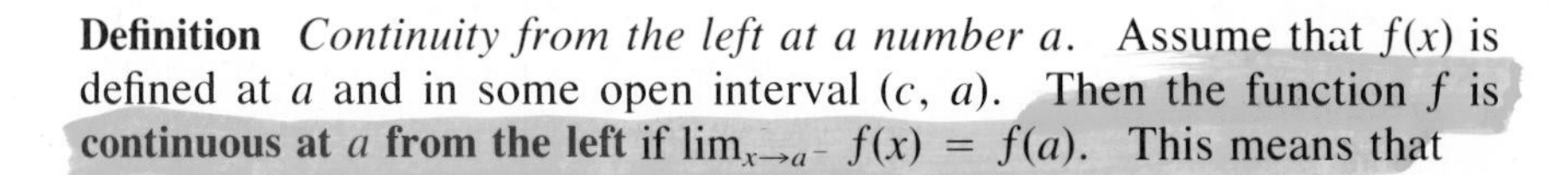

Definition *Continuity from the left at a number a.* Assume that $f(x)$ is defined at a and in some open interval (c, a). Then the function f is **continuous at a from the left** if $\lim_{x \to a^-} f(x) = f(a)$. This means that

1 $\lim_{x \to a^-} f(x)$ exists and

2 that limit is $f(a)$.

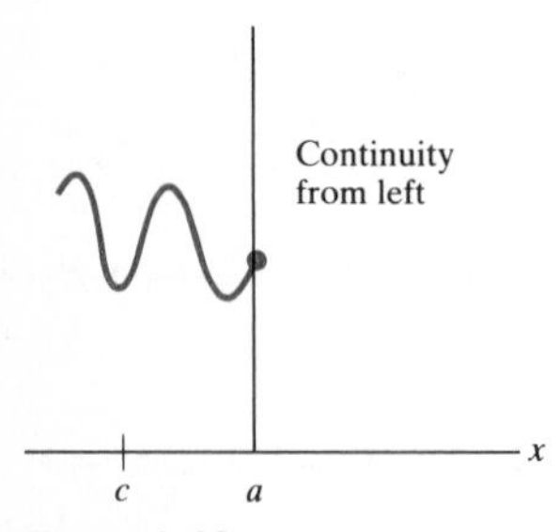

Figure 2.28

Figure 2.28 illustrates this type of continuity.

The next definition applies if the function is defined in some open interval that includes the number a. It essentially combines the first two definitions.

Definition *Continuity at a number a.* Assume that $f(x)$ is defined in some open interval (b, c) that contains the number a. Then the function f is **continuous at a** if $\lim_{x \to a} f(x) = f(a)$. This means that

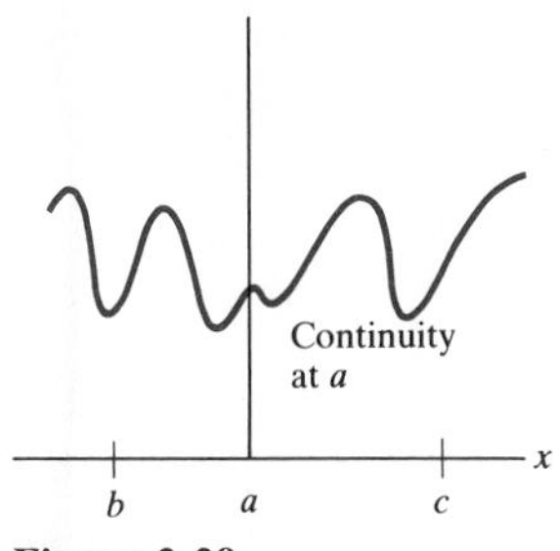

Figure 2.29

1. $\lim_{x \to a} f(x)$ exists and
2. that limit is $f(a)$.

This third definition amounts to asking that the function be continuous both from the right and from the left at a. It is illustrated by Fig. 2.29.

Some examples will show how the definitions are applied and why all three definitions have to be made.

EXAMPLE 1 Let $f(x)$ be the largest integer that is less than or equal to x, denoted $[x]$. Graph f and determine at which numbers it is continuous.

SOLUTION Begin with a table:

The function $[x]$ is also denoted $\lfloor x \rfloor$, and called the "floor" of x.

x	0	0.5	0.9	1.0	1.5	1.9
$f(x) = [x]$	0	0	0	1	1	1

An application of the greatest integer function: How old are you?

As the table suggests,

$$f(x) = 0 \quad \text{for} \quad 0 \le x < 1 \qquad \text{and} \qquad f(x) = 1 \quad \text{for} \quad 1 \le x < 2.$$

The graph of f, which consists of horizontal line segments, is shown in Fig. 2.30.

Is f continuous at $a = 1$? Inspection of the graph shows that

$$\lim_{x \to 1^+} f(x) = 1 \qquad \text{and} \qquad \lim_{x \to 1^-} f(x) = 0.$$

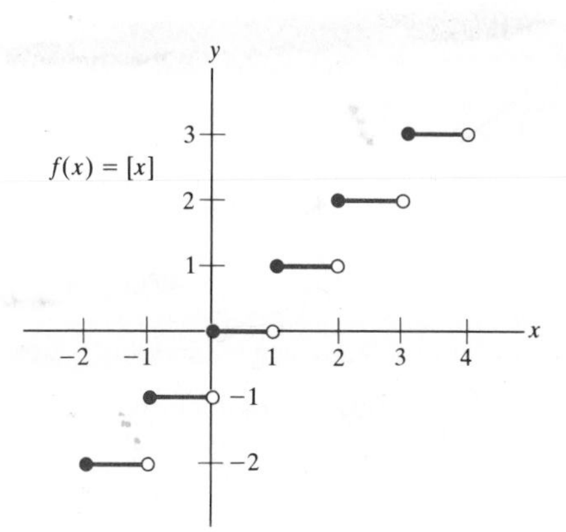

Figure 2.30

Since these limits are not equal, $\lim_{x \to 1} f(x)$ does not exist. No matter what $f(1)$ is, there is no chance for the function to be continuous at $a = 1$.

Similar reasoning shows that f is not continuous at any integer a.

Is f continuous at $a = \frac{1}{2}$? Inspection of the graph shows that as x approaches $\frac{1}{2}$, $f(x)$ approaches 0. That is,

$$\lim_{x \to 1/2} f(x) = 0.$$

Moreover,

$$f(\tfrac{1}{2}) = 0.$$

Since $\lim_{x \to 1/2} f(x)$ exists and equals $f(\frac{1}{2})$, f is continuous at $a = \frac{1}{2}$.

Similar reasoning shows that f is continuous at any number a that is not an integer.

Incidentally, at $a = 1$ (or at any integer) f is continuous from the right but not continuous from the left. ■

EXAMPLE 2 Let $f(x) = x^2$ for all x. Show that f is continuous at $a = 3$.

SOLUTION As $x \to 3$, $f(x) = x^2$ approaches 9; that is,

$$\lim_{x \to 3} x^2 = 9.$$

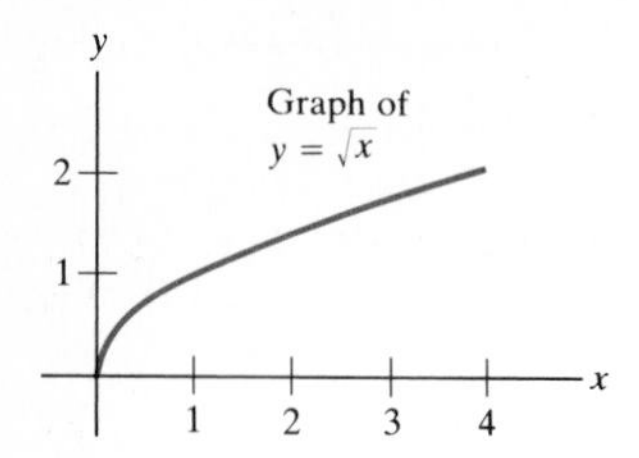

Figure 2.31

Next, compute $f(3)$, which is 3^2, or 9. Since $\lim_{x \to 3} f(x)$ exists and equals $f(3)$, f is continuous at 3. (In fact, f is continuous at each real number.) ■

EXAMPLE 3 Let $f(x) = \sqrt{x}$ for $x \geq 0$. Show that f is continuous from the right at $a = 0$.

SOLUTION As the graph of $f(x) = \sqrt{x}$ in Fig. 2.31 reminds us, the domain of f does not contain an open interval around 0. It is meaningful to speak of "continuity from the right" at 0 but not of "continuity from the left."

Since $\sqrt{x}$ approaches 0 as x approaches 0, $\lim_{x \to 0^+} f(x) = 0$. Is this limit the same as $f(0)$? Since $f(0) = \sqrt{0} = 0$, the answer is "yes." In short, f is continuous from the right at 0. ■

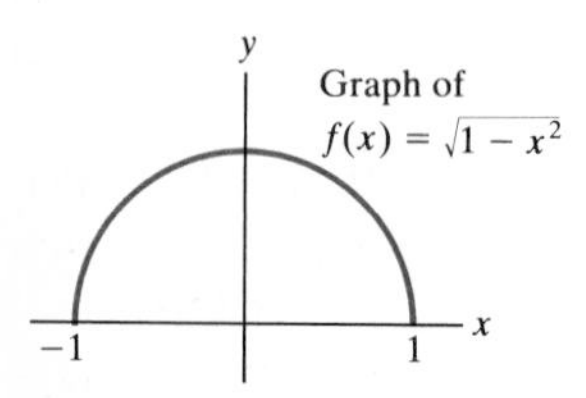

Figure 2.32

All three forms of continuity must be called upon in the case of $f(x) = \sqrt{1 - x^2}$, whose domain is $[-1, 1]$ and whose graph is a semicircle, as shown in Fig. 2.32. At $a = -1$, $f(x) = \sqrt{1 - x^2}$ is continuous from the right. At $a = 1$ it is continuous from the left. At any number a in the interval $(-1, 1)$ it is continuous.

We will want to call the functions x^2, $\sqrt{x}$, and $\sqrt{1 - x^2}$ "continuous." The greatest integer function $[x]$ will not be continuous. The following definitions define the notion of "continuous function"; they depend on the type of domain of the function.

Definition *Continuous function.* Let f be a function whose domain is the x axis or is made up of open intervals. Then f is a **continuous function** if it is continuous at each number a in its domain.

Thus x^2 is a continuous function. So is $1/x$, whose domain consists of the intervals $(-\infty, 0)$ and $(0, \infty)$. Although this function explodes at 0, this does not prevent it from being a continuous function. *The key to being continuous is that the function is continuous at each number in its domain.* The number 0 is not in the domain of $1/x$.

Continuous functions on closed intervals

Only a slight modification of the definition is necessary to cover functions whose domains involve closed intervals. We will say that a function whose domain is the closed interval $[a, b]$ is *continuous* if it is continuous at each point in the open interval (a, b), continuous from the right at a, and continuous from the left at b. Thus $\sqrt{1 - x^2}$ is continuous on the interval $[-1, 1]$.

In a similar spirit, we say that a function with domain $[a, \infty)$ is continuous if it is continuous at each point in (a, ∞) and continuous from the right at a. Thus $\sqrt{x}$ is a continuous function. A similar definition covers functions whose domains are of the form $(-\infty, b]$.

Examples of continuous functions

Many of the functions met in algebra and trigonometry are continuous. For instance, 2^x, $\sqrt{x}$, $\sin x$, $\tan x$, and any polynomial are continuous. So is any rational function (the quotient of two polynomials). (This is proved in Appendix G.) Moreover, algebraic combinations of continuous functions are continuous. For example, since x^3 and $\sin x$ are continuous, so are $x^3 + \sin x$, $x^3 - \sin x$, and $x^3 \sin x$. The function $x^3/\sin x$, which is not defined when $\sin x = 0$, is continuous on its domain. The following definitions are needed to make these statements general.

Definition *Sum, difference, product, and quotient of functions.* Let f and g be two functions. The functions, $f + g$, $f - g$, fg, and f/g are defined as follows.

$$(f + g)(x) = f(x) + g(x) \quad \text{for } x \text{ in the domains of both } f \text{ and } g.$$

$$(f - g)(x) = f(x) - g(x) \quad \text{for } x \text{ in the domains of both } f \text{ and } g.$$

$$(fg)(x) = f(x)g(x) \quad \text{for } x \text{ in the domains of both } f \text{ and } g.$$

$$\left(\frac{f}{g}\right)(x) = \frac{f(x)}{g(x)} \quad \text{for } x \text{ in the domains of both } f \text{ and } g,\ g(x) \neq 0.$$

If f and g are defined at least in an open interval that includes the number a and if f and g are continuous at a, then so are $f + g$, $f - g$, and fg. Moreover, if $g(a) \neq 0$, f/g is also continuous at a. (Proofs of these statements are to be found in Appendix G.)

A function obtained by the composition of continuous functions is also continuous. That is, if the function g is continuous at a and the function f is continuous at $g(a)$, then the composition, $f \circ g$, is continuous at a. For instance, the function $\sqrt[3]{1 + x^2}$ is continuous since both the polynomial $1 + x^2$ and the cube root function are continuous.

A useful property of a continuous function

Note carefully the role of continuity in the following example.

EXAMPLE 4 Find $\lim_{x\to 0} \sqrt[3]{\dfrac{\sin x}{x}}$.

SOLUTION As $x \to 0$, $(\sin x)/x \to 1$. Moreover the cube root function is continuous. Therefore,

The limit can go inside.

$$\lim_{x\to 0} \sqrt[3]{\frac{\sin x}{x}} = \sqrt[3]{\lim_{x\to 0} \frac{\sin x}{x}} = \sqrt[3]{1} = 1. \quad ■$$

Example 4 generalizes, as follows. Let f be a continuous function. If g is some other function for which $\lim_{x\to a} g(x)$ exists and is in the domain of f and $g(x)$ is in the domain of f for x near a, then

For continuous f, "f" and "lim" can be switched.

$$\lim_{x\to a} f(g(x)) = f\left(\lim_{x\to a} g(x)\right).$$

In Example 4, $f(x) = \sqrt[3]{x}$ and $g(x) = (\sin x)/x$. See also the discussion after Example 7 in Sec. 2.2.

The next example shows that the graph of a function can have a sharp corner even though the function is continuous.

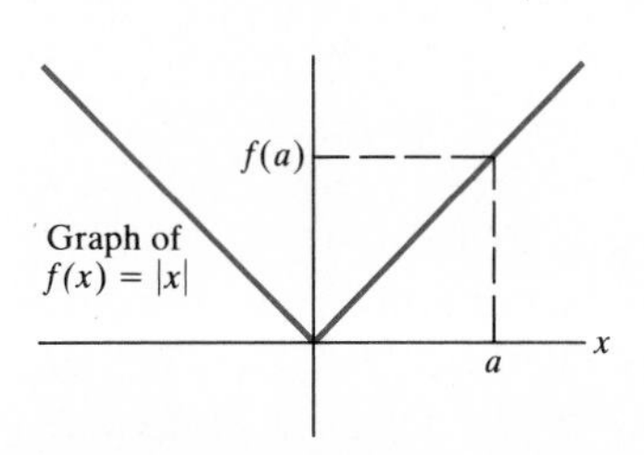

Figure 2.33

EXAMPLE 5 Determine at which numbers a the absolute-value function $f(x) = |x|$ is continuous.

SOLUTION The graph of $f(x) = |x|$ is shown in Fig. 2.33. For each number a, inspection of the graph shows that $\lim_{x\to a} f(x)$ exists and equals $f(a)$. So f is continuous at each number a.

Since f is continuous at each number a, it is a continuous function. ■

The next example explores continuity and limits in terms of the graph of a function.

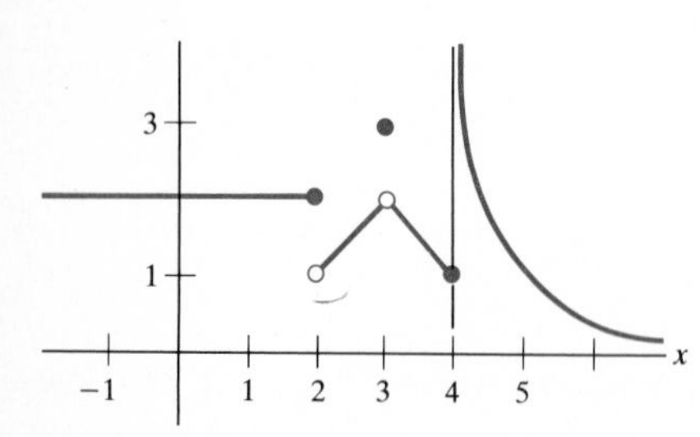

Figure 2.34

EXAMPLE 6 Figure 2.34 is the graph of a certain function whose domain is the x axis. (Obviously, the entire graph cannot be shown.)
(*a*) At which a does $\lim_{x\to a} f(x)$ not exist?
(*b*) At which a is f not continuous?

SOLUTION
(*a*) Consider $a = 2$. We have

$$\lim_{x\to 2^-} f(x) = 2 \qquad \text{and} \qquad \lim_{x\to 2^+} f(x) = 1.$$

Since these limits are not equal, $\lim_{x\to 2} f(x)$ does not exist.

Consider $a = 4$. Since $\lim_{x\to 4^+} f(x) = \infty$, $\lim_{x\to 4^+} f(x)$ does not exist. Hence $\lim_{x\to 4} f(x)$ does not exist.

At $a = 3$, $\lim_{x\to 3^-} f(x) = 2$ and $\lim_{x\to 3^+} f(x) = 2$. Therefore $\lim_{x\to 3} f(x)$ exists and equals 2. The fact that $f(3) = 3$ has no influence on whether $\lim_{x\to 3} f(x)$ exists.

At all a other than 2 and 4, $\lim_{x\to a} f(x)$ exists.

(*b*) Since $\lim_{x\to 2} f(x)$ and $\lim_{x\to 4} f(x)$ do not exist, f cannot be continuous at 2 and 4. Furthermore, since $\lim_{x\to 3} f(x)$ does not equal $f(3)$, f is not continuous at 3. However, f is continuous at all numbers a other than 2, 3, and 4. ■

As Example 6 shows, a function whose domain is the x axis can fail to be continuous at a given number a for either of two reasons: First, $\lim_{x\to a} f(x)$ might not exist. Second, when $\lim_{x\to a} f(x)$ does exist, $f(a)$ might not equal that limit.

EXERCISES FOR SEC. 2.5: CONTINUOUS FUNCTIONS

In each of Exercises 1 to 8 there is a graph of a function f defined on $[0, 1]$. In each case answer the question on the basis of the graph. *Note:* Only if $\lim_{x\to a} f(x) = L$, L a real number, is the limit said to exist.

1 (See Fig. 2.35.)
(*a*) Does $\lim_{x\to 1/2} f(x)$ exist? If so, evaluate it.
(*b*) Is f continuous at $\frac{1}{2}$?

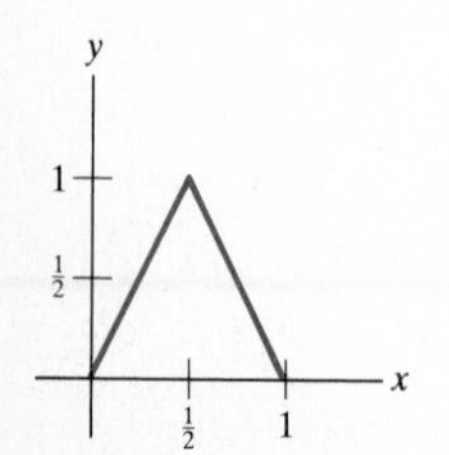

Figure 2.35

2 (See Fig. 2.36.)
(*a*) Does $\lim_{x\to 1/2} f(x)$ exist? If so, evaluate it.
(*b*) Is f continuous at $\frac{1}{2}$?

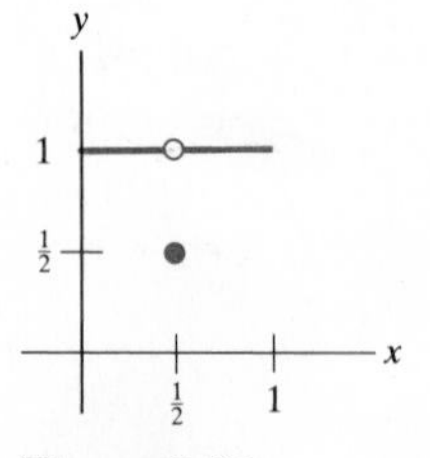

Figure 2.36

3 (See Fig. 2.37.)
(*a*) Does $\lim_{x \to 1/2} f(x)$ exist? If so, evaluate it.
(*b*) Is f continuous at $\frac{1}{2}$?

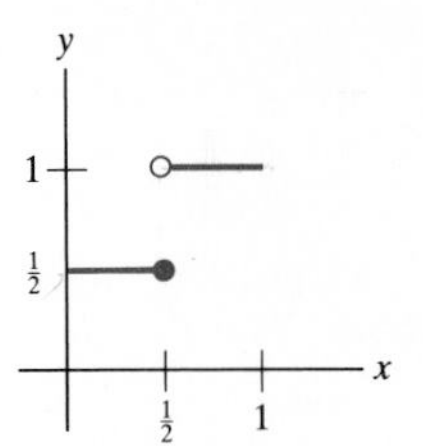

Figure 2.37

4 (See Fig. 2.38.)
(*a*) Does $\lim_{x \to 1^-} f(x)$ exist? If so, evaluate it.
(*b*) Is f continuous at 1?

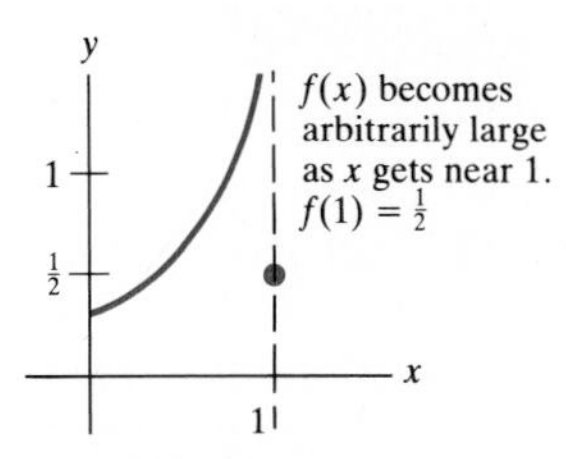

Figure 2.38

5 (See Fig. 2.39.)
(*a*) Does $\lim_{x \to 0^+} f(x)$ exist? If so, evaluate it.
(*b*) Is f continuous at 0?

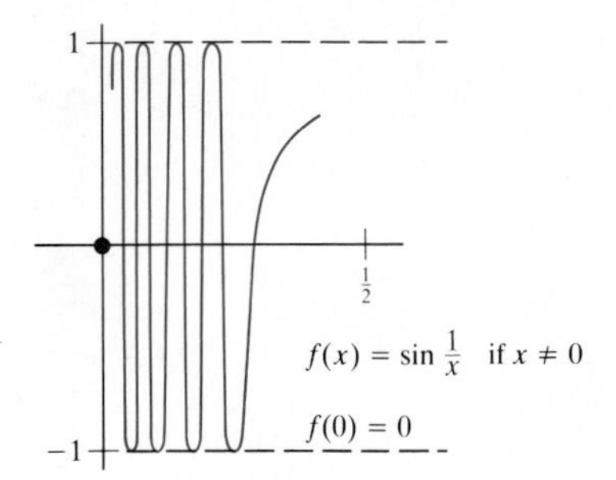

The graph oscillates infinitely often between the lines $y = 1$ and $y = -1$.

Figure 2.39

6 (See Fig. 2.40.)
(*a*) Does $\lim_{x \to 0^+} f(x)$ exist? If so, evaluate it.
(*b*) Is f continuous at 0?

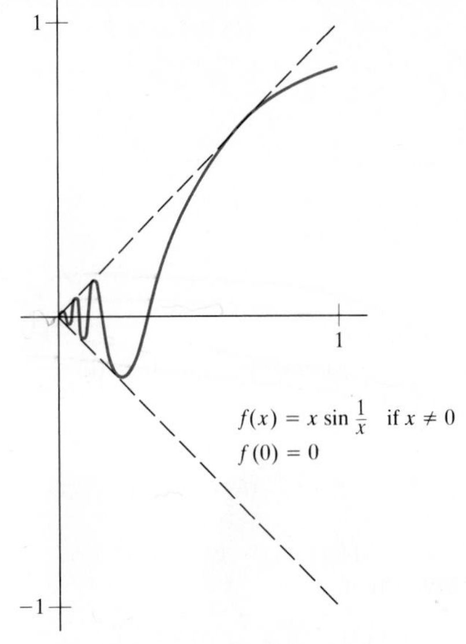

The graph oscillates infinitely often between the dashed lines.

Figure 2.40

7 (See Fig. 2.41.)
(*a*) Does $\lim_{x \to 1/4} f(x)$ exist?
(*b*) Is f continuous at $\frac{1}{4}$?
(*c*) Does $\lim_{x \to 3/4} f(x)$ exist?
(*d*) Is f continuous at $\frac{3}{4}$?
(*e*) Is f continuous at $\frac{1}{2}$?
(*f*) Is f continuous at 0?

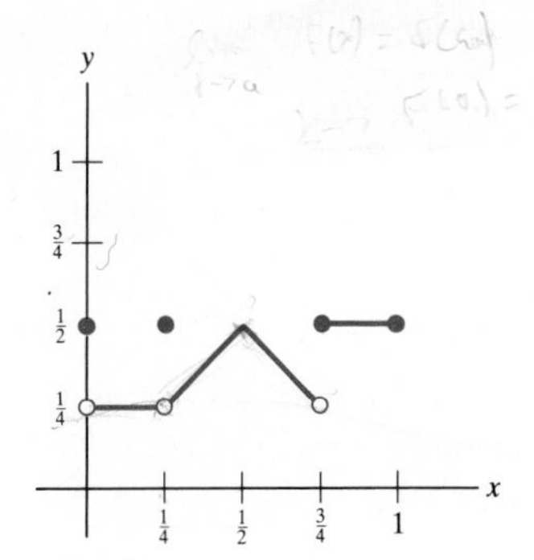

Figure 2.41

8 (See Fig. 2.42.)
(*a*) Does $\lim_{x \to 1/2} f(x)$ exist?
(*b*) Is f continuous at $\frac{1}{2}$?

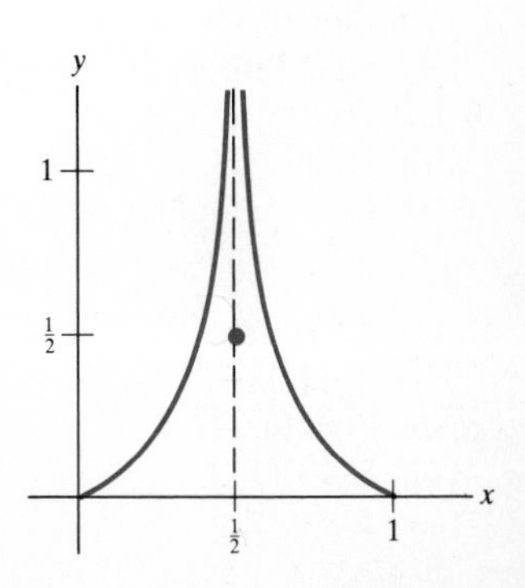

Figure 2.42

9 Let $f(x)$ equal the least integer that is greater than or equal to x. For instance, $f(3) = 3$, $f(3.4) = 4$, $f(3.9) = 4$. (This function is sometimes denoted $\lceil x \rceil$ and called the "ceiling" of x.)
(*a*) Graph f.
(*b*) Does $\lim_{x\to 4^-} f(x)$ exist? If so, what is it?
(*c*) Does $\lim_{x\to 4^+} f(x)$ exist? If so, what is it?
(*d*) Does $\lim_{x\to 4} f(x)$ exist? If so, what is it?
(*e*) Is f continuous at 4?
(*f*) Where is f continuous?
(*g*) Where is f not continuous?

10 Let f be the "nearest integer, with rounding down" function. That is,

$$f(x) = \begin{cases} \text{the integer nearest to } x \text{ if } x \text{ is not midway between two consecutive integers,} \\ x - \tfrac{1}{2} \text{ if } x \text{ is midway between two consecutive integers.} \end{cases}$$

For instance, $f(1.4) = 1$, $f(1.5) = 1$, and $f(1.6) = 2$.
(*a*) Graph f.
(*b*) Does $\lim_{x\to 3.5^-} f(x)$ exist? If so, evaluate it.
(*c*) Does $\lim_{x\to 3.5^+} f(x)$ exist? If so, evaluate it.
(*d*) Does $\lim_{x\to 3.5} f(x)$ exist? If so, evaluate it.
(*e*) Is f continuous at 3.5?
(*f*) Where is f continuous?
(*g*) Where is f not continuous?

11 Let $f(x) = (1 - \cos x)/x$ for $x \neq 0$. Is it possible to define $f(0)$ in such a way that f is continuous throughout the x axis?

12 Let $f(x) = (x^3 - 1)/(x - 1)$ if $x \neq 1$. Is it possible to define $f(1)$ in such a way that f is continuous throughout the x axis?

13 Let $f(x) = 2 - x$ if $x < 1$ and let $f(x) = x^2$ if $x > 1$.
(*a*) Graph f.
(*b*) Can $f(1)$ be defined in such a way that f is continuous through the x axis?

14 Let $f(x) = x^2$ for $x < 1$ and let $f(x) = 2x$ for $x > 1$.
(*a*) Graph f.
(*b*) Can $f(1)$ be defined in such a way that f is continuous throughout the x axis?

15 Let $f(x) = 0$ for $x < 1$ and let $f(x) = (x - 1)^2$ for $x > 1$.
(*a*) Graph f.
(*b*) Can $f(1)$ be defined in such a way that f is continuous throughout the x axis?

16 Let $f(x) = x + |x|$.
(*a*) Graph f.
(*b*) Is f continuous at 0?

■

17 Let $f(x) = 2^{1/x}$ for $x \neq 0$.
(*a*) Find $\lim_{x\to\infty} f(x)$.
(*b*) Find $\lim_{x\to-\infty} f(x)$.
(*c*) Does $\lim_{x\to 0^+} f(x)$ exist?
(*d*) Does $\lim_{x\to 0^-} f(x)$ exist?
(*e*) Graph f, incorporating the information in parts (*a*) to (*d*).
(*f*) Is it possible to define $f(0)$ in such a way that f is continuous throughout the x axis?

18 Let $f(x) = 2^{-1/x^2}$ for $x \neq 0$.
(*a*) Find $\lim_{x\to\infty} f(x)$.
(*b*) Find $\lim_{x\to 0^+} f(x)$.
(*c*) Graph f.
(*d*) Is it possible to define $f(0)$ in such a way that f is continuous throughout the x axis?

19 (*a*) Graph $f(x) = (\sin x)/x$.
(*b*) Is it possible to define $f(0)$ in such a way that f is continuous throughout the x axis?

20 Let $f(x) = x$ for rational x and let $f(x) = x^2$ for irrational x.
(*a*) Sketch the graph of f. (Use a dotted curve to indicate a curve from which points are missing.)
(*b*) At which inputs, if any, is f continuous?

21 Let $f(x) = x^2$ for rational x and let $f(x) = x^3$ for irrational x.
(*a*) Sketch the graph of f. (Use a dotted curve to indicate a curve from which points are missing.)
(*b*) At which inputs, if any, is f continuous?

22 Let f be a continuous function defined for all x. Assume that $f(x) = 0$ when x is rational. Explain why $f(x) = 0$ when x is irrational as well.

23 Let f and g be continuous functions defined for all x. Assume that $f(x) = g(x)$ for all rational x. Deduce that $f(x) = g(x)$ for all real numbers x. (*Hint:* Exercise 22 may be of use.)

24 This exercise determines all continuous functions f such that $f(x + y) = f(x)f(y)$ for all real numbers x and y and such that the values of f are always positive.
(*a*) Let b be a fixed positive number. Let $f(x) = b^x$. Check that $f(x + y) = f(x)f(y)$. [Part (*b*) will show that there are no other functions that satisfy the stated conditions.]
(*b*) Assume that f is a continuous function such that $f(x) > 0$ for all x and $f(x + y) = f(x)f(y)$ for all x and y. Let $f(1) = c$.
(1) Show that $f(n) = c^n$ for any positive integer n.
(2) Show that $f(0) = 1$.
(3) Show that $f(n) = c^n$ for any negative integer n.
(4) Show that $f(1/n) = \sqrt[n]{c}$ for any positive integer n.
(5) Show that $f(m/n) = (\sqrt[n]{c})^m$ for any integer m and positive integer n.
(6) By (5), $f(x) = c^x$ for any rational number x. Assuming that f is continuous and that the exponential function c^x is continuous, deduce that $f(x) = c^x$ for all real numbers x. (*Hint:* See Exercise 23.)

25 Let f be a continuous function whose domain is the x

axis and which has the property that

$$f(x + y) = f(x) + f(y)$$

for all numbers x and y. This exercise shows that f must be of the form $f(x) = cx$ for some constant c. [The function cx does satisfy the equation, $f(x + y) = f(x) + f(y)$, since $c(x + y) = cx + cy$.]

(*a*) Let $f(1) = c$. Show that $f(2) = 2c$.
(*b*) Show that $f(0) = 0$.
(*c*) Show that $f(-1) = -c$.
(*d*) Show that for any positive integer n, $f(n) = cn$.
(*e*) Show that for any negative integer n, $f(n) = cn$.
(*f*) Show that $f(\frac{1}{2}) = c/2$.
(*g*) Show that for any nonzero integer n, $f(1/n) = c/n$.
(*h*) Show that for any integer m and positive integer n, $f(m/n) = c(m/n)$.
(*i*) Show that for any irrational number x, $f(x) = cx$. (This is where the continuity of f enters. See Exercise 23.) Parts (*h*) and (*i*) together complete the solution.

26 (Calculator) *The reason* 0^0 *is not defined.* It might be hoped that if the positive number b and the number x are both close to 0, then b^x might be close to some fixed number. If that were so, it would suggest a definition of 0^0. The following computations using the y^x-key show that when b and x are both near 0, b^x is not near any specific number. Compute (*a*) $(0.001)^{0.001}$ and (*b*) $(0.0000001)^{0.1}$. Consider also $(0.001)^0$ and $0^{0.001}$.

2.6 THE MAXIMUM-VALUE THEOREM AND THE INTERMEDIATE-VALUE THEOREM

Continuous functions have two properties of particular importance in calculus: the "maximum-value" property and the "intermediate-value" property. Both are quite plausible, and a glance at the graph of a "typical" continuous function may persuade us that they are true of all continuous functions. No proofs will be offered; they depend on the precise definitions of limits given in Secs. 2.7 and 2.8 and are part of an advanced calculus course.

The first theorem asserts that a function that is continuous throughout the closed interval $[a, b]$ takes on a largest value somewhere in the interval. (It also takes on a smallest value.)

Maximum-Value and Minimum-Value Theorem Let f be continuous throughout the closed interval $[a, b]$. Then there is at least one number in $[a, b]$ at which f takes on a maximum value. That is, for some number c in $[a, b]$,

$$f(c) \geq f(x) \qquad \text{for all } x \text{ in } [a, b].$$

Similarly, f takes on a minimum value somewhere in the interval. ■

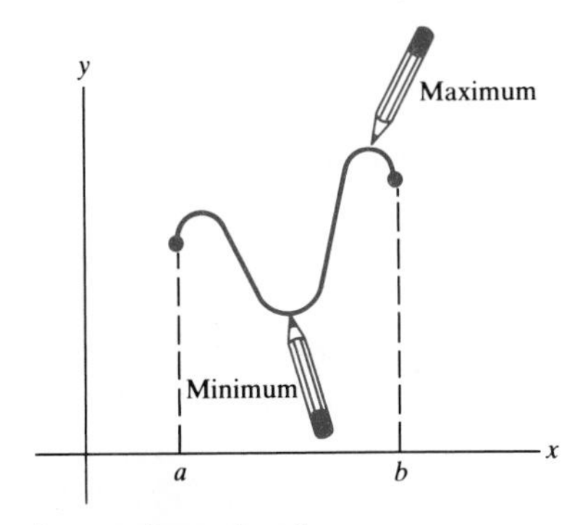

As a pencil runs along the graph of a continuous function from one point to another, it passes through at least one maximum point and at least one minimum point.

Figure 2.43

To persuade yourself that this theorem is plausible, imagine sketching the graph of a continuous function. As your pencil moves along the graph from some point on the graph to some other point on the graph, it passes through a highest point and also through a lowest point. (See Fig. 2.43.)

The maximum-value theorem guarantees that a maximum value exists, but it does *not tell how* to find it. The problem of finding the maximum value (and minimum value) is discussed in Chap. 4.

EXAMPLE 1 Let $f(x) = \cos x$ and $[a, b] = [0, 3\pi]$. Find all numbers in $[0, 3\pi]$ at which f takes on a maximum value. Also find all numbers in $[0, 3\pi]$ at which f takes on a minimum value.

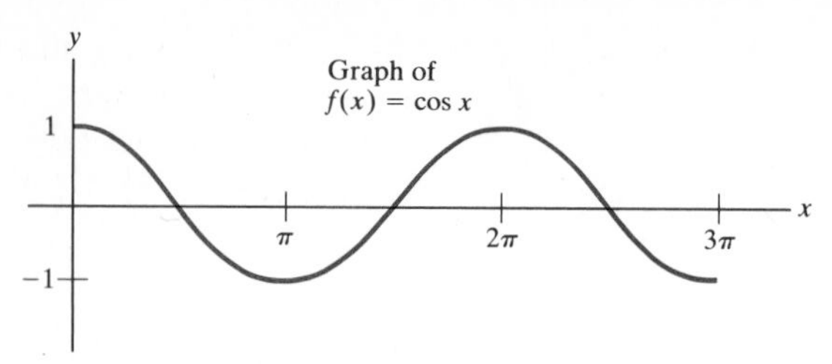

Figure 2.44

SOLUTION Figure 2.44 is a graph of $\cos x$ for x in $[0, 3\pi]$. Inspection of the graph shows that the maximum value of $\cos x$ for $0 \le x \le 3\pi$ is 1, and it is attained when $x = 0$ and when $x = 2\pi$. The minimum value is -1, which is attained when $x = \pi$ and when $x = 3\pi$. ■

Extreme values

"Extrema" is the plural of "extremum."

The maximum and minimum values of a function are frequently called its **extreme values** or **extrema**. Thus the extreme values of $\cos x$ for x in $[0, 3\pi]$ are 1 and -1.

To apply the maximum-value theorem, we must know that the function is continuous and the interval is closed (that is, contains its endpoints). The next three examples show that if either of these assumptions is deleted, the conclusion no longer need hold. In Examples 2 and 3 the interval is not closed; in Example 4 the function is not continuous.

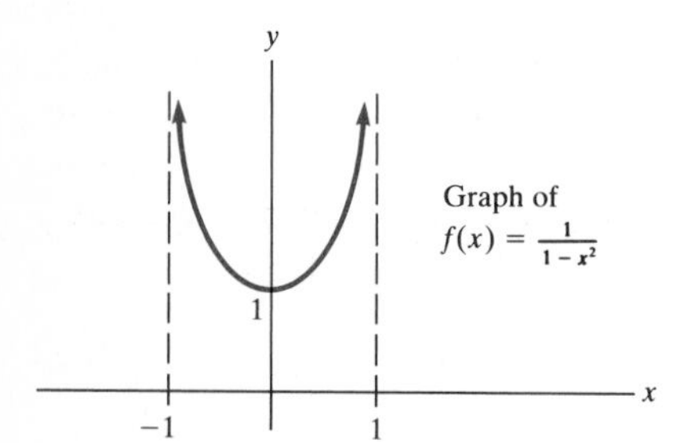

Figure 2.45

EXAMPLE 2 Let $f(x) = 1/(1 - x^2)$ and let (a, b) be the open interval $(-1, 1)$. Show that f does not have a maximum value for x in (a, b).

SOLUTION For x near 1, $f(x)$ gets arbitrarily large since the denominator $1 - x^2$ is close to 0. The graph of f, for x in $(-1, 1)$, is shown in Fig. 2.45. This function is continuous throughout the open interval $(-1, 1)$, but there is no number c in $(-1, 1)$ at which f has a maximum value. However, f has a minimum value, $f(0) = 1$. ■

EXAMPLE 3 Let $f(x) = 1/x$ and let (a, b) be the open interval $(0, 1)$. Show that f does not have a maximum value in (a, b).

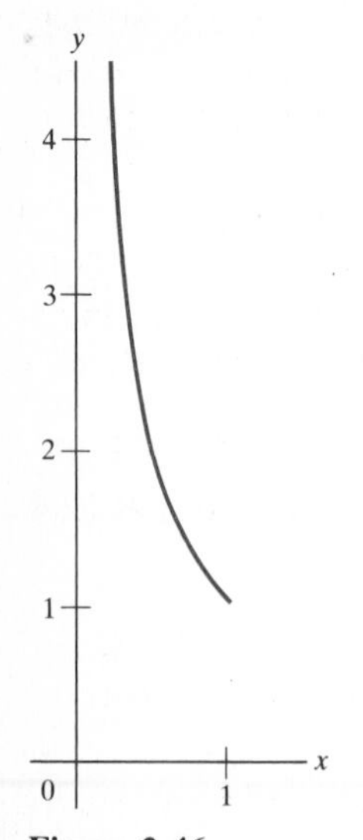

Figure 2.46

SOLUTION Figure 2.46 shows the pertinent part of the graph of $1/x$. Since $\lim_{x \to 0^+} 1/x = \infty$, the function has no maximum value.

Moreover, the function has no minimum value for x in $(0, 1)$. It does take on values arbitrarily close to 1 for inputs that are close to 1, but there is no number in the open interval $(0, 1)$ at which $f(x)$ is equal to 1. ■

In the next example the interval is closed, but the function is not continuous.

EXAMPLE 4 Let $f(x) = x$ if x is not an integer and let $f(x) = 0$ if x is an integer. Show that f does not assume a maximum value on the interval $[0, 1]$.

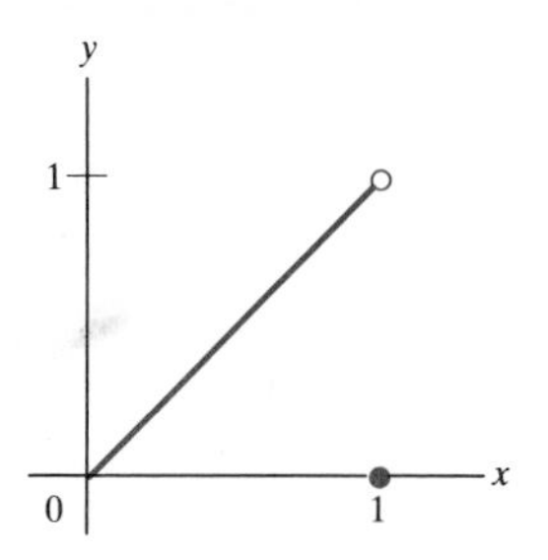

Figure 2.47

SOLUTION First graph f for x in $[0, 1]$, as in Fig. 2.47. Inspection of the graph shows that f does not take on a maximum value at any number in $[0, 1]$. As $x \to 1$ from the left, $f(x) \to 1$, but $f(x)$ does not attain the value 1. (Incidentally, f takes on its minimum value 0 for the interval $[0, 1]$ at two places, at $x = 0$ and at $x = 1$.) ■

The next theorem says that a function which is continuous throughout an interval takes on all values between any two of its values.

Intermediate-Value Theorem Let f be continuous throughout the closed interval $[a, b]$. Let m be any number between $f(a)$ and $f(b)$. [That is, $f(a) \leq m \leq f(b)$ if $f(a) \leq f(b)$, or $f(a) \geq m \geq f(b)$ if $f(a) \geq f(b)$.] Then there is at least one number c in $[a, b]$ such that $f(c) = m$. ■

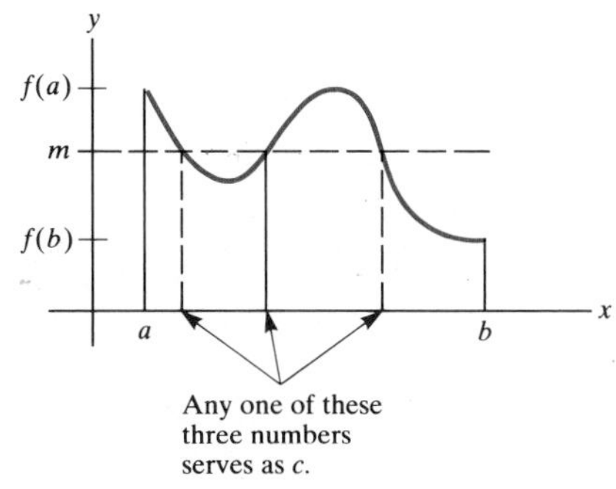

Figure 2.48

In ordinary English, the intermediate-value theorem reads: A continuous function defined on $[a, b]$ takes on all values between $f(a)$ and $f(b)$. Pictorially, it asserts that a horizontal line of height m must meet the graph of f at least once if m is between $f(a)$ and $f(b)$, as shown in Fig. 2.48. In other words, when you move a pencil along the graph of a continuous function from one height to another, the pencil passes through all intermediate heights.

Even though the theorem guarantees the existence of c, it does *not tell how* to find it. To find c, we must solve an equation, namely, $f(c) = m$.

EXAMPLE 5 Use the intermediate-value theorem to show that the equation $2x^3 + x^2 - x + 1 = 5$ has a solution in the interval $[1, 2]$.

SOLUTION Let $P(x) = 2x^3 + x^2 - x + 1$. Then

$$P(1) = 2 \cdot 1^3 + 1^2 - 1 + 1 = 3$$

and

$$P(2) = 2 \cdot 2^3 + 2^2 - 2 + 1 = 19.$$

The intermediate-value theorem can help guarantee a solution.

Since P is continuous and 5 is between $P(1) = 3$ and $P(2) = 19$, we may apply the intermediate-value theorem to P in the case $a = 1$, $b = 2$, and $m = 5$. Thus there is at least one number c between 1 and 2 such that $P(c) = 5$. This completes the answer.

(To get a more accurate estimate for a number c such that $P(c) = 5$, find a shorter interval for which the intermediate-value theorem can be applied. For instance, $P(1.2) \approx 4.7$ and $P(1.3) \approx 5.8$. By the intermediate-value theorem, there is a number c in $[1.2, 1.3]$ such that $P(c) = 5$.) ■

EXAMPLE 6 Show that the equation $x^5 - 2x^2 + x + 11 = 0$ has at least one real root.

SOLUTION For x large and positive the polynomial $P(x) = x^5 - 2x^2 + x + 11$ is positive [since $\lim_{x\to\infty} P(x) = \infty$]. Thus there is a number b such that $P(b) > 0$. Similarly, for x negative and of large absolute value, $P(x)$ is negative [since $\lim_{x\to-\infty} P(x) = -\infty$]. Select a number a such that $P(a) < 0$.

The number 0 is between $P(a)$ and $P(b)$. Since P is continuous on the interval $[a, b]$, there is a number c in $[a, b]$ such that $P(c) = 0$. This number c is a real solution to the equation $x^5 - 2x^2 + x + 11 = 0$. ■

Any polynomial of odd degree has a real root.

Note that the argument in Example 6 applies to any polynomial of odd degree. The argument does not hold for polynomials of even degree, since the equations $x^2 + 1 = 0$, $x^4 + 1 = 0$, $x^6 + 1 = 0$, and so on have no real solutions.

EXAMPLE 7 Use the intermediate-value theorem to show that there is a number c in $[1, 2]$ such that $4 - c = 2^c$.

SOLUTION Introduce the function $f(x) = (4 - x) - 2^x$, the difference of $4 - x$ and 2^x, and show that there is a number c in $[1, 2]$ such that $f(c) = 0$.

Observe that

$$f(1) = 4 - 1 - 2^1 = 1,$$

$$f(2) = 4 - 2 - 2^2 = -2,$$

and that f is continuous.

Since 0 is between $f(1)$ and $f(2)$, we may apply the intermediate-value theorem with $a = 1$, $b = 2$, and $m = 0$. The theorem assures us that there is at least one number c in $[1, 2]$ such that $f(c) = 0$; that is, we have $4 - c - 2^c = 0$. From this it follows that $4 - c = 2^c$. ■

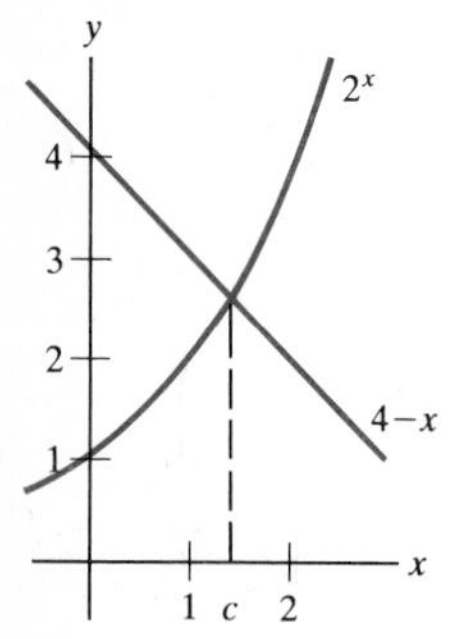

Figure 2.49

In Example 7 the intermediate-value theorem does not tell what c is. The graphs of $4 - x$ and 2^x in Fig. 2.49 suggest that c is unique and about 1.4. Further arithmetic, done on a calculator, shows that $c \approx 1.39$.

From Examples 6 and 7 we make two observations:

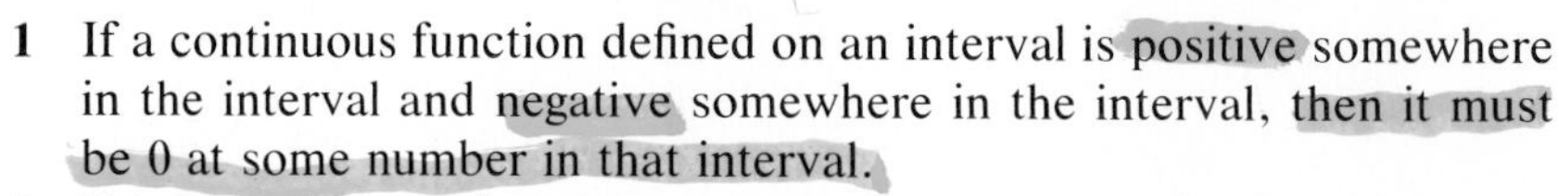

1. If a continuous function defined on an interval is positive somewhere in the interval and negative somewhere in the interval, then it must be 0 at some number in that interval.
2. To show that two functions are equal at some number in an interval, show that their difference is 0 at some number in the interval.

EXERCISES FOR SEC. 2.6: THE MAXIMUM-VALUE THEOREM AND THE INTERMEDIATE-VALUE THEOREM

1. Does the function $(x^3 + x^4)/(1 + 5x^2 + x^6)$ have (*a*) a maximum value for x in $[1, 4]$? (*b*) a minimum value for x in $[1, 4]$?
2. Does the function $2^x - x^3 + x^5$ have (*a*) a maximum value for x in $[-3, 10]$? (*b*) a minimum value for x in $[-3, 10]$?
3. Does the function x^3 have a maximum value for x in (*a*) $[2, 4]$? (*b*) $[-3, 5]$? (*c*) $(1, 6)$?
4. Does the function x^4 have a minimum value for x in (*a*) $[-5, 6]$? (*b*) $(-2, 4)$? (*c*) $(3, 7)$? (*d*) $(-4, 4)$?
5. Does the function $2 - x^2$ have (*a*) a maximum value for x in $(-1, 1)$? (*b*) a minimum value for x in $(-1, 1)$?
6. Does the function $2 + x^2$ have (*a*) a maximum value for x in $(-1, 1)$? (*b*) a minimum value for x in $(-1, 1)$?

7 Show that the equation $x^5 + 3x^4 + x - 2 = 0$ has at least one root in the interval $[0, 1]$.

8 Show that the equation $x^5 - 2x^3 + x^2 - 3x + 1 = 0$ has at least one root in the interval $[1, 2]$.

In Exercises 9 to 15 verify the intermediate-value theorem for the specified function f, the interval $[a, b]$, and the indicated value m. Find all c's in each case.

9 Function $3x + 5$; interval $[1, 2]$; $m = 10$.

10 Function $x^2 - 2x$; interval $[-1, 4]$; $m = 5$.

11 Function $\sin x$; interval $[\pi/2, 11\pi/2]$; $m = 0$.

12 Function $\cos x$; interval $[0, 5\pi]$; $m = 0$.

13 Function $\cos x$; interval $[0, 5\pi]$; $m = \frac{1}{2}$.

14 Function 2^x; interval $[0, 3]$; $m = 4$.

15 Function $x^3 - x$; interval $[-2, 2]$; $m = 0$.

16 Show that the equation $2^x - 3x = 0$ has a solution in the interval $[0, 1]$.

17 Does the equation $x + \sin x = 1$ have a solution?

18 Does the equation $x^3 = 2^x$ have a solution?

19 Use the intermediate-value theorem to show that the equation $3x^3 + 11x^2 - 5x = 2$ has a solution.

20 Let $f(x) = 1/x$, $a = -1$, $b = 1$, $m = 0$. Note that $f(a) \le 0 \le f(b)$. Is there at least one c in $[a, b]$ such that $f(c) = 0$? If so, find c; if not, does this imply that the intermediate-value theorem is sometimes false?

■

21 Let f and g be two continuous functions defined at least on the interval $[a, b]$. Assume that $f(a) < g(a)$ and that $f(b) > g(b)$. Prove that there is a number c in (a, b) such that $f(c) = g(c)$. *Hint:* Apply the intermediate-value theorem to the function h defined by $h(x) = f(x) - g(x)$.

22 Let $P(x) = a_n x^n + a_{n-1} x^{n-1} + \cdots + a_0$ be a polynomial of odd degree n and with lead coefficient a_n positive. Show that there is at least one real number r such that $P(r) = 0$.

23 (This continues Exercise 22.) The *factor theorem* from algebra asserts that the number r is a root of the polynomial $P(x)$ if and only if $x - r$ is a factor of $P(x)$. For instance, 2 is a root of the polynomial $x^2 - 3x + 2$ and $x - 2$ is a factor of the polynomial: $x^2 - 3x + 2 = (x - 2)(x - 1)$. This is reviewed in Appendix C.

(*a*) Use the factor theorem and Exercise 22 to show that every polynomial of odd degree has a factor of degree 1.

(*b*) Show that none of the polynomials $x^2 + 1$, $x^4 + 1$, or $x^{100} + 1$ has a first-degree factor.

(*c*) Check $x^4 + 1 = (x^2 + \sqrt{2}\,x + 1)(x^2 - \sqrt{2}\,x + 1)$. (It can be shown using complex numbers that every polynomial is the product of polynomials of degrees at most 2.)

In Exercises 24 to 26 "completing the square," which is reviewed in Appendix C, may come in handy.

24 What is the smallest possible value of $x^2 - 4x + 5$ for all inputs x?

25 Find two positive numbers whose sum is 4 and whose product is as large as possible. *Hint:* Call one of the numbers x. What is the other?

26 Which rectangle of perimeter 1 meter has the largest area?

A set in the plane bounded by a curve is *convex* if for any two points P and Q in the set the line segment joining them also lies in the set. (See Fig. 2.50.)

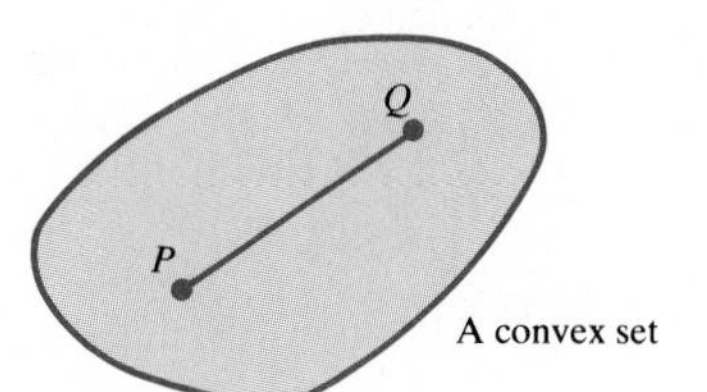

Figure 2.50

Circles, triangles, and parallelograms are convex sets. The quadrilateral shown in Fig. 2.51 is not convex. Convex sets will be referred to in the following exercises and occasionally in the exercises of later chapters.

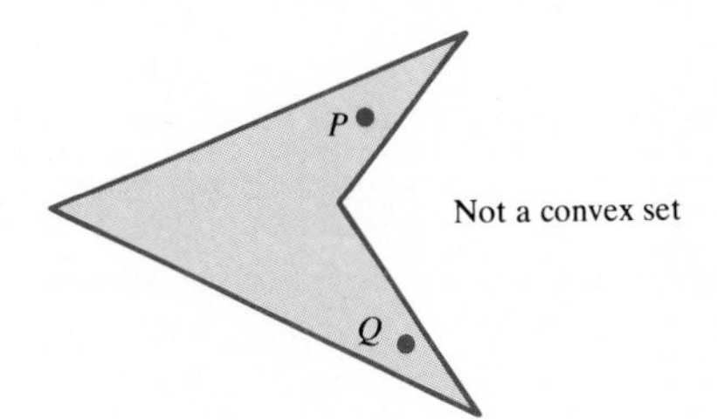

Figure 2.51

In Exercises 27 to 31 you will need to define various functions geometrically. *You may assume that they are continuous.*

27 Let L be a line in the plane and let K be a convex set. Show that there is a line parallel to L that cuts K into two pieces of equal area. *Hint:* Consider all lines parallel to L that meet K and notice how they divide K. Apply the intermediate-value theorem to an appropriate function.

28 Let P be a point in the plane and let K be a convex set. Is there is a line through P that cuts K into two pieces of equal area?

29 Let K_1 and K_2 be two convex sets in the plane. Is there a line that simultaneously cuts K_1 into two pieces of equal area and cuts K_2 into two pieces of equal area? (This is known as the "two pancakes" question.)

30 Let K be a convex set in the plane. Show that there

is a line that simultaneously cuts K into two pieces of equal area and cuts the boundary of K into two pieces of equal length.

31 Let K be a convex set. Show that there are two perpendicular lines that cut K into four pieces of equal area. (It is not known whether it is always possible to find two perpendicular lines that divide K into four pieces whose areas are $\frac{1}{8}$, $\frac{1}{8}$, $\frac{3}{8}$, and $\frac{3}{8}$ of the area of K, with the parts of equal area sharing an edge, as in Fig. 2.52.)

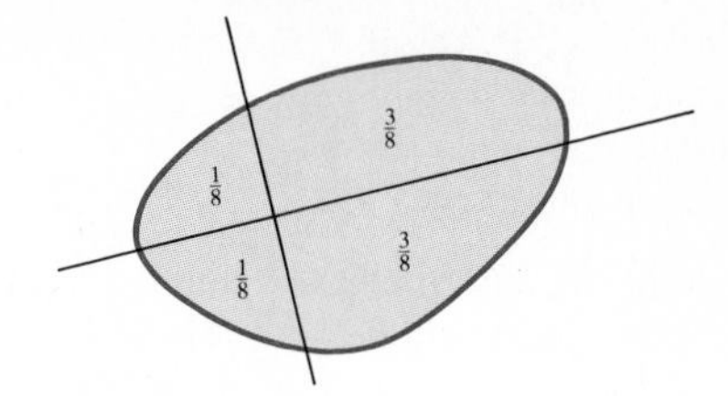

Figure 2.52

2.7 PRECISE DEFINITIONS OF "$\lim_{x\to\infty} f(x) = \infty$" AND "$\lim_{x\to\infty} f(x) = L$"

In the definitions of the limits considered in Secs. 2.1 and 2.2 appear such phrases as "x approaches a," "$f(x)$ approaches a specific number," "as x gets larger," and "$f(x)$ becomes and remains arbitrarily large." Such phrases, although appealing to the intuition and conveying the sense of a limit, are not precise. The definitions seem to suggest moving objects and call to mind the motion of a pencil point as it traces out the graph of a function.

This informal approach was adequate during the early development of calculus, from Leibniz and Newton in the seventeenth century through the Bernoullis, Euler, and Gauss in the eighteenth and early nineteenth centuries. But by the mid-nineteenth century, mathematicians, facing more complicated functions and more difficult theorems, no longer could depend solely on intuition. They realized that glancing at a graph was no longer adequate to understand the behavior of functions—especially if theorems covering a broad class of functions were needed.

It was Weierstrass who developed, in the period 1841–1856, a way to define limits without any hint of motion or of pencils tracing out graphs. His approach, on which he lectured after joining the faculty at the University of Berlin in 1859, has since been followed by pure and applied mathematicians throughout the world. Even an undergraduate advanced calculus course depends on Weierstrass's approach.

In this section we examine how Weierstrass would define the concepts:

$$\lim_{x\to\infty} f(x) = \infty \qquad \text{and} \qquad \lim_{x\to\infty} f(x) = L.$$

Throughout, "f" refers to a numerical function. In the next section we consider "$\lim_{x\to a} f(x) = L$."

Recall the definition of "$\lim_{x\to\infty} f(x) = \infty$" given in the table in Sec. 2.2.

Informal

Informal definition of $\lim_{x\to\infty} f(x) = \infty$:

$f(x)$ is defined for all x beyond some number and, as x gets large through positive values, $f(x)$ becomes and remains arbitrarily large and positive.

To take us part way to the precise definition, let us reword the informal definition, paraphrasing it in the following definition, which is still informal.

Reworded

Reworded informal definition of $\lim_{x\to\infty} f(x) = \infty$ [assume that $f(x)$ is defined for all x greater than some number c]:

If x is sufficiently large and positive, then $f(x)$ is necessarily large and positive.

The precise definition parallels the reworded definition.

Precise

> *Precise definition* of $\lim_{x\to\infty} f(x) = \infty$ [assume that $f(x)$ is defined for all x greater than some number c]:
>
> For each number E there is a number D such that for all $x > D$ it is true that
>
> $$f(x) > E.$$

The "challenge and reply" approach to limits

Think of the number E as a challenge and D as the reply. The *larger* E is, the *larger* D must usually be. Only if a number D (which depends on E) can be found for *every* number E can we make the claim that "$\lim_{x\to\infty} f(x) = \infty$."

To picture the idea behind the precise definition, consider the graph in Fig. 2.53 of a function f for which $\lim_{x\to\infty} f(x) = \infty$. For each possible choice of a horizontal line, say, at height E, if you are far enough to the right on the graph of f, you stay above that line. That is, there is a number D such that if $x > D$, then $f(x) > E$, as illustrated in Fig. 2.54.

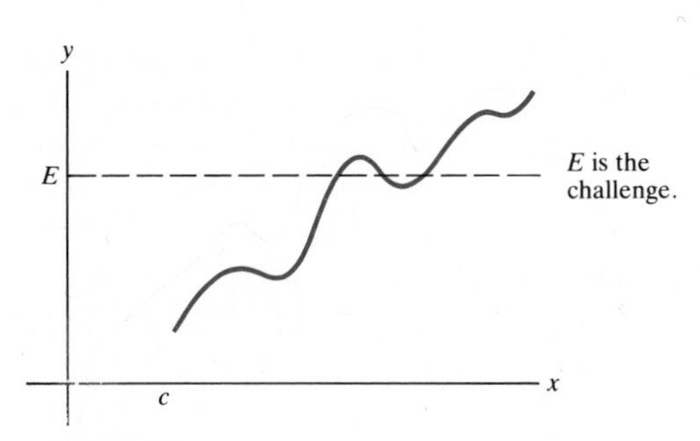

Figure 2.53

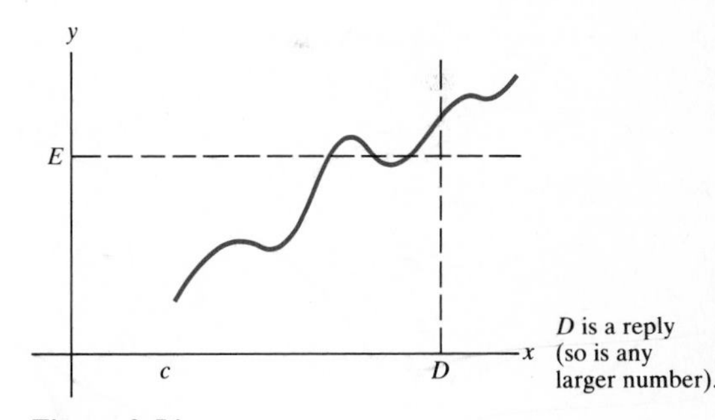

Figure 2.54

Examples 1 and 2 illustrate how the precise definition is used.

EXAMPLE 1 Using the precise definition, show that $\lim_{x\to\infty} 2x = \infty$.

SOLUTION Let E be any number. We must show that there is a number D such that whenever $x > D$, it follows that $2x > E$. (For example, if $E = 100$, then $D = 50$ would do. It is indeed the case that if $x > 50$, then $2x > 100$.) The number D will depend on E.

Now, the inequality $2x > E$ is equivalent to

$$x > \frac{E}{2}.$$

In other words, if $x > E/2$, then $2x > E$. So $D = E/2$ suffices. That is, for $x > D$ $(= E/2)$, $2x > E$. We conclude immediately that

$$\lim_{x \to \infty} 2x = \infty. \quad \blacksquare$$

The response D is not unique.

In Example 1 a formula was provided for a suitable D in terms of E, namely, $D = E/2$. For instance, when $E = 1000$, $D = 500$ suffices. In fact, any larger value of D also is suitable. If $x > 600$, it is still the case that $2x > 1000$ (since then $2x > 1200$). If one value of D is a satisfactory response to a given challenge E, then any larger value of D also is a satisfactory response.

The graph of $f(x) = \frac{x}{2} + \sin x$ is shown in Fig. 2.55.

EXAMPLE 2 Using the precise definition, show that $\lim_{x \to \infty} \left(\frac{x}{2} + \sin x\right) = \infty$.

SOLUTION Let E be any number. We must exhibit a number D, depending on E, such that $x > D$ forces

$$\frac{x}{2} + \sin x > E. \tag{1}$$

Now, $\sin x \geq -1$ for all x. So, if we can force

$$\frac{x}{2} > E + 1, \tag{2}$$

then it will follow that

$$\frac{x}{2} + \sin x > E.$$

Inequality (2) is equivalent to

$$x > 2(E + 1).$$

Thus $D = 2(E + 1)$ will suffice. That is,

$$\text{If } x > 2(E + 1), \qquad \text{then } \frac{x}{2} + \sin x > E.$$

To verify this assertion, we must check that $D = 2(E + 1)$ is a satisfactory reply to E. Assume that $x > 2(E + 1)$. Then

$$\frac{x}{2} > E + 1$$

and

$$\sin x \geq -1.$$

If $a > b$ and $c \geq d$, then $a + c > b + d$.

Adding these last two inequalities gives

$$\frac{x}{2} + \sin x > (E + 1) - 1,$$

or simply,

$$\frac{x}{2} + \sin x > E,$$

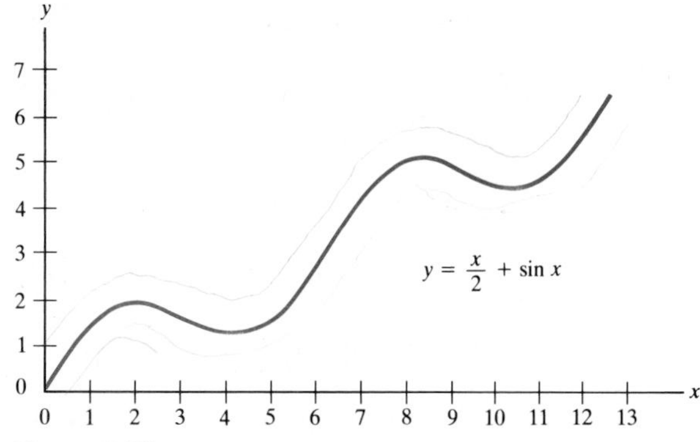

Figure 2.55

which is inequality (1).

Thus we are permitted to assert that

$$\lim_{x\to\infty}\left(\frac{x}{2} + \sin x\right) = \infty. \quad \blacksquare$$

The graph of the function examined in Example 2 appears in Fig. 2.55. Note that the function is not increasing. Nevertheless, as x increases, the function does tend to become and remain large, despite the small dips downward.

Next, recall the definition of "$\lim_{x\to\infty} f(x) = L$" given in the table of Sec. 2.2. L refers to a fixed number.

Informal

Informal definition of $\lim_{x\to\infty} f(x) = L$ [assume that $f(x)$ is defined for all x beyond some number c]:

As x gets large through positive values, $f(x)$ approaches L.

Again we reword this definition before offering the precise definition.

Reworded

Reworded informal definition of $\lim_{x\to\infty} f(x) = L$ [assume that there is a number c such that $f(x)$ is defined for all $x > c$]:

If x is sufficiently large and positive, then $f(x)$ is necessarily near L.

Again, the precise definition parallels the reworded informal definition. In order to make precise the phrase "$f(x)$ is necessarily near L," we shall use the absolute value of $f(x) - L$ to measure the distance from $f(x)$ to L. The following definition says that "if x is large enough, then $|f(x) - L|$ is as small as we please."

Precise

Precise definition of $\lim_{x\to\infty} f(x) = L$ [assume that $f(x)$ is defined for all x beyond some number c]:

For each positive number ϵ, there is a number D such that for all $x > D$ it is true that

$$|f(x) - L| < \epsilon.$$

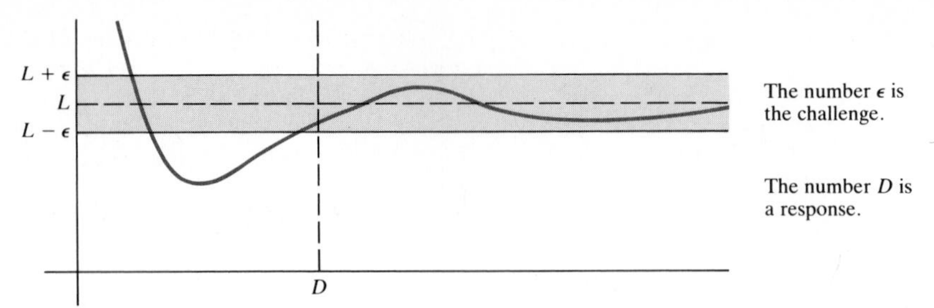

Figure 2.56

ϵ (epsilon) is the Greek letter corresponding to the English letter e.

The positive number ϵ is the challenge, and D is a response. The smaller ϵ is, the larger D usually must be chosen. The geometric meaning of the precise definition of $\lim_{x\to\infty} f(x) = L$ is shown in Fig. 2.56.

The narrower the band, the larger D must usually be.

Draw two lines parallel to the x axis, one of height $L + \epsilon$ and one of height $L - \epsilon$. They are the two edges of an endless band of width 2ϵ. Assume that for each positive ϵ, a number D can be found, depending on ϵ, such that the part of the graph to the right of $x = D$ lies within the band. Then we say that "as x approaches ∞, $f(x)$ approaches L" and write

$$\lim_{x\to\infty} f(x) = L.$$

EXAMPLE 3 Use the precise definition of "$\lim_{x\to\infty} f(x) = L$," to show that

$$\lim_{x\to\infty}\left(1 + \frac{1}{x}\right) = 1.$$

SOLUTION Here $f(x) = 1 + 1/x$, which is defined for all $x > 0$. The number L is 1. We must show that for each positive number ϵ, however small, there is a number D such that, for all $x > D$,

$$\left|\left(1 + \frac{1}{x}\right) - 1\right| < \epsilon. \tag{3}$$

Inequality (3) reduces to

$$\left|\frac{1}{x}\right| < \epsilon.$$

Since we shall consider only $x > 0$, this inequality is equivalent to

$$\frac{1}{x} < \epsilon. \tag{4}$$

Multiplying inequality (4) by the positive number x yields the equivalent inequality

$$1 < \epsilon x. \tag{5}$$

Division of inequality (5) by the positive number ϵ yields

$$\frac{1}{\epsilon} < x \qquad \text{or} \qquad x > \frac{1}{\epsilon}.$$

These steps are reversible. This shows that $D = 1/\epsilon$ is a suitable reply to the challenge ϵ. If $x > 1/\epsilon$, then

$$\left|\left(1 + \frac{1}{x}\right) - 1\right| < \epsilon.$$

According to the precise definition of "$\lim_{x\to\infty} f(x) = L$," we may conclude that

$$\lim_{x\to\infty}\left(1 + \frac{1}{x}\right) = 1. \quad \blacksquare$$

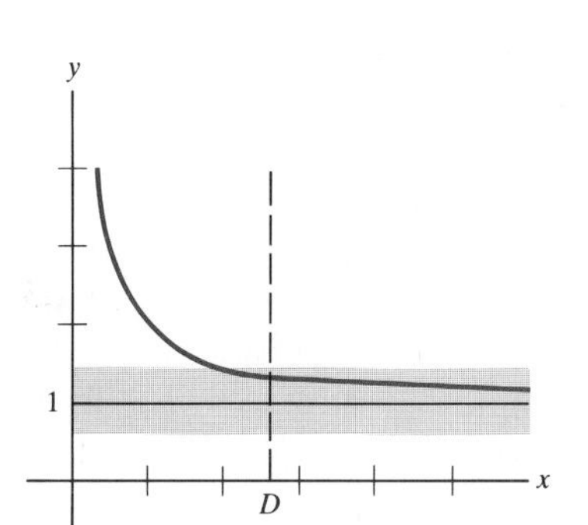

Figure 2.57

The graph of $f(x) = 1 + 1/x$, shown in Fig. 2.57, reinforces the argument. It seems plausible that no matter how narrow a band someone may place around the line $y = 1$, it will always be possible to find a number D such that the part of the graph to the right of $x = D$ stays within that band. In Fig. 2.57 the typical band is shown shaded.

The precise definitions also can be used to show that some claim about an alleged limit is false. The next example illustrates how this is done.

EXAMPLE 4 Show that the claim that $\lim_{x\to\infty} \sin x = 0$ is false.

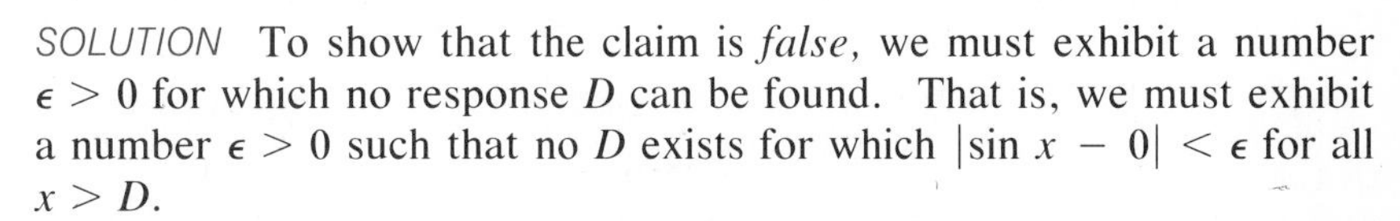

SOLUTION To show that the claim is *false*, we must exhibit a number $\epsilon > 0$ for which no response D can be found. That is, we must exhibit a number $\epsilon > 0$ such that no D exists for which $|\sin x - 0| < \epsilon$ for all $x > D$.

Recall that $\sin(\pi/2) = 1$ and that $\sin x = 1$ whenever $x = \pi/2 + 2n\pi$ for any integer n. This means that there are arbitrarily large values of x for which $\sin x = 1$. This suggests how to exhibit an $\epsilon > 0$ for which no response D can be found. Simply pick ϵ to be some positive number less than or equal to 1. For instance, $\epsilon = 0.7$ will do.

For any number D there is always a number $x^* > D$ such that we have $\sin x^* = 1$. This means that $|\sin x^* - 0| = 1 > 0.7$. Hence no response can be found for $\epsilon = 0.7$. Thus the claim that $\lim_{x\to\infty} \sin x = 0$ is false. $\blacksquare$

EXERCISES FOR SEC. 2.7: PRECISE DEFINITIONS OF $\lim_{x\to\infty} f(x) = \infty$ AND $\lim_{x\to\infty} f(x) = L$

1 Let $f(x) = 3x$.
- (*a*) Find a number D such that, for $x > D$, it follows that $f(x) > 600$.
- (*b*) Find another number D such that, for $x > D$, it follows that $f(x) > 600$.
- (*c*) What is the smallest number D such that, for $x > D$, it follows that $f(x) > 600$?

2 Let $f(x) = 4x$.
- (*a*) Find a number D such that, for $x > D$, it follows that $f(x) > 1000$.
- (*b*) Find another number D such that, for $x > D$, it follows that $f(x) > 1000$.
- (*c*) What is the smallest number D such that, for $x > D$, it follows that $f(x) > 1000$?

3 Let $f(x) = 5x$. Find a number D such that, for all $x > D$, (*a*) $f(x) > 2000$, (*b*) $f(x) > 10{,}000$.

4 Let $f(x) = 6x$. Find a number D such that, for all $x > D$, (*a*) $f(x) > 1200$, (*b*) $f(x) > 1800$.

In Exercises 5 to 12 use the precise definition of the assertion "$\lim_{x\to\infty} f(x) = \infty$" to establish the given limits.

5 $\lim_{x\to\infty} 3x = \infty$ **6** $\lim_{x\to\infty} 4x = \infty$

7 $\lim_{x\to\infty} (x + 5) = \infty$ **8** $\lim_{x\to\infty} (x - 600) = \infty$

9 $\lim_{x\to\infty} (2x + 4) = \infty$

10 $\lim_{x\to\infty} (3x - 1200) = \infty$

11 $\lim_{x\to\infty} (4x + 100 \cos x) = \infty$

12 $\lim_{x\to\infty} (2x - 300 \cos x) = \infty$

13 Let $f(x) = x^2$.
(*a*) Find a number D such that, for $x > D$, $f(x) > 100$.
(*b*) Let E be any nonnegative number. Find a number D such that, for $x > D$, it follows that $f(x) > E$.
(*c*) Let E be any negative number. Find a number D such that, for $x > D$, it follows that $f(x) > E$.
(*d*) Using the precise definition of "$\lim_{x\to\infty} f(x) = \infty$," show that $\lim_{x\to\infty} x^2 = \infty$.

14 Using the precise definition of "$\lim_{x\to\infty} f(x) = \infty$," show that $\lim_{x\to\infty} x^3 = \infty$.

Exercises 15 to 22 concern the precise definition of "$\lim_{x\to\infty} f(x) = L$."

15 Let $f(x) = 3 + 1/x$ if $x \neq 0$.
(*a*) Find a number D such that, for $x > D$, it follows that $|f(x) - 3| < \frac{1}{10}$.
(*b*) Find another number D such that, for $x > D$, it follows that $|f(x) - 3| < \frac{1}{10}$.
(*c*) What is the smallest number D such that, for $x > D$, it follows that $|f(x) - 3| < \frac{1}{10}$?
(*d*) Using the precise definition of "$\lim_{x\to\infty} f(x) = L$," show that $\lim_{x\to\infty} (3 + 1/x) = 3$.

16 Let $f(x) = 2/x$ if $x \neq 0$.
(*a*) Find a number D such that, for $x > D$, it follows that $|f(x) - 0| < \frac{1}{100}$.
(*b*) Find another number D such that, for $x > D$, it follows that $|f(x) - 0| < \frac{1}{100}$.
(*c*) What is the smallest number D such that, for $x > D$, it follows that $|f(x) - 0| < \frac{1}{100}$?
(*d*) Using the precise definition of "$\lim_{x\to\infty} f(x) = L$," show that $\lim_{x\to\infty} 2/x = 0$.

In Exercises 17 to 22 use the precise definition of "$\lim_{x\to\infty} f(x) = L$" to establish the given limits.

17 $\lim_{x\to\infty} [(\sin x)/x] = 0$ (*Hint:* $|\sin x| \leq 1$ for all x.)

18 $\lim_{x\to\infty} [(x + \cos x)/x] = 1$

19 $\lim_{x\to\infty} 4/x^2 = 0$

20 $\lim_{x\to\infty} [(2x + 3)/x] = 2$

21 $\lim_{x\to\infty} [1/(x - 100)] = 0$

22 $\lim_{x\to\infty} [(2x + 10)/(3x - 5)] = \frac{2}{3}$

■

23 Using the precise definition of "$\lim_{x\to\infty} f(x) = \infty$," show that the claim that

$$\lim_{x\to\infty} \frac{x}{x+1} = \infty$$

is false.

24 Using the precise definition of "$\lim_{x\to\infty} f(x) = L$," show that the claim that $\lim_{x\to\infty} \sin x = \frac{1}{2}$ is false.

25 Using the precise definition of "$\lim_{x\to\infty} f(x) = L$," show that the claim that $\lim_{x\to\infty} 3x = 6$ is false.

26 Using the precise definition of "$\lim_{x\to\infty} f(x) = L$," show that for every number L the assertion "$\lim_{x\to\infty} 2x = L$" is false.

In Exercises 27 to 30 develop precise definitions of the given limits. Phrase your definitions in terms of a challenge number E or ϵ and a reply D. Show the geometric meaning of your definition on a graph.

27 $\lim_{x\to\infty} f(x) = -\infty$

28 $\lim_{x\to-\infty} f(x) = \infty$

29 $\lim_{x\to-\infty} f(x) = -\infty$

30 $\lim_{x\to-\infty} f(x) = L$

31 Let $f(x) = 5$ for all x. (*a*) Using the precise definition of "$\lim_{x\to\infty} f(x) = L$," show that $\lim_{x\to\infty} f(x) = 5$. (*b*) Using the precise definition of "$\lim_{x\to-\infty} f(x) = L$," show that $\lim_{x\to-\infty} f(x) = 5$.

2.8 PRECISE DEFINITION OF "$\lim_{x\to a} f(x) = L$"

Recall the informal definition given in Sec. 2.2.

Informal definition of $\lim_{x\to a} f(x) = L$:
Let f be a function and a some fixed number. Assume that the domain of f contains open intervals (c, a) and (a, b) for some number $c < a$ and some number $b > a$.

If, as x approaches a, either from the left or from the right, $f(x)$ approaches a specific number L, then L is called the *limit* of $f(x)$ as x approaches a. This is written

$$\lim_{x\to a} f(x) = L.$$

Keep in mind that a need not be in the domain of f. Even if a happens to be in the domain of f, the value $f(a)$ plays no role in determining whether $\lim_{x\to a} f(x) = L$.

Reworded informal definition of $\lim_{x\to a} f(x) = L$ [assume that $f(x)$ is defined for all x in some intervals (c, a) and (a, b)]:

If x is sufficiently close to a but not equal to a, then $f(x)$ is necessarily near L.

The "ϵ, δ" definition of "$\lim_{x\to a} f(x) = L$"

The precise definition parallels the reworded informal definition. The letter δ that appears in it is the lower case Greek "delta," equivalent to the English letter d.

> *Precise definition of* $\lim_{x\to a} f(x) = L$ [assume that $f(x)$ is defined in some intervals (c, a) and (a, b)]:
>
> For each positive number ϵ there is a positive number δ such that for all x that satisfy the inequality
>
> $$0 < |x - a| < \delta$$
>
> it is true that
>
> $$|f(x) - L| < \epsilon.$$

The meaning of $0 < |x - a| < \delta$

The inequality $0 < |x - a|$ that appears in the definition is just a fancy way of saying "x is not a." The inequality $|x - a| < \delta$ asserts that x is within a distance δ of a. The two inequalities may be combined as the single statement $0 < |x - a| < \delta$, which describes the open interval $(a - \delta, a + \delta)$ from which a is deleted. This deletion is made since the value $f(a)$ plays no role in the definition of $\lim_{x\to a} f(x)$.

Once again ϵ is the challenge. The response is δ. Usually, the smaller ϵ is, the smaller δ will have to be.

The geometric significance of the precise definition of "$\lim_{x\to a} f(x) = L$" is shown in Fig. 2.58. The narrow horizontal band of width 2ϵ is again the challenge. The response is a sufficiently narrow vertical band, of width 2δ, such that the part of the graph within that vertical band (except perhaps at $x = a$) also lies in the challenging horizontal band of width 2ϵ. In Fig. 2.59 the vertical band shown is not narrow enough to meet the challenge of the horizontal band shown. But the vertical band shown in Fig. 2.60 is sufficiently narrow.

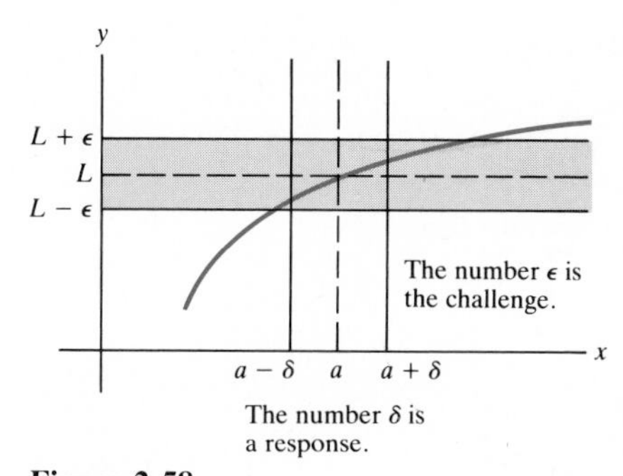

Figure 2.58

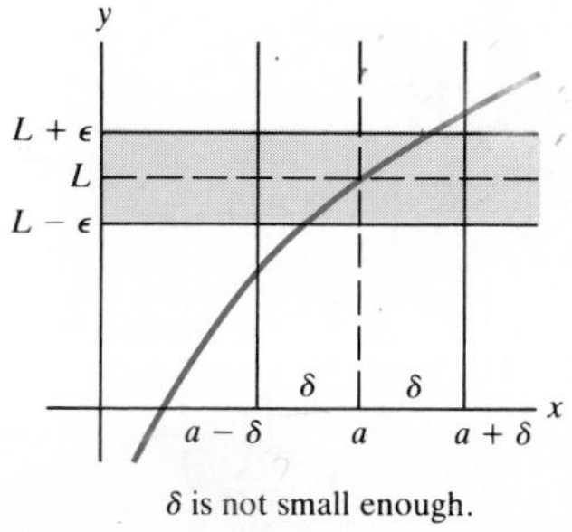

δ is not small enough.

Figure 2.59

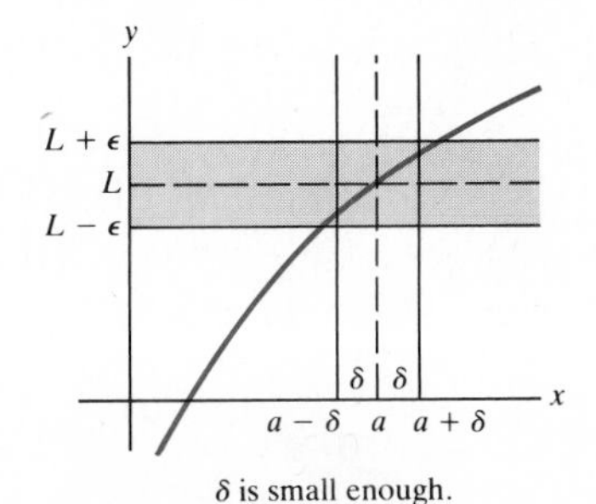

δ is small enough.

Figure 2.60

Assume that for each positive number ϵ it is possible to find a positive number δ such that the parts of the graph between $x = a - \delta$ and $x = a$ and between $x = a$ and $x = a + \delta$ lie within the given horizontal band. Then we say that "as x approaches a, $f(x)$ approaches L." The narrower the horizontal band around the line $y = L$, the smaller δ usually must be.

EXAMPLE 1 Use the precise definition of "$\lim_{x\to a} f(x) = L$" to show that $\lim_{x\to 0} x^2 = 0$.

SOLUTION In this case $a = 0$ and $L = 0$. Let ϵ be a positive number. We wish to find a positive number δ such that for $0 < |x - 0| < \delta$ it follows that $|x^2 - 0| < \epsilon$.

Since $|x|^2 = |x^2|$, we are asking, "for which x is $|x|^2 < \epsilon$?" This inequality is satisfied when

$$|x| < \sqrt{\epsilon}.$$

In other words, when $|x| < \sqrt{\epsilon}$, it follows that $|x^2 - 0| < \epsilon$. Thus $\delta = \sqrt{\epsilon}$ suffices.

(For instance, when $\epsilon = 1$, $\delta = \sqrt{1} = 1$ is a suitable response. When $\epsilon = 0.01$, $\delta = 0.1$ suffices.) ■

EXAMPLE 2 Use the precise definition of "$\lim_{x\to a} f(x) = L$" to show that $\lim_{x\to 2} (3x + 5) = 11$.

SOLUTION Here $a = 2$ and $L = 11$. Let ϵ be a positive number. We wish to find a number $\delta > 0$ such that for $0 < |x - 2| < \delta$ it follows that $|(3x + 5) - 11| < \epsilon$.

So let us find out for which x it is true that $|(3x + 5) - 11| < \epsilon$. This inequality is equivalent to

$$|3x - 6| < \epsilon$$

or

$$3|x - 2| < \epsilon$$

or

$$|x - 2| < \frac{\epsilon}{3}.$$

Thus $\delta = \epsilon/3$ is an adequate response. If $0 < |x - 2| < \epsilon/3$, then $|(3x + 5) - 11| < \epsilon$. ■

The algebra of finding a response δ can be much more involved for other functions, such as $f(x) = ax^2 + bx + c$.

EXERCISES FOR SEC. 2.8: PRECISE DEFINITION OF "$\lim_{x\to a} f(x) = L$"

In Exercises 1 to 4 using the precise definition of "$\lim_{x\to a} f(x) = L$," establish the assertions.

1 $\lim_{x\to 0} \dfrac{x^2}{4} = 0$

2 $\lim_{x\to 0} 4x^2 = 0$

3 $\lim_{x\to 1} (3x + 5) = 8$

4 $\lim_{x\to 1} \dfrac{5x + 3}{4} = 2$

5 Give an example of a number $\delta > 0$ such that $|x^2 - 4| < 1$ if $0 < |x - 2| < \delta$.

6 Give an example of a number $\delta > 0$ such that $|x^2 + x - 2| < 0.5$ if $0 < |x - 1| < \delta$.

■

7 (*a*) Show that, if $0 < \delta < 1$ and $|x - 3| < \delta$, then $|x^2 - 9| < 7\delta$.
(*b*) From (*a*) deduce that $\lim_{x\to 3} x^2 = 9$.

8 (*a*) Show that, if $0 < \delta < 1$ and $|x - 4| < \delta$, then

$$|\sqrt{x} - 2| < \frac{\delta}{\sqrt{3} + 2}.$$

(*Hint:* Rationalize $\sqrt{x} - 2$.)
(*b*) From (*a*) deduce that $\lim_{x\to 4} \sqrt{x} = 2$.

9 (*a*) Show that, if $0 < \delta < 1$ and $|x - 3| < \delta$, then $|x^2 + 5x - 24| < 12\delta$. (*Hint:* First factor $x^2 + 5x - 24$.)
(*b*) From (*a*) deduce that $\lim_{x\to 3} (x^2 + 5x) = 24$.

10 (*a*) Show that, if $0 < \delta < 1$ and $|x - 2| < \delta$, then

$$\left|\frac{1}{x} - \frac{1}{2}\right| < \frac{\delta}{2}.$$

(*b*) From (*a*) deduce that $\lim_{x\to 2} 1/x = \frac{1}{2}$.

In Exercises 11 to 16 develop precise definitions of the given limits. Phrase your definitions in terms of a challenge, E or ϵ, and a response, δ.

11 $\lim_{x\to a^+} f(x) = L$ 12 $\lim_{x\to a^-} f(x) = L$

13 $\lim_{x\to a} f(x) = \infty$ 14 $\lim_{x\to a} f(x) = -\infty$

15 $\lim_{x\to a^+} f(x) = \infty$ 16 $\lim_{x\to a^-} f(x) = \infty$

17 Let $f(x) = 9x^2$.
(*a*) Find $\delta > 0$ such that, for $0 < |x - 0| < \delta$ it follows that $|9x^2 - 0| < \frac{1}{100}$.
(*b*) Let ϵ be any positive number. Find a positive number δ such that, for $0 < |x - 0| < \delta$ it follows that $|9x^2 - 0| < \epsilon$.
(*c*) Show that $\lim_{x\to 0} 9x^2 = 0$.

18 Let $f(x) = x^3$.
(*a*) Find $\delta > 0$ such that for $0 < |x - 0| < \delta$ it follows that $|x^3 - 0| < \frac{1}{1000}$.
(*b*) Show that $\lim_{x\to 0} x^3 = 0$.

19 Show that the assertion "$\lim_{x\to 2} 3x = 5$" is false. To do this, it is necessary to exhibit a positive number ϵ such that there is no response number $\delta > 0$. (*Hint:* Draw a picture.)

20 Use the precise definition of "$\lim_{x\to a} f(x) = L$" to show that $\lim_{x\to 5} x^2 = 25$.

2.S SUMMARY

The two central ideas of this chapter are limits and continuous functions. Limits were used to find asymptotes, as an aid in graphing. Two limits needed in the next chapter were determined:

$$\lim_{\theta\to 0} \frac{\sin\theta}{\theta} = 1 \quad \text{and} \quad \lim_{\theta\to 0} \frac{1 - \cos\theta}{\theta} = 0.$$

The definition of a continuous function is phrased in terms of behavior of the function at and near each number in an interval. Although the definition of continuity depends on the definition of limit, the easiest way to think of a function that is continuous throughout an interval is that its graph is a curve that can be drawn without lifting pencil from paper.

Two properties of continuous functions, known as the maximum-value property and the intermediate-value property, were discussed. They will be used often in later chapters.

The final two sections concerned precise definitions of limits.

Vocabulary and Symbols

limit $\lim_{x\to a} f(x)$, $\lim_{x\to\infty} f(x)$, etc.
right-hand limit $\lim_{x\to a^+} f(x)$
left-hand limit $\lim_{x\to a^-} f(x)$
constant function
polynomial function
rational function
asymptote (vertical, horizontal, tilted)
greatest integer less than or equal to x, $[x]$
continuous at a
continuous on an interval
continuous
sum, difference, product, quotient of functions $f + g$, $f - g$, fg, f/g
maximum value, minimum value, extreme value, extremum
maximum-value theorem
intermediate-value theorem

Key Facts

For the definitions of the various limits, such as

$$\lim_{x \to a} f(x) = L, \qquad \lim_{x \to \infty} f(x) = L, \qquad \text{and} \qquad \lim_{x \to -\infty} f(x) = L,$$

see Secs. 2.1 and 2.2. (Precise definitions are given in Secs. 2.7 and 2.8, which are not covered in this summary.)

PROPERTIES OF LIMITS

If $\lim_{x \to a} f(x)$ and $\lim_{x \to a} g(x)$ both exist, then

$$\lim_{x \to a} (f(x) + g(x)) = \lim_{x \to a} f(x) + \lim_{x \to a} g(x)$$

$$\lim_{x \to a} (f(x) - g(x)) = \lim_{x \to a} f(x) - \lim_{x \to a} g(x)$$

$$\lim_{x \to a} f(x)g(x) = \lim_{x \to a} f(x) \lim_{x \to a} g(x)$$

$$\lim_{x \to a} f(x)/g(x) = \lim_{x \to a} f(x)/\lim_{x \to a} g(x), \qquad \text{if } \lim_{x \to a} g(x) \neq 0$$

$$\lim_{x \to a} f(x)^{g(x)} = \left(\lim_{x \to a} f(x)\right)^{\lim_{x \to a} g(x)}, \qquad \text{if } \lim_{x \to a} f(x) > 0.$$

Limits of Rational Functions

$$\lim_{x \to \infty} \frac{ax^n + \cdots}{bx^m + \cdots} = \lim_{x \to \infty} \frac{ax^n}{bx^m}$$

(The degree of the numerator is n and the degree of the denominator is m.)

Consequently,

$$\lim_{x \to \infty} \frac{ax^n + \cdots}{bx^m + \cdots} = \begin{cases} a/b & \text{if } m = n \\ 0 & \text{if } n < m \\ \infty \text{ or } -\infty & \text{if } n > m \text{ (depending on the signs of } a \text{ and } b). \end{cases}$$

Similar assertions hold for $x \to -\infty$.

CONTINUITY

Let $f(x)$ be defined for all x. Then f is continuous at $x = a$ if

$$\lim_{x \to a} f(x) = f(a).$$

This means that

1 $\lim_{x \to a} f(x)$ exists.

2 $\lim_{x \to a} f(x)$ equals $f(a)$.

A similar definition holds if f, though not defined for all x, is defined at least on some open interval that includes the number a.

For the definitions of "continuous from the right" and "continuous from the left" see Sec. 2.5.

A function that is continuous at each point of an open interval is said to be continuous on that interval. Similar definitions cover functions whose domains are closed intervals.

If f and g are continuous on the same interval, so are $f + g$, $f - g$, and fg. Moreover, f/g is continuous wherever $g(x)$ is not 0. The composition of continuous functions is continuous.

If f is continuous and $\lim_{x\to a} g(x)$ exists and is in the domain of f, then

$$\lim_{x\to a} f(g(x)) = f\left(\lim_{x\to a} g(x)\right).$$

("It is legal to move 'lim' past 'f' if f is continuous.")

> **MAXIMUM-VALUE THEOREM**
>
> Let f be continuous throughout the closed interval $[a, b]$. Then there is at least one number in $[a, b]$ at which f takes on a maximum value. That is, for some number c in $[a, b]$, $f(c) \geq f(x)$, for all x in $[a, b]$.

A corresponding minimum-value theorem also holds.

> **INTERMEDIATE-VALUE THEOREM**
>
> Let f be continuous throughout the closed interval $[a, b]$. Let m be any number between $f(a)$ and $f(b)$. Then there is at least one number c in $[a, b]$ such that $f(c) = m$.

In particular, if f is continuous throughout $[a, b]$ and if one of $f(a)$ and $f(b)$ is negative and the other is positive, then there is a number c in $[a, b]$ such that $f(c) = 0$.

With the aid of the preceding fact, it was shown that a polynomial of odd degree has at least one real root. In other words, the graph of a polynomial of odd degree always crosses the x axis at least once.

GUIDE QUIZ ON CHAP. 2: LIMITS AND CONTINUOUS FUNCTIONS

1 Define "$\lim_{x\to a} f(x) = L$." (The informal definition, not the one in Sec. 2.8.)

2 Define "$\lim_{x\to\infty} f(x) = L$." (The informal definition, not the one in Sec. 2.7.)

3 Figure 2.61 is the graph of a function f whose domain is $[1, 5]$.

(*a*) Does $\lim_{x\to 2} f(x)$ exist?
(*b*) Is f continuous at 2?
(*c*) Does $\lim_{x\to 3} f(x)$ exist?
(*d*) Does $\lim_{x\to 5} f(x)$ exist?
(*e*) Is f continuous on $[1, 5]$?
(*f*) Is f continuous on $(3, 5)$?

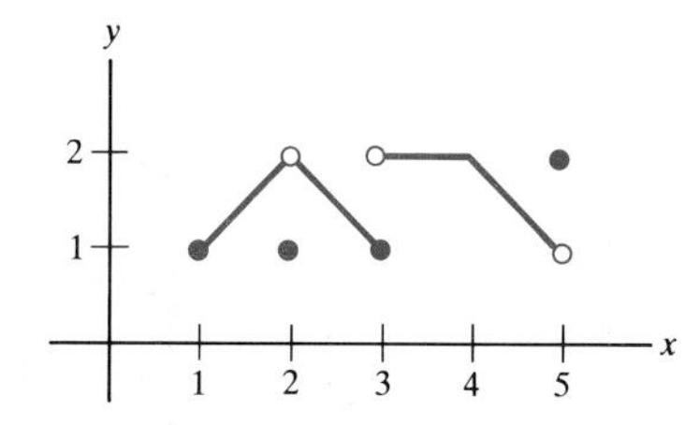

Figure 2.61

4 Examine the following limits:

(*a*) $\lim_{x\to 1} (x^2 + 5x)$ (*b*) $\lim_{x\to\infty} \dfrac{3x^4 - 100x + 3}{5x^4 + 7x - 1}$

(c) $\lim_{x\to 0} \frac{3x^4 - 100x + 3}{5x^4 + 7x - 1}$ (d) $\lim_{x\to -\infty} \frac{500x^3 - x^2 - 5}{x^4 + x}$

(e) $\lim_{t\to 0} \frac{\sin 3t}{6t}$ (f) $\lim_{x\to -\infty} \frac{-6x^5 + 4x}{x^2 + x + 5}$

(g) $\lim_{x\to\infty} 2^{-x}$ (h) $\lim_{x\to 0} \frac{x^3 + 8}{x + 2}$

(i) $\lim_{x\to -2} \frac{x^3 + 8}{x + 2}$ (j) $\lim_{x\to 0} \sin \frac{1}{x}$

(k) $\lim_{x\to\infty} \sin x$ (l) $\lim_{x\to\infty} \frac{1 + 3\cos x}{x^2}$

(m) $\lim_{x\to\infty} (\sqrt{4x^2 + 5x} - \sqrt{4x^2 + x})$

(n) $\lim_{x\to 16} \frac{\sqrt{x} - 4}{x - 16}$

5 If $\lim_{x\to a} f(x) = 3$ and $\lim_{x\to a} g(x) = 4$, what, if anything, can be said about
(a) $\lim_{x\to a} f(x)g(x)$? (b) $\lim_{x\to a} f(x)/g(x)$?
(c) $\lim_{x\to a} [f(x) + g(x)]$? (d) $\lim_{x\to a} [f(x) - 3]/[g(x) - 4]$?
(e) $\lim_{x\to a} [f(x) - 3]^{g(x)}$?

6 If $\lim_{x\to a} f(x) = 0$ and $\lim_{x\to a} g(x) = \infty$, what, if anything, can be said about
(a) $\lim_{x\to a} [f(x) + g(x)]$? (b) $\lim_{x\to a} f(x)/g(x)$?
(c) $\lim_{x\to a} f(x)^{g(x)}$? (d) $\lim_{x\to a} [2 + f(x)]^{g(x)}$?
(e) $\lim_{x\to a} f(x)g(x)$?

7 (a) State the assumptions in the maximum-value theorem.
(b) State the conclusion of the maximum-value theorem.

8 (a) State the assumptions in the intermediate-value theorem.
(b) State the conclusion of the intermediate-value theorem.

REVIEW EXERCISES FOR CHAP. 2: LIMITS AND CONTINUOUS FUNCTIONS

In Exercises 1 to 36 examine the limits. Evaluate those which exist. Determine those which do not exist and, among these, the ones that are infinite.

1 $\lim_{x\to 1} \frac{x^3 + 1}{x^2 + 1}$ **2** $\lim_{x\to 1} \frac{x^3 - 1}{x^2 - 1}$

3 $\lim_{x\to 2} \frac{x^4 - 16}{x^3 - 8}$ **4** $\lim_{x\to 0} \frac{x^4 - 16}{x^3 - 8}$

5 $\lim_{x\to\infty} \frac{x^7 - x^2 + 1}{2x^7 + x^3 + 300}$ **6** $\lim_{x\to -\infty} \frac{x^9 + 6x + 3}{x^{10} - x - 1}$

7 $\lim_{x\to -\infty} \frac{x^3 + 1}{x^2 + 1}$ **8** $\lim_{x\to -\infty} \frac{x^4 + x^2 + 1}{3x^2 + 4}$

9 $\lim_{x\to 4} \frac{\sqrt{x} - 2}{x - 4}$ **10** $\lim_{x\to 81} \frac{x - 81}{\sqrt{x} - 9}$

11 $\lim_{x\to\infty} (\sqrt{x^2 + 2x + 3} - \sqrt{x^2 - 2x + 3})$

12 $\lim_{x\to\infty} (\sqrt{2x^2} - \sqrt{2x^2 - 6x})$

13 $\lim_{x\to 1^+} \frac{1}{x - 1}$ **14** $\lim_{x\to 1^-} \frac{1}{x - 1}$

In Exercises 15 and 16, $[x]$ denotes the "greatest integer" function.

15 $\lim_{x\to 3^-} [2x]$ **16** $\lim_{x\to 3^+} [2x]$

17 $\lim_{x\to 0^+} 2^{1/x}$ **18** $\lim_{x\to 0^-} 2^{1/x}$

19 $\lim_{x\to\infty} 2^{1/x}$ **20** $\lim_{x\to -\infty} 2^{1/x}$

21 $\lim_{x\to\infty} \frac{(x + 1)(x + 2)}{(x + 3)(x + 4)}$ **22** $\lim_{x\to -\infty} \frac{(x + 1)^{100}}{(2x + 50)^{100}}$

23 $\lim_{x\to\pi/2} \frac{\cos x}{1 + \sin x}$ **24** $\lim_{x\to\pi/2} \frac{\cos x}{1 - \sin x}$

25 $\lim_{x\to 0} \frac{\sin x}{3x}$ **26** $\lim_{x\to\infty} \frac{\sin x}{3x}$

27 $\lim_{x\to\pi/2^+} \cos x$ **28** $\lim_{x\to\pi/2^+} \sec x$

29 $\lim_{x\to 0^-} \sin x$ **30** $\lim_{x\to 0^-} \csc x$

31 $\lim_{x\to\infty} \sin \frac{1}{x}$ **32** $\lim_{x\to\infty} x \sin \frac{1}{x}$

33 $\lim_{x\to\pi/4} x^2 \cos x$ **34** $\lim_{x\to\infty} x^2 \cos x$

35 $\lim_{\theta\to\infty} (\cos^2\theta + \sin^2\theta)$ **36** $\lim_{\theta\to\infty} (\cos^2\theta - \sin^2\theta)$

In Exercises 37 to 42 exhibit specific functions f and g that meet all three conditions. (The answers are not unique.)

37 $\lim_{x\to 0} f(x) = 0$, $\lim_{x\to 0} g(x) = 0$, and $\lim_{x\to 0} f(x)/g(x) = 5$

38 $\lim_{x\to\infty} f(x) = 0$, $\lim_{x\to\infty} g(x) = \infty$, and $\lim_{x\to\infty} f(x)g(x) = 20$

39 $\lim_{x\to\infty} f(x) = 0$, $\lim_{x\to\infty} g(x) = \infty$, and $\lim_{x\to\infty} f(x)g(x) = \infty$

40 $\lim_{x\to\infty} f(x) = \infty$, $\lim_{x\to\infty} g(x) = \infty$, and $\lim_{x\to\infty} [f(x) - g(x)] = 3$

41 $\lim_{x\to\infty} f(x) = \infty$, $\lim_{x\to\infty} g(x) = \infty$, and $\lim_{x\to\infty} [f(x) - g(x)] = \infty$

42 $\lim_{x\to\infty} f(x) = \infty$, $\lim_{x\to\infty} g(x) = \infty$, and $\lim_{x\to\infty} f(x)/g(x) = \infty$

43 Does $x + \sin x$ have a maximum value for x in (*a*) $[0, 100]$? (*b*) $[0, \infty)$?

44 Does $x^3 + x + 1$ have a minimum value for x in (*a*) $[-100, 5]$? (*b*) $(-\infty, 5]$?

45 Does $1/(1 + x^2)$ have (*a*) a maximum value for x in $(-1, 1)$? (*b*) a minimum value for x in $(-1, 1)$?

46 Does $1/x^3$ have a maximum value for x in
(*a*) $[2, 100]$? (*b*) $[2, \infty)$?
A minimum value for x in
(*c*) $[2, 100]$? (*d*) $[2, \infty)$?

47 Show that the equation $x^5 = 2^x$ has a solution (*a*) less than 2, and (*b*) greater than 2.

48 Show that the equation $x^3 - 2x^2 - 3x + 1 = 0$ has a solution (*a*) less than 0, (*b*) in $[0, 2]$, and (*c*) larger than 2.

■

49 (*a*) Does $(\sin x - \cos x)^2$ have a maximum value? If so, find it.
(*b*) Does $(\sin x - \cos x)^2$ have a minimum value? If so, find it.

50 Assume that $\lim_{x\to 3} f(x) = 0$ and $\lim_{x\to 3} g(x) = 0$. What, if anything, can be said about
(*a*) $\lim_{x\to 3} [f(x) - g(x)]$? (*b*) $\lim_{x\to 3} \sin f(x)$?
(*c*) $\lim_{x\to 3} \cos f(x)$? (*d*) $\lim_{x\to 3} f(x)g(x)$?
(*e*) $\lim_{x\to 3} [f(x)]^3/g(x)$?

In each of Exercises 51 to 53 verify the intermediate-value theorem for the indicated function, closed interval, and value.

51 2^x, $[1, 8]$, $m = 4$

52 $\sin x$, $[0, 9\pi/2]$, $m = \frac{1}{2}$

53 $\tan x$, $[0, \pi/3]$, $m = 1$

54 Find $\lim_{x\to 0} \dfrac{\tan x - \sin x}{x}$.

55 Find $\lim_{x\to 0} \dfrac{\tan x - \sin x}{x^2}$.

56 Let $f(x) = \sin(1/x)$ if $x \neq 0$. Is it possible to define $f(0)$ in such a way that f is continuous throughout the x axis?

57 Let $f(x) = x \sin(1/x)$ if $x \neq 0$. Is it possible to define $f(0)$ in such a way that f is continuous throughout the x axis?

58 Let f be a continuous function defined on the x axis. Assume that $f(f(f(x))) = x$ for all x. Prove that $f(x) = x$ for all x. [*Hint:* Show that f is either an increasing function or a decreasing function.]

59 Let f be a continuous function such that $f(x)$ is in $[0, 1]$ when x is in $[0, 1]$. Prove that there is at least one number c in $[0, 1]$ such that $f(c) = c$. [*Hint:* Consider the function g given by $g(x) = f(x) - x$.]

60 Let f be a continuous function such that $f(f(x)) = x$ for all x. Prove that there is at least one number c such that $f(c) = c$.

61 If f is a function, then by a **chord** of f we shall mean a line segment whose ends are on the graph of f. Now let f be continuous throughout $[0, 1]$, and let $f(0) = f(1) = 0$.
(*a*) Explain why there is a horizontal chord of f of length $\frac{1}{2}$.
(*b*) Explain why there is a horizontal chord of f of length $1/n$, where $n = 1, 2, 3, 4, \ldots$.
(*c*) Must there exist a horizontal chord of f of length $\frac{2}{3}$?
(*d*) What is the answer to (*c*) if we also demand that $f(x) \geq 0$ for all x in $[0, 1]$?

62 Examine (*a*) $\lim_{x\to 0^+} (4^x + 3^x)^{1/x}$, (*b*) $\lim_{x\to 0^-} (4^x + 3^x)^{1/x}$, and (*c*) $\lim_{x\to\infty} (4^x + 3^x)^{1/x}$.

63 Let K be a bounded convex set and P any point inside K. (*a*) Is there always a chord through P that is divided by P into two pieces one of which is twice as long as the other? (*b*) Is there always a chord through P that is divided by P into pieces of equal length.

64 Let K be a bounded convex set. Is there always a circumscribing (*a*) right triangle? (*b*) equilateral triangle? (*c*) rectangle? (*d*) square?

3 The Derivative

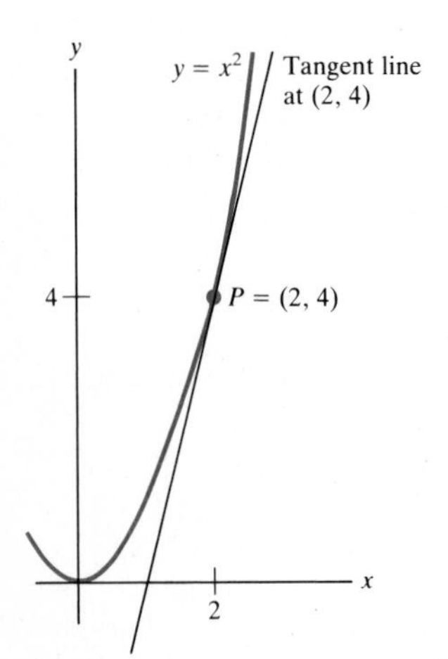

Figure 3.1

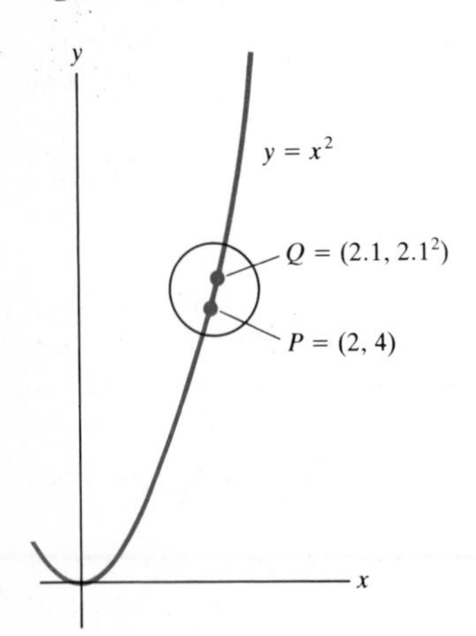

Figure 3.2

This chapter introduces one of the most important concepts of calculus, the derivative. The approach uses geometric and physical illustrations, but a few examples and exercises show that the derivative has far more varied applications.

3.1 FOUR PROBLEMS WITH ONE THEME

This section discusses four problems which at first glance may seem unrelated. The first one concerns the slope of a tangent line to a curve. The second involves velocity. The final two concern magnification and density. A little arithmetic will quickly show that they are all just different versions of one mathematical idea.

PROBLEM 1 *Slope.* What is the slope of the tangent line to the graph of $y = x^2$ at the point $P = (2, 4)$, as shown in Fig. 3.1?

For the present, by the **tangent line** to a curve at a point P on the curve shall be meant the line through P that has the "same direction" as the curve at P. This will be made precise in the next section.

SOLUTION We need two distinct points to determine a line, but the only point we have on the tangent line at (2, 4) is the point (2, 4) itself. To get around this difficulty, we will choose a point Q on the parabola $y = x^2$, near P, and compute the slope of the line through P and Q. Such a line is called a *secant*. For instance, choose $Q = (2.1, 2.1^2)$ and compute the slope of the line through P and Q as shown in Figs. 3.2 and 3.3.

$$\frac{2.1^2 - 2^2}{2.1 - 2} = \frac{4.41 - 4}{0.1} = \frac{0.41}{0.1} = 4.1.$$

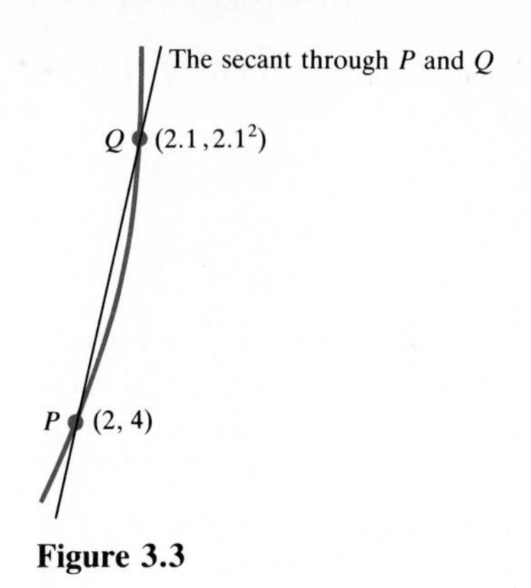

Figure 3.3

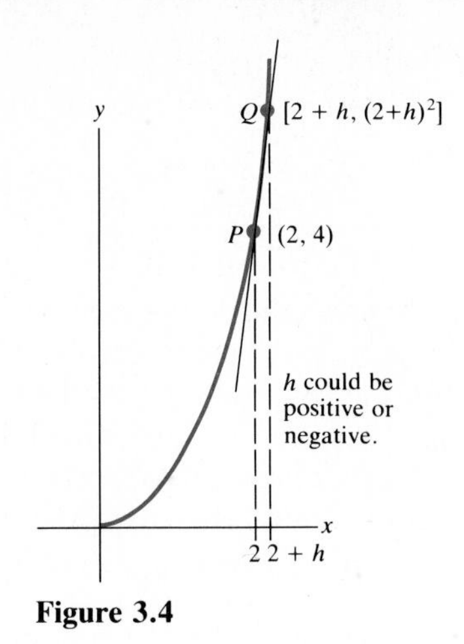

Figure 3.4

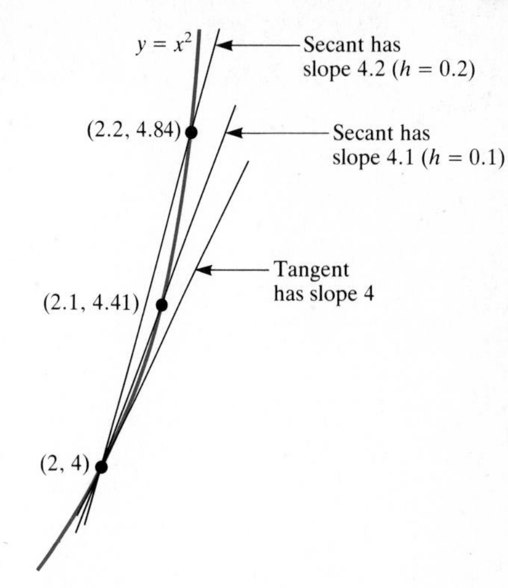

Figure 3.5

Thus an estimate of the slope of the tangent line is 4.1. Note that in making this estimate there was no need to draw the curve.

To obtain a better estimate, we could repeat the process using the line through $P = (2, 4)$ and $Q = (2.01, 2.01^2)$. Rather than do this, it is simpler to consider a *typical* point Q. That is, consider the line through $P = (2, 4)$ and $Q = (2 + h, (2 + h)^2)$ when h is small, either positive or negative. (See Fig. 3.4.) This line has slope

$$\frac{(2 + h)^2 - 2^2}{h}.$$

h could be > 0 *or* < 0.

To find out what happens to this quotient as h gets closer to 0, apply the techniques of limits. We have

$$\lim_{h\to 0} \frac{(2 + h)^2 - 2^2}{h} = \lim_{h\to 0} \frac{4 + 4h + h^2 - 4}{h} = \lim_{h\to 0} (4 + h) = 4.$$

Observe the use of limits.

Thus the tangent line at (2, 4) has slope 4.

Figure 3.5 indicates the idea of the solution by showing how secant lines approximate the tangent line. It suggests the blowup of a small part of the curve, $y = x^2$.

■

PROBLEM 2 *Speed.* A rock initially at rest falls $16t^2$ feet in t seconds. What is its speed after 2 seconds?

SOLUTION If the rock moves a distance of D feet in an interval of time of t seconds, we know what is meant by its average speed during that time, namely, the quotient D/t feet per second. We will use this idea to deal with the much more abstract idea of "speed at a given time," the so-called instantaneous speed. In the next section the notion of speed at a given instant will be made precise.

For practice, make an estimate by finding the average speed of the rock during a short time interval, say from 2 to 2.01 seconds. At the start of this interval the rock has fallen $16(2^2) = 64$ feet. By the end it has fallen

$16(2.01)^2 = 16(4.0401) = 64.6416$ feet. So during 0.01 second it fell 0.6416 feet. Its average speed during this time interval is

$$\frac{0.6416}{0.01} = 64.16 \text{ feet per second,}$$

an estimate of the speed at time $t = 2$ seconds.

Figure 3.6 shows the position of the falling rock from time $t = 2$ seconds to time $t = 2.1$ seconds at intervals of 0.02 second. Note that in the first 0.02 second the rock falls 1.2864 feet and in the last 0.02 second it falls 1.3376 feet. Clearly, it is speeding up.

Although we will keep $h > 0$, estimates could just as well be made with $h < 0$.

Rather than making another estimate with the aid of a still shorter interval of time, let us consider the typical time interval from 2 to $2 + h$ seconds, $h > 0$. During this short time of h seconds the rock travels $16(2 + h)^2 - 16 \cdot 2^2 = 16[(2 + h)^2 - 2^2]$ feet, as shown in Fig. 3.7. The average speed of the rock during this period is

$$\frac{16[(2 + h)^2 - 2^2]}{h} \text{ feet per second.}$$

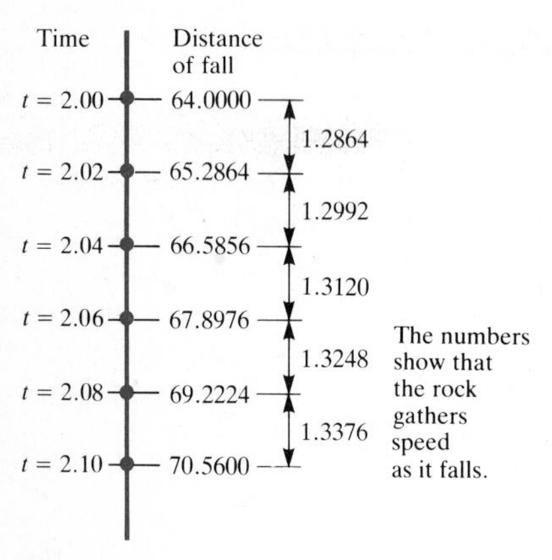

Figure 3.6

When h is close to 0, what happens to this average speed? It approaches

$$\lim_{h \to 0} \frac{16[(2 + h)^2 - 2^2]}{h} = 16 \lim_{h \to 0} \frac{(2 + h)^2 - 2^2}{h}$$
$$= 16 \lim_{h \to 0} (4 + h) = 16 \cdot 4 = 64 \text{ feet per second.} \blacksquare$$

Even though Problems 1 and 2 seem unrelated at first, their solutions turn out to be practically identical: The slope in Problem 1 is approximated by the quotient

$$\frac{(2 + h)^2 - 2^2}{h},$$

and the speed in Problem 2 is approximated by the quotient

$$\frac{16(2 + h)^2 - 16 \cdot 2^2}{h} = 16 \frac{(2 + h)^2 - 2^2}{h}.$$

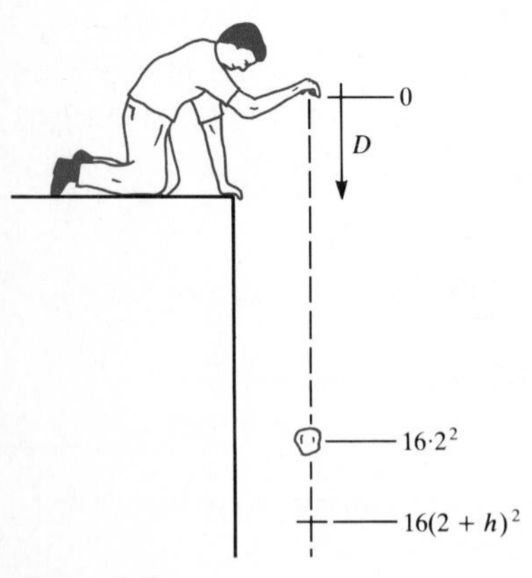

Figure 3.7

The only difference between the solutions is that the second quotient has an extra factor, 16.

The third problem concerns magnification, a concept that occurs in everyday life. For instance, photographs can be blown up or reduced in size, with each part magnified or shrunk by the same factor. However, magnification may vary from point to point, as with a curved mirror. If the projection of an interval of length L is an interval of length L^*, we say that the interval on the photograph is magnified by the factor L^*/L.

As with slope and speed, we treat "magnification at a point" intuitively. The next section will make it precise.

PROBLEM 3 *Magnification.* A light, two lines (a slide and a screen), and a complicated lens are placed as in Fig. 3.8. This arrangement projects the point on the bottom line, whose coordinate is x, to the point on the

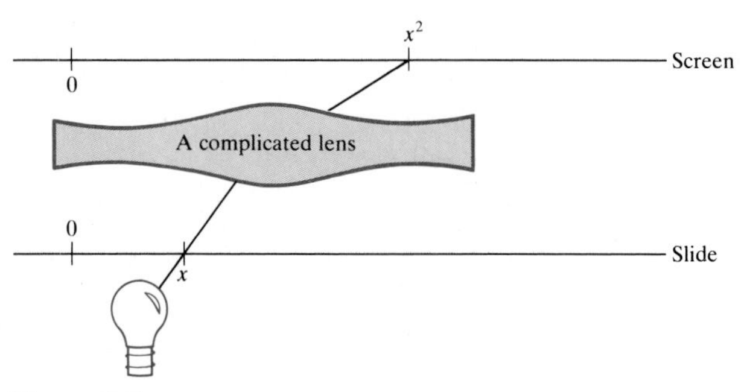

Figure 3.8

top line, whose coordinate is x^2. For example, 2 is projected onto 4, and 3 onto 9. The projection of the interval [2, 3] is [4, 9], which is five times as long. The magnification of the interval [2, 3] is said to be 5. The projection of the interval $[0, \frac{1}{3}]$ is $[0, \frac{1}{9}]$, which is only one-third as long. In this case, the interval is magnified by a factor of $\frac{1}{3}$. For large x, the lens magnifies to a great extent; for x near 0, the lens markedly reduces. What is the "magnification at $x = 2$"?

SOLUTION The lens projects the point having the coordinate 2 onto the point having the coordinate 2^2. More concisely, the image of 2 is $2^2 = 4$. The image of 3 is $3^2 = 9$; the image of 5 is $5^2 = 25$; and so on. Let us join some sample points to their images by straight lines, as in Fig. 3.9. This diagram shows that the interval [2, 3] on the slide is magnified to become the interval [4, 9] on the screen, a fivefold magnification. Similarly, [3, 4] on the slide has as its image on the screen [9, 16], a sevenfold magnification. The magnifying power of the lens increases from left to right.

To estimate the magnification at 2 on the slide, examine the projection of a small interval in the vicinity of 2. Let us see what the image of [2, 2.1] is on the screen. Since the image of 2 is 2^2 and the image of 2.1 is 2.1^2, the image of the interval [2, 2.1] of length 0.1 is the interval $[2^2, 2.1^2]$ of length

$$2.1^2 - 2^2 = 0.41.$$

The magnifying factor over the interval [2, 2.1] is

$$\frac{0.41}{0.1} = 4.1.$$

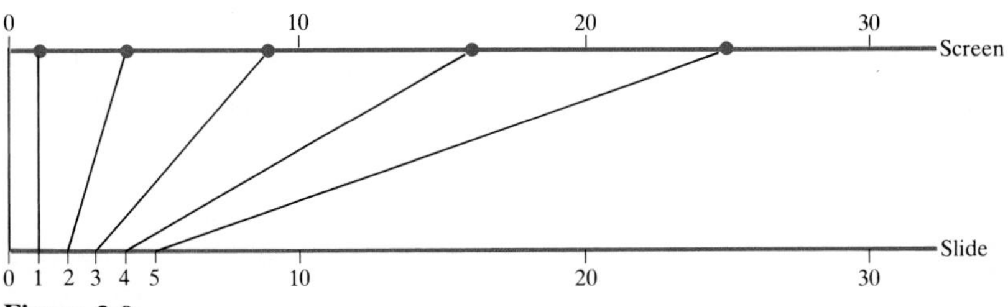

Figure 3.9

This number, 4.1, is an estimate of the magnification at $x = 2$.

You have probably guessed the next step. Rather than go on and consider the magnification of another specific interval, such as the interval [2, 2.01], we go directly to a typical interval with 2 as its left end.

For convenience we use $h > 0$. Estimates could be made with $h < 0$ also.

The image of the interval $[2, 2 + h]$, where h is greater than 0, is the interval $[2^2, (2 + h)^2]$. Since $[2, 2 + h]$ has length h and $[2^2, (2 + h)^2]$ has length $(2 + h)^2 - 2^2$, the magnification of the interval $[2, 2 + h]$ is

$$\frac{(2 + h)^2 - 2^2}{h} = 4 + h, \qquad h > 0.$$

As already observed, when h approaches 0, this quotient approaches 4. Thus the magnification at 2 is 4. In terms of limits,

$$\text{Magnification at } 2 = \lim_{h \to 0} \frac{(2 + h)^2 - 2^2}{h} = 4. \quad ■$$

The next problem is concerned with density, which is a measure of the heaviness of a material. Density is defined as the quotient

Density is mass divided by volume.

$$\frac{\text{Total mass}}{\text{Total volume}}.$$

Water has a density of 1 gram per cubic centimeter, while the density of lead is 11.3 grams per cubic centimeter. Air has a density of only 0.0013 gram per cubic centimeter. The density of an object may vary from point to point. For instance, the density of matter near the center of the earth is much greater than that near the surface. In fact, the *average* density of the earth is 5.5 grams per cubic centimeter, more than five times that of water.

The idea of density provides a concrete analog of several mathematical ideas and will be referred to frequently in later chapters. The next problem concerns a string of varying density. This density will be considered in terms of grams per linear centimeter rather than grams per cubic centimeter. The matter is imagined as a continuous distribution, not composed of isolated molecules.

Later chapters also will be concerned with density of matter distributed along a curve, in a flat object, and in a solid. The notion of "density at a point" will be made precise in the next section. For the moment, we deal with it intuitively.

PROBLEM 4 *Density*. The mass of the left-hand x centimeters of a nonhomogeneous string 10 centimeters long is x^2 grams, as shown in Fig. 3.10. For instance, the left half has a mass of 25 grams, while the whole string has a mass of 100 grams. Clearly, the right half is denser than the

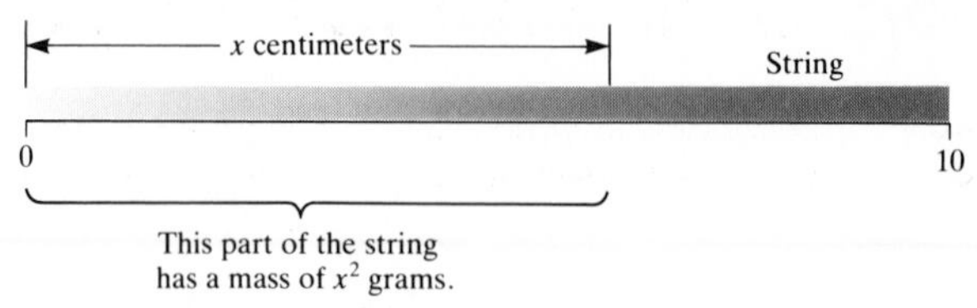

Figure 3.10

2.1

0 1 2 3 4 5 6 7 8 9 10

Figure 3.11

left half. What is the density, in grams per centimeter, of the material at $x = 2$?

SOLUTION To estimate the density of the string 2 centimeters from its left end, examine the mass of the material in the interval [2, 2.1]. (See Fig. 3.11.)

The material in the interval [2, 2.1] has a mass of $2.1^2 - 2^2$ grams, which equals $4.41 - 4 = 0.41$ gram. Thus the average density for this interval is $0.41/0.1 = 4.1$ grams per centimeter.

Rather than make another estimate, consider the density in the typical small interval $[2, 2 + h]$. The mass in this interval is

$$(2 + h)^2 - 2^2 \text{ grams.}$$

The interval has length h centimeters. Thus the density of matter in this interval is

$$\frac{(2 + h)^2 - 2^2}{h} = 4 + h \text{ grams per centimeter.}$$

As h approaches 0, this quotient approaches 4, and we say that the density 2 centimeters from the left end of the string is 4 grams per centimeter.

In terms of limits,

$$\text{Density at } 2 = \lim_{h \to 0} \frac{(2 + h)^2 - 2^2}{h} = 4. \quad \blacksquare$$

From a mathematical point of view, the problems of finding the slope of the tangent line, the speed of the rock, the magnification of the lens, and the density along the string are the same. Each leads to the same type of quotient as an estimate. In each case, the behavior of this quotient is studied as h approaches 0. In each case, the answer is a limit.

The underlying mathematical theme is explored in the next section, which introduces the derivative and will enable us to give precise definitions of "slope," "speed," "magnification," and "density" when these vary from point to point.

EXERCISES FOR SEC. 3.1: FOUR PROBLEMS WITH ONE THEME

Exercises 1 to 8 concern slope. In each case use the technique of Problem 1 to find the slope of the tangent line to the curve at the point.

1 $y = x^2$ at the point $(3, 3^2) = (3, 9)$

2 $y = x^2$ at the point $(\frac{1}{2}, (\frac{1}{2})^2) = (\frac{1}{2}, \frac{1}{4})$

3 $y = x^2$ at the point $(-2, (-2)^2) = (-2, 4)$

4 $y = x^2$ at the point $(1, 1^2) = (1, 1)$

5 $y = x^3$ at $(2, 2^3) = (2, 8)$ [*Hint:* Recall that $(a + b)^3 = a^3 + 3a^2b + 3ab^2 + b^3$.]

6 $y = x^3$ at $(1, 1^3) = (1, 1)$

7 (*a*) $y = x^2$ at (0, 0)
(*b*) Sketch the graph of $y = x^2$ and the tangent line at (0, 0).
8 (*a*) $y = x^3$ at (0, 0)
(*b*) Sketch the graph of $y = x^3$ and the tangent line at (0, 0). [Be especially careful when sketching the graph near (0, 0).]

In Exercises 9 to 12 use the method of Problem 2 to find the speed of the rock after

9 3 seconds
10 $\frac{1}{2}$ second
11 1 second
12 $\frac{1}{4}$ second

13 A certain object travels t^3 feet in the first t seconds.
(*a*) How far does it travel during the time interval from 2 to 2.1 seconds?
(*b*) What is its average speed during that time interval?
(*c*) Let h be any positive number. Find the average speed of the object from time 2 to time $2 + h$ seconds.
(*d*) Find the speed of the object at time 2 seconds by letting h approach 0 in part (*c*).

14 A certain object travels t^3 feet in the first t seconds.
(*a*) Find its average speed during the time interval from 3 to 3.01 seconds.
(*b*) Find its average speed during the time interval from 3 to $3 + h$ seconds, $h > 0$.
(*c*) By letting h approach 0 in part (*b*), find the speed of the object at time 3 seconds.

In the slope problem the nearby point Q was always pictured as being to the right of P. The point Q could just as well have been chosen to the left of P. Exercises 15 and 16 illustrate this case.

15 Consider the parabola $y = x^2$.
(*a*) Find the slope of the line through $P = (2, 4)$ and $Q = (1.9, 1.9^2)$.
(*b*) Find the slope of the line through $P = (2, 4)$ and $Q = (1.99, 1.99^2)$.
(*c*) Find the slope of the line through $P = (2, 4)$ and $Q = (2 + h, (2 + h)^2)$, where $h < 0$.
(*d*) Show that as h approaches 0, the slope in (*c*) approaches 4.

16 Consider the curve $y = x^3$.
(*a*) Find the slope of the line through $P = (2, 2^3)$ and $Q = (1.9, 1.9^3)$.
(*b*) Find the slope of the line through $P = (2, 2^3)$ and $Q = (2 + h, (2 + h)^3)$, where $h < 0$.
(*c*) Show that as h approaches 0, the slope in (*b*) approaches 12.

The next two exercises are intended to emphasize the limitation of graphs in finding the slope of a tangent line.

17 (*a*) Draw the curve $y = x^2$ as carefully as you can.
(*b*) Draw as carefully as you can the tangent line at (4, 16).
(*c*) Using a ruler or the scale on your graph, estimate the slope of the line you drew in (*b*).
(*d*) Find the slope of the line through (4, 16) and the nearby point $(4.01, 4.01^2)$.
(*e*) Find the slope of the line through (4, 16) and the nearby point $(3.99, 3.99^2)$.
(*f*) How does your result in (*c*) compare with those in (*d*) and (*e*)?

18 (*a*) Draw the curve $y = x^2$ as carefully as you can.
(*b*) Draw by eye the tangent line at $(-1, 1)$.
(*c*) What is the slope of the line you drew in (*b*)?
(*d*) Examining the appropriate quotients, show that the tangent line at $(-1, 1)$ has slope -2.

Exercises 19 to 22 concern magnification.

19 By what factor does the lens in Problem 3 magnify the interval (*a*) [1, 1.1]? (*b*) [1, 1.01]? (*c*) [1, 1.001]? (*d*) Find the magnification at 1.

20 By what factor does the lens in Problem 3 magnify the interval (*a*) [3, 3.1]? (*b*) [3, 3.01]? (*c*) [3, 3.001]? (*d*) Find the magnification at 3.

21 By what factor does the lens in Problem 3 magnify the interval (*a*) [0.49, 0.5]? (*b*) [0.499, 0.5]? (*c*) Find the magnification at 0.5 by examining the magnification of intervals of the form $[0.5 + h, 0.5]$, where $h < 0$.

22 By what factor does the lens in Problem 3 magnify the interval (*a*) [1.49, 1.5]? (*b*) [1.499, 1.5]? (*c*) Find the magnification at 1.5 by examining the typical interval of the form $[1.5 + h, 1.5]$, where $h < 0$.

Exercises 23 and 24 concern density.

23 The left x centimeters of a string have a mass of x^2 grams.
(*a*) What is the mass in the interval [3, 3.01]?
(*b*) Using the interval [3, 3.01], estimate the density at 3.
(*c*) Using the interval [2.99, 3], estimate the density at 3.
(*d*) By considering intervals of the form $[3, 3 + h]$, $h > 0$, find the density at the point 3 centimeters from the left end.
(*e*) By considering intervals of the form $[3 + h, 3]$, $h < 0$, find the density at the point 3 centimeters from the left end.

24 The left x centimeters of a string have a mass of x^2 grams.
(*a*) What is the mass in the interval [2, 2.01]?
(*b*) Using the interval [2, 2.01], estimate the density at 2.
(*c*) Using the interval [1.99, 2], estimate the density at 2.
(*d*) By considering intervals of the form $[2, 2 + h]$, $h > 0$, find the density at the point 2 centimeters from the left end.
(*e*) By considering intervals of the form $[2 + h, 2]$, $h < 0$, find the density at the point 2 centimeters from the left end.

25 Find an equation of the tangent line to $y = x^2$ at the point $(-1, 1)$.

26 Find an equation of the tangent line to $y = x^3$ at the point $(\frac{1}{2}, \frac{1}{8})$.

■

The next two exercises show that the idea common to the four problems in this section also appears in biology and economics.

27 A certain bacterial culture has a mass of t^2 grams after t minutes of growth.
(*a*) How much does it grow during the time interval [2, 2.01]?
(*b*) What is its rate of growth during the time interval [2, 2.01]?
(*c*) What is its rate of growth when $t = 2$?

28 A thriving business has a profit of t^2 million dollars in its first t years. Thus from time $t = 3$ to time $t = 3.5$ (the first half of its fourth year) it has a profit of $(3.5)^2 - 3^2$ million dollars, giving an annual rate of

$$\frac{(3.5)^2 - 3^2}{0.5} = 6.5 \text{ million dollars per year.}$$

(*a*) What is its annual rate of profit during the time interval [3, 3.1]?
(*b*) What is its annual rate of profit during the time interval [3, 3.01]?
(*c*) What is its annual rate of profit after 3 years?

29 (*a*) Graph the curve $y = 2x^2 + x$.
(*b*) By eye, draw the tangent line to the curve at the point (1, 3). Using a ruler, estimate the slope of the tangent line.
(*c*) Sketch the line that passes through the point (1, 3) and the point $[1 + h, 2(1 + h)^2 + (1 + h)]$.
(*d*) Find the slope of the line in (*c*).
(*e*) Letting x get closer and closer to 1, find the slope of the tangent line at (1, 3). How close was your estimate in (*b*)?

30 An object travels $2t^2 + t$ feet in t seconds.
(*a*) Find its average speed during the interval of time $[1, 1 + h]$, where h is greater than 0.
(*b*) Letting h get closer and closer to 0, find the speed at time 1.

31 Find the slope of the tangent line to the curve $y = x^2$ of Problem 1 at the typical point $P = (x, x^2)$. To do this, consider the slope of the line through P and the nearby point $Q = (x + h, (x + h)^2)$ and let h approach 0.

32 Find the speed of the falling rock of Problem 2 at any time t. To do this, consider the average speed during the time interval $[t, t + h]$ and then let h approach 0.

33 Find the magnification of the lens in Problem 3 at the typical point x by considering the magnification of the short interval $[x, x + h]$, where $h > 0$, and then let h approach 0.

34 Find the density of the string in Problem 4 at a typical point x centimeters from the left end. To do this, consider the mass in a short interval $[x, x + h]$, where $h > 0$, and let h approach 0.

35 (*a*) Sketch the curve $y = x^3 - x^2$.
(*b*) Using the method of the nearby point, find the slope of the tangent line to the curve at the typical point $(x, x^3 - x^2)$.
(*c*) Find all points on the curve where the tangent line is horizontal.
(*d*) Find all points where the tangent line has slope 1.

36 Answer the same questions as in Exercise 35 for the curve $y = x^3 - x$.

37 Does the tangent line to the curve $y = x^2$ at the point (1, 1) pass through the point (6, 12)?

38 An astronaut is traveling from left to right along the curve $y = x^2$. When she shuts off the engine, she will fly off along the line tangent to the curve at the point where she is at that moment. At what point should she shut off the engine in order to reach the point
(*a*) (4, 9)? (*b*) (4, −9)?

39 See Exercises 36 and 38. Where can an astronaut who is traveling from left to right along $y = x^3 - x$ shut off the engine and pass through the point (2, 2)?

3.2 THE DERIVATIVE

The solution of the slope problem in Sec. 3.1 (as well as those of the magnification problem and the density problem) led to the limit

$$\lim_{h \to 0} \frac{(2 + h)^2 - 2^2}{h}.$$

The speed problem involved a similar limit,

$$\lim_{h \to 0} \frac{16(2 + h)^2 - 16 \cdot 2^2}{h}.$$

These limits arose from the particular formulas x^2 and $16t^2$ that had been picked. In each case we formed a **difference quotient**,

$$\frac{\text{Difference in outputs}}{\text{Difference in inputs}},$$

and examined its limit as the change in the inputs was made smaller and smaller.

The whole procedure can be carried out for functions other than x^2 and $16t^2$ and at numbers other than 2.

The four problems in Sec. 3.1 had one theme in common.

The underlying common theme of the four problems in Sec. 3.1 is the important mathematical concept, the **derivative** of a numerical function, which will now be defined.

In the following definition x is fixed and h approaches 0.

Definition *The derivative of a function at the number x.* Let f be a function that is defined at least in some open interval that contains the number x. If

$$\lim_{h \to 0} \frac{f(x + h) - f(x)}{h}$$

The f' notation

exists, it is called the **derivative of f at x** and is denoted $f'(x)$. The function is said to be **differentiable** at x.

The derivative at an endpoint

If the function f is defined only to the right of x, in an interval of the form $[x, b)$, then in the definition of the derivative "$h \to 0$" would be replaced by "$h \to 0^+$." The function is then said to be "differentiable on the right." A similar stipulation is made if f is defined only in an interval of the form $(a, x]$, and the function is said to be "differentiable on the left."

The numerator, $f(x + h) - f(x)$, is the change, or difference, in the outputs; the denominator, h, is the change in the inputs. (See Fig. 3.12.) Keep in mind that $x + h$ can be either to the right or left of x. Similarly, $f(x + h)$ can be either larger or smaller than $f(x)$.

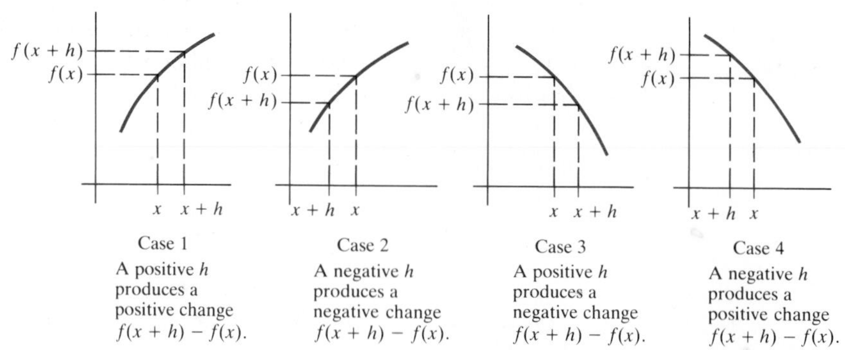

Figure 3.12

A few examples will illustrate the concept of the derivative.

EXAMPLE 1 Find the derivative of the squaring function at the number 2.

SOLUTION In this case, $f(x) = x^2$ for any input x. By definition, the derivative of this function at 2 is

$$\lim_{h\to 0} \frac{(2 + h)^2 - 2^2}{h} = \lim_{h\to 0} (4 + h) = 4.$$

We say that "the derivative of the function x^2 at 2 is 4." ■

The next example determines the derivative of the squaring function at any input, not just at 2.

EXAMPLE 2 Find the derivative of the function x^2 at any number x.

SOLUTION By definition, the derivative at x is

$$\lim_{h\to 0} \frac{(x + h)^2 - x^2}{h} = \lim_{h\to 0} \frac{x^2 + 2hx + h^2 - x^2}{h}$$

$$= \lim_{h\to 0} \frac{2hx + h^2}{h} = \lim_{h\to 0} (2x + h) = 2x.$$

The derivative of the squaring function at x is $2x$. ■

That the derivative of the function x^2 is the function $2x$ is denoted

$$(x^2)' = 2x.$$

This notation is convenient when dealing with a specific function. [*Warning:* Don't replace x by a specific number in this notation. For instance, do not write that $(3^2)'$ equals $2 \cdot 3$. This is not correct.]

The result in Example 2 can be interpreted in terms of each of the four problems in Sec. 3.1. For example, we now know from Example 2 that the slope of the tangent line to the parabola $y = x^2$ at the point (x, x^2) is $2x$. In particular, the slope of the tangent line at $(2, 2^2)$ is $2 \cdot 2 = 4$, a result found in Sec. 3.1. Also, according to the formula for the derivative, $(x^2)' = 2x$, the slope of the tangent line to $y = x^2$ at $(-2, (-2)^2)$ is $2 \cdot (-2) = -4$ and at $(0, 0)$ is $2 \cdot 0 = 0$. A glance at Fig. 3.13 shows that these are reasonable results. The derivative of x^2 is a function. It assigns to the number x the slope of the tangent line to the parabola $y = x^2$ at the point (x, x^2).

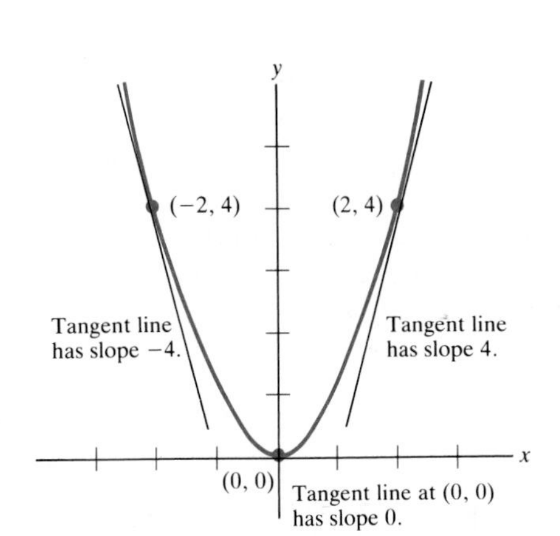

Figure 3.13

The next two examples illustrate the idea of the derivative with functions other than x^2.

EXAMPLE 3 Find $f'(x)$ if $f(x) = x^3$.

Keep in mind that x is fixed and that $h \to 0$.

SOLUTION In this case, $f(x + h) = (x + h)^3$ and $f(x) = x^3$. The derivative of the function at x is therefore

$$\lim_{h\to 0} \frac{(x + h)^3 - x^3}{h} = \lim_{h\to 0} \frac{x^3 + 3x^2h + 3xh^2 + h^3 - x^3}{h}$$

$$= \lim_{h\to 0} \frac{3x^2h + 3xh^2 + h^3}{h} = \lim_{h\to 0} (3x^2 + 3xh + h^2) = 3x^2.$$

The derivative of x^3 at x is $3x^2$. ■

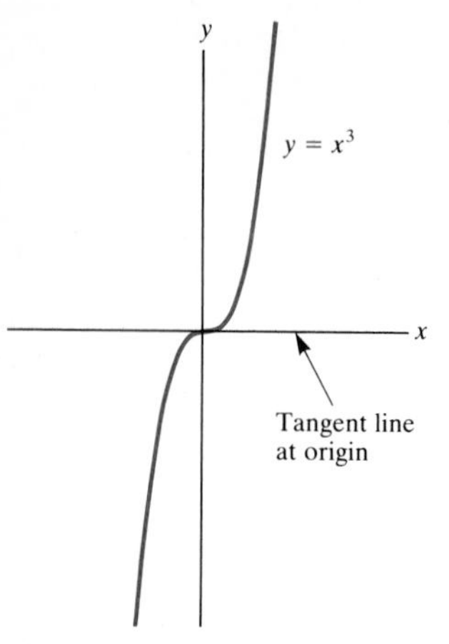

Figure 3.14

In view of Example 3, we may say that "the derivative of the function x^3 is the function $3x^2$" and write $(x^3)' = 3x^2$.

Example 3 tells us, for instance, that the slope of the graph of $y = x^3$ is never negative (since x^2 is never negative). Moreover, at $x = 0$ the slope is $3 \cdot 0^2 = 0$. Thus the tangent line to the curve $y = x^3$ at the origin is horizontal, as is shown in Fig. 3.14.

It may seem strange that a tangent line can *cross the curve,* as this tangent line does. However, the basic property of a tangent line is that it indicates the direction of a curve at a point. In high school geometry, where only tangent lines to circles are considered, the tangent line never crosses the curve.

Example 3 can also be interpreted in terms of the speed of a moving object. If an object moves x^3 feet in the first x seconds, its speed after x seconds is $3x^2$ feet per second.

EXAMPLE 4 Find the derivative of the square root function $f(x) = \sqrt{x}$.

SOLUTION Since the domain of $\sqrt{x}$ contains no negative numbers, assume that $x \geq 0$. If $x > 0$, then, by definition of the derivative,

When finding a derivative, be sure to write "lim" at each step.

$$f'(x) = \lim_{h\to 0} \frac{\sqrt{x+h} - \sqrt{x}}{h}$$

$$= \lim_{h\to 0} \left(\frac{\sqrt{x+h} - \sqrt{x}}{h}\right)\left(\frac{\sqrt{x+h} + \sqrt{x}}{\sqrt{x+h} + \sqrt{x}}\right)$$

$$= \lim_{h\to 0} \frac{x+h-x}{h(\sqrt{x+h} + \sqrt{x})}$$

$$= \lim_{h\to 0} \frac{1}{\sqrt{x+h} + \sqrt{x}}$$

$$= \frac{1}{2\sqrt{x}}.$$

Memorize $(\sqrt{x})' = \dfrac{1}{2\sqrt{x}}$.

If $x = 0$, the limit we are considering is

$$\lim_{h\to 0} \frac{1}{\sqrt{0+h} + \sqrt{0}},$$

which does not exist. We say that the derivative of $\sqrt{x}$ does not exist at 0. ■

According to Example 4, $(\sqrt{x})' = 1/(2\sqrt{x})$. Is this result reasonable? It says that when x is large, the slope of the tangent line at $(x, \sqrt{x})$ is near 0 [since $1/(2\sqrt{x})$ is near 0]. Let us draw the graph and see. First we make a brief table, as shown in the margin. With the aid of these six points, the graph can be sketched. (See Fig. 3.15.) For points far to the right on the graph the tangent line is indeed almost horizontal, as the formula $1/(2\sqrt{x})$ suggests. When x is near 0, the derivative $1/(2\sqrt{x})$ is large. The graph gets steeper and steeper near $x = 0$.

x	0	1	4	9	16	25
$\sqrt{x}$	0	1	2	3	4	5

In Examples 2 and 3 it was shown that

$$(x^2)' = 2x \quad \text{and} \quad (x^3)' = 3x^2.$$

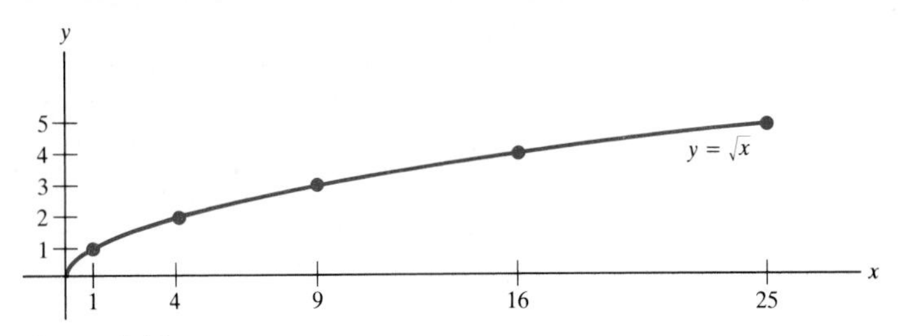

Figure 3.15

Both are special cases of the following theorem.

Theorem 1 For each positive integer n,

$$(x^n)' = nx^{n-1}.$$

Proof It is necessary to find what happens to the quotient

$$\frac{(x + h)^n - x^n}{h}$$

when $h \to 0$.

When you multiply out $(x + h)(x + h) \cdots (x + h)$ (n factors), you get x^n once and $x^{n-1}h$ n times; all other terms you get have h^2 as a factor. Thus

$$\frac{(x + h)^n - x^n}{h} = \frac{x^n + nx^{n-1}h + (\text{terms that involve } h^2) - x^n}{h}$$

$$= nx^{n-1} + \text{terms involving } h.$$

Why do the "terms involving h" disappear?

Hence $$\lim_{h\to 0} \frac{(x + h)^n - x^n}{h} = \lim_{h\to 0} (nx^{n-1} + \text{terms involving } h)$$
$$= nx^{n-1}.$$

Consequently, the derivative of x^n is nx^{n-1}. ■

Direct application of this theorem yields, for instance:

The derivative of x^4 is $4x^{4-1} = 4x^3$.

The derivative of x^3 is $3x^{3-1} = 3x^2$.

The derivative of x^2 is $2x^{2-1} = 2x$.

The derivative of x^1 is $1x^0 = 1$ (in agreement with the fact that the line given by the formula $y = x$ has slope 1).

The next theorem generalizes the fact that $(\sqrt{x})' = 1/(2\sqrt{x})$, which can be written

$$(x^{1/2})' = \tfrac{1}{2}x^{1/2-1}.$$

Theorem 2 For each positive integer n,

$$(x^{1/n})' = \frac{1}{n}x^{1/n-1}$$

(for those x at which both $x^{1/n}$ and $x^{1/n-1}$ are defined).

Proof We must examine

$$\lim_{h\to 0} \frac{(x+h)^{1/n} - x^{1/n}}{h}.$$

Check the identity by multiplying out the right side for $n = 2, 3, 4$.

As in Example 4, we rationalize the numerator, using the algebraic identity

$$d^n - c^n = (d-c)\underbrace{(d^{n-1} + d^{n-2}c + \cdots + dc^{n-2} + c^{n-1})}_{n \text{ summands}},$$

as follows. Let $d = (x+h)^{1/n}$ and $c = x^{1/n}$. Then

$$\begin{aligned}
&\frac{(x+h)^{1/n} - x^{1/n}}{h} \\
&= \frac{(x+h)^{1/n} - x^{1/n}}{h} \frac{[(x+h)^{1/n}]^{n-1} + [(x+h)^{1/n}]^{n-2}x^{1/n} + \cdots + (x^{1/n})^{n-1}}{[(x+h)^{1/n}]^{n-1} + [(x+h)^{1/n}]^{n-2}x^{1/n} + \cdots + (x^{1/n})^{n-1}} \\
&= \frac{[(x+h)^{1/n}]^n - (x^{1/n})^n}{h} \frac{1}{[(x+h)^{1/n}]^{n-1} + [(x+h)^{1/n}]^{n-2}x^{1/n} + \cdots + (x^{1/n})^{n-1}} \\
&= \frac{x+h-x}{h} \frac{1}{\underbrace{[(x+h)^{1/n}]^{n-1} + [(x+h)^{1/n}]^{n-2}x^{1/n} + \cdots + (x^{1/n})^{n-1}}_{n \text{ summands}}}.
\end{aligned}$$

$\dfrac{x+h-x}{h} = 1.$

Thus
$$\lim_{h\to 0} \frac{(x+h)^{1/n} - x^{1/n}}{h} = \frac{1}{(x^{1/n})^{n-1} + (x^{1/n})^{n-1} + \cdots + (x^{1/n})^{n-1}}$$
$$= \frac{1}{nx^{(n-1)/n}} = \frac{1}{n}x^{1/n-1}. \quad \blacksquare$$

The derivative of x^a is ax^{a-1} for fixed rational exponent a.

Theorems 1 and 2 both fit into the same pattern: to find the derivative of x^a, for a fixed exponent a, lower the exponent by 1 and multiply by the original exponent. In Sec. 3.6 this rule will be extended to all rational exponents a and in Chap. 6 to all irrational exponents as well. At this point the rule is established only when the exponent is a positive integer or the reciprocal of a positive integer.

EXAMPLE 5 Use Theorem 2 to find $(\sqrt[3]{x})'$ at $x = 8$.

SOLUTION $\sqrt[3]{x} = x^{1/3}$. Thus

$$(\sqrt[3]{x})' = (x^{1/3})' = \tfrac{1}{3}x^{1/3-1}$$
$$= \tfrac{1}{3}x^{-2/3} = \frac{1}{3}\frac{1}{x^{2/3}}.$$

In particular, at $x = 8$, the derivative is

$$\frac{1}{3}\cdot\frac{1}{8^{2/3}} = \frac{1}{3}\cdot\frac{1}{4} = \frac{1}{12}. \quad \blacksquare$$

Now that we have the concept of the derivative, we are in a position to define **tangent line**, **speed**, **magnification**, and **density**, terms used only intuitively until now. These definitions are suggested by the similarity of the computations made in the four problems in Sec. 3.1.

The slope of a nonvertical line equals the quotient $(y_2 - y_1)/(x_2 - x_1)$, where $P_1 = (x_1, y_1)$ and $P_2 = (x_2, y_2)$ are any two distinct points on the line. Now it is possible to define the slope of a curve at a point on the curve.

(In all five definitions it is assumed that the derivative exists.)

Definition *Slope of a curve.* The **slope** of the graph of the function f at $(x, f(x))$ is the derivative of f at x.

Definition *Tangent line to a curve.* The **tangent line** to the graph of the function f at the point $P = (x, y)$ is the line through P that has a slope equal to the derivative of f at x.

Definition *Velocity and speed of a particle moving on a line.* The **velocity** at time t of an object whose position on a line at time t is given by $f(t)$ is the derivative of f at time t. The **speed** of the particle is the absolute value of the velocity.

Note the distinction between velocity and speed. Velocity can be negative; speed is either positive or 0.

Definition *Magnification of a linear projector.* The **magnification** at x of a lens that projects the point x of one line onto the point $f(x)$ of another line is the derivative of f at x.

Definition *Density of material.* The **density** at x of material distributed along a line in such a way that the left-hand x centimeters have a mass of $f(x)$ grams is equal to the derivative of f at x.

Slope, velocity, magnification, and density are just interpretations, or applications, of the derivative. But biology, economics, chemistry, engineering, physics, computer models, and management use the derivative both to describe the concept "the rate at which some quantity is changing" and as a device for calculating that rate of change. The derivative itself is a purely mathematical concept; it is a special limit formed in a certain way from a function:

$$\lim_{h \to 0} \frac{f(x + h) - f(x)}{h}.$$

EXERCISES FOR SEC. 3.2: THE DERIVATIVE

In Exercises 1 to 16 use the definition of the derivative to find the derivatives of the given functions.

1 x^4 **2** x^5

3 $2x$ **4** $5x$

5 $x^2 + 3$ **6** $4x^2 + 5$

7 $-5x^2 + 4x$ **8** $2x^3 - 4x^2 + 5$

9 $7\sqrt{x}$ **10** $x^2 + 3\sqrt{x}$

11 $1/x$ **12** $1/(x + 1)$

13 $1/x^2$ **14** $\sqrt{x} + 5/x$

15 $3 - 6/x$ **16** $1/x^3$

In Exercises 17 to 24, use Theorems 1 and 2 to find the derivatives of the given functions at the given numbers.

17 x^4 at -1 **18** x^4 at $\frac{1}{2}$

19 x^5 at a **20** x^5 at $\sqrt{2}$

21 $\sqrt[3]{t}$ at -8

22 $\sqrt[3]{t}$ at -1

23 $\sqrt[4]{x}$ at 16

24 $\sqrt[4]{x}$ at 81

■

25 Let $f(x) = x^4$. (*a*) What is the slope of the line joining $(1, 1)$ to $(1.1, 1.1^4)$? (*b*) What is the slope of the tangent line to the curve at the point $(1, 1)$?

26 An object travels t^4 feet in the first t seconds.
(*a*) What is its average velocity from time $t = 2$ to time $t = 2.01$?
(*b*) What is its average velocity from time $t = 1.99$ to time $t = 2$?
(*c*) What is its velocity at time $t = 2$?

27 A lens projects x on the slide to x^4 on the screen.
(*a*) How much does it magnify the interval $[1, 1.01]$?
(*b*) What is its magnification at $x = 1$?

28 The left x centimeters of a string have a mass of x^3 grams.
(*a*) What is the average density of the interval $[2, 2.01]$?
(*b*) What is the average density of the interval $[1.99, 2.01]$?
(*c*) What is the density at $x = 2$?

29 (*a*) Show that the tangent line to the curve $y = x^3$ at $(1, 1)$ passes through $(2, 4)$.
(*b*) Use (*a*) to draw the tangent to the curve $y = x^3$ at $(1, 1)$. (It is not necessary to draw the curve.)

30 (*a*) Using the definition of the derivative, find the derivative of $x + x^2$.
(*b*) Graph $y = x + x^2$.
(*c*) For which x is the derivative in (*a*) equal to 0?
(*d*) Using (*c*), find at which point on the graph of $y = x + x^2$ the slope is 0.
(*e*) In view of (*d*), what do you think is the smallest possible value of $x + x^2$?

31 Figure 3.16 shows the graph of a function f whose domain is the interval $[0, 4]$. [Solid dots indicate $f(1)$ and $f(2)$.]

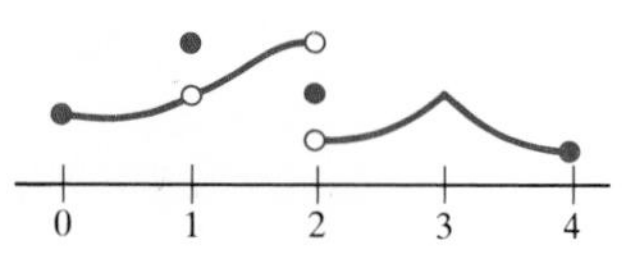

Figure 3.16

(*a*) For which numbers a does $\lim_{x\to a} f(x)$ *not* exist?
(*b*) At which numbers a is f not continuous?
(*c*) At which numbers a is f not differentiable?

3.3 THE DERIVATIVE AND CONTINUITY; ANTIDERIVATIVES

After presenting another notation for the difference quotient

$$\frac{f(x + h) - f(x)}{h}$$

and the derivative $f'(x)$, this section shows the relation between "having a derivative" and "being continuous." It concludes by introducing the notion of an "antiderivative."

Δ is the capital Greek letter corresponding to the English D.

It is also common to give the difference or change h the name Δx ("delta x"). The difference quotient then takes the form

$$\frac{f(x + \Delta x) - f(x)}{\Delta x},$$

Δx is not a product.

and the derivative is defined as

$$f'(x) = \lim_{\Delta x \to 0} \frac{f(x + \Delta x) - f(x)}{\Delta x}.$$

Furthermore, the difference in the outputs is often named Δf or Δy:

$$f(x + \Delta x) - f(x) = \Delta f$$

and so

$$f(x + \Delta x) = f(x) + \Delta f.$$

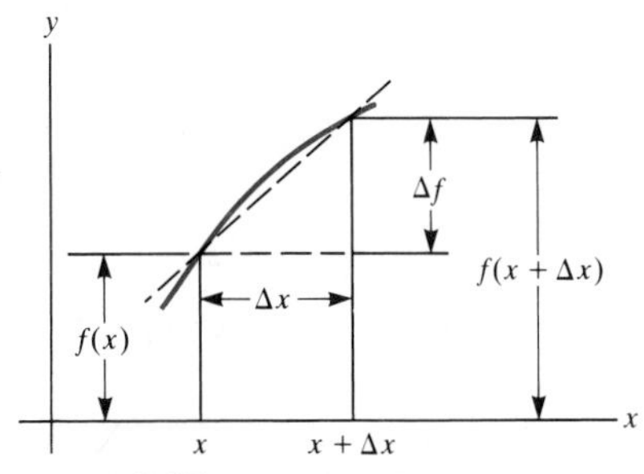

Figure 3.17

The latter equation says that "the value of the function at $x + \Delta x$ is equal to the value of the function at x plus the change in the function." With Δx denoting the change in the inputs and Δf denoting the change in the outputs, we have

$$f'(x) = \lim_{\Delta x \to 0} \frac{\Delta f}{\Delta x}.$$

Figure 3.17 illustrates the Δ notation for the difference quotient.

EXAMPLE 1 Find $(x^2)'$ using the Δ notation.

SOLUTION By the definition of the derivative, the derivative of the squaring function at x is

$$\lim_{\Delta x \to 0} \frac{(x + \Delta x)^2 - x^2}{\Delta x}.$$

Since $$(x + \Delta x)^2 = x^2 + 2x \cdot \Delta x + (\Delta x)^2,$$

the limit equals $$\lim_{\Delta x \to 0} \frac{x^2 + 2x \cdot \Delta x + (\Delta x)^2 - x^2}{\Delta x}.$$

Then we have

$$\begin{aligned} (x^2)' &= \lim_{\Delta x \to 0} \frac{2x\,\Delta x + (\Delta x)^2}{\Delta x} && \text{algebra} \\ &= \lim_{\Delta x \to 0} \frac{\Delta x(2x + \Delta x)}{\Delta x} && \text{factoring} \\ &= \lim_{\Delta x \to 0} (2x + \Delta x) && \text{canceling } \Delta x \neq 0 \\ &= 2x. \end{aligned}$$

So the derivative of x^2 is $2x$, in agreement with the result in Example 2 of the preceding section. ■

It would be of value to compare the two different calculations of $(x^2)'$.

Notations for the Derivative The derivative $f'(x)$ is also commonly denoted

$$\frac{df}{dx} \quad \text{or} \quad D(f).$$

$\frac{dy}{dx}$ is the notation used by Leibniz.

If $f(x)$ is denoted y, the derivative is also denoted

$$\frac{dy}{dx} \quad \text{or} \quad D(y).$$

For instance, $$\frac{d(x^3)}{dx} = 3x^2 \quad \text{or} \quad D(x^3) = 3x^2.$$

If the formula for the function is long, it is customary to write

$$\frac{d}{dx}(f(x)) \qquad \text{instead of} \qquad \frac{d(f(x))}{dx}.$$

For instance, the notation

$$\frac{d}{dx}(x^3 - x^2 + 6x + \sqrt{x})$$

is clearer than
$$\frac{d(x^3 - x^2 + 6x + \sqrt{x})}{dx}.$$

If the variable is denoted by some letter other than x, such as t (for time), we write, for instance,

$$\frac{d(t^3)}{dt} = 3t^2, \qquad \frac{d}{dt}(t^3) = 3t^2, \qquad \text{and} \qquad D(t^3) = 3t^2.$$

dy and dx are without meaning at this point.

Keep in mind that in the notations df/dx and dy/dx, the symbols df, dy, and dx have no meaning by themselves. The symbol dy/dx should be thought of as a single entity, just like the numeral 8, which we do not think of as formed of two 0's.

The dot notation

In the study of motion, Newton's dot notation is often used. If x is a function of time t, then $\dot{x}$ denotes the derivative dx/dt.

In the next theorem we will be concerned with $f'(a)$, the derivative at a number a. This is simply

$$\lim_{h \to 0} \frac{f(a + h) - f(a)}{h}.$$

Now, if we let $x = a + h$, hence $h = x - a$, this limit reads

$$\lim_{x \to a} \frac{f(x) - f(a)}{x - a}.$$

In some expositions this is used as the definition of the derivative at a, $f'(a)$.

If f is differentiable at each number x in some interval, it is said to be **differentiable** throughout that interval.

A very small piece of the graph of a differentiable function looks almost like a straight line, as shown in Fig. 3.18. In this sense, the differentiable functions are even "better" than the continuous functions. The following theorem shows that a differentiable function is necessarily continuous. However, a function can be continuous without being differentiable, as Example 2 will show.

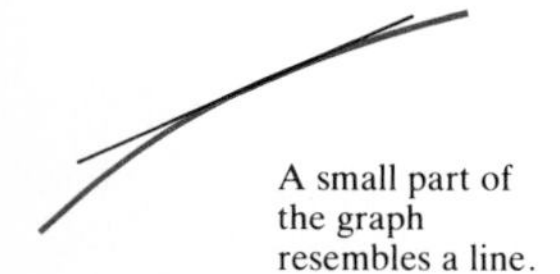

Figure 3.18

Theorem If f is differentiable at a, then it is continuous at a.

Proof We must show that $\lim_{x \to a} f(x)$ exists and equals $f(a)$. To accomplish both goals at one time, we will show that

$$\lim_{x \to a} (f(x) - f(a)) = 0.$$

For then
$$\lim_{x \to a} f(x) = \lim_{x \to a} [(f(x) - f(a)) + f(a)]$$
$$= 0 + f(a) = f(a).$$

To begin, note that for $x \neq a$,

$$f(x) - f(a) = \frac{f(x) - f(a)}{x - a}(x - a).$$

Thus,
$$\lim_{x \to a} (f(x) - f(a)) = \lim_{x \to a} \frac{f(x) - f(a)}{x - a}(x - a)$$

$$= \lim_{x \to a} \frac{f(x) - f(a)}{x - a} \lim_{x \to a} (x - a) \qquad \text{(since both limits exist)}$$

$$= f'(a) \cdot 0 = 0.$$

Thus f is continuous at a. ■

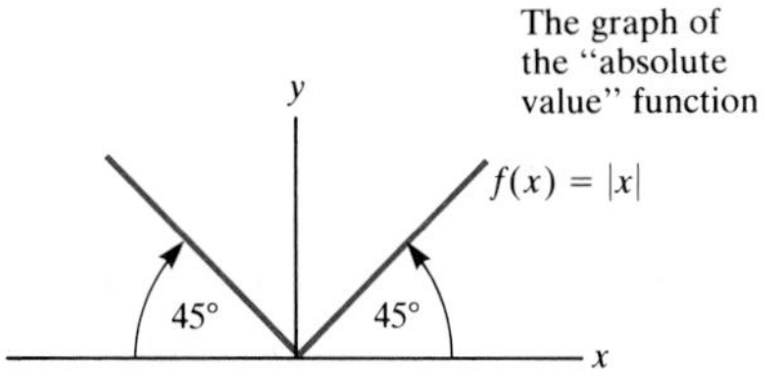

The graph of the "absolute value" function

Figure 3.19

The converse of this theorem is not true. The function $|x|$ is continuous at 0, but, as the graph in Fig. 3.19 suggests, it is not differentiable at 0. This is demonstrated in Example 2.

EXAMPLE 2 Show that the absolute-value function $f(x) = |x|$ is not differentiable at 0.

SOLUTION Since $\lim_{x \to 0} |x| = 0 = f(0)$, the function $f(x) = |x|$ is continuous at $x = 0$. To show that f is not differentiable at $x = 0$, we must show that

$$\lim_{\Delta x \to 0} \frac{|\Delta x| - |0|}{\Delta x - 0}$$

does not exist. To accomplish this, we will examine the right- and left-hand limits at 0.

As $\Delta x \to 0$ from the right, Δx is positive. Thus

$$\lim_{\Delta x \to 0^+} \frac{|\Delta x| - |0|}{\Delta x - 0} = \lim_{\Delta x \to 0^+} \frac{\Delta x - 0}{\Delta x - 0} = \lim_{\Delta x \to 0^+} \frac{\Delta x}{\Delta x}$$

$$= \lim_{\Delta x \to 0^+} 1 = 1.$$

As $\Delta x \to 0$ from the left, Δx is negative. Thus

$$\lim_{\Delta x \to 0^-} \frac{|\Delta x| - |0|}{\Delta x - 0} = \lim_{\Delta x \to 0^-} \frac{-\Delta x - 0}{\Delta x - 0} = \lim_{\Delta x \to 0^-} \frac{-\Delta x}{\Delta x}$$

$$= \lim_{\Delta x \to 0^-} (-1) = -1.$$

Since the left-hand limit at 0 does not equal the right-hand limit at 0,

$$\lim_{\Delta x \to 0} \frac{|\Delta x| - |0|}{\Delta x - 0}$$

does not exist. Thus $|x|$ is not differentiable at 0. ■

If f and F are two functions and f is the derivative of F, then F is called an **antiderivative** of f. For instance, since $(x^3)' = 3x^2$, the function x^3 is an antiderivative of $3x^2$. This concept will play an important role beginning with Chap. 5. As more derivatives are computed later in this chapter, practice in finding antiderivatives will be provided. Techniques for finding antiderivatives are given in Chap. 7.

$x^3 + 2001$ *is also an antiderivative of* $3x^2$, *as can be checked.*

EXERCISES FOR SEC. 3.3: THE DERIVATIVE AND CONTINUITY; ANTIDERIVATIVES

In Exercises 1 to 10 use the Δ notation to find the given derivatives.

1 $\dfrac{d(x^3)}{dx}$

2 $\dfrac{d(x^4)}{dx}$

3 $\dfrac{d(\sqrt{x})}{dx}$

4 $\dfrac{d(3\sqrt{x})}{dx}$

5 $\dfrac{d(5x^2)}{dx}$

6 $\dfrac{d}{dx}(5x^2 + 3x + 2)$

7 $D\left(\dfrac{3}{x}\right)$

8 $D\left(\dfrac{5}{x^2}\right)$

9 $\dfrac{d}{dx}\left(\dfrac{3}{x} - 4x + 2\right)$

10 $\dfrac{d}{dx}(x^3 - 5x + 1982)$

11 Let $f(x) = x^2$. Find Δf if (*a*) $x = 1$ and $\Delta x = 0.1$, (*b*) $x = 3$ and $\Delta x = -0.1$.

12 Let $f(x) = 1/x$. Find Δf if (*a*) $x = 2$ and $\Delta x = \frac{1}{5}$, (*b*) $x = 2$ and $\Delta x = -\frac{1}{8}$.

13 Using the Δ notation, find the derivatives of
(*a*) $6x + 3$,
(*b*) $6x + 7$,
(*c*) $6x - 273$,
(*d*) $6x + C$ for any constant C.
(*e*) Give five different antiderivatives of the constant function $f(x) = 6$.

14 (*a*) Using the Δ notation, find the derivative of the function $x^3/3 + C$ for any constant C.
(*b*) How many different antiderivatives does the function x^2 have?

■

15 Using the definition of the derivative, find the derivative of (*a*) x^4, (*b*) $17x^4$, and (*c*) kx^4 for any constant k. (*d*) Give an example of an antiderivative for the function x^3.

16 (*a*) Find the derivative of kx^5 for any constant k, using the definition of the derivative.
(*b*) Give an example of an antiderivative for the function x^4.

17 (*a*) Let k and C be constants. Find the derivative of $kx^6 + C$, using the Δ notation.
(*b*) Give three different antiderivatives of x^5.

18 Figure 3.20 is the graph of a hypothetical function f.

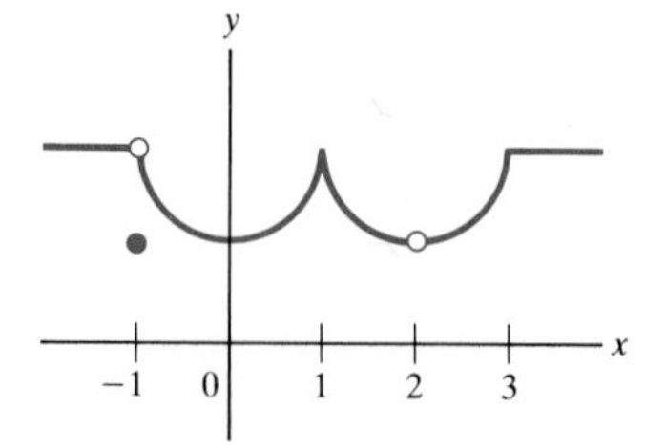

Figure 3.20

(*a*) For which numbers a does $\lim_{x \to a} f(x)$ exist, but f is not continuous at a?
(*b*) For which numbers a is f continuous at a, but not differentiable at a?

19 Answer the same questions as in Exercise 18 for Figure 3.21.

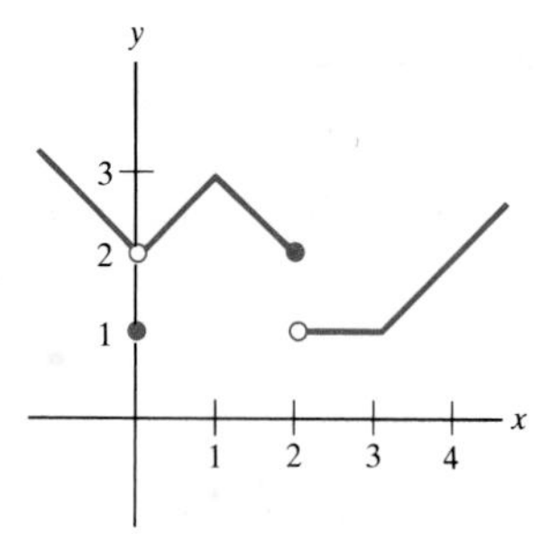

Figure 3.21

20 Answer the same questions as in Exercise 18 for Figure 3.22.

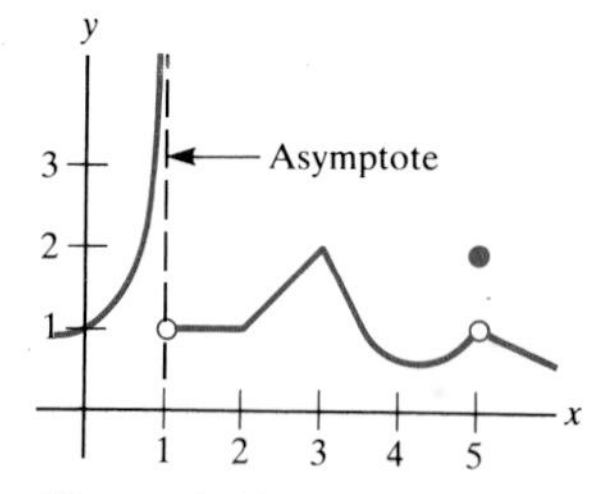

Figure 3.22

21 Let $f(x) = x^3$.
(*a*) Graph f.
(*b*) On the graph show x, $x + \Delta x$, Δx, $f(x)$, $f(x + \Delta x)$, and Δf for $x = 2$ and $\Delta x = 0.3$.

In Exercises 22 to 25 use the definition of the derivative to find:

22 $\dfrac{d}{dx}(2x^3 - 3x^2 + 4x - 5)$

23 $D\left(\dfrac{1}{3x + 5}\right)$

24 $D(\sin x)$. [*Hint:* Use the identity for $\sin (A + B)$.]

25 $D(\cos x)$.

26 (Calculator) Let $f(x) = 2^x$. (*a*) Compute the quotient $[f(0 + \Delta x) - f(0)]/\Delta x$ for $\Delta x = 0.1, 0.01, -0.1$, and -0.01. (*b*) What would be your estimate of $(2^x)'$ at $x = 0$.

27 Give an antiderivative of (*a*) $2x$, (*b*) x, (*c*) 7, (*d*) x^2.

28 Give an antiderivative of (*a*) $4x^3$, (*b*) x^3, (*c*) x^4, (*d*) $1/\sqrt{x}$.

3.4 THE DERIVATIVES OF THE SUM, DIFFERENCE, PRODUCT, AND QUOTIENT

This section develops methods for finding derivatives of functions, or what is called **differentiating** functions. With these methods it will be a routine matter to find, for instance, the derivative of

$$\frac{(1 + \sqrt{x})(x^3 + 1)}{x^2 + 5x + 3}$$

without going back to the definition of the derivative and (at great effort) finding the limit of a difference quotient.

Before developing the methods, it will be useful to find the derivative of any constant function.

Theorem 1 The derivative of a constant function is 0; in symbols,

$$(c)' = 0 \qquad \text{or} \qquad \frac{dc}{dx} = 0 \qquad \text{or} \qquad D(c) = 0.$$

Proof Let c be a fixed number and let f be the constant function, $f(x) = c$ for all inputs x. Then

$$f'(x) = \lim_{\Delta x \to 0} \frac{f(x + \Delta x) - f(x)}{\Delta x}.$$

Since the function f has the same output c for all inputs,

$$f(x + \Delta x) = c \qquad \text{and} \qquad f(x) = c.$$

Thus

$$\begin{aligned} f'(x) &= \lim_{\Delta x \to 0} \frac{c - c}{\Delta x} = \lim_{\Delta x \to 0} \frac{0}{\Delta x} \\ &= \lim_{\Delta x \to 0} 0 \qquad (\text{since } \Delta x \neq 0) \\ &= 0. \end{aligned}$$

This shows that the derivative of any constant function is 0 for all x. ■

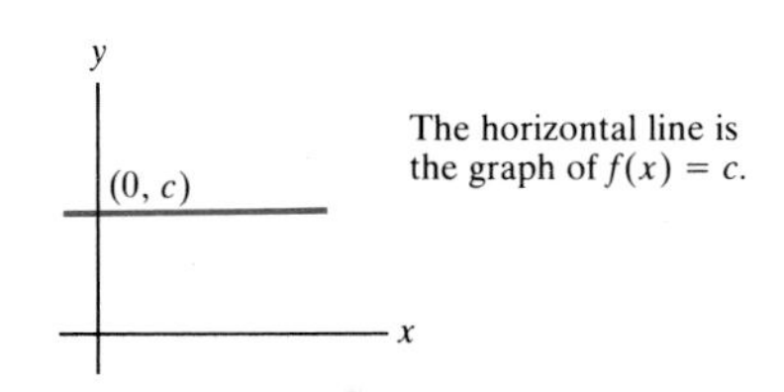

The horizontal line is the graph of $f(x) = c$.

Figure 3.23

From two points of view, Theorem 1 is no surprise: Since the graph of $f(x) = c$ is a horizontal line, it coincides with each of its tangent lines, as can be seen in Fig. 3.23. Also, if we think of x as time and $f(x)$ as the position of a particle, Theorem 1 implies that a stationary particle has zero velocity.

EXAMPLE 1 Differentiate the function π^3.

SOLUTION The number π^3, in this case, is short for the constant function

$$f(x) = \pi^3$$

$c' = 0$ no matter how fancy the constant c may be.

for all x. Since f is constant, $f'(x) = 0$. Thus

$$(\pi^3)' = 0. \quad \blacksquare$$

The derivatives of $f + g$ and $f - g$

The next theorem asserts that if the functions f and g have derivatives at a certain number x, then so does their sum $f + g$, and

$$\frac{d}{dx}(f + g) = \frac{df}{dx} + \frac{dg}{dx}.$$

In other words, "the derivative of the sum is the sum of the derivatives." A similar formula holds for the derivative of $f - g$.

Theorem 2 If f and g are differentiable functions, then so is $f + g$. Its derivative is given by the formula

$$(f + g)' = f' + g'.$$

Similarly, $$(f - g)' = f' - g'.$$

Proof Give the function $f + g$ the name u. That is,

$$u(x) = f(x) + g(x).$$

Then $$u(x + \Delta x) = f(x + \Delta x) + g(x + \Delta x).$$

So,

$$\begin{aligned}\Delta u &= u(x + \Delta x) - u(x)\\ &= [f(x + \Delta x) + g(x + \Delta x)] - [f(x) + g(x)]\\ &= [f(x + \Delta x) - f(x)] + [g(x + \Delta x) - g(x)]\\ &= \Delta f + \Delta g.\end{aligned}$$

Thus

$$\begin{aligned}u'(x) &= \lim_{\Delta x \to 0} \frac{\Delta u}{\Delta x} = \lim_{\Delta x \to 0} \frac{\Delta f + \Delta g}{\Delta x}\\ &= \lim_{\Delta x \to 0}\left(\frac{\Delta f}{\Delta x} + \frac{\Delta g}{\Delta x}\right) = \lim_{\Delta x \to 0} \frac{\Delta f}{\Delta x} + \lim_{\Delta x \to 0} \frac{\Delta g}{\Delta x}\\ &= f'(x) + g'(x).\end{aligned}$$

Hence $f + g$ is differentiable, and

$$(f + g)' = f' + g'.$$

A similar argument applies to $f - g$. $\blacksquare$

Theorem 2 extends to any finite number of differentiable functions. For example,

$$(f + g + h)' = f' + g' + h'$$

and $$(f - g + h)' = f' - g' + h'.$$

EXAMPLE 2 Using Theorem 2, differentiate $x^2 + x^3$.

SOLUTION $\frac{d}{dx}(x^2 + x^3) = \frac{d}{dx}(x^2) + \frac{d}{dx}(x^3)$ Theorem 2

$= 2x + 3x^2$ $D(x^n) = nx^{n-1}$. ■

EXAMPLE 3 Differentiate $x^4 - \sqrt{x} + 5$.

SOLUTION $\frac{d}{dx}(x^4 - \sqrt{x} + 5) = \frac{d}{dx}(x^4) - \frac{d}{dx}(\sqrt{x}) + \frac{d}{dx}(5)$ Theorem 2

$$= 4x^3 - \frac{1}{2\sqrt{x}} + 0 = 4x^3 - \frac{1}{2\sqrt{x}}.$$ ■

The derivative of fg

The following theorem concerning the derivative of the product of two functions may be surprising, for it turns out that the derivative of the product is *not* the product of the derivatives. The formula is more complicated than that for the derivative of the sum; hence it is a little harder to apply. (It asserts that the derivative of the product is "the first function times the derivative of the second plus the second function times the derivative of the first.")

Theorem 3 If f and g are differentiable functions, then so is fg. Its derivative is given by the formula

The product rule

$$(fg)' = fg' + gf'.$$

Proof Call the function fg simply u. That is,

$$u(x) = f(x)g(x).$$

Then $$u(x + \Delta x) = f(x + \Delta x)g(x + \Delta x).$$

Rather than subtract directly, first write

$$f(x + \Delta x) = f(x) + \Delta f \quad \text{and} \quad g(x + \Delta x) = g(x) + \Delta g.$$

Then
$$\begin{aligned} u(x + \Delta x) &= [f(x) + \Delta f][g(x) + \Delta g] \\ &= f(x)g(x) + f(x)\,\Delta g + g(x)\,\Delta f + \Delta f\,\Delta g. \end{aligned}$$

Hence
$$\begin{aligned} \Delta u &= u(x + \Delta x) - u(x) \\ &= f(x)g(x) + f(x)\,\Delta g + g(x)\,\Delta f + \Delta f\,\Delta g - f(x)g(x) \\ &= f(x)\,\Delta g + g(x)\,\Delta f + \Delta f\,\Delta g \end{aligned}$$

and
$$\frac{\Delta u}{\Delta x} = f(x)\frac{\Delta g}{\Delta x} + g(x)\frac{\Delta f}{\Delta x} + \Delta f\,\frac{\Delta g}{\Delta x}.$$

As $\Delta x \to 0$, $\Delta g/\Delta x \to g'(x)$, $\Delta f/\Delta x \to f'(x)$, and, because f is differentiable (hence continuous), $\Delta f \to 0$. It follows that

$$\lim_{\Delta x \to 0} \frac{\Delta u}{\Delta x} = f(x)g'(x) + g(x)f'(x) + 0 \cdot g'(x).$$

The formula for $(fg)'$ was discovered by Leibniz in 1676. (His first guess was wrong.)

Therefore, u is differentiable and

$$u' = fg' + gf'.$$ ■

EXAMPLE 4 Find $\frac{d}{dx}[(x^2 + x^3)(x^4 - \sqrt{x} + 5)]$.

SOLUTION (Note that in Examples 2 and 3 the derivatives of both factors, $x^2 + x^3$ and $x^4 - \sqrt{x} + 5$, were found.)

By Theorem 3,

$$\frac{d}{dx}[(x^2 + x^3)(x^4 - \sqrt{x} + 5)]$$

$$= (x^2 + x^3)\frac{d}{dx}(x^4 - \sqrt{x} + 5) + (x^4 - \sqrt{x} + 5)\frac{d}{dx}(x^2 + x^3)$$

$$= (x^2 + x^3)\left(4x^3 - \frac{1}{2\sqrt{x}}\right) + (x^4 - \sqrt{x} + 5)(2x + 3x^2). \blacksquare$$

A special case of the formula $(fg)' = fg' + gf'$ occurs so frequently that it is singled out in Theorem 4.

The derivative of cf

Theorem 4 If c is a constant function and f is a differentiable function, then cf is differentiable and its derivative is given by formula

$$(cf)' = cf'.$$

Proof By Theorem 3,

$$(cf)' = cf' + c'f \qquad \text{derivative of a product}$$

$$= cf' + 0 \cdot f \qquad \text{derivative of constant is 0}$$

$$= cf'. \blacksquare$$

In other notations for the derivative, Theorem 4 is expressed as

$$\frac{d(cf)}{dx} = c\frac{df}{dx} \quad \text{and} \quad D(cf) = cD(f).$$

A constant factor can go past the derivative symbol.

Theorem 4 asserts that "it is legal to move a constant factor outside the derivative symbol."

EXAMPLE 5 Find $D(6x^3)$.

SOLUTION

$$D(6x^3) = 6D(x^3) \qquad \text{6 is constant}$$

$$= 6 \cdot 3x^2 \qquad D(x^n) = nx^{n-1}$$

$$= 18x^2. \blacksquare$$

EXAMPLE 6 Find $D(x^5/11)$.

SOLUTION

$$D\left(\frac{x^5}{11}\right) = D(\tfrac{1}{11}x^5) = \tfrac{1}{11}D(x^5)$$

$$= \tfrac{1}{11}(5x^4) = \tfrac{5}{11}x^4. \blacksquare$$

$\frac{d}{dx}\left(\frac{f}{c}\right) = \frac{1}{c}\frac{d}{dx}(f).$

Example 6 generalizes to the fact that for a nonzero constant c,

$$\left(\frac{f}{c}\right)' = \frac{f'}{c}.$$

The formula for the derivative of the product extends to the product of several differentiable functions. For instance,

$$(fgh)' = f'gh + fg'h + fgh'.$$

In each summand only one derivative appears. (See Exercise 49.)

EXAMPLE 7 Differentiate $\sqrt{x}\,(x^2 - 2)(2x^3 + 1)$.

SOLUTION By the preceding remark,

$$\begin{aligned}[\sqrt{x}\,(x^2 - 2)(2x^3 + 1)]' &= (\sqrt{x})'(x^2 - 2)(2x^3 + 1) + \sqrt{x}(x^2 - 2)'(2x^3 + 1) + \sqrt{x}(x^2 - 2)(2x^3 + 1)' \\ &= \frac{1}{2\sqrt{x}}(x^2 - 2)(2x^3 + 1) + \sqrt{x}(2x)(2x^3 + 1) + \sqrt{x}(x^2 - 2)(6x^2). \quad \blacksquare\end{aligned}$$

Any polynomial can be differentiated by the methods already developed, as Example 8 illustrates.

EXAMPLE 8 Differentiate $6x^8 - x^3 + 5x^2 + \pi^3$.

Differentiate a polynomial "term by term."

SOLUTION

$$\begin{aligned}(6x^8 - x^3 + 5x^2 + \pi^3)' &= (6x^8)' - (x^3)' + (5x^2)' + (\pi^3)' \\ &= 6 \cdot 8x^7 - 3x^2 + 5 \cdot 2x + 0 \\ &= 48x^7 - 3x^2 + 10x. \quad \blacksquare\end{aligned}$$

The derivative of f/g

It will next be shown that if the functions f and g are differentiable at a number x, and if $g(x) \neq 0$, then f/g is differentiable at x. The formula for $(f/g)'$ is a bit messy; a suggestion for remembering it is given after the proof.

Theorem 5 If f and g are differentiable functions, then so is f/g and

The quotient rule

$$\left(\frac{f}{g}\right)' = \frac{gf' - fg'}{g^2} \quad \text{[where } g(x) \text{ is not 0].}$$

Proof Denote the quotient function f/g by u. That is,

$$u(x) = \frac{f(x)}{g(x)}.$$

Then

$$u(x + \Delta x) = \frac{f(x + \Delta x)}{g(x + \Delta x)}.$$

[Since we consider only values of x such that $g(x) \neq 0$ and g is continuous, for Δx sufficiently small, $g(x + \Delta x) \neq 0$.]

Before computing $\lim_{\Delta x \to 0} (\Delta u/\Delta x)$, express the numerator Δu as simply as possible in terms of $f(x)$, Δf, $g(x)$, and Δg:

$$\begin{aligned}
\Delta u &= u(x+\Delta x) - u(x) && \text{by definition of } \Delta u \\
&= \frac{f(x+\Delta x)}{g(x+\Delta x)} - \frac{f(x)}{g(x)} && \text{by definition of the function } u \\
&= \frac{f(x)+\Delta f}{g(x)+\Delta g} - \frac{f(x)}{g(x)} && \text{by definition of } \Delta f \text{ and } \Delta g \\
&= \frac{g(x)[f(x)+\Delta f] - f(x)[g(x)+\Delta g]}{[g(x)+\Delta g]g(x)} && \text{placing over common denominator} \\
&= \frac{g(x)f(x) + g(x)\,\Delta f - f(x)g(x) - f(x)\,\Delta g}{[g(x)+\Delta g]g(x)} && \text{multiplying out} \\
&= \frac{g(x)\,\Delta f - f(x)\,\Delta g}{[g(x)+\Delta g]g(x)} && \text{simplifying.}
\end{aligned}$$

After this simplification, $u'(x)$ can be found as follows:

$$\begin{aligned}
u'(x) &= \lim_{\Delta x\to 0}\frac{\Delta u}{\Delta x} = \lim_{\Delta x\to 0}\frac{\dfrac{g(x)\,\Delta f - f(x)\,\Delta g}{[g(x)+\Delta g]g(x)}}{\Delta x} \\
&= \lim_{\Delta x\to 0}\frac{g(x)\,\Delta f - f(x)\,\Delta g}{[g(x)+\Delta g]g(x)\,\Delta x} && \text{algebra: } \frac{\left(\dfrac{a}{b}\right)}{c} = \frac{a}{bc} \\
&= \lim_{\Delta x\to 0}\frac{g(x)\dfrac{\Delta f}{\Delta x} - f(x)\dfrac{\Delta g}{\Delta x}}{[g(x)+\Delta g]g(x)} && \text{algebra: } \frac{ab-cd}{ef} = \frac{a\dfrac{b}{f} - c\dfrac{d}{f}}{e} \\
&= \frac{g(x)f'(x) - f(x)g'(x)}{g(x)g(x)} && \text{taking limits.}
\end{aligned}$$

This establishes the formula for $(f/g)'$. ■

A suggestion for using the formula for the derivative of a quotient

Word of advice: When using the formula for $(f/g)'$, first write down the parts where g^2 and g appear:

$$\frac{g\qquad\qquad}{g^2}.$$

In this way you will get the denominator correct and have a good start on the numerator. You may then go on to complete the numerator, *remembering that it has a minus sign:*

$$\left(\frac{f}{g}\right)' = \frac{gf' - fg'}{g^2}.$$

EXAMPLE 9 Compute $[x^2/(x^3+1)]'$, showing each step in detail.

SOLUTION

Step 1 $$\left(\frac{x^2}{x^3+1}\right)' = \frac{(x^3+1)\cdots}{(x^3+1)^2}$$ write denominator and start numerator

Step 2 $$= \frac{(x^3+1)(x^2)' - (x^2)(x^3+1)'}{(x^3+1)^2}$$ complete numerator, remembering minus sign

Step 3 $\quad = \dfrac{(x^3 + 1)(2x) - x^2(3x^2)}{(x^3 + 1)^2}$ compute derivatives

Step 4 $\quad = \dfrac{2x^4 + 2x - 3x^4}{(x^3 + 1)^2}$ algebra

Step 5 $\quad = \dfrac{2x - x^4}{(x^3 + 1)^2}$ algebra: collecting. ■

As Example 9 illustrates, the techniques for differentiating polynomials and quotients suffice to differentiate any rational function.

The next example uses the formulas for the derivatives of the product and the quotient.

EXAMPLE 10 Differentiate $[(x^3 + 1)\sqrt{x}]/x^2$, where $x > 0$.

SOLUTION

$$D\left[\frac{(x^3 + 1)\sqrt{x}}{x^2}\right]$$

$$= \frac{x^2D[(x^3 + 1)\sqrt{x}] - (x^3 + 1)\sqrt{x}D(x^2)}{(x^2)^2} \qquad D\left(\frac{f}{g}\right)$$

$$= \frac{x^2[(x^3 + 1)D(\sqrt{x}) + \sqrt{x}D(x^3 + 1)] - (x^3 + 1)\sqrt{x}(2x)}{x^4} \qquad D(fg)$$

$$= \frac{x^2\left[(x^3 + 1)\dfrac{1}{2\sqrt{x}} + \sqrt{x}(3x^2)\right] - (x^3 + 1)\sqrt{x}(2x)}{x^4} \qquad D(x^n).$$

The result can be simplified a little by algebra. A factor x can be canceled in numerator and denominator. Also, coefficients can be placed at the front of the terms. This gives

This can be simplified further, but there is no need to here.

$$D\left[\frac{(x^3 + 1)\sqrt{x}}{x^2}\right] = \frac{x[(x^3 + 1)/(2\sqrt{x}) + 3\sqrt{x}x^2] - 2\sqrt{x}(x^3 + 1)}{x^3}. \quad ■$$

The following corollary is just a special case of the formula for $(f/g)'$. Since it is needed often, it is worth memorizing.

Corollary 1

$$\left(\frac{1}{g}\right)' = -\frac{g'}{g^2} \qquad \text{where } g(x) \text{ is not } 0.$$

The derivative of 1/g

Proof

$$\left(\frac{1}{g}\right)' = \frac{g \cdot (1)' - 1 \cdot g'}{g^2} \qquad \text{derivative of a quotient}$$

$$= \frac{g \cdot 0 - g'}{g^2} \qquad \text{derivative of a constant}$$

$$= -\frac{g'}{g^2}. \quad ■$$

EXAMPLE 11 Find $D\left(\dfrac{1}{2x^3 + x + 5}\right)$.

SOLUTION By the formula for $(1/g)'$,

$$D\left(\frac{1}{2x^3 + x + 5}\right) = \frac{-D(2x^3 + x + 5)}{(2x^3 + x + 5)^2}$$

$$= \frac{-(6x^2 + 1)}{(2x^3 + x + 5)^2}. \quad ■$$

In Sec. 3.2 it was shown that $D(x^n) = nx^{n-1}$ when n is a positive integer. The next corollary shows that the same formula holds when n is a negative integer.

Corollary 2 If n is a negative integer, $n = -1, -2, -3, \ldots$, then

$$(x^n)' = nx^{n-1}.$$

Proof Let $n = -m$, where m is a positive integer. Then

$$(x^n)' = (x^{-m})'$$

$$= \left(\frac{1}{x^m}\right)'$$

$$= \frac{-(x^m)'}{(x^m)^2} \qquad D\left(\frac{1}{f}\right) = \frac{-f'}{f^2}$$

$$= \frac{-mx^{m-1}}{x^{2m}} \qquad D(x^m) = mx^{m-1} \text{ for } m \text{ a positive integer}$$

$$= -mx^{-m-1}$$

$$= nx^{n-1} \qquad n = -m. \quad ■$$

EXAMPLE 12 Use Corollary 2 to find the derivative of $1/x$.

SOLUTION

$$\frac{d(1/x)}{dx} = \frac{d(x^{-1})}{dx}$$

$$= (-1)x^{-1-1} \qquad D(x^n) = nx^{n-1} \text{ for any integer } n$$

$$= -x^{-2}.$$

Thus

$$\frac{d(1/x)}{dx} = \frac{-1}{x^2}. \quad ■$$

EXAMPLE 13 Find $D(1/x^3)$.

SOLUTION $D\left(\dfrac{1}{x^3}\right) = D(x^{-3}) = -3x^{-3-1} = -3x^{-4} = \dfrac{-3}{x^4}$. ■

EXAMPLE 14 Differentiate $(1/2t)^5$.

SOLUTION $D\left[\left(\frac{1}{2t}\right)^5\right] = D\left(\frac{1}{32t^5}\right) = D\left(\frac{1}{32}\cdot\frac{1}{t^5}\right)$

$$= \frac{1}{32}D(t^{-5}) = \frac{1}{32}(-5)t^{-6} = -\frac{5}{32}\frac{1}{t^6}. \quad ■$$

The derivative of x^n in case $n = 0$ is 0 since $x^0 = 1$ for $x \neq 0$. This agrees with the formula nx^{n-1}. Thus $(x^n)' = nx^{n-1}$ for any integer.

EXAMPLE 15 Find an equation of the line tangent to the curve $y = x^3/3 + x^2 - 2\sqrt{x}$ at the point $(1, -\frac{2}{3})$.

SOLUTION First find the slope of the tangent line at $(1, -\frac{2}{3})$; in other words, find the derivative of $x^3/3 + x^2 - 2\sqrt{x}$ at $x = 1$. The derivative at x is

$$\frac{dy}{dx} = \frac{3x^2}{3} + 2x - 2\frac{1}{2\sqrt{x}} = x^2 + 2x - \frac{1}{\sqrt{x}}.$$

When $x = 1$, the derivative is 2. By the point-slope formula (see Sec. B.2), an equation of the tangent line is

$$\frac{y - (-\frac{2}{3})}{x - 1} = 2,$$

which can be put in the standard form, $2x - y - \frac{8}{3} = 0$. ■

SUMMARY OF THIS SECTION

$$(f + g)' = f' + g' \qquad (f - g)' = f' - g'$$

$$(fg)' = fg' + gf' \qquad \left(\frac{f}{g}\right)' = \frac{gf' - fg'}{g^2} \qquad \left(\frac{1}{g}\right)' = \frac{-g'}{g^2}$$

$$c' = 0 \qquad (cf)' = cf' \qquad \left(\frac{f}{c}\right)' = \frac{f'}{c} \qquad (c \text{ denotes a constant function.})$$

$$(x^n)' = nx^{n-1} \qquad \text{for any integer } n$$

The derivative of a polynomial is the sum of the derivatives of its terms.

EXERCISES FOR SEC. 3.4: THE DERIVATIVES OF THE SUM, DIFFERENCE, PRODUCT, AND QUOTIENT

In Exercises 1 to 34 differentiate with the aid of formulas, *not* by using the definition of the derivative.

1 $x^5 - 2x^2 + 3$

2 $x^3 + 5x^2 + 2$

3 $2x^4 - 6x^2 + 5x + 2$

4 $x^4 - x - \sqrt{2}$

5 $(x^2 + 3x + 1)(x^3 - 2x)$

6 $(5x^7 - x^2 + 4x + 2)(3x^2 - 7)$

7 $\dfrac{3x^4 - x^2 + 5x + 2}{7}$

8 $\dfrac{2x^3 + 3x + \pi^2}{10}$

9 $5\sqrt{x}$

10 $7\sqrt[3]{x}$

11 $12/x$

12 $-5/x^3$

13 $\frac{3 + x}{3 + x^2}$

14 $\frac{1 + x}{2 - x}$

15 $\frac{t^2 - 3t + 1}{t^3 + 1}$

16 $\frac{s^3 + 2s}{5s^2 + s + 2}$

17 $(1 + \sqrt{x})(x^3)$

18 $\left(\frac{2}{x} + \sqrt[3]{x}\right)(x^3 + 1)$

19 $(2x)^3$

20 $\left(\frac{3}{x}\right)^5$

21 $1 - \frac{1}{x} + \frac{1}{x^2}$

22 $x^2 + x + \frac{1}{x} + \frac{1}{x^2}$

23 $\frac{1}{x^3 + 2x + 1}$

24 $\frac{1}{x + \sqrt{x}}$

25 $\frac{(x^3 + x + 1)(x^2 - 1)}{5x^2 + 3}$

26 $\frac{(x^3 + 1)(5 + \sqrt[3]{x})}{x^5}$

27 $(2x + 1)^2$ [*Hint:* Write it as $(2x + 1)(2x + 1)$.]

28 $(3w^2 - 2w + 5)^2$ [*Hint:* Write it as the product $(3w^2 - 2w + 5)(3w^2 - 2w + 5)$.]

29 $\frac{1 + (1/x)}{1 - (1/x)}$

30 $\left(x + \frac{2}{x}\right)(x^3 + 6x + 1)$

31 $\frac{1}{\sqrt{x}}$

32 $\frac{1}{\sqrt[3]{x}}$

33 $\frac{(x^2 + 3x + 1)\sqrt{x}}{x^3 - 5x + 2}$

34 $\frac{\sqrt[3]{x}\sqrt{x}}{x^3 + 1}$

35 Give two antiderivatives for each of these functions: (*a*) x^3, (*b*) x, (*c*) $1/x^2$, (*d*) $1/x^3$.

36 Give two antiderivatives for each of these functions: (*a*) $1/\sqrt{x}$, (*b*) $3x^2 + 3x + 1$, (*c*) $1/\sqrt[3]{x^2}$, (*d*) $4x^5$.

■

In each of Exercises 37 to 39 find the slope of the given curve at the point with the given x coordinate.

37 $y = x^3 - x^2 + 2x$, at $x = 1$

38 $y = 1/(2x + 1)$, at $x = 2$

39 $y = \sqrt{x}(x^2 + 2)$, at $x = 4$

In each of Exercises 40 to 42 the distance an object travels in the first t seconds is given by the formula. Find the velocity and the speed (= absolute value of velocity) at the given time t.

40 $2t^4 + t^3 + 2t$, $t = 1$

41 $5\sqrt{t}$, $t = 9$

42 $\sqrt[4]{t/9}$, $t = 16$

In each of Exercises 43 and 44 a lens projects x on the linear slide to a point on the linear screen given by the formula. Find the magnification at the given point x.

43 $\sqrt[3]{x}$, $x = 8$

44 $4x^2 + x + 2$, $x = 2$

In each of Exercises 45 and 46 the mass of the left x centimeters of a string is given by the indicated formula. Find the density at the point x.

45 $x\sqrt[3]{x}$, $x = 8$

46 $x/(x + 1)$, $x = 2$

47 Show that if f, g, and h are differentiable functions, then $(f + g + h)' = f' + g' + h'$. *Hint:* Use Theorem 2 twice after writing $f + g + h$ as $(f + g) + h$.

48 Using the definition of the derivative, prove that $(f - g)' = f' - g'$.

49 Show that if f, g, and h are differentiable functions, then $(fgh)' = f'gh + fg'h + fgh'$. *Hint:* First write fgh as $(fg)h$. Then use the formula for the derivative of the product of two functions.

50 Let f be a differentiable function. (*a*) Show that $D(f^2) = 2ff'$. (*b*) Using Exercise 49, show that $D(f^3) = 3f^2f'$.

51 Obtain the formula $(1/g)' = -g'/g^2$, in Corollary 1, by introducing the function $f(x) = 1/g(x)$ and using the definition of the derivative to find f'.

52 (Economics: elasticity) Economists try to forecast the impact a change in price will have on demand. For instance, if the price of a gallon of gasoline is raised 10 cents, either by the oil company or by the government (in the form of a tax), how much petroleum will be conserved? To deal with such problems economists use the concept of the **elasticity of demand**, which will now be defined. Let $y = f(x)$ be the demand for a product as a function of the price x, that is, the amount that will be bought at the price x in a given time. Assume that y is a differentiable function of x.

(*a*) Is the derivative y' in general positive or is it negative?

(*b*) What are the dimensions of y' if y is measured in gallons and x is measured in cents?

(*c*) Why is the ratio $(\Delta y/y)/(\Delta x/x)$ called a "dimensionless quantity"? (*Note:* If gasoline is measured in liters instead of gallons, this ratio does not change.)

(*d*) The **elasticity** of demand at the price x is defined as

$$\epsilon = \lim_{\Delta x \to 0} \frac{\Delta y/y}{\Delta x/x}.$$

Show that
$$\epsilon = \left(\frac{x}{y}\right)y'.$$

(*e*) Estimate ϵ if a 1 percent increase in the price causes a 2 percent decrease in demand.

(*f*) Estimate ϵ if a 2 percent increase in price causes a 1 percent decrease in demand.

(*g*) If $|\epsilon| > 1$, the demand is called **elastic**; if $|\epsilon| < 1$, it is called **inelastic**. Why?

(*h*) Show that $y = x^{-3}$ has a constant elasticity (that is, the elasticity is independent of x).

53 Is there a line that is simultaneously tangent to the curves $y = x^2$ and $y = -x^2 + 2x - 2$? (First sketch the curves.) If so, find all such lines.

54 (*a*) Find an equation of the tangent line to the curve $y = 2x^2 - 3x + 1$ at $(1, 0)$.

(*b*) Find an equation of the line through $(1, 0)$ perpendicular to the tangent line in (*a*). (Section B.2 discusses the slope of a line perpendicular to a given line.)

(*c*) At what points does the line in (*b*) intersect the curve?

55 Figure 3.24 shows the typical tangent line to the curve $y = 1/x$ at a point P. Show that the area of the triangle OAB is constant, independent of the choice of point P.

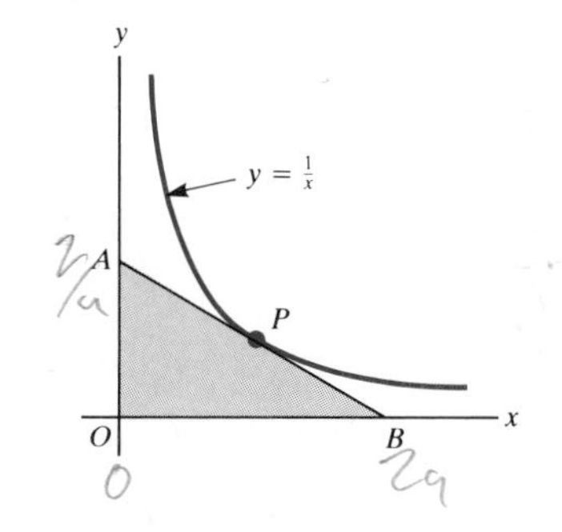

Figure 3.24

56 (*a*) Find an equation of the tangent line to the curve $y = x^3$ at the point (a, a^3).

(*b*) Does the tangent line in (*a*) always meet the curve at another point than (a, a^3)? If not, when does it?

57 There are four points on the curve $y = x^4 - 8x^2$ where the associated tangent line passes through $(-\frac{11}{3}, 49)$. Find the x coordinates of these points. (*Hint:* Two are integers. See Appendix C for advice in searching for rational roots of a polynomial.)

58 (*Cubic splines*) Scientists and engineers frequently need to construct differentiable functions to fit data. **Cubic splines** is one way of devising such functions. Given the points $P_0 = (x_0, y_0)$, $P_1 = (x_1, y_1)$, . . . , $P_n = (x_n, y_n)$, where $x_i < x_{i+1}$, one can construct a cubic polynomial on each interval $[x_i, x_{i+1}]$ such that the function pieced together from them passes through each point P_i, is differentiable, and its derivative has preassigned values at P_0 and P_n.

Let $P_0 = (-1, 1)$, $P_1 = (0, 1)$, and $P_2 = (1, 2)$. Find the cubic spline curve passing through P_0, P_1, and P_2 with first derivative equal to 1 at P_0 and -1 at P_2 and to (*a*) 0 at P_1, and (*b*) 2 at P_1. [*Hint:* Find $f(x) = ax^3 + bx^2 + cx + d$ that satisfies the conditions on the first interval, expressing the coefficients in terms of c. Similarly, use $g(x) = Ax^3 + Bx^2 + Cx + D$ on the second interval. Note that any value of the derivative at P_1 can be preassigned.]

3.5 THE DERIVATIVES OF THE TRIGONOMETRIC FUNCTIONS

The derivatives of the six trigonometric functions—sin x, cos x, tan x, sec x, csc x, and cot x—will be found in this section.

The derivatives of sin x *and* cos x

In order to find $\frac{d}{dx}(\sin x)$ and $\frac{d}{dx}(\cos x)$, it will be necessary to make use of the limits

$$\lim_{\Delta x \to 0} \frac{\sin \Delta x}{\Delta x} = 1 \quad \text{and} \quad \lim_{\Delta x \to 0} \frac{1 - \cos \Delta x}{\Delta x} = 0,$$

which were found in Sec. 2.4.

Theorem 1 The derivative of the sine function is the cosine function; symbolically,

$$(\sin x)' = \cos x \quad \text{or} \quad \frac{d}{dx}(\sin x) = \cos x.$$

Proof The derivative at x of a function f is defined as

$$\lim_{\Delta x \to 0} \frac{f(x + \Delta x) - f(x)}{\Delta x}.$$

In this case, f is the function "sine," and the limit under consideration is

$$\lim_{\Delta x \to 0} \frac{\sin (x + \Delta x) - \sin x}{\Delta x}.$$

Keep in mind that x is fixed while $\Delta x \to 0$. As $\Delta x \to 0$, the numerator approaches

$$\sin x - \sin x = 0,$$

When finding a derivative, we run into "zero-over-zero."

while the denominator Δx also approaches 0. Since the expression 0/0 is meaningless, it is necessary to change the form of the quotient

$$\frac{\sin (x + \Delta x) - \sin x}{\Delta x}$$

before letting Δx approach 0.

Let us use the trigonometric identity

$$\sin (A + B) = \sin A \cos B + \cos A \sin B$$

in the case $A = x$ and $B = \Delta x$, obtaining

$$\sin (x + \Delta x) = \sin x \cos \Delta x + \cos x \sin \Delta x.$$

Then the numerator, $\sin (x + \Delta x) - \sin x$, takes the form

$$\begin{aligned} \sin x \cos \Delta x + \cos x \sin \Delta x - \sin x &= \sin x (\cos \Delta x - 1) + \cos x \sin \Delta x \\ &= -\sin x (1 - \cos \Delta x) + \cos x \sin \Delta x. \end{aligned}$$

Therefore,

$$\begin{aligned} \lim_{\Delta x \to 0} \frac{\sin (x + \Delta x) - \sin x}{\Delta x} &= \lim_{\Delta x \to 0} \frac{-\sin x (1 - \cos \Delta x) + \cos x \sin \Delta x}{\Delta x} \\ &= \lim_{\Delta x \to 0} \left(-\sin x \frac{1 - \cos \Delta x}{\Delta x} + \cos x \frac{\sin \Delta x}{\Delta x} \right) \\ &= (-\sin x)(0) + (\cos x)(1) = \cos x. \end{aligned}$$

In short, the derivative of the sine function is the cosine function. This concludes the proof. ■

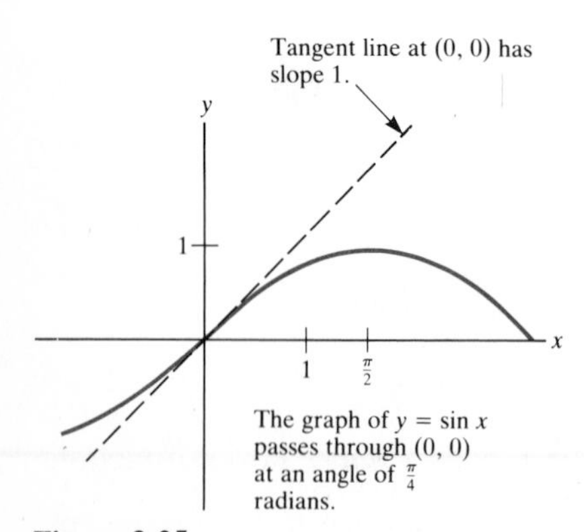

Figure 3.25

The formula obtained in Theorem 1 provides interesting information about the graph of $y = \sin x$. Since

$$\frac{d}{dx} (\sin x) = \cos x,$$

the derivative of the sine function when $x = 0$ is cos 0, which is 1. This implies that the slope of the curve $y = \sin x$, when $x = 0$, is 1. Consequently, the graph of $y = \sin x$ passes through the origin at an angle of $\pi/4$ radians (45°). See Fig. 3.25.

Theorem 2 The derivative of the cosine function is the negative of the sine function; symbolically,

$$(\cos x)' = -\sin x. \quad ■$$

We omit the proof, which is similar to that of Theorem 1. It makes use of the trigonometric identity

$$\cos (A + B) = \cos A \cos B - \sin A \sin B,$$

and is left as Exercise 19.

Recall that

$$\sec x = \frac{1}{\cos x} \quad \text{and} \quad \tan x = \frac{\sin x}{\cos x},$$

$$\csc x = \frac{1}{\sin x} \quad \text{and} \quad \cot x = \frac{\cos x}{\sin x}.$$

In the next two theorems the derivatives of these four functions are obtained.

The derivatives of sec x *and* csc x

Theorem 3 $(\sec x)' = \sec x \tan x$ and $(\csc x)' = -\csc x \cot x$.

Proof

$$(\sec x)' = \left(\frac{1}{\cos x}\right)' = \frac{-(\cos x)'}{\cos^2 x} \quad \text{using} \quad \left(\frac{1}{g}\right)' = \frac{-g'}{g^2}$$

$$= \frac{-(-\sin x)}{\cos^2 x}$$

$$= \frac{\sin x}{\cos^2 x}$$

$$= \frac{1}{\cos x} \frac{\sin x}{\cos x}$$

$$= \sec x \tan x.$$

Thus $(\sec x)' = \sec x \tan x$. The derivative of $\csc x$ is obtained similarly. ■

The derivatives of tan x *and* cot x

Theorem 4 $(\tan x)' = \sec^2 x$ and $(\cot x)' = -\csc^2 x$.

Proof

$$(\tan x)' = \left(\frac{\sin x}{\cos x}\right)'$$

$$= \frac{\cos x (\sin x)' - \sin x (\cos x)'}{\cos^2 x}$$

$$= \frac{\cos x \cos x - \sin x (-\sin x)}{\cos^2 x}$$

$$= \frac{\cos^2 x + \sin^2 x}{\cos^2 x} = \frac{1}{\cos^2 x} = \sec^2 x.$$

The derivative of cot x is obtained similarly. ■

This table summarizes the six formulas. Note that the derivatives of the cosine, cosecant, and cotangent have minus signs.

The derivatives of the cofunctions (cosine, cotangent, cosecant) have minus signs.

TABLE OF TRIGONOMETRIC DERIVATIVES

$(\sin x)' = \cos x$	$(\cos x)' = -\sin x$
$(\tan x)' = \sec^2 x$	$(\cot x)' = -\csc^2 x$
$(\sec x)' = \sec x \tan x$	$(\csc x)' = -\csc x \cot x$

EXAMPLE 1 Differentiate $x - \sin x \cos x$.

SOLUTION

$$\begin{aligned} D(x - \sin x \cos x) &= D(x) - D(\sin x \cos x) \\ &= 1 - [\sin x\, D(\cos x) + \cos x\, D(\sin x)] \\ &= 1 - [\sin x\,(-\sin x) + \cos x\,(\cos x)] \\ &= 1 + \sin^2 x - \cos^2 x. \end{aligned}$$

Since $1 - \cos^2 x = \sin^2 x$, the last expression can be simplified to $\sin^2 x + \sin^2 x$, or simply $2 \sin^2 x$. ■

The next section will present a shortcut for differentiating $\sin 2x$. The derivative is *not* $\cos 2x$. The next example uses a trigonometric identity to find $(\sin 2x)'$.

EXAMPLE 2 Differentiate $\sin 2x$.

SOLUTION $\sin 2x = 2 \sin x \cos x.$

Thus

$$\begin{aligned} D(\sin 2x) &= D(2 \sin x \cos x) \\ &= 2D(\sin x \cos x) \\ &= 2\,[\sin x\, D(\cos x) + \cos x\, D(\sin x)] \\ &= 2\,[\sin x\,(-\sin x) + \cos x\,(\cos x)] \\ &= 2\,(\cos^2 x - \sin^2 x) = 2 \cos 2x. \end{aligned}$$

The derivative of $\sin 2x$ *is* $2 \cos 2x$, *not* $\cos 2x$.

In short, the derivative of $\sin 2x$ is $2 \cos 2x$. ■

EXAMPLE 3 Find $D(x^3 \sec x)$.

SOLUTION

$$\begin{aligned} D(x^3 \sec x) &= x^3 D(\sec x) + \sec x\, D(x^3) \\ &= x^3 \sec x \tan x + \sec x\,(3x^2), \end{aligned}$$

which is usually written $x^3 \sec x \tan x + 3x^2 \sec x$ for clarity. ■

EXAMPLE 4 Find $\dfrac{d}{dx}\left(\dfrac{\csc x}{\sqrt{x}}\right)$.

SOLUTION
$$\frac{d}{dx}\left(\frac{\csc x}{\sqrt{x}}\right) = \frac{\sqrt{x}\left[\dfrac{d}{dx}(\csc x)\right] - \csc x\left[\dfrac{d}{dx}(\sqrt{x})\right]}{(\sqrt{x})^2}$$

$$= \frac{\sqrt{x}\,(-\csc x \cot x) - \csc x \dfrac{1}{2\sqrt{x}}}{x}.$$

Since $\sqrt{x}/x = 1/\sqrt{x}$ and $x\sqrt{x} = x^{3/2}$, this can be simplified a little to

$$\frac{-\csc x \cot x}{\sqrt{x}} - \frac{\csc x}{2x^{3/2}}. \quad ■$$

Why Radian Measure Is Used in Calculus

Throughout this section angles are measured in radians, as is customary in calculus. If we measured angles in degrees instead, the formulas for the derivatives of the trigonometric functions would be more complicated. Each formula would have an extra factor, $\pi/180$, as we will now show.

In Sec. 2.4 it was shown that when angles are measured in radians,

$$\lim_{\theta \to 0} \frac{\sin \theta}{\theta} = 1.$$

When angles are measured in degrees, this limit is not 1. Let Sin θ denote the sine of an angle of θ degrees. The following table suggests that the limit is much smaller (angles measured in degrees; data to four significant figures):

θ	10	5	1	0.1
Sin θ	0.1736	0.08716	0.01745	0.001745
$\dfrac{\text{Sin } \theta}{\theta}$	0.01736	0.01743	0.01745	0.01745

The data suggest that $\lim_{\theta \to 0}$ (Sin θ)/θ is about 0.01745. As you will see when doing Exercise 30, the limit is precisely $\pi/180$. If you then go through this section and carry through the steps that determined the derivatives of the trigonometric functions, you will see that an extra factor of $\pi/180$ must be put into each formula. Clearly, the use of radians as the measure of angles gives the simplest formulas for trigonometric derivatives.

EXERCISES FOR SEC. 3.5: THE DERIVATIVES OF THE TRIGONOMETRIC FUNCTIONS

In each of Exercises 1 to 8 differentiate the function and simplify your answer.

1 $\sin x - x\cos x$

2 $\cos x + x\sin x$

3 $2x\sin x + 2\cos x - x^2\cos x$

4 $3x^2\sin x - 6\sin x - x^3\cos x + 6x\cos x$

5 $\tan x - x$

6 $-\cot x - x$

7 $2x\cos x - 2\sin x + x^2\sin x$

8 $(3x^2 - 6)\cos x + (x^2 - 6)\sin x$

In Exercises 9 to 18 differentiate the functions.

9 $\dfrac{1 + \sin x}{\cos x}$

10 $\dfrac{1 - \sin x}{\cos x}$

11 $\dfrac{1 + 3\sec x}{\tan x}$

12 $x^3\sec x$

13 $\dfrac{\csc x}{\sqrt[3]{x}}$

14 $3\csc x + 2\tan x$

15 $\sin x\tan x$

16 $x^2\cos x\cot x$

17 $\dfrac{\cot x}{1 + x^2}$

18 $\dfrac{x}{\sec x}$

■

19 Using the identity for $\cos(A + B)$, show that the derivative of $\cos x$ is $-\sin x$.

20 Differentiate $\cos 2x$.

21 Find the slope of the curve $y = \cos x$ at the point for which x is (*a*) 0, (*b*) $\pi/6$, (*c*) $\pi/4$, (*d*) $\pi/3$, and (*e*) $\pi/2$.

22 (*a*) Find the slope of the curve $y = \tan x$ when $x = \pi/4$.
(*b*) Using (*a*), estimate the angle that the tangent line to the curve $y = \tan x$ at $(\pi/4, 1)$ makes with the x axis. This angle is called the "angle of inclination" of the line.

23 A mass bobbing up and down on the end of a spring has the y coordinate $y = 3\sin t$ centimeters at time t seconds. (*a*) How high does it go? (*b*) How low? (*c*) What is its velocity when $t = 0$ and when $t = \pi$? (*d*) What is its speed when $t = 0$ and when $t = \pi$?

24 The height of the ocean surface above (or below) mean sea level is, say, $y = 2\sin t$ feet at t hours.
(*a*) Find the rate at which the tide is rising or falling at time t.
(*b*) Is the surface rising more rapidly at low tide or when it is at mean sea level?

25 At what angle does the graph of $y = \tan x$ cross the x axis?

26 What is the angle of inclination of the tangent line to the curve $y = \sin x$ at the point $(\pi, 0)$?

27 Give two antiderivatives for each of these functions: (*a*) $\cos x$, (*b*) $5\cos x$, (*c*) $\sin x$, and (*d*) $-3\sin x$.

28 Using the definition of the derivative and the identity for $\sin(A + B)$, show that $(\sin 7x)' = 7\cos 7x$.

29 Using the definition of the derivative and the identity for $\cos(A + B)$, show that $D(\cos 11x) = -11\sin 11x$.

30 Let $\operatorname{Sin}\theta$ denote the sine of an angle of θ degrees. Since an angle of θ degrees is the same as an angle of $\pi\theta/180$ radians, $\operatorname{Sin}\theta = \sin(\pi\theta/180)$, where $\sin x$ denotes the sine of an angle of x radians. Deduce that $(\operatorname{Sin}\theta)' = (\pi/180)\operatorname{Cos}\theta$, where $\operatorname{Cos}\theta$ denotes the cosine of an angle of θ degrees. *Hint:* First find $\lim_{\theta\to 0}(\operatorname{Sin}\theta)/\theta$.

31 Find $D(\csc\theta)$, using Theorems 1 and 2.

32 Find $D(\cot\theta)$, using Theorems 1 and 2.

3.6 COMPOSITE FUNCTIONS AND THE CHAIN RULE

Composite functions were discussed in Sec. 1.2.

The differentiation techniques obtained so far do not enable us to differentiate such functions as

$$(1 + 2x)^{100}, \qquad \sqrt{1 + x^2}, \qquad \text{or} \qquad \sin x^3.$$

We could differentiate $(1 + 2x)^{100}$, but only with great effort, by first expanding $(1 + 2x)^{100}$ to form a polynomial of degree 100 and then differentiating that polynomial. This section develops a shortcut for differentiating composite functions, such as $(1 + 2x)^{100}$, $\sqrt{1 + x^2}$, and $\sin x^3$, which are built up from simpler functions by composition.

If f and g are differentiable functions, is the composite function $f \circ g$ also differentiable? If so, what is its derivative? More concretely: If

$y = f(u)$ and $u = g(x)$, then y is a function of x. How can we find dy/dx?

Take the simple case, $y = 3u$ and $u = 2x$. Hence $y = 6x$. In this case,

$$\frac{dy}{du} = 3 \qquad \frac{du}{dx} = 2 \qquad \text{and} \qquad \frac{dy}{dx} = 6.$$

So dy/dx is the product of the derivatives dy/du and du/dx. This observation suggests the all-important **chain rule**, which will be proved at the end of this section after several examples show how it is used.

An easily remembered form of the chain rule

THE CHAIN RULE (informal statement)

If y is a differentiable function of u and u is a differentiable function of x, then y is a differentiable function of x and

$$\frac{dy}{dx} = \frac{dy}{du}\frac{du}{dx}.$$

The equation $\frac{dy}{dx} = \frac{dy}{du}\frac{du}{dx}$

is read as "derivative of y with respect to x equals derivative of y with respect to u times derivative of u with respect to x."

As we have already remarked, the notation "dy/dx" is not a fraction, but rather a *notation* for the derivative of y with respect to x. The chain rule is a statement about derivatives, not about fractions, and we should not think of the "du" as "canceling out."

In the D notation, the chain rule reads

The chain rule in D notation

$$D_x(y) = D_u(y)D_x(u).$$

The subscripts are needed to indicate the appropriate variable. $D_x(y)$ is the derivative of y with respect to x, that is, $\lim_{\Delta x \to 0} (\Delta y/\Delta x)$. $D_u(y)$ is the derivative of y with respect to u, that is, $\lim_{\Delta u \to 0} (\Delta y/\Delta u)$.

Let $h(x) = f(g(x))$ be the composite function of f and g. In function notation, the chain rule is

$$h'(x) = f'(g(x))g'(x).$$

In practical terms, this means "differentiate the 'outside function,' plug in the 'inside function,' and then multiply by the derivative of the inside function."

It may help to picture two projectors, one following the other, that is, the g projector followed by the f projector, as in Fig. 3.26. The first magnification is by the factor $g'(x)$; the second is by the factor $f'(g(x))$. The effect of the successive magnifications is their product, $f'(g(x))g'(x)$.

How to Use the Chain Rule Some examples will show the power of the chain rule, which is the technique most often used when computing derivatives. At first, each calculation will be displayed in full detail. How-

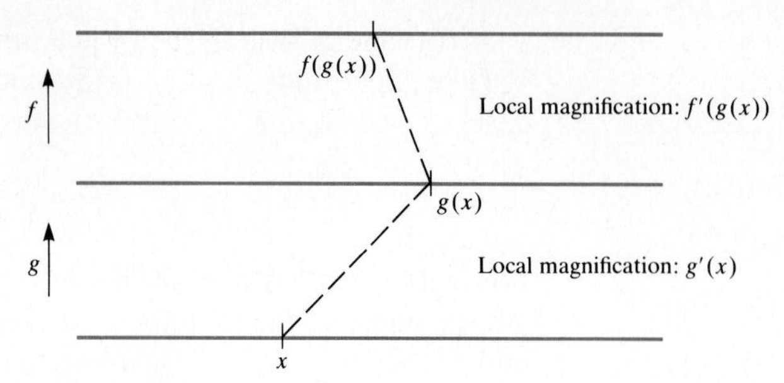

Figure 3.26

ever, Example 6 indicates how the calculations are done after sufficient practice.

EXAMPLE 1 Differentiate $\sqrt{1 + x^2}$.

SOLUTION $y = \sqrt{1 + x^2}$ is a composite function,

$$y = \sqrt{u} \quad \text{where} \quad u = 1 + x^2.$$

By the chain rule,

$$\frac{dy}{dx} = \frac{dy}{du} \cdot \frac{du}{dx},$$

The inside function is $1 + x^2$. The outside function is $\sqrt{u}$.

or

$$\frac{d}{dx}(\sqrt{1 + x^2}) = \frac{d}{du}(\sqrt{u}) \cdot \frac{d}{dx}(1 + x^2)$$

$$= \frac{1}{2\sqrt{u}} \cdot 2x = \frac{x}{\sqrt{u}}$$

$$= \frac{x}{\sqrt{1 + x^2}}.$$

In short,

$$\frac{d}{dx}(\sqrt{1 + x^2}) = \frac{x}{\sqrt{1 + x^2}}. \quad ■$$

EXAMPLE 2 Differentiate $\sin x^3$.

SOLUTION $y = \sin x^3$ can be expressed as

$$y = \sin u \quad \text{where} \quad u = x^3.$$

By the chain rule,

$$\frac{d}{dx}(\sin x^3) = \frac{d}{du}(\sin u) \cdot \frac{d}{dx}(x^3)$$

$$= \cos u \cdot 3x^2 = 3x^2 \cos u$$

Note that the answer is not $\cos x^3$.

$$= 3x^2 \cos x^3. \quad ■$$

Differentiation of $\sin^3 x$ is quite different from Example 2. In this case, $y = u^3$, where $u = \sin x$. So

$$\frac{d}{dx}(\sin^3 x) = \frac{d(u^3)}{du} \cdot \frac{d}{dx}(\sin x)$$
$$= 3u^2 \cos x$$
$$= 3 \sin^2 x \cos x.$$

EXAMPLE 3 Differentiate $(1 + 2x)^{100}$.

SOLUTION $y = (1 + 2x)^{100}$ is the composition of

$$y = u^{100} \quad \text{and} \quad u = 1 + 2x.$$

By the chain rule,

$$\frac{d}{dx}[(1 + 2x)^{100}] = \frac{d}{du}(u^{100}) \cdot \frac{d}{dx}(1 + 2x)$$
$$= 100u^{99} \cdot 2 = 200(1 + 2x)^{99}. \quad \blacksquare$$

Note that the answer is not $100(1 + 2x)^{99}$.

The chain rule extends to a function built up as the composition of three or more functions. For instance, if

$$y = f(u), \quad u = g(v), \quad \text{and} \quad v = h(x),$$

then y is a function of x and it can be shown that

$$\frac{dy}{dx} = \frac{dy}{du}\frac{du}{dv}\frac{dv}{dx}.$$

An extended form of the chain rule

The next example applies this fact.

EXAMPLE 4 Differentiate $\sqrt{(1 + x^2)^5}$.

SOLUTION $y = \sqrt{(1 + x^2)^5}$ can be expressed as

$$y = \sqrt{u}, \quad u = v^5, \quad \text{and} \quad v = 1 + x^2.$$

Then
$$\frac{dy}{dx} = \frac{dy}{du}\frac{du}{dv}\frac{dv}{dx}$$

or
$$\frac{d}{dx}\left(\sqrt{(1 + x^2)^5}\right) = \frac{d}{du}(\sqrt{u})\frac{d}{dv}(v^5)\frac{d}{dx}(1 + x^2)$$
$$= \frac{1}{2\sqrt{u}} \cdot 5v^4 \cdot 2x = \frac{5v^4x}{\sqrt{u}}$$
$$= \frac{5(1 + x^2)^4x}{\sqrt{v^5}} = \frac{5(1 + x^2)^4x}{\sqrt{(1 + x^2)^5}}$$
$$= 5x(1 + x^2)^{3/2}. \quad \blacksquare$$

EXAMPLE 5 Differentiate $\sin^2 3x$.

SOLUTION $y = \sin^2 3x$ can be written as

$$y = u^2, \quad u = \sin v, \quad \text{and} \quad v = 3x.$$

Thus
$$\frac{d}{dx}(\sin^2 3x) = \frac{d}{du}(u^2)\frac{d}{dv}(\sin v)\frac{d}{dx}(3x)$$
$$= (2u)(\cos v)(3) = 6u \cos v$$
$$= 6 \sin v \cos v = 6 \sin 3x \cos 3x. \blacksquare$$

EXAMPLE 6 Compute $\frac{d}{dx}(x^2 \sin^5 2x)$.

SOLUTION First of all, by the formula for the derivative of the product,

$$\frac{d}{dx}(x^2 \sin^5 2x) = x^2 \frac{d}{dx}(\sin^5 2x) + \sin^5 2x \frac{d}{dx}(x^2).$$

The chain rule is needed for computing

$$\frac{d}{dx}(\sin^5 2x).$$

Without all the details (that is, introduction of the letters u and v and exhibition of the various functions in detail), the computation looks like this:

$$\frac{d}{dx}(\sin^5 2x) = (5 \sin^4 2x)(\cos 2x)(2)$$
$$= 10 \sin^4 2x \cos 2x.$$

Thus
$$\frac{d}{dx}(x^2 \sin^5 2x) = x^2(10 \sin^4 2x \cos 2x) + (\sin^5 2x)(2x)$$
$$= 10x^2 \sin^4 2x \cos 2x + 2x \sin^5 2x. \blacksquare$$

As these examples suggest, the chain rule is one of the most frequently used tools in the computation of derivatives.

The table below records a few special cases of the chain rule. They are used so often that they are worth memorizing. In each case u is a differentiable function of x.

y	$\frac{dy}{dx}$
u^n	$nu^{n-1}\frac{du}{dx}$
$\sin u$	$\cos u \frac{du}{dx}$
$\cos u$	$-\sin u \frac{du}{dx}$

When first working with the chain rule it is safest to write down every step, showing the various functions with the aid of the letters u, v, and so on. However, with practice, it will not be necessary to record every detail.

The next example will be used later.

EXAMPLE 7 Let y be a differentiable function of x. Then y^2 is also a differentiable function of x. Express its derivative with respect to x, $D_x(y^2)$, in terms of the derivative of y with respect to x, $D_x(y)$.

SOLUTION Denote y^2 by w. Then

$$w = y^2 \qquad \text{where } y \text{ is a function of } x.$$

By the chain rule,

$$D_x(w) = D_y(w)D_x(y)$$
$$= 2yD_x(y).$$

In short,

$$D_x(y^2) = 2yD_x(y). \quad ■$$

In Secs. 3.2 and 3.3 it was shown that $D(x^n) = nx^{n-1}$ for any integer and $D(x^{1/n}) = (1/n)x^{1/n-1}$ for any positive integer. These are special cases of the formula for the derivative of x^r for any rational exponent r. This formula will now be obtained with the aid of the chain rule.

The derivative of x^r

Theorem 1 Let r be a rational number. Then $D(x^r) = rx^{r-1}$.

Proof Since r is rational, it can be written as m/n for some integers m and n, with n positive. Let $y = x^r = x^{m/n} = (x^{1/n})^m$. Then y is a composite function:

$$y = u^m \qquad \text{where} \qquad u = x^{1/n}.$$

By the chain rule,

$$\frac{dy}{dx} = \frac{d(u^m)}{du} \cdot \frac{d(x^{1/n})}{dx}$$

$$= mu^{m-1} \cdot \frac{1}{n} x^{(1/n)-1} \qquad \text{known formulas}$$

$$= m(x^{1/n})^{m-1} \cdot \frac{1}{n} x^{(1/n)-1}$$

$$= mx^{(m-1)/n} \cdot \frac{1}{n} x^{(1/n)-1} \qquad \text{power of a power}$$

$$= \frac{m}{n} x^{(m-1)/n+(1/n)-1} \qquad \text{basic law of exponents}$$

$$= \frac{m}{n} x^{(m/n)-1} = rx^{r-1}. \quad ■$$

EXAMPLE 8 Use the formula for $D(x^r)$ to differentiate $\sqrt[3]{x}\sqrt{x}$.

SOLUTION

$$\sqrt[3]{x}\sqrt{x} = x^{1/3}x^{1/2} = x^{1/3+1/2} = x^{5/6}.$$

Then

$$D(x^{5/6}) = \frac{5}{6} x^{(5/6)-1} = \frac{5}{6} x^{-1/6} = \frac{5}{6} \frac{1}{\sqrt[6]{x}}. \quad ■$$

Proof of the Chain Rule

Theorem 2 *The chain rule.* If f and g are differentiable functions, then for all x such that $g(x)$ is in the domain of f, $h = f \circ g$ is differentiable at x, and

$$h'(x) = f'(g(x)) \cdot g'(x).$$

More briefly, if $y = f(u)$ and $u = g(x)$, then $y = h(x)$ and

$$\frac{dy}{dx} = \frac{dy}{du}\frac{du}{dx}.$$

Proof We will prove the theorem with the extra assumption $g'(x)$ is not 0 at the particular number x under consideration. Since $g'(x) = \lim_{\Delta x \to 0} \Delta u/\Delta x$, this implies that when Δx is sufficiently small, Δu is not 0. This fact will be used in the argument.

To examine $h'(x)$ it is necessary to go back to the definition of the derivative,

$$h'(x) = \lim_{\Delta x \to 0} \frac{\Delta y}{\Delta x}.$$

The computation will involve Δx, $\Delta u = \Delta g$, and Δy, shown in Fig. 3.27. That is, Δx, which is not 0, determines a number Δu, the change in u,

$$\Delta u = g(x + \Delta x) - g(x),$$

and a number Δy, the change in y,

$$\Delta y = h(x + \Delta x) - h(x).$$

We are using the theorem on page 90 to show that g is continuous.

It is important to note that since g is differentiable, $\Delta u \to 0$ as $\Delta x \to 0$. However, it could happen that Δu is 0, even though Δx is not 0. But the assumption we made about $g'(x)$ assures us that when Δx is small, Δu is *not* 0.

Now for the short proof:

$$h'(x) = \lim_{\Delta x \to 0} \frac{\Delta y}{\Delta x}$$

$$= \lim_{\Delta x \to 0} \frac{\Delta y}{\Delta u}\frac{\Delta u}{\Delta x} \qquad \text{algebra (It's ``okay'' to divide by } \Delta u\text{, since it is not 0.)}$$

$$= \lim_{\Delta x \to 0} \frac{\Delta y}{\Delta u} \lim_{\Delta x \to 0} \frac{\Delta u}{\Delta x} \qquad \text{since both limits exist}$$

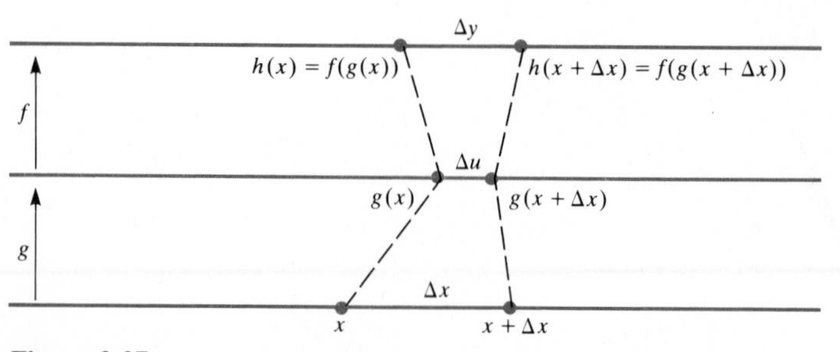

Figure 3.27

$$= \lim_{\Delta u \to 0} \frac{\Delta y}{\Delta u} \lim_{\Delta x \to 0} \frac{\Delta u}{\Delta x} \qquad \text{since } \Delta u \to 0 \text{ as } \Delta x \to 0$$

$$= \frac{dy}{du}\frac{du}{dx}.$$

[The special case, where $g'(x) = 0$, is discussed in Exercise 62.] This concludes the proof. ■

DIFFERENTIATION BY MACHINE

We calculate derivatives by a small number of rules obtained in this chapter and in Chap. 6. These rules may be programmed into computers, which can then do our differentiation for us. Such programs do "symbolic" rather than "numerical" mathematics. One of the most powerful symbolic programs is MACSYMA, developed for mainframe computers at the Massachusetts Institute of Technology. However, even microcomputers are now powerful enough to do many calculus computations, including differentiation.

Microsoft muMath is a microcomputer program that can handle derivatives and other calculus procedures. For example, entering the command DIF(COS x^2, x) results in the answer −2xSIN(x^2), where x^2 represents the square of x.

Although severely limited by comparison with as powerful a mainframe program as MACSYMA, muMath has no trouble with differentiation. It is available for Apple II systems with Z80 (CP/M) cards, CP/M computers, and MS-DOS or PC-DOS microcomputers (IBM PC or PC-compatible). Microsoft muMath is licensed from the Soft Warehouse and is obtainable (for $300 as of September 1986) either from Microsoft Corporation, 16011 NE 36th Way, Redmond, WA 98073-9717, or The Soft Warehouse, P.O. Box 11174, Honolulu, Hawaii 96828-0174. A similar program called PowerMath is available for the Apple Macintosh (for $99.95 as of September 1986) from Brainpower, 24009 Ventura Boulevard, Calabasas, Calif. 91320; its calculus functions, however, are limited to polynomials.

As microprocessors become more powerful, we may expect to see versions of MACSYMA and other mainframe symbolic programs running on microcomputers. Several colleges already offer such programs through remote terminals and minicomputer systems.

EXERCISES FOR SEC. 3.6: COMPOSITE FUNCTIONS AND THE CHAIN RULE

Exercises 1 to 8 concern the notion of a composite function. In each case show that the function is a composition of other functions by introducing the necessary symbols, such as u and v, and describing the functions.

1 $(x^3 + x^2 - 2)^{50}$ 2 $(\sqrt{x} + 1)^{10}$

3 $\sqrt{x + 3}$ 4 $\sqrt[3]{1 + x^2}$

5 $\sin 2x$ 6 $\sin^2 x$

7 $\cos^3 2x$ 8 $\sqrt[3]{(1 + 2x)^{50}}$

In Exercises 9 to 38 differentiate the functions.

9 $(2x^3 - x)^{40}$ 10 $(x^4 - 6)^{100}$

11 $\left(1 + \frac{3}{x}\right)^4$ 12 $\left(1 - \frac{2}{x^2}\right)^5$

13 $\sqrt{x^3 + x + 2}$ 14 $\sqrt[3]{x^3 + 8}$

15 $\sin 3x$ 16 $(1 + 4x)^{10}$

17 $\frac{x^3 \tan (1/x)}{1 + x^2}$ 18 $\sqrt{1 + x^2} \cot^3 5x$

19 $(2x + 1)^5(3x + 1)^7$ 20 $x^5(x^2 + 2)^3 \cos^3 5x$

21 $x^2 \sin^5 3x$ 22 $x^3 \cos^2 3x \sin^2 2x$

23 $\frac{1}{(2x + 3)^5}$ [*Shortcut:* First write it as $(2x + 3)^{-5}$.]

24 $\frac{1}{(3x - 5)^7}$ 25 $\frac{(2x + 1)^3}{(3x + 1)^4}$

26 $\dfrac{x^2}{(x^2+1)^3}$

27 $\dfrac{(x^3-1)^5 \cot^3 5x}{x}$

28 $\dfrac{\sqrt{1-x^2}\sec^2\sqrt{x}}{1+x}$

29 $\left(\dfrac{1+2x}{1+3x}\right)^4$

30 $\left(\dfrac{x^2+3x+5}{2x-1}\right)^5$

31 $\tan^2\sqrt{x}$

32 $(2x+1)^3 \cot^5 x^3$

33 $\dfrac{1}{\sqrt{1-x^2}}$

34 $\dfrac{x}{\sqrt{1-x^2}}$

35 $\sqrt{5(3x-2)^4+1}$

36 $\sqrt[3]{(2x+1)^2+x}$

37 $\cos 3x \sin 4x$

38 $\sec 2x \tan 5x$

Exercises 39 to 44 concern the formula $D(x^r) = rx^{r-1}$. In each case use it when finding the derivative of the indicated function.

39 $x^{7/5}$

40 $\sqrt[3]{x^4}$

41 $(x^3+2)^{7/9}$

42 $\sqrt[3]{(x^2+x)^2}$

43 $\sqrt[3]{x} \cdot x^3$

44 $\sqrt[3]{(2x+1)^2}$

In each of Exercises 45 to 48 y is a differentiable function of x; express the indicated derivative in terms of y and dy/dx.

45 $\dfrac{d}{dx}(y^5)$

46 $\dfrac{d}{dx}\left(\dfrac{1}{y}\right)$

47 $\dfrac{d}{dx}(\sin y)$

48 $\dfrac{d}{dx}(\cos y^2)$

49 Differentiate (*a*) $\sin(\sin 3x)$ and (*b*) $\cos^2(\tan 5x)$.

In each of Exercises 50 to 53 give an example of an antiderivative of the given function.

50 $(2x+1)^4$

51 $\cos 3x$

52 $x^{4/3}$

53 $\sin 2x$

In Exercises 54 to 58 differentiate the given functions and simplify your answers as far as possible.

54 $\dfrac{2(9x-2)}{135}\sqrt{(3x+1)^3}$

55 $-\frac{1}{3}\cos 3x + \frac{1}{9}\cos^3 3x$

56 $\dfrac{3x}{8} - \dfrac{3\sin 10x}{80} - \dfrac{\sin^3 5x \cos 5x}{20}$

57 $5\sqrt{4+\tan^2 2x}$

58 $\dfrac{x}{x+\sin^3 5x}$

■

59 Find an equation of the tangent line to the curve

(*a*) $y = \sin x$ at the point $\left(\dfrac{\pi}{4}, \dfrac{\sqrt{2}}{2}\right)$.

(*b*) $y = \tan 2x$ at the point $\left(\dfrac{\pi}{6}, \sqrt{3}\right)$.

60 Differentiation can obtain new trigonometric identities from old ones. What identity do you obtain if you differentiate both sides of

(*a*) $\sin 2x = 2 \sin x \cos x$?

(*b*) $\sin(x+c) = \sin x \cos c + \cos x \sin c$, where c is a constant?

61 Combine the formula for $D(1/x)$ with the chain rule to show that $D(1/g) = -g'/g^2$ for any differentiable function g.

62 This exercise outlines the proof of the chain rule when $g'(x) = 0$.

(*a*) Show that in this case it is sufficient to prove that

$$\lim_{\Delta x \to 0} \frac{\Delta y}{\Delta x} = 0.$$

(*b*) For some values of Δx, we have $\Delta u \neq 0$; for others, we have $\Delta u = 0$. Show that as $\Delta x \to 0$ through values of the first type, then $\Delta y/\Delta x \to 0$. [*Hint:* Write $\Delta y/\Delta x = (\Delta y/\Delta u)(\Delta u/\Delta x)$.]

(*c*) Show that, when Δx is of the second type, Δy is 0; hence

$$\frac{\Delta y}{\Delta x} = 0.$$

Thus, as $\Delta x \to 0$ through values of the second type, $\Delta y/\Delta x \to 0$.

(*d*) Combine (*b*) and (*c*) to show that as $\Delta x \to 0$, $\Delta y/\Delta x \to 0$.

3.S SUMMARY

This chapter defined the derivative of a function as

$$\lim_{h\to 0} \frac{f(x+h) - f(x)}{h} \quad \text{or} \quad \lim_{\Delta x\to 0} \frac{f(x+\Delta x) - f(x)}{\Delta x}$$

if this limit exists. Informally, the derivative is

$$\text{The limit of } \frac{\text{difference in outputs}}{\text{difference in inputs}},$$

as the difference in inputs approaches 0.

The derivative measures how quickly the value of a function changes. If a slight change in the input causes a relatively large change in the output, the derivative will be large. Most of the chapter was spent developing techniques for computing derivatives without having to return to the definition of the derivative and calculate a (perhaps horrendous) limit each time you want to know the derivative of some function. These labor-saving methods are listed later in this summary.

The derivative was motivated by slope, velocity, magnification, and density—concepts drawn from geometry or the physical world. The derivative has many more applications, and we will cite four more. But whenever the rate at which some quantity changes is studied, a derivative will surely enter the picture.

Biology

Let $P(t)$ be a differentiable function that estimates the size of a population at time t. Then the derivative $P'(t)$ tells how fast the population is increasing [if $P'(t) > 0$] or decreasing [if $P'(t) < 0$] at time t.

Physiology

Let $Q(t)$ be the amount of blood, in cubic centimeters, that flows through an artery during the first t seconds of an observation. Then the derivative $Q'(t)$ is the rate, in cubic centimeters per second, at which blood flows through the artery at time t.

Economics

Let $C(x)$ be the cost in dollars of producing x refrigerators. [In reality x is an integer; in economic theory and practice it is convenient to assume that $C(x)$ is defined for all real numbers in some interval and that $C(x)$ is a differentiable function.] The derivative $C'(x)$ is called the **marginal cost**. This marginal cost, as we will now show, is roughly the cost of producing the $(x + 1)$st refrigerator. The actual cost of producing refrigerator number $x + 1$ is the cost of producing the first $x + 1$ refrigerators less the cost of producing the first x refrigerators. So the cost of producing the $(x + 1)$st refrigerator is $C(x + 1) - C(x)$, which equals

$$\frac{C(x+1) - C(x)}{1},$$

which, by the definition of the derivative, is an approximation of $C'(x)$. Or, looked at the opposite way, $C'(x)$ is an approximation of the ratio $[C(x + 1) - C(x)]/1$, the cost of the $(x + 1)$st refrigerator.

Similarly, if $R(x)$ is the total revenue received for x refrigerators, then the derivative $R'(x)$ is called the **marginal revenue**, which can be thought of as the extra revenue obtained by selling the $(x + 1)$st refrigerator.

Energy

Let $Q(t)$ be the total amount of crude oil in the earth at time t, measured in barrels. (One barrel holds 42 gallons.) The derivative $Q'(t)$ tells how fast $Q(t)$ changes. If no new reserves are being formed, then $Q'(t)$ is negative, approximately $-50{,}000{,}000$ barrels per day. Estimates of $Q(t)$ for $t = 1980$ vary but are on the order of $2 \cdot 10^{12}$ barrels (two trillion barrels). If $Q'(t)$ remains constant, all known and conjectured reserves would be used up in about a century.

Predictions of the rate at which petroleum—or any other natural resource—will be used depend on estimates of derivatives.

The following table is worth careful study. The bottom row describes the derivative, which is the underlying mathematical concept, free of any particular interpretation. Each of the other lines describe one of its many *applications* or *interpretations*.

If we interpret x as	*and $f(x)$ as*	*then* $\frac{f(x + \Delta x) - f(x)}{\Delta x}$ *is*	*and, as Δx approaches 0, the quotient approaches*
The abscissa of a point in the plane	The ordinate of that point	The slope of a certain line	The slope of a tangent line
Time	The location of a particle moving on a line	An average velocity over a time interval	The velocity at time x
A point on a linear slide	Its projection on a linear screen	An average magnification	The magnification at x
A location on a non-uniform string	The mass from 0 to x	An average density	The density at x
Time	Mass of a bacterial culture at time x	An average growth rate over a time interval	The growth rate at time x
Time	Total profit up to time x	An average rate of profit over a time interval	The rate of profit at time x
Just a number: the input	A number depending on x: the output	A quotient: the change in the output divided by the change in the input	The derivative evaluated at x (the rate of change of the function with respect to x)

Vocabulary and Symbols

difference quotient $\quad \frac{f(x + \Delta x) - f(x)}{\Delta x}$

derivative $\quad f', \frac{d}{dx}(f), \frac{df}{dx}, \frac{dy}{dx}, D(f)$, and the dot notation $\quad \dot{x} = \frac{dx}{dt}$

differentiable
velocity and speed
magnification
density
tangent to a curve
antiderivative
chain rule
Δx (change in input)
Δf (change in output)
slope of a curve

Key Facts

The derivative is defined as a limit:

$$\lim_{\Delta x \to 0} \frac{f(x + \Delta x) - f(x)}{\Delta x}$$

if the limit exists.

If $f'(x)$ exists at a particular number x, then f is said to be differentiable at that number. A function that is differentiable throughout an interval is said to be differentiable on that interval. Most functions met in applications are differentiable throughout their domains with perhaps the exception of a few isolated points. For instance, $\sqrt{x}$ is differentiable throughout $(0, \infty)$ but not at 0.

Wherever a function is differentiable it is necessarily continuous. However, a function can be continuous at a number yet not be differentiable there. For example $|x|$ is continuous at 0 but not differentiable there. (In 1831, Bolzano constructed a function that is continuous everywhere but differentiable nowhere!)

The computational formulas and techniques obtained in this chapter are recorded in the following two tables.

FORMULAS FOR DERIVATIVES

f	f'	*Remark*
c	0	Constant function.
x	1	
x^r	rx^{r-1}	Rational r fixed.
$\sqrt{x}$	$\frac{1}{2\sqrt{x}}$	Although a special case of x^r, worth memorizing.
$\frac{1}{x}$	$\frac{-1}{x^2}$	Although a special case of x^r, worth memorizing.
$a_n x^n + \cdots + a_1 x + a_0$	$na_n x^{n-1} + \cdots + a_1$	Differentiate polynomials term by term.
$\sin x$	$\cos x$	
$\cos x$	$-\sin x$	Remember that the derivatives of the "co" functions have the minus sign.
$\tan x$	$\sec^2 x$	
$\cot x$	$-\csc^2 x$	
$\sec x$	$\sec x \tan x$	
$\csc x$	$-\csc x \cot x$	

Combining these formulas with the chain rule shows, for instance, that

$$\frac{d}{dx}((u(x))^r) = r(u(x))^{r-1}u'(x)$$

and

$$\frac{d}{dx}(\sin u(x)) = (\cos u(x))u'(x),$$

where $u(x)$ is a differentiable function of x.

TECHNIQUES OF DIFFERENTIATION

$$(cf)' = cf' \qquad \left(\frac{f}{c}\right)' = \frac{f'}{c} \qquad (c \text{ constant})$$

$$(f+g)' = f' + g' \qquad (f-g)' = f' - g'$$

$$(fg)' = fg' + gf' \qquad \left(\frac{f}{g}\right)' = \frac{gf' - fg'}{g^2} \qquad \left(\frac{1}{g}\right)' = -\frac{g'}{g^2}$$

And, most important, most often used, the chain rule: If $y = f(u)$ and $u = g(x)$, then

$$\frac{dy}{dx} = \frac{dy}{du}\frac{du}{dx}.$$

GUIDE QUIZ ON CHAP. 3: THE DERIVATIVE

1 Define the derivative.

2 Use the definition of the derivative to compute

(a) $\frac{d}{dx}(5x^3 - 2x + 2)$ (b) $\frac{d}{dx}\left(\frac{5}{3x+2} + 6x\right)$

(c) $\frac{d}{dx}(3 \sin 2x)$ (d) $\frac{d}{dx}(x^{-2})$

3 Using formulas developed in the chapter, differentiate
(a) $5\sqrt{x}$ (b) $x^2\sqrt{3 - 2x^2}$ (c) $\cos 5x$
(d) $(1 + x^2)^{3/4}$ (e) $\sqrt[3]{\tan 6x}$ (f) $x^3 \sin 5x$

(g) $\frac{1}{\sqrt{2x+1}}$ (h) $(2x^5 - x^3)^{-4}$ (i) $\sqrt[3]{x^3 - 3}$

(j) $\frac{2x^3 + 2}{3x + 1}$ (k) $\frac{1}{5x^2 + 1}$ (l) $\frac{1}{(3x+2)^{10}}$

(m) $(1 + 2x)^5 x^3 \sec 3x$ (n) $\csc \sqrt{x}$
(o) $(1 + 3 \cot 4x)^{-2}$

4 On a sketch of the graph of a typical function f,
(a) show the line whose slope is $[f(x + h) - f(x)]/h$;
(b) show the tangent line at the point $(x, f(x))$.

5 (a) Sketch the graph of $y = 3x^2 + 5x + 6$.
(b) By inspection of your graph, estimate the x coordinate of the point where the tangent line is horizontal.
(c) Using the derivative, solve (b) precisely.

6 (a) Without sketching the graph of $y = x^4$, draw the line that is tangent to it at the point $(\frac{1}{2}, \frac{1}{16})$.
(b) Find an equation for this line.

7 (a) Graph $y = x^3 - 12x$ with the aid of this table.

x	-2	-1	0	1	2	3
$x^3 - 12x$						

(b) Evaluate $(x^3 - 12x)'$.
(c) Find all x such that the derivative in (b) has the value 0.
(d) At what points on the graph in (a) is the tangent line horizontal? Specify both the x and y coordinates.

8 Let f be a differentiable function and let h be a positive number. Interpret $f(x + h) - f(x)$, h, and their quotient $[f(x + h) - f(x)]/h$ if (a) $f(x)$ is the height of a rocket x seconds after lift-off; (b) $f(x)$ is the number of bacteria in a bacterial culture at time x; (c) $f(x)$ is the mass of the left-hand x centimeters of a rod; (d) $f(x)$ is the position of the image on the linear screen of the point x on the linear slide.

9 A bug is wandering on the x axis. At time t seconds it is at the point $x = t^2 - 2t$. Assume that distance is measured in meters.
(a) What is the bug's velocity at time t?
(b) What is the bug's velocity at $t = \frac{1}{4}$?
(c) What is the bug's speed when $t = \frac{1}{4}$?
(d) Is the bug moving to the right or to the left when $t = \frac{1}{4}$?

10 Give an antiderivative for (a) x^2, (b) $1/x^2$, (c) $\sin 3x$, (d) $3x^2 + 4x + 5$, (e) $\sec^2 x$, and (f) $\sin x \cos x$.

REVIEW EXERCISES FOR CHAP. 3: THE DERIVATIVE

In Exercises 1 to 6 use the definition of the derivative to differentiate the given functions.

1 $5x^3$ 2 $\sqrt{3x}$ 3 $1/(x + 3)$
4 $(2x + 1)^2$ 5 $\cos 3x$ 6 $\sin 5x$

In Exercises 7 to 36 find the derivatives of the given functions.

7 $2x^5 + x^3 - x$ 8 $t^4 - 5t^2 + 2$

9 $\frac{x^2}{4x + 1}$ 10 $\frac{(3x + 1)^4}{(2x - 1)^2}$

11 $\sqrt{3x^2 + 2x + 4}$ 12 $\sqrt{5x^2 - x}$
13 $\sqrt[3]{(2t - 1)^2}$ 14 $(t^2 + 1)^{3/4}$
15 $\sin^2 5x$ 16 $\cos^3 7x$

17 $\frac{(5x + 1)^4}{7}$ 18 $\frac{(3x - 2)^{-5}}{11}$

19 $x \sin 3x$ 20 $x^2 \cos 4x$

21 $\tan^2 \sqrt[3]{1 + 2x}$ 22 $\left(\frac{\sin 2x}{1 + \tan 3x}\right)^3$

23 $\frac{x^3 \cos 2x}{1 + x^2}$ 24 $\frac{x^2 \sin 5x}{(2x + 1)^3}$

25 $\sqrt[3]{1 + \sqrt{x^2 + 3}}$ 26 $\sqrt[4]{x^3 + \sqrt[3]{x} + \sin^2 x}$

27 $\sqrt[3]{(\cot 5x)^7}$ 28 $\sqrt[5]{(\csc 3x)^{11}}$
29 $\sqrt{8x + 3}$ 30 $\sqrt{5x - 1}$

31 $\dfrac{x^2}{x^3 + 1}$

32 $\dfrac{x^3 + 1}{x^2}$

33 $[(x^2 + 3x)^4 + x]^{-5/7}$

34 $\left(\dfrac{3x + 1}{2x + 1}\right)^4$

35 $\dfrac{x\sqrt{2x + 1}\cos^2 6x}{5}$

36 $\dfrac{5(1 + x^2)^3}{x\sqrt{2x + 1}}$

In Exercises 37 to 40 the number a is a constant. In each case differentiate the given function and simplify your answer.

37 $\dfrac{1}{a^2}\sin ax - \dfrac{1}{a}x\cos ax$

38 $\dfrac{x^2}{4} - \dfrac{x\sin 2ax}{4a} - \dfrac{\cos 2ax}{8a^2}$

39 $\dfrac{x}{2} - \dfrac{\sin 2ax}{4a}$

40 $-\dfrac{1}{a}\cos ax + \dfrac{1}{3a}\cos^3 ax$

41 The height of a ball thrown straight up is $64t - 16t^2$ feet after t seconds.
(*a*) Show that its velocity after t seconds is $64 - 32t$ feet per second.
(*b*) What is its velocity when $t = 0$? $t = 1$? $t = 2$? $t = 3$?
(*c*) What is its speed when $t = 0$? $t = 1$? $t = 2$? $t = 3$?
(*d*) For what values of t is the ball rising? falling?

42 In the study of the seepage of irrigation water into soil, equations such as $y = \sqrt{t}$ are sometimes used. The equation says that the water penetrates $\sqrt{t}$ feet in t hours.
(*a*) What is the physical significance of the derivative $1/(2\sqrt{t})$?
(*b*) What does (*a*) say about the rate at which water penetrates the soil when t is large?

43 Find an equation of the tangent line to the curve $y = x^3 - 2x^2$ at $(1, -1)$.

44 Find an equation of the tangent line to the curve $y = 2x^4 - 6x^2 + 8$ at $(2, 16)$.

45 (*a*) The left-hand x centimeters of a rod have a mass of $3x^4$ grams. What is its density at $x = 1$?
(*b*) Devise a magnification problem mathematically equivalent to (*a*).
(*c*) Devise a velocity problem mathematically equivalent to (*a*).

46 The left-hand x centimeters of a string have a mass of $\sqrt{x}$ grams. What is its density when x is (*a*) $\frac{1}{4}$? (*b*) 1? (*c*) Is its density defined at $x = 0$?

47 A snail crawls $\sqrt{t}$ feet in t seconds. What is its speed when t is (*a*) $\frac{1}{9}$? (*b*) 1? (*c*) 4? (*d*) 9?

48 Sketch a graph of $y = x^3$.
(*a*) Why can the tangent line to this graph at $(0, 0)$ *not* be defined as "the line through $(0, 0)$ that meets the graph just once"?
(*b*) Why can the tangent line to this graph at $(1, 1)$ *not* be defined as "the line through $(1, 1)$ that meets the graph just once"?
(*c*) How is the tangent line at any point on the graph defined?

49 (Contributed by David G. Mead) (*a*) Sketch the curves $y = x^2 + 1$ and $y = -x^2$. (*b*) Find equations of the lines L that are tangent to both curves simultaneously.

50 After t hours a certain bacterial population has a mass of $500 + t^3$ grams. Find the rate at which it grows (in grams per hour) when (*a*) $t = 0$, (*b*) $t = 1$, and (*c*) $t = 2$.

51 A lens projects the point x on the x axis onto the point x^3 on a linear screen.
(*a*) How much does it magnify the interval $[2, 2.1]$?
(*b*) How much does it magnify the interval $[1.9, 2]$?
(*c*) What is its magnification at 2?

52 Let f be the function whose value at x is $4x^2$.
(*a*) Compute $[f(2.1) - f(2)]/0.1$.
(*b*) What is the interpretation of the quotient in (*a*) if $f(x)$ denotes the total profit of a firm (in millions of dollars) in its first x years?
(*c*) What is the interpretation of the quotient in (*a*) if $f(x)$ denotes the ordinate in the graph of the parabola $y = 4x^2$?
(*d*) What is the interpretation of the quotient in (*a*) if $f(x)$ is the distance a particle moves in the first x seconds?

53 It costs a certain firm $C(x) = 1000 + 5x + x^2/200$ dollars to produce x calculators, for $x \le 400$.
(*a*) How much does it cost to produce 0 calculators? (This represents start-up costs, which are independent of the number produced.)
(*b*) What is the marginal cost $C'(x)$?
(*c*) What is the marginal cost when $x = 10$?
(*d*) Compute $C(11) - C(10)$, the cost of producing the eleventh calculator.

54 (*a*) If the function f records the trade-in value of a car (dependent on its age), then we may think of the derivative f' as ________ .
(*b*) When is the derivative of (*a*) negative? positive? Which is the more usual case?

55 (*Economics*) A certain growing business firm makes a profit of t^2 million dollars in its first t years.
(*a*) How much profit does it make during its third year, that is, from time $t = 2$ to time $t = 3$?
(*b*) How much profit does it make from time $t = 2$ to time $t = 2.5$ (a duration of half a year)?
(*c*) Using (*b*), show that its average rate of profit from time $t = 2$ to time $t = 2.5$ is 4.5 million dollars per year.

(*d*) Find its "rate of profit at time $t = 2$" by considering short intervals of time from 2 to t, $t > 2$, and letting t approach 2.

56 (*Biology*) A certain increasing bacterial population has a mass of t^2 grams after t hours.
(*a*) By how many grams does the population increase from time $t = 3$ hours to time $t = 4$ hours?
(*b*) By how many grams does it increase from time $t = 3$ hours to $t = 3.01$ hours?
(*c*) By how many grams does the population increase from time $t = 3$ hours to time t hours, where t is larger than 3?
(*d*) Using (*c*), show that the average rate of growth from time 3 to time t, $t > 3$, is $3 + t$ grams per hour. As t approaches 3, the average growth rate approaches 6 grams per hour, which is called "the growth rate at time 3 hours."

57 In each of these functions, y denotes a differentiable function of x. Express the derivative of each with respect to x in terms of y and dy/dx.
(*a*) y^3 (*b*) $\cos y$ (*c*) $1/y$

58 Let $f(t)$ be the height in miles of the cloud top above burst height t minutes after the explosion of a 1-megaton nuclear bomb. Figure 3.28 is a graph of this function. (Note that the vertical and horizontal scales are different.) Assume that the cloud is not dispersed.
(*a*) What is the physical meaning of $f'(t)$?
(*b*) As t increases, what happens to $f'(t)$?
(*c*) As t increases, what happens to $f(t)$?
(*d*) Estimate how rapidly the cloud is rising at the time of explosion.
(*e*) Estimate how rapidly the cloud is rising 1 minute after the explosion.
In (*d*) and (*e*) make the estimate by drawing a tangent line. The estimate will be in miles per minute.

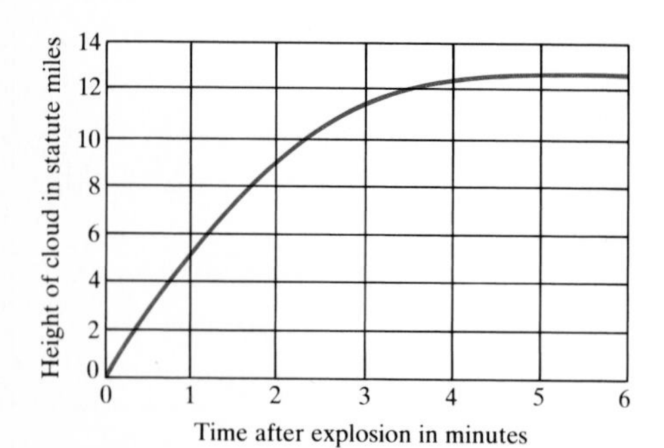

Height of cloud top above burst height at various times after a 1-megaton explosion for a moderately low air burst.

Figure 3.28

59 Give two antiderivatives for each of these functions:
(*a*) $4x^3$ (*b*) x^3
(*c*) $x^4 + x^3 + \cos x$ (*d*) $x^3 + \sin x$
(*e*) $(x^2 + 1)^2$ *Hint:* For (*e*), first expand $(x^2 + 1)^2$.

60 (*a*) Does $D(x^2 + x^3)$ equal $D(x^2) + D(x^3)$?
(*b*) Does $D(x^2x^3)$ equal $D(x^2)D(x^3)$?
(*c*) Does $D(x^2 - x^3)$ equal $D(x^2) - D(x^3)$?
(*d*) Does $D\left(\frac{x^2}{x^3}\right)$ equal $\frac{D(x^2)}{D(x^3)}$?

61 Figure 3.29 shows the graph of a function f. (*a*) At which numbers a does $\lim_{x\to a} f(x)$ not exist? (*b*) At which numbers a does $\lim_{x\to a} f(x)$ exist yet f is not continuous at a? (*c*) Where is f continuous but not differentiable?

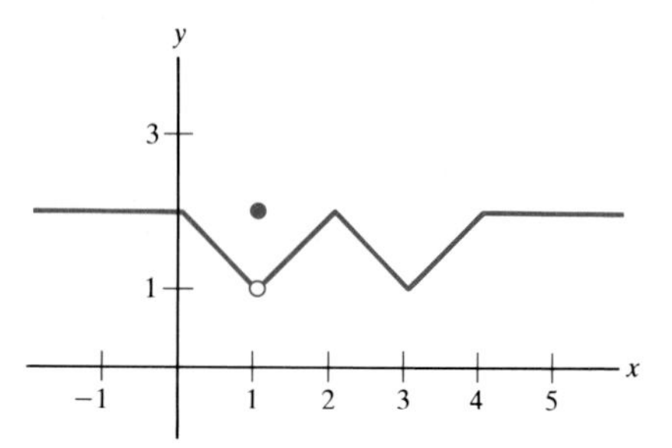

Figure 3.29

■

The next exercise briefly introduces a concept (that will later be treated in detail) which might be met early in an elementary physics or chemistry class.

62 In the case of a function of more than one variable we may differentiate with respect to any one of the variables, treating the others as constants. For instance, the derivative of x^3y^2z with respect to x is $3x^2y^2z$; the derivative of x^3y^2z with respect to y is $2x^3yz$; the derivative with respect to z is x^3y^2. Such derivatives are called **partial derivatives** and are denoted with the symbol ∂ (called "del") as follows:

$$\frac{\partial}{\partial x}(x^3y^2z) = 3x^2y^2z,$$

$$\frac{\partial}{\partial y}(x^3y^2z) = 2x^3yz,$$

$$\frac{\partial}{\partial z}(x^3y^2z) = x^3y^2.$$

Compute the following partial derivatives:

(*a*) $\frac{\partial}{\partial x}\left(\frac{x^2z}{y}\right)$ (*b*) $\frac{\partial}{\partial y}\left(\frac{x^2z}{y}\right)$ (*c*) $\frac{\partial}{\partial z}\left(\frac{x^2z}{y}\right)$

(*d*) $\frac{\partial}{\partial x}(\cos 3x \sin 4y)$ (*e*) $\frac{\partial}{\partial y}(\cos 3x \sin 4y)$

63 Tell what is wrong with this alleged proof that $2 = 1$: Observe that $x^2 = x \cdot x = x + x + \cdots + x$ (x times).

Differentiation with respect to x yields the equation $2x = 1 + 1 + \cdots + 1$ (x 1s). Thus $2x = x$. Setting $x = 1$ shows that $2 = 1$.

64 Define $f(x)$ to be $x^2 \sin (1/x)$ if $x \neq 0$, and $f(0)$ to be 0.
(*a*) Show that f has a derivative at 0, namely, 0.

Hint: Investigate $\displaystyle\lim_{\Delta x \to 0} \frac{f(\Delta x) - f(0)}{\Delta x}$.

(*b*) Show that f has a derivative at $x \neq 0$.
(*c*) Show that the derivative of f is not continuous at $x = 0$.

65 Let $V(r) = 4\pi r^3/3$, the volume of a sphere of radius r. Let $S(r) = 4\pi r^2$, the surface area of a sphere of radius r.
(*a*) Show that $V'(r) = S(r)$.
(*b*) With the aid of a picture showing concentric spheres of radii r and $r + \Delta r$, explain why the result in (*a*) is plausible.

Express the limits in Exercises 66 and 67 as derivatives, and evaluate them.

66 $\displaystyle\lim_{w \to 2} \frac{(1 + w^2)^3 - 125}{w - 2}$

67 $\displaystyle\lim_{\Delta x \to 0} \frac{\sin \sqrt{3 + \Delta x} - \sin \sqrt{3}}{\Delta x}$

68 Find all continuous functions f, whose domains are the x axis, such that $f(0) = 5$ and, for each x, $f(x)$ is an integer.

69 (*a*) Draw a freehand curve indicating a typical function f.
(*b*) Label on it the three points $P_0 = (x, f(x))$, $P_1 = (x + h, f(x + h))$, and $P_2 = (x - h, f(x - h))$.
(*c*) Show that the slope of the line through P_1 and P_2 is

$$\frac{f(x + h) - f(x - h)}{2h}.$$

(*d*) For a differentiable function, what do you think is the value of

$$\lim_{h \to 0} \frac{f(x + h) - f(x - h)}{2h}?$$

(*e*) Compute the limit in (*d*) if $f(x) = x^3$.

70 Let f and g be differentiable functions. Show that

(*a*) $\displaystyle\frac{(fg)'}{fg} = \frac{f'}{f} + \frac{g'}{g}$

(*b*) $\displaystyle\frac{(f/g)'}{f/g} = \frac{f'}{f} - \frac{g'}{g}$

(It is assumed that the denominators are not 0.)

(*c*) Generalize (*a*) to three functions, f, g, and h.

4 Applications of the Derivative

This chapter applies the derivative to graphing, to the study of motion, to finding maximum and minimum values of a function, and to estimating the change in output of a function when the input changes by a small amount.

4.1 ROLLE'S THEOREM AND THE MEAN-VALUE THEOREM

This section presents theorems that are the basis for many applications of the derivative. It also argues for their plausibility and illustrates them with specific functions. The proofs, deferred to the end of the section, show how each theorem follows from its predecessor.

Let f be a differentiable function defined at least on the closed interval $[a, b]$. Because it is differentiable, it is necessarily continuous. As mentioned in Sec. 2.6, the function f must therefore take on a maximum value for some number c in $[a, b]$. That is, for some number c in $[a, b]$,

$$f(c) \geq f(x)$$

for all x in $[a, b]$. What can be said about $f'(c)$, the derivative at c?

First, if c is neither a nor b, that is, c is in the open interval (a, b), the maximum would appear as in Fig. 4.1. It seems likely that a tangent to the graph at $(c, f(c))$ would be parallel to the x axis, in which case

$$f'(c) = 0.$$

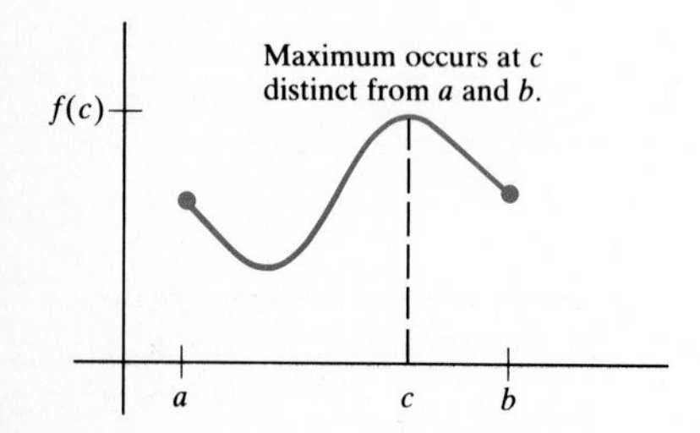

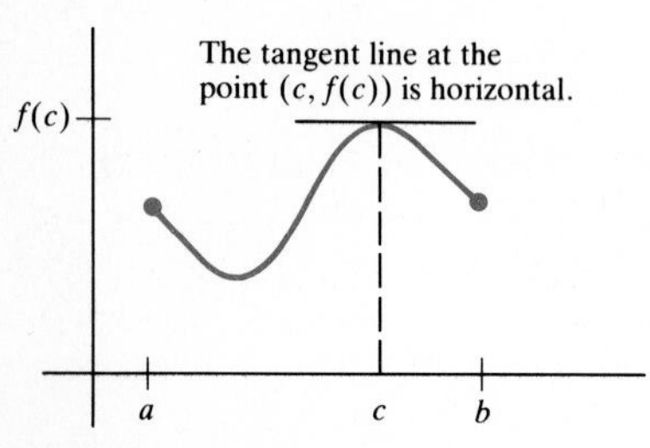

Figure 4.1

If, instead, the maximum occurs at an endpoint of the interval, at a or b, as the graph in Fig. 4.2 illustrates, the derivative at such a point need not be 0. In this graph, the maximum occurs at b, where the derivative is not 0.

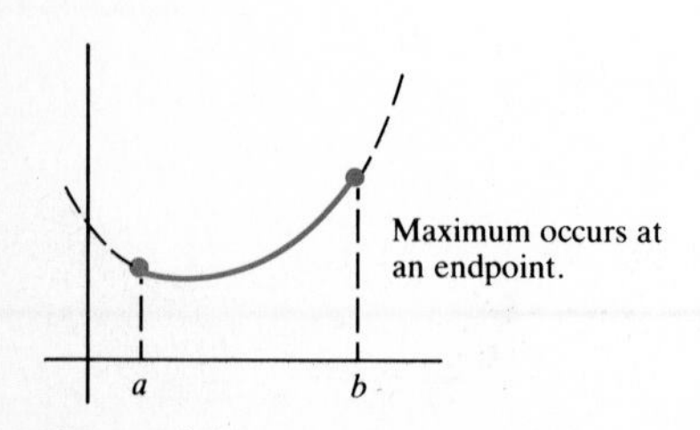

Figure 4.2

The case in which the maximum (or minimum) occurs away from the ends of the interval, that is, in the interior of the interval, is so important that we state it as a theorem.

THEOREM OF THE INTERIOR EXTREMUM

Let f be a function defined at least on the open interval (a, b). If f takes on an extreme value at a number c in this interval and if $f'(c)$ exists, then

$$f'(c) = 0.$$

This theorem will be exploited beginning in Sec. 4.2 to find maximum and minimum values of a function. If an extreme value occurs within an open interval and the derivative exists there, the derivative must be 0 there. In the present section it will be used to establish Rolle's theorem. To motivate Rolle's theorem, we introduce the notion of a chord of a graph.

Michel Rolle was a seventeenth-century mathematician.

Definition A line segment joining two points on the graph of a function f is called a **chord** of f.

Assume that a certain differentiable function f has a chord parallel to the x axis, as in Fig. 4.3. It seems reasonable that the graph will then have at least one horizontal tangent line. (In the case shown, there are three such lines tangent to the graph, as is indicated in Fig. 4.4.) This is the substance of the next theorem.

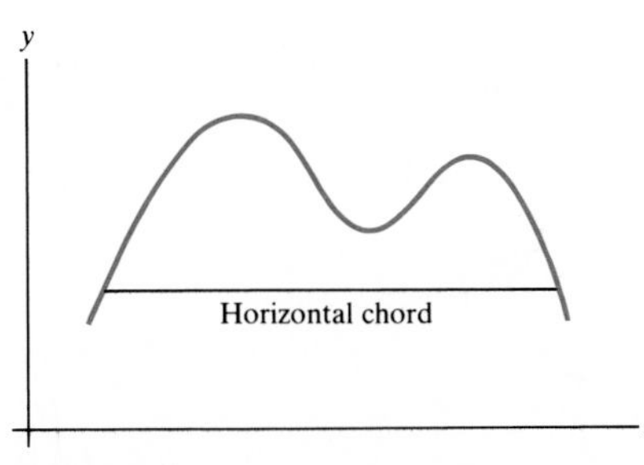

Figure 4.3

Rolle's Theorem Let f be a continuous function on the closed interval $[a, b]$ and have a derivative at all x in the open interval (a, b). If $f(a) = f(b)$, then there is at least one number c in (a, b) such that $f'(c) = 0$. ■

EXAMPLE 1 Verify Rolle's theorem for the case $f(x) = \cos x$ and $[a, b] = [\pi, 5\pi]$.

SOLUTION Note that $f(\pi) = -1 = f(5\pi)$. Since $\cos x$ is differentiable for all x, it is continuous on $[\pi, 5\pi]$ and differentiable on $(\pi, 5\pi)$. According to Rolle's theorem, there must be at least one number c in $(\pi, 5\pi)$ for which $(\cos x)'$ is 0. Now, $(\cos x)' = -\sin x$. Thus there should be at least one solution of the equation

$$-\sin x = 0$$

in the open interval $(\pi, 5\pi)$. As can be checked, the equation has three such solutions, namely, 2π, 3π, and 4π. ■

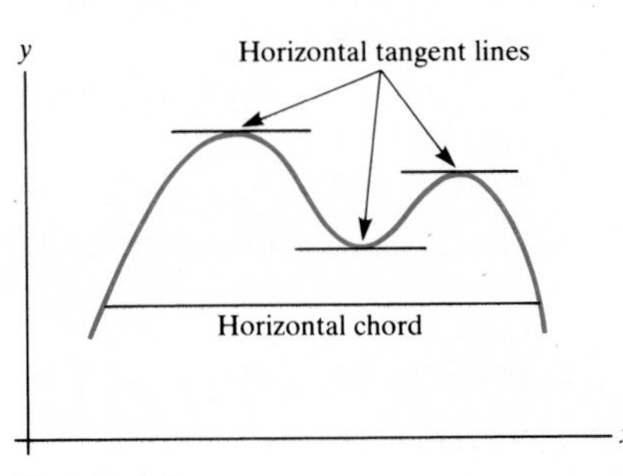

Figure 4.4

The next example has a slight twist, since the function is not differentiable at the ends of the interval $[a, b]$.

EXAMPLE 2 Verify Rolle's theorem for the case $f(x) = \sqrt{1 - x^2}$ and $[a, b] = [-1, 1]$.

SOLUTION Observe that $f(-1) = 0 = f(1)$, that f is continuous, and that $f'(x) = -x/\sqrt{1 - x^2}$, which is defined for all x in $(-1, 1)$. Rolle's theorem then guarantees that there is at least one number c in $(-1, 1)$ such that $f'(c) = 0$. We can find c by setting $f'(c) = 0$:

$$\frac{-c}{\sqrt{1-c^2}} = 0.$$

Thus $c = 0$ (and this happens to be unique). ■

The next example shows that it is necessary to assume in Rolle's theorem that f is differentiable throughout (a, b).

EXAMPLE 3 Can Rolle's theorem be applied to $f(x) = |x|$ in the interval $[a, b] = [-2, 2]$?

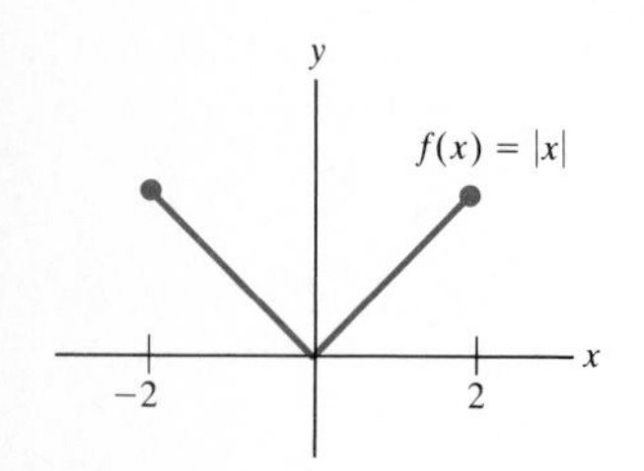

Figure 4.5

SOLUTION First, $f(-2) = 2 = f(2)$; second, f is continuous. The function, however, fails to be differentiable at 0. (See Example 2 in Sec. 3.3.) Not all the hypotheses of Rolle's theorem hold; thus there is no need for the conclusion to hold. Indeed, there is no number c such that $f'(c) = 0$, as a glance at the graph of f in Fig. 4.5 shows. ■

EXAMPLE 4 Use Rolle's theorem to determine how many real roots there are for the equation

$$x^3 - 6x^2 + 15x + 3 = 0.$$

SOLUTION Since $f(x) = x^3 - 6x^2 + 15x + 3$ is a polynomial of odd degree, there is at least one real number r such that $f(r) = 0$. (Recall the argument in Sec. 2.6 based on the intermediate-value theorem.) Could there be another root, s? If so, by Rolle's theorem, there would be a number c (between r and s) at which $f'(c) = 0$.

To check, we compute the derivative of $f(x)$ and see if it is ever equal to 0. We have $f'(x) = 3x^2 - 12x + 15$.

To find when $f'(x)$ is 0, we solve the equation $3x^2 - 12x + 15 = 0$ by the quadratic formula, obtaining

$$x = \frac{12 \pm \sqrt{144 - 180}}{6}.$$

Since $144 - 180$ is negative, the equation has no real roots. Since $f'(x)$ is never 0, it follows that the polynomial $x^3 - 6x^2 + 15x + 3$ has only one real root. ■

Rolle's theorem asserts that if the graph of a function has a horizontal chord, then it has a tangent line parallel to that chord. The mean-value theorem is a generalization of Rolle's theorem, since it concerns any chord of f, not just horizontal chords.

In geometric terms, the theorem asserts that if you draw a chord for the graph of a well-behaved function (as in Fig. 4.6), then somewhere above or below that chord the graph has at least one tangent line parallel to the chord.

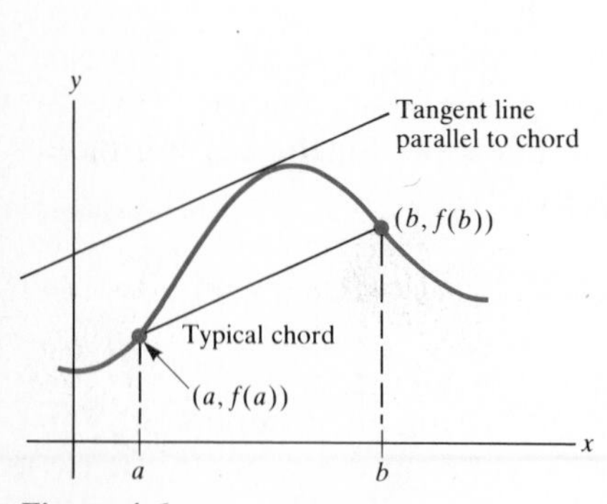

Figure 4.6

Let us translate this geometric statement into the language of functions. Call the ends of the chord $(a, f(a))$ and $(b, f(b))$. The slope of the chord is then

$$\frac{f(b) - f(a)}{b - a},$$

while the slope of the tangent line at a typical point $(x, f(x))$ on the graph is $f'(x)$. The mean-value theorem then asserts that there is at least one number c in the open interval (a, b) such that

$$f'(c) = \frac{f(b) - f(a)}{b - a}.$$

Mean-Value Theorem Let f be a continuous function on the closed interval $[a, b]$ and have a derivative at every x in the open interval (a, b). Then there is at least one number c in the open interval (a, b) such that

$$f'(c) = \frac{f(b) - f(a)}{b - a}. \quad ■$$

EXAMPLE 5 Verify the mean-value theorem for $f(x) = 2x^3 - 8x + 1$, $a = 1$, and $b = 3$.

SOLUTION

$$f(a) = f(1) = 2 \cdot 1^3 - 8(1) + 1 = -5$$

and

$$f(b) = f(3) = 2 \cdot 3^3 - 8(3) + 1 = 31.$$

According to the mean-value theorem, there is at least one number c between $a = 1$ and $b = 3$ such that

$$f'(c) = \frac{31 - (-5)}{3 - 1} = \frac{36}{2} = 18.$$

Let us find c explicitly. Since $f'(x) = 6x^2 - 8$, we need to solve the equation

$$6x^2 - 8 = 18,$$

that is,

$$6x^2 = 26$$

or

$$x^2 = \tfrac{26}{6}.$$

The solutions are $\sqrt{\tfrac{13}{3}}$ and $-\sqrt{\tfrac{13}{3}}$. But only $\sqrt{\tfrac{13}{3}}$ is in $(1, 3)$. Hence there is only one number, namely, $\sqrt{\tfrac{13}{3}}$, that serves as the c whose existence is guaranteed by the mean-value theorem. ■

The interpretation of the derivative as slope suggested the mean-value theorem. What does the mean-value theorem say when the derivative is interpreted, say, as velocity? This question is considered in Example 6.

EXAMPLE 6 A car moving on the x axis has the x coordinate $f(t)$ at time t. At time a its position is $f(a)$. At some later time b its position is $f(b)$. What does the mean-value theorem assert for this car?

SOLUTION The quotient

$$\frac{f(b) - f(a)}{b - a}$$

equals $$\frac{\text{Change in position}}{\text{Change in time}},$$

or "average velocity" for the interval of time $[a, b]$. The mean-value theorem asserts that at some time during this period the velocity of the car must equal its average velocity. To be specific, if a car travels 210 miles in 3 hours, then at some time its speedometer must read 70 miles per hour. ■

There are several ways of writing the mean-value theorem. For example, the equation

$$f'(c) = \frac{f(b) - f(a)}{b - a}$$

is equivalent to $$f(b) - f(a) = (b - a)f'(c),$$

hence to $$f(b) = f(a) + (b - a)f'(c).$$

In this form, the mean-value theorem asserts that $f(b)$ is equal to $f(a)$ plus a quantity that involves the derivative f'. The following important corollaries exploit this alternative view of the mean-value theorem.

Corollary 1 If the derivative of a function is 0 throughout an interval, then the function is constant throughout that interval.

Proof Let s and t be any two numbers in the interval and let the function be denoted by f. To prove the corollary, it suffices to prove that $f(t) = f(s)$.

By the mean-value theorem there is a number c between s and t such that

$$f(t) = f(s) + (t - s)f'(c).$$

But $f'(c) = 0$, since $f'(x)$ is 0 for all x in the given interval. Hence

$$f(t) = f(s) + (t - s)(0),$$

which proves that $$f(t) = f(s). \quad ■$$

When Corollary 1 is interpreted in terms of motion, it is quite plausible. It asserts that if a particle has zero velocity for a period of time, then it does not move during that time.

EXAMPLE 7 Use Corollary 1 to show that $f(x) = \cos^2 3x + \sin^2 3x$ is a constant. Find the constant.

SOLUTION $f'(x) = -6 \cos 3x \sin 3x + 6 \sin 3x \cos 3x = 0$. Corollary 1 says that f is constant. To find the constant, just evaluate f at some specific number, say at 0. We have $f(0) = \cos^2 (3 \cdot 0) + \sin^2 (3 \cdot 0) = \cos^2 0 + \sin^2 0 = 1$. Thus

$$\cos^2 3x + \sin^2 3x = 1$$

for all x. This should be no surprise since, by the pythagorean theorem, $\cos^2 \theta + \sin^2 \theta = 1$. ■

Corollary 2 If two functions have the same derivatives throughout an interval, then they differ by a constant. That is, if $f'(x) = g'(x)$ for all x in an interval, then there is a constant C such that $f(x) = g(x) + C$.

Proof Define a third function h by the equation

$$h(x) = f(x) - g(x).$$

Then
$$h'(x) = f'(x) - g'(x) = 0.$$

Since the derivative of h is 0, Corollary 1 implies that h is constant, that is, $h(x) = C$ for some fixed number C. Thus

$$f(x) - g(x) = C \quad \text{or} \quad f(x) = g(x) + C,$$

and the corollary is proved. ■

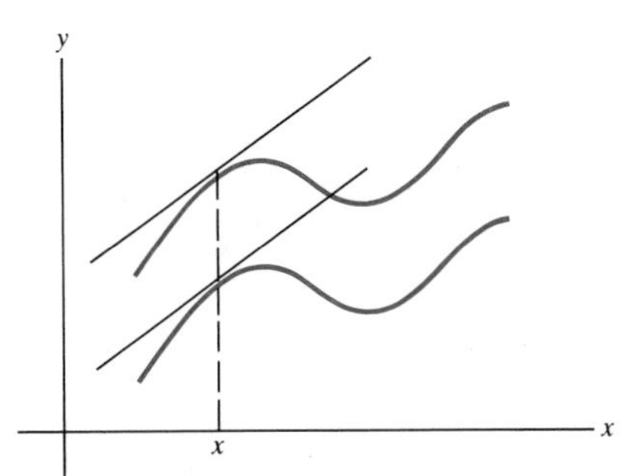

If two graphs have parallel tangent lines at all pairs of points with the same x coordinate, then one graph is obtainable from the other by raising or lowering it.

Figure 4.7

Is Corollary 2 plausible when the derivative is interpreted as slope? In this case, the corollary asserts that if the graphs of two functions have the property that their tangent lines at points with the same x coordinate are parallel, then one graph can be obtained from the other by raising (or lowering) it by an amount C. If you sketch two such graphs (as in Fig. 4.7), you will see that the corollary is reasonable.

EXAMPLE 8 What functions have a derivative equal to $2x$ everywhere?

SOLUTION One such function is x^2; another is $x^2 + 25$. For any constant C, $D(x^2 + C) = 2x$. Are there any other possibilities? Corollary 2 tells us there are not, for if f is a function such that $f'(x) = 2x$, then $f'(x) = (x^2)'$ for all x. Thus the functions f and x^2 differ by a constant, say C, that is,

$$f(x) = x^2 + C.$$

The only functions whose derivatives are $2x$ are of the form $x^2 + C$. ■

Antiderivatives are unique up to a constant difference C.

Example 8 shows that every antiderivative of the function $2x$ must be of the form $x^2 + C$ for some constant C. More generally, if $F(x)$ is a particular antiderivative of the function $f(x)$ on an interval, then any other antiderivative of $f(x)$ there must be of the form $F(x) + C$ for some constant C.

Corollary 1 asserts that if $f'(x) = 0$ for all x, then f is constant. What can be said about f if $f'(x)$ is *positive* for all x? In terms of the graph of f, this assumption implies that all the tangent lines slope upward. It is reasonable to expect that as we move from left to right on the graph in Fig. 4.8, the y coordinate increases, that is, the function f is increasing. In Corollary 3 this is proved.

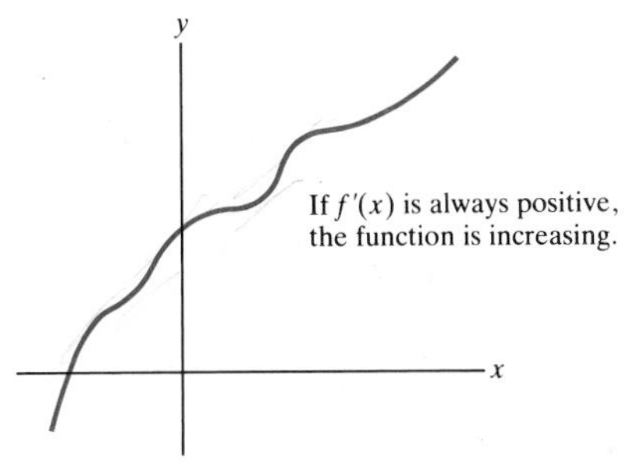

Figure 4.8

Corollary 3 If f is continuous on $[a, b]$ and has a positive derivative on the open interval (a, b), then f is increasing on the interval $[a, b]$. If f is continuous on $[a, b]$ and has a negative derivative on the open interval (a, b), then f is decreasing on the interval $[a, b]$.

Proof We prove the "increasing" case. Take two numbers x_1 and x_2

such that

$$a \le x_2 < x_1 \le b.$$

By the mean-value theorem, there is some number c between x_1 and x_2 such that

$$f(x_1) = f(x_2) + (x_1 - x_2)f'(c).$$

Since $x_1 - x_2$ is positive, and since $f'(c)$ is assumed to be positive, it follows that

$$(x_1 - x_2)f'(c) > 0.$$

Thus $f(x_1) > f(x_2)$, and the corollary is proved. (The "decreasing" case is proved similarly.) ■

As an illustration, consider $3x + 2 \sin x$; its derivative is $3 + 2 \cos x$. Since $\cos x \ge -1$, $3 + 2 \cos x$ is positive. Thus $3x + 2 \sin x$ is an increasing function.

Proofs of the Three Theorems

Proof of the Interior-Extremum Theorem We will prove this for the case of a maximum, the other case being similar. Assume that $f(c)$ is the maximum value of f on $[a, b]$, that c is in (a, b), and that $f'(c)$ exists. It is to be shown that $f'(c) = 0$. We shall prove that $f'(c) \le 0$ and that $f'(c) \ge 0$. From this it will follow that $f'(c)$ must be 0.

Consider the quotient

$$\frac{f(c + \Delta x) - f(c)}{\Delta x}$$

used in defining $f'(c)$. Take Δx so small that $c + \Delta x$ is in the interval $[a, b]$. Since $f(c)$ is the maximum value of $f(x)$ for x in $[a, b]$,

$$f(c + \Delta x) \le f(c).$$

Hence

$$f(c + \Delta x) - f(c) \le 0.$$

Therefore, when Δx is positive,

$$\frac{f(c + \Delta x) - f(c)}{\Delta x}$$

is negative or 0. Consequently, as $\Delta x \to 0$ through positive values,

$$\frac{f(c + \Delta x) - f(c)}{\Delta x},$$

being negative or 0, cannot approach a positive number. Thus

$$f'(c) = \lim_{\Delta x \to 0} \frac{f(c + \Delta x) - f(c)}{\Delta x} \le 0.$$

If, on the other hand, Δx is negative, then the denominator of

$$\frac{f(c + \Delta x) - f(c)}{\Delta x}$$

is negative, and the numerator is still ≤ 0. Hence, for negative Δx,

$$\frac{f(c + \Delta x) - f(c)}{\Delta x} \geq 0$$

(the quotient of two negative numbers being positive). Thus as $\Delta x \to 0$ through negative values, the quotient approaches a number ≥ 0. Hence $f'(c) \geq 0$.

Since $0 \leq f'(c) \leq 0$, $f'(c)$ must be 0, and the theorem is proved. ■

Proof of Rolle's Theorem Since f is continuous, it has a maximum value M and a minimum value m for x in $[a, b]$. Certainly $m \leq M$.

If $m = M$, f is constant and $f'(x) = 0$ for all x in $[a, b]$. Then any number x in (a, b) will serve as the desired number c.

If $m < M$, then the minimum and maximum cannot both occur at the ends of the interval a and b, since $f(a) = f(b)$. One of them, at least, occurs at a number c, $a < c < b$. And, by the interior-extremum theorem, at that c, $f'(c)$ is 0. This proves Rolle's theorem. ■

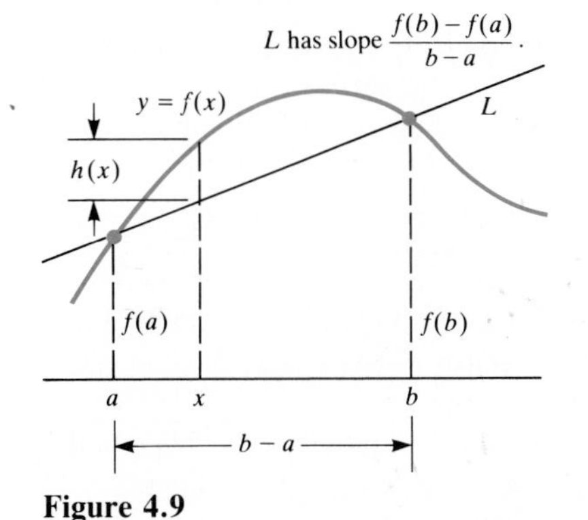

Figure 4.9

Proof of the Mean-Value Theorem We shall prove the theorem by introducing a function to which Rolle's theorem can be applied. The chord through $(a, f(a))$ and $(b, f(b))$ is part of a line L whose equation is, let us say, $y = g(x)$. Let $h(x) = f(x) - g(x)$, which represents the difference between the graph of f and the line L for a given x. It is clear from Fig. 4.9 that $h(a) = h(b)$, since both equal 0.

By Rolle's theorem, there is at least one number c in the open interval (a, b) such that $h'(c) = 0$. But $h'(c) = f'(c) - g'(c)$. Since $y = g(x)$ is the equation of the line through $(a, f(a))$ and $(b, f(b))$, $g'(x)$ is the slope of the line, that is,

$$g'(x) = \frac{f(b) - f(a)}{b - a}$$

for all x.

Hence $$0 = h'(c) = f'(c) - \frac{f(b) - f(a)}{b - a}.$$

In short, $$f'(c) = \frac{f(b) - f(a)}{b - a},$$

and the mean-value theorem is proved. ■

EXERCISES FOR SEC. 4.1: ROLLE'S THEOREM AND THE MEAN-VALUE THEOREM

Exercises 1 and 2 concern the interior extremum.

1 Consider the function $f(x) = x^2$ only for x in $[-1, 2]$.
   (a) Graph the function $f(x)$ for x in $[-1, 2]$.
   (b) What is the maximum value of $f(x)$ for x in the interval $[-1, 2]$?
   (c) Does $f'(x)$ exist at the maximum?
   (d) Does $f'(x)$ equal 0 at the maximum?
   (e) Does $f'(x)$ equal 0 at the minimum?

2 Consider the function $f(x) = \sin x$ for x in $[0, \pi]$.
   (a) Graph the function $f(x)$ for x in $[0, \pi]$.
   (b) Does $f'(x)$ equal 0 at the maximum value of $f(x)$ for x in $[0, \pi]$?
   (c) Does $f'(x)$ equal 0 where $f(x)$ has a minimum for x in $[0, \pi]$?

Exercises 3 to 6 concern Rolle's theorem.

3 (*a*) Graph $f(x) = x^{2/3}$ for x in $[-1, 1]$.
(*b*) Show that $f(-1) = f(1)$.
(*c*) Is there a number c in $(-1, 1)$ such that $f'(c) = 0$?
(*d*) Why does this function not contradict Rolle's theorem?

4 (*a*) Graph $f(x) = 1/x^2$ for x in $[-1, 1]$.
(*b*) Show that $f(-1) = f(1)$.
(*c*) Is there a number c in $(-1, 1)$ such that $f'(c) = 0$?
(*d*) Why does this function not contradict Rolle's theorem?

In each of Exercises 5 and 6, verify that the given function satisfies the hypotheses of Rolle's theorem. Find all numbers c that satisfy the conclusion of the theorem.

5 $x^4 - 2x^2 + 1$ and $[-2, 2]$

6 $\sin x + \cos x$ and $[0, 4\pi]$

In Exercises 7 to 10, find explicitly all values of c which satisfy the mean-value theorem for the given functions and intervals.

7 $\cos x$ and $[0, 6\pi]$

8 $\sin x$ and $[0, 5\pi]$

9 $x^3 + 3x^2 + 3x$ and $[1, 2]$

10 $x^3 + x^2 - x$ and $[0, 1]$

11 (*a*) Differentiate $\sec^2 x$ and $\tan^2 x$.
(*b*) The derivatives in (*a*) are equal. Corollary 2 then asserts that there exists a constant C such that $\sec^2 x = \tan^2 x + C$. Find that constant.

12 (*a*) Differentiate $\csc^2 x$ and $\cot^2 x$.
(*b*) The derivatives in (*a*) are equal. Find the constant C (promised by Corollary 2) such that $\csc^2 x = \cot^2 x + C$.

In Exercises 13 to 16 the function f is differentiable at all real numbers. What can be said about the number of solutions of the equation $f(x) = 3$?

13 $f'(x) < 0$ for all x.

14 $f'(x) < 0$ for $x < 5$; $f'(x) > 0$ for $x > 5$.

15 $f'(x) > 0$ for $x < -2$; $f'(x) < 0$ for x in $(-2, 1)$; $f'(x) > 0$ for $x > 1$.

16 $f'(x) = 0$ only at $x = 4$ and $x = 6$. Assume $f'(x)$ is continuous.

17 Over which intervals is the function $3x^3 - 6x^2 + 4x$ increasing?

18 Over which intervals is the function $2x^3 - 9x^2 + 12x$ increasing?

In Exercises 19 and 20 find all antiderivatives of the given functions.

19 (*a*) $8x^3$ (*b*) $\sin 2x$ (*c*) $1/x^2$ (*d*) $1/(x + 1)^3$

20 (*a*) $10x^4$ (*b*) $\sec^2 3x$ (*c*) $x^3 + 5x^2 + 1$ (*d*) $(2x + 1)^{10}$

■

21 Which of the corollaries to the mean-value theorem implies that (*a*) if two cars on a straight road have the same velocity at every instant, they remain a fixed distance apart? (*b*) if all the tangents to a curve are horizontal, the curve is a horizontal straight line? Explain in each case.

22 Assume that f has a derivative for all x, that $f(3) = 7$, and that $f(8) = 17$. What can we conclude about f' at some number?

23 Consider the function f given by the formula $f(x) = x^3 - 3x$.
(*a*) At which numbers x is $f'(x) = 0$?
(*b*) Use the theorem of the interior extremum to show that the maximum value of $x^3 - 3x$ for x in $[1, 5]$ occurs either at 1 or at 5.
(*c*) What is the maximum value of $x^3 - 3x$ for x in $[1, 5]$?

24 At time t seconds a thrown ball has the height $f(t) = -16t^2 + 32t + 40$ feet.
(*a*) Show that after 2 seconds it returns to its initial height $f(0)$.
(*b*) What does Rolle's theorem imply about the velocity of the ball?
(*c*) Verify Rolle's theorem in this case by computing the numbers c which it asserts exist.

The answers to Exercises 25 and 26 should be phrased in colloquial English. (Section 3.1 introduced the concepts of density and magnification.)

25 State the mean-value theorem in terms of density and mass. [Let x be the distance from the left end of a string and $f(x)$ the mass of the string from 0 to x.] When stated in these terms, does the mean-value theorem seem reasonable?

26 State the mean-value theorem in terms of a slide and a screen. [Let x denote the position on the (linear) slide and $f(x)$ denote the position of the image on the screen.] In optical terms, what does the mean-value theorem say?

27 Differentiate for practice:

(*a*) $\sqrt{1 - x^2}\sin 3x$ (*b*) $\dfrac{\sqrt[3]{x}}{x^2 + 1}$ (*c*) $\tan \dfrac{1}{(2x + 1)^2}$

28 Let $f(x) = (x^2 - 4)/(x + \frac{1}{2})$. Observe that $f(2) = f(-2)$.
(*a*) What does Rolle's theorem say about f?
(*b*) For which values of x is $f'(x) = 0$?

29 Can a polynomial of degree 100 have more than 100 real roots? (Use results in this section to explain your answer.)

30 For which values of the constant k is the function $7x + k \sin 2x$ always increasing?

31 Consider the function $f(x) = x^3 + ax^2 + c$. Show that if $a < 0$ and $c > 0$, then f has exactly one negative root.

32 Let f have a derivative for all x.
(*a*) Is every chord of the graph of f parallel to some tangent to the graph of f?
(*b*) Is every tangent to the graph of f parallel to some chord of the graph of f?

33 A driver covered the 400 miles of Highway 5 from Los Angeles to Sacramento in 6 hours. He bragged to a Highway Patrol officer. The officer (a student of calculus) promptly cited him for speeding. What theorem did he use to justify giving the driver a ticket?

34 Let r be a rational number, $0 < r < 1$, and let x be a positive real number.
(*a*) (Calculator) Which is larger, $(1 + x)^r$ or $1 + rx$? Experiment for some choices of r and x, with some x very small and some large.
(*b*) Make a conjecture and prove it.

35 The same as Exercise 34, except $r > 1$.

36 Is there a differentiable function f whose domain is the x axis such that f is increasing and yet the derivative is *not* positive for all x?

37 Let f and g be two functions differentiable on (a, b) and continuous on $[a, b]$. Assume that $f(a) = g(a)$ and that $f'(x) < g'(x)$ for all x in (a, b). Prove the inequality $f(b) < g(b)$.

38 (See Exercise 37.)
(*a*) Show that $\tan x > x$ if $0 < x < \pi/2$.
(*b*) What does (*a*) say about certain lengths related to the unit circle?

39 Let $f(x) = 2x^5 - 10x + 5$.
(*a*) Show that $f'(x)$ is positive for x in $(1, \infty)$ and in $(-\infty, -1)$ but negative for x in $(-1, 1)$.
(*b*) Compute $f(-1)$ and $f(1)$ and sketch a graph of f. (Don't try to find its x intercepts.)
(*c*) Using (*a*) and (*b*), show that the equation $2x^5 - 10x + 5 = 0$ has exactly three real roots.

Remark: It is proved in advanced algebra that none of the roots of the equation in (*c*) can be expressed in terms of combinations of square roots, cube roots, fourth roots, fifth roots, ... However, the roots of polynomials of degree 2, 3, or 4 can be expressed in terms of such roots. (The quadratic formula takes care of the case of degree 2. There is no corresponding formula for degree 5; no such formula can *ever* be found.)

40 (*a*) Recall the definition of $g(x)$ in the proof of the mean-value theorem, and show that

$$g(x) = f(a) + \frac{x - a}{b - a}[f(b) - f(a)].$$

(*b*) Using (*a*), show that

$$g'(x) = \frac{f(b) - f(a)}{b - a}.$$

41 Is this proposed proof of the mean-value theorem correct? *Proof:* Tilt the x and y axes until the x axis is parallel to the given chord. The chord is now "horizontal," and we may apply Rolle's theorem.

42 Which polynomials of degree at most 3 are one-to-one functions?

43 In *Surely You're Joking, Mr. Feynman,* Norton, New York, 1985, Caltech physicist and Nobel laureate Richard P. Feynman writes

> I often liked to play tricks on people when I was at MIT. One time, in mechanical drawing class, some joker picked up a French curve (a piece of plastic for drawing smooth curves—a curly, funny-looking thing) and said, "I wonder if the curves on that thing have some special formula?"
>
> I thought for a moment and said, "Sure they do. The curves are very special curves. Lemme show ya," and I picked up my French curve and began to turn it slowly. "The French curve is made so that at the lowest point on each curve, no matter how you turn it, the tangent is horizontal."
>
> All the guys in the class were holding their French curve up at different angles, holding their pencil up to it at the lowest point and laying it along, and discovering that, sure enough, the tangent is horizontal.

How was Feynman playing a trick on his classmates?

4.2 USING THE DERIVATIVE AND LIMITS WHEN GRAPHING A FUNCTION

The primitive and inefficient way to graph a function is to make a table of values, plot many points, and draw a curve through the points (hoping that the chosen points adequately represent the function). Chapter 2 refined the technique somewhat. As mentioned in Sec. B.1 of Appendix B, the x and y *intercepts* are of aid in graphing, for they tell where the graph meets the x and y axes. Furthermore, horizontal and vertical *asymptotes* were discussed; they can be of use in sketching the graph for large $|x|$ and also near a number where the function becomes infinite (usually because a denominator is 0). For instance, the line $x = 1$ is a vertical asymptote of $1/(x - 1)$; the line $y = 0$ is a horizontal asymptote

The x and y intercepts

of the same curve. The line $x = \pi/2$ is a vertical asymptote of the curve $y = \tan x$.

This section shows how to use the derivative and limits to help graph a function. Of particular interest will be these questions:

Where is the derivative equal to 0?

Where is the derivative positive? Negative?

How does the function behave for $|x|$ large?

The answers will tell a good deal about the general shape of a particular graph; it will then not be necessary to plot so many specific points on the graph.

First, a few helpful definitions.

Definition *Critical number and critical point.* A number c at which $f'(c) = 0$ is called a **critical number** for the function f. The corresponding point $(c, f(c))$ on the graph of f is a **critical point** on that graph.

Definition *Relative maximum (local maximum).* The function f has a **relative maximum** (or **local maximum**) at the number c if there is an open interval (a, b) around c such that $f(c) \geq f(x)$ for all x in (a, b) that lie in the domain of f. A **local** or **relative minimum** is defined analogously.

Definition *Global maximum.* The function f has a **global maximum** (or **absolute maximum**) at the number c if $f(c) \geq f(x)$ for all x in the domain of f. A **global minimum** is defined analogously.

Note that a global maximum is necessarily a local maximum as well. A local maximum is like the summit of a single mountain; a global maximum corresponds to Mount Everest.

Figure 4.10 illustrates the notions of critical point, local maximum, global maximum, local minimum, and global minimum in the graph of a hypothetical function. Any given function may have none of these, or some, or all.

By the theorem of the interior extremum in Sec. 4.1, there is a close relation between a local maximum (or minimum) and critical points for a

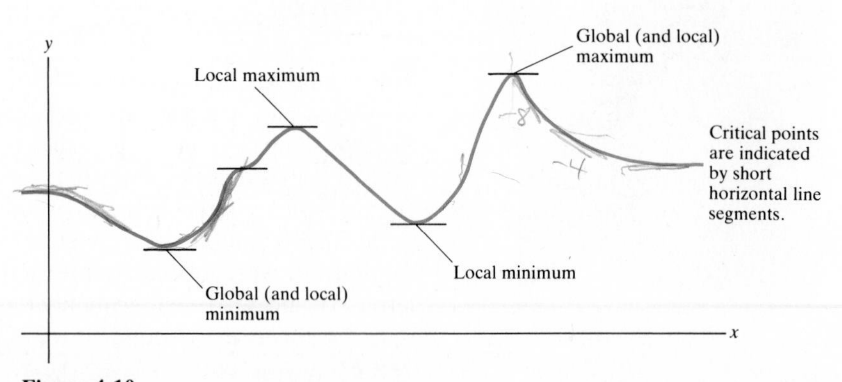

Figure 4.10

differentiable function. If a local maximum occurs at a number c that lies within some open interval within the domain of f, then $f'(c) = 0$. This means that c is a critical number. However, a critical point need not be a local extremum. This is illustrated by the function x^3, whose derivative $3x^2$ is 0 at 0. Thus $c = 0$ is a critical number of the function x^3. A glance at Fig. 4.11 shows that x^3 has neither a local maximum nor a local minimum at 0.

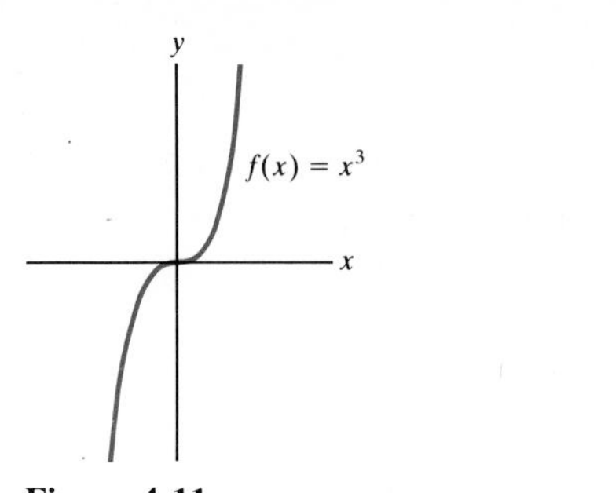

Figure 4.11

To determine whether a function has a local extremum at c, it is not enough to know that $f'(c) = 0$. It is also important to know how the derivative behaves for inputs near c.

A function may not be differentiable at a local extremum. For instance, consider $f(x) = x^{2/3}$, which is graphed in Fig. 4.12. Clearly it has a local minimum at $x = 0$. However, $f'(x) = \frac{2}{3}x^{-1/3}$ is not defined at $x = 0$. [This is an unusual situation, but it should be kept in mind as a possibility when dealing with functions that are not differentiable throughout the domain of interest. The curve $y = x^{2/3}$ is said to have a cusp at (0, 0).]

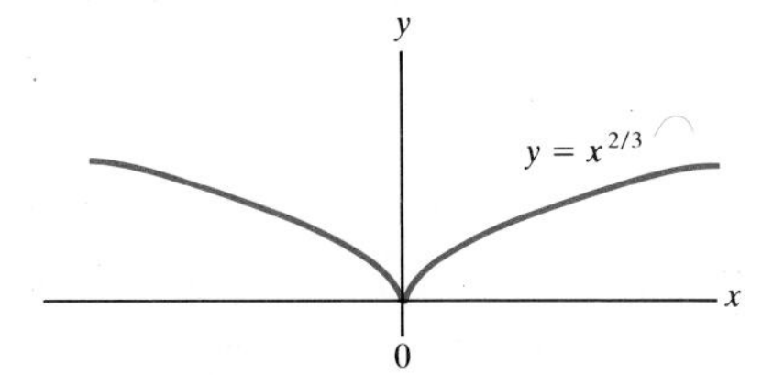

Figure 4.12

The following test for local maximum or local minimum is an immediate consequence of the fact that when the derivative is positive the function increases and when it is negative it decreases.

FIRST-DERIVATIVE TEST FOR LOCAL MAXIMUM AT $x = c$

Let f be a function and let c be a number in its domain. Assume that numbers a and b exist such that $a < c < b$ and

1. f is continuous on the open interval (a, b).
2. f is differentiable on the open interval (a, b), except possibly at c.
3. $f'(x)$ is positive for all $x < c$ in the interval and is negative for all $x > c$ in the interval.

Then f has a **local maximum** at c.

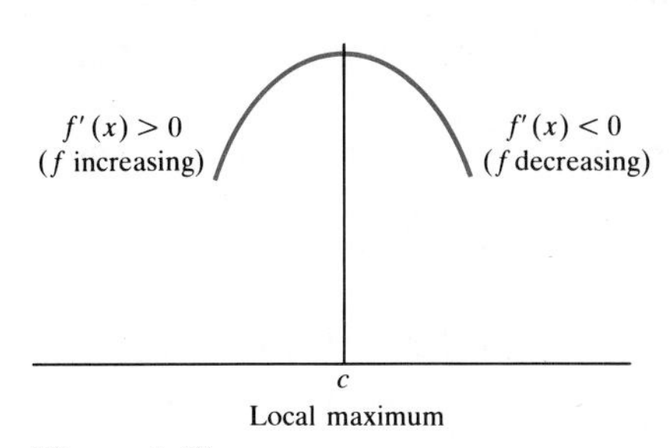

Figure 4.13

A similar test, with "positive" and "negative" interchanged, holds for a **local minimum**. (See Figs. 4.13 and 4.14.)

Informally, the derivative test says, "If the derivative changes sign at c, then the function has either a local minimum or a local maximum." To decide which it is, just make a crude sketch of the graph near $(c, f(c))$ to show on which side of c the function is increasing and on which side it is decreasing.

Note that in the derivative test there is no assumption that the derivative exists at $x = c$. Thus the test applies to $f(x) = |x|$, which has a local minimum at $x = 0$, where the derivative changes from -1 to 1. However, $f(x) = |x|$ is not differentiable at 0, as shown in Example 2 of Sec. 3.3.

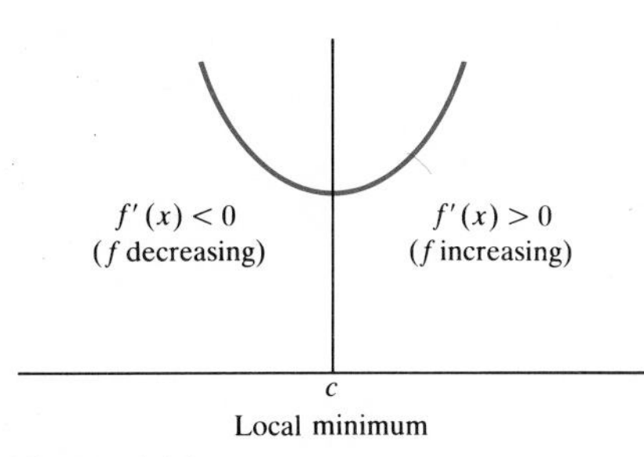

Figure 4.14

EXAMPLE 1 Graph $f(x) = (2x - 5)/(x - 1)$.

SOLUTION Note that at $x = 1$ the function is not defined.

Next observe that $f(0) = 5$ and that $f(x) = 0$ only when the numerator,

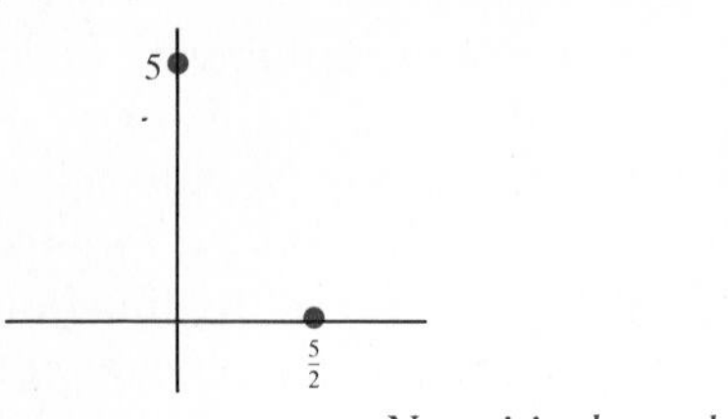

Figure 4.15

$2x - 5$, is 0. Thus the y intercept is 5 and the only x intercept is $x = \frac{5}{2}$, the solution of the equation $2x - 5 = 0$. This is recorded in Fig. 4.15.

Next, *determine the critical numbers of f*. To do this, compute $f'(x)$:

$$f'(x) = D\left(\frac{2x - 5}{x - 1}\right) = \frac{(x - 1) \cdot 2 - (2x - 5) \cdot 1}{(x - 1)^2} = \frac{3}{(x - 1)^2}.$$

No critical numbers

Since the numerator is never 0, there are no critical numbers.

Where is the function increasing? Decreasing? Since the derivative is $3/(x - 1)^2$, it is positive throughout the domain of the function. The function is always increasing.

Always increasing

How does the function behave when $|x|$ is large? We have

$$\lim_{x \to \infty} \frac{2x - 5}{x - 1} = \lim_{x \to \infty} \frac{2 - 5/x}{1 - 1/x} = \frac{2}{1} = 2; \qquad \text{similarly } \lim_{x \to -\infty} \frac{2x - 5}{x - 1} = 2.$$

Horizontal asymptote

Thus the line $y = 2$ is an asymptote of the graph both far to the right and far to the left. Since the function is *increasing*, the graph, for $|x|$ large, resembles the sketch in Fig. 4.16.

Are there any vertical asymptotes? In other words, are there any numbers a near which the function becomes arbitrarily large? Since

$$f(x) = \frac{2x - 5}{x - 1},$$

Vertical asymptote

and $x - 1$ is small when x is near 1, near $a = 1$ the function "blows up." If x is near 1 and right of 1, the numerator is near $2 \cdot 1 - 5 = -3$, hence negative, and the denominator is a small positive number. Thus

$$\lim_{x \to 1^+} \frac{2x - 5}{x - 1} = -\infty; \qquad \text{similarly, } \lim_{x \to 1^-} \frac{2x - 5}{x - 1} = \infty.$$

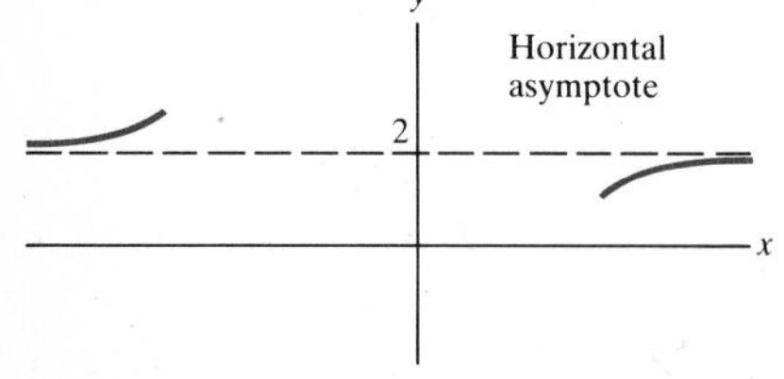

Figure 4.16

The line $x = 1$ is a vertical asymptote.

With this information, the graph can be completed. It is shown in Fig. 4.17.

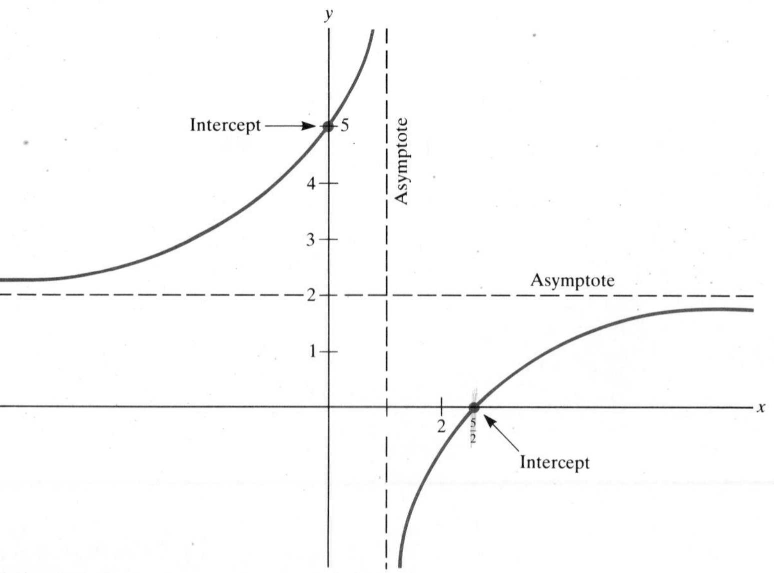

Figure 4.17

■

The next example illustrates the graphing of a polynomial.

EXAMPLE 2 Sketch the graph of $f(x) = 2x^3 - 3x^2 - 12x$.

SOLUTION Note that $f(x)$, being a polynomial, is defined for all x.

Intercepts

Since $f(0) = 2 \cdot 0^3 - 3 \cdot 0^2 - 12 \cdot 0 = 0$, the y intercept is 0. To find the x intercepts it is necessary to solve the equation $f(x) = 0$. In the case of this function, the equation can be solved easily:

$$2x^3 - 3x^2 - 12x = 0$$

so

$$x(2x^2 - 3x - 12) = 0.$$

Either $x = 0$ or $2x^2 - 3x - 12 = 0$. The latter equation can be solved by the quadratic formula:

$$x = \frac{-(-3) \pm \sqrt{(-3)^2 - 4(2)(-12)}}{2 \cdot 2} = \frac{3 \pm \sqrt{9 + 96}}{4} = \frac{3 \pm \sqrt{105}}{4}.$$

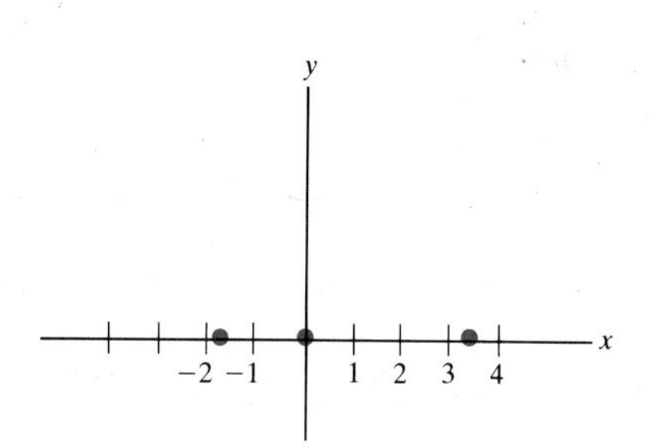

Figure 4.18

These two solutions are approximately -1.8 and 3.3.

The intercepts are recorded in Fig. 4.18.

Critical numbers

When is $f'(x) = 0$? We have

$$f'(x) = 6x^2 - 6x - 12$$
$$= 6(x^2 - x - 2)$$
$$= 6(x - 2)(x + 1).$$

Thus $f'(x) = 0$ when $6(x - 2)(x - 1) = 0$, that is, when $x = 2$ or $x = -1$. At these critical numbers, the function has the values

$$f(2) = 2 \cdot 2^3 - 3 \cdot 2^2 - 12 \cdot 2 = -20$$

and

$$f(-1) = 2(-1)^3 - 3(-1)^2 - 12(-1) = 7.$$

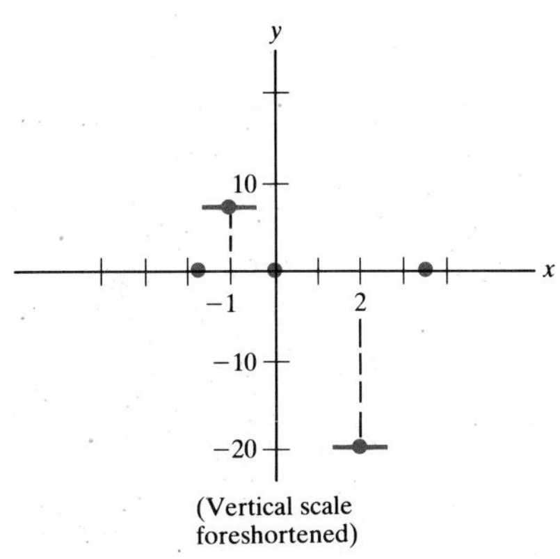

Figure 4.19

Figure 4.19 records the data gathered so far. The short segments indicate horizontal tangents.

Next, examine the sign of $f'(x)$ to determine where the function is increasing and where it is decreasing. Recall that $f'(x) = 6(x - 2)(x + 1)$ and use the accompanying chart as an aid.

	$x < -1$	-1	$-1 < x < 2$	2	$x > 2$
Sign of $x - 2$	− − −		− − −		+ + +
Sign of $x + 1$	− − −		+ + +		+ + +
Sign of $6(x - 2)(x + 1)$	+ + +		− − −		+ + +

Where increasing or decreasing

Thus the function is increasing for $x < -1$ and for $x > 2$; it is decreasing for $-1 < x < 2$. The information gathered so far is recorded in Fig. 4.20.

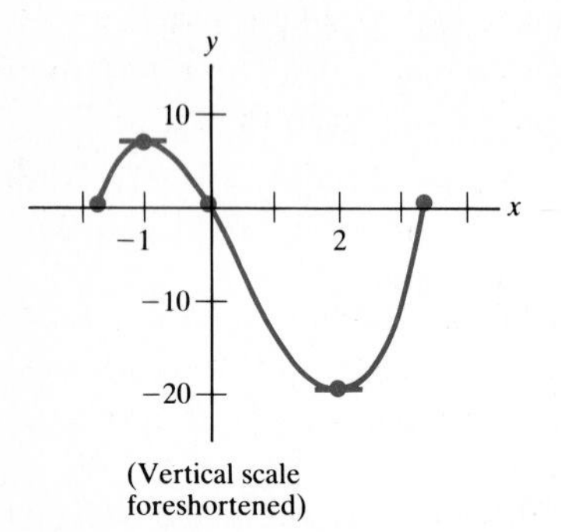

Figure 4.20

Finally, consider the behavior of $f(x) = 2x^3 - 3x^2 - 12x$ when $|x|$ is large. Since

$$\lim_{x\to\infty} f(x) = \lim_{x\to\infty} (2x^3 - 3x^2 - 12x)$$

$$= \lim_{x\to\infty} x^3\left(2 - \frac{3}{x} - \frac{12}{x^2}\right) = \infty,$$

the graph does not have a horizontal asymptote as $x \to \infty$. Similar reasoning shows that it has no horizontal asymptote as $x \to -\infty$. In fact,

$$\lim_{x\to-\infty} f(x) = -\infty.$$

With this last information the curve can be sketched. The graph (with the y axis compressed) appears in Fig. 4.21.

There is a local maximum at $x = -1$, a local minimum at $x = 2$, but no global maximum or minimum.

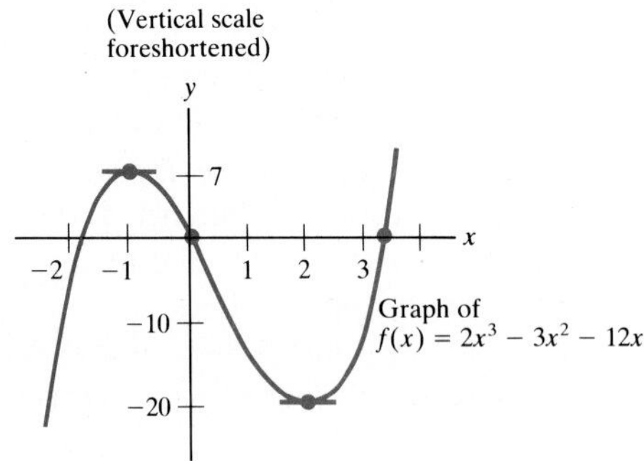

The graph crosses the x axis at $x = 0$,

$\dfrac{3 - \sqrt{105}}{4} \approx -1.8$, and $\dfrac{3 + \sqrt{105}}{4} \approx 3.3$

Figure 4.21 ■

EXAMPLE 3 Graph $f(x) = 3x^4 - 4x^3$. Discuss relative maxima and minima.

SOLUTION To find the intercepts note that $f(0) = 0$ and $3x^4 - 4x^3 = 0$ when $x^3(3x - 4) = 0$, that is, when $x = 0$ or $x = \frac{4}{3}$. The derivative is

$$f'(x) = 12x^3 - 12x^2 = 12x^2(x - 1).$$

The critical numbers are the solutions of the equation

$$12x^2(x - 1) = 0,$$

namely, 0 and 1.

How does the sign of $f'(x) = 12x^2(x - 1)$ behave when x is near 0? For $x < 0$, $12x^2$ is positive and $x - 1$ is negative; hence $12x^2(x - 1)$ is negative. For $0 < x < 1$, $12x^2$ is positive and $x - 1$ is still negative. Thus the sign of $f'(x)$ does *not* change as x passes through 0. In fact,

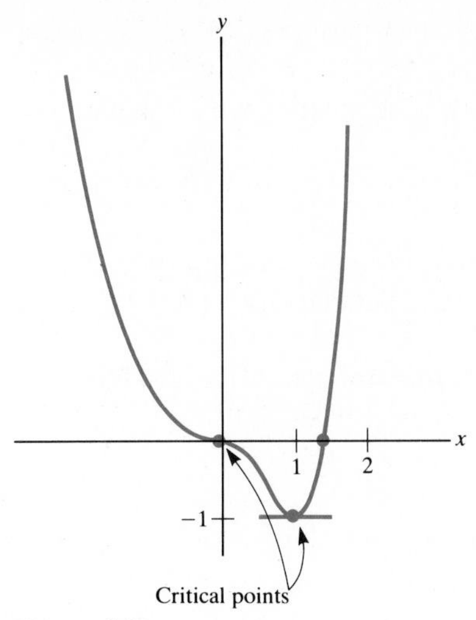

Figure 4.22

since $f'(x)$ remains negative (except at 0), the function f is decreasing for $x \le 1$. Thus there is no relative maximum or minimum at $x = 0$.

How does the sign of $f'(x) = 12x^2(x - 1)$ behave when x is near 1? The factor $12x^2$ remains positive, but $x - 1$ changes sign from negative to positive. Hence at $x = 1$ the function has a local minimum.

Writing
$$f(x) = 3x^4 - 4x^3 = x^4\left(3 - \frac{4}{x}\right)$$
shows that when $|x|$ is large, $f(x)$ behaves like $3x^4$ (since $4/x$ is then near 0). Since $3x^4$ becomes arbitrarily large when x is large, the function has no global maximum. The graph in Fig. 4.22 shows the x intercepts and the critical points. Note that at $x = 1$ a global minimum occurs. ■

The Maximum over a Closed Interval

In many applied problems we are interested in the behavior of a differentiable function just over some closed interval $[a, b]$. Such a function will have a global maximum for that interval by the maximum-value theorem of Sec. 2.6. That maximum can occur either at an endpoint—a or b—or else at some number c in the open interval (a, b). In the latter case, c must be a critical number, for $f'(c) = 0$ by the interior-maximum theorem of Sec. 4.1.

Figures 4.23 and 4.24 show some of the ways in which a relative or global maximum or minimum can occur for a function considered only on a closed interval $[a, b]$.

Global maximum
Relative maximum
Relative minimum
The derivative is 0 at these two numbers.

The global maximum occurs at an end. (The derivative need not be 0.)

Figure 4.23

The major point to keep in mind is that the maximum value of a function f that is differentiable on a closed interval occurs

1 At an endpoint of the interval, or
2 At a critical number [where $f'(x) = 0$]

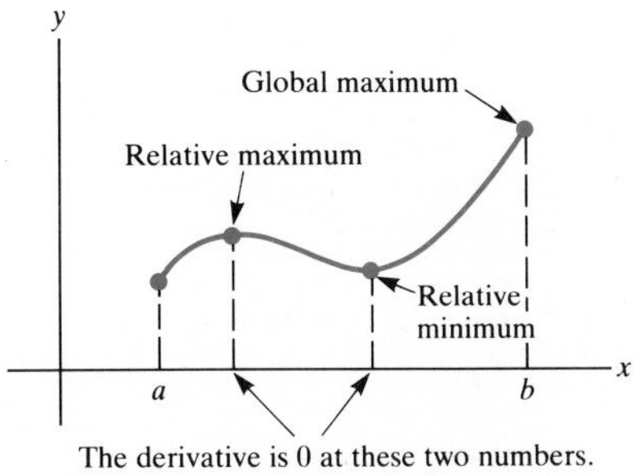

EXAMPLE 4 Find the maximum value of $f(x) = x^3 - 3x^2 + 3x$ for x in $[0, 2]$.

SOLUTION First compute f at the ends of the interval, 0 and 2:
$$f(0) = 0 \quad \text{and} \quad f(2) = 2.$$
Next, compute $f'(x)$, which is $3x^2 - 6x + 3$. When is $f'(x) = 0$? When
$$3x^2 - 6x + 3 = 0,$$
or
$$3(x^2 - 2x + 1) = 0,$$
or
$$3(x - 1)^2 = 0.$$
Thus 1 is the only critical number, and it lies in the interval $[0, 2]$.

The maximum of f must therefore occur either at an endpoint of the interval (at 0 or 2) or at the only critical number, 1. It is necessary to calculate $f(1)$ to determine where the maximum occurs:
$$f(1) = 1^3 - 3 \cdot 1^2 + 3 \cdot 1 = 1.$$

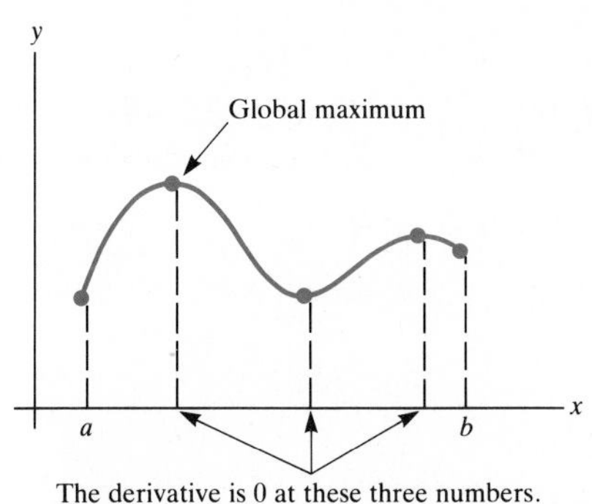

The global maximum occurs at a number other than a or b. (The derivative is 0.)

Figure 4.24

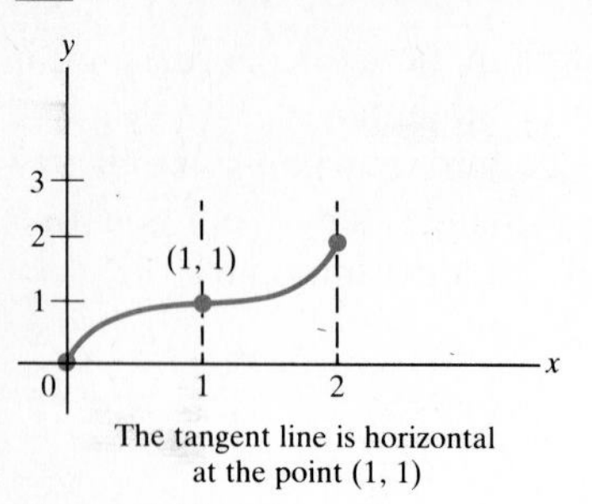

The tangent line is horizontal at the point (1, 1)

Figure 4.25

Since $f(0) = 0$, $f(2) = 2$, and $f(1) = 1$, the maximum value is 2, occurring at the endpoint 2.

Now that the problem is solved, it may be instructive to sketch the graph of the function. Since

$$f'(x) = 3(x - 1)^2$$

is positive for all x other than 1, the function is increasing. Figure 4.25 shows how the graph looks. Observe that the minimum occurs at 0. ■

The following flowchart shows how to maximize a differentiable function on a closed interval.

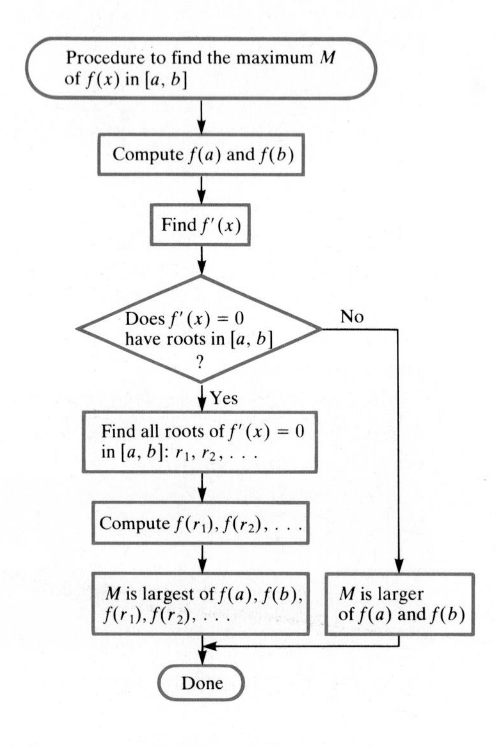

EXERCISES FOR SEC. 4.2: USING THE DERIVATIVE AND LIMITS WHEN GRAPHING A FUNCTION

In each of Exercises 1 to 8 find all critical numbers of the given function and use the first-derivative test to determine whether a local maximum, a local minimum, or neither, occurs there.

1 x^5

2 x^6

3 $(x - 1)^3$

4 $(x - 1)^4$

5 $3x^4 + x^3$

6 $2x^3 + 3x^2$

7 $x \sin x + \cos x$

8 $x \cos x - \sin x$

In Exercises 9 to 24 graph the given functions, showing any intercepts, asymptotes, critical points, or local or global extrema.

9 $x^3 - 3x^2 + 3x$

10 $x^4 - 4x^3 + 4x^2$

11 $x^4 - 4x + 3$

12 $x^5 + 5x$

13 $x^2 - 6x + 5$

14 $2x^2 + 3x + 5$

15 $x^4 + 2x^3 - 3x^2$

16 $2x^3 + 3x^2 - 6x$

17 $\dfrac{3x + 1}{3x - 1}$

18 $\dfrac{x}{x + 1}$

19 $\dfrac{x}{x^2 + 1}$

20 $\dfrac{x}{x^2 - 1}$

21 $\dfrac{1}{2x^2 - x}$

22 $\dfrac{1}{x^2 - 3x + 2}$

23 $\dfrac{x^2 + 3}{x^2 - 4}$

24 $\dfrac{\sqrt{x^2 + 1}}{x}$

Exercises 25 to 32 concern functions whose domains are restricted to closed intervals. In each find the maximum and the minimum value for the given function over the given interval.

25 $x^2 - x^4$; $[0, 1]$

26 $4x - x^2$; $[0, 5]$

27 $4x - x^2$; $[0, 1]$

28 $2x^2 - 5x$; $[-1, 1]$

29 $x^3 - 2x^2 + 5x$; $[-1, 3]$

30 $x/(x^2 + 1)$; $[0, 3]$

31 $x^2 + x^4$; $[0, 1]$

32 $(x + 1)/\sqrt{x^2 + 1}$; $[0, 3]$

■

In Exercises 33 to 41 graph the functions.

33 $\dfrac{x - 1}{x^2 + 1}$

34 $x + \dfrac{1}{x}$

35 $\dfrac{\sin x}{1 + 2\cos x}$

36 $\dfrac{\sqrt{x^2 - 1}}{x}$

37 $\dfrac{1}{(x - 1)^2(x - 2)}$

38 $\dfrac{3x^2 + 5}{x^2 - 1}$

39 $2x^{1/3} + x^{4/3}$

40 $\dfrac{3x^2 + 5}{x^2 + 1}$

41 $\sqrt{3}\sin x + \cos x$.

42 Let f and g be polynomials without a common root.
 (*a*) Show that if the degree of g is odd, the graph of f/g has a vertical asymptote.
 (*b*) Show that if f and g have the same degree, the graph of f/g has a horizontal asymptote.
 (*c*) Show that if the degree of f is less than the degree of g, the graph of f/g has a horizontal asymptote.

4.3 HIGHER DERIVATIVES AND THE BINOMIAL THEOREM

Velocity is the rate at which distance changes. The rate at which velocity changes is called **acceleration**. Thus if $y = f(t)$ denotes position on a line at time t, then the derivative $dy/dt = \dot{y}$ equals the velocity, and the derivative of the derivative, that is,

$$\frac{d}{dt}\left(\frac{dy}{dt}\right)$$

equals the acceleration.

The second derivative

The derivative of the derivative of a function $y = f(x)$ is called the **second derivative** of the function. It is denoted

$$\frac{d^2y}{dx^2},\quad D^2y,\quad y'',\quad f'',\quad D^2f,\quad f^{(2)},\quad f^{(2)}(x),\quad \text{or}\quad \frac{d^2f}{dx^2}.$$

If $y = f(t)$, where t denotes time, the second derivative d^2y/dt^2 is also denoted $\ddot{y}$.

For instance, if $y = x^3$,

$$\frac{dy}{dx} = 3x^2 \quad\text{and}\quad \frac{d^2y}{dx^2} = 6x.$$

Other ways of denoting the second derivative of this function are

$$D^2(x^3) = 6x,\qquad \frac{d^2(x^3)}{dx^2} = 6x,\qquad\text{and}\qquad (x^3)'' = 6x.$$

y	$\dfrac{dy}{dx}$	$\dfrac{d^2y}{dx^2}$
x^3	$3x^2$	$6x$
$\dfrac{1}{x}$	$\dfrac{-1}{x^2}$	$\dfrac{2}{x^3}$
$\sin 5x$	$5\cos 5x$	$-25\sin 5x$

The table in the margin lists dy/dx, the first derivative, and d^2y/dx^2, the second derivative, for a few functions.

Most functions f met in applications of calculus can be differentiated repeatedly in the sense that Df exists, the derivative of Df, namely, D^2f, exists, the derivative of D^2f exists, and so on. The derivative of the second derivative,

$$\frac{d}{dx}\left(\frac{d^2y}{dx^2}\right),$$

is called the **third derivative** and is denoted many ways, such as

$$\frac{d^3y}{dx^3}, \quad D^3y, \quad y''', \quad f''', \quad f^{(3)}, \quad f^{(3)}(x), \quad \text{or} \quad \frac{d^3f}{dx^3}.$$

The **fourth derivative** $f^{(4)}(x)$ is defined as the derivative of the third derivative and is represented by similar notations. Similarly, $f^{(n)}(x)$ is defined for $n = 5, 6, \ldots$. The derivatives $f^{(n)}(x)$ for $n \geq 2$ are called the **higher derivatives** of f. (The first derivative is also denoted $f^{(1)}(x)$.)

EXAMPLE 1 Compute $f^{(n)}(x)$ if $f(x) = x^3 - 2x^2 + x + 5$ and n is a positive integer.

SOLUTION

$$f^{(1)}(x) = \frac{df}{dx} = 3x^2 - 4x + 1$$

$$f^{(2)}(x) = \frac{d}{dx}(f^{(1)}(x)) = 6x - 4$$

$$f^{(3)}(x) = \frac{d}{dx}(f^{(2)}(x)) = 6$$

$$f^{(4)}(x) = 0 \qquad \text{since } f^{(3)}(x) \text{ is constant.}$$

Since $f^{(4)}(x)$ is constant, its derivative, $f^{(5)}(x)$, is 0 for all x. Similarly, $f^{(6)}(x) = 0$, $f^{(7)}(x) = 0$, and so on. ■

As Example 1 may suggest, for any polynomial $f(x)$ of degree at most 3, $f^{(n)}(x) = 0$ for all integers $n \geq 4$. The next example is quite different.

EXAMPLE 2 Compute $f^{(n)}(x)$ if $f(x) = \cos x$.

SOLUTION

$$f^{(1)}(x) = \frac{d}{dx}(\cos x) = -\sin x$$

$$f^{(2)}(x) = \frac{d}{dx}(f^{(1)}(x)) = \frac{d}{dx}(-\sin x) = -\cos x$$

$$f^{(3)}(x) = \frac{d}{dx}(-\cos x) = \sin x$$

$$f^{(4)}(x) = \frac{d}{dx}(\sin x) = \cos x$$

$$f^{(5)}(x) = \frac{d}{dx}(\cos x) = -\sin x$$

$$f^{(6)}(x) = \frac{d}{dx}(-\sin x) = -\cos x$$

Note that $f^{(4)}(x) = f(x)$, $f^{(5)}(x) = f^{(1)}(x)$, and so on. The higher derivatives repeat every four steps. ■

It follows from Example 2 that it is not possible to find a polynomial $P(x)$ such that $\cos x = P(x)$ for all x in an interval (a, b).

Of the higher derivatives, the second derivative is the most commonly used in physical applications. Derivatives higher than the second are important in the study of the error in approximating a quantity by some algorithms (for instance, in approximating some function by a polynomial).

The higher derivatives can be used to obtain the **binomial theorem**, which asserts that for a positive integer n,

$$(1 + x)^n = 1 + nx + \frac{n(n-1)}{2!}x^2 + \cdots + \frac{n(n-1)\cdots(n-k+1)}{k!}x^k + \cdots + \frac{n(n-1)\cdots 1}{n!}x^n.$$

The symbol $k!$ stands for the product of the integers from 1 through k. For instance, $4! = 4 \cdot 3 \cdot 2 \cdot 1 = 24$. Also, $0!$ is defined to be 1. The coefficient of x^k can also be written

$$\frac{n!}{k!(n-k)!}.$$

EXAMPLE 3 Use the higher derivatives to show that

$$(1 + x)^4 = 1 + 4x + \frac{4 \cdot 3}{2!}x^2 + \frac{4 \cdot 3 \cdot 2}{3!}x^3 + \frac{4 \cdot 3 \cdot 2}{4!}x^4.$$

SOLUTION When $(1 + x)^4$ is multiplied out, it yields a polynomial of degree 4. Thus there are constants a_0, a_1, a_2, a_3, and a_4 such that

$$a_0 + a_1x + a_2x^2 + a_3x^3 + a_4x^4 = (1 + x)^4 \tag{1}$$

for all x. Taking successive higher derivatives of both sides of Eq. (1), we obtain

$$a_1 + 2a_2x + 3a_3x^2 + 4a_4x^3 = 4(1 + x)^3 \tag{2}$$

$$2a_2 + 3 \cdot 2a_3x + 4 \cdot 3a_4x^2 = 4 \cdot 3(1 + x)^2 \tag{3}$$

$$3 \cdot 2a_3 + 4 \cdot 3 \cdot 2a_4x = 4 \cdot 3 \cdot 2(1 + x) \tag{4}$$

$$4 \cdot 3 \cdot 2a_4 = 4 \cdot 3 \cdot 2 \tag{5}$$

Plugging $x = 0$ into Eq. (1), we obtain $a_0 = 1$. Plugging $x = 0$ into Eq. (2) shows that $a_1 = 4$. Similarly replacing x by 0 in Eqs. (3), (4), and (5) yields

$$a_2 = \frac{4 \cdot 3}{2} = \frac{4 \cdot 3}{2!}, \qquad a_3 = \frac{4 \cdot 3 \cdot 2}{3!}, \qquad \text{and} \qquad a_4 = \frac{4 \cdot 3 \cdot 2}{4!} = 1.$$

This establishes the binomial theorem for $n = 4$. Evaluation of the coefficients a_i, $i = 0, 1, \ldots, 4$ then shows that

$$(1 + x)^4 = 1 + 4x + 6x^2 + 4x^3 + x^4.$$

A similar argument works for $(1 + x)^n$. ■

EXERCISES FOR SEC. 4.3: HIGHER DERIVATIVES AND THE BINOMIAL THEOREM

In Exercises 1 to 8 compute $f^{(1)}$, $f^{(2)}$, $f^{(3)}$, and $f^{(4)}$ for the given functions f.

1 $x^2 - x + 5$ 2 $2x^2 - 5x$ 3 x^3
4 $x^4 + 5x^2$ 5 $\sin 2x$ 6 $\cos 3x$
7 $1/x^2$ 8 $1/(x + 1)$

In Exercises 9 and 10 find the second derivative.

9 (*a*) $\tan 2x$ (*b*) $\sqrt{1 + 2x}$ (*c*) $\cot \sqrt{x}$
10 (*a*) $\csc 3x$ (*b*) $(1 + 3x)^{10}$ (*c*) $\sec x^2$

■

11 Is there a function f such that $f(0) = -2$, $f(1) = 1$, and $f''(x) = 0$ for all x? If so, how many such functions are there?
12 Find all functions $f(x)$ such that $f^{(3)}(x) = 0$ for all x.
13 Find all functions $f(x)$ such that $f^{(2)}(x) = 2 \sin 3x$.
14 Let a, b, c, and d be constants such that $a^2 \leq 8b/3$. Show that the equation $x^4 + ax^3 + bx^2 + cx + d = 0$ has at most three distinct real roots.
15 Let $f(x) = (1 + x)^3 = a_0 + a_1x + a_2x^2 + a_3x^3$, where the constants a_0, a_1, a_2, and a_3 are to be determined.
(*a*) Compute $f(0)$, $f^{(1)}(0)$, $f^{(2)}(0)$, and $f^{(3)}(0)$, using the formula $f(x) = (1 + x)^3$.
(*b*) Compute the quantities in (*a*), using the formula $f(x) = a_0 + a_1x + a_2x^2 + a_3x^3$.
(*c*) Comparing the results in (*a*) and (*b*), show that $(1 + x)^3 = 1 + 3x + 3x^2 + x^3$.
16 Let $f(x) = (1 + x)^n = a_0 + a_1x + a_2x^2 + \cdots + a_nx^n$, the polynomial obtained when $(1 + x)^n$ is multiplied out. Compute $f^{(k)}(0)$ in two different ways, as in the preceding exercise, to obtain a formula for a_k and thus establish the binomial theorem in general.
17 From the binomial expansion for $(1 + x)^4$, deduce the binomial expansion of $(a + b)^4$.
18 (*a*) Show that when x is small, a good estimate of $(1 + x)^n$ is $1 + nx$.
(*b*) What is a good estimate of the error when using $1 + nx$ as an estimate of $(1 + x)^n$?
(*c*) Illustrate (*a*) and (*b*) with $n = 6$ and $x = 0.01$.

4.4 CONCAVITY AND THE SECOND DERIVATIVE

Whether the first derivative is positive, negative, or zero tells a good deal about a function and its graph. This section will explore the geometric significance of the second derivative being positive, negative, or zero. The following section will show how the second derivative is used in the study of motion.

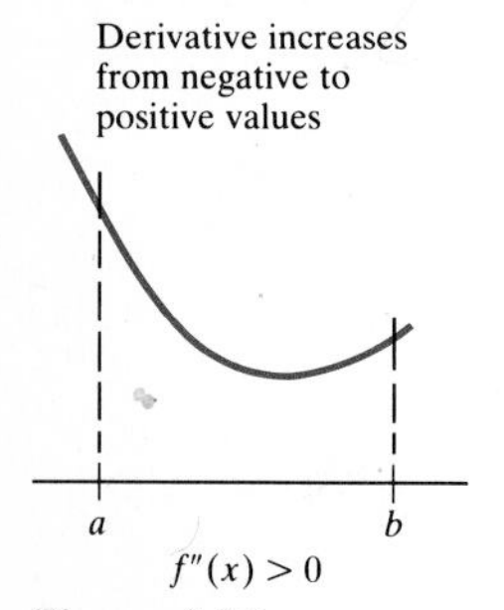

Figure 4.26

Concave Upward and Concave Downward Assume that $f''(x)$ is positive for all x in the open interval (a, b). Since f'' is the derivative of f', it follows that f' is an increasing function throughout the interval (a, b). In other words, as x increases, the slope of the graph of $y = f(x)$ increases as we move from left to right on that part of the graph corresponding to the interval (a, b). The slope may increase from negative to positive values, as in Fig. 4.26. Or the slope may be positive throughout (a, b) and increasing, as in Fig. 4.27. Or the slope may be negative throughout (a, b) and increasing, as in Fig. 4.28.

As you drive along such a graph from left to right, your car keeps turning to the left.

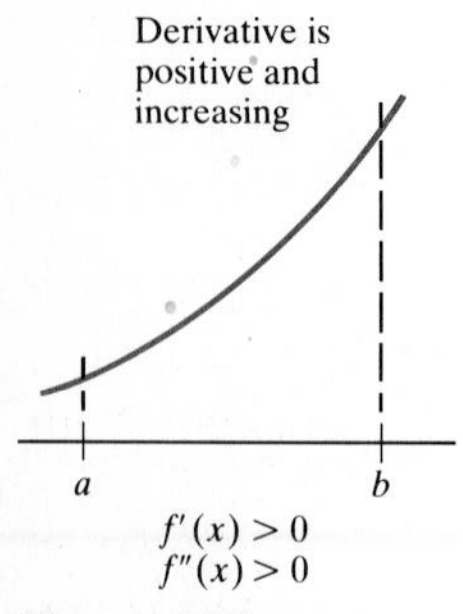

Figure 4.27

Definition *Concave upward.* A function f whose first derivative is increasing throughout the open interval (a, b) is called **concave upward** in that interval.

Note that when a function is concave upward, it is shaped like part of a cup (Concave *UP*ward).

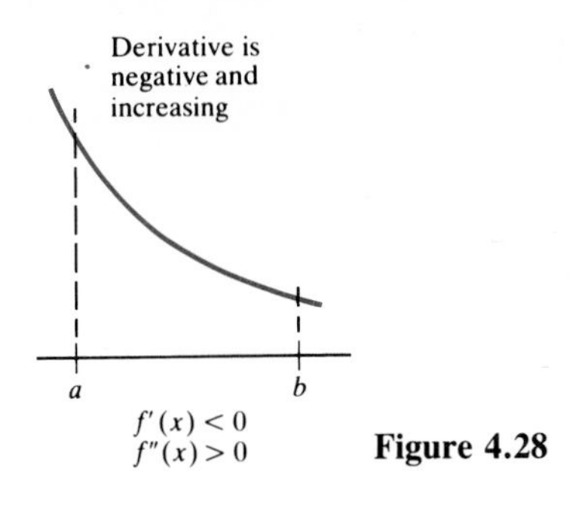

Figure 4.28

It can be proved that where a curve is concave upward it lies *above its tangent lines* and *below its chords,* as shown in Fig. 4.29. (See Exercises 36 to 38.)

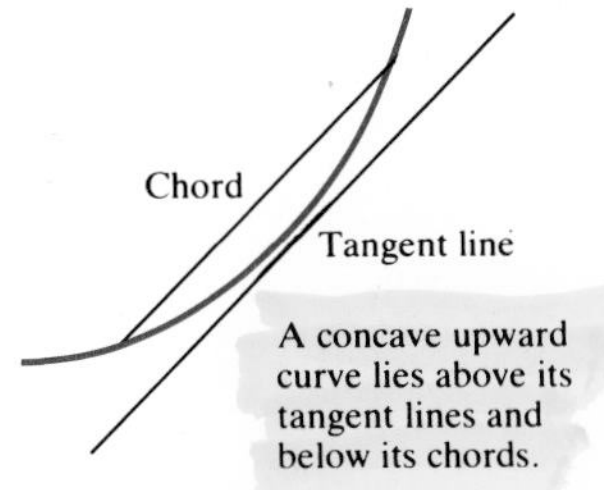

Figure 4.29

As was observed, in an interval where $f''(x)$ is positive, the function $f'(x)$ is increasing, and so the function f is concave upward. However, if a function is concave upward, $f''(x)$ is not necessarily positive. For instance, $y = x^4$ is concave upward over any interval, since the derivative $4x^3$ is increasing. The second derivative $12x^2$ is not always positive; at $x = 0$ it is 0.

If, on the other hand, $f''(x)$ is negative throughout (a, b), then f' is a decreasing function and the graph of f looks like part of the curve in Fig. 4.30.

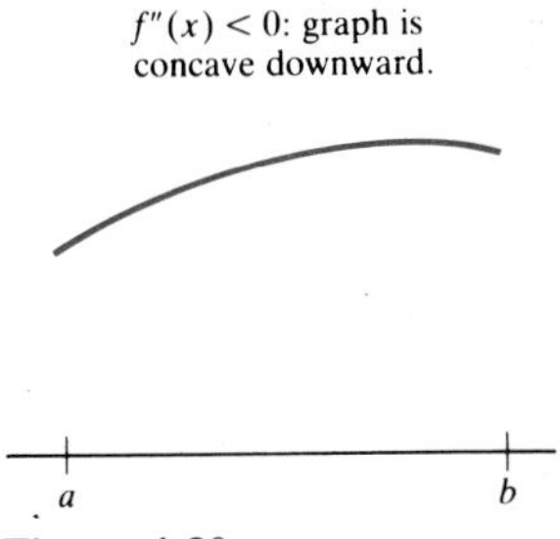

Figure 4.30

Definition *Concave downward.* A function f whose first derivative is decreasing throughout an open interval (a, b) is called **concave** ~~**upward**~~ downward in that interval.

Where a function is concave downward, it lies *below its tangent lines* and *above its chords.*

EXAMPLE 1 Where is the graph of $f(x) = x^3$ concave upward? Concave downward?

SOLUTION First compute the second derivative. Since $D(x^3) = 3x^2$, $D^2(x^3) = 6x$.

Clearly $6x$ is positive for all positive x and negative for all negative x. The graph, shown in Fig. 4.31, is concave upward if $x > 0$ and concave downward if $x < 0$. Note that the sense of concavity changes at $x = 0$. When you drive along this curve from left to right, your car turns to the right until you pass through (0, 0). Then it starts turning to the left. ■

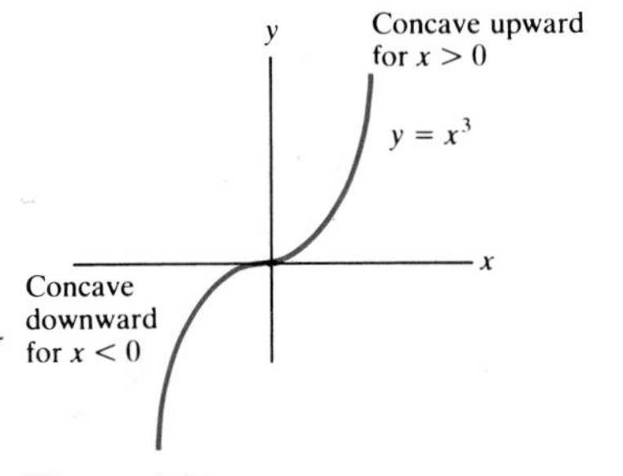

Figure 4.31

EXAMPLE 2 Consider the function $f(x) = \sin x$ for x in $[0, 2\pi]$. Where is the graph concave upward? Concave downward?

SOLUTION

$$f(x) = \sin x,$$
$$f'(x) = \cos x,$$
$$f''(x) = -\sin x.$$

The second derivative, $-\sin x$, is negative for $0 < x < \pi$. It is positive for $\pi < x < 2\pi$. Therefore, the graph is concave downward for x in $(0, \pi)$ and concave upward for x in $(\pi, 2\pi)$, as shown in Fig. 4.32. ■

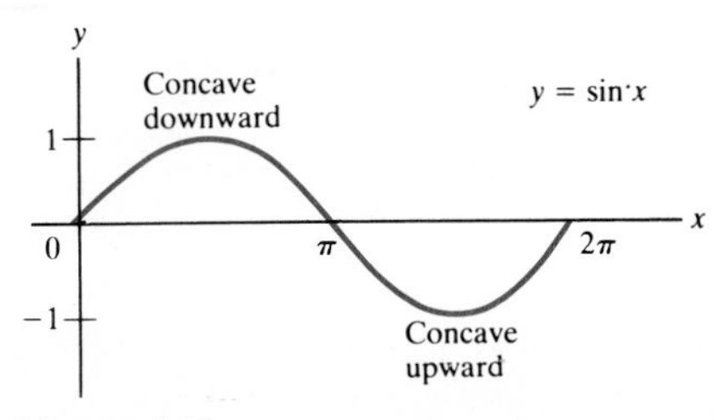

Figure 4.32

The sense of concavity is a useful tool in sketching the graph of a function. Of special interest in Examples 1 and 2 is the presence of a point on the graph where the sense of concavity changes. Such a point is called an **inflection point**.

Definition *Inflection point and inflection number.* Let f be a function and let a be a number. Assume that there are numbers b and c such that $b < a < c$ and

1 f is continuous on the open interval (b, c).
2 f is concave upward in the interval (b, a) and concave downward in the interval (a, c), or vice versa.

Then the point $(a, f(a))$ is called an **inflection point** or **point of inflection**. The number a is called an **inflection number**.

Observe that if the second derivative changes sign at the number a, then a is an inflection number.

If the second derivative exists at an inflection point, it must be 0. But there can be an inflection point even if f'' is not defined there, as shown by the next example, which is closely related to Example 1.

EXAMPLE 3 Examine the concavity of $y = x^{1/3}$.

SOLUTION Here

$$y' = \tfrac{1}{3}x^{-2/3} \qquad \text{and} \qquad y'' = \frac{1}{3}\cdot\frac{-2}{3}x^{-5/3}.$$

Neither y' nor y'' is defined at 0; however, the sign of y'' changes at 0. When x is negative, y'' is positive; when x is positive, y'' is negative. The concavity switches from upward to downward at $x = 0$. The graph is shown in Fig. 4.33. ■

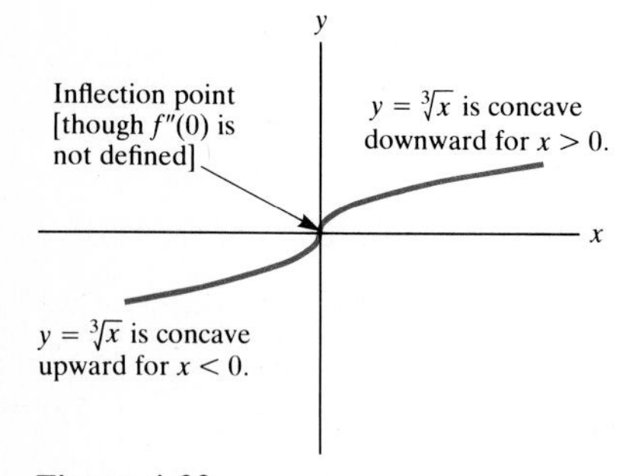

Figure 4.33

When graphing a function f, find where $f(x) = 0$, where $f'(x) = 0$, and where $f''(x) = 0$ if the solutions are easy. Determine where $f'(x)$ is positive and where it is negative. Determine also where $f''(x)$ is positive and where it is negative. The following table contrasts the interpretations of the signs of f, f', and f''. (It is assumed that f, f', and f'' are continuous.)

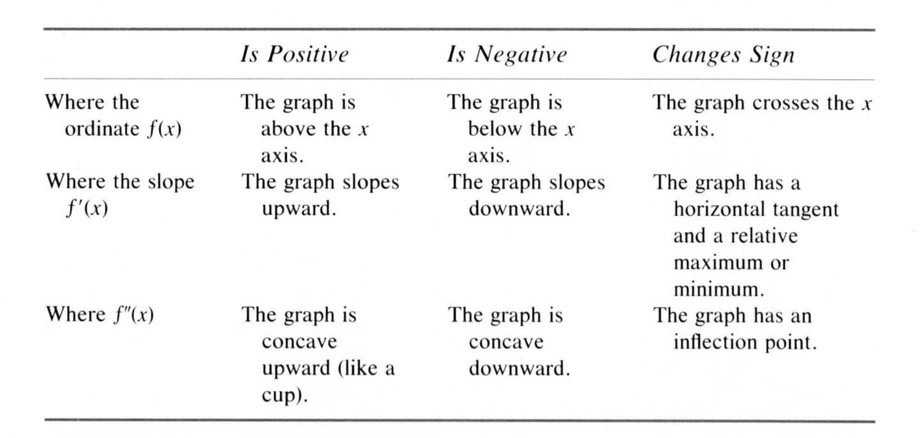

	Is Positive	*Is Negative*	*Changes Sign*
Where the ordinate $f(x)$	The graph is above the x axis.	The graph is below the x axis.	The graph crosses the x axis.
Where the slope $f'(x)$	The graph slopes upward.	The graph slopes downward.	The graph has a horizontal tangent and a relative maximum or minimum.
Where $f''(x)$	The graph is concave upward (like a cup).	The graph is concave downward.	The graph has an inflection point.

Keep in mind that the graph can have an inflection point at a, even though the second derivative is not defined at a (Example 3). Similarly, a graph can have a maximum or minimum at a, even though the first derivative is not defined at a. [Consider $f(x) = |x|$ at $a = 0$.]

EXAMPLE 4 Graph $f(x) = 2x^3 - 3x^2$, showing intercepts, critical points, and inflection points.

SOLUTION To find x intercepts, set $f(x) = 0$:

$$2x^3 - 3x^2 = 0$$

$$x^2(2x - 3) = 0$$

$$x^2 = 0 \quad \text{or} \quad 2x - 3 = 0.$$

Intercepts

Thus the x intercepts are $x = 0$ and $x = \frac{3}{2}$. The y intercept is simply $f(0)$, which is 0.

To find critical numbers, set $f'(x) = 0$:

$$f'(x) = (2x^3 - 3x^2)' = 6x^2 - 6x = 6x(x - 1) = 0;$$

thus $$x = 0 \quad \text{or} \quad x - 1 = 0.$$

Critical numbers

There are two critical numbers, $x = 0$ and $x = 1$.

To find inflection numbers, determine where the sign of $f''(x)$ changes. Since the second derivative exists everywhere, $f''(x)$ will exist and be 0 at an inflection number. So set $f''(x) = 0$ and check whether the second derivative changes sign at any of the solutions. We have

$$f''(x) = (6x^2 - 6x)' = 12x - 6.$$

The equation $12x - 6 = 0$ has only one solution:

$$x = \tfrac{1}{2}.$$

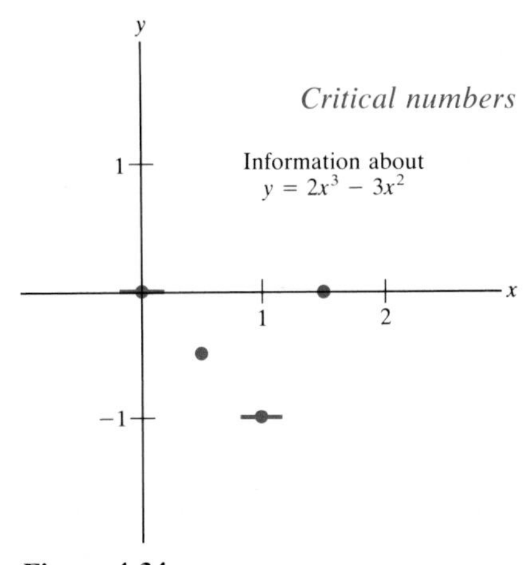

Figure 4.34

Inflection numbers

But is $x = \frac{1}{2}$ an inflection number? To check, we must see whether the second derivative changes sign at $\frac{1}{2}$. The second derivative is $12x - 6 = 12(x - \frac{1}{2})$. For $x > \frac{1}{2}$, the second derivative is positive; for $x < \frac{1}{2}$, the second derivative is negative. Thus $x = \frac{1}{2}$ is an inflection number.

The inputs of interest are 0 (which is an x intercept and a critical number), $\frac{3}{2}$ (which is an x intercept), 1 (which is a critical number), and $\frac{1}{2}$ (which is an inflection number). Compute the outputs for these inputs in order to plot the key points on the graph.

x	0	$\frac{1}{2}$	1	$\frac{3}{2}$
$f(x) = 2x^3 - 3x^2$	0	$-\frac{1}{2}$	-1	0

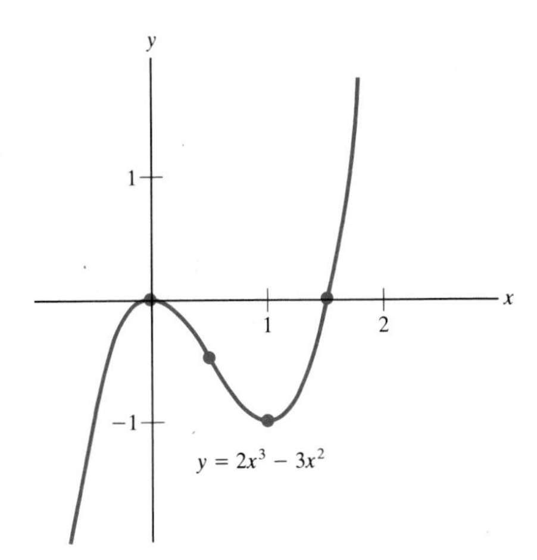

Figure 4.35

Figure 4.34 displays the information gathered so far. Noting that $\lim_{x\to\infty} (2x^3 - 3x^2) = \infty$ and that $\lim_{x\to-\infty} (2x^3 - 3x^2) = -\infty$, we complete the graph freehand, as shown in Fig. 4.35. [Try completing it yourself; if you don't have local extrema at (0, 0) and (1, −1), then you will force there to be more critical numbers.] ■

The Second Derivative and Local Extrema The second derivative is also useful in testing whether at a critical number there is a relative maximum or relative minimum. For instance, let a be a critical number for the function f and assume that $f''(a)$ happens to be negative. If f'' is continuous in some open interval that contains a, then $f''(a)$ remains negative for a suitably small open interval that contains a. This means that the graph of f is concave downward near $(a, f(a))$, hence lies below its tangent lines. In particular, it lies below the horizontal tangent line at the critical point $(a, f(a))$, as illustrated in Fig. 4.36. Thus the function has a **relative maximum** at the critical number a. This observation suggests the following test for a relative maximum or minimum.

The second-derivative test for extrema

Theorem *Second-derivative test for relative maximum or minimum.* Let f be a function such that $f'(x)$ is defined at least on some open interval containing the number a. Assume that $f''(a)$ is defined. If

$$f'(a) = 0 \quad \text{and} \quad f''(a) < 0,$$

then f has a local maximum at a.

Similarly, if

$$f'(a) = 0 \quad \text{and} \quad f''(a) > 0,$$

then f has a local minimum at a.

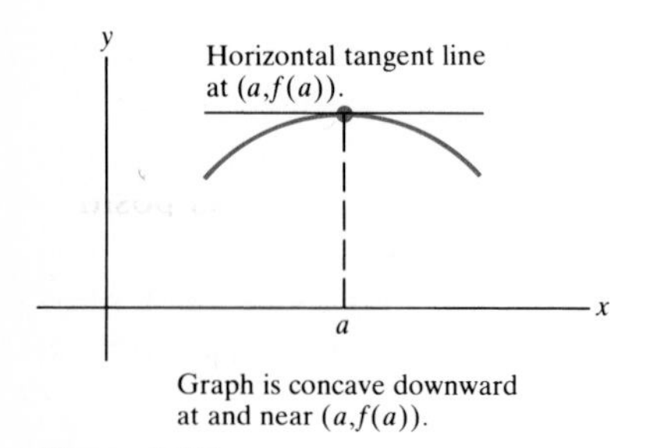

Figure 4.36

Proof Assume that $f'(a) = 0$ and that $f''(a)$ is negative. Then for x sufficiently close to a, the difference quotient

$$\frac{f'(x) - f'(a)}{x - a}$$

must be negative. Since $f'(a) = 0$,

$$\frac{f'(x)}{x - a}$$

is negative.

When x is greater than a, the denominator $x - a$ is positive; thus $f'(x)$ is negative. When x is less than a, $x - a$ is negative; thus $f'(x)$ is positive.

Since the sign of $f'(x)$ changes from positive to negative at a as x increases, there is a local maximum at $x = a$.

A similar argument shows that if $f''(a)$ is positive at a critical number, then the function f has a local minimum at a. ■

EXAMPLE 5 Find all local extrema of the function $f(x) = x^4 - 2x^3$.

SOLUTION Since f is differentiable throughout its domain, any local extremum can occur only at a critical number. So begin by finding the critical numbers, as follows:

$$f'(x) = (x^4 - 2x^3)' = 4x^3 - 6x^2 = x^2(4x - 6).$$

Setting $f'(x) = 0$ gives $x^2 = 0$ or $4x - 6 = 0$. The critical numbers are therefore

$$x = 0 \quad \text{and} \quad x = \tfrac{3}{2}.$$

Now use the second derivative to determine whether either of these corresponds to a local extremum.

The second derivative is

$$f''(x) = (4x^3 - 6x^2)' = 12x^2 - 12x.$$

At $x = \frac{3}{2}$ we have

$$f''(\tfrac{3}{2}) = 12(\tfrac{3}{2})^2 - 12(\tfrac{3}{2}) = 27 - 18 = 9,$$

which is positive. Since $f'(\frac{3}{2}) = 0$ and $f''(\frac{3}{2}) > 0$, f has a local minimum at $x = \frac{3}{2}$.

How about the other critical number, $x = 0$? In this case,

$$f''(0) = 12 \cdot 0^2 - 12 \cdot 0 = 0.$$

Since $f''(0) = 0$, the second-derivative test tells us nothing about the critical number 0. Instead, we must resort to the first-derivative test and examine the sign of $f'(x) = 4x^3 - 6x^2 = x^2(4x - 6)$ for x near 0. For x sufficiently near 0, whether to the right of 0 or to the left, x^2 is positive and $4x - 6$ is negative. Thus $f'(x)$ is negative for x near 0. Since f is a decreasing function near 0, it has neither a local maximum nor a local minimum at 0. [As may be checked, there happens to be an inflection point at (0, 0).] The function is graphed in Fig. 4.37.

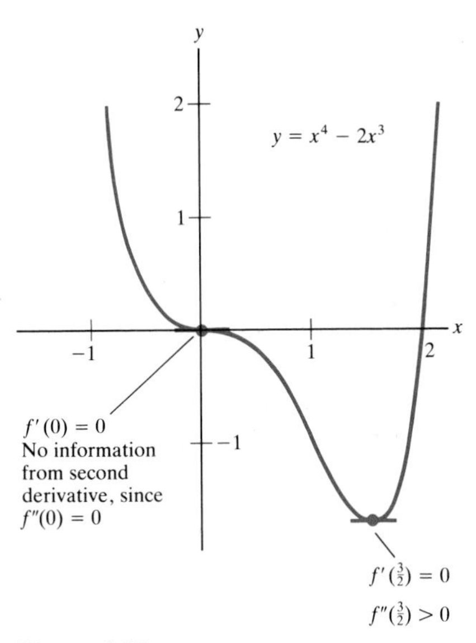

Figure 4.37 ■

Summary of This Section This section, which examined the geometric significance of the second derivative, introduced the concepts of **concave upward**, **concave downward**, **inflection point**, and **inflection number**. It also developed a second-derivative test for a local maximum [$f'(a) = 0$ and $f''(a)$ negative] and for a local minimum [$f'(a) = 0$ and $f''(a)$ positive]. If $f'(a) = 0$ and $f''(a) = 0$, no conclusion can be drawn without further analysis. Perhaps there is an inflection point (as occurred in Example 5); perhaps there is a local minimum (this is the case with the function $y = x^4$ at 0, as may be checked by graphing); perhaps there is a local maximum (as is the case with the function $y = -x^4$).

The following table shows the steps in graphing a function. Of course, not all these steps will apply for each function. For instance, there may not be any intercepts or asymptotes. A critical number can be examined by either the first-derivative test or the second-derivative test to determine whether it gives a relative maximum or relative minimum; which method is preferable depends on the function.

General Procedure for Graphing a Function

	Calculations	*Geometric Meaning*
Domain	1. Find where $f(x)$ is defined.	Find horizontal extent of graph.
Intercepts	2. Find $f(0)$ and the values of x for which $f(x) = 0$.	Find where graph crosses the axes.
Critical numbers	3. Find where $f'(x) = 0$.	Find where tangent line is horizontal.
	4. Compute $f(x)$ at all critical numbers.	Data needed for critical points.
Increasing, decreasing	5. Find the values of x for which $f'(x)$ is positive and those for which $f'(x)$ is negative.	Find where graph goes up and where it goes down as pencil moves to the right.
Horizontal asymptotes	6. Find $\lim_{x\to\infty} f(x)$ and $\lim_{x\to-\infty} f(x)$.	Find horizontal asymptotes or general behavior when $\lvert x\rvert$ is large.
Vertical asymptotes	7. Find the values of a where $\lim_{x\to a^+} f(x)$ or $\lim_{x\to a^-} f(x)$ is infinite.	Find vertical asymptotes (where the graph "blows up").
Concavity and inflection points	8. Find the values of x for which $f''(x)$ is positive and those for which $f''(x)$ is negative. Note where it changes sign.	Find where the graph is concave upward and where it is concave downward. Note inflection points.
	9. Sketch the graph, showing intercepts, critical points, asymptotes, local and global maxima and minima, and inflection points.	

EXERCISES FOR SEC. 4.4: CONCAVITY AND THE SECOND DERIVATIVE

In Exercises 1 to 18 graph the functions, showing any relative maxima, relative minima, and inflection points.

1 $x^3 - x^2$

2 $x^3 + x^2$

3 $x^4 + 2x^3$

4 $x^3 - 3x^2$

5 $x^4 - 4x^3$

6 $3x^5 - 5x^4$

7 $\dfrac{1}{1 + x^2}$

8 $\dfrac{1}{1 + x^4}$

9 $x^3 - 6x^2 - 15x$

10 $\dfrac{x^2}{2} + \dfrac{1}{x}$

11 $\dfrac{1}{1 + x^3}$

12 $\dfrac{1}{1 + x^5}$

13 $\tan x$

14 $\sin x + \sqrt{3}\cos x$

15 $\dfrac{1}{1 + 3x^2}$

16 $\dfrac{x}{1 + 3x^2}$

17 $(x - 1)^4$

18 $(x - 1)^5$

In each of Exercises 19 to 26 sketch the general appearance of the graph of the given function near (1, 1) on the basis of the information given. Assume that f, f', and f'' are continuous.

19 $f(1) = 1$, $f'(1) = 0$, $f''(1) = 1$

20 $f(1) = 1$, $f'(1) = 0$, $f''(1) = -1$

21 $f(1) = 1$, $f'(1) = 0$, $f''(1) = 0$ (Sketch four possibilities.)

22 $f(1) = 1$, $f'(1) = 0$, $f''(1) = 0$, $f''(x) < 0$ for $x < 1$ and $f''(x) > 0$ for $x > 1$

23 $f(1) = 1$, $f'(1) = 0$, $f''(1) = 0$ and $f''(x) < 0$ for x near 1

24 $f(1) = 1$, $f'(1) = 1$, $f''(1) = -1$

25 $f(1) = 1$, $f'(1) = 1$, $f''(1) = 0$ and $f''(x) < 0$ for $x < 1$ and $f''(x) > 0$ for $x > 1$

26 $f(1) = 1$, $f'(1) = 1$, $f''(1) = 0$ and $f''(x) > 0$ for x near 1

■

27 Let $f(x) = ax^2 + bx + c$, where a, b, and c are constants, $a \neq 0$. Show that f has no inflection points.

28 Let $f(x) = ax^3 + bx^2 + cx + d$, where a, b, c, and d are constants, $a \neq 0$. Show that f has exactly one inflection point.

29 Figure 4.38 appeared in "Energy Use in the United States Food System," by John S. Steinhart and Carol E. Steinhart, in *Perspectives on Energy*, edited by Lon C. Ruedisili and Morris W. Firebaugh, Oxford, New York, 1975. The graph shows farm output as a function of energy input.

(*a*) What is the practical significance of the fact that the function has a positive derivative?

(*b*) What is the practical significance of the inflection point?

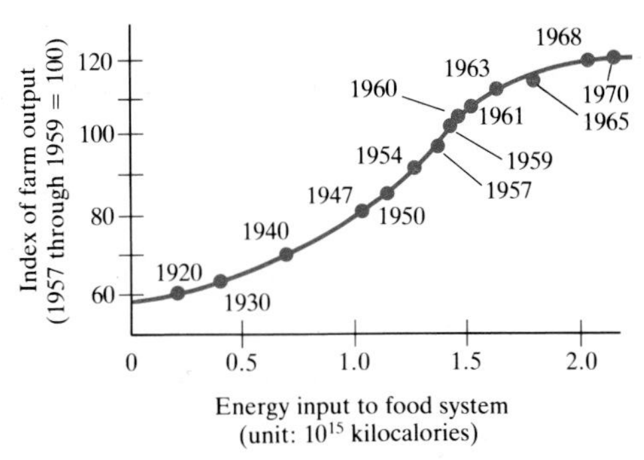

Farm output as a function of energy input to the United States food system, 1920 through 1970.

Figure 4.38

30 (Contributed by David Hayes.) Let f be a function that is continuous for all x and differentiable for all x other than 0. Figure 4.39 is the graph of its derivative $f'(x)$ as a function of x.

(*a*) Answer the following questions about f (*not* about f'). Where is f increasing? decreasing? concave upward? concave downward? What are the critical numbers? Where do any relative maxima or relative minima occur?

(*b*) Assuming that $f(0) = 1$, graph a hypothetical function f that satisfies the conditions given.

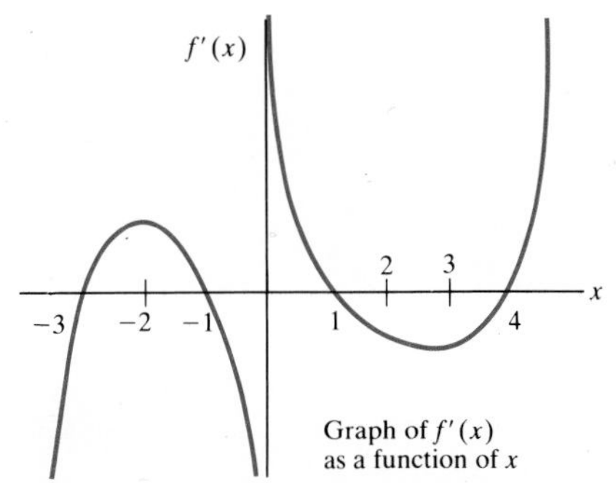

Figure 4.39

31 In the theory of **inhibited growth** it is assumed that the growing quantity y approaches some limiting size M. Specifically, one assumes that the rate of growth is proportional both to the amount present and to the amount left to grow:

$$\frac{dy}{dt} = ky(M - y).$$

Prove that the graph of y as a function of time has an inflection point when the amount y is exactly half the amount M.

32 Let f be a function such that $f''(x) = (x - 1)(x - 2)$.

(*a*) For which x is f concave upward?

(*b*) For which x is f concave downward?

(*c*) List its inflection numbers.

(*d*) Find a specific function f whose second derivative is $(x - 1)(x - 2)$.

33 Graph $y = 2(x - 1)^{5/3} + 5(x - 1)^{2/3}$, paying particular attention to points where the derivative does not exist.

34 A certain function $y = f(x)$ has the property that

$$y' = \sin y + 2y + x.$$

Show that at a critical number the function has a local minimum.

35 Sketch the graph of a function f such that for all x:

(*a*) $f(x) > 0$, $f'(x) > 0$, $f''(x) > 0$.

(*b*) $f'(x) < 0$, $f''(x) < 0$.

(*c*) Can there be a function such that for all x, $f(x) > 0$, $f'(x) < 0$, and $f''(x) < 0$? Explain.

36 Let f be a function such that $f(0) = 0 = f(1)$ and $f''(x) \geq 0$ for all x in $[0, 1]$.

(*a*) Using a sketch, explain why $f(x) \leq 0$ for all x in $[0, 1]$.

(*b*) Without a sketch prove that $f(x) \leq 0$ for all x in $[0, 1]$.

37 (See Exercise 36.) Prove that if f is a function such that $f''(x) > 0$ for all x, then the graph of $y = f(x)$ lies below its chords; i.e.,

$$f(ax_1 + (1 - a)x_2) < af(x_1) + (1 - a)f(x_2)$$

for any a in $(0, 1)$, and for any x_1 and x_2, $x_1 \neq x_2$.

38 Prove (without using a picture) that where the graph of f is concave upward it lies above its tangent lines.

39 Prove (without referring to a picture) that if the graph of f lies above its tangent lines for all x in $[a, b]$, then $f''(x) \geq 0$ for all x in $[a, b]$. [The case $y = x^4$ shows that we should not try to prove that $f''(x) > 0$.]

40 Does every polynomial of even degree $n \geq 2$ have at least one critical point? at least one inflection point? a global maximum or minimum?

41 Does every polynomial of odd degree $n \geq 3$ have at least one critical point? a global maximum or minimum?

Exercises 42 and 43 are related.

42 Let f be a function whose second derivative is of the

form $f^{(2)}(x) = (x - a)^k g(x)$, where k is a positive integer, a is a fixed number, and g is a continuous function such that $g(a) \neq 0$. (*a*) Show that if k is odd, then a is an inflection number. (*b*) Show that if k is even, then a is not an inflection number.

43 Explain why a polynomial of odd degree at least 3 always has at least one inflection point. [*Suggestion:* Let f be the polynomial and let $a_1, a_2, \ldots, a_j$ be the roots of its second derivative. Then $f^{(2)}(x)$ can be written as $(x - a_1)^{k1}(x - a_2)^{k2} \cdots (x - a_j)^{kj}g(x)$, where g is a polynomial with no real roots.]

4.5 MOTION AND THE SECOND DERIVATIVE

In this section the second derivative will be used in the study of acceleration.

EXAMPLE 1 A falling rock drops $16t^2$ feet in the first t seconds. Find its velocity and acceleration.

SOLUTION Place the y axis in the usual position, with 0 at the beginning of the fall and the part with positive values above 0, as in Fig. 4.40. At time t the object has the y coordinate

$$y = -16t^2.$$

The velocity is $(-16t^2)' = -32t$ feet per second, and the acceleration is $(-32t)' = -32$ feet per second per second. The velocity changes at a constant rate. That is, the acceleration is constant. ■

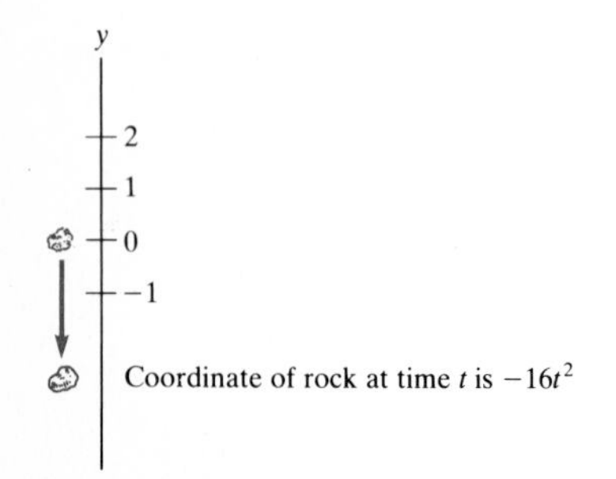

Figure 4.40

The second derivative represents acceleration in other contexts, as the next example shows.

EXAMPLE 2 Translate this news item into calculus: "The latest unemployment figures can be read as bearing out the forecast that the recession is nearing its peak. Though unemployment continues to increase, it is doing so at a slower rate than before."

SOLUTION Let y be a differentiable function of t that approximates the number of people unemployed at time t. As time changes, so does y: $y = f(t)$. The rate of change in unemployment is the derivative, dy/dt. The news that "unemployment continues to increase" is recorded by the inequality

$$\frac{dy}{dt} > 0.$$

There is optimism in the article. The rate of increase, dy/dt, is itself declining ("unemployment continues to increase . . . at a *slower rate* than before"). The function dy/dt is decreasing. Thus its derivative

$$\frac{d}{dt}\left(\frac{dy}{dt}\right)$$

is negative:

$$\frac{d^2y}{dt^2} < 0.$$

Economics need not be a dismal science. If the first derivative is bleak, maybe the second derivative is promising.

In short, the bad news is that dy/dt is positive. But there is good news: d^2y/dt^2 is negative.

The promise that "the recession is nearing its peak" amounts to the prediction that soon dy/dt will be 0 and then switch sign to become negative. In short, a local maximum in the graph of $y = f(t)$ appears in the economists' crystal ball. ■

Knowing the initial position and initial velocity of a moving object and its acceleration at all times, we can calculate its position at all times, as we see in the next two examples, where acceleration is constant.

EXAMPLE 3 In the simplest motion, no forces act on a moving particle. Assume that a particle is moving on the x axis and no forces act on it. Let its location at time t seconds be $x = f(t)$ feet. If at time $t = 0$, $x = 3$ feet and the velocity is 5 feet per second, determine $f(t)$.

SOLUTION The assumption that no force operates on the particle means that $d^2x/dt^2 = 0$. Call the velocity v. Then

$$\frac{dv}{dt} = \frac{d^2x}{dt^2} = 0.$$

Now v is itself a function of time. Since its derivative is 0, v must be constant:

$$v(t) = C$$

for some constant C. Since $v(0) = 5$, the constant C must be 5.

To find the position x as a function of time, note that

$$\frac{dx}{dt} = 5.$$

This equation implies that x must be of the form

$$x = 5t + K$$

for some constant K. Now, when $t = 0$, $x = 3$. Thus $K = 3$. In short, at any time t seconds, the particle is at $x = 5t + 3$ feet. ■

The next example concerns the case in which the acceleration is constant, but not zero.

EXAMPLE 4 A ball is thrown straight up, with a speed of 64 feet per second, from a cliff 96 feet above the ground. Where is the ball t seconds later? When does it reach its maximum height? How high above the ground does the ball rise? When does the ball hit the ground? Assume that there is no air resistance and that the acceleration due to gravity is constant.

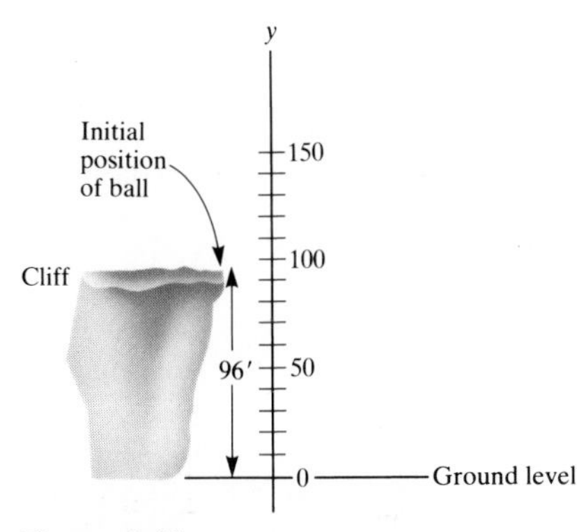

Figure 4.41

SOLUTION Introduce a vertical coordinate axis to describe the position of the ball. It is more natural to call it the y axis, and so velocity is dy/dt and acceleration is d^2y/dt^2. Place the origin at ground level and let the positive part of the y axis be above the ground, as in Fig. 4.41.

At time $t = 0$, the velocity $dy/dt = 64$, since the ball is thrown up at a speed of 64 feet per second. (If it had been thrown down, dy/dt would be -64.) As time increases, dy/dt decreases from 64 to 0 (when the ball reaches the top of its path and begins its descent) and continues to decrease through negative values as the ball falls down to the ground. Since v is decreasing, the acceleration dv/dt is negative. The (constant) value of dv/dt, obtained from experiments, is approximately -32 feet per second per second.

From the equation

$$\frac{dv}{dt} = -32,$$

Velocity is an antiderivative of acceleration.

it follows that

$$v = -32t + C,$$

where C is some constant. To find C, recall that $v = 64$ when $t = 0$. Thus

$$64 = -32 \cdot 0 + C,$$

and $C = 64$. Hence $v = -32t + 64$ for any time t until the ball hits the ground. Now $v = dy/dt$, so

$$\frac{dy}{dt} = -32t + 64.$$

Position is an antiderivative of velocity.

This equation implies that

$$y = -16t^2 + 64t + K,$$

where K is a constant. To find K, make use of the fact that $y = 96$ when $t = 0$. Thus

$$96 = -16 \cdot 0^2 + 64 \cdot 0 + K,$$

and $K = 96$.

We have obtained a complete description of the position of the ball at any time t while it is in the air:

$$y = -16t^2 + 64t + 96.$$

This, together with $v = -32t + 64$, provides answers to many questions about the ball's flight.

Maximum height

When does it reach its maximum height? When $v = 0$; that is, when $-32t + 64 = 0$, or when $t = 2$ seconds.

How high above the ground does the ball rise? Simply compute y when $t = 2$. This gives $-16 \cdot 2^2 + 64 \cdot 2 + 96 = 160$ feet.

Hitting the ground

When does the ball hit the ground? When $y = 0$. Find t such that

$$y = -16t^2 + 64t + 96 = 0.$$

Division by -16 yields the simpler equation $t^2 - 4t - 6 = 0$, which has the solutions

$$t = \frac{4 \pm \sqrt{16 + 24}}{2} = 2 \pm \sqrt{10}.$$

Since $2 - \sqrt{10}$ is negative and the ball cannot hit the ground before it is thrown, the physically meaningful solution is $2 + \sqrt{10}$. The ball lands $2 + \sqrt{10}$ seconds after it is thrown; it is in the air for about 5.2 seconds.

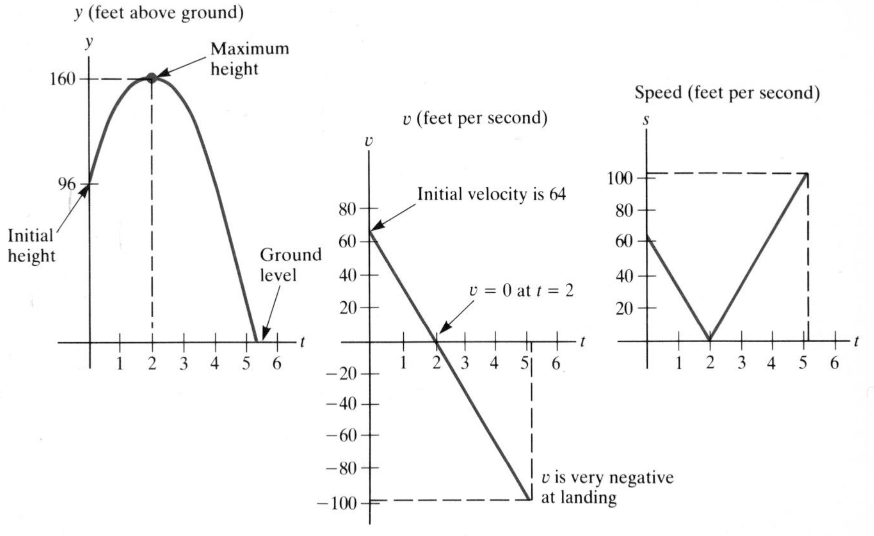

Figure 4.42

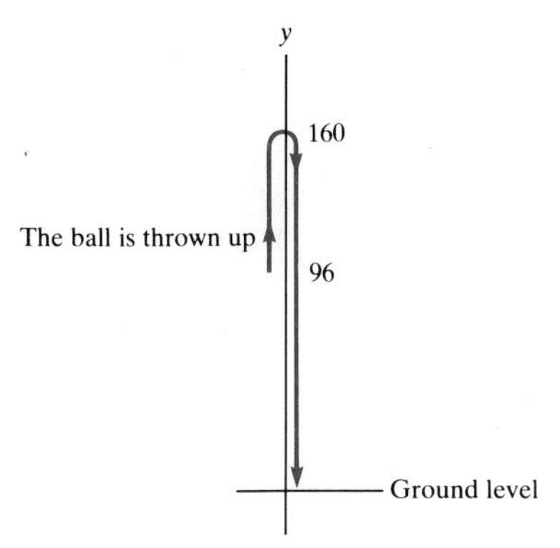

Figure 4.43

The graphs of y, v, and speed, as functions of time, provide another perspective on the motion of the ball, as shown in Fig. 4.42. Of course, the actual path of the ball is restricted to a vertical line and looks somewhat as pictured in Fig. 4.43. ■

Reasoning like that in Example 4 establishes the following description of motion in which the acceleration is constant.

MOTION UNDER CONSTANT ACCELERATION

Assume that a particle moving on the y axis has a constant acceleration a at any time. Assume that at time $t = 0$ it has an initial velocity v_0 and has the initial y coordinate y_0. Then at any time $t \geq 0$ its y coordinate is

$$y = \frac{a}{2}t^2 + v_0t + y_0.$$

Example 4 is the special case, $a = -32$, $v_0 = 64$, and $y_0 = 96$.

The motion of a mass attached to a spring illustrates nonconstant acceleration. Consider a mass m free to move horizontally and attached to the spring shown in Fig. 4.44. (Assume there is no friction.)

If you pull the mass to point B, the spring will draw it back to point C. The velocity of the mass carries it past point C, to A. Since the spring is then compressed, it propels the mass back toward point C again. The mass continues to oscillate back and forth between points A and B. If you start the mass at point C, the midpoint of segment AB, where the spring is neither stretched nor compressed, then the mass will simply stay at this point.

Introduce an x axis through the line AB, with 0 at the rest position of

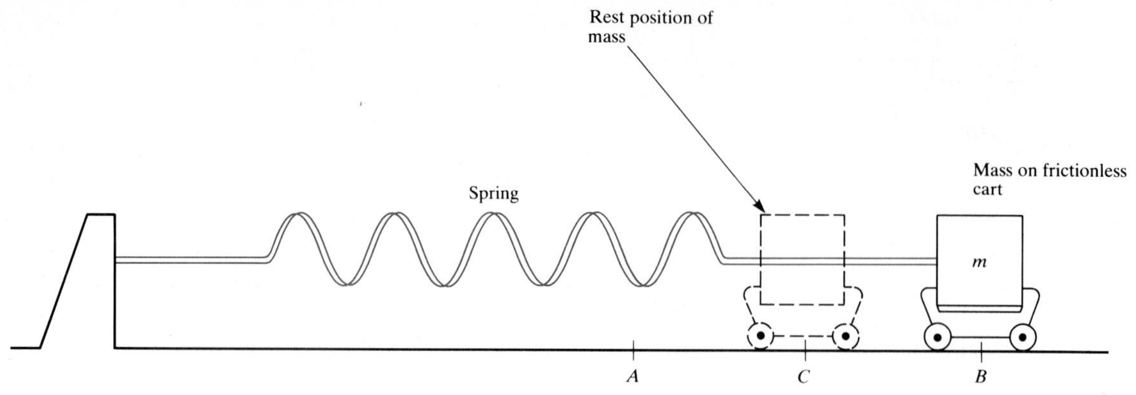

Figure 4.44

the mass, point C. Pull the mass away from the rest point and release it. At time t seconds it will have the x coordinate $x = x(t)$. When the mass is to the right of 0 [$x(t) > 0$], the spring pulls on the mass, tending to *decrease* the velocity dx/dt. The larger $x(t)$, the larger is this tendency to decrease the velocity. Using Hooke's law, we can show that the rate at which the velocity dx/dt changes is proportional to the displacement $x(t)$. We have, then, in mathematical terms,

$$\frac{d}{dt}\left(\frac{dx}{dt}\right) \text{ is proportional to } x(t).$$

This means that there is a *positive* constant c such that

$$\frac{d^2x}{dt^2} = -cx. \tag{1}$$

Motion that satisfies Eq. (1) is called **simple harmonic motion.** It is easy to verify that for any constants A and k, the function $x(t) = A \cos(\sqrt{c}t + k)$ satisfies Eq. (1). To check this statement, let's differentiate $A \cos(\sqrt{c}t + k)$ twice:

$$x(t) = A \cos(\sqrt{c}t + k),$$

$$\frac{dx}{dt} = -A\sqrt{c} \sin(\sqrt{c}t + k),$$

$$\frac{d^2x}{dt^2} = -A\,c \cos(\sqrt{c}t + k).$$

This shows that $$\frac{d^2x}{dt^2} = -cx.$$

It can be shown that the only functions that satisfy Eq. (1) are of the form described.

The force of gravity depends on distance

In contrast to the assumption in Example 4, the acceleration due to gravity is not constant, but varies inversely as the square of the distance from the center of the earth. (However, it is almost constant if the particle moves only small distances—such as a few miles up; so treating it as constant in practical engineering is justified.) We must take this into consideration when determining the **escape velocity**, the minimum speed

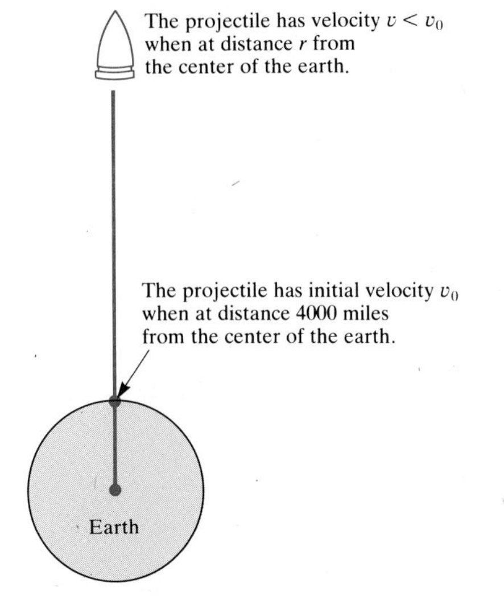

Figure 4.45

a payload must have in order to "coast to infinity" rather than fall back to earth.

The mathematical basis for determining the escape velocity was developed in Newton's *Principia*, published in 1687. There he investigated not only the inverse square law, but also other laws of attraction, such as the inverse cube, which are not known to occur in nature.

We begin by studying the motion of a projectile fired with an initial velocity of v_0 miles per second from the surface of the earth. (Assume all launches are straight up.) For this purpose, introduce a coordinate system whose origin is at the center of the earth. Let r denote the distance of the projectile from the center of the earth. (See Fig. 4.45.)

The velocity of the projectile is defined as

$$v = \frac{dr}{dt},$$

and its acceleration as

$$\frac{d^2r}{dt^2} = \frac{dv}{dt}.$$

This follows from Newton's second law.

Assume that the acceleration due to gravity is proportional to the force of gravity on the particle; that is,

$$\text{Acceleration} = \frac{dv}{dt} = \frac{-K}{r^2}, \tag{2}$$

where K is some positive constant. [The negative sign in Eq. (2) reminds us that gravity slows the projectile down.]

Before analyzing Eq. (2) further, we determine K. At the surface of the earth, where $r = 4000$ miles, the acceleration due to gravity is -32 feet per second per second, which is approximately -0.006 mile per second per second. Thus K satisfies the equation

$$-0.006 = \frac{-K}{4000^2},$$

and

$$K = (4000)^2(0.006). \tag{3}$$

Now return to Eq. (2), which links velocity, time, and distance. It is possible to eliminate time from Eq. (2) by using the chain rule:

$$\frac{dv}{dt} = \frac{dv}{dr}\frac{dr}{dt} = \frac{dv}{dr}v.$$

So Eq. (2) can be written

$$v\frac{dv}{dr} = \frac{-K}{r^2}, \tag{4}$$

an equation linking velocity and distance. Observe that Eq. (4) is equivalent to the equation

$$\frac{d}{dr}\left(\frac{v^2}{2}\right) = \frac{d}{dr}\left(\frac{K}{r}\right).$$

Since the functions $v^2/2$ and K/r have the same derivative with respect to r, they differ by a constant. Thus

$$\frac{v^2}{2} = \frac{K}{r} + C, \tag{5}$$

where C is constant. To determine C, again use information available at the surface of the earth, namely, $v = v_0$ when $r = 4000$. From Eq. (5) it follows that

$$\frac{v_0^2}{2} = \frac{K}{4000} + C,$$

and

$$C = \frac{v_0^2}{2} - \frac{K}{4000}. \tag{6}$$

Combining Eqs. (3), (5), and (6) yields

$$\begin{aligned} \frac{v^2}{2} &= \frac{K}{r} + \left(\frac{v_0^2}{2} - \frac{K}{4000}\right) \\ &= \frac{v_0^2}{2} + K\left(\frac{1}{r} - \frac{1}{4000}\right) \\ &= \frac{v_0^2}{2} + (4000)^2(0.006)\left(\frac{1}{r} - \frac{1}{4000}\right). \end{aligned}$$

Hence

$$v^2 = v_0^2 + (4000)^2(0.012)\left(\frac{1}{r} - \frac{1}{4000}\right). \tag{7}$$

Equation (7) describes v as a function of r. If v in Eq. (7) is never 0, that is, if the payload never reaches a maximum distance from the earth, then the payload will not fall back to the earth. Thus, by Eq. (7), if v_0 is such that the equation

$$0 = v_0^2 + (4000)^2(0.012)\left(\frac{1}{r} - \frac{1}{4000}\right) \tag{8}$$

has no positive solution r, then v_0 is large enough to send the payload on an endless journey. To find such v_0, rewrite Eq. (8) as

$$(4000)(0.012) - v_0^2 = (4000)^2(0.012)\left(\frac{1}{r}\right). \tag{9}$$

If the left side of Eq. (9) is greater than 0, then there is a solution for r, and the payload will reach a maximum distance; if the left side of Eq. (9) is less than or equal to 0, then there is no solution for r. The smallest v_0 that satisfies the inequality $(4000)(0.012) - v_0^2 \le 0$ is

$$v_0 = \sqrt{(4000)(0.012)} = \sqrt{48} \approx 6.93 \text{ miles per second.}$$

The escape velocity is 6.93 miles per second, which is about 25,000 miles per hour.

Escape velocity is $\sqrt{2}$ times orbit velocity.

In Sec. 14.5 it will be shown that the velocity necessary to maintain an object in a circular orbit at the surface of the earth is about 4.90 miles per second, which is approximately 18,000 miles per hour. This orbit velocity equals the escape velocity divided by $\sqrt{2}$.

EXERCISES FOR SEC. 4.5: MOTION AND THE SECOND DERIVATIVE

1 Translate into calculus the following news report about the leaning tower of Pisa. "The tower's angle from the vertical was increasing more rapidly." *Suggestion:* Let $\theta = f(t)$ be the angle of deviation from the vertical at time t.

Incidentally, the tower, begun in 1174 and completed

in 1350, is 179 feet tall and leans about 14 feet from the vertical. Each day it leans, on the average, another $\frac{1}{5000}$ inch.

2 Translate this news headline into calculus: "Gasoline prices increase more slowly." *Suggestion:* Let $G(t)$ be the price of a gallon of gasoline at time t.

Exercises 3 to 6 concern Example 4.

3 (*a*) How long after the ball in Example 4 is thrown does it pass by the top of the cliff?
(*b*) What are its speed and velocity then?

4 If the ball in Example 4 had simply been dropped from the cliff, what would y be as a function of time? How long would the ball fall?

5 In view of the result of Exercise 4 interpret physically each of the three terms on the right side of the formula $y = -16t^2 + 64t + 96$.

6 Let $y = f(t)$ describe the motion on the y axis of an object whose acceleration has the constant value a. Show that

$$y = \frac{a}{2}t^2 + v_0 t + y_0,$$

where v_0 is the velocity when $t = 0$, and y_0 is the position when $t = 0$.

7 At time $t = 0$ a particle is at $y = 3$ feet and has a velocity of -3 feet per second; it has a constant acceleration of 6 feet per second per second. Find its position at any time t.

8 At time $t = 0$ a particle is at $y = 10$ feet and has a velocity of 8 feet per second; it has a constant acceleration of -8 feet per second per second. (*a*) Find its position at any time t. (*b*) What is its maximum y coordinate?

9 At time $t = 0$ a particle is at $y = 0$ and has a velocity of 0 feet per second. Find its position at any time t if its acceleration is always -32 feet per second per second.

10 At time $t = 0$ a particle is at $y = -4$ feet and has a velocity of 6 feet per second; it has a constant acceleration of -32 feet per second per second. (*a*) Find its position at any time t. (*b*) What is its largest y coordinate?

■

11 In harmonic motion, $x(t) = A\cos(\sqrt{c}t + k)$.
(*a*) What is the maximum displacement from 0?
(*b*) What is the maximum speed? Where is the mass then?
(*c*) What is the minimum speed? Where is the mass then?

12 A car accelerates with constant acceleration from 0 (rest) to 60 miles per hour in 15 seconds. How far does it travel in this period? Be sure to do your computations either all in seconds or all in hours; for instance, 60 miles per hour is 88 feet per second.

13 A mass at the end of a spring oscillates. At time t seconds its position (relative to its position at rest) is $y = 6\sin t$ inches.
(*a*) Graph y as a function of t.
(*b*) What is the maximum displacement of the mass from its rest position?
(*c*) Show that its acceleration is proportional to its displacement y.
(*d*) Where is it when its speed is maximum?
(*e*) Where is it when the absolute value of its acceleration is maximum?

14 Show that a ball thrown straight up from the ground takes as long to rise as to fall back to its initial position. How does the velocity with which it strikes the ground compare with its initial velocity? Consider the same question for its speed.

15 Let $y = (t - 1)^{2/3}$. Show that y satisfies the "differential equation"

$$\frac{d^2y}{dt^2} = -\frac{2}{9}\frac{1}{y^2}.$$

This differential equation says that the acceleration of y is inversely proportional to the square of y. It describes the motion of an object "coasting to infinity" away from the earth. Note that its velocity approaches 0.

Exercises 16 to 21 relate to the launch of the payload.

16 If a payload is launched with the velocity of 7 miles per second, which is greater than its escape velocity, what happens to its velocity far out in its journey? In other words, determine $\lim_{r\to\infty} v$.

17 If we launch a payload with a speed of 6 miles per second, how far will it go from the center of the earth?

18 At what speed must we launch a payload if it is to reach the moon, 240,000 miles from the center of the earth? (Disregard the gravitational field of the moon.)

19 When a payload is launched with precisely the escape velocity, what happens to its velocity far out in its journey? In other words, determine $\lim_{r\to\infty} v$.

20 (Disregard air resistance.) In order to propel an object 100 miles straight up, what must the launching velocity be if it is assumed that (*a*) the force of gravity varies as in the discussion of escape velocity? (*b*) the force of gravity is constant?

21 Could it happen that a projectile shot straight out from the earth neither returns nor travels to "infinity" but approaches a certain finite limiting position? (Disregard other gravitational fields than the earth's.)

22 Let $x(t) = A\sin(\sqrt{c}t) + B\cos(\sqrt{c}t)$. Does this function satisfy Eq. (1)?

4.6 APPLIED MAXIMUM AND MINIMUM PROBLEMS

One of the most important applications of calculus is obtaining the most efficient design of a product. Frequently the problem of minimizing cost or maximizing the volume of a certain object reduces to minimizing or maximizing some function $f(x)$. In that case, the methods developed in Secs. 4.2 and 4.4 may be called on. They are the use of critical points, the first-derivative test, and the second-derivative test. Recall that when maximizing or minimizing a function over a closed interval it is essential to consider also the values of the function at the endpoints.

The five examples that follow are typical. The only novelty is the challenge of how to translate each problem into the terminology of functions.

EXAMPLE 1 If we cut four congruent squares out of the corners of a square piece of cardboard 12 inches on each side, we can fold up the four remaining flaps to obtain a tray without a top. What size squares should be cut in order to maximize the volume of the tray?

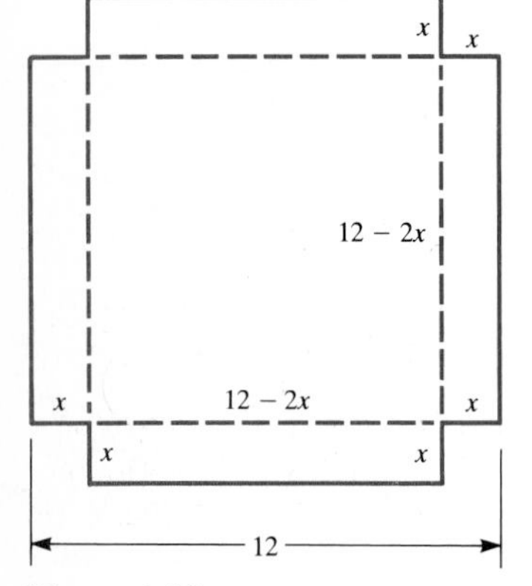

Figure 4.46

SOLUTION Let us remove squares of side x, as shown in Figs. 4.46 and 4.47. Folding on the dotted lines, we obtain a tray of volume

$$V(x) = (12 - 2x)^2(x) = 4x^3 - 48x^2 + 144x.$$

Since each side of the cardboard square has length 12 inches, the only values of x which make sense are those in the closed interval $[0, 6]$. Thus we wish to find the number x in $[0, 6]$ that maximizes $V(x)$.

Notice that $V(x) = (12 - 2x)^2(x)$ is small when x is near 0 (that is, when we try to economize by making the height of the tray small) and small when x is near 6 (that is, when we try to economize by making the base small). We have a "two-influence" problem; to find the best balance between them, we use calculus.

The maximum value of $V(x)$ for x in $[0, 6]$ occurs either at 0, 6, or a critical number [where $V'(x) = 0$]. Now, $V(0) = 0$ (the tray has height 0), and $V(6) = 0$ (the tray has a base of area 0). These are minimum values for the volume, certainly not the maximum volume, so the maximum must occur at some critical number in $(0, 6)$.

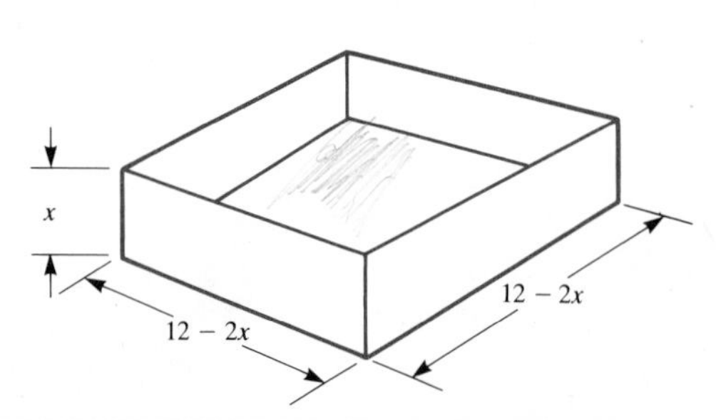

Figure 4.47

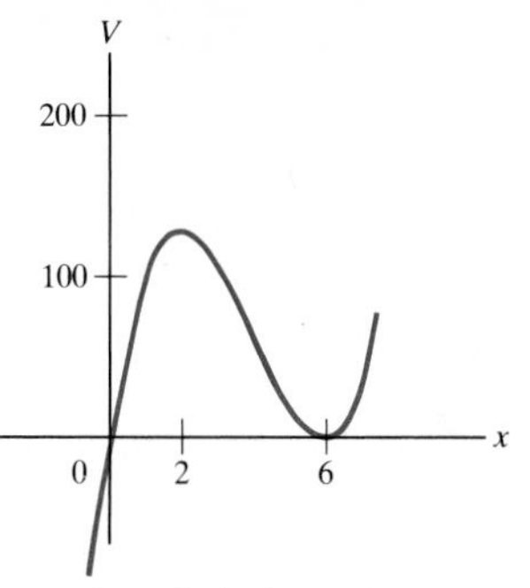

Only values of x in the portion above [0, 6] correspond to physically realizable trays

Figure 4.48

Next compute $V'(x)$:

$$\begin{aligned} V'(x) &= (4x^3 - 48x^2 + 144x)' \\ &= 12x^2 - 96x + 144 \\ &= 12(x^2 - 8x + 12) \\ &= 12(x - 6)(x - 2). \end{aligned}$$

The equation $12(x - 6)(x - 2) = 0$ has two roots, namely, 2 and 6. The critical numbers are 2 and 6. As already remarked, the maximum does not occur at 0 or 6. Hence it occurs at $x = 2$. When $x = 2$, the volume is

$$\begin{aligned} V(2) &= (12 - 2 \cdot 2)^2(2) = 8^2 \cdot 2 \\ &= 128 \text{ cubic inches.} \end{aligned}$$

This is the largest possible volume and is obtained when the length of the cut is 2 inches.

As a matter of interest, let us graph the function V, showing its behavior for all x, not just for values of x significant to the problem. Note in Fig. 4.48 that at $x = 6$ the tangent is horizontal. ■

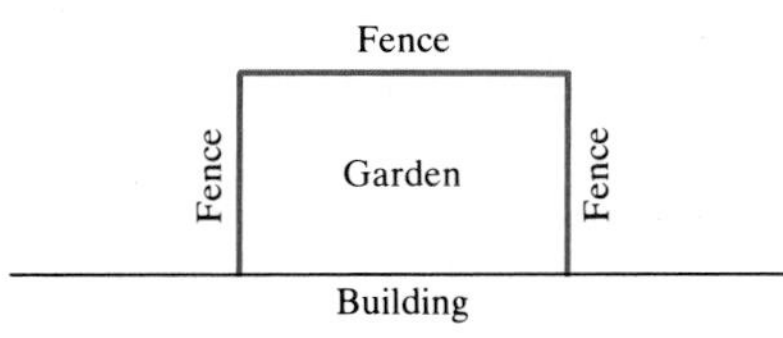

The three sides of the garden not along the building total 100 feet.

Figure 4.49

EXAMPLE 2 A couple have enough wire to construct 100 feet of fence. They wish to use it to form three sides of a rectangular garden, one side of which is along a building, as shown in Fig. 4.49. What shape garden should they choose in order to enclose the largest possible area?

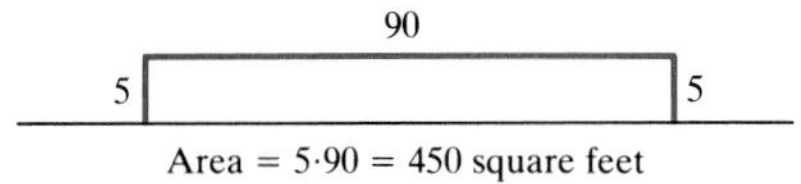

Area = 5·90 = 450 square feet

Figure 4.50

SOLUTION Figures 4.50 to 4.52 show some possible ways of laying out the 100 feet of fence. For convenience, let x denote the length of the side of the garden that is perpendicular to the building and y the length of the side parallel to the building. Since 100 feet of fencing is available, $2x + y = 100$. When $x = 5$, the area is 450 square feet. When x has increased to 20, y has decreased to 60 feet, and the area is 1200 square feet. When $x = 40$, $y = 20$, and the area is only 800 square feet. It is not immediately clear how to choose x to maximize the area of the garden. What makes the problem interesting is that when you increase one dimension of the rectangle, the other automatically decreases. The area, which is the product of the two dimensions, is subject to two opposing forces, one causing it to increase, the other to decrease. This type of problem is easily solved with the aid of the derivative.

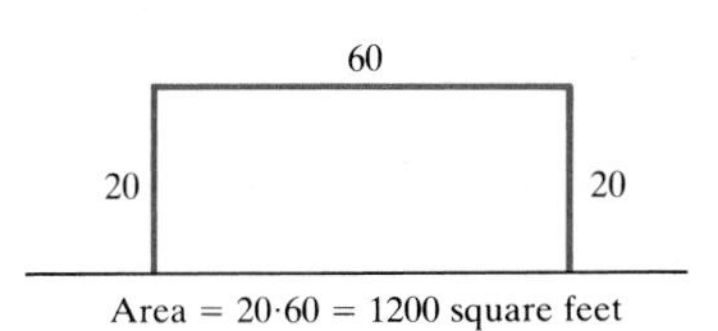

Area = 20·60 = 1200 square feet

Figure 4.51

First of all, express the area A of the garden in terms of x and y:

$$A = xy.$$

Then use the equation $100 = 2x + y$ to express y in terms of x:

$$y = 100 - 2x.$$

Thus the area A is

$$A = x(100 - 2x) \text{ square feet.}$$

(See Fig. 4.53.)

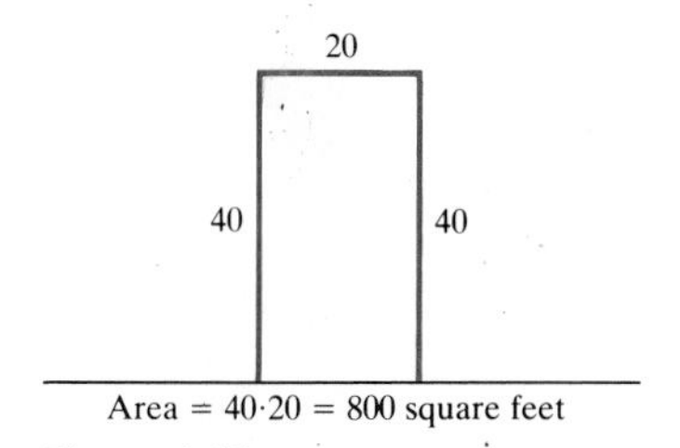

Area = 40·20 = 800 square feet

Figure 4.52

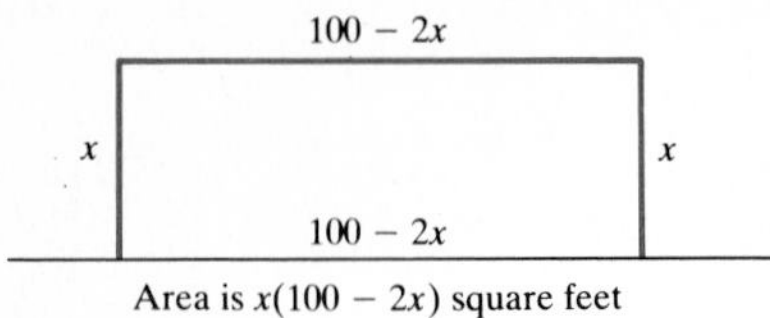

Figure 4.53

Clearly, $0 \leq x \leq 50$. Thus the problem now has become: Maximize $f(x) = x(100 - 2x)$ for x in $[0, 50]$.

In this case, $f(x) = 100x - 2x^2$; hence $f'(x) = 100 - 4x$. Set the derivative equal to 0:

$$0 = 100 - 4x \quad \text{or} \quad 4x = 100.$$

Hence
$$x = 25.$$

Thus 25 is the only critical number for the function. The maximum of f occurs either at 25 or at one of the ends of the interval, 0 or 50. Now,

$$f(0) = 0[100 - 2 \cdot 0] = 0,$$

$$f(50) = 50[100 - 2 \cdot 50] = 0,$$

$$f(25) = 25[100 - 2 \cdot 25] = 1250.$$

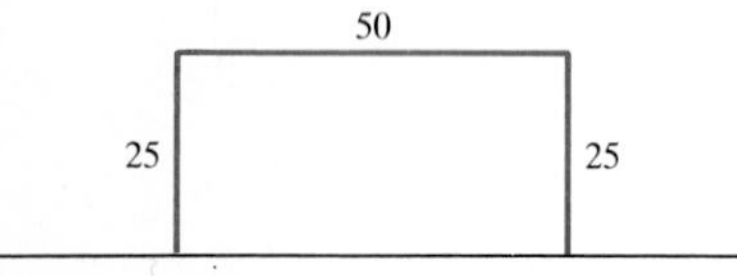

Figure 4.54

Thus the maximum possible area is 1250 square feet, and the fence should be laid out as shown in Fig. 4.54. ■

Examples 1 and 2 illustrate the general procedure for solving applied maximum (or minimum) problems.

PROCEDURE FOR FINDING A MAXIMUM

1. Draw and label the necessary pictures.
2. Name the various quantities in the problem by letters, such as x, y, A, V.
3. Identify the variable that is to be maximized.
4. Express the quantity to be maximized in terms of one or more other quantities.
5. By eliminating variables, express the quantity to be maximized as a function of one variable.
6. Maximize the function obtained in step 5 by examining its derivative.

Example 3 also illustrates this procedure, as applied instead to a minimization problem.

EXAMPLE 3 Of all the tin cans that enclose a volume of 100 cubic inches, which requires the least metal?

SOLUTION Denote the radius of a can of volume 100 cubic inches by r and its height by h. The can may be flat or tall. If the can is flat, the side uses little metal, but then the top and bottom bases are large. If the can is shaped like a mailing tube, then the two bases require little metal, but the curved side requires a great deal of metal. (See Fig. 4.55.) What is the ideal compromise between these two extremes?

The surface area S of the can is given by

$$S = 2\pi r^2 + 2\pi rh, \tag{1}$$

which accounts for the two circular bases and the side. Since the amount of metal in the can is proportional to S, it suffices to minimize S.

In the tin can under consideration, the radius and height are related by

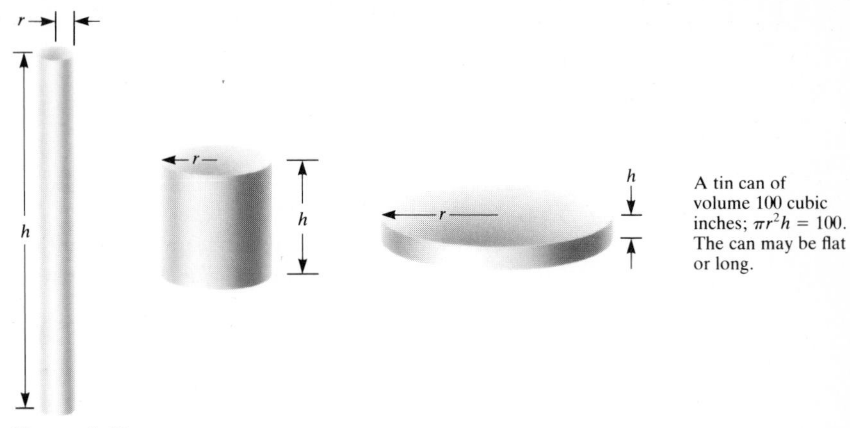

A tin can of volume 100 cubic inches; $\pi r^2 h = 100$. The can may be flat or long.

Figure 4.55

the constraint

$$\pi r^2 h = 100. \tag{2}$$

In order to express S as a function of one variable, use Eq. (2) to eliminate either r or h. Choosing to eliminate h, we solve Eq. (2) for h:

$$h = \frac{100}{\pi r^2}.$$

Substitution into Eq. (1) yields

$$S = 2\pi r^2 + 2\pi r \frac{100}{\pi r^2} \quad \text{or} \quad S = 2\pi r^2 + \frac{200}{r}. \tag{3}$$

Equation (3) expresses S as a function of just one variable, r. The domain of this function for our purposes is $(0, \infty)$, since the tin can has a positive radius.

Compute dS/dr:

$$\frac{dS}{dr} = 4\pi r - \frac{200}{r^2} = \frac{4\pi r^3 - 200}{r^2}. \tag{4}$$

This derivative is 0 only when

$$4\pi r^3 = 200, \tag{5}$$

that is, when

$$r = \sqrt[3]{\frac{50}{\pi}}.$$

Thus $r = \sqrt[3]{50/\pi}$ is the only critical number. Does it in fact provide a minimum?

Using the second derivative to test for global minimum

First, let us check by the second-derivative test. Differentiation of Eq. (4) yields

$$\frac{d^2S}{dr^2} = 4\pi + \frac{400}{r^3},$$

which is positive for all $r > 0$: the graph is concave upward for all $r > 0$. Hence $r = \sqrt[3]{50/\pi}$ provides not only a local minimum but a global minimum.

Using the first derivative to test for global minimum

The first derivative will also enable us to reach the same conclusion. Recall that

$$\frac{dS}{dr} = \frac{4\pi r^3 - 200}{r^2}.$$

At the critical number, the numerator is 0. If r is less than the critical number, the numerator, hence the quotient, is *negative*. If r is larger than the critical number, the quotient is positive. Thus the function decreases for $0 < r < \sqrt[3]{50/\pi}$ and increases for $r > \sqrt[3]{50/\pi}$. Thus the critical number indeed provides an absolute or global minimum. ■

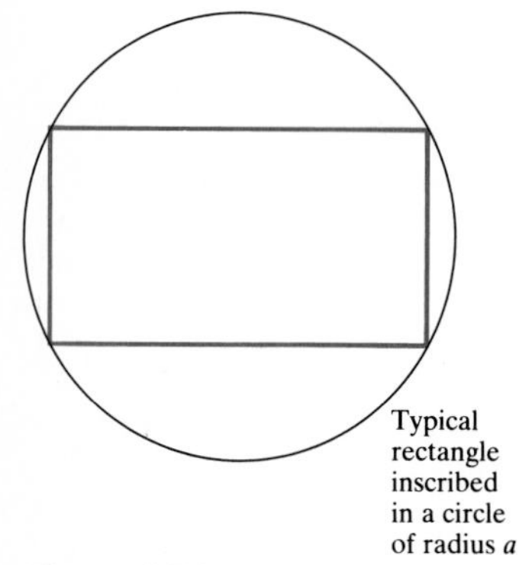

Figure 4.56

EXAMPLE 4 Find the dimensions of the rectangle of largest area that can be inscribed in a circle of radius a.

SOLUTION A circle of radius a and a typical inscribed rectangle are shown in Fig. 4.56. Label the lengths of the sides w and h, as shown in Fig. 4.57. The area of the typical rectangle is

$$A = hw. \tag{6}$$

But h and w are linked to each other. (If h is small, w is large; if h is large, w is small.)

In order to express the formula $A = hw$ as a function of one variable, either h or w, it is necessary to find the relation between them. The relation must express the fact that the rectangle is inscribed in a circle of radius a.

To find the "missing equation" it may be necessary to draw the "missing line."

The distance from any point P on the circle to its center C is a, as shown in Fig. 4.58. The radius CP is the hypotenuse of a right triangle whose two legs have lengths $w/2$ and $h/2$, as in Fig. 4.59. The pythagorean theorem then provides an equation linking w and h:

$$\left(\frac{w}{2}\right)^2 + \left(\frac{h}{2}\right)^2 = a^2,$$

or, equivalently,

$$w^2 + h^2 = 4a^2. \tag{7}$$

Solving Eq. (7) for h, say, gives

$$h = \sqrt{4a^2 - w^2}. \tag{8}$$

Taken together, Eqs. (6) and (8) show how the area of the inscribed

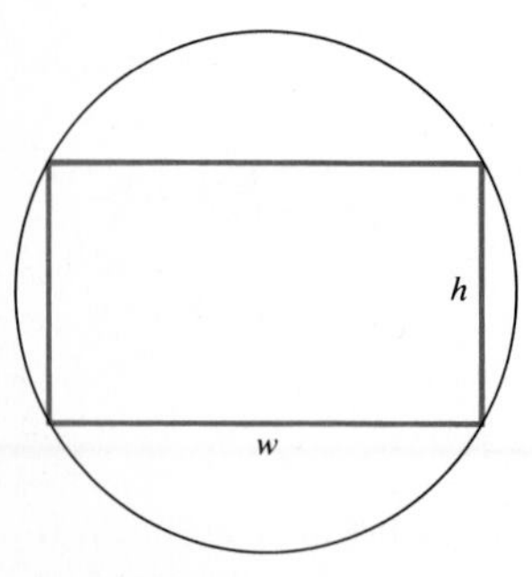

Figure 4.57

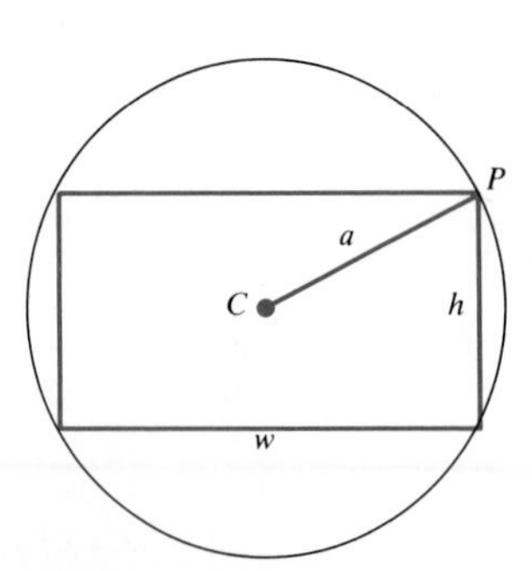

Figure 4.58

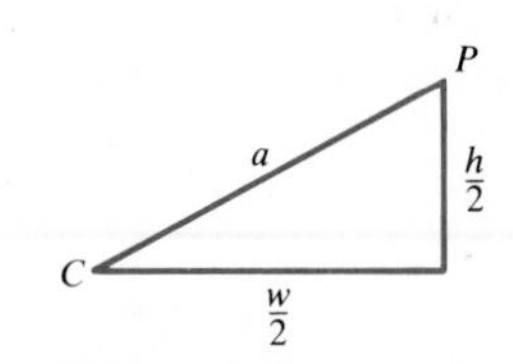

Figure 4.59

rectangle varies with w:

$$A(w) = hw = w\sqrt{4a^2 - w^2}. \tag{9}$$

The geometry is done; the calculus begins. $A(w)$ is defined and continuous for w in $[0, 2a]$. Thus $A(w)$ has a maximum either at 0, at $2a$, or at a critical number.

To find any critical numbers, differentiate $A(w)$:

$$\begin{aligned} A'(w) &= w \cdot \frac{1}{2} \cdot \frac{-2w}{\sqrt{4a^2 - w^2}} + \sqrt{4a^2 - w^2} \\ &= \frac{-w^2}{\sqrt{4a^2 - w^2}} + \sqrt{4a^2 - w^2} \\ &= \frac{-w^2 + 4a^2 - w^2}{\sqrt{4a^2 - w^2}} \\ &= \frac{4a^2 - 2w^2}{\sqrt{4a^2 - w^2}}. \end{aligned} \tag{10}$$

For $A'(w)$ to be 0, the numerator of Eq. (10) must be 0:

$$\begin{aligned} 4a^2 - 2w^2 &= 0 \\ 2w^2 &= 4a^2 \\ w^2 &= 2a^2 \\ w &= \sqrt{2}a. \end{aligned}$$

(The negative square root of $2a^2$ lies outside the interval $[0, 2a]$.)

Thus the maximum value of $A(w)$ is the maximum of the three numbers $A(0)$, $A(2a)$, and $A(\sqrt{2}a)$. A quick computation using Eq. (9) or a glance at Fig. 4.58 shows that $A(0) = 0$ and $A(2a) = 0$. Thus the maximum must occur at $w = \sqrt{2}a$.

By Eq. (8), the corresponding h is

$$h = \sqrt{4a^2 - (\sqrt{2}a)^2} = \sqrt{4a^2 - 2a^2} = \sqrt{2a^2} = \sqrt{2}a.$$

The rectangle of largest area inscribed in the circle is a square. ■

We could also solve this problem by maximizing $[A(w)]^2$. This approach avoids square roots.

EXERCISES FOR SEC. 4.6: APPLIED MAXIMUM AND MINIMUM PROBLEMS

(Formulas for various volumes and areas are listed inside the back cover.)

Exercises 1 to 4 are related to Example 1. In each case find the length of the cut that maximizes the volume of the tray. The dimensions of the cardboard are given.

1 5 inches by 5 inches

2 5 inches by 7 inches

3 4 inches by 8 inches

4 6 inches by 10 inches

5 Solve Example 2, expressing A in terms of y instead of x.

6 Solve Example 2 if there is instead enough wire to construct 160 feet of fence.

7 Solve Example 3, expressing S in terms of h instead of r.

8 Of all cylindrical tin cans without a top that contain 100 cubic inches, which requires the least material?

9 Of all enclosed rectangular boxes with square bases that have a volume of 1000 cubic inches, which uses the least material?

10 Of all topless rectangular boxes with square bases that have a volume of 1000 cubic inches, which uses the least material?

11 Solve Example 4, but express the area of the rectangle

not in terms of w and h, but in terms of the angle θ, shown in Fig. 4.60.

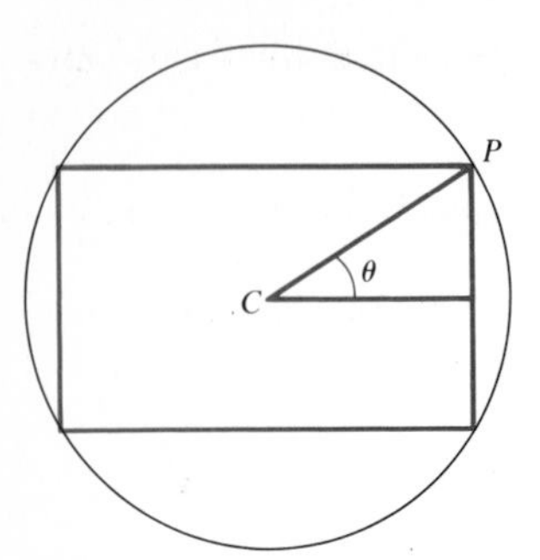

Figure 4.60

12 Find the dimensions of the rectangle of largest perimeter that can be inscribed in a circle of radius a.

13 Show that of all rectangles of a given perimeter, the square has the largest area. *Suggestion:* Call the fixed perimeter p and keep in mind that it is constant.

14 Show that of all rectangles of a given area, the square has the shortest perimeter. *Suggestion:* Call the fixed area A and keep in mind that it is constant.

15 A rancher wants to construct a rectangular corral. He also wants to divide the corral by a fence parallel to one of the sides. He has 240 feet of fence. What are the dimensions of the corral of largest area he can enclose?

16 A river has a 45° turn, as indicated in Fig. 4.61. A rancher wants to construct a corral bounded on two sides by the river and on two sides by 1 mile of fence ABC, as shown. Find the dimensions of the corral of largest area.

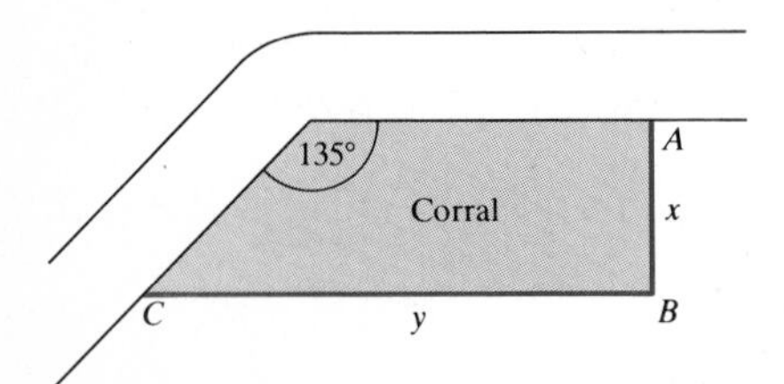

Figure 4.61

17 (*a*) How should one choose two nonnegative numbers whose sum is 1 in order to maximize the sum of their squares?

(*b*) To minimize the sum of their squares?

18 How should one choose two nonnegative numbers whose sum is 1 in order to maximize the product of the square of one of them and the cube of the other?

19 An irrigation channel made of concrete is to have a cross section in the form of an isosceles trapezoid, three of whose sides are 4 feet long. See Fig. 4.62. How should the trapezoid be shaped if it is to have the maximum possible area? Consider the area as a function of x and solve.

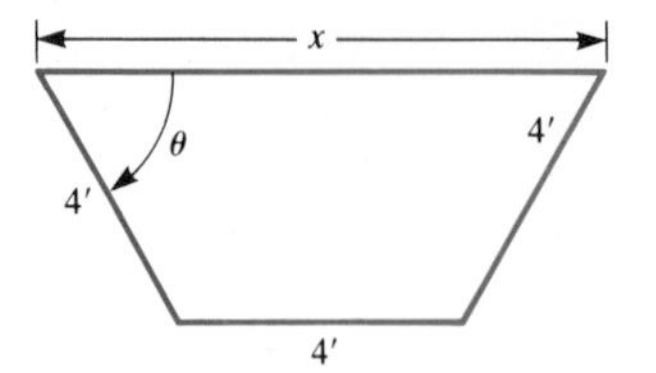

Figure 4.62

20 (*a*) Solve Exercise 19, expressing the area as a function of θ instead of x.

(*b*) Do the answers in (*a*) and Exercise 19 agree?

In Exercises 21 to 24 use the fact that the combined length and girth (distance around) of a package to be sent through the mails cannot exceed 108 inches.

21 Find the dimensions of the right circular cylinder of largest volume that can be sent through the mail.

22 Find the dimensions of the right circular cylinder of largest surface area that can be sent through the mail.

23 Find the dimensions of the rectangular box with square base of largest volume that can be sent through the mail.

24 Find the dimensions of the rectangular box with square base of largest surface area that can be sent through the mail.

■

Exercises 25 to 30 concern "minimal cost" problems.

25 A cylindrical can is to be made to hold 100 cubic inches. The material for its top and bottom costs twice as much per square inch as the material for its side. Find the radius and height of the most economical can. *Warning:* This is not the same as Example 3.

(*a*) Would you expect the most economical can in this problem to be taller or shorter than the solution to Example 3? (Use common sense, not calculus.)

(*b*) For convenience, call the cost of 1 square inch of the material for the side k cents. Thus the cost of 1 square inch of the material for the top and bottom is $2k$ cents. (The precise value of k will not affect the answer.) Show that a can of radius r and height h costs

$$C = 4k\pi r^2 + 2k\pi rh \qquad \text{cents.}$$

(*c*) Find r that minimizes the function C in (*b*). Keep in mind during any differentiation that k is constant.

(*d*) Find the corresponding h.

26 A rectangular box with a square base is to hold 100 cubic inches. Material for the sides costs twice as much per square inch as the material for the top and bottom.
(*a*) If the base has side x and the height is y, what does the box cost?
(*b*) Find the dimensions of the most economical box.

27 A rectangular box with a square base is to hold 100 cubic inches. Material for the top of the box costs 2 cents per square inch; material for the sides costs 3 cents per square inch; material for the bottom costs 5 cents per square inch. Find the dimensions of the most economical box.

28 The cost of operating a certain truck (for gasoline, oil, and depreciation) is $(15 + s/9)$ cents per mile when it travels at a speed of s miles per hour. A truck driver earns \$9 per hour. What is the most economical speed at which to operate the truck during a 600-mile trip?
(*a*) If you considered only the truck, would you want s to be small or large?
(*b*) If you considered only the expense of the driver's wages, would you want s to be small or large?
(*c*) Express cost as a function of s and solve. (Be sure to put the costs all in terms of cents or all in terms of dollars.)
(*d*) Would the answer be different for a 1000-mile trip?

29 A government contractor who is removing earth from a large excavation can route trucks over either of two roads. There are 10,000 cubic yards of earth to move. Each truck holds 10 cubic yards. On one road the cost per truck load is $1 + 2x^2$ cents, when x trucks use that road; the function records the cost of congestion. On the other road the cost is $2 + x^2$ cents per truckload when x trucks use that road. How many trucks should be dispatched to each of the two roads?

30 On one side of a river 1 mile wide is an electric power station; on the other side, s miles upstream, is a factory. (See Fig. 4.63.) It costs 3 dollars per foot to run cable over land and 5 dollars per foot under water. What is the most economical way to run cable from the station to the factory?
(*a*) Using no calculus, what do you think would be (approximately) the best route if s were very small? if s were very large?
(*b*) Solve with the aid of calculus, and draw the routes for $s = \frac{1}{2}, \frac{3}{4}$, 1, and 2.
(*c*) Solve for arbitrary s.
Warning: Minimizing the length of cable is not the same as minimizing its cost.

31 (From *Dynamics of Airplanes,* by John E. Younger and Baldwin M. Woods, Wiley, New York.) "Recalling that

$$I = A \cos^2 \theta + C \sin^2 \theta - 2E \cos \theta \sin \theta,$$

we wish to find θ when I is a maximum or a minimum." Show that at an extremum of I,

$$\tan 2\theta = \frac{2E}{C - A}. \qquad \text{(Assume that } A \neq C.)$$

32 (From *University Physics,* p. 46, by Alvin Hudson and Rex Nelson, Harcourt-Brace-Jovanovich, New York, 1982.) "By differentiating the equation for the horizontal range,

$$R = \frac{v_0^2 \sin 2\theta}{g},$$

show that the initial elevation angle θ for maximum range is 45°." In the formula for R, v_0 and g are constants. (R is the horizontal distance a baseball covers if you throw it at an angle θ with speed v_0. Air resistance is disregarded.)
(*a*) Using calculus, show that the maximum range occurs when $\theta = 45°$.
(*b*) Solve the same problem without calculus.

33 Fencing is to be added to an existing wall of length 20 feet, as shown in Fig. 4.64. How should the extra fence be added to maximize the area of the enclosed rectangle if the additional fence is
(*a*) 40 feet long?
(*b*) 80 feet long?
(*c*) 60 feet long?

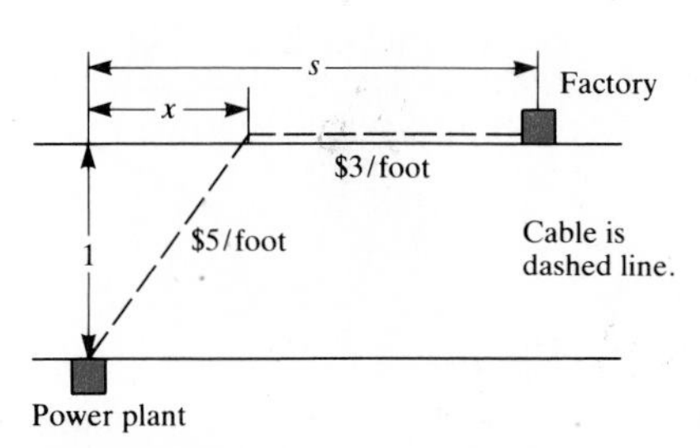

Figure 4.63

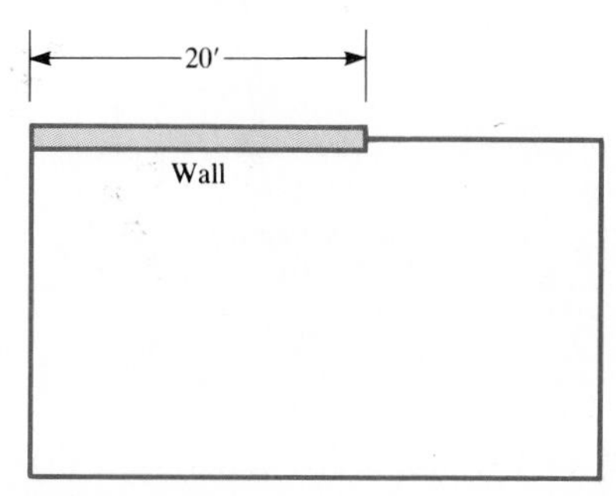

Figure 4.64

34 Let A and B be constants. Find the maximum and minimum values of $A \cos t + B \sin t$.

35 A spider at corner S of a cube of side 1 inch wishes to capture a fly at the opposite corner F. (See Fig. 4.65.) The spider, who must walk on the surface of the solid cube, wishes to find the shortest path.
(*a*) Find a shortest path with the aid of calculus.
(*b*) Find a shortest path without calculus.

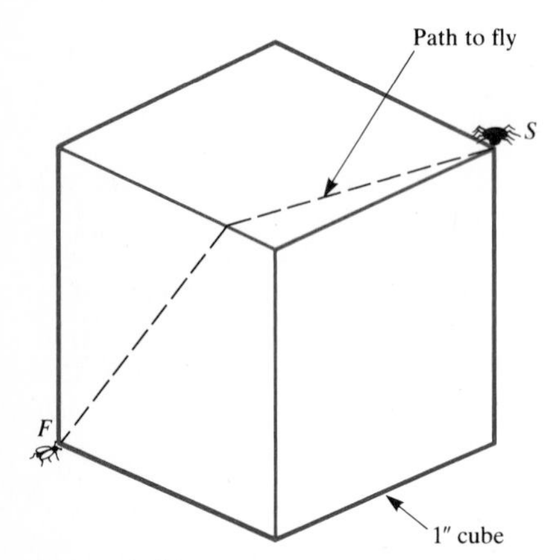

Figure 4.65

36 A ladder of length b leans against a wall of height a, $a < b$. What is the maximal horizontal distance that the ladder can extend beyond the wall if its base rests on the horizontal ground?

37 A woman can walk 3 miles per hour on grass and 5 miles per hour on sidewalk. She wishes to walk from point A to point B, shown in Fig. 4.66 in the least time. What route should she follow if s is (*a*) $\frac{1}{2}$? (*b*) $\frac{3}{4}$? (*c*) 1?

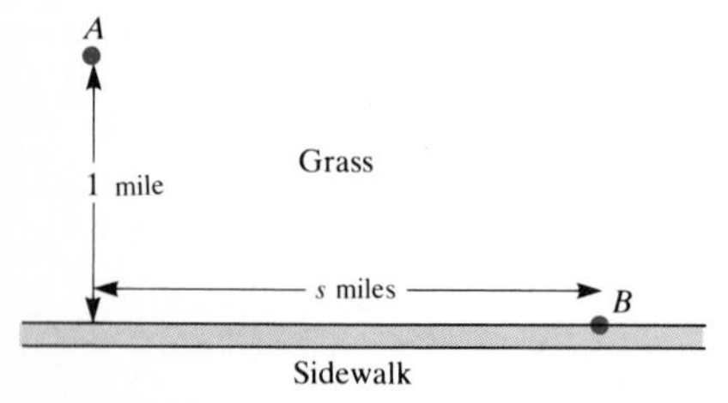

Figure 4.66

38 The potential energy in a diatomic molecule is given by the formula

$$U(r) = U_0\left[\left(\frac{r_0}{r}\right)^{12} - 2\left(\frac{r_0}{r}\right)^{6}\right],$$

where U_0 and r_0 are constants and r is the distance between the atoms. For which value of r is $U(r)$ a minimum?

39 What are the dimensions of the right circular cylinder of largest volume that can be inscribed in a sphere of radius a?

40 The stiffness of a rectangular beam is proportional to the product of the width and the cube of the height of its cross section. What shape beam should be cut from a log in the form of a right circular cylinder of radius r in order to maximize its stiffness?

41 A rectangular box-shaped house is to have a square floor. Three times as much heat per square foot enters through the roof as through the walls. What shape should the house be if it is to enclose a given volume and minimize heat entry? (Assume no heat enters through the floor.)

42 (See Fig. 4.67.) Find the coordinates of the points $P = (x, y)$, with $y \leq 1$, on the parabola $y = x^2$, that (*a*) minimize $\overline{PA}^2 + \overline{PB}^2$, (*b*) maximize $\overline{PA}^2 + \overline{PB}^2$.

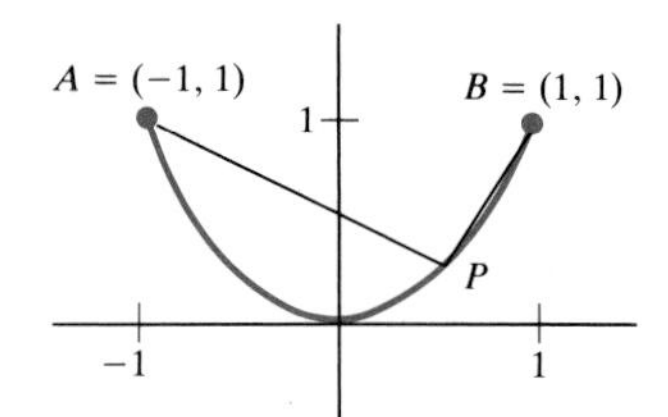

Figure 4.67

43 The speed of traffic through the Lincoln Tunnel in New York City depends on the density of the traffic. Let S be the speed in miles per hour and let D be the density in vehicles per mile. The relation between S and D was seen to be approximated closely by the formula

$$S = 42 - \frac{D}{3},$$

for $D \leq 100$.
(*a*) Express in terms of S and D the total number of vehicles that enter the tunnel in an hour.
(*b*) What value of D will maximize the flow in (*a*)?

44 When a tract of timber is to be logged, a main logging road is built from which small roads branch off as feeders. The question of how many feeders to build arises in practice. If too many are built, the cost of construction would be prohibitive. If too few are built, the time spent moving the logs to the roads would be prohibitive. The formula for total cost,

$$y = \frac{CS}{4} + \frac{R}{VS},$$

is used in a logger's manual to find how many feeder roads are to be built. R, C, and V are known constants: R is the cost of road at "unit spacing"; C is the cost of moving a log a unit distance; V is the value of timber per acre. S denotes the distance between the regularly spaced feeder roads. (See Fig. 4.68.) Thus the cost y is a function of S, and the object is to find that value of S that minimizes y. The manual says, "To find the desired S set the two summands equal to each other and solve:

$$\frac{CS}{4} = \frac{R}{VS}."$$

Show that the method is valid.

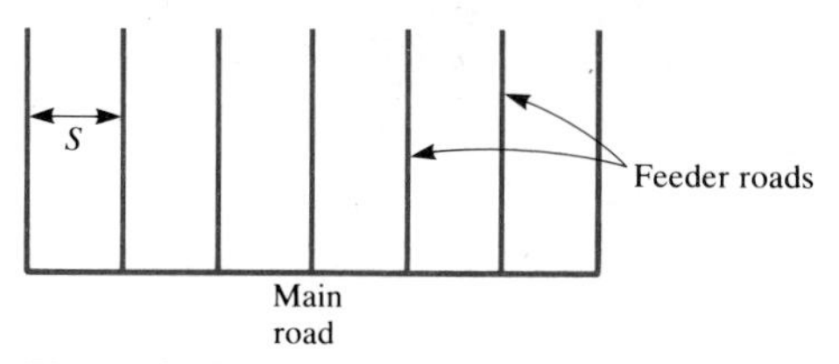

Figure 4.68

45 A delivery service is deciding how many warehouses to set up in a large city. The warehouses will serve similarly shaped regions of equal area A and, let us assume, an equal number of people.

(*a*) Why would transportation costs per item presumably be proportional to $\sqrt{A}$?

(*b*) Assuming that the warehouse cost per item is inversely proportional to A, show that C, the cost of transportation and storage per item, is of the form $t\sqrt{A} + w/A$, where t and w are appropriate constants.

(*c*) Show that C is a minimum when $A = (2w/t)^{2/3}$.

46 A pipe of length b is carried down a long corridor of width $a < b$ and then around corner C. (See Fig. 4.69.) During the turn y starts out at 0, reaches a maximum, and then returns to 0. (Try this with a short stick.) Find that maximum in terms of a and b. *Suggestion:* Express y in terms of a, b, and θ; θ is a variable, while a and b are constants.

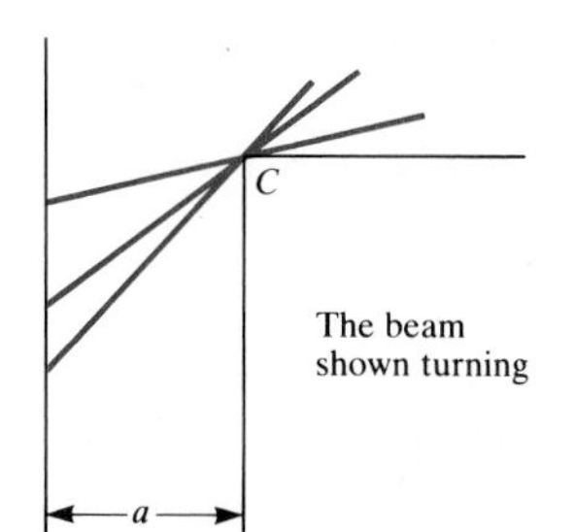

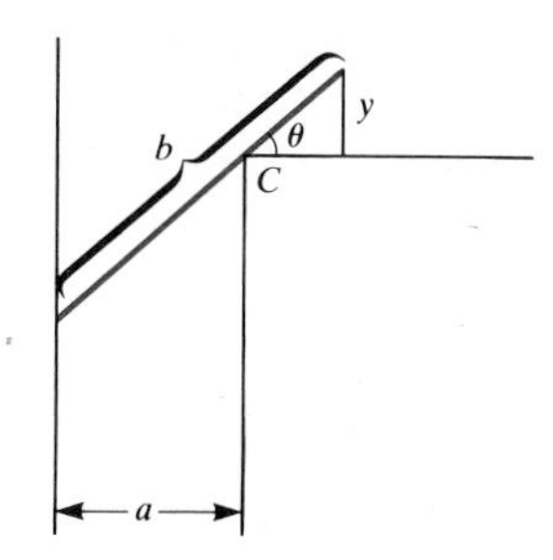

Figure 4.69

47 Figure 4.70 shows two corridors meeting at a right angle. One has width 8; the other, width 27. Find the length of the longest pipe that can be carried horizontally from one hall, around the corner and into the other hall. *Suggestion:* Do Exercise 46 first.

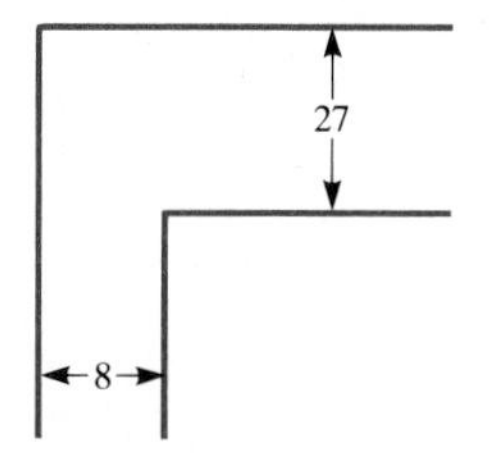

Figure 4.70

48 Two houses, A and B, are a distance p apart. They are distances q and r, respectively, from a straight road, and on the same side of the road. Find the length of the shortest path that goes from A to the road, and then on to the other house B.

(*a*) Use calculus.

(*b*) Use only elementary geometry. *Hint:* Introduce an imaginary house C such that the midpoint of B and C is on the road and the segment BC is perpendicular to the road; that is, "reflect" B across the road to become C.

49 The base of a painting on a wall is a feet above the eye of an observer, as shown in Fig. 4.71. The vertical side of the painting is b feet long. How far from the wall should the observer stand to maximize the angle that the painting subtends? *Hint:* It is more convenient to maximize $\tan\theta$ than θ itself. Recall that

$$\tan(A - B) = \frac{\tan A - \tan B}{1 + \tan A \tan B}.$$

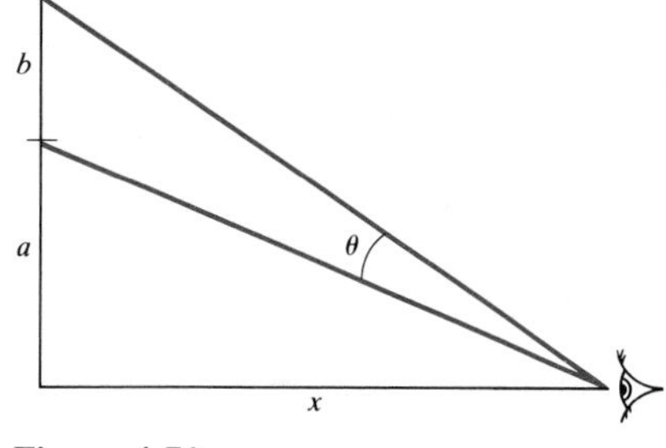

Figure 4.71

50 Find the point P on the x axis such that the angle APB in Fig. 4.72 is maximal. *Suggestion:* Note hint in Exercise 49.

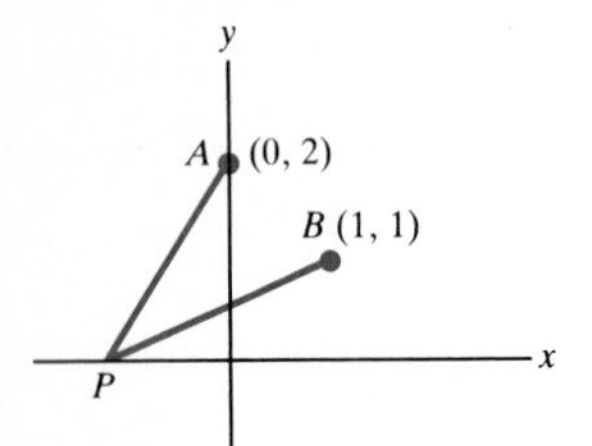

Figure 4.72

51 (*Economics*) Let p denote the price of some commodity and y the number sold at that price. To be concrete, assume that $y = 250 - p$ for $0 \le p \le 250$. Assume that it costs the producer $100 + 10y$ dollars to manufacture y units. What price p should the producer choose in order to maximize total profit, that is, "revenue minus cost"?

52 (*Leibniz on light*) A ray of light travels from point A to point B in Fig. 4.73 in minimal time. The point A is in one medium, such as air or a vacuum. The point B is in another medium, such as water or glass. In the first medium light travels at velocity v_1 and in the second at velocity v_2. The media are separated by line L. Show that for the path APB of minimal time,

$$\frac{\sin \alpha}{v_1} = \frac{\sin \beta}{v_2}.$$

Leibniz solved this problem with calculus in a paper published in 1684. (The result is called **Snell's law of refraction**)

Leibniz then wrote, "other very learned men have sought in many devious ways what someone versed in this calculus can accomplish in these lines as by magic." (See C. H. Edwards Jr., *The Historical Development of the Calculus*, p. 259, Springer-Verlag, New York, 1979.)

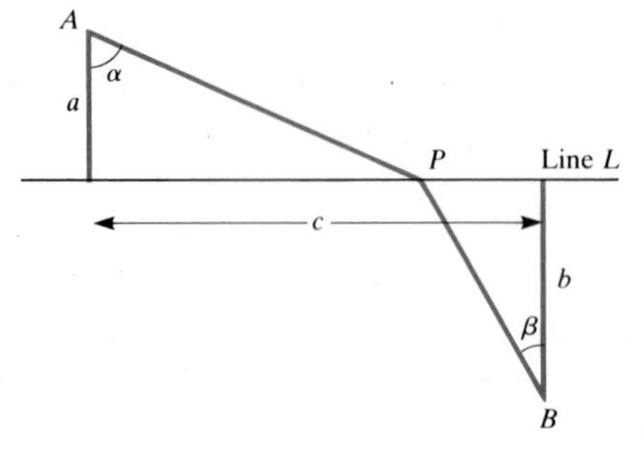

Figure 4.73

53 The following calculation occurs in an article by Manfred Kochen, "On Determining Optimum Size of New Cities": The net utility to the total client-centered system is

$$U = \frac{RLv}{A} n^{1/2} - nK - \frac{ALc}{v} n^{-1/2}.$$

All symbols except U and n are constant; n is a measure of decentralization. Regarding U as a differentiable function of n, we can determine when $dU/dn = 0$. This occurs when

$$\frac{RLv}{2A} n^{-1/2} - K + \frac{ALc}{2v} n^{-3/2} = 0.$$

This is a cubic equation for $n^{-1/2}$.
(*a*) Check that the differentiation is correct.
(*b*) Of what cubic polynomial is $n^{-1/2}$ a root?

4.7 IMPLICIT DIFFERENTIATION

Sometimes a function $y = f(x)$ is given indirectly by an equation that relates x and y. For instance, consider the equation

$$x^2 + y^2 = 25. \tag{1}$$

This equation can be solved for y: $y^2 = 25 - x^2$; so either

$$y = \sqrt{25 - x^2} \quad \text{or} \quad y = -\sqrt{25 - x^2}.$$

There are thus two continuous functions that satisfy Eq. (1).

Equation (1) is said to describe the function $y = f(x)$ **implicitly**. The equations

$$y = \sqrt{25 - x^2} \quad \text{and} \quad y = -\sqrt{25 - x^2}$$

describe the function $y = f(x)$ **explicitly**.

It is possible to differentiate a function given implicitly without having to solve for the function and express it explicitly. An example will illustrate the method, which is simply to differentiate both sides of the equation that defines the function implicitly. This procedure is called **implicit differentiation**.

EXAMPLE 1 Let $y = f(x)$ be the continuous function that satisfies the equation

$$x^2 + y^2 = 25$$

such that $y = 4$ when $x = 3$. Find dy/dx when $x = 3$ and $y = 4$.

SOLUTION Differentiating both sides of the equation

$$x^2 + y^2 = 25$$

with respect to x yields

Note that the chain rule is used to find $\frac{d(y^2)}{dx}$. See Example 7 in Sec. 3.6.

$$\frac{d}{dx}(x^2 + y^2) = \frac{d}{dx}(25),$$

$$2x + 2y\frac{dy}{dx} = 0.$$

Hence

$$x + y\frac{dy}{dx} = 0.$$

In particular, when $x = 3$ and $y = 4$,

$$3 + 4\frac{dy}{dx} = 0,$$

and therefore,

$$\frac{dy}{dx} = -\frac{3}{4}.$$

The problem could also be solved by differentiating $\sqrt{25 - x^2}$. But the algebra involved is more complicated, since it is necessary to differentiate a square root. ■

In the next example implicit differentiation is the only way to find the derivative, for in this case there is no formula expressible in terms of trigonometric and algebraic functions giving y explicitly in terms of x.

EXAMPLE 2 Assume that the equation

$$2xy + \pi \sin y = 2\pi$$

defines a function $y = f(x)$. Find dy/dx when $x = 1$ and $y = \pi/2$. (Note that $x = 1$ and $y = \pi/2$ satisfy the equation.)

SOLUTION Implicit differentiation yields

$$\frac{d}{dx}(2xy + \pi \sin y) = \frac{d(2\pi)}{dx},$$

$$2\left(x\frac{dy}{dx} + y\frac{dx}{dx}\right) + \pi(\cos y)\frac{dy}{dx} = 0,$$

or
$$2x\frac{dy}{dx} + 2y + \pi(\cos y)\frac{dy}{dx} = 0.$$

For $x = 1$ and $y = \pi/2$, this last equation becomes

$$2 \cdot 1\frac{dy}{dx} + 2\frac{\pi}{2} + \pi\left(\cos\frac{\pi}{2}\right)\frac{dy}{dx} = 0$$

or
$$2\frac{dy}{dx} + 2\frac{\pi}{2} = 0.$$

Hence
$$\frac{dy}{dx} = -\frac{\pi}{2}. \quad \blacksquare$$

Implicit Differentiation and Extrema

Example 3 of Sec. 4.6 answered the question, "Of all the tin cans that enclose a volume of 100 cubic inches, which requires the least metal?" The radius of the most economical can is $\sqrt[3]{50/\pi}$. From this and the fact that its volume is 100 cubic inches, its height could be calculated. In the next example implicit differentiation is used to answer the same question. Not only will the algebra be simpler than before, but the answer will provide more information, since also the general shape—the proportion between height and radius—is revealed. Before reading the next example, it would be instructive to read over the solution in Sec. 4.6.

EXAMPLE 3 Of all the tin cans that enclose a volume of 100 cubic inches, which requires the least metal?

SOLUTION The height h and radius r of any can of volume 100 cubic inches are related by the equation

$$\pi r^2 h = 100. \tag{2}$$

The surface area S of the can is

$$S = 2\pi r^2 + 2\pi rh. \tag{3}$$

Consider h, and hence S, as functions of r. However, *it is not necessary to find these functions explicitly.*

Differentiation of Eqs. (2) and (3) with respect to r yields

All we use about 100 is that it is a constant.

$$\pi\left(r^2\frac{dh}{dr} + 2rh\right) = \frac{d(100)}{dr} = 0 \tag{4}$$

and
$$\frac{dS}{dr} = 4\pi r + 2\pi\left(r\frac{dh}{dr} + h\right). \tag{5}$$

Since when S is a minimum, $dS/dr = 0$, we have

$$0 = 4\pi r + 2\pi\left(r\frac{dh}{dr} + h\right). \tag{6}$$

Equations (4) and (6) yield, with a little algebra, a relation between h and r, as follows:

Factoring πr out of Eq. (4) and 2π out of Eq. (6) shows that

$$r\frac{dh}{dr} + 2h = 0 \quad \text{and} \quad 2r + r\frac{dh}{dr} + h = 0. \tag{7}$$

Elimination of dh/dr from Eqs. (7) yields

$$2r + r\left(\frac{-2h}{r}\right) + h = 0,$$

which simplifies to

$$2r = h. \qquad (8)$$

Equation (8) asserts that the height of the most economical can is the same as its diameter. Moreover, this is the ideal shape, no matter what the prescribed volume happens to be. [Equation (4) follows from Eq. (2) merely because 100 is constant.]

The specific dimensions of the most economical can are found by combining the equations

$$2r = h \qquad (8)$$

and

$$\pi r^2 h = 100. \qquad (2)$$

Elimination of h from these two equations shows that

$$\pi r^2(2r) = 100 \quad \text{or} \quad r^3 = \frac{50}{\pi}.$$

Hence

$$r = \sqrt[3]{\frac{50}{\pi}} \quad \text{and} \quad h = 2r = 2\sqrt[3]{\frac{50}{\pi}}. \quad ■$$

As in the case of Example 3, implicit differentiation finds the proportions of a general solution before finding the exact values of the variables. Often it is the proportion, rather than the (perhaps messier) explicit values, that gives more insight into the answer. For instance, Eq. (8) tells that the diameter equals the height for the most economical can.

The procedure illustrated in Example 3 is quite general. It may be of use when maximizing (or minimizing) a quantity that at first is expressed as a function of two variables which are linked by an equation. The equation that links them is called the **constraint**. In Example 3, the constraint is $\pi r^2 h = 100$.

The constraint

General procedure for using implicit differentiation in an applied maximum problem

HOW TO USE IMPLICIT DIFFERENTIATION IN AN EXTREMUM PROBLEM

1. Name the various quantities in the problem by letters, such as x, y, A, V.
2. Identify which quantity is to be maximized (or minimized).
3. Express the quantity to be maximized (or minimized) in terms of other quantities, such as x and y.
4. Obtain an equation relating x and y. (This equation is called a constraint.)
5. Differentiate implicitly both the constraint and the expression to be maximized (or minimized), interpreting all the various quantities to be functions of x (or, perhaps, of y).
6. Set the derivative of the expression to be maximized (or minimized) equal to 0 and combine with the derivative of the constraint to obtain an equation relating x and y at a maximum (or minimum).
7. Step 6 gives only a relation or proportion between x and y at an extremum. If the explicit values of x and y are desired, find them by using the fact that x and y also satisfy the constraint.

Warning: Sometimes an extremum occurs where a derivative, such as dy/dx, is not defined. (Exercise 29 illustrates this possibility.)

EXERCISES FOR SEC. 4.7: IMPLICIT DIFFERENTIATION

In Exercises 1 to 4 find dy/dx at the indicated values of x and y in two ways: explicitly (solving for y first) and implicitly.

1 $xy = 4$ at $(1, 4)$ 2 $x^2 - y^2 = 3$ at $(2, 1)$
3 $x^2y + xy^2 = 12$ at $(3, 1)$ 4 $x^2 + y^2 = 100$ at $(6, -8)$

In Exercises 5 to 8 find dy/dx at the given points by implicit differentiation.

5 $\frac{2xy}{\pi} + \sin y = 2$ at $(1, \pi/2)$

6 $2y^3 + 4xy + x^2 = 7$ at $(1, 1)$
7 $x^5 + y^3x + yx^2 + y^5 = 4$ at $(1, 1)$
8 $x + \tan xy = 2$ at $(1, \pi/4)$
9 Solve Example 3 by implicit differentiation, but differentiate Eqs. (2) and (3) with respect to h instead of r.
10 What is the shape of the cylindrical can of largest volume that can be constructed with a given surface area? Do not find the radius and height of the largest can; find the ratio between them. *Suggestion:* Call the surface area S and keep in mind that it is constant.

In Exercises 11 to 16 solve by implicit differentiation:

11 Example 2 of Sec. 4.6. 12 Example 4 of Sec. 4.6.
13 Exercise 13 of Sec. 4.6. 14 Exercise 14 of Sec. 4.6.
15 Exercise 23 of Sec. 4.6. 16 Exercise 24 of Sec. 4.6.

In Exercises 17 to 20 find dy/dx at a general point (x, y) on the given curve.

17 $xy^3 + \tan(x + y) = 1$
18 $\sec(x + 2y) + \cos(x - 2y) + y = 2$
19 $-7x^2 + 48xy + 7y^2 = 25$
20 $\sin^3(xy) + \cos(x + y) + x = 1$

■

Exercise 21 shows how to find y'' if y is given implicitly.

21 Assume that $y(x)$ is a differentiable function of x and that $x^3y + y^4 = 2$. Assume that $y(1) = 1$. Find $y''(1)$, following these steps.
(*a*) Show that $x^3y' + 3x^2y + 4y^3y' = 0$.
(*b*) Use (*a*) to find $y'(1)$.
(*c*) Differentiate the equation in (*a*) to show that

$$x^3y'' + 6x^2y' + 6xy + 4y^3y'' + 12y^2(y')^2 = 0.$$

(*d*) Use the equation in (*c*) to find $y''(1)$. [*Hint:* $y(1)$ and $y'(1)$ are known.]

22 Find $y''(1)$ if $y(1) = 2$ and $x^5 + xy + y^5 = 35$.
23 Find $y'(1)$ and $y''(1)$ if $y(1) = 0$ and $\sin y = x - x^3$.
24 Find $y''(2)$ if $y(2) = 1$ and $x^3 + x^2y - xy^3 = 10$.
25 Use implicit differentiation to find the highest and lowest points on the ellipse $x^2 + xy + y^2 = 12$.
26 Does the tangent line to the curve $x^3 + xy^2 + x^3y^5 = 3$ at the point $(1, 1)$ pass through the point $(-2, 3)$?

Exercises 27 and 28 obtain by implicit differentiation the formulas for differentiating $x^{1/n}$ and $x^{m/n}$ with the assumption that they are differentiable functions.

27 Let n be a positive integer. Assume that $y = x^{1/n}$ is a differentiable function of x. From the equation $y^n = x$ deduce by implicit differentiation that $y' = (1/n)x^{1/n-1}$.
28 Let m be a nonzero integer and n a positive integer. Assume that $y = x^{m/n}$ is a differentiable function of x. From the equation $y^n = x^m$ deduce by implicit differentiation that $y' = (m/n)x^{m/n-1}$.
29 (*a*) What difficulty arises when you use implicit differentiation to maximize $x^2 + y^2$ subject to $x^2 + 4y^2 = 16$.
(*b*) Show that a maximum occurs where dy/dx is not defined. What is the maximum of $x^2 + y^2$ subject to $x^2 + 4y^2 = 16$?
(*c*) The problem can be viewed geometrically as "Maximize $x^2 + y^2$ for points on the ellipse $x^2 + 4y^2 = 16$." Sketch the ellipse and interpret (*b*) in terms of it.

4.8 THE DIFFERENTIAL

The applied sciences are greatly concerned with the errors that may occur in measurements taken in the field or the laboratory. For this reason, some freshman laboratory manuals have several pages on estimating errors. For instance, if you measure an angle θ with an "uncertainty" or error of perhaps as much as 5°, what error may there be in $\sin\theta$? This type of question can be answered with the aid of derivatives.

Let $y = f(x)$ be a differentiable function. Then by the definition of a derivative, $\Delta y/\Delta x$ is a good approximation of $f'(x)$ when Δx is small. But we may look at this same fact in a different way. When Δx is small, the derivative $f'(x)$ is a good estimate of $\Delta y/\Delta x$, an assertion written

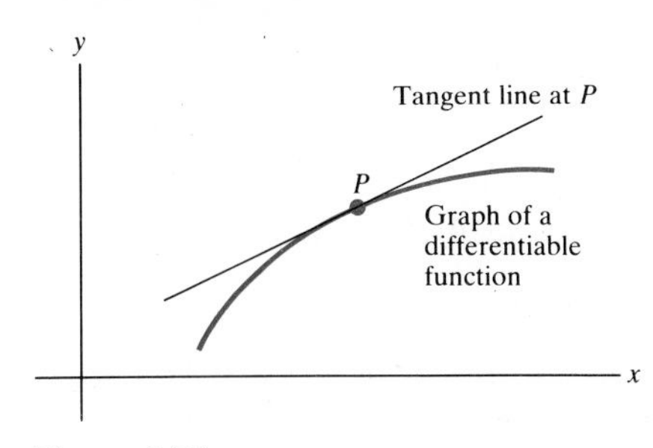

Figure 4.74

$$\frac{\Delta y}{\Delta x} \approx f'(x) \qquad (\Delta x \text{ "small"}).$$

Multiplying both sides by Δx gives us

$$\Delta y \approx f'(x)\,\Delta x \qquad (\Delta x \text{ "small"}). \tag{1}$$

Equation (1) asserts that $f'(x)\Delta x$ *is a good estimate of* Δy *when* Δx *is small.*

Let's check this claim with the function $f(x) = \sqrt{x}$ at $x = 25$. Since $f'(x) = 1/(2\sqrt{x})$, we have $f'(25) = \frac{1}{10}$. Now compute the change

$$\Delta y = f(25 + \Delta x) - f(25) = \sqrt{25 + \Delta x} - \sqrt{25} = \sqrt{25 + \Delta x} - 5.$$

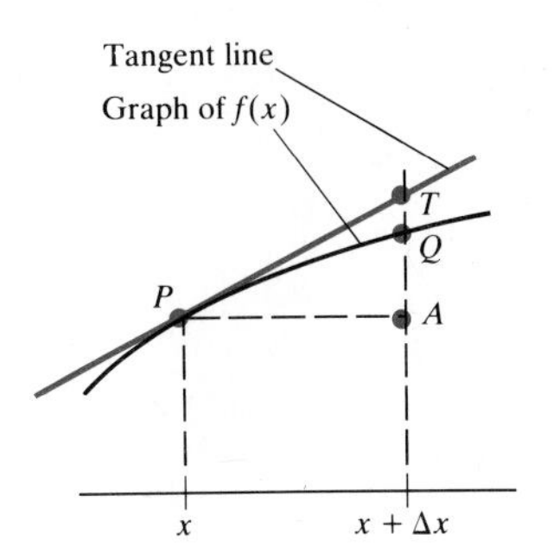

Figure 4.75

Then Eq. (1) becomes

$$\sqrt{25 + \Delta x} - 5 \approx \tfrac{1}{10}\,\Delta x. \tag{2}$$

When $\Delta x = 0.1$, the right side of Eq. (2) is 0.01 and the left side is $\sqrt{25.1} - 5$, which to six decimal places is $5.009990 - 5 = 0.009990$. Clearly, the difference between $f'(x)\,\Delta x$ and Δy in this case is quite small.

The expression $f'(x)\,\Delta x$ is of both practical and theoretical interest. Because of its importance, it is given a name.

Definition Let $y = f(x)$ be a differentiable function. Then $f'(x)\,\Delta x$ is called the **differential** of f and is denoted df (or dy):

$$df = f'(x)\,\Delta x \qquad (dy = f'(x)\,\Delta x).$$

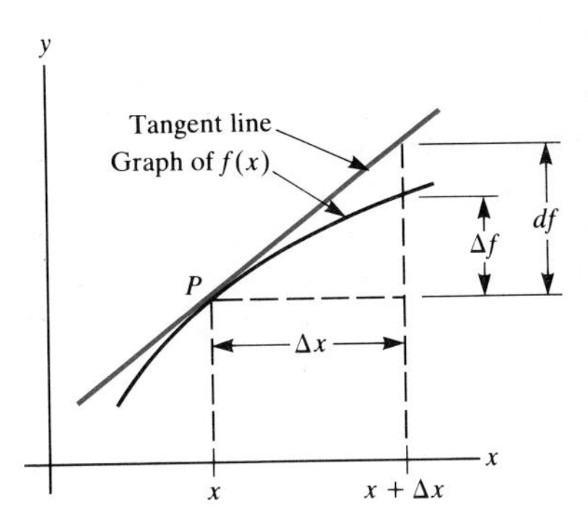

Figure 4.76

For a given function f, the value of df depends on both x and Δx. So df is actually a function of two variables. As we saw, df is a good approximation of Δf when Δx is small.

Although we introduced the differential as an estimate of change or "error," it can also be viewed geometrically. A very short piece of the graph around a point P, of a differentiable function, looks straight and closely resembles a short segment of the tangent line to the graph at P. (See Fig. 4.74.)

In Fig. 4.75 we saw a blowup of a very small portion of Fig. 4.74 near P. In Fig. 4.75, $\overline{AQ} = \Delta y$, the change in the function f as the input changes from x to $x + \Delta x$. The length of segment AT represents the vertical change along the tangent line. Since $\overline{AT}/\Delta x$ is the slope of the tangent line,

$$\frac{\overline{AT}}{\Delta x} = f'(x).$$

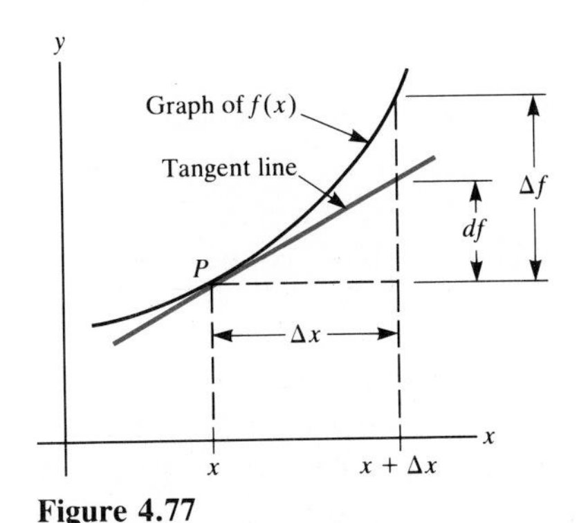

Figure 4.77

Hence
$$\overline{AT} = f'(x)\,\Delta x.$$

Thus *the differential* $f'(x)\,\Delta x$ *represents vertical change along the tangent line*. (This is often taken as the definition of the differential.)

Figure 4.76 shows Δx and $df = dy$. If the tangent lies below the graph, then the diagram appears as shown in Fig. 4.77 (and df underestimates Δf).

The next example shows how the differential is used in estimating possible induced errors in laboratory experiments.

EXAMPLE 1 If there is a 5° uncertainty in measuring an angle θ, how much uncertainty is there in $\cos \theta$?

SOLUTION First, translate the error in measuring the angle into radians, by the proportion

$$\frac{\text{Number of radians}}{\pi} = \frac{\text{Number of degrees}}{180}$$

An error of 5° corresponds to an error of $5\pi/180$ radians. Let $f(\theta) = \cos \theta$. Then

$$df = -\sin \theta \, \Delta\theta. \tag{3}$$

Since the error in measuring θ may be as large as $5\pi/180$ radians, we have $|\Delta\theta| = 5\pi/180$. By Eq. (3),

$$|df| = |\sin \theta| \frac{5\pi}{180}. \tag{4}$$

So the uncertainty in $\cos \theta$ is about $|\sin \theta|\,(5\pi/180)$. If $\theta = 60°$, this uncertainty is $(\sqrt{3}/2)\,(5\pi/180) \approx 0.076$. If $\theta = 90°$, this uncertainty is $1(5\pi/180) \approx 0.087$. ■

The differential can be used to estimate the value of a function at the input $x + \Delta x$ in terms of information at x. If we know the value of the function at x, $f(x)$, and the derivative, $f'(x)$, we have

$$f(x + \Delta x) = f(x) + \Delta f \approx f(x) + df.$$

So we have the **approximation formula**,

$$f(x + \Delta x) \approx f(x) + f'(x)\,\Delta x. \tag{5}$$

EXAMPLE 2 Use the approximation formula to estimate $\sqrt[3]{29}$.

SOLUTION Because we wish to estimate $\sqrt[3]{29}$, we introduce the cube root function $f(x) = \sqrt[3]{x}$. We know the exact value of $\sqrt[3]{x}$ when $x = 27$, which is near 29, so we use Eq. (5) in the form

$$\sqrt[3]{27 + 2} \approx \sqrt[3]{27} + f'(27)\,(29 - 27). \tag{6}$$

Since $f(x) = x^{1/3}$, $f'(x) = \frac{1}{3}x^{-2/3}$. Thus Eq. (6) becomes

$$\sqrt[3]{29} \approx 3 + \frac{1}{3(27)^{2/3}}\,(2)$$

or

$$\sqrt[3]{29} \approx 3 + \frac{2}{27} \approx 3 + 0.0741 = 3.0741.$$

As may be checked on a calculator, $\sqrt[3]{29} \approx 3.0723$, rounded to four decimals. ■

The method used in Example 2 amounts to the following general procedure:

How to use a differential to estimate an output of a function

TO ESTIMATE $f(b)$

1. Find a number a near b at which $f(a)$ and $f'(a)$ are easy to calculate.
2. Find $\Delta x = b - a$. (Δx may be positive or negative.)
3. Compute $f(a) + f'(a)\,\Delta x$. This is an estimate of $f(b)$. In short $f(b) \approx f(a) + (b - a)f'(a)$.

EXAMPLE 3 Use a differential to estimate $\sqrt{67}$.

SOLUTION The object is to estimate the value of the square root function $f(x) = \sqrt{x}$ at the input $x = 67$. (Here $b = 67$.)

In this case, $f(64)$ is known. (So use $a = 64$.) We have

$$f(64) = \sqrt{64} = 8 \qquad \text{and} \qquad f'(64) = \frac{1}{2\sqrt{64}} = \frac{1}{16}.$$

Since $67 = 64 + 3$, Δx is 3. Therefore,

$$\begin{aligned}\sqrt{67} = f(64 + 3) &\approx f(64) + df \\ &= f(64) + f'(64)(3) \\ &= 8 + \tfrac{1}{16}(3) = 8.1875.\end{aligned}$$

Thus

$$\sqrt{67} \approx 8.1875.$$

(A calculator shows that to four decimal places, $\sqrt{67} \approx 8.1854$. So the estimate obtained by the differential is not far off.) ■

If the derivative of a function is known, so is its differential. For example,

$$d(\tan x) = \sec^2 x\,\Delta x,$$

$$d(x^5) = 5x^4\,\Delta x,$$

and

$$d(x) = 1\,\Delta x = \Delta x.$$

$dx = \Delta x$.

Notice that $d(x) = \Delta x$. For this reason it is customary to write Δx also as dx. The differential of f, then, is also written as

$$df = f'(x)\,dx \qquad \text{or} \qquad dy = f'(x)\,dx.$$

Thus we can also write that

$$d(\tan x) = \sec^2 x\,dx \qquad \text{and} \qquad d(x^5) = 5x^4\,dx.$$

The origin of the symbol $\frac{dy}{dx}$

The symbols dy and dx now have meaning individually. It is meaningful to divide both sides of the equation

$$dy = f'(x)\,dx$$

by dx, obtaining

$$dy \div dx = f'(x).$$

This is the origin of the symbol dy/dx for the derivative. It goes back to Leibniz at the end of the seventeenth century when dx denoted a number "vanishingly small," blasted by Bishop Berkeley in 1734 as "a ghost of a departed quantity." Now, however, dx denotes a number and dy is defined as $f'(x)\,dx$.

Incidentally, the logician Abraham Robinson in 1960 vindicated these "ghosts" by enlarging the real number system to one that includes numbers both "infinitely small" and "infinitely large."

EXAMPLE 4 The side of a cube is measured with an error of at most 1 percent. What percent error may this cause in calculating the volume of the cube?

SOLUTION Let x be the length of a side of the cube and V its volume. Let dx denote the possible error in measuring x. The relative error

$$\frac{dx}{x}$$

is at most 0.01 in absolute value. That is, $|dx|/x \leq 0.01$.

Estimating relative error

The differential dV is an estimate of the actual error in calculating the volume. Thus

$$\frac{dV}{V}$$

is an estimate of the relative error in the volume.

Since $$dV = d(x^3) = 3x^2\,dx,$$

it follows that $$\frac{dV}{V} = \frac{3x^2\,dx}{x^3} = 3\,\frac{dx}{x}.$$

Therefore, the relative error in the volume is about three times the relative error in measuring the side, hence at most about 3 percent. (In fact, it can be as much as 3.0301 percent.) ■

We have used the differential to estimate a change (or "error" or "uncertainty") Δf and also the value $f(x + \Delta x)$ if $f(x)$ is known, namely, $f(x + \Delta x) \approx f(x) + df$. But such estimates are of little value unless we have some idea of how accurate they generally are.

Let us look at the size of the discrepancy between Δf and df in a particular example, $f(x) = x^3$ for x near 2. In this case, $f'(x) = 3x^2$, so $df = 3x^2\,\Delta x$. When $x = 2$, $df = 12\,\Delta x$. The following table compares df and $\Delta f = (2 + \Delta x)^3 - 2^3$ for $\Delta x = 1, 0.1, 0.01, 0.001$, and -0.001.

Δx	$df = 12\,\Delta x$	$\Delta f = (2 + \Delta x)^3 - 8$	$\lvert\Delta f - df\rvert$
1	12	19	7
0.1	1.2	1.261	0.061
0.01	0.12	0.120601	0.000601
0.001	0.012	0.012006001	0.000006001
−0.001	−0.012	−0.011994001	0.000005999

Each time we cut our "error" Δx by a factor of 10, the difference between df and Δf is cut by a factor of about 100, the square of 10. This illustrates a general principle, which is developed in "numerical analysis": the discrepancy between df and Δf is usually proportional to $(\Delta x)^2$. (See Exercise 42.)

However, we can show that the differential df is a good estimate of Δf in the sense that their ratio is near 1 when Δx is small. This is proved in the following theorem.

Theorem Let f be differentiable at a number x and assume that $f'(x) \neq 0$. Then

$$\lim_{\Delta x \to 0} \frac{\Delta f}{df} = 1.$$

Proof Consider the quotient

$$\frac{\Delta f}{df} = \frac{\Delta f}{f'(x)\,\Delta x} = \frac{\Delta f}{\Delta x}\frac{1}{f'(x)}.$$

[Since $f'(x) \neq 0$, division is permissible.] Thus

$$\lim_{\Delta x \to 0} \frac{\Delta f}{df} = \lim_{\Delta x \to 0} \frac{\Delta f}{\Delta x}\frac{1}{f'(x)} = f'(x)\frac{1}{f'(x)} = 1.$$

This concludes the proof. ■

EXERCISES FOR SEC. 4.8: THE DIFFERENTIAL

In Exercises 1 to 6 compute dy and Δy for the given functions and values of x and dx and represent them on graphs of the functions.

1 x^2 at $x = 1$ and $dx = 0.3$
2 x^3 at $x = \frac{1}{2}$ and $dx = 0.1$
3 $\sqrt{x}$ at $x = 9$ and $dx = -2$
4 $\sqrt[3]{x}$ at $x = 27$ and $dx = -4$
5 $\tan x$ at $x = \pi/6$ and $dx = \pi/12$
6 $\sin x$ at $x = \pi/3$ and $dx = -\pi/12$

In Exercises 7 to 10 use the approximation formula to establish the given estimates for small values of h.

7 $\dfrac{1}{1+h} \approx 1 - h$
8 $(1+h)^{10} \approx 1 + 10h$
9 $\sqrt{1+h} \approx 1 + \dfrac{h}{2}$
10 $\sqrt[3]{1+h} \approx 1 + \dfrac{h}{3}$

In Exercises 11 to 18 calculate the differentials, expressing them in terms of x and dx.

11 $d(1/x^3)$
12 $d(\sqrt{1+x^2})$
13 $d(\sin 2x)$
14 $d(\cos 5x)$
15 $d(\csc x)$
16 $d(\tan x^3)$
17 $d\left(\dfrac{\cot 5x}{x}\right)$
18 $d(x^3 \sec^2 5x)$

In Exercises 19 to 34 use differentials to estimate the given quantities.

19 $\sqrt{98}$
20 $\sqrt{103}$
21 $\sqrt[3]{25}$
22 $\sqrt[3]{28}$
23 $\tan\left(\dfrac{\pi}{4} - 0.01\right)$
24 $\tan\left(\dfrac{\pi}{4} + 0.01\right)$
25 $\sin\left(\dfrac{\pi}{3} - 0.02\right)$
26 $\sin\left(\dfrac{\pi}{3} + 0.02\right)$
27 $\sin 0.13$
28 $\sin 0.3$
29 $1/4.03$
30 $\sqrt{15.7}$
31 $\sin 32°$ (*Warning:* First translate into radians.)
32 $\sin 2°$
33 $\cos 28°$
34 $\tan 3°$
35 The period T of a pendulum is proportional to the square root of its length l. That is, there is a constant k such that $T = k\sqrt{l}$. If the length is measured with an error of at most p percent, use differentials to estimate the possible error in calculating the period.
36 The side of a square is measured with an error of at most 5 percent. Estimate the largest percent error this may induce in the measurement of the area.

■

37 Let $f(x) = x^2$, the area of a square of side x, shown in Fig. 4.78.
(*a*) Compute df and Δf in terms of x and Δx.
(*b*) In the square in the diagram, shade the part whose area is Δf.
(*c*) Shade the part of the square in (*b*) whose area is df.

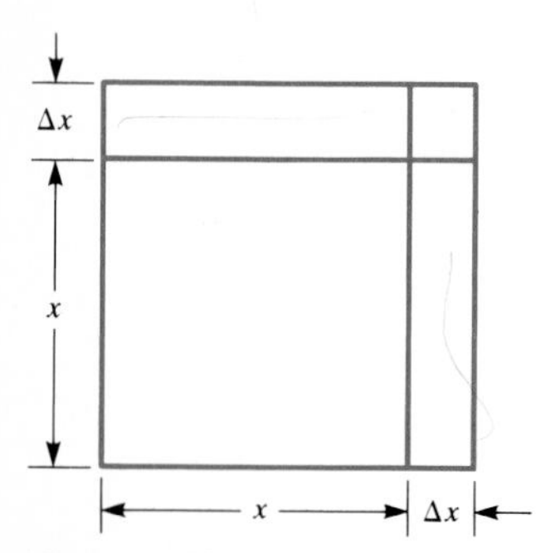

Figure 4.78

38 (See Exercise 37.) Let $f(x) = x^3$, the volume of a cube of side x.
(*a*) Compute df and Δf in terms of x and Δx.
(*b*) Draw a diagram analogous to Fig. 4.78 showing cubes of sides x and $x + \Delta x$.
(*c*) Indicate in the diagram in (*b*) the part whose volume is df.

39 Prove that if f and g are two differentiable functions, then
(*a*) $d(f - g) = df - dg$
(*b*) $d(fg) = f\,dg + g\,df$
(*c*) $d\left(\dfrac{f}{g}\right) = \dfrac{g\,df - f\,dg}{g^2}$.

40 Let y be a differentiable function of u and u a differentiable function of x. Then $dy = D_u(y)\,du$ and $du = D_x(u)\,dx$. But y is a composite function of x, and one writes $dy = D_x(y)\,dx$. Show that the two values of dy are equal.

41 (*a*) Estimate $\sin\theta$, $\cos\theta$, and $\tan\theta$ for small θ using differentials.
(*b*) Explain why $\cos\theta \approx 1 - (\theta^2/2)$ is a better estimate for $\cos\theta$ than the one in (*a*). [*Hint:* Recall the identity $(1 - \cos\theta)(1 + \cos\theta) = \sin^2\theta$.]

42 [This exercise shows why df generally differs from Δf by an amount proportional to $(dx)^2$.] Let f be differentiable and have a continuous second derivative. Then $f(b) = f(a) + f'(a)(b - a) + R$. (Here $b - a$ plays the role of dx.) We shall express the "discrepancy" R in terms of f''. (Note that $R = \Delta f - df$.) We accomplish this by introducing a new function $g(x) = f(a) + f'(a)(x - a) + k(x - a)^2$, where the constant k is chosen so that $g(b) = f(b)$.
(*a*) Show that $g(a) = f(a)$.
(*b*) Deduce that there is a number c_1, between a and b, such that $g'(c_1) = f'(c_1)$. (*Hint:* Consider $g - f$.)
(*c*) Show that $g'(a) = f'(a)$.
(*d*) Deduce that there is a number c_2, between a and c_1, such that $g''(c_2) = f''(c_2)$.
(*e*) From (*d*) conclude that $R = f''(c_2)(b - a)^2/2$. Since $dx = b - a$, this shows that the discrepancy is on the order of $(dx)^2$.

4.S SUMMARY

This chapter applied the derivative to graphs, motion, extrema, and estimates.

Section 4.1 provided the foundation for the chapter. The theorem of the interior extremum showed that there is a close tie between a maximum or minimum value and a derivative being equal to 0, although the two conditions are not equivalent. The mean-value theorem, which followed from Rolle's theorem, showed that a function whose derivative is positive is increasing and that two functions that have the same derivative differ only by a constant. The diagram on the next page shows how the theorems in Sec. 4.1 fit together.

Section 4.2 described a procedure for graphing functions which involves intercepts, asymptotes, and the first derivative.

The higher derivatives were introduced in Sec. 4.3 and were used to show how to obtain the binomial theorem for positive-integer exponents.

In Sec. 4.4, the notions of concave upward, concave downward, inflection point, and inflection number were introduced. (In advanced texts, "concave upward" is often called "convex" and "concave downward" is called "concave.")

Section 4.5 discussed the use of the second derivative to represent acceleration. Motion under constant acceleration was described completely. (The position turns out to be a second-degree polynomial in time t.) Harmonic motion was also discussed.

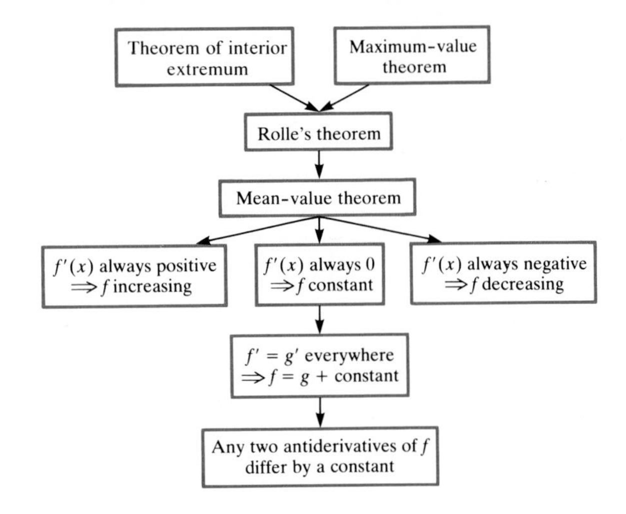

Section 4.6 applied the techniques for finding extrema to such problems as finding a maximum volume or a minimal cost.

Implicit differentiation and its use in certain extremum problems was the subject of Sec. 4.7.

Section 4.8 introduced the differential $dy = f'(x)\,dx$. It is a function of x and dx that records change along a tangent line.

Vocabulary and Symbols

theorem of the interior extremum
chord of f
Rolle's theorem
mean-value theorem
increasing (decreasing) function
critical number
critical point
relative (local) maximum or minimum
global maximum or minimum
first-derivative test for local extremum
second derivative d^2y/dx^2, y'', f'', $f^{(2)}$, $D^2(y)$
higher derivatives d^ny/dx^n, $f^{(n)}$, $D^n(f)$
acceleration
concave upward, concave downward
inflection number
inflection point
second-derivative test for local extremum
implicit function
implicit differentiation
constraint
differential dy, df

Key Facts

THEOREM OF THE INTERIOR EXTREMUM

("Maximum" case) Let f be defined at least on (a, b). Let c be a number in (a, b) such that $f(c) \geq f(x)$ for all x in (a, b). If f is differentiable at c, then $f'(c) = 0$.

The "minimum" case is similar.

ROLLE'S THEOREM

Let f be continuous on $[a, b]$ and differentiable on (a, b). If $f(a) = f(b)$, then there is at least one number c in (a, b) such that $f'(c) = 0$.

Informally, Rolle's theorem asserts that if a graph of a differentiable function has a horizontal chord, then it has a horizontal tangent line.

MEAN-VALUE THEOREM

Let f be continuous on $[a, b]$ and differentiable on (a, b). Then there is at least one number c in (a, b) such that

$$f'(c) = \frac{f(b) - f(a)}{b - a}.$$

Informally, the mean-value theorem asserts that for any chord on the graph of a differentiable function, there is a tangent line parallel to it.

The conclusion of the mean-value theorem may also be written as

$$f(b) = f(a) + f'(c)(b - a),$$

for some number c in (a, b).

INFORMATION PROVIDED BY f' AND f''

Where f' is positive, f is increasing.
Where f'' is positive, f is concave upward.
Where f' is negative, f is decreasing.
Where f'' is negative, f is concave downward.
Where $f' = 0$, f may have an extremum.
Where $f'' = 0$, f may have an inflection point.
First-derivative test for local maximum at c: $f'(c) = 0$ and f' changes from positive to negative at c.
Second-derivative test for local maximum at c: $f'(c) = 0$ and $f''(c)$ is negative.
First-derivative test for local minimum at c: $f'(c) = 0$ and f' changes from negative to positive at c.
Second-derivative test for local minimum at c: $f'(c) = 0$ and $f''(c)$ is positive.
[If $f'(c)$ or $f''(c)$ does not exist, it is best to study the behavior of $f(x)$ and $f'(x)$ for x near c.]

Two functions, defined over the same interval, with equal derivatives, differ by a constant. [From this it follows, for instance, that if $F'(x) = 2x$, then $F(x)$ must be of the form $x^2 + C$ for some constant C.] This fact will be of use in the next chapter where antiderivatives will be needed.

The second derivative records acceleration, the rate of change of velocity. If the acceleration is constant, the function is given by the formula

$$y = \frac{a}{2}t^2 + v_0t + y_0,$$

where a = acceleration, v_0 = initial velocity, and y_0 = initial position.

In harmonic motion, the acceleration is proportional to the displacement, $d^2x/dt^2 = -cx$, $c > 0$. Such motion is oscillating and given by the formula $x = A \cos(\sqrt{c}t + k)$ for some constants A and k.

The following table summarizes the procedure for graphing f in terms of questions about f, f', and f'':

How to graph using f, f', f''
What are the intercepts?
What are the critical numbers?
Where is the function increasing? decreasing?
Are there any local maxima or minima?
Where is the second derivative positive? negative? zero?
Where is the curve concave upward? concave downward?
Are there any inflection points?
Are there any vertical or horizontal asymptotes?
What happens when $\|x\| \to \infty$?

(See also the table in Sec. 4.4.)

To find the derivative of a function given implicitly, differentiate the defining equation, remembering that the chain rule may be needed. Differentiation of the resulting equation will then give an equation for the second derivative.

The quantity $df = f'(x)\,\Delta x$ [or $f'(x)dx$], the change along a tangent line, is an estimate of Δf, the change along the curve. When Δx is small, Δf and df will be small and their ratio will be near 1.

If you know $f(a)$ and $f'(a)$ and if b is near a, then you can estimate $f(b)$ by

$$f(b) \approx f(a) + f'(a)(b - a).$$

This amounts to the same thing as the approximations

$$f(x + \Delta x) \approx f(x) + f'(x)\,\Delta x \qquad \text{or} \qquad f(x + \Delta x) \approx f(x) + df.$$

GUIDE QUIZ ON CHAP. 4: APPLICATIONS OF THE DERIVATIVE

1 (*a*) State all the assumptions in Rolle's theorem.
(*b*) State the conclusion of Rolle's theorem.

2 (*a*) State all the assumptions in the mean-value theorem.
(*b*) State the conclusion of the mean-value theorem.

3 What does each of the following imply about the graph of a function?
(*a*) As you move from left to right, $f(x)$ changes sign at a from positive to negative.
(*b*) As you move from left to right, $f'(x)$ changes sign at a from positive to negative.
(*c*) As you move from left to right, $f''(x)$ changes sign at a from positive to negative.

4 Use the mean-value theorem to show that a function whose derivative is zero throughout an interval is constant.

5 (*a*) Prove that $\tan x - x$ is an increasing function of x, when $0 \le x \le \pi/2$.
(*b*) Deduce that $\tan x > x$ for x in $(0, \pi/2)$.
(*c*) From (*b*) obtain the inequality $x \cos x - \sin x < 0$, if x is in $(0, \pi/2)$.
(*d*) Prove that $(\sin x)/x$ is a decreasing function for x in $(0, \pi/2)$.

6 (*a*) Describe all functions whose derivatives equal $\sin 3x$.
(*b*) How are you sure that you have found all possibilities in (*a*)?

7 How should one choose two nonnegative numbers whose

sum is 1 in order to minimize the sum of the square of one and the cube of the other?

8 A track of a certain length L is to be laid out in the shape of two semicircles at the ends of a rectangle, as shown in Fig. 4.79. Find the relative proportion of the radius of the circle r and the length of the straight section x if the track is to enclose a maximum area. Discuss also the case of minimum area.

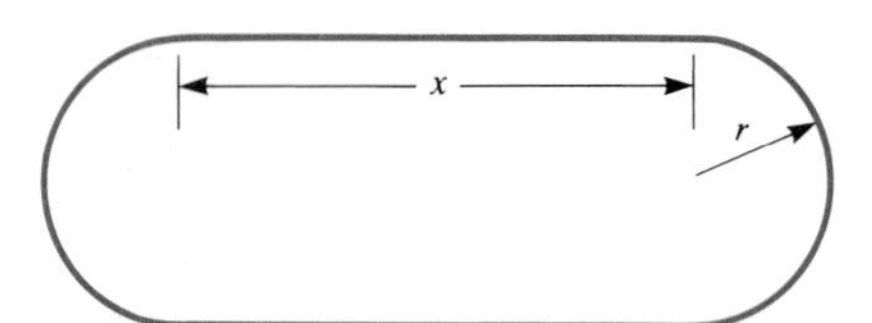

Figure 4.79

9 Using differentials, show that a good estimate (*a*) of $\sqrt{25 + dx}$ is $5 + dx/10$; (*b*) of $\sqrt{a^2 + dx}$ is $a + dx/2a$ (if a is positive). (*c*) Use the formula in (*b*) to estimate $\sqrt{65.6}$.

10 Graph $1/(x^2 - 3x + 2)$.

11 Find $y'(0)$ and $y''(0)$ if $y(0) = 2$ and $y^3 + x^2y + x^3 = 8$.

12 Give all functions whose derivatives are
(*a*) $x^3 + 2x^2$ (*b*) $1/x^3$
(*c*) $5 \sin 2x$ (*d*) $\dfrac{x}{\sqrt{1 + x^2}}$

13 Use differentials to show that for small x, $\tan x \approx x$ (angle in radians).

14 Graph $f(x) = 3x^4 - 16x^3 + 24x^2$.

15 Find the fourth derivative of
(*a*) $2x^5 - 1/x$ (*b*) $\cos 2x$
(*c*) $17x^3 - 5x + 2$ (*d*) $\sqrt{x}$

16 (*a*) How do you go about finding the global maximum of a differentiable function on a closed interval?
(*b*) Describe two different tests for a relative (local) minimum at a critical point.

17 Using higher derivatives, obtain the binomial theorem for $(1 + x)^5$.

18 Graph $(x^3 + 1)/(x^2 + 1)$, showing any asymptotes.

REVIEW EXERCISES FOR CHAP. 4: APPLICATIONS OF THE DERIVATIVE

1 Show that the equation $x^5 + 2x^3 - 2 = 0$ has exactly one solution in the interval $[0, 1]$. (Why does it have at least one? Why is there at most one?)

2 Show that the equation $3 \tan x + x^3 = 2$ has exactly one solution in the interval $[0, \pi/4]$.

3 Let $f(x) = 1/x$.
(*a*) Show that $f'(x)$ is negative for all x in the domain of f.
(*b*) If $x_1 > x_2$, is $f(x_1) < f(x_2)$?

4 A rancher wishes to fence in a rectangular pasture 1 square mile in area, one side of which is along a road. The cost of fencing along the road is higher and equals 5 dollars a foot. The fencing for the other three sides costs 3 dollars a foot. What is the shape of the most economical pasture?

5 Graph $y = \dfrac{1}{x^2} + \dfrac{1}{x - 1}$.

6 Translate the following excerpt from a news article into the terminology of calculus:

> With all the downward pressure on the economy, the first signs of a slowing of inflation seem to be appearing. Some sensitive commodity price indexes are down; the overall wholesale price index is rising at a slightly slower rate.

7 (*a*) Graph $y = \sqrt{x}$ for $0 \le x \le 5$.
(*b*) Compute dy for $x = 4$ and $dx = 1$.
(*c*) Compute Δy for $x = 4$ and $\Delta x = 1$.
(*d*) Using the graph in (*a*), show dy and Δy.

8 Fill in this table:

Interpretation of $f(x)$	*Interpretation of* $f'(x)$
The y coordinate in a graph of $y = f(x)$	
Total distance traveled up to time x	
Projection by lens of point x	
Size of population at time x	
Total mass of left x centimeters of string	
Velocity at time x	

9 Explain why each of these two proposed definitions of a tangent line is inadequate:
(*a*) A line L is tangent to a curve at a point P if L meets the curve only at P.
(*b*) A line L is tangent to a curve at a point P if L meets the curve at P and does not cross the curve at P.

10 A window is made of a rectangle and an equilateral triangle, as shown in Fig. 4.80. What should the dimensions be to maximize the area of the window if its perimeter is prescribed?

Figure 4.80

11 A wire of length L is to be cut into two pieces. One piece will be shaped into an equilateral triangle and the other into a square. How should the wire be cut in order to
(*a*) minimize the sum of the areas of the triangle and square?
(*b*) maximize the sum of the areas?

12 A square foot of glass is to be melted into two shapes. Some of it will be the thin surface of a cube. The rest will be the thin surface of a sphere. How much of the glass should be used for the cube and how much for the sphere if
(*a*) their total volume is to be a minimum?
(*b*) their total volume is to be a maximum?

13 If f is defined for all x, $f(0) = 0$, and $f'(x) \geq 1$ for all x, what is the most that can be said about $f(3)$? Explain.

14 (*a*) Using differentials, show that $\sqrt[3]{8 + h} \approx 2 + h/12$ when h is small.
(*b*) What is the percent error when $h = 1$? $h = -1$?

15 For what value of the exponent a is the function $y = x^a$ a solution to the differential equation

$$\frac{dy}{dx} = -y^2?$$

16 Find all functions f such that

$$\frac{d^2f}{dx^2} = x.$$

17 Graph $y = \sqrt{x}/(1 + x)$.

18 Graph $y = x^4 - 12x^3 + 54x^2$.

19 Show that the equation $2x^7 + 3x^5 + 6x + 10 = 0$ has exactly one real solution.

20 For each of the following give an example of a function whose domain is the x axis, such that
(*a*) f has a global minimum at $x = 1$, but 1 is not a critical number.
(*b*) f has an inflection point at $(0, 0)$, but $f''(0)$ is not defined.
(*c*) $f'(2) = 0$, but f does not have a local extremum at $x = 2$.
(*d*) $f''(2) = 0$, but 2 is not an inflection number.

21 Differentiate for practice:

(*a*) $\dfrac{2x^3 - x}{x + 2}$ (*b*) $x^5\sqrt{1 + 3x}$ (*c*) $\dfrac{(2x - 1)^5}{7}$

(*d*) $\sin^4 \sqrt{x}$ (*e*) $\cos (1/x^3)$ (*f*) $\tan \sqrt{1 - x^2}$

22 A rectangular box with a square base is to be constructed. Material for the top and bottom costs a cents per square inch and material for the sides costs b cents per square inch.
(*a*) For a given cost, what shape has the largest volume? (Express your answer in terms of the ratio between height and dimension of base.)
(*b*) For a given volume, what shape is most economical?

23 What is the minimum slope of $y = x^3 - 9x^2 + 15x$?

24 Use a differential to estimate each of the following for small h.

(*a*) $\sec\left(\dfrac{\pi}{3} + h\right)$ (*b*) $\sqrt[3]{1 + h^2}$ (*c*) $\dfrac{1}{(1 - h)^2}$

25 Use a differential to estimate
(*a*) $(1.002)^5$ (*b*) $(0.996)^3$

26 Find $y'(1)$ if $y(1) = 1$ and $\tan\left(\dfrac{\pi}{4}xy\right) + y^3 + x = 3$.

27 Find a tilted asymptote to the graph of $y = \sqrt{x^2 + 2x}$ and graph the function.

28 The derivative of a certain function f is 5 when x is 2.
(*a*) If $f(x)$ is the distance in feet that a rocket travels in x seconds, about how far does it travel from $x = 2$ to $x = 2.1$ seconds?
(*b*) If $f(x)$ is the projection of x on a slide, about how long is the projection of the interval $[2, 2.1]$?
(*c*) If $f(x)$ is the depth in feet that water penetrates the soil in the first x hours, how much does the water penetrate in the 6 minutes from 2 to 2.1 hours?

29 A certain function $y = f(x)$ has the property that

$$\frac{dy}{dx} = 3y^2.$$

Show that

$$\frac{d^2y}{dx^2} = 18y^3.$$

30 If dy/dx is proportional to x^2, show that d^2y/dx^2 is proportional to x.

31 If dy/dx is proportional to y^2, show that d^2y/dx^2 is proportional to y^3.

32 Using differentials, show that

$$\sin\left(\frac{\pi}{6} + h\right) \approx \frac{1}{2} + \frac{\sqrt{3}}{2}h \qquad \text{for small } h.$$

33 Show that if $dy/dx = 3y^4$, then $d^2y/dx^2 = 36y^7$.

34 Find the maximum value of $\sin^2 \theta \cos \theta$.

■

35 What are the dimensions of the rectangle of largest area that can be inscribed in the ellipse $x^2/a^2 + y^2/b^2 = 1$? (Assume that the sides of the rectangle are parallel to the axes.)

36 Show that the equation $x^5 - 6x + 3 = 0$ has exactly three real roots.

37 Of all squares that can be inscribed in a square of side a what is the side of the one of smallest area?

38 What point on the parabola $y = x^2$ is closest to the point $(3, 0)$?

39 Let f be a differentiable function and let A be a point not on the graph of f. Show that if B is the point on the graph closest to A, then the segment AB is perpendicular to the tangent line to the curve at B.

40 Find the volume of the largest right circular cone that can be inscribed in a sphere of radius a.

41 In each case below decide if there is a function that meets all the conditions. If there is, sketch the graph of such a function. If there is none, indicate why not.
(*a*) $f'(x) > 0$ and $f''(x) < 0$ for all x
(*b*) $f(x) > 0$ and $f''(x) < 0$ for all x
(*c*) $f(x) > 0$ and $f'(x) > 0$ for all x

42 (*a*) The area of a circle of radius r is πr^2. Use a differential to estimate the change in the area when the radius changes from r to $r + dr$.
(*b*) The circumference of a circle of radius r is $2\pi r$. Explain why $2\pi r$ appears in the answer to (*a*).

43 The graph of a certain function is shown in Fig. 4.81. List the x coordinates of (*a*) relative maxima, (*b*) relative minima, (*c*) critical points, (*d*) global maximum, (*e*) global minimum.

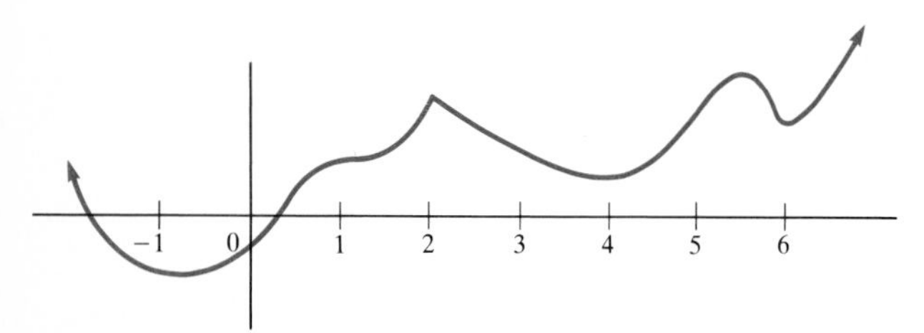

Figure 4.81

44 How should one choose two positive numbers whose product is 2 in order to minimize the sum of their squares?

45 The left-hand x centimeters of a string 12 centimeters long has a mass of $18x^2 - x^3$ grams.
(*a*) What is its density x centimeters from the left-hand end?
(*b*) Where is its density greatest?

46 Of all right circular cones with fixed volume V, which shape has the least surface area, including the area of the base? (The area of the curved part of a cone of slant height l and radius r is πrl.)

47 Of all right circular cones with fixed surface area A (including the area of the base), which shape has the largest volume?

48 What point on the line $y = 3x + 7$ is closest to the origin? (Instead of minimizing the distance, it is much more convenient to minimize the square of the distance. Doing so avoids square roots.)

49 (*a*) What is the maximum value of the function $y = 3 \sin t + 4 \cos t$?
(*b*) What is the maximum value of the function $y = A \sin kt + B \cos kt$, where A, B, and k are constants, $k \neq 0$?

50 Let $f(x) = (x - 1)^n(x - 2)$, where n is an integer, $n \geq 2$.
(*a*) Show that $x = 1$ is a critical number.
(*b*) For which values of n will $x = 1$ provide a relative maximum? a relative minimum? neither?

51 Let p and q be constants.
(*a*) Show that, if $p > 0$, the equation $x^3 + px + q = 0$ has exactly one real root.
(*b*) Show that, if $4p^3 + 27q^2 < 0$, the cubic equation $x^3 + px + q = 0$ has three distinct real roots.

52 (*a*) Sketch the graph of $y = 1/x$.
(*b*) Estimate by eye the point (x_0, y_0) on the graph closest to $(3, 1)$.
(*c*) Show that x_0 is a solution of the equation $x^4 - 3x^3 + x - 1 = 0$.
(*d*) Show that the equation in (*c*) has a root between 2.9 and 3.

53 Show that, if $f'(x) \neq 0$, then dy is a good approximation to Δy when dx is small, in the sense that

$$\lim_{dx \to 0} \frac{\Delta y - dy}{dx} = 0.$$

54 A swimmer stands at a point A on the bank of a circular pond of diameter 200 feet. He wishes to reach the diametrically opposite point B by swimming to some point P on the bank and walking the arc PB along the bank. If he swims 100 feet per minute and walks 200 feet per minute, to what point P should he swim in order to reach B in the shortest possible time?

55 Let f be differentiable everywhere. Assume that $f'(a) = 0$ and $f'(b) = 1$. Must there be a number c, with $a < c < b$, such that $f'(c) = \frac{1}{2}$? *Warning:* f' need not be continuous.

56 Each of Figs. 4.82 to 4.85 is the graph of the velocity $v(t)$ of a particle moving on a line. In each case sketch the general shape of the graph of acceleration as a function of time.

(*a*)

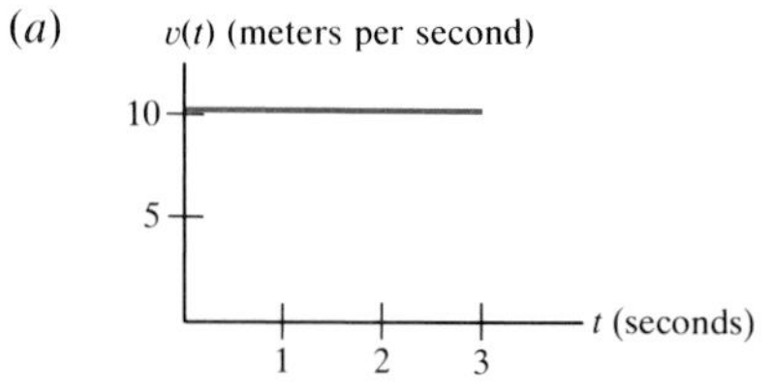

Figure 4.82

(*b*)

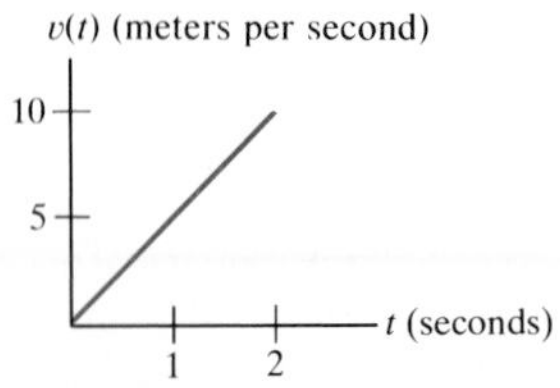

Figure 4.83

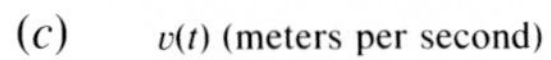

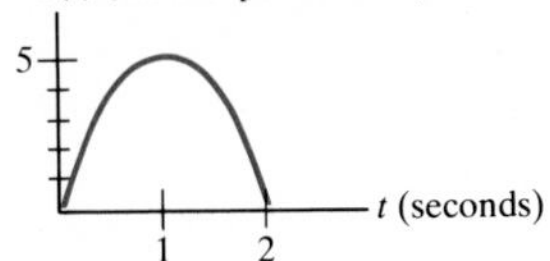

Figure 4.84

(*d*) $v(t)$ (meters per second)

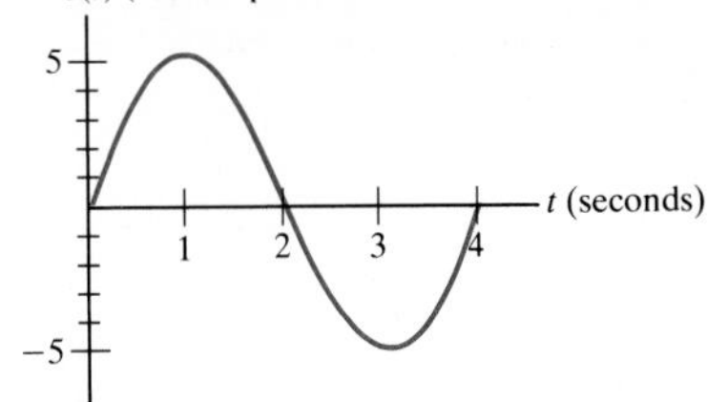

Figure 4.85

5 The Definite Integral

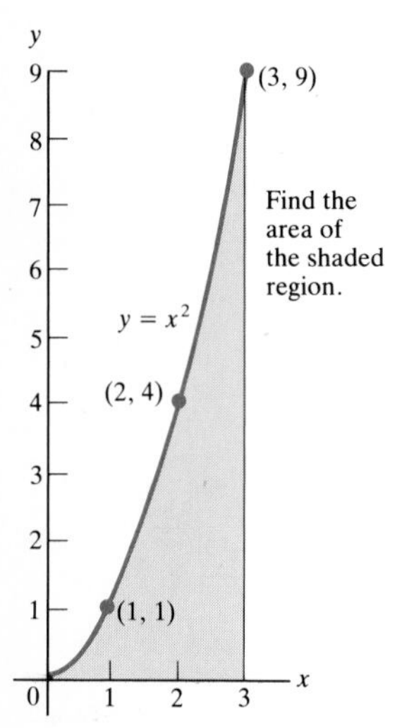

Figure 5.1

Chapters 3 and 4 were concerned with the derivative, which gives local information, such as the slope at a particular point on a curve or the velocity at a particular time. The present chapter introduces the second major concept of calculus, the definite integral. In contrast to the derivative, the definite integral gives overall global information, such as the area under a curve.

The derivative turns out to be the main tool for evaluating definite integrals.

5.1 ESTIMATES IN FOUR PROBLEMS

Just as Chap. 3 introduced the derivative by four problems, this chapter introduces the definite integral by four problems. At first glance these problems may seem unrelated, but by the end of the section it will be clear that they represent one basic problem in various guises.

An Area Problem

PROBLEM 1 Find the area of the region bounded by the curve $y = x^2$, the x axis, and the vertical line $x = 3$, as shown in Fig. 5.1.

An estimate for the area under $y = x^2$

An estimate of the area can be made using a staircase of six rectangles, as shown in Fig. 5.2.

First break the interval from 0 to 3 into six smaller intervals, each of length $\frac{1}{2}$. Then above each small interval draw the rectangle whose height is that of the curve $y = x^2$ above the midpoint of that interval. The total area of the six rectangles is easily computed. This is equal to

$$(\tfrac{1}{4})^2(\tfrac{1}{2}) + (\tfrac{3}{4})^2(\tfrac{1}{2}) + (\tfrac{5}{4})^2(\tfrac{1}{2}) + (\tfrac{7}{4})^2(\tfrac{1}{2}) + (\tfrac{9}{4})^2(\tfrac{1}{2}) + (\tfrac{11}{4})^2(\tfrac{1}{2}),$$

which reduces to $\frac{286}{32}$ or 8.9375.

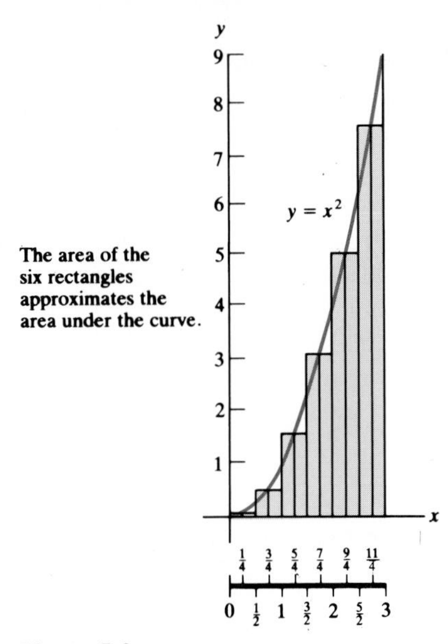

The area of the six rectangles approximates the area under the curve.

Figure 5.2

We have *not* computed the area. The preceding computation provides only an *estimate,* 8.9375, of the area. ■

A Total Mass Problem

PROBLEM 2 A thin nonuniform string 3 centimeters long is made of a material that is very light near one end and very heavy near the other end. In fact, at a distance of x centimeters from the left end it has a density of $5x^2$ grams per centimeter. Find the mass of the string, as shown in Fig. 5.3.

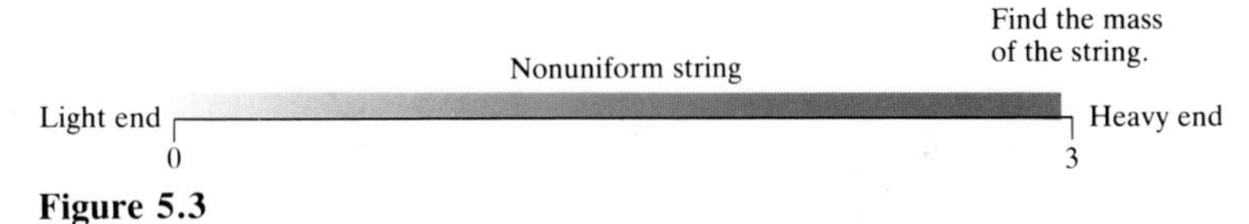

Figure 5.3

Let us cut the string into six sections of equal length, as shown in Fig. 5.4.

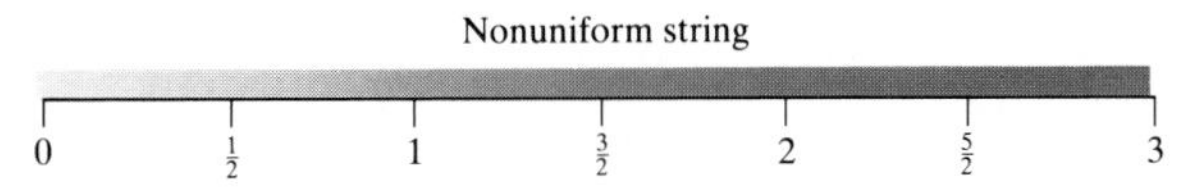

Figure 5.4

The density of the string in each of the six pieces varies less than it does over the whole length of the string. If the density were constant in a section, we would have

$$\text{Density} = \frac{\text{mass}}{\text{length}},$$

which we could rewrite as

$$\text{Mass} = \text{density} \cdot \text{length}.$$

Recall that density at a point is defined in Sec. 3.2.

We shall treat each of the six sections as having throughout a constant density equal to that at its midpoint. To obtain an estimate of the mass of each of the six sections, let us multiply the density at the midpoint of the section by the length of the section.

The left section has a density of $5(\frac{1}{4})^2$ grams per centimeter at its midpoint, $\frac{1}{4}$, and thus has a mass of about $5(\frac{1}{4})^2(\frac{1}{2})$ gram. The next section, from $\frac{1}{2}$ to 1, has a density of $5(\frac{3}{4})^2$ at its midpoint and thus has a mass of about $5(\frac{3}{4})^2(\frac{1}{2})$ grams. An estimate of the mass of each of the four other sections can be made similarly. An estimate of the total mass of the nonuniform string is then the sum

An estimate for the mass of the nonuniform string

$$5(\tfrac{1}{4})^2(\tfrac{1}{2}) + 5(\tfrac{3}{4})^2(\tfrac{1}{2}) + 5(\tfrac{5}{4})^2(\tfrac{1}{2}) + 5(\tfrac{7}{4})^2(\tfrac{1}{2}) + 5(\tfrac{9}{4})^2(\tfrac{1}{2}) + 5(\tfrac{11}{4})^2(\tfrac{1}{2}).$$

This sum is five times the sum in Problem 1 and hence equals 5(8.9375), a little less than 45 grams. More important is the similarity in form between this sum and the sum used in the first problem. ■

A Total Distance Problem

PROBLEM 3 An engineer drives a car whose clock and speedometer work, but whose odometer (mileage recorder) is broken. On a 3-hour trip out of a congested city into the countryside she begins at a snail's pace and, as the traffic thins, gradually speeds up. Indeed, she notices that after traveling t hours her speed is $8t^2$ miles per hour. Thus after the first $\frac{1}{2}$ hour she is crawling along at 2 miles per hour, but after 3 hours she is traveling at 72 miles per hour. How far does the engineer travel in 3 hours?

An estimate for the total distance

The speed during the 3-hour trip varies from 0 to 72 miles per hour. During shorter time intervals such a wide fluctuation will not occur.

As in the first two problems, cut the 3 hours of the trip into six equal intervals, each $\frac{1}{2}$ hour long, and use them to make an estimate of the total distance covered. Represent time by a line segment cut into six parts of equal length, as in Fig. 5.5.

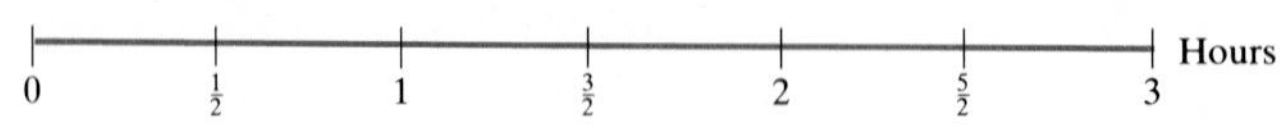

Figure 5.5

To estimate the distance the engineer travels in the first $\frac{1}{2}$ hour, multiply her speed at $\frac{1}{4}$ hour by the duration of the first interval of time, $\frac{1}{2}$ hour. Since her speed at time t is $8t^2$, after $\frac{1}{4}$ hour her speed is $8(\frac{1}{4})^2$ miles per hour. Thus during the first $\frac{1}{2}$ hour the engineer travels about $8(\frac{1}{4})^2(\frac{1}{2})$ mile. During the second $\frac{1}{2}$ hour she travels about $8(\frac{3}{4})^2(\frac{1}{2})$ miles.

Making similar estimates for each of the other $\frac{1}{2}$-hour periods, we obtain this estimate for the length of the trip:

$$8(\tfrac{1}{4})^2(\tfrac{1}{2}) + 8(\tfrac{3}{4})^2(\tfrac{1}{2}) + 8(\tfrac{5}{4})^2(\tfrac{1}{2}) + 8(\tfrac{7}{4})^2(\tfrac{1}{2}) + 8(\tfrac{9}{4})^2(\tfrac{1}{2}) + 8(\tfrac{11}{4})^2(\tfrac{1}{2}).$$

This sum is eight times the sum in Problem 1; hence it equals

$$8 \cdot (8.9375) = 71.5 \text{ miles.}$$

Keep in mind that this is only an estimate of the length of the trip. ■

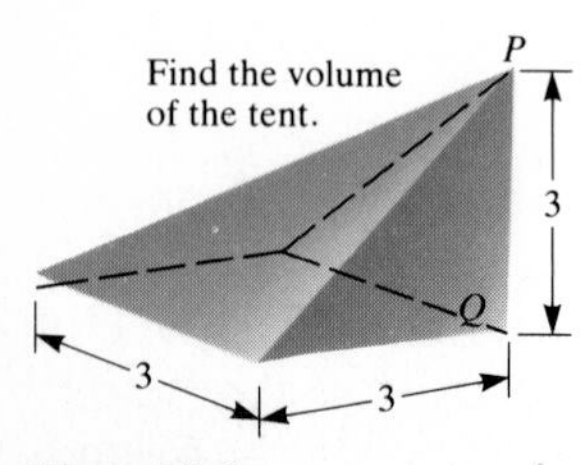

Figure 5.6

A Volume Problem

PROBLEM 4 Find the volume inside a tent with a square floor of side 3 feet, whose pole, 3 feet long, rises above a corner of the floor. The tent is shown in Fig. 5.6. It can also be thought of as the surface obtained when the piece of paper shown in Fig. 5.7 is folded along the dotted lines and the free edges taped in such a way that A, B, C, and D come together (to become P).

An estimate for the volume

Observe that the cross section of the tent made by any plane parallel to the base is a square, as shown in Fig. 5.8.

This time we cut a vertical line, representing the pole, into six sections of equal length. Then we approximate each slab by a flat rectangular box $\frac{1}{2}$ foot high. The cross section of the smallest box is obtained by passing a horizontal plane through the midpoint of the highest of the six sections. The remaining five boxes are determined in a similar manner, as shown in Figs. 5.9 and 5.10. As the side view of the six boxes shows, the square cross section of the top box has a side equal to $\frac{1}{4}$ foot, the box below it has a side equal to $\frac{3}{4}$ foot, and so on until we reach the bottom box, whose side is $\frac{11}{4}$ feet.

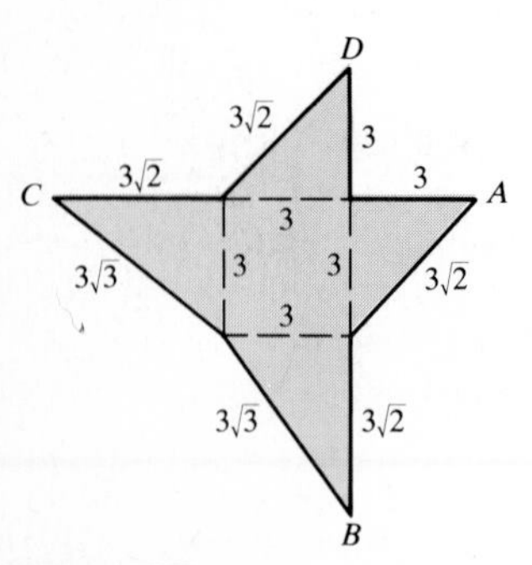

Figure 5.7

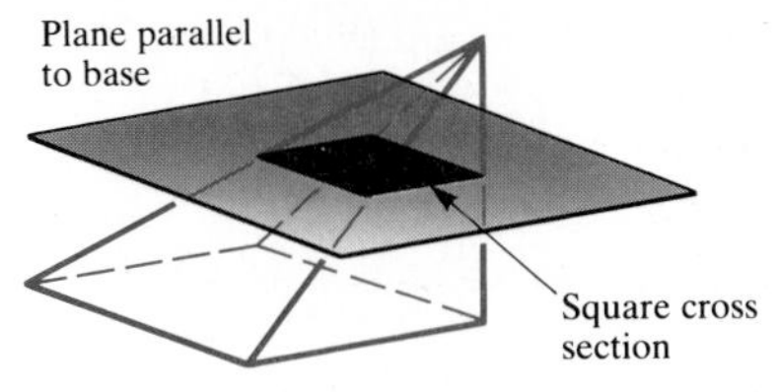

Figure 5.8

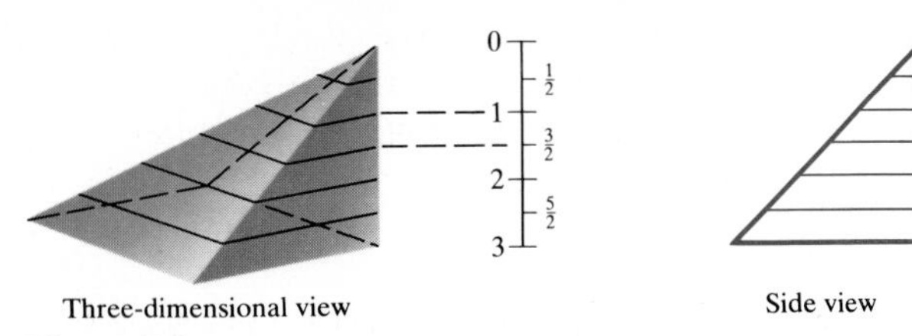

Figure 5.9

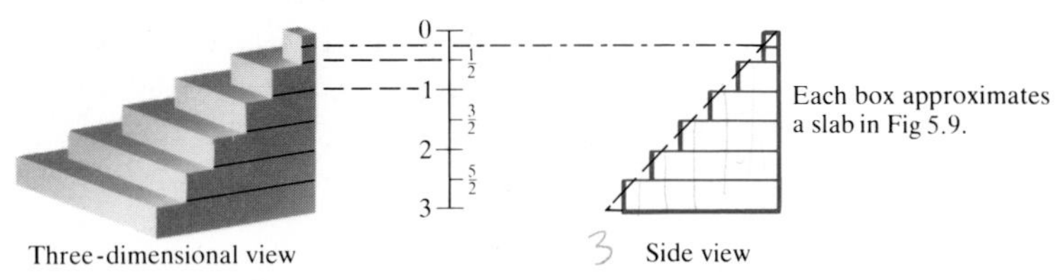

Figure 5.10

Since the volume of a box is just the area of its base times its height, the total volume of the six boxes is

$$\left(\tfrac{1}{4}\right)^2\left(\tfrac{1}{2}\right) + \left(\tfrac{3}{4}\right)^2\left(\tfrac{1}{2}\right) + \left(\tfrac{5}{4}\right)^2\left(\tfrac{1}{2}\right) + \left(\tfrac{7}{4}\right)^2\left(\tfrac{1}{2}\right) + \left(\tfrac{9}{4}\right)^2\left(\tfrac{1}{2}\right) + \left(\tfrac{11}{4}\right)^2\left(\tfrac{1}{2}\right) \text{ cubic feet.}$$

This is the same sum that was met in estimating the area under the curve $y = x^2$. Thus the volume of the tent is estimated as 8.9375 cubic feet. ■

None of the four problems is yet solved; in each case, all we have is an estimate. In Sec. 5.3 the precise answers will be found.

EXERCISES FOR SEC. 5.1: ESTIMATES IN FOUR PROBLEMS

To show the similarity of the four problems, the interval [0, 3] has been cut into six sections each time and the midpoint of each section used to determine the cross section, density, speed, or area. Of course, we are free to cut the interval into more or fewer sections and to use a point other than the midpoint in each section. Furthermore, there is no need to restrict the sections to be of equal length. Exercises 1 to 5 concern other estimates for the same problems.

1 (*a*) Estimate the area in Problem 1 by using three sections, each of length 1, and the midpoint each time.
(*b*) Estimate the same area by using the same three sections, but now use the y coordinate of the point on $y = x^2$ above the right end of each section to determine the rectangle.
(*c*) Draw the three rectangles used in (*b*). Is their total area more or less than the area under the curve?
(*d*) Proceed as in (*b*) and (*c*), but use the point on $y = x^2$ above the left endpoint of each section.
(*e*) Using information gathered in (*b*) and (*d*), complete this sentence: The area in Problem 1 is certainly less than ________ but larger than ________ .

2 Cutting the interval from 0 to 3 into five sections of equal length, estimate the area in Problem 1 by finding the sum of the areas of five rectangles whose heights are determined by (*a*) midpoints, (*b*) right endpoints, and (*c*) left endpoints. (*d*) Using information gathered in (*b*) and (*c*), complete this sentence: The area in Problem 1 is certainly less than ________ but larger than ________ .

3 Estimate the mass of the string in Problem 2 by cutting it into five sections of equal length. For an estimate of the mass of each of these sections use the density at (*a*) the midpoint of each section, (*b*) the right endpoint, and (*c*) the left endpoint. (*d*) On the basis of (*b*) and (*c*), the mass in Problem 2 is less than ________ but larger than ________ .

4 Cutting the interval of 3 hours into five periods of $\frac{3}{5}$ hour each, estimate the length of the engineer's trip in Problem 3. For the approximate velocity in each period use

the speedometer reading at (*a*) the middle of the period, (*b*) the end of the period, and (*c*) the beginning of the period. (*d*) In view of (*b*) and (*c*), the length of the trip is less than ________ but larger than ________.

5 Make an estimate for each of the four problems, using in each case the accompanying division into four sections. As the points where the cross section, density, or velocity is computed, use $\frac{1}{2}$, $\frac{3}{2}$, 2, and $\frac{14}{5}$ (one of these is in each of the four sections). See Fig. 5.11.

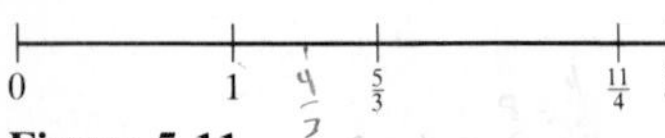

Figure 5.11

6 Estimate the area between the curve $y = x^3$, the x axis, and the vertical line $x = 6$ using a division into (*a*) three sections of equal length and midpoints; (*b*) six sections of equal length and midpoints; (*c*) six sections of equal length and left endpoints; (*d*) six sections of equal length and right endpoints.

7 (Calculator) Estimate the volume of the tent in Problem 4 by cutting [0, 3] into 10 sections of equal length and taking cross sections at
(*a*) left endpoints, (*b*) right endpoints.

8 (Calculator) Estimate the area under $y = x^2$ and above [0, 1] by cutting [0, 1] into ten sections of equal length and using (*a*) left endpoints, (*b*) right endpoints.

■

9 A business which now shows no profit is to increase its profit flow gradually in the next 3 years until it reaches a rate of 9 million dollars per year. At the end of the first half year the rate is to be $\frac{1}{4}$ million dollars per year; at the end of 2 years, 4 million dollars per year. In general, at the end of t years, where t is any number between 0 and 3, the rate of profit is to be t^2 million dollars per year. Estimate the total profit during the next 3 years if the plan is successful. Use six intervals of equal length and midpoints.

10 Estimate the area under $y = x^2$ and directly above the interval [1, 5] by the midpoint method with the aid of a partition of [1, 5] into (*a*) four sections of equal length, (*b*) eight sections of equal length.

11 A right circular cone has a height of 3 feet and a radius of 3 feet, as shown in Fig. 5.12. Estimate its volume by the sum of the volumes of six cylindrical slabs, just as we estimated the volume of the tent with the aid of six rectangular slabs. In particular, (*a*) show with the aid of a diagram how the same sections and midpoints we used determine six cylinders, and (*b*) compute their total volume.

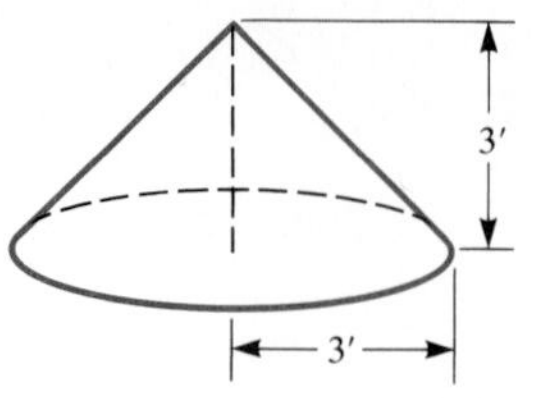

Right circular cone of height 3 feet and radius 3 feet

Figure 5.12

12 (Calculator) Estimate the area of the region under the curve $y = \sin x$ and above the interval $[0, \pi/2]$, cutting the interval as shown in Fig. 5.13 and using (*a*) left endpoints, (*b*) right endpoints. All but the last section are of the same size.

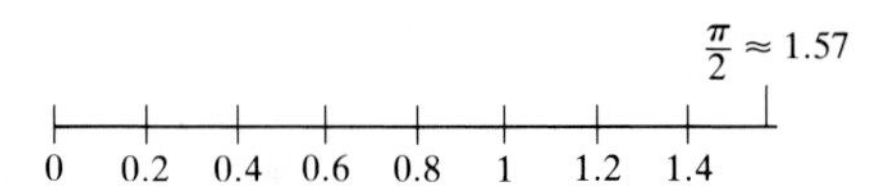

Figure 5.13

13 Estimate the area of the region under $y = 1/x$ and above [1, 2], using five sections of equal length and (*a*) left endpoints, (*b*) right endpoints.

14 Differentiate for practice:

(*a*) $(1 + x^2)^{4/3}$

(*b*) $\dfrac{(1 + x^3)\sin 3x}{\sqrt[3]{5x}}$

(*c*) $\dfrac{3x}{8} + \dfrac{3x \sin 4x}{32} + \dfrac{\cos^3 2x \sin 2x}{8}$

(*d*) $\dfrac{3}{8(2x + 3)^2} - \dfrac{1}{4(2x + 3)}$

(*e*) $\dfrac{\cos^3 2x}{6} - \dfrac{\cos 2x}{2}$

(*f*) $x^3 \sqrt{x^2 - 1} \tan 5x$

15 Give an example of a function F whose derivative is
(*a*) $(x + 2)^3$ (*b*) $(x^2 + 1)^2$ (*c*) $x \sin x^2$
(*d*) $x^3 + \dfrac{1}{x^3}$ (*e*) $\dfrac{1}{\sqrt{x}}$

16 The kinetic energy of an object, for example, a bullet or car, of mass m and speed v is defined as $mv^2/2$ ergs.

(Here mass is measured in grams and speed in centimeters per second.) Now, in a certain machine a uniform rod 3 centimeters long and weighing 32 grams rotates once per second around one of its ends. Estimate the kinetic energy of this rod by cutting it into six sections, each $\frac{1}{2}$ centimeter long, and taking as the "speed of a section" the speed of its midpoint.

17 Draw an accurate graph of $y = x^2$ and the six rectangles with heights equal to the ordinates of the curve at the midpoints that we used to estimate the area. Does each of these rectangles underestimate or overestimate the area under $y = x^2$ and above the base of the rectangle? (Form your opinion on the basis of your drawing.)

18 This exercise concerns the areas of regions under the curve $1/x$.

(*a*) Estimate the area under $y = 1/x$ and above $[1, 2]$, using five sections of equal length and left endpoints.

(*b*) Estimate the area under $y = 1/x$ and above $[3, 6]$, using five sections of equal length and left endpoints.

(*c*) The answers to (*a*) and (*b*) are the same. Would they be the same if you used 100 sections of equal length (instead of 5)? Explain.

(*d*) Write a short paragraph explaining why the area under $y = 1/x$ above $[1, 2]$ equals the area under $y = 1/x$ above $[3, 6]$.

(*e*) Let a and b be numbers greater than 1. Explain why the area under $1/x$ and above $[1, a]$ equals the area under $1/x$ and above $[b, ab]$.

(*f*) For $t > 1$, let $G(t)$ equal the area under the curve $1/x$ and above $[1, t]$. Show that, for a and b greater than 1, $G(ab) = G(a) + G(b)$.

(*g*) What function f studied in precalculus resembles the function G in that $f(ab) = f(a) + f(b)$?

5.2 SUMMATION NOTATION AND APPROXIMATING SUMS

In Sec. 5.1 sums of a particular type were formed and computed. Such sums play an essential role in the theory of the definite integral developed in this chapter and in applications in several later chapters. Since such sums will be needed often, let us introduce a convenient notation for them, the so-called sigma notation, named after the Greek letter Σ, which corresponds to the *s* of sum. The sigma notation is useful in dealing with sums in which the summands all have the same general form.

$\sum$ notation or summation notation

Definition *Sigma notation.* Let $a_1, a_2, \ldots, a_n$ be n numbers. The sum $a_1 + a_2 + \cdots + a_n$ will be denoted in **sigma notation** by the symbol $\sum_{i=1}^{n} a_i$ or $\Sigma_{i=1}^{n} a_i$, which is read as "the sum of a sub i as i goes from 1 to n."

In the sigma notation, the formula for the typical summand is given, as is a description of where the summation starts and ends.

EXAMPLE 1 Write the sum $1^2 + 2^2 + 3^2 + 4^2$ in the sigma notation.

SOLUTION Since the ith summand is the square of i and the summation extends from $i = 1$ to $i = 4$, we have

$$1^2 + 2^2 + 3^2 + 4^2 = \sum_{i=1}^{4} i^2.$$

Simple arithmetic shows that the sum is equal to 30. ■

EXAMPLE 2 Compute $\Sigma_{i=1}^{3} 2^i$.

SOLUTION This is short for the sum $2^1 + 2^2 + 2^3$, which is $2 + 4 + 8$, or 14. ■

In the definition of the sigma notation, the letter i (for "index") was used. Any letter, such as j or k, would do just as well. Such an index is sometimes called a **summation index** or **dummy index**

EXAMPLE 3 Compute $\displaystyle\sum_{j=1}^{4} \frac{1}{j}$.

SOLUTION This is short for $\frac{1}{1} + \frac{1}{2} + \frac{1}{3} + \frac{1}{4}$, which is approximately 2.083. ■

Had Example 3 read "Compute $\Sigma_{k=1}^{4} \dfrac{1}{k}$," the result would be the same:

$$\sum_{k=1}^{4} \frac{1}{k} = \frac{1}{1} + \frac{1}{2} + \frac{1}{3} + \frac{1}{4}.$$

The particular letter used to indicate the form of the typical summand is of no special importance.

The summation notation has two properties which will be of use in coming chapters. First of all, if c is a fixed number, then

$$\sum_{i=1}^{n} ca_i = ca_1 + ca_2 + \cdots + ca_n$$

$$= c(a_1 + a_2 + \cdots + a_n) = c \sum_{i=1}^{n} a_i.$$

$\displaystyle\sum_{i=1}^{n} ca_i = c \sum_{i=1}^{n} a_i$

Thus
$$\sum_{i=1}^{n} ca_i = c \sum_{i=1}^{n} a_i.$$

This distributive rule is read as "a constant factor can be moved past Σ."

Second,

$$\sum_{i=1}^{n} (a_i + b_i) = (a_1 + b_1) + (a_2 + b_2) + \cdots + (a_n + b_n)$$

$$= (a_1 + a_2 + \cdots + a_n) + (b_1 + b_2 + \cdots + b_n)$$

$\displaystyle\sum_{i=1}^{n} (a_i + b_i) = \sum_{i=1}^{n} a_i + \sum_{i=1}^{n} b_i.$

$$= \sum_{i=1}^{n} a_i + \sum_{i=1}^{n} b_i.$$

This is a direct consequence of the rules of algebra.

EXAMPLE 4 Compute $\displaystyle\sum_{i=1}^{4} \left(i^2 + \frac{1}{i}\right)$.

SOLUTION This may be rewritten as

$$\sum_{i=1}^{4} i^2 + \sum_{i=1}^{4} \frac{1}{i}.$$

By Examples 1 and 3, the sum, to three decimals, is $30 + 2.083 = 32.083$. ■

EXAMPLE 5 What is the value of $\Sigma_{i=1}^{5} 3$?

SOLUTION In this case, $a_i = 3$ for each index i. Each summand has the value 3. Thus

$$\sum_{i=1}^{5} 3 = 3 + 3 + 3 + 3 + 3 = 15.$$

$\sum_{i=1}^{n} c = cn.$

More generally, if c is a fixed number not depending on i, then $\Sigma_{i=1}^{n} c = cn$. ■

The next example shows how to interpret the sigma notation when the index does not start at 1.

EXAMPLE 6 Compute $\Sigma_{i=2}^{6} 5i$ (read as "the sum of $5i$ as i goes from 2 to 6").

SOLUTION This is short for $5 \cdot 2 + 5 \cdot 3 + 5 \cdot 4 + 5 \cdot 5 + 5 \cdot 6$, which equals

$$5(2 + 3 + 4 + 5 + 6),$$

or 100. ■

Another useful fact to note is that

Breaking a sum into a "front" end and a "back" end

$$\sum_{i=1}^{m} a_i + \sum_{i=m+1}^{n} a_i = \sum_{i=1}^{n} a_i, \qquad 1 \le m < n.$$

For instance,

$$\sum_{i=1}^{4} i^2 + \sum_{i=5}^{10} i^2 = \sum_{i=1}^{10} i^2.$$

EXAMPLE 7 Let b_0, b_1, b_2, b_3 be four numbers. Form the three differences

$$a_1 = b_1 - b_0, \qquad a_2 = b_2 - b_1, \qquad a_3 = b_3 - b_2,$$

and compute

$$\Sigma_{i=1}^{3} a_i.$$

SOLUTION

$$\sum_{i=1}^{3} a_i = a_1 + a_2 + a_3 = (b_1 - b_0) + (b_2 - b_1) + (b_3 - b_2).$$

Cancellations of b_1 and $-b_1$ and of b_2 and $-b_2$ show that

$$a_1 + a_2 + a_3 = b_3 - b_0. \quad ■$$

Example 7 can easily be generalized from four to any finite list of numbers. If $b_0, b_1, \ldots, b_n$ are $n + 1$ numbers and $a_i = b_i - b_{i-1}$, $i = 1, \ldots, n$, then

$$\sum_{i=1}^{n} a_i = \sum_{i=1}^{n} (b_i - b_{i-1}) = b_n - b_0.$$

Telescoping sums In short, the sum $\Sigma_{i=1}^{n} (b_i - b_{i-1})$ "telescopes" to $b_n - b_0$.

As an application of telescoping sums we obtain short formulas for the sums $1 + 2 + 3 + \cdots + n$ and $1^2 + 2^2 + 3^2 + \cdots + n^2$.

Consider the telescoping sum $\Sigma_{i=1}^{n} [(i + 1)^2 - i^2]$. [In this case, $b_i = (i + 1)^2$.] We have the equation

$$\sum_{i=1}^{n} [(i + 1)^2 - i^2] = (n + 1)^2 - 1. \tag{1}$$

On the other hand, by the binomial theorem, we know that

$$(i + 1)^2 - i^2 = i^2 + 2i + 1 - i^2 = 2i + 1.$$

Thus Eq. (1) can be written

$$\sum_{i=1}^{n} (2i + 1) = (n + 1)^2 - 1,$$

which simplifies to

$$2 \sum_{i=1}^{n} i + n = n^2 + 2n.$$

It follows that

Short formula for $\sum_{i=1}^{n} i$

$$\sum_{i=1}^{n} i = \frac{n^2}{2} + \frac{n}{2} = \frac{n(n + 1)}{2}.$$

To obtain a short formula for $\Sigma_{i=1}^{n} i^2$, start with the telescoping sum

$$\sum_{i=1}^{n} [(i + 1)^3 - i^3] = (n + 1)^3 - 1. \tag{2}$$

By the binomial theorem, $(i + 1)^3 = i^3 + 3i^2 + 3i + 1$. Hence Eq. (2) becomes

$$\sum_{i=1}^{n} (i^3 + 3i^2 + 3i + 1 - i^3) = n^3 + 3n^2 + 3n$$

or

$$3 \sum_{i=1}^{n} i^2 + 3 \sum_{i=1}^{n} i + \sum_{i=1}^{n} 1 = n^3 + 3n^2 + 3n. \tag{3}$$

Since $\Sigma_{i=1}^{n} 1 = n$ and, as was just shown, $\Sigma_{i=1}^{n} i = (n^2/2) + (n/2)$, Eq. (3) provides this equation for $\Sigma_{i=1}^{n} i^2$:

$$3 \sum_{i=1}^{n} i^2 + 3 \frac{n^2}{2} + 3 \frac{n}{2} + n = n^3 + 3n^2 + 3n,$$

from which it follows that

Short formula for $\sum_{i=1}^{n} i^2$

$$\sum_{i=1}^{n} i^2 = \frac{n^3}{3} + \frac{n^2}{2} + \frac{n}{6} = \frac{n(n + 1)(2n + 1)}{6}. \tag{4}$$

This formula will be needed in Sec. 5.3. (See Exercise 29 for an application to computer programming.)

Sigma Notation for the Approximating Sums The sums used in Sec. 5.1 to approximate the area, mass, distance, or volume were all made the

$$\sum_{i=1}^{N} a_i + \sum_{i=N+1}^{M} a_i = \sum_{i=1}^{m} a_i$$

same way. Consider, for instance, how an approximating sum for the area under x^2 and above the interval $[a, b]$ is formed.

First, the interval $[a, b]$ is partitioned into smaller sections, perhaps all of equal length, perhaps not. There could be any finite number of sections. Say that there are n sections. These sections are determined by choosing $n - 1$ numbers in (a, b), $x_1, x_2, \ldots, x_{n-1}$,

$$a < x_1 < x_2 < \cdots < x_{n-1} < b.$$

For convenience, introduce

$$x_0 = a \qquad \text{and} \qquad x_n = b.$$

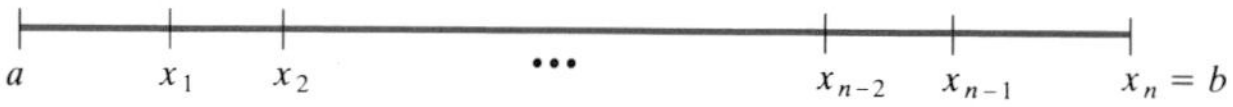

The partition: $x_0 = a$ x_1 x_2 ... x_{n-2} x_{n-1} $x_n = b$

Figure 5.14

Typical section has ends x_{i-1} and x_i.

x_{i-1} x_i

Figure 5.15

The ith section, $i = 1, 2, \ldots, n$ has the left endpoint x_{i-1} and the right endpoint x_i, as shown in Fig. 5.14. The typical section is shown in Fig. 5.15. For instance, the first section is $[x_0, x_1]$, which is $[a, x_1]$, the second section is $[x_1, x_2]$, . . . , and the nth section is $[x_{n-1}, x_n]$, which is $[x_{n-1}, b]$. The length of the ith section is $x_i - x_{i-1}$, which is often denoted Δx_i.

The ith section is $[x_{i-1}, x_i]$.

$\Delta x_i = x_i - x_{i-1}$.

After the partition or division into n sections is formed, a number is selected in each section at which to evaluate x^2. The number chosen in $[x_{i-1}, x_i]$ could be its left endpoint x_{i-1}, its right endpoint x_i, its midpoint $(x_{i-1} + x_i)/2$, or any point whatsoever in the section. To allow for the most general possible choice, denote the number chosen in the ith section by c_i. The number c_i is called a **sampling number**. It is shown in Fig. 5.16.

c_i *is sampling number.*

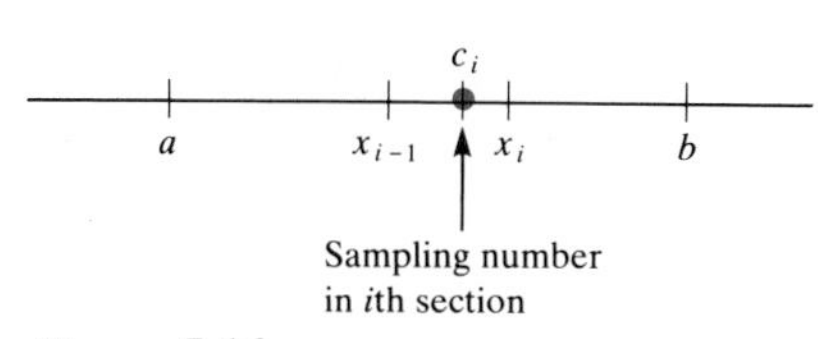

Figure 5.16

The next step is to evaluate the function x^2 at each c_i and form the sum with n summands:

$$c_1^2(x_1 - x_0) + c_2^2(x_2 - x_1) + \cdots + c_i^2(x_i - x_{i-1}) + \cdots + c_n^2(x_n - x_{n-1}). \tag{5}$$

It takes a long time to write down the sum (5) or to read it aloud. The Σ notation compresses it to the simple expression,

$$\sum_{i=1}^{n} c_i^2(x_i - x_{i-1}), \tag{6}$$

which is read, "the sum from 1 to n of $c_i^2(x_i - x_{i-1})$." If the length $x_i - x_{i-1}$ is denoted Δx_i, Eq. (6) reduces to

$$\sum_{i=1}^{n} c_i^2 \, \Delta x_i.$$

It must be kept in mind that $x_0 = a$ and $x_n = b$.

EXAMPLE 8 Write in sigma notation the typical approximating sum for the area of the region under $f(x) = x^2$ and above $[2, 7]$ using right endpoints as sampling numbers.

SOLUTION In this case, $a = 2$, $b = 7$, and $c_i = x_i$. The typical sum is

$$\sum_{i=1}^{n} x_i^2(x_i - x_{i-1})$$

or

$$\sum_{i=1}^{n} x_i^2 \Delta x_i.$$

It is to be understood that

$$2 = x_0 < x_1 < \cdots < x_{n-1} < x_n = 7. \quad \blacksquare$$

But x^2 is just one possible function. Similar approximating sums can be formed for any other function, such as x^3, $1/x$, or $\sin x$. The typical approximating sum for any function $f(x)$ and any interval $[a, b]$ in the domain of the function is formed just like Eq. (5). First, a partition of $[a, b]$ is determined:

$$a = x_0 < x_1 < \cdots < x_{n-1} < x_n = b.$$

Then a sampling number c_i is picked in the ith section, $i = 1, 2, \ldots, n$. The function $f(x)$ is evaluated at each c_i, and finally the approximating sum is formed:

$$f(c_1)(x_1 - x_0) + \cdots + f(c_i)(x_i - x_{i-1}) + \cdots + f(c_n)(x_n - x_{n-1}).$$

In Σ notation this reduces to

$$\sum_{i=1}^{n} f(c_i)(x_i - x_{i-1})$$

or

$$\sum_{i=1}^{n} f(c_i) \Delta x_i.$$

An approximating sum is also called a Riemann sum.

Such an approximating sum is also called a **Riemann sum** in honor of the nineteenth-century mathematician, Georg Riemann, who made many fundamental contributions to various branches of mathematics.

EXERCISES FOR SEC. 5.2: SUMMATION NOTATION AND APPROXIMATING SUMS

In Exercises 1 to 4 evaluate the sums.

1 (*a*) $\sum_{i=1}^{3} i$ (*b*) $\sum_{i=1}^{4} 2i$ (*c*) $\sum_{d=1}^{3} d^2$

2 (*a*) $\sum_{i=2}^{4} i^2$ (*b*) $\sum_{j=2}^{4} j^2$ (*c*) $\sum_{i=1}^{3} (i^2 + i)$

3 (*a*) $\sum_{i=1}^{4} 1^i$ (*b*) $\sum_{k=2}^{6} (-1)^k$ (*c*) $\sum_{j=1}^{150} 3$

4 (*a*) $\sum_{i=3}^{5} \frac{1}{i}$ (*b*) $\sum_{i=0}^{4} \cos 2\pi i$ (*c*) $\sum_{i=1}^{3} 2^{-i}$

In Exercises 5 to 8 write in the sigma notation. (Do not evaluate.)

5 (*a*) $1 + 2 + 2^2 + 2^3 + \cdots + 2^{100}$
(*b*) $x^3 + x^4 + x^5 + x^6 + x^7$
(*c*) $\frac{1}{3} + \frac{1}{4} + \cdots + \frac{1}{102}$

6 (*a*) $\frac{1}{2} + \frac{1}{3} + \cdots + \frac{1}{100}$
(*b*) $\frac{1}{3} + \frac{1}{5} + \frac{1}{7} + \frac{1}{9} + \frac{1}{11}$
(*c*) $\frac{1}{1^2} + \frac{1}{3^2} + \frac{1}{5^2} + \cdots + \frac{1}{101^2}$

7 (*a*) $x_0^2(x_1 - x_0) + x_1^2(x_2 - x_1) + x_2^2(x_3 - x_2)$
(*b*) $x_1^2(x_1 - x_0) + x_2^2(x_2 - x_1) + x_3^2(x_3 - x_2)$

8 (*a*) $8t_0^2(t_1 - t_0) + 8t_1^2(t_2 - t_1) + \cdots + 8t_{99}^2(t_{100} - t_{99})$
(*b*) $8t_1^2(t_1 - t_0) + 8t_2^2(t_2 - t_1) + \cdots + 8t_n^2(t_n - t_{n-1})$

In Exercises 9 and 10 evaluate the telescoping sums.

9 (*a*) $\sum_{i=1}^{100} (2^i - 2^{i-1})$ (*b*) $\sum_{i=2}^{100} \left(\frac{1}{i} - \frac{1}{i-1}\right)$

(*c*) $\sum_{i=1}^{50} \left(\frac{1}{2i+1} - \frac{1}{2(i-1)+1}\right)$

10 (*a*) $\sum_{i=1}^{100} \left(\frac{x_i^3}{3} - \frac{x_{i-1}^3}{3}\right)$ (*b*) $\sum_{i=5}^{70} \left(\frac{1}{x_i} - \frac{1}{x_{i-1}}\right)$

11 Writing out each sum in longhand, show that

(*a*) $\sum_{i=1}^{3} a_i = \sum_{j=1}^{3} a_j = \sum_{k=1}^{3} a_k$

(*b*) $\sum_{i=1}^{3} (a_i + 4) = 12 + \sum_{j=2}^{4} a_{j-1}$

12 Writing out each sum in longhand, show that

(*a*) $\sum_{i=1}^{3} (a_i - b_i) = \sum_{i=1}^{3} a_i - \sum_{i=1}^{3} b_i$

(*b*) $\sum_{i=1}^{2} a_i b_i$ is *not* always equal to $\sum_{i=1}^{2} a_i \cdot \sum_{i=1}^{2} b_i$.

(*c*) $\sum_{i=1}^{3} \left(\sum_{j=1}^{3} b_j\right) a_i = \sum_{j=1}^{3} \left(\sum_{i=1}^{3} a_i\right) b_j$

In Exercises 13 to 18 evaluate the approximating sum $\Sigma_{i=1}^{n} f(c_i)(x_i - x_{i-1})$. In each case the interval is [1, 3] and the partition consists of four sections of equal length. Express answers to two decimal places.

13 $f(x) = 3x$, $c_i = x_i$
14 $f(x) = 3x$, $c_i = x_{i-1}$
15 $f(x) = 5$, $c_i = x_i$
16 $f(x) = x^3 + x$, $c_i = x_i$
17 $f(x) = 1/x$, $c_1 = 1.25$, $c_2 = 1.8$, $c_3 = 2.2$, $c_4 = 3$
18 $f(x) = 2^x$, $c_i = x_{i-1}$

19 Evaluate (*a*) $\sum_{i=1}^{100} i$; (*b*) $\sum_{j=1}^{1000} 3j$.

20 Evaluate (*a*) $\sum_{i=1}^{100} i^2$; (*b*) $\sum_{i=1}^{n} (2i^2 + 3i + 6)$.

■

21 Using the method by which Eq. (4) was obtained, show that

$$\sum_{i=1}^{n} i^3 = \frac{n^4}{4} + \frac{n^3}{2} + \frac{n^2}{4}.$$

22 Using the method by which Eq. (4) was obtained, together with the result in Exercise 21, obtain a formula for $\Sigma_{i=1}^{n} i^4$.

23 Let $f(x) = 11$ for all x. Consider the partition $x_0 = a < x_1 < \cdots < x_n = b$ and let c_i be in $[x_{i-1}, x_i]$. Evaluate $\Sigma_{i=1}^{n} f(c_i)(x_i - x_{i-1})$.

24 Write in Σ notation the typical approximating sum for the area of the region under x^3 and above [1, 7] in which the sampling numbers c_i are (*a*) left endpoints, (*b*) right endpoints, (*c*) midpoints.

In Exercises 25 to 27 evaluate $\Sigma_{i=1}^{n} f(c_i)(x_i - x_{i-1})$ for the given data.

25 $f(x) = \sqrt{x}$, $x_0 = 1$, $x_1 = 3$, $x_2 = 5$, $c_1 = 1$, $c_2 = 4$ $(n = 2)$

26 $f(x) = \sqrt[3]{x}$, $x_0 = 0$, $x_1 = 1$, $x_2 = 4$, $x_3 = 10$, $c_1 = 0$, $c_2 = 1$, $c_3 = 8$ $(n = 3)$

27 $f(x) = 1/x$, $x_0 = 1$, $x_1 = 1.25$, $x_2 = 1.5$, $x_3 = 1.75$, $x_4 = 2$, $c_1 = 1$, $c_2 = 1.25$, $c_3 = 1.6$, $c_4 = 2$ $(n = 4)$

28 Write the expression

$$c^{n-1} + c^{n-2}d + c^{n-3}d^2 + \cdots + d^{n-1}$$

in the sigma notation.

29 A computer program designed to evaluate a certain function f takes kn^2 seconds to evaluate $f(n)$, where k is a constant. It takes 3 seconds to evaluate $f(1000)$. About how long will it take to evaluate $f(n)$ for all n from 1 to 1,000,000?

30 (Calculator) Let $S_n = \Sigma_{i=1}^{n} 1/[i(i+1)]$.

(*a*) Compute S_2, S_3, S_4, and S_5 to at least four decimal places.

(*b*) What do you think happens to S_n as $n \to \infty$?

31 (Calculator) Let $S_n = \Sigma_{i=n}^{2n} 1/i$. For example,

$$S_3 = \tfrac{1}{3} + \tfrac{1}{4} + \tfrac{1}{5} + \tfrac{1}{6}.$$

(*a*) Compute S_1, S_2, S_3, and S_4 to at least three decimal places.

(*b*) What do you think happens to S_n as $n \to \infty$?

(*c*) Show that $S_n > \frac{1}{2}$ for all positive integers n.

5.3 THE DEFINITE INTEGRAL

There are two main concepts in calculus: the derivative and the definite integral. This section defines the definite integral and uses it to solve the four problems in Sec. 5.1. In Sec. 5.1 there were only *estimates* of the four quantities; now we will obtain their exact values. Sums of the form

$$\sum_{i=1}^{n} f(c_i)(x_i - x_{i-1})$$

were used to estimate certain quantities such as area, mass, distance, and volume. The larger n is and the shorter the sections $[x_{i-1}, x_i]$ are, the closer we would expect these approximating sums to be to the quantity we are trying to find. We are really interested in *what happens to these approximating sums as all the sections in the partition are chosen smaller and smaller.* This leads to the notion of the definite integral of a function over an interval, which will be defined after we introduce a measure of the "fineness" of a partition.

Definition *Mesh.* The **mesh** of a partition is the length of the longest section (or sections) in the partition.

For instance, the partition used in Sec. 5.1 has mesh equal to $\frac{1}{2}$.

Definition *The definite integral of a function f over an interval* $[a, b]$. If f is a function defined on $[a, b]$ and the sums $\Sigma_{i=1}^{n} f(c_i)(x_i - x_{i-1})$ approach a certain number as the mesh of partitions of $[a, b]$ shrinks toward 0 (no matter how the sampling number c_i is chosen in $[x_{i-1}, x_i]$), that certain number is called the **definite integral of f over** $[a, b]$.

Notation for the definite integral: $\int_a^b f(x)\,dx$

The definite integral is also called the **definite integral of f from a to b** and the **integral of f from a to b**. This number is denoted $\int_a^b f(x)\,dx$. The symbol $\int$ comes from the letter s of sum; the dx traditionally suggests a small section of the x axis and will be more meaningful and useful later. It is important to realize that area, mass, distance traveled, and volume are merely applications of the definite integral. (It is a mistake to link the definite integral too closely with one of its applications, just as it narrows our understanding of the number 2 to link it always with the idea of two fingers.) The definite integral $\int_a^b f(x)\,dx$ is also called the **Riemann integral** in honor of the mathematician who defined it.

Slope, velocity, magnification, and density are particular interpretations or applications of the derivative, which is a purely mathematical concept defined as a limit:

$$\text{Derivative of } f \text{ at } x = \lim_{\Delta x \to 0} \frac{f(x + \Delta x) - f(x)}{\Delta x}.$$

Similarly, area, total distance, mass, and volume are just particular interpretations of the definite integral, which is also defined as a limit:

$$\text{Definite integral of } f \text{ over } [a, b] = \lim_{\text{mesh}\to 0} \sum_{i=1}^{n} f(c_i)(x_i - x_{i-1}).$$

In advanced calculus it is proved that if f is continuous, then

Recall that $x_0 = a$ and $x_n = b$.

$$\lim_{\text{mesh}\to 0} \sum_{i=1}^{n} f(c_i)(x_i - x_{i-1})$$

exists; that is, a continuous function always has a definite integral. For emphasis, we record this fact, an important result in advanced calculus, as a theorem.

Theorem *Existence of the definite integral.* Let f be a continuous function defined on $[a, b]$. Then the approximating sums

$$\sum_{i=1}^{n} f(c_i)(x_i - x_{i-1})$$

approach a single number as the mesh of the partition of $[a, b]$ approaches 0. Hence $\int_a^b f(x)\,dx$ exists. ■

To bring the definition down to earth, let us use it to evaluate the definite integral of a constant function.

EXAMPLE Let f be the function whose value at any number x is 4; that is, f is the constant function given by the formula $f(x) = 4$. Use only the definition of the definite integral to compute

$$\int_1^3 f(x)\,dx.$$

SOLUTION In this case, a typical partition has $x_0 = 1$ and $x_n = 3$. The approximating sum

$$\sum_{i=1}^{n} f(c_i)(x_i - x_{i-1})$$

becomes

$$\sum_{i=1}^{n} 4(x_i - x_{i-1})$$

since, no matter how the sampling number c_i is chosen, $f(c_i) = 4$. Now

$$\sum_{i=1}^{n} 4(x_i - x_{i-1}) = 4\sum_{i=1}^{n} (x_i - x_{i-1}) = 4(x_n - x_0),$$

since the sum is telescoping. Since $x_n = 3$ and $x_0 = 1$, it follows that all approximating sums have the same value, namely,

$$4(3 - 1) = 8.$$

It does not matter whether the mesh is small or where the c_i are picked in each section. Thus, as the mesh approaches 0,

$$\sum_{i=1}^{n} f(c_i)(x_i - x_{i-1})$$

approaches 8. Indeed, the sums are always 8. Thus

$$\int_1^3 4\,dx = 8. \quad ■$$

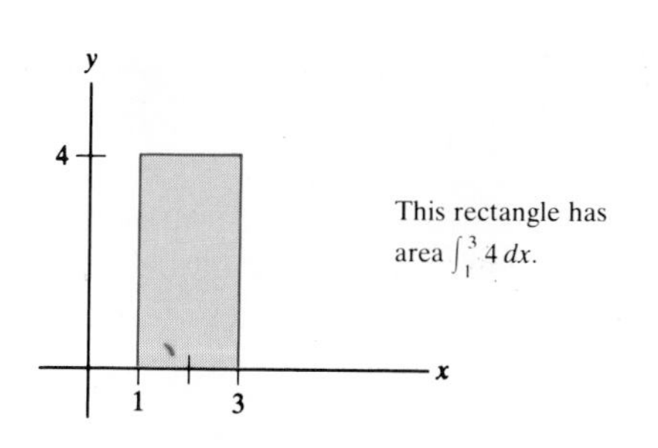

This rectangle has area $\int_1^3 4\,dx$.

We could have guessed the value of $\int_1^3 4\,dx$ by interpreting the definite integral as an area. To do so, draw a rectangle of height 4 and base coinciding with the interval [1, 3]. (See Fig. 5.17.) Since the area of a rectangle is its base times its height, it follows again that $\int_1^3 4\,dx = 8$.

Similar reasoning shows that for any constant function that has the fixed value c,

$$\int_a^b c\,dx = c(b - a).$$

Although area is the most intuitive of the interpretations, physical scientists, if they want to think of the definite integral concretely, should think of it as giving total mass if we know the density everywhere. This interpretation carries through easily to higher dimensions; the area interpretation does not.

It is the concept of the definite integral that links the four problems of Sec. 5.1, which are summarized below.

PROBLEM 1 The area under the curve $y = x^2$ and above $[0, 3]$ equals the definite integral $\int_0^3 x^2\,dx$.

PROBLEM 2 The mass of the string whose density is $5x^2$ grams per centimeter equals the definite integral $\int_0^3 5x^2\,dx$.

PROBLEM 3 The distance traveled by the engineer whose speed is $8t^2$ miles per hour at time t equals the definite integral $\int_0^3 8t^2\,dt$ (the t reminding us of time).

PROBLEM 4 The volume of the tent equals the definite integral $\int_0^3 x^2\,dx$.

It is somewhat satisfying to have reduced all four problems to one problem, that of evaluating the definite integral of x^2 from 0 to 3, $\int_0^3 x^2\,dx$. Once it is found, all four problems are solved.

But it is one thing to define a certain limit; it is quite a different matter to evaluate it. Recall that the derivative was defined in Sec. 3.2, but it took several sections to develop techniques for computing it. Just as we resorted to algebra to differentiate x^2, x^3, and $\sqrt{x}$ in Sec. 3.2, we will use algebra to compute $\int_0^3 x^2\,dx$.

Evaluating $\int_a^b x^2\,dx$

It takes no more work to evaluate $\int_0^b x^2\,dx$ than $\int_0^3 x^2\,dx$, so let us evaluate $\int_0^b x^2\,dx$. Although the reasoning will be free of any particular interpretation, the definite integral $\int_0^b x^2\,dx$ can be thought of as the area of the region under the curve $y = x^2$ and above the interval $[0, b]$, as shown in Fig. 5.18.

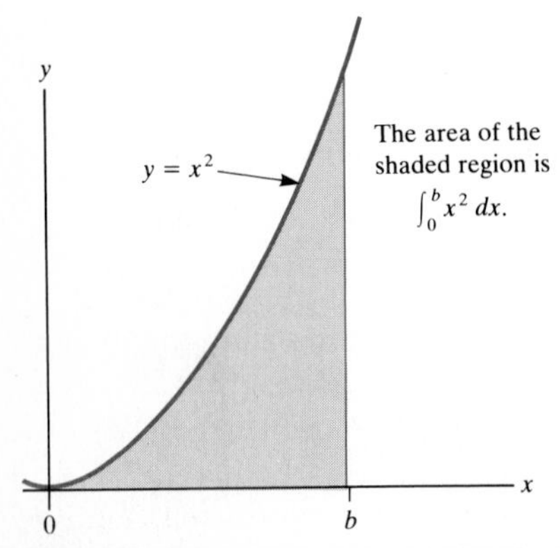

Figure 5.18

We assume that $\int_0^b x^2\,dx$ exists; it remains only to examine certain approximating sums to find that limit. For ease of computation, we use partitions in which all the sections have the same length. As sampling points, we use right-hand endpoints.

So we let n be a positive integer and partition $[0, b]$ by the numbers

$$0, \frac{b}{n}, \frac{2b}{n}, \ldots, \frac{ib}{n}, \ldots, \frac{nb}{n} = b,$$

as shown in Fig. 5.19. Each section has length b/n. The ith interval is $[(i-1)b/n, ib/n]$; its right-hand endpoint is ib/n. The partition has $x_i = ib/n$, and $x_i - x_{i-1} = b/n$. The sampling point is $c_i = ib/n$. The function is x^2. Thus

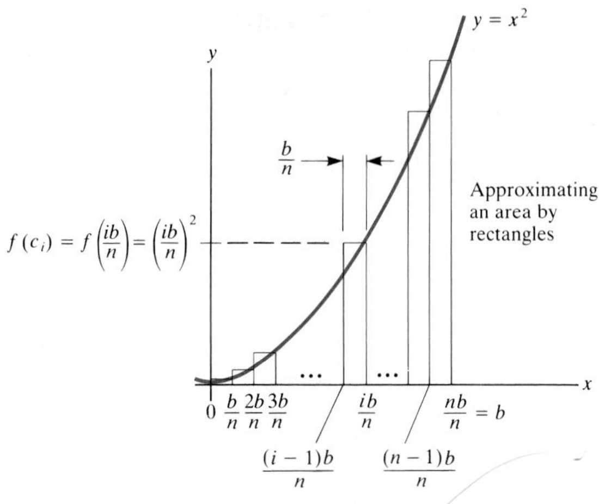

Figure 5.19

$$\sum_{i=1}^{n} f(c_i)(x_i - x_{i-1}) = \sum_{i=1}^{n} \left(\frac{ib}{n}\right)^2 \frac{b}{n}$$

$\sum_{i=1}^{n} i^2$ was found in Sec. 5.2.

$$= \frac{b}{n}\frac{b^2}{n^2}\sum_{i=1}^{n} i^2$$

$$= \frac{b^3}{n^3}\left(\frac{n^3}{3} + \frac{n^2}{2} + \frac{n}{6}\right)$$

$$= \frac{b^3}{3} + \frac{b^3}{2n} + \frac{b^3}{6n^2}$$

As $n \to \infty$, this approaches $b^3/3$. We conclude that

$$\int_0^b x^2\, dx = \frac{b^3}{3}. \tag{1}$$

In particular, when $b = 3$, we conclude from Eq. (1) that $\int_0^3 x^2\, dx = 3^3/3 = 9$.

Precise answers to the four problems

From this result we immediately conclude that:

The area under $y = x^2$ and above $[0, 3]$ is 9 square units.

The mass of the string in Problem 2 of Sec. 5.1 is $5 \cdot 9 = 45$ grams.

The engineer in Problem 3 of Sec. 5.1 travels $8 \cdot 9 = 72$ miles.

The volume of the tent in Problem 4 of Sec. 5.1 is 9 cubic feet.

The same technique that showed that $\int_0^b x^2\, dx = b^3/3$ can be used to show that

$$\int_a^b x^2\, dx = \frac{b^3}{3} - \frac{a^3}{3},$$

where $0 \le a \le b$. Of course, you might have already guessed this by considering it in terms of area. A sketch shows that the area from a to

b is equal to the difference between the area from 0 to b and the area from 0 to a. So we expect that

$$\int_a^b x^2\,dx = \int_0^b x^2\,dx - \int_0^a x^2\,dx$$
$$= \frac{b^3}{3} - \frac{a^3}{3}.$$

Exercise 5 presents a formal demonstration that does not depend on areas or sketches.

This is a good time to summarize the four main applications of the definite integral in full generality. Each has already been illustrated by one of the four problems in Sec. 5.1.

Area of a Plane Region as a Definite Integral Let S be some region in the plane whose area is to be found. Let L be a line in the plane which will be considered to be the x axis. Each line in the plane and perpendicular to L meets S in what shall be called a cross section. (If the line misses S, the cross section is empty.) See Figs. 5.20 to 5.22. Let the coordinate

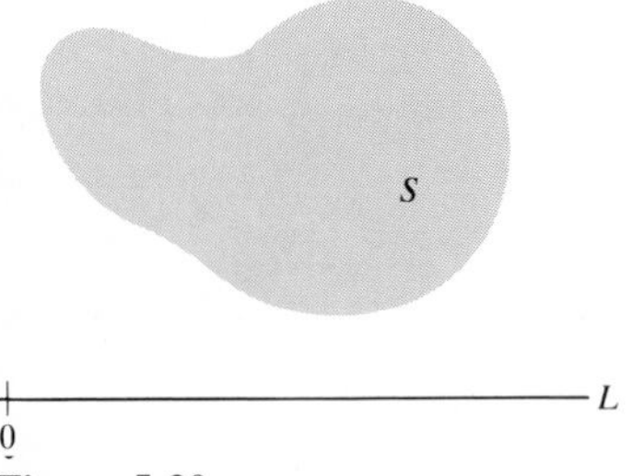

Figure 5.20

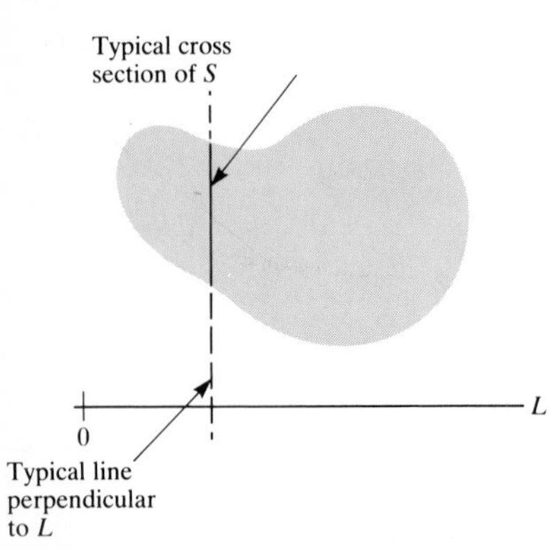

Figure 5.21

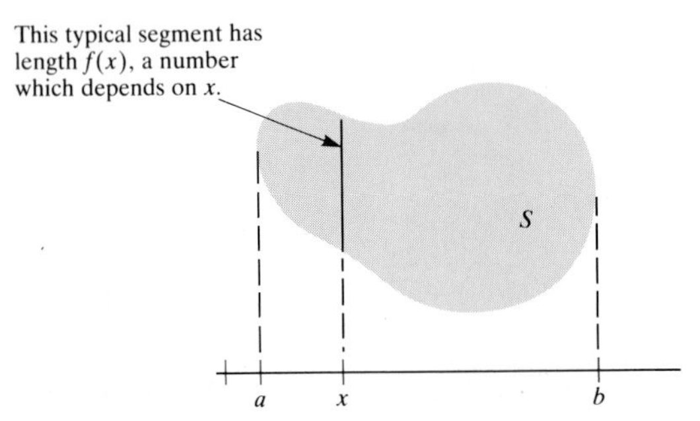

Figure 5.22

on L where the typical line meets L be x. The length of the typical cross section is denoted by $f(x)$. Assume that the lines that are perpendicular to L and that meet S intersect L in an interval whose ends are a and b. [In Problem 1 of Sec. 5.1, L is the x axis, $f(x) = x^2$, $a = 0$, and $b = 3$.]

To estimate the area of S, proceed just as for the region under $y = x^2$. First cut the interval $[a, b]$ into n sections by means of the numbers $x_0 = a, x_1, x_2, \ldots, x_n = b$, as in Fig. 5.23.

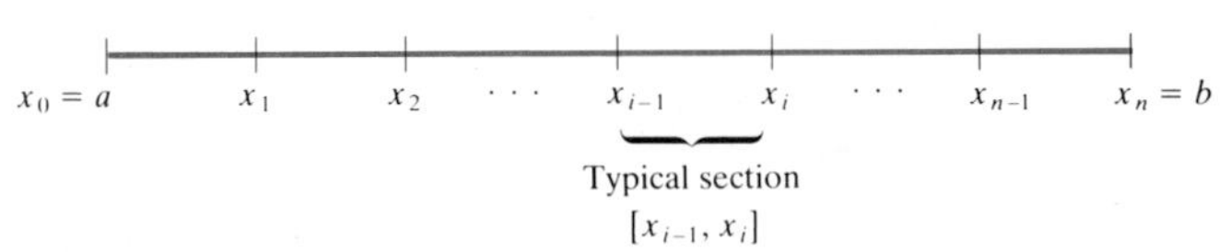

Figure 5.23

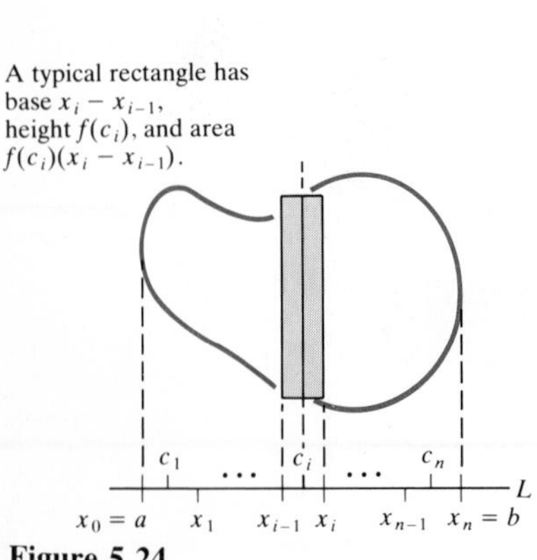

Figure 5.24

In each of these sections, select a number at random. In the section $[x_0, x_1]$ select c_1, in $[x_1, x_2]$ select c_2, and so on. Simply stated, in the ith interval, $[x_{i-1}, x_i]$, select c_i.

With this choice of x_i's and c_i's, form a set of rectangles whose typical member is shown in Fig. 5.24. Then $\Sigma_{i=1}^n f(c_i)(x_i - x_{i-1})$ is an estimate of the area of S. As the lengths $x_i - x_{i-1}$ are chosen to be smaller and smaller, we would expect that these sums tend toward the area of S.

But, by the definition of the definite integral, the sums

$$\sum_{i=1}^{n} f(c_i)(x_i - x_{i-1})$$

approach

$$\int_a^b f(x)\,dx$$

as the mesh → 0. Thus

Area of $S = \int_a^b f(x)\,dx$, where $f(x)$ is the length of a cross section of S.

In short, "area is the definite integral of cross-sectional length." In practical terms, this tells us that if we can compute definite integrals, then we can compute areas.

The Mass of a String as a Definite Integral A string is made of a material whose density may vary from point to point. (Such a string is called nonuniform or **nonhomogeneous**.) How would its total mass be computed if its density at each point is known?

First, place the string somewhere on the x axis and denote by $f(x)$ its density at x. The string occupies an interval $[a, b]$ on the axis, as in Fig. 5.25. Then cut the string into n sections $[x_{i-1}, x_i]$, $i = 1, 2, \ldots, n$,

Figure 5.25

as in Fig. 5.26. In each section the density is almost constant. So the mass of the ith section is approximately

$$f(c_i)(x_i - x_{i-1}),$$

where c_i is some point in $[x_{i-1}, x_i]$. (See Fig. 5.27.) Thus we see that

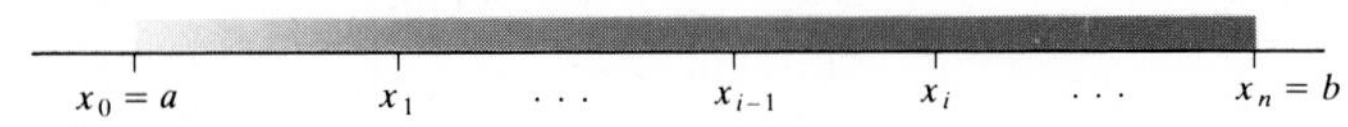

Figure 5.26

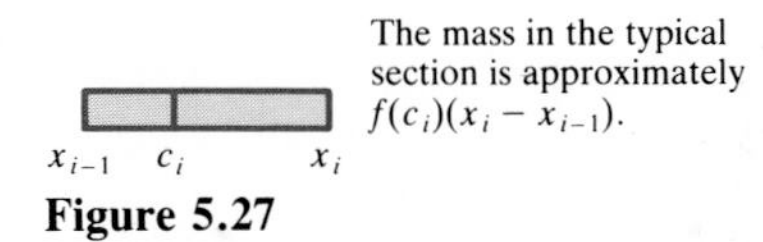

Figure 5.27

The mass in the typical section is approximately $f(c_i)(x_i - x_{i-1})$.

$\Sigma_{i=1}^{n} f(c_i)(x_i - x_{i-1})$ is an estimate of the total mass. And, what is more important, it seems plausible that as the mesh of the partition approaches 0, the sum $\Sigma_{i=1}^{n} f(c_i)(x_i - x_{i-1})$ approaches the mass of the string. [The case when $a = 0$, $b = 3$, and $f(x) = 5x^2$ is Problem 2 in Sec. 5.1.] But, by the definition of the definite integral, the sums

$$\sum_{i=1}^{n} f(c_i)(x_i - x_{i-1})$$

approach

$$\int_a^b f(x)\,dx$$

as the mesh → 0. Thus

$$\text{Mass} = \int_a^b f(x)\,dx, \text{ where } f(x) \text{ is the density at } x.$$

In short, "mass is the definite integral of density."

The Distance Traveled as a Definite Integral An engineer takes a trip that begins at time a and ends at time b. Imagine that at any time t during the trip her velocity is $f(t)$, depending on the time t. How far does she travel? [The case in which $a = 0$, $b = 3$, $f(t) = 8t^2$ is Problem 3 in Sec. 5.1.]

First, cut the time interval $[a, b]$ into smaller intervals by a partition and estimate the trip's length by summing the estimates of the distance the engineer travels during each of the time intervals. (See Fig. 5.28.)

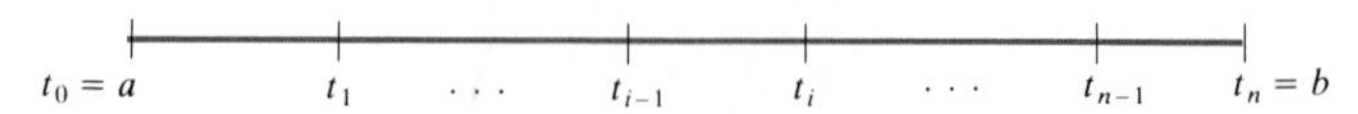

Figure 5.28

During a small interval of time, the velocity changes little. We thus expect to obtain a reasonable estimate of the distance covered during the ith time interval $[t_{i-1}, t_i]$ by observing the speedometer reading at some instant T_i in that interval, $f(T_i)$, and computing the product $f(T_i)(t_i - t_{i-1})$. Thus $\Sigma_{i=1}^n f(T_i)(t_i - t_{i-1})$ is an estimate of the length of the trip. Moreover, as the mesh of the partition approaches zero, the sum $\Sigma_{i=1}^n f(T_i)(t_i - t_{i-1})$ approaches the length of the trip.

Since these sums also approach the definite integral $\int_a^b f(t)\,dt$, we have

$$\text{Total distance} = \int_a^b f(t)\,dt, \text{ where } f(t) \text{ is the velocity at time } t.$$

In short, "distance is the definite integral of velocity."

The Volume of a Solid Region as a Definite Integral Suppose that we wish to compute the volume of a solid S, and we happen to know the area $A(x)$ of each cross section made by planes in a fixed direction (see Fig. 5.29).

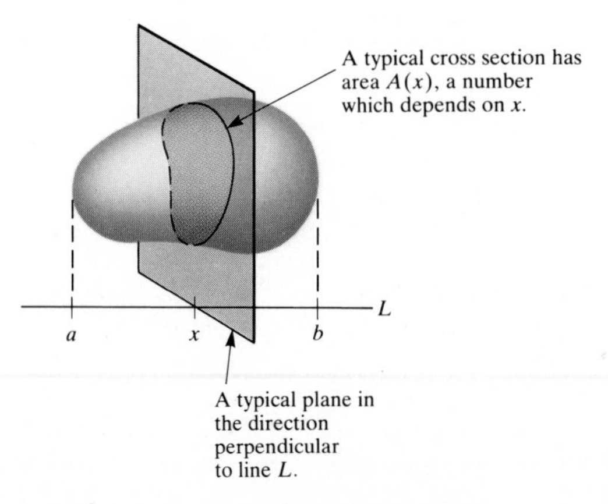

Figure 5.29

A typical slab is an irregular cylinder of cross-sectional area $A(c_i)$ and thickness $x_i - x_{i-1}$.

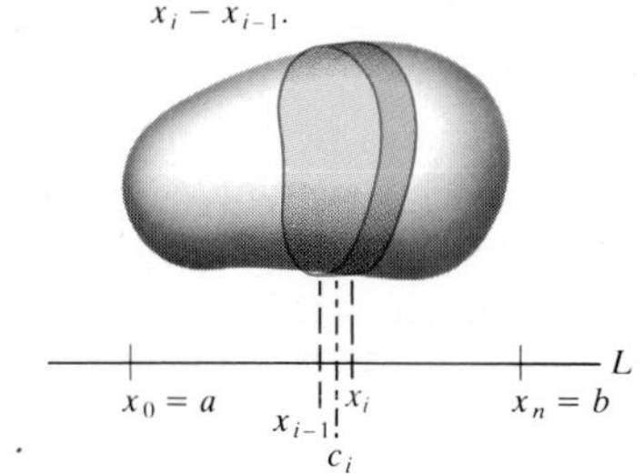

Figure 5.30

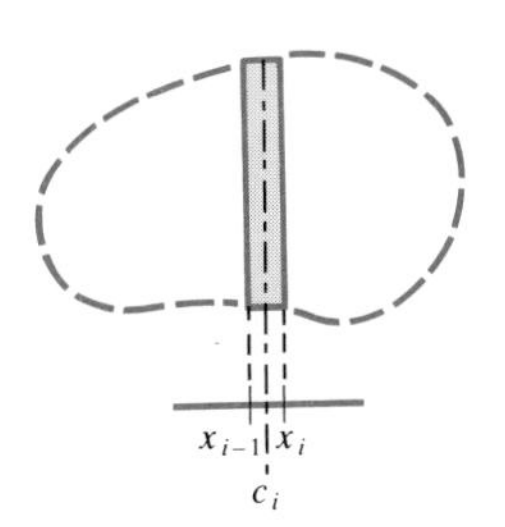

Figure 5.31

[In the case of the tent in Problem 4 of Sec. 5.1, $a = 0$, $b = 3$, and $A(x) = x^2$.]

Every partition of $[a, b]$ and selection of c's provides an estimate of the volume of S, the sum of the volumes of slabs. A typical slab is shown in perspective and side views in Figs. 5.30 and 5.31.

Thus $\Sigma_{i=1}^{n} A(c_i)(x_i - x_{i-1})$ is an estimate of the volume of the solid S. As the mesh of the partition shrinks, the slabs become thin and the sum of their volumes becomes a more and more accurate estimate of the volume of S. As the mesh of the partition approaches 0, we see that the sum $\Sigma_{i=1}^{n} A(c_i)(x_i - x_{i-1})$ approaches the volume of S.

From this it follows that

$$\text{Volume of } S = \int_a^b A(x)\,dx, \text{ where } A(x) \text{ is the cross-sectional area at } x.$$

In short, "volume is the definite integral of cross-sectional area."

The following table shows the similarities of these four general types of problems. To emphasize these similarities, all the functions, whether cross-sectional length, density, velocity, or cross-sectional area, are denoted by the same symbol $f(x)$.

Spend some time examining this table. The concepts it describes will be used often.

$f(x)$	$\sum_{i=1}^{n} f(c_i)(x_i - x_{i-1})$	$\int_a^b f(x)\,dx$
Variable length of cross section of set in plane	Approximation to area of the set in the plane	The area of the set in the plane
Variable density of string	Approximation to mass of the string	The mass of the string
Variable velocity	Approximation to the distance traveled	The distance traveled
Variable area of cross section of a solid	Approximation to the volume of the solid	The volume of the solid

Underlying these four applications is one purely mathematical concept, the definite integral, $\int_a^b f(x)\,dx$. The definite integral is defined as a certain limit; it is a number. Chapter 7 develops methods for computing this limit for many functions. It is essential to keep the definition of $\int_a^b f(x)\,dx$ clear. Otherwise you may eventually get the definite integral confused with the methods for computing it.

EXERCISES FOR SEC. 5.3: THE DEFINITE INTEGRAL

1 Find the mesh of each of these partitions of [1, 6]:
(a) $x_0 = 1, x_1 = 2, x_2 = 3, x_3 = 4, x_4 = 5, x_5 = 6$
(b) $x_0 = 1, x_1 = 3, x_2 = 5, x_3 = 6$
(c) $x_0 = 1, x_1 = 4, x_2 = 4.5, x_3 = 5, x_4 = 5.5, x_5 = 6$

2 Find the mesh of each of these partitions of [2, 6]:
(a) $x_0 = 2, x_1 = 4, x_2 = 6$
(b) $x_0 = 2, x_1 = 3, x_2 = 6$
(c) $x_0 = 2, x_1 = 2.5, x_2 = 3, x_3 = 3.5, x_4 = 4, x_5 = 5, x_6 = 6$

3 Find the area under the curve $y = x^2$ and above the interval (a) [0, 5]; (b) [0, 4]; (c) [4, 5].

4 Figure 5.32 shows the curve $y = x^2$. What is the ratio between the shaded area under the curve and the area of the rectangle $ABCD$?

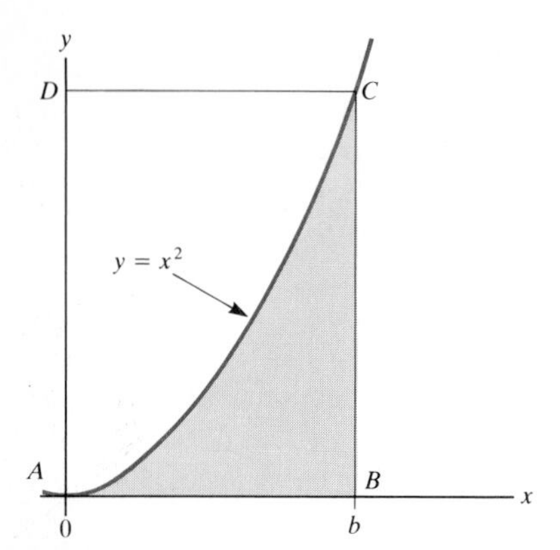

Figure 5.32

5 (This exercise obtains $\int_a^b x^2\,dx$.) Let n be a positive integer. Partition $[a, b]$ into n sections of equal length using the numbers $x_0, x_1, \ldots, x_n$.
(*a*) What is x_0?
(*b*) What is x_1?
(*c*) What is x_2? x_i?
(*d*) Using the right-hand endpoint of each section as sampling point, compute $\sum_{i=1}^n x_i^2 (x_i - x_{i-1})$.
(*e*) Use (*d*) to find $\int_a^b x^2\,dx$.

6 The same as Exercise 5, except use the left-hand endpoint of each section as sampling point.

7 A rocket moving with a varying speed travels at $f(t)$ miles per second at time t. Let $t_0, \ldots, t_n$ be a partition of $[a, b]$, and let $T_1, \ldots, T_n$ be sampling numbers. What is the physical interpretation of
(*a*) $t_i - t_{i-1}$? (*b*) $f(T_i)$? (*c*) $f(T_i)(t_i - t_{i-1})$?
(*d*) $\sum_{i=1}^{n} f(T_i)(t_i - t_{i-1})$? (*e*) $\int_a^b f(t)\,dt$?

8 A string occupying the interval $[a, b]$ has the density $f(x)$ grams per centimeter at x. Let $x_0, \ldots, x_n$ be a partition of $[a, b]$, and let $c_1, \ldots, c_n$ be sampling numbers. What is the physical interpretation of
(*a*) $x_i - x_{i-1}$? (*b*) $f(c_i)$?
(*c*) $f(c_i)(x_i - x_{i-1})$? (*d*) $\sum_{i=1}^{n} f(c_i)(x_i - x_{i-1})$?
(*e*) $\int_a^b f(x)\,dx$?

9 Water is flowing into a lake at the rate of $f(t)$ gallons per second at time t. Answer the five questions in Exercise 7 for this interpretation of the function f, using the same partition and sampling numbers as in Exercise 7.

10 A business firm has a flow of income at the rate of $f(t)$ dollars per year at time t. Consider f to be a continuous function. [Thus in a short period of time of Δt years near time t it earns approximately $f(t)\,\Delta t$ dollars.] Answer the five questions in Exercise 7 for this interpretation of the function, using the same partition and sampling numbers as in Exercise 7.

11 The density of a string is x^2 grams per centimeter at the point x. Find the mass in the interval $[1, 3]$.

12 An object is moving with a velocity of t^2 feet per second at time t seconds. How far does it travel during the time interval $[2, 5]$?

Exercises 13 to 18 concern the definite integral, free of any particular interpretation.

13 Estimate $\int_1^3 (1/x)\,dx$, using a partition into four sections of equal length and, as sampling points, (*a*) left endpoints, (*b*) right endpoints.

14 Estimate $\int_0^3 2^x\,dx$, using a partition into three sections of equal length and, as sampling points, (*a*) left endpoints, (*b*) right endpoints.

15 Estimate $\int_0^1 x^3\,dx$, using a partition into five sections of equal length and, as sampling points, (*a*) left endpoints, (*b*) right endpoints.

16 (Calculator) Estimate $\int_0^1 x^3\,dx$, using a partition into 10 sections of equal length and left endpoints as sampling points.

17 (Calculator) Estimate $\int_0^{\pi/2} \sin x\,dx$, using the partition $x_0 = 0$, $x_1 = \pi/6$, $x_2 = \pi/4$, $x_3 = \pi/3$, and $x_4 = \pi/2$ and, as sampling points, (*a*) left endpoints, (*b*) right endpoints. Express the answers to two decimal places.

18 (Calculator) Estimate $\int_0^1 \sqrt{x}\,dx$, using a partition into 10 sections of equal length and, as sampling points, (*a*) left endpoints, (*b*) right endpoints, (*c*) midpoints.

19 (Exercise 21 of Sec. 5.2 provides a formula for $\sum_{i=1}^n i^3$.)
(*a*) Using the method of this section to find $\int_0^b x^2\,dx$, find $\int_0^b x^3\,dx$.
(*b*) Find the area under the curve $y = x^3$ and above the interval $[1, 2]$.

20 In Exercise 22 of Sec. 5.2 it was shown that

$$\sum_{i=1}^n i^4 = \frac{n^5}{5} + \frac{n^4}{2} + \frac{n^3}{3} - \frac{n}{30}$$

(*a*) Obtain a formula for $\sum_{i=1}^n i^5$.
(*b*) Using the method of this section, find $\int_0^b x^5\,dx$.

21 Estimate $\int_1^3 (1/x^2)\,dx$, using a partition into four sections of equal length and left endpoints as sampling points.

22 Estimate $\int_1^4 (2x + 1)\,dx$, using a partition into three sections of equal length and right endpoints as sampling points.

23 Figure 5.33 shows the graph of $y = x$ and the region below it and above the interval $[a, b]$, $0 \le a < b$. Using

the formula for the area of a trapezoid (see inside back cover), show that $\int_a^b x\,dx = b^2/2 - a^2/2$.

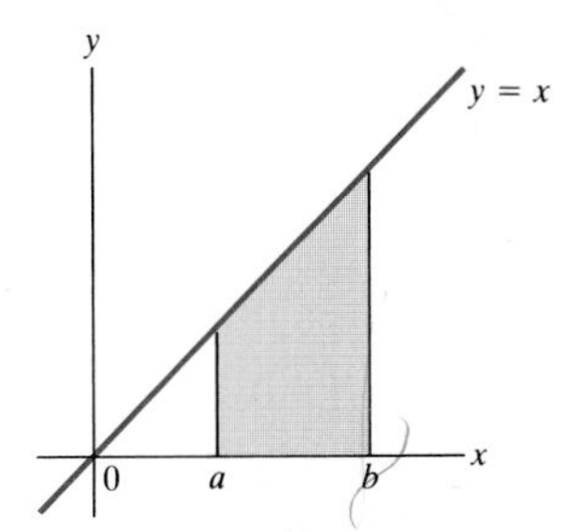

Figure 5.33

24 (See Exercise 23.)

(*a*) Set up an appropriate definite integral $\int_a^b f(x)\,dx$ which equals the volume of the headlight in Fig. 5.34 whose cross section by a typical plane perpendicular to the x axis at x is a circle whose radius is $\sqrt{x/\pi}$.

(*b*) Evaluate the definite integral in (*a*) with the aid of Exercise 23.

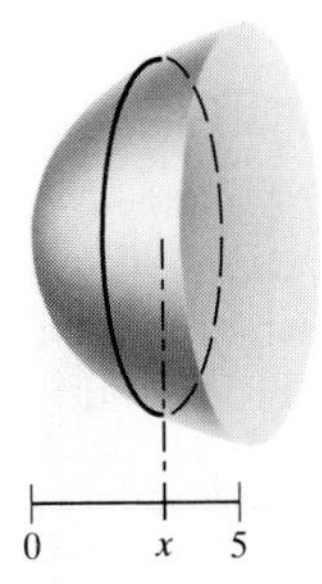

Figure 5.34

25 By considering Fig. 5.35, in particular the area of region ACD, show that $\int_0^a \sqrt{x}\,dx = \frac{2}{3}a^{3/2}$.

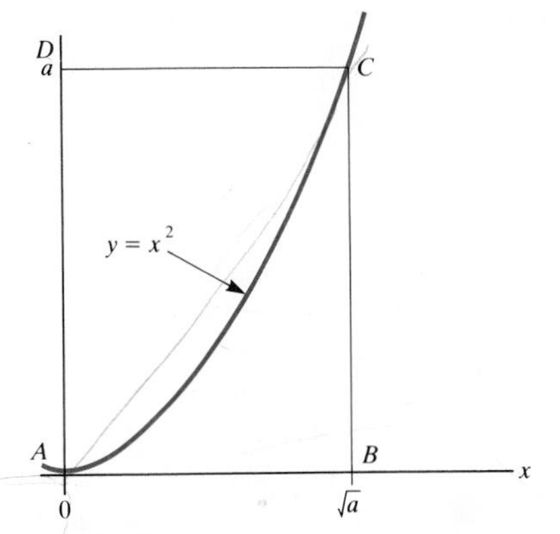

Figure 5.35

26 Show that the volume of a right circular cone of radius a and height h is $\pi a^2h/3$. (*Suggestion:* First show that a cross section by a plane perpendicular to the axis of the cone and a distance x from the vertex is a circle of radius ax/h.)

■

27 (*a*) Sketch a graph of $y = 1/(1 + x^2)$ for $0 \le x \le 1$.

(*b*) Let A be the area under the graph in (*a*) and above $[0, 1]$. Show that $A < 1$.

(*c*) Use elementary geometry to show $\frac{3}{4} < A$.

(*d*) Use a partition of $[0, 1]$ into five sections to obtain lower and upper estimates of A. It will be shown that $\int_0^1 1/(1 + x^2)\,dx = \pi/4$, hence that $A \approx 0.7854$.

The next two exercises concern a property of the function $1/x$ first noticed in the seventeenth century. (See Exercise 18 in Sec. 5.1.)

28 (*a*) Write out the typical approximating sum for the area under the curve $y = 1/x$ and above the interval $[1, 2]$, using left endpoints as the c_i and the typical partition $x_0 = 1, x_1, \ldots, x_n = 2$.

(*b*) Show that $3x_0, 3x_1, \ldots, 3x_n$ is a partition of the interval $[3, 6]$. Show that if the left endpoints are used to form the approximating sum for the area under $y = 1/x$ and above $[3, 6]$, then this sum has the same value as the one obtained in (*a*).

(*c*) Show that the area under the curve $y = 1/x$ and above $[1, 2]$ equals the area under the curve $y = 1/x$ and above $[3, 6]$.

29 (See Exercise 28.)

(*a*) Show that if $A(t)$ is the area under the curve $y = 1/x$ and above $[1, t]$, then $A(2) = A(6) - A(3)$.

(*b*) Show that if $x > 1$ and $y > 1$, then $A(x)$ is equal to $A(xy) - A(y)$.

(*c*) By (*b*), $A(xy) = A(x) + A(y)$ for x and y greater than 1. What famous functions f have the property that $f(xy) = f(x) + f(y)$ for all positive x and y?

Exercises 30 and 31 are related.

30 Let f be increasing on the interval $[a, b]$. Let $x_0, x_1, \ldots, x_n$ be a partition of $[a, b]$ of mesh p. Show that the largest possible approximating sum for $\int_a^b f(x)\,dx$ formed with this partition differs from the smallest possible approximating sum for $\int_a^b f(x)\,dx$ formed with the same partition by at most $[f(b) - f(a)]p$.

31 Let f be an increasing function on $[0, 2]$. Assume that $f(0) = 3$ and $f(2) = 8$. You are going to estimate $\int_0^2 f(x)\,dx$ by an approximating sum involving n sections of equal length. How large should you choose n in order to be sure the error is at most 0.001?

32 If you make three tents out of paper with the pattern given in Sec. 5.1, you will be able to fit them together easily to form a cube. If you have more geometric intuition than time, you might prefer to see that this is so by examining Fig. 5.36. Use this information to find the volume of the tent directly. (This trick solves all four problems in Sec. 5.1, since the answer to one determines the answers to the other three.)

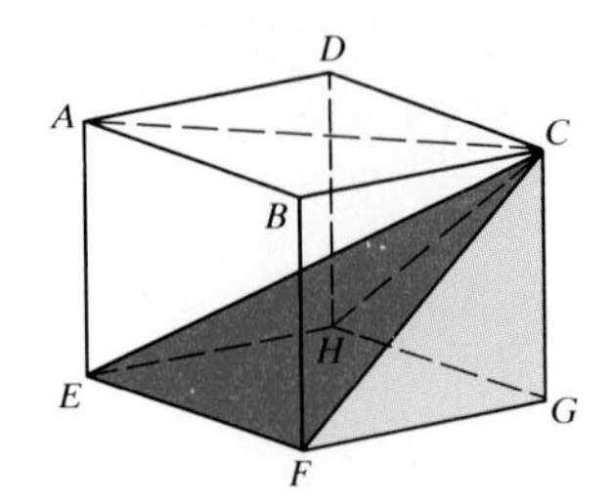

Figure 5.36

5.4 THE FUNDAMENTAL THEOREMS OF CALCULUS

This section shows that there is an intimate connection between the definite integral and the derivative. This relationship, expressed in the fundamental theorems of calculus, provides a tool for computing many, but not all, definite integrals without having to form a single approximating sum. The argument will be intuitive. Section 5.6 approaches the fundamental theorems of calculus from a purely mathematical point of view.

The First Fundamental Theorem of Calculus

In Chap. 3 it was shown that *velocity is the derivative of the distance*. In Sec. 5.3 it was shown that *the definite integral of velocity is the change in distance*. These two facts suggest that there is a close relation between derivatives and definite integrals. In order to express this relation mathematically, let us introduce some mathematical symbols to describe these observations about velocity and change in distance.

For convenience, assume velocity positive (so particle moves to the right).

Let x denote time, and let $F(x)$ denote the coordinate of a particle moving on a line. Then the velocity at time x is the derivative $F'(x)$. The change in distance of the moving particle from time a to time b is

$$\text{Final coordinate} - \text{initial coordinate} = F(b) - F(a).$$

The assertion "the definite integral of velocity is the change in distance" now reads mathematically

$$\int_a^b F'(x)\,dx = F(b) - F(a).$$

This equation generalizes the statement "rate · time = distance," which is valid for a particle moving at a constant speed.

This physical argument suggests that there is a purely mathematical theorem in the background. Moreover, computational evidence from Sec. 5.3 also points in the same direction for the particular function $F(x) = x^3/3$. It was shown in Sec. 5.3 that

$$\int_a^b x^2\,dx = \frac{b^3}{3} - \frac{a^3}{3}, \qquad 0 \le a < b.$$

If $F(x) = x^3/3$, then $F'(x) = x^2$, and the equation

$$\int_a^b x^2\,dx = \frac{b^3}{3} - \frac{a^3}{3}$$

now reads

$$\int_a^b F'(x)\,dx = F(b) - F(a).$$

Thus two separate lines of reasoning both suggest the general and purely mathematical result:

This equation links the derivative and the definite integral.

$$\int_a^b F'(x)\,dx = F(b) - F(a),$$

that is, "the definite integral of the derivative of a function over an interval is simply the difference in the values of the function at the ends of the interval."

As it stands, the conjecture is not quite correct. A function F may have such a wild derivative F' that $\int_a^b F'(x)\,dx$ does not exist; that is, the approximating sums do not approach a single number as the mesh of the partition approaches 0. But, as mentioned in Sec. 5.3, it is proved in advanced calculus that if a function f is continuous on the interval $[a, b]$, then $\int_a^b f(x)\,dx$ *does* exist.

The observations about velocity and $\int_a^b x^2\,dx$ suggest the following theorem, whose proof will be presented in Sec. 5.6.

First Fundamental Theorem of Calculus If f is continuous on $[a, b]$ and if F is an antiderivative of f, then

$$\int_a^b f(x)\,dx = F(b) - F(a). \quad \blacksquare$$

The symbol f is introduced for F' with a view toward applying this theorem to the computation of definite integrals. It says, "If you want to compute $\int_a^b f(x)\,dx$, search for a function F whose derivative is f, that is,

$$F' = f.$$

Then $\int_a^b f(x)\,dx$ is just $F(b) - F(a)$."

The following examples will exhibit the power and the limitations of the fundamental theorem of calculus. (Generally, the adjective "first" is omitted.)

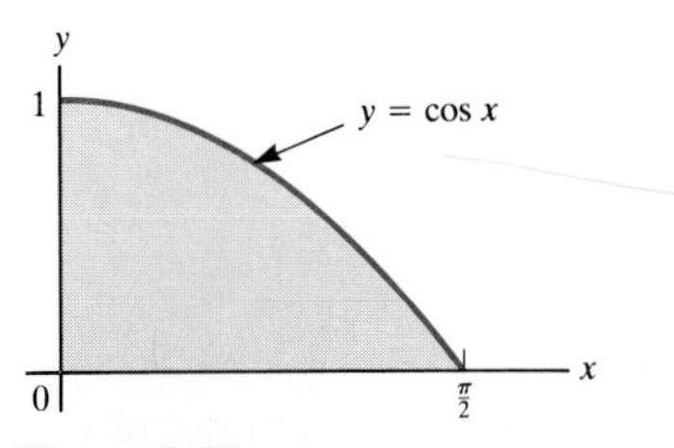

Figure 5.37

EXAMPLE 1 Find the area of the region under the curve $y = \cos x$, above the x axis, and between $x = 0$ and $x = \pi/2$. (See Fig. 5.37.)

SOLUTION As was shown in Sec. 5.3, area is the definite integral of the cross-sectional length. In this case,

$$\text{Area} = \int_0^{\pi/2} \cos x\,dx.$$

The fundamental theorem of calculus asserts that if we can find a function F such that

This shows why an antiderivative is so useful.

$$F'(x) = \cos x,$$

then the definite integral can be evaluated easily, as $F(\pi/2) - F(0)$. Now, in Chap. 3 it was shown that the derivative of the sine function is the cosine function. So let

$$F(x) = \sin x.$$

The fundamental theorem of calculus then says

$$\int_0^{\pi/2} \cos x \, dx = F\left(\frac{\pi}{2}\right) - F(0)$$

$$= \sin\frac{\pi}{2} - \sin 0 = 1 - 0 = 1. \quad ■$$

When we evaluate $\int_0^{\pi} \cos x \, dx$, we obtain $\sin \pi - \sin 0 = 0 - 0 = 0$. What does this say about areas? Inspection of Fig. 5.38 shows what is happening.

For x in $[\pi/2, \pi]$, $\cos x$ is negative and the curve $y = \cos x$ lies *below* the x axis. If we interpret the corresponding area as negative, then we see that it cancels with the area from 0 to $\pi/2$. So let us agree that when we say "$\int_a^b f(x)\,dx$ represents area under the curve $y = f(x)$," we mean that it represents the area between the curve and the x axis, with area below the x axis taken as negative.

"Algebraic" area

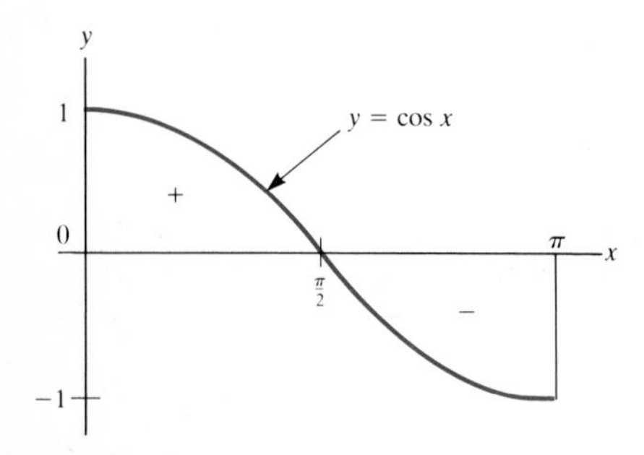

Figure 5.38

EXAMPLE 2 In Sec. 5.3 it was noted that

$$\int_a^b x^2 \, dx = \frac{b^3}{3} - \frac{a^3}{3},$$

when both a and b are nonnegative and $a < b$. What does the fundamental theorem of calculus say about $\int_a^b x^2 \, dx$ for any a and b, $a < b$?

SOLUTION The function $F(x) = x^3/3$ is an antiderivative of x^2. According to the fundamental theorem of calculus,

$$\int_a^b x^2 \, dx = F(b) - F(a) = \frac{b^3}{3} - \frac{a^3}{3}.$$

This holds even if a or b is negative, provided that a is less than b. ■

The next two examples form a fable whose moral should be remembered.

EXAMPLE 3 Compute $\int_0^{\pi/4} x \cos x \, dx$.

SOLUTION To apply the fundamental theorem of calculus, it is necessary to find an antiderivative of $x \cos x$. It happens that the derivative of the function $x \sin x + \cos x$ equals $x \cos x$. (Chapter 7 presents some methods for finding an antiderivative, when it is possible to do so.) The fundamental theorem of calculus asserts that if

$$F(x) = x \sin x + \cos x,$$

then $$\int_0^{\pi/4} x \cos x \, dx \underset{\text{FTC}}{=} F\left(\frac{\pi}{4}\right) - F(0)$$

The letters FTC under the equality sign record the use of the fundamental theorem of calculus.

$$= \left(\frac{\pi}{4}\sin\frac{\pi}{4} + \cos\frac{\pi}{4}\right) - (0 \cdot \sin 0 + \cos 0)$$

$$= \left(\frac{\pi}{4}\cdot\frac{\sqrt{2}}{2} + \frac{\sqrt{2}}{2}\right) - (0 \cdot 0 + 1)$$

$$= \frac{\pi\sqrt{2}}{8} + \frac{\sqrt{2}}{2} - 1. \quad \blacksquare$$

EXAMPLE 4 Compute $\int_0^{\pi/4} x \tan x \, dx$.

ATTEMPT AT SOLUTION To apply the fundamental theorem of calculus, it is necessary to find a function F such that

$$F'(x) = x \tan x.$$

Elementary functions

As will be shown, there is such a function F. However, mathematicians have proved that F is *not an elementary function*. That is, F is not expressible in terms of polynomials, logarithms, exponentials, trigonometric functions, or any composition of these functions. We are therefore blocked, for the fundamental theorem of calculus is of use in computing $\int_a^b f(x)\,dx$ only if f is "nice" enough to be the derivative of an elementary function. ■

A moral about antiderivatives

The moral of these last two examples is this: It is not easy to tell by glancing at f whether the desired F is elementary. After all, $x \tan x$ looks no more complicated than $x \cos x$, yet it is not the derivative of an elementary function, while $x \cos x$ is.

As another example, $\cos\sqrt{x}$ looks more complicated than $\cos x^2$. Yet it turns out that $\cos\sqrt{x}$ is the derivative of an elementary function, while $\cos x^2$ is not. [It is not hard to check that $(2\sqrt{x}\sin\sqrt{x} + 2\cos\sqrt{x})'$ is $\cos\sqrt{x}$.]

Every time that we differentiate an elementary function we get an elementary function. For instance, $(\sqrt{x})' = 1/(2\sqrt{x})$, $(x^3 - 3x^2)' = 3x^2 - 6x$, and $(\tan x)' = \sec^2 x$. But if you start with an elementary function f and search for an elementary function F whose derivative is to be f, you may be frustrated—not because it may be hard to find F—but because no such F exists.

The Second Fundamental Theorem of Calculus

There is a second theorem, closely related to the first fundamental theorem of calculus, which is also called the fundamental theorem of calculus. It describes the connection between the derivative and the definite integral in a different way from the first fundamental theorem of calculus.

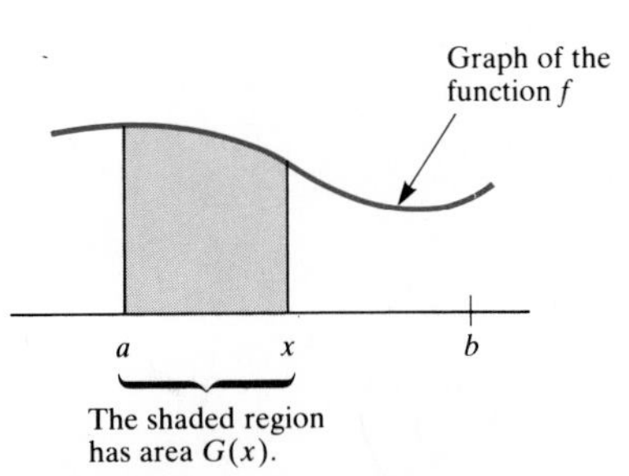

Figure 5.39

Let f be a continuous function such that $f(x)$ is positive for x in $[a, b]$. For x in $[a, b]$, let $G(x)$ be the area of the region under the graph of f and above the interval $[a, x]$, as shown in Fig. 5.39. Let Δx be a small positive number. Then $G(x + \Delta x)$ is the area under the graph of f and above the interval $[a, x + \Delta x]$. Consequently,

$$\Delta G = G(x + \Delta x) - G(x)$$

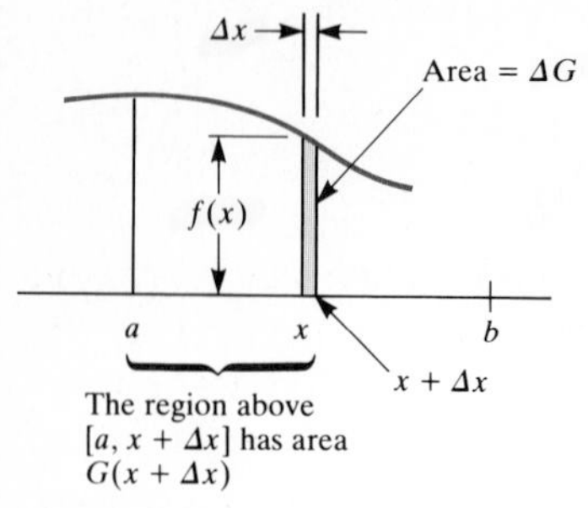

Figure 5.40

is the area of the narrow strip shaded in Fig. 5.40. When Δx is small, the narrow shaded strip above $[x, x + \Delta x]$ resembles a rectangle of base Δx and height $f(x)$, with area $f(x)\,\Delta x$. Therefore, it seems reasonable that when Δx is small,

$$\frac{\Delta G}{\Delta x} \text{ is approximately } f(x).$$

In short, it seems plausible that

$$\lim_{\Delta x \to 0} \frac{\Delta G}{\Delta x} = f(x).$$

Briefly,
$$G'(x) = f(x).$$

Now
$$G(x) = \int_a^x f(t)\,dt,$$

since area is the definite integral of the cross-sectional length.

Considerations of area thus lead us to suspect that the equation

$$\frac{d}{dx}\left(\int_a^x f(t)\,dt\right) = f(x)$$

holds for definite integrals in general. This equation says that "the derivative of the definite integral of f with respect to the right end coordinate of the interval is simply f evaluated at that coordinate." This is the substance of the second fundamental theorem of calculus.

Second Fundamental Theorem of Calculus Let f be continuous on an open interval containing the interval $[a, b]$. Let

$$G(x) = \int_a^x f(t)\,dt$$

for $a \le x \le b$. Then G is differentiable on $[a, b]$ and its derivative is f; that is,

$$G'(x) = f(x). \quad \blacksquare$$

As a consequence of this theorem, every continuous function is the derivative of some function. This is stated as a corollary for emphasis.

Corollary Let f be continuous on an interval $[a, b]$. Then f is the derivative of some function.

Proof Let $F(x) = \int_a^x f(t)\,dt$. Then, by the second fundamental theorem,

$$F'(x) = f(x);$$

that is,
$$\frac{d}{dx}\left(\int_a^x f(t)\,dt\right) = f(x).$$

This proves the corollary. $\blacksquare$

EXAMPLE 5 Find a function whose derivative is $\sin x^2$.

SOLUTION The function F defined by

$$F(x) = \int_0^x \sin t^2 \, dt$$

is such a function. ■

The information obtained in Example 5 is of no use in trying to exploit the fundamental theorem of calculus to compute, say, $\int_0^{\pi/4} \sin x^2 \, dx$. Although there is a function whose derivative is $\sin x^2$, there is no *elementary* function whose derivative is $\sin x^2$; hence other means must be used to compute $\int_0^{\pi/4} \sin x^2 \, dx$. (The proof that there is no such elementary function is complicated and is given in graduate-level courses.)

EXAMPLE 6 Differentiate the function $f(x) = \int_0^{x^2} \sqrt{1 + t^2} \, dt$.

SOLUTION The upper limit of integration is x^2, not x. The chain rule will be needed. Let

$$y = \int_0^{x^2} \sqrt{1 + t^2} \, dt.$$

How to differentiate a definite integral if upper limit is a function of x

Then

$$y = \int_0^u \sqrt{1 + t^2} \, dt, \qquad \text{where } u = x^2.$$

By the second fundamental theorem of calculus,

$$\frac{dy}{du} = \sqrt{1 + u^2}.$$

The chain rule then says that

$$\begin{aligned} \frac{dy}{dx} = \frac{dy}{du} \cdot \frac{du}{dx} &= \sqrt{1 + u^2} \cdot 2x \\ &= \sqrt{1 + (x^2)^2} \cdot 2x \\ &= 2x\sqrt{1 + x^4}. \quad ■ \end{aligned}$$

We have not given mathematical proofs of the fundamental theorems of calculus, only plausibility arguments. The first we motivated by considering velocity; the second, by area under a curve. In Sec. 5.6 mathematical arguments are given that are free of any particular application of the definite integral.

EXERCISES FOR SEC. 5.4: THE FUNDAMENTAL THEOREMS OF CALCULUS

1 Let k be a constant, and let a be a fixed rational number, $a \neq -1$. Show by differentiation that $kx^{a+1}/(a + 1)$ is an antiderivative of kx^a.

2 Use the formula in Exercise 1 to find an antiderivative of
(*a*) $5x^3$ (*b*) $-4x^2$ (*c*) x^{-4} (*d*) $5/x^3$
(*e*) $\sqrt{x}$ (*f*) $1/\sqrt{x}$ (*g*) $\sqrt[3]{x}$ (*h*) $1/x^2$

In Exercises 3 to 12 use the fundamental theorem of calculus to evaluate the given quantities.

3 The area of the region under the curve $3x^2$ and above $[1, 4]$.

4 The area of the region under the curve $1/x^2$ and above $[2, 3]$.

5 The area of the region under the curve $6x^4$ and above $[-1, 1]$.

6 The area of the region under the curve $\sqrt{x}$ and above $[25, 36]$.

7 The distance an object travels from time $t = 1$ second

to time $t = 2$ seconds, if its speed at time t seconds is t^5 feet per second.

8 The distance an object travels from time $t = 1$ second to time $t = 8$ seconds if its speed at time t seconds is $7\sqrt[3]{t}$ feet per second.

9 The total mass of a string in the section $[1, 2]$ if its density at x is $5x^3$ grams per centimeter.

10 The total mass of a string in the section $[\frac{1}{4}, 1]$ if its density at x is $4\sqrt{x}$ grams per centimeter.

11 The volume of a solid located between a plane at $x = 1$ and a plane located at $x = 5$ if the cross-sectional area of the solid by a plane corresponding to x is $6x^3$ square centimeters. (Assume that the planes are all perpendicular to the x axis.)

12 Like Exercise 11, except that the typical cross-sectional area is $1/x^3$ instead of $6x^3$.

In Exercises 13 to 20 use the fundamental theorem of calculus to evaluate the definite integrals.

13 $\int_1^2 x^3\,dx$ **14** $\int_{-1}^1 x^3\,dx$

15 $\int_0^3 6x\,dx$ **16** $\int_1^4 \frac{1}{x^2}\,dx$

17 $\int_4^9 5\sqrt{x}\,dx$ **18** $\int_1^9 \frac{1}{\sqrt{x}}\,dx$

19 $\int_1^8 \sqrt[3]{x^2}\,dx$ **20** $\int_2^3 \frac{4}{x^3}\,dx$

Exercises 21 to 26 concern the second fundamental theorem of calculus. In Exercises 21 and 22 find the derivative of the given function f of x, $x > 1$, in two ways. First evaluate the function with the aid of the first fundamental theorem of calculus, and then differentiate the result. In the second approach, use the second fundamental theorem of calculus.

21 $\int_1^x t^4\,dt$ **22** $\int_1^x \frac{1}{t^2}\,dt$

In Exercises 23 to 26 use the second fundamental theorem of calculus to differentiate the given functions of x.

23 $\int_1^x t^5\,dt$ **24** $\int_1^{\sqrt{x}} \frac{1}{t^3}\,dt$

25 $\int_0^{\sin x} \sqrt{1 + t^3}\,dt$ **26** $\int_{-1}^{\sqrt{x}} 3^t\,dt$

■

27 A plane at a distance x from the center of a sphere of radius r, $0 \le x \le r$, meets the sphere in a circle. (See Fig. 5.41.)

(a) Show that the radius of the circle is $\sqrt{r^2 - x^2}$.
(b) Show that the area of the circle is $\pi r^2 - \pi x^2$.
(c) Using the fundamental theorem, find the volume of the sphere.

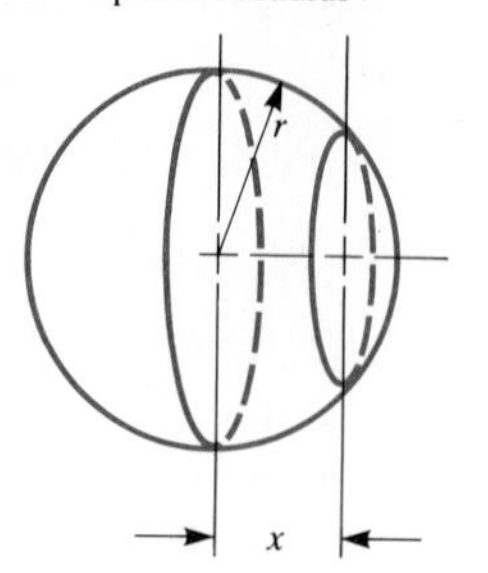

Figure 5.41

28 We interpreted $f(x)$ as velocity and $F(x)$ as distance in order to motivate the first fundamental theorem of calculus. Instead, let $f(x)$ be density and $F(x)$ be mass along a wire. Specifically, let $F(x)$ be the mass in grams of the left-hand x centimeters of a wire. Then the density of the wire at x is $F'(x)$, which we denote $f(x)$.

(a) Express the mass of the section $[a, b]$ in terms of F.
(b) Express the same mass in terms of f.
(c) Compare (a) and (b) to obtain another piece of evidence in favor of the first fundamental theorem of calculus.

Exercises 29 to 34 concern the volume of the solid obtained by revolving a region in the xy plane around the x axis.

29 Let $y = f(x)$ be nonnegative for x in $[a, b]$. The region below $y = f(x)$ and above $[a, b]$ is revolved around the x axis to form a "solid of revolution," as shown in Fig. 5.42.

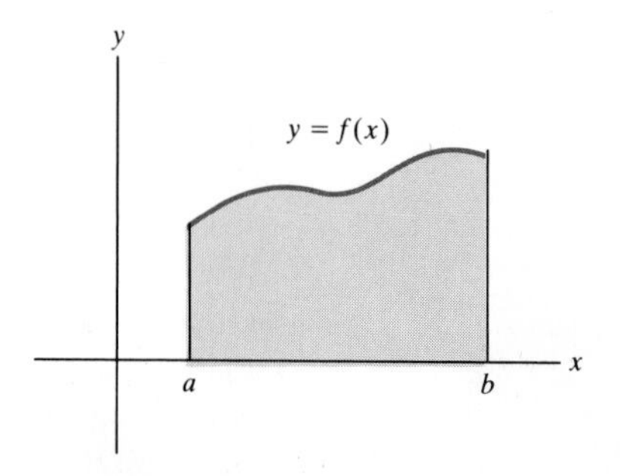

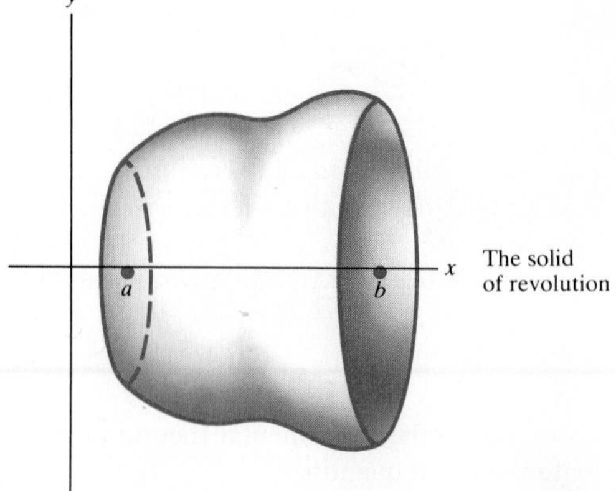

Figure 5.42

(*a*) What is the area of the cross section of the solid made by a plane perpendicular to the x axis and passing through the point on the x axis with coordinate x?

(*b*) Using (*a*), obtain a definite integral that equals the volume of the solid.

In Exercises 30 to 34 a solid of revolution is formed by revolving around the x axis the region under the graph of the given function and above the given interval. Find its volume using the formula developed in Exercise 29.

30 x^3, $[1, 2]$ **31** $x^{3/2}$, $[0, 1]$

32 $1/x$, $[1, 2]$ **33** $x + 1$, $[2, 3]$

34 $5x$, $[0, 1]$

35 In Example 1, $F(x) = \sin x$ was used to compute $\int_0^{\pi/2} \cos x \, dx$.

(*a*) Show that the function $5 + \sin x$ also has a derivative equal to $\cos x$.

(*b*) Use $F(x) = 5 + \sin x$ to evaluate $\int_0^{\pi/2} \cos x \, dx$.

36 There is no elementary function whose derivative is $\sqrt{\sin x}$. Consider the curve $y = \sqrt{\sin x}$ for x in the interval $[0, \pi]$.

(*a*) Set up a definite integral for the area of the region under the curve and above $[0, \pi]$.

(*b*) Set up a definite integral for the volume of the solid obtained by revolving the region in (*a*) around the x axis. (*Hint:* See Exercise 29.)

(*c*) Show that the fundamental theorem of calculus is useful in finding the volume in (*b*) but not in finding the area in (*a*).

In each of Exercises 37 and 38 there are three definite integrals. Two are easy to evaluate with the aid of the fundamental theorem of calculus, while the other, like $\int_0^{\pi/4} x \tan x \, dx$, cannot be evaluated by it. Decide which integrals can be evaluated by the fundamental theorem and evaluate them.

37 (*a*) $\displaystyle\int_0^1 \sqrt[3]{x + x^2} \, dx$ (*b*) $\displaystyle\int_0^1 \sqrt[3]{1 + x} \, dx$

(*c*) $\displaystyle\int_0^1 \frac{1}{\sqrt[3]{1 + x}} \, dx$

38 (*a*) $\displaystyle\int_0^1 x\sqrt{x^2 + 1} \, dx$ (*b*) $\displaystyle\int_0^1 x^2\sqrt[3]{x^3 + 1} \, dx$

(*c*) $\displaystyle\int_0^1 \sqrt[3]{x^3 + 1} \, dx$

(*Hint:* Try a power of the expression under the radical.)

5.5 PROPERTIES OF THE ANTIDERIVATIVE AND THE DEFINITE INTEGRAL

This section describes some of the important properties of the antiderivative and the definite integral and gives a notation for the antiderivative. Frequently, an antiderivative is called an **integral** or **indefinite integral.** There is a danger that "integral" will become confused with "definite integral." It should be kept in mind that the definite integral $\int_a^b f(x) \, dx$ is defined as a number, a limit of certain sums, while an integral or antiderivative is a function. Sometimes, in application, both the definite integral and the indefinite integral are called integrals; it takes a clear mind and mastery of the definitions to keep the ideas separate. The table of integrals in a mathematical handbook is primarily a table of antiderivatives (functions); it is usually followed by a short section that lists the values of a few common definite integrals (numbers).

Inside the front cover of this book is a short table of antiderivatives or "indefinite integrals."

Notation An antiderivative of f is denoted $\displaystyle\int f(x) \, dx$.

Thus if $F' = f$, we write $F = \int f(x) \, dx$ and say "F is an antiderivative of f" or, for convenience, "$F(x)$ is an antiderivative of $f(x)$." For instance, we say "x^3 is an antiderivative of $3x^2$." Note that $x^3 = \int 3x^2 \, dx$ and $x^3 + 1 = \int 3x^2 \, dx$. This does *not* imply that $x^3 = x^3 + 1$.

For any constant C, $(x^3 + C)' = 3x^2$. Thus

$$x^3 + C = \int 3x^2 \, dx.$$

There are no other antiderivatives of $3x^2$. If you have found one anti-

THE LONG ROAD TO CALCULUS

The first five chapters have presented the foundations of calculus in this order: functions, limits and continuity, the derivative, the definite integral, and the fundamental theorem that joins the last two. This bears little relation to the order in which these concepts were actually developed. Nor can we sense in this approach, which follows the standard calculus syllabus, the long struggle that culminated in the creation of calculus.

The origins of calculus go back over 2000 years to the work of the Greeks on areas and tangents. Archimedes (287–212 B.C.) found the area of a section of a parabola, an accomplishment that amounts in our terms to evaluating $\int_0^b x^2\,dx$. He also found the area of an ellipse and both the surface area and the volume of a sphere. Apollonius (around 260–200 B.C.) wrote about tangents to ellipses, parabolas, and hyperbolas, and Archimedes discussed the tangents to a certain spiral-shaped curve. Little did they suspect that the "area" and "tangent" problems were to converge many centuries later.

With the collapse of the Greek world, symbolized by the Emperor Justinian's closing in A.D. 529 of Plato's Academy, which had survived for a thousand years, it was the Arab world that preserved the works of Greek mathematicians. In its liberal atmosphere, Arab, Christian, and Jewish scholars worked together, translating and commenting on the old writings, occasionally adding their own embellishments. For instance, Alhazen (A.D. 965–1039) computed volumes of certain solids, in essence evaluating $\int_0^b x^3\,dx$ and $\int_0^b x^4\,dx$.

It was not until the seventeenth century that several ideas came together to form calculus. In 1637, both Descartes (1596–1650) and Fermat (1601–1665) introduced analytic geometry. Descartes examined a given curve with the aid of algebra, while Fermat took the opposite tack, exploring the geometry hidden in a given equation. For instance, Fermat showed that the graph of $ax^2 + bxy + cy^2 + dx + ey + f = 0$ is always an ellipse, hyperbola, parabola, or one of their degenerate forms.

In this same period, Cavalieri (1598–1647) found the area under the curve $y = x^n$ for $n = 1, 2, 3, \ldots, 9$ by a method the length of whose computations grew rapidly as the exponent increased. Stopping at $n = 9$, he conjectured that the pattern would continue for larger exponents. In the next 20 years, several mathematicians justified his guess. So, even the calculation of the area under $y = x^n$ for a positive integer n, which we take for granted, represented a hard-won triumph.

"What about the other exponents?" we may wonder. Before 1665 there were no other exponents. Nevertheless, it was possible to work with the function which we denote $y = x^{p/q}$ for positive integers p and q by describing it as the function y such that $y^q = x^p$. Wallis (1616–1703) found the area under this curve by a method that smacks more of magic than of mathematics. However, Fermat obtained the same result with the aid of an infinite geometric series.

The problem of determining tangents to curves was also in vogue in the first half of the seventeenth century. Descartes showed how to find a line perpendicular to a curve at a point P (by constructing a circle that meets the curve only at P); the tangent was then the line through P perpendicular to that line. Fermat found tangents in a way similar to ours and applied it to maximum-minimum problems.

The stage was set for the union of the "tangent" and "area" techniques. Indeed, Barrow (1630–1677), Newton's teacher at Cambridge, obtained a result equivalent to the fundamental theorem of calculus, but it was not expressed in a useful form.

Newton (1642–1727) arrived in Cambridge in 1661, and during the two years 1665–1666, which he spent at his family's farm to avoid the plague, he developed the essentials of calculus—recognizing that finding tangents and calculating areas are inverse processes. The first integral table ever compiled is to be found in one of his manuscripts of this period. But Newton did not publish his results at that time, perhaps because of the depression in the book trade after the Great Fire of London in 1665. During those two remarkable years he also introduced negative and fractional exponents, thus demonstrating that such diverse operations as multiplying a number by itself several times, taking its reciprocal, and finding a root of some power of that number are just

derivative F for a function f, then any other antiderivative of f is of the form $F(x) + C$ for some constant C. This is established in the following theorem.

Theorem 1 If F and G are both antiderivatives of f on an interval $[a, b]$, then there is a constant C such that

$$F(x) = G(x) + C.$$

special cases of a single general exponential function a^x, where x is a positive integer, -1, or a fraction.

Independently, however, Leibniz (1646–1716) also invented calculus. A lawyer, diplomat, and philosopher, for whom mathematics was a serious avocation, Leibniz established his version in the years 1673–1676, publishing his researches in 1684 and 1686, well before Newton's first publication in 1711. To Leibniz we owe the notations dx and dy, the terms "differential calculus" and "integral calculus," the integral sign, and the word "function." Newton's notation survives only in the symbol $\dot{x}$ for differentiation with respect to time, which is still used in physics.

It was to take two more centuries before calculus reached its present state of precision and rigor. The notion of a function gradually evolved from "curve" to "formula" to any rule that assigns one quantity to another. The great calculus text of Euler, published in 1748, emphasized the function concept by including not even one graph.

In several texts of the 1820s, Cauchy (1789–1857) defined "limit" and "continuous function" much as we do today. He also gave a definition of the definite integral, which with a slight change by Riemann (1826–1866) in 1854 became the definition standard today. So by the mid-nineteenth century the discoveries of Newton and Leibniz were put on a solid foundation.

In 1833, Liouville (1809–1882) demonstrated that the fundamental theorem could not be used to evaluate integrals of all elementary functions. In fact, he showed that the only values of the constant k for which $\int \sqrt{1-x^2}\,\sqrt{1-kx^2}\,dx$ is elementary are 0 and 1.

Still some basic questions remained, such as "What do we mean by area?" (For instance, does the set of points situated within some square and having both coordinates rational have an area? If so, what is this area?) It was as recently as 1887 that Peano (1858–1932) gave a precise definition of area—that quantity which earlier mathematicians had treated as intuitively given.

The history of calculus therefore consists of three periods. First, there was the long stretch when there was no hint that the tangent and area problems were related. Then came the discovery of their intimate connection and the exploitation of this relation from the end of the seventeenth century through the eighteenth century. This was followed by a century in which the loose ends were tied up.

The twentieth century has seen calculus applied in many new areas, for it is the natural language for dealing with continuous processes, such as change with time. In this century mathematicians have also obtained some of the deepest theoretical results about its foundations. Calculus is definitely alive and well and still growing.

References

E. T. Bell, *Men of Mathematics,* Simon and Schuster, New York, 1937. (In particular, chap. 6 on Newton and chap. 7 on Leibniz.)

Carl B. Boyer, *The History of the Calculus and Its Conceptual Development,* Dover, New York, 1959.

Carl B. Boyer, *A History of Mathematics,* Wiley, New York, 1968.

M. J. Crowe, *A History of Vector Analysis,* Notre Dame, 1967. (In particular, chap. 5.)

C. H. Edwards, Jr., *The Historical Development of the Calculus,* Springer-Verlag, New York, 1979. (In particular, chap. 8, "The Calculus According to Newton," and chap. 9 "The Calculus According to Leibniz.")

Morris Kline, *Mathematical Thought from Ancient to Modern Times,* Oxford, New York, 1972. (In particular, chap. 17.)

Pronunciation

Descartes	"Day-CART"
Fermat	"Fair-MA"
Leibniz	"LIBE-nits"
Euler	"OIL-er"
Cauchy	"KOH-shee"
Riemann	"REE-mahn"
Liouville	"LEE-oo-veel"
Peano	"Pay-AHN-oh"

Proof The functions F and G have the same derivative f. By Corollary 2 in Sec. 4.1, they must differ by a constant. This proves the theorem. ■

In tables C is usually omitted.

When writing down an antiderivative, it is best to add the constant C. (It will be needed in the study of differential equations.) For example,

$$\int 5\,dx = 5x + C,$$

$$\int x^3\,dx = \frac{x^4}{4} + C,$$

and

$$\int \sin 2x\,dx = -\frac{\cos 2x}{2} + C.$$

Any property of derivatives implies a corresponding property of antiderivatives. The next theorem records three of the most important properties of antiderivatives.

Theorem 2 Assume that f and g are functions with antiderivatives $\int f(x)\,dx$ and $\int g(x)\,dx$. Then the following hold:

Properties of antiderivatives

(a) $\int cf(x)\,dx = c\int f(x)\,dx$ for any constant c.
(b) $\int (f(x) + g(x))\,dx = \int f(x)\,dx + \int g(x)\,dx$.
(c) $\int (f(x) - g(x))\,dx = \int f(x)\,dx - \int g(x)\,dx$.

Proof

(*a*) It is necessary to show that the derivative of $c\int f(x)\,dx$ is $cf(x)$. The differentiation follows:

$$\frac{d}{dx}\left(c\int f(x)\,dx\right) = c\,\frac{d}{dx}\left(\int f(x)\,dx\right)$$ A constant moves past the derivative symbol.

$$= cf(x)$$ Definition of $\int f(x)\,dx$.

(*b*) It is necessary to show that the derivative of $\int f(x)\,dx + \int g(x)\,dx$ is $f(x) + g(x)$:

$$\frac{d}{dx}\left(\int f(x)\,dx + \int g(x)\,dx\right) = \frac{d}{dx}\left(\int f(x)\,dx\right) + \frac{d}{dx}\left(\int g(x)\,dx\right)$$

$$= f(x) + g(x).$$

(*c*) The proof of property (*c*) is similar to that of property (*b*). ■

The last two parts of Theorem 2 extend to any finite number of functions. For instance,

$$\int (f(x) - g(x) + h(x))\,dx = \int f(x)\,dx - \int g(x)\,dx + \int h(x)\,dx.$$

Theorem 3 Let a be a rational number other than -1. Then

$$\int x^a\,dx = \frac{x^{a+1}}{a+1} + C.$$

Proof

$$\left(\frac{x^{a+1}}{a+1} + C\right)' = \frac{(a+1)x^{a+1-1}}{a+1} = x^a. \quad ■$$

EXAMPLE 1 Find $\int (2x^5 - 3x^2 + 4)\,dx$.

SOLUTION

$$\int (2x^5 - 3x^2 + 4)\,dx = \int 2x^5\,dx - \int 3x^2\,dx + \int 4\,dx$$ Theorem 2

$$= 2\int x^5\,dx - 3\int x^2\,dx + \int 4\,dx \qquad \text{Theorem 2}$$

A single constant is enough.

$$= 2\frac{x^6}{6} - 3\frac{x^3}{3} + 4x + C \qquad \text{Theorem 3}$$

$$= \frac{x^6}{3} - x^3 + 4x + C. \quad \blacksquare$$

As Example 1 illustrates, an antiderivative of any polynomial is again a polynomial.

Notation $F(b) - F(a)$ is abbreviated to $F(x)\Big|_a^b$.

EXAMPLE 2 Evaluate $\int_1^2 \frac{1}{x^2}\,dx$ by the fundamental theorem of calculus.

SOLUTION $\int_1^2 \frac{1}{x^2}\,dx = \frac{-1}{x}\Big|_1^2 = \frac{-1}{2} - \frac{-1}{1} = \frac{-1}{2} + 1 = \frac{1}{2}.$ ■

The fundamental theorem of calculus asserts that

$$\underbrace{\int_1^2 \frac{1}{x^2}\,dx}_{\text{The definite integral; a limit of sums.}} = \underbrace{\int \frac{1}{x^2}\,dx\Big|_1^2}_{\text{The difference between an antiderivative evaluated at 2 and at 1}}.$$

The symbols on the right and left of the equal sign are so similar that it is tempting to think that the equation is obvious or says nothing whatever.
Beware: This compact equation is in fact a terse statement of the (first) fundamental theorem of calculus.

Definition *Integrand.* In the definite integral $\int_a^b f(x)\,dx$ and in the antiderivative $\int f(x)\,dx$, $f(x)$ is called the **integrand**.

The related processes of computing $\int_a^b f(x)\,dx$ and of finding an antiderivative $\int f(x)\,dx$ are both called **integrating** $f(x)$. Thus integration refers to two separate but related problems: computing a number $\int_a^b f(x)\,dx$ or finding a function $\int f(x)\,dx$. The fundamental theorem of calculus states that the second process may be of use in computing $\int_a^b f(x)\,dx$.

Properties of the Definite Integral

In the notation for the definite integral $\int_a^b f(x)\,dx$, b is larger than a. It will be useful to be able to speak of "the definite integral from a to b" even if b is less than or equal to a. The following definitions meet this need and will be used in the next section in the proofs of the two fundamental theorems of calculus.

Definition *The integral from a to b, where b is less than a.* If b is less than a, then

$$\int_a^b f(x)\,dx = -\int_b^a f(x)\,dx.$$

EXAMPLE 3 Compute $\int_3^0 x^2\,dx$, the integral from 3 to 0 of x^2.

SOLUTION The symbol $\int_3^0 x^2\,dx$ is defined as $-\int_0^3 x^2\,dx$. As was shown in Sec. 5.3, $\int_0^3 x^2\,dx = 9$. Thus

$$\int_3^0 x^2\,dx = -9. \quad \blacksquare$$

Definition *The integral from a to a.* $\int_a^a f(x)\,dx = 0.$

Remark: The definite integral is defined with the aid of partitions. Rather than permit partitions to have sections of length 0, it is simpler just to make the preceding definition.

The point of making these two definitions is that now the symbol $\int_a^b f(x)\,dx$ is defined for any numbers a and b and any continuous function f. It is no longer necessary that a be less than b.

The definite integral has several properties, some of which will be used in this section and some later in the text.

Properties of the Definite Integral Let f and g be continuous functions, and let c be a constant. Then

1 $\int_a^b cf(x)\,dx = c\int_a^b f(x)\,dx.$

2 $\int_a^b (f(x) + g(x))\,dx = \int_a^b f(x)\,dx + \int_a^b g(x)\,dx.$

3 $\int_a^b (f(x) - g(x))\,dx = \int_a^b f(x)\,dx - \int_a^b g(x)\,dx.$

4 If $f(x) \geq 0$ for all x in $[a, b]$, $a < b$, then

$$\int_a^b f(x)\,dx \geq 0.$$

5 If $f(x) \geq g(x)$ for all x in $[a, b]$, $a < b$, then

$$\int_a^b f(x)\,dx \geq \int_a^b g(x)\,dx.$$

6 If a, b, and c are numbers, then

$$\int_a^c f(x)\,dx + \int_c^b f(x)\,dx = \int_a^b f(x)\,dx.$$

7 If m and M are numbers and $m \leq f(x) \leq M$ for all x between a and b, then

$$m(b-a) \leq \int_a^b f(x)\,dx \leq M(b-a) \qquad \text{if } a < b,$$

and

$$m(b-a) \geq \int_a^b f(x)\,dx \geq M(b-a) \qquad \text{if } b < a.$$

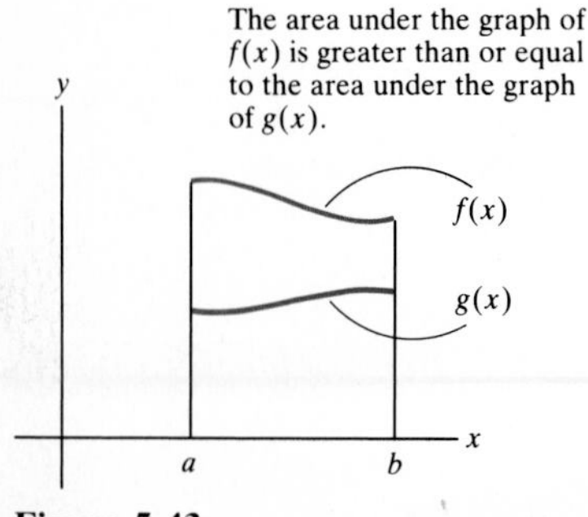

Figure 5.43

Interpretation of the definite integral as an area (for positive integrands) makes these seven assertions plausible. For instance, property 5 amounts then to the assertion that if one plane region contains another, then it has at least as large an area as the region it contains. (See Fig. 5.43.)

In the case that $a < c < b$ and $f(x)$ assumes only positive values, property 6 asserts that the area of the region below the graph of f and above the interval $[a, b]$ is the sum of the areas of the regions below the graph and above the smaller intervals $[a, c]$ and $[c, b]$. This is certainly plausible.

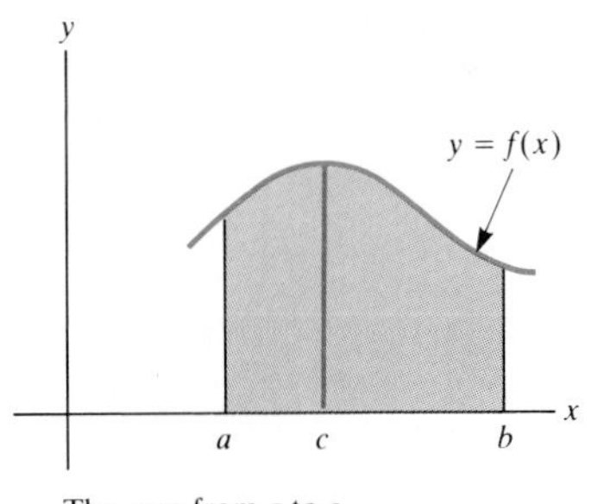

The area from a to c plus the area from c to b is equal to the area from a to b.

Figure 5.44

Figure 5.44 expresses property 6 geometrically.

The inequalities in property 7 compare the area under $f(x)$ with the areas of two rectangles, one of height M and one of height m. (See Fig. 5.45.)

Properties 1 to 7 can be proved by examining approximating sums and seeing how they behave as the mesh approaches 0. (See Exercise 55, for instance.)

A mean-value theorem for definite integrals follows from these properties.

Mean-Value Theorem for Definite Integrals Let a and b be numbers, and let f be a continuous function defined for x between a and b. Then there is a number c between a and b such that

$$\int_a^b f(x)\,dx = f(c)(b - a).$$

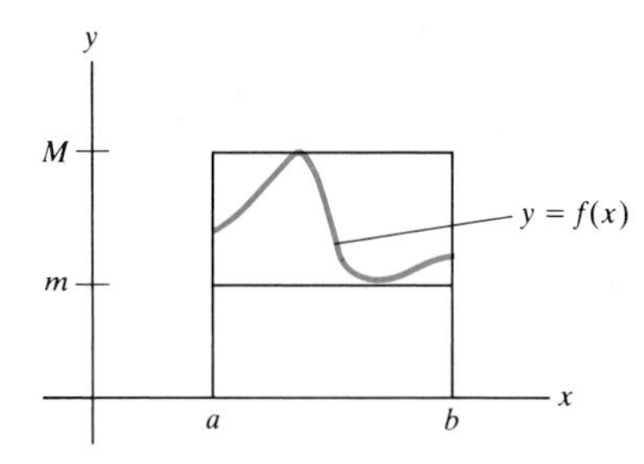

Figure 5.45

Proof Consider the case $a < b$. Let M be the maximum and m the minimum of $f(x)$ for x in $[a, b]$. [Recall the maximum- (and minimum-) value theorem of Sec. 2.6.] By property 7,

$$m \le \frac{\int_a^b f(x)\,dx}{b - a} \le M.$$

By the intermediate-value theorem of Sec. 2.6, there is a number c in $[a, b]$ such that

$$f(c) = \frac{\int_a^b f(x)\,dx}{b - a},$$

and the theorem is proved. (The case $b < a$ can be obtained from the case $a < b$.) ■

The area of the rectangle is the same as the area of the region below the curve.

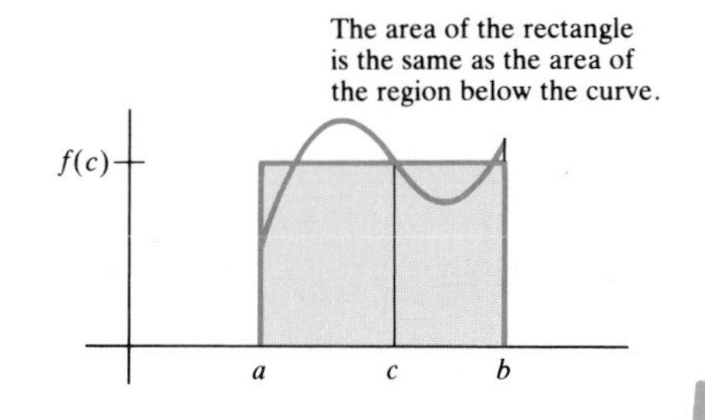

If $f(x)$ is positive, $\int_a^b f(x)\,dx$ can be interpreted as the area of the region below $y = f(x)$ and above $[a, b]$. The mean-value theorem then asserts that there is a rectangle whose base is $[a, b]$ and whose height is $f(c)$ that has the same area as the region for some c in $[a, b]$. This rectangle is shown in Fig. 5.46. This number $f(c)$ is called the "average of" $f(x)$ over the interval $[a, b]$.

Definition Let $f(x)$ be defined on the interval $[a, b]$. Assume that $\int_a^b f(x)\,dx$ exists. Then the quotient

$$\frac{\int_a^b f(x)\,dx}{b - a}$$

is called the **average** of the function on this interval.

EXAMPLE 4 Find the average of $\sin x$ for x in (*a*) $[0, \pi/2]$, (*b*) $[0, 2\pi]$.

SOLUTION

(*a*) The average for the interval $[0, \pi/2]$ is

$$\frac{\int_0^{\pi/2} \sin x\,dx}{(\pi/2) - 0} = \frac{-\cos x\,\big|_0^{\pi/2}}{\pi/2} = \frac{1}{\pi/2} = \frac{2}{\pi} \approx 0.64.$$

(*b*) The average for the interval $[0, 2\pi]$ is

$$\frac{\int_0^{2\pi} \sin x\, dx}{2\pi - 0} = \frac{-\cos x \,\big|_0^{2\pi}}{2\pi} = \frac{0}{2\pi} = 0.$$

EXERCISES FOR SEC. 5.5: PROPERTIES OF THE ANTIDERIVATIVE AND THE DEFINITE INTEGRAL

In Exercises 1 to 10 evaluate the antiderivatives, adding in each case the constant C. Check each answer by differentiating it.

1 $\int (2x - x^3 + x^5)\, dx$ **2** $\int \left(6x^2 + \frac{1}{\sqrt{x}}\right) dx$

3 $\int \left(3x + 4 \sin 2x - \frac{1}{4x^2}\right) dx$

4 $\int (\sin 2x + \cos 4x)\, dx$ **5** $\int \sqrt{2x + 5}\, dx$

6 $\int \frac{1}{(2x + 1)^3}\, dx$ **7** $\int \tan 2x \sec 2x\, dx$

8 $\int \csc^2 3x\, dx$ **9** $\int (1 - \cos 2x)\, dx$

10 $\int \sin^2 x\, dx$ (*Hint:* Use a trigonometric identity.)

11 Compute: (*a*) $\int x^2\, dx \Big|_1^2$, (*b*) $\int \frac{1}{(x+1)^2}\, dx \Big|_1^2$, (*c*) $\int \sin 3x\, dx \Big|_1^2$.

12 Compute: (*a*) $\int \cos x\, dx \Big|_0^{\pi/4}$, (*b*) $\int \sec^2 x\, dx \Big|_0^1$, (*c*) $\int \frac{1}{\sqrt[3]{x}}\, dx \Big|_1^2$.

13 Compute:

(*a*) $x^2 \Big|_1^2$ (*b*) $x^2 \Big|_{-1}^1$

(*c*) $\sin x \Big|_0^{\pi}$ (*d*) $\cos x \Big|_0^{\pi}$

14 Find $\int 1\, dx$. (This is usually written $\int dx$.)

15 (*a*) Is $\int x^2\, dx$ a function or is it a number?
(*b*) Is $\int x^2\, dx \big|_1^3$ a function or is it a number?
(*c*) Is $\int_1^3 x^2\, dx$ a function or is it a number?

16 (*a*) Which of these two numbers is defined as a limit of sums:

$$\int x^2\, dx \Big|_1^3 \quad \text{or} \quad \int_1^3 x^2\, dx?$$

(*b*) Why are the two numbers in (*a*) equal?

17 True or false: (*a*) Every elementary function has an elementary derivative. (*b*) Every elementary function has an elementary antiderivative.

18 True or false: (*a*) $\sin x^2$ has an elementary antiderivative. (*b*) $\sin x^2$ has an antiderivative.

19 Find dy/dx if (*a*) $y = \int \sin(x^2)\, dx$, (*b*) $y = 3x + \int_{-2}^{3} \sin x^2\, dx$, (*c*) $y = \int_{-2}^{x} \sin t^2\, dt$.

20 (*a*) Compute $\frac{d}{dx}\left(\int_4^{x^2} \frac{\sqrt{1+u^2}}{2\sqrt{u}}\, du\right)$ for $x > 0$.

(*b*) Compute $\frac{d}{dx}\left(\int_2^{x} \sqrt{1+t^4}\, dt\right)$ for $x > 0$.

(*c*) In view of (*a*) and (*b*), there is a constant C such that

$$\int_2^x \sqrt{1+t^4}\, dt = \int_4^{x^2} \frac{\sqrt{1+u^2}}{2\sqrt{u}}\, du + C.$$

Find C.

In Exercises 21 and 22 verify the equations quoted from a table of antiderivatives (integrals). Just differentiate each of the alleged antiderivatives and see whether you obtain the integrand. The number a is constant in each case.

21 $\int x^2 \sin ax\, dx = \frac{2x}{a^2} \sin ax + \frac{2}{a^3} \cos ax - \frac{x^2}{a} \cos ax + C.$

22 $\int x \sin^2 ax\, dx = \frac{x^2}{4} - \frac{x \sin 2ax}{4a} - \frac{\cos 2ax}{8a^2} + C.$

In Exercises 23 to 26 evaluate the expressions.

23 $\int_0^0 2^{x^2}\, dx$ **24** $\int_1^0 x^2\, dx$

25 $\int_2^1 (12x^3 - 2x)\, dx$ **26** $\int_3^2 (x+1)^{-2}\, dx$

In Exercises 27 to 30 differentiate $f(x)$.

27 $f(x) = \int_x^{17} \sin^3 t\, dt$
(*Hint:* First rewrite it as $-\int_{17}^{x} \sin^3 t\, dt$.)

28 $f(x) = \int_{x^2}^{17} \sin^3 2t\, dt$

29 $f(x) = \int_{2x}^{3x} t \tan t \, dt$
(*Hint:* First rewrite it as $\int_{2x}^{0} t \tan t \, dt + \int_{0}^{3x} t \tan t \, dt$.)

30 $f(x) = \int_{x^2}^{x^3} \sqrt[3]{1 + t^2} \, dt$

In Exercises 31 to 34 find d^2y/dx^2.

31 $y = \int_{1984}^{x} \sin t^2 \, dt$

32 $y = \int_{1776}^{x} t^2 \sqrt{1 + 2t} \, dt$

33 $y = \int_{x}^{1492} \frac{t^3 \sin 2t}{\sqrt{1 + 3t}} \, dt$

34 $y = \int_{x^2}^{1865} t^2 \sqrt[3]{1 + 2t} \cos 3t \, dt$

In Exercises 35 to 38 find at least one number c whose existence is guaranteed by the mean-value theorem for integrals.

35 $\int_{0}^{4\pi} \sin x \, dx$

36 $\int_{1}^{3} x^2 \, dx$

37 $\int_{1}^{3} x^{-3} \, dx$

38 $\int_{0}^{1} (x - x^2) \, dx$

■

In Exercises 39 to 42 find the minimum, maximum, and average value of the given function on the given interval.

39 $f(x) = x^2$; $[1, 3]$

40 $f(x) = \sec^2 x$; $[0, \pi/4]$

41 $f(x) = 1/x^2$; $[1, 3]$

42 $f(x) = \sin 2x$; $[0, \pi/2]$

43 Give an example of a function f such that $f(4) = 0$ and $f'(x) = \sqrt[3]{1 + x^2}$.

44 How often should a machine be overhauled? This depends on the rate $f(t)$ at which it depreciates and the cost A of overhaul. Denote the time interval between overhauls by T.
(*a*) Explain why you would like to minimize $g(T) = [A + \int_0^T f(t) \, dt]/T$.
(*b*) Find dg/dT.
(*c*) Show that when $dg/dT = 0$, $f(T) = g(T)$.
(*d*) Is this reasonable?

45 An unmanned satellite automatically reports its speed every minute. If a graph is drawn showing speed as a function of time during the flight, what is the physical interpretation of the area under the curve and above the time axis? Explain.

46 As a stone is lowered into water, we record the volume of water it displaces. When x inches are submerged, the stone displaces $V(x)$ cubic inches of water. How can we find the area of the cross section of the stone made by the plane of the surface of the water when it is submerged to a depth of x inches? Assume we know $V(x)$ for all x.

47 A man whose jeep has a vertical windshield drives a mile through a vertical rain consisting of drops that are uniformly distributed and falling at a constant rate. (See Fig. 5.47.) Should he go slow or fast in order to minimize the amount of rain that strikes the windshield?

Figure 5.47

48 Let f be a continuous function that has a derivative nowhere. (There are such functions.) Construct a function that has a derivative everywhere but a second derivative nowhere.

49 Let $v(t)$ be the velocity at time t of an object moving on a straight line. The velocity may be positive or negative.
(*a*) If you graph $v(t)$ for t in $[a, b]$, what is the physical meaning of $\int_a^b v(t) \, dt$?
(*b*) What is the physical meaning of the slope of the graph?

50 A number is **dyadic** if it can be expressed as the quotient of two integers m/n, where n is a power of 2. (These are the fractions into which an inch is usually divided.) Between any two numbers lies an infinite set of dyadic numbers and also an infinite set of numbers that are not dyadic. With this background, we shall define a function f that does *not* have a definite integral over the interval $[0, 1]$ as follows:

$$\text{Let } f(x) = \begin{cases} 0 & \text{if } x \text{ is dyadic;} \\ 3 & \text{if } x \text{ is not dyadic.} \end{cases}$$

(*a*) Show that for any partition of $[0, 1]$ it is possible to choose sampling numbers c_i such that

$$\sum_{i=1}^{n} f(c_i)(x_i - x_{i-1}) = 3.$$

(*b*) Show that for any partition of $[0, 1]$ it is possible to choose sampling numbers c_i such that

$$\sum_{i=1}^{n} f(c_i)(x_i - x_{i-1}) = 0.$$

(*c*) Why does f not have a definite integral over the interval $[0, 1]$?

51 Show that the mean-value theorem for definite integrals of continuous functions is a consequence of the mean-value theorem for differentiable functions stated in Sec. 4.1.

52 Prove that if f is an increasing function on the interval $[a, b]$ and $f = F'$, then $\int_a^b f(x)\,dx$ exists and equals $F(b) - F(a)$. (Do not assume the fundamental theorem of calculus.)

53 Prove that if g is differentiable and $|g'(x)| < k$ for all x in $[a, b]$, then the function $f(x) = g(x) + kx$ is increasing on $[a, b]$.

54 (See Exercises 52 and 53.) Prove that if f is a differentiable function whose derivative f' is bounded between two fixed numbers over the interval $[a, b]$ and $f = F'$, then $\int_a^b f(x)\,dx$ exists and equals $F(b) - F(a)$. (Do not assume that a continuous function has a definite integral.)

55 Justify property 7 for definite integrals by showing that every approximating sum for $\int_a^b f(x)\,dx$ is $\le M(b - a)$ and $\ge m(b - a)$, if $a < b$.

56 (Optimal replacement) The following argument appears in *Optimal Replacement Policy* by D. W. Jorgenson, J. J. McCall, and R. Radner, pp. 92–93, Rand McNally, Chicago, 1967:

The average cost per unit good time, $V(N)$, is

$$V(N) = \frac{N + K}{\int_0^N R(t)\,dt}, \qquad (*)$$

where K is a constant, "the imputed down time," and N is the time of replacement. To determine the optimum preparedness maintenance policy, $V(N)$ is minimized with respect to replacement age N. Differentiation of (*) with respect to N yields the condition

$$\frac{dV}{dN} = \frac{\int_0^N R(t)\,dt - (N + K)R(N)}{[\int_0^N R(t)\,dt]^2} = 0.$$

Verify that the differentiation is correct.

5.6 PROOFS OF THE TWO FUNDAMENTAL THEOREMS OF CALCULUS

Geometric and physical intuition suggested the two fundamental theorems of calculus. It may be of value to give a mathematical proof for these two theorems, a proof which is independent of any particular interpretation of the definite integral. This will offer a chance to think of the definite integral as a purely mathematical concept.

The proof of the second fundamental theorem will be given first.

Proof of the Second Fundamental Theorem The second fundamental theorem asserts that the derivative of $G(x) = \int_a^x f(t)\,dt$ is $f(x)$. (See Sec. 5.4 for its full statement.)

It is necessary to study the behavior of the quotient

$$\frac{\Delta G}{\Delta x} = \frac{G(x + \Delta x) - G(x)}{\Delta x}$$

as $\Delta x \to 0$ (x is fixed). The number Δx may be positive or negative.

We have
$$\Delta G = \int_a^{x+\Delta x} f(t)\,dt - \int_a^x f(t)\,dt.$$

By property 6 in Sec. 5.5,

$$\int_a^{x+\Delta x} f(t)\,dt = \int_a^x f(t)\,dt + \int_x^{x+\Delta x} f(t)\,dt.$$

Hence
$$\Delta G = \int_a^x f(t)\,dt + \int_x^{x+\Delta x} f(t)\,dt - \int_a^x f(t)\,dt$$

$$= \int_x^{x+\Delta x} f(t)\,dt.$$

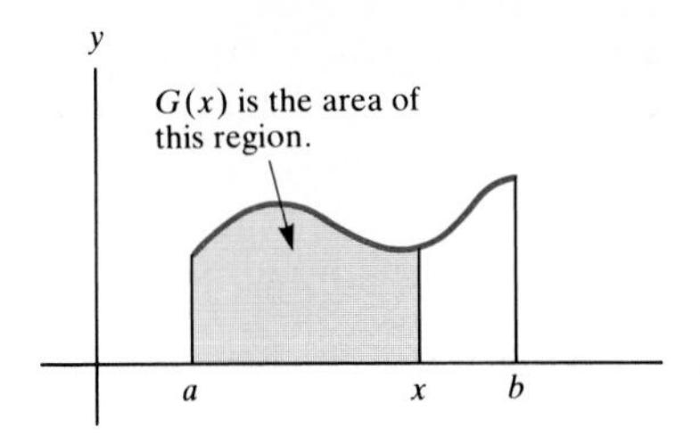

Figure 5.48

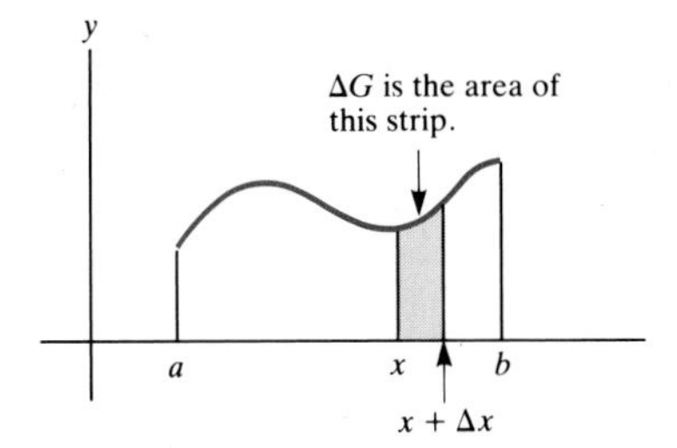

Figure 5.49

Now, by the mean-value theorem for definite integrals,

$$\Delta G = \int_x^{x+\Delta x} f(t)\,dt = f(c)((x + \Delta x) - x) \qquad \text{for some } c \text{ between } x \text{ and } x + \Delta x$$

$$= f(c)\,\Delta x.$$

Thus

$$\frac{\Delta G}{\Delta x} = f(c)$$

for some number c between x and $x + \Delta x$. (Note that c is not constant but depends on the choice of Δx and the fixed number x.)

Now it will be possible to compute

$$\lim_{\Delta x \to 0} \frac{\Delta G}{\Delta x}$$

because, by the above reasoning,

$$\lim_{\Delta x \to 0} \frac{\Delta G}{\Delta x} = \lim_{\Delta x \to 0} f(c).$$

Since c is between x and $x + \Delta x$, as $\Delta x \to 0$, c must approach x.

Since f is continuous at x and c is between x and $x + \Delta x$, it follows that

$$\lim_{\Delta x \to 0} f(c) = f(x).$$

This proves that G is differentiable and that $G' = f$. ■

Remark: It may be illuminating to follow the various steps in this proof if we interpret "definite integral" as "area." If $f(x)$ is positive for all x in $[a, b]$, then $G(x)$ can be thought of as the area under the curve $y = f(t)$ from a to x. Then for $\Delta x > 0$, $G(x + \Delta x) - G(x) = \Delta G$ represents the area of a narrow strip above the interval $[x, x + \Delta x]$. Choose c in such a way that the rectangle with base Δx and height $f(c)$ has an area equal to the area of the shaded strip in Fig. 5.49. Then $\Delta G/\Delta x = f(c)$, the height of the rectangle chosen as described. As Δx approaches 0, $f(c)$ approaches $f(x)$, since f is continuous. See Figs. 5.48 to 5.50.

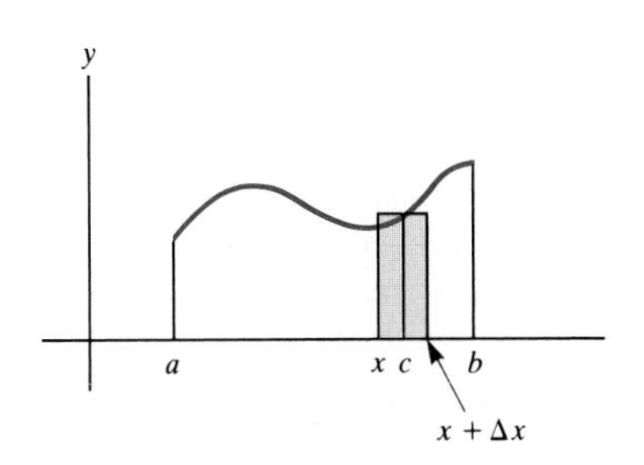

The area of the rectangle of base Δx and height $f(c)$ equals the area of the shaded strip.

Figure 5.50

EXAMPLE Let f be continuous. Find $\dfrac{d}{dx}\left(\int_x^b f(t)\,dt\right)$. In other words, differentiate the definite integral with respect to the *lower* limit of integration.

SOLUTION To do this, write $\int_x^b f(t)\,dt = -\int_b^x f(t)\,dt$. Thus the derivative equals

$$\frac{d}{dx}\left(-\int_b^x f(t)\,dt\right) = -\frac{d}{dx}\left(\int_b^x f(t)\,dt\right) = -f(x).$$

Note the minus sign. ■

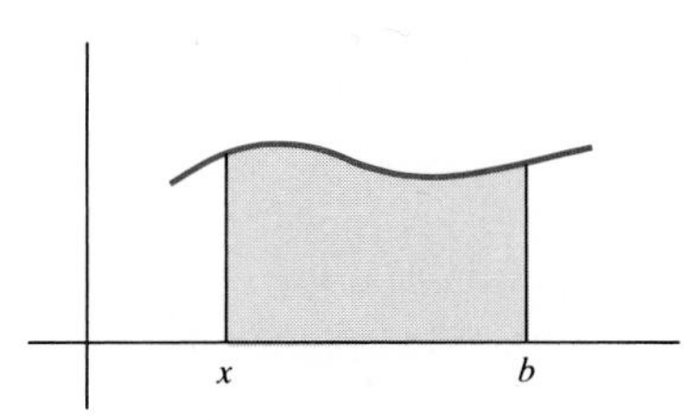

As x increases, the area of the shaded region decreases.

Figure 5.51

Remark: The minus sign in the answer in the example makes sense if $f(x)$ is positive, $x < b$, and $\int_x^b f(t)\,dt$ is thought of as area. As x increases, the area *decreases,* and its derivative should be negative. (See Fig. 5.51.)

Now let us turn to the proof of the first fundamental theorem.

Proof of the First Fundamental Theorem The first fundamental theorem, which asserts that if $f = F'$, then $\int_a^b f(x)\,dx = F(b) - F(a)$, follows from the second, as will now be shown. (See Sec. 5.4 for its complete statement.)

Let $G(x) = \int_a^x f(t)\,dt$. By the second fundamental theorem,

$$G'(x) = f(x).$$

Since it is assumed that $F'(x) = f(x)$, it follows that the functions G and F have the same derivative. Thus they differ by a constant (see Corollary 2, Sec. 4.1):

$$G(x) = F(x) + C, \qquad C \text{ constant.}$$

Then
$$G(b) - G(a) = [F(b) + C] - [F(a) + C]$$
$$= F(b) - F(a).$$

But
$$G(a) = \int_a^a f(x)\,dx = 0,$$

and
$$G(b) = \int_a^b f(x)\,dx.$$

Hence
$$G(b) - G(a) = \int_a^b f(x)\,dx.$$

Consequently,
$$\int_a^b f(x)\,dx = G(b) - G(a) = F(b) - F(a),$$

and the theorem is proved. ■

EXERCISES FOR SEC. 5.6: PROOFS OF THE TWO FUNDAMENTAL THEOREMS OF CALCULUS

1 The first fundamental theorem can be proved directly, without referring to the second fundamental theorem. Assume that f is continuous, $f = F'$, and $\int_a^b f(x)\,dx$ exists. The steps are outlined as follows:

(*a*) Given x_{i-1} and x_i, $x_{i-1} < x_i$, in $[a, b]$, show that there is a number c_i in $[x_{i-1}, x_i]$ such that

$$F(x_i) - F(x_{i-1}) = F'(c_i)(x_i - x_{i-1}).$$

(*b*) Given x_{i-1} and x_i in $[a, b]$, show that there is a number c_i in $[x_{i-1}, x_i]$ such that

$$f(c_i)(x_i - x_{i-1}) = F(x_i) - F(x_{i-1}).$$

(*c*) Let $x_0 = a, x_1, \ldots, x_n = b$ determine a partition of $[a, b]$ into n sections. Show that, if the sampling numbers c_i are chosen as in (*b*), then

$$\sum_{i=1}^{n} f(c_i)(x_i - x_{i-1}) = F(b) - F(a).$$

(*d*) Use (*c*) to show that

$$\lim_{\text{mesh}\to 0} \sum_{i=1}^{n} f(c_i)(x_i - x_{i-1}) = F(b) - F(a)$$

[even if the c_i are not chosen as in (*b*)]. This proves the first fundamental theorem directly.

2 Assume that the function f is defined for all x and has a continuous derivative. Assume that $f(0) = 0$ and that $0 < f'(x) \le 1$.

(*a*) Prove that

$$\left[\int_0^1 f(x)\,dx\right]^2 \ge \int_0^1 [f(x)]^3\,dx.$$

(*Hint:* Prove a more general result, namely that the inequality holds when the upper limit of integration 1 is replaced by t, $0 \le t \le 1$.)

(*b*) Give an example where equality occurs in (*a*).

3 Letting $f(x)$ denote the velocity of an object at time x, interpret each step of the proof of the second fundamental theorem in terms of velocity and distance.

4 (*a*) What is meant by an "elementary function"?

(*b*) Give an example of an elementary function whose antiderivative is not elementary.

(*c*) Give an example of a function that is not elementary, yet has an elementary derivative.

5 Find

(a) $\lim_{\Delta x \to 0} \frac{\int_2^{5+\Delta x} \sin x^2\, dx - \int_2^5 \sin x^2\, dx}{\Delta x}$

(b) $\frac{d^2}{dx^2}\left(\int_0^{x^2} \frac{dt}{\sqrt{1 - 5t^3}}\right)^2$

5.S SUMMARY

Four problems led to the concept of the definite integral: area of the region under a curve, total mass of a string of varying density, total distance traveled when the speed is varying, and volume of a certain tent.

All four problems required the same procedure: choosing partitions of some interval, sampling numbers, and then forming approximating sums:

$$\sum_{i=1}^{n} f(c_i)(x_i - x_{i-1}).$$

In Sec. 5.1 the interval was [0, 3], and the function f was the squaring function.

The typical sum

$$\sum_{i=1}^{n} f(c_i)(x_i - x_{i-1})$$

is *not* the definite integral, any more than $[f(x + \Delta x) - f(x)]/\Delta x$ is the derivative. A definite integral $\int_a^b f(x)\, dx$ is the limit of the approximating sums as their mesh is chosen smaller and smaller:

$$\int_a^b f(x)\, dx = \lim_{\text{mesh}\to 0} \sum_{i=1}^{n} f(c_i)(x_i - x_{i-1}).$$

The following table shows at a glance why the definite integral is related to mass, area, volume, and distance. Study this table carefully. It records the core ideas of the chapter.

Function	*Interpretation of typical summand*	*Approximating sum*	*Definite integral*	*Meaning of definite integral*
Density	$f(c_i)(x_i - x_{i-1})$ is estimate of mass in $[x_{i-1}, x_i]$.	$\sum_{i=1}^{n} f(c_i)(x_i - x_{i-1})$	$\int_a^b f(x)\, dx$	Mass
Length of cross section of a plane region by a line	$f(c_i)(x_i - x_{i-1})$ is area of an approximating rectangle.	$\sum_{i=1}^{n} f(c_i)(x_i - x_{i-1})$	$\int_a^b f(x)\, dx$	Area
Area of cross section of a solid by a plane	$f(c_i)(x_i - x_{i-1})$ is volume of a thin approximating slab.	$\sum_{i=1}^{n} f(c_i)(x_i - x_{i-1})$	$\int_a^b f(x)\, dx$	Volume
Speed	$f(T_i)(t_i - t_{i-1})$ is estimate of distance covered from time t_{i-1} to time t_i.	$\sum_{i=1}^{n} f(T_i)(t_i - t_{i-1})$	$\int_a^b f(t)\, dt$	Distance
Just a function (no application in mind)	$f(c_i)(x_i - x_{i-1})$ is just a product of two numbers.	$\sum_{i=1}^{n} f(c_i)(x_i - x_{i-1})$	$\int_a^b f(x)\, dx$	Just a number (no application in mind)

The close tie between the derivative and the definite integral is expressed in the two fundamental theorems of calculus:

FIRST FUNDAMENTAL THEOREM

If f is continuous on the interval $[a, b]$ and if F is an antiderivative of f, then

$$\int_a^b f(x)\,dx = F(b) - F(a).$$

SECOND FUNDAMENTAL THEOREM

Let f be continuous on an open interval containing the interval $[a, b]$. Then, for $a \le x \le b$,

$$\frac{d}{dx}\left(\int_a^x f(t)\,dt\right) = f(x).$$

The first fundamental theorem is also called the fundamental theorem and is abbreviated by the letters FTC. It provides a tool for computing many definite integrals. If an antiderivative of f is elementary, then FTC is of use. But there are elementary functions, for instance, $\sqrt{1 + x^3}$, which are not derivatives of elementary functions. In these cases, it may be necessary to estimate the definite integral, say, by an approximating sum.

Any function whose derivative is the function f is called an **antiderivative** of f and is denoted $\int f(x)\,dx$. Any two antiderivatives of a function over an interval differ by a constant.

The second fundamental theorem implies that every continuous function is the derivative of some function. More specifically, it tells the rate of change of a definite integral as the interval over which the integral is computed is changed.

The FTC is *not* a theorem about area or mass. It is a theorem about the limit of the sums $\sum_{i=1}^{n} f(c_i)(x_i - x_{i-1})$. In many applications, it is first shown that a certain quantity (area, distance, volume, mass, etc.) is estimated by sums of that type, and then the FTC is called on. However, it may or may not be of use.

In the first five chapters we have covered the core of calculus: limits, derivatives, definite integrals, antiderivatives, and the relations between them. The remaining chapters present further applications and generalizations of these ideas.

Vocabulary and Symbols

sigma notation $\sum_{i=1}^{n} a_i$

partition

section of a partition

sampling number c_i

mesh

definite integral $\int_a^b f(x)\,dx$

telescoping sum $\sum_{i=1}^{n} (b_i - b_{i-1}) = b_n - b_0$

approximating sum $\sum_{i=1}^{n} f(c_i)(x_i - x_{i-1})$

first fundamental theorem of calculus = fundamental theorem of calculus = first fundamental theorem = fundamental theorem = FTC.

$$\int_a^b f(x)\,dx = -\int_b^a f(x)\,dx \quad \text{if } b < a; \qquad \int_a^a f(x)\,dx = 0$$

second fundamental theorem of calculus = second fundamental theorem

elementary function
integral
antiderivative $\int f(x)\,dx$
average of a function
indefinite integral
integrand
$F(x)\Big|_a^b = F(b) - F(a)$

Key Facts

If f is continuous, $\int_a^b f(x)\,dx$ exists.

$$\int x^a\,dx = \frac{x^{a+1}}{a+1} + C,\ a \neq -1.$$

PROPERTIES OF ANTIDERIVATIVES

1. $\int cf(x)\,dx = c\int f(x)\,dx \qquad (c \text{ constant}).$
2. $\int (f(x) + g(x))\,dx = \int f(x)\,dx + \int g(x)\,dx.$
3. $\int (f(x) - g(x))\,dx = \int f(x)\,dx - \int g(x)\,dx.$

PROPERTIES OF DEFINITE INTEGRALS

1. $\int_a^b cf(x)\,dx = c\int_a^b f(x)\,dx \qquad (c \text{ constant}).$
2. $\int_a^b (f(x) + g(x))\,dx = \int_a^b f(x)\,dx + \int_a^b g(x)\,dx.$
3. $\int_a^b (f(x) - g(x))\,dx = \int_a^b f(x)\,dx - \int_a^b g(x)\,dx.$
4. If $f(x) \geq 0$, then $\int_a^b f(x)\,dx \geq 0 \qquad (a < b).$
5. If $f(x) \geq g(x)$, then $\int_a^b f(x)\,dx \geq \int_a^b g(x)\,dx \qquad (a < b).$
6. $\int_a^b f(x)\,dx = \int_a^c f(x)\,dx + \int_c^b f(x)\,dx.$
7. If m and M are constants, $m \leq f(x) \leq M$, then

$$m(b - a) \leq \int_a^b f(x)\,dx \leq M(b - a) \qquad (a < b).$$

MEAN-VALUE THEOREM FOR INTEGRALS

Let a and b be numbers and let f be a continuous function defined for x between a and b. Then there is a number c between a and b such that

$$\int_a^b f(x)\,dx = f(c)(b - a).$$

Although there are formulas for computing some definite integrals, do not forget that a definite integral is a limit of sums. There are two reasons for keeping this fundamental concept clear:

1. In many applications in science the concept of the definite integral is more important than its use as a computational tool.
2. Many definite integrals cannot be evaluated by a formula. Some of the more important of these have been tabulated to several decimal places and published in handbooks of mathematical tables.

To emphasize that the fundamental theorem of calculus does not dispose of all definite integrals, here is a little assortment of elementary functions whose integrals are not elementary:

$$\sqrt{x}\sqrt[n]{1+x}, \qquad \sqrt[n]{1+x^2} \quad \text{for } n = 3, 4, 5, \ldots$$

$$\sqrt{1+x^n} \quad \text{for } n = 3, 4, 5, \ldots$$

$$\frac{\sin x}{x}, \qquad x \tan x, \qquad \frac{2^x}{x}, \qquad 2^{x^2}$$

Even when an elementary antiderivative exists, given the easy access to calculators and computers, we may prefer to get a good estimate of the definite integral by calculating an approximating sum rather than by struggling to find an antiderivative. (Programmable calculators can provide estimates of definite integrals accurate to several decimal places.)

GUIDE QUIZ ON CHAP. 5: THE DEFINITE INTEGRAL

1 Define (*a*) $\int_a^b f(x)\,dx$, (*b*) $\int f(x)\,dx$.

2 Estimate $\int_0^3 dx/(1 + x^3)$ by using an approximating sum with three sections of equal length and, as sampling points, (*a*) left endpoints, (*b*) right endpoints.

3 (*a*) Differentiate $1/(2x + 3)$.
(*b*) Find the area under the curve $y = 1/(2x + 3)^2$ and above the interval $[0, 1]$.

4 The curve $y = 1/x$ is rotated around the x axis. Find the volume of the solid enclosed by the resulting surface, between $x = 1$ and $x = 4$. (See Fig. 5.52.)

5 What is the average of $\sec^2 2x$ for x in $[\pi/12, \pi/8]$?

6 Find $D^2(\int_{x^2}^{x^3} \cos 3t\,dt)$.

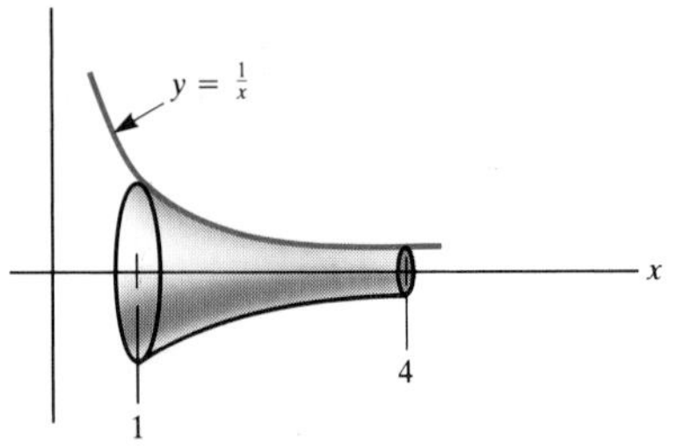

Figure 5.52

REVIEW EXERCISES FOR CHAP. 5: THE DEFINITE INTEGRAL

In each of Exercises 1 to 6 find the area of the region under the given curve and above the given interval.

1 $y = 2x^3$; $[1, 2]$

2 $y = 6x^2 + 10x^4$; $[-1, 2]$

3 $y = \sin 3x$; $[0, \pi/6]$

4 $y = 3\cos 2x$; $[\pi/6, \pi/4]$

5 $y = \dfrac{1}{x^3}$; $[2, 3]$

6 $y = \dfrac{1}{(x+1)^4}$; $[0, 1]$

In each of Exercises 7 to 16 give a formula for the antiderivatives of the given function (including the constant C).

7 $\sec^2 x$

8 $\sec^2 3x$

9 $\sec x \tan x$

10 $5 \sec 3x \tan 3x$

11 $4 \csc x \cot x$

12 $-\csc 5x \cot 5x$

13 $(x^3 + 1)^2$ (*Suggestion:* First multiply it out.)

14 $\left(x + \dfrac{1}{x}\right)^2$

15 $100x^{19}$

16 $\dfrac{24}{x^3}$

17 (*a*) Differentiate $(x^3 + 1)^6$.
(*b*) Find $\int (x^3 + 1)^5 x^2\,dx$.

18 (*a*) Differentiate $\sin x^3$.
(*b*) Find $\int x^2 \cos x^3\,dx$.

19 Write these sums without using the sigma notation:

(*a*) $\sum_{j=1}^{3} d^j$ (*b*) $\sum_{k=1}^{4} x^k$

(*c*) $\sum_{i=0}^{3} i2^{-i}$ (*d*) $\sum_{i=2}^{5} \dfrac{i+1}{i}$

(*e*) $\sum_{i=2}^{4} \left(\dfrac{1}{i} - \dfrac{1}{i+1}\right)$ (*f*) $\sum_{i=1}^{4} \sin \dfrac{\pi i}{4}$.

20 Write these sums without sigma notation and then evaluate:

(*a*) $\sum_{i=1}^{100} (2^i - 2^{i-1})$ (*b*) $\sum_{i=0}^{100} (2^{i+1} - 2^i)$

(*c*) $\sum_{i=1}^{100} \left(\dfrac{1}{i} - \dfrac{1}{i+1}\right)$

■

21 Let n be a positive integer and f a function.

(*a*) Show that $\sum_{i=1}^{n} f(i/n)(1/n)$ is an approximating sum for the definite integral $\int_0^1 f(x)\,dx$.

(*b*) What is the length of the ith section of the partition in (*a*)?

(*c*) What is the mesh of the partition?

(*d*) Where does the sampling number c_i lie in the ith section?

22 Explain why $\frac{1}{100}\sum_{i=1}^{100} f(i/100)$ is an estimate of $\int_0^1 f(x)\,dx$.

23 What definite integrals are estimated by the following sums?

(*a*) $\displaystyle\sum_{i=1}^{200}\left(\frac{i}{100}\right)^3\frac{1}{100}$ (*b*) $\displaystyle\sum_{i=1}^{100}\left(\frac{i-1}{100}\right)^4\frac{1}{100}$

(*c*) $\displaystyle\sum_{i=101}^{300}\left(\frac{i}{100}\right)^5\frac{1}{100}$

24 (See Exercise 23.)

(*a*) Show that $(1/n^3)\sum_{i=1}^{n} i^2$ is an approximation of $\int_0^1 x^2\,dx$.

(*b*) Compute the sum in (*a*) when $n = 4$.

(*c*) Compute $\int_0^1 x^2\,dx$.

(*d*) Find $\lim_{n\to\infty} (1/n^3)\sum_{i=1}^{n} i^2$.

25 Find $\lim_{\Delta x\to 0} (\int_3^{3+\Delta x} \sin\sqrt{t}\,dt/\Delta x)$.

26 Let f be a function such that $\int_0^x f(t)\,dt = [f(x)]^2$ for $x \geq 0$. Assume that $f(x) > 0$ for $x > 0$.

(*a*) Find $f(0)$. (*b*) Find $f(x)$ for $x > 0$.

27 A particle moves on a line in such a way that its time-average of velocity over any interval of time $[a, b]$ is the same as its velocity at time $(a + b)/2$. Prove that the velocity $v(t)$ must be of the form $ct + d$ for appropriate constants c and d. *Hint:* Begin by differentiating the relation $\int_a^b v(t)\,dt = [v((a + b)/2)](b - a)$ with respect to b and with respect to a.

28 A particle moves on a line in such a way that the time-average of velocity over any interval of time $[a, b]$ is equal to the average of its velocities at the beginning and the end of the interval of time. Prove that the velocity $v(t)$ must be of the form $ct + d$ for appropriate constants c and d.

6 Topics in Differential Calculus

This chapter obtains the derivatives of the logarithmic, exponential, and inverse trigonometric functions. In addition, it describes further applications of the derivative.

6.1 REVIEW OF LOGARITHMS

Consider the question

$$3^? = 9;$$

read as "3 raised to what power equals 9?" The answer, whatever its numerical value might be, is called "the logarithm of 9 to the base 3." Since

$$3^2 = 9,$$

we say that "the logarithm of 9 to the base 3 is 2." The general definition of logarithm follows:

Definition *Logarithm.* If b and c are positive numbers, $b \neq 1$, and

$$b^x = c,$$

then the number x is the **logarithm** of c to the base b and is written

Remember: A logarithm is an exponent!

$$\log_b c.$$

Any exponential equation $b^x = c$ may be translated into a logarithmic equation $x = \log_b c$, just as any English statement may be translated into French. This table illustrates some of these translations. Read it over several times, perhaps aloud, until you can, when covering a column, fill in the correct translation of the other column.

Exponential Language	*Logarithmic Language*
$3^2 = 9$	$\log_3 9 = 2$
$7^0 = 1$	$\log_7 1 = 0$
$10^3 = 1{,}000$	$\log_{10} 1{,}000 = 3$
$10^{-2} = 0.01$	$\log_{10} 0.01 = -2$
$9^{1/2} = 3$	$\log_9 3 = \frac{1}{2}$
$8^{2/3} = 4$	$\log_8 4 = \frac{2}{3}$
$8^{-1} = \frac{1}{8}$	$\log_8 \frac{1}{8} = -1$
$5^1 = 5$	$\log_5 5 = 1$
$b^1 = b$	$\log_b b = 1$

Since $b^x = c$ is equivalent to $x = \log_b c$, it follows that

$$b^{\log_b c} = c.$$

$b^{\log_b c} = c.$

The equality $b^{\log_b c} = c$ is not deep; it just restates the definition of a logarithm.

EXAMPLE 1 Find $\log_5 125$.

SOLUTION Look for an answer to the question

"5 to what power equals 125?"

or, equivalently, for a solution of the equation

$$5^x = 125.$$

Since $5^3 = 125$, the answer is 3; that is,

$$\log_5 125 = 3 \quad \blacksquare$$

EXAMPLE 2 Find $\log_{10} \sqrt{10}$.

SOLUTION By the definition of $\log_{10} \sqrt{10}$,

$$10^{\log_{10}\sqrt{10}} = \sqrt{10}.$$

Now,

$$10^{1/2} = \sqrt{10}.$$

Thus

$$\log_{10} \sqrt{10} = \tfrac{1}{2}.$$

In words, "the power to which we must raise 10 to get $\sqrt{10}$ is $\frac{1}{2}$." As a calculator shows, $\sqrt{10} \approx 3.162$. Thus

$$\log_{10} 3.162 \approx \tfrac{1}{2} = 0.5. \quad \blacksquare$$

x	$\log_{10} x$
100	2
10	1
1	0
0.1	-1
0.01	-2

In order to get an idea of the logarithm as a function, consider logarithms to the familiar base of 10:

$$y = \log_{10} x.$$

Begin with a table, as shown in the margin. We must restrict ourselves to $x > 0$. A negative number, such as -1, cannot have a logarithm, since there is no power of 10 that equals -1. Similarly, 0 does not have a

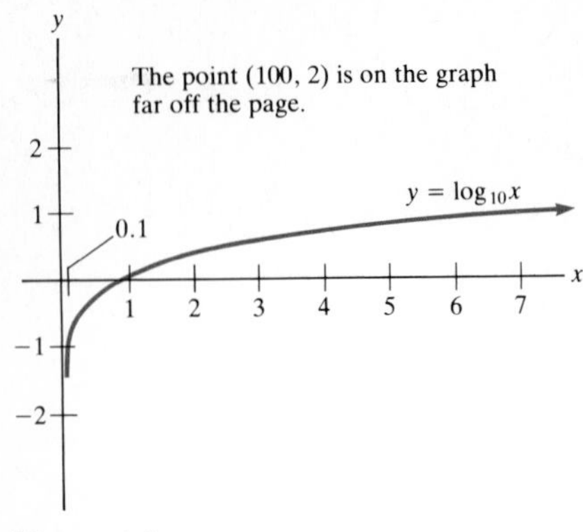

Figure 6.1

logarithm since the equation $10^x = 0$ has no solution. The domain of the function "log to the base 10" consists of the positive real numbers.

With the aid of the five points in the table, the graph is easy to sketch and is shown in Fig. 6.1. The graph lies to the right of the y axis. Far to the right it rises slowly; not until x reaches 100 does the y coordinate reach 2. Furthermore, $\log_{10} 1{,}000 = 3$ and $\log_{10} 10^n = n$ for any positive integer n. Although $\log_{10} x$ grows slowly, it does become arbitrarily large as $x \to \infty$: we have $\lim_{x\to\infty} \log_{10} x = \infty$.

To see how $\log_{10} x$ behaves when x is a small positive number, note that $\log_{10} 0.01 = -2$, $\log_{10} 0.001 = -3$, and $\log_{10} 10^{-n} = -n$ for any positive integer n. We see that $\lim_{x\to 0^+} \log_{10} x = -\infty$. Keeping in mind that $\log_{10} x$ is an increasing function, we conclude that for $x > 0$, $\log_{10} x$ takes on all values, positive and negative. Similarly, for any base $b > 1$, $\lim_{x\to\infty} \log_b x = \infty$ and $\lim_{x\to 0^+} \log_b x = -\infty$.

Logarithms to the base 10 are called **common logarithms**. Many calculators have a $\log_{10}$-key, usually labeled "log." Such a key replaces logarithm tables that filled books, the first of which was published by John Napier in 1614.

Properties of Logarithms Since each exponential equation $b^x = c$ translates into the corresponding logarithmic equation $x = \log_b c$, every property of exponentials must carry over to some property of logarithms. For instance, the information that

$$b^0 = 1$$

translates, in the language of logarithms, to

$$\log_b 1 = 0.$$

The logarithm of 1 in any base b is 0.

The equation
$$b^1 = b$$

amounts to saying that the logarithm of b (the base) in the base b is 1:

$$\log_b b = 1.$$

Thus $\log_{10} 10 = 1$ and $\log_{2.718}(2.718) = 1$.

The following table lists the fundamental properties of exponential functions together with the corresponding properties of logarithms.

Exponents	*Logarithms*
$b^0 = 1$	$\log_b 1 = 0$
$b^{1/2} = \sqrt{b}$	$\log_b \sqrt{b} = \frac{1}{2}$
$b^1 = b$	$\log_b b = 1$
$b^{x+y} = b^x b^y$	$\log_b cd = \log_b c + \log_b d$
$b^{-x} = \dfrac{1}{b^x}$	$\log_b \left(\dfrac{1}{c}\right) = -\log_b c$
$b^{x-y} = \dfrac{b^x}{b^y}$	$\log_b \left(\dfrac{c}{d}\right) = \log_b c - \log_b d$
$(b^x)^y = b^{xy}$	$\log_b c^m = m \log_b c$

Of these identities for logarithms, the most fundamental is

$$\log_b cd = \log_b c + \log_b d, \tag{1}$$

which asserts that the log of the product is the sum of the logs of the factors. The proof is instructive. (Proofs of the others are left as exercises.)

Theorem 1 For any positive numbers c and d and for any base b,

The log of the product

$$\log_b cd = \log_b c + \log_b d.$$

Proof By the definition of the logarithm as an exponent,

$$c = b^{\log_b c} \quad \text{and} \quad d = b^{\log_b d}. \tag{2}$$

Thus

$$cd = b^{\log_b c} b^{\log_b d}.$$

By the basic law of exponents ($b^x b^y = b^{x+y}$),

$$b^{\log_b c} b^{\log_b d} = b^{\log_b c + \log_b d} \tag{3}$$

Combining Eqs. (2) and (3) shows that

$$cd = b^{\log_b c + \log_b d}.$$

So the exponent to which b must be raised to get cd is

$$\log_b c + \log_b d.$$

In other words, the logarithm of cd to the base b is $\log_b c + \log_b d$. ■

EXAMPLE 3 Use the last line in the preceding table to evaluate $\log_9 (3^7)$ and $\log_5 \sqrt[3]{25^2}$.

SOLUTION

$$\log_9 (3^7) = 7 \log_9 3 = 7(\tfrac{1}{2}) = \tfrac{7}{2}.$$

$$\log_5 \sqrt[3]{25^2} = \log_5 (25)^{2/3} = \tfrac{2}{3} \log_5 25 = (\tfrac{2}{3})2 = \tfrac{4}{3}. \quad ■$$

The next example shows how logarithms can be used to solve equations in which the unknown appears in an exponent.

EXAMPLE 4 Find x if $5 \cdot 3^x \cdot 7^{2x} = 2$.

SOLUTION First rewrite the equation as

$$3^x \cdot 7^{2x} = \tfrac{2}{5} = 0.4,$$

and then take logarithms to the base 10 of both sides:

$$\log_{10} (3^x 7^{2x}) = \log_{10} 0.4$$

$$\log_{10} 3^x + \log_{10} 7^{2x} = \log_{10} 0.4 \qquad \text{log of a product}$$

$$x \log_{10} 3 + 2x \log_{10} 7 = \log_{10} 0.4 \qquad \text{log of an exponential}$$

$$x = \frac{\log_{10} 0.4}{\log_{10} 3 + 2 \log_{10} 7} \qquad \text{solving for } x$$

$$x \approx \frac{-0.3979}{0.4771 + 2(0.8451)} \qquad \text{calculator}$$

$$x \approx -0.1836. \quad ■$$

A calculator or computer program may make $\log_{10} x$ immediately available. In that case, how could you find logarithms with a different base, say, $\log_2 7$? The next example answers this question.

EXAMPLE 5 Express $\log_2 7$ in terms of common logarithms (base 10 logarithms).

SOLUTION Since a logarithm is an exponent, we begin with the equation

$$2^{\log_2 7} = 7.$$

Thus
$$\log_{10} (2^{\log_2 7}) = \log_{10} 7. \tag{4}$$

By the rule for the logarithm of a power [$\log_b (c^m) = m \log_b c$], transform Eq. (4) into

$$\log_2 7 \log_{10} 2 = \log_{10} 7,$$

and find that
$$\log_2 7 = \frac{\log_{10} 7}{\log_{10} 2}.$$

(A calculator then shows that $\log_{10} 7 \approx 0.8451$, $\log_{10} 2 \approx 0.3010$, and so $\log_2 7 \approx 2.807$. This answer is reasonable, since $\log_2 8 = 3$.) ■

The argument in Example 5 shows that for two different bases, b and c,

$$\log_b x = \log_b c \cdot \log_c x.$$

Since $\log_b c$ is a constant, this equation tells us that $\log_b x$ is proportional to $\log_c x$ and the constant of proportionality is $\log_b c$. (For instance, $\log_3 x = \log_3 9 \log_9 x = 2 \log_9 x$.) So for $b, c > 1$, the graphs of $y = \log_b x$ and $y = \log_c x$ have the same general shape: the first is obtained from the second by an expansion parallel to the y axis by a constant factor, $\log_b c$. (See Fig. 6.2.)

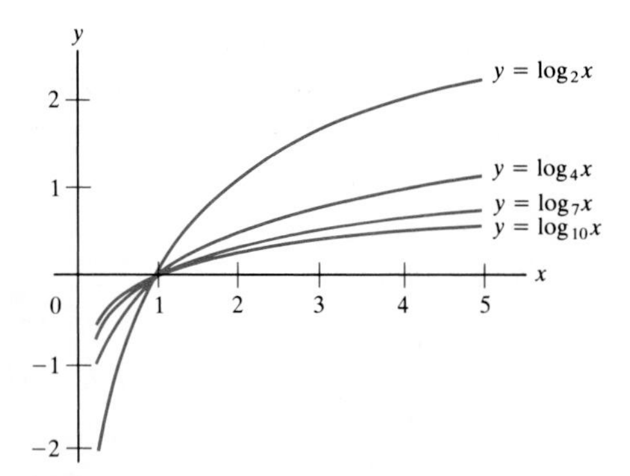

Figure 6.2

EXAMPLE 6 As every computer buff knows, 2^{10} is approximately 1,000. (It is actually 1,024.) Use this information to estimate $\log_{10} 2$.

SOLUTION Start with the approximation

$$2^{10} \approx 10^3,$$

and take $\log_{10}$ of both sides, getting

$$10 \log_{10} 2 \approx 3.$$

Hence

$$\log_{10} 2 \approx 0.3.$$

(To three decimal places, $\log_{10} 2 \approx 0.301$.) ■

Sometimes the formula for $\log f(x)$ may be much simpler than the formula for the function $f(x)$ itself. Example 7 illustrates this fact, which will come in handy later to simplify the computation of certain derivatives.

EXAMPLE 7 Let $f(x) = \sqrt[3]{\dfrac{(1 + x^2)^5 \sin^7 5x}{x^{2x}}}$. Express $\log_{10} f(x)$ as simply as possible.

SOLUTION

$$\log_{10} f(x) = \tfrac{1}{3} \log_{10} \left[\frac{(1 + x^2)^5 \sin^7 5x}{x^{2x}}\right] \qquad \text{log of a power}$$

$$= \tfrac{1}{3}\{\log_{10} [(1 + x^2)^5 \sin^7 5x] - \log_{10} x^{2x}\} \qquad \text{log of a quotient}$$

$$= \tfrac{1}{3}[\log_{10} (1 + x^2)^5 + \log_{10} \sin^7 5x - \log_{10} x^{2x}] \qquad \text{log of a product}$$

$$= \tfrac{1}{3}[5 \log_{10} (1 + x^2) + 7 \log_{10} \sin 5x - 2x \log_{10} x]. \qquad \text{log of a power}$$

And finally,

$$\log_{10} f(x) = \tfrac{5}{3} \log_{10} (1 + x^2) + \tfrac{7}{3} \log_{10} \sin 5x - \tfrac{2}{3}x \log_{10} x.$$

Simplification removed the radical sign, quotient, and two powers from the formula for $f(x)$. ■

As mentioned earlier, for $b > 1$, $\log_b x \to \infty$ as $x \to \infty$, but very slowly. In fact, it grows much more slowly than any fixed power of x, x^a, $a > 0$, as is illustrated by the following table, which compares $\log_2 x$ with $x^{0.1}$:

x	1	2	2^5	2^{10}	2^{100}	2^{200}
$\log_2 x$	0	1	5	10	100	200
$x^{0.1}$	1	1.072	1.414	2	1,024	1,048,576

For large x, $\log_2 x$ is much smaller than $\sqrt[10]{x}$.

Even though at $x = 2^{10}$, $\log_2 x$ is larger than $x^{0.1}$, for large x, the power $x^{0.1}$ takes a commanding lead. By the method in Sec. 6.10 it can be shown that for $b > 1$ and $a > 0$,

$$\lim_{x \to \infty} \frac{\log_b x}{x^a} = 0.$$

EXERCISES FOR SEC. 6.1: REVIEW OF LOGARITHMS

1 Translate these equations into the language of logarithms.
(*a*) $2^5 = 32$ (*b*) $3^4 = 81$
(*c*) $10^{-3} = 0.001$ (*d*) $5^0 = 1$
(*e*) $1{,}000^{1/3} = 10$ (*f*) $49^{1/2} = 7$

2 Translate these equations into the language of logarithms.
(*a*) $8^{2/3} = 4$ (*b*) $10^3 = 1{,}000$
(*c*) $10^{-4} = 0.0001$ (*d*) $3^0 = 1$
(*e*) $10^{1/2} = \sqrt{10}$ (*f*) $(\frac{1}{2})^{-2} = 4$

3 (*a*) Fill in this table:

x	$\frac{1}{9}$	$\frac{1}{3}$	1	3	9
$\log_3 x$					

(*b*) Plot the five points in (*a*) and graph $y = \log_3 x$.

4 (*a*) Fill in this table:

x	$\frac{1}{16}$	$\frac{1}{4}$	1	2	4	8	16
$\log_4 x$							

(*b*) Plot the seven points in (*a*) and graph $y = \log_4 x$.

5 Translate these equations into the language of exponents.
(*a*) $\log_2 7 = x$ (*b*) $\log_5 2 = s$
(*c*) $\log_3 \frac{1}{3} = -1$ (*d*) $\log_7 49 = 2$

6 Translate these equations into the language of exponents.
(*a*) $\log_{10} 1{,}000 = 3$ (*b*) $\log_5 \frac{1}{25} = -2$
(*c*) $\log_{1/2} (\frac{1}{4}) = 2$ (*d*) $\log_{64} 128 = \frac{7}{6}$

7 Evaluate
(*a*) $2^{\log_2 16}$ (*b*) $2^{\log_2 (1/2)}$ (*c*) $2^{\log_2 7}$ (*d*) $2^{\log_2 g}$

8 Evaluate
(*a*) $10^{\log_{10} 100}$ (*b*) $10^{\log_{10} 0.01}$
(*c*) $10^{\log_{10} 7}$ (*d*) $10^{\log_{10} p}$

9 Evaluate
(*a*) $\log_3 \sqrt{3}$ (*b*) $\log_3 (3^5)$ (*c*) $\log_3 (\frac{1}{27})$

10 If $\log_4 A = 2.1$, evaluate
(*a*) $\log_4 A^2$ (*b*) $\log_4 (1/A)$ (*c*) $\log_4 16A$

In each of Exercises 11 to 14 solve for x.

11 $2 \cdot 3^x = 7$ **12** $3 \cdot 5^x = 6^x$
13 $3^{5x} = 2^{7x}$ **14** $10^{2x}3^{2x} = 5$

15 (*a*) Evaluate $\log_2 8$, $\log_8 2$, and then their product.
(*b*) Show that, for any two bases a and b,

$$(\log_a b) \times (\log_b a) = 1.$$

Suggestion: Start with $a^{\log_a b} = b$ and take $\log_b$ of both sides.

16 If $\log_3 5 = a$, what is $\log_5 3$?

17 Assume that $\log_{10} 2 \approx 0.30$ and $\log_{10} 3 \approx 0.48$. From this information estimate
(*a*) $\log_{10} 4$ (*b*) $\log_{10} 5$
(*c*) $\log_{10} 6$ (*d*) $\log_{10} 8$
(*e*) $\log_{10} 9$ (*f*) $\log_{10} 1.5$
(*g*) $\log_{10} 1.2$ (*h*) $\log_{10} 1.33$
(*i*) $\log_{10} 20$ (*j*) $\log_{10} 200$
(*k*) $\log_{10} 0.006$

18 Assume that $\log_{10} 2 \approx 0.30$ and $\log_{10} 3 \approx 0.48$. From this information, obtain estimates for
(*a*) $\log_{10} \sqrt{2}$ (*b*) $\log_{10} 0.5$
(*c*) $\log_{10} \frac{2}{3}$ (*d*) $\log_{10} \sqrt[3]{3}$
(*e*) $\log_{10} 18$ (*f*) $\log_{10} 12$
(*g*) $\log_{10} 0.75$ (*h*) $\log_{10} 7.5$
(*i*) $\log_{10} (1/7.5)$ (*j*) $\log_{10} 0.075$
(*k*) $\log_{10} (30\sqrt[3]{2^5})$ (*l*) $\log_{10} \frac{9}{32}$

In Exercises 19 and 20 use logarithms to the base 10 to calculate the given logarithms to two decimal places.

19 (*a*) $\log_3 5$ (*b*) $\log_2 3$
20 (*a*) $\log_{1/2} 3$ (*b*) $\log_7 \frac{1}{2}$

21 (*a*) Express $\log_{1/2} x$ in terms of $\log_2 x$.
(*b*) Sketch the graphs of $y = \log_{1/2} x$ and $y = \log_2 x$.
(*c*) How is one graph in (*b*) obtainable from the other?

22 Let $b > 1$.
(*a*) How are the graphs of $\log_b x$ and $\log_{1/b} x$ related?
(*b*) How are the graphs of $\log_b x$ and $\log_{b^2} x$ related?

23 From the fact that $(b^x)^y = b^{xy}$, deduce that $\log_b c^m = m \log_b c$. (*Hint:* Write c as $b^{\log_b c}$ and consider c^m.)

24 From the fact that $b^{x-y} = b^x/b^y$, deduce that $\log_b (c/d) = \log_b c - \log_b d$. (*Hint:* Write c as $b^{\log_b c}$ and d as $b^{\log_b d}$.)

In Exercises 25 to 28 express $\log_{10} f(x)$ as simply as possible for the given $f(x)$.

25 $f(x) = \dfrac{\cos^7 x \sqrt{(x^2 + 5)^3}}{4 + \tan^2 x}$

26 $f(x) = \sqrt{(1 + x^2)^5(3 + x)^4\sqrt{1 + 2x}}$

27 $f(x) = (x\sqrt{2} + \cos x)^{x^2}$ **28** $f(x) = \sqrt{\dfrac{x(1 + x)}{\sqrt{(1 + 2x)^3}}}$

■

29 Find $\log_2 [\log_2 (\log_2 2^{1.024})]$.

30 Is $\log_2 (c + d)$ ever equal to $\log_2 c + \log_2 d$? Explain your answer.

31 (The slide rule) Logarithms are the basis for the design of the slide rule, a device for multiplying or dividing two numbers, to three significant figures. Slide rules were common from the early part of the seventeenth century to their recent eclipse by the hand-held calcu-

Figure 6.3

lator. A slide rule consists of two sticks (or circular disks), one of which is fixed and the other is free to slide next to it. A scale is introduced on each stick by placing the number N at a distance $\log_{10} N$ inches from the left end of the stick. (Any base would do as well as 10.) Thus 1 is placed at the left end of each stick and 10 at $\log_{10} 10 = 1$ inch from the left end. The scale of the top stick is at the bottom edge; the scale of the lower stick is at the top edge. Each stick has the scale shown in Fig. 6.3.

(*a*) Make two such scales on paper or cardboard.

(*b*) Use them to multiply 4 times 25, as follows. Place the 1 of the lower stick at the 4 of the upper stick. Above the 25 of the lower stick appears the product. Check that this works with your two sticks.

(*c*) Explain precisely why the slide rule works.

(*d*) How would you use it to divide two numbers?

32 In the Richter scale, the intensity M of an earthquake is related to the energy E of the earthquake by the formula $\log_{10} E = 11.4 + 1.5\,M$. ($E$ is measured in ergs.)

(*a*) If one earthquake has a thousand times the energy of another, how much larger is its Richter rating M?

(*b*) What is the ratio of the energy of the San Francisco earthquake of 1906 ($M = 8.3$) with that of the Eureka earthquake of 1980 ($M = 7$)?

(*c*) What is the Richter rating of a 10 megaton H-bomb, that is, of an H-bomb whose energy is equivalent to that in 10 million tons of TNT? (One ton of TNT contains 4.2×10^6 ergs.)

For a fuller discussion of the Richter scale, see the *Encyclopaedia Britannica, Macropaedia,* vol. 6, pp. 71–72.

33 Prove that $\log_3 2$ is irrational. (*Hint:* Assume that it is rational, that is, equal to m/n for some integers m and n, and obtain a contradiction.)

34 If $0 < b < 1$, examine (*a*) $\lim_{x\to\infty} \log_b x$, and (*b*) $\lim_{x\to 0^+} \log_b x$. (*Hint:* Express $\log_b x$ in terms of $\log_{1/b} x$.)

35 As of October of 1985, the largest known prime was $2^{216{,}091} - 1$.

(*a*) When written in decimal notation, how many digits will it have?

(*b*) How many pages of this book would be needed to print it? (One page can hold about 6,400 digits.)

6.2 THE NUMBER e

This section describes a number that is as important in calculus as π is in the study of the circle. We introduce this number, which is always denoted e, by an example involving compound interest.

Say that you put A dollars in an account at the beginning of some interest period. This period could be 1 year or 6 months or 3 months or an even shorter time period. The bank agrees to pay an interest rate r for money left on deposit for that period. In practice, r is around 6 percent $=$ 0.06 per year, or 0.03 per half year, etc. During the time period the value of the account would increase by rA dollars and amount to a total of

$$A + rA = (1 + r)A \text{ dollars.}$$

So, to find the final amount, just multiply the initial amount by $1 + r$:

$$\begin{pmatrix}\text{Amount at end of}\\ \text{an interest period}\end{pmatrix} = (1 + r)(\text{initial amount}). \qquad (1)$$

Now assume that the interest rate is very generous, 100 percent per year, that is, $r = 1$ for the year. One dollar left in the bank a year would grow to

$$(1 + 1)1 = 2 \text{ dollars},$$

by formula (1).

If instead the bank compounds interest twice a year, the interest rate for each of the 6-month periods would be 50 percent, or $r = 0.50$, or simply $r = \frac{1}{2}$. At the end of the first 6 months an initial 1 dollar would grow to

$$(1 + \tfrac{1}{2})1 = 1 + \tfrac{1}{2} \text{ dollars},$$

again by formula (1). During the second 6 months this amount, $1 + \frac{1}{2}$, would grow to

$$(1 + \tfrac{1}{2})(1 + \tfrac{1}{2}) = (1 + \tfrac{1}{2})^2 = 2.25 \text{ dollars},$$

again by formula (1).

A competing bank offers to compound interest 3 times a year, at 4-month intervals. Thus $r = \frac{1}{3}$. At the end of the first 4 months 1 dollar would grow to

$$(1 + \tfrac{1}{3})1 = 1 + \tfrac{1}{3} \text{ dollars}.$$

During the second period of 4 months, this amount, $1 + \frac{1}{3}$ dollars, left on deposit, grows to

$$(1 + \tfrac{1}{3})(1 + \tfrac{1}{3}) = (1 + \tfrac{1}{3})^2 \text{ dollars}.$$

During the final 4 months, the amount $A = (1 + \frac{1}{3})^2$ grows to

$$(1 + \tfrac{1}{3})A = (1 + \tfrac{1}{3})(1 + \tfrac{1}{3})^2 = (1 + \tfrac{1}{3})^3 \approx 2.37 \text{ dollars}.$$

Other bankers join in, one compounding 4 times a year (quarterly), another daily, and another hourly. When interest is compounded 4 times a year, a deposit of 1 dollar would grow to

$$(1 + \tfrac{1}{4})^4 \approx 2.44 \text{ dollars}$$

in a year. If interest is deposited n times a year, 1 dollar would grow to

$$\left(1 + \frac{1}{n}\right)^n \text{ dollars}$$

in a year. The following table shows the amount in the account at the end of the year if 1 dollar is compounded n times at an annual interest rate of 100 percent.

	Simple Interest	*Semi-annually*	*Every 4 Months*	*Quarterly*	*Monthly*	*Daily*	*Hourly*
Number of times compounded (n)	1	2	3	4	12	365	8,760
Value at end of year (rounded off)	\$2	\$2.25	\$2.37037	\$2.44141	\$2.61304	\$2.71457	\$2.71813

The more often the interest is computed, the more the account grows. However, no matter how frequently interest is compounded—even if every second—the account does not get arbitrarily large in one year.

It turns out that it gets arbitrarily close to a certain amount, which, to five decimal places, is \$2.71828. This brings us to the definition of the number e.

Definition *The number e.*

e is not a repeating decimal. The next digit is 4.

$$e = \lim_{n \to \infty} \left(1 + \frac{1}{n}\right)^n \approx 2.718281828.$$

Observe that for large n the expression $(1 + 1/n)^n$ is of the form

$$(1 + \text{small number})^{\text{reciprocal of same small number}}.$$

So we may consider

$$(1 + x)^{1/x}$$

when x is near 0, even if x is not of the form $1/n$, that is, is not the reciprocal of an integer. It can be shown that $\lim_{x \to 0} (1 + x)^{1/x}$ exists and equals e:

$$\lim_{x \to 0} (1 + x)^{1/x} = e. \qquad (2)$$

Often Eq. (2) is taken as the definition of e. It is this expression for e that will be used in the next section, where we will find the derivatives of the logarithm functions.

We *assume* that the limit in Eq. (2) exists, since a proof that it does exist would amount to a big detour. (See Exercises 17 and 18 and Appendix H.) Note that there are two conflicting influences on $(1 + x)^{1/x}$ when x is near 0. For simplicity, consider only $x > 0$. First, the base gets near 1, so there is a chance that $(1 + x)^{1/x}$ gets near 1. But the exponent $1/x$ gets arbitrarily large, so there is also a chance that $(1 + x)^{1/x}$ gets large, since the base $1 + x$ is larger than 1. It turns out that the force pushing $(1 + x)^{1/x}$ toward 1 is not strong enough to do that, but it does manage to keep $(1 + x)^{1/x}$ from exceeding $e \approx 2.71828$.

From the fact that $\lim_{x \to 0} (1 + x)^{1/x} = e$, we can obtain other closely related limits. For instance, $\lim_{h \to 0} (1 + 2h)^{1/2h} = e$. (Note that $2h \to 0$ and the exponent is the reciprocal of $2h$.)

EXAMPLE Find $\lim_{h \to 0} (1 + 2h)^{1/h}$.

SOLUTION The expression $(1 + 2h)^{1/h}$ is not of the form

$$(1 + \text{small number})^{\text{reciprocal of same small number}}$$

since $1/h$ is not the reciprocal of $2h$. A little algebra gets around this obstacle:

$$\begin{aligned}
\lim_{h \to 0} (1 + 2h)^{1/h} &= \lim_{h \to 0} (1 + 2h)^{2/2h} \\
&= \lim_{h \to 0} [(1 + 2h)^{1/2h}]^2 && (b^c)^d = b^{cd} \\
&= \left[\lim_{h \to 0} (1 + 2h)^{1/2h}\right]^2 && x^2 \text{ is continuous} \\
&= e^2. \quad \blacksquare
\end{aligned}$$

The number e was introduced by considering a bank that pays 100 percent interest per year. However, e is useful in analyzing compound interest even when the rate is more modest, say, 6 percent per year. Reasoning similar to the case for 100 percent interest shows that 1 dollar grows to

$$\left(1 + \frac{0.06}{n}\right)^n \text{ dollars} \tag{3}$$

if interest is compounded n times a year. To find out what happens to (3) as $n \to \infty$, rewrite it in such a way that the reciprocal of $0.06/n$ appears as an exponent:

$$\left(1 + \frac{0.06}{n}\right)^n = \left(1 + \frac{0.06}{n}\right)^{(n/0.06)0.06}$$
$$= \left[\left(1 + \frac{0.06}{n}\right)^{n/0.06}\right]^{0.06}.$$

Thus, as $n \to \infty$, $0.06/n \to 0$, and we obtain

$$\lim_{n\to\infty} \left(1 + \frac{0.06}{n}\right)^n = e^{0.06}.$$

Finding $e^{0.06}$ on a calculator with an e^x-key, we see that $e^{0.06} \approx 1.06184$. Thus so-called continuous compounding is not much more rewarding than simple interest in which 1 dollar grows to \$1.06. [Compounded at 6-month intervals, 1 dollar grows to $(\$1.03)^2 = \$1.0609$.]

More generally, when the annual interest rate is r (usually in the range from 0.05 to 0.20) and compounding is done n times per year, 1 dollar grows to $(1 + r/n)^n$ dollars at the end of one year. By arguments similar to those already used, it then can be shown that

$$\lim_{n\to\infty} \left(1 + \frac{r}{n}\right)^n = e^r.$$

The formula for "continuous compounding"

So, in continuous compounding, where the annual rate of interest is r, 1 dollar grows to e^r dollars in one year.

Leonhard Euler (pronounced "oiler"), the great Swiss mathematician, introduced the number e (and named it) in his calculus text *Introductio in analisin Infinitorum,* vol. 1, 1748, p. 90, in these words:

inventam, $a = 1 + \frac{1}{1} + \frac{1}{1.2} + \frac{1}{1.2.3} + \frac{1}{1.2.3.4} + \&c.$, qui termini, ſi in fractiones decimales convertantur atque actu addantur, præbebunt hunc valorem pro a = 2,71828182845904523536028. cujus ultima adhuc nota veritati eſt conſentanea. Quod ſi jam ex hac baſi Logarithmi conſtruantur, ii vocari ſolent Logarithmi *naturales* ſeu *hyperbolici*, quoniam quadratura hyperbolæ per iſtiuſmodi Logarithmos exprimi poteſt. Ponamus autem brevitatis gratia pro numero hoc 2,718281828459 &c. conſtanter litteram e, quæ ergo denotabit baſin Logarithmorum naturalium ſeu hyperbolicorum,

Here it is defined as the limit as $n \to \infty$ of sums of the form

$$1 + \frac{1}{1} + \frac{1}{2!} + \frac{1}{3!} + \frac{1}{4!} + \cdots + \frac{1}{n!}.$$

(See Exercise 16.)

The number e appears often in places where you would not expect it. For example, imagine that you write letters to n friends, address the n envelopes, and put the letters randomly in the envelopes, one to an envelope. The probability that all the letters are in wrong envelopes is approximately $1/e$. As $n \to \infty$, the probability approaches $1/e$.

As another example, the nth prime number is approximately equal to $n \log_e n$. If P_n denotes the nth prime number, then $\lim_{n\to\infty} P_n/(n \log_e n) = 1$. The 100th prime number is 541, and $541/(100 \log_e 100) \approx 1.17$. The 664,699th prime number is 10,006,721; the quotient $P_n/(n \log_e n) = 10{,}006{,}721/(664{,}699 \log_e 664{,}699) \approx 1.12$.

EXERCISES FOR SEC. 6.2: THE NUMBER e

In Exercises 1 to 6 evaluate the limits.

1 $\lim_{t\to 0} (1 + t)^{1000}$ **2** $\lim_{x\to\infty} 1.001^x$

3 $\lim_{h\to 0} (1 + 3h)^{1/4h}$ **4** $\lim_{h\to 0} (1 - h)^{1/h}$

5 $\lim_{\Delta x\to 0} \left(1 + \frac{\Delta x}{x}\right)^{x/\Delta x}$, $x \neq 0$ is fixed

6 $\lim_{n\to\infty} \left(1 + \frac{3}{n}\right)^{n/2}$

7 What do you think happens to $(1 + x)^{1/x}$ as $x \to -1$ from the right? Do some calculations first.

8 What do you think happens to $(1 + x)^{1/x}$ as $x \to \infty$? Do some calculations first.

In the text we considered an account that started with 1 dollar and a time period of 1 year. Exercises 9 and 10 generalize this to an arbitrary initial amount and time period.

9 An amount of A dollars is deposited in a bank that pays an interest rate of r percent per year. How much will be in the account at the end of t years if the interest is compounded exactly n times during this period at intervals of t/n years?

10 (See Exercise 9.) Let $A(n)$ denote the amount in the account in Exercise 9 at the end of t years if the compounding is done n times. Find $\lim_{n\to\infty} A(n)$.

11 A bank pays an annual interest rate of 50 percent. Assume that \$1,000 is deposited at the beginning of the year. How much will there be in the account at the end of the year if interest is (*a*) simple (not compounded), (*b*) compounded every 6 months, (*c*) compounded monthly, (*d*) compounded daily, (*e*) compounded continuously?

12 Like Exercise 11, but with an interest rate of 8 percent per year.

■

13 (Calculator)
(*a*) Compute $(2^h - 1)/h$ for $h = 0.1, 0.01$, and -0.01.
(*b*) On the basis of (*a*), estimate the derivative of 2^x at 0.
(*c*) Similarly, estimate the derivative of 3^x at 0.
(*d*) Similarly, estimate the derivative of e^x at 0.

14 Show that if you knew the derivative of b^x at 0, then you would know it for all x.

15 On the basis of Exercises 7 and 8 and the definition of e, sketch the graph of $y = (1 + x)^{1/x}$ for $x > -1$, $x \neq 0$.

16 (*a*) Show that

$$(1 + 1/n)^n < \frac{1}{0!} + \frac{1}{1!} + \frac{1}{2!} + \frac{1}{3!} + \cdots + \frac{1}{n!}.$$

(*b*) Compute $\frac{1}{0!} + \frac{1}{1!} + \cdots + \frac{1}{6!}$. (*Hint:* Recall that $0! = 1$.)

17 Using approximating sums and the definition of a definite integral, show that

(*a*) $\int_1^2 dx/x < 1$ (*b*) $\int_1^3 dx/x > 1$

18 Pretend that you have *never heard of* e *and* $\lim_{h\to 0} (1 + h)^{1/h}$. This exercise outlines an argument using areas that $\lim_{h\to 0} (1 + h)^{1/h}$ exists. For simplicity, consider only rational $h > 0$.

(*a*) Sketch the curves $y = 1/x$ and $y = 1/x^{1-h}$ relative to the same axes. Note that for $x > 1$ the second curve lies above the first curve.

(*b*) For a given number h (rational) find the number $A(h)$ such that

$$\int_1^{A(h)} \frac{dx}{x^{1-h}} = 1.$$

(*c*) Using Exercise 17, show that there is a number B such that $\int_1^B dx/x = 1$. Note that $2 < B < 3$.

(*d*) Using (*a*), (*b*), and (*c*), show that, for $h > 0$, $(1 + h)^{1/h} < B$.

(*e*) Why would you expect, on the basis of geometric intuition, that $\lim_{h \to 0} (1 + h)^{1/h} = B$. This number B is, of course, the number e. (A similar argument works for $h < 0$.)

6.3 THE DERIVATIVES OF THE LOGARITHMIC FUNCTIONS

If a is rational and *not equal to* -1, then

$$\int x^a \, dx = \frac{x^{a+1}}{a+1} + C.$$

The formula works, for instance, when $a = -1.01$:

$$\begin{aligned} \int x^{-1.01} \, dx &= \frac{x^{-1.01+1}}{-1.01 + 1} + C \\ &= \frac{x^{-0.01}}{-0.01} + C \\ &= -100x^{-0.01} + C. \end{aligned}$$

But the formula breaks down when $a = -1$.

In view of the fundamental theorem of calculus, it would be helpful to have a formula for $\int x^a \, dx$ when $a = -1$, that is, a formula for

$$\int \frac{dx}{x}.$$

The function $\frac{1}{x}$ leads us to study logarithms.

It turns out that an antiderivative of $1/x$ is $\log_e x$. In order to show this, we will determine the derivative of $\log_b x$ for any base b and then consider the special case $b = e$.

In the proof of Theorem 1 it will be assumed that the function $\log_b x$ is continuous.

Theorem 1 The derivative of the function $\log_b x$ is

$$\frac{\log_b e}{x}$$

for all positive numbers x.

Proof It is necessary to compute

$$\lim_{\Delta x \to 0} \frac{\log_b (x + \Delta x) - \log_b x}{\Delta x}.$$

Recall that $f'(x) = \lim_{\Delta x \to 0} \frac{f(x + \Delta x) - f(x)}{\Delta x}$, *by definition of the derivative.*

Before letting $\Delta x \to 0$, rewrite the difference quotient,

$$\frac{\log_b (x + \Delta x) - \log_b x}{\Delta x},$$

using algebra and properties of logarithms, as follows:

$$\frac{\log_b (x + \Delta x) - \log_b x}{\Delta x}$$

$= \dfrac{\log_b \left(\dfrac{x + \Delta x}{x}\right)}{\Delta x}$	$\log_b c - \log_b d = \log_b (c/d)$
$= \dfrac{1}{\Delta x} \log_b \left(1 + \dfrac{\Delta x}{x}\right)$	algebra
$= \log_b \left(1 + \dfrac{\Delta x}{x}\right)^{1/\Delta x}$	$\log_b c^m = m \log_b c$
$= \log_b \left[\left(1 + \dfrac{\Delta x}{x}\right)^{x/\Delta x}\right]^{1/x}$	power of a power
$= \dfrac{1}{x} \log_b \left(1 + \dfrac{\Delta x}{x}\right)^{x/\Delta x}$	$\log_b c^m = m \log_b c.$

After these manipulations, it is easy to take limits:

$$\lim_{\Delta x \to 0} \frac{\log_b (x + \Delta x) - \log_b x}{\Delta x}$$

$= \lim_{\Delta x \to 0} \dfrac{1}{x} \log_b \left(1 + \dfrac{\Delta x}{x}\right)^{x/\Delta x}$	
$= \dfrac{1}{x} \lim_{\Delta x \to 0} \log_b \left(1 + \dfrac{\Delta x}{x}\right)^{x/\Delta x}$	$1/x$ is fixed
$= \dfrac{1}{x} \log_b \left[\lim_{\Delta x \to 0} \left(1 + \dfrac{\Delta x}{x}\right)^{x/\Delta x}\right]$	$\log_b$ is continuous
$= \dfrac{1}{x} \log_b e$	$x/\Delta x$ is the reciprocal of $(\Delta x)/x$.

Thus $\log_b$ has a derivative

$$(\log_b x)' = \frac{\log_b e}{x}. \quad ■$$

In particular,

$$(\log_{10} x)' = \frac{\log_{10} e}{x}.$$

As a table of logarithms or a calculator shows, $\log_{10} e \approx 0.434$. Thus

$$(\log_{10} x)' \approx \frac{0.434}{x}. \tag{1}$$

It is interesting to compare Eq. (1) with the slope of the graph of $\log_{10} x$ shown in Fig. 6.1. According to Eq. (1), the slope of the graph is always positive. The slope approaches 0 as $x \to \infty$; the slope approaches ∞ as $x \to 0$ through positive values. These conclusions are consistent with the general appearance of the graph.

Which is the best of all possible bases b to use? More precisely, for which base b does the formula

$$\frac{\log_b e}{x}$$

take its simplest form? Certainly not $b = 10$. It would be nice to choose the base b in such a way that

$$\log_b e = 1;$$

Why e is used as a base for logarithms

that is, b^1 must equal e. In this case, b is e. The best of all bases to use for logarithms is e. The derivative of the $\log_e$ function is given by

$$\frac{d}{dx}(\log_e x) = \frac{\log_e e}{x} = \frac{1}{x}.$$

In this case, there is no constant, such as 0.434, to memorize.

The natural logarithm

Scientific calculators usually have a ln-*key* ($\log_e$) *and a* log-*key* ($\log_{10}$).

For this reason, the base e is preferred in calculus. We shall write $\log_e x$ as **ln *x***, the **natural logarithm** of x. (Only for purposes of arithmetic, such as multiplying with the aid of logarithms, is base 10 preferable.) Most handbooks of mathematical tables include tables of $\log_{10} x$ (common logarithm) and $\ln x$ (natural logarithm).

A simple equation summarizes much of this section:

$$\boxed{\frac{d}{dx}(\ln x) = \frac{1}{x} \qquad x > 0.}$$

Math texts often use log *x to denote the natural logarithm.*

It is well worth memorizing the following statement: *The derivative of the natural logarithm function* ln *x is the reciprocal function* $1/x$.

EXAMPLE 1 Find the area under the curve $y = 1/x$ and above the interval [1, 100].

SOLUTION The area equals the definite integral

$$\int_1^{100} \frac{1}{x}\, dx.$$

By the fundamental theorem of calculus,

$$\int_1^{100} \frac{1}{x}\, dx = \ln x \Big|_1^{100} = \ln 100 - \ln 1$$

$$= \ln 100 \approx 4.605. \quad ■$$

EXAMPLE 2 Find $[\ln (x^2 + 1)]'$.

SOLUTION Let $y = \ln (x^2 + 1)$. Then $y = \ln u$, where $u = x^2 + 1$. By the chain rule,

$$\frac{dy}{dx} = \frac{dy}{du}\frac{du}{dx}$$

$$= \frac{d}{du}(\ln u)\frac{d}{dx}(x^2 + 1)$$

$$= \frac{1}{u} \cdot 2x = \frac{2x}{x^2 + 1}.$$

Thus the derivative of $\ln (x^2 + 1)$ is $2x/(x^2 + 1)$. ■

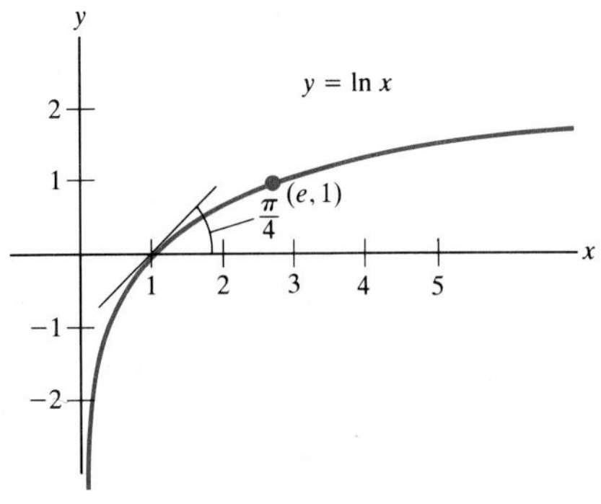

Figure 6.4

EXAMPLE 3 Graph $y = \ln x$. At what angle does it cross the x axis?

SOLUTION The graph is shown in Fig. 6.4. Note that at $x = e \approx 2.718$, the y coordinate is 1. The graph crosses the x axis at (1, 0). Since the derivative of $\ln x$ is $1/x$, the slope of the curve $y = \ln x$ at $x = 1$ is $1/1$, which is 1. Since the slope of the tangent line at (1, 0) is 1, the tangent makes an angle of $\pi/4$ radian (45°) with the x axis. ■

Since $(\ln x)' = 1/x$,

$$\int \frac{1}{x}\,dx = \ln x + C. \tag{2}$$

However, Eq. (2) makes no sense if x is negative, since only positive numbers have logarithms. The next theorem provides an antiderivative for $1/x$, even where x is negative.

Theorem 2

$$\int \frac{1}{x}\,dx = \ln |x| + C \qquad \text{for } x > 0 \text{ or for } x < 0. \tag{3}$$

Proof It is necessary to show that the derivative of $\ln |x|$ is $1/x$. For positive x this has already been done. Now consider x negative.

For $x < 0$, $|x| = -x$. Thus, for negative x,

$$\begin{aligned}\frac{d}{dx}(\ln |x|) &= \frac{d}{dx}(\ln(-x)) \\ &= \frac{1}{(-x)}\frac{d(-x)}{dx} \qquad \text{by chain rule, as in Example 2} \\ &= \frac{1}{-x}(-1) = \frac{1}{x}.\end{aligned}$$

This completes the proof. ■

EXAMPLE 4 Compute $\displaystyle\int_{-3}^{-1} \frac{dx}{x}$.

SOLUTION

$$\begin{aligned}\int_{-3}^{-1} \frac{dx}{x} &\underset{\text{FTC}}{=} \ln |x| \Big|_{-3}^{-1} \qquad \text{by Theorem 2} \\ &= \ln|-1| - \ln|-3| \\ &= \ln 1 - \ln 3 = 0 - \ln 3 = -\ln 3. \ ■\end{aligned}$$

Careful: $\displaystyle\int_a^b \frac{dx}{x}$ *does not exist if* $a < 0 < b$.

The next theorem shows that the logarithm function enables us to integrate many functions besides $1/x$.

Theorem 3 Let $f(x)$ be a differentiable function. Then, if $f(x) \neq 0$,

$$\int \frac{f'(x)}{f(x)}\,dx = \ln |f(x)| + C.$$

Proof Let $y = \ln|f(x)|$. Then $y = \ln|u|$, where $u = f(x)$. By the chain rule and Theorem 2,

$$\frac{dy}{dx} = \frac{1}{u} f'(x) = \frac{f'(x)}{f(x)}. \quad ■$$

How to integrate $\frac{f'(x)}{f(x)}$

With the aid of the formula in Theorem 3 we can integrate the quotient of two functions if the numerator is exactly the derivative of the denominator. In fact, if the numerator is a constant times the derivative of the denominator, we can still use the formula, as the next example shows.

EXAMPLE 5 Compute $\int \frac{x^2\,dx}{x^3 + 1}$.

SOLUTION The numerator is not quite the derivative of the denominator. However, the numerator is $\frac{1}{3}$ times the derivative of the denominator; that is,

$$x^2 = \tfrac{1}{3} \cdot 3x^2.$$

With this observation, the integration of $x^2/(x^3 + 1)$ is quick:

$$\int \frac{x^2\,dx}{x^3 + 1} = \int \frac{(\frac{1}{3})3x^2\,dx}{x^3 + 1}$$

$$= \frac{1}{3}\int \frac{3x^2\,dx}{x^3 + 1} \qquad \int cf(x)\,dx = c\int f(x)\,dx \ (c \text{ constant})$$

$$= \tfrac{1}{3}\ln|x^3 + 1| + C. \quad ■$$

The next example presents a special case of implicit differentiation called **logarithmic differentiation**. This is a method for differentiating a function whose logarithm is simpler than the function itself.

EXAMPLE 6 Differentiate $y = \frac{\sqrt[3]{x}\,\sqrt{(1 + x^2)^3}}{x^{4/5}}$.

SOLUTION Rather than compute dy/dx directly, take logarithms of both sides of the equation first, obtaining

$$\ln y = \tfrac{1}{3}\ln x + \tfrac{3}{2}\ln(1 + x^2) - \tfrac{4}{5}\ln x.$$

Then differentiate this equation implicitly:

$$\frac{1}{y}\frac{dy}{dx} = \frac{1}{3x} + \frac{3}{2}\frac{2x}{1 + x^2} - \frac{4}{5}\frac{1}{x}.$$

Solving for dy/dx yields

$$\frac{dy}{dx} = y\left(\frac{1}{3x} + \frac{3x}{1 + x^2} - \frac{4}{5x}\right)$$

$$= \frac{\sqrt[3]{x}\sqrt{(1 + x^2)^3}}{x^{4/5}}\left(\frac{1}{3x} + \frac{3x}{1 + x^2} - \frac{4}{5x}\right).$$

The reader is invited to find dy/dx directly from the explicit formula for y. Doing so will show the advantage of logarithmic differentiation. ■

The following table records derivatives and antiderivatives involving logarithms.

Derivative	*Antiderivative*
$(\ln x)' = \dfrac{1}{x}$	$\int \dfrac{dx}{x} = \ln x + C,\ x > 0$
$(\ln \lvert x \rvert)' = \dfrac{1}{x}$	$\int \dfrac{dx}{x} = \ln \lvert x \rvert + C,\ x \neq 0$
$[\ln f(x)]' = \dfrac{f'(x)}{f(x)}$	$\int \dfrac{f'(x)}{f(x)}\,dx = \ln f(x) + C,\ f(x) > 0$
$[\ln \lvert f(x) \rvert]' = \dfrac{f'(x)}{f(x)}$	$\int \dfrac{f'(x)}{f(x)}\,dx = \ln \lvert f(x) \rvert + C,\ f(x) \neq 0$
$(\log_b x)' = \dfrac{\log_b e}{x}$	Not needed as antiderivative

EXAMPLE 7 Graph $y = (\ln x)/x$, showing intercepts, critical points, inflection points, extrema, and asymptotes.

SOLUTION The domain of the function $(\ln x)/x$ is the interval $(0, \infty)$. Recall from Sec. 6.1 that $\lim_{x\to\infty} (\ln x)/x = 0$. Hence the x axis is an asymptote. Moreover, when x is a small positive number, $\ln x$ is a negative number of large absolute value. Thus

$$\lim_{x\to 0^+} \frac{\ln x}{x} = \lim_{x\to 0^+} (\ln x)\left(\frac{1}{x}\right) = -\infty,$$

and so the y axis is also an asymptote.

Next examine $D[(\ln x)/x]$ and $D^2[(\ln x)/x]$. We have

$$D\left(\frac{\ln x}{x}\right) = \frac{1 - \ln x}{x^2} \quad \text{and} \quad D^2\left(\frac{\ln x}{x}\right) = \frac{-3 + 2\ln x}{x^3}.$$

The first derivative is 0 when $1 - \ln x = 0$, that is, when $x = e$. It is positive for $0 < x < e$ and negative for $x > e$. Hence at $x = e$ the curve has a global maximum.

The second derivative is 0 when $-3 + 2 \ln x = 0$, that is, when $\ln x = \frac{3}{2}$, $x = e^{3/2}$. It is negative for $0 < x < e^{3/2}$ and positive for $x > e^{3/2}$. Hence the curve has an inflection point when $x = e^{3/2}$.

All this information is incorporated in Fig. 6.5. ■

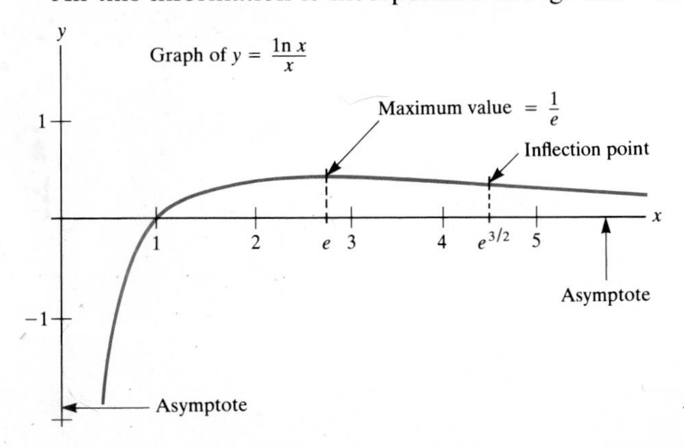

Figure 6.5

EXERCISES FOR SEC. 6.3: THE DERIVATIVES OF THE LOGARITHMIC FUNCTIONS

In Exercises 1 to 8 differentiate the functions.

1 $\ln (1 + x^2)$

2 $x^3 \ln (1 + x^2)$

3 $\dfrac{\ln x}{x}$, $x > 0$

4 $(\ln x)^3$, $x > 0$

5 $\sin 4x \ln 3x$, $x > 0$

6 $\sec 5x \ln (1 + x^4)$

7 $[\ln (\sin x)]^2$, $0 < x < \pi$

8 $\sin [\ln (1 + x^6)]$

In Exercises 9 to 14 differentiate and simplify your answers.

9 $\ln |2x + 3|$

10 $\dfrac{x}{3} - \dfrac{1}{9} \ln |3x + 1|$

11 $\dfrac{2}{25(5x + 2)} + \dfrac{1}{25} \ln |5x + 2|$

12 $x + 3 - 6 \ln |x + 3| - \dfrac{9}{x + 3}$

13 $\ln |x + \sqrt{x^2 - 5}|$

14 $\ln |x + \sqrt{x^2 + 1}|$

In Exercises 15 to 20 first simplify by using laws of logarithms; then differentiate.

15 $\dfrac{1}{5} \ln \dfrac{x}{3x + 5}$, $x > 0$

16 $-\dfrac{1}{3x} + \dfrac{5}{9} \ln \left| \dfrac{3x + 5}{x} \right|$

17 $\dfrac{1}{10} \ln \left| \dfrac{5 + x}{5 - x} \right|$

18 $\sqrt{x^2 + 1} \ln \dfrac{\sqrt{x^2 + 1} - 1}{x}$

19 $\ln [(x^2 + 1)^3 (x^5 + 1)^4]$

20 $\ln \dfrac{\sqrt{2x + 1}\sqrt[3]{3x + 2}}{(x^2 + 1)^5}$

In Exercises 21 to 24 differentiate by logarithmic differentiation.

21 $(1 + 3x)^5 (\sin 3x)^6$

22 $\sqrt{1 + x^2}\, \sqrt[3]{(1 + \cos 3x)^5}$

23 $\dfrac{(\sec 4x)^{5/3} \sin^3 2x}{\sqrt{x}}$

24 $\dfrac{(\cot 5x)^3}{\sqrt[4]{x}\, (x^3 + 2)^{5/2}}$

In Exercises 25 and 26 find the indicated antiderivatives.

25 (*a*) $\displaystyle\int \frac{5\, dx}{5x + 1}$ (*b*) $\displaystyle\int \frac{x\, dx}{x^2 + 5}$

(*c*) $\displaystyle\int \frac{\cos x\, dx}{\sin x}$ (*d*) $\displaystyle\int \frac{(1/x)\, dx}{\ln x}$

26 (*a*) $\displaystyle\int \frac{dx}{3x + 2}$

(*b*) $\displaystyle\int \frac{\sin x\, dx}{\cos x}$

(*c*) $\displaystyle\int \frac{(6x + 1)\, dx}{3x^2 + x + 5}$

In Exercises 27 and 28 graph the functions as in Example 7.

27 $y = (\ln x)/x^2$

28 $y = (\ln x)/x^3$

29 Find the area under $y = (\ln x)/x$ and above $[e, e^2]$. [*Hint:* First find the derivative of $(\ln x)^2$.]

30 Find the area under $y = x/(1 + x^2)$ and above $[2, 5]$.

31 Differentiate $\log_{10} \sqrt[3]{x}$.

32 Differentiate $\log_2((x^2 + 1)^3 \sin 3x)$.

■

33 Assuming that f and g are differentiable functions with positive values, obtain by logarithmic differentiation the formulas for (*a*) $(fg)'$, (*b*) $(f/g)'$.

34 Let n be an integer larger than 1. The area of the shaded region in Fig. 6.6 is equal to

$$\frac{1}{1} + \frac{1}{2} + \frac{1}{3} + \cdots + \frac{1}{n - 1} - \ln n.$$

As $n \to \infty$, the area of the shaded region approaches a number, denoted γ (gamma), called **Euler's constant.** To three decimal places, $\gamma \approx 0.577$. It is not known whether γ is rational. Using geometric intuition (no calculus), show that γ is (*a*) less than 1, (*b*) greater than $\frac{1}{2}$.

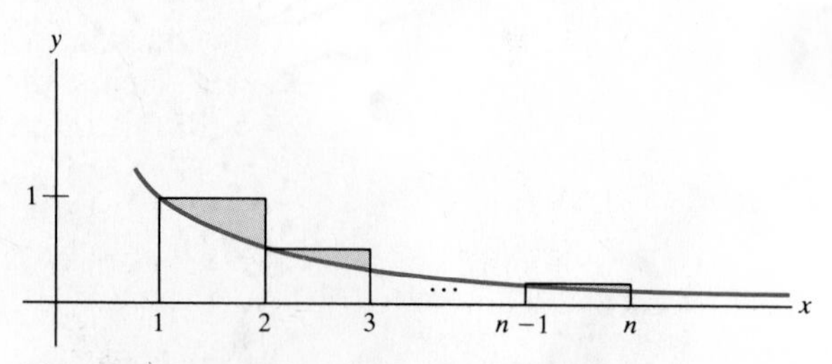

Figure 6.6

35 (*a*) Find the area of the region under the curve $y = 1/x$ and above $[1, b]$.

(*b*) Using the result in (*a*), show that the area of the region under $y = 1/x$ and above $[1, \infty)$ is infinite.

(*c*) The region below $y = 1/x$ and above $[1, \infty)$ is revolved around the x axis. Find the volume of that part of the resulting solid situated to the left of $x = b$, $b > 1$. (See Exercise 29 of Sec. 5.4.)

(*d*) Show that the volume of the unbounded solid in (*c*) is finite.

(*e*) In view of (*b*) it is impossible to paint the region in (*b*) with a finite amount of paint. However, in view of (*d*), we could fill the unbounded solid of revolution with a finite amount of paint, then dip the region in (*b*) into the paint. That would paint the region in (*b*) with a finite amount of paint. What is wrong?

6.4 THE DERIVATIVE OF b^x

See Appendix D for a review of exponents.

In this section we obtain the derivative of the exponential function b^x for any base $b > 0$.

If $y = b^x$, then $x = \log_b y$. This means that the exponential function $y = b^x$ and the logarithm function $x = \log_b y$ for $b \neq 1$ are inverses of each other.

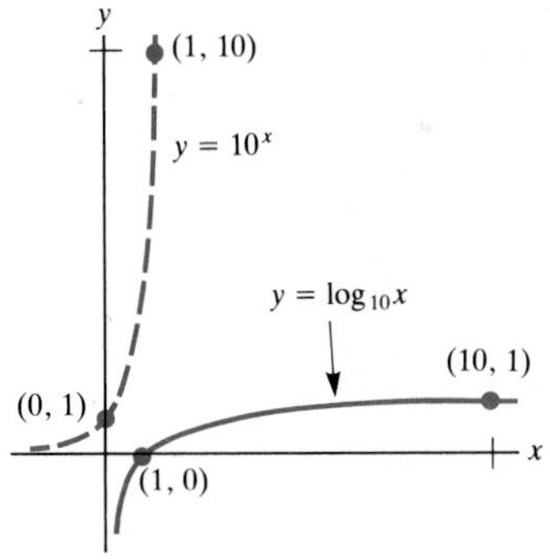

EXAMPLE 1 Graph $y = 10^x$ and its inverse function $y = \log_{10} x$ on the same axes.

SOLUTION In order to graph the logarithm and exponential functions, it is advisable to prepare brief tables first:

x	10	1	0.1
$\log_{10} x$	1	0	-1

x	1	0	-1
10^x	10	1	0.1

Graph the functions with the aid of the tables, as in Fig. 6.7. ■

Figure 6.7

We will obtain the derivative of b^x by exploiting the derivative of its inverse function, $\log_b x$. For this reason, let us see how the derivative of a function f is related to the derivative of its inverse function g.

Assume that the function f is differentiable on the interval $[a, b]$ and that its derivative is always positive. Then f is an increasing function, hence one-to-one. Its graph is indicated in Fig. 6.8.

The graph of its inverse function g is the mirror image of that for f in the line $y = x$. (Recall Sec. 1.3.) Since f is differentiable, its graph locally resembles a line. Since the graph of g is a copy of that of f, we would expect g to be a differentiable function as well. This expectation is formally justified in an advanced calculus course, but we will assume it without proof in the following theorem.

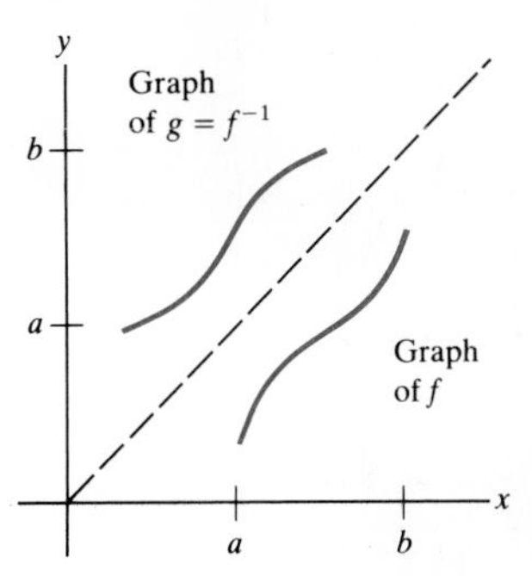

Figure 6.8

Theorem 1 Let f be a differentiable function with domain $[a, b]$ and range $[c, d]$. Assume that $f'(x)$ is positive for all x in $[a, b]$. Then f is a one-to-one function and its inverse function g from $[c, d]$ to $[a, b]$ is differentiable. The same conclusion holds if $f'(x)$ is negative for all x in $[a, b]$. ■

If f satisfies the condition in the theorem, then we may write

$$x = g(y) \qquad \text{and} \qquad y = f(x).$$

Thus $x = g(f(x))$, and x may be viewed as a composite function of x.

The chain rule then tells us that

$$\frac{dx}{dx} = \frac{dx}{dy}\frac{dy}{dx} \quad \text{or} \quad 1 = \frac{dx}{dy}\frac{dy}{dx}.$$

This equation, which relates the derivatives of a function and its inverse, will be used in this section to find the derivative of b^x and in the next to differentiate the inverse trigonometric functions. [In more detail it asserts that $1 = g'(f(x))f'(x)$; the derivative of g is evaluated at $f(x)$, while the derivative of f is evaluated at x.]

Theorem 2 *The derivative of* e^x. Let $y = e^x$ be the exponential function with base e. Then

$$\frac{d(e^x)}{dx} = e^x.$$

Proof Let $y = e^x$. This function is the inverse of the natural logarithmic function, that is,

$$y = e^x \quad \text{is equivalent to} \quad x = \ln y.$$

Since the derivative dx/dy exists, the derivative dy/dx exists, wherever dx/dy is not 0.

Since $$x = \ln y,$$

we have $$1 = \frac{dx}{dx} = \frac{dx}{dy} \cdot \frac{dy}{dx},$$

or $$1 = \frac{d}{dy}(\ln y) \cdot \frac{dy}{dx}.$$

Hence $$1 = \frac{1}{y} \cdot \frac{dy}{dx}.$$

Thus $$\frac{dy}{dx} = y = e^x.$$

e^x is equal to its own derivative!

In short, $$\frac{dy}{dx} = e^x,$$

and the theorem is proved. ■

EXAMPLE 2 Find the derivative of e^{3x}.

SOLUTION Let $y = e^{3x}$. Then $y = e^u$, where $u = 3x$.

Thus
$$\begin{aligned}\frac{dy}{dx} &= \frac{dy}{du}\frac{du}{dx} && \text{chain rule}\\ &= \frac{d(e^u)}{du}\frac{d(3x)}{dx}\\ &= e^u \cdot 3 && \text{Theorem 2}\\ &= e^{3x} \cdot 3\\ &= 3e^{3x}.\end{aligned}$$ ■

The formula for the derivative of e^x is quite simple. Let us next compute the derivative of 10^x.

EXAMPLE 3 Find the derivative of 10^x.

SOLUTION Write 10 as a power of e:

$$10 = e^{\ln 10}.$$

Then
$$10^x = (e^{\ln 10})^x.$$

By the "power-of-a-power" rule,

$$(e^{\ln 10})^x = e^{(\ln 10)x}.$$

Thus
$$10^x = e^{(\ln 10)x}.$$

Since $\ln 10$ is a constant, this problem is similar to Example 2.

Let $y = e^{(\ln 10)x}$. This can be written as $y = e^u$, where $u = (\ln 10)x$.

Then
$$\frac{dy}{dx} = \frac{d(e^u)}{du}\frac{d}{dx}[(\ln 10)x] \qquad \text{chain rule}$$
$$= e^u \cdot \ln 10 \qquad \text{Theorem 2}$$
$$= e^{(\ln 10)x} \cdot \ln 10$$
$$= 10^x \cdot \ln 10 = (\ln 10) \cdot 10^x.$$

Note that $\dfrac{d(10^x)}{dx} \approx (2.3)10^x$.

Thus
$$\frac{d(10^x)}{dx} = (\ln 10) \cdot 10^x. \quad ■$$

Similarly, for any base b,

$$\frac{d(b^x)}{dx} = (\ln b)b^x.$$

Why e is the ideal base for an exponential function

(Since it is easy to forget the coefficient $\ln b$, it may be safer to write b^x as $e^{x \ln b}$ and differentiate than to memorize this formula.) The coefficient $\ln b$ equals 1 only if the base b is chosen to be e; then there is nothing to remember, except $(e^x)' = e^x$. That is why e is the most convenient base for an exponential function in calculus.

EXAMPLE 4 Differentiate $f(x) = \dfrac{e^{ax}}{a^3}(a^2x^2 - 2ax + 2)$, where a is a nonzero constant.

SOLUTION First bring a^3, which is constant, to the front of the formula by writing the function as

$$f(x) = \frac{1}{a^3}e^{ax}(a^2x^2 - 2ax + 2).$$

Now differentiate:

$$f'(x) = \frac{1}{a^3}[e^{ax}(a^2x^2 - 2ax + 2)]' \qquad \text{derivative of constant times a function}$$

$$= \frac{1}{a^3}[e^{ax}(a^2 \cdot 2x - 2a) + (a^2x^2 - 2ax + 2)ae^{ax}]$$ derivative of product, chain rule, Theorem 2

$$= \frac{1}{a^3}[(2a^2x - 2a + a^3x^2 - 2a^2x + 2a)e^{ax}]$$

$$= x^2e^{ax}$$ after canceling. ■

The next theorem generalizes the formula $(x^a)' = ax^{a-1}$ from rational a to any real number a.

Theorem 3 Let a be a fixed real number. Then for $x > 0$,

$$\frac{d(x^a)}{dx} = ax^{a-1}.$$

Proof Let $y = x^a$. Since $x = e^{\ln x}$,

$$y = (e^{\ln x})^a.$$

Hence, by the "power-of-a-power" rule,

$$y = e^{a \ln x}.$$

This can be written as $y = e^u$, where $u = a \ln x$.

Hence $$\frac{dy}{dx} = \frac{d(e^u)}{du}\frac{d}{dx}(a \ln x)$$ chain rule

$$= e^u \frac{a}{x} = e^{a \ln x}\frac{a}{x}$$

$$= x^a \frac{a}{x} = ax^{a-1}.$$

This proves the theorem. ■

Incidentally, this provides a second proof that $(x^a)' = ax^{a-1}$ for positive x and rational a.

For example, $d(x^\pi)/dx = \pi x^{\pi-1}$ and $(x^{\sqrt{2}})' = \sqrt{2}(x^{\sqrt{2}-1})$.

At this point we can differentiate x^a (the exponent is fixed and the base varies) and b^x (the base is fixed and the exponent varies). What if both the base and the exponent vary? For instance, how would we differentiate x^x? One way is to rewrite the base as $e^{\ln x}$. Then

$$x^x = (e^{\ln x})^x = e^{x \ln x};$$

since $e^{x \ln x}$ has a fixed base, it can be differentiated by the chain rule, as e^{3x} was differentiated in Example 2. However, a simpler way is to use logarithmic differentiation, as shown in Example 5.

EXAMPLE 5 Use logarithmic differentiation to find $(x^x)'$.

SOLUTION Let $y = x^x$. We wish to find y'.

Begin by taking logarithms,

$$\ln y = \ln (x^x)$$

or $$\ln y = x \ln x$$ $\ln c^m = m \ln c$.

Then differentiate this last equation with respect to x, obtaining:

$$\frac{1}{y}y' = (x \ln x)'$$

$$= x \cdot \frac{1}{x} + (\ln x)$$

$$= 1 + \ln x.$$

Solve for y': $$y' = y(1 + \ln x).$$

Thus $$y' = x^x(1 + \ln x). \blacksquare$$

How to differentiate $[f(x)]^{g(x)}$

To differentiate $y = f(x)^{g(x)}$, differentiate $\ln y = \ln[f(x)^{g(x)}] = g(x) \ln f(x)$ implicitly and solve for y'.

For any fixed exponent $a > 0$ and fixed base $b > 1$, both x^a and b^x get arbitrarily large as $x \to \infty$. However, the exponential function b^x eventually grows much more rapidly. The following table compares x^3 and 1.1^x for a few values of x.

Watch 1.1^x catch up.

x	1	10	100	1,000
x^3	1	1,000	1,000,000	10^9
1.1^x	1.1	2.5937	13,780.6	2.4699×10^{41}

The data suggest that $\lim_{x\to\infty} x^3/1.1^x = 0$. In fact, for $a > 0$ and $b > 1$, we have

$$\lim_{x\to\infty} \frac{x^a}{b^x} = 0.$$

(This can be proved with the aid of the binomial theorem. See Exercises 48 and 49.)

EXAMPLE 6 Graph $y = \dfrac{x}{e^x}$.

SOLUTION The domain is the entire x axis; $x = 0$ provides an intercept. Since $\lim_{x\to\infty} x/e^x = 0$, the x axis is an asymptote. Since $\lim_{x\to-\infty} e^x = 0$, we have $\lim_{x\to-\infty} x/e^x = -\infty$.

Asymptotes

Next examine dy/dx, which equals $(1 - x)/e^x$. There is a critical point when $x = 1$. When $x < 1$, dy/dx is positive, and when $x > 1$, dy/dx is negative. Hence a global maximum occurs when $x = 1$.

Critical point

The second derivative of x/e^x is $(x - 2)/e^x$, which changes sign at $x = 2$; hence there is an inflection point there. The graph is concave downward for $x < 2$ and concave upward for $x > 2$.

Inflection point

This information is incorporated in Fig. 6.9.

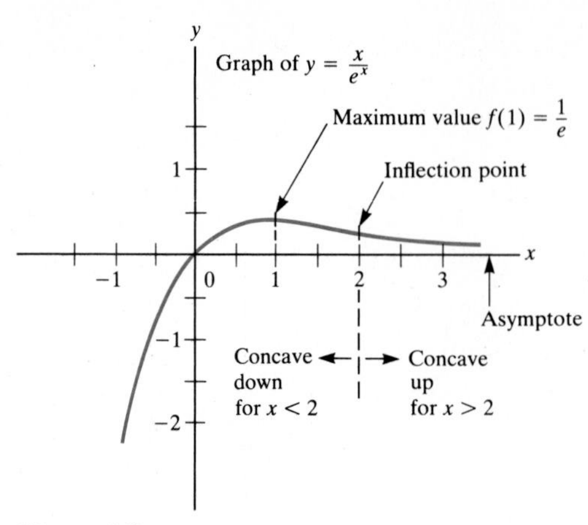

Figure 6.9

■

EXERCISES FOR SEC. 6.4: THE DERIVATIVE OF b^x

In Exercises 1 to 18 differentiate the given functions.

1 e^{x^2}　　2 xe^{-4x}

3 x^2e^{2x}　　4 $(\sin 2x)(\sin e^{-x})$

5 2^{-x^2}　　6 $3^{\sqrt{x}}$

7 $x^{(x^2)}$　　8 $(2 + \cos x)^{\sin x}$

9 $e^{\ln 3x}$　　10 $\ln e^{x^2}$

11 $x^{\tan 3x}$　　12 $(\tan \sqrt{x})e^{-x}$

13 $\dfrac{e^{-4x} \cos 5x}{1 + e^x}$　　14 $\dfrac{10^{x^2} \cot 5x}{\ln (1 + x^2)}$

15 $x^{\sqrt{3}} (\sin 3x)e^{x^2}$　　16 $\dfrac{x^\pi \tan e^x}{2^x}$

17 $\ln(x + \sqrt{1 + e^{3x}})$　　18 $\ln e^{\cos 3\theta}$

In Exercises 19 to 22 differentiate and simplify. (The numbers a, b, and c are positive constants.)

19 $\dfrac{e^{ax}(ax - 1)}{a^2}$

20 $\dfrac{xb^{ax}}{a \ln b} - \dfrac{b^{ax}}{a^2 (\ln b)^2}$

21 $\dfrac{e^{ax}}{a^2 + b^2}(a \sin bx - b \cos bx)$

22 $\dfrac{1}{ac} \ln (b + ce^{ax})$

In each of Exercises 23 to 26 a region in the plane is described. (*a*) Find its area. (*b*) Find the volume of the solid formed by revolving the region around the x axis. (See Exercise 29 in Sec. 5.4.)

23 Under e^{3x}, above $[1, 5]$.

24 Under $5e^{-2x}$, above $[0, \ln 2]$.

25 Under 10^x, above $[0, 3]$.

26 Under 2^{-x}, above $[-4, 1]$.

In Exercises 27 to 34 use a differential to estimate the given quantities. Assume x small. You may also assume $\ln 10 \approx 2.30$, $e \approx 2.72$, and $\log_{10} e \approx 0.43$.

27 e^x　　28 $e^{1.1}$

29 10^x　　30 $10^{1.1}$

31 $\ln (1 + x)$　　32 $\ln 1.1$

33 $\log_{10}(1 + x)$　　34 $\log_{10} 0.98$

In each of Exercises 35 to 40 find (*a*) intercepts, (*b*) critical points, (*c*) local maxima or minima, (*d*) inflection points, and (*e*) asymptotes. (*f*) Graph the function.

35 $f(x) = xe^{-x}$　　36 $f(x) = x^2e^{-x}$

37 $f(x) = x^3e^{-x}$　　38 $f(x) = x\,2^{-x}$

39 $f(x) = (x - x^2)e^{-x}$　　40 $f(x) = xe^x$

■

41 Using results of this section and the definition of the derivative, evaluate:

(*a*) $\lim_{h \to 0} \dfrac{e^h - 1}{h}$　　(*b*) $\lim_{x \to 1} \dfrac{2^x - 2}{x - 1}$　　(*c*) $\lim_{h \to 0} \dfrac{10^h - 1}{h}$

42 Let $f(x) = 5 + (x - x^2)e^x$.

(*a*) Show that $f(0) = f(1)$.

(*b*) Find all numbers c in $(0, 1)$ whose existence is guaranteed by Rolle's theorem, stated in Sec. 4.1.

43 Let A and k be constants. Show that the derivative of Ae^{kx} is proportional to Ae^{kx}.

44 (*a*) Graph $y = e^x$ and $y = -x$ on the same axes.

(*b*) Using the graphs in (*a*), graph $y = e^x - x$.
(*c*) Find all points on the graph in (*b*) where the tangent line is horizontal.

45 (Calculator) The formula $y = e^{-t} \sin t$ describes a decaying alternating current. Consider t in $[0, 4\pi]$.
(*a*) Fill in this table and plot the resulting points.

t	0	$\frac{\pi}{2}$	π	$\frac{3\pi}{2}$	2π	$\frac{5\pi}{2}$	3π	$\frac{7\pi}{2}$	4π
$e^{-t} \sin t$									

(*b*) Graph $y = e^{-t} \sin t$ for t in $[0, 2\pi]$.
(*c*) Find all points on the graph in (*b*) where the tangent line is horizontal.

46 (*a*) Show that there is a number $x > 1$ such that $x^3 = 1.01^x$.
(*b*) Could there be two values of $x > 1$ such that $x^3 = 1.01^x$?

47 In planning a dam, an engineer estimates that for a dam of height h meters, the cost will be $ae^{h/20}$ dollars, the volume of the lake created bh^3 cubic meters, and the surface area of the lake ch^2 square meters, where a, b, and c are constants.
(*a*) What height maximizes the volume-to-cost ratio?
(*b*) What height maximizes the area-to-cost ratio?

48 This exercise shows that if $b > 1$, then $\lim_{x\to\infty} x/b^x = 0$.
(*a*) Show that for x sufficiently large, x/b^x is a decreasing function.
(*b*) Write $b = 1 + c$, $c > 0$. Using the binomial theorem for $n > 2$, show that

$$b^n > 1 + nc + \frac{n(n-1)}{2}c^2 \qquad \text{if } n > 2.$$

(*c*) From (*b*) deduce that

$$\lim_{n\to\infty} \frac{n}{b^n} = 0.$$

(*d*) From (*a*) and (*c*) deduce that $\lim_{x\to\infty} x/b^x = 0$.

49 (This continues Exercise 48.) Show that if $n > 3$ and $b > 1$, then

$$b^n > 1 + nc + \frac{n(n-1)}{2}c^2 + \frac{n(n-1)(n-2)}{3!}c^3.$$

Then, modeling your argument on Exercise 48, show that $\lim_{x\to\infty} x^2/b^x = 0$.

A similar argument shows that for any $b > 1$ and any positive a, $\lim_{x\to\infty} x^a/b^x = 0$. (The case $a \le 0$ is trivial. Why?)

50 When doing this exercise disregard all work done in this section. Pretend that you are back in Chap. 3, where the derivative was defined. Let f be the exponential function given by the formula $f(x) = 10^x$.
(*a*) Copy and complete this table:

x	-2	-1	0	1
10^x	0.01			10

(*b*) Graph the function for x in $[-2, 1]$. (The same scale should be used for both axes.)
(*c*) Using a ruler, draw what you think would be the tangent line at $(0, 1)$.
(*d*) Using a ruler, estimate the slope of the line you drew in (*c*).
(*e*) Show that the derivative of 10^x at $x = 0$ is

$$\lim_{h\to 0} \frac{10^h - 1}{h}.$$

(*f*) The limit in (*e*) is far from obvious. What does (*d*) suggest as an estimate of this limit?
(*g*) Let the limit in (*e*) be denoted c. Show that $(10^x)' = c10^x$.

51 (*a*) Show that $y = x^5 + 3x$ is a one-to-one function.
(*b*) Does it have an inverse function?
(*c*) Can you find a formula for the inverse function?

6.5 THE DERIVATIVES OF THE INVERSE TRIGONOMETRIC FUNCTIONS

The derivative of $\frac{1}{2} \ln (1 + x^2)$ is $x/(1 + x^2)$. However, up to this point we have no function whose derivative is simply $1/(1 + x^2)$. Nor do we have any function whose derivative is $1/\sqrt{1 - x^2}$ or $\sqrt{1 - x^2}$. Such functions, which are needed in integral calculus, will be obtained in this section. Surprisingly, they turn out to involve trigonometric functions.

arctan x and Its Derivative

First, we will provide a function whose derivative is $1/(1 + x^2)$.

Consider the function $y = \tan x$ in the open interval $-\pi/2 < x < \pi/2$.

As x increases in this interval, $\tan x$ increases. (See Fig. 6.10.) Thus the function $\tan x$ is one-to-one if the domain is restricted to be between $-\pi/2$ and $\pi/2$. Note that as $x \to \pi/2$ or $x \to -\pi/2$, $|\tan x|$ gets very large.

The graph of the inverse function g is obtained by folding the graph of $y = \tan x$ across the line $y = x$. As x gets large, note in Fig. 6.11 that $g(x) \to \pi/2$.

The inverse of the tangent function is called the **arctangent function** and is written $\arctan x$ or $\tan^{-1} x$. [This is not the reciprocal of $\tan x$, which is written $\cot x$, $1/\tan x$, or $(\tan x)^{-1}$ to avoid confusion.] As an example, since $\tan \pi/4 = 1$,

Remember: $\arctan x = \tan^{-1} x$ *is an angle measured in radians, a dimensionless number.*

$$\arctan 1 = \frac{\pi}{4} \quad \text{or} \quad \tan^{-1} 1 = \frac{\pi}{4}.$$

Observe that the *domain of the arctan function* is the entire x axis and that when $|x|$ is large, $\arctan x$ is near $\pi/2$ or $-\pi/2$. The *range of the arctan function* is the open interval $(-\pi/2, \pi/2)$.

It is frequently useful to picture the tangent and arctangent functions in terms of the unit circle, as shown in Figs. 6.12 and 6.13.

Graph of the tangent function for x in $(-\frac{\pi}{2}, \frac{\pi}{2})$

Figure 6.10

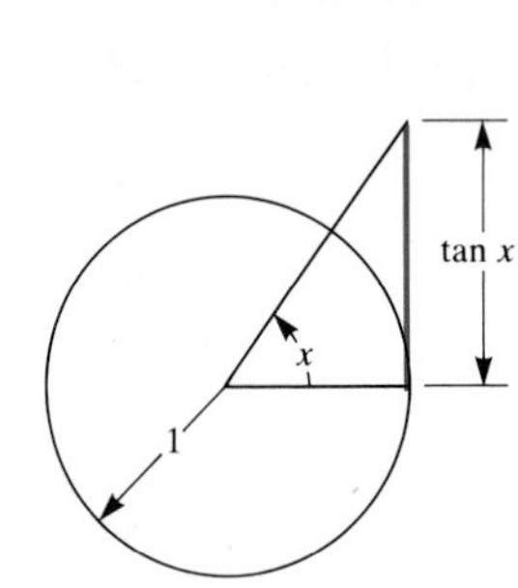

Figure 6.12

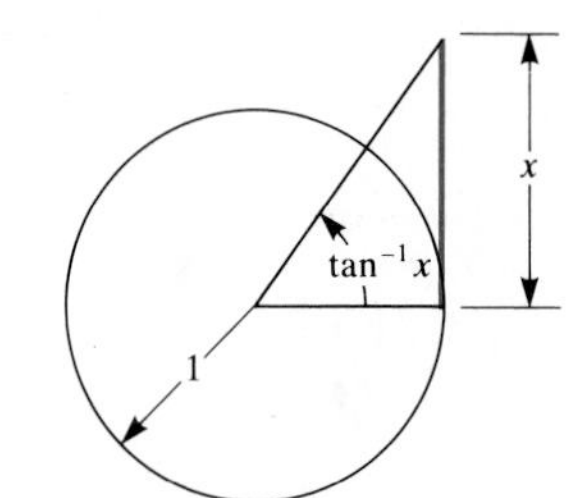

Figure 6.13

Graph of the inverse of the tangent function

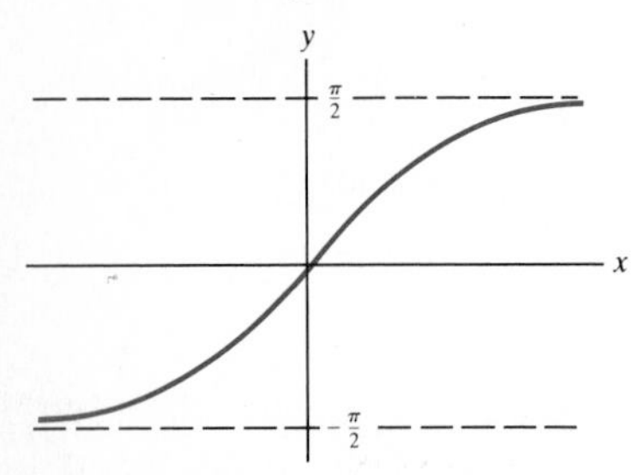

Figure 6.11

The inverse tangent function is easily evaluated on most scientific calculators. To get a feel for this function, fill in the following table, remembering to set the calculator for radians (not degrees).

x	$\tan^{-1} x$
−1000	
−100	
−10	
−1	
0	
1	
10	
100	
1000	

Theorem 1 obtains the derivative of $\tan^{-1} x$. The proof makes use of the identity $\sec^2 \theta = 1 + \tan^2 \theta$. (To obtain this identity, divide both sides of $\cos^2 \theta + \sin^2 \theta = 1$ by $\cos^2 \theta$.)

Theorem 1 $\dfrac{d}{dx}(\tan^{-1} x) = \dfrac{1}{1 + x^2}$.

Proof Let $y = \tan^{-1} x$. The problem is to find dy/dx. By the definition of the inverse tangent function, $x = \tan y$. Note that $dx/dy = \sec^2 y$.
As in the preceding section,

$$\frac{dx}{dx} = \frac{dx}{dy} \cdot \frac{dy}{dx},$$

or

$$1 = \sec^2 y \frac{dy}{dx}.$$

Hence

$$\frac{dy}{dx} = \frac{1}{\sec^2 y} = \frac{1}{1 + \tan^2 y} = \frac{1}{1 + x^2}.$$

This completes the proof. ■

Memorize the formula

$$(\tan^{-1} x)' = \frac{1}{1 + x^2}.$$

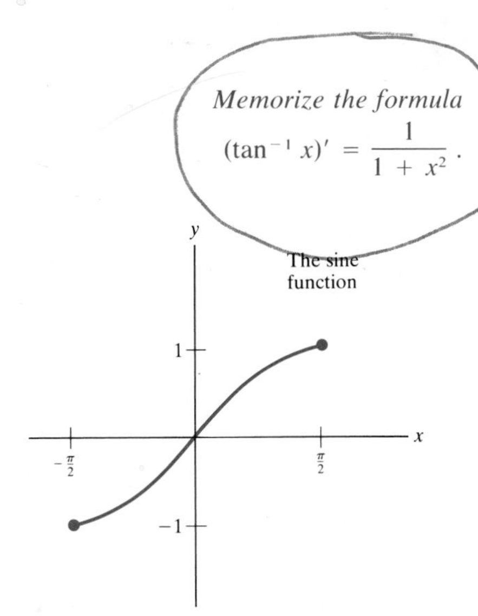

Figure 6.14

EXAMPLE 1 Find $\frac{d}{dx}(\tan^{-1} \sqrt{x})$.

SOLUTION Theorem 1 and the chain rule are needed. Let

$$y = \tan^{-1} \sqrt{x}$$

Then

$$y = \tan^{-1} u, \qquad \text{where } u = \sqrt{x}.$$

Thus

$$\frac{dy}{dx} = \frac{dy}{du} \cdot \frac{du}{dx}$$

$$= \frac{d}{du}(\tan^{-1} u) \frac{d}{dx}(\sqrt{x})$$

$$= \frac{1}{1 + u^2} \cdot \frac{1}{2\sqrt{x}} = \frac{1}{1 + x} \cdot \frac{1}{2\sqrt{x}}$$

$$= \frac{1}{2\sqrt{x}(1 + x)}. \quad ■$$

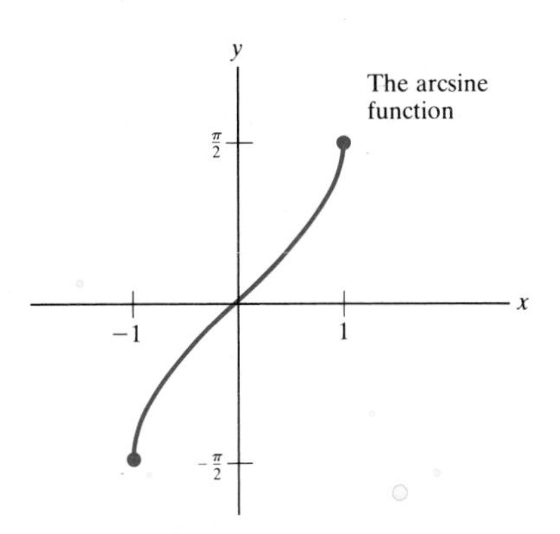

Figure 6.15

arcsin x and Its Derivative

We turn next to the inverse of the sine function.

The sine function is not one-to-one. For instance, $\sin(\pi/4) = \sqrt{2}/2 = \sin(3\pi/4)$. However, if the domain is restricted to $[-\pi/2, \pi/2]$, a one-to-one function results. The function $y = \sin x$ increases from -1 to 1 as x goes from $-\pi/2$ to $\pi/2$.

The inverse function is called the **arcsine function** and is usually written $\arcsin x$ or $\sin^{-1} x$. Since $\sin \pi/2 = 1$, we have, for instance,

$$\arcsin 1 = \frac{\pi}{2} \qquad \text{or} \qquad \sin^{-1} 1 = \frac{\pi}{2}.$$

Both these latter equations say "the angle whose sine is 1 is $\pi/2$." For more practice with $\sin^{-1} x$, use a calculator to fill in the table in the margin.

x	$\sin^{-1} x$
-1	
-0.6	
-0.2	
0	
0.2	
0.6	
1	

We graph the sine and arcsine functions in Figs. 6.14 and 6.15.

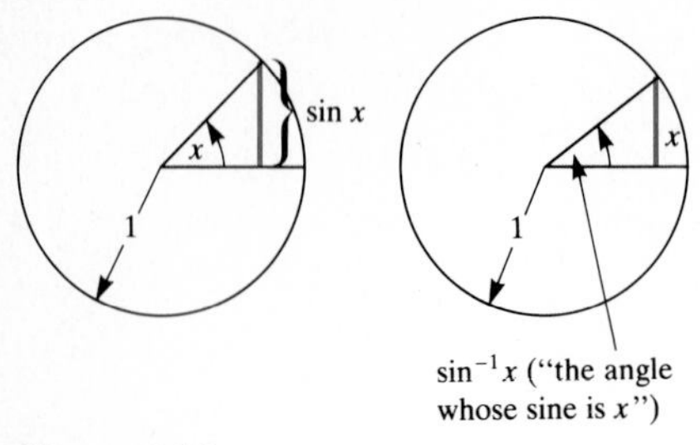

Figure 6.16

It is also useful to visualize these two functions in terms of the unit circle; they are depicted in Fig. 6.16. Note that

$$-\frac{\pi}{2} \leq \arcsin x \leq \frac{\pi}{2}.$$

Note also that the domain of the arcsine function is $[-1, 1]$ and its range is $[-\pi/2, \pi/2]$.

The proof of the next theorem is similar to that of Theorem 1.

Theorem 2 $\dfrac{d}{dx}(\sin^{-1} x) = \dfrac{1}{\sqrt{1-x^2}}$.

Proof Let $y = \sin^{-1} x$; hence $x = \sin y$. Thus

$$\frac{dx}{dx} = \frac{d}{dx}(\sin y)$$

$$= \frac{d}{dy}(\sin y)\frac{dy}{dx}$$

or

$$1 = \cos y \frac{dy}{dx}.$$

Hence

$$\frac{dy}{dx} = \frac{1}{\cos y}.$$

To express $\cos y$ in terms of x, draw the unit circle and indicate on it that $x = \sin y$. Inspection of Fig. 6.17 shows that $\cos y$ is positive (since $-\pi/2 \leq y \leq \pi/2$) and

$$\cos^2 y + x^2 = 1.$$

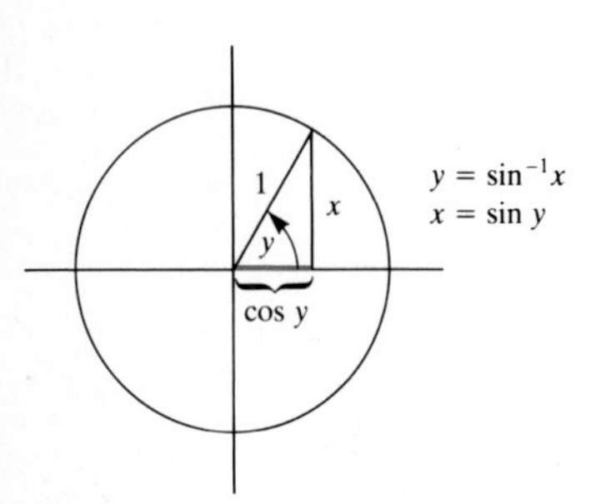

Figure 6.17

Thus $\cos y = \sqrt{1-x^2}$, the positive square root, and dy/dx takes the form

$$\frac{dy}{dx} = \frac{1}{\sqrt{1-x^2}}.$$

This proves that

$$\frac{d}{dx}(\sin^{-1} x) = \frac{1}{\sqrt{1-x^2}}. \quad \blacksquare$$

EXAMPLE 2 Find the derivative of $\sin^{-1}\dfrac{3x}{4}$.

SOLUTION Let $y = \sin^{-1}\dfrac{3x}{4}$. This is a composite function, with

$$y = \sin^{-1} u, \quad \text{where} \quad u = \frac{3x}{4}.$$

Now

$$\frac{dy}{du} = \frac{1}{\sqrt{1-u^2}} \quad \text{and} \quad \frac{du}{dx} = \frac{3}{4}.$$

Thus

$$\frac{dy}{dx} = \frac{dy}{du}\cdot\frac{du}{dx}$$

$$= \frac{1}{\sqrt{1-u^2}} \cdot \frac{3}{4} = \frac{1}{\sqrt{1-(3x/4)^2}} \cdot \frac{3}{4}$$

$$= \frac{1}{\sqrt{1-9x^2/16}} \cdot \frac{3}{4}$$

$$= \frac{\sqrt{16}}{\sqrt{16-9x^2}} \cdot \frac{3}{4}$$

$$= \frac{3}{\sqrt{16-9x^2}}. \quad ■$$

EXAMPLE 3 **Differentiate $x\sqrt{1-x^2} + \sin^{-1} x$.**

SOLUTION

$$\frac{d}{dx}(x\sqrt{1-x^2} + \sin^{-1} x) = \frac{d}{dx}(x\sqrt{1-x^2}) + \frac{d}{dx}(\sin^{-1} x)$$

$$= x\frac{d}{dx}(\sqrt{1-x^2}) + \sqrt{1-x^2}\frac{dx}{dx} + \frac{1}{\sqrt{1-x^2}}$$

$$= x \cdot \frac{1}{2} \cdot \frac{-2x}{\sqrt{1-x^2}} + \sqrt{1-x^2} + \frac{1}{\sqrt{1-x^2}}$$

$$= \frac{-x^2}{\sqrt{1-x^2}} + \sqrt{1-x^2} + \frac{1}{\sqrt{1-x^2}}$$

$$= \frac{1-x^2}{\sqrt{1-x^2}} + \sqrt{1-x^2}$$

$$= \sqrt{1-x^2} + \sqrt{1-x^2}$$

$$= 2\sqrt{1-x^2}. \quad ■$$

Note that Example 3 provides an antiderivative for $2\sqrt{1-x^2}$, that is,

$$\int 2\sqrt{1-x^2}\,dx = x\sqrt{1-x^2} + \sin^{-1} x + C.$$

Why inverse trig functions are important

The integrals of many algebraic functions involve inverse trigonometric functions. It is this fact that makes the inverse trigonometric functions important in calculus.

arccos x and Its Derivative

The function $\cos x$ is decreasing for x in $[0, \pi]$. Thus it has an inverse function, which assigns to each number in $[-1, 1]$ a number in $[0, \pi]$. The inverse function is called the **arccosine function**, denoted $\arccos x$ or $\cos^{-1} x$.

For instance, what is

$$\cos^{-1} \tfrac{1}{2}?$$

It is the angle between 0 and π whose cosine is $\frac{1}{2}$. That angle is $\pi/3$. Thus

$$\cos^{-1} \frac{1}{2} = \frac{\pi}{3} \approx 1.05.$$

x	$\cos^{-1} x$
-1	
-0.5	
-0.2	
0	
0.2	
0.5	
1	

For practice with $\cos^{-1} x$ you might fill in the table in the margin using a calculator. The graph of $y = \cos^{-1} x$ is shown in Fig. 6.18.

The domain of the arccosine function is $[-1, 1]$, and its range is $[0, \pi]$. Thus it has the same domain as the arcsine function but a different range. Why?

Reasoning like that in the proof of Theorem 2 shows that

$$(\cos^{-1} x)' = \frac{-1}{\sqrt{1 - x^2}}.$$

(The negative sign reflects the fact that $\cos^{-1} x$ is a decreasing function.)

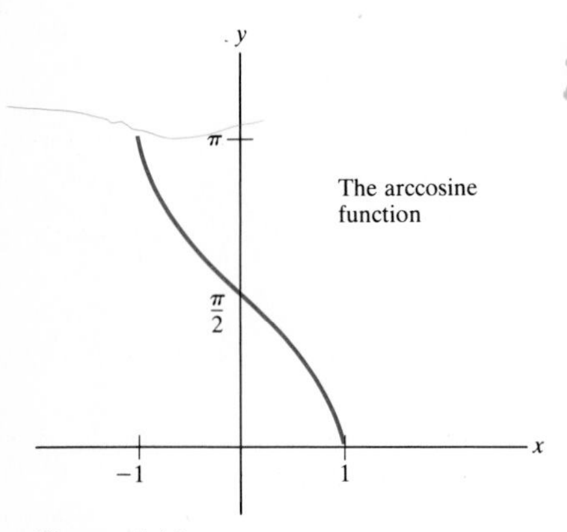

Figure 6.18

Theorem 3 $\dfrac{d}{dx}(\cos^{-1} x) = \dfrac{-1}{\sqrt{1 - x^2}}$. ■

Since the derivative of $\cos^{-1} x$ differs from the derivative of $\sin^{-1} x$ only by the constant factor -1, it is not needed for finding antiderivatives. In most integral tables, $\sin^{-1} x$ is used in preference to $\cos^{-1} x$.

arcsec x and Its Derivative

After $\tan^{-1} x$ and $\sin^{-1} x$, the next most important function for the computation of antiderivatives is the inverse of the secant function. Recall that $\sec x = 1/\cos x$; it is graphed in Appendix E. Since $|\sec x| \geq 1$, the inverse of the secant function is defined only for inputs of absolute value ≥ 1. For $|x| \geq 1$, define $\sec^{-1} x$ to be that angle in $[0, \pi]$ whose secant is x. If the secant of an angle is x, then the cosine of that angle is $1/x$. Thus

x	$\sec^{-1} x$
-10	
-2	
-1	
1	
2	
10	

$$\sec^{-1} x = \cos^{-1} \frac{1}{x}.$$

This formula, which is sometimes taken as a definition of the inverse secant function, makes it possible to compute $\sec^{-1} x$ on a calculator. For practice, you might fill in the table in the margin. The graph of $\sec^{-1} x$ is shown in Fig. 6.19.

Observe that the domain of the arcsecant function consists of all x such that $|x| > 1$. The range consists of all numbers from 0 to π except $\pi/2$.

To compute arcsecants, express their inverse nature in words. For instance, to evaluate $y = \sec^{-1} 2$, reason as follows:

$$\begin{aligned} y &= \text{angle whose secant is } 2 \\ &= \text{angle whose cosine is } \tfrac{1}{2} \\ &= \frac{\pi}{3}. \end{aligned}$$

Thus

$$\sec^{-1} 2 = \frac{\pi}{3}.$$

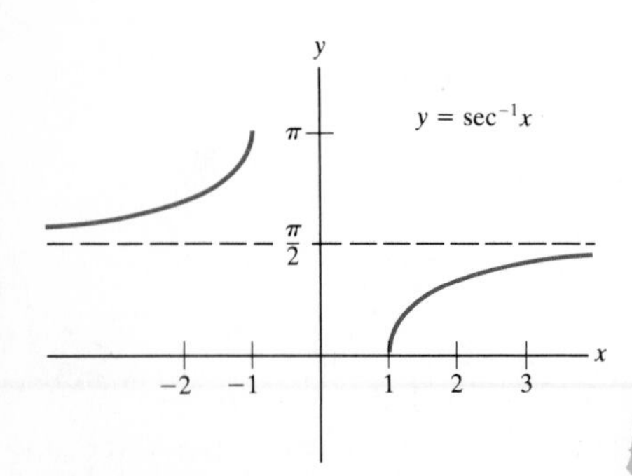

Figure 6.19

Theorem 4 $\dfrac{d}{dx}(\sec^{-1} x) = \dfrac{1}{|x|\sqrt{x^2 - 1}}$, where $|x| > 1$.

Proof Let $$y = \sec^{-1} x.$$

Then $$x = \sec y,$$

and $$\frac{dx}{dx} = \frac{dx}{dy} \cdot \frac{dy}{dx}$$

or $$1 = \sec y \tan y \frac{dy}{dx}.$$

Hence $$\frac{dy}{dx} = \frac{1}{\sec y \tan y} = \frac{1}{x \tan y}.$$

All that remains is to express $\tan y$ in terms of x.

Since $x^2 = \sec^2 y = 1 + \tan^2 y$, it follows that

$$\tan y = \pm\sqrt{x^2 - 1}, \qquad \text{where } |x| > 1.$$

Which sign is to be chosen? If $x > 1$, $y = \sec^{-1} x$ is in the range $(0, \pi/2)$; thus $\tan y$ is positive. If $x < -1$, $y = \sec^{-1} x$ is in the range $(\pi/2, \pi)$; thus $\tan y$ is negative. But in both cases $x \tan y$ is positive. Thus

$$\frac{dy}{dx} = \frac{1}{|x| \sqrt{x^2 - 1}}.$$

(Note that this derivative is positive, in agreement with the graph of $y = \sec^{-1} x$ on page 264; its tangent lines slope upward.) ■

EXAMPLE 4 Differentiate $y = \sec^{-1} 5x$.

SOLUTION The chain rule is required. Here

$$y = \sec^{-1} u, \qquad \text{where } u = 5x.$$

Thus $$\frac{dy}{dx} = \frac{1}{|u| \sqrt{u^2 - 1}} \cdot 5 = \frac{1}{|5x| \sqrt{25x^2 - 1}} \cdot 5$$

$$= \frac{1}{|5|\,|x| \sqrt{25x^2 - 1}} \cdot 5 = \frac{1}{|x| \sqrt{25x^2 - 1}}. \quad ■$$

The inverses of the remaining two trigonometric functions, $\cot x$ and $\csc x$, will not be needed. For the record, they may be defined as follows:

$$\cot^{-1} x = \frac{\pi}{2} - \tan^{-1} x \qquad \text{all } x$$

$$\csc^{-1} x = \sin^{-1} \frac{1}{x}, \qquad |x| \geq 1.$$

Their derivatives are given by the formulas

$$(\cot^{-1} x)' = \frac{-1}{1 + x^2} \qquad \text{and} \qquad (\csc^{-1} x)' = \frac{-1}{|x| \sqrt{x^2 - 1}},$$

which add nothing of value to our tools for integration.

With this section, the roster of functions needed in calculus is completed. There are the **algebraic functions**: polynomials in x, rational functions, and functions $y = f(x)$ defined implicitly as roots of an equation of the form $a_0(x) + a_1(x)y + \cdots + a_n(x)y^n = 0$, where the $a_i(x)$ are polynomials in x. For instance, the function $y = x^{2/3}$ is algebraic because it satisfies the equation $x^3 - y^2 = 0$.

Any function that can be obtained by algebraic means from polynomials, logarithms, exponentials, the six trigonometric functions, and their inverses is called an **elementary function**. An elementary function that is not algebraic is called **transcendental**. The function

$$\frac{\sqrt{1 + x^3} - \sin^{-1}(e^{-x^2})}{x^2 \tan(\ln 3x)}$$

is elementary (and transcendental). The formulas for differentiation assure us that its derivative is also an elementary function. However, as we saw in Chap. 5, an antiderivative of an elementary function need not be elementary.

Is the derivative of a transcendental function necessarily transcendental?

The key points in this section are the definitions of $\tan^{-1} x$, $\sin^{-1} x$, and $\sec^{-1} x$ and the calculations of their derivatives.

Function	*Derivative*	*Domain*
$\tan^{-1} x$	$\frac{1}{1 + x^2}$	All real numbers
$\sin^{-1} x$	$\frac{1}{\sqrt{1 - x^2}}$	$[-1, 1]$
$\sec^{-1} x$	$\frac{1}{\lvert x\rvert \sqrt{x^2 - 1}}$	All x such that $\lvert x\rvert \geq 1$

With each derivative comes a corresponding antiderivative:

$$\int \frac{dx}{1 + x^2} = \tan^{-1} x + C,$$

$$\int \frac{dx}{\sqrt{1 - x^2}} = \sin^{-1} x + C, \qquad |x| < 1,$$

$$\int \frac{dx}{x\sqrt{x^2 - 1}} = \sec^{-1} x + C, \qquad x > 1.$$

If x is positive, the absolute value sign can be omitted.

EXERCISES FOR SEC. 6.5: DERIVATIVES OF THE INVERSE TRIGONOMETRIC FUNCTIONS

Exercises 1 to 16 concern the three inverse functions, $\tan^{-1} x$, $\sin^{-1} x$, and $\sec^{-1} x$.

1 Draw a circle of radius 10 centimeters and use it, a centimeter ruler, and a protractor to estimate
(*a*) $\tan^{-1} 1.5$ (*b*) $\tan^{-1} 0.7$ (*c*) $\tan^{-1}(-1.2)$
(*d*) $\sin^{-1} 0.4$ (*e*) $\sin^{-1}(-0.5)$ (*f*) $\sin^{-1} 0.8$
If your protractor reads degrees, turn your answer into radians by dividing by 57 (an approximation of $180/\pi$).

2 Evaluate
(*a*) $\sin^{-1} 1$ (*b*) $\tan^{-1} 1$
(*c*) $\sin^{-1}(-\sqrt{3}/2)$ (*d*) $\tan^{-1}(-\sqrt{3})$
(*e*) $\sec^{-1} \sqrt{2}$

3 Use a calculator to find
(*a*) $\tan^{-1} 3$ (*b*) $\tan^{-1}(-2)$
(*c*) $\sin^{-1} 0.4$ (*d*) $\sec^{-1} 3$

4 What happens when you try to find arcsin 2 on your calculator? Why?

5 Which of these are *meaningless*?

(*a*) $\cos^{-1} 1.5$ (*b*) $\sec^{-1} 1.5$

(*c*) $\tan^{-1} 1.5$ (*d*) $\sec^{-1} 0.3$

In Exercises 6 to 8 use a calculator to fill in the tables (angles in radians). Then use the data to graph the functions.

6

x	-20	-10	-1	0	1	10	20
$\tan^{-1} x$							

7

x	-1	-0.8	-0.6	-0.4	-0.2	0	0.2	0.4	0.6	0.8	1
$\sin^{-1} x$											

8

x	-20	-10	-1	1	10	20
$\sec^{-1} x$						

[Use the fact that $\sec^{-1} x = \cos^{-1}(1/x)$.]

Exercises 9 and 10 obtain relations between inverse trigonometric functions. They depend on the right triangle shown in Fig. 6.20 and the fact that $\sin\theta = b/c$, $\cos\theta = a/c$, and $\tan\theta = b/a$.

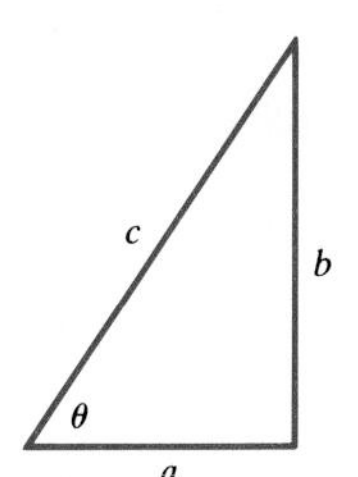

Figure 6.20

9 (See Fig. 6.21.) Show that for $x \geq 0$,

$$\sin^{-1}\frac{x}{\sqrt{x^2+1}} = \tan^{-1} x.$$

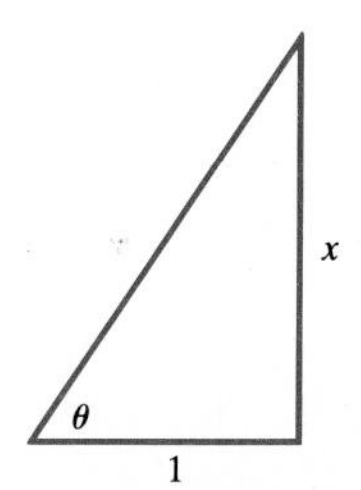

Figure 6.21

10 (See Fig. 6.22.)

(*a*) Express $\cos\theta$ in terms of x.

(*b*) Express $\sec\theta$ in terms of x.

(*c*) Verify that for $x \geq 1$, $\sec^{-1} x = \cos^{-1}(1/x)$.

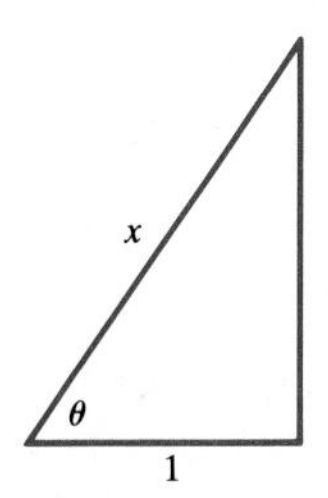

Figure 6.22

In Exercises 11 to 16 evaluate the expressions without recourse to a calculator. A sketch of the unit circle or of an appropriate right triangle may help.

11 $\sin(\tan^{-1} 1)$

12 $\tan(\sec^{-1} 2)$

13 $\tan[\sin^{-1}(-\sqrt{2}/2)]$

14 $\tan[\sin^{-1}(\sqrt{3}/2)]$

15 $\sin(\sin^{-1} 0.3)$

16 $\sin(\tan^{-1} 0)$

Exercises 17 to 46 differentiate the given functions.

17 $\sin^{-1} 5x$

18 $\tan^{-1} 3x$

19 $\sec^{-1} 3x$

20 $\sin^{-1} e^{-x}$

21 $\tan^{-1}\sqrt[3]{x}$

22 $-\dfrac{1}{3}\sin^{-1}\dfrac{3}{x}$

23 $x^2 \sec^{-1}\sqrt{x}$

24 $\dfrac{1}{\sin^{-1} 2x}$

25 $\sin 3x \sin^{-1} 3x$

26 $x^3 \tan^{-1} 2x$

27 $\dfrac{x \sec^{-1} 3x}{e^{2x}}$

28 $\arcsin(2x - 3)$

29 $\arctan\sqrt{x}$

30 $\operatorname{arcsec}\sqrt{x}$

31 $\ln \sec^{-1}\sqrt{x}$

32 $\ln[(\sin^{-1} 5x)^2]$

33 $\dfrac{x}{\tan^{-1} 10^x}$

34 $10^{\sec^{-1} 2x}$

35 $\sin^{-1} x - \sqrt{1 - x^2}$

36 $2^x \cdot \log_3 x \cdot \sec 3x$

37 $(\tan^{-1} 2x)^3$

38 $(\sin^{-1}\sqrt{x} - 1)^4$

39 $\dfrac{x}{2}\sqrt{2 - x^2} + \sin^{-1}\dfrac{x}{\sqrt{2}}$

40 $\sqrt{3x^2 - 1} - \tan^{-1}\sqrt{3x^2 - 1}$

41 $\frac{2}{3}\sec^{-1}\sqrt{3x^5}$

42 $\dfrac{1}{2}\left[(x - 3)\sqrt{6x - x^2} + 9\sin^{-1}\dfrac{x - 3}{3}\right]$

43 $\sqrt{1 + x}\sqrt{2 - x} - 3\sin^{-1}\sqrt{\dfrac{2 - x}{3}}$

44 $x\sin^{-1} 3x + \frac{1}{3}\sqrt{1 - 9x^2}$

45 $x(\sin^{-1} 2x)^2 - 2x + \sqrt{1 - 4x^2}\sin^{-1} 2x$

46 $x\tan^{-1} 5x - \frac{1}{10}\ln(1 + 25x^2)$

■

In Exercises 47 to 49 differentiate the given functions. Note that quite different functions may have very similar derivatives.

47 (*a*) $\ln(x + \sqrt{x^2 - 9})$ (*b*) $\sin^{-1}\frac{x}{3}$

48 (*a*) $-\frac{1}{5}\ln\frac{5 + \sqrt{25 - x^2}}{x}$ (*b*) $-\frac{1}{5}\sin^{-1}\frac{5}{x}$

49 (*a*) $\ln\frac{\sqrt{2x^2 + 1} - 1}{x}$ (*b*) $\sec^{-1} x\sqrt{2}$

50 (*a*) Show that

$$\int \frac{dx}{\sqrt{1 - a^2x^2}} = \frac{1}{a}\sin^{-1} ax + C \qquad (a > 0).$$

Use (*a*) to find

(*b*) $\int \frac{dx}{\sqrt{1 - 25x^2}}$ (*c*) $\int \frac{dx}{\sqrt{1 - 3x^2}}$

51 (*a*) Show that

$$\int \frac{dx}{\sqrt{a^2 - x^2}} = \sin^{-1}\frac{x}{a} + C \qquad (a > 0).$$

Use (*a*) to find

(*b*) $\int \frac{dx}{\sqrt{25 - x^2}}$ (*c*) $\int \frac{dx}{\sqrt{5 - x^2}}$

The next two exercises offer an interesting contrast.

52 (*a*) Sketch $y = 1/(1 + x^2)$.
(*b*) Find the area under the curve in (*a*) and above $[0, b]$, $b > 0$.
(*c*) Would you say that the area under the curve and above $[0, \infty)$ is finite or infinite? If finite, what is it?

53 (*a*) Sketch $y = 1/(1 + x)$ for $x \geq 0$.
(*b*) Find the area under the curve in (*a*) and above $[0, b]$, $b > 0$.
(*c*) Would you say that the area under the curve and above $[0, \infty)$ is finite or infinite? If finite, what is it?

54 Consider the curve $y = \sqrt{1 - x^2}$ for x in $[0, 1]$.
(*a*) Sketch the curve. (*Suggestion:* Show it is part of a circle.)
(*b*) Show that

$$\int \sqrt{1 - x^2}\, dx = \frac{x\sqrt{1 - x^2}}{2} + \frac{\sin^{-1} x}{2} + C.$$

(*c*) Find the area below $y = \sqrt{1 - x^2}$ and above $[0, \frac{1}{2}]$.
(*d*) Find the area below $y = \sqrt{1 - x^2}$ and above $[0, 1]$.

Exercises 55 to 57 are related.

55 (*a*) Compute the differential $d(\tan^{-1} x)$.
(*b*) Use (*a*) to estimate $\tan^{-1} 1.1$. (*Suggestion:* $\tan^{-1} 1$ is known. Use $\pi \approx 3.14$.)

56 (*a*) Compute the differential $d(\sin^{-1} x)$.
(*b*) Use (*a*) to estimate $\sin^{-1} 0.47$.

57 (*a*) Compute the differential $d(\sec^{-1} x)$.
(*b*) Use (*a*) to estimate $\sec^{-1} 2.08$.

58 Find the area below $y = 1/(x\sqrt{x^2 - 1})$ and above $[\sqrt{2}, 2]$.

59 (*a*) Differentiate $\tan^{-1} ax$, where a is a constant.
(*b*) Find the area under $y = 1/(1 + 3x^2)$ and above $[0, 1]$.

60 Show that $\tan^{-1}\frac{1}{2} + \tan^{-1}\frac{1}{3} = \pi/4$. *Hint:* Use a trigonometric identity.

61 (*a*) Show that $\sin^{-1}(x/\sqrt{x^2 + 1}) = \tan^{-1} x$ by differentiating both sides.
(*b*) Show it by a picture.

62 Show that $\sec^{-1} x = \cos^{-1}(1/x)$ by differentiating both sides.

63 Show that $\cos^{-1}(-x) = \pi - \cos^{-1} x$ by differentiating both sides.

64 The segment AB in Fig. 6.23 has fixed length a. As it is moved to the right, does the angle θ that it subtends decrease?

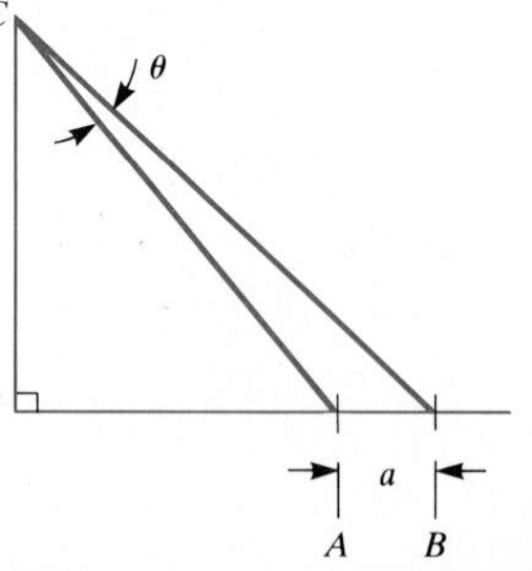

Figure 6.23

6.6 RELATED RATES

Sometimes the rate at which one quantity is changing is known and we wish to find the rate at which some related quantity is changing. Example 1 is typical and indicates a general method of attacking them.

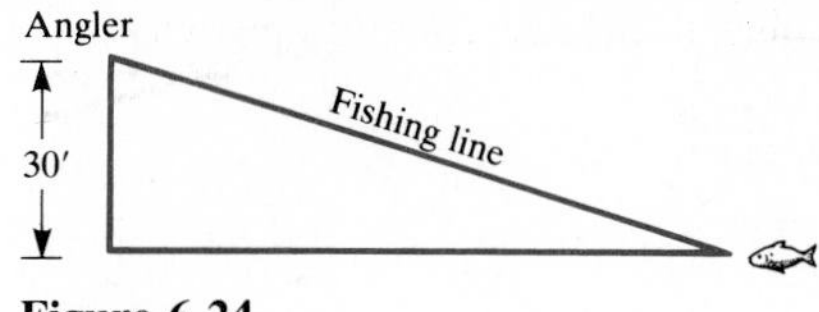

Figure 6.24

EXAMPLE 1 An angler has a fish at the end of his line, which is reeled in at 2 feet per second from a bridge 30 feet above the water. At what speed is the fish moving through the water when the amount of line out is 50 feet? 31 feet? Assume the fish is at the surface of the water. (See Fig. 6.24.)

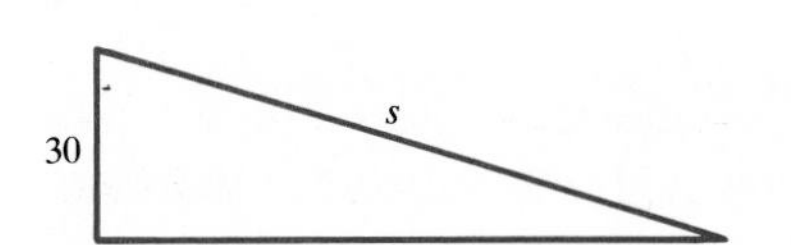

Figure 6.25

SOLUTION Let s be the length of the line and x the horizontal distance of the fish from the bridge. (See Fig. 6.25.)

Since the line is reeled in at the rate of 2 feet per second,

$$\frac{ds}{dt} = -2.$$

The rate at which the fish moves through the water is given by the derivative, dx/dt. The problem is to find dx/dt when $s = 50$ and also when $s = 31$.

The quantities x and s are related by the equation given by the pythagorean theorem:

$$x^2 + 30^2 = s^2.$$

Both x and s are functions of time t. Thus both sides of the equation may be differentiated with respect to t, yielding

$$\frac{d(x^2)}{dt} + \frac{d(30^2)}{dt} = \frac{d(s^2)}{dt}$$

or

$$2x\frac{dx}{dt} + 0 = 2s\frac{ds}{dt}.$$

Hence

$$x\frac{dx}{dt} = s\frac{ds}{dt}.$$

This last equation provides the tool for answering the questions.

Since $ds/dt = -2$,

$$x\frac{dx}{dt} = s(-2).$$

Hence

$$\frac{dx}{dt} = \frac{-2s}{x}.$$

When $s = 50$,

$$x^2 + 30^2 = 50^2,$$

from which it follows that $x = 40$. Thus when 50 feet of line are out, the speed is

$$\frac{2s}{x} = \frac{2 \cdot 50}{40} = 2.5 \text{ feet per second.}$$

When $s = 31$,

$$x^2 + 30^2 = 31^2.$$

Hence

$$x = \sqrt{31^2 - 30^2} = \sqrt{961 - 900} = \sqrt{61}.$$

Thus when 31 feet of line are out, the fish is moving at the speed of

$$\frac{2s}{x} = \frac{2 \cdot 31}{\sqrt{61}} = \frac{62}{\sqrt{61}} \approx 7.9 \text{ feet per second.} \quad \blacksquare$$

General procedure for finding related rates

The method used in Example 1 applies to many related rate problems. This is the general procedure, broken into three steps:

Warning: *Differentiate,* then *substitute the specific numbers for the variables. If you reversed the order, you would just be differentiating constants.*

1 Find an equation relating the varying quantities.
2 Differentiate both sides of the equation implicitly with respect to time or some other appropriate variable.
3 Use the equation obtained in step 2 to determine the unknown rate from the given rates. It is at this point that you substitute numbers for the variables and known rates of change.

In Example 1 it would be a tactical mistake to indicate in Fig. 6.25 that the horizontal leg of the triangle is 50 feet long, for if one leg is 30 feet and the other is 50 feet, the triangle is determined; there is nothing left free to vary with time. It is safest to label all the lengths or quantities that can change with letters x, y, s, and so on, even if not all are needed in the solution. Only after you finish differentiating do you determine what the rates are at a specified value of the variable.

EXAMPLE 2 A woman on the ground is watching a jet through a telescope as it approaches at a speed of 10 miles per minute at an altitude of 7 miles. At what rate is the angle of the telescope changing when the horizontal distance of the jet from the woman is 24 miles? When the jet is directly above the woman?

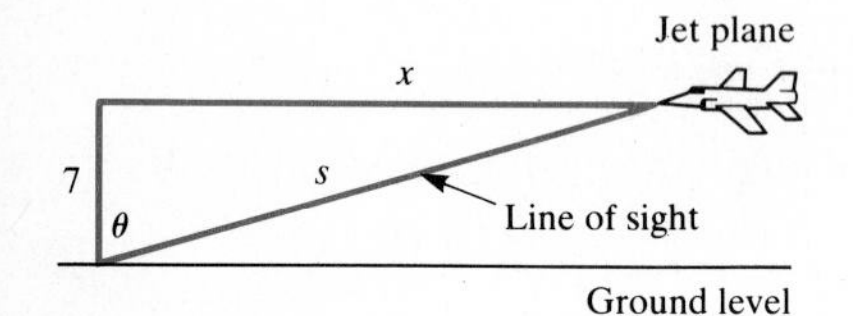

Figure 6.26

SOLUTION To begin, sketch a diagram and label the parts that may be of interest, as has been done in Fig. 6.26. Observe that

$$\frac{dx}{dt} = -10 \text{ miles per minute.}$$

The rate at which θ changes, $d\theta/dt$, is to be found.

Step 1 consists of finding an equation relating θ and x. One such equation is

$$\theta = \tan^{-1}\frac{x}{7}.$$

Step 2 is to differentiate this equation with respect to time:

$$\frac{d\theta}{dt} = \frac{1}{1 + (x/7)^2}\frac{dx/dt}{7}.$$

Since $dx/dt = -10$,

$$\frac{d\theta}{dt} = \frac{1}{1 + x^2/49} \cdot \frac{-10}{7} = \frac{-70}{49 + x^2} \text{ radians per minute.}$$

Finally, we replace the variable x by 24, obtaining, when $x = 24$,

$$\frac{d\theta}{dt} = \frac{-70}{49 + 24^2} = \frac{-70}{625} \text{ radians per minute.}$$

(This is about $-6°$ per minute.) When the jet is directly above the woman, $x = 0$, and we obtain $d\theta/dt = -70/49$ radians per minute (which is about $-82°$ per minute.) ■

The method described in Example 1 for determining unknown rates from known ones extends to finding an unknown acceleration. Just differentiate another time. Example 3 illustrates the procedure.

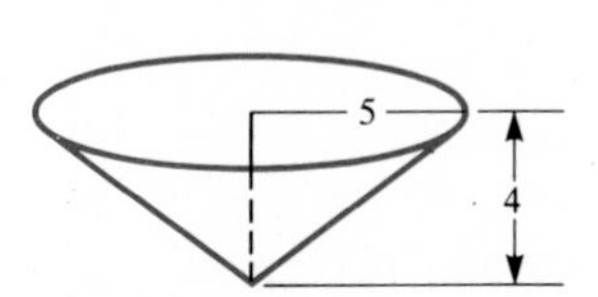

Figure 6.27

EXAMPLE 3 Water flows into a conical tank at the constant rate of 3 cubic meters per second. The radius of the cone is 5 meters and its height is 4 meters. Let $h(t)$ represent the height of the water above the bottom of the cone at time t. Find dh/dt (the rate at which the water is rising in the tank) and d^2h/dt^2, when the tank is filled to a height of 2 meters. (See Fig. 6.27.)

SOLUTION Let $V(t)$ be the volume of water in the tank at time t. The data imply that

$$\frac{dV}{dt} = 3,$$

and hence
$$\frac{d^2V}{dt^2} = 0.$$

To find dh/dt and d^2h/dt^2, first obtain an equation relating V and h.

When the tank is filled to the height h, the water forms a cone of height h and radius r. (See Fig. 6.28.) By similar triangles,

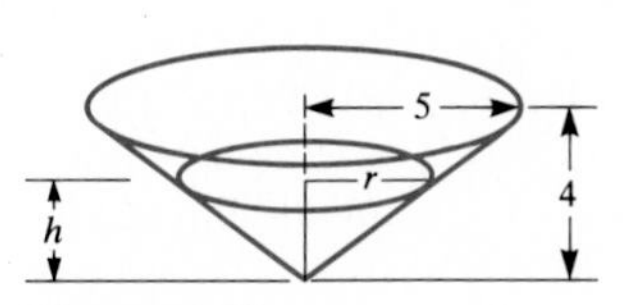

Figure 6.28

$$\frac{r}{h} = \frac{5}{4} \quad \text{or} \quad r = \frac{5h}{4}.$$

Thus
$$V = \tfrac{1}{3}\pi r^2 h = \tfrac{1}{3}\pi(\tfrac{5}{4}h)^2 h = \tfrac{25}{48}\pi h^3.$$

The equation relating V and h is

$$V = \frac{25\pi}{48} h^3. \tag{1}$$

From here on, the procedure is automatic: Just differentiate as often as needed.

Differentiating once (using the chain rule) yields

$$\frac{dV}{dt} = \frac{25\pi}{48}\frac{d(h^3)}{dh}\frac{dh}{dt} = \frac{25\pi}{16} h^2 \frac{dh}{dt}. \tag{2}$$

Since $dV/dt = 3$ all the time, and $h = 2$ at the moment of interest, substitute these values into Eq. (2), obtaining

$$3 = \frac{25\pi}{16} 2^2 \frac{dh}{dt}.$$

Hence
$$\frac{dh}{dt} = \frac{12}{25\pi} \text{ meters per second}$$

when $h = 2$.

Do not differentiate $\frac{dh}{dt} = \frac{12}{25\pi}$.

To find d^2h/dt^2, differentiate Eq. (2), obtaining

$$\frac{d^2V}{dt^2} = \frac{25\pi}{16}\left(h^2\frac{d^2h}{dt^2} + 2h\frac{dh}{dt}\frac{dh}{dt}\right). \tag{3}$$

Since $d^2V/dt^2 = 0$ all the time, and, when $h = 2$, $dh/dt = 12/(25\pi)$, Eq. (3) implies that

$$0 = \frac{25\pi}{16}\left[2^2\frac{d^2h}{dt^2} + 2\cdot 2\left(\frac{12}{25\pi}\right)^2\right]. \tag{4}$$

Solving Eq. (4) for d^2h/dt^2 shows that

$$\frac{d^2h}{dt^2} = \frac{-144}{625\pi^2} \text{ meters per second per second.}$$

Since d^2h/dt^2 is negative, the rate at which the water rises in the tank is slowing down. In general, the higher the water, the slower it rises. Even though V changes at a constant rate, h does not. ■

EXERCISES FOR SEC. 6.6: RELATED RATES

Exercises 1 and 2 are related to Example 1.

1 How fast is the fish moving through the water when it is 1 foot horizontally from the bridge?

2 The angler in Example 1 decides to let the line out as the fish swims away. The fish swims away at a constant speed of 5 feet per second relative to the water. How fast is the angler paying out his line when the horizontal distance from the bridge to the fish is (*a*) 1 foot? (*b*) 100 feet?

3 A 10-foot ladder is leaning against a wall. If a person pulls the base of the ladder away from the wall at the rate of 1 foot per second, how fast is the top going down the wall when the base of the ladder is (*a*) 6 feet from the wall? (*b*) 8 feet from the wall? (*c*) 9 feet from the wall?

4 A kite is flying at a height of 300 feet in a horizontal wind. When 500 feet of string are out, the kite is pulling the string out at a rate of 20 feet per second. What is the wind velocity? (Assume the string remains straight.)

5 A beachcomber walks 2 miles per hour along the shore as the beam from a rotating light 3 miles offshore follows him. (See Fig. 6.29.)

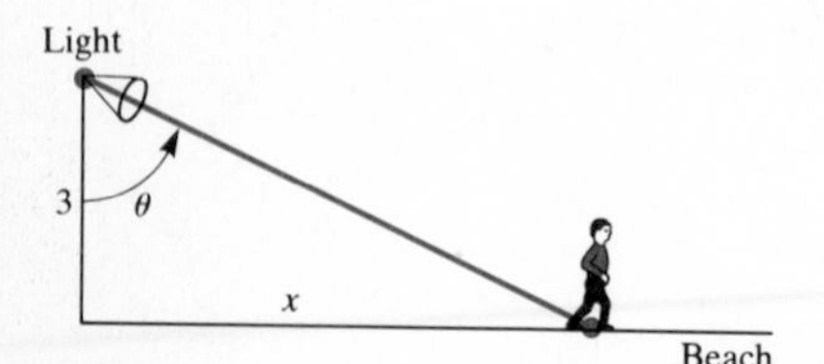

Figure 6.29

(*a*) Intuitively, what do you think happens to the rate at which the light rotates as the beachcomber walks further and further along the shore away from the lighthouse?

(*b*) Letting x describe the distance of the beachcomber from the point on the shore nearest the light and θ the angle of the light, obtain an equation relating θ and x.

(*c*) With the aid of (*b*), show that $d\theta/dt = 6/(9 + x^2)$ (radians per hour).

(*d*) Does the formula in (*c*) agree with your guess in (*a*)?

6 A man 6 feet tall walks at the rate of 5 feet per second away from a lamp that is 20 feet high. At what rate is his shadow lengthening when he is (*a*) 10 feet from the lamp? (*b*) 100 feet from the lamp?

7 The length of a rectangle is increasing at the rate of 7 feet per second, and the width is decreasing at the rate of 3 feet per second. When the length is 12 feet and the width is 5 feet, find the rate of change of (*a*) the area, (*b*) the perimeter, (*c*) the length of the diagonal.

8 A shrinking spherical balloon loses air at the rate of 1 cubic inch per second. At what rate is its radius changing when the radius is (*a*) 2 inches? (*b*) 1 inch? (The volume V of a sphere of radius r is $4\pi r^3/3$.)

9 Bulldozers are moving earth at the rate of 1,000 cubic yards per hour onto a conically shaped hill whose height remains equal to its radius. At what rate is the height of the hill increasing when the hill is (*a*) 20 yards high? (*b*) 100 yards high? (The volume of a cone of radius r and height h is $\pi r^2h/3$.)

10 The lengths of the two legs of a right triangle depend on time. One leg, whose length is x, increases at the rate of 5 feet per second, while the other, of length y, decreases at the rate of 6 feet per second. At what rate is the hypotenuse changing when $x = 3$ feet and $y = 4$ feet? Is the hypotenuse increasing or decreasing then?

11 Two sides of a triangle and their included angle are changing with respect to time. The angle increases at the rate of 1 radian per second, one side increases at the rate of 3 feet per second, and the other side decreases at the rate of 2 feet per second. Find the rate at which the area is changing when the angle is $\pi/4$, the first side is 4 feet long, and the second side is 5 feet long. Is the area decreasing or increasing then?

12 A large spherical balloon is being inflated at the rate of 100 cubic feet per minute. At what rate is the radius increasing when the radius is (*a*) 10 feet? (*b*) 20 feet?

Exercises 13 to 16 concern acceleration. The notation $\dot{x}$ for dx/dt, $\dot{\theta}$ for $d\theta/dt$, $\ddot{x}$ for d^2x/dt^2, and $\ddot{\theta}$ for $d^2\theta/dt^2$ is common in physics.

13 What is the acceleration of the fish described in Example 1 when the length of line is (*a*) 300 feet? (*b*) 31 feet?

14 Find $\ddot{\theta}$ in Example 2 when the horizontal distance from the jet is (*a*) 7 miles, (*b*) 1 mile.

15 A particle moves on the parabola $y = x^2$ in such a way that $\dot{x} = 3$ throughout the journey. Find formulas for (*a*) $\dot{y}$ and (*b*) $\ddot{y}$.

16 Call one acute angle of a right triangle θ. The adjacent leg has length x and the opposite leg has length y.

(*a*) Obtain an equation relating x, y, and θ.

(*b*) Obtain an equation involving $\dot{x}$, $\dot{y}$, and $\dot{\theta}$ (and other variables).

(*c*) Obtain an equation involving $\ddot{x}$, $\ddot{y}$, and $\ddot{\theta}$ (and other variables).

■

17 A two-piece extension ladder leaning against a wall is collapsing at the rate of 2 feet per second at the same time as its foot is moving away from the wall at the rate of 3 feet per second. How fast is the top of the ladder moving down the wall when it is 8 feet from the ground and the foot is 6 feet from the wall? (See Fig. 6.30.)

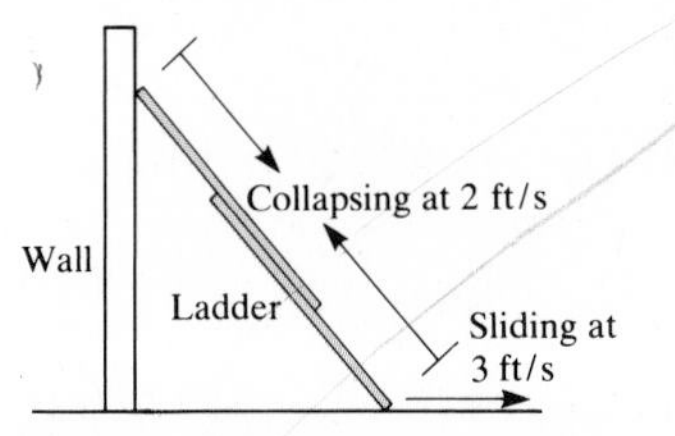

Figure 6.30

18 The atmospheric pressure at an altitude of x kilometers is approximately $1{,}000(0.88)^x$ millibars. A rocket is rising at the rate of 5 kilometers per second vertically. At what rate is the atmospheric pressure changing (in millibars per second) when the altitude of the rocket is (*a*) 1 kilometer? (*b*) 50 kilometers?

19 A woman is walking on a bridge that is 20 feet above a river as a boat passes directly under the center of the bridge (at a right angle to the bridge) at 10 feet per second. At that moment the woman is 50 feet from the center and approaching it at the rate of 5 feet per second. (*a*) At what rate is the distance between the boat and woman changing at that moment? (*b*) Is the rate at which they are approaching or separating increasing or is it decreasing?

20 A spherical raindrop evaporates at a rate proportional to its surface area. Show that the radius shrinks at a constant rate.

21 A couple is on a Ferris wheel when the sun is directly overhead. The diameter of the wheel is 50 feet, and its speed is 0.1 revolution per second. (*a*) What is the speed of their shadows on the ground when they are at a two-o'clock position? (*b*) A one-o'clock position? (*c*) Show that the shadow is moving its fastest when they are at the top or bottom, and its slowest when they are at the three-o'clock position.

22 Water is flowing into a hemispherical kettle of radius 5 feet at the constant rate of 1 cubic foot per minute.

(*a*) At what rate is the top surface of the water rising when its height above the bottom of the kettle is 3 feet? 4 feet? 5 feet?

(*b*) If $h(t)$ is the depth in feet at time t, find $\ddot{h}$ when $h = 3$, 4, and 5.

23 A man in a hot-air balloon is ascending at the rate of 10 feet per second. How fast is the distance from the balloon to the horizon (that is, the distance the man can see) increasing when the balloon is 1,000 feet high? Assume that the earth is a ball of radius 4,000 miles. (See Fig. 6.31.)

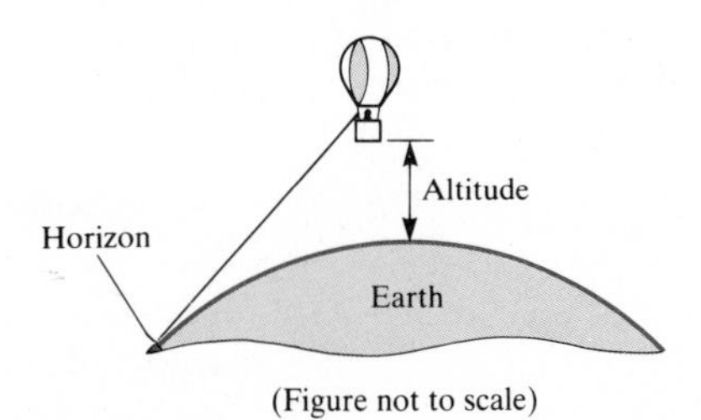

Figure 6.31

6.7 SEPARABLE DIFFERENTIAL EQUATIONS

The abbreviation for "differential equation" is D.E.

An equation that involves one or more of the derivatives of a function is called a **differential equation**. For example, Sec. 4.5 examined the differential equation for motion under constant acceleration,

$$\frac{d^2y}{dt^2} = a \qquad (a \text{ constant}), \tag{1}$$

and the differential equation of harmonic motion,

$$\frac{d^2x}{dt^2} = -cx.$$

Finding an antiderivative $F(x)$ for a function $f(x)$ amounts to solving the differential equation

$$\frac{dF}{dx} = f(x) \tag{2}$$

for the unknown function $F(x)$.

Solution of a D.E.

A **solution** of a differential equation is any function that satisfies the equation. To **solve** a differential equation means to find all its solutions. In Sec. 4.5 it was shown that the most general solution of Eq. (1) is

$$y = \frac{at^2}{2} + v_0t + y_0,$$

A solution of a D.E. is a function, not a number.

where v_0 and y_0 are constants. The most general solution of

$$\frac{dF}{dx} = x^2 \tag{3}$$

is

$$F(x) = \frac{x^3}{3} + C,$$

where C represents an arbitrary constant.

A differential equation can be much more complicated than Eq. (1) or (3), as the differential equation

$$\left(\frac{d^2y}{dx^2}\right)^2 + 3x\frac{dy}{dx} + \sin x = \frac{d^3y}{dx^3} \tag{4}$$

Order of a D.E.

illustrates. The **order** of a differential equation is the highest order of the derivatives that appear in it. Thus Eq. (1) is of order 2, Eqs. (2) and (3) are of order 1, and Eq. (4) is of order 3.

This section examines a special and important type of first-order differential equation, called **separable**. After showing how to solve it, we will apply it in Sec. 6.8 to the study of natural growth and decay and to inhibited growth.

A **separable differential equation** is one that can be written in the form

Separable D.E.

$$\frac{dy}{dx} = \frac{f(x)}{g(y)}, \tag{5}$$

where $f(x)$ and $g(y)$ are differentiable functions. Such an equation can be solved by *separating the variables*, that is, bringing all the x's to one side and all the y's to the other side to obtain the following equation in differentials:

$$g(y)\,dy = f(x)\,dx. \tag{6}$$

This is solved by integrating both sides:

$$\int g(y)\,dy = \int f(x)\,dx + C. \tag{7}$$

Some examples will illustrate the technique.

EXAMPLE 1 Solve $\dfrac{dy}{dx} = \dfrac{2x}{3y}$ $(y > 0)$.

SOLUTION Separating the variables, we obtain

$$3y\,dy = 2x\,dx.$$

Thus

$$\int 3y\,dy = \int 2x\,dx + C$$

or

$$\frac{3y^2}{2} = x^2 + C. \tag{8}$$

Equation (8) determines y as a function of x implicitly. Each choice of C produces a solution. ■

EXAMPLE 2 Solve the differential equation

$$\frac{dy}{dx} = \frac{2y}{x} \qquad (x, y > 0). \tag{9}$$

SOLUTION At first glance the equation does not appear to be of the form in Eq. (5). However, it can be rewritten in the form

$$\frac{dy}{dx} = \frac{(1/x)}{(1/2y)},$$

so it has the form of a separable differential equation. Separation of the variables is not hard:

$$\frac{dy}{dx} = \frac{2y}{x} \qquad \frac{dy}{2y} = \frac{dx}{x}.$$

Hence

$$\int \frac{dy}{2y} = \int \frac{dx}{x} + C$$

or

$$\tfrac{1}{2}\ln y = \ln x + C. \tag{10}$$

Since x, y assumed > 0, $\ln|x| = \ln x$, $\ln|y| = \ln y$.

In this case, let us solve for y explicitly:

$$\ln y = 2\ln x + 2C$$

$$y = e^{2\ln x + 2C} \quad \text{definition of natural logarithm}$$

$$= e^{2\ln x}e^{2C} \quad \text{basic law of exponents}$$

$$= (e^{\ln x})^2 e^{2C} \quad \text{power of a power}$$

$$= x^2e^{2C}.$$

Since e^{2C} is an arbitrary positive constant, call it k. Thus the most general solution of Eq. (9) is

$$y = kx^2. \tag{11}$$

As a check on this solution, see if $y = kx^2$ satisfies Eq. (9):

$$\frac{d(kx^2)}{dx} \overset{?}{=} \frac{2(kx^2)}{x}$$

$$2kx \overset{?}{=} \frac{2kx^2}{x}.$$

Yes, it checks. ■

The solution of a separable differential equation (in fact, any first-order differential equation) will generally involve one arbitrary constant. Each choice of that constant determines a specific function that satisfies the differential equation.

The Differential Equations of Natural Growth and Decay The next example treats a differential equation that is important in the study of growth and decay. It arises in such diverse areas as biology, ecology, physics, chemistry, and economic forecasting. Its applications will be illustrated in the next section.

EXAMPLE 3 Solve the differential equation

$$\frac{dy}{dx} = ky \qquad (y > 0), \tag{12}$$

where k is a nonzero constant.

SOLUTION Separation of the variables yields

$$\frac{dy}{y} = k\,dx$$

$$\int \frac{dy}{y} = \int k\,dx + C$$

$$\ln y = kx + C$$

$$y = e^{kx+C}$$

$$y = e^C e^{kx}.$$

Denote the arbitrary positive constant e^C by the letter A. Then

$$y = Ae^{kx}. \tag{13}$$

The most general solution of $dy/dx = ky$ is $y = Ae^{kx}$. ■

Example 4 solves a differential equation that arises in the study of bounded or **inhibited growth**, which will be applied in the next section. At one point in the solution, the algebraic identity

$$\frac{M}{y(M - y)} = \frac{1}{y} + \frac{1}{M - y}$$

will be needed. (Check this identity before reading Example 4.)

EXAMPLE 4 Solve the differential equation

$$\frac{dy}{dx} = ky\left(1 - \frac{y}{M}\right), \tag{14}$$

where k and M are positive constants and $0 < y < M$.

SOLUTION Separate the variables and integrate:

$$\frac{dy}{y[1 - (y/M)]} = k\,dx$$

$$\int \frac{dy}{y[1 - (y/M)]} = \int k\,dx + C$$

$$\int \frac{M\,dy}{y(M - y)} = \int k\,dx + C \qquad \text{algebra}$$

$$\int \left(\frac{1}{y} + \frac{1}{M - y}\right) dy = \int k\,dx + C \qquad \text{algebraic identity}$$

$$\int \frac{dy}{y} + \int \frac{dy}{M - y} = \int k\,dx + C$$

$$\ln y - \ln (M - y) = kx + C \qquad 0 < y < M, \text{ so } \ln|y| = \ln y \text{ and } \ln|M - y| = \ln (M - y)$$

$$\ln\left(\frac{y}{M - y}\right) = kx + C$$

$$\frac{y}{M - y} = e^{kx+C} \qquad \text{definition of natural logarithm}$$

$$\frac{M - y}{y} = e^{-kx-C} \qquad \text{taking reciprocals}$$

$$\frac{M - y}{y} = e^{-C}e^{-kx} \qquad \text{basic law of exponents}$$

$$= ae^{-kx} \qquad \text{setting } e^{-C} = a$$

$$M - y = ae^{-kx}y$$

$$M = (1 + ae^{-kx})y$$

Finally,

$$y = \frac{M}{1 + ae^{-kx}}. \quad ■ \tag{15}$$

EXERCISES FOR SEC. 6.7: SEPARABLE DIFFERENTIAL EQUATIONS

In Exercises 1 to 6 check by substitution that the given functions satisfy the given differential equations.

1 $y = Ae^{kx}$; $dy/dx = ky$.

2 $y = ke^{x^2/2}$; $dy/dx = xy$.

3 $y = e^{-3x}$; $d^2y/dx^2 = 9y$.

4 $y = Ae^{-3x} + Be^{3x}$; $d^2y/dx^2 = 9y$ (A and B constants).

5 $y = A \cos 3x + B \sin 3x$; $d^2y/dx^2 = -9y$.

6 $y = M/(1 + ae^{-kx})$; $dy/dx = ky(1 - y/M)$. (This checks the result in Example 4.)

In Exercises 7 to 20 solve the differential equations.

7 $dy/dx = x^3/y^4$

8 $dy/dx = y^4/x^3$

9 $dy/dx = y/x$

10 $dy/dx = e^y/x$

11 $dy/dx = \sqrt{1 - y^2}/(1 + x^2)$

12 $dy/dx = (y + 4)/(x\sqrt{x^2 - 1})$, $x > 1$, $y > 0$

13 $x + y^2 \dfrac{dy}{dx} = 0$

14 $x + y \dfrac{dy}{dx} = 3$

15 $y \dfrac{dy}{dx} - (1 + y^2)x^2 = 0$

16 $(3 + x)e^{2y} \dfrac{dy}{dx} = 1$

17 $\dfrac{1}{\sqrt{y^2 - 1}} \dfrac{dy}{dx} = yx^2$

18 $\dfrac{(1 + \sqrt[3]{1 + 2y})}{\cos 3x} \dfrac{dy}{dx} = 5$

19 $\dfrac{dy}{dx} = \dfrac{\sqrt{1 - y^2}}{\sqrt{3 - x}}$, $x < 3$, $|y| < 1$

20 $(1 + 2y)^3 \dfrac{dy}{dx} = \sec^2 3x$

■

21 The differential equation

$$L \frac{di}{dt} + Ri = E$$

occurs in the study of electric circuits. L, R, and E are constants that describe the inductance, resistance, and voltage, i is the current, and t is time. Assume that di/dt is positive. Then $E - Ri$ is positive as well.
(*a*) Solve for di/dt in terms of i.
(*b*) The equation in (*a*) is separable. Solve it and express the answer in terms of R, L, E, and i_0, the initial current.

22 Find all functions $y = f(t)$ such that

$$\frac{dy}{dt} = k(y - A),$$

where k and A are constants. For negative k, this is Newton's law of cooling; y is the temperature of some heated object at time t. The room temperature is A. The differential equation $dy/dt = k(y - A)$ says, "The object cools at a rate proportional to the difference between its temperature and the room temperature."

23 A company is founded with a capital investment A. The plan is to have its rate of investment proportional to its total investment at any time. Let $f(t)$ denote the rate of investment at time t.
(*a*) Show that there is a constant k such that $f(t) = k[A + \int_0^t f(x)\, dx]$ for any $t \geq 0$.
(*b*) Find a formula for f.

24 A particle moving through a liquid meets a "drag" proportional to the velocity; that is, its acceleration is proportional to its velocity. Let x denote its position and v its velocity at time t.
(*a*) Show that there is a positive constant k such that

$$\frac{dv}{dt} = -kv.$$

(*b*) Show that there is a constant A such that

$$v = Ae^{-kt}.$$

(*c*) Show that there is a constant B such that

$$x = -\frac{1}{k} Ae^{-kt} + B.$$

(*d*) How far does the particle travel as t goes from 0 to ∞?

25 Consider the differential equation

$$\frac{dy}{dx} + ky = Q,$$

where k is a constant and Q is a function of x.
(*a*) Multiply both sides of the equation by e^{kx} and show that

$$\frac{d}{dx}(e^{kx}y) = e^{kx}Q.$$

(*b*) Deduce that $y = e^{-kx} \int e^{kx}Q\, dx$ for some antiderivative of $e^{kx}Q$.
(*c*) Using (*b*), solve the equation $dy/dx + ky = \sin ax$, where k and a are constants. (*Hint:* Use formula (60) in the integral table in front of the book.) This type of equation appears in the study of alternating currents.

6.8 NATURAL AND INHIBITED GROWTH

The change in the size of the world's population is determined by two basic variables: the birth rate and the death rate. If they are equal, the size of the population remains constant. But if one is larger than the other, the population either grows or shrinks. For example, in the United States in 1984 there were 3,690,000 births and 2,046,000 deaths. (In addition, there were 523,000 legal immigrants.)

The birth rate and death rate are determined by different variables, and therefore, it would be sheer coincidence if they were equal. With improved agricultural techniques, medical care, and preventive medicine, the death rate is well below the birth rate.

Consequently, for the past 2 centuries the world population has in-

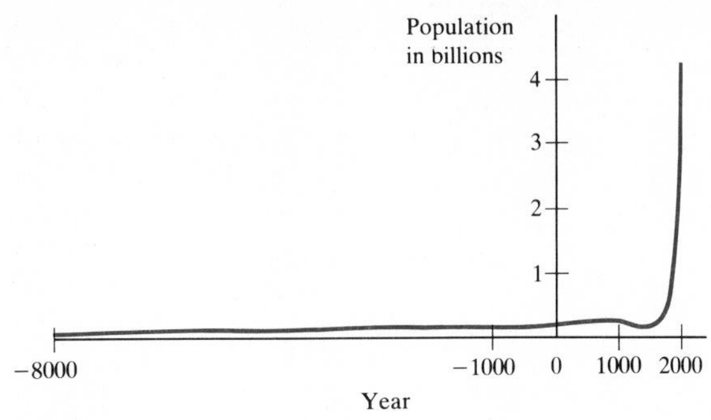

Figure 6.32

creased dramatically, as shown in Fig. 6.32. The size of the world population in 1650 is estimated to have been about 0.5 billion; in 1750 about 0.7 billion; in 1850 about 1.1 billion; in 1980 about 4.5 billion. The population has been growing at an accelerating pace. If back in 1850 you had moved every human being to China, then that total number would just match the present population of China.

In 1986 the world population reached 5 billion.

It took a million years for the world population to reach 1 billion. It reached 2 billion in another 120 years, 3 billion in another 32 years, and 4 billion in just 15 more years.

What will the population be in the year 2000 if the present rate of growth continues? To answer this question we must describe the size of the population mathematically.

Let $P(t)$ denote the size of the population at time t. Actually, $P(t)$ is an integer, and the graph of P has "jumps" whenever someone is born or dies. However, assume that P is a "smooth" (differentiable) function that approximates the size of the population.

The derivative P' then records the rate of change of the population. If social, medical, and technological factors remain constant, then it is reasonable to expect the rate of growth $P'(t)$ to be proportional to the size of the population $P(t)$: A large population will produce more babies in a year than a small population. More precisely, there is a fixed number k, independent of time, such that

The D.E. of natural growth and decay

$$P'(t) = kP(t). \tag{1}$$

This is the differential equation of **natural growth** (or **decay**, if k is negative). The constant k is called the **growth constant**.

To forecast the population, it is necessary to solve Eq. (1). But Eq. (1) is just the differential equation in Example 3 of the preceding section in different symbols. By that example,

$$P(t) = Ae^{kt}. \tag{2}$$

What is the meaning of the constant A? To find out, set $t = 0$ in Eq. (2), obtaining

$$P(0) = Ae^{k \cdot 0} \qquad \text{or} \qquad P(0) = A.$$

The meaning of A

Thus A is the amount or size of the population at the "initial time" $t = 0$.

To find the meaning of k, first let us interpret e^k. Let

$$b = e^k.$$

Then Eq. (2) takes the simpler form

$$P(t) = Ab^t. \tag{3}$$

Thus the function P at time $t + 1$, one unit after t, has the value

$$P(t + 1) = Ab^{t+1},$$

and

$$\frac{P(t + 1)}{P(t)} = \frac{Ab^{t+1}}{Ab^t} = b.$$

Thus

$$P(t + 1) = bP(t).$$

The meaning of b

The constant b is the ratio between the population at time $t + 1$ and the population at time t.

In 1986 the world population was growing at the relative rate of 1.8 percent (= 0.018) per year. Thus $b = 1 + 0.018 = 1.018$. But $b = e^k$. So

$$k = \ln b = \ln 1.018$$
$$\approx 0.0178.$$

Note that k, the growth constant, is fairly close to the growth rate per unit time, 1.8 percent, or 0.018, per year.

The relative growth rate per unit time, r

In describing growing populations newspapers report the relative growth per year, which is usually in the 1 to 3 percent range. If this number is denoted r, then

$$b = 1 + r.$$

Use a differential to show that for small r, $\ln(1 + r) \approx r$.

Since r is small, $k = \ln(1 + r) \approx r$. (See Exercise 1.) Thus for small r, the growth constant k is approximately the same as the observed relative growth rate per unit time, r, that is,

$$k \approx r \qquad \text{(if } r \text{ is small)}.$$

For small r, $k \approx r$ and $b \approx 1 + k$.

Or to put it another way,

$$b = e^k \approx 1 + k.$$

EXAMPLE 1 The population of the United States in 1986 was estimated to be about 241 million and to be growing at 0.9 percent per year.
(a) Estimate the population in the year 2000.
(b) When will the population double?

SOLUTION In this case, the relative growth rate per unit time r is 0.009. Thus

$$b = 1.009.$$

Introduce a time scale with $t = 0$ corresponding to the year 1986. Let $P(t)$ be the population of the United States at time t, that is, in the year $1986 + t$. Since $P(0) = 241$ million,

$$P(t) = 241(1.009)^t \text{ million}$$

at time t.

(*a*) To find the population in the year 2000, which is 14 years after the time $t = 0$, compute $P(14)$:

$$P(14) = 241(1.009)^{14} \text{ million}$$
$$\approx 241(1.1336) \text{ million} \qquad \text{(use the } y^x\text{-key on your calculator)}$$
$$\approx 273 \text{ million.}$$

So if the growth rate continues at 0.9 percent per year, the population will be 273 million in the year 2000.

(*b*) To find when the population will double, solve the equation

$$2 \cdot 241 = 241(1.009)^t$$

or

$$2 = 1.009^t. \tag{4}$$

To solve Eq. (4), take the natural log of both sides:

$$\ln 2 = t \ln 1.009.$$

Since population figures are not precise, this estimate is good enough.

Since

$$\ln 1.009 \approx 0.009,$$
$$\ln 2 \approx (0.009)t$$

or

$$t \approx \frac{\ln 2}{0.009} \approx \frac{0.69}{0.009} \approx 77 \text{ years.}$$

Thus in about 77 years the population would double. ■

The doubling time, t_2

The time it takes for a quantity growing in accord with the formula $P(t) = Ab^t$ to double is called its **doubling time**. As in Example 1, the doubling time, denoted t_2, is given by the formula

$$t_2 = \frac{\ln 2}{\ln b}.$$

(We solve the equation $2A = Ab^t$.) But if the growth rate per unit time, r, is small, then $\ln b \approx r$, and

$$t_2 \approx \frac{\ln 2}{r} \approx \frac{0.69}{r}. \tag{5}$$

An estimate for t_2 when r is small

Equation (5) gives a quick way to estimate doubling time. For instance, in 1986 the less developed countries had an annual growth rate $r \approx 2.2$ percent $= 0.022$. Therefore, their populations may double in

$$t_2 \approx \frac{0.69}{0.022} \approx 31 \text{ years.}$$

Natural decay

A substance can decay at a rate proportional to the amount present. In this case, the same differential equation still holds,

$$P'(t) = kP(t),$$

but $P'(t)$ is negative, and therefore, k is negative. The next example illustrates this type of decay.

EXAMPLE 2 Carbon 14 (chemical symbol ^{14}C), one of the isotopes of carbon, is radioactive and decays at a rate proportional to the amount present. In about 5730 years only half the material is left. This is expressed by saying that its **half-life** is 5730 years. The half-life is denoted by $t_{1/2}$. Find the constant k in the formula Ae^{kt} in this case.

Half-life $t_{1/2}$

SOLUTION Let $P(t) = Ae^{kt}$ be the amount present after t years. If A is the initial amount, then

$$P(5730) = \frac{A}{2}.$$

Thus
$$Ae^{k \cdot 5730} = \frac{A}{2},$$

or
$$e^{5730k} = \tfrac{1}{2}.$$

Thus
$$5730k = \ln \tfrac{1}{2} = -\ln 2 \approx -0.69.$$

Solving for k, we have
$$k \approx \frac{-0.69}{5730} \approx -0.00012.$$

Thus
$$P(t) \approx Ae^{-0.00012t}.$$

Exercise 20 shows how this formula may be used to determine the age of fossils and other once-living things. ■

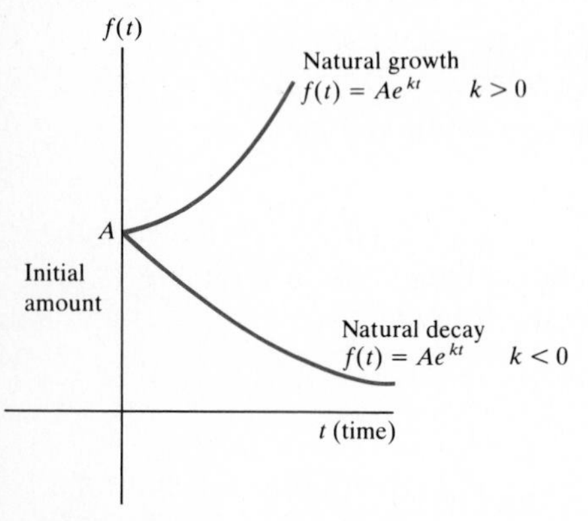

Figure 6.33

The graphs of natural growth and natural decay look quite different. The first gets arbitrarily steep as time goes on; the second approaches the t axis and becomes almost horizontal. (See Fig. 6.33.)

If a quantity is increasing subject to natural growth, its graph may quickly shoot off an ordinary size piece of paper. The next example describes a special way of graphing an equation of the form $y = Ae^{kt}$.

EXAMPLE 3 Let $y = Ae^{kt}$. Let $Y = \ln y$. Show that the graph of Y, considered as a function of t, is a straight line.

SOLUTION
$$\begin{aligned} Y &= \ln y \\ &= \ln Ae^{kt} \\ &= \ln A + \ln e^{kt} \\ &= \ln A + kt. \end{aligned}$$

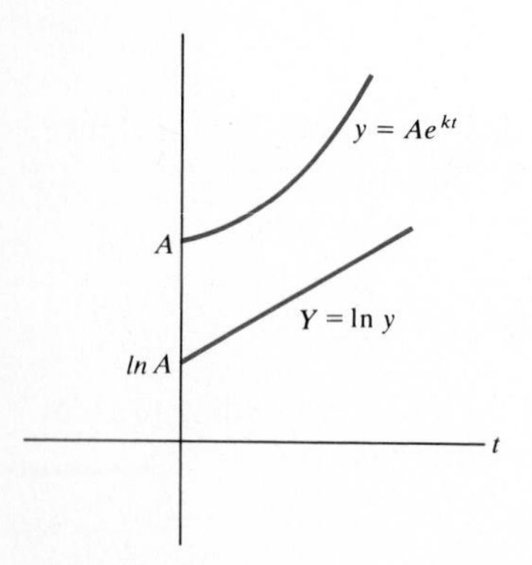

Figure 6.34

The graph is thus a straight line with Y intercept $\ln A$ and slope k. The diagram in Fig. 6.34 contrasts the graphs of y and $Y = \ln y$ as functions of t. ■

Special graph paper called **semilog**, based on the observation in Example 3, is available. It enables one to graph $Y = \ln y$ directly without having to compute logarithms. On the vertical axis the number y is placed a distance from the x axis proportional to $\ln y$. (See Exercise 23.)

Inhibited Growth In many cases of growth there is obviously a finite upper bound M which the population size must approach. It is reasonable to assume (or to take as a model) that

$$\frac{dP}{dt} = kP(t)\left(1 - \frac{P(t)}{M}\right). \tag{6}$$

This equation asserts that the population $P(t)$ grows at a rate proportional both to itself and to the fraction left to grow. Thus as $P(t)$ approaches its limiting size M, the fraction

$$1 - \frac{P(t)}{M}$$

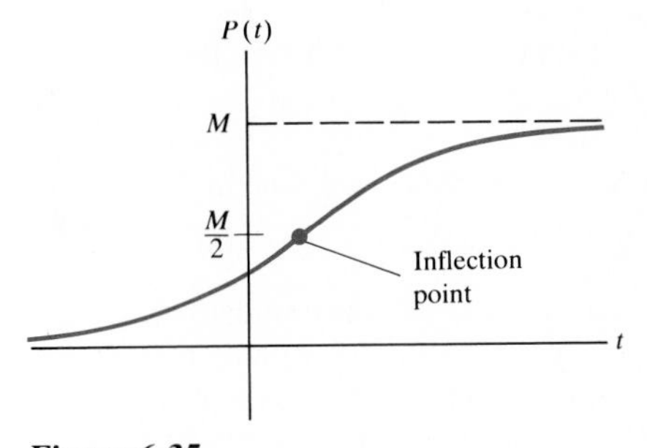

Figure 6.35

approaches 0 and acts as a damper in Eq. (6), decreasing the rate of change dP/dt.

Growth satisfying Eq. (6) is called **logistic**, **inhibited**, or **sigmoidal** (the last because its graph, shown in Fig. 6.35, is S-shaped). It was first proposed in the 1840s by P. F. Verhulst.

In 1913, Tor Carlson plotted the growth of brewer's yeast in a test tube and found that the curve fit the logistic model. Since then, many cases of growth have been found to fit that model.

By Example 4 in the preceding section,

$$P(t) = \frac{M}{1 + ae^{-kt}}. \tag{7}$$

The logistic curve is symmetric with respect to its inflection point.

Note that as $t \to \infty$, $e^{-kt} \to 0$, so $P(t) \to M$, as desired. The logistic curve has an inflection point at height $M/2$ (see Exercise 21). Therefore, if experimental data indicated an inflection point at a height h, then the ultimate population limit could be conjectured as $2h$.

EXAMPLE 4 What is the significance of the constant a in Eq. (7)?

SOLUTION Setting $t = 0$ in Eq. (7) yields

$$P(0) = \frac{M}{1 + a}.$$

Thus

$$(1 + a)P(0) = M,$$

$$1 + a = \frac{M}{P(0)},$$

and

$$a = \frac{M}{P(0)} - 1.$$

The logistic growth function expressed in terms of initial and ultimate populations

The constant a is one less than the ratio between the limiting population M and the initial population $P(0)$. Thus

$$P(t) = \frac{M}{1 + \{[M/P(0)] - 1\}e^{-kt}}. \quad ■$$

EXERCISES FOR SEC. 6.8: NATURAL AND INHIBITED GROWTH

1 Use differentials to show that for k small, (*a*) $\ln(1 + k) \approx k$ and (*b*) $e^k \approx 1 + k$.

2 When the relative growth rate r is small, estimate (*a*) t_2 for $r > 0$ and (*b*) $t_{1/2}$ for $r < 0$.

3 A substance is decaying at the rate of 2 percent per year. (*a*) What is b? (*b*) Find k to four decimal places.

4 A population is growing at the rate of 2 percent per year. (*a*) What is b? (*b*) Find k to four decimal places.

5 The amount of a certain growing substance increases at the rate of 10 percent per hour. Find (*a*) r, (*b*) b, (*c*) k, (*d*) t_2.

6 A quantity is increasing according to the law of natural growth. The amount present at time $t = 0$ is A. It will double when $t = 10$.
(*a*) Express the amount in the form Ae^{kt} for suitable k.
(*b*) Express the amount in the form Ab^t for suitable b.

7 The mass of a certain bacterial culture after t hours is $10 \cdot 3^t$ grams.
(*a*) What is the initial amount?
(*b*) What is the growth constant k?
(*c*) What is the percent increase in any period of 1 hour?

8 Let $f(t) = 3 \cdot 2^t$.
(*a*) Solve the equation $f(t) = 12$.
(*b*) Solve the equation $f(t) = 5$.
(*c*) Find k such that $f(t) = 3e^{kt}$.

9 In 1985 the world population was about 4.9 billion and was increasing at the rate of 1.8 percent per year. If it continues to grow at that rate, when will it (*a*) double? (*b*) quadruple? (*c*) reach 100 billion?

10 The population of Latin America has a doubling time of 27 years. Estimate the percent it grows per year.

11 In 1986 the United States population was 241 million and increasing at the rate of 0.9 percent per year and the population of Mexico was 80 million and increasing at the rate of 3.0 percent per year. When will the two populations be the same size if they continue to grow at the same rates?

12 A bacterial culture grows from 100 to 400 grams in 10 hours according to the law of natural growth.
(*a*) How much was present after 3 hours?
(*b*) How long will it take the mass to double? quadruple? triple?

13 At 1 P.M. a bacterial culture weighed 100 grams. At 4:30 P.M. it weighed 250 grams. Assuming that it grows at a rate proportional to the amount present, find (*a*) at what time it will grow to 400 grams, (*b*) its growth constant, and (*c*) the growth rate per hour.

14 The population of the Soviet Union in 1986 was 280 million and growing at the relative rate of 0.8 percent per year and the population of the United States was 241 million and growing at the relative rate of 0.9 percent per year. When will the population of the United States equal that of the Soviet Union?

15 See Example 2. How much carbon 14 remains after (*a*) 11,460 years? (*b*) 2,000 years?

16 The half-life of radium is about 1,600 years.
(*a*) From this, find k in the expression of Ae^{kt}.
(*b*) How long does it take 75 percent of the radium to disintegrate?
(*c*) Solve (*b*) without using calculus.
(*d*) How long will it take for 90 percent of the radium to disintegrate?
(*e*) Without calculus, show that the answer to (*d*) is between 4,800 and 6,400 years.

17 A bacterial culture grows at a rate proportional to the amount present. From 9 to 11 A.M. it increases from 100 to 200 grams. What will be its weight at noon?

18 A disintegrating radioactive substance decreases from 12 to 11 grams in 1 day. Find its half-life.

19 A radioactive substance disintegrates at the instantaneous rate of 0.05 gram per day when its mass is 10 grams.
(*a*) How much of the substance will remain after t days if the initial amount is A?
(*b*) What is its half-life?

20 If the carbon 14 concentration in the carbon from a plant or piece of wood of unknown age is half that of the carbon 14 concentration in a present-day live specimen, then it is about 5,730 years old. Show that if A_c and A_u are the radioactivities of samples prepared from contemporary and from undated materials, respectively, then the age of the undated material is about $t = 8{,}300 \ln (A_c/A_u)$. (This method is dependable up to an age of about 70,000 years.) See Radiocarbon Dating in *Encyclopedia of Science and Technology,* 5th ed., McGraw-Hill, New York, 1982.

■

21 Show that at the inflection point of the logistic curve the y coordinate is $M/2$. Use Eq. (6) and differentiate.

22 (Doomsday equation) A differential equation of the form $dP/dt = kP^{1.01}$ is called a **doomsday equation**. The rate of growth is just slightly higher than that for natural growth.
(*a*) Solve the equation.
(*b*) Show that there is a finite number t_1 such that $\lim_{t \to t_1^-} P(t) = \infty$. Naturally, t_1 is doomsday.

23 Figure 6.36 has a logarithmic vertical axis. Since only one axis is logarithmic, this coordinate system is called **semilogarithmic**.

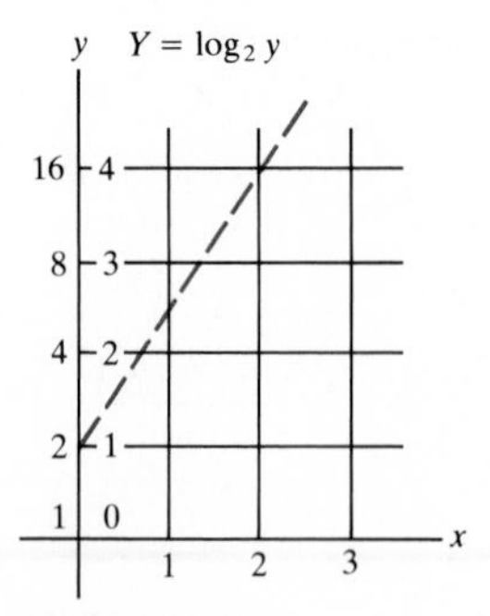

Figure 6.36

(*a*) By inspection of the graph, find Y as a function of x.
(*b*) From (*a*) find y as a function of x.
(*c*) When you graph $y = Ae^{kt}$ on semilog paper, what kind of curve results?

24 Figure 6.37 shows semilogarithmic graph paper. Graph $y = 3 \cdot 2^x$ on it. (For clarity, some horizontal lines corresponding to integer coordinates are omitted.)

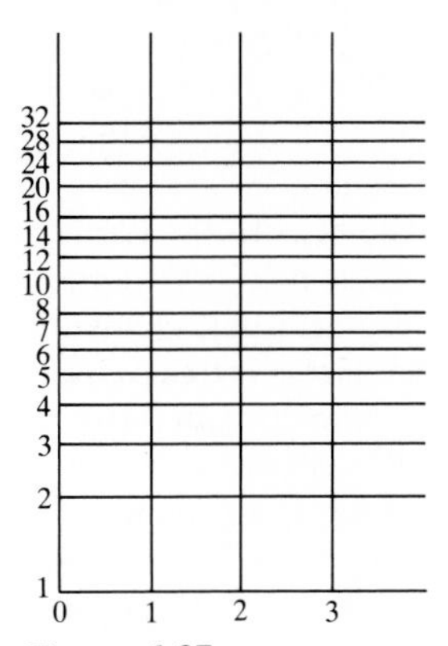

Figure 6.37

25 Semilogarithmic graph paper is available in bookstores where engineering forms are sold. Figure 6.38 is part of one such sheet, which is $8\frac{1}{2}$ by 11 inches. For this exercise either copy Fig. 6.38 or buy a piece of semilog paper.

The population of the world in 1950 was 2.5 billion; in 1970, 3.7 billion; in 1980, 4.5 billion.
(*a*) Plot these data on semilog paper like that in Fig. 6.38. (They should lie on a straight line.)
(*b*) Draw a line through the three points in (*a*) and use it to estimate the population in the year 2000.
(*c*) Using the line drawn in (*b*), estimate when the world population will reach 10 billion.

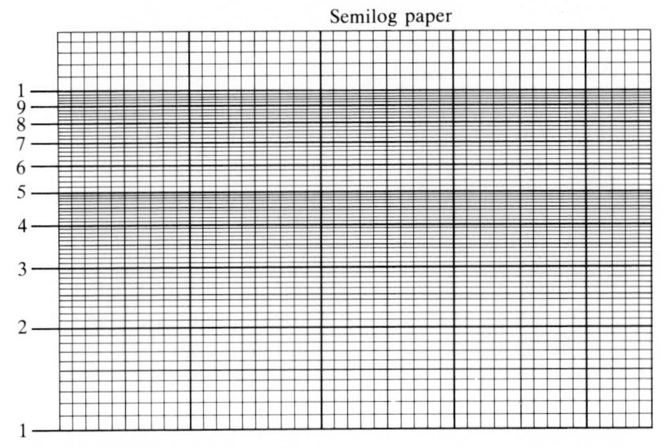

Figure 6.38

26 Find all functions $f(x)$ such that $f(x) = 3\int_0^x f(t)\,dt$.

27 Let $I(x)$ be the intensity of sunlight at a depth of x meters in the ocean. As x increases, $I(x)$ decreases.
(*a*) Why is it reasonable to assume that there is a constant k (negative) such that $\Delta I \approx kI(x)\,\Delta x$ for small Δx?
(*b*) Deduce that $I(x) = I(0)e^{kx}$, where $I(0)$ is the intensity of sunlight at the surface. Incidentally, sunlight at a depth of 1 meter is only $\frac{1}{4}$ as intense as at the surface.

28 A salesman, trying to persuade a tycoon to invest in Standard Coagulated Mutual Fund, shows him the accompanying graph which records the value of a similar investment made in the fund in 1950. "Look! In the first 5 years the investment increased \$1,000," the salesman observed, "but in the past 5 years it increased by \$2,000. It's really improving. Look at the slope of the graph from 1970 to 1975, which you can see clearly in Fig. 6.39."

The tycoon replied, "Hogwash; in order to present an unbiased graph, you should use semilog paper. Though your graph is steeper from 1970 to 1975, in fact, the rate of return is less than from 1950 to 1955. Indeed, that was your best period."
(*a*) If the percentage return on the accumulated investment remains the same over each 5-year period as the first 5-year period, sketch the graph.
(*b*) Explain the tycoon's reasoning.

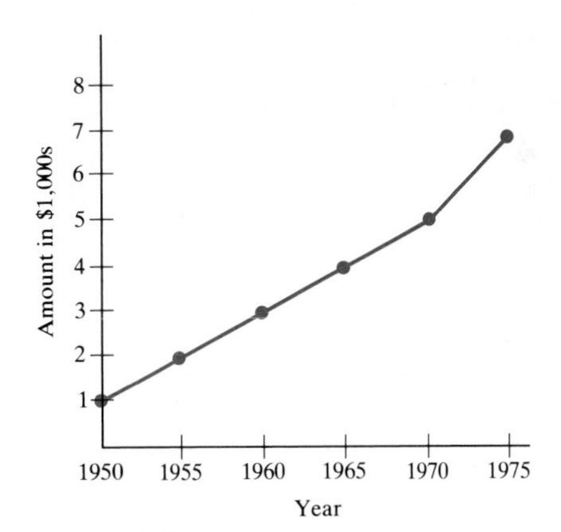

Figure 6.39

29 A certain fish increases in number at a rate proportional to the size of the population. In addition, it is being harvested at a constant rate. Let $P(t)$ be the size of the fish population at time t.
(*a*) Show that there are positive constants h and k such that for small Δt, $\Delta P \approx kP\,\Delta t - h\,\Delta t$.
(*b*) Find a formula for $P(t)$ in terms of $P(0)$, h, and k.

30 This newspaper article illustrates the rapidity of exponential growth:

U.S. HIT FOR 200-YR. DEBT

LAS VEGAS—An autograph dealer is demanding the U.S. government pay off a 200-year-old note. At seven percent interest, the debt amounts to $14 billion.

It was issued to Haym Salomon on March 27, 1782, by finance chief Robert Morse in return for a $30,000 loan.

The note is payable to the bearer—but the statute of limitations ran out on it some 150 years ago.

(*a*) Is the figure of $14 billion correct?
(*b*) What interest rate would be required to produce an account of $14 billion if interest were compounded once a year?
(*c*) Answer (*b*) for continuous compounding.

31 From a newspaper article:

'Rule of 72'

I've been hearing bankers and investment advisers talk about something called the "rule of 72." Could you explain what it means? — B. H.

How quickly would you like to double your money? That's what the "rule of 72" will tell you. To find out how fast your money will double at any given interest rate or yield, simply divide that yield into 72. This will tell you how many years doubling will take.

Let's say you have a long-term certificate of deposit paying 12 percent. At that rate your money would double in six years. A money-market fund paying 10 percent would take 7.2 years to double your investment.

What is the exact rule for computing doubling time if interest is added (*a*) once a year? (*b*) continuously?

32 The following situations are all mathematically the same:
(1) A drug is administered in a dose of A grams to a patient and gradually leaves the system through excretion.
(2) Initially there is an amount A of smoke in a room. The air conditioner is turned on and gradually the smoke is removed. (Assume that the smoke is always thoroughly mixed.)
(3) Initially there is an amount A of some pollutant in a lake, when further dumping of toxic materials is prohibited. The rate at which water enters the lake equals the rate at which it leaves. (Assume the pollutant is thoroughly mixed.)

In each case, let $P(t)$ be the initial amount present (whether drug, smoke, or pollutant).
(*a*) Why is it reasonable to assume that there is a constant k such that for small intervals of time, Δt, $\Delta P \approx kP(t)\Delta t$?
(*b*) From (*a*) deduce that $P(t) = Ae^{kt}$.
(*c*) Is k positive or negative?

Exercises 33 to 37 are based on D. L. Meadows, D. H. Meadows, J. Randers, and W. W. Behrens III, *The Limits to Growth*, Universe Books, New York, 1972. All the time intervals are approximations.

33 Let $Y(t)$ be the amount of some natural resource that is consumed from time $t = 0$ to time $t > 0$. For our purposes, $t = 0$ corresponds to the year 1970. Let $c(t)$ be the rate at which that resource is being consumed. Since population and industry are growing, $c(t)$ is an increasing function. Assume that $c(t)$ is increasing exponentially; that is, there are constants A and k such that $c(t) = Ae^{kt}$. Show that $Y(t) = A(e^{kt} - 1)/k$.

34 The amounts of such natural resources as aluminum, natural gas, and petroleum are finite. The formula obtained in Exercise 33 enables one to estimate how long a given resource will last subject to various assumptions.

Let R be the amount of a given resource remaining at time $t = 0$.
(*a*) Show that if the rate of consumption remains constant, then the resource will last R/A years. This is called the **static index** and is denoted by s.
(*b*) Show that if the rate of consumption continues to grow exponentially, the resource will last

$$\frac{\ln(ks + 1)}{k} \quad \text{years.}$$

This is called the **exponential index**. The letter s denotes the static index, defined in (*a*).

35 The static index of aluminum is 100 years (known global reserves are 1.17×10^9 tons). The projected growth rate for consumption is 6.4 percent a year. Thus $c(t) = A(1.064)^t$. However, $e^{0.064}$ is a very good approximation to 1.064 and is used in the cited reference to obtain the formula $c(t) = Ae^{0.064t}$.
(*a*) Show that the exponential index is 31 years (approximately). Show that if the global reserves are five times as large as quoted, the static index will be 500 years and the exponential index will be 55 years.
(*b*) Show that if $k = 0.032$ instead of 0.064, the static index is unaffected but the exponential index will be 45 years. (Assume known global reserves.)

36 The known global reserves of natural gas are 1.14×10^{15} cubic feet. The static index is 38 years. If consumption continues to increase at 4.7 percent per year, show that the exponential index (*a*) is 22 years, (*b*) is about 50 years if reserves are 5 times as large.

37 The known global reserves of petroleum are 455×10^8 barrels. The static index is 31 years. If the rate of consumption continues to increase at 3.9 percent per year, show that the exponential index is 20 years and, if reserves are five times as large, about 51 years.

38 (*a*) Assume that k_1 and k_2 are constants and that y_1 and y_2 are functions such that $dy_1/dx = k_1y_1$ and $dy_2/dx = k_2y_2$. Without solving for y_1 and y_2 explicitly, show that $d(y_1y_2)/dx = (k_1 + k_2)y_1y_2$.

(*b*) If energy consumption per person in the United States increases at the rate of 1 percent per year and the population grows at the rate of 0.9 percent per year, at what rate does the energy consumption in the nation increase? *Hint:* The answer is not 1.9 percent.

39 If the population of the Western Hemisphere is growing exponentially and the population of the Eastern Hemisphere is growing exponentially, does it follow that the population of the world is growing exponentially?

6.9 L'HÔPITAL'S RULE

Pronounced "Low-pee-tal's rule"

The problem of finding a limit has arisen in graphing a curve and will appear often in later chapters. Fortunately, there are some general techniques for computing a wide variety of limits. This section discusses one of the most important of these methods, l'Hôpital's rule, which concerns the limit of a quotient of two functions.

If f and g are functions and a is a number such that

$$\lim_{x \to a} f(x) = 2 \quad \text{and} \quad \lim_{x \to a} g(x) = 3,$$

then

$$\lim_{x \to a} \frac{f(x)}{g(x)} = \frac{2}{3}.$$

This problem presents no difficulty; no more information is needed about the functions f and g. But if

$$\lim_{x \to a} f(x) = 0 \quad \text{and} \quad \lim_{x \to a} g(x) = 0,$$

then finding

$$\lim_{x \to a} \frac{f(x)}{g(x)}$$

may present a serious problem because the rule for the limit of a quotient stated in Sec. 2.2 does not apply when $\lim_{x \to a} g(x) = 0$. For instance, Sec. 2.4 was dedicated to showing that

$$\lim_{\theta \to 0} \frac{\sin \theta}{\theta} = 1.$$

Here $f(\theta) = \sin \theta \to 0$ and $g(\theta) = \theta \to 0$ as $\theta \to 0$. The quotient $f(\theta)/g(\theta)$ approaches 1. In that same section it was proved that

$$\lim_{\theta \to 0} \frac{1 - \cos \theta}{\theta} = 0.$$

In this second limit, the numerator rushes toward 0 so much faster than the denominator that the quotient approaches 0. These two examples serve to point out that if you know *only* that

$$\lim_{x \to a} f(x) = 0 \quad \text{and} \quad \lim_{x \to a} g(x) = 0,$$

then you do not have enough information to determine

$$\lim_{x \to a} \frac{f(x)}{g(x)}.$$

In such cases, simply plugging in the limits of the numerator and denominator produces $\frac{0}{0}$, a meaningless expression traditionally called an "indeterminate form."

Theorem 1 describes a general technique for dealing with the troublesome quotient

$$\frac{f(x)}{g(x)}$$

when $\qquad f(x) \to 0 \quad$ and $\quad g(x) \to 0.$

It is known as the **zero-over-zero case** of l'Hôpital's rule.

Theorem 1 *L'Hôpital's rule (zero-over-zero case).* Let a be a number and let f and g be differentiable over some open interval (a, b). Assume also that $g'(x)$ is not 0 for any x in that interval. If

$$\lim_{x \to a^+} f(x) = 0, \qquad \lim_{x \to a^+} g(x) = 0,$$

and

$$\lim_{x \to a^+} \frac{f'(x)}{g'(x)} = L,$$

then

$$\lim_{x \to a^+} \frac{f(x)}{g(x)} = L. \quad ■$$

Similar rules hold for $x \to a^-$, $x \to a$, $x \to \infty$, and $x \to -\infty$. Of course, corresponding changes must be made in the hypotheses.

Before worrying about *why* this theorem is true, we illustrate its use by an example.

EXAMPLE 1 Find $\displaystyle\lim_{x \to 1^+} \frac{x^5 - 1}{x^3 - 1}$.

SOLUTION In this case,

$$a = 1, \qquad f(x) = x^5 - 1, \qquad \text{and} \qquad g(x) = x^3 - 1.$$

All the assumptions of l'Hôpital's rule are satisfied. In particular,

$$\lim_{x \to 1^+} (x^5 - 1) = 0 \qquad \text{and} \qquad \lim_{x \to 1^+} (x^3 - 1) = 0.$$

According to l'Hôpital's rule,

$$\lim_{x \to 1^+} \frac{x^5 - 1}{x^3 - 1} = \lim_{x \to 1^+} \frac{(x^5 - 1)'}{(x^3 - 1)'},$$

if the latter limit exists. Now,

We do not differentiate the quotient; we differentiate the numerator and denominator separately.

$$\lim_{x \to 1^+} \frac{(x^5 - 1)'}{(x^3 - 1)'} = \lim_{x \to 1^+} \frac{5x^4}{3x^2} \qquad \text{differentiation of numerator and denominator}$$

$$= \lim_{x \to 1^+} \tfrac{5}{3}x^2 \qquad \text{algebra}$$

$$= \tfrac{5}{3}.$$

This can also be solved by factoring $x^5 - 1$ and $x^3 - 1$.

Thus

$$\lim_{x \to 1^+} \frac{x^5 - 1}{x^3 - 1} = \frac{5}{3}. \quad ■$$

A complete proof of Theorem 1 may be found in Exercises 177 and 178 of Sec. 6.S. Let us pause long enough here to make the theorem plausible.

Argument for a special case of Theorem 1

To do so, consider the *special case* where f, f', g, and g' are all continuous throughout an open interval containing a. Assume that $g'(x) \neq 0$ throughout the interval. Since we have $\lim_{x\to a^+} f(x) = 0$ and $\lim_{x\to a^+} g(x) = 0$, it follows by continuity that $f(a) = 0$ and $g(a) = 0$. Now assume that $\lim f'(x)/g'(x) = L$. Then

$$\begin{aligned}
\lim_{x\to a^+} \frac{f(x)}{g(x)} &= \lim_{x\to a^+} \frac{f(x) - f(a)}{g(x) - g(a)} && \text{since } f(a) = 0 \text{ and } g(a) = 0 \\
&= \lim_{x\to a^+} \frac{\dfrac{f(x) - f(a)}{x - a}}{\dfrac{g(x) - g(a)}{x - a}} && \text{algebra} \\
&= \frac{\lim\limits_{x\to a^+} \dfrac{f(x) - f(a)}{x - a}}{\lim\limits_{x\to a^+} \dfrac{g(x) - g(a)}{x - a}} && \text{limit of quotients} \\
&= \frac{f'(a)}{g'(a)} && \text{by definition of } f'(a) \text{ and } g'(a) \\
&= \frac{\lim\limits_{x\to a^+} f'(x)}{\lim\limits_{x\to a^+} g'(x)} && f' \text{ and } g' \text{ are continuous} \\
&= \lim_{x\to a^+} \frac{f'(x)}{g'(x)} && \text{``limit of quotient'' property} \\
&= L && \text{by assumption.}
\end{aligned}$$

Consequently,

$$\lim_{x\to a^+} \frac{f(x)}{g(x)} = L.$$

Sometimes it may be necessary to apply l'Hôpital's rule more than once, as in the next example.

EXAMPLE 2 Find $\lim\limits_{x\to 0} \dfrac{\sin x - x}{x^3}$.

SOLUTION As $x \to 0$, both numerator and denominator approach 0. By l'Hôpital's rule,

$$\lim_{x\to 0} \frac{\sin x - x}{x^3} = \lim_{x\to 0} \frac{\cos x - 1}{3x^2}.$$

Repeated application of l'Hôpital's rule

But as $x \to 0$, both $\cos x - 1 \to 0$ and $3x^2 \to 0$. So use l'Hôpital's rule again:

$$\lim_{x\to 0} \frac{\cos x - 1}{3x^2} = \lim_{x\to 0} \frac{-\sin x}{6x}.$$

Or recall from Sec. 2.4 that $\lim\limits_{x\to 0} \dfrac{\sin x}{x} = 1.$

Both $\sin x$ and $6x$ approach 0 as $x \to 0$. Use l'Hôpital's rule yet another time:

$$\lim_{x\to 0} \frac{-\sin x}{6x} = \lim_{x\to 0} \frac{-\cos x}{6} = -\tfrac{1}{6}.$$

So after three applications of l'Hôpital's rule we find that

$$\lim_{x \to 0} \frac{\sin x - x}{x^3} = -\frac{1}{6}. \quad \blacksquare$$

Sometimes a limit may be simplified before l'Hôpital's rule is applied. For instance, consider

$$\lim_{x \to 0} \frac{(\sin x - x) \cos^5 x}{x^3}.$$

Since $\lim_{x \to 0} \cos^5 x = 1$, we have

$$\lim_{x \to 0} \frac{(\sin x - x) \cos^5 x}{x^3} = \left(\lim_{x \to 0} \frac{\sin x - x}{x^3} \right) \cdot 1,$$

which, by Example 2, is $-\frac{1}{6}$. This shortcut saves a lot of work, as may be checked by finding the limit using l'Hôpital's rule without separating $\lim_{x \to 0} \cos^5 x$.

Theorem 1 concerns the problem of finding the limit of $f(x)/g(x)$ when both $f(x)$ and $g(x)$ approach 0, the zero-over-zero case of l'Hôpital's rule. But a similar problem arises when both $f(x)$ and $g(x)$ get arbitrarily large as $x \to a$ or as $x \to \infty$. The behavior of the quotient $f(x)/g(x)$ will be influenced by how rapidly $f(x)$ and $g(x)$ become large.

For example,

$$\begin{aligned} \lim_{x \to \infty} \frac{x^2 + 5x + 2}{x^3 + x - 1} &= \lim_{x \to \infty} \frac{x^2[1 + (5/x) + (2/x^2)]}{x^3[1 + (1/x^2) - (1/x^3)]} \\ &= \lim_{x \to \infty} \frac{1}{x} \left[\frac{1 + (5/x) + (2/x^2)}{1 + (1/x^2) - (1/x^3)} \right] \\ &= 0 \cdot 1 = 0. \end{aligned}$$

On the other hand,

$$\begin{aligned} \lim_{x \to \infty} \frac{4x + 1}{2x} &= \lim_{x \to \infty} \left(2 + \frac{1}{2x} \right) \\ &= 2. \end{aligned}$$

In this case, the numerator is increasing about twice as rapidly as the denominator.

Another indeterminate limit

The next theorem presents a form of l'Hôpital's rule that covers the case in which $f(x) \to \infty$ and $g(x) \to \infty$. It is called the **infinity-over-infinity case** of l'Hôpital's rule.

Theorem 2 *L'Hôpital's rule (infinity-over-infinity case).* Let f and g be defined and differentiable for all x larger than some fixed number. Then, if

L'Hôpital's rule for the infinity-over-infinity case

$$\lim_{x \to \infty} f(x) = \infty, \qquad \lim_{x \to \infty} g(x) = \infty, \qquad \text{and} \qquad \lim_{x \to \infty} \frac{f'(x)}{g'(x)} = L,$$

it follows that

$$\lim_{x \to \infty} \frac{f(x)}{g(x)} = L.$$

A similar result holds for $x \to a$, $x \to a^-$, $x \to a^+$, or $x \to -\infty$. Moreover,

$\lim_{x\to\infty} f(x)$ and $\lim_{x\to\infty} g(x)$ could both be $-\infty$, or one could be ∞ and the other $-\infty$. ■

Figure 6.40

The proof of this is left to an advanced calculus course. However, it is easy to see why it is plausible. Imagine that $f(t)$ and $g(t)$ describe the locations on the x axis of two cars at time t. Call the cars the f-car and the g-car. See Fig. 6.40. Their velocities are therefore $f'(t)$ and $g'(t)$. These two cars are on endless journeys. But let us assume that as time $t \to \infty$ the f-car tends to travel at a speed closer and closer to L times the speed of the g-car. That is, assume that

$$\lim_{t\to\infty} \frac{f'(t)}{g'(t)} = L.$$

No matter how the two cars move in the short run, it seems reasonable that in the *long run* the f-car will tend to travel about L times as far as the g-car; that is,

$$\lim_{t\to\infty} \frac{f(t)}{g(t)} = L.$$

Although not a proof, this argument does provide a perspective for a rigorous proof.

EXAMPLE 3 Use l'Hôpital's rule to find $\lim_{x\to\infty} \dfrac{x}{e^x}$.

SOLUTION By Theorem 2,

$$\lim_{x\to\infty} \frac{x}{e^x} = \lim_{x\to\infty} \frac{x'}{(e^x)'}$$

$$= \lim_{x\to\infty} \frac{1}{e^x} = 0. \quad ■$$

EXAMPLE 4 Find $\lim_{x\to\infty} \dfrac{x^3}{2^x}$.

SOLUTION Since both numerator and denominator approach ∞ as $x \to \infty$, l'Hôpital's rule may be applied. It asserts that

$$\lim_{x\to\infty} \frac{x^3}{2^x} = \lim_{x\to\infty} \frac{3x^2}{2^x \ln 2} \quad \text{if the latter limit exists}$$

$$= \lim_{x\to\infty} \frac{6x}{2^x (\ln 2)^2} \quad \text{using the rule again}$$

$$= \lim_{x\to\infty} \frac{6}{2^x (\ln 2)^3} \quad \text{again l'Hôpital's rule}$$

$$= 0.$$

Thus as $x \to \infty$, 2^x grows much faster than x^3. ■

As in Example 4, l'Hôpital's rule in the infinity-over-infinity case can be used to show that for $a > 0$ and $b > 1$, $\lim_{x\to\infty} x^a/b^x = 0$. (This limit was mentioned in Sec. 6.4.) Similarly, it can be used to show that $\lim_{x\to\infty} (\log_b x)/x^a = 0$, a limit discussed in Sec. 6.1.

The next example conveys a warning.

EXAMPLE 5 Find

$$\lim_{x\to\infty} \frac{x - \cos x}{x}. \tag{1}$$

SOLUTION Both numerator and denominator approach ∞ as $x \to \infty$. Trying l'Hôpital's rule, we obtain

$$\lim_{x\to\infty} \frac{x - \cos x}{x} = \lim_{x\to\infty} \frac{1 + \sin x}{1}.$$

But $\lim_{x\to\infty} (1 + \sin x)$ does not exist.

L'Hôpital's rule may fail to provide an answer.

What can we conclude about (1)? Nothing at all. L'Hôpital's rule says that if $\lim_{x\to\infty} f'/g'$ exists, then $\lim_{x\to\infty} f/g$ exists. It says nothing about the case when $\lim_{x\to\infty} f'/g'$ does not exist.

It is not difficult to evaluate (1) directly, as follows:

$$\begin{aligned} \lim_{x\to\infty} \frac{x - \cos x}{x} &= \lim_{x\to\infty} \left(1 - \frac{\cos x}{x}\right) && \text{algebra} \\ &= 1 - 0 && \text{since } |\cos x| \le 1 \\ &= 1. \quad \blacksquare \end{aligned}$$

Transforming Some Limits So L'Hôpital's Rule Applies Many limit problems can be transformed to limits to which l'Hôpital's rule applies. For instance, the problem of finding

The zero-times-infinity case

$$\lim_{x\to 0^+} x \ln x$$

does not seem to be related to l'Hôpital's rule, since it does not involve the quotient of two functions. As $x \to 0^+$, one factor, x, approaches 0 and the other factor, $\ln x$, approaches $-\infty$. It is not obvious how their product, $x \ln x$, behaves as $x \to 0^+$. But a little algebraic manipulation transforms it into a problem to which l'Hôpital's rule applies, as the next example shows.

EXAMPLE 6 Find $\lim_{x\to 0^+} x \ln x$.

SOLUTION Rewrite $x \ln x$ as a quotient, $(\ln x)/(1/x)$. Let $f(x) = \ln x$ and $g(x) = 1/x$. Note that

$$\lim_{x\to 0^+} \ln x = -\infty \quad \text{and} \quad \lim_{x\to 0^+} \frac{1}{x} = \infty.$$

A case of Theorem 2, with $x \to 0^+$, asserts that

$$\begin{aligned} \lim_{x\to 0^+} \frac{\ln x}{1/x} &= \lim_{x\to 0^+} \frac{1/x}{-1/x^2} \\ &= \lim_{x\to 0} (-x) \\ &= 0. \end{aligned}$$

Thus $$\lim_{x \to 0^+} \frac{\ln x}{1/x} = 0,$$

from which it follows that $\lim_{x \to 0^+} x \ln x = 0$. (The factor x, which approaches 0, dominates the factor $\ln x$, which gets arbitrarily large in absolute value.) ■

The remaining examples illustrate other limits that can be found by first relating them to limit problems to which l'Hôpital's rule applies.

Try this on your calculator first.

EXAMPLE 7 Find $\lim_{x \to 0^+} x^x$.

The zero-to-the-zero case

SOLUTION Since this limit involves an exponential, not a quotient, it does not fit directly into l'Hôpital's rule. But a little algebraic manipulation will change the problem to one covered by l'Hôpital's rule.

Let $$y = x^x.$$

Then $$\ln y = \ln x^x = x \ln x.$$

By Example 6,

$$\lim_{x \to 0^+} \ln y = 0.$$

Thus $$\lim_{x \to 0^+} y = \lim_{x \to 0} e^{\ln y}.$$

If $\ln y \to 0$, *then* $y \to 1$.

Since $\ln y \to 0$ as $x \to 0^+$, $e^{\ln y} \to 1$ as $x \to 0^+$. (This depends on the continuity of e^x. In short, $\lim_{x \to 0^+} x^x = 1$. ■

The following table shows how some limits not of the zero-over-zero or infinity-over-infinity forms can be brought into those forms:

Form	*Name*	*Method*
$f(x)g(x)$; $f(x) \to 0$, $g(x) \to \infty$	Zero-times-infinity	Rewrite as $\dfrac{g(x)}{1/f(x)}$.
$f(x)^{g(x)}$; $f(x) \to 1$, $g(x) \to \infty$ $f(x)^{g(x)}$; $f(x) \to 0$, $g(x) \to 0$	One-to-the-infinity Zero-to-the-zero	Write $y = f(x)^{g(x)}$; take $\ln y$, find limit of $\ln y$ and then the limit of $y = e^{\ln y}$.

EXERCISES FOR SEC. 6.9: L'HÔPITAL'S RULE

In Exercises 1 to 12 check that l'Hôpital's rule applies and use it to find the limits.

1 $\lim_{x\to 2} \frac{x^3 - 8}{x^2 - 4}$

2 $\lim_{x\to 1} \frac{x^7 - 1}{x^3 - 1}$

3 $\lim_{x\to 0} \frac{\sin 3x}{\sin 2x}$

4 $\lim_{x\to 0} \frac{\sin x^2}{(\sin x)^2}$

5 $\lim_{x\to\infty} \frac{x^3}{e^x}$

6 $\lim_{x\to\infty} \frac{x^5}{3^x}$

7 $\lim_{x\to 0} \frac{1 - \cos x}{x^2}$

8 $\lim_{x\to 0} \frac{\sin x - x}{(\sin x)^3}$

9 $\lim_{x\to 0} \frac{\tan 3x}{\ln (1 + x)}$

10 $\lim_{x\to 1} \frac{\cos (\pi x/2)}{\ln x}$

11 $\lim_{x\to\infty} \frac{(\ln x)^2}{x}$

12 $\lim_{x\to 0} \frac{\sin^{-1} x}{e^{2x} - 1}$

In each of Exercises 13 to 18 transform the problem into one to which l'Hôpital's rule applies; then find the limit.

13 $\lim_{x\to 0} (1 - 2x)^{1/x}$

14 $\lim_{x\to 0} (1 + \sin 2x)^{\csc x}$

15 $\lim_{x\to 0^+} (\sin x)^{(e^x - 1)}$

16 $\lim_{x\to 0^+} x^2 \ln x$

17 $\lim_{x\to 0^+} (\tan x)^{\tan 2x}$

18 $\lim_{x\to 0^+} (e^x - 1) \ln x$

In Exercises 19 to 46 find the limits. Use any method. *Warning:* l'Hôpital's rule, carelessly applied, may give a wrong answer or no answer.

19 $\lim_{x\to\infty} \frac{2^x}{3^x}$

20 $\lim_{x\to\infty} \frac{2^x + x}{3^x}$

21 $\lim_{x\to\infty} \frac{\log_2 x}{\log_3 x}$

22 $\lim_{x\to 1} \frac{\log_2 x}{\log_3 x}$

23 $\lim_{x\to\infty} \left(\frac{1}{x} - \frac{1}{\sin x}\right)$

24 $\lim_{x\to\infty} (\sqrt{x^2 + 3} - \sqrt{x^2 + 4x})$

25 $\lim_{x\to\infty} \frac{x^2 + 3 \cos 5x}{x^2 - 2 \sin 4x}$

26 $\lim_{x\to\infty} \frac{e^x - 1/x}{e^x + 1/x}$

27 $\lim_{x\to 0} \frac{3x^3 + x^2 - x}{5x^3 + x^2 + x}$

28 $\lim_{x\to\infty} \frac{3x^3 + x^2 - x}{5x^3 + x^2 + x}$

29 $\lim_{x\to\infty} \frac{\sin x}{4 + \sin x}$

30 $\lim_{x\to\infty} 5 \sin 3x$

31 $\lim_{x\to 1^+} (x - 1) \ln (x - 1)$

32 $\lim_{x\to\pi/2} \frac{\tan x}{x - (\pi/2)}$

33 $\lim_{x\to 0} (\cos x)^{1/x}$

34 $\lim_{x\to 0^+} x^{1/x}$

35 $\lim_{x\to\infty} \frac{\sin 2x}{\sin 3x}$

36 $\lim_{x\to 1} \frac{x^2 - 1}{x^3 - 1}$

37 $\lim_{x\to 0} \frac{xe^x(1 + x)^3}{e^x - 1}$

38 $\lim_{x\to 0} \frac{xe^x \cos^2 6x}{e^{2x} - 1}$

39 $\lim_{x\to 0} (\csc x - \cot x)$

40 $\lim_{x\to 0} \frac{\csc x - \cot x}{\sin x}$

41 $\lim_{x\to 0} \frac{5^x - 3^x}{\sin x}$

42 $\lim_{x\to 0} \frac{\tan^5 x - \tan^3 x}{1 - \cos x}$

43 $\lim_{x\to 2} \frac{x^3 + 8}{x^2 + 5}$

44 $\lim_{x\to\pi/4} \frac{\sin 5x}{\sin 3x}$

45 $\lim_{x\to 0} \left(\frac{1}{1 - \cos x} - \frac{2}{x^2}\right)$

46 $\lim_{x\to 0} \frac{\sin^{-1} x}{\tan^{-1} 2x}$

■

47 In Fig. 6.41 the unit circle is centered at O, BQ is a vertical tangent line, and the length of BP is the same as the length of BQ. What happens to the point E as $Q \to B$?

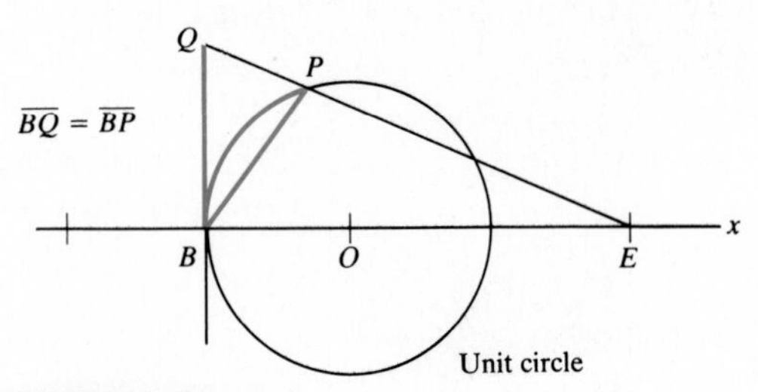

Figure 6.41

48 In Fig. 6.42 the unit circle is centered at the origin, BQ is a vertical tangent line, and the length of BQ is the same as the arc length $\widehat{BP}$. Prove that the x coordinate of R approaches -2 as $P \to B$.

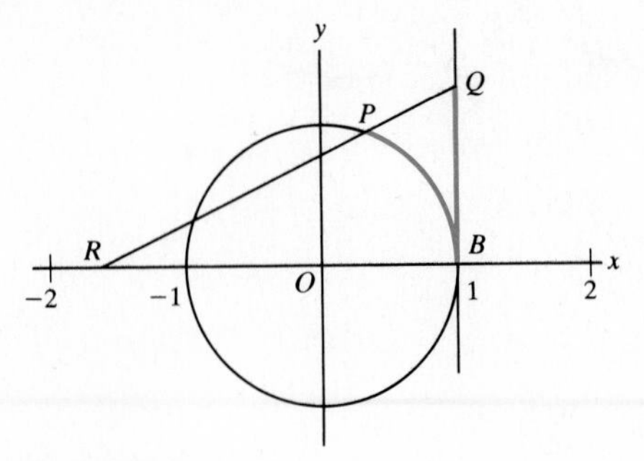

Figure 6.42

49 *Warning:* As Albert Einstein observed, "Common sense is the deposit of prejudice laid down in the mind before the age of 18." Exercise 20 of Sec. 2.4 asked the reader to guess a certain limit. Now that limit will be computed. In Fig. 6.43, which shows a circle, let $f(\theta)$ = area of triangle ABC and let $g(\theta)$ = area of the shaded region formed by deleting triangle OAC from the sector OBC. Clearly, $0 < f(\theta) < g(\theta)$.
(*a*) What would you guess is the value of $\lim_{\theta\to 0} f(\theta)/g(\theta)$?
(*b*) Find $\lim_{\theta\to 0} f(\theta)/g(\theta)$.

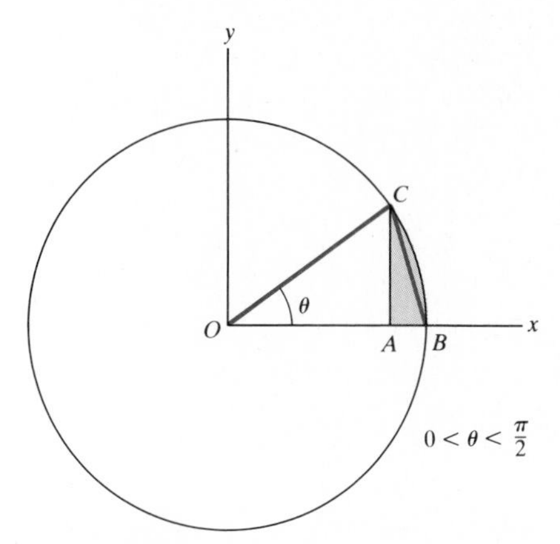

Figure 6.43

50 Figure 6.44 shows a triangle ABC and a shaded region cut from the parabola $y = x^2$ by a horizontal line. Find the limit, as $x \to 0$, of the ratio between the area of the triangle and the area of the shaded region.

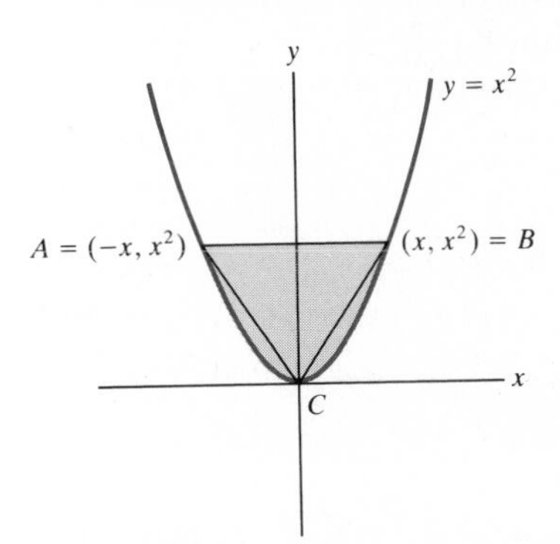

Figure 6.44

51 (*Economics*) In Eugene Silberberg, *The Structure of Economics,* McGraw-Hill, New York, 1978, this argument appears:

Consider the production function

$$y = k[\alpha x_1^{-\rho} + (1 - \alpha)x_2^{-\rho}]^{-1/\rho},$$

where k, α, x_1, x_2 are positive constants and $\alpha < 1$. Taking limits as $\rho \to 0^+$, we find that

$$\lim_{\rho\to 0^+} y = kx_1^{\alpha}x_2^{1-\alpha},$$

which is the Cobb-Douglas function, as expected.

Fill in the details.

52 Linus proposes this proof for Theorem 1: "Since

$$\lim_{x\to a^+} f(x) = 0 \qquad \text{and} \qquad \lim_{x\to a^+} g(x) = 0,$$

I will define $f(a) = 0$ and $g(a) = 0$. Next I consider $x > a$ but near a. I now have continuous functions f and g defined on the closed interval $[a, x]$ and differentiable on the open interval (a, x). So, using the mean-value theorem, I conclude that there is a number c, $a < c < x$, such that

$$\frac{f(x) - f(a)}{x - a} = f'(c) \qquad \text{and} \qquad \frac{g(x) - g(a)}{x - a} = g'(c).$$

Since $f(a) = 0$ and $g(a) = 0$, these equations tell me that

$$f(x) = (x - a)f'(c) \qquad \text{and} \qquad g(x) = (x - a)g'(c).$$

Thus
$$\frac{f(x)}{g(x)} = \frac{f'(c)}{g'(c)}.$$

Hence
$$\lim_{x\to a^+} \frac{f(x)}{g(x)} = \lim_{x\to a^+} \frac{f'(c)}{g'(c)} = L."$$

Alas, Linus made one error. What is it?

53 Find $\displaystyle\lim_{x\to 0} \left(\frac{1 + 2^x}{2}\right)^{1/x}$.

54 In R. P. Feynman, *Lectures on Physics,* Addison-Wesley, Reading, Mass., 1963, this remark appears: "Here is the quantitative answer of what is right instead of kT. This expression

$$\frac{\hbar\omega}{e^{\hbar\omega/kT} - 1}$$

should, of course, approach kT as $\omega \to 0$ . . . See if you can prove that it does—learn how to do the mathematics." Do the mathematics.

55 Show that for any polynomial $P(x)$ of degree at least one, $\lim_{x\to\infty} (\ln x)/P(x) = 0$.

56 Graph $y = (1 + x)^{1/x}$ for $x > -1$, $x \neq 0$, showing (*a*) where y is decreasing, (*b*) asymptotes, and (*c*) behavior of y for x near 0.

In Exercises 57 to 60 determine which of the following limits can be found from the given information about the functions. Give their values. Which cannot? In those cases, give examples that show the limits are not determined.

(*a*) $\lim_{x\to a} f(x)g(x)$ (*b*) $\lim_{x\to a} [f(x) + g(x)]$

(*c*) $\lim_{x\to a} \dfrac{f(x)}{g(x)}$ (*d*) $\lim_{x\to a} \dfrac{f(x) - 1}{g(x)}$

(*e*) $\lim_{x\to a} \dfrac{f(x) - 1}{g(x) - 1}$ (*f*) $\lim_{x\to a} [1 - f(x)]^{g(x)}$

(*g*) $\lim_{x\to a} [1 - f(x)]^{1-g(x)}$

57 $\lim_{x\to a} f(x) = 0$, $\lim_{x\to a} g(x) = 0$

58 $\lim_{x\to a} f(x) = 0$, $\lim_{x\to a} g(x) = 1$

59 $\lim_{x\to a} f(x) = 0$, $\lim_{x\to a} g(x) = \infty$

60 $\lim_{x\to a} f(x) = 1$, $\lim_{x\to a} g(x) = \infty$

61 (*a*) Recalling that $e = \lim_{x\to 0} (1 + x)^{1/x}$, what would you expect to be the value of

$$\lim_{x\to\infty} \frac{e^x}{\left[\left(1 + \dfrac{1}{x}\right)^x\right]^x} ?$$

(*b*) Find the limit in (*a*).

In Exercises 62 to 64 find the limits. Try l'Hôpital's rule first, but it may not help.

62 $\lim_{x\to 0} \dfrac{\int_1^{1+x} e^{t^2}\,dt}{\int_2^{2+x} e^{t^2}\,dt}$

63 $\lim_{x\to\infty} \dfrac{e^{\sqrt{x}}}{e^x}$

64 $\lim_{x\to\infty} \dfrac{x + \sqrt{x}}{x - \sqrt{x}}$

65 Using l'Hôpital's rule or otherwise, show that $\lim_{x\to\infty} x^a/b^x = 0$ for any a and for $b > 1$.

66 Using l'Hôpital's rule or otherwise, show that $\lim_{x\to\infty} (\log_b x)/x^a = 0$ for any base b and for $a > 0$.

67 Find $\lim_{x\to 0} [(\sin x)/x]^{1/x^2}$.

6.10 NEWTON'S METHOD FOR SOLVING AN EQUATION

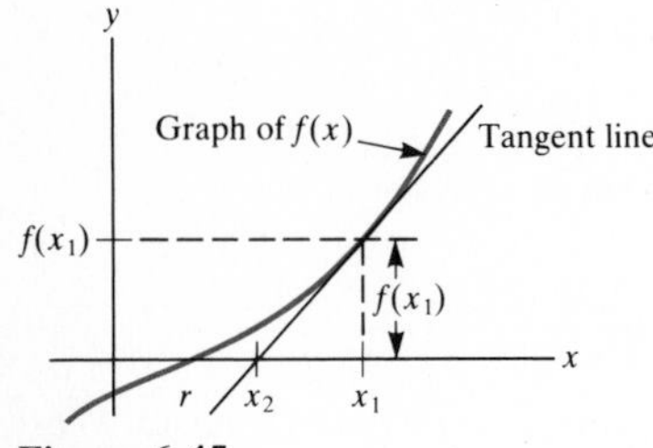

Figure 6.45

Suppose that we wish to estimate a solution (or root) r of an equation $f(x) = 0$. If a first guess is, say, x_1, then Fig. 6.45 suggests that a better estimate of r may be x_2, the point at which the tangent line at $(x_1, f(x_1))$ crosses the x axis. To find x_2 explicitly, observe that the slope of the tangent line at $(x_1, f(x_1))$ is $f'(x_1)$ and is also $f(x_1)/(x_1 - x_2)$. Thus

$$f'(x_1) = \frac{f(x_1)}{x_1 - x_2}.$$

Solving this equation for x_2 yields

Newton's recursive formula for estimating a root of $f(x) = 0$

$$x_2 = x_1 - \frac{f(x_1)}{f'(x_1)}, \tag{1}$$

which is meaningful if $f'(x_1)$ is not 0.

Equation (1) is the basis of Newton's method for estimating a root of an equation. Generally, it is applied several times to increase accuracy.

EXAMPLE 1 Use Newton's method to estimate the square root of 3, that is, the positive root of the equation $x^2 - 3 = 0$.

SOLUTION Here $f(x) = x^2 - 3$ and $f'(x) = 2x$. According to Eq. (1), if the first guess is x_1, then the next estimate x_2 should be

$$x_2 = x_1 - \frac{f(x_1)}{f'(x_1)} = x_1 - \frac{x_1{}^21 - 3}{2x_1} = \frac{x_1 + 3/x_1}{2}.$$

If $x_1 = 2$, say, then

$$x_2 = \frac{2 + \frac{3}{2}}{2} = 1.75.$$

For a better estimate of $\sqrt{3}$, repeat and use 1.75 instead of 2. Thus

$$x_3 = \frac{x_2 + 3/x_2}{2} = \frac{1.75 + 3/1.75}{2} \approx 1.73214$$

is the third estimate, to five decimals. One more repetition of the process yields (to five decimals) $x_4 \approx 1.73205$, which is quite close to $\sqrt{3}$, whose decimal expansion begins 1.7320508. ■

In fact, x_4 agrees with $\sqrt{3}$ to seven decimals.

Since the recursive process represented by Newton's method is of practical use and easily programmed, it is important to know under what circumstances $|x_i - r|$ approaches 0 as $i \to \infty$.

Several questions come to mind: Does Newton's method always work? That is, does it always produce a sequence that approaches a root? If not, why not? When does it surely work? And if it does, how rapidly does the sequence of estimates $x_1, x_2, x_3, \ldots$ approach the root r?

The method can fail because the function may have no root. (See Fig. 6.46.) Or there may be a root, but the behavior of the function at the first estimate x_1 causes the second estimate x_2 to be far from the root. (See Fig. 6.47.)

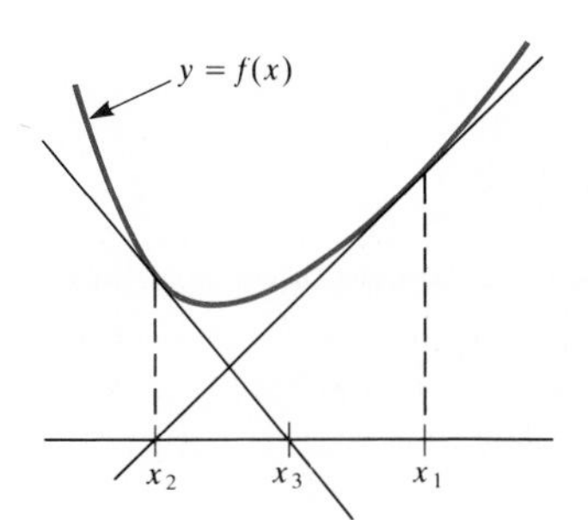

Figure 6.46

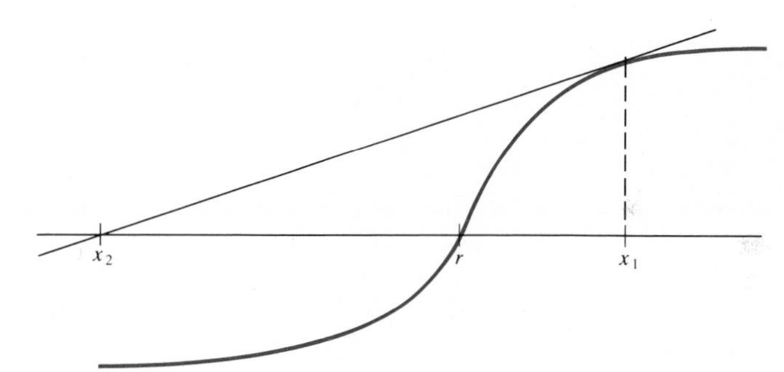

Figure 6.47

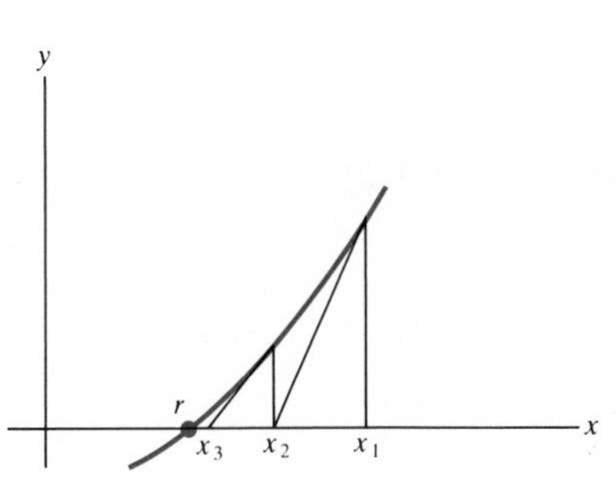

Figure 6.48

However, it appears to work in the situation shown in Fig. 6.48. (The successive estimates remain in the interval $[r, x_1]$ and get closer to r.) If the graph of f happens to be a nonhorizontal straight line, a quick sketch shows that no matter what choice of x_1 is made, the number x_2 is exactly the root r. In other words, if $f''(x)$ is identically 0, Newton's method is perfectly accurate. It is therefore reasonable to expect that the accuracy of Newton's method is influenced by $f''(x)$. [When $f''(x)$ is small, the method is probably more accurate.] On the other hand, if $f'(x_1)$ is near 0, the tangent line at $(x_1, f(x_1))$ is nearly horizontal and may depart a great deal from the graph of f by the time it crosses the x axis. Hence $f'(x)$ also should influence the accuracy. [When $f'(x)$ is large, the method is probably more accurate.]

The following theorem shows that, if $|f''(x)|$ is not too large nor $|f'(x)|$ too small, then $|x_i - r|$ does approach 0 as $i \to \infty$. Its proof is sketched in Exercise 22.

Theorem Let r be a root of $f(x) = 0$ and x_i an estimate of r such that $f'(x_i)$ is not 0. Let

$$x_{i+1} = x_i - \frac{f(x_i)}{f'(x_i)}.$$

If f' and f'' are continuous and M is a number such that

$$\left|\frac{f''(x)}{f'(t)}\right| \le M$$

for all x and t in the interval from x_i to r, then

$$|x_{i+1} - r| \le \frac{M}{2}|x_i - r|^2. \quad ■ \tag{2}$$

If $f''(x)$, $f'(x)$, and $f(x)$ are positive from $x = r$ to $x = x_1 > r$, then, as Fig. 6.48 shows, $x_1 > x_2 > x_3 > \cdots > r$. This means that the successive estimates $x_2, x_3, \ldots$ are situated between the initial estimate x_1 and the root r. Hence the hypotheses of the theorem apply to x_i for all i.

In this case, how swiftly does the decreasing sequence $x_1, x_2, x_3, \ldots$ approach r? Notice that if x_1 is close to r, say $|x_1 - r| \le 0.1$, then

$$|x_2 - r| \le \frac{M}{2}(0.1)^2 = \frac{M}{2}(0.01).$$

Thus, if M is not too large, x_2 is a much better approximation to r than is x_1. For instance, if $M = 2$, then

$$|x_2 - r| \le 0.01$$

and

$$|x_3 - r| \le \frac{M}{2}(x_2 - r)^2 \le 0.0001$$

and so on. Hence, if x_1 is an estimate of r accurate to one decimal place, then x_2 is accurate to two decimal places, x_3 is accurate to four decimal places, and so on. The number of decimal places of accuracy tends to double at each step of the Newton recursion. For instance, the Newton recursion formula for $\sqrt{10}$ (≈ 3.162278) is

$$x_{i+1} = \frac{x_i + 10/x_i}{2}.$$

The following table shows the results of the recursive process when the initial estimate x_1 is 3:

Step	*Estimate*	*Correct Digits*	*Number of Correct Decimal Digits*
1	$x_1 = 3$	3	0
2	$x_2 = 3.166667$	3.16	2
3	$x_3 = 3.162281$	3.1622	4

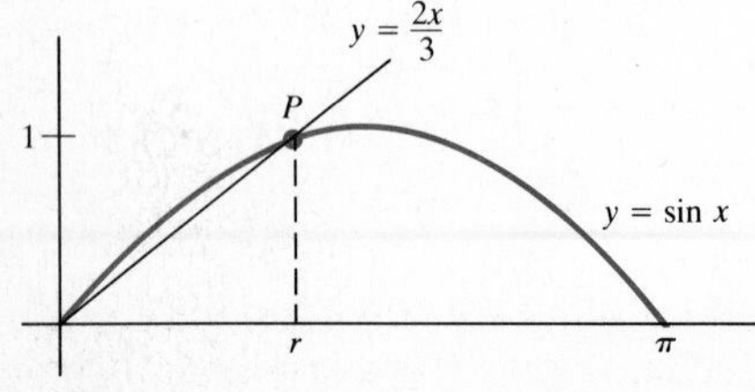

Figure 6.49

EXAMPLE 2 The line $y = 2x/3$ crosses the curve $y = \sin x$ at a point P, whose x coordinate r is between 0 and π, as shown in Fig. 6.49. The number r is a solution of the equation $2x/3 = \sin x$, since the graphs have equal y coordinates at $x = r$. Use Newton's method to approximate r.

SOLUTION A glance at the graph in Fig. 6.49 suggests that r is approximately 1.5. To obtain a better estimate, note that r is a root of the equation

$$f(x) = \sin x - \frac{2x}{3} = 0.$$

Since $f'(x) = \cos x - \frac{2}{3}$, Newton's method provides this second estimate of r:

$$x_2 = 1.5 - \frac{f(1.5)}{f'(1.5)} = 1.5 - \frac{\sin 1.5 - 2(1.5)/3}{\cos 1.5 - \frac{2}{3}}.$$

Angle in radians, not degrees

Computing sin 1.5 and cos 1.5, we find that

$$x_2 \approx 1.5 - \frac{0.9975 - 1}{0.0707 - 0.6667} = 1.5 - \frac{0.0025}{0.5960} \approx 1.496.$$

Incidentally, to three decimals, $r = 1.496$. ■

EXERCISES FOR SEC. 6.10: NEWTON'S METHOD FOR SOLVING AN EQUATION

1 Let a be a positive number. Show that the Newton recursion formula for estimating $\sqrt{a}$ is given by

$$x_{i+1} = \frac{x_i + a/x_i}{2}.$$

2 Use the formula of Exercise 1 to estimate $\sqrt{15}$. Choose $x_1 = 4$ and compute x_2 and x_3 to three decimals.

3 Use the formula of Exercise 1 to estimate $\sqrt{19}$. Choose $x_1 = 4$ and compute x_2 and x_3 to three decimals.

4 In estimating $\sqrt{3}$, an electronic computer began with $x_1 = 50$. What does Newton's method give for x_2, x_3, and x_4?

5 (*a*) Show that Newton's method gives this recursion formula for estimating $\sqrt[3]{7}$:

$$x_{i+1} = \frac{2}{3}x_i + \frac{7}{3x_i^2}.$$

(*b*) Let $x_1 = 1$, and compute x_2 and x_3.
(*c*) Let $x_1 = 2$, and compute x_2 and x_3.

6 Let $f(x) = x^4 + x - 19$.
(*a*) Show that $f(2) < 0 < f(3)$ and that f must thus have a root r between 2 and 3.
(*b*) Apply Newton's method, starting with $x_1 = 2$. Compute x_2 and x_3.

7 Let $f(x) = x^5 + x - 1$.
(*a*) Show that there is exactly one root of the equation $f(x) = 0$ in the interval $[0, 1]$. (Examine f'.)
(*b*) Using $x_1 = \frac{1}{2}$ as a first estimate, apply Newton's method to find a second estimate x_2.

8 Let $f(x) = 2x^3 - x^2 - 2$.
(*a*) Show that there is exactly one root of the equation $f(x) = 0$ in the interval $[1, 2]$.
(*b*) Using $x_1 = \frac{3}{2}$ as a first estimate, apply Newton's method to find a second estimate x_2.

9 (*a*) Graph $y = e^x$ and $y = x + 2$ relative to the same axes.
(*b*) With the aid of (*a*), estimate roots of the equation $e^x - x - 2 = 0$.
(*c*) Use Newton's method and a calculator to estimate the roots to two-decimal accuracy.

10 (*a*) Show that there is a number r in $[0, 1]$ such that $\ln(1 + r) = 1 - r$.
(*b*) Show that there is only one such number r in the interval $[0, 1]$.
(*c*) Use Newton's method with $x_1 = 0.5$ to find x_2, a closer approximation to r.

11 (*a*) Graph $y = \ln x$ and $y = \sin x$ relative to the same axes.
(*b*) With the aid of the graphs in (*a*), estimate the x coordinate of the point such that $\sin x - \ln x = 0$.
(*c*) Using the estimate in (*b*) as x_1, find another estimate x_2 by Newton's method.

12 (*a*) Graph $y = x \sin x$ for x in $[0, \pi]$.
(*b*) Using the first and second derivatives, show that it has a unique relative maximum in the interval $[0, \pi]$.
(*c*) Show that the maximum value of $x \sin x$ occurs when $x \cos x + \sin x = 0$.
(*d*) Use Newton's method, with $x_1 = \pi/2$, to find an estimate x_2 for a root of $x \cos x + \sin x = 0$.
(*e*) Use Newton's method again to find x_3.

13 (*a*) Graph $y = e^x$ and $y = \tan x$ relative to the same axes.
(*b*) Show that the equation $e^x - \tan x = 0$ has a solution between 0 and $\pi/2$.
(*c*) Choose x_1, an estimate of the solution in (*b*), on the basis of the graph in (*a*). Then determine x_2 by Newton's method.

14 (*a*) Show that the equation $3x + \sin x - e^x = 0$ has a root between 0 and 1.

(*b*) Starting with $x_1 = 0.5$, compute x_2 and x_3, the estimates of the root in (*a*) by Newton's recursion.

Exercises 15 and 16 show that care should be taken in applying Newton's method.

15 Let $f(x) = 2x^3 - 4x + 1$.

(*a*) Show that there must be a root r of $f(x) = 0$ in $[0, 1]$.

(*b*) Take $x_1 = 1$, and apply Newton's method to obtain x_2 and x_3, estimates of r.

(*c*) Graph f, and show what is happening in the sequence of estimates.

16 Let $f(x) = x^2 + 1$.

(*a*) Using Newton's method with $x_1 = 2$, compute x_2, x_3, x_4, and x_5 to two decimal places.

(*b*) Using the graph of f, show geometrically what is happening in (*a*).

(*c*) Using Newton's method with $x_1 = \sqrt{3}/3$, compute x_2 and x_3. What happens to x_n as $n \to \infty$?

(*d*) What happens when you use Newton's method, starting with $x_1 = 1$?

■

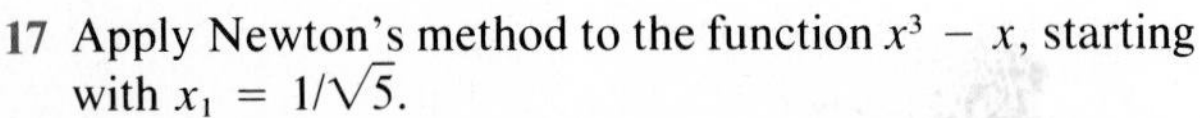

17 Apply Newton's method to the function $x^3 - x$, starting with $x_1 = 1/\sqrt{5}$.

(*a*) Compute x_2 and x_3 exactly (not as decimal approximations).

(*b*) Graph $x^3 - x$ and explain why Newton's method fails in this case.

In Exercises 18 to 20 (Figs. 6.50 to 6.52) use Newton's method to estimate θ (to two decimal places). Angles are in radians. Also show that there is only one answer if $0 < \theta < \pi/2$.

18

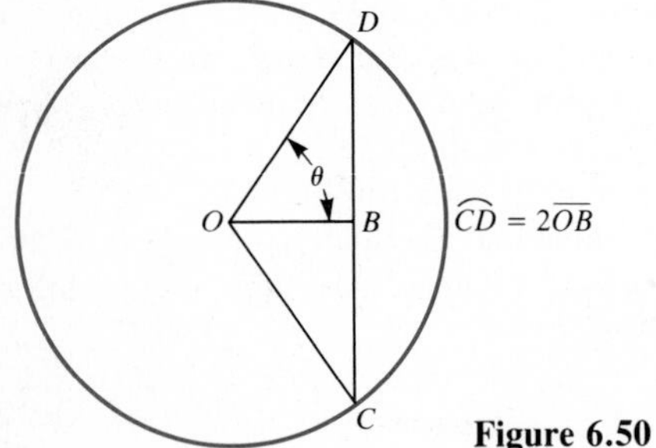

Figure 6.50

19

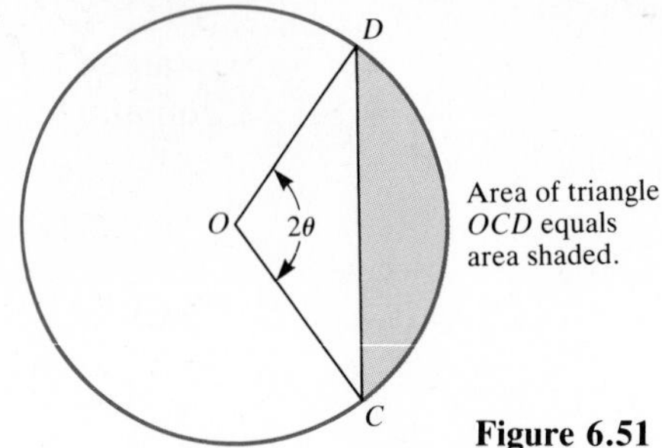

Figure 6.51

20

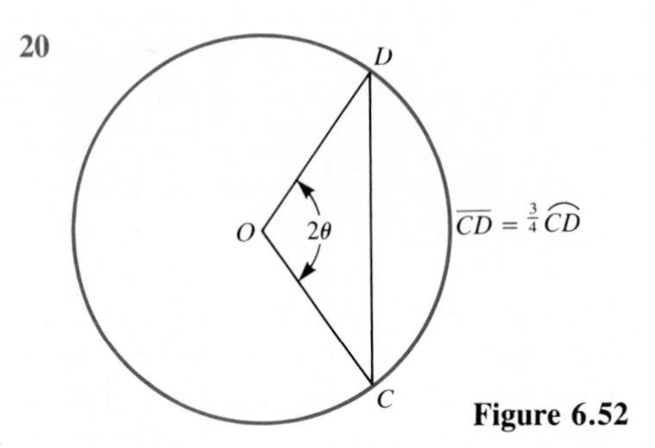

Figure 6.52

21 The equation $x \tan x = 1$ occurs in the theory of vibrations.

(*a*) How many roots does it have in $[0, \pi/2]$?

(*b*) Find them to two decimal places.

22 This exercise justifies inequality (2). In Exercise 42 of Sec. 4.8 it was shown that if f'' is continuous, then there is a number c_2 in $[a, b]$ such that

$$f(b) = f(a) + f'(a)(b - a) + \frac{f''(c_2)}{2}(b - a)^2.$$

(*a*) Replace b by r and a by x_i to show that

$$x_{i+1} - r = \frac{(r - x_i)^2}{2}\frac{f''(c_2)}{f'(x_i)}.$$

(*b*) Deduce inequality (2).

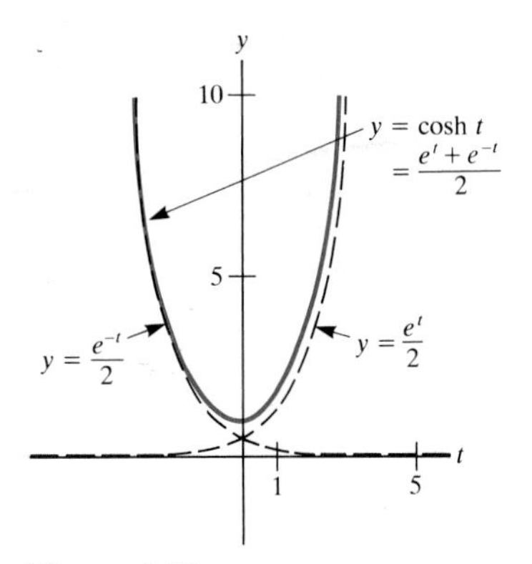

Figure 6.53

6.11 THE HYPERBOLIC FUNCTIONS AND THEIR INVERSES

Certain combinations of the exponential functions e^x and e^{-x} occur often enough in differential equations and engineering—for instance, in the study of electric transmission and suspension cables—to be given names. This section defines these so-called **hyperbolic functions** and obtains their basic properties. Since the letter x will be needed later for another purpose, we will use the letter t when writing the two preceding exponentials, namely, e^t and e^{-t}.

Definition *The hyperbolic cosine.* Let t be a real number. The **hyperbolic cosine** of t, denoted $\cosh t$, is given by the formula

Pronounced as written, "cosh," rhyming with "gosh"

$$\cosh t = \frac{e^t + e^{-t}}{2}.$$

To graph $\cosh t$, note first that

$$\cosh(-t) = \frac{e^{-t} + e^{-(-t)}}{2} = \frac{e^{-t} + e^{t}}{2} = \cosh t.$$

Since $\cosh(-t) = \cosh t$, the cosh function is even, and so its graph is symmetric with respect to the vertical axis. Furthermore, $\cosh t$ is the sum of two terms:

$$\cosh t = \frac{e^t}{2} + \frac{e^{-t}}{2}.$$

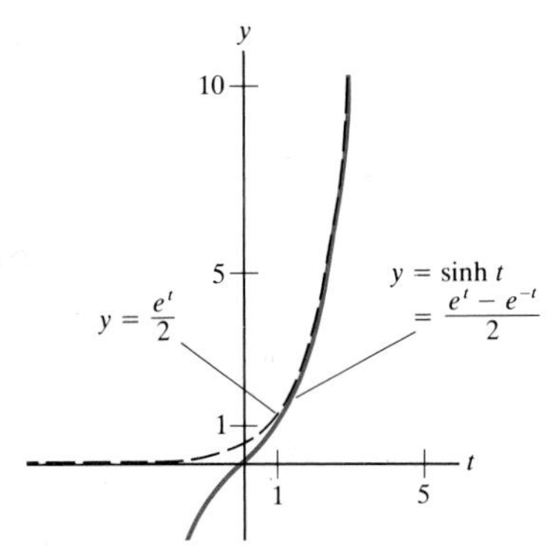

Figure 6.54

As $t \to \infty$, the second term, $e^{-t}/2$, approaches 0. Thus for $t > 0$ and large, the graph of $\cosh t$ is just a little above that of $e^t/2$. This information, together with the fact that $\cosh 0 = (e^0 + e^{-0})/2 = 1$, is the basis for Fig. 6.53.

For $|t| \to \infty$, the graph of $y = \cosh t$ is asymptotic to the graph of $y = e^t/2$ or $y = e^{-t}/2$.

The curve $y = \cosh t$ in Fig. 6.53 is called a **catenary** (from the Latin *catena*, meaning "chain"). A chain or rope, suspended from its ends, forms a curve that is part of a catenary.

The other hyperbolic function used in practice is defined as follows.

Definition *The hyperbolic sine.* Let t be a real number. The **hyperbolic sine** of t, denoted $\sinh t$, is given by the formula

"sinh" is pronounced "sinch," rhyming with "pinch."

$$\sinh t = \frac{e^t - e^{-t}}{2}.$$

sinh t is an odd function.

It is a simple matter to check that $\sinh 0 = 0$ and $\sinh(-t) = -\sinh t$, so that the graph of $\sinh t$ is symmetric with respect to the origin. Moreover, it lies below the graph of $e^t/2$. However, as $t \to \infty$, the two graphs approach each other since $e^{-t}/2 \to 0$ as $t \to \infty$. Figure 6.54 shows the graph of $\sinh t$.

Note the contrast between $\sinh t$ and $\sin t$. As t becomes large, the hyperbolic sine becomes large, $\lim_{t\to\infty} \sinh t = \infty$ and $\lim_{t\to-\infty} \sinh t = -\infty$. There is a similar contrast between $\cosh t$ and $\cos t$. While the trigonometric functions are periodic, the hyperbolic functions are not.

Example 1 shows why the functions $(e^t + e^{-t})/2$ and $(e^t - e^{-t})/2$ are called **hyperbolic**.

EXAMPLE 1 Show that for any real number t the point with coordinates

$$x = \cosh t, \qquad y = \sinh t$$

Why they are called hyperbolic functions

lies on the hyperbola $\qquad x^2 - y^2 = 1.$

SOLUTION Compute $\cosh^2 t - \sinh^2 t$ and see whether it equals 1. We have

$$\begin{aligned}\cosh^2 t - \sinh^2 t &= \left(\frac{e^t + e^{-t}}{2}\right)^2 - \left(\frac{e^t - e^{-t}}{2}\right)^2 \\ &= \frac{e^{2t} + 2e^te^{-t} + e^{-2t}}{4} - \frac{e^{2t} - 2e^te^{-t} + e^{-2t}}{4} \\ &= \frac{2+2}{4} \qquad \text{cancellation} \\ &= 1.\end{aligned}$$

$\cosh^2 t - \sinh^2 t = 1$

Observe that since $\cosh t \geq 1$, the point $(\cosh t, \sinh t)$ is on the right half of the hyperbola $x^2 - y^2 = 1$, as shown in Figure 6.55.

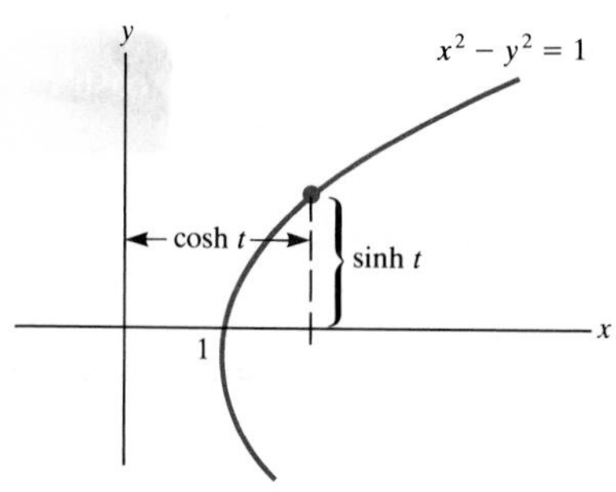

Figure 6.55 ■

[Since $(\cos\theta, \sin\theta)$ lies on the circle $x^2 + y^2 = 1$, the trigonometric functions are called circular functions.]

Derivatives of the Hyperbolic Functions

The four other hyperbolic functions, namely, the hyperbolic tangent, the hyperbolic secant, the hyperbolic cotangent, and the hyperbolic cosecant, are defined as follows:

These are defined like the circular functions, tan, sec, cot, csc.

$$\tanh t = \frac{\sinh t}{\cosh t} \qquad \operatorname{sech} t = \frac{1}{\cosh t} \qquad \coth t = \frac{\cosh t}{\sinh t} \qquad \operatorname{csch} t = \frac{1}{\sinh t}$$

Each can be expressed in terms of exponentials. For instance,

$$\tanh t = \frac{(e^t - e^{-t})/2}{(e^t + e^{-t})/2} = \frac{e^t - e^{-t}}{e^t + e^{-t}}.$$

As $t \to \infty$, $e^t \to \infty$ and $e^{-t} \to 0$. Thus $\lim_{t\to\infty} \tanh t = 1$. Similarly, $\lim_{t\to-\infty} \tanh t = -1$. Figure 6.56 is a graph of $y = \tanh t$.

$-1 < \tanh t < 1$, so tanh is bounded, and cosh and sinh are unbounded (the exact opposite of their trig analogs).

The derivatives of the six hyperbolic functions can be computed directly. For instance,

$$(\cosh t)' = \frac{(e^t + e^{-t})'}{2} = \frac{e^t - e^{-t}}{2} = \sinh t.$$

Function	*Derivative*
cosh t	sinh t
sinh t	cosh t
tanh t	$\operatorname{sech}^2 t$
coth t	$-\operatorname{csch}^2 t$
sech t	$-\operatorname{sech} t \tanh t$
csch t	$-\operatorname{csch} t \coth t$

The table in the margin lists the six derivatives. Notice that the formulas, except for the minus signs, are like those for the derivatives of the trigonometric functions.

Inverse hyperbolic functions appear on some calculators and in integral tables. Just as the hyperbolic functions are expressed in terms of the exponential function, each inverse hyperbolic function can be expressed in terms of a logarithm. They provide useful antiderivatives as well as solutions to some differential equations.

Consider the inverse of sinh t first. Since sinh t is increasing, it is one-to-one; there is no need to restrict its domain. To find its inverse, it is necessary to solve the equation

$$x = \sinh t$$

Finding the inverse of the hyperbolic sine

for t as a function of x. The steps are straightforward:

$$x = \frac{e^t - e^{-t}}{2},$$

$$2x = e^t - \frac{1}{e^t},$$

$$2xe^t = (e^t)^2 - 1,$$

or

$$(e^t)^2 - 2xe^t - 1 = 0. \tag{1}$$

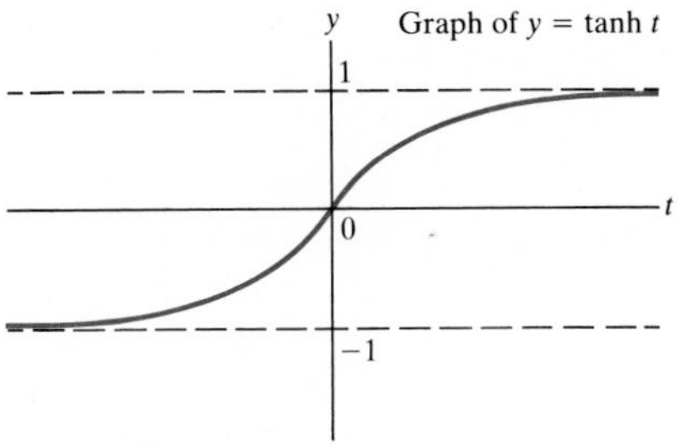

Graph of $y = \tanh t$

Figure 6.56

Equation (1) is quadratic in the unknown e^t. By the quadratic formula,

$$e^t = \frac{2x \pm \sqrt{(2x)^2 + 4}}{2}$$
$$= x \pm \sqrt{x^2 + 1}.$$

Since $e^t > 0$ and $\sqrt{x^2 + 1} > x$, the plus sign is kept and the minus sign is rejected. Thus

$$e^t = x + \sqrt{x^2 + 1} \quad \text{and} \quad t = \ln(x + \sqrt{x^2 + 1}).$$

Consequently, the inverse of the function sinh t is given by the formula

Formula for $\sinh^{-1} x$

$$\sinh^{-1} x = \ln(x + \sqrt{x^2 + 1}).$$

Computation of $\tanh^{-1} x$ is a little different. Since the derivative of $\tanh t$ is $\operatorname{sech}^2 t$, the function $\tanh t$ is increasing and has an inverse. However, $|\tanh t| < 1$, and so the inverse function will be defined only for $|x| < 1$.

We find the inverse of $x = \tanh t$ as follows:

$$x = \tanh t = \frac{e^t - e^{-t}}{e^t + e^{-t}} = \frac{e^t - (1/e^t)}{e^t + (1/e^t)} = \frac{e^{2t} - 1}{e^{2t} + 1}.$$

Thus
$$xe^{2t} + x = e^{2t} - 1$$
$$1 + x = e^{2t}(1 - x)$$
$$e^{2t} = \frac{1 + x}{1 - x}$$
$$2t = \ln\left(\frac{1 + x}{1 - x}\right)$$
$$t = \frac{1}{2}\ln\left(\frac{1 + x}{1 - x}\right).$$

Formula for $\tanh^{-1} x$

Thus
$$\tanh^{-1} x = \frac{1}{2}\ln\left(\frac{1 + x}{1 - x}\right) \qquad |x| < 1.$$

Inverses of the other four hyperbolic functions are computed similarly. The functions $\cosh^{-1} x$ and $\text{sech}^{-1} x$ are chosen to be positive. Their formulas are included in the following table:

The derivatives are found by differentiating the formulas.

$\cosh x$ *and* $\text{sech}\, x$ *are one-to-one for* $x \geq 0$

Function	*Formula*	*Derivative*	*Domain*
$\cosh^{-1} x$	$\ln(x + \sqrt{x^2 - 1})$	$\frac{1}{\sqrt{x^2 - 1}}$	$x \geq 1$
$\sinh^{-1} x$	$\ln(x + \sqrt{x^2 + 1})$	$\frac{1}{\sqrt{x^2 + 1}}$	x axis
$\tanh^{-1} x$	$\frac{1}{2}\ln\left(\frac{1 + x}{1 - x}\right)$	$\frac{1}{1 - x^2}$	$\|x\| < 1$
$\coth^{-1} x$	$\frac{1}{2}\ln\left(\frac{x + 1}{x - 1}\right)$	$\frac{1}{1 - x^2}$	$\|x\| > 1$
$\text{sech}^{-1} x$	$\ln\left(\frac{1 + \sqrt{1 - x^2}}{x}\right)$	$\frac{-1}{x\sqrt{1 - x^2}}$	$0 < x \leq 1$
$\text{csch}^{-1} x$	$\ln\left(\frac{1}{x} + \sqrt{1 + \frac{1}{x^2}}\right)$	$\frac{-1}{\|x\|\sqrt{1 + x^2}}$	$x \neq 0$

EXAMPLE 2 Verify the formula
$$\int \frac{dx}{a^2 - x^2} = \frac{1}{a}\tanh^{-1}\frac{x}{a} + C \qquad a > 0,\ |x| < a,$$
included in many integral tables.

SOLUTION Let

$$y = \frac{1}{a}\tanh^{-1}\frac{x}{a} + C$$
$$= \frac{1}{a}\frac{1}{2}\ln\left(\frac{1 + x/a}{1 - x/a}\right) + C \qquad \text{from preceding table}$$
$$= \frac{1}{2a}\ln\left(\frac{a + x}{a - x}\right) + C \qquad \text{algebra}$$
$$= \frac{1}{2a}[\ln(a + x) - \ln(a - x)] + C. \qquad [\ln(A/B) = \ln A - \ln B]$$

Differentiation then yields

$$y' = \frac{1}{2a}\left(\frac{1}{a+x} + \frac{1}{a-x}\right) = \frac{1}{2a} \cdot \frac{2a}{a^2 - x^2} = \frac{1}{a^2 - x^2}. \quad ■$$

EXERCISES FOR SEC. 6.11: THE HYPERBOLIC FUNCTIONS AND THEIR INVERSES

Exercises 1 to 14 concern the hyperbolic functions.

1 Show that $\cosh t + \sinh t = e^t$.

2 Show that $\cosh t - \sinh t = e^{-t}$.

In Exercises 3 to 6 differentiate the given functions and express the derivatives in terms of hyperbolic functions.

3 $\sinh t$ **4** $\tanh t$

5 $\operatorname{sech} t$ **6** $\coth t$

In Exercises 7 to 14 differentiate the functions. Be sure to use the chain rule.

7 $\cosh 3x$ **8** $\sinh 5x$

9 $\tanh \sqrt{x}$ **10** $\operatorname{sech} (\ln x)$

11 $e^{3x} \sinh x$ **12** $\dfrac{(\tanh 2x)(\operatorname{sech} 3x)}{\sqrt{1 + 2x}}$

13 $(\cosh 4x)(\coth 5x)(\operatorname{csch} x^2)$

14 $\dfrac{(\coth 5x)^{5/2}(\tanh 3x)^{1/3}}{3x + 4 \cos x}$ (Call it y and use logarithmic differentiation.)

Exercises 15 to 22 concern the inverse hyperbolic functions. In Exercises 15 to 18 obtain the given formulas using the technique illustrated in the text.

15 $\cosh^{-1} x = \ln (x + \sqrt{x^2 - 1}),\ x \geq 1$

16 $\operatorname{sech}^{-1} x = \ln \left(\dfrac{1 + \sqrt{1 - x^2}}{x}\right),\ 0 < x \leq 1$

17 $\coth^{-1} x = \dfrac{1}{2} \ln \left(\dfrac{x + 1}{x - 1}\right),\ |x| > 1$

18 $\operatorname{csch}^{-1} x = \ln \left(\dfrac{1}{x} + \sqrt{1 + \dfrac{1}{x^2}}\right),\ x \neq 0$

In Exercises 19 to 22 verify the integration formulas by differentiation (as in Example 2).

19 $\displaystyle\int \frac{dx}{\sqrt{x^2 - 1}} = \cosh^{-1} x + C,\ x > 1$

20 $\displaystyle\int \frac{dx}{\sqrt{x^2 + 1}} = \sinh^{-1} x + C$

21 $\displaystyle\int \frac{dx}{x\sqrt{1 - x^2}} = -\operatorname{sech}^{-1} x + C,\ 0 < x < 1$

22 $\displaystyle\int \frac{dx}{x\sqrt{1 + x^2}} = -\operatorname{csch}^{-1} x + C,\ x > 0$

In Exercises 23 to 28 use the definitions of the hyperbolic functions to verify the given identities.

23 (*a*) $\cosh (x + y) = \cosh x \cosh y + \sinh x \sinh y$
(*b*) $\sinh (x + y) = \sinh x \cosh y + \cosh x \sinh y$

24 $\tanh (x + y) = \dfrac{\tanh x + \tanh y}{1 + \tanh x \tanh y}$

25 (*a*) $\cosh (x - y) = \cosh x \cosh y - \sinh x \sinh y$
(*b*) $\sinh (x - y) = \sinh x \cosh y - \cosh x \sinh y$

26 (*a*) $\cosh 2x = \cosh^2 x + \sinh^2 x$
(*b*) $\sinh 2x = 2 \sinh x \cosh x$

27 (*a*) $2 \sinh^2 (x/2) = \cosh x - 1$
(*b*) $2 \cosh^2 (x/2) = \cosh x + 1$

28 $\operatorname{sech}^2 x + \tanh^2 x = 1$

29 At what angle does the graph of $y = \tanh x$ cross the x axis?

30 At what angle does the graph of $y = \sinh x$ cross the x axis?

■

31 (Calculator)
(*a*) Compute $\cosh t$ and $\sinh t$ for $t = -3, -2, -1, 0, 1, 2$, and 3.
(*b*) Plot the seven points $(\cosh t, \sinh t)$ given by (*a*). They should lie on the hyperbola $x^2 - y^2 = 1$.

32 (Calculator)
(*a*) Compute $\cosh t$ and $e^t/2$ for $t = 0, 1, 2, 3$, and 4.
(*b*) Using the data in (*a*), graph $\cosh t$ and $e^t/2$ relative to the same axes.

33 (Calculator)
(*a*) Compute $\tanh x$ for $x = 0, 1, 2$, and 3.
(*b*) Using the data in (*a*) and the fact that $\tanh (-x) = -\tanh x$, graph $y = \tanh x$.

34 Some integral tables contain the formulas

$$\int \frac{dx}{\sqrt{ax + b}\sqrt{cx + d}} = \begin{cases} \dfrac{2}{\sqrt{-ac}} \tan^{-1} \sqrt{\dfrac{-c(ax + b)}{a(cx + d)}} & a > 0,\ c < 0 \\[2ex] \dfrac{2}{\sqrt{ac}} \tanh^{-1} \sqrt{\dfrac{c(ax + b)}{a(cx + d)}} & a, c > 0 \end{cases}$$

The first formula is used if a and c have opposite signs, the second if they have the same signs. Check each of the formulas by differentiation.

35 One of the applications of hyperbolic functions is to the study of motion in which the resistance of the medium is proportional to the square of the velocity. Suppose that a body starts from rest and falls x meters in t seconds. Let g (a constant) be the acceleration due to gravity. It can be shown that there is a constant $V > 0$ such that

$$x = \frac{V^2}{g} \ln \cosh \frac{gt}{V}.$$

(*a*) Find the velocity $v(t) = dx/dt$ as a function of t.
(*b*) Show that $\lim_{t\to\infty} v(t) = V$.
(*c*) Compute the acceleration dv/dt as a function of t.
(*d*) Show that the acceleration equals $g - g(v/V)^2$.
(*e*) What is the limit of the acceleration as $t \to \infty$?

36 Two particles repel each other with a force proportional to the distance x between them. There is thus a positive constant k such that

$$\frac{d^2x}{dt^2} = kx.$$

(*a*) Show that, for any constants A and B, the function $x = A \cosh \sqrt{k}t + B \sinh \sqrt{k}t$ satisfies the given differential equation.
(*b*) If at time $t = 0$ the particles are a distance a apart and motionless, show that $A = a$ and $B = 0$. (Thus $x = a \cosh \sqrt{k}t$.)

37 Find the inflection points on the curve $y = \tanh x$.

6.S SUMMARY

This chapter completed the task of finding the derivatives of all the elementary functions. It opened with a review of logarithms ("a logarithm is an exponent"), defined e, and obtained the derivatives of the logarithm functions.

Then, with the aid of the notion of an inverse function, it found the derivative of b^x and the inverse trigonometric functions.

Separable differential equations were solved and applied to natural growth and decay and also inhibited growth.

Then l'Hôpital's rule was described. It provides a tool for dealing with many limits that are of the form $\lim_{x\to a} f(x)/g(x)$, where $f(x)$ and $g(x)$ both approach 0 or both approach ∞. Some other limits can be transformed to the type covered by l'Hôpital's rule.

Newton's method for estimating a root, $x_{i+1} = x_i - f(x_i)/f'(x_i)$, was discussed. The final section treated the hyperbolic functions, which will not be reviewed in this summary.

Vocabulary and Symbols

$e = \lim_{n\to\infty} (1 + 1/n)^n$
$\quad = \lim_{x\to 0} (1 + x)^{1/x} \approx 2.718$
$\ln x$ (= $\log_e x$)
one-to-one
inverse trigonometric function
 arcsin x, $\sin^{-1} x$, etc.
elementary function
algebraic function
transcendental function
differential equation D.E.
solution of a differential equation
order of a differential equation
separable differential equation
natural growth or decay
 $dP/dt = kP$
relative rate of growth per unit time r
growth constant k
 (negative k for decay)
doubling time t_2
half-life $t_{1/2}$
inhibited growth
 $dP/dt = kP(1 - P/M)$
l'Hôpital's rule
Newton's formula

Key Facts

The following table lists the derivatives found in Chap. 3 and in this chapter:

Formulas

Remember where the minus signs go.

Function	*Derivative*	*Function*	*Derivative*
x^a	ax^{a-1}	e^x	e^x
$[u(x)]^a$	$a[u(x)]^{a-1}u'(x)$	b^x	$(\ln b)\, b^x$
$\sqrt{x}$	$\dfrac{1}{2\sqrt{x}}$	$\sin^{-1} x$	$\dfrac{1}{\sqrt{1-x^2}}$
$\sin x$	$\cos x$	$\tan^{-1} x$	$\dfrac{1}{1+x^2}$
$\cos x$	$-\sin x$		
$\tan x$	$\sec^2 x$		
$\cot x$	$-\csc^2 x$	$\sec^{-1} x$	$\dfrac{1}{\lvert x\rvert \sqrt{x^2-1}}$
$\sec x$	$\sec x \tan x$		
$\csc x$	$-\csc x \cot x$		
$\ln x$	$\dfrac{1}{x},\ x > 0$		
$\ln \lvert x\rvert$	$\dfrac{1}{x},\ x \neq 0$		

It is not necessary to memorize the formulas for the derivatives of $\cos^{-1}$, $\cot^{-1}$, and $\csc^{-1}$. (They are obtained by putting minus signs in front of the last three formulas in the list.)

In a related-rate problem, find an equation linking the variables and then differentiate implicitly with respect to time. To find acceleration, differentiate again.

A separable D.E. has the form

$$\frac{dy}{dx} = \frac{g(x)}{h(y)}.$$

[It may also have the form $dy/dx = g(x)h(y)$ or $dy/dx = h(y)/g(x)$.] The equation is solved by bringing all x's to one side and all y's to the other (separating the variables), then integrating:

$$\int h(y)\, dy = \int g(x)\, dx + C.$$

L'Hôpital's rule concerns the behavior of $f(x)/g(x)$ in case numerator and denominator both approach 0 or both approach ∞. If

$$\lim_{x\to a} \frac{f'(x)}{g'(x)} \text{ exists,}$$

then

$$\lim_{x\to a} \frac{f(x)}{g(x)} = \lim_{x\to a} \frac{f'(x)}{g'(x)}.$$

The forms "zero times infinity," "one to the infinity," and "zero to the zero" can be reduced to l'Hôpital's form. In the last two cases, take logs, find the limits of the logs, and then be sure to complete the problem by finding the original limits.

Newton's method provides a recursive estimate of a root r of a function f, namely, $x_{i+1} = x_i - f(x_i)/f'(x_i)$. Keep in mind that there are times when the process breaks down.

GUIDE QUIZ ON CHAP. 6: TOPICS IN DIFFERENTIAL CALCULUS

1 Differentiate:

(a) $e^{\sin^{-1} 3x}$ (b) $\sin[(\tan^{-1} 4x)^2]$

(c) $(x^3 + 5x)x^{\sqrt{x}}$ (d) $\dfrac{\tan 5^x}{\sec^{-1} 2x}$

(e) $\left\{\dfrac{x^{-3}[\ln(x^2+1)]^8}{\sqrt[3]{\cos^2 2x}}\right\}^5$

2 Differentiate and simplify your answer.

(a) $\dfrac{e^{ax}}{a^2+b^2}(a\cos bx + b\sin bx)$

(b) $x\sin^{-1} ax + \dfrac{1}{a}\sqrt{1-a^2x^2}$

(c) $x\tan^{-1} ax - \dfrac{1}{2a}\ln(1+a^2x^2)$

(d) $x\sec^{-1} ax - \dfrac{1}{a}\ln(ax + \sqrt{a^2x^2-1})$, $ax > 0$

3 The quantities x and y, which are differentiable functions of t, are related by the equation

$$xy + e^y = e + 1.$$

When $t = 0$, $x = 1$, $y = 1$, $\dot{x} = 2$, and $\ddot{x} = 3$. Find $\dot{y}$ and $\ddot{y}$ when $t = 0$.

4 A load of concrete M hangs from a rope which passes over a pulley B. (See Fig. 6.57.) A construction worker at C pulls the rope as he walks away at 5 feet per second. The level of the pulley is 10 feet higher than his hand. At what rate is the load rising if the length $\overline{BC}$ is (a) 15 feet? (b) 100 feet?

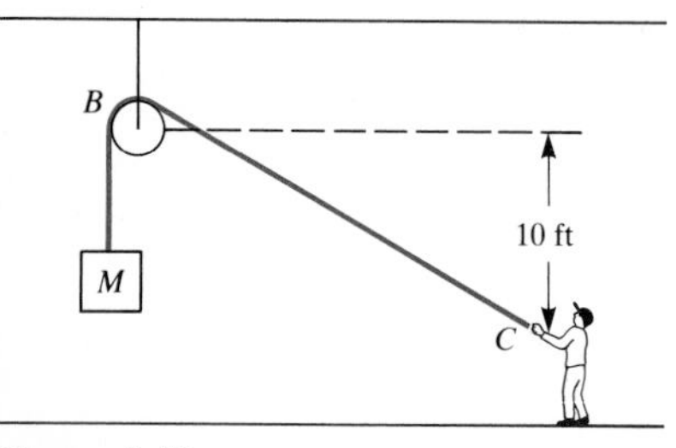

Figure 6.57

5 Solve: (a) $\dfrac{dy}{dx} = e^{-2y}x^3$ (b) $\dfrac{dy}{dx} = \dfrac{4y^2+1}{y}$

6 Radon has a half-life of 3.825 days. How long does it take for radon to diminish to only 10 percent of its original amount?

7 Examine these limits:

(a) $\lim_{x\to 0} \dfrac{\sin 2x}{\tan^{-1} 3x}$ (b) $\lim_{x\to\infty} \dfrac{\sin 2x}{\tan^{-1} 3x}$

(c) $\lim_{x\to(\pi/2)^-} (\sec x - \tan x)$ (d) $\lim_{x\to 0} (1 - \cos 2x)^x$

(e) $\lim_{x\to(\pi/2)^-} \dfrac{\int_0^x \tan^2\theta\, d\theta}{\tan x}$

8 Arrange these functions in order of increasing size for large x: x^3, $\ln x$, 1.001^x, 2^x, $\log_{10} x$.

9 Let $f(x) = (\ln x)/x$.

(a) What is the domain of f?

(b) Find any x intercepts of f.

(c) Find any critical numbers of f.

(d) Use the second-derivative test to find whether there is a local maximum or minimum at any critical number.

(e) For which x is f increasing? Decreasing?

(f) Does f have a global minimum? A global maximum?

(g) Find $\lim_{x\to 0^+} f(x)$.

(h) Find $\lim_{x\to\infty} f(x)$.

(i) Graph f.

10 (a) What is meant by "e"?

(b) What is e to three decimal places?

11 Assume that a calculator has a *log*-key ($\log_{10}$) but no *ln*-key. How could you use the *log*-key to find $\ln 7$?

REVIEW EXERCISES FOR CHAP. 6: TOPICS IN DIFFERENTIAL CALCULUS

The next chapter, which treats the problem of finding an antiderivative, assumes a mastery of the formal computation of derivatives. Do as many of the first 86 exercises as you can find time for.

In Exercises 1 to 86 differentiate and simplify. (Assume that the input for any logarithm is positive.)

1 $\sqrt{1+x^3}$ 2 5^{x^2}

3 $\sqrt{x}$ 4 $\dfrac{1}{\sqrt{x}}$

5 $\cos^2 3x$ 6 $\sin^{-1} 3x$

7 $\sqrt{x^3}$ 8 $\dfrac{1}{3}\tan^{-1}\dfrac{x}{3}$

9 $\sqrt{\sin x}$

10 $\dfrac{\cos 5x}{x^2}$

11 $\cot x^2$

12 $\dfrac{1}{x^3}$

13 $\dfrac{e^{x^2}}{2x}$

14 $x^2 \sin 3x$

15 $x^{5/6} \sin^{-1} x$

16 $e^{\sqrt{x}}$

17 $x^2 e^{3x}$

18 $\sin^2 2x$

19 $\ln (\sec 3x + \tan 3x)$

20 $\dfrac{1}{5} \tan^{-1} \dfrac{x}{5}$

21 $\cos \sqrt{x}$

22 $e^{-x} \tan x^2$

23 $\ln (\sec x + \tan x)$

24 $3 \cot 5x + 5 \csc 3x$

25 $\dfrac{1}{\sqrt{6 + 3x^2}}$

26 $\ln (\sin 2x)$

27 $\sqrt{\tfrac{2}{15} (5x + 7)^3}$

28 $x \cos 5x + \sin 5x$

29 $\dfrac{x}{3} - \dfrac{4}{9} \ln (3x + 4)$

30 $\dfrac{\sqrt{4 - 9x^2}}{x}$

31 $(1 + x^2)^5 \sin 3x$

32 $\cos [\log_{10} (3x + 1)]$

33 $(2x^3 - 2x + 5)^4$

34 $(\sin 3x)^5$

35 $(1 + 2x)^5 \cos 3x$

36 $\dfrac{\sin 3x}{(x^2 + 1)^5}$

37 $\cos 3x \sin 4x$

38 $\tan \sqrt{x}$

39 $\csc 3x^2$

40 $\dfrac{1}{\sqrt{1 - x^2}}$

41 $x \left(\dfrac{x^2}{1 + x}\right)^3$

42 $\left(\dfrac{1 + 2x}{1 + 3x}\right)^4$

43 $x \sec 3x$

44 $[\ln (x^2 + 1)]^3$

45 $\dfrac{(x^3 - 1)^3 (x^{10} + 1)^4}{(2x + 1)^5}$

46 $\dfrac{x}{\sqrt{1 - x^2}}$

47 $\ln (x + \sqrt{x^2 + 1})$

48 $\dfrac{\sin^3 2x}{x^2 + x}$

49 $e^{-x} \tan^{-1} x^2$

50 $\log_{10} (x^2 + 1)$

51 $2e^{\sqrt{x}} (\sqrt{x} - 1)$

52 $\sin^{-1} \dfrac{x}{5}$

53 $\dfrac{\ln x}{x^2}$

54 $(x^2 + 1) \sin 2x$

55 $\dfrac{\sin^2 x}{\cos x}$

56 $x^5 - 2x + \ln (2x + 3)$

57 $(\sec^{-1} 3x) x^2 \ln (1 + x^2)$

58 $\sin^3 (1 + x^2)$

59 $x - 2 \ln (x - 1) + \dfrac{1}{x + 1}$

60 $\ln \sqrt{\dfrac{1 + x^2}{1 + x^3}}$

61 $\ln \left(\dfrac{1}{6x^2 + 3x + 1}\right)$

62 $\ln (\sqrt{4 + x}\, \sqrt[3]{x^2 + 1})$

63 $\ln \left[\dfrac{(5x + 1)^3 (6x + 1)^2}{(2x + 1)^4}\right]$

64 $\dfrac{1}{8} \left[\ln (2x + 1) + \dfrac{2}{2x + 1} - \dfrac{1}{2(2x + 1)^2}\right]$

65 $\dfrac{1}{2} \left(x\sqrt{9 - x^2} + 9 \sin^{-1} \dfrac{x}{3}\right)$

66 $\dfrac{1}{8} \left[2x + 3 - 6 \ln (2x + 3) - \dfrac{9}{2x + 3}\right]$

67 $\tan 3x - 3x$

68 $\dfrac{1}{\sqrt{6}} \tan^{-1} \left(x \sqrt{\dfrac{2}{3}}\right)$

69 $\ln (x + \sqrt{x^2 + 25})$

70 $\dfrac{e^x (\sin 2x - 2 \cos 2x)}{5}$

71 $x \sin^{-1} x + \sqrt{1 - x^2}$

72 $x \tan^{-1} x - \tfrac{1}{2} \ln (1 + x^2)$

73 $\tfrac{1}{3} \ln (\tan 3x + \sec 3x)$

74 $\tfrac{1}{6} \tan^2 3x + \tfrac{1}{3} \ln \cos 3x$

75 $-\tfrac{1}{3} \cos 3x + \tfrac{1}{9} \cos^3 3x$

76 $\tfrac{1}{8} e^{-2x} (4x^2 + 4x + 2)$

77 $e^{3x} \sin^2 2x \tan x$

78 $\dfrac{1}{3} \ln \left(\dfrac{\sqrt{2x + 3} - 3}{\sqrt{2x + 3} + 3}\right)$

79 $\dfrac{x}{2} \sqrt{4x^2 + 3} + \dfrac{3}{4} \ln (2x + \sqrt{4x^2 + 3})$

80 $\sqrt{1 + \sqrt[3]{x}}$

81 $\dfrac{x^3 (x^4 - x + 3)}{(x + 1)^2}$

82 $\dfrac{1}{1 + \csc 5x}$

83 $(1 + 3x)^{x^2}$

84 $2^x 5^{x^2} 7^{x^3}$

85 $\dfrac{x \ln x}{(1 + x^2)^5}$

86 $x^3 \cot^3 \sqrt{x^3}$

87 (*a*) Fill in this table for the function $\log_2 x$:

x	$\frac{1}{8}$	$\frac{1}{4}$	$\frac{1}{2}$	1	2	4	8
$\log_2 x$							

(*b*) Plot the seven points in (*a*) and graph $\log_2 x$.
(*c*) What is $\lim_{x \to \infty} \log_2 x$?
(*d*) What is $\lim_{x \to 0^+} \log_2 x$?

88 (*a*) Use a differential to show that for small x, $\ln (1 + x) \approx x$.
(*b*) Use (*a*) to estimate ln 1.1.

89 (*a*) Use a differential to show that for small x, $\log_{10}(1 + x) \approx 0.434\,x$.
(*b*) Use (*a*) to estimate $\log_{10} 1.05$.

In Exercises 90 to 100 check the equations by differentiation. The letters a and b denote constants.

90 $\int \ln ax\,dx = x \ln ax - x + C$

91 $\int x \ln ax\,dx = \frac{x^2}{2} \ln ax - \frac{x^2}{4} + C$

92 $\int x^2 \ln ax\,dx = \frac{x^3}{3} \ln ax - \frac{x^3}{9} + C$

93 $\int \frac{dx}{x \ln ax} = \ln(\ln ax) + C$

94 $\int \frac{dx}{\sin x} = \ln|\csc x - \cot x| + C$

95 $\int \tan ax\,dx = -\frac{1}{a} \ln|\cos ax| + C$

96 $\int \tan^3 ax\,dx = \frac{1}{2a} \tan^2 ax + \frac{1}{a} \ln|\cos ax| + C$

97 $\int \frac{dx}{\sin ax \cos ax} = \frac{1}{a} \ln|\tan ax| + C$

98 $\int \frac{dx}{\sqrt{ax^2 + b}} = \frac{1}{\sqrt{a}} \ln|x\sqrt{a} + \sqrt{ax^2 + b}| + C \quad (a > 0)$

99 $\int \frac{x\,dx}{(ax + b)^2} = \frac{b}{a^2(ax + b)} + \frac{1}{a^2} \ln|ax + b| + C$

100 $\int (\ln ax)^2\,dx = x(\ln ax)^2 - 2x \ln ax + 2x + C$

101 (*a*) Differentiate $\ln\left|\frac{1 + x}{1 - x}\right|$.
(*b*) Find the area of the region under the curve $y = 1/(1 - x^2)$ and above $[0, \frac{1}{2}]$.

102 Find the area of the region under the curve $y = x^3/(x^4 + 5)$ and above $[1, 2]$.

103 Differentiate and then simplify your answer:

$$\ln\left|\frac{\sqrt{ax + b} - b}{\sqrt{ax + b} + b}\right|.$$

104 Show that, for b and $c > 1$, $\log_b x/\log_c x$ is constant by differentiating it.

105 Find these integrals:

(*a*) $\int \frac{3x^2 + 1}{x^3 + x - 6}\,dx$ (*b*) $\int \frac{\cos 2x}{\sin 2x}\,dx$

(*c*) $\int \frac{dx}{5x + 3}$ (*d*) $\int \frac{dx}{(5x + 3)^2}$

106 (Calculator) (*a*) Evaluate $(1 - 2h)^{1/h}$ for $h = 0.01$. (*b*) Find $\lim_{h \to 0} (1 - 2h)^{1/h}$.

107 Differentiate:

(*a*) $\ln\left[\frac{\sqrt[3]{\tan 4x}\,(1 - 2x)^5}{\sqrt[3]{(1 + 3x)^2}}\right]$

(*b*) $\frac{(1 + x^2)^3 \sqrt{1 + x}}{\sin 3x}$

108 The graph in Fig. 6.58 shows the tangent line to the curve $y = \ln x$ at a point (x_0, y_0). Prove that AB has length 1, independent of the choice of (x_0, y_0).

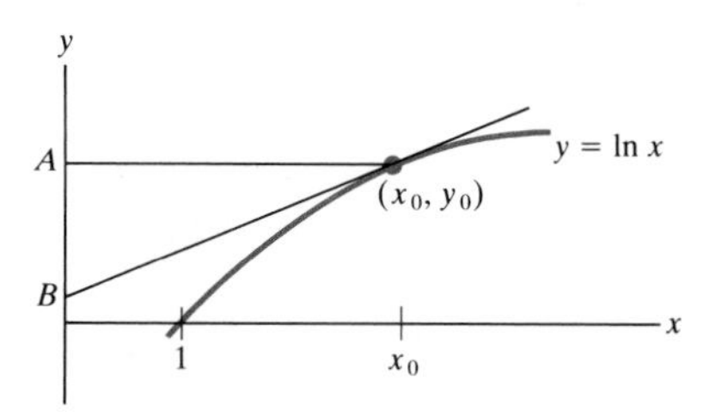

Figure 6.58

109 Find (*a*) $\lim_{x \to 1} \log_x 2$, (*b*) $\log_{x \to \infty} \log_x 2$. *Hint:* What is the relation between $\log_a b$ and $\log_b a$?

110 The **information content** or **entropy** of a binary source (such as a telegraph that transmits dots and dashes) whose two values occur with probabilities p and $1 - p$ is defined as $H(p) = -p \ln p - (1 - p) \ln(1 - p)$, where $0 < p < 1$. Show that H has a maximum at $p = \frac{1}{2}$. The practical significance of this result is that for maximum flow of information per unit time, the two values should, in the long run, appear in equal proportions.

111 (See Exercise 110.) Let p be fixed so that $0 < p < 1$. Define $M(q) = -p \ln q - (1 - p) \ln(1 - q)$. Show that $H(p) \le M(q)$ for $0 < q < 1$ and that equality holds if and only if $p = q$.

In each of Exercises 112 to 114 evaluate the limit by first showing that it is the derivative of a certain function at a specific number.

112 $\lim_{x \to 3} \frac{\ln(1 + 2x) - \ln 7}{x - 3}$

113 $\lim_{x \to 0} \frac{\ln(2 + x) - \ln 2}{x}$

114 $\lim_{x \to 0} \frac{\ln(1 + x)}{x}$

115 Graph $y = \ln|x|$.

116 Compute:

(*a*) $\int \frac{dx}{(3x + 2)^2}$ (*b*) $\int \frac{dx}{\sqrt{3x + 2}}$ (*c*) $\int \frac{dx}{3x + 2}$

117 Find the coordinates of the point P on the curve $y = \ln x$ such that the line through $(0, 0)$ and P is tangent to the curve.

118 Use logarithmic differentiation, as in Example 6 of Sec. 6.3, to find the derivative of

(*a*) $\dfrac{x^{3/5}(1+2x)^4 \sin^3 2x}{\tan^2 5x}$ (*b*) $\dfrac{x^3}{\sqrt[3]{x^3+x^2\cos 4x}}$

119 Find $y'(0)$ if $y = f(x)$ is a function satisfying the equation

$$\ln(1+y) + xy = \ln 2.$$

120 Does the graph of $y = x^4 - 4\ln x$ have (*a*) any local maxima? (*b*) local minima? (*c*) inflection points?

In Exercises 121 to 159 examine the limits.

121 $\lim_{x\to\infty} x^{1/\log_2 x}$

122 $\lim_{x\to\pi/4} \dfrac{\sin x - \sqrt{2}/2}{x - \pi/4}$

123 $\lim_{h\to 0} \dfrac{e^{3+h} - e^3}{h}$

124 $\lim_{x\to 1} \dfrac{\sin \pi x}{x-1}$

125 $\lim_{x\to 0} \dfrac{\cos\sqrt{x} - 1}{\tan x}$

126 $\lim_{x\to\infty} 2^x e^{-x}$

127 $\lim_{x\to 0} (1+2x^2)^{1/x^2}$

128 $\lim_{x\to 0} (1+3x)^{1/x}$

129 $\lim_{x\to-\infty} \dfrac{e^x - e^{-x}}{e^x + e^{-x}}$

130 $\lim_{x\to\infty} \dfrac{x^2+5}{2x^2+6x}$

131 $\lim_{x\to 0} \dfrac{1-\cos x}{x + \tan x}$

132 $\lim_{x\to 0^+} (\sin x)^{\sin x}$

133 $\lim_{x\to 0} \dfrac{\sin 2x}{e^{3x} - 1}$

134 $\lim_{x\to\infty} \dfrac{(x^2+1)^5}{e^x}$

135 $\lim_{x\to\infty} \dfrac{3x - \sin x}{x + \sqrt{x}}$

136 $\lim_{x\to 0} \dfrac{xe^x}{e^x - 1}$

137 $\lim_{x\to 2} \dfrac{x^3 - 8}{x^2 - 4}$

138 $\lim_{x\to 0} \dfrac{\sin x^2}{x \sin x}$

139 $\lim_{x\to\pi/2} \dfrac{\sin x}{1+\cos x}$

140 $\lim_{x\to 1} \dfrac{e^x + 1}{e^x - 1}$

141 $\lim_{x\to 2} \dfrac{x^3+8}{x^2+2}$

142 $\lim_{x\to\pi/4} \dfrac{\cos x^2}{\cos x}$

143 $\lim_{x\to 3^+} \dfrac{\ln(x-3)}{x-3}$

144 $\lim_{x\to 0^+} \dfrac{1-\cos x}{x - \tan x}$

145 $\lim_{x\to\infty} \dfrac{\cos x}{x}$

146 $\lim_{x\to 2} \dfrac{5^x + 3^x}{x}$

147 $\lim_{x\to 0} \dfrac{5^x - 3^x}{x}$

148 $\lim_{x\to\infty} [(x^2 - 2x)^{1/2} - x]$

149 $\lim_{x\to 0} (1+3x)^{2/x}$

150 $\lim_{x\to\infty} \dfrac{e^{-x}}{x^2}$

151 $\lim_{x\to\infty} \dfrac{\ln(x^2+1)}{\ln(x^2+8)}$

152 $\lim_{x\to\infty} e^{-x}\ln x$

153 $\lim_{x\to 0^+} \sin x \ln x$

154 $\lim_{x\to\pi/2^-} (1 - \sin x)^{\tan x}$

155 $\lim_{x\to\infty} (2^x - x^{10})^{1/x}$

156 $\lim_{x\to 0^+} \left(\dfrac{\sin x}{x}\right)^{1/x}$

157 $\lim_{x\to\infty} \dfrac{(x^3 - x^2 - 4x)^{1/3}}{x}$

158 $\lim_{x\to\pi/4} \dfrac{\sin x}{x}$

159 $\lim_{h\to 0} \dfrac{e^{3+h} - e^3}{1-h}$

160 Graph $y = xe^{-x}/(x+1)$, showing intercepts, critical points, relative maxima and minima, and asymptotes.

161 (*a*) Why does calculus use radian measure?
(*b*) Why does calculus use the base e for logarithms?
(*c*) Why does calculus use the base e for exponentials?

162 Using the e^x-key, the *ln*-key and the × (multiplication) but not the y^x-key, how would you calculate 3^{80} on a calculator?

163 In which cases below is it possible to determine $\lim_{x\to a} f(x)^{g(x)}$ without further information about the functions?
(*a*) $\lim_{x\to a} f(x) = 0$; $\lim_{x\to a} g(x) = 7$
(*b*) $\lim_{x\to a} f(x) = 2$; $\lim_{x\to a} g(x) = 0$
(*c*) $\lim_{x\to a} f(x) = 0$; $\lim_{x\to a} g(x) = 0$
(*d*) $\lim_{x\to a} f(x) = 0$; $\lim_{x\to a} g(x) = \infty$
(*e*) $\lim_{x\to a} f(x) = \infty$; $\lim_{x\to a} g(x) = 0$
(*f*) $\lim_{x\to a} f(x) = \infty$; $\lim_{x\to a} g(x) = -\infty$

164 In which cases below is it possible to determine $\lim_{x\to a} f(x)/g(x)$ without further information about the functions?
(*a*) $\lim_{x\to a} f(x) = 0$; $\lim_{x\to a} g(x) = \infty$
(*b*) $\lim_{x\to a} f(x) = 0$; $\lim_{x\to a} g(x) = 1$
(*c*) $\lim_{x\to a} f(x) = 0$; $\lim_{x\to a} g(x) = 0$
(*d*) $\lim_{x\to a} f(x) = \infty$; $\lim_{x\to a} g(x) = -\infty$

165 (*a*) State the assumptions in the zero-over-zero case of l'Hôpital's rule.
(*b*) State the conclusion.

166 Prove that

(*a*) $(\sin^{-1} x)' = \dfrac{1}{\sqrt{1-x^2}}$ (*b*) $(e^x)' = e^x$

(*c*) $(\tan^{-1} x)' = \dfrac{1}{1+x^2}$

167 What is the inverse of each of these functions?
(*a*) $\ln x$ (*b*) e^x (*c*) x^3
(*d*) $3x$ (*e*) $\sqrt[3]{x}$ (*f*) $\sin^{-1} x$

168 (*a*) Let $f(x) = 5^{7x}6^{8x+3}$. Show that $f'(x)$ is proportional to $f(x)$.
(*b*) Does (*a*) contradict the theorem that asserts that

the only functions whose derivatives are proportional to the functions are of the form Ae^{kx}?

169 The graph in Fig. 6.59 shows the tangent line to the curve $y = e^x$ at a point (x_0, y_0). Find the length of the segment AB.

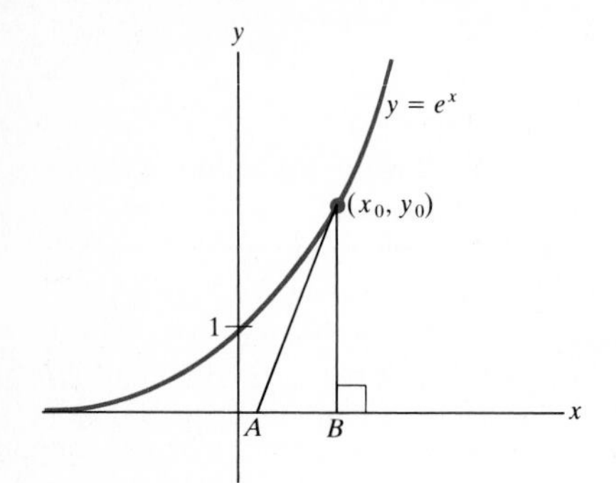

Figure 6.59

■

170 Find all positive integer solutions of the equation $x^y = y^x$ where $x \neq y$. (*Hint:* First take logarithms of both sides of the equation.)

171 Let $f(x) = (1 + x)^{1/x}$ for $x > -1$, $x \neq 0$. Let $f(0) = e$.

(*a*) Show that f is continuous at $x = 0$.

(*b*) Show that f is differentiable at $x = 0$. What is $f'(0)$?

172 Newton computed ln 0.8, ln 0.9, ln 1.1, and ln 1.2. Then, using these values and laws of logarithms, he found ln 2, ln 3, and ln 5. How could that be done?

173 Find $\lim_{x\to\infty} \left[\dfrac{a^x - 1}{(a - 1)\,x}\right]^{1/x}$. The answer depends on the constant $a > 0$, $a \neq 1$.

174 It was once conjectured that the speed of a ball falling from rest is proportional to the distance s that it drops.

(*a*) Show that, if this conjecture were correct, s would grow exponentially as a function of time.

(*b*) With the aid of (*a*) show that the speed would also grow exponentially.

(*c*) Recalling that the initial speed is 0, show that (*b*) leads to an absurd conclusion.

In fact, ds/dt is proportional to time t rather than to distance s.

175 According to a Chinese riddle, the lilies in a pond doubled in number each day. At the end of a month of 30 days the lilies finally covered the entire pond. On what day was exactly half the pond covered?

176 Find the limit of $(1^x + 2^x + 3^x)^{1/x}$ as (*a*) $x \to 0$, (*b*) $x \to \infty$, and (*c*) $x \to -\infty$.

177 The proof of Theorem 1 in Sec. 6.9, to be outlined in Exercise 178, depends on the following generalized mean-value theorem.

Generalized mean-value theorem. Let f and g be two functions that are continuous on $[a, b]$ and differentiable on (a, b). Furthermore, assume that $g'(x)$ is never 0 for x in (a, b). Then there is a number c in (a, b) such that

$$\frac{f(b) - f(a)}{g(b) - g(a)} = \frac{f'(c)}{g'(c)}.$$

(*a*) During a given time interval one car travels twice as far as another car. Use the generalized mean-value theorem to show that there is at least one instant when the first car is traveling exactly twice as fast as the second car.

(*b*) To prove the generalized mean-value theorem, introduce a function h:

$$h(x) = f(x) - f(a) - \frac{f(b) - f(a)}{g(b) - g(a)}[g(x) - g(a)].$$

Show that $h(b) = 0$ and $h(a) = 0$. Then apply Rolle's theorem to h.

Remark: The function h is geometrically quite similar to the function h used in the proof of the mean-value theorem in Sec. 4.1. It is easy to check that $h(x)$ is the vertical distance between the point $(f(x), g(x))$ and the line through $(f(a), g(a))$ and $(f(b), g(b))$.

178 This exercise proves Theorem 1 of Sec. 6.9, l'Hôpital's rule in the zero-over-zero case. Assume the hypotheses of that theorem. Define $f(a) = 0$ and $g(a) = 0$, so that f and g are continuous at a. Note that

$$\frac{f(x)}{g(x)} = \frac{f(x) - f(a)}{g(x) - g(a)},$$

and apply the generalized mean-value theorem from Exercise 177.

179 Let f and g be functions defined on some open interval. Assume that they have only positive values and that they are differentiable and possess second derivatives which are not 0 in the interval considered. Let $F(x) = \ln f(x)$ and let $G(x) = \ln g(x)$.

(*a*) If F is concave upward, must f be concave upward?

(*b*) If f is concave upward, must F be concave upward?

(*c*) If f and g are concave upward, must $f + g$ be concave upward?

(*d*) If f and g are concave upward, must fg be concave upward?

(*e*) If F and G are concave upward, must $\ln fg$ be concave upward?

(*f*) If F and G are concave upward, must $\ln(f + g)$ be concave upward?

In each case explain your answer.

180 If
$$\lim_{t\to\infty} f(t) = \infty = \lim_{t\to\infty} g(t)$$
and
$$\lim_{t\to\infty} \frac{\ln f(t)}{\ln g(t)} = 1,$$
must
$$\lim_{t\to\infty} \frac{f(t)}{g(t)} = 1?$$

181 If
$$\lim_{t\to\infty} f(t) = \infty = \lim_{t\to\infty} g(t)$$
and
$$\lim_{t\to\infty} \frac{f(t)}{g(t)} = 3,$$
what can be said about
$$\lim_{t\to\infty} \frac{\ln f(t)}{\ln g(t)}?$$
(Do not assume that f and g are differentiable.)

182 Give an example of a pair of functions f and g such that we have $\lim_{x\to 0} f(x) = 1$, $\lim_{x\to 0} g(x) = \infty$, and $\lim_{x\to 0} f(x)^{g(x)} = 2$.

183 Let f be a function such that f, f', and f'' are continuous, $f(x) \geq 0$, $f(0) = 0$, $f'(0) = 0$, and $f''(0) > 0$. The graph of f is indicated in Fig. 6.60. Find the limit as $x \to 0^+$ of the quotient
$$\frac{\text{Area under curve and above } [0, x]}{\text{Area of triangle } OAP}.$$

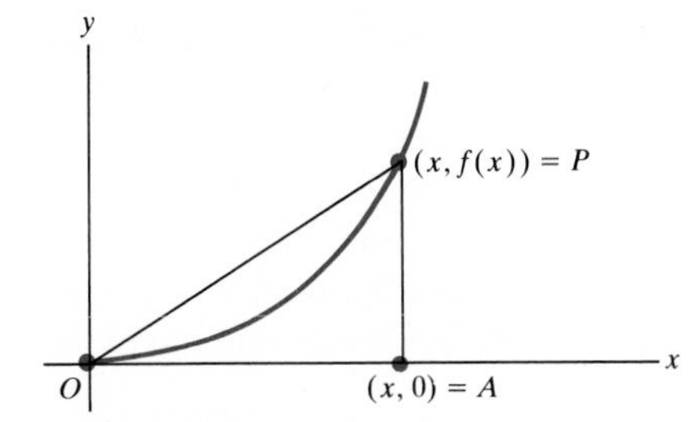

Figure 6.60

184 (*a*) Show that the equation $2x^5 - 10x + 5 = 0$ has precisely three real roots.

(*b*) Show that one of the roots in (*a*) is between 0 and 1.

(*c*) Use Newton's method to estimate the root in (*b*) to two-decimal accuracy.

Galois, early in the nineteenth century, proved that the root in (*c*) cannot be expressed in terms of square roots, cube roots, fourth roots, fifth roots, etc. [His theorem applies to any polynomial with integer coefficients such that (i) its degree is a prime $p \geq 5$, (ii) it is not the product of two polynomials of lower degree with integer coefficients, and (iii) exactly $p - 2$ of its roots are real.]

185 Show that $x^3 - 3x - 3$ has exactly one real root and use Newton's method to estimate it to two decimal places.

186 Show that $x^3 + x - 6$ has exactly one real root and use Newton's method to estimate it to two decimal places.

187 Figure 6.61 shows the graph of the velocity $v(t)$ of a particle moving along the x axis.

(*a*) Approximately how far does the particle move during the time interval [0, 1]? During the time interval [1, 2]? During the time interval [2, 4]?

(*b*) Let $x(t)$ be its coordinate at time t. Assume that $x(0) = 0$. Graph $x(t)$ as a function of t.

(*c*) Graph the acceleration as a function of t.

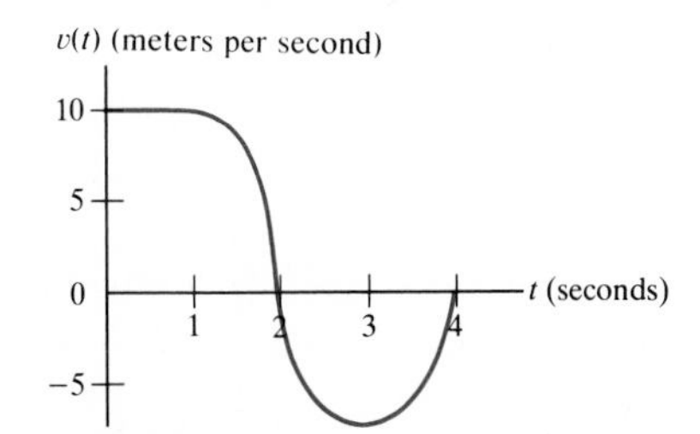

Figure 6.61

188 This table records the velocity $v(t)$ in meters per second of an object moving on the x axis during its first second of motion:

t	0	0.1	0.2	0.3	0.4	0.5	0.6	0.7	0.8	0.9	1.0
$v(t)$	0	2	2.5	2	3	3.1	3.2	3.5	3.2	3	3.1

(*a*) Sketch a rough graph of velocity as a function of time.

(*b*) Sketch a rough graph of acceleration as a function of time.

(*c*) Let $x(t)$ be the position of the object at time t. Assuming that $x(0) = 0$, sketch a rough graph of $x(t)$ as a function of t.

7 Computing Antiderivatives

The problem of computing antiderivatives differs from that of computing derivatives in two important aspects. First of all, some elementary functions, such as e^{x^2}, do not have elementary antiderivatives. Second, a slight change in the form of an integrand can cause a great change in the form of its antiderivative; for instance,

$$\int \frac{1}{x^2+1}\,dx = \tan^{-1} x + C \qquad \text{and} \qquad \int \frac{x}{x^2+1}\,dx = \tfrac{1}{2}\ln(x^2+1) + C.$$

A few moments of browsing through a table of antiderivatives (usually called a **table of integrals**) will yield many such examples. A short table of integrals may be found on the inside front cover of this book.

To be convenient, a table of antiderivatives should be short; it should not try to anticipate every antiderivative that may arise in practice. Sometimes it is necessary to transform a problem into one listed in the table or else to solve it without the aid of the table. It will often be quicker to use these techniques than to thumb through the pages of the table.

Programs for antidifferentiation on microcomputers are discussed in Sec. 7.S. They too have their limitations, and it is still important to acquire some skill in finding antiderivatives.

7.1 THE SUBSTITUTION TECHNIQUE

This section describes the substitution technique that changes the form of an integral, preferably to that of an easier integral. Before describing this technique, we collect some basic facts about integrals in order to have a supply of functions that can be "integrated at a glance."

Every formula for a derivative provides a corresponding formula for an antiderivative or integral. For instance, since $(x^3/3)' = x^2$, it follows that

$$\int x^2\,dx = \frac{x^3}{3} + C.$$

The following miniature integral table lists a few formulas that should be memorized. Each can be checked by differentiation.

$$\int x^a\,dx = \frac{x^{a+1}}{a+1} + C \qquad \text{for } a \neq -1$$

$$\int \frac{1}{x}\,dx = \ln|x| + C$$

$$\int e^x\,dx = e^x + C$$

$$\int \sin x\,dx = -\cos x + C \qquad \text{(Remember the } -.\text{)}$$

$$\int \cos x\,dx = \sin x + C$$

$$\int \frac{1}{\sqrt{1-x^2}}\,dx = \sin^{-1} x + C$$

$$\int \frac{1}{1+x^2}\,dx = \tan^{-1} x + C$$

$$\int \frac{1}{|x|\sqrt{x^2-1}}\,dx = \sec^{-1} x + C$$

Furthermore,

$$\int \frac{f'}{f}\,dx = \ln|f| + C.$$

A constant can be moved past the integral sign.

If c is constant, then

$$\int cf(x)\,dx = c\int f(x)\,dx \qquad \text{and} \qquad \int \frac{f(x)}{c}\,dx = \frac{1}{c}\int f(x)\,dx.$$

An antiderivative of a sum of functions can be found by adding antiderivatives of the functions.

For two functions f and g,

$$\int [f(x) + g(x)]\,dx = \int f(x)dx + \int g(x)\,dx.$$

Similarly,
$$\int [f(x) - g(x)]\,dx = \int f(x)\,dx - \int g(x)\,dx.$$

A few examples will show how these formulas serve to integrate many functions.

Antiderivative of a polynomial

EXAMPLE 1 Find $\int (2x^4 - 3x + 2)\,dx$.

SOLUTION
$$\begin{aligned}\int (2x^4 - 3x + 2)\,dx &= \int 2x^4\,dx - \int 3x\,dx + \int 2\,dx \\ &= 2\int x^4\,dx - 3\int x\,dx + 2\int 1\,dx \\ &= 2\frac{x^5}{5} - 3\frac{x^2}{2} + 2x + C. \quad \blacksquare\end{aligned}$$

One constant of integration is enough.

Antiderivative of f'/f

EXAMPLE 2 Find $\int \frac{4x^3}{x^4 + 1}\, dx$.

SOLUTION The numerator is precisely the derivative of the denominator. Hence

$$\int \frac{4x^3}{x^4 + 1}\, dx = \ln |x^4 + 1| + C.$$

Since $x^4 + 1$ is always positive, the absolute-value sign is not needed, and

$$\int \frac{4x^3}{x^4 + 1}\, dx = \ln (x^4 + 1) + C. \quad \blacksquare$$

EXAMPLE 3 Find $\int \sqrt{x}\, dx$.

Antiderivative of x^a

SOLUTION

$$\int \sqrt{x}\, dx = \int x^{1/2}\, dx = \frac{x^{1/2+1}}{\frac{1}{2} + 1} + C$$

$$= \tfrac{2}{3}x^{3/2} + C = \tfrac{2}{3}(\sqrt{x})^3 + C. \quad \blacksquare$$

EXAMPLE 4 Find $\int \frac{1}{x^3}\, dx$.

SOLUTION

$$\int \frac{1}{x}\, dx = \int x^{-3}\, dx = \frac{x^{-3+1}}{-3 + 1} + C$$

$$= -\tfrac{1}{2}x^{-2} + C = -\frac{1}{2x^2} + C. \quad \blacksquare$$

EXAMPLE 5 Find $\int \left(3 \cos x - 4 \sin x + \frac{1}{x^2}\right) dx$.

SOLUTION

$$\int \left(3 \cos x - 4 \sin x + \frac{1}{x^2}\right) dx = 3 \int \cos x\, dx - 4 \int \sin x\, dx + \int \frac{1}{x^2}\, dx$$

$$= 3 \sin x + 4 \cos x - \frac{1}{x} + C. \quad \blacksquare$$

EXAMPLE 6 Find $\int \frac{x}{1 + x^2}\, dx$.

Multiplying the integrand by a constant

SOLUTION If the numerator were $2x$, then the numerator would be the derivative of the denominator and the antiderivative would be $\ln (1 + x^2)$. But the numerator can be multiplied by 2 if we simultaneously divide by 2:

$$\int \frac{x}{1 + x^2}\, dx = \frac{1}{2} \int \frac{2x}{1 + x^2}\, dx.$$

This step depends on the fact that a constant can be moved past the integral sign:

$$\frac{1}{2}\int \frac{2x}{1+x^2}\,dx = \frac{1}{2}\cdot 2\int \frac{x}{1+x^2}\,dx = \int \frac{x}{1+x^2}\,dx.$$

Thus
$$\int \frac{x}{1+x^2}\,dx = \frac{1}{2}\int \frac{2x}{1+x^2}\,dx$$
$$= \tfrac{1}{2}\ln(1+x^2) + C. \quad \blacksquare$$

Since $1 + x^2 > 0$, the absolute value is not needed in $\ln(1 + x^2)$.

Notation In Examples 2, 4, and 6 the integrand was separated from the dx. This was done for emphasis. Usually the dx is combined with the integrand:

$$\int \frac{4x^3}{x^4+1}\,dx \quad \text{is written} \quad \int \frac{4x^3\,dx}{x^4+1}$$

and
$$\int 1\,dx \quad \text{is written} \quad \int dx.$$

Now we introduce the most commonly used technique of integration, the **substitution method**. Several examples will illustrate the technique, which involves a change of variable. The proof that it works will be given later, in Theorem 1.

EXAMPLE 7 Find $\int (\sin x^2)\, 2x\, dx$.

SOLUTION Note that $2x$ is the derivative of x^2. Introduce $u = x^2$. Then

$$du = 2x\,dx \quad \text{and} \quad \int (\sin x^2)\,2x\,dx = \int \sin u\,du.$$

Now it is easy to find $\int \sin u\,du$:

$$\int \sin u\,du = -\cos u + C.$$

Replacing u by x^2 in $-\cos u$ yields $-\cos x^2$. Thus

$$\int (\sin x^2)\,2x\,dx = -\cos x^2 + C.$$

This answer can be checked by differentiation (using the chain rule):

$$\frac{d}{dx}(-\cos x^2 + C) = \sin x^2 \frac{d}{dx}(x^2) + 0$$
$$= (\sin x^2)\,2x. \quad \blacksquare$$

EXAMPLE 8 Find $\int e^{x^5} 5x^4\,dx$.

SOLUTION Introduce $u = x^5$. Then $du = 5x^4\,dx$ and

$$\int e^{x^5}5x^4\,dx = \int e^u\,du$$
$$= e^u + C$$
$$= e^{x^5} + C. \quad \blacksquare$$

EXAMPLE 9 Find $\int \sin^2 \theta \cos \theta \, d\theta$.

SOLUTION Note that $\cos \theta$ is the derivative of $\sin \theta$, and introduce $u = \sin \theta$. Hence

$$du = \cos \theta \, d\theta.$$

Then
$$\int \underbrace{\sin^2 \theta}_{u^2} \underbrace{\cos \theta \, d\theta}_{du} = \int u^2 \, du$$
$$= \frac{u^3}{3} + C$$
$$= \frac{\sin^3 \theta}{3} + C. \quad \blacksquare$$

EXAMPLE 10 Find $\int (1 + x^3)^5 x^2 \, dx$.

SOLUTION The derivative of $1 + x^3$ is $3x^2$, which differs from the x^2 in the integrand only by the constant factor 3. So let $u = 1 + x^3$. Hence

$$du = 3x^2 \, dx \quad \text{and} \quad \frac{du}{3} = x^2 \, dx.$$

Then
$$\int (1 + x^3)^5 x^2 \, dx = \int u^5 \frac{du}{3}$$
$$= \frac{1}{3} \int u^5 \, du = \frac{1}{3} \frac{u^6}{6} + C$$
$$= \frac{(1 + x^3)^6}{18} + C.$$

It would be instructive to check this answer by differentiation. $\blacksquare$

In Example 10 note that if the x^2 were not present in the integrand, the substitution method would not work. To find $\int (1 + x^3)^5 \, dx$, it would be necessary to multiply out $(1 + x^3)^5$ first, a most unpleasant chore.

EXAMPLE 11 Compare the problems of finding these antiderivatives: $\int \frac{dx}{\sqrt{1 + x^3}}$ and $\int \frac{x^2 \, dx}{\sqrt{1 + x^3}}$.

SOLUTION It turns out that the first antiderivative is *not* an elementary function, whereas the second is easy, because the x^2 is present.

Since x^2 differs from the derivative of $1 + x^3$ only by a constant factor 3, use the substitution $u = 1 + x^3$. Hence

$$du = 3x^2 \, dx \quad \text{and} \quad \frac{du}{3} = x^2 \, dx.$$

Thus
$$\int \frac{x^2 \, dx}{\sqrt{1 + x^3}} = \int \frac{1}{\sqrt{u}} \frac{du}{3} = \frac{1}{3} \int \frac{du}{\sqrt{u}} = \frac{1}{3} \int u^{-1/2} \, du$$
$$= \frac{1}{3} \frac{u^{1/2}}{\frac{1}{2}} + C = \tfrac{2}{3}(1 + x^3)^{1/2} + C$$
$$= \tfrac{2}{3}\sqrt{1 + x^3} + C. \quad \blacksquare$$

In the next example, the choice of substitution is suggested not by something that can serve as du, but rather by the desire to simplify a denominator.

EXAMPLE 12 Find $\int_2^4 \frac{x^2 + 1}{(2x - 3)^2}\, dx$.

To evaluate $\int \frac{P(x)}{(ax + b)^n}\, dx$, where $P(x)$ is a polynomial, let $u = ax + b$.

SOLUTION Since the denominator complicates the problem, try the substitution $u = 2x - 3$. We then have

$$du = 2\, dx, \qquad dx = \frac{du}{2}, \qquad \text{and} \qquad x = \frac{u + 3}{2}.$$

Thus
$$\int \frac{x^2 + 1}{(2x - 3)^2}\, dx = \int \frac{[(u + 3)/2]^2 + 1}{u^2} \frac{du}{2}$$
$$= \int \frac{u^2 + 6u + 13}{8u^2}\, du = \int \left(\frac{1}{8} + \frac{3}{4u} + \frac{13}{8u^2}\right) du$$
$$= \frac{u}{8} + \frac{3}{4} \ln |u| - \frac{13}{8u} + C$$
$$= \frac{2x - 3}{8} + \frac{3}{4} \ln |2x - 3| - \frac{13}{8(2x - 3)} + C.$$

Consequently,
$$\int_2^4 \frac{x^2 + 1}{(2x - 3)^2}\, dx = \left[\frac{2x - 3}{8} + \frac{3}{4} \ln |2x - 3| - \frac{13}{8(2x - 3)} + C\right]\Bigg|_2^4$$
$$= \left(\tfrac{5}{8} + \tfrac{3}{4} \ln 5 - \tfrac{13}{40} + C\right) - \left(\tfrac{1}{8} + \tfrac{3}{4} \ln 1 - \tfrac{13}{8} + C\right)$$
$$= \tfrac{9}{5} + \tfrac{3}{4} \ln 5. \quad \blacksquare$$

As mentioned in the marginal note, the substitution in Example 12, of the type $u = ax + b$, is often useful. Example 13 is another application of this type of substitution.

EXAMPLE 13 Integral tables include a formula for $\int \frac{dx}{(ax + b)^n}$, $n \neq 1$. Use a substitution to find the formula.

SOLUTION Let $u = ax + b$. Then

$$du = a\, dx \qquad \text{and} \qquad dx = \frac{du}{a}.$$

Thus
$$\int \frac{dx}{(ax + b)^n} = \int \frac{du/a}{u^n} = \frac{1}{a} \int u^{-n}\, du$$
$$= \frac{1}{a} \frac{u^{-n+1}}{(-n + 1)} + C = \frac{(ax + b)^{-n+1}}{a(-n + 1)} + C$$
$$= \frac{1}{a(-n + 1)(ax + b)^{n-1}} + C. \quad \blacksquare$$

A substitution is worth trying in two cases:

When to substitute

1 The integrand can be written in the form of a product of a special type:

function of $u(x)$	$\times$	derivative of $u(x)$

for some function $u(x)$.

2 The integrand becomes simpler when a part of it is denoted $u(x)$.

The substitution technique, or "change of variables," extends to definite integrals, $\int_a^b f(x)\,dx$, with one important proviso: When making the substitution from x to u, be sure to replace the interval $[a, b]$ by the interval whose endpoints are $u(a)$ and $u(b)$. An example will illustrate the necessary change in the limits of integration. (It should be compared with Example 12.) The technique is justified in Theorem 2.

EXAMPLE 14 Transform the definite integral $\int_2^4 \frac{x^2 + 1}{(2x - 3)^2}\,dx$ using the substitution $u = 2x - 3$. Then evaluate the new definite integral.

SOLUTION Let $u = 2x - 3$. Then $du = 2\,dx$, $dx = du/2$, $x = (u + 3)/2$, and as x goes from 2 to 4, u goes from

$$2 \cdot 2 - 3 = 1 \quad \text{to} \quad 2 \cdot 4 - 3 = 5.$$

This is the last you see of x in the problem.

Thus
$$\int_2^4 \frac{x^2 + 1}{(2x - 3)^2}\,dx = \int_1^5 \frac{[(u + 3)/2]^2 + 1}{u^2}\frac{du}{2} \qquad \text{here } \int_2^4 \text{ is replaced by } \int_1^5$$

$$= \int_1^5 \left(\frac{1}{8} + \frac{3}{4u} + \frac{13}{8u^2}\right) du$$

u is positive, so $\ln |u| = \ln u$.

$$= \left(\frac{u}{8} + \frac{3}{4}\ln u - \frac{13}{8u}\right)\Bigg|_1^5 \qquad \text{omit } C\text{, since it will cancel}$$

Compare with Example 12.

$$= \left(\frac{5}{8} + \frac{3}{4}\ln 5 - \frac{13}{8 \cdot 5}\right) - \left(\frac{1}{8} + \frac{3}{4}\ln 1 - \frac{13}{8}\right)$$

$$= \tfrac{9}{5} + \tfrac{3}{4}\ln 5. \quad \blacksquare$$

In each example the method is basically the same. In order to apply the substitution technique to find

$$\int f(x)dx,$$

look for a function $u = h(x)$ such that

$$f(x) = g(h(x))h'(x),$$

for some function g, or more simply,

$$f(x)\,dx = g(u)\,du.$$

Then find an antiderivative of g, $\int g(u)\,du$, and replace u by $h(x)$ in this antiderivative.

However, when evaluating $\int_a^b f(x)\,dx$ replace a and b by $u(a)$ and $u(b)$ and carry out all computations in terms of u. It is not necessary to go back and work with the letter x.

It is to be hoped that the problem of finding $\int g(u)\,du$ is easier than that of finding $\int f(x)\,dx$. If it is not, try another substitution or one of the methods presented in the rest of the chapter. It is important to keep in mind that there is no simple routine method for antidifferentiation of elementary functions. Practice in integration pays off in the quick recognition of which technique is most promising.

Finally, we justify the substitution technique.

Theorem 1 *The substitution method.* Let $g(u)$ be a continuous function and let $h(x)$ be a differentiable function. Assume that $G(u)$ is an antiderivative of $g(u)$. Then $G(h(x))$ is an antiderivative of $g(h(x))h'(x)$. [That is, if $G(u) = \int g(u)\,du$, then $G(h(x)) = \int g(h(x))\,h'(x)\,dx$.]

Proof By the definition of an antiderivative, we must show that the derivative of $G(h(x))$ is in fact $g(h(x))h'(x)$. To do this, let $y = G(h(x))$; that is, $y = G(u)$, where $u = h(x)$. By the chain rule,

The chain rule is the basis of the substitution technique.

$$\frac{dy}{dx} = \frac{dy}{du}\frac{du}{dx} = G'(u)h'(x) = g(u)h'(x) = g(h(x))h'(x).$$

This proves the theorem. ■

Theorem 2 *Substitution in the definite integral.* Let f be a continuous function on the interval $[a, b]$, $u = h(x)$ be a differentiable function on the same interval, and g be a continuous function such that

$$f(x)\,dx = g(u)\,du;$$

that is,

$$f(x) = g(h(x))h'(x).$$

Then

$$\int_a^b f(x)\,dx = \int_{h(a)}^{h(b)} g(u)\,du.$$

Proof Let $G(u)$ be an antiderivative of $g(u)$. By Theorem 1, $G(h(x))$ is an antiderivative of $f(x)$. Thus

$$\int_a^b f(x)\,dx \underset{\text{FTC}}{=} G(h(x))\Big|_a^b = G(h(b)) - G(h(a)).$$

But

$$\int_{h(a)}^{h(b)} g(u)\,du \underset{\text{FTC}}{=} G(h(b)) - G(h(a)).$$

This proves that substitution in the definite integral is valid. ■

EXERCISES FOR SEC. 7.1: THE SUBSTITUTION TECHNIQUE

In Exercises 1 to 24 compute the antiderivatives. Do *not* use the substitution technique.

1 $\int 5x^3\,dx$

2 $\int \frac{dx}{6x^3}$

3 $\int x^{1/3}\,dx$

4 $\int \sqrt[3]{x^2}\,dx$

5 $\int \frac{3}{\sqrt{x}}\,dx$

6 $\int \frac{1}{\sqrt[4]{x}}\,dx$

7 $\int 5e^{-2x}\,dx$

8 $\int \frac{5}{1 + x^2}\,dx$

9 $\int \frac{6\,dx}{|x|\sqrt{x^2 - 1}}$

10 $\int \frac{5\,dx}{\sqrt{1 - x^2}}$

11 $\int \frac{x^3}{1 + x^4}\,dx$

12 $\int \frac{e^x}{1 + e^x}\,dx$

13 $\int \frac{\sin x}{1 + \cos x}\,dx$

14 $\int \frac{dx}{1 + 3x}$

15 $\int \frac{1 + 2x}{x + x^2}\,dx$

16 $\int \frac{1 + 2x}{1 + x^2}\,dx$

17 $\int (x^2 + 3)^2\,dx$

18 $\int (1 + e^x)^2\,dx$

19 $\int (1 + 3x)x^2\,dx$

20 $\int \left(\sqrt{x} - \frac{2}{x} + \frac{x}{2}\right) dx$

21 $\int x^2 \sqrt{x}\,dx$

22 $\int 6 \sin x\,dx$

23 $\int \frac{1 + \sqrt{x}}{x}\,dx$

24 $\int \frac{(\sqrt{x} + 1)^2}{\sqrt[3]{x}}\,dx$

In Exercises 25 to 38 *use the given substitutions* to find the antiderivatives or definite integrals.

25 $\int (1 + 3x)^5\, 3\,dx$; $u = 1 + 3x$

26 $\int e^{\sin\theta} \cos\theta\,d\theta$; $u = \sin\theta$

27 $\int_0^1 \frac{x}{\sqrt{1 + x^2}}\,dx$; $u = 1 + x^2$

28 $\int_{\sqrt{8}}^{\sqrt{15}} \sqrt{1 + x^2}\,x\,dx$; $u = 1 + x^2$

29 $\int \sin 2x\,dx$; $u = 2x$

30 $\int \frac{e^{2x}}{(1 + e^{2x})^2}\,dx$; $u = 1 + e^{2x}$

31 $\int_{-1}^{2} e^{3x}\,dx$; $u = 3x$

32 $\int_2^3 \frac{e^{1/x}}{x^2}\,dx$; $u = \frac{1}{x}$

33 $\int \frac{1}{\sqrt{1 - 9x^2}}\,dx$; $u = 3x$

34 $\int \frac{t\,dt}{\sqrt{2 - 5t^2}}$; $u = 2 - 5t^2$

35 $\int_{\pi/6}^{\pi/4} \tan\theta \sec^2\theta\,d\theta$; $u = \tan\theta$

36 $\int_{\pi^2/16}^{\pi^2/4} \frac{\sin\sqrt{x}}{\sqrt{x}}\,dx$; $u = \sqrt{x}$

37 $\int \frac{(\ln x)^4}{x}\,dx$; $u = \ln x$

38 $\int \frac{\sin(\ln x)}{x}\,dx$; $u = \ln x$

In Exercises 39 to 60 use appropriate substitutions to find the antiderivatives.

39 $\int (1 - x^2)^5 x\,dx$

40 $\int \frac{x\,dx}{(x^2 + 1)^3}$

41 $\int \sqrt[3]{1 + x^2}\,x\,dx$

42 $\int \frac{\sin\theta}{\cos^2\theta}\,d\theta$

43 $\int \frac{e^{\sqrt{t}}}{\sqrt{t}}\,dt$

44 $\int e^x \sin e^x\,dx$

45 $\int \sin 3\theta\,d\theta$

46 $\int \frac{dx}{\sqrt{2x + 5}}$

47 $\int (x - 3)^{5/2}\,dx$

48 $\int \frac{dx}{(4x + 3)^3}$

49 $\int \frac{2x + 3}{x^2 + 3x + 2}\,dx$

50 $\int \frac{2x + 3}{(x^2 + 3x + 5)^4}\,dx$

51 $\int e^{2x}\,dx$

52 $\int \frac{dx}{\sqrt{x}(1 + \sqrt{x})^3}$

53 $\int x^4 \sin x^5\,dx$

54 $\int \frac{\cos(\ln x)\,dx}{x}$

55 $\int \frac{x}{1 + x^4}\,dx$

56 $\int \frac{x^3}{1 + x^4}\,dx$

57 $\int \frac{x\,dx}{1 + x}$

58 $\int \frac{x}{\sqrt{1 - x^4}}\,dx$

59 $\int \frac{\ln 3x\,dx}{x}$

60 $\int \frac{\ln x^2\,dx}{x}$

In Exercises 61 to 66 use a substitution to transform the definite integral to a more convenient definite integral. Be sure to change the limits of integration. (Do not evaluate the definite integral.)

61 $\int_1^2 e^{x^3} x^2\,dx$

62 $\int_{\pi/2}^{\pi} \sin^3\theta \cos\theta\,d\theta$

63 $\int_0^1 \frac{x^2 - 3}{(x + 1)^4}\,dx$

64 $\int_1^2 \frac{x^3 - x}{(3x + 1)^3}\,dx$

65 $\int_1^e \frac{(\ln x)^3}{x}\,dx$

66 $\int_0^{\pi/2} \cos^5\theta \sin\theta\,d\theta$

■

In Exercises 67 to 70 use a substitution to evaluate the integral.

67 $\int \frac{x^2\,dx}{ax+b}$, $a \neq 0$

68 $\int \frac{x\,dx}{(ax+b)^2}$, $a \neq 0$

69 $\int \frac{x^2\,dx}{(ax+b)^2}$, $a \neq 0$

70 $\int x(ax+b)^n\,dx$, $n \neq -1, -2$

71 Jack (using the substitution $u = \cos\theta$) claims that $\int 2\cos\theta\sin\theta\,d\theta = -\cos^2\theta$, while Jill (using the substitution $u = \sin\theta$) claims that the answer is $\sin^2\theta$. Who is right?

72 Jill says, "$\int_0^\pi \cos^2\theta\,d\theta$ is obviously positive." Jack claims, "No, it's zero. Just make the substitution $u = \sin\theta$; hence $du = \cos\theta\,d\theta$. Then I get

$$\int_0^\pi \cos^2\theta\,d\theta = \int_0^\pi \cos\theta\cos\theta\,d\theta$$
$$= \int_0^0 \sqrt{1-u^2}\,du = 0.$$

Simple."

(a) Who is right? What is the mistake?

(b) Use the identity $\cos^2\theta = (1+\cos 2\theta)/2$ to evaluate the integral without substitution.

73 Jill asserts that $\int_{-2}^1 2x^2\,dx$ is obviously positive. "After all, the integrand is never negative and $-2 < 1$." "You're wrong again," Jack replies, "It's negative. Here are my computations. Let $u = x^2$; hence $du = 2x\,dx$. Then

$$\int_{-2}^1 2x^2\,dx = \int_{-2}^1 x \cdot 2x\,dx$$
$$= \int_4^1 \sqrt{u}\,du = -\int_1^4 \sqrt{u}\,du,$$

which is obviously negative." Who is right?

74 Let f be a continuous function. Show that $\int_0^2 f(x)\,dx = \int_0^1 [f(x) + f(x+1)]\,dx$.

7.2 INTEGRATION BY PARTS

Just as the chain rule is the basis for integration by substitution, the formula for the derivative of a product is the basis for integration by parts.

Theorem *Integration by parts.* If u and v are differentiable functions and $\int vu'\,dx$ is an antiderivative of vu', then

$$uv - \int vu'\,dx$$

is an antiderivative of uv'. In symbols,

$$\int uv'\,dx = uv - \int vu'\,dx,$$

or, in the notation of differentials,

This differential form is the most useful.

$$\int u\,dv = uv - \int v\,du.$$

Proof Differentiate

$$uv - \int vu'\,dx$$

to see if the result is uv'. We have

$$\left(uv - \int vu'\,dx\right)' = (uv)' - \left(\int vu'\,dx\right)' \qquad \text{derivative of difference}$$

$$= (uv)' - vu' \qquad \text{definition of } \int vu'\,dx$$

$$= uv' + vu' - vu' \qquad \text{derivative of product}$$

$$= uv'.$$

Thus $uv - \int vu'\,dx$ is an antiderivative of uv', as was to be shown. ■

EXAMPLE 1 Find $\int xe^x\,dx$.

SOLUTION To use the formula

$$\int u\,dv = uv - \int v\,du,$$

it is necessary to write $\int xe^x\,dx$

in the form $\int u\,dv$.

(The resulting $\int v\,du$, it is hoped, is easier to find than the original integral $\int u\,dv$.)

The integrand is so simple that there is not much choice. Try

$$u = x \quad \text{and} \quad dv = e^x\,dx;$$

that is, break up the integrand this way:

$$\int \underbrace{x}_{u}\,\underbrace{e^x\,dx}_{dv}.$$

Then find du and v. Since $u = x$, it follows that $du = dx$. Since $dv = e^x\,dx$, we choose $v = e^x$. (Of course, v could be $e^x + C$ for any constant C, but choose the simplest v whose derivative is e^x.) Applying integration by parts yields

$$\int \underbrace{x}_{u}\,\underbrace{e^x\,dx}_{dv} = \underbrace{x}_{u}\,\underbrace{e^x}_{v} - \int \underbrace{e^x}_{v}\,\underbrace{dx}_{du}.$$

Is $\int v\,du$ easier than the original integral, $\int u\,dv$? Yes;

$$\int e^x\,dx = e^x.$$

Save the constant of integration till the end.

Hence

$$\int xe^x\,dx = xe^x - e^x + C.$$

The reader may check this by differentiation. ■

EXAMPLE 2 Use integration by parts to find $\int (\sin x)x\,dx$.

SOLUTION Two ways to break up $(\sin x)x\,dx$ come to mind:

$$\text{(A)} \quad \underbrace{\sin x}_{u}\,\underbrace{x\,dx}_{dv}$$

and

$$\text{(B)} \quad \underbrace{x}_{u}\,\underbrace{\sin x\,dx}_{dv}.$$

Let us see what each method gives.

Method A.

$$\int \underbrace{\sin x}_{u} \underbrace{x\,dx}_{dv} = \underbrace{\sin x}_{u} \underbrace{\frac{x^2}{2}}_{v} - \int \underbrace{\frac{x^2}{2}}_{v} \underbrace{\cos x\,dx}_{du}$$

$$u = \sin x \qquad du = \cos x\,dx$$

$$dv = x\,dx \qquad v = \frac{x^2}{2}$$

The new integrand is harder than the one we started with. Although $\cos x$ is not harder than $\sin x$, $(x^2/2)\cos x$ is surely harder than $x \sin x$, for the exponent has increased from 1 to 2. This route being fruitless, we try method B.

Method B. For convenience, first rewrite the integrand, $(\sin x)x$, as $x \sin x$. Then we have

$$\int \underbrace{x}_{u} \underbrace{\sin x\,dx}_{dv} = \underbrace{x}_{u} \underbrace{(-\cos x)}_{v} - \int \underbrace{(-\cos x)}_{v} \underbrace{dx}_{du}$$

$$u = x \qquad du = dx$$

$$dv = \sin x\,dx \qquad v = -\cos x$$

This time $\int v\,du$ is easier than $\int u\,dv$; the exponent of x went from 1 down to 0. To finish method B, we have to evaluate $\int \cos x\,dx$. All told,

$$\int x \sin x\,dx = -x\cos x + \int \cos x\,dx$$
$$= -x\cos x + \sin x + C. \quad ■$$

EXAMPLE 3 Find $\int x \ln x\,dx$.

SOLUTION Setting $dv = \ln x\,dx$ is not a wise move, since $v = \int \ln x\,dx$ is not immediately apparent. But setting $u = \ln x$ is promising because $du = d(\ln x) = dx/x$ is much easier than $\ln x$. This second approach goes through smoothly:

That v is messier than dv is no threat in this case.

$$\int x \ln x\,dx = \int \underbrace{\ln x}_{u} \underbrace{x\,dx}_{dv} = \underbrace{\ln x}_{u} \underbrace{\frac{x^2}{2}}_{v} - \int \underbrace{\frac{x^2}{2}}_{v} \underbrace{\frac{dx}{x}}_{du}$$

$$= \frac{x^2 \ln x}{2} - \int \frac{x\,dx}{2}$$

$$= \frac{x^2 \ln x}{2} - \frac{x^2}{4} + C.$$

The result may be checked by differentiation. ■

Some words of advice

The key to applying integration by parts is the labeling of u and dv. Usually three conditions should be met:

1 v can be found by integrating and should not be too messy.
2 du should not be messier than u.
3 $\int v\,du$ should be easier than the original $\int u\,dv$.

For instance, $\int x^2 e^x\,dx$, with $u = x^2$ and $dv = e^x\,dx$, meets these criteria. In this case, $v = e^x$ can be found and is not too messy; $du = d(x^2) = 2x\,dx$ is easier than u; and as Example 4 shows, $\int v\,du$ is indeed easier than $\int u\,dv$.

EXAMPLE 4 Find $\displaystyle\int x^2 e^x\,dx$.

SOLUTION

$$\int \underbrace{x^2}_{u}\,\underbrace{e^x\,dx}_{dv} = \underbrace{x^2}_{u}\,\underbrace{e^x}_{v} - \int \underbrace{e^x}_{v}\,\underbrace{2x\,dx}_{du} = x^2e^x - 2\int xe^x\,dx$$

$$u = x^2 \qquad du = 2x\,dx$$

$$dv = e^x\,dx \qquad v = e^x$$

By Example 1,

$$\int xe^x\,dx = xe^x - e^x.$$

Check the answer by differentiation.

Thus

$$\int x^2e^x\,dx = x^2e^x - 2[xe^x - e^x] + C$$
$$= x^2e^x - 2xe^x + 2e^x + C. \quad ■$$

Integration by parts, with $u = x^3$, could be used to express $\int x^3e^x\,dx$ in terms of $\int x^2e^x\,dx$. Another integration by parts, with $u = x^2$, then expresses $\int x^2e^x\,dx$ in terms of $\int xe^x\,dx$, as was done in Example 4. Each time, integration by parts lowers the exponent by 1.

The idea behind this applies to integrals of the form $\int P(x)g(x)\,dx$, where $P(x)$ is a polynomial and $g(x)$ is a function—such as $\sin x$, $\cos x$, or e^x—that can be repeatedly integrated. Let $u = P(x)$ and $dv = g(x)\,dx$.

The next example shows how to integrate any inverse trigonometric function.

EXAMPLE 5 Find $\displaystyle\int \tan^{-1} x\,dx$.

SOLUTION Recall that the derivative of $\tan^{-1} x$ is $1/(1 + x^2)$, a much simpler function than $\tan^{-1} x$. This suggests the following approach:

$$\int \underbrace{\tan^{-1} x}_{u}\,\underbrace{dx}_{dv} = \underbrace{(\tan^{-1} x)}_{u}\,\underbrace{x}_{v} - \int \underbrace{x}_{v}\,\underbrace{\frac{dx}{1 + x^2}}_{du} \qquad \left(du = \frac{dx}{1 + x^2},\quad v = x\right)$$

Integrating an inverse function by parts

$$= x\tan^{-1} x - \int \frac{x}{1 + x^2}\,dx.$$

It is easy to compute

$$\int \frac{x}{1 + x^2}\,dx,$$

since the numerator is a constant times the derivative of the denominator:

$$\int \frac{x\,dx}{1 + x^2} = \frac{1}{2}\int \frac{2x}{1 + x^2}\,dx = \tfrac{1}{2}\ln(1 + x^2).$$

Check by differentiation. Hence $$\int \tan^{-1} x\,dx = x\tan^{-1} x - \tfrac{1}{2}\ln(1+x^2) + C. \quad \blacksquare$$

Evaluation by integration by parts of a definite integral $\int_a^b f(x)\,dx$, where $f(x) = u(x)v'(x)$, takes the form

$$\begin{aligned}\int_a^b f(x)\,dx &= \int_a^b u\,dv = uv\Big|_a^b - \int_a^b v\,du \\ &= u(b)v(b) - u(a)v(a) - \int_a^b v(x)u'(x)\,dx.\end{aligned}$$

For instance, by Example 5,

$$\begin{aligned}\int_0^1 \tan^{-1} x\,dx &= x\tan^{-1} x\Big|_0^1 - \int_0^1 \frac{x}{1+x^2}\,dx \\ &= x\tan^{-1} x\Big|_0^1 - \frac{1}{2}\ln(1+x^2)\Big|_0^1 \\ &= 1\tan^{-1} 1 - 0\tan^{-1} 0 - \tfrac{1}{2}\ln(1+1^2) + \tfrac{1}{2}\ln(1+0^2) \\ &= \frac{\pi}{4} - \frac{1}{2}\ln 2.\end{aligned}$$

EXAMPLE 6 Find $\int e^x \cos x\,dx$.

SOLUTION Proceed as follows:

$$\int \underbrace{e^x}_{u}\underbrace{\cos x\,dx}_{dv} = \underbrace{e^x}_{u}\underbrace{\sin x}_{v} - \int \underbrace{\sin x}_{v}\underbrace{e^x\,dx}_{du}$$

$$du = e^x\,dx \qquad v = \sin x.$$

Repeated integration by parts

It may seem that nothing useful has been accomplished; $\cos x$ is replaced by $\sin x$. But watch closely as the new integral is treated by an integration by parts. Capital letters U and V, instead of u and v, are used to distinguish this computation from the preceding one.

$$\begin{aligned}\int \underbrace{e^x}_{U}\underbrace{\sin x\,dx}_{dV} &= \underbrace{e^x}_{U}\underbrace{(-\cos x)}_{V} - \int \underbrace{(-\cos x)}_{V}\underbrace{e^x\,dx}_{dU} \qquad (dU = e^x\,dx, \quad V = -\cos x) \\ &= -e^x\cos x + \int e^x \cos x\,dx.\end{aligned}$$

Combining the two yields

$$\begin{aligned}\int e^x\cos x\,dx &= e^x\sin x - \left(-e^x\cos x + \int e^x\cos x\,dx\right) \\ &= e^x(\sin x + \cos x) - \int e^x \cos x\,dx. \qquad (1)\end{aligned}$$

It might seem that we could rewrite this as

$$2\int e^x\cos x\,dx = e^x(\sin x + \cos x) \qquad (2)$$

and conclude that

$$\int e^x\cos x\,dx = \tfrac{1}{2}e^x(\sin x + \cos x).$$

This cannot be correct, since there is no constant of integration. In fact, the antiderivatives which appear on the left and right sides of Eq. (1) need not be equal, but could differ by a constant K. Hence, instead of Eq. (2) we have

$$2\int e^x \cos x\, dx = e^x(\sin x + \cos x) + K,$$

so

$$\int e^x \cos x\, dx = \tfrac{1}{2}e^x(\sin x + \cos x) + \frac{K}{2}.$$

Giving the constant $K/2$ the name "C," we then have

$$\int e^x \cos x\, dx = \tfrac{1}{2}e^x(\sin x + \cos x) + C. \quad \blacksquare$$

Reduction or recursion formulas

EXAMPLE 7 Many formulas in a table of integrals express the integral of a function that involves the nth power of some expression in terms of the integral of a function that involves the $(n-1)$th or lower power of the same expression. These are **reduction formulas**. Usually they are obtained by an integration by parts. For instance, derive the formula

$$\int \sin^n x\, dx = -\frac{\sin^{n-1} x \cos x}{n} + \frac{n-1}{n}\int \sin^{n-2} x\, dx, \qquad n \geq 2.$$

SOLUTION First write $\int \sin^n x\, dx$ as $\int \sin^{n-1} x \sin x\, dx$. Then let $u = \sin^{n-1} x$ and $dv = \sin x\, dx$. Thus

$$du = (n-1)\sin^{n-2} x \cos x\, dx \qquad \text{and} \qquad v = -\cos x.$$

Integration by parts yields

$$\int \underbrace{\sin^{n-1} x}_{u}\, \underbrace{\sin x\, dx}_{dv} = \underbrace{(\sin^{n-1} x)}_{u}\, \underbrace{(-\cos x)}_{v} - \int \underbrace{(-\cos x)}_{v}\, \underbrace{(n-1)\sin^{n-2} x \cos x\, dx}_{du}.$$

But the integral on the right of the preceding equation is equal to

$$\begin{aligned}-\int (n-1)\cos^2 x \sin^{n-2} x\, dx &= -(n-1)\int (1-\sin^2 x)\sin^{n-2} x\, dx \\ &= -(n-1)\int \sin^{n-2} x\, dx + (n-1)\int \sin^n x\, dx.\end{aligned}$$

Thus

$$\int \sin^n x\, dx = -\sin^{n-1} x \cos x + (n-1)\int \sin^{n-2} x\, dx - (n-1)\int \sin^n x\, dx.$$

Rather than being dismayed by the reappearance of $\int \sin^n x\, dx$, collect like terms:

$$n\int \sin^n x\, dx = -\sin^{n-1} x \cos x + (n-1)\int \sin^{n-2} x\, dx,$$

from which the quoted formula follows. $\blacksquare$

Integration by parts is used for more purposes than finding antiderivatives. In Sec. 10.8 it will be applied to analyze the error in a certain type of estimate. It is also of importance in differential equations. The next example illustrates the versatility of integration by parts.

EXAMPLE 8 Find $\lim_{a\to\infty} \int_0^1 e^{x^2} \cos ax\, dx$.

SOLUTION Letting $u = e^{x^2}$ and $dv = \cos ax\, dx$, we have

$$\int_0^1 e^{x^2} \cos ax\, dx = \frac{e^{x^2} \sin ax}{a}\Bigg|_0^1 - \int_0^1 \frac{2xe^{x^2} \sin ax\, dx}{a}$$

$$= \frac{e \sin a}{a} - \frac{2}{a}\int_0^1 xe^{x^2} \sin ax\, dx.$$

Now, for x in $[0, 1]$,

$$-e \le xe^{x^2} \sin ax \le e.$$

Hence, $$-e = \int_0^1 -e\, dx \le \int_0^1 xe^{x^2} \sin ax\, dx \le \int_0^1 e\, dx = e.$$

Consequently, $$\left|\int_0^1 xe^{x^2} \sin ax\, dx\right| \le e$$

and $$\lim_{a\to\infty}\left(\frac{e \sin a}{a} - \frac{1}{a}\int_0^1 2xe^{x^2} \sin ax\, dx\right) = 0.$$

Exercise 46 explores this further. Thus the limit of $\int_0^1 e^{x^2} \cos ax\, dx$ as $a \to \infty$ is 0. ■

EXERCISES FOR SEC. 7.2: INTEGRATION BY PARTS

In Exercises 1 to 22 evaluate the integrals by integration by parts.

1 $\int xe^{2x}\, dx$

2 $\int (x + 3)e^{-x}\, dx$

3 $\int x \sin 2x\, dx$

4 $\int (x + 3) \cos 2x\, dx$

5 $\int x \ln 3x\, dx$

6 $\int (2x + 1) \ln x\, dx$

7 $\int_1^2 x^2e^{-x}\, dx$

8 $\int_0^1 x^2e^{2x}\, dx$

9 $\int_0^1 \sin^{-1} x\, dx$

10 $\int_0^{1/2} \tan^{-1} 2x\, dx$

11 $\int x^2 \ln x\, dx$

12 $\int x^3 \ln x\, dx$

13 $\int x(3x + 5)^{10}\, dx$

14 $\int (x + 2)(2x - 1)^{50}\, dx$

15 $\int_2^3 (\ln x)^2\, dx$

16 $\int_2^3 (\ln x)^3\, dx$

17 $\int_1^e \frac{\ln x\, dx}{x^2}$

18 $\int_e^{e^2} \frac{\ln x\, dx}{x^3}$

19 $\int e^x \sin x\, dx$

20 $\int e^{-2x} \sin 3x\, dx$

21 $\int \frac{\ln (1 + x^2)\, dx}{x^2}$

22 $\int x \ln (x^2)\, dx$

■

23 Obtain this recursion formula, which is usually to be found in a table of integrals:

$$\int \sin^n ax\, dx$$
$$= -\frac{\sin^{n-1} ax \cos ax}{na} + \frac{n-1}{n}\int \sin^{n-2} ax\, dx.$$

24 Obtain this recursion formula ($m, n \ge 0$):

$$\int \sin^m x \cos^n x\, dx$$
$$= -\frac{\sin^{m-1} x \cos^{n+1} x}{m+n} + \frac{m-1}{m+n}\int \sin^{m-2} x \cos^n x\, dx.$$

25 Find $\int \ln (x + 1)\, dx$ using
(a) $u = \ln (x + 1)$, $dv = dx$, $v = x$
(b) $u = \ln (x + 1)$, $dv = dx$, $v = x + 1$
(c) Which is easier?

In Exercises 26 to 28 obtain recursion formulas for the integrals.

26 $\int x^ne^{ax}\, dx$, n an integer > 0, a a nonzero constant

27 $\int (\ln x)^n\, dx$, n an integer > 0

28 $\int x^n \sin x\, dx$, n an integer > 0

In Exercises 29 to 32 find the integrals. In each case a substitution is required before integration by parts can be used. In Exercises 31 and 32 the notation exp (u) is used for e^u. This notation is often used for clarity.

29 $\int \sin \sqrt{x}\, dx$ **30** $\int \sin \sqrt[3]{x}\, dx$

31 $\int \exp(\sqrt{x})\, dx$ **32** $\int \exp(\sqrt[3]{x})\, dx$

33 Use the recursion in Example 7 to find

(*a*) $\int \sin^2 x\, dx$ (*b*) $\int \sin^4 x\, dx$ (*c*) $\int \sin^6 x\, dx$

34 Use the recursion in Example 7 to find

(*a*) $\int \sin^3 x\, dx$ (*b*) $\int \sin^5 x\, dx$

35 $\int [(\sin x)/x]\, dx$ is not elementary. Deduce that $\int \cos x \ln x\, dx$ is not elementary.

36 $\int x \tan x\, dx$ is not elementary. Deduce that $\int (x/\cos x)^2\, dx$ is not elementary.

In Exercises 37 to 40 find the integrals two ways: (*a*) by substitution, (*b*) by integration by parts.

37 $\int x\sqrt{3x+7}\, dx$ **38** $\int \frac{x\, dx}{\sqrt{2x+7}}$

39 $\int x(ax+b)^3\, dx$ **40** $\int \frac{x\, dx}{ax+b}$, $a \neq 0$

41 Let I_n denote $\int_0^{\pi/2} \sin^n \theta\, d\theta$, where n is a nonnegative integer.

(*a*) Evaluate I_0 and I_1.

(*b*) Using the recursion in Example 7, show that

$$I_n = \frac{n-1}{n} I_{n-2}, \quad \text{for } n \geq 2.$$

(*c*) Use (*b*) to evaluate I_2 and I_3.

(*d*) Use (*b*) to evaluate I_4 and I_5.

(*e*) Find a formula for I_n, n odd.

(*f*) Find a formula for I_n, n even.

(*g*) Explain why $\int_0^{\pi/2} \cos^n \theta\, d\theta = \int_0^{\pi/2} \sin^n \theta\, d\theta$.

42 Show that

$$\int_0^1 xf''(x)\, dx = 3$$

for every function $f(x)$ that satisfies the following conditions: (*i*) $f(x)$ is defined for all x, (*ii*) $f''(x)$ is continuous, (*iii*) $f(0) = f(1)$, (*iv*) $f'(1) = 3$.

Exercises 43 and 44 are related.

43 In a certain race, a car starts from rest and ends at rest, having traveled 1 mile in 1 minute. Let $v(t)$ be its velocity at time t and $a(t)$ be its acceleration at time t. Show that

(*a*) $\int_0^1 v(t)\, dt = 1$ (*b*) $\int_0^1 a(t)\, dt = 0$

(*c*) $\int_0^1 ta(t)\, dt = -1$

44 (Continuation of Exercise 43.)

(*a*) Show that at some time t we have $|a(t)| > 4$.

(*b*) Show graphically [drawing $v(t)$ as a function of time] that a race can be driven as in Exercise 43, but with $|a(t)| \leq 4.1$ for all t.

Exercise 45 shows how integration by parts can be used to study the approximation of a function by a polynomial.

45 Let f have derivatives of all orders.

(*a*) Explain why $f(b) = f(0) + \int_0^b f'(x)\, dx$.

(*b*) Using an integration by parts on the definite integral in (*a*), with $u = f'(x)$ and $dv = dx$, show that

$$f(b) = f(0) + f'(0)b + \int_0^b f^{(2)}(x)(b-x)\, dx.$$

Hint: Use $v = x - b$.

(*c*) Similarly, show that

$$f(b) = f(0) + f'(0)b + \frac{f^{(2)}(0)}{2} b^2 + \frac{1}{2}\int_0^b f^{(3)}(x)(b-x)^2\, dx.$$

(*d*) Check that (*c*) is correct for any quadratic polynomial $f(x) = Ax^2 + Bx + C$.

(*e*) Use another integration by parts on the formula in (*c*) to obtain the "next formula."

46 (See Example 8.)

(*a*) Describe the general appearance of the graph of $y = \cos ax$, when a is large.

(*b*) Describe the general appearance of the graph of $y = e^{x^2} \cos ax$, when a is large.

(*c*) In view of (*b*), is the result in Example 8 reasonable?

47 Let f have a continuous derivative for x in $[a, b]$. Examine $\lim_{c\to\infty} \int_a^b f(x) \sin cx\, dx$.

48 Evaluate $\int_1^2 x^4 e^{2x}\, dx$.

Evaluate $\int_0^1 (x^4 + 2x) \sin 3x\, dx$.

7.3 HOW TO INTEGRATE CERTAIN RATIONAL FUNCTIONS

This section shows how to compute

$ax^2 + bx + c$ is irreducible.

$$\int \frac{dx}{(ax+b)^n}, \quad \int \frac{dx}{(ax^2+bx+c)^n}, \quad \text{and} \quad \int \frac{x\,dx}{(ax^2+bx+c)^n}.$$

(The polynomial $ax^2 + bx + c$ is assumed to be irreducible, that is, not the product of two polynomials of degree 1.) These three types of integrals will play a basic role in Sec. 7.4.

How to tell when $ax^2 + bx + c$ is irreducible

To determine whether $ax^2 + bx + c$ is irreducible, compute its discriminant, $b^2 - 4ac$. If it is negative, the polynomial is irreducible. If it is 0 or positive, the polynomial is reducible, that is, the product of first-degree factors. (See Exercise 43.)

How to Compute $\int \frac{dx}{(ax+b)^n}$

As was shown in Example 13 of Sec. 7.1, the computation of $\int dx/(ax+b)^n$ can be accomplished by the substitution $u = ax + b$. For instance,

$$\begin{aligned}\int \frac{dx}{(3x+2)^5} &= \int u^{-5}\frac{du}{3} \\ &= \frac{1}{3}\frac{u^{-4}}{-4} + C \\ &= \frac{-1}{12}\frac{1}{(3x+2)^4} + C.\end{aligned} \qquad \left\{\begin{aligned} u &= 3x+2 \\ du &= 3\,dx \\ \frac{du}{3} &= dx\end{aligned}\right\}$$

Also, with the same u,

$$\int \frac{dx}{3x+2} = \int \frac{du/3}{u} = \frac{1}{3}\int \frac{du}{u} = \frac{1}{3}\ln|u| + C = \frac{1}{3}\ln|3x+2| + C.$$

The integral $\int dx/(3x + 2)$ could also be evaluated by noticing that the numerator is almost the derivative of the denominator.

For any polynomial $P(x)$ the integral

$$\int \frac{P(x)\,dx}{(ax+b)^n} \tag{1}$$

can be computed by making the substitution $u = ax + b$. (Example 12 of Sec. 7.1 illustrates the computations.) Thus a rational function of the form (1) is not hard to integrate.

Some examples will show how to compute $\int dx/(ax^2 + bx + c)$, where $ax^2 + bx + c$ is irreducible.

EXAMPLE 1 Find $\int \frac{dx}{4x^2+1}$.

SOLUTION This resembles

$$\int \frac{dx}{x^2+1} = \tan^{-1} x + C.$$

For this reason, make a substitution so that $u^2 = 4x^2$. To do this, let $u = 2x$; hence

$u = -2x$ would work too.

$$du = 2\,dx \quad \text{and} \quad \frac{du}{2} = dx.$$

Then
$$\int \frac{dx}{4x^2 + 1} = \int \frac{1}{u^2 + 1}\frac{du}{2} = \frac{1}{2}\int \frac{du}{u^2 + 1} = \frac{1}{2}\tan^{-1} u + C$$
$$= \tfrac{1}{2}\tan^{-1} 2x + C. \quad ■$$

EXAMPLE 2 Find $\int \frac{dx}{4x^2 + 9}$.

SOLUTION Again the motivation is provided by the fact that

$$\int \frac{dx}{x^2 + 1} = \tan^{-1} x + C.$$

This time choose u such that $9u^2 = 4x^2$. This substitution is suggested by the equation

$$\frac{1}{4x^2 + 9} = \frac{1}{9u^2 + 9} = \frac{1}{9}\frac{1}{u^2 + 1}.$$

So choose u such that $3u = 2x$; hence

$$3\,du = 2\,dx \quad \text{and} \quad \tfrac{3}{2}\,du = dx.$$

Thus
$$\int \frac{dx}{4x^2 + 9} = \int \frac{1}{9u^2 + 9}\frac{3}{2}\,du = \frac{3}{18}\int \frac{du}{u^2 + 1}$$
$$= \frac{1}{6}\tan^{-1} u + C = \frac{1}{6}\tan^{-1}\frac{2x}{3} + C.$$

(Note that only at the end is it necessary to solve for u; $u = 2x/3$.) ■

Had Example 2 asked for the definite integral $\int_0^2 dx/(4x^2 + 9)$, the same substitution $u = \frac{2}{3}x$ would apply, giving

$$\int_0^2 \frac{dx}{4x^2 + 9} = \frac{3}{18}\int_0^{4/3} \frac{du}{u^2 + 1} = \frac{1}{6}\tan^{-1}\frac{4}{3} - \frac{1}{6}\tan^{-1} 0.$$

The next example uses "completing the square," an algebraic technique described in Appendix C.

EXAMPLE 3 Find $\int \frac{dx}{x^2 + 4x + 13}$.

SOLUTION (Since $4^2 - 4 \cdot 1 \cdot 13$ is negative, $x^2 + 4x + 13$ is irreducible.) Begin by completing the square in the denominator:

Completing the square

$$x^2 + 4x + 13 = x^2 + 4x + 2^2 + 13 - 2^2 = (x + 2)^2 + 9.$$

Thus
$$\int \frac{dx}{x^2 + 4x + 13} = \int \frac{dx}{(x + 2)^2 + 9},$$

an integral reminiscent of those in Examples 1 and 2.

To complete the integration, introduce a function u such that

$$9u^2 = (x + 2)^2.$$

To do this, let $3u = x + 2$; that is, $u = (x + 2)/3$. It follows that $3\,du = dx$. Thus

$$\int \frac{dx}{(x+2)^2 + 9} = \int \frac{3\,du}{9u^2 + 9} = \frac{3}{9}\int \frac{du}{u^2 + 1} = \frac{1}{3}\tan^{-1} u + C.$$

Consequently,

It is instructive to check this by differentiation.

$$\int \frac{dx}{x^2 + 4x + 13} = \frac{1}{3}\tan^{-1}\frac{x+2}{3} + C. \quad \blacksquare$$

In the next example the coefficient of x^2 is not 1; completing the square involves a little more algebra.

EXAMPLE 4 Find $\displaystyle\int \frac{dx}{4x^2 + 8x + 13}$.

SOLUTION First, complete the square in the denominator, $4x^2 + 8x + 13$, as follows:

$$\begin{aligned} 4x^2 + 8x + 13 &= 4(x^2 + 2x) + 13 \\ &= 4[x^2 + 2x + (\tfrac{2}{2})^2] + 13 - 4(\tfrac{2}{2})^2 \\ &= 4(x+1)^2 + 9. \end{aligned}$$

The integral now reads $\displaystyle\int \frac{dx}{4(x+1)^2 + 9}$,

which resembles Example 2.

Choose a substitution such that $9u^2 = 4(x + 1)^2$. To do this, choose u so that $3u = 2(x + 1)$; that is, $u = 2(x + 1)/3$. Consequently,

$$3\,du = 2\,dx \qquad \text{and} \qquad dx = \frac{3\,du}{2}.$$

The substitution yields

$$\begin{aligned} \int \frac{dx}{4(x+1)^2 + 9} &= \int \frac{\frac{3}{2}\,du}{9u^2 + 9} \\ &= \frac{3}{2}\cdot\frac{1}{9}\int \frac{du}{u^2 + 1} \\ &= \tfrac{1}{6}\tan^{-1} u + C. \end{aligned}$$

Thus $$\int \frac{dx}{4x^2 + 8x + 13} = \frac{1}{6}\tan^{-1}\frac{2(x+1)}{3} + C,$$

a result that the skeptical may check by differentiation. $\blacksquare$

As these examples show, to compute

$$\int \frac{dx}{ax^2 + bx + c} \qquad (b^2 - 4ac < 0),$$

complete the square and then make a substitution. The integral will involve an arctangent.

The integral $\int dx/(ax^2 + bx + c)^n$, $n > 1$, is computed by a recursive formula to be found in Exercise 44 and in integral tables.

The computation of $\int x\,dx/(ax^2 + bx + c)$ *can be reduced to that of* $\int dx/(ax^2 + bx + c)$, as is shown in Example 5.

EXAMPLE 5 Find $\displaystyle\int \frac{x\,dx}{4x^2 + 8x + 13}$.

SOLUTION If the numerator were $8x + 8$, it would be the derivative of the denominator. The problem would then be covered by the formula

$$\int \frac{f'}{f}\,dx = \ln|f| + C.$$

This prompts the following maneuver:

Now recall Example 4.

$4x^2 + 8x + 13$ is always positive. In fact, it equals $4(x + 1)^2 + 9$, which is never less than 9.

$$\begin{aligned}\int \frac{x\,dx}{4x^2 + 8x + 13} &= \frac{1}{8}\int \frac{8x\,dx}{4x^2 + 8x + 13}\\ &= \frac{1}{8}\left(\int \frac{8x + 8}{4x^2 + 8x + 13}\,dx - \int \frac{8}{4x^2 + 8x + 13}\,dx\right)\\ &= \frac{1}{8}\left[\ln(4x^2 + 8x + 13) - \frac{8}{6}\tan^{-1}\frac{2(x + 1)}{3}\right] + C.\end{aligned}$$

(The result of Example 4 was used.) ■

Summary of This Section

Integrand	*Method of Integration*
$\dfrac{1}{(ax + b)^n}$	Substitute $u = ax + b$.
$\dfrac{1}{ax^2 + c}$, $a, c > 0$	Substitute so $cu^2 = ax^2$: $u = \sqrt{\dfrac{a}{c}}\,x$.
$\dfrac{1}{ax^2 + bx + c}$, $b^2 - 4ac < 0$	Complete the square, then substitute.
$\dfrac{x}{ax^2 + bx + c}$, $b^2 - 4ac < 0$	First write as $\dfrac{1}{2a}\cdot\dfrac{(2ax + b) - b}{ax^2 + bx + c}$, then break into two parts.

Note that, for any constants c_1 and d_1,

$$\int \frac{c_1x + d_1}{ax^2 + bx + c}\,dx = \int \frac{c_1x}{ax^2 + bx + c}\,dx + \int \frac{d_1}{ax^2 + bx + c}\,dx.$$

This will be needed in the next section.

EXERCISES FOR SEC. 7.3: HOW TO INTEGRATE CERTAIN RATIONAL FUNCTIONS

Compute the integrals in Exercises 1 to 36.

1 $\int \frac{dx}{3x - 4}$

2 $\int \frac{2\,dx}{3x + 6}$

3 $\int \frac{5\,dx}{(2x + 7)^2}$

4 $\int \frac{dx}{(4x + 1)^3}$

5 $\int \frac{dx}{x^2 + 9}$

6 $\int \frac{dx}{9x^2 + 1}$

7 $\int \frac{x\,dx}{x^2 + 9}$

8 $\int \frac{x\,dx}{x^2 + 2}$

9 $\int \frac{2x + 3}{x^2 + 9}\,dx$

10 $\int \frac{3x - 5}{x^2 + 9}\,dx$

11 $\int \frac{dx}{16x^2 + 25}$

12 $\int \frac{dx}{9x^2 + 4}$

13 $\int \frac{x\,dx}{16x^2 + 25}$

14 $\int \frac{x\,dx}{9x^2 + 4}$

15 $\int \frac{x + 2}{9x^2 + 4}\,dx$

16 $\int \frac{2x - 1}{9x^2 + 4}\,dx$

17 $\int \frac{dx}{2x^2 + 3}$

18 $\int \frac{x\,dx}{2x^2 + 3}$

19 $\int \frac{dx}{x^2 + 2x + 3}$

20 $\int \frac{dx}{x^2 + 2x + 5}$

21 $\int \frac{dx}{x^2 - 2x + 3}$

22 $\int \frac{x\,dx}{x^2 - 2x + 3}$

23 $\int \frac{dx}{2x^2 + x + 3}$

24 $\int \frac{dx}{3x^2 - 12x + 13}$

25 $\int \frac{dx}{x^2 + 4x + 7}$

26 $\int \frac{dx}{x^2 + 4x + 9}$

27 $\int \frac{dx}{2x^2 + 4x + 7}$

28 $\int \frac{dx}{2x^2 + 6x + 5}$

29 $\int \frac{2x\,dx}{x^2 + 2x + 3}$

30 $\int \frac{2x\,dx}{x^2 + 2x + 5}$

31 $\int \frac{3x\,dx}{5x^2 + 3x + 2}$

32 $\int \frac{x\,dx}{5x^2 - 3x + 2}$

33 $\int \frac{x + 1}{x^2 + x + 1}\,dx$

34 $\int \frac{x + 3}{x^2 + x + 1}\,dx$

35 $\int \frac{3x + 5}{3x^2 + 2x + 1}\,dx$

36 $\int \frac{x + 5}{2x^2 + 3x + 5}\,dx$

■

In Exercises 37 and 38 determine which polynomials are irreducible and which are reducible. If a polynomial is reducible, write it as the product of first-degree factors.

37 (*a*) $x^2 - 9$ (*b*) $x^2 - 5$ (*c*) $x^2 + 9$ (*d*) $x^2 + 3x + 2$ (*e*) $x^2 + 6x + 9$ (*f*) $2x^2 + 3x + 2$ (*g*) $x^2 + 5x + 2$

38 (*a*) $x^2 - 4$ (*b*) $x^2 - 3$ (*c*) $x^2 + 3$ (*d*) $2x^2 + 3x + 1$ (*e*) $2x^2 + 3x + 7$ (*f*) $2x^2 + 3x - 7$ (*g*) $49x^2 + 25$

39 Compute $\int dx/(ax^2 + c)$, $a, c > 0$.

40 Compute $\int x\,dx/(ax^2 + c)$, $a, c > 0$.

41 Let $ax^2 + bx + c$ be reducible.

(*a*) Verify that

$$\int \frac{dx}{ax^2 + bx + c} = \frac{1}{\sqrt{b^2 - 4ac}} \ln \left| \frac{2ax + b - \sqrt{b^2 - 4ac}}{2ax + b + \sqrt{b^2 - 4ac}} \right| + C, \qquad b^2 > 4ac.$$

(*b*) Obtain the formula in (*a*) by the methods of this section.

42 Let $ax^2 + bx + c$ be irreducible.

(*a*) Verify that

$$\int \frac{dx}{ax^2 + bx + c} = \frac{2}{\sqrt{4ac - b^2}} \tan^{-1} \frac{2ax + b}{\sqrt{4ac - b^2}} + C, \qquad b^2 < 4ac.$$

(*b*) Obtain the formula in (*a*) by the methods of this section.

43 (*a*) Show that if r_1 and r_2 are the roots of the equation $ax^2 + bx + c = 0$, then $a(x - r_1)(x - r_2) = ax^2 + bx + c$. (*Suggestion:* Use the quadratic formula.)

(*b*) Show that if $ax^2 + bx + c = a(x - r_1)(x - r_2)$, then the numbers r_1 and r_2 are roots of the equation $ax^2 + bx + c = 0$.

(*c*) Deduce from (*a*) and (*b*) that $ax^2 + bx + c$ is irreducible exactly when $b^2 - 4ac < 0$.

44 Let $ax^2 + bx + c$ be irreducible. Verify the following recursion formula by differentiating the right side of the equation:

$$\int \frac{dx}{(ax^2 + bx + c)^{n+1}} = \frac{2ax + b}{n(4ac - b^2)(ax^2 + bx + c)^n} + \frac{2(2n - 1)a}{n(4ac - b^2)} \int \frac{dx}{(ax^2 + bx + c)^n}.$$

45 Let $ax^2 + bx + c$ be irreducible. Verify the following reduction formula by differentiating the right side of the equation:

$$\int \frac{x\,dx}{(ax^2 + bx + c)^{n+1}} = \frac{-(2c + bx)}{n(4ac - b^2)(ax^2 + bx + c)^n} - \frac{b(2n - 1)}{n(4ac - b^2)} \int \frac{dx}{(ax^2 + bx + c)^n}.$$

46 Use the identity in Exercise 44 and the result of Example 1 to find

$$\int \frac{dx}{(4x^2 + 1)^2}.$$

47 Use the identity in Exercise 45 and the result of Example 4 to find

$$\int \frac{x\,dx}{(4x^2 + 8x + 13)^2}.$$

7.4 INTEGRATION OF RATIONAL FUNCTIONS BY PARTIAL FRACTIONS

The algebraic technique known as **partial fractions** makes it possible to integrate any rational function. For instance, later in this section it will be shown how to compute the integral

$$\int \frac{x^4 + x^3 - 3x + 5}{x^3 + 2x^2 + 2x + 1}\,dx. \tag{1}$$

[No integral table lists a form that covers (1).] The technique of partial fractions is also used in differential equations.

This section, which is purely algebraic, depends on this result from advanced algebra: Every rational function can be expressed as a sum of a polynomial (which may be 0) and constant multiples of the three types of functions met in Sec. 7.3:

$ax^2 + bx + c$ is assumed to be irreducible. If not, the second and third types can be expressed in terms of the first type, $\left(\text{e.g., } \frac{1}{x^2 - 1} = \frac{\frac{1}{2}}{x - 1} - \frac{\frac{1}{2}}{x + 1}\right).$

$$\frac{1}{(ax + b)^n}, \qquad \frac{1}{(ax^2 + bx + c)^n}, \qquad \text{and} \qquad \frac{x}{(ax^2 + bx + c)^n}. \tag{2}$$

(Moreover, the representation is unique.) Partial fractions were employed in Example 4 of Sec. 6.7 without proper credit when use was made of the identity

$$\frac{M}{y(M - y)} = \frac{1}{y} + \frac{1}{M - y}.$$

Since any polynomial and each of the three types of rational functions in (2) can be integrated, any rational function can be integrated. The only new question of interest is "What is the method for expressing a rational function as a sum of these four types of simpler functions?" A general method is presented in this section. The resulting expression is called the **partial-fraction representation** of the rational function.

To express A/B, where A and B are polynomials, as the sum of partial fractions, follow these steps:

1. Make degree of numerator less than degree of denominator.

Step 1 If the degree of A is *equal to or greater than* the degree of B, divide B into A to obtain a quotient and a remainder: $A = QB + R$, where the degree of R is less than the degree of B or else $R = 0$. Then

$$\frac{A}{B} = Q + \frac{R}{B}.$$

Apply the remaining steps to R/B.

EXAMPLE 1 If $\dfrac{A}{B} = \dfrac{3x^3 + x}{x^2 + 3x + 5}$, carry out step 1.

SOLUTION Since the degree of the numerator is *not* less than the degree of the denominator, carry out a long division:

$$\begin{array}{r|rrrrl}
 & & 3x & & -\ 9 & \text{quotient} \\ \cline{2-5}
x^2 + 3x + 5 & 3x^3 & +\ 0x^2 & +\ x & +\ 0 & \\
 & 3x^3 & +\ 9x^2 & +\ 15x & & \\ \cline{2-5}
 & & -\ 9x^2 & -\ 14x & +\ 0 & \\
 & & -\ 9x^2 & -\ 27x & -\ 45 & \\ \cline{3-5}
 & & & 13x & +\ 45 & \text{remainder}
\end{array}$$

Thus
$$\frac{3x^3 + x}{x^2 + 3x + 5} = 3x - 9 + \frac{13x + 45}{x^2 + 3x + 5}.$$

(To check, just multiply both sides by $x^2 + 3x + 5$.) ■

Similarly, in the case

$$\frac{3x^2 + x}{x^2 + 3x + 5},$$

a division would be carried out first.

EXAMPLE 2 Carry out step 1 on

$$\frac{A}{B} = \frac{4x^3 + x + 1}{x + 3}.$$

SOLUTION

$$\begin{array}{r|rrrrl}
 & 4x^2 & -\ 12x & +\ 37 & & \text{quotient} \\ \cline{2-5}
x + 3 & 4x^3 & +\ 0x^2 & +\ x & +\ 1 & \\
 & 4x^3 & +\ 12x^2 & & & \\ \cline{2-3}
 & & -12x^2 & +\ x & & \\
 & & -12x^2 & -\ 36x & & \\ \cline{3-4}
 & & & 37x & +\ 1 & \\
 & & & 37x & +\ 111 & \\ \cline{4-5}
 & & & & -\ 110 & \text{remainder}
\end{array}$$

Hence
$$\frac{4x^3 + x + 1}{x + 3} = 4x^2 - 12x + 37 + \frac{-110}{x + 3}.$$

(Note that the right side of this equation can be integrated easily. An antiderivative is $4x^3/3 - 6x^2 + 37x - 110 \ln |x + 3|$.) ■

2. Factor denominator.

Step 2 If the degree of A is *less* than the degree of B, then express B as the product of polynomials of degree 1 or 2, where the second-degree factors are *irreducible*. (It can be proved that this is possible.) *No irreducible factor should simply be a constant times another irreducible factor.*

3. List summands of form $\dfrac{k_i}{(px+q)^i}$.

Step 3 If $px + q$ appears exactly n times in the factorization of B, form the sum

$$\frac{k_1}{px+q} + \frac{k_2}{(px+q)^2} + \cdots + \frac{k_n}{(px+q)^n},$$

where the constants $k_1, k_2, \ldots, k_n$ are to be determined later.

4. List summands of form $\dfrac{c_jx+d_j}{(ax^2+bx+c)^j}$.

Step 4 If $ax^2 + bx + c$ appears exactly m times in the factorization of B, then form the sum

$$\frac{c_1x+d_1}{ax^2+bx+c} + \frac{c_2x+d_2}{(ax^2+bx+c)^2} + \cdots + \frac{c_mx+d_m}{(ax^2+bx+c)^m},$$

where the constants $c_1, c_2, \ldots, c_m$ and $d_1, d_2, \ldots, d_m$ are to be determined later.

5. Find constants k_i, c_j, d_j.

Step 5 Determine the appropriate k's, c's, and d's defined in steps 3 and 4, such that A/B is equal to the sum of all the terms formed in steps 3 and 4 for all factors of B defined in step 2.

Remark: The rational function

$$\frac{c_jx+d_j}{(ax^2+bx+c)^j}$$

equals

$$\frac{c_jx}{(ax^2+bx+c)^j} + \frac{d_j}{(ax^2+bx+c)^j},$$

the sum of two functions that can be integrated by the methods of the preceding section.

EXAMPLE 3 Indicate the form of the partial-fraction representation of

$$\frac{A}{B} = \frac{2x^2+3x+3}{(x+1)^3}.$$

SOLUTION The denominator, if multiplied out, is a polynomial of degree larger than the degree of the numerator. In this case, the denominator has only one first-degree factor, $x + 1$ (repeated three times), and no quadratic factors. Only step 3 applies, so that we write

$$\frac{2x^2+3x+3}{(x+1)^3} = \frac{k_1}{x+1} + \frac{k_2}{(x+1)^2} + \frac{k_3}{(x+1)^3}.$$

The constants k_1, k_2, and k_3 can be found by methods discussed later in this section. ■

EXAMPLE 4 What does the partial-fraction decomposition of $\frac{A}{B} = \frac{2x^3 - 6x^2 + 2}{(x + 1)(x^2 + x + 1)^2}$ look like?

SOLUTION The denominator has degree 5 (if multiplied out), while the numerator has degree 3. Thus step 1 does not apply. The irreducible factors of B are $x + 1$, which appears only once, and $x^2 + x + 1$, which appears to the second power. Both steps 3 and 4 apply, so that we write

$$\frac{2x^3 - 6x^2 + 2}{(x + 1)(x^2 + x + 1)^2} = \frac{k_1}{x + 1} + \frac{c_1x + d_1}{x^2 + x + 1} + \frac{c_2x + d_2}{(x^2 + x + 1)^2}. \quad ■$$

The next example shows how to find the promised constants in a simpler case, where there are fewer constants.

EXAMPLE 5 Express $\frac{4x - 7}{x^2 - 3x + 2}$ as the sum of partial fractions.

SOLUTION Beginning at step 2, factor $x^2 - 3x + 2$ as $(x - 1)(x - 2)$. The denominator B in this case has only first-degree factors; hence step 4 will not apply. Since both $x - 1$ and $x - 2$ appear only once in the factorization, we have $n = 1$ in each case for step 3.

According to step 3, constants k_1 and k_2 exist such that

$$\frac{4x - 7}{x^2 - 3x + 2} = \frac{k_1}{x - 1} + \frac{k_2}{x - 2}. \tag{3}$$

To find k_1 and k_2, multiply both sides of Eq. (3) by $x^2 - 3x + 2$, obtaining

$$4x - 7 = k_1(x - 2) + k_2(x - 1). \tag{4}$$

The method of substituting specific values

Equation (4) holds for all x, since it is an algebraic identity. Thus it holds when x is replaced by any specific number. To find two equations for the two unknowns k_1 and k_2, we replace the x in Eq. (4) by two numbers. Since $x - 1$ vanishes for x equal to 1, and since $x - 2$ vanishes for x equal to 2, it is convenient to replace x by 1 and by 2. This gives

$$4 \cdot 1 - 7 = k_1(1 - 2) + k_2 \cdot 0 \qquad \text{setting } x = 1 \text{ in Eq. (4);}$$

$$4 \cdot 2 - 7 = k_1 \cdot 0 + k_2(2 - 1) \qquad \text{setting } x = 2 \text{ in Eq. (4).}$$

These equations simplify to

$$-3 = -k_1 \qquad \text{and} \qquad 1 = k_2.$$

Hence $k_1 = 3$ and $k_2 = 1$, and

$$\frac{4x - 7}{x^2 - 3x + 2} = \frac{3}{x - 1} + \frac{1}{x - 2}. \quad ■$$

In Example 5 the unknown constants were found by substituting specific values for x. There is another way, called **equating coefficients**. It depends on the fact that if two polynomials are equal for all x, then corresponding coefficients must be equal. To illustrate this technique, return to Eq. (4).

According to Eq. (4),

$$4x - 7 = k_1(x - 2) + k_2(x - 1)$$

or

$$4x - 7 = (k_1 + k_2)x - (2k_1 + k_2).$$

Comparing corresponding coefficients, we conclude that

$$4 = k_1 + k_2 \qquad \text{comparing coefficients of } x;$$

and

$$-7 = -2k_1 - k_2 \qquad \text{comparing constant terms.}$$

These simultaneous equations may be solved for k_1 and k_2. Adding them gives

$$-3 = -k_1.$$

Hence $k_1 = 3$. Substituting $k_1 = 3$ into $4 = k_1 + k_2$ gives $k_2 = 1$.

EXAMPLE 6 Express $\dfrac{x^2 + 7x + 1}{(x + 2)^2(2x + 1)}$ as the sum of partial fractions.

SOLUTION The degree of the numerator is less than the degree of the denominator (which is 3). Hence step 1 does not apply. Since the denominator is already factored (a common occurrence in practice), step 2 is done. We shall do steps 3 and 5 simultaneously; step 4 does not apply, since the denominator B has no second-degree irreducible factors.

Since $x + 2$ appears twice in the factorization and $2x + 1$ appears once,

$$\frac{x^2 + 7x + 1}{(x + 2)^2(2x + 1)} = \frac{k_1}{x + 2} + \frac{k_2}{(x + 2)^2} + \frac{l}{2x + 1}. \tag{5}$$

To find the constants k_1, k_2, and l, remove the denominators in Eq. (5) by multiplying both sides by $(x + 2)^2(2x + 1)$:

$$x^2 + 7x + 1 = k_1(x + 2)(2x + 1) + k_2(2x + 1) + l(x + 2)^2. \tag{6}$$

Three equations are needed to find the three unknowns k_1, k_2, and l. To obtain them, replace x in Eq. (6) by three different numbers in turn. Since $x + 2 = 0$ when $x = -2$ and $2x + 1 = 0$ when $x = -\frac{1}{2}$, replace x by -2 and then by $-\frac{1}{2}$. To obtain a third equation, use $x = 0$. Thus

$$-9 = -3k_2 \qquad \text{setting } x = -2 \text{ in Eq. (6);}$$

$$-\tfrac{9}{4} = (\tfrac{9}{4})l \qquad \text{setting } x = -\tfrac{1}{2} \text{ in Eq. (6);}$$

$$1 = 2k_1 + k_2 + 4l \qquad \text{setting } x = 0 \text{ in Eq. (6).}$$

Thus $k_2 = 3$, $l = -1$, and finally, $k_1 = 1$. Replacing k_1, k_2, and l in Eq. (5) yields

$$\frac{x^2 + 7x + 1}{(x + 2)^2(2x + 1)} = \frac{1}{x + 2} + \frac{3}{(x + 2)^2} - \frac{1}{2x + 1}. \quad ■$$

This is (1) in the opening paragraph.

EXAMPLE 7 Express $\dfrac{x^4 + x^3 - 3x + 5}{(x + 1)(x^2 + x + 1)}$ as the sum of partial fractions.

SOLUTION Since the degree of the numerator, 4, is at least as large as the degree of the denominator, 3, step 1 is applicable. Divide by the denominator, $(x + 1)(x^2 + x + 1) = x^3 + 2x^2 + 2x + 1$, as follows:

$$\begin{array}{r|rrrrrl}
 & & & & x & -\ 1 & \quad\text{quotient} \\
\cline{2-6}
x^3 + 2x^2 + 2x + 1 & x^4 + & x^3 + & 0x^2 & -\ 3x & +\ 5 & \\
 & x^4 + & 2x^3 + & 2x^2 & +\ x & & \\
\cline{2-6}
 & & -\ x^3 - & 2x^2 & -\ 4x & +\ 5 & \\
 & & -\ x^3 - & 2x^2 & -\ 2x & -\ 1 & \\
\cline{3-6}
 & & & & -\ 2x & +\ 6 & \quad\text{remainder}
\end{array}$$

Hence, $$\frac{x^4 + x^3 - 3x + 5}{(x + 1)(x^2 + x + 1)} = x - 1 + \frac{-2x + 6}{(x + 1)(x^2 + x + 1)}. \tag{7}$$

Next represent $$\frac{-2x + 6}{(x + 1)(x^2 + x + 1)}$$

as a sum of partial quotients in accordance with steps 3 and 4. Since $x + 1$ and $x^2 + x + 1$ are irreducible, we seek constants k_1, c_1, and d_1 such that

$$\frac{-2x + 6}{(x + 1)(x^2 + x + 1)} = \frac{k_1}{x + 1} + \frac{c_1x + d_1}{x^2 + x + 1}. \tag{8}$$

To find k_1, c_1, and d_1, multiply Eq. (8) by $(x + 1)(x^2 + x + 1)$, obtaining

$$-2x + 6 = k_1(x^2 + x + 1) + (c_1x + d_1)(x + 1). \tag{9}$$

Let $x = -1$ (the root of $x + 1 = 0$); then let $x = 0$ and $x = 1$, which are easy numbers to work with, arriving at

$$\begin{aligned}
8 &= k_1 && \text{setting } x = -1 \text{ in Eq. (9);} \\
6 &= k_1 + d_1 && \text{setting } x = 0 \text{ in Eq. (9);} \\
4 &= 3k_1 + 2c_1 + 2d_1 && \text{setting } x = 1 \text{ in Eq. (9).}
\end{aligned}$$

The first equation yields $k_1 = 8$, the second yields $d_1 = -2$, and the third yields $c_1 = -8$.

Thus Eq. (8) takes the form

$$\frac{-2x + 6}{(x + 1)(x^2 + x + 1)} = \frac{8}{x + 1} - \frac{8x + 2}{x^2 + x + 1}. \tag{10}$$

Combining Eqs. (7) and (10) shows that

$$\frac{x^4 + x^3 - 3x + 5}{(x + 1)(x^2 + x + 1)} = x - 1 + \frac{8}{x + 1} - \frac{8x + 2}{x^2 + x + 1}. \quad \blacksquare \tag{11}$$

No integrations have been carried out in this section, which is purely algebraic. This section, together with the preceding one, shows how to integrate any rational function.

This flowchart describes the procedure for representing a rational function as a sum of partial fractions.

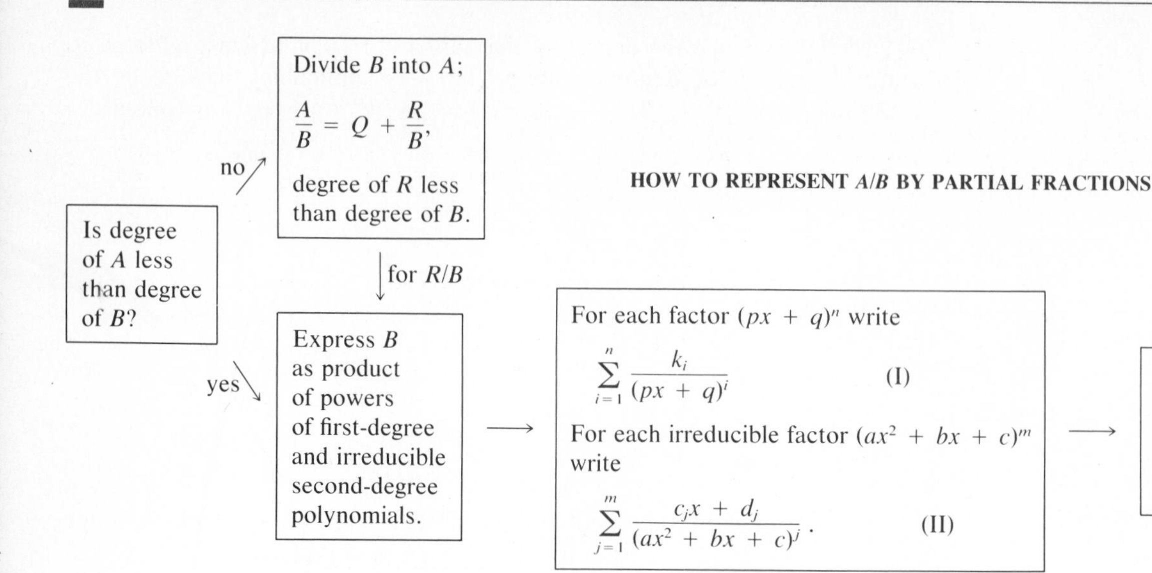

Useful Facts The number of unknown constants equals the degree of B. (Check that you have the right number.)

If $b^2 - 4ac < 0$, $ax^2 + bx + c$ is irreducible. If $b^2 - 4ac \geq 0$, $ax^2 + bx + c$ is reducible.

EXERCISES FOR SEC. 7.4: INTEGRATION OF RATIONAL FUNCTIONS BY PARTIAL FRACTIONS

In Exercises 1 to 12 indicate the form of the partial-fraction representation of the rational function listed, but do *not* find the constants k_i, c_j, and d_j.

1 $\frac{x + 3}{(x + 1)(x + 2)}$

2 $\frac{5}{(x - 1)(x + 3)}$

3 $\frac{1}{x^2 - 4}$

4 $\frac{x + 3}{x^2 - 4}$

5 $\frac{x^2 + 3x + 1}{(x + 1)^3}$

6 $\frac{5x + 6x^2}{(x + 1)^3}$

7 $\frac{x^2 + x + 2}{x^2 - 1}$

8 $\frac{x^3}{x + 1}$

9 $\frac{x^3}{(x - 1)(x + 2)}$

10 $\frac{x + 3}{(x - 1)^2(x + 2)^2}$

11 $\frac{x^4 + 3x^2}{(x^2 + x + 1)^3}$

12 $\frac{x^7 - 1}{(x^2 + x + 1)^3(x + 1)^2}$

In Exercises 13 to 28 express the rational functions in terms of partial fractions.

13 $\frac{x^2 + x + 3}{x + 1}$

14 $\frac{x}{x - 3}$

15 $\frac{2x^2 + 2x + 3}{2x + 1}$

16 $\frac{3x - 3}{x^2 - 9}$

17 $\frac{-x}{(x + 1)(x + 2)}$

18 $\frac{4x + 1}{x(x + 1)}$

19 $\frac{x^2 - 3x - 1}{x(x + 1)^2}$

20 $\frac{4x^3 - 4x^2 + x + 1}{x^2(x - 1)^2}$

21 $\frac{2x^2 + 3x + 3}{(x + 1)^3}$

22 $\frac{2x^3 - 6x^2 + 2}{(x + 1)(x^2 + x + 1)^2}$

23 $\frac{x + 4}{(x + 1)^2}$

24 $\frac{2x^2 + 9x + 12}{(x + 2)(x + 3)^2}$

25 $\dfrac{x^3}{x^2 + 3x + 5}$ **26** $\dfrac{x^3}{x^2 - x - 6}$

27 $\dfrac{x^4 + x^3 + 4x^2 + 1}{x(x^2 + x + 1)}$ **28** $\dfrac{3x^3 + x^2 - 3}{x^2(x^2 + 3x + 3)}$

■

In Exercises 29 to 36 compute the integral of the function in the cited exercise.

29 Exercise 17. **30** Exercise 18.
31 Exercise 19. **32** Exercise 20.
33 Exercise 25. **34** Exercise 26.
35 Exercise 27. **36** Exercise 28.

37 Compute

$$\int \frac{3x^5 + 2x^3 - x}{(x^2 - 1)(x^2 + 1)}\, dx.$$

38 (*a*) If a is a constant other than 0, what is the partial-fraction decomposition of $1/(a^2 - x^2)$?
(*b*) Using (*a*), find $\int dx/(a^2 - x^2)$.

39 Find the area under the curve $y = 1/(x^3 - x)$ and above $[2, 3]$.

40 Find the area under the curve $y = 1/(x^3 + x)$ and above $[2, 3]$.

41 Solve the differential equation

$$\frac{dy}{dx} = \frac{y^2 - 1}{x^2(x + 1)}.$$

42 Since the only polynomials that are irreducible over the real numbers are the first-degree polynomials and the irreducible second-degree polynomials, the polynomial $x^4 + 1$ must be reducible. Factor it. [*Suggestion:* Find constants a and b such that $x^4 + 1 = (x^2 + ax + 1)(x^2 + bx + 1)$.]

43 Compute as easily as possible

(*a*) $\displaystyle\int \frac{x^3\, dx}{x^4 + 1}$ (*b*) $\displaystyle\int \frac{x\, dx}{x^4 + 1}$ (*c*) $\displaystyle\int \frac{dx}{x^4 + 1}$

44 (*a*) Write $x^4 + x^2 + 1$ as the product of irreducible polynomials of second degree.
(*b*) Compute $\displaystyle\int \frac{dx}{x^4 + x^2 + 1}$.

45 There is a partial-fraction representation of rational numbers. The denominators, instead of being irreducible polynomials or their powers, are prime numbers or their powers. (An integer ≥ 2 is prime if it is not the product of positive integers smaller than itself.) A partial fraction has the form k/p^n, where p is prime, n is an integer ≥ 1, and k is an integer, $|k| < p$. Each rational number, A/B, is the sum of an integer (perhaps 0) and partial fractions of the form k/p^n, where p^n divides B. For instance,

$$\frac{19}{24} = \frac{0}{2} + \frac{1}{2^2} + \frac{-1}{2^3} + \frac{2}{3} \qquad \text{(note: } 24 = 2^3 \cdot 3\text{);}$$

$$\frac{37}{10} = 3 + \frac{1}{2} + \frac{1}{5} \qquad \text{(note: } 10 = 2 \cdot 5\text{).}$$

Find a partial-fraction representation of (*a*) $\frac{5}{6}$, (*b*) $\frac{4}{15}$, (*c*) $\frac{19}{15}$, (*d*) $\frac{7}{27}$.

46 (This continues Exercise 45.)
(*a*) Find integers x and y such that

$$\frac{1}{77} = \frac{x}{11} + \frac{y}{7},$$

and $|x| < 11$, $|y| < 7$.
(*b*) Find another such pair.
As (*a*) and (*b*) show, the partial-fraction representation of rational numbers is not unique. That it is unique for rational functions is far from obvious. It is sometimes proved to be unique in an upper-division or graduate algebra course.

47 This exercise describes a technique for finding the partial-fraction representation of $A(x)/B(x)$ if the degree of $A(x)$ is less than the degree of $B(x)$ and $B(x)$ is the product of distinct first-degree factors, $B(x) = (x - q_1)(x - q_2) \cdots (x - q_n)$.
Write

$$\frac{A(x)}{(x - q_1)(x - q_2) \cdots (x - q_n)} = \frac{k_1}{x - q_1} + \frac{k_2}{x - q_2} + \cdots + \frac{k_n}{x - q_n}. \tag{12}$$

(*a*) Show why

$$k_1 = \frac{A(q_1)}{(q_1 - q_2)(q_1 - q_3) \cdots (q_1 - q_n)}.$$

Hint: Multiply both sides of Eq. (12) by $x - q_1$.
(*b*) You get k_1 by covering $x - q_1$ with your thumb and plugging q_1 into what remains on the left side of Eq. (12). How do you find k_2? k_3?
(*c*) Use this technique to represent by partial fractions

$$\frac{x^2 + x + 3}{x(x - 1)(x - 2)}.$$

(*d*) Use this technique to represent

$$\frac{x + 5}{(2x + 1)(x + 3)}.$$

[First write $2x + 1$ as $2(x + \frac{1}{2})$.]

7.5 HOW TO INTEGRATE POWERS OF TRIGONOMETRIC FUNCTIONS

This section shows how to integrate certain products of powers of the six trigonometric functions, $\cos\theta$, $\sin\theta$, $\tan\theta$, $\sec\theta$, $\csc\theta$, and $\cot\theta$. Since we have $\tan\theta = \sin\theta/\cos\theta$, $\sec\theta = 1/\cos\theta$, $\csc\theta = 1/\sin\theta$, and $\cot\theta = \cos\theta/\sin\theta$, any such product can be expressed in the form $\cos^m\theta \sin^n\theta$ for some integers m and n, positive, zero, or negative. *The methods in this section thus show how to compute* $\int \cos^m\theta \sin^n\theta\, d\theta$ *for certain convenient combinations of* m *and* n. (The method in the next section shows that $\int \cos^m\theta \sin^n\theta\, d\theta$ is always elementary.) Typical of the integrals that can be evaluated by these methods are

$$\int \sin^2\theta\, d\theta, \qquad \int \sin^5\theta\, d\theta, \qquad \int \cos^3\theta \sin^4\theta\, d\theta,$$

$$\int \tan^5\theta \sec^3\theta\, d\theta, \qquad \text{and} \qquad \int \sec\theta\, d\theta.$$

The methods are summarized in the table at the end of this section.

Consider the integral

$$\int \cos^m\theta \sin^n\theta\, d\theta, \tag{1}$$

where m and n are *nonnegative integers*. *The technique for integrating* (1) *varies with* m *and* n.

If $m = 1$, say, and $n \geq 1$, then (1) becomes

$$\int \cos\theta \sin^n\theta\, d\theta. \tag{2}$$

The substitution $u = \sin\theta$ turns (2) into the easy integral

$$\int u^n\, du.$$

If $m = 0$, then (1) takes the form $\int \sin^n\theta\, d\theta$. A recursive formula was developed in Example 7 of Sec. 7.2 which expressed this integral in terms of $\int \sin^{n-2}\theta\, d\theta$. However, there are other approaches available for the cases $n = 2$ and n odd, as illustrated by Examples 1 and 2.

EXAMPLE 1 Find $\int \sin^2\theta\, d\theta$ by using the identity $\sin^2\theta = \dfrac{1-\cos 2\theta}{2}$.

SOLUTION

How to find $\int \sin^2\theta\, d\theta$

$$\int \sin^2\theta\, d\theta = \int \frac{1-\cos 2\theta}{2}\, d\theta = \int \frac{1}{2}\, d\theta - \int \frac{\cos 2\theta}{2}\, d\theta = \frac{\theta}{2} - \frac{\sin 2\theta}{4} + C$$

$$= \frac{\theta}{2} - \frac{2\sin\theta\cos\theta}{4} + C = \frac{\theta}{2} - \frac{\sin\theta\cos\theta}{2} + C. \quad ■$$

The particular definite integrals

$$\int_0^{\pi/2} \sin^2\theta\, d\theta \qquad \text{and} \qquad \int_0^{\pi/2} \cos^2\theta\, d\theta$$

occur so frequently that the following memory device is worth pointing out. A quick sketch of the graphs of $\sin^2\theta$ and $\cos^2\theta$ shows that the two definite integrals are equal. Also, since $\sin^2\theta + \cos^2\theta = 1$,

$$\int_0^{\pi/2} \sin^2\theta\, d\theta + \int_0^{\pi/2} \cos^2\theta\, d\theta = \int_0^{\pi/2} 1\, d\theta = \frac{\pi}{2}.$$

How to remember $\int_0^{\pi/2} \sin^2\theta\, d\theta$ and $\int_0^{\pi/2} \cos^2\theta\, d\theta$

Hence

$$\int_0^{\pi/2} \sin^2\theta\, d\theta = \frac{\pi}{4} = \int_0^{\pi/2} \cos^2\theta\, d\theta.$$

In the next example a quick way is shown to find $\int \sin^n\theta\, d\theta$ if n is an *odd* positive integer.

EXAMPLE 2 Find $\displaystyle\int \sin^5\theta\, d\theta$.

SOLUTION Recall that

$$d(\cos\theta) = -\sin\theta\, d\theta.$$

Thus

$$\int \sin^5\theta\, d\theta = \int \sin^4\theta \sin\theta\, d\theta$$
$$= -\int (1 - \cos^2\theta)^2\, d(\cos\theta).$$

Letting $u = \cos\theta$, we obtain that

$$\int \sin^5\theta\, d\theta = -\int (1 - u^2)^2\, du$$
$$= -\int (1 - 2u^2 + u^4)\, du$$
$$= -\left(u - \frac{2u^3}{3} + \frac{u^5}{5}\right) + C$$
$$= -\cos\theta + \frac{2\cos^3\theta}{3} - \frac{\cos^5\theta}{5} + C. \quad \blacksquare$$

$\cos^m \sin^{odd}$

More generally, to find $\int \cos^m\theta \sin^n\theta\, d\theta$, where m and n are nonnegative integers and n is odd, pair one $\sin\theta$ with $d\theta$ to form $\sin\theta\, d\theta = -d(\cos\theta)$, and use the identity $\sin^2\theta = 1 - \cos^2\theta$ together with the substitution $u = \cos\theta$. The new integrand will be a polynomial in u. A similar approach works on $\int \cos^m\theta \sin^n\theta\, d\theta$ if m is odd, as is illustrated by Example 3.

$\cos^{odd} \sin^n$

EXAMPLE 3 Find $\displaystyle\int \cos^3\theta \sin^4\theta\, d\theta$.

SOLUTION Pair one $\cos\theta$ with $d\theta$ to form

$$\cos\theta\, d\theta = d(\sin\theta),$$

which suggests the substitution $u = \sin\theta$:

$$\int \cos^3\theta \sin^4\theta\, d\theta = \int \cos^2\theta \sin^4\theta\,(\cos\theta\, d\theta)$$
$$= \int (1 - \sin^2\theta)\sin^4\theta\, d(\sin\theta)$$
$$= \int (1 - u^2)u^4\, du = \int (u^4 - u^6)\, du$$
$$= \frac{u^5}{5} - \frac{u^7}{7} + C = \frac{\sin^5\theta}{5} - \frac{\sin^7\theta}{7} + C. \quad \blacksquare$$

If both m and n are even in $\int \cos^m \theta \sin^n \theta \, d\theta$, the method of Example 3 does not apply. It then helps to use the identities

$$\cos^2 \theta = \frac{1 + \cos 2\theta}{2} \quad \text{and} \quad \sin^2 \theta = \frac{1 - \cos 2\theta}{2}.$$

Example 1 is an instance of this approach. Example 4 is a more involved case.

EXAMPLE 4 Find $\int_0^{\pi/4} \cos^2 \theta \sin^4 \theta \, d\theta$.

SOLUTION First find an antiderivative of $\cos^2 \theta \sin^4 \theta$:

$$\begin{aligned}
\int \cos^2 \theta \sin^4 \theta \, d\theta &= \int \cos^2 \theta \, (\sin^2 \theta)^2 \, d\theta \\
&= \int \frac{1 + \cos 2\theta}{2} \left(\frac{1 - \cos 2\theta}{2} \right)^2 d\theta \\
&= \frac{1}{8} \int (1 + \cos 2\theta)(1 - 2 \cos 2\theta + \cos^2 2\theta) \, d\theta \\
&= \frac{1}{8} \int (1 - \cos 2\theta - \cos^2 2\theta + \cos^3 2\theta) \, d\theta. \qquad (3)
\end{aligned}$$

The four summands in (3) can be integrated separately:

C will be added at the end.

$$\int 1 \, d\theta = \theta, \qquad \int -\cos 2\theta \, d\theta = \frac{-\sin 2\theta}{2},$$

$$\int -\cos^2 2\theta \, d\theta = \int -\frac{1 + \cos 4\theta}{2} \, d\theta = -\frac{\theta}{2} - \frac{\sin 4\theta}{8},$$

$$\begin{aligned}
\int \cos^3 2\theta \, d\theta = \int \cos^2 2\theta \cos 2\theta \, d\theta &= \int (1 - \sin^2 2\theta) \cos 2\theta \, d\theta \\
&= \int (\cos 2\theta - \sin^2 2\theta \cos 2\theta) \, d\theta \\
&= \frac{\sin 2\theta}{2} - \frac{\sin^3 2\theta}{6}.
\end{aligned}$$

Thus the right side of (3) equals

$$\frac{1}{8} \left(\theta - \frac{\sin 2\theta}{2} - \frac{\theta}{2} - \frac{\sin 4\theta}{8} + \frac{\sin 2\theta}{2} - \frac{\sin^3 2\theta}{6} \right) + C,$$

or

$$\frac{1}{8} \left(\frac{\theta}{2} - \frac{\sin 4\theta}{8} - \frac{\sin^3 2\theta}{6} \right) + C.$$

Consequently,

$$\begin{aligned}
\int_0^{\pi/4} \cos^2 \theta \sin^4 \theta \, d\theta &= \frac{1}{8} \left(\frac{\theta}{2} - \frac{\sin 4\theta}{8} - \frac{\sin^3 2\theta}{6} \right) \Bigg|_0^{\pi/4} \\
&= \frac{1}{8} \left[\left(\frac{\pi}{8} - \frac{\sin \pi}{8} - \frac{\sin^3 (\pi/2)}{6} \right) - \left(0 - \frac{\sin 0}{8} - \frac{\sin^3 0}{6} \right) \right] \\
&= \frac{1}{8} \left(\frac{\pi}{8} - \frac{1}{6} \right). \qquad \blacksquare
\end{aligned}$$

How to Integrate $\tan^m \theta \sec^n \theta$

$\int \tan^m\theta \sec^n\theta \, d\theta$ when m is odd or n is even

Recall that $d(\tan \theta) = \sec^2 \theta \, d\theta$ and $d(\sec \theta) = \sec \theta \tan \theta \, d\theta$. These formulas facilitate the computation of $\int \tan^m \theta \sec^n \theta \, d\theta$, m and n non-negative integers, when m is odd or n is even. When m is odd, form $\sec \theta \tan \theta \, d\theta$; when n is even, form $\sec^2 \theta \, d\theta$.

EXAMPLE 5 Find $\int \tan^5 \theta \sec^3 \theta \, d\theta$.

SOLUTION Recall that

$$d(\sec \theta) = \sec \theta \tan \theta \, d\theta,$$

and that

$$\tan^2 \theta = \sec^2 \theta - 1.$$

Let $u = \sec \theta$ and $du = \sec \theta \tan \theta \, d\theta$.

How to deal with $\tan^{odd} \sec^n$

Then

$$\begin{aligned}\int \tan^5 \theta \sec^3 \theta \, d\theta &= \int \tan^4 \theta \sec^2 \theta (\sec \theta \tan \theta \, d\theta) \\ &= \int (\sec^2 \theta - 1)^2 \sec^2 \theta (\sec \theta \tan \theta \, d\theta) \\ &= \int (u^2 - 1)^2 u^2 \, du \\ &= \int (u^6 - 2u^4 + u^2) \, du \\ &= \frac{u^7}{7} - \frac{2u^5}{5} + \frac{u^3}{3} + C \\ &= \frac{\sec^7 \theta}{7} - \frac{2 \sec^5 \theta}{5} + \frac{\sec^3 \theta}{3} + C. \quad \blacksquare\end{aligned}$$

Incidentally, since $\tan^5 \theta \sec^3 \theta = \cos^{-8} \theta \sin^5 \theta$, Example 5 can be interpreted as another case of integrating $\cos^m \theta \sin^n \theta$.

EXAMPLE 6 Find $\int \tan^6 \theta \sec^4 \theta \, d\theta$.

SOLUTION Recall that

$$d(\tan \theta) = \sec^2 \theta \, d\theta.$$

So pair $\sec^2 \theta$ with $d\theta$ to form $\sec^2 \theta \, d\theta$. This suggests the substitution

How to deal with $\tan^m \sec^{even}$

$$u = \tan \theta \qquad du = \sec^2 \theta \, d\theta.$$

Recall also that

$$\sec^2 \theta = \tan^2 \theta + 1.$$

Then

$$\begin{aligned}\int \tan^6 \theta \sec^4 \theta \, d\theta &= \int \tan^6 \theta \sec^2 \theta \sec^2 \theta \, d\theta \\ &= \int u^6(u^2 + 1) \, du \\ &= \int (u^8 + u^6) \, du \\ &= \frac{u^9}{9} + \frac{u^7}{7} + C \\ &= \frac{\tan^9 \theta}{9} + \frac{\tan^7 \theta}{7} + C. \quad \blacksquare\end{aligned}$$

A slightly different trick disposes of $\int \tan^n \theta \, d\theta$, as the next example illustrates.

EXAMPLE 7 Obtain a recursion formula for $\int \tan^n \theta \, d\theta$.

SOLUTION Keep in mind that $(\tan \theta)' = \sec^2 \theta$ and that $\tan^2 \theta = \sec^2 \theta - 1$:

A recursion for $\int \tan^n \theta \, d\theta$

$$\begin{aligned}
\int \tan^n \theta \, d\theta &= \int \tan^{n-2} \theta \tan^2 \theta \, d\theta \\
&= \int \tan^{n-2} \theta (\sec^2 \theta - 1) \, d\theta \\
&= \int \tan^{n-2} \theta \sec^2 \theta \, d\theta - \int \tan^{n-2} \theta \, d\theta \\
&= \int u^{n-2} \, du - \int \tan^{n-2} \theta \, d\theta \qquad (u = \tan \theta) \\
&= \frac{\tan^{n-1} \theta}{n-1} - \int \tan^{n-2} \theta \, d\theta.
\end{aligned}$$

Repeated application of this recursion eventually produces $\int \tan^1 \theta \, d\theta$ or $\int \tan^0 \theta \, d\theta$. Both are easily computed:

$\int \tan \theta \, d\theta = -\ln |\cos \theta| + C.$

$$\int \tan \theta \, d\theta = \int \frac{\sin \theta \, d\theta}{\cos \theta} = -\ln |\cos \theta| + C$$

and

$$\int \tan^0 \theta \, d\theta = \int 1 \, d\theta = \theta + C. \quad \blacksquare$$

EXAMPLE 8 Obtain a recursion formula for $\int \sec^n \theta \, d\theta$.

SOLUTION Write $\sec^n \theta$ as $\sec^{n-2} \theta \sec^2 \theta$ and use integration by parts:

$$\begin{aligned}
\int \underbrace{\sec^{n-2} \theta}_{u} \underbrace{\sec^2 \theta \, d\theta}_{dv} &= \underbrace{\sec^{n-2} \theta}_{u} \underbrace{\tan \theta}_{v} - \int \underbrace{\tan \theta}_{v} \underbrace{(n-2) \sec^{n-2} \theta \tan \theta \, d\theta}_{du} \\
&= \sec^{n-2} \theta \tan \theta - (n-2) \int \sec^{n-2} \theta \tan^2 \theta \, d\theta \\
&= \sec^{n-2} \theta \tan \theta - (n-2) \int \sec^{n-2} \theta (\sec^2 \theta - 1) \, d\theta \\
&= \sec^{n-2} \theta \tan \theta - (n-2) \int \sec^n \theta \, d\theta + (n-2) \int \sec^{n-2} \theta \, d\theta.
\end{aligned}$$

Collecting $\int \sec^n \theta \, d\theta$, we obtain

$$(n-1) \int \sec^n \theta \, d\theta = \sec^{n-2} \theta \tan \theta + (n-2) \int \sec^{n-2} \theta \, d\theta,$$

and therefore,

A recursion for $\int \sec^n \theta \, d\theta$

$$\int \sec^n \theta \, d\theta = \frac{\sec^{n-2} \theta \tan \theta}{n-1} + \frac{n-2}{n-1} \int \sec^{n-2} \theta \, d\theta. \quad \blacksquare$$

How to Integrate $\sec \theta$

A century before the invention of calculus, the cartographer Gerhardus Mercator had to estimate $\int_0^\alpha \sec \theta \, d\theta$ in order to determine where to place the lines of latitude on his maps. On a Mercator map, a straight line corresponds to a voyage with a constant compass heading, a property of great use to navigators. (See Exercise 44 for the geometry.) Henry Bond, in 1645, while scrutinizing a table of ln tan $(\alpha/2 + \pi/4)$, conjectured that $\int_0^\alpha \sec\theta\, d\theta = \ln\tan(\alpha/2 + \pi/4)$, but he offered no proof. In 1666, Nicolaus Mercator (no relation to Gerhardus) offered the royalties on one of his inventions to the mathematician who could prove that Bond's conjecture was right. Within two years James Gregory provided the missing proof, well before the tools of calculus were available.

This integral will be evaluated three ways: twice in the next example and once in Exercise 19 of the next section.

EXAMPLE 9 Find $\int \sec\theta\, d\theta$.

SOLUTION *Method A:*

$$\int \sec\theta\, d\theta = \int \frac{\sec\theta(\sec\theta + \tan\theta)}{\sec\theta + \tan\theta}\, d\theta$$

$$= \int \frac{\sec^2\theta + \sec\theta\tan\theta}{\sec\theta + \tan\theta}\, d\theta$$

$$= \ln|\sec\theta + \tan\theta| + C.$$

To see the motivation, differentiate the answer.

While this is the shortest method, it does seem artificial. The next method may seem a little less contrived.

Method B:

$$\int \sec\theta\, d\theta = \int \frac{1}{\cos\theta}\, d\theta$$

$$= \int \frac{\cos\theta}{\cos^2\theta}\, d\theta$$

$$= \int \frac{\cos\theta}{1 - \sin^2\theta}\, d\theta.$$

The substitution $u = \sin\theta$ and $du = \cos\theta\, d\theta$ transforms this last integral into the integral of a rational function:

$$\int \frac{du}{1 - u^2} = \frac{1}{2}\int \left(\frac{1}{1 + u} + \frac{1}{1 - u}\right) du$$

$$= \tfrac{1}{2}[\ln(1 + u) - \ln(1 - u)] + C$$

$$= \frac{1}{2}\ln\frac{1 + u}{1 - u} + C.$$

Both $1 + u$ and $1 - u$ are positive, since $u = \sin\theta$.

Since $u = \sin\theta$, $\quad \dfrac{1}{2}\ln\dfrac{1 + u}{1 - u} = \dfrac{1}{2}\ln\dfrac{1 + \sin\theta}{1 - \sin\theta}.$

Note that $\sec\theta + \tan\theta$ can be negative, but $(1 + \sin\theta)/(1 - \sin\theta)$ cannot be.

The reader may check that this equals $\ln|\sec\theta + \tan\theta|$ by showing that

$$\frac{1 + \sin\theta}{1 - \sin\theta} = (\sec\theta + \tan\theta)^2.$$

Neither method gives Bond's conjecture, $\ln\tan(\theta/2 + \pi/4)$. That formula will be obtained directly and in a fairly straightforward way in the next section. However, it is an amusing exercise to show that $\tan(\theta/2 + \pi/4) = \sec\theta + \tan\theta$. From this Bond's conjecture follows immediately. ■

Up to this point, the angles have been equal in the various trigonometric functions that appeared in an integrand. However, the integrals of $\sin mx \sin nx$, $\sin mx \cos nx$, and $\cos mx \sin nx$, where the angles are not equal, are of importance and can be computed with the aid of these identities:

$$\sin A \sin B = \tfrac{1}{2}\cos(A-B) - \tfrac{1}{2}\cos(A+B);$$
$$\sin A \cos B = \tfrac{1}{2}\sin(A+B) + \tfrac{1}{2}\sin(A-B);$$
$$\cos A \cos B = \tfrac{1}{2}\cos(A-B) + \tfrac{1}{2}\cos(A+B).$$

These identities can be checked by using the well-known identities for $\sin(A \pm B)$ and $\cos(A \pm B)$.

EXAMPLE 10 Find $\int_0^{\pi/4} \sin 3x \sin 2x\, dx$.

SOLUTION

$$\begin{aligned}\int_0^{\pi/4} \sin 3x \sin 2x\, dx &= \int_0^{\pi/4}\left(\frac{1}{2}\cos x - \frac{1}{2}\cos 5x\right)dx \\ &= \left(\frac{1}{2}\sin x - \frac{1}{10}\sin 5x\right)\Bigg|_0^{\pi/4} \\ &= \left(\frac{\sqrt{2}}{4} + \frac{\sqrt{2}}{20}\right) - \left(\frac{0}{2} - \frac{0}{10}\right) \\ &= \frac{3\sqrt{2}}{10}. \quad \blacksquare\end{aligned}$$

The following table summarizes the techniques discussed and similar ones for other powers of trigonometric functions:

Integrand	*Technique*
$\sin^2\theta$	Write $\sin^2\theta$ as $\dfrac{1-\cos 2\theta}{2}$.
$\cos^2\theta$	Write $\cos^2\theta$ as $\dfrac{1+\cos 2\theta}{2}$.
$\sin^n\theta$ (n odd)	Write $\sin^n\theta\, d\theta = \sin^{n-1}\theta\,(\sin\theta\, d\theta)$ and use $u = \cos\theta$; hence $1 - u^2 = \sin^2\theta$.
$\cos^m\theta \sin^n\theta$ (n odd)	Write $\cos^m\theta \sin^n\theta\, d\theta = \cos^m\theta \sin^{n-1}\theta\,(\sin\theta\, d\theta)$ and use $u = \cos\theta$; hence $1 - u^2 = \sin^2\theta$.
$\cos^m\theta \sin^n\theta$ (m odd)	Write $\cos^m\theta \sin^n\theta\, d\theta = \cos^{m-1}\theta \sin^n\theta\,(\cos\theta\, d\theta)$ and use $u = \sin\theta$; hence $1 - u^2 = \cos^2\theta$.
$\cos^m\theta \sin^n\theta$ (m and n positive even integers)	Replace $\cos^2\theta$ by $\dfrac{1+\cos 2\theta}{2}$ and $\sin^2\theta$ by $\dfrac{1-\cos 2\theta}{2}$.
$\tan^m\theta \sec^n\theta$ ($n \ge 2$ even)	Write $\tan^m\theta \sec^n\theta\, d\theta$ as $\tan^m\theta \sec^{n-2}\theta\,(\sec^2\theta\, d\theta)$ and use $u = \tan\theta$; hence $1 + u^2 = \sec^2\theta$.
$\tan^m\theta \sec^n\theta$ (m odd)	Write $\tan^m\theta \sec^n\theta\, d\theta$ as $\tan^{m-1}\theta \sec^{n-1}\theta\,(\tan\theta \sec\theta\, d\theta)$ and use $u = \sec\theta$; hence $u^2 - 1 = \tan^2\theta$.
$\tan^n\theta$ ($n \ge 2$)	Write $\tan^n\theta = \tan^{n-2}\theta \tan^2\theta = \tan^{n-2}\theta \sec^2\theta - \tan^{n-2}\theta$ and repeat.
$\tan\theta$	$\int \tan\theta\, d\theta = -\ln\lvert\cos\theta\rvert + C$.

Integrand	*Technique*
$\sec\theta$	$\int \sec\theta\, d\theta = \ln\lvert\sec\theta + \tan\theta\rvert + C$.
$\cot^m\theta\csc^n\theta$ ($n \ge 2$ even)	Write $\cot^m\theta\csc^n\theta$ as $\cot^m\theta\csc^{n-2}\theta\,(\csc^2\theta\, d\theta)$ and use $u = \cot\theta$; hence $1 + u^2 = \csc^2\theta$.
$\cot^m\theta\csc^n\theta$ (m odd)	Write $\cot^m\theta\csc^n\theta\, d\theta$ as $\cot^{m-1}\theta\csc^{n-1}\theta\,(\cot\theta\csc\theta\, d\theta)$ and use $u = \csc\theta$; hence $u^2 - 1 = \cot^2\theta$.
$\tan^m\theta\sec^n\theta$ (m even, n even)	Replace $\sec^2\theta$ by $\tan^2\theta + 1$.
$\sec^n\theta$ ($n \ge 2$)	Recursion in Example 8.
$\sin mx \sin nx$ $\sin mx \cos nx$ $\cos mx \sin nx$	Use trigonometric identities for $\sin A \sin B$, $\sin A \cos B$, and $\cos A \cos B$.

EXERCISES FOR SEC. 7.5: HOW TO INTEGRATE POWERS OF TRIGONOMETRIC FUNCTIONS

In Exercises 1 to 36 find the integrals.

1 $\int \sin^2 2\theta\, d\theta$

2 $\int \cos^2 3\theta\, d\theta$

3 $\int_{\pi/4}^{\pi/2} \sin^2 4\theta\, d\theta$

4 $\int_0^{\pi/10} \cos^2 5\theta\, d\theta$

5 $\int \sin^5 2\theta\, d\theta$

6 $\int \sin^7 \theta\, d\theta$

7 $\int_0^{\pi/2} \cos^3 x\, dx$

8 $\int_0^{\pi/2} \cos^5 x\, dx$

9 $\int \sin^3\theta \cos^4\theta\, d\theta$

10 $\int \cos^5\theta \sin^2\theta\, d\theta$

11 $\int \cos^3\theta \sin^3\theta\, d\theta$

12 $\int \cos^2\theta \sin^5\theta\, d\theta$

13 $\int_{\pi/6}^{\pi/4} \cos^4 x\, dx$

14 $\int_0^{\pi} \sin^4 x\, dx$

15 $\int \tan^4\theta \sec^2\theta\, d\theta$

16 $\int \tan^4\theta \sec^4\theta\, d\theta$

17 $\int \tan^3\theta \sec\theta\, d\theta$

18 $\int \tan\theta \sec^3\theta\, d\theta$

19 $\int \tan^3\theta\, d\theta$

20 $\int \tan^4\theta\, d\theta$

21 $\int_0^{\pi/4} \sec^4\theta\, d\theta$

22 $\int_0^{\pi/4} \sec^5\theta\, d\theta$

23 $\int \sin^2\theta \cos^3\theta\, d\theta$

24 $\int \tan^5\theta \sec^3\theta\, d\theta$

25 $\int \cot\theta\, d\theta$

26 $\int \cot^3\theta\, d\theta$

27 $\int \sec 3\theta\, d\theta$

28 $\int \cot^3\theta \csc^5\theta\, d\theta$

29 $\int \cos^5\theta \sin^3\theta\, d\theta$

30 $\int \cos^4\theta \sin^4\theta\, d\theta$

31 $\int \cot^3\theta \csc^4\theta\, d\theta$

32 $\int (\sin\theta + 2\cos\theta)^2\, d\theta$

33 $\int \cot^4\theta \csc^4\theta\, d\theta$

34 $\int (\tan\theta + 2\cot\theta)^2\, d\theta$

35 $\int \sin^3\theta \tan^2\theta\, d\theta$

36 $\int \frac{\sin^3\theta}{\cos^5\theta}\, d\theta$

■

In Exercise 37 to 40 evaluate the definite integrals. The following formula, obtained in Exercise 41 of Sec. 7.2, may be of aid:

$$\int_0^{\pi/2} \sin^n\theta\, d\theta = \begin{cases} \dfrac{1\cdot 3\cdot 5\cdots(n-1)}{2\cdot 4\cdot 6\cdots n}\dfrac{\pi}{2} & n \text{ even} \\ \dfrac{2\cdot 4\cdot 6\cdots(n-1)}{3\cdot 5\cdot 7\cdots n} & n \text{ odd} \end{cases}$$

37 $\int_0^{\pi/2} \cos^5\theta\, d\theta$

$\left(\textit{Hint: } \text{Why is } \int_0^{\pi/2} \cos^n\theta\, d\theta = \int_0^{\pi/2} \sin^n\theta\, d\theta?\right)$

38 $\int_0^{\pi/2} \sin^6\theta\, d\theta$

39 $\int_0^{2\pi} \sin^6\theta\, d\theta$

40 $\int_0^{2\pi} \cos^7 \theta \, d\theta$

41 Find $\int \sin^2 \theta \cos^2 \theta \, d\theta$ by use of the identity $\sin 2\theta = 2 \sin \theta \cos \theta$.

42 Compute $\int \frac{d\theta}{1 + \cos \theta}$ with the aid of the identity

$$\cos^2 \frac{\theta}{2} = \frac{1 + \cos \theta}{2}.$$

43 Show that $\tan\left(\frac{\theta}{2} + \frac{\pi}{4}\right) = \sec \theta + \tan \theta$ for $0 \le \theta < \pi/2$.

44 In a Mercator map the meridians (vertical lines of longitude) are spaced at equal distances, but the lines of latitude (horizontal lines) are not. Instead they are placed so that "locally the map shrinks horizontal and vertical distances by the same factor." Figure 7.1 shows the spherical earth, with radius 1 for convenience and a corresponding section of the map.

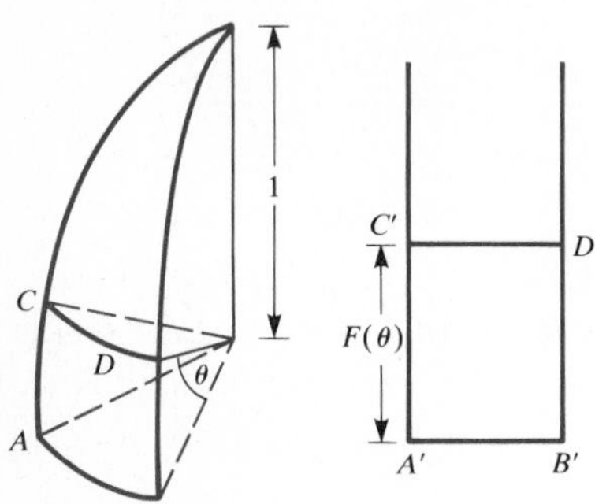

A', *B'*, *C'*, and *D'* are the images on the map of the points *A*, *B*, *C*, and *D* on the sphere of radius 1.

The arc $\overset{\frown}{CD}$ has latitude θ.

Figure 7.1

(*a*) Show that if $\overline{A'B'} = AB$, then $\overline{C'D'} = \sec \theta \, CD$, where AB and CD are the lengths of the arcs AB and CD, respectively.

(*b*) Let $F(\theta)$ be the distance on the map from $A'B'$ to $C'D'$. In order that the map preserve locally the ratios between vertical and horizontal distances, why should $F'(\theta) = \sec \theta$? *Hint:* Review magnification in Secs. 3.1 and 3.2.

(*c*) Deduce that $F(\alpha) = \int_0^\alpha \sec \theta \, d\theta$.

(*d*) In a certain Mercator map the distance from the equator to the 30° latitude line is 3 inches. How far is it from the 30° latitude line to the 60° latitude line?

(*e*) Can a Mercator map of the whole world be drawn on a finite piece of paper?

For an extensive discussion of Mercator's map, see Philip M. Tuchinsky, *Mercator's World Map and the Calculus*, UMAP Module 206, EDC, UMAP, 55 Chapel St., Newton, Mass. 02160.

45 Find $\int \csc \theta \, d\theta$.

46 Show that

$$\frac{1}{2} \ln \frac{1 + \sin \theta}{1 - \sin \theta} = \ln |\sec \theta + \tan \theta|.$$

47 An arbitrary periodic sound wave can be approximated by a sum of simpler sound waves that correspond to pure pitches. This suggests representing a function $y = f(x)$ as the sum of cosine and sine functions:

$$f(x) = \frac{a_0}{2} + \sum_{k=1}^{n} a_k \cos kx + \sum_{k=1}^{n} b_k \sin kx,$$

for an integer n and some constants $a_0, a_1, \ldots, a_n, b_1, \ldots, b_n$. Show that if the representation is valid, then (*a*) $a_m = (1/\pi) \int_0^{2\pi} f(x) \cos mx \, dx$, and (*b*) $b_m = (1/\pi) \int_0^{2\pi} f(x) \sin mx \, dx$. [*Hint:* For (*a*), evaluate $\int_0^{2\pi} f(x) \cos mx \, dx$.] (In the theory of Fourier series such representations, with n replaced by ∞, are studied.)

48 When a voltage V is introduced across a resistance R, the power dissipated is V^2/R. If the voltage varies with time but R is constant, then the average power during the time interval $[a, b]$ is defined as

$$\frac{1}{b - a} \frac{1}{R} \int_a^b [V(t)]^2 \, dt.$$

In case of sinusoidal voltage, $V(t) = M \sin t$, where M is the maximum voltage.

(*a*) Show that the average power over the time interval $[0, \pi/2]$ is $\frac{1}{2}M^2/R$.

(*b*) Show that the constant voltage $M/\sqrt{2}$ would produce the same power.

When we say we have 110 volts in our homes, we are referring to $M/\sqrt{2}$. The maximum voltage is actually $\sqrt{2}(110) \approx 154$ volts.

The **root mean square** (RMS) of a function $f(x)$ is defined as the square root of $\int_a^b [f(x)]^2 \, dx/(b - a)$. A stereo system is often described as "producing 50 watts RMS," even though this is a simple average, with no squares and no square roots.

7.6 HOW TO INTEGRATE ANY RATIONAL FUNCTION OF sin θ AND cos θ

This section shows how to integrate a function such as

$$\frac{\sin^3\theta + 5\cos\theta\sin\theta}{3 + 4\cos\theta - 7\sin^2\theta}.$$

Some definitions will be needed before describing the method.

A **polynomial in x and y** is a sum of terms of the form ax^iy^j, where i and j are nonnegative integers and a is a real number. For instance, $1 + x - 2xy + y^2$ and $x^3y - 3xy + x^2y^3$ are polynomials in x and y. The quotient of two such polynomials is called a **rational function of x and y** and is denoted $R(x, y)$. Thus

$$\frac{1 + x - 2xy + y^2}{x^3y - 3xy + x^2y^3}$$

is a rational function of x and y.

$R(\cos\theta, \sin\theta)$

If, in $R(x, y)$, x is replaced by $\cos\theta$ and y by $\sin\theta$, we obtain a **rational function of $\cos\theta$ and $\sin\theta$**. Thus

$$\frac{1 + \cos\theta - 2\cos\theta\sin\theta + \sin^2\theta}{\cos^3\theta\sin\theta - 3\cos\theta\sin\theta + \cos^2\theta\sin^3\theta}$$

is a rational function of $\cos\theta$ and $\sin\theta$.

The technique described in this section—a particular substitution—reduces the integration of any rational function of $\cos\theta$ and $\sin\theta$ to the integration of a rational function of u. The latter can be accomplished by partial fractions.

Description of the Method The method depends on the fact that $\cos\theta$ and $\sin\theta$ can both be expressed as rational functions of $\tan(\theta/2)$.

Consider $-\pi < \theta < \pi$ and let $u = \tan(\theta/2)$. The right triangle in Fig. 7.2 shows the geometry of this substitution for $u \geq 0$, and Fig. 7.3 shows the case when u is negative.

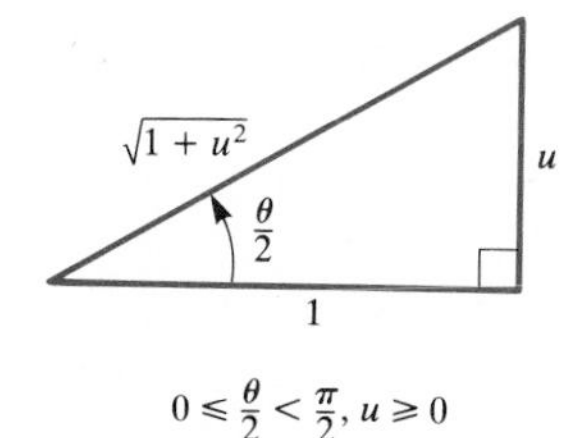

$0 \leq \frac{\theta}{2} < \frac{\pi}{2}, u \geq 0$

Figure 7.2

As Figs. 7.2 and 7.3 show,

$$\cos\frac{\theta}{2} = \frac{1}{\sqrt{1+u^2}} \quad \text{and} \quad \sin\frac{\theta}{2} = \frac{u}{\sqrt{1+u^2}}.$$

Thus $\quad \cos\theta = \cos^2\frac{\theta}{2} - \sin^2\frac{\theta}{2} = \left(\frac{1}{\sqrt{1+u^2}}\right)^2 - \left(\frac{u}{\sqrt{1+u^2}}\right)^2 = \frac{1-u^2}{1+u^2}$

and $\quad \sin\theta = 2\sin\frac{\theta}{2}\cos\frac{\theta}{2} = 2\left(\frac{u}{\sqrt{1+u^2}}\right)\left(\frac{1}{\sqrt{1+u^2}}\right) = \frac{2u}{1+u^2}.$

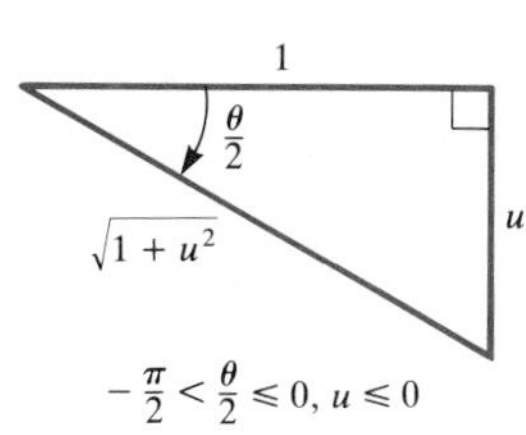

$-\frac{\pi}{2} < \frac{\theta}{2} \leq 0, u \leq 0$

Figure 7.3

In addition, $d\theta$ can be expressed easily in terms of u and du, as will now be shown. Since $u = \tan(\theta/2)$,

$$du = \frac{1}{2}\sec^2\frac{\theta}{2}\,d\theta = \frac{1}{2}\left(1 + \tan^2\frac{\theta}{2}\right)d\theta = \tfrac{1}{2}(1 + u^2)\,d\theta.$$

Thus

$$d\theta = \frac{2\,du}{1+u^2}.$$

The substitution $$u = \tan \frac{\theta}{2}$$

thus leads to the formulas

$$\sin \theta = \frac{2u}{1 + u^2}, \qquad \cos \theta = \frac{1 - u^2}{1 + u^2}, \qquad \text{and} \qquad d\theta = \frac{2\, du}{1 + u^2}. \tag{1}$$

Although always applicable, this method is not always the most convenient one to use.

This substitution transforms any integral of a rational function of $\cos \theta$ and $\sin \theta$ into an integral of a rational function of u. The resulting rational function can then be integrated by the method of partial fractions.

EXAMPLE 1 Find $\displaystyle\int \frac{d\theta}{1 - \sin \theta}$.

SOLUTION The substitution (1) transforms the integral to

$$\int \frac{\dfrac{2\, du}{1 + u^2}}{1 - \dfrac{2u}{1 + u^2}} = \int \frac{2\, du}{1 - 2u + u^2} = \int \frac{2\, du}{(1 - u)^2}$$

$$= \frac{2}{1 - u} + C = \frac{2}{1 - \tan (\theta/2)} + C.$$

The identity $\tan (\theta/2) = \sin \theta/(1 + \cos \theta)$ could be used to express the answer in terms of $\sin \theta$ and $\cos \theta$. ■

EXAMPLE 2 Compute

$$\int_0^{\pi/2} \frac{d\theta}{4 \sin \theta + 3 \cos \theta}. \tag{2}$$

SOLUTION As θ goes from 0 to $\pi/2$, $u = \tan (\theta/2)$ goes from 0 to 1. The substitution (1) transforms (2) into

$$\int_0^1 \frac{\dfrac{2\, du}{1 + u^2}}{4\left(\dfrac{2u}{1 + u^2}\right) + 3\left(\dfrac{1 - u^2}{1 + u^2}\right)} = \int_0^1 \frac{2\, du}{8u + 3(1 - u^2)}$$

$$= \int_0^1 \frac{2\, du}{8u + 3 - 3u^2} = \int_0^1 \frac{2\, du}{(3u + 1)(3 - u)}$$

$$= \int_0^1 \left(\frac{\frac{3}{5}}{3u + 1} + \frac{\frac{1}{5}}{3 - u}\right) du \qquad \text{partial fractions}$$

$$= [\tfrac{1}{5} \ln (3u + 1) - \tfrac{1}{5} \ln (3 - u)]\Big|_0^1$$

$$= (\tfrac{1}{5} \ln 4 - \tfrac{1}{5} \ln 2) - (\tfrac{1}{5} \ln 1 - \tfrac{1}{5} \ln 3)$$

$$= \frac{\ln 4 - \ln 2 + \ln 3}{5} = \frac{\ln 6}{5}. \quad ■$$

Remarks on the Method The substitution (1) transforms any rational function of the six trigonometric functions into a rational function of u.

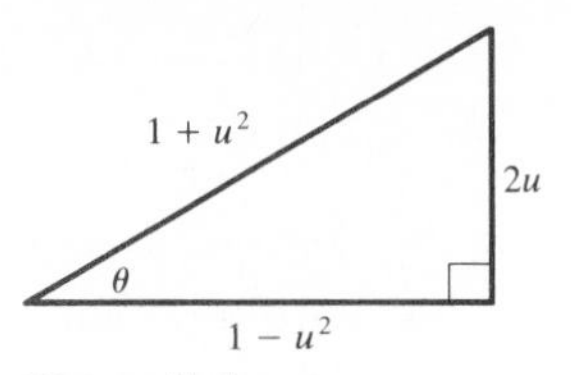

Figure 7.4

To apply (1), first express tan θ, sec θ, cot θ, or csc θ in terms of cos θ and sin θ.

Figure 7.4 may be of use in remembering the substitution. Just remember that $(1 - u^2)^2 + (2u)^2 = (1 + u^2)^2$, an identity that can be checked easily. Inspection of Fig. 7.4 shows that

$$\cos\theta = \frac{1 - u^2}{1 + u^2} \quad \text{and} \quad \sin\theta = \frac{2u}{1 + u^2}.$$

The formula for $d\theta$ resembles that for sin θ:

$$d\theta = \frac{2\,du}{1 + u^2}.$$

Keep in mind that the substitution $u = \tan(\theta/2)$ is called on only when easier ways, such as those in the preceding section, do not work.

EXERCISES FOR SEC. 7.6: HOW TO INTEGRATE ANY RATIONAL FUNCTION OF sin θ AND cos θ

In Exercises 1 to 4 use the substitution (1) to transform the integrands to rational functions of u. Do *not* evaluate the resulting integrals.

1 $\displaystyle\int \frac{\sin\theta + \cos\theta}{1 + 2\cos\theta}\,d\theta$

2 $\displaystyle\int \frac{\sin^2\theta\,d\theta}{\sin\theta + \cos\theta}$

3 $\displaystyle\int \frac{3 + \tan\theta}{2 + \tan\theta}\,d\theta$

4 $\displaystyle\int \frac{\sec^2\theta\,d\theta}{1 + 4\sin\theta}$

In Exercises 5 to 16 evaluate the integrals.

5 $\displaystyle\int \frac{d\theta}{4\cos\theta + 3\sin\theta}$

6 $\displaystyle\int \frac{d\theta}{4\cos\theta - 3\sin\theta}$

7 $\displaystyle\int_0^{\pi/4} \frac{d\theta}{1 + 2\cos\theta - \sin\theta}$

8 $\displaystyle\int_0^{\pi/2} \frac{d\theta}{1 + 3\cos\theta + \sin\theta}$

9 $\displaystyle\int \frac{d\theta}{5\cos\theta + 5\sin\theta + 1}$

10 $\displaystyle\int \frac{d\theta}{5 + 3\cos\theta + 4\sin\theta}$

11 $\displaystyle\int \frac{\sin\theta + 2\cos\theta}{1 + \cos\theta}\,d\theta$

12 $\displaystyle\int \frac{3\sin\theta - \cos\theta}{1 + \cos\theta}\,d\theta$

13 $\displaystyle\int \frac{\sin\theta\,d\theta}{4 - 3\tan\theta}$

14 $\displaystyle\int \frac{\sin\theta\,d\theta}{3 + 4\tan\theta}$

15 $\displaystyle\int \frac{d\theta}{4 + 5\cos\theta}$

16 $\displaystyle\int \frac{d\theta}{8 + 17\cos\theta}$

■

17 Find

$$\int \frac{\sin\theta\,d\theta}{26\cos\theta + 4\sin\theta + 4\sin\theta\cos\theta + 12\sin^2\theta + 26}.$$

18 Find $\displaystyle\int \frac{3 + \cos\theta}{2 - \cos\theta}\,d\theta.$

19 In Sec. 7.5 $\int \sec\theta\,d\theta$ was computed in two ways.

(*a*) Use substitution (1) to show that

$$\int \sec\theta\,d\theta = \ln\left(\frac{1 + \tan\theta/2}{1 - \tan\theta/2}\right) + C.$$

(*b*) Use the identity

$$\tan(A + B) = \frac{\tan A + \tan B}{1 - \tan A \tan B}$$

to show that the answer in (*a*) is the same as $\ln\tan(\theta/2 + \pi/4)$, which was Bond's conjecture.

20 Let $p(u)$, $q(u)$, and $r(u)$ be three polynomials such that $[p(u)]^2 + [q(u)]^2 = [r(u)]^2$. Make a change of variable by

$$\cos\theta = \frac{p(u)}{r(u)} \quad \text{and} \quad \sin\theta = \frac{q(u)}{r(u)}.$$

Show that $\displaystyle d\theta = \frac{rq' - qr'}{prr'}\,dr.$

This section was based on the fact that

$$(1 - u^2)^2 + (2u)^2 = (1 + u^2)^2.$$

Substitution (1) may therefore be considered independently of the angle $\theta/2$. See Alan H. Schoenfeld, The Curious Substitution $z = \tan(\theta/2)$ and the Pythagorean Theorem, *American Mathematical Monthly*, vol. 84, pp. 370–372, May 1977.

7.7 HOW TO INTEGRATE RATIONAL FUNCTIONS OF x AND $\sqrt{a^2 - x^2}$, $\sqrt{a^2 + x^2}$, $\sqrt{x^2 - a^2}$, OR $\sqrt[n]{ax + b}$

The first part of this section describes trigonometric substitutions that turn certain rational functions of quantities that involve square roots into rational functions of $\sin\theta$ and $\cos\theta$; these can be integrated by the method of Sec. 7.6. The second part describes an algebraic substitution that reduces a rational function of x and $\sqrt[n]{ax + b}$ to a rational function of a single variable, which can then be treated by the technique of partial fractions.

Three Trigonometric Substitutions

A rational function of x and $\sqrt{a^2 - x^2}$, $\sqrt{a^2 + x^2}$, or $\sqrt{x^2 - a^2}$ can be integrated by using a trigonometric substitution. If the integrand is a rational function of x and

How to integrate $R(x, \sqrt{a^2 - x^2})$, $R(x, \sqrt{a^2 + x^2})$, $R(x, \sqrt{x^2 - a^2})$

Case 1 $\sqrt{a^2 - x^2}$; let $x = a\sin\theta$ $\quad \left(a > 0, -\dfrac{\pi}{2} \le \theta \le \dfrac{\pi}{2}\right)$.

Case 2 $\sqrt{a^2 + x^2}$; let $x = a\tan\theta$ $\quad \left(a > 0, -\dfrac{\pi}{2} < \theta < \dfrac{\pi}{2}\right)$.

Case 3 $\sqrt{x^2 - a^2}$; let $x = a\sec\theta$ $\quad \left(a > 0, 0 \le \theta \le \pi, \theta \ne \dfrac{\pi}{2}\right)$.

The motivation behind this general procedure is quite simple. Consider case 1, for instance. If you replace x in $\sqrt{a^2 - x^2}$ by $a\sin\theta$, you obtain

How to make the square root sign in $\sqrt{a^2 - x^2}$ *disappear*

$$\begin{aligned}\sqrt{a^2 - x^2} = \sqrt{a^2 - (a\sin\theta)^2} &= \sqrt{a^2(1 - \sin^2\theta)} \\ &= \sqrt{a^2\cos^2\theta} \\ &= a\cos\theta.\end{aligned}$$

(Keep in mind that a and $\cos\theta$ are positive.) The important thing is that *the square root sign disappears.*

Case 3 raises a fine point. We have $a > 0$. However, whenever x is negative, θ is a second-quadrant angle, so $\tan\theta$ is *negative*. In that case,

$$\begin{aligned}\sqrt{x^2 - a^2} &= \sqrt{(a\sec\theta)^2 - a^2} \\ &= a\sqrt{\sec^2\theta - 1} \\ &= a\sqrt{\tan^2\theta} \\ &= a(-\tan\theta) \qquad \text{since } -\tan\theta \text{ is positive.}\end{aligned}$$

If $c < 0$, $\sqrt{c^2} = -c$.

In the examples and exercises involving case 3 it will be assumed that x varies through nonnegative values, so that θ remains in the first quadrant and $\sqrt{\sec^2\theta - 1} = \tan\theta$.

Note that for $\sqrt{a^2 - x^2}$ to be meaningful, $|x|$ must be no larger than a. On the other hand, for $\sqrt{x^2 - a^2}$ to be meaningful, $|x|$ must be at least as large as a. The quantity $\sqrt{a^2 + x^2}$ is meaningful for all values of x.

EXAMPLE 1 Make an appropriate substitution to remove the square root sign in each of these integrals. Do not evaluate.

(a) $\displaystyle\int \frac{x^3\,dx}{3 + \sqrt{16 - x^2}}$ (b) $\displaystyle\int \frac{x^5\,dx}{1 + \sqrt{3 + x^2}}$ (c) $\displaystyle\int \frac{(x^2 + 2)\,dx}{x + \sqrt{x^2 - 9}}$

SOLUTION

(a) In this case, $a = 4$ and the substitution $x = 4 \sin \theta$ is appropriate:

$$\int \frac{x^3\,dx}{3 + \sqrt{16 - x^2}} = \int \frac{(4 \sin \theta)^3 4 \cos \theta\,d\theta}{3 + \sqrt{16 - (4 \sin \theta)^2}} \qquad dx = d(4 \sin \theta)$$

$$= \int \frac{256 \sin^3 \theta \cos \theta\,d\theta}{3 + \sqrt{16 - 16 \sin^2 \theta}}$$

$$= \int \frac{256 \sin^3 \theta \cos \theta\,d\theta}{3 + 4\sqrt{1 - \sin^2 \theta}}$$

$$= \int \frac{256 \sin^3 \theta \cos \theta\,d\theta}{3 + 4 \cos \theta}.$$

Since $-\frac{\pi}{2} \leq \theta \leq \frac{\pi}{2}$, $\cos \theta$ is not negative, so $\cos \theta = \sqrt{1 - \sin^2 \theta}$.

(b) In this case, $a^2 = 3$, so $a = \sqrt{3}$. The substitution $x = \sqrt{3} \tan \theta$ is appropriate. We then have

$$\int \frac{x^5\,dx}{1 + \sqrt{3 + x^2}} = \int \frac{(\sqrt{3} \tan \theta)^5\,d(\sqrt{3} \tan \theta)}{1 + \sqrt{3 + (\sqrt{3} \tan \theta)^2}}$$

$$= \int \frac{27 \tan^5 \theta \sec^2 \theta\,d\theta}{1 + \sqrt{3 + 3 \tan^2 \theta}}$$

$$= \int \frac{27 \tan^5 \theta \sec^2 \theta\,d\theta}{1 + \sqrt{3}\sqrt{1 + \tan^2 \theta}}$$

$$= \int \frac{27 \tan^5 \theta \sec^2 \theta\,d\theta}{1 + \sqrt{3} \sec \theta}.$$

(c) In this case, $a = 3$, and the substitution $x = 3 \sec \theta$ is appropriate. Then

$$\int \frac{(x^2 + 2)\,dx}{x + \sqrt{x^2 - 9}} = \int \frac{((3 \sec \theta)^2 + 2)\,d(3 \sec \theta)}{3 \sec \theta + \sqrt{(3 \sec \theta)^2 - 9}}$$

$$= \int \frac{(9 \sec^2 \theta + 2)\,3 \sec \theta \tan \theta\,d\theta}{3 \sec \theta + 3\sqrt{\sec^2 \theta - 1}}$$

$$= \int \frac{(9 \sec^2 \theta + 2) \sec \theta \tan \theta\,d\theta}{\sec \theta + \tan \theta}. \qquad ■$$

EXAMPLE 2 Find $\int x^3 \sqrt{16 - x^2}\,dx$.

SOLUTION Let $x = 4 \sin \theta$; hence $dx = 4 \cos \theta\,d\theta$. Then

$$\int x^3\sqrt{16 - x^2}\,dx = \int (4 \sin \theta)^3\sqrt{16 - (4 \sin \theta)^2}\,4 \cos \theta\,d\theta$$

$$= 4^4 \int \sin^3 \theta \sqrt{16 - 16 \sin^2 \theta} \cos \theta\,d\theta$$

$$= 4^5 \int \sin^3 \theta \sqrt{1 - \sin^2 \theta} \cos \theta \, d\theta$$

$$= 4^5 \int \sin^3 \theta \cos^2 \theta \, d\theta$$

$$= 4^5 \int \sin^2 \theta \cos^2 \theta \sin \theta \, d\theta$$

$$= 4^5 \int (1 - \cos^2 \theta) \cos^2 \theta \sin \theta \, d\theta$$

$$= 4^5 \int (1 - u^2)u^2(-du) \qquad u = \cos \theta$$

$$= 4^5 \int (u^4 - u^2) \, du$$

$$= 4^5 \left(\frac{u^5}{5} - \frac{u^3}{3} \right) + C$$

$$= 4^5 \left(\frac{\cos^5 \theta}{5} - \frac{\cos^3 \theta}{3} \right) + C. \tag{1}$$

To express the answer in terms of x, express $\cos \theta$ in terms of x. To do this, use Fig. 7.5, which records the relation $x = 4 \sin \theta$ (or $\sin \theta = x/4$).

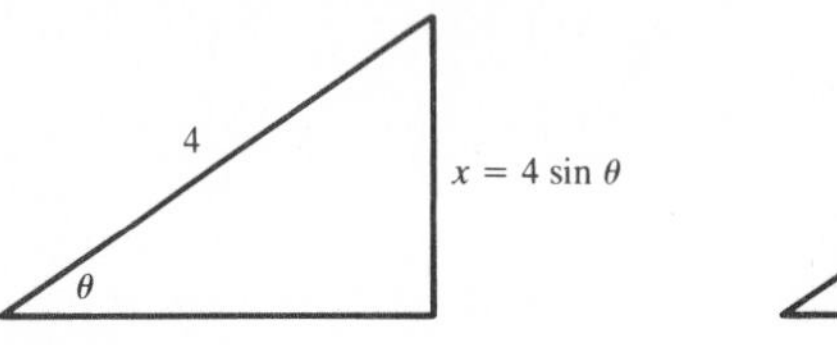

Figure 7.5

The third side has length $\sqrt{16 - x^2}$, by the pythagorean theorem. Thus

Figure 7.5 shows the case $x > 0$; draw the case $x < 0$.

$$\cos \theta = \frac{\sqrt{16 - x^2}}{4},$$

and (1) becomes $$4^5 \left[\frac{1}{5} \left(\frac{\sqrt{16 - x^2}}{4} \right)^5 - \frac{1}{3} \left(\frac{\sqrt{16 - x^2}}{4} \right)^3 \right] + C$$

$$= \frac{(\sqrt{16 - x^2})^5}{5} - 16 \frac{(\sqrt{16 - x^2})^3}{3} + C. \quad ■ \tag{2}$$

Had Example 2 asked instead for the value of $\int_{-2}^{2} x^3 \sqrt{16 - x^2} \, dx$, the solution would have been much simpler. Since the integrand is an odd function, $\int_{-2}^{2} x^3 \sqrt{16 - x^2} \, dx = 0$.

EXAMPLE 3 Compute $\int \sqrt{1 + x^2} \, dx$.

SOLUTION The identity $\sec \theta = \sqrt{1 + \tan^2 \theta}$ suggests the substitution described in case 2:

$$x = \tan \theta.$$

Hence $$dx = \sec^2 \theta \, d\theta.$$

(See Fig. 7.6 for the geometry of this substitution.)

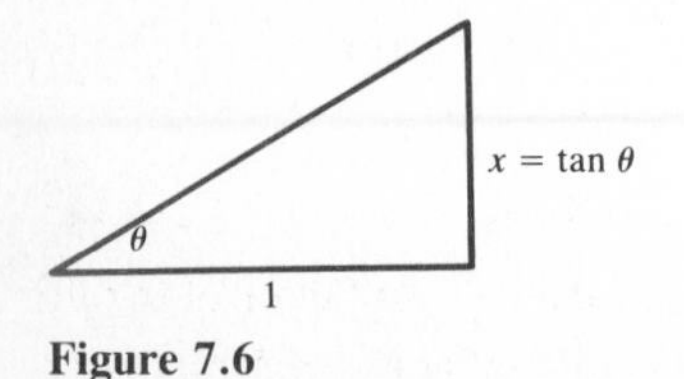

Figure 7.6

Thus $$\int \sqrt{1 + x^2} \, dx = \int \sec \theta \sec^2 \theta \, d\theta = \int \sec^3 \theta \, d\theta.$$

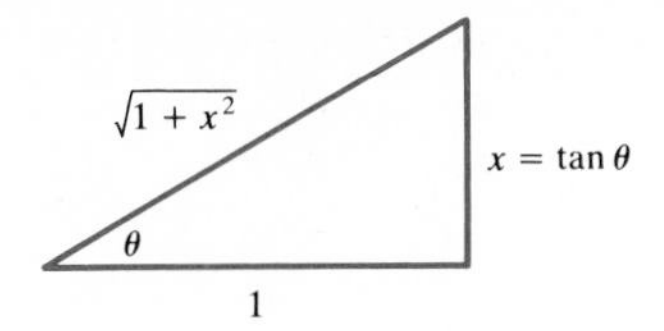

Figure 7.7

By Examples 8 and 9 in Sec. 7.5,

$$\int \sec^3\theta\,d\theta = \frac{\sec\theta\tan\theta}{2} + \frac{1}{2}\ln|\sec\theta + \tan\theta| + C.$$

To express the antiderivative just obtained in terms of x rather than θ, it is necessary to express $\tan\theta$ and $\sec\theta$ in terms of x. Starting with the definition $x = \tan\theta$, find $\sec\theta$ by means of the relation $\sec\theta = \sqrt{1 + \tan^2\theta} = \sqrt{1 + x^2}$, as in Fig. 7.7. Thus

$$\int \sqrt{1 + x^2}\,dx = \frac{x\sqrt{1 + x^2}}{2} + \frac{1}{2}\ln(\sqrt{1 + x^2} + x) + C. \quad \blacksquare$$

EXAMPLE 4 Compute $\displaystyle\int_4^5 \frac{dx}{\sqrt{x^2 - 9}}$.

SOLUTION Let $x = 3\sec\theta$; hence $dx = 3\sec\theta\tan\theta\,d\theta$. (See Fig. 7.8.) Thus, letting $\alpha = \sec^{-1}(\frac{4}{3})$ and $\beta = \sec^{-1}(\frac{5}{3})$, we obtain

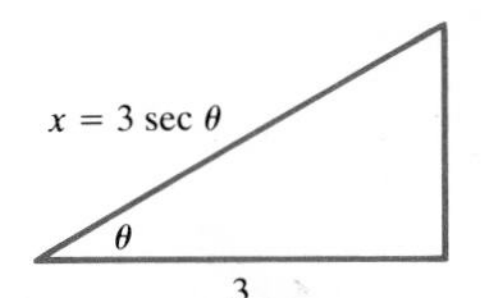

Figure 7.8

$$\begin{aligned}
\int_4^5 \frac{dx}{\sqrt{x^2 - 9}} &= \int_\alpha^\beta \frac{3\sec\theta\tan\theta\,d\theta}{\sqrt{9\sec^2\theta - 9}} \\
&= \int_\alpha^\beta \frac{\sec\theta\tan\theta\,d\theta}{\tan\theta} \\
&= \int_\alpha^{\beta\alpha} \sec\theta\,d\theta \\
&= \ln|\sec\theta + \tan\theta|\,\Big|_{\sec^{-1}(4/3)}^{\sec^{-1}(5/3)} \qquad \text{by Example 9 of Sec. 7.5} \\
&= \ln\left(\frac{5}{3} + \frac{4}{3}\right) - \ln\left(\frac{4}{3} + \frac{\sqrt{7}}{3}\right) \qquad \text{using Fig. 7.9 to find } \tan\theta \text{ at the limits of integration} \\
&= \ln 3 - \ln\left(\frac{4 + \sqrt{7}}{3}\right) \\
&= \ln\left(\frac{9}{4 + \sqrt{7}}\right). \quad \blacksquare
\end{aligned}$$

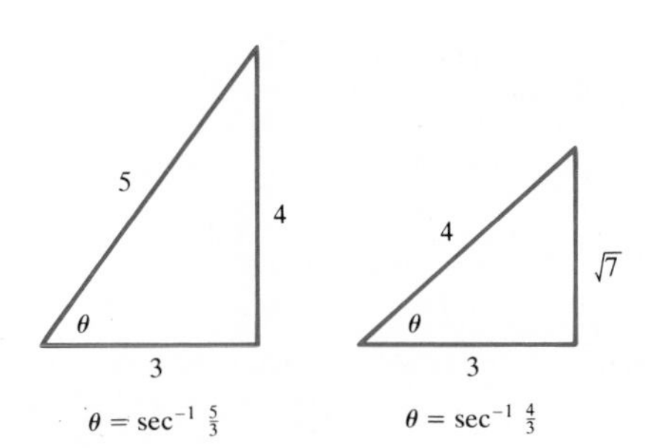

Figure 7.9

A rational function of x and $\sqrt{a^2 - b^2x^2}$, $\sqrt{a^2x^2 - b^2}$, or $\sqrt{a^2 + b^2x^2}$ can be treated by a similar procedure. For instance, to deal with a rational function of x and $\sqrt{a^2 - b^2x^2}$, make the substitution suggested by

$$b^2x^2 = a^2\sin^2\theta,$$

that is,

$$bx = a\sin\theta,$$

or

$$x = \frac{a}{b}\sin\theta \quad \text{and} \quad dx = \frac{a}{b}\cos\theta\,d\theta.$$

Then

$$\begin{aligned}
\sqrt{a^2 - b^2x^2} &= \sqrt{a^2 - a^2\sin^2\theta} \\
&= a\cos\theta.
\end{aligned}$$

A rational function of x and $a^2 - b^2x^2$ is a rational function of x and therefore can be integrated by partial fractions. But in some special cases a trigonometric substitution may provide another solution, as Example 5 illustrates.

EXAMPLE 5 Transform $\int \frac{dx}{(4 - 9x^2)^2}$ into a trigonometric integral.

Since $9x^2 < 4$, $|\sin\theta| = \frac{|3x|}{2}$ is indeed ≤ 1.

SOLUTION Assume that $9x^2 < 4$. Let $9x^2 = 4\sin^2\theta$, choosing $3x = 2\sin\theta$; hence $dx = \frac{2}{3}\cos\theta\, d\theta$.

Thus
$$\int \frac{dx}{(4 - 9x^2)^2} = \int \frac{\frac{2}{3}\cos\theta\, d\theta}{(4 - 4\sin^2\theta)^2} = \frac{2}{3}\int \frac{\cos\theta\, d\theta}{16(1 - \sin^2\theta)^2}$$
$$= \frac{1}{24}\int \frac{\cos\theta\, d\theta}{\cos^4\theta} = \frac{1}{24}\int \frac{1}{\cos^3\theta}\, d\theta.$$

Consequently,
$$\int \frac{dx}{(4 - 9x^2)^2} = \frac{1}{24}\int \sec^3\theta\, d\theta.$$

Recall that $|\sec\theta| \geq 1$.

If $9x^2 > 4$, then the substitution $9x^2 = 4\sec^2\theta$ is suggested, with the choice $3x = 2\sec\theta$. ■

An Algebraic Substitution

Let n be a positive integer. *Any rational function of x and $\sqrt[n]{ax + b}$ can be transformed into a rational function of u by the substitution*

$$u = \sqrt[n]{ax + b}$$

and thus can be integrated by partial fractions.

If $u = \sqrt[n]{ax + b}$, then $u^n = ax + b$. From this it follows that

$$x = \frac{u^n - b}{a} \quad \text{and} \quad dx = \frac{nu^{n-1}\, du}{a}.$$

Two examples illustrate the method.

EXAMPLE 6 Find $\int x^2 \sqrt[3]{4x + 5}\, dx$.

SOLUTION Let $u = \sqrt[3]{4x + 5}$. Then

$$u^3 = 4x + 5, \quad x = \frac{u^3 - 5}{4}, \quad \text{and} \quad dx = \frac{3u^2\, du}{4}.$$

Consequently,

$$\int x^2 \sqrt[3]{4x + 5}\, dx$$
$$= \int \left(\frac{u^3 - 5}{4}\right)^2 \cdot u \cdot \frac{3u^2\, du}{4} = \frac{3}{64}\int u^3(u^3 - 5)^2\, du$$
$$= \frac{3}{64}\int u^3(u^6 - 10u^3 + 25)\, du = \frac{3}{64}\int (u^9 - 10u^6 + 25u^3)\, du$$
$$= \frac{3}{64}\left(\frac{u^{10}}{10} - \frac{10u^7}{7} + \frac{25u^4}{4}\right) + C$$
$$= \frac{3}{64}\left[\frac{(\sqrt[3]{4x + 5})^{10}}{10} - \frac{10}{7}(\sqrt[3]{4x + 5})^7 + \frac{25}{4}(\sqrt[3]{4x + 5})^4\right] + C. \quad ■$$

EXAMPLE 7 Express $\int \frac{\sqrt{2x+1}}{1+\sqrt[3]{2x+1}}\,dx$ as the integral of a rational function of a single variable.

SOLUTION At first glance this problem does not seem to fit into the form "integral of a rational function of x and $\sqrt[n]{ax+b}$." However, both $\sqrt{2x+1}$ and $\sqrt[3]{2x+1}$ are powers of $\sqrt[6]{2x+1}$:

$$\sqrt{2x+1} = (2x+1)^{1/2} = (2x+1)^{3/6} = (\sqrt[6]{2x+1})^3$$

and $$\sqrt[3]{2x+1} = (2x+1)^{1/3} = (2x+1)^{2/6} = (\sqrt[6]{2x+1})^2.$$

This suggests the substitution $u = \sqrt[6]{2x+1}$; hence

$$u^6 = 2x+1, \qquad x = \frac{u^6-1}{2}, \qquad \text{and} \qquad dx = \frac{6u^5\,du}{2} = 3u^5\,du.$$

Consequently, $$\int \frac{\sqrt{2x+1}}{1+\sqrt[3]{2x+1}}\,dx = \int \frac{u^3(3u^5\,du)}{1+u^2},$$

an integral computable by partial fractions. ■

EXERCISES FOR SEC. 7.7: HOW TO INTEGRATE RATIONAL FUNCTIONS OF x AND $\sqrt{a^2 - x^2}$, $\sqrt{a^2 + x^2}$, $\sqrt{x^2 - a^2}$, OR $\sqrt[n]{ax + b}$

In Exercises 1 to 4 state an appropriate substitution for x (do not evaluate).

1 $\int x^3\sqrt{4+x^2}\,dx$ **2** $\int x^5(2+3x^2)^{5/2}\,dx$

3 $\int \frac{1+(5-x^2)^{3/2}}{x}\,dx$ **4** $\int x(x^2-5)^{7/2}\,dx$

In Exercises 5 to 8 transform the definite integral by an appropriate substitution (do not evaluate).

5 $\int_1^3 \frac{x^3\,dx}{\sqrt{x^2+2}+(\sqrt{x^2+2})^3}$

6 $\int_0^1 x^5(4-3x^2)^{10}\,dx$

7 $\int_2^3 \frac{x+\sqrt{3x^2-4}}{x^3}\,dx$ **8** $\int_0^1 x^7(5-3x^2)^4\,dx$

In Exercises 9 to 24 find the integrals.

9 $\int_1^2 \sqrt{4-x^2}\,dx$ **10** $\int_{-2/\sqrt{3}}^{2/\sqrt{3}} \sqrt{3-x^2}\,dx$

11 $\int_4^5 \frac{x^2\,dx}{\sqrt{x^2-9}}$ **12** $\int_{3/2}^{3\sqrt{3}/2} \frac{x^2\,dx}{\sqrt{9-x^2}}$

13 $\int \frac{dx}{\sqrt{9+x^2}}$ **14** $\int \frac{dx}{\sqrt{2+x^2}}$

15 $\int \frac{dx}{\sqrt{x}+3}$ **16** $\int \frac{x\,dx}{\sqrt{x}+3}$

17 $\int_1^{16} \frac{dx}{\sqrt[4]{x}+\sqrt{x}}$ **18** $\int \frac{dx}{\sqrt[3]{x}+\sqrt{x}}$

19 $\int \frac{\sqrt{2x+1}\,dx}{x}$ **20** $\int x^2\sqrt{2x+1}\,dx$

21 $\int x\sqrt[3]{3x+2}\,dx$ **22** $\int x(3x+2)^{5/3}\,dx$

23 $\int \frac{\sqrt{x}+3}{\sqrt{x}-2}\,dx$ **24** $\int \frac{dx}{(x-2)\sqrt{x+2}}$

■

In Exercises 25 to 32 find the integrals. (a is a positive constant.)

25 $\int \sqrt{a^2-x^2}\,dx$ **26** $\int \sqrt{x^2-a^2}\,dx$

27 $\int \sqrt{a^2 + x^2}\, dx$

28 $\int \frac{dx}{\sqrt{a^2 - x^2}}$

29 $\int \frac{dx}{\sqrt{25x^2 - 16}}$

30 $\int_{\sqrt{2}}^{2} \sqrt{x^2 - 1}\, dx$

31 $\int_0^{1/2} x^3 \sqrt{1 - x^2}\, dx$

32 $\int \frac{\sqrt{4 + x^2}}{x}\, dx$

33 Show that any rational function of $\sqrt{x}$ and $\sqrt{x + 1}$ can be integrated by using the substitution $x = \tan^2 \theta$.

34 Show that any rational function of x, $\sqrt{x + a}$, and $\sqrt{x + b}$ can be integrated by introducing the substitution defined by

$$x + a = \frac{1}{4}(a - b)\left(t + \frac{1}{t}\right)^2;$$

hence
$$x + b = \frac{1}{4}(a - b)\left(t - \frac{1}{t}\right)^2.$$

However, it is not the case that every rational function of $\sqrt{x + a}$, $\sqrt{x + b}$, and $\sqrt{x + c}$ has an elementary integral. For instance,

$$\int \frac{1}{\sqrt{x}\sqrt{x - 1}\sqrt{x + 1}}\, dx = \int \frac{1}{\sqrt{x^3 - x}}\, dx$$

is not an elementary function.

35 Show that every rational function of x and $\sqrt[n]{\frac{ax + b}{cx + d}}$ has an elementary antiderivative.

7.8 WHAT TO DO IN THE FACE OF AN INTEGRAL

Since the exercises in each section of this chapter are focused on the techniques of that section, it is usually clear what technique to use on a given integral. But what if an integral is met "in the wild," where there is no clue how to evaluate it? This section suggests what to do in this typical situation.

The more integrals you compute, the more quickly you will be able to choose an appropriate technique. There is no substitute for practice. (This section and the summary section offer many more exercises.)

There are only a few techniques on which to draw:

Substitution	Sec. 7.1
Parts	Sec. 7.2
Partial fractions	Secs. 7.3 and 7.4
Powers of trigonometric functions	Sec. 7.5
$R(\sin\theta, \cos\theta)$	Sec. 7.6
$R(x, \sqrt{a^2 - x^2})$, $R(x, \sqrt{a^2 + x^2})$, $R(x, \sqrt{x^2 - a^2})$	Sec. 7.7
$R(x, \sqrt[n]{ax + b})$	Sec. 7.7
$\int \frac{f'(x)\, dx}{f(x)} = \ln \lvert f(x) \rvert + C$ (not to be forgotten)	Sec. 7.1

The following recommendations for a strategy of integration are based on my own experience and *Integration, Getting It All Together,* by Alan H. Schoenfeld, U 203–205, EDC, UMAP, 55 Chapel St., Newton, Mass. 02160.

One possible general strategy consists of answering these questions.

Questions	*Examples*
Can the integrand be simplified by some algebraic or trigonometric identity?	$\int \frac{3 + x^2}{x} dx = \int \frac{3\,dx}{x} + \int \frac{x^2\,dx}{x}$ $\int \sin^2 \theta\, d\theta = \int \frac{1 - \cos 2\theta}{2} d\theta$
Will a substitution simplify the integrand? In particular, can the integrand be written as the product [function of $u(x)$] × [$u'(x)$] ?	$\int (1 + x^4)^5 x^3\,dx$ $= \frac{1}{4}\int (1 + x^4)^5 4x^3\,dx$ $= \frac{1}{4}\int u^5\,du \quad (u = 1 + x^4)$ $\int \frac{3x^2}{1 + x^3} dx = \int \frac{du}{u} = \ln \lvert u \rvert$ $(u = 1 + x^3)$
Would integration by parts help? Can I write the problem as $\int u\,dv$ in such a way that $\int v\,du$ is easier?	$\int x^4 e^x\,dx = x^4 e^x - \int e^x 4x^3\,dx$ (which reduces exponent of x)
Is the integrand a rational function of x?	$\int \frac{dx}{x^3 - 1} = \int \left(\frac{Ax + B}{x^2 + x + 1} + \frac{C}{x - 1}\right) dx$ $\left(\text{but do } \int \frac{x^2\,dx}{x^3 - 1} \text{ by the ``}f'/f\text{ method''}\right)$
Is the integrand of a special form?	$\int \sin^5 \theta \cos^2 \theta\, d\theta$ $\int \frac{3 + 2\cos\theta}{5 + 22\sin\theta} d\theta$ $\int x^3 \sqrt{9 + x^2}\,dx$ $\int \frac{\sqrt[3]{x}\,dx}{1 + (\sqrt[3]{x})^5}$

The following examples illustrate the strategy.

EXAMPLE 1 $\displaystyle\int \frac{x\,dx}{x^2 - 9}$.

DISCUSSION The substitutions $x = 3 \sec \theta$ or $x = 3 \sin \theta$ would work, depending on whether $|x| > 3$ or $|x| < 3$. However, notice that $x\,dx$ is almost the differential of $x^2 - 9$. This suggests the substitution $u = x^2 - 9$, $du = 2x\,dx$:

$$\int \frac{x\,dx}{x^2 - 9} = \int \frac{du/2}{u} = \frac{1}{2}\int u^{-1}\,du$$

$$= \tfrac{1}{2} \ln |u| + C = \tfrac{1}{2} \ln |x^2 - 9| + C. \quad \blacksquare$$

EXAMPLE 2 $\displaystyle\int \frac{x\,dx}{1 + x^4}$.

DISCUSSION Since the integrand is a rational function of x, partial fractions would work. This requires factoring $x^4 + 1$ and then representing $x/(1 + x^4)$ as a sum of partial fractions. With some struggle it can be found that

$$x^4 + 1 = (x^2 + \sqrt{2}\,x + 1)(x^2 - \sqrt{2}\,x + 1).$$

Then constants A, B, C, and D will have to be found such that

$$\frac{x}{1 + x^4} = \frac{Ax + B}{x^2 + \sqrt{2}x + 1} + \frac{Cx + D}{x^2 - \sqrt{2}x + 1}.$$

The method would work but would certainly be tedious.

Try another attack. The numerator x is almost the derivative of x^2. The substitution $u = x^2$ is at least worth testing:

$$u = x^2, \qquad du = 2x\,dx,$$

$$\int \frac{x\,dx}{1 + x^4} = \int \frac{du/2}{1 + u^2},$$

which is easy. The answer is $\frac{1}{2}\tan^{-1} u + C = \frac{1}{2}\tan^{-1} x^2 + C$. ■

EXAMPLE 3 $\displaystyle\int \frac{\sin^3\theta\,d\theta}{\cos^4\theta}$.

DISCUSSION Since this a rational function of $\sin\theta$ and $\cos\theta$, the substitution $u = \tan(\theta/2)$ would work, but maybe there is an easier way.

The fact that $\sin\theta$ appears to an odd power suggests using $\sin\theta\,d\theta$ to form a differential. That is, introduce $u = \cos\theta$, $du = -\sin\theta\,d\theta$. Then

$$\begin{aligned}
\int \frac{\sin^3\theta\,d\theta}{\cos^4\theta} &= \int \frac{\sin^2\theta\sin\theta\,d\theta}{\cos^4\theta} = \int \frac{(1 - \cos^2\theta)\sin\theta\,d\theta}{\cos^4\theta} \\
&= \int \frac{(1 - u^2)(-du)}{u^4} \\
&= \int \frac{(u^2 - 1)\,du}{u^4} \\
&= \int (u^{-2} - u^{-4})\,du = \frac{u^{-1}}{-1} + \frac{u^{-3}}{3} + C \\
&= -\sec\theta + \tfrac{1}{3}\sec^3\theta + C.
\end{aligned}$$ ■

EXAMPLE 4 $\displaystyle\int \frac{1 + x}{1 + x^2}\,dx$.

DISCUSSION This is a rational function of x, but partial fractions will not help, since the integrand is already in its partial-fraction representation.

The numerator is not the derivative of the denominator, but it comes close enough to persuade us to break the integrand into two summands:

$$\int \frac{1 + x}{1 + x^2}\,dx = \int \frac{dx}{1 + x^2} + \int \frac{x\,dx}{1 + x^2}.$$

Both the latter integrals can be done by sight. The first is $\tan^{-1} x + C$, and the second is $\frac{1}{2} \ln (1 + x^2) + C$. ■

EXAMPLE 5 $\displaystyle\int \frac{e^{2x}}{1 + e^x}\, dx.$

DISCUSSION At first glance, this integral looks so peculiar that it may not even be elementary. However, e^x is a fairly simple function, with $d(e^x) = e^x\, dx$. This suggests trying the substitution $u = e^x$ and seeing what happens:

$u = 1 + e^x$ is even easier

$$u = e^x, \qquad du = e^x\, dx.$$

Thus

$$dx = \frac{du}{e^x} = \frac{du}{u}.$$

But what will be done to e^{2x}? Recalling that $e^{2x} = (e^x)^2$, we anticipate no problem:

$$\int \frac{e^{2x}}{1 + e^x}\, dx = \int \frac{u^2}{1 + u}\frac{du}{u} = \int \frac{u\, du}{1 + u},$$

which can be integrated quickly:

$$\int \frac{u\, du}{1 + u} = \int \frac{u + 1 - 1}{1 + u}\, du = \int \left(1 - \frac{1}{1 + u}\right) du$$

$$= u - \ln |1 + u| + C.$$

$$= e^x - \ln (1 + e^x) + C.$$

The same substitution could have been done more elegantly:

$$\int \frac{e^{2x}}{1 + e^x}\, dx = \int \frac{e^x(e^x\, dx)}{1 + e^x} = \int \frac{u\, du}{1 + u}. \quad ■$$

EXAMPLE 6 $\displaystyle\int \frac{x^3\, dx}{(1 - x^2)^5}.$

DISCUSSION Partial fractions would work, but the denominator, when factored, would be $(1 + x)^5(1 - x)^5$. There would be 10 unknown constants to find. Look for an easier approach.

Since the denominator is the obstacle, try $u = x^2$ or $u = 1 - x^2$, to see if the integrand gets simpler. Let us examine what happens in each case. Try $u = x^2$ first. Assume that we are interested only in getting an antiderivative for positive x, $x = \sqrt{u}$:

If x were negative, we would use $x = -\sqrt{u}$.

$$u = x^2, \qquad du = 2x\, dx, \qquad dx = \frac{du}{2x} = \frac{du}{2\sqrt{u}}.$$

Then

$$\int \frac{x^3\, dx}{(1 - x^2)^5} = \int \frac{u^{3/2}}{(1 - u)^5}\frac{du}{2\sqrt{u}} = \frac{1}{2}\int \frac{u\, du}{(1 - u)^5}.$$

The same substitution could be carried out as follows:

$$\int \frac{x^3\, dx}{(1 - x^2)^5} = \int \frac{x^2 x\, dx}{(1 - x^2)^5} = \int \frac{u(du/2)}{(1 - u)^5} = \frac{1}{2}\int \frac{u\, du}{(1 - u)^5}.$$

The substitution $v = 1 - u$ then results in an easy integral, as the reader may check. (The two substitutions $u = x^2$ and $v = 1 - u$ are equivalent to the single substitution $v = 1 - x^2$.)

So let us try $u = 1 - x^2$, $du = -2x\,dx$; thus

$$\int \frac{x^3\,dx}{(1 - x^2)^5} = \int \frac{x^2(x\,dx)}{(1 - x^2)^5} = \int \frac{(1 - u)(-du/2)}{u^5} = \int \frac{1}{2}(u^{-4} - u^{-5})\,du,$$

an integral that can be computed without further substitution. So $u = 1 - x^2$ is quicker than $u = x^2$.

Assume $|x| \leq 1$.

Since $1 - x^2 = (\sqrt{1 - x^2})^2$, it might be amusing to try the substitution $x = \sin\theta$, $dx = \cos\theta\,d\theta$ instead; thus

$$\int \frac{x^3\,dx}{(1 - x^2)^5} = \int \frac{\sin^3\theta\cos\theta\,d\theta}{(1 - \sin^2\theta)^5} = \int \frac{\sin^3\theta\cos\theta\,d\theta}{\cos^{10}\theta}$$

$$= \int \frac{\sin^3\theta\,d\theta}{\cos^9\theta}$$

$$= \int \frac{\sin^2\theta\sin\theta\,d\theta}{\cos^9\theta}$$

$$= \int \frac{(1 - \cos^2\theta)\sin\theta\,d\theta}{\cos^9\theta}$$

$$= \int \frac{(1 - u^2)(-du)}{u^9} \qquad (u = \cos\theta),$$

which is not a difficult integral.

Of the three methods, the substitution $u = 1 - x^2$ is the most efficient. ■

EXAMPLE 7 $\displaystyle\int \frac{\ln x\,dx}{x}$.

DISCUSSION Integration by parts, with $u = \ln x$ and $dv = dx/x$, may come to mind. In that case, $du = dx/x$ and $v = \ln x$; thus

$$\int \underbrace{\ln x}_{u}\,\underbrace{\frac{dx}{x}}_{dv} = \underbrace{(\ln x)}_{u}\,\underbrace{(\ln x)}_{v} - \int \underbrace{\ln x}_{v}\,\underbrace{\frac{dx}{x}}_{du}.$$

Bringing the $\int \ln x\,dx/x$ all to one side produces the equation

$$2\int \ln x\,\frac{dx}{x} = (\ln x)^2,$$

from which it follows that

$$\int \ln x\,\frac{dx}{x} = \frac{(\ln x)^2}{2} + C.$$

The method worked, but it is not the easiest one to use. Since $1/x$ is the derivative of $\ln x$, we could have used the substitution $u = \ln x$, $du = dx/x$; thus

$$\int \frac{\ln x\,dx}{x} = \int u\,du = \frac{u^2}{2} + C = \frac{(\ln x)^2}{2} + C. \quad ■$$

EXAMPLE 8 $\int x^3 e^{x^2}\, dx.$

If we can raise an exponent, we should be able to lower it.

DISCUSSION Integration by parts may come to mind, since if $u = x^3$, then $du = 3x^2\, dx$ is simpler. However, dv must then be $e^{x^2}\, dx$ and force v to be nonelementary. This is a dead end.

So try integration by parts with $u = e^{x^2}$ and $dv = x^3\, dx$. What will $v\, du$ be? We have $v = x^4/4$ and $du = 2x\, e^{x^2}\, dx$. Thus $v\, du = \frac{1}{2} x^5 e^{x^2}\, dx$, which is worse than the original $u\, dv$. The exponent of x has been raised by 2, from 3 to 5.

This time try $u = x^2$ and $dv = x\, e^{x^2}\, dx$; thus $du = 2x\, dx$ and $v = e^{x^2}/2$. Integration by parts yields

$$\int x^3 e^{x^2}\, dx = \int \underbrace{x^2}_{u}\, \underbrace{x\, e^{x^2}\, dx}_{dv} = \underbrace{x^2}_{u}\, \underbrace{\frac{e^{x^2}}{2}}_{v} - \int \underbrace{\frac{e^{x^2}}{2}}_{v}\, \underbrace{2x\, dx}_{du}$$

$$= \frac{x^2 e^{x^2}}{2} - \frac{e^{x^2}}{2} + C.$$

Another approach is to use the substitution $u = x^2$ followed by an integration by parts. ■

EXAMPLE 9 $\displaystyle\int \frac{1 - \sin\theta}{\theta + \cos\theta}\, d\theta.$

DISCUSSION Because θ appears by itself, the integrand is not a rational function of $\sin\theta$ and $\cos\theta$. However, the numerator is the derivative of the denominator, so the integral is $\ln|\theta + \cos\theta| + C$. ■

EXAMPLE 10 $\displaystyle\int \frac{1 - \sin\theta}{\cos\theta}\, d\theta.$

DISCUSSION The substitution $u = \tan(\theta/2)$ would work, but there is an easier approach. Break the integrand into two summands:

$$\int \frac{1 - \sin\theta}{\cos\theta}\, d\theta = \int \left(\frac{1}{\cos\theta} - \frac{\sin\theta}{\cos\theta}\right) d\theta$$

$$= \int (\sec\theta - \tan\theta)\, d\theta$$

$$= \ln|\sec\theta + \tan\theta| + \ln|\cos\theta| + C.$$

Since $\ln A + \ln B = \ln AB$, the answer can be simplified to

$$\ln(|\sec\theta + \tan\theta|\, |\cos\theta|) + C.$$

But $\sec\theta \cos\theta = 1$ and $\tan\theta\cos\theta = \sin\theta$. The answer becomes even simpler:

$$\int \frac{1 - \sin\theta}{\cos\theta}\, d\theta = \ln(1 + \sin\theta) + C. \quad ■$$

EXERCISES FOR SEC. 7.8: WHAT TO DO IN THE FACE OF AN INTEGRAL

All the integrals in Exercises 1 to 78 are elementary. In each case, list the technique or techniques that would be of use. If there is a preferred technique, state what it is. Do *not* evaluate the integrals. (For practice in evaluating integrals, see Sec. 7.S, Exercises 33 to 174.)

1 $\int \frac{1+x}{x^2}\,dx$

2 $\int \frac{x^2}{1+x}\,dx$

3 $\int \frac{dx}{x^2+x^3}$

4 $\int \frac{x+1}{x^2+x^3}\,dx$

5 $\int \tan^{-1} 2x\,dx$

6 $\int \sin^{-1} 2x\,dx$

7 $\int x^{10}e^x\,dx$

8 $\int \frac{\ln x}{x^2}\,dx$

9 $\int \frac{\sec^2\theta\,d\theta}{\tan\theta}$

10 $\int \frac{\tan\theta\,d\theta}{\sin^2\theta}$

11 $\int \frac{x^3}{\sqrt[3]{x+2}}\,dx$

12 $\int \frac{x^2}{\sqrt[3]{x^3+2}}\,dx$

13 $\int \frac{\cos^3\theta\,d\theta}{\sqrt[3]{\sin\theta}}$

14 $\int \sqrt{\cos\theta}\,\sin\theta\,d\theta$

15 $\int \tan^2\theta\,d\theta$

16 $\int \frac{d\theta}{\sec^2\theta}$

17 $\int \frac{x}{(\sqrt{9-x^2})^5}\,dx$

18 $\int \frac{dx}{(\sqrt{9-x^2})^5}$

19 $\int e^{\sqrt{x}}\,dx$

20 $\int \sin\sqrt{x}\,dx$

21 $\int \frac{dx}{(x^2+x+1)^5}$

22 $\int \frac{2x+1}{(x^2+x+1)^5}\,dx$

23 $\int \frac{dx}{(x^2-4x+3)^2}$

24 $\int \frac{(x-2)\,dx}{(x^2-4x+3)^2}$

25 $\int \frac{x^5}{x+1}\,dx$

26 $\int \frac{x+1}{x^5}\,dx$

27 $\int \frac{e^{3x}\,dx}{1+e^x+e^{2x}}$

28 $\int \frac{\ln x}{x(1+\ln x)}\,dx$

29 $\int \frac{dx}{x\sqrt{x^2+9}}$

30 $\int \frac{x\,dx}{\sqrt{x^2+9}}$

31 $\int \frac{dx}{(3+\sin x)^2}$

32 $\int \frac{\cos x\,dx}{(3+\sin x)^2}$

33 $\int \ln(e^x)\,dx$

34 $\int \ln(\sqrt[3]{x})\,dx$

35 $\int \tan^6 x\sec x\,dx$

36 $\int \sec^6 x\tan x\,dx$

37 $\int \frac{x^4-1}{x+2}\,dx$

38 $\int \frac{x+2}{x^4-1}\,dx$

39 $\int \frac{dx}{\sqrt{x}(3+\sqrt{x})^2}$

40 $\int \frac{dx}{(3+\sqrt{x})^3}$

41 $\int (1+\tan\theta)^3\sec^2\theta\,d\theta$

42 $\int (1+\tan\theta)^3\sec\theta\,d\theta$

43 $\int \frac{e^x+e^{-x}}{e^x-e^{-x}}\,dx$

44 $\int \frac{e^{2x}+1}{e^x-e^{-x}}\,dx$

45 $\int \frac{(x+3)(\sqrt{x+2}+1)}{\sqrt{x+2}-1}\,dx$

46 $\int \frac{(\sqrt[3]{x+2}-1)\,dx}{\sqrt{x+2}+1}$

47 $\int \frac{dx}{x^2-9}$

48 $\int (x^2-9)^{10}\,dx$

49 $\int \frac{x+7}{(3x+2)^{10}}\,dx$

50 $\int \frac{x^3\,dx}{(3x+2)^7}$

51 $\int \frac{2^x+3^x}{4^x}\,dx$

52 $\int \frac{2^x}{1+2^x}\,dx$

53 $\int \frac{(x+\sin^{-1}x)\,dx}{\sqrt{1-x^2}}$

54 $\int \frac{x+\tan^{-1}x}{1+x^2}\,dx$

55 $\int x^3\sqrt{1+x^2}\,dx$

56 $\int x(1+x^2)^{3/2}\,dx$

57 $\int \frac{x\,dx}{\sqrt{x^2-1}}$

58 $\int \frac{x^3}{\sqrt{x^2-1}}\,dx$

59 $\int \sec^2 x\tan^2 x\,dx$

60 $\int \sec^3 x\tan^3 x\,dx$

61 $\int \cos^4\theta\sin^6\theta\,d\theta$

62 $\int \cos^3\theta\sin^6\theta\,d\theta$

63 $\int \sin^6\theta\,d\theta$

64 $\int \sin^5\theta\,d\theta$

65 $\int \frac{dx}{(x^2 - 9)^{3/2}}$

66 $\int \frac{x\,dx}{(x^2 - 9)^{3/2}}$

67 $\int \frac{\tan^{-1} x}{1 + x^2}\,dx$

68 $\int \frac{\tan^{-1} x}{x^2}\,dx$

69 $\int \frac{\sin (\ln x)}{x}\,dx$

70 $\int \cos x \ln (\sin x)\,dx$

71 $\int \frac{x\,dx}{\sqrt{x^2 + 4}}$

72 $\int \frac{dx}{\sqrt{x^2 + 4}}$

73 $\int \frac{dx}{x^2 + x + 5}$

74 $\int \frac{x\,dx}{x^2 + x + 5}$

75 $\int \frac{x + 3}{(x + 1)^5}\,dx$

76 $\int \frac{x^5 + x + \sqrt{x}}{x^3}\,dx$

77 $\int (x^2 + 9)^{10}\,x\,dx$

78 $\int (x^2 + 9)^{10}\,x^3\,dx$

■

In each of Exercises 79 to 81, (*a*) state for which two integers $n \geq 1$ you can evaluate the integral, and (*b*) evaluate it. (*Warning:* For most n, the integral is not elementary. Base your answer on your experience with integrals.)

79 $\int \sqrt{1 + x^n}\,dx$

80 $\int (1 + x^2)^{1/n}\,dx$

81 $\int (1 + x)^{1/n}\sqrt{1 - x}\,dx$

82 Find $\int \frac{dx}{\sqrt{x + 2} - \sqrt{x - 2}}$.

83 Find $\int \sqrt{1 - \cos x}\,dx$.

84 Find $\int \frac{\sec^2 x\,dx}{\sqrt{10 - \sec^2 x}}$.

85 Find $\int \frac{dx}{x(x^{20} + 1)}$.

7.9 ESTIMATES OF DEFINITE INTEGRALS

This section presents ways of estimating a definite integral $\int_a^b f(x)\,dx$. There are several reasons for wanting to make such an estimate. First, the integrand $f(x)$ may not have an elementary antiderivative; thus the fundamental theorem of calculus could not be used (for example, $\int_0^1 e^{x^2}\,dx$). Second, even though the antiderivative is elementary, it may be tedious to compute [for example, $\int_0^1 1/(1 + x^5)\,dx$]. Third, the values of the integrand $f(x)$ may be known only at a few values of x (for example, temperature may be recorded hourly).

The definite integral $\int_a^b f(x)\,dx$ is, by definition, a limit of sums of the form

$$\sum_{i=1}^{n} f(c_i)(x_i - x_{i-1}). \qquad (1)$$

Any such sum consequently provides an estimate of $\int_a^b f(x)\,dx$. However, the two methods described in this section, the trapezoidal method and Simpson's method, generally provide much better estimates for the same amount of arithmetic.

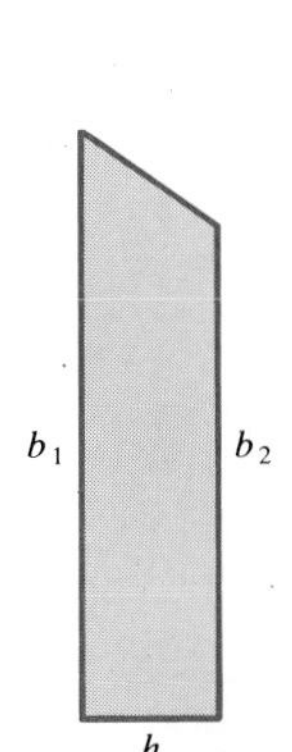

Area = $\frac{(b_1 + b_2)h}{2}$

Figure 7.10

The sum (1) can be thought of as a sum of areas of rectangles. In the **trapezoidal method**, trapezoids are used instead of rectangles. Recall that the area of a trapezoid of height h and bases b_1 and b_2 is $(b_1 + b_2)h/2$. (See Fig. 7.10.)

The trapezoidal method

Let n be a positive integer. Divide the interval $[a, b]$ into n sections of equal length $h = (b - a)/n$ with

$$x_0 = a, \quad x_1 = a + h, \quad x_2 = a + 2h, \quad \ldots \quad, x_n = b.$$

The sum

$$\frac{f(x_0) + f(x_1)}{2} \cdot h + \frac{f(x_1) + f(x_2)}{2} \cdot h + \cdots + \frac{f(x_{n-1}) + f(x_n)}{2} \cdot h$$

In the trapezoidal method, $h = \frac{b-a}{n}$ and f is computed at $n + 1$ inputs.

is the **trapezoidal estimate** of $\int_a^b f(x)\,dx$. It is usually written

$$\boxed{\frac{h}{2}[f(x_0) + 2f(x_1) + 2f(x_2) + \cdots + 2f(x_{n-1}) + f(x_n)].} \tag{2}$$

Note that $f(x_0)$ and $f(x_n)$ have coefficient 1, while all the other $f(x_i)$'s have coefficient 2. This is due to the double counting of the edges common to two trapezoids.

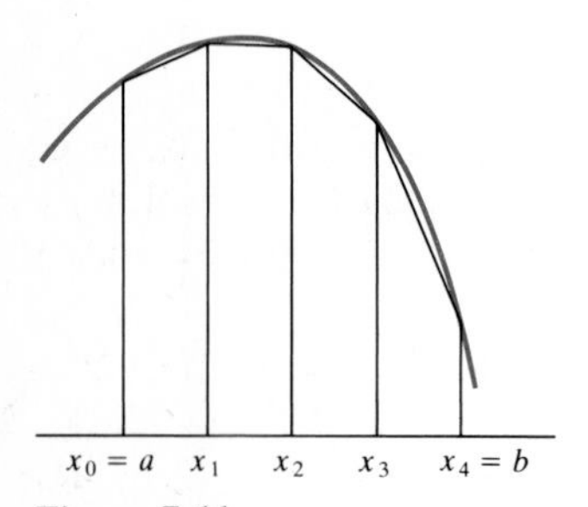

Figure 7.11

The diagram in Fig. 7.11 illustrates the trapezoidal approximation for the case $n = 4$. Note that if f is concave downward, the trapezoidal approximation underestimates $\int_a^b f(x)\,dx$. If f is a linear function, the trapezoidal method, of course, gives the integral exactly.

EXAMPLE 1 Use the trapezoidal method with $n = 4$ to estimate

$$\int_0^1 \frac{dx}{1 + x^2}.$$

SOLUTION In this case, $a = 0$, $b = 1$, and $n = 4$, so $h = (1 - 0)/4 = \frac{1}{4}$. The successive coefficients in the trapezoidal estimate (2) are 1, 2, 2, 2, and 1.

The trapezoidal estimate is

$$\frac{h}{2}\left[f(0) + 2f\left(\frac{1}{4}\right) + 2f\left(\frac{2}{4}\right) + 2f\left(\frac{3}{4}\right) + f(1)\right].$$

Now $h/2 = \frac{1}{4}/2 = \frac{1}{8}$. To compute the sum in brackets make a table:

x_i	$f(x_i)$	*Coefficient*	*Summand*	*Decimal Form*
0	$\frac{1}{1 + 0^2}$	1	$1 \cdot \frac{1}{1 + 0}$	1.0000000
$\frac{1}{4}$	$\frac{1}{1 + (\frac{1}{4})^2}$	2	$2 \cdot \frac{1}{1 + \frac{1}{16}}$	1.8823529
$\frac{2}{4}$	$\frac{1}{1 + (\frac{2}{4})^2}$	2	$2 \cdot \frac{1}{1 + \frac{1}{4}}$	1.6000000
$\frac{3}{4}$	$\frac{1}{1 + (\frac{3}{4})^2}$	2	$2 \cdot \frac{1}{1 + \frac{9}{16}}$	1.2800000
$\frac{4}{4}$	$\frac{1}{1 + (\frac{4}{4})^2}$	1	$1 \cdot \frac{1}{1 + 1}$	0.5000000

The trapezoidal sum is therefore approximately

$$\tfrac{1}{8}(1 + 1.8823529 + 1.6 + 1.28 + 0.5) = \tfrac{1}{8}(6.2623529) \approx 0.782794.$$

Thus

$$\int_0^1 \frac{dx}{1 + x^2} \approx 0.782794. \quad ■$$

The integral in Example 1 can be evaluated by the fundamental theorem of calculus. It equals $\tan^{-1} 1 - \tan^{-1} 0 = \pi/4 \approx 0.785398$. The trap-

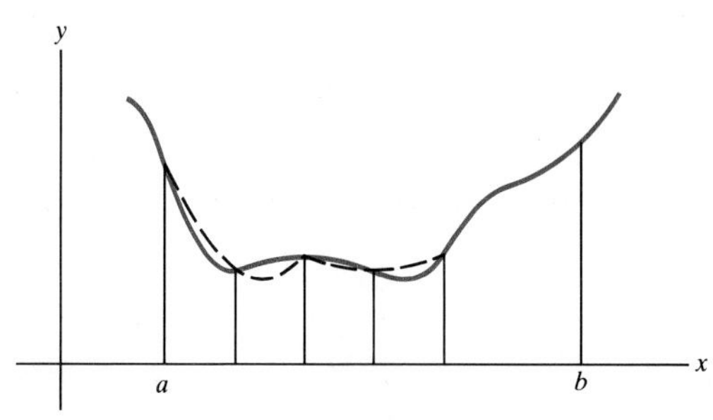

The dashed lines are parts of parabolas

Figure 7.12

ezoidal estimate is correct to two decimal places.

Simpson's method uses parabolas.

In the trapezoidal method a curve is approximated by lines. In **Simpson's method** a curve is approximated by parabolas. (See Fig. 7.12.) As Exercise 25 shows, Simpson's method is exact if $f(x)$ is a polynomial of degree at most 3.

In Simpson's method n is even

In Simpson's method the interval $[a, b]$ is divided into an *even* number of sections.

Divide the interval $[a, b]$ into n sections of equal length $h = (b - a)/n$ with

$$x_0 = a, \quad x_1 = a + h, \quad x_2 = a + 2h, \quad \ldots \quad, x_n = b.$$

The sum

Note that $f(x_0)$ and $f(x_n)$ have coefficient 1, while the coefficients of the other $f(x_i)$'s alternate 4, 2, 4, 2, . . . , 2, 4.

$$\boxed{\frac{h}{3}[f(x_0) + 4f(x_1) + 2f(x_2) + 4f(x_3) + \cdots + 2f(x_{n-2}) + 4f(x_{n-1}) + f(x_n)]} \tag{3}$$

is the **Simpson estimate** of $\int_a^b f(x)\,dx$.

EXAMPLE 2 Use Simpson's method with $n = 4$ to estimate $\displaystyle\int_0^1 \frac{dx}{1 + x^2}$.

SOLUTION Again $h = \frac{1}{4}$. Simpson's formula (3) takes the form

$$\frac{\frac{1}{4}}{3}[f(0) + 4f(\tfrac{1}{4}) + 2f(\tfrac{2}{4}) + 4f(\tfrac{3}{4}) + f(1)].$$

x_i	$f(x_i)$	*Coefficient*	*Summand*	*Decimal Form*
0	$\frac{1}{1 + 0^2}$	1	$1 \cdot \frac{1}{1 + 0}$	1.0000000
$\frac{1}{4}$	$\frac{1}{1 + (\frac{1}{4})^2}$	4	$4 \cdot \frac{1}{1 + \frac{1}{16}}$	3.7647059
$\frac{2}{4}$	$\frac{1}{1 + (\frac{2}{4})^2}$	2	$2 \cdot \frac{1}{1 + \frac{1}{4}}$	1.6000000
$\frac{3}{4}$	$\frac{1}{1 + (\frac{3}{4})^2}$	4	$4 \cdot \frac{1}{1 + \frac{9}{16}}$	2.5600000
$\frac{4}{4}$	$\frac{1}{1 + (\frac{4}{4})^2}$	1	$1 \cdot \frac{1}{1 + 1}$	0.5000000

The Simpson approximation of $\int_0^1 dx/(1 + x^2)$ is therefore

$$\tfrac{1}{12}(1 + 3.7647059 + 1.6 + 2.56 + 0.5) = \tfrac{1}{12}(9.4247059) \approx 0.785392.$$

Thus $$\int_0^1 \frac{dx}{1 + x^2} \approx 0.785392. \quad ■$$

Since $\int_0^1 dx/(1 + x^2) \approx 0.785398$, Simpson's estimate is correct to five decimal places. Simpson's method usually provides a much more accurate estimate of an integral than the trapezoidal estimate for the same amount of arithmetic. (As in Examples 1 and 2, each uses the same number of points at which to evaluate the function; the difference is in the weights given the values of the function at those points.)

The accuracy of the trapezoidal and Simpson methods

The trapezoidal estimate is exact for a polynomial of the form $f(x) = a + bx$, since the region below its graph is a trapezoid. Now, such a polynomial can be thought of as a function whose second derivative is 0 for all x. Therefore, it is reasonable to expect that the accuracy of the trapezoidal method is influenced by the second derivative of f. It can be shown that if M_2 is the maximum value of $|f^{(2)}(x)|$ for x in $[a, b]$, then the error in using the trapezoidal approximation (2) is at most

The error in the trapezoidal method

$$\frac{(b - a)M_2h^2}{12}.$$

The factor h^2 is the key indicator of the accuracy. It suggests that cutting h in half will cut the error by a factor of 4.

As mentioned earlier, Simpson's estimate is exact for any polynomial of the form $f(x) = a + bx + cx^2 + dx^3$, that is, for any function whose fourth derivative is identically 0. The error in using Simpson's method for other functions is measured by the fourth derivative. Let M_4 be the maximum value of $|f^{(4)}(x)|$ for x in $[a, b]$. Then the error in using Simpson's formula (3) in estimating $\int_a^b f(x)\, dx$ is at most

The error in the Simpson method

$$\frac{(b - a)M_4h^4}{180}.$$

Since $h^4 \to 0$ faster than $h^2 \to 0$ as $h \to 0$, we would expect Simpson's method to be generally more accurate than the trapezoidal method. Cutting h in half would tend to cut the error by a factor of 16.

EXERCISES FOR SEC. 7.9: ESTIMATES OF DEFINITE INTEGRALS

Exercises 1 to 6 concern the trapezoidal estimate. In Exercises 1 to 4: (*a*) estimate the given integrals by the trapezoidal method, using the given values of n; (*b*) Evaluate the integrals and calculate the absolute values of the errors in the estimates.

1 $\int_0^1 x^2\, dx$, $n = 3$

2 $\int_0^1 x^2\, dx$, $n = 4$

3 $\int_1^5 \frac{dx}{x}$, $n = 2$

4 $\int_1^7 \frac{dx}{x}$, $n = 6$

5 Estimate $\int_0^1 dx/(1 + x^3)$ using the trapezoidal method with $n = 4$.

6 Estimate $\int_0^1 dx/(1 + x^4)$ using the trapezoidal method with $n = 5$.

In Exercises 7 to 12 estimate the given integrals by Simpson's method using the given values of n.

7 $\int_1^5 \frac{dx}{x}$, $n = 2$

8 $\int_1^5 \frac{dx}{x}$, $n = 4$

9 $\int_1^7 \frac{dx}{x}$, $n = 2$

10 $\int_1^7 \frac{dx}{x}$, $n = 6$

11 $\int_0^1 \frac{dx}{1 + x^3}$, $n = 4$

12 $\int_0^1 \frac{dx}{1 + x^3}$, $n = 8$

In Exercises 13 and 14, using the given n, estimate the given integrals by (*a*) the trapezoidal method, (*b*) Simpson's method.

13 $\int_1^2 \frac{e^x\, dx}{x}$, $n = 6$

14 $\int_1^5 e^{-x^2}\, dx$, $n = 6$

■

The "right-point" estimate of $\int_a^b f(x)\, dx$ is obtained as follows: Select a positive integer n and divide $[a, b]$ into n sections of equal length $h = (b - a)/n$. The points of subdivision are $x_0 < x_1 < \cdots < x_n$, with $x_0 = a$ and $x_n = b$. The right-point estimate is the approximating sum

$$h[f(x_1) + f(x_2) + \cdots + f(x_n)].$$

The "left-point" estimate is defined similarly; it is given by $h[f(x_0) + f(x_1) + \cdots + f(x_{n-1})]$.

15 Show that if $f(a) = f(b)$, the left-point, right-point, and trapezoidal estimates for a given value of h are the same.

16 Show that for a given n the average of the left-point estimate and the right-point estimate equals the trapezoidal estimate.

The next two exercises present cases in which the bounds of maximum error are actually assumed.

17 Show that if the trapezoidal method with $n = 1$ is used to estimate $\int_0^1 x^2\, dx$, the error equals $(b - a)M_2h^2/12$, where $a = 0$, $b = 1$, $h = 1$, and M_2 is the maximum value of $|d^2(x^2)/dx^2|$ for x in $[0, 1]$.

18 Show that if Simpson's method with $n = 2$ is used to estimate $\int_0^1 x^4\, dx$, the error equals $(b - a)M_4h^4/180$, where $a = 0$, $b = 1$, $h = \frac{1}{2}$, and M_4 is the maximum value of $|d^4(x^4)/dx^4|$ for x in $[0, 1]$.

19 (*a*) Compute $\int_0^1 dx/(1 + x^3)$ using partial fractions. *Hint:* $1 + x^3 = (1 + x)(1 - x + x^2)$.

(*b*) Estimate $\int_0^1 dx/(1 + x^3)$ using Simpson's method with $n = 6$.

(*c*) Compute the absolute value of the error of the estimate made in (*b*).

Exercises 20 to 25 describe the geometric motivation of Simpson's method.

20 Let $f(x) = Ax^2 + Bx + C$. Show that

$$\int_{-h}^{h} f(x)\, dx = \frac{h}{3}[f(-h) + 4f(0) + f(h)].$$

Hint: Just compute both sides.

21 Let f be a function. Show that there is a parabola $y = Ax^2 + Bx + C$ that passes through the three points $(-h, f(-h))$, $(0, f(0))$, and $(h, f(h))$. (See Fig. 7.13.)

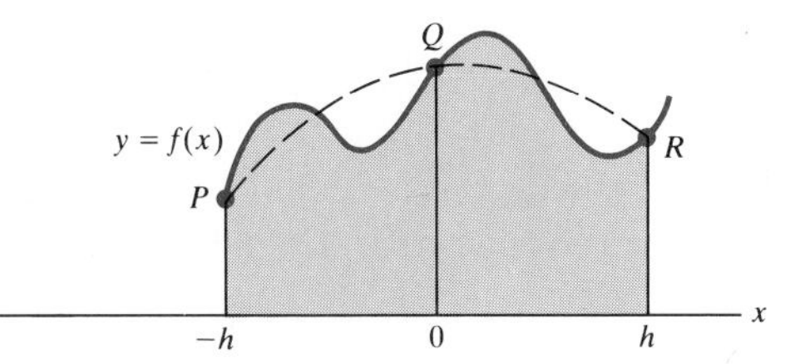

The dashed graph is a parabola, $y = Ax^2 + Bx + C$, through P, Q, and R. The area of the region below the parabola is precisely

$$\frac{h}{3}[f(-h) + 4f(0) + f(h)]$$

and is an approximation of the area of the shaded region.

Figure 7.13

22 The equation in Exercise 20 (which was known to the Greeks) is called the **prismoidal formula** Use it to compute the volume of a sphere of radius a.

23 Let $f(x) = Ax^2 + Bx + C$. Show that

$$\int_{c-h}^{c+h} f(x)\, dx = \frac{h}{3}[f(c - h) + 4f(c) + f(c + h)].$$

Hint: Use the substitution $x = c + t$ to reduce this to Exercise 20.

24 First, $[a, b]$ is divided into n sections (n even), which are grouped into $n/2$ pairs of adjacent sections. Over each pair the function is approximated by the parabola that passes through the three points of the graph with x coordinates equal to those which determine the two sections of the pair. (See Fig. 7.12.) The integral of this quadratic function is used as an estimate of the integral of f over each pair of adjacent sections. Show that when these $n/2$ separate estimates are added, Simpson's formula results.

25 Since Simpson's method was designed to be exact when $f(x) = Ax^2 + Bx + C$, one would expect the error associated with it to involve $f^{(3)}(x)$. By a quirk of good fortune, Simpson's method happens to be exact even when $f(x)$ is a *cubic*, $Ax^3 + Bx^2 + Cx + D$. This suggests that the error involves $f^{(4)}(x)$, not $f^{(3)}(x)$.

(*a*) Show that if $f(x) = x^3$,

$$\int_{-h}^{h} f(x)\, dx = \frac{h}{3}[f(-h) + 4f(0) + f(h)].$$

(*b*) Show that Simpson's estimate is exact for cubics.

26 There are many other methods for estimating definite integrals. Some old methods, which had been of only theoretical interest because of their messy arithmetic,

have, with the advent of computers, assumed practical importance. This exercise illustrates the simplest of the so-called Gaussian quadrature formulas. For simplicity, consider only integrals over $[-1, 1]$.
(*a*) Show that

$$\int_{-1}^{1} f(x)\,dx = f\left(\frac{-1}{\sqrt{3}}\right) + f\left(\frac{1}{\sqrt{3}}\right)$$

for $f(x) = 1, x, x^2$, and x^3.
(*b*) Let a and b be two numbers, $-1 \le a < b \le 1$, such that

$$\int_{-1}^{1} f(x)\,dx = f(a) + f(b)$$

for $f(x) = 1, x, x^2$, and x^3. Show that $a = -1/\sqrt{3}$ and $b = 1/\sqrt{3}$.
(*c*) Show that the approximation $\int_{-1}^{1} f(x)\,dx \approx f(-1/\sqrt{3}) + f(1/\sqrt{3})$ has no error when f is a polynomial of degree at most 3.

Part (*c*) suggests that the error in this method involves $f^{(4)}(x)$. It is proved in numerical analysis that the absolute value of the error is at most $M_4/135$, where M_4 is the maximum of $|f^{(4)}(x)|$ for x in $[-1, 1]$. Incidentally, the estimate

$$\int_{-1}^{1} f(x)\,dx \approx \tfrac{5}{9}f(-\sqrt{\tfrac{3}{5}}) + \tfrac{8}{9}f(0) + \tfrac{5}{9}f(\sqrt{\tfrac{3}{5}})$$

is exact for polynomials of degree at most 5. Gaussian quadrature is discussed in Anthony Ralston and Philip Rabinowitz, *A First Course in Numerical Analysis,* 2d ed., McGraw-Hill, New York, 1978, pp. 98–101.

27 Use the first formula in Exercise 26(*c*) to estimate $\int_{-1}^{1} dx/(1 + x^2)$.

28 Let f be a function such that $|f^{(2)}(x)| \le 10$ and $|f^{(4)}(x)| \le 50$ for all x in $[1, 5]$. If $\int_1^5 f(x)\,dx$ is to be estimated with an error of at most 0.01, how small must h be in (*a*) the trapezoidal approximation? (*b*) Simpson's approximation?

29 Consider the integral $\int_0^4 dx/(1 + x^4)$.
(*a*) Estimate the integral by the trapezoidal method, $n = 4$.
(*b*) Estimate the integral by Simpson's method, $n = 4$.
(*c*) Evaluate the integral, making use of the fact that $x^4 + 1 = (x^2 + \sqrt{2}x + 1)(x^2 - \sqrt{2}x + 1)$. After expressing the integrand in terms of partial fractions, use an integral table.

30 Let T be the trapezoidal estimate of $\int_a^b f(x)\,dx$, using $x_0 = a, x_1, \ldots, x_n = b$. Let M be the "midpoint estimate," $\sum_{i=1}^{n} f(c_i)(x_i - x_{i-1})$, where $c_i = (x_{i-1} + x_i)/2$. Let S be Simpson's estimate using the $2n + 1$ points $x_0, c_1, x_1, c_2, x_2, \ldots, c_n, x_n$. Show that

$$S = \tfrac{2}{3}M + \tfrac{1}{3}T.$$

31 This table shows the temperature $f(t)$ as a function of time.

Time	1	2	3	4	5	6	7
Temperature	81	75	80	83	78	70	60

(*a*) Use Simpson's method to estimate $\int_1^7 f(t)\,dt$.
(*b*) Use the result in (*a*) to estimate the average temperature.

32 Estimate the area of the region below $y = \cos x^2$ and above $[0, 1]$, using $n = 6$ and (*a*) the trapezoidal formula, (*b*) Simpson's formula.

7.S SUMMARY

Method	*Description*
Substitution	Introduce $u = h(x)$. If $f(x)\,dx = g(u)\,du$, then $\int f(x)\,dx = \int g(u)\,du$.
Substitution in the definite integral	If, in the above substitution, $u = A$ when $x = a$ and $u = B$ when $x = b$, then $\int_a^b f(x)\,dx = \int_A^B g(u)\,du$.
Table of integrals	Obtain and become familiar with a table of integrals. Substitution, together with integral tables, will usually be adequate.
Integration by parts	$\int u\,dv = uv - \int v\,du$. Choose u and v so $u\,dv = f(x)\,dx$ and $\int v\,du$ is easier than $\int u\,dv$.
Partial fractions (applies to any rational function of x)	This is an algebraic method. Write the integrand as a sum of a polynomial (if the degree of the numerator is greater than or equal to the degree of the denominator) plus terms of the type $\dfrac{k_i}{(ax+b)^i}$ and $\dfrac{c_jx+d_j}{(ax^2+bx+c)^j}$. The number of unknown constants is the same as the degree of the denominator. A table of integrals treats the integrals of these two types. (For the first type, use the substitution $u = ax + b$; for the second, complete the square.)
To integrate certain powers of trigonometric functions	There are many special cases. For instance, rewrite $\int \cos^k\theta \sin^{2n+1}\theta\,d\theta$ as $\int \cos^k\theta\,(1-\cos^2\theta)^n \sin\theta\,d\theta$ and make the substitution $u = \cos\theta$. Keep in mind that $d(\sin\theta) = \cos\theta\,d\theta$; $d(\cos\theta) = -\sin\theta\,d\theta$; $d(\sec\theta) = \sec\theta\tan\theta\,d\theta$; $d(\tan\theta) = \sec^2\theta\,d\theta$. Also, use trigonometric identities, such as $\sin^2\theta = (1-\cos 2\theta)/2$ and $\cos^2\theta = (1+\cos 2\theta)/2$.
To integrate any rational function of $\cos\theta$ and $\sin\theta$	Let $u = \tan(\theta/2)$. Then $\cos\theta = \dfrac{1-u^2}{1+u^2}$, $\sin\theta = \dfrac{2u}{1+u^2}$, $d\theta = \dfrac{2\,du}{1+u^2}$, and the new integrand is a rational function of u.
To integrate rational functions of x and one of $\sqrt{a^2-x^2}$, $\sqrt{a^2+x^2}$, $\sqrt{x^2-a^2}$, a^2-x^2, a^2+x^2	For $\sqrt{a^2-x^2}$ or a^2-x^2, let $x = a\sin\theta$. For $\sqrt{a^2+x^2}$ or a^2+x^2, let $x = a\tan\theta$. For $\sqrt{x^2-a^2}$, let $x = a\sec\theta$. (Recall the two right triangles shown below.) [Triangle: hypotenuse $\sec\theta$, adjacent side 1, opposite side $\tan\theta$, angle θ] [Triangle: hypotenuse 1, adjacent side $\cos\theta$, opposite side $\sin\theta$, angle θ]
To integrate rational functions of x and $\sqrt[n]{ax+b}$	Let $u = \sqrt[n]{ax+b}$, hence $u^n = ax+b$, $nu^{n-1}\,du = a\,dx$, and $x = (u^n-b)/a$. The new integrand is a rational function of u.

The fundamental theorem of calculus, proved in Chap. 5, raised the problem of finding antiderivatives. Now, some very simple and important functions do not have elementary antiderivatives; for instance,

$$\int \frac{\sin x\,dx}{x}, \quad \int e^{x^2}\,dx, \quad \int \frac{dx}{\ln x}, \quad \int x\tan x\,dx,$$

$$\int \frac{\ln x}{x+1}\,dx, \quad \int \sqrt{1-\frac{\sin^2 x}{4}}\,dx, \quad \text{and} \quad \int \sqrt[3]{x-x^2}\,dx$$

are not elementary. If the definite integral $\int_0^1 e^{x^2}\,dx$ is needed, an estimate must be made, for example, by an approximating sum (after all, a definite integral is defined as a limit of such sums), the trapezoidal method, or Simpson's method. The "elliptic integral" $\int_0^{\pi/2} \sqrt{1 - k^2 \sin^2 x}\, dx$, frequently used in engineering, is tabulated for various values of k to four decimal places in most handbooks.

Some definite integrals over intervals of the form $[-a, a]$ can be simplified before evaluation. If $f(x)$ is an even function, then $\int_{-a}^{a} f(x)\, dx = 2 \int_0^a f(x)\, dx$; if it is an odd function, then $\int_{-a}^{a} f(x)\, dx = 0$. (For instance, $\int_{-1}^{1} xe^{x^4}\, dx = 0$.)

TWO WAYS TO ESTIMATE $\int_a^b f(x)\, dx$

Method	*Formula*	*Weights*	*Bound on Error*
Trapezoidal	$\frac{h}{2}[f(x_0) + 2f(x_1) + 2f(x_2) + \cdots + 2f(x_{n-1}) + f(x_n)]$	1, 2, 2, 2, . . . , 2, 1	$\frac{(b - a)M_2h^2}{12}$
Simpson's	$\frac{h}{3}[f(x_0) + 4f(x_1) + 2f(x_2) + \cdots + 4f(x_{n-1}) + f(x_n)]$	1, 4, 2, 4, 2, . . . , 2, 4, 1	$\frac{(b - a)M_4h^4}{180}$

(M_k is maximum of $|f^{(k)}(x)|$ for $a \le x \le b$.) In Simpson's method the number n must be even. In the trapezoidal formula $h/2$ is a factor; in Simpson's method $h/3$ is a factor. In both methods $h = (b - a)/n$. Also in both, the inputs x_i are evenly spaced with successive x_i's a distance h apart: $x_i = a + ih$, $i = 0, 1, 2, \ldots, n$.

ANTIDIFFERENTIATION BY MACHINE

As we have seen in this chapter, the rules for antidifferentiation are not nearly as complete or direct as those for differentiation. Consequently, computer programs have much greater difficulty producing integrals than in computing derivatives.

For example, muMath (see Sec. 3.S) will toss back in essentially unchanged form any integral it cannot do. The antidifferentiation command INT(1/(x^5 + 1), x) results in INT(1/(1 + x^5), x). However, muMath can find an antiderivative for xe^x: INT(x #E^x, x) yields #E^x x − #E^x, where #E is e and ^ represents exponentiation. Occasionally, the user can circumvent the limitations of muMath's antidifferentiation algorithms by first modifying the integrand. For example, muMath has the PARFRAC command to attempt partial-fraction decomposition of algebraic expressions. While muMath cannot calculate INT(1/((x − 1)*(x + 1)*(x + 2)), x), if one first assigns the variable F to the expression PARFRAC(1/((x − 1)*(x + 1)*(x + 2)), x), muMath can easily compute INT(F, x).

The muMath package includes the muSimp programming language in which muMath is written. A user willing to learn muSimp can add new features to muMath.

When muMath encounters a function with a nonelementary antiderivative, it will return the integral unchanged. However, since muMath cannot recognize all elementary cases, its failure to compute an antiderivative does not show that the result is nonelementary. Students with access to MACSYMA, muMath, or a similar program will still often have to rely on themselves rather than on computers.

DEFINITE INTEGRATION BY MACHINE

By use of the fundamental theorem of calculus, computer programs that compute antiderivatives may also compute definite integrals. For example, in Microsoft muMath, the cube function may be integrated over the interval $[a, b]$ with the following command: DEFINT(x^3, x, a, b), which yields (−a^4 + b^4)/4, as expected. If specific numbers are given for the limits of integration, muMath will return a numeric answer.

If a numeric answer is all that is desired, many scientific pocket calculators now offer definite integration through approximation techniques. Although any programmable calculator can be used to apply Simpson's rule, preprogrammed algorithms of even greater power are available on advanced calculators, such as those from Hewlett-Packard and Texas Instruments. (The Hewlett-Packard algorithm does not use equally spaced points and is a business secret.) After you define the function to be integrated, and its limits of integration, the calculator iteratively computes approximate answers until attaining a specific number of decimal places.

GUIDE QUIZ ON CHAP. 7: COMPUTING ANTIDERIVATIVES

In Problems 1 to 20 evaluate the integrals.

1 $\int_1^2 \frac{x^3\,dx}{1+x^4}$

2 $\int \sqrt{4-9x^2}\,dx$

3 $\int \frac{dx}{x^4-1}$

4 $\int \tan^5 2x \sec^2 2x\,dx$

5 $\int \frac{x^4\,dx}{x^4-1}$

6 $\int_0^{4/3} \frac{dx}{\sqrt{9x^2+16}}$

7 $\int \frac{dx}{2\sqrt{x}-\sqrt[4]{x}}$

8 $\int \frac{dx}{(x-3)\sqrt{x+3}}$

9 $\int \frac{dx}{3-x^2}$

10 $\int \sin^5 2x\,dx$

11 $\int e^x \cos 2x\,dx$

12 $\int \frac{dx}{\sin^5 3x}$

13 $\int \frac{x^3\,dx}{1+x^8}$

14 $\int \frac{dx}{x\sqrt{4+x^2}}$

15 $\int \frac{x^2\,dx}{\sqrt{x^2-9}}$

16 $\int (9-x^2)^{3/2}\,dx$

17 $\int \frac{dx}{3+\cos x}$

18 $\int \frac{dx}{x^2\sqrt{x^2+25}}$

19 $\int \frac{x^4-\sqrt{x}}{x^3}\,dx$

20 $\int \sin 3x \cos 5x\,dx$

21 (*a*) Describe the trapezoidal method.
(*b*) For what type of functions $f(x)$ does it give the exact value of $\int_a^b f(x)\,dx$?
(*c*) What is the general bound on the error?
(*d*) If you reduce the width of the trapezoids by a factor of 10, what would you expect to happen to the error?

22 (*a*) Describe Simpson's method.
(*b*) For what type of functions does it give the exact value of $\int_a^b f(x)\,dx$?
(*c*) What is the general bound on the error?
(*d*) If you reduce the width of the subdivision by a factor of 10, what would you expect to happen to the error?

REVIEW EXERCISES FOR CHAP. 7: COMPUTING ANTIDERIVATIVES

1 (*a*) By an appropriate substitution, transform this definite integral into a simpler definite integral:

$$\int_0^{\pi/2} \sqrt{(1+\cos\theta)^3}\,\sin\theta\,d\theta.$$

(*b*) Evaluate the new definite integral in (*a*).

2 Two of these antiderivatives are elementary functions; evaluate them.

(*a*) $\int \ln x\,dx$ (*b*) $\int \frac{\ln x\,dx}{x}$ (*c*) $\int \frac{dx}{\ln x}$

3 Evaluate

(*a*) $\int_1^2 (1+x^3)^2\,dx$ (*b*) $\int_1^2 (1+x^3)^2 x^2\,dx$

4 Compute with the aid of a table of integrals:

(*a*) $\int \frac{e^x\,dx}{5e^{2x}-3}$ (*b*) $\int \frac{dx}{\sqrt{x^2-3}}$

5 Compute

(*a*) $\int \frac{dx}{x^3}$ (*b*) $\int \frac{dx}{\sqrt{x+1}}$ (*c*) $\int \frac{e^x\,dx}{1+5e^x}$

6 Compute

$$\int \frac{5x^4-5x^3+10x^2-8x+4}{(x^2+1)(x-1)}\,dx$$

7 Compute

$$\int \frac{x^3\,dx}{(1+x^2)^4}$$

in two different ways:
(*a*) By the substitution $u = 1 + x^2$.
(*b*) By the substitution $x = \tan\theta$.

8 Transform the definite integral

$$\int_0^3 \frac{x^3}{\sqrt{x+1}}\,dx$$

to another definite integral in two different ways (and evaluate):
(*a*) By the substitution $u = x + 1$.
(*b*) By the substitution $u = \sqrt{x + 1}$.

9 Compute $\int x^2 \ln(1 + x)\,dx$ (*a*) without an integral table; (*b*) with an integral table.

10 Find $\int \frac{x\,dx}{\sqrt{9x^4 + 16}}$ (*a*) without an integral table; (*b*) with an integral table.

11 Compute $\int \frac{\sin\theta\,d\theta}{1 + \sin^2\theta}$
(*a*) by using the substitution that applies to any rational function of $\cos\theta$ and $\sin\theta$;
(*b*) by writing $\sin^2\theta$ as $1 - \cos^2\theta$ and using the substitution $u = \cos\theta$.

12 (*a*) Without an integral table, evaluate

$$\int \sin^5\theta\,d\theta \quad \text{and} \quad \int \tan^6\theta\,d\theta.$$

(*b*) Evaluate them with an integral table.

13 Two of these three antiderivatives are elementary. Find them.
(*a*) $\int \sqrt{1 - 4\sin^2\theta}\,d\theta$ (*b*) $\int \sqrt{4 - 4\sin^2\theta}\,d\theta$
(*c*) $\int \sqrt{1 + \cos\theta}\,d\theta$

14 (*a*) Transform the definite integral

$$\int_{-1}^{4} \frac{x + 2}{\sqrt{x + 3}}\,dx$$

into another definite integral using the substitution $u = x + 3$.
(*b*) Evaluate the integral obtained in (*a*).

15 The fundamental theorem can be used to evaluate one of these definite integrals, but not the other. Evaluate one of them.

(*a*) $\int_0^1 \sqrt[3]{x\sqrt{x}}\,dx$ (*b*) $\int_0^1 \sqrt[3]{1 - x\sqrt{x}}\,dx$

16 Verify that the following factorizations into irreducible polynomials are correct.
(*a*) $x^3 - 1 = (x - 1)(x^2 + x + 1)$
(*b*) $x^4 - 1 = (x - 1)(x + 1)(x^2 + 1)$
(*c*) $x^3 + 1 = (x + 1)(x^2 - x + 1)$
(*d*) $x^4 + 1 = (x^2 + \sqrt{2}x + 1)(x^2 - \sqrt{2}x + 1)$

In Exercises 17 to 24 express as a sum of partial fractions. (Do not integrate.) Exercise 16 may be helpful.

17 $\frac{2x^2 + 3x + 1}{x^3 - 1}$ **18** $\frac{x^4 + 2x^2 - 2x + 2}{x^3 - 1}$

19 $\frac{2x - 1}{x^3 + 1}$

20 $\frac{x^4 + 3x^3 - 2x^2 + 3x - 1}{x^4 - 1}$

21 $\frac{x^3 - (1 + \sqrt{2})x^2 + (1 - \sqrt{2})x - 1}{x^4 + 1}$

22 $\frac{2x + 5}{x^2 + 3x + 2}$

23 $\frac{5x^3 + 11x^2 + 6x + 1}{x^2 + x}$ **24** $\frac{5x^3 + 6x^2 + 8x + 5}{(x^2 + 1)(x + 1)}$

25 For which values of the nonnegative integers m and n are the following integrations comparatively simple?

(*a*) $\int \sin^m x\,dx$ (*b*) $\int \sec^n x\,dx$

(*c*) $\int \sin^m x \cos^n x\,dx$ (*d*) $\int \sec^m x \tan^n x\,dx$

(*e*) $\int \cot^m x \csc^n x\,dx$

26 (*a*) Develop the reduction formula relating

$$\int \sin^n x\,dx \quad \text{to} \quad \int \sin^{n-2} x\,dx.$$

(*b*) If n is odd, what technique may be used for finding $\int \sin^n x\,dx$?

In each of Exercises 27 to 32 use an appropriate substitution to obtain an integrand that is a rational function of a single variable or of trigonometric functions. Do not evaluate.

27 $\int \frac{(\sqrt{4 - x^2})^3 + 1}{[(4 - x^2)^3 + 5]\sqrt{4 - x^2}}\,dx$

28 $\int \frac{x + \sqrt[3]{x - 2}}{x^2 - \sqrt[3]{x - 2}}\,dx$

29 $\int \frac{(x^2 - 5)^7}{x^2 + 3 + \sqrt{x^2 - 5}}\,dx$

30 $\int \frac{\cos^2\theta + \sin\theta}{1 - \sin\theta\cos\theta}\,d\theta$

31 $\int \frac{3\tan^2\theta + \sec\theta + 1}{2 + \tan\theta + \cos\theta}\,d\theta$

32 $\int \frac{(4 + x^2)^{1/2}}{5 + (4 + x^2)^{3/2}}\,dx$

In Exercises 33 to 174 find the integrals.

33 $\int \frac{\cos x\,dx}{\sin^3 x - 8}$ **34** $\int \frac{dx}{\sqrt{2 + \sqrt{x}}}$

35 $\int \frac{\sqrt{x^2 + 1}}{x^4}\,dx$ **36** $\int \frac{\sin x\,dx}{1 + 3\cos^2 x}$

37 $\int \frac{\sin x\, dx}{3 + \cos x}$

38 $\int x\sqrt{x^4 - 1}\, dx$

39 $\int x^2\sqrt{x^3 - 1}\, dx$

40 $\int \sin \sqrt{x}\, dx$

41 $\int \frac{dx}{(4 + x^2)^2}$

42 $\int (\sqrt[3]{x} + \sqrt[3]{x + 1})\, dx$

43 $\int \sin^2 3x \cos^2 3x\, dx$

44 $\int \sin^3 3x \cos^2 3x\, dx$

45 $\int \tan^4 3\theta\, d\theta$

46 $\int \frac{x^2\, dx}{x^4 - 1}$

47 $\int \frac{x^4 + x^2 + 1}{x^3}\, dx$

48 $\int \frac{3\, dx}{\sqrt{1 - 5x^2}}$

49 $\int 10^x\, dx$

50 $\int \frac{x^3}{(x^4 + 1)^3}\, dx$

51 $\int \frac{x\, dx}{(x^4 + 1)^2}$

52 $\int \cos^3 x \sin^2 x\, dx$

53 $\int \cos^2 x\, dx$

54 $\int x\sqrt{x + 4}\, dx$

55 $\int x\sqrt{x^2 + 4}\, dx$

56 $\int \frac{x + 2}{x^2 + 1}\, dx$

57 $\int \frac{x^2\, dx}{1 + x^6}$

58 $\int \sqrt[3]{4x + 7}\, dx$

59 $\int x^2 \sin x^3\, dx$

60 $\int \frac{\ln x^4}{x}\, dx$

61 $\int x^4 \ln x\, dx$

62 $\int \frac{\tan^{-1} 3x}{1 + 9x^2}\, dx$

63 $\int \frac{e^{\sqrt{x}}}{\sqrt{x}}\, dx$

64 $\int \sin (\ln x)\, dx$

65 $\int \ln (x^3 - 1)\, dx$

66 $\int \tan x\, dx$

67 $\int \frac{x\, dx}{\sqrt{(x^2 + 1)^3}}$

68 $\int \frac{2 + \sqrt[3]{x}}{x}\, dx$

69 $\int \frac{dx}{\sqrt{(x + 1)^3}}$

70 $\int \frac{2x + 3}{x^2 + 3x + 5}\, dx$

71 $\int \frac{3\, dx}{x^2 + 4x + 5}$

72 $\int \frac{3\, dx}{x^2 + 4x - 5}$

73 $\int \frac{x\, dx}{1 + \sqrt[3]{x}}$

74 $\int \ln \sqrt{2x - 1}\, dx$

75 $\int \frac{x^7\, dx}{\sqrt{x^2 + 1}}$

76 $\int x^3 \tan^{-1} x\, dx$

77 $\int \frac{\tan^{-1} x}{x^2}\, dx$

78 $\int \frac{dx}{x^3 + 4x}$

79 $\int e^x \sin 3x\, dx$

80 $\int \sqrt{\frac{1}{x^2} + \frac{1}{x^4}}\, dx$

81 $\int \frac{dx}{(4 - x^2)^{3/2}}$

82 $\int x^{1/4}(1 + x^{1/5})\, dx$

83 $\int \frac{x\, dx}{x^4 - 2x^2 - 3}$

84 $\int \sin^{-1} \sqrt[3]{x}\, dx$

85 $\int \frac{x^2\, dx}{\sqrt[3]{x - 1}}$

86 $\int \ln (4 + x^2)\, dx$

87 $\int \frac{\sqrt{x^2 + 4}}{x}\, dx$

88 $\int \sqrt{\tan \theta} \sec^2 \theta\, d\theta$

89 $\int \sec^5 \theta \tan \theta\, d\theta$

90 $\int \tan^6 \theta\, d\theta$

91 $\int \frac{dx}{x\sqrt{x^2 + 9}}$

92 $\int (e^x + 1)^2\, dx$

93 $\int \frac{(1 - x)^2}{\sqrt[3]{x}}\, dx$

94 $\int (1 + \sqrt{x})x\, dx$

95 $\int \sin^2 2x \cos x\, dx$

96 $\int (e^{2x})^3 e^x\, dx$

97 $\int \left(e^x - \frac{1}{e^x}\right)^2 dx$

98 $\int \frac{dx}{(\sqrt{x} + 1)(\sqrt{x})}$

99 $\int x \sin^{-1} x^2\, dx$

100 $\int x \sin^{-1} x\, dx$

101 $\int \frac{dx}{e^{2x} + 5e^x}$

102 $\int \frac{e^x\, dx}{1 - 6e^x + 9e^{2x}}$

103 $\int (2x + 1)\sqrt{3x + 2}\, dx$

104 $\int \frac{2x^3 + 1}{x^3 - 4x^2}\, dx$

105 $\int \frac{x^2\, dx}{(x - 1)^3}$

106 $\int \frac{dx}{\sqrt{9 + x^2}}$

107 $\int \frac{e^x + 1}{e^x - 1}\, dx$

108 $\int \frac{dx}{4x^2 + 1}$

109 $\int (1 + 3x^2)^2\, dx$

110 $\int \frac{x\, dx}{x^3 + 1}$

111 $\int \frac{x^3\, dx}{x^3 + 1}$

112 $\int \frac{x^2\, dx}{\sqrt{2x + 1}}$

113 $\int \frac{dx}{\sqrt{2x + 1}}$

114 $\int (x + \sin x)^2\, dx$

115 $\int \frac{x\,dx}{x^4 - 3x^2 - 2}$

116 $\int \frac{x^3\,dx}{x^4 - 1}$

117 $\int \frac{e^x\,dx}{1 + e^{2x}}$

118 $\int \frac{dx}{x^2 + 5x - 6}$

119 $\int \frac{dx}{x^2 + 5x + 6}$

120 $\int \frac{x\,dx}{2x^2 + 5x + 6}$

121 $\int \frac{4x + 10}{x^2 + 5x + 6}\,dx$

122 $\int \sqrt{4x^2 + 1}\,dx$

123 $\int \frac{dx}{2x^2 + 5x + 6}$

124 $\int \sqrt{-4x^2 + 1}\,dx$

125 $\int \frac{dx}{2x^2 + 5x - 6}$

126 $\int \frac{dx}{2 + 3\sin x}$

127 $\int \frac{dx}{\sin^2 x}$

128 $\int \frac{dx}{3 + 2\sin x}$

129 $\int \frac{dx}{\sin^4 x}$

130 $\int \ln(x^2 + 5)\,dx$

131 $\int \frac{dx}{\sqrt{x - 2\sqrt{3 - x}}}$

132 $\int x^3 e^{-5x}\,dx$

133 $\int \sqrt{(1 + 2x)(1 - 2x)}\,dx$

134 $\int x \sin 3x\,dx$

135 $\int \frac{2x\,dx}{\sqrt{x^2 + 1}}$

136 $\int \frac{2\,dx}{\sqrt{x^2 + 1}}$

137 $\int \frac{x^4 + 4x^3 + 6x^2 + 4x - 3}{x^4 - 1}\,dx$

138 $\int \frac{x^3 + 6x^2 + 11x + 5}{(x + 2)^2(x + 1)}\,dx$

139 $\int \frac{-3x^2 - 11x - 11}{(x + 2)^2(x + 1)}\,dx$

140 $\int \frac{x^2 - 3x}{(x + 1)(x - 1)^2}\,dx$

141 $\int \frac{12x^2 + 2x + 3}{4x^3 + x}\,dx$

142 $\int \frac{6x^3 + 2x + \sqrt{3}}{1 + 3x^2}\,dx$

143 $\int \frac{-6x^3 - 13x - 3\sqrt{3}}{1 - 3x^2}\,dx$

144 $\int \frac{x\,dx}{\sqrt{1 - 9x^2}}$

145 $\int \frac{dx}{\sqrt{1 - 9x^2}}$

146 $\int \frac{dx}{x\sqrt{3x^2 - 5}}$

147 $\int \frac{dx}{(3x^2 + 2)^{3/2}}$

148 $\int \frac{dx}{\sin 5x}$

149 $\int \frac{dx}{\cos 4x}$

150 $\int \frac{x^2\,dx}{1 + 3x^3 + 2x^6}$

151 $\int e^x \sin^2 x\,dx$

152 $\int x^3\sqrt{1 - 3x^2}\,dx$

153 $\int \sqrt{1 + \sqrt{1 + \sqrt{x}}}\,dx$

154 $\int \frac{x^3\,dx}{1 - 4x^2}$

155 $\int e^{\sqrt[4]{x}}\,dx$

156 $\int \sec^4 x\,dx$

157 $\int \frac{dx}{\cos^3 x}$

158 $\int \cos^3 x\,dx$

159 $\int x^2 \ln(x^3 + 1)\,dx$

160 $\int \frac{\ln x + \sqrt{x}}{x}\,dx$

161 $\int \frac{(3 + x^2)^2\,dx}{x}$

162 $\int \frac{dx}{e^x}$

163 $\int \frac{(1 + 3\cos x)^2\,dx}{\sin x}$

164 $\int \frac{3 + \cos\theta}{2 - \cos\theta}\,d\theta$

165 $\int (e^{2x} + 1)e^{-x}\,dx$

166 $\int \sqrt{9x^2 - 4}\,dx$

167 $\int \frac{2x^2 + 4x + 3}{x^3 + 2x^2 + 3x}\,dx$

168 $\int \frac{dx}{x^4 + 3x^2 + 1}$

169 $\int \frac{\sec^2\theta\,d\theta}{\sqrt{\sec^2\theta - 1}}$

170 $\int \ln(2x + x^2)\,dx$

171 $\int x \sin^2 x\,dx$

172 $\int \frac{dx}{1 + 2e^{3x}}$

173 $\int x \tan^2 x\,dx$

174 $\int \sqrt{1 + \cos 3\theta}\,d\theta$

175 Compute

(*a*) $\int \frac{dx}{x^2 + 4x + 3}$

(*b*) $\int \frac{dx}{x^2 + 4x + 4}$

(*c*) $\int \frac{dx}{x^2 + 4x + 5}$

(*d*) $\int \frac{dx}{x^2 + 4x - 2}$

176 Compute $\int \frac{x^3\,dx}{(x-1)^2}$

(*a*) Using partial fractions
(*b*) Using the substitution $u = x - 1$
(*c*) Which method is easier?

177 Compute

(*a*) $\int \sec^5 x\,dx$ (*b*) $\int \sec^5 x \tan x\,dx$

(*c*) $\int \frac{\sin x}{(\cos x)^3}\,dx$

178 (*a*) Compute $\int \frac{x^{2/3}\,dx}{x+1}$.

(*b*) What does a table of integrals say about

$$\int \frac{x^{2/3}\,dx}{x+1}?$$

179 Find $\int x\sqrt{(1-x^2)^5}\,dx$ using the substitutions (*a*) $u = x^2$, (*b*) $u = 1 - x^2$, (*c*) $x = \sin\theta$.

180 Compute $\int x\sqrt[3]{x+1}\,dx$ using

(*a*) The substitution $u = \sqrt[3]{x+1}$
(*b*) The substitution $u = x + 1$

181 Transform $\int (x^2/\sqrt{1+x})\,dx$ by each of the substitutions (*a*) $u = \sqrt{1+x}$, (*b*) $u = 1 + x$, (*c*) $x = \tan^2\theta$. (*d*) Solve the easiest of the resulting problems.

In Exercises 182 to 185 evaluate the integrals.

182 $\int_0^1 (e^x+1)^3 e^x\,dx$ **183** $\int_0^1 (x^4+1)^5 x^3\,dx$

184 $\int_1^e \frac{\sqrt{\ln x}}{x}\,dx$ **185** $\int_0^{\pi/2} \frac{\cos\theta\,d\theta}{\sqrt{1+\sin\theta}}$

■

186 Estimate $\int_0^3 f(x)\,dx$ if it is known that $f(0) = 10$, $f(0.5) = 13$, $f(1) = 14$, $f(1.5) = 16$, $f(2) = 18$, $f(2.5) = 10$, $f(3) = 6$ by (*a*) the trapezoidal method, (*b*) Simpson's method.

187 Transform the problem of finding $\int (x^3/\sqrt{1+x^2})\,dx$ to a different problem, using (*a*) integration by parts with $dv = (x\,dx)/\sqrt{1+x^2}$, (*b*) the substitution $x = \tan\theta$, (*c*) the substitution $u = \sqrt{1+x^2}$.

188 Two of these three integrals are elementary. Evaluate them.

(*a*) $\int \sin^2 x\,dx$ (*b*) $\int \sin\sqrt{x}\,dx$ (*c*) $\int \sin x^2\,dx$

189 Compute $\int x\sqrt{1+x}\,dx$ in three ways:

(*a*) Let $u = \sqrt{1+x}$.
(*b*) Let $x = \tan^2\theta$.
(*c*) By parts, with $u = x$, $dv = \sqrt{1+x}\,dx$.

190 Only one of the functions $\sqrt{x}\sqrt[3]{1-x}$ and $\sqrt{1-x}\sqrt[3]{1-x}$ has an elementary antiderivative F. Find F.

191 Consider the problem of finding the area under $y = e^{x^2}$ from $x = 0$ to $x = 1$.

(*a*) Why is the FTC useless in determining this area?
(*b*) Estimate $\int_0^1 e^{x^2}\,dx$ by utilizing a partition of [0, 1] into the five sections [0, 0.2], [0.2, 0.4], [0.4, 0.6], [0.6, 0.8], [0.8, 1], and as sampling numbers $c_1 = 0.1$, $c_2 = 0.3$, $c_3 = 0.5$, $c_4 = 0.7$, and $c_5 = 0.9$.

192 Assuming that $\int (e^x/x)\,dx$ is not elementary (a theorem of Liouville), prove that $\int 1/\ln x\,dx$ is not elementary.

193 Evaluate $\int_{-1}^{1} xe^{x^4}\,dx$.

194 The factor theorem asserts that if r is a root of a polynomial $P(x)$, then $x - r$ is a divisor of $P(x)$. Use this theorem to factor each of the following polynomials into irreducible factors:

(*a*) $x^3 - 1$ (*b*) $x^3 + 1$ (*c*) $x^3 - 5$
(*d*) $x^3 + 8$ (*e*) $x^4 - 1$ (*f*) $x^3 - 3x + 2$

195 (See Exercise 194.) Represent in partial fractions:

(*a*) $\frac{x}{x^3 - 1}$ (*b*) $\frac{1}{x^3 - 3x + 2}$,

(*c*) $\frac{x^5}{x^3 + 1}$ (*d*) $\frac{1}{x^3 + 8}$.

196 Show these integrals are elementary: (*a*) $\int x^{1/3}(1+x)^{5/3}\,dx$, (*b*) $\int \sqrt[4]{x}(1+x)^3\,dx$. (*Hint:* Let $x = 1/t$.)

197 One integral table lists the antiderivative of $\int (\sqrt{x^2+a^2}/x)\,dx$ as

$$\sqrt{x^2+a^2} - a\ln\left(\frac{a+\sqrt{x^2+a^2}}{x}\right),$$

while another lists it as

$$\sqrt{x^2+a^2} + a\ln\left(\frac{\sqrt{x^2+a^2}-a}{x}\right).$$

Is there an error in one?

198 One of these integrals is elementary; the other is not. Evaluate the one that is elementary.

(*a*) $\int \ln(\cos x)\,dx$ (*b*) $\int \cos(\ln x)\,dx$

199 (*a*) Explain why $\int x^m e^x\,dx$ is an elementary function for any positive integer m.
(*b*) Explain why $\int x^m(\ln x)^n\,dx$ is an elementary function for any positive integers m and n.

200 (*a*) Prove the trigonometric identity

$$\sin mx \sin nx = \tfrac{1}{2}\{\cos[(m-n)x] - \cos[(m+n)x]\}$$

by expanding the right side.
(*b*) Use it to compute $\int \sin 2x \sin 3x\,dx$.

201 From the fact that $\int x\tan x\,dx$ is not elementary, deduce that the following are not elementary:

(a) $\int x^2 \sec^2 x\, dx$ (b) $\int x^2 \tan^2 x\, dx$

(c) $\int \frac{x^2\, dx}{1 + \cos x}$

202 Three of these six antiderivatives are elementary. Compute them.

(a) $\int x \cos x\, dx$ (b) $\int \frac{\cos x}{x}\, dx$ (c) $\int \frac{x\, dx}{\ln x}$

(d) $\int \frac{\ln x^2}{x}\, dx$ (e) $\int \sqrt{x - 1}\sqrt{x}\sqrt{x + 1}\, dx$

(f) $\int \sqrt{x - 1}\, \sqrt{x + 1}\, x\, dx$

203 From the fact that $\int (\sin x)/x\, dx$ is not elementary, deduce that the following are not elementary:

(a) $\int (\cos^2 x)/x^2\, dx$ (b) $\int (\sin^2 x)/x^2\, dx$

(c) $\int \sin e^x\, dx$ (d) $\int \cos x \ln x\, dx$

Exercise 204 is the basis of Exercises 205 to 214.

204 Let p and q be rational numbers. Prove that $\int x^p(1 - x)^q\, dx$ is an elementary function (a) if p is an integer (*Hint:* If $q = s/t$, let $1 - x = v^t$), (b) if q is an integer, (c) if $p + q$ is an integer.

Chebyshev proved that these are the only cases for which the antiderivative in question is elementary. In particular,

$$\int \sqrt{x}\sqrt[3]{1 - x}\, dx \quad \text{and} \quad \int \sqrt[3]{x - x^2}\, dx$$

are not elementary. Chebyshev's theorem also holds for $\int x^p(1 + x)^q\, dx$.

205 Deduce from Exercise 204 that $\int \sqrt{1 - x^3}\, dx$ is not elementary.

206 Deduce from Exercise 204 that $\int (1 - x^n)^{1/m}\, dx$, where m and n are positive integers, is elementary if and only if $m = 1$, $n = 1$, or $m = 2 = n$.

207 Deduce from Exercise 204 that $\int \sqrt{\sin x}\, dx$ is not elementary. *Hint:* Let $u = \sin^2 x$.

208 Deduce from Exercise 204 that $\int \sin^a x\, dx$, where a is rational, is elementary if and only if a is an integer.

209 Deduce from Exercise 204 that $\int \sin^p x \cos^q x\, dx$, where p and q are rational, is elementary if and only if p or q is an odd integer or $p + q$ is an even integer.

210 Deduce from Exercise 209 that $\int \sec^p x \tan^q x\, dx$, where p and q are rational, is elementary only if $p + q$ or q is odd, or if p is even.

211 (a) Deduce from Exercise 204 that $\int (x/\sqrt{1 + x^n})\, dx$, where n is a positive integer, is elementary only when $n = 1, 2$, or 4.
(b) Evaluate the integral for $n = 1, 2$, and 4.

212 (a) Deduce from Exercise 204 that $\int (x^2/\sqrt{1 + x^n})\, dx$, where n is a positive integer, is elementary only when $n = 1, 2, 3$, or 6.
(b) Evaluate the integral for $n = 1, 2, 3$, and 6.

213 (a) Using Exercise 204, determine for which positive integers n the integral $\int (x^n/\sqrt{1 + x^4})\, dx$ is elementary.
(b) Evaluate the integral for $n = 3$ and $n = 5$.

214 Liouville proved that if f and g are rational functions and if $\int e^{f(x)} g(x)\, dx$ is an elementary function, then $\int e^{f(x)} g(x)\, dx$ can be expressed in the form $e^{f(x)} w(x)$, where $w(x)$ is a rational function. With the aid of this result, prove that $\int e^x/x\, dx$ is not an elementary function. *Hint:* Assume $[e^x(p/q)]' = e^x/x$, where p and q are relatively prime polynomials. Write $q = x^i r$, where $i \geq 0$ and x does not divide the polynomial r.

215 This is an excerpt from an applied chemistry text:

> The time for the depth D of the water to fall from 10 to 5 feet equals
>
> $$\int_5^{10} \frac{dD}{\sqrt{D - 4} + 4\sqrt{D - 2}}.$$
>
> The integral is evaluated by first rationalizing the denominator and then. . . .

Evaluate the integral. (Use an integral table.)

216 The following is an excerpt from an engineering text:

> The last equation may be written
>
> $$\theta = \int \frac{c\, dr}{r\sqrt{r^6 - c^2}}$$
>
> where c is a constant. The integral is easily evaluated by the substitution. . . .

(a) What substitution did the text recommend?
(b) Using Exercise 204, determine for which positive integers, n, $\int (c/(r^n \sqrt{r^6 - c^2}))\, dr$ is elementary.

217 The following is a quote from an article on the management of energy resources:

> Let $u(t)$ be the rate at which water flows through the turbines at time t. Then $U(t) = \int_0^t u(s)\, ds$ is the total flow during the time $[0, t]$. Let $z(t)$ represent the rate of demand for electric energy at time t, as measured in equivalent water flow. Then the cost incurred during this period of time is $\int_0^t c(z(s) - u(s))\, ds$, where c is a cost function for imported energy.

Explain why these interpretations of the integrals make sense.

218 If you are going to estimate $\int_0^1 e^{x^2}\, dx$ by the trapezoidal method, how small should you choose the subdivision length h to be sure that the error is less than 0.01?

219 Like the preceding exercise, but for Simpson's method.

8 Applications of the Definite Integral

This chapter presents various geometric and physical applications of the definite integral. It combines the ideas of Chap. 5 with the techniques of Chap. 7.

8.1 COMPUTING AREA BY PARALLEL CROSS SECTIONS

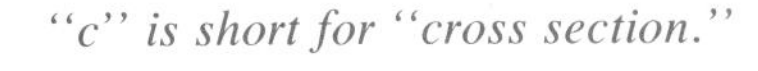

In Sec. 5.3 it was shown that the area of a plane region is equal to the integral of its cross-sectional length. Let $c(x)$ denote the length of the intersection with the given region of the vertical line through $(x, 0)$. (See Fig. 8.1.) Then the area of the region is equal to $\int_a^b c(x)\,dx$, where a and b are shown in Fig. 8.1. Note that x need not refer to the x axis of the xy plane; it may refer to any conveniently chosen line in the plane. It may even refer to the y axis; in this case, the cross-sectional length would be denoted $c(y)$.

To compute an area:

1 Find a, b and the cross-sectional length $c(x)$.
2 Evaluate $\int_a^b c(x)\,dx$ by the fundamental theorem of calculus if $\int c(x)\,dx$ is elementary.

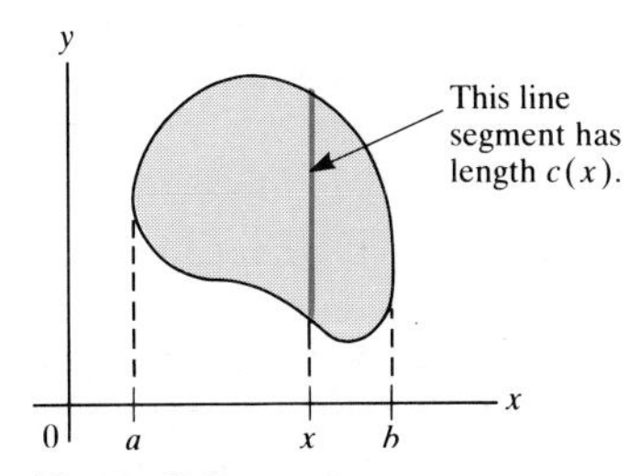

Figure 8.1

Chapter 7 showed how to accomplish step 2. The present section is concerned primarily with step 1, how to find the cross-sectional length $c(x)$.

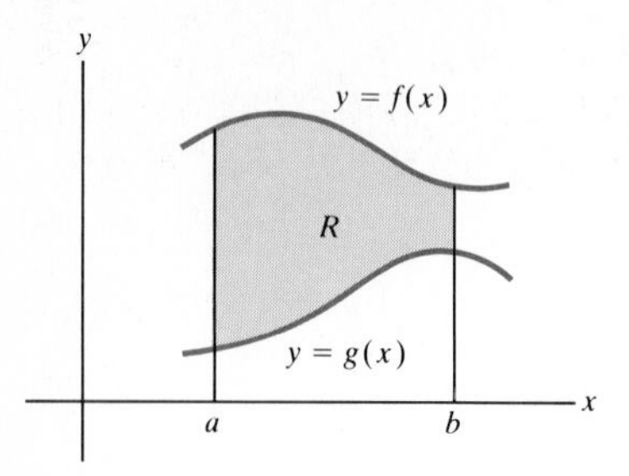

Figure 8.2

How to Find the Area between Two Curves

Let f and g be two continuous functions such that $f(x) \geq g(x)$ for all x in the interval $[a, b]$. Let R be the region between the curve $y = f(x)$ and the curve $y = g(x)$ for x in $[a, b]$, as shown in Fig. 8.2. In these circumstances, the length of the cross section of R made by a line perpendicular to the x axis can be computed in terms of f and g. Inspection of Fig. 8.3 shows that

$$c(x) = f(x) - g(x).$$

In short, to find $c(x)$, subtract the smaller value $g(x)$ from the larger value $f(x)$. The area of R is then given by

Area between two curves

$$\text{Area} = \int_a^b [f(x) - g(x)]\, dx.$$

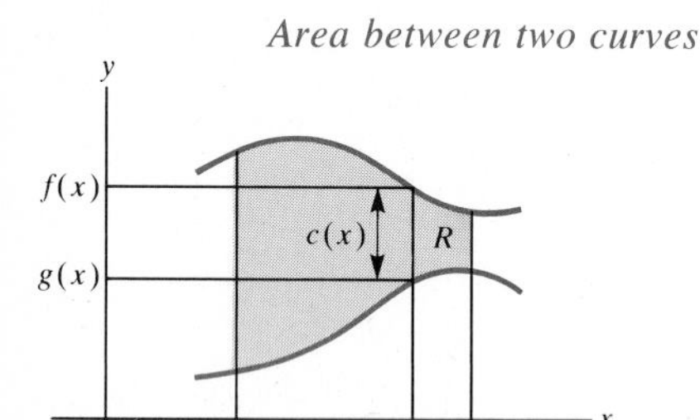

Figure 8.3

EXAMPLE 1 Find the area of the region shown in Fig. 8.4. The region is bounded by the curve $y = x^2$, the line $y = -\frac{3}{2}x$, and the line $x = 2$.

SOLUTION In this case, $f(x) = x^2$ and $g(x) = -\frac{3}{2}x$. For x in $[0, 2]$, the cross-sectional length is $c(x) = x^2 - (-\frac{3}{2}x)$. Thus the area of the region is

$$\int_0^2 \left[x^2 - \left(-\frac{3x}{2}\right)\right] dx = \int_0^2 \left(x^2 + \frac{3x}{2}\right) dx.$$

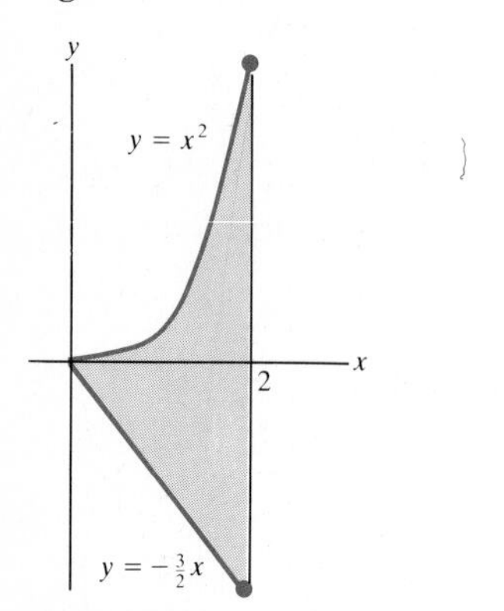

Figure 8.4

This definite integral can be evaluated by the fundamental theorem of calculus:

$$\int_0^2 \left(x^2 + \frac{3x}{2}\right) dx = \left(\frac{x^3}{3} + \frac{3x^2}{4}\right)\Bigg|_0^2$$

$$= \left(\frac{2^3}{3} + \frac{3 \cdot 2^2}{4}\right) - \left(\frac{0^3}{3} + \frac{3 \cdot 0^2}{4}\right) = \frac{17}{3}. \quad ■$$

EXAMPLE 2 Find the area of the region in Fig. 8.4, but this time use cross sections parallel to the x axis.

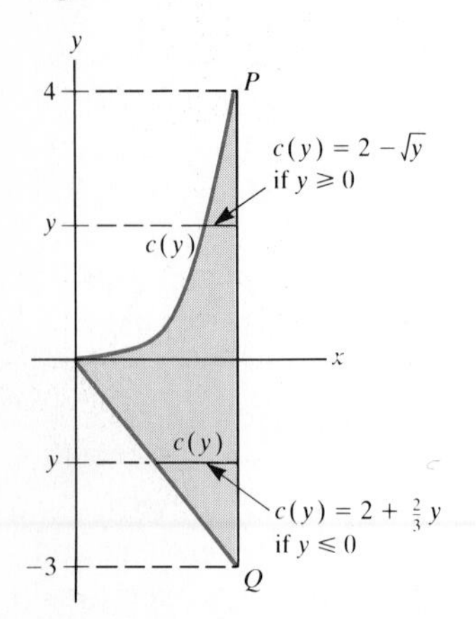

Figure 8.5

SOLUTION Since the cross-sectional length is to be expressed in terms of y, first express the equations of the curves bounding the region in terms of y. The curve $y = x^2$ may be written as $x = \sqrt{y}$, since we are interested only in positive x. The curve $y = -\frac{3}{2}x$ can be expressed as $x = -\frac{2}{3}y$ by solving for x in terms of y. The line $x = 2$ also bounds the region. (See Fig. 8.5.)

For each number y in a certain interval $[a, b]$ that will be determined, the line with y coordinate equal to y meets the region in Fig. 8.5 in a line segment of length $c(y)$. It is necessary to determine the numbers a and b as well as the formula for $c(y)$.

The point P in Fig. 8.5 lies on the parabola $y = x^2$ (or $x = \sqrt{y}$) and has the x coordinate 2. Thus $P = (2, 2^2) = (2, 4)$. The point Q lies on

y varies from the y coordinate of Q to the y coordinate of P.

the line $y = -\frac{3}{2}x$ and has x coordinate 2. Thus $Q = (2, -\frac{3}{2} \cdot 2) = (2, -3)$. Consequently, a cross section of the region is determined for each number y in the interval $[-3, 4]$. The area of the region is therefore

$$\int_{-3}^{4} c(y)\, dy. \tag{1}$$

Next, a formula for $c(y)$ must be found. For $0 \le y \le 4$, the cross section is determined by the line $x = 2$ and the parabola $x = \sqrt{y}$. Thus for $0 \le y \le 4$,

$$c(y) = 2 - \sqrt{y},$$

the larger minus the smaller. (See Fig. 8.5.) For $-3 \le y \le 0$, the cross section is determined by the line $x = 2$ and the line $x = -\frac{2}{3}y$. Thus for $-3 \le y \le 0$,

$$c(y) = 2 - \left(-\frac{2}{3}y\right) = 2 + \frac{2y}{3}.$$

The integral (1) breaks into two separate integrals, each of which is easily evaluated by the fundamental theorem of calculus:

$$\begin{aligned}
\int_{-3}^{0} c(y)\, dy + \int_{0}^{4} c(y)\, dy &= \int_{-3}^{0} \left(2 + \frac{2y}{3}\right) dy + \int_{0}^{4} (2 - \sqrt{y})\, dy \\
&= \left(2y + \frac{y^2}{3}\right)\bigg|_{-3}^{0} + \left(2y - \frac{2}{3}y^{3/2}\right)\bigg|_{0}^{4} \\
&= \left(2 \cdot 0 + \frac{0^2}{3}\right) - \left(2(-3) + \frac{(-3)^2}{3}\right) \\
&\quad + \left(2 \cdot 4 - \frac{2}{3}4^{3/2}\right) - \left(2 \cdot 0 - \frac{2}{3}0^{3/2}\right) \\
&= 0 - (-3) + \frac{8}{3} - 0 = \frac{17}{3}. \quad \blacksquare
\end{aligned}$$

Choose the direction for cross sections to make life easy.

Example 1 needed only one integral, but Example 2 needed two. Moreover, in Example 2 the formula for the cross-sectional length when $0 \le y \le 4$ involved $\sqrt{y}$, which is a little harder to integrate than x^2, which appeared in the corresponding formula in Example 1. Although both approaches to finding the area of the region in Fig. 8.4 are valid, the one with cross sections parallel to the y axis is more convenient.

How to Find Cross Sections Geometrically

Perhaps a region is described not by formulas but geometrically, maybe as a circle, triangle, trapezoid, or some other polygon. Two geometric facts are often of use in finding the cross-sectional length in these circumstances:

The Pythagorean Theorem In a right triangle whose legs have lengths a and b and whose hypotenuse has length c, $c^2 = a^2 + b^2$.

Corresponding Parts of Similar Triangles are Proportional If a, b,

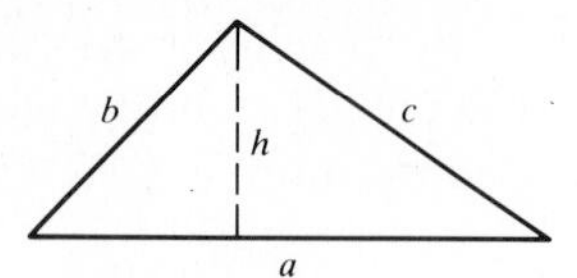

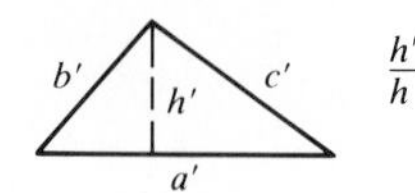

Figure 8.6

and c are the lengths of the sides of one triangle and a', b', and c' are the lengths of the corresponding sides of a similar triangle, then $a'/a = b'/b = c'/c$.

In addition, corresponding altitudes of the two triangles are also in the same proportion. (See Fig. 8.6.)

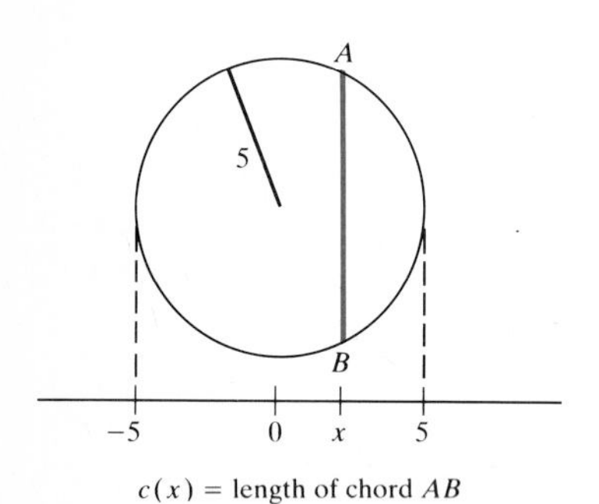

$c(x)$ = length of chord AB

Figure 8.7

EXAMPLE 3 Find the cross-sectional length $c(x)$ if R is a circle of radius 5. Set up a definite integral for the area of R.

SOLUTION We are free to put the axis anywhere in the plane. Place it in such a way that its origin is below the center of the circle. This provides the simplest formula for $c(x)$, the length of typical chord AB perpendicular to the x axis. This is illustrated in Fig. 8.7.

Note that $a = -5$ and $b = 5$. To get a feel for $c(x)$ in this case, note by glancing at the diagram that

$$c(-5) = 0, \quad c(0) = 10 \text{ (the circle's diameter)}, \quad \text{and} \quad c(5) = 0.$$

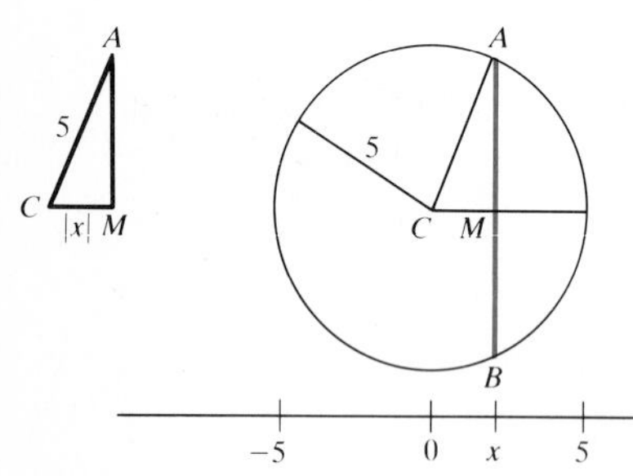

Figure 8.8

To find $c(x)$ for any x in the interval $[-5, 5]$, draw the line through the center of the circle and parallel to the x axis. It meets the segment AB at a point M. Call the center of the circle C. Also draw the segment AC, a radius of the circle, as in Fig. 8.8. Then

$$c(x) = \overline{AB} = 2\overline{AM}.$$

To find $\overline{AM}$, use the right triangle ACM. One side CM has length $|x|$, while the hypotenuse has length 5. Hence $5^2 = |x|^2 + \overline{AM}^2$. Since $|x|^2 = x^2$, this equation gives a simple formula for $\overline{AM}^2$:

$$\overline{AM}^2 = 5^2 - x^2 = 25 - x^2.$$

Hence

$$\overline{AM} = \sqrt{25 - x^2},$$

and, since $c(x) = 2\overline{AM}$,

$$c(x) = 2\sqrt{25 - x^2}.$$

Thus the area of the circle is

$$\int_{-5}^{5} 2\sqrt{25 - x^2}\, dx. \tag{2}$$

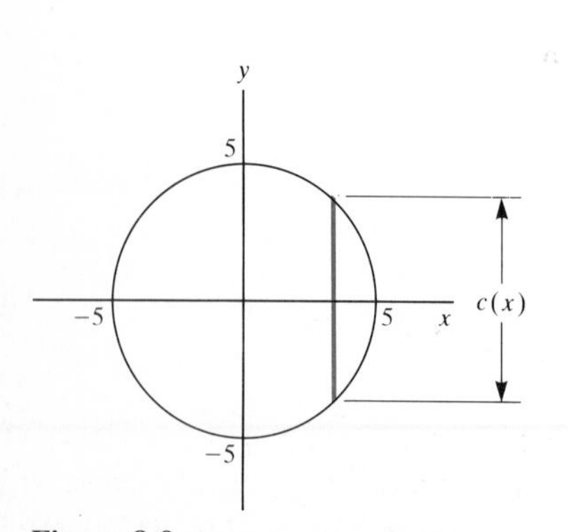

Figure 8.9

We could also set up a definite integral for the area of a circle of radius 5 by first finding the equation of the circle relative to a convenient choice of axes. If the origin of an xy coordinate system is placed at the center of the circle, as in Fig. 8.9, then the circle has the equation

$$x^2 + y^2 = 25.$$

The top half of the circle is the graph of $y = \sqrt{25 - x^2}$, and the bottom half is the graph of $y = -\sqrt{25 - x^2}$. Hence the cross section is $c(x) =$

$\sqrt{25 - x^2} - (-\sqrt{25 - x^2}) = 2\sqrt{25 - x^2}$. The area of the disk is thus

$$\int_{-5}^{5} 2\sqrt{25 - x^2}\, dx,$$

the same integral we had before.

EXAMPLE 4 Set up a definite integral for the area of the region above the parabola $y = x^2$ and below the line through (2, 0) and (0, 1) shown in Fig. 8.10.

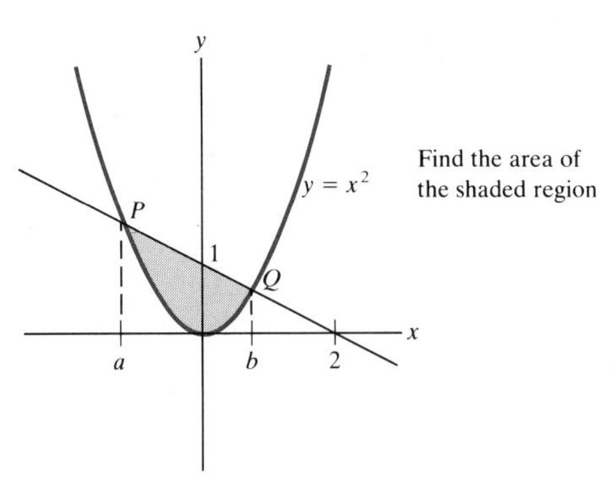

Figure 8.10

SOLUTION Since the x intercept of the line is 2 and the y intercept is 1, an equation for the line is

$$\frac{x}{2} + \frac{y}{1} = 1. \tag{3}$$

Hence $y = 1 - x/2$. The length $c(x)$ of a cross section of the region taken parallel to the y axis is therefore

$$c(x) = \left(1 - \frac{x}{2}\right) - x^2 = 1 - \frac{x}{2} - x^2.$$

To find the interval $[a, b]$ of integration, we must find the x coordinates of the points P and Q in Fig. 8.10. For these values of x,

$$x^2 = 1 - \frac{x}{2},$$

so

$$2x^2 + x - 2 = 0. \tag{4}$$

The solutions of Eq. (4) are

$$x = \frac{-1 \pm \sqrt{17}}{4}.$$

Hence

$$\text{Area} = \int_{(-1-\sqrt{17})/4}^{(-1+\sqrt{17})/4} \left(1 - \frac{x}{2} - x^2\right) dx. \quad \blacksquare \tag{5}$$

The formula for cross-sectional length $c(x)$ or $c(y)$ depends on where we choose to place the origin of the x axis or y axis. A prudent choice may simplify the formula for the cross-sectional length, as the next example illustrates.

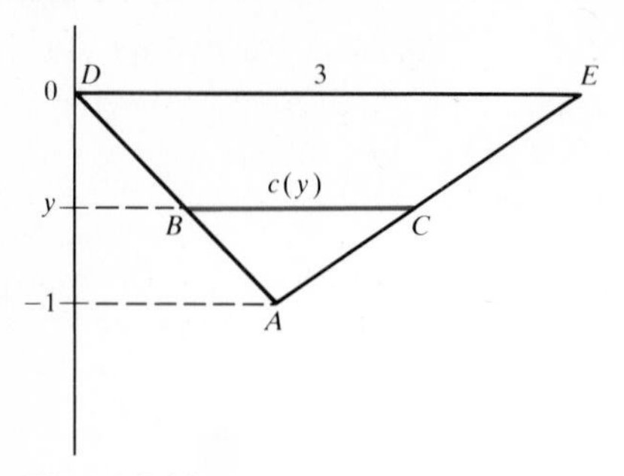

Figure 8.11

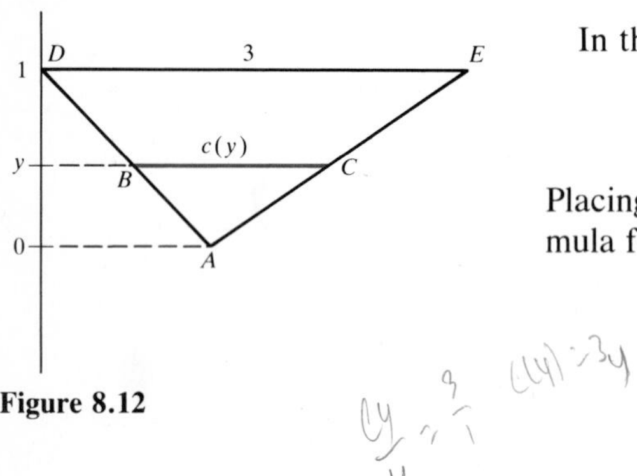

Figure 8.12

EXAMPLE 5 Find the formula for $c(y)$ as shown in Figs. 8.11 and 8.12.

SOLUTION In the first figure, ΔABC is similar to ΔADE. Since corresponding parts of similar triangles are proportional, the sides BC and DE are in the same ratio as the corresponding altitudes perpendicular to them; that is,

$$\frac{\overline{BC}}{\overline{DE}} = \frac{y - (-1)}{0 - (-1)}.$$

Hence $$\frac{c(y)}{3} = y + 1 \quad \text{and} \quad c(y) = 3(y + 1).$$

In the case of Fig. 8.12, we have, again by similar triangles,

$$\frac{c(y)}{3} = \frac{y}{1} \quad \text{and} \quad c(y) = 3y.$$

Placing the y coordinate system as in Fig. 8.12 provides the simpler formula for cross-sectional length. ■

EXERCISES FOR SEC. 8.1: COMPUTING AREA BY PARALLEL CROSS SECTIONS

In each of Exercises 1 to 8 find the area of the region between the two curves and above (or below) the given interval.

1 $y = x^2$ and $y = x^3$; $[0, 1]$
2 $y = x^2$ and $y = x^3$; $[1, 2]$
3 $y = x^2$ and $y = \sqrt{x}$; $[0, 1]$
4 $y = x^3$ and $y = \sqrt[3]{x}$; $[1, 2]$
5 $y = x^3$ and $y = -x$; $[1, 2]$
6 $y = 1 + x$ and $y = \ln x$; $[1, e]$
7 $y = \sin x$ and $y = \cos x$; $[0, \pi/4]$
8 $y = \sin x$ and $y = \cos x$; $[\pi/2, \pi]$

In Exercises 9 to 16 sketch the finite regions bounded by the given curves. Then find their areas by (*a*) vertical cross sections, (*b*) horizontal cross sections.

9 $y = x^2$ and $y = 3x - 2$
10 $y = 2x^2$ and $y = x + 1$
11 $y = 4x$ and $y = 2x^2$
12 $y = x^2$ and $y = 4$
13 $y = 1/x^2$, $y = 0$, $x = 1$, $x = 3$
14 $x = y^2$ and $x = 3y - 2$
15 $y = \sin x$, $y = 0$, $x = \pi/2$ $(x \leq \pi/2)$
16 $y = \tan x$, $y = 0$, $x = \pi/4$ (Consider only $x \geq 0$.)
17 Evaluate integral (2) in Example 3.
18 Find the area of the region in Example 4 by (*a*) evaluating the integral in Eq. (5), (*b*) first expressing the area in terms of cross sections parallel to the x axis and then evaluating the resulting integrals.

■

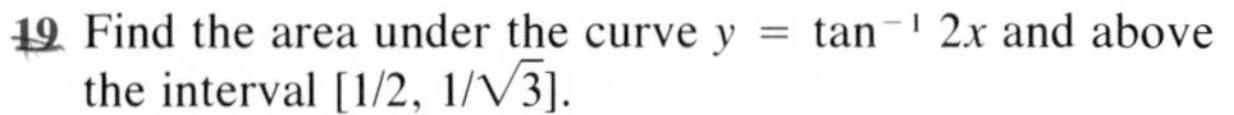

19 Find the area under the curve $y = \tan^{-1} 2x$ and above the interval $[1/2, 1/\sqrt{3}]$.
20 (*a*) Find the area $A(b)$ under the curve $y = e^{-x}\cos^2 x$ and above the interval $[0, b]$.
(*b*) Find $\lim_{b\to\infty} A(b)$.
21 Find the area of the region in the first quadrant below $y = -7x + 29$ and above the portion of $y = 8/(x^2 - 8)$ that lies in the first quadrant.
22 Find the area of the region between $y = \cos^{10} x$ and $y = \cos^{11} x$ for x in $[0, \pi/2]$. (Recall Exercise 41 of Sec. 7.2.)
23 Find the area of the region below $y = 10^x$ and above $y = \log_{10} x$ for x in $[1, 10]$.
24 Find the area under the curve $y = x/(x^2 + 5x + 6)$ and above the interval $[1, 2]$.
25 Find the area of the region below $y = (2x + 1)/(x^2 + x)$ and above the interval $[2, 3]$.
26 Find the area of the region that lies above the curve $y = x^2 - 3x + 2$ and below the x axis.
27 Find the area of the region under the curve $y = x^2/(x^3 + x^2 + x + 1)$ and above the interval $[1, 2]$.

28 Find the area of the region under $y = \sin x$ and above $[0, \pi/2]$ using cross sections parallel to (*a*) the y axis, (*b*) the x axis.

29 Find the vertical cross-sectional length $c(x)$ if R is the shaded region shown in Fig. 8.13.

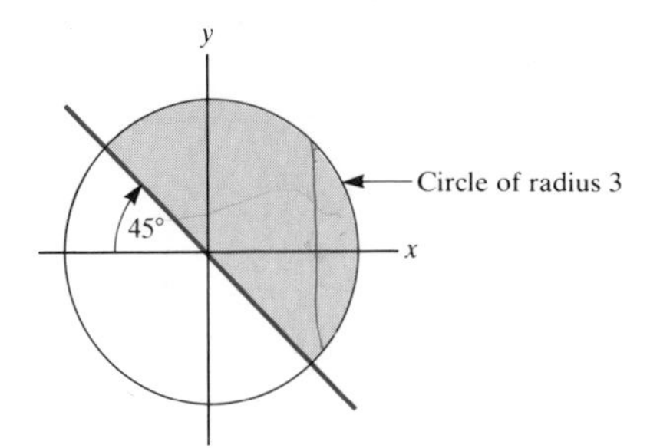

Figure 8.13

30 Find the horizontal cross-sectional length $c(y)$ for the region in Exercise 29.

In Exercises 31 to 33 $c(x)$ and $c(y)$ are distinct functions.

31 (*a*) Draw the region bordered by $y = \sin x$ and the x axis for x in $[0, \pi]$.
(*b*) Find $c(x)$, the vertical cross-sectional length.
(*c*) Find $c(y)$, the horizontal cross-sectional length.

32 (*a*) Draw the region bordered by $y = x/2$, $y = x - 1$, and the x axis.
(*b*) Find $c(x)$, the vertical cross-sectional length.
(*c*) Find $c(y)$, the horizontal cross-sectional length.

33 (*a*) Draw the region bordered by $y = \ln x$, $y = x/e$, and the x axis.
(*b*) Find $c(x)$, the vertical cross-sectional length.
(*c*) Find $c(y)$, the horizontal cross-sectional length.

34 What fraction of the rectangle whose vertices are $(0, 0)$, $(a, 0)$, (a, a^4), and $(0, a^4)$, with a positive, is occupied by the region under the curve $y = x^4$ and above $[0, a]$?

35 Find the area of the shaded region in Fig. 8.14 (*a*) without calculus, (*b*) using vertical cross sections, (*c*) using horizontal cross sections. [*Suggestion* for (*a*): Use the formula for the area of a sector.]

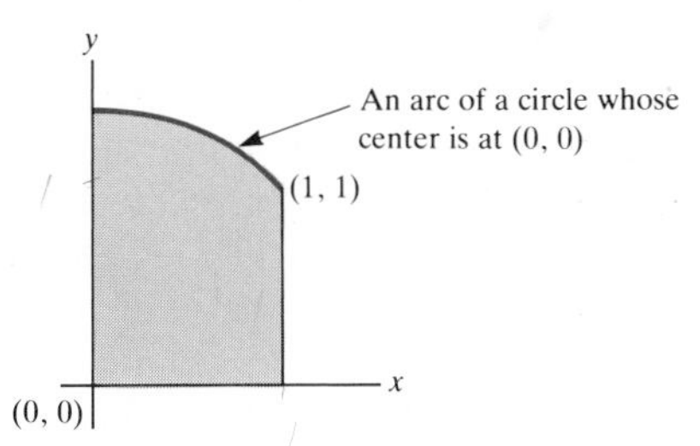

Figure 8.14

36 Find the area between the curves $y^2 = x$ and $y = x - 2$, using (*a*) horizontal cross sections, (*b*) vertical cross sections.

37 Sketch the region common to two circles of radius 1 whose centers are a distance 1 apart. Find the area of this region, using (*a*) vertical cross sections, (*b*) horizontal cross sections, (*c*) elementary geometry, but no calculus.

38 (*a*) Draw the region R inside the ellipse

$$\frac{x^2}{4} + \frac{y^2}{9} = 1.$$

(*b*) Find $c(x)$, the vertical cross-sectional length.
(*c*) Find $c(y)$, the horizontal cross-sectional length.
(*d*) Find the area of R.

39 (*a*) Find the area of the region under the curve $y = \sin x$ and above $[0, \pi/2]$.
(*b*) Why is there a line parallel to the x axis that cuts this region into two regions of equal area?
(*c*) Use Newton's method to estimate a in the equation $y = a$ of that line to two decimal places.

40 Figure 8.15 shows a right triangle ABC.
(*a*) Find equations for the lines parallel to each edge, AC, BC, and AB, that cut the triangle into two pieces of equal area.
(*b*) Are the three lines in (*a*) concurrent; that is, do they meet at a single point?

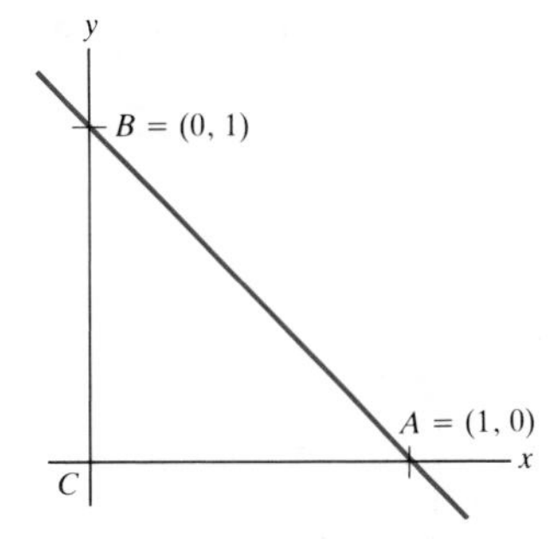

Figure 8.15

41 Show that the shaded area in Fig. 8.16 on page 390 is two-thirds the area of the parallelogram $ABCD$. This is an illustration of a theorem of Archimedes concerning sectors of parabolas.

42 Let f be an increasing function with $f(0) = 0$, and assume that it has an elementary antiderivative. Then f^{-1} is an increasing function, and $f^{-1}(0) = 0$. Prove that if f^{-1} is elementary, then it also has an elementary antiderivative. *Hint:* See Fig. 8.17 on page 390.

43 Find the area of the region bounded by $x = y^2$ and $x = 3y - 2y^2$.

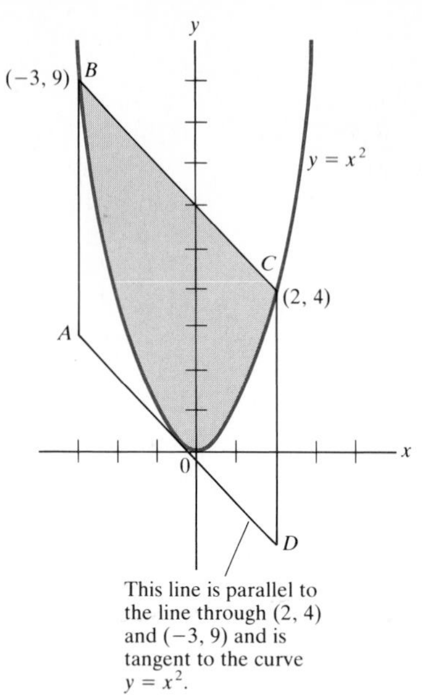

Figure 8.16

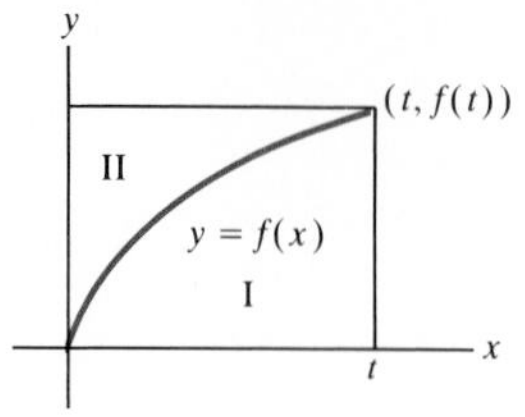

Figure 8.17

8.2 COMPUTING VOLUME BY PARALLEL CROSS SECTIONS

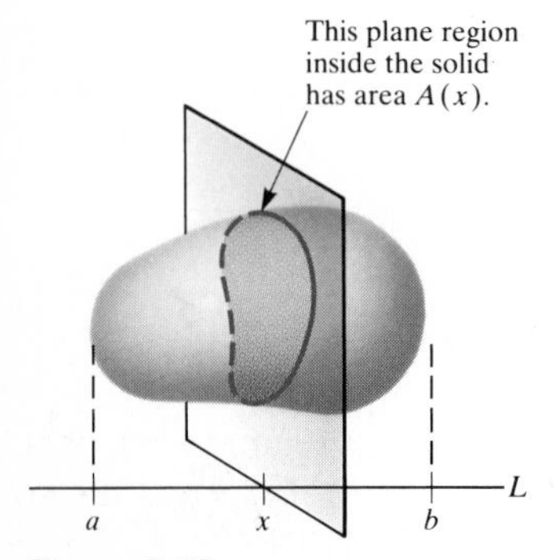

Figure 8.18

In Sec. 5.3 it was shown that the volume V of a spatial region, a "solid," can be expressed as a definite integral of cross-sectional area $A(x)$,

$$V = \int_a^b A(x)\,dx,$$

as shown in Fig. 8.18. So to find the volume of some solid, follow these steps:

1 Choose a line L to serve as an x axis.
2 For each plane perpendicular to that axis, find the area of the cross section of the solid made by the plane. Call this area $A(x)$.
3 Determine the limits of integration, a and b, for the region.
4 Evaluate the definite integral $\int_a^b A(x)\,dx$.

Most of the effort is usually spent in finding the integrand $A(x)$.

In addition to the pythagorean theorem and the properties of similar triangles, formulas for the areas of familiar plane figures may be needed. (See inside back cover.) Also keep in mind that if corresponding dimensions of similar figures have the ratio k, then their areas have the ratio k^2; that is, the area is proportional to the square of the lengths of corresponding line segments.

A few examples will show what may be involved in finding the cross-sectional area $A(x)$.

EXAMPLE 1 Find the volume of the solid whose base is the region bounded by the x axis and the arch of the curve $y = \sin x$ from $x = 0$ to $x = \pi$

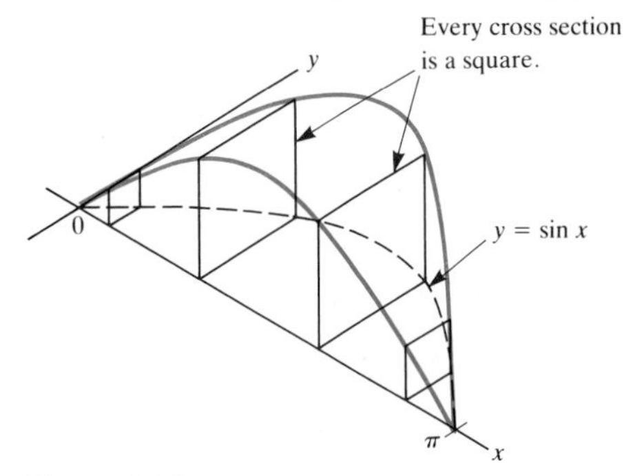

Figure 8.19

and for which each plane section perpendicular to the x axis is a square whose base lies in the region. The solid is shown in Fig. 8.19.

SOLUTION Since the cross section is a square whose side is $\sin x$, its area is $\sin^2 x$. Thus

$$A(x) = \sin^2 x.$$

The definite integral for the volume is therefore

$$\int_0^{\pi} \sin^2 x \, dx.$$

Think of the graph of $y = \sin^2 x$ from $x = 0$ to $x = \pi$.

To evaluate this integral, note that

$$\int_0^{\pi} \sin^2 x \, dx = 2 \int_0^{\pi/2} \sin^2 x \, dx.$$

By the short cut in Sec. 7.5,

$$\int_0^{\pi/2} \sin^2 x \, dx = \frac{\pi}{4}.$$

Thus the volume of the solid is $2(\pi/4) = \pi/2$. ■

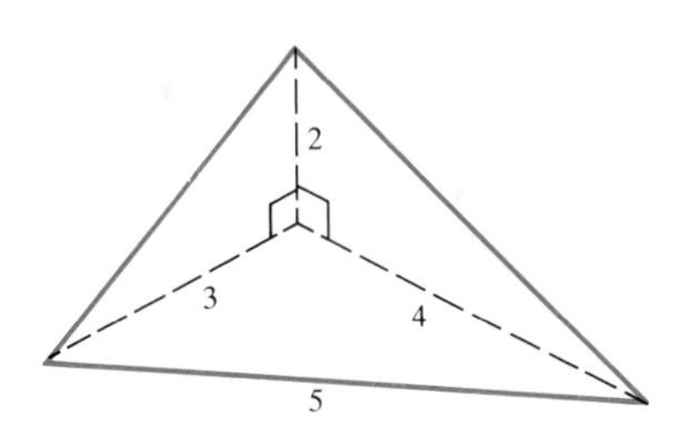

Figure 8.20

EXAMPLE 2 Find the volume of a solid triangular pyramid whose base is a right triangle of sides 3, 4, and 5. The altitude of the pyramid is above the vertex of the right angle and has length 2. (See Fig. 8.20.)

SOLUTION There are three convenient directions in which to define cross sections by planes, namely, parallel to each of the three right-triangular faces. Choose, say, planes parallel to the base, as shown in Fig. 8.21.

Introduce an x axis perpendicular to the base triangle and with origin in the plane of that triangle. We will find the cross-sectional area $A(x)$ in two ways.

The typical cross section is a triangle T. Its area is $A(x) = \frac{1}{2}bh$, where b and h are shown in Fig. 8.21. By similar triangles,

$$\frac{b}{3} = \frac{2-x}{2} \quad \text{and} \quad \frac{h}{4} = \frac{2-x}{2}.$$

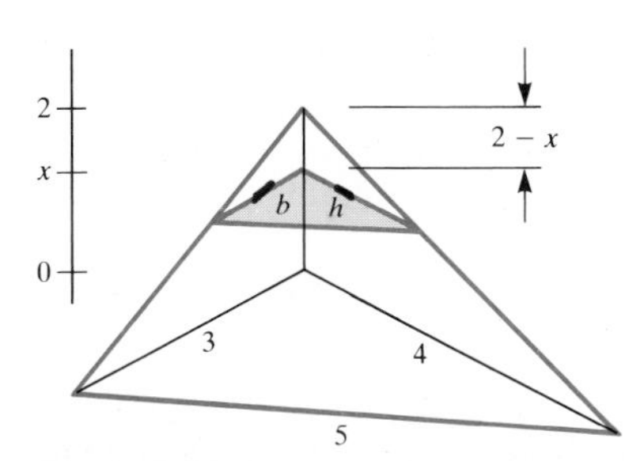

Figure 8.21

Thus $b = \frac{3}{2}(2 - x)$, $h = 2(2 - x)$, and

$$A(x) = \tfrac{1}{2}bh = (\tfrac{1}{2})(\tfrac{3}{2})(2 - x)2(2 - x) = \tfrac{3}{2}(2 - x)^2.$$

The same formula for the area of T can also be obtained by comparing T to the base triangle, which has area $\frac{1}{2} \cdot 3 \cdot 4 = 6$. Recall that the areas of similar figures have the ratio k^2, where k is the ratio of their corresponding sides. The corresponding sides of lengths b and 3 have the ratio

$$k = \frac{b}{3} = \frac{2 - x}{2},$$

so

$$\frac{\text{Area of } T}{\text{Area of base}} = \left(\frac{2 - x}{2}\right)^2$$

or

$$\frac{A(x)}{6} = \left(\frac{2 - x}{2}\right)^2.$$

Solving this last equation for $A(x)$ gives

$$A(x) = \tfrac{3}{2}(2 - x)^2.$$

Thus the volume of the pyramid is

$$\int_0^2 \tfrac{3}{2}(2 - x)^2\, dx = -\frac{3}{2}\frac{(2 - x)^3}{3}\Bigg|_0^2 = \left[-\frac{3}{2}\frac{(2 - 2)^3}{3}\right] - \left[-\frac{3}{2}\frac{(2 - 0)^3}{3}\right]$$

$$= 0 + 4 = 4.$$

The volume of the pyramid is 4 cubic units. ■

EXAMPLE 3 Find the volume of a sphere of radius a.

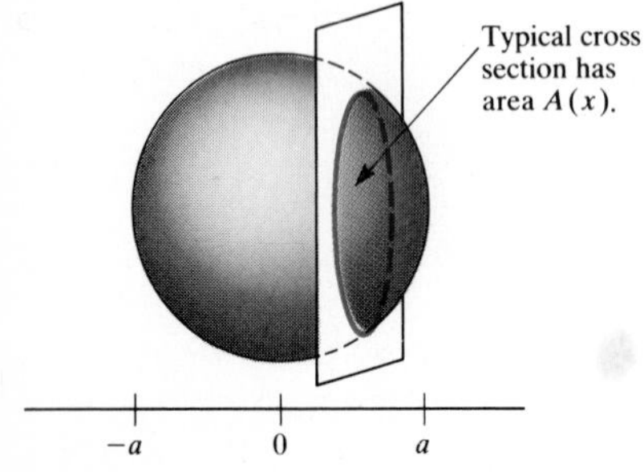

Figure 8.22

SOLUTION Place an x axis in such a way that its origin is beneath the center of the sphere, as in Fig. 8.22. The typical cross section is a circle. Since the area of a circle of radius r is πr^2, all that remains is to find r^2 in terms of x.

To accomplish this, draw a side view of the sphere and the typical cross section, showing r and x clearly, as in Fig. 8.23.

By the pythagorean theorem,

$$a^2 = |x|^2 + r^2 = x^2 + r^2$$

and

$$r^2 = a^2 - x^2.$$

Consequently,

$$A(x) = \pi r^2 = \pi(a^2 - x^2).$$

The volume of the sphere of radius a is therefore

$$\int_{-a}^{a} \pi(a^2 - x^2)\, dx.$$

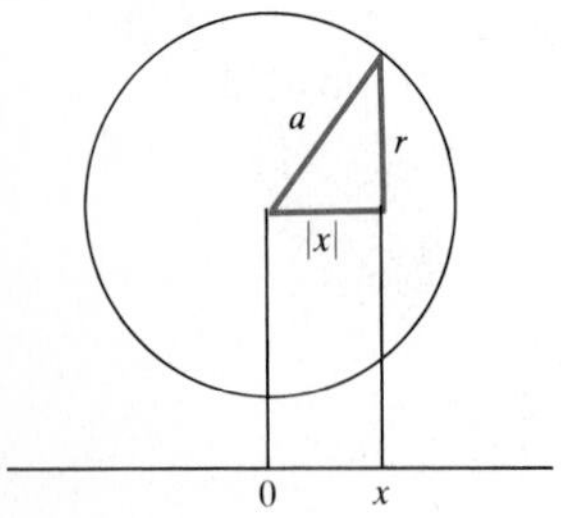

Figure 8.23

By the fundamental theorem of calculus,

$$\int_{-a}^{a} \pi(a^2 - x^2)\, dx = \pi\left(a^2x - \frac{x^3}{3}\right)\Bigg|_{-a}^{a}$$

$$= \pi\left(a^3 - \frac{a^3}{3}\right) - \pi\left(a^2(-a) - \frac{(-a)^3}{3}\right)$$

$$= \tfrac{4}{3}\pi a^3. \quad ■$$

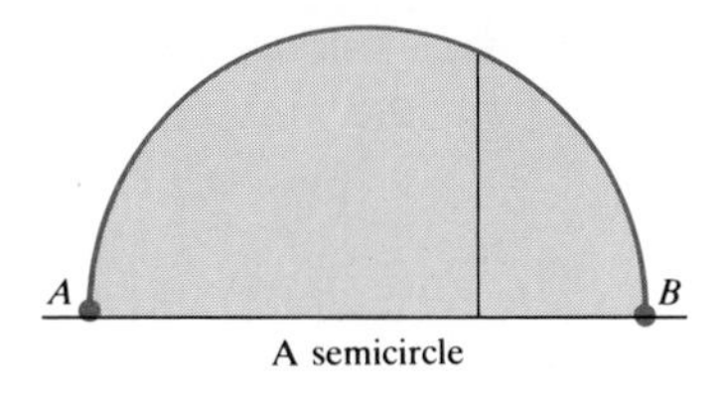

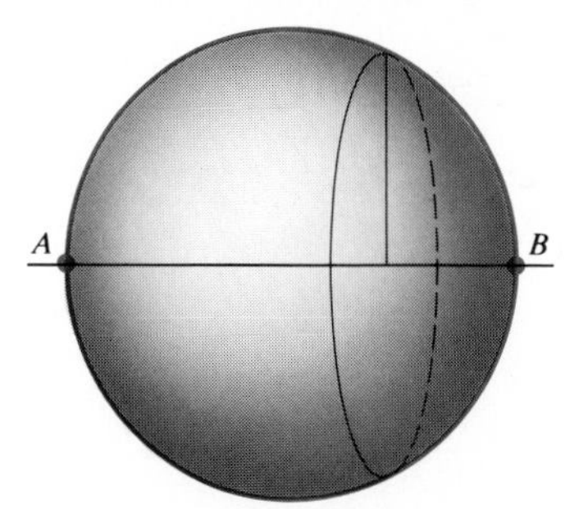

The sphere obtained by revolving the semicircle around AB.

Figure 8.24

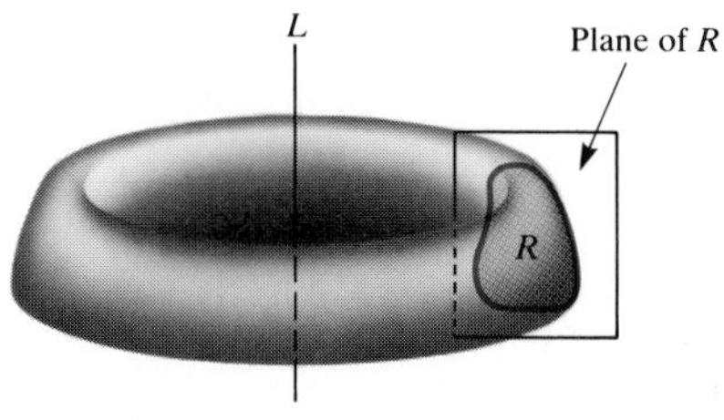

Figure 8.25

The sphere can be viewed as the solid obtained by revolving the semicircular region shown in Fig. 8.24 about its diameter AB. This is a special case of a "solid of revolution," which we will describe.

Let R be a region in the plane and L a line in the plane. (See Fig. 8.25.) Assume that L does not meet R at all or that L meets R only at points of the boundary. The solid formed by revolving R about L is called a **solid of revolution**. If L does not meet R, the solid is shaped like a ring. (If R is a disk, the solid is shaped like a doughnut.)

Let us see how to compute the volume of a solid of revolution when R is the region under the curve $y = f(x)$ and above the interval $[a, b]$ and L is the x axis. (See Figs. 8.26 and 8.27.)

To find the volume, first find the area $A(x)$ of a typical cross section made by a plane perpendicular to the x axis corresponding to the coordinate x. This cross section is a disk of radius $f(x)$, as shown in Fig. 8.28.

Thus $A(x) = \pi[f(x)]^2$. Since the volume of a solid is the integral of its cross-sectional area, we conclude that

Formula for volume of solid of revolution formed by revolving region below a curve

$$\text{Volume of the solid in Fig. 8.27} = \int_a^b \pi[f(x)]^2\,dx. \tag{1}$$

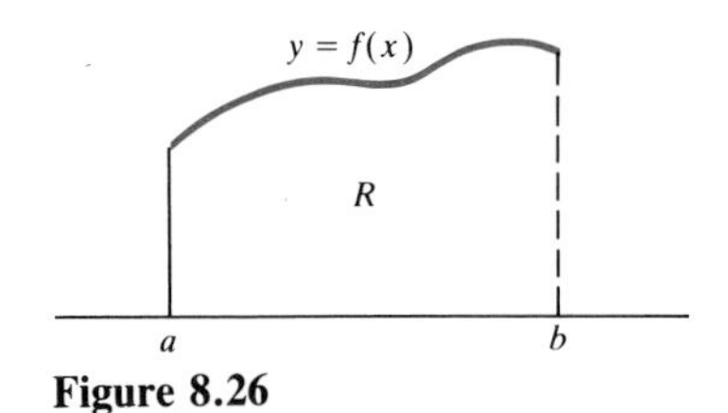

Figure 8.26

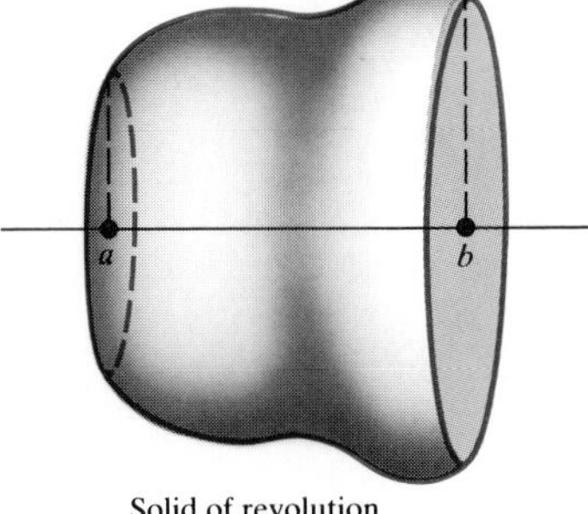

Solid of revolution obtained by revolving R.

Figure 8.27

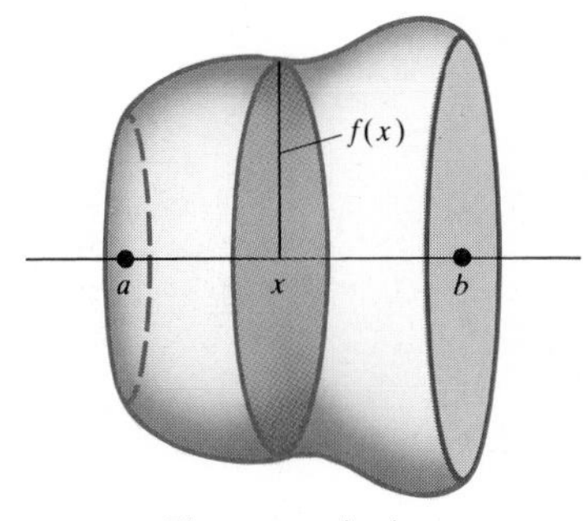

The cross section is a disk of radius $f(x)$.

Figure 8.28

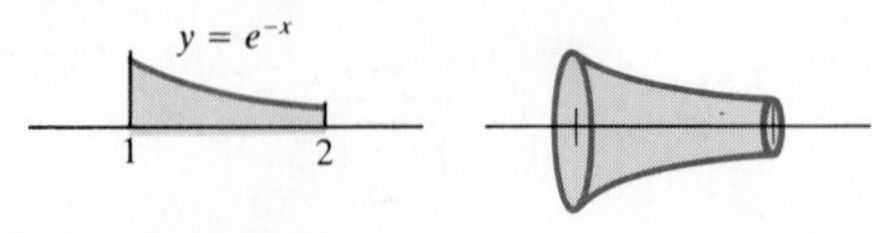

Figure 8.29

EXAMPLE 4 The region under $y = e^{-x}$ and above [1, 2] is revolved around the x axis. Find the volume of the resulting solid of revolution. (See Fig. 8.29.)

SOLUTION By formula (1), the volume is

$$V = \int_1^2 \pi(e^{-x})^2\,dx = \int_1^2 \pi e^{-2x}\,dx$$

$$\underset{\text{FTC}}{=} \left.\frac{\pi e^{-2x}}{-2}\right|_1^2 = \frac{\pi}{-2}e^{-4} - \frac{\pi}{-2}e^{-2}$$

$$= \frac{\pi}{2}(e^{-2} - e^{-4}). \quad \blacksquare$$

Revolving the region between two curves

A similar approach works for finding the volume of the solid of revolution formed when the region between two curves is revolved around the x axis.

Let $y = f(x)$ and $y = g(x)$ be two continuous functions such that $f(x) \geq g(x) \geq 0$ for x in the interval $[a, b]$. Let R be the region bounded by the curves $y = f(x)$ and $y = g(x)$ and above $[a, b]$. The region R is revolved around the x axis to form a solid of revolution. What is the volume of this solid? (See Fig. 8.30.)

The solid may have a hole.

The "washer" method

The cross section of the solid in Fig. 8.30 made by a plane perpendicular to the x axis is a ring, as shown in Fig. 8.31. The ring is bounded by a circle of radius $f(x)$ and a circle of radius $g(x)$. Thus its area $A(x)$ is

This is not $\pi[f(x) - g(x)]^2$.

$$\pi[f(x)]^2 - \pi[g(x)]^2,$$

or

$$A(x) = \pi\{[f(x)]^2 - [g(x)]^2\}.$$

Consequently,

Formula for volume of solid of revolution formed by revolving region between two curves

$$\boxed{\text{Volume of solid in Fig. 8.30} = \int_a^b \pi\{[f(x)]^2 - [g(x)]^2\}\,dx.} \quad (2)$$

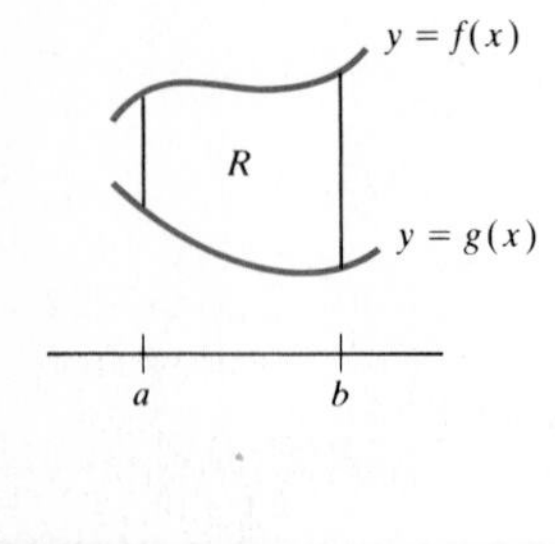

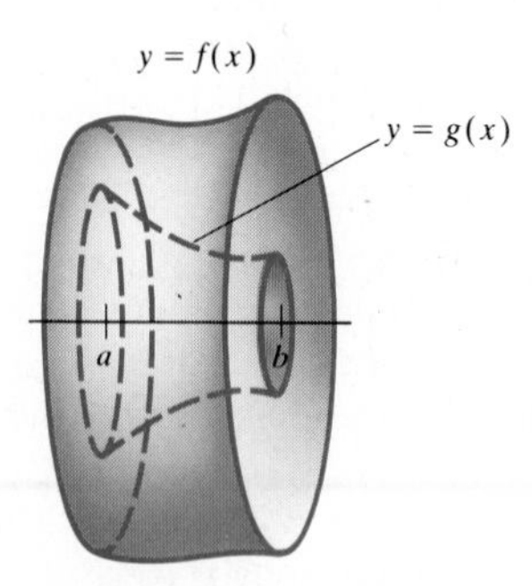

Figure 8.30

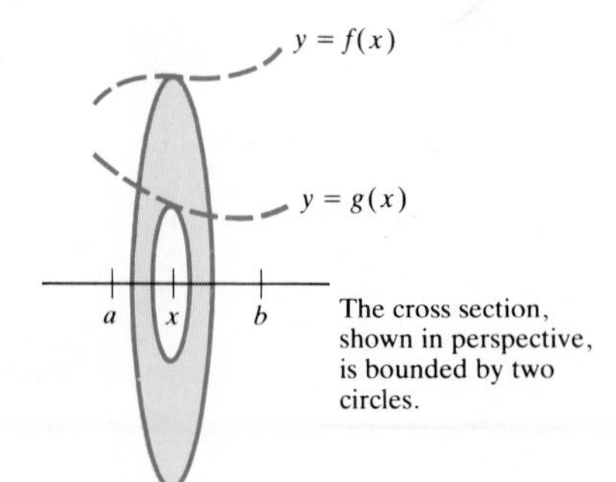

Figure 8.31

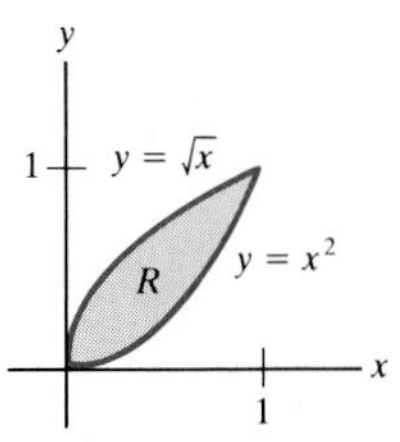

Figure 8.32

EXAMPLE 5 The region between $y = \sqrt{x}$ and $y = x^2$ is revolved around the x axis. Find the volume of the solid of revolution produced.

SOLUTION Figure 8.32 shows the plane region R being revolved. The region R lies above the interval $[0, 1]$ and between the curves $y = \sqrt{x}$ and $y = x^2$. By formula (2), the volume of the solid of revolution is

$$\int_0^1 \pi[(\sqrt{x})^2 - (x^2)^2]\,dx = \int_0^1 \pi(x - x^4)\,dx$$

$$\underset{\text{FTC}}{=} \pi\left(\frac{x^2}{2} - \frac{x^5}{5}\right)\Bigg|_0^1 = \pi(\tfrac{1}{2} - \tfrac{1}{5}) - \pi(\tfrac{0}{2} - \tfrac{0}{5})$$

$$= \frac{3\pi}{10}. \quad \blacksquare$$

EXERCISES FOR SEC. 8.2: COMPUTING VOLUME BY PARALLEL CROSS SECTIONS

In each of Exercises 1 to 4, (*a*) draw the solid, (*b*) set up an integral for its volume, (*c*) find the volume.

1 The base of the solid is the region in the xy plane bounded by $y = x^2$, the x axis, and the line $x = 1$. A cross section by a plane perpendicular to the x axis is a square one side of which is in the base.

2 The base of the solid is the region in the xy plane bounded by $y = x^2$ and $y = x$. A cross section by a plane perpendicular to the x axis is a square one side of which is in the base.

3 The base of the solid is the region in the xy plane bounded by $y = x^2$ and $y = x^3$. A cross section by a plane perpendicular to the y axis is a square one side of which is in the base.

4 The base of the solid is the region in the xy plane bounded by $y = e^x$, $x = 1$, $x = 2$, and the x axis. A cross section by a plane perpendicular to the x axis is an equilateral triangle one side of which is in the base.

5 Solve Example 2 if the origin of the x axis, instead of being in the plane of the base triangle, is level with the top vertex and the positive part of the x axis is downward.

6 Solve Example 2 using cross sections parallel to the triangle with sides of lengths 2 and 4.

7 Solve Example 2 using cross sections parallel to the triangle with sides of lengths 2 and 3.

8 Find the volume of the triangular pyramid shown in Fig. 8.33.

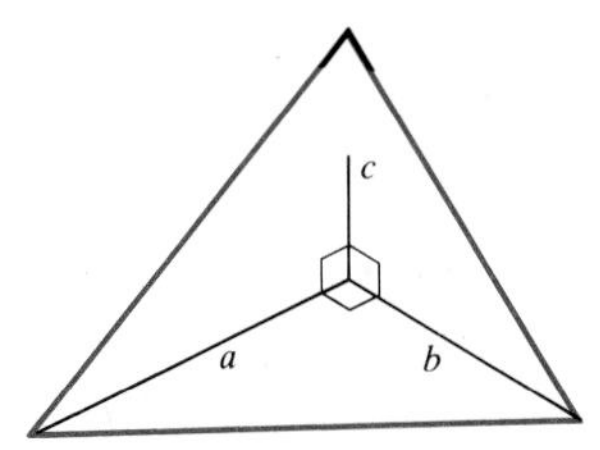

Figure 8.33

Exercises 9 to 14 concern volumes of solids of revolution. In each of them a region R in the plane is revolved around the x axis to produce a solid of revolution. In each case, (*a*) draw the region, (*b*) draw the solid of revolution, (*c*) draw the typical cross section (a disk or a ring), (*d*) set up a definite integral for the volume, and (*e*) evaluate the integral.

9 R is bounded by $y = \sqrt{x}$, the x axis, $x = 1$, and $x = 2$.

10 R is bounded by $y = 1/\sqrt{1 + x^2}$, the x axis, $x = 0$, and $x = 1$.

11 R is bounded by $y = x^2$ and $y = x^3$.

12 R is bounded by $y = 1/\sqrt{x}$, $y = 1/x$, $x = 1$, and $x = 2$.

13 R is bounded by $y = \tan x$, $y = \sin x$, $x = 0$, and $x = \pi/4$.

14 R is bounded by $y = \sec x$, $y = \cos x$, $x = \pi/6$, and $x = \pi/3$.

15 A right circular cone has height h and radius a. Consider cross sections parallel to the base. Introduce an x axis such that $x = 0$ corresponds to the vertex of the cone and $x = h$ corresponds to the base.
(*a*) Draw the cone, the x axis, and the typical cross section.
(*b*) Compute $A(x)$, the typical cross-sectional area for planes perpendicular to the x axis.
(*c*) Set up a definite integral for the volume.
(*d*) Find the volume.

16 A lumberjack saws a wedge out of a cylindrical tree of radius a. His first cut is parallel to the ground and stops at the axis of the tree. His second cut makes an angle θ with the first cut and meets it along a diameter. Draw the solid.

17 (See Exercise 16.) Place the x axis in such a way that the cross sections of the wedge in Exercise 16 are triangles. (*Continued on next page.*)

(*a*) Draw the typical cross section.
(*b*) Find the area of a typical cross section made by a plane at a distance x from the axis of the tree.
(*c*) Find the volume of the wedge.

18 (See Exercise 16.) Place the x axis in such a way that the sections of the wedge in Exercise 16 are rectangles.
(*a*) Draw a typical cross section.
(*b*) Find its area if it is made by a plane at a distance x from the axis of the tree.
(*c*) Find the volume of the wedge.

19 A drill of radius 3 inches bores a hole through a sphere of radius 5 inches, passing symmetrically through the center of the sphere.
(*a*) Draw the part of the sphere removed by the drill.
(*b*) Find $A(x)$, the area of a cross section of the region in (*a*) made by a plane perpendicular to the axis of the drill and at a distance x from the center of the sphere.
(*c*) Find the volume removed.

Exercise 20 provides an easy way to remember the volume of a cone or pyramid.

20 A solid is formed in the following manner. A plane region R and a point P not in that plane are given. The solid consists of all line segments joining P to points in R. If R has area A and P is a distance h from the plane of R, show that the volume of the solid is $Ah/3$. (See Fig. 8.34.)

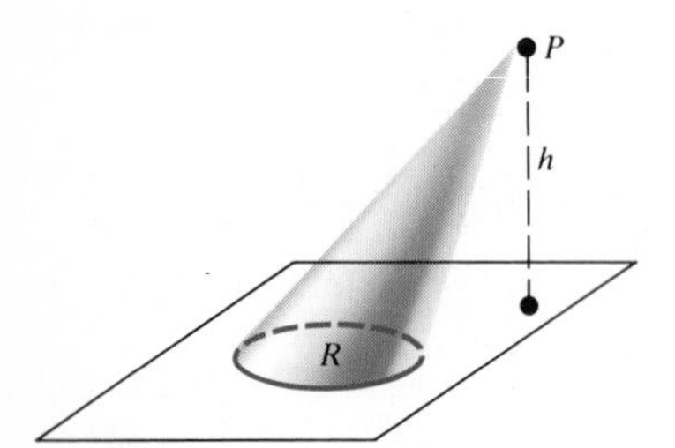

Figure 8.34

21 The base of a solid is a disk of radius 3. Each plane perpendicular to a given diameter meets the solid in a square, one side of which is in the base of the solid. (See Fig. 8.35.) Find its volume.

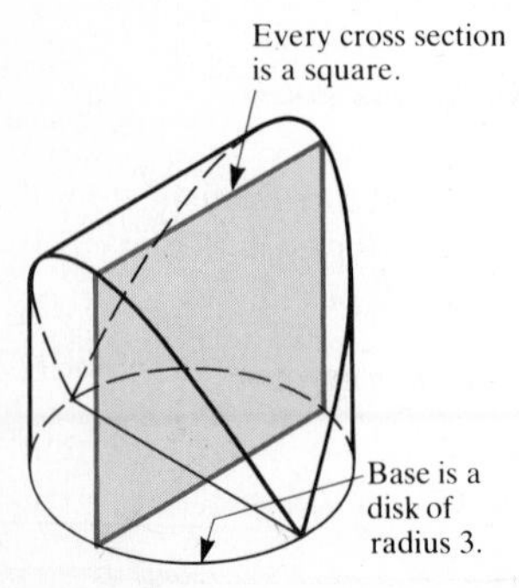

Figure 8.35

22 The base of a solid S is the region bounded by $y = 1 - x^2$ and $y = 1 - x^4$. Cross sections of S by planes that are perpendicular to the x axis are squares. Find the volume of S.

23 Find the volume of a solid if each of its cross sections perpendicular to the x axis, for $0 \leq x \leq 1$, is a circle, one of whose diameters is a line segment from the curve $y = \sqrt{x}$ to the line $y = x$.

24 Find the volume of the solid whose base is the disk of radius 5 and whose cross sections perpendicular to the x axis are equilateral triangles. (See Fig. 8.36.)

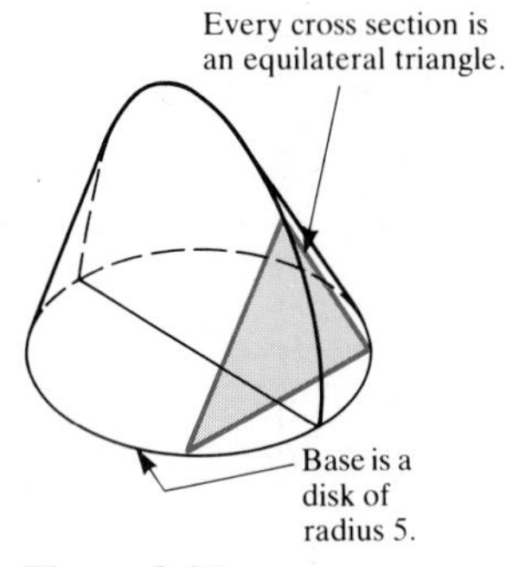

Figure 8.36

25 What fraction of the volume of a sphere is contained between parallel planes that trisect the diameter to which they are perpendicular?

26 A solid is formed by revolving the region below $y = e^{2x} \sin 3x$ and above $[0, \pi/3]$ around the x axis. Find its volume.

■

Exercises 27 to 31 offer practice in drawing cross sections of solid regions.

27 A right circular cylindrical glass full of water is tilted until the amount remaining covers exactly half the base, as shown in Fig. 8.37. (It helps if you do this with a glass of water.)
(*a*) In what direction can you take parallel cross sections of the water that are parts of circles?
(*b*) Sketch a typical such cross section.

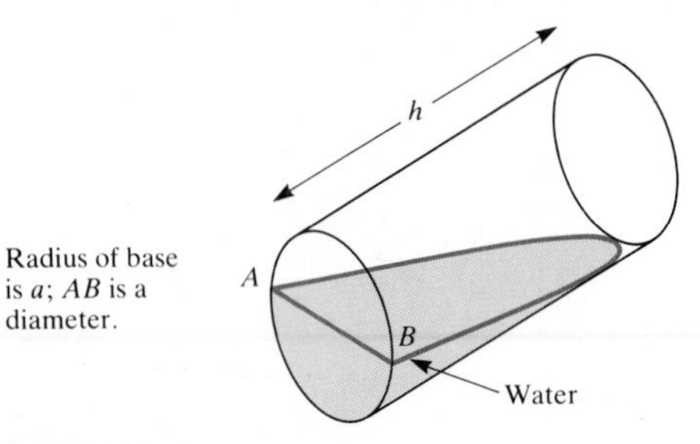

Radius of base is a; AB is a diameter.

Figure 8.37

28 (See Fig. 8.37.)
(*a*) In what direction can you take parallel cross sections that are triangles?
(*b*) Sketch a typical such cross section.

29 (See Fig. 8.37.)
(*a*) In what direction can you take parallel cross sections that are rectangles?
(*b*) Sketch a typical such cross section.

30 Draw a cross section of a solid cube by a plane that is (*a*) a square, (*b*) an equilateral triangle, (*c*) a five-sided polygon, (*d*) a regular hexagon. [*Hint* for (*d*): The vertices of the hexagon are midpoints of edges of the cube.]

31 Figure 8.38 indicates an unbounded solid right circular cone. Draw a cross section that is bounded by (*a*) a circle, (*b*) an ellipse (but not a circle), (*c*) a parabola, (*d*) one-half of a hyperbola.

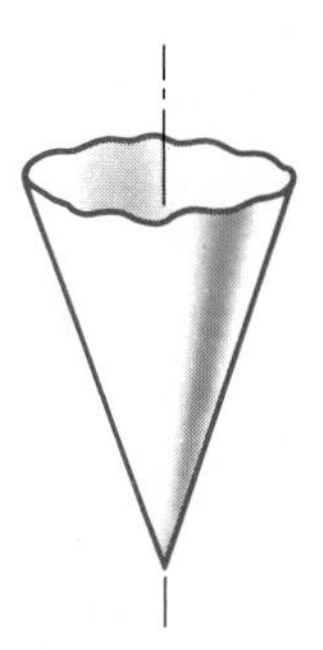

Figure 8.38

32 Find the volume of one octant of the region common to two right circular cylinders of radius 1 whose axes intersect at right angles, as shown in Fig. 8.39.

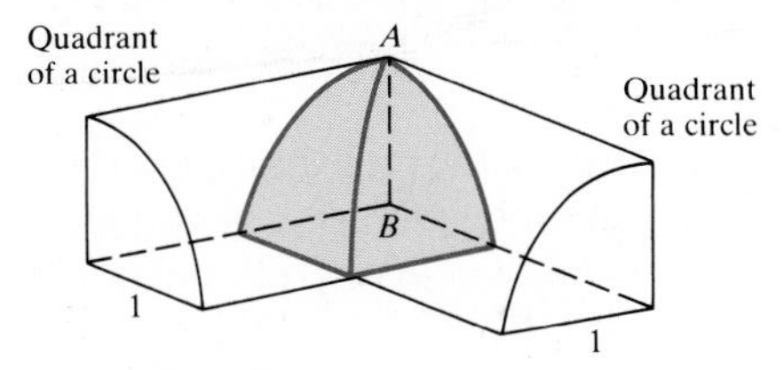

Figure 8.39

33 (Contributed by Steve Abell) A lead sinker is to be made by revolving a section of a circular disk around the line AB, as shown in Fig. 8.40. The length of the sinker is to be 3 times its maximum radius a, as shown in Fig. 8.40. The volume of the sinker is to be 3 cubic inches. Find the radius of the disk and the maximum radius of the sinker.

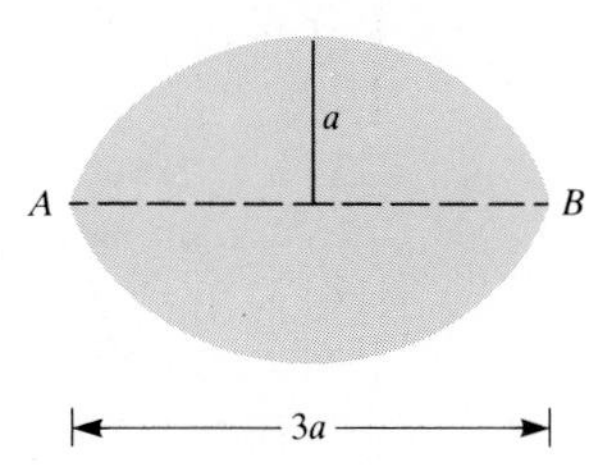

Figure 8.40

34 The glass in Exercise 27 has radius a and height h. Find the volume of water using the triangular sections of Exercise 28.

35 Find the volume of water described in Exercise 34 using rectangular cross sections. (See Exercise 29.)

Exercises 36 to 38 concern a right circular cylindrical glass of water tilted so that the water just covers the base. (See Fig. 8.41.) The glass has radius a and height h. In each of Exercises 36 to 38, find the volume of the water by the indicated method.

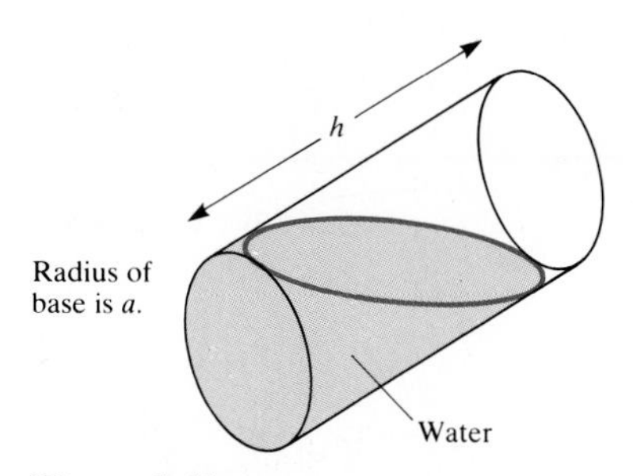

Figure 8.41

36 Parallel cross sections that are rectangles.
37 Parallel cross sections that are trapezoids.
38 Geometric intuition (no calculus).

8.3 HOW TO SET UP A DEFINITE INTEGRAL

This section presents an informal shortcut for setting up a definite integral to evaluate some quantity. First, the formal and informal approaches are contrasted in the case of setting up the definite integral for area. Then the informal approach will be illustrated as commonly applied in a variety of fields.

The Formal Approach Recall how the formula $A = \int_a^b f(x)\,dx$ was obtained in Sec. 5.3. The interval $[a, b]$ was partitioned by the numbers $x_0 < x_1 < x_2 < \cdots < x_n$ with $x_0 = a$ and $x_n = b$. A sampling number was chosen in each section $[x_{i-1}, x_i]$. For convenience, let us use x_{i-1} as that sampling number. The sum

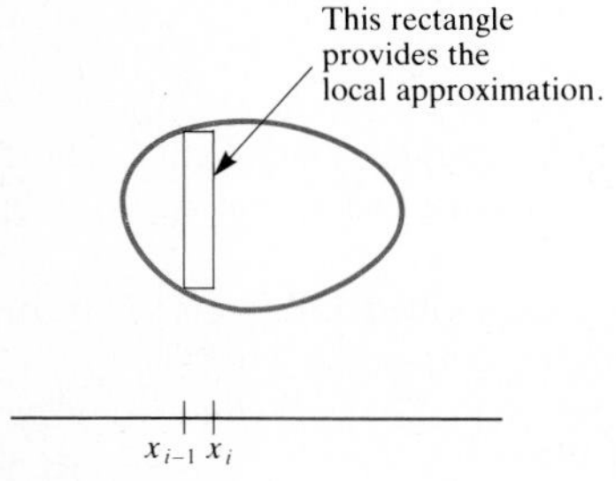

Figure 8.42

$$\sum_{i=1}^{n} f(x_{i-1})(x_i - x_{i-1}) \tag{1}$$

is then formed. As the mesh of the partition approaches 0, the sum (1) approaches the area of the region under consideration. But, by the definition of the definite integral, the sum (1) approaches

$$\int_a^b f(x)\,dx$$

as the mesh of the partition approaches 0. Thus

$$\text{Area} = \int_a^b f(x)\,dx. \tag{2}$$

That is the "formal" approach to obtain the formula (2). Now consider the "informal" approach, which is just a shorthand for the formal approach.

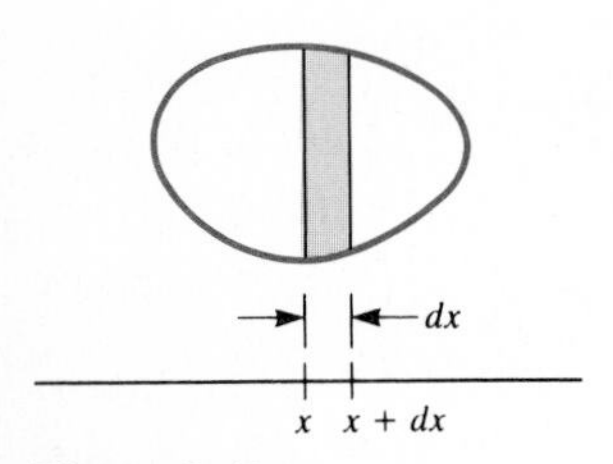

Figure 8.43

The Informal Approach The heart of the formal approach is the **local estimate** $f(x_{i-1})(x_i - x_{i-1})$, the area of a rectangle of height $f(x_{i-1})$ and width $x_i - x_{i-1}$, which is shown in Fig. 8.42.

In the informal approach attention is focused on the **local approximation**. No mention is made of the partition or the sampling numbers or the mesh approaching 0. We illustrate this shorthand approach by obtaining formula (2) informally.

Consider a small positive number dx. What would be a good estimate of the area of the region corresponding to the short interval $[x, x + dx]$ of width dx shown in Fig. 8.43? The area of the rectangle of width dx and height $f(x)$ shown in Fig. 8.44 would seem to be a plausible estimate. The area of this thin rectangle is

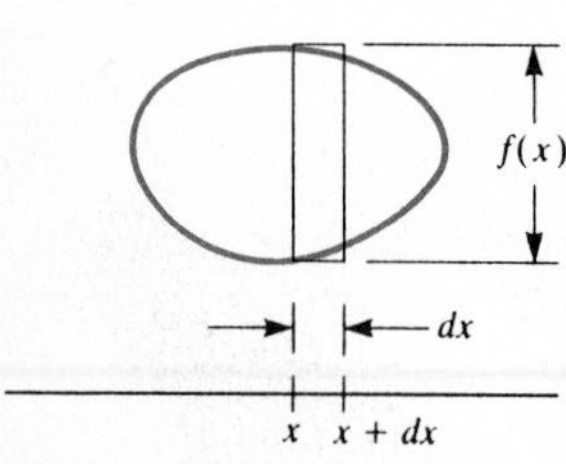

Figure 8.44

$$f(x)\,dx. \tag{3}$$

Without further ado, we then write

$$\text{Area} = \int_a^b f(x)\,dx, \tag{4}$$

which is formula (2). The leap from the local approximation (3) to the definite integral (4) omits many steps of the formal approach. This informal approach is the shorthand commonly used in applications of calculus. It should be emphasized that it is only an abbreviation of the formal approach, which deals with approximating sums.

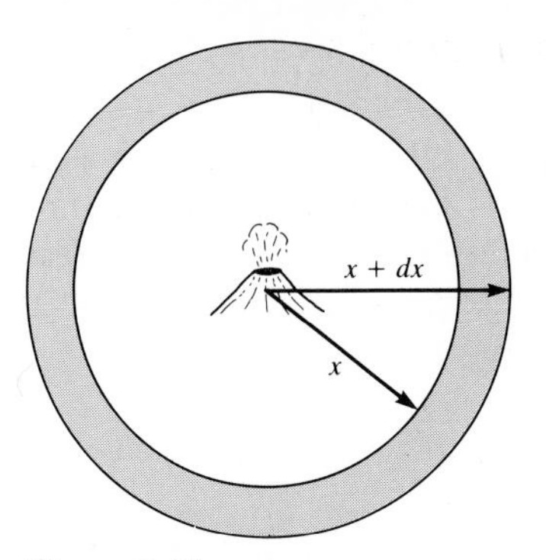

Figure 8.45

EXAMPLE 1 (Volcanic ash settling) After the explosion of a volcano, ash gradually settles from the atmosphere and falls on the ground. Assume that the depth of the ash at a distance x feet from the volcano is Ae^{-kx} feet, where A and k are positive constants. Set up a definite integral for the total volume of ash that falls within a distance b of the volcano.

SOLUTION First estimate the volume of ash that falls on a very narrow ring of width dx and inner radius x centered at the volcano. (See Fig. 8.45.) This estimate can be made since the depth of the ash depends only on the distance from the volcano. On this ring the depth is almost constant.

The area of this ring is approximately that of a rectangle of length $2\pi x$ and width dx. So the area of the ring is approximately

$$2\pi x\ dx.$$

[Exercise 1 shows that its area is $2\pi x\ dx + \pi(dx)^2$.]

Although the depth of the ash on this narrow ring is not constant, it does not vary much. A good estimate of the depth throughout the ring is Ae^{-kx}. Thus the volume of ash that falls on the typical ring of inner radius x and outer radius $x + dx$ is approximately

The local approximation

$$Ae^{-kx}\ 2\pi x\ dx \text{ cubic feet.} \tag{5}$$

Once we have the key local estimate (5), we immediately write down the definite integral for the total volume of ash that falls within a distance b of the volcano:

$$\text{Total volume} = \int_0^b Ae^{-kx}\ 2\pi x\ dx.$$

(The limits of integration must be determined just as in the formal approach.) This completes the informal setting up of the definite integral. (It could be evaluated by integration by parts, but this is not our concern at this point.) ■

Definition of kinetic energy

The second example of the informal approach to setting up definite integrals concerns kinetic energy. The kinetic energy associated with an object of mass m grams and speed v centimeters per second is defined as $mv^2/2$ ergs. If the various parts of the object are not all moving at the same speed, an integral may be needed to express the total kinetic energy.

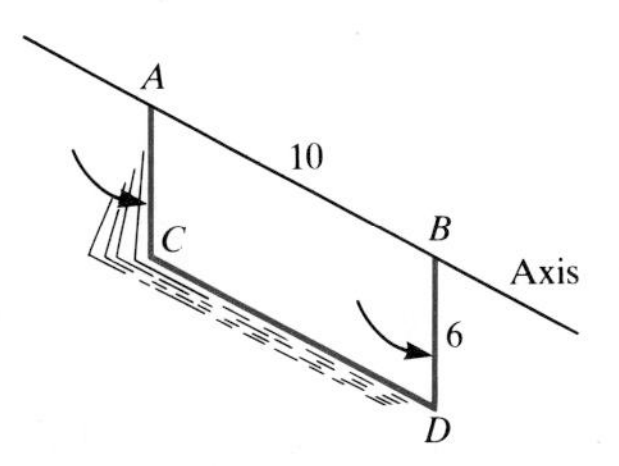

Figure 8.46

EXAMPLE 2 A thin rectangular piece of sheet metal is spinning around one of its longer edges 3 times per second, as shown in Fig. 8.46. The length of its shorter edge is 6 centimeters and the length of its longer edge is 10 centimeters. The density of the sheet metal is 4 grams per square centimeter. Find the kinetic energy of the spinning rectangle.

SOLUTION To find the total kinetic energy of the rotating piece of sheet

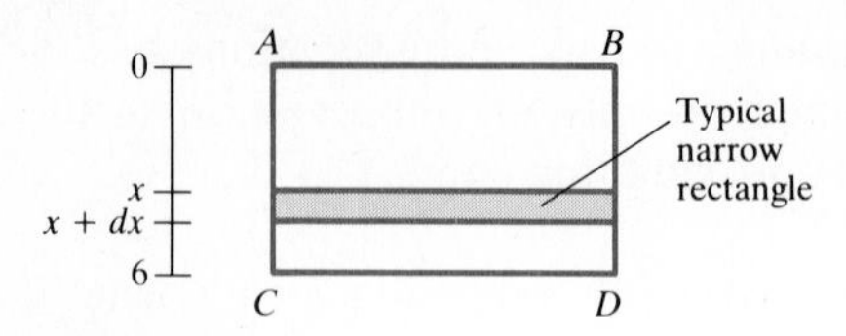

Figure 8.47

metal, imagine it divided into narrow rectangles of length 10 centimeters and width dx centimeters parallel to the edge AB; a typical one is shown in Fig. 8.47. (Introduce an x axis parallel to edge AC with the origin corresponding to A.) Since all points of this typical narrow rectangle move at roughly the same speed, we will be able to estimate its kinetic energy. That estimate will provide the key local approximation of the informal approach to setting up a definite integral.

First of all, the mass of the typical rectangle is

$$4 \cdot 10\, dx \text{ grams,}$$

since its area is 10 dx square centimeters and the density is 4 grams per square centimeter.

Second, we must estimate its speed. The narrow rectangle is spun 3 times per second around a circle of radius x. In 1 second each point in it covers a distance of about

$$3 \cdot 2\pi x = 6\pi x \text{ centimeters.}$$

Consequently, the speed of the typical rectangle is

$$6\pi x \text{ centimeters per second.}$$

The local estimate of the kinetic energy associated with the typical rectangle is therefore

$$\tfrac{1}{2} \underbrace{40\, dx}_{\text{mass}} \underbrace{(6\pi x)^2}_{\substack{\text{speed}\\ \text{squared}}} \quad \text{ergs,}$$

The local approximation

or simply
$$720\pi^2 x^2\, dx \text{ ergs.} \tag{6}$$

Having obtained the local estimate (6), we jump directly to the definite integral and conclude that

$$\text{Total energy of spinning rectangle} = \int_0^6 720\pi^2 x^2\, dx \text{ ergs.} \quad \blacksquare$$

The next example concerns water pressure.

The force exerted by water

The pressure of water at a depth of h feet is $62.4h$ pounds per square foot. This pressure exerts a force on a flat horizontal surface submerged at the depth of h feet equal to the product of the pressure and the area of the surface.

Consider, for instance, the horizontal floor of a swimming pool which is 8 feet deep. If the floor has an area of 800 square feet, then the total force of the water on the floor is

$$\underbrace{8(62.4)}_{\text{pressure}} \underbrace{(800)}_{\text{area}} \quad \text{pounds.}$$

The pressure is exerted in all directions.

The water exerts a pressure on the walls of the pool as well. In fact, at a given depth the pressure is the same in all directions. (A submerged swimmer does not avoid the pressure against her eardrums by turning her head.) But computing the force against the wall is harder than computing the force against the floor, since the pressure is not constant along the wall. A definite integral is required, as shown, informally, in Example 3.

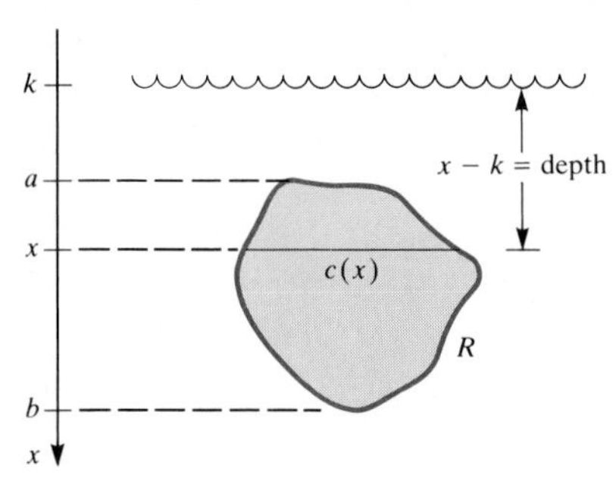

Figure 8.48

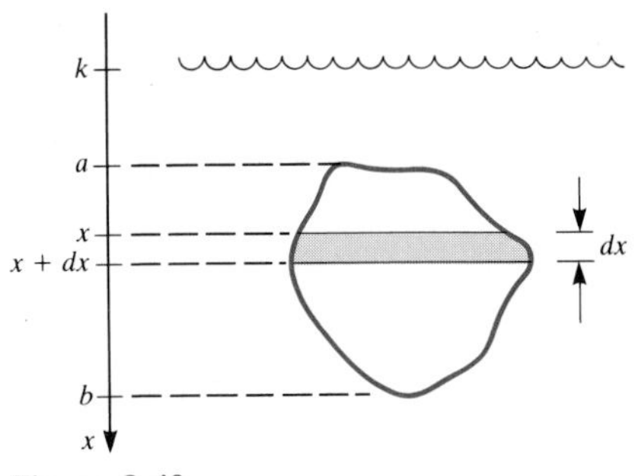

Figure 8.49

EXAMPLE 3 Let R be part of the submerged vertical wall of a pool. Set up a definite integral for the total force against the surface R.

SOLUTION Begin by introducing a vertical x axis whose positive part is directed downward. Let the x coordinate of the surface of the water be k. We could choose $k = 0$, but this is not always the best choice. Define a, b, and the cross-sectional function for R as usual. (See Fig. 8.48.)

Consider the force against a typical narrow strip of R bounded by horizontal lines corresponding to the x coordinates x and $x + dx$, where dx is a small positive number, as shown in Fig. 8.49. The depth of this strip is approximately $x - k$ feet.

The pressure of the water throughout this strip is approximately $62.4x$ pounds per square foot. The area of the strip is approximately that of a rectangle of height dx and length $c(x)$. Thus the force against this narrow strip is approximately

The local approximation

$$\underbrace{62.4(x - k)}_{\text{pressure}}\,\underbrace{c(x)\,dx}_{\text{area}} \qquad \text{pounds.}$$

Having obtained a local estimate of the force, we then, using the informal approach, conclude that

This formula will be referred to in later sections.

$$\boxed{\text{Total force against } R = \int_a^b 62.4(x - k)c(x)\,dx \text{ pounds.}} \tag{7}$$

■

Further illustrations of the informal style of setting up definite integrals are found in the exercises.

EXERCISES FOR SEC. 8.3: HOW TO SET UP A DEFINITE INTEGRAL

1 In Example 1 the area of the ring with inner radius x and outer radius $x + dx$ was informally estimated to be approximately $2\pi x\,dx$.
  (a) Using the formula for the area of a circle, show that the area of the ring is $2\pi x\,dx + \pi(dx)^2$.
  (b) Show that the ring has the same area as a trapezoid of height dx and bases of lengths $2\pi x$ and $2\pi(x + dx)$.

2 The following analysis of primitive agriculture is taken from *Is There an Optimum Level of Population?* edited by S. Fred Singer, McGraw-Hill, New York, 1971:

> Consider a circular range of radius a with the home

base of production at the center. Let $G(r)$ denote the density of foodstuffs (in calories per square meter) at radius r meters from the home base. Then the total number of calories produced in the range is given by the definite integral ______ .

Using the informal approach, set up the definite integral that appeared in the blank.

3 The depth of rain at a distance r feet from the center of a storm is $g(r)$ feet.

(*a*) Estimate the total volume of rain that falls between a distance r feet and a distance $r + dr$ feet from the center of the storm. (Assume that dr is a small positive number.)

(*b*) Using (*a*), set up a definite integral for the total volume of rain that falls between 1,000 and 2,000 feet from the center of the storm.

Exercises 4 and 5 concern Example 2, kinetic energy.

4 A circular piece of metal of radius 7 centimeters has a density of 3 grams per square centimeter. It rotates 5 times per second around an axis perpendicular to the circle and passing through the center of the circle.

(*a*) Devise a local approximation for the kinetic energy of a narrow ring in the circle.

(*b*) With the aid of (*a*), set up a definite integral for the kinetic energy of the rotating metal.

(*c*) Evaluate the integral in (*b*).

5 The piece of sheet metal in Example 2 is rotated around the line midway between the edges AB and CD at the rate of 5 revolutions per second.

(*a*) Using the informal approach, obtain a local approximation for the kinetic energy of a narrow strip of the metal.

(*b*) Using (*a*), set up a definite integral for the kinetic energy of the piece of sheet metal.

(*c*) Evaluate the integral in (*b*).

Exercises 6 to 9 concern Example 3. Find the total force against the submerged vertical surfaces shown.

6

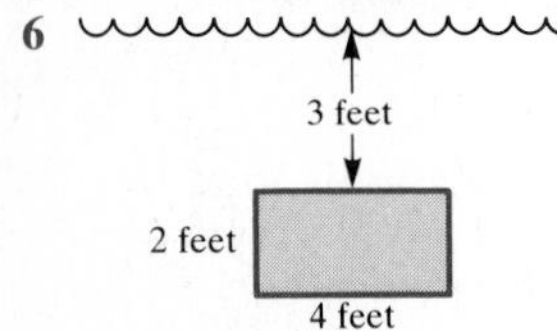

Figure 8.50

7

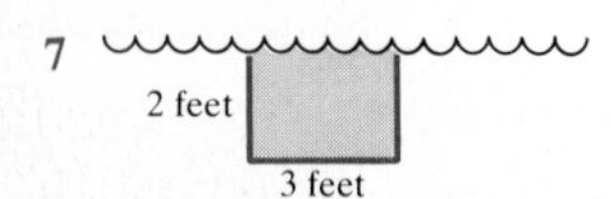

Figure 8.51

8

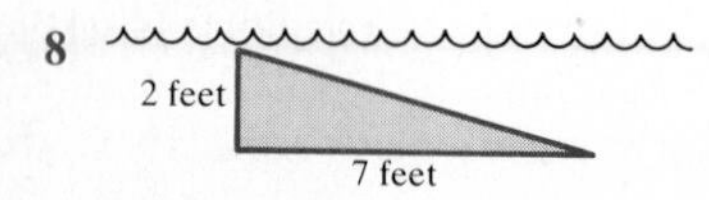

Figure 8.52

9

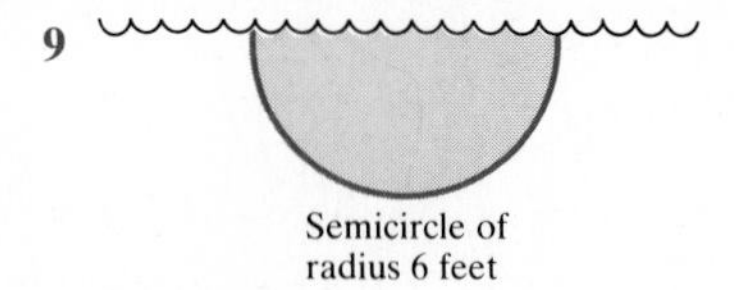

Figure 8.53

In Exercises 10 to 14 use the formula developed in Example 3 to find the total force against the shaded regions shown in Figs. 8.54 to 8.58. Choose the origin of the x axis conveniently and keep in mind that the positive part of the x axis points down.

10

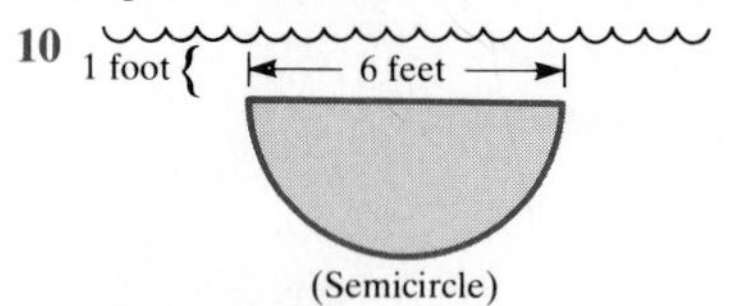

Figure 8.54

11 **12**

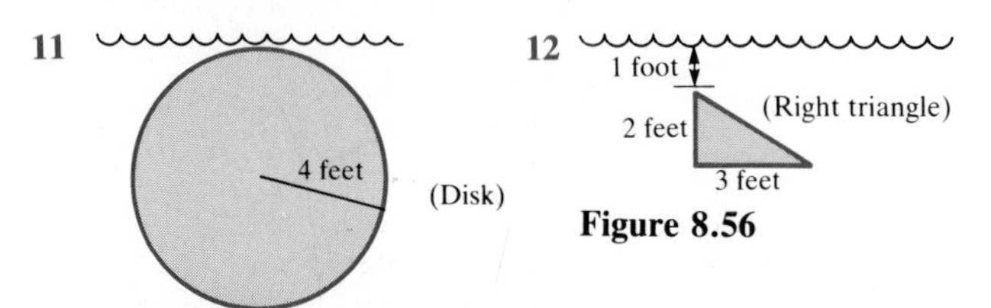

Figure 8.55

Figure 8.56

13

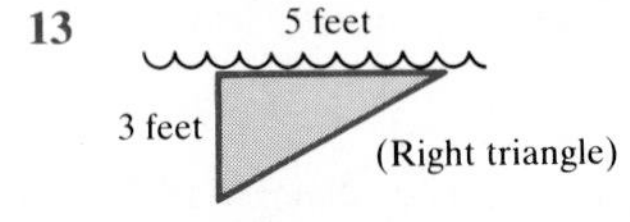

Figure 8.57

14

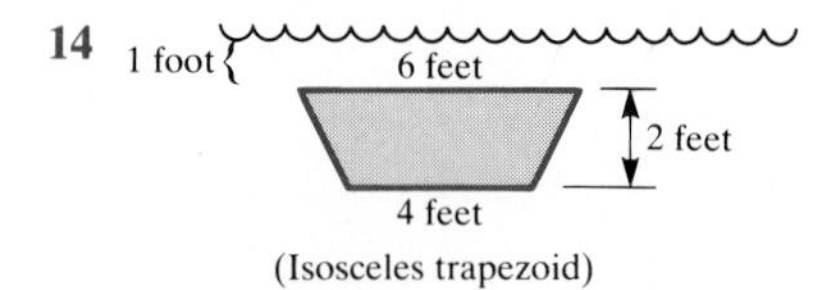

Figure 8.58

■

15 (*Poiseuille's law of blood flow*) A fluid flowing through a pipe does not all move at the same velocity. The velocity of any part of the fluid depends on its distance from the center of the pipe. The fluid at the center of the pipe moves fastest, whereas the fluid near the wall of the pipe moves slowest. Assume that the velocity of the fluid at a distance x centimeters from the axis of the pipe is $g(x)$ centimeters per second.

(*a*) Estimate informally the flow of fluid (in cubic centimeters per second) through a thin ring of inner radius r and outer radius $r + dr$ centimeters centered at the axis of the pipe and perpendicular to the axis.

(*b*) Using (*a*), set up a definite integral for the flow (in cubic centimeters per second) of liquid through the pipe. (Let the radius of the pipe be b centimeters.)

(*c*) Poiseuille (1797–1869), when studying the flow of blood through arteries, used the function $g(r) = k(b^2 - r^2)$, where k is a constant. Show that in this case the flow of blood through an artery is proportional to the fourth power of the radius of the artery.

16 (*Kinetic energy*) The density of a rod x centimeters from its left end is $g(x)$ grams per centimeter. The rod has a length of b centimeters. The rod is spun around its left end 7 times per second.

(*a*) Estimate the mass of the rod in the section that is between x and $x + dx$ centimeters from the left end. (Assume that dx is small.)

(*b*) Estimate the kinetic energy of the mass in (*a*).

(*c*) Set up a definite integral for the kinetic energy of the rotating rod.

17 Obtain the formula $V = \int_a^b A(x)\,dx$ informally.

18 At time t hours, $0 \le t \le 24$, a firm uses electricity at the rate of $e(t)$ kilowatts. The rate schedule indicates that the cost per kilowatt-hour at time t is $c(t)$ dollars. Assume that both e and c are continuous functions.

(*a*) Estimate the cost of electricity consumed between times t and $t + dt$, where dt is a small positive number.

(*b*) Using (*a*), set up a definite integral for the total cost of electricity for the 24-hour period.

19 (*Present value*) The **present value** of a promise to pay one dollar t years from now is $g(t)$ dollars.

(*a*) What is $g(0)$?

(*b*) Why is it reasonable to assume that $g(t) \le 1$ and that g is a decreasing function of t?

(*c*) What is the present value of a promise to pay q dollars t years from now?

(*d*) Assume that an investment made now will result in an income flow at the rate of $f(t)$ dollars per year t years from now. (Assume that f is a continuous function.) Estimate informally the present value of the income to be earned between time t and time $t + dt$, where dt is a small positive number.

(*e*) On the basis of the local estimate made in (*d*), set up a definite integral for the present value of all the income to be earned from now to time b years in the future.

20 (*Population*) Let the number of females in a certain population in the age range from x years to $x + dx$ years, where dx is a small positive number, be approximately $f(x)\,dx$. Assume that, on average, women of age x at the beginning of a calendar year produce $m(x)$ offspring during the year. Assume that both f and m are continuous functions.

(*a*) What definite integral represents the number of women between ages a and b years?

(*b*) What definite integral represents the total number of offspring during the calendar year produced by women whose ages at the beginning of the calendar year were between a and b years?

21 Find the force of the water against the rectangle, shown in Fig. 8.59, inclined at an angle of 30° to the vertical and whose top edge lies on the water surface. Use a definite integral (*a*) in which the interval of integration is vertical; (*b*) in which the interval of integration is inclined at 30° to the vertical; (*c*) in which the interval of integration is horizontal. In each case, draw a neat picture that shows the interval $[a, b]$ of integration and compute the integrand carefully.

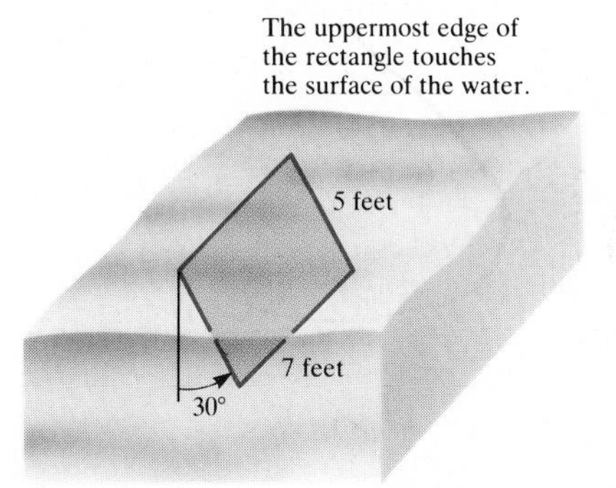

Figure 8.59

The **moment of inertia** about the line L of a point mass m at a distance x from L is defined to be x^2m. In Exercises 22 to 24 use this definition to devise an appropriate local estimate. In each case, treat all the mass between a distance x and a distance $x + dx$ from the line L as being at distance x from L. In each case, assume that the object is homogeneous; that is, its density is constant.

22 Find the moment of inertia of a solid right circular cylinder of mass M, radius a, and height h about its axis.

23 Find the moment of inertia of a ball of mass M and radius a about a diameter.

24 Find the moment of inertia of a right circular cone of mass M, radius a, and height h about its axis.

25 Let $F(t)$ be the fraction of ball bearings that wear out during the first t hours of use. Thus $F(0) = 0$.

(*a*) As t increases, what would you think happens to $F(t)$?

(*b*) Show that during the short interval of time $[t, t + dt]$, the fraction of ball bearings that wear out is approximately $F'(t)\,dt$. (Assume F is differentiable.)

(*c*) Assume all wear out in at most 1,000 hours. What is $F(1{,}000)$?

(*d*) Using the assumptions in (*b*) and (*c*), devise a definite integral for the average life of the ball bearings.

26 (*Beware of intuition.*) Consider the following argument: "Approximate the surface area of the sphere of radius a shown in Fig. 8.60 as follows. To approximate the surface area between x and $x + dx$, let us try using the area of the narrow curved part of the cylinder used to approximate the volume between x and $x + dx$. (This part is shaded in Fig. 8.60.) This local approximation can be pictured (when unrolled and laid flat) as a rectangle of width dx and length $2\pi r$. The surface area of a sphere is $\int_{-a}^{a} 2\pi r\,dx = 4\pi \int_0^a \sqrt{a^2 - x^2}\,dx$. But $\int_0^a \sqrt{a^2 - x^2}\,dx = \pi a^2/4$, since it equals the area of a quadrant of a disk. Hence the area of the sphere is $\pi^2 a^2$." This does not agree with the correct value, $4\pi a^2$, which was discovered by Archimedes in the third century B.C. What is wrong with the argument?

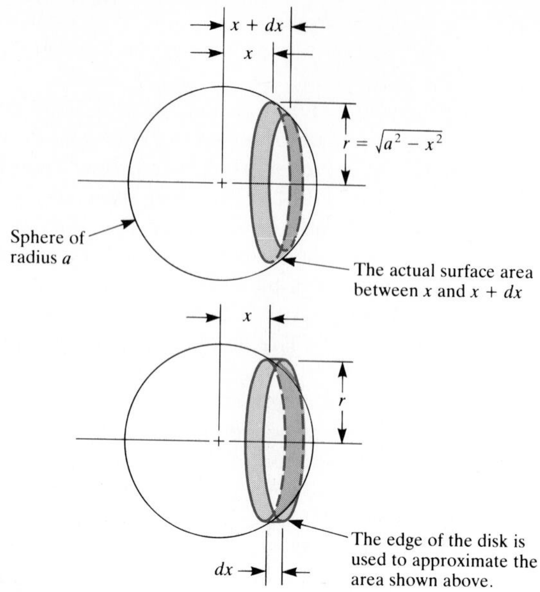

Figure 8.60

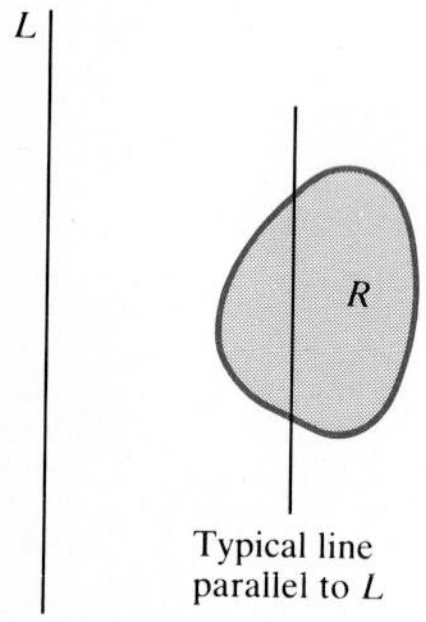

L

R

Typical line parallel to L

Figure 8.61

The line L can be in any direction in the plane.

8.4 THE SHELL TECHNIQUE

In the method of parallel cross sections, the solid of revolution is approximated by very thin disks or washers. But there is a completely different way of viewing a solid of revolution. The solid may also be approximated by concentric hollow pipes (or shells).

Consider a region R in a plane and a line L in that plane. Assume that R lies on one side of L. In addition, assume that each line in the plane of R which is parallel to L meets R in a line segment or a point, as in Fig. 8.61. The solid of revolution formed by revolving R around L is shown in Fig. 8.62. Approximate R by rectangles, as shown in Fig. 8.63. Then the volume of the solid in Fig. 8.62 is approximated by the volume of the solid of revolution formed by revolving around L the region formed by the rectangles in Fig. 8.63. When the rectangles shown in Fig. 8.63 are revolved around the line L, each one sweeps out a cylindrical shell. One of these shells or pipes is shown in Fig. 8.64.

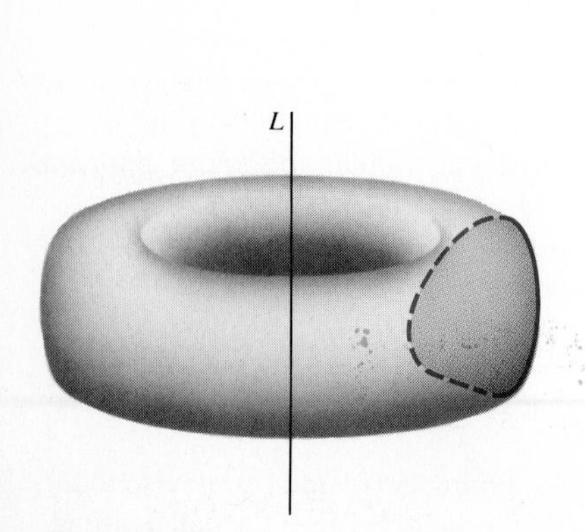

Figure 8.62

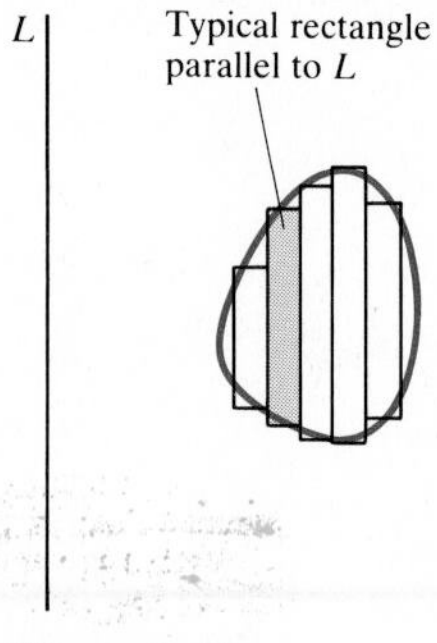

Figure 8.63

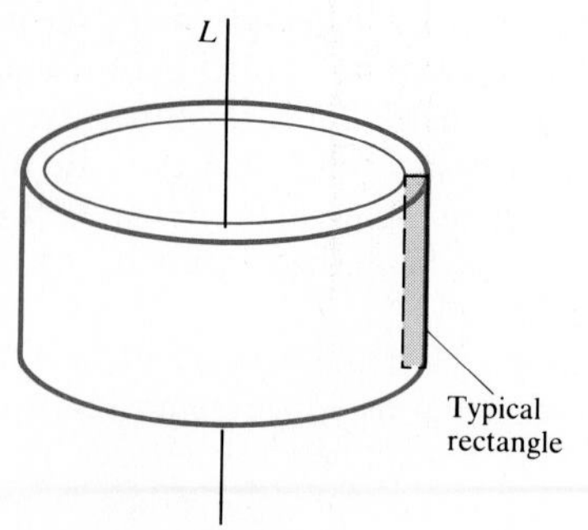

This is the shell formed by revolving the "typical rectangle" in Fig. 8.63 around L.

Figure 8.64

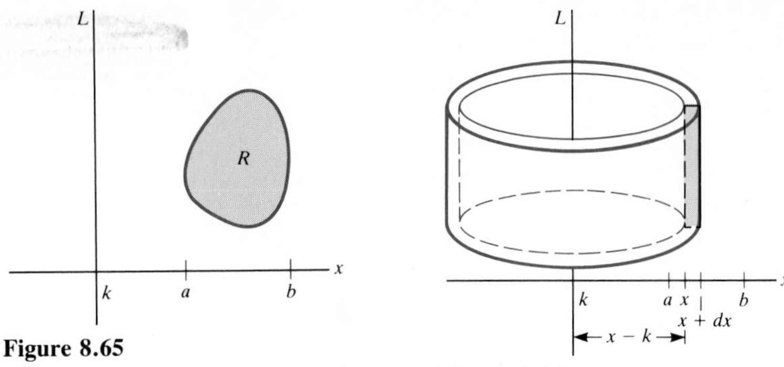

Figure 8.65

Figure 8.66

These observations suggest a way of forming a definite integral for the volume of the solid of revolution formed by revolving R around L. The informal approach introduced in Sec. 8.3 will be used.

Introduce an x axis in the plane of R and perpendicular to L. Assume that L lies to the left of R and cuts the x axis at k and that R lies above the interval $[a, b]$ as in Fig. 8.65.

Estimate the volume of the solid of revolution formed by revolving around L the part of R between those lines parallel to L which meet the x axis at x and $x + dx$. (Assume that dx is a small positive number. See Fig. 8.66.)

To estimate the volume of the shell or tube in Fig. 8.66, begin by letting $c(x)$ be the length of the cross section of R made by a line parallel to L and meeting the coordinate axis at x. Imagine cutting the tube along a direction parallel to L and then laying the tube flat. When laid flat, the tube will resemble a thin slab of thickness dx, width $c(x)$, and length $2\pi(x - k)$, as shown in Fig. 8.67. The volume of the tube, therefore, is presumably about

This is the key local estimate.

$$2\pi(x - k)c(x)\, dx. \tag{1}$$

The radius is $x - k$, as in Fig. 8.66, so circumference is $2\pi(x - k)$.

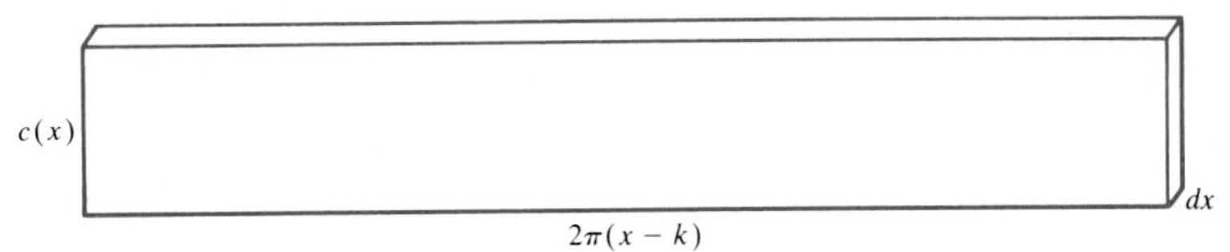

Figure 8.67

With the aid of this local estimate (1), we can then conclude that the volume of the solid of revolution is

Formula for volume of a solid of revolution by shells

$$\text{Volume} = \int_a^b 2\pi(x - k)c(x)\, dx. \tag{2}$$

This is the formula for computing volumes by the **shell technique**. If $x - k$ is denoted $R(x)$, the "radius of the shell," then

A shorter version of the formula for volume by shells

$$\text{Volume} = \int_a^b 2\pi R(x)c(x)\,dx. \tag{3}$$

These two diagrams show the essence of the shell technique.

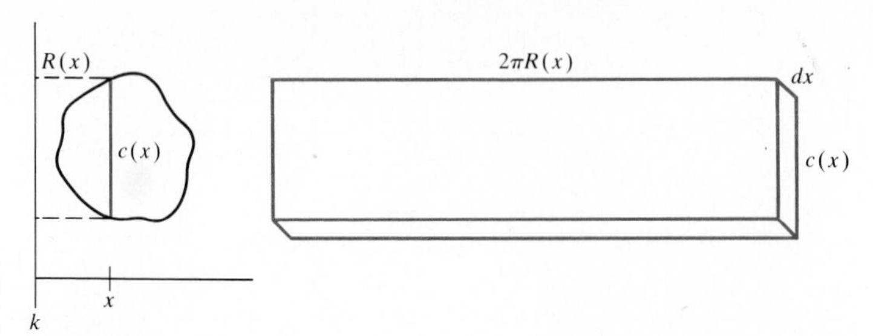

Figure 8.68

Memory aid: The expression $2\pi R(x)c(x)\,dx$ is the volume of a flat box of height dx, width $c(x)$, and length $2\pi R(x)$, the circumference of a circle of radius $R(x)$. Figure 8.68 gives the shell technique at a glance.

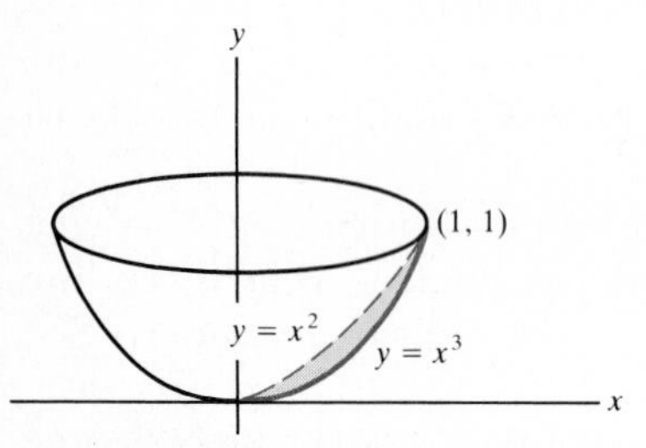

The region below $y = x^2$ and above $y = x^3$ is revolved around the y axis.

Figure 8.69

EXAMPLE 1 The region between the curves $y = x^2$ and $y = x^3$ is revolved around the y axis. Find the volume of the solid of revolution produced. (It is shaped like a bowl, as shown in Fig. 8.69.)

SOLUTION Use the given x axis to set up the integral $\int_a^b 2\pi R(x)c(x)\,dx$. Since the y axis is the line L about which the region is revolved, $R(x) = x$. Also, $c(x) = x^2 - x^3$. As may be checked, the interval of integration is [0,1]. Thus

$$\begin{aligned}\text{Volume} &= \int_0^1 2\pi x(x^2 - x^3)\,dx \\ &= \int_0^1 2\pi(x^3 - x^4)\,dx \\ &\underset{\text{FTC}}{=} 2\pi\left(\frac{x^4}{4} - \frac{x^5}{5}\right)\Bigg|_0^1 \\ &= 2\pi\left[\left(\frac{1^4}{4} - \frac{1^5}{5}\right) - \left(\frac{0^4}{4} - \frac{0^5}{5}\right)\right] \\ &= 2\pi\left(\frac{1}{4} - \frac{1}{5}\right) \\ &= \frac{2\pi}{20} = \frac{\pi}{10}. \quad \blacksquare\end{aligned}$$

EXAMPLE 2 The triangle whose vertices are (0, 0), (2, 0), and (2, 1) is revolved around the x axis. Find the volume of the resulting cone. (See Fig. 8.70.)

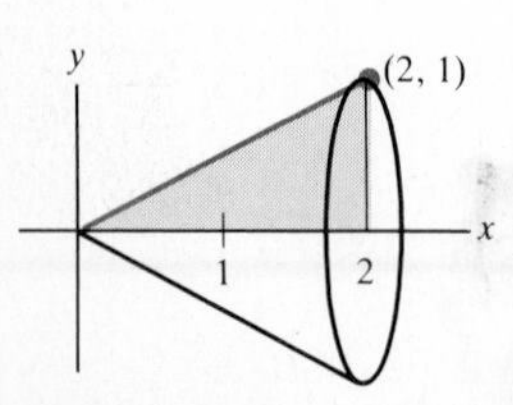

Figure 8.70

SOLUTION In this case, the shells are formed by partitions of the y axis, not the x axis. (See Fig. 8.71.) The formula for the volume, written with y replacing x, is

$$\int_a^b 2\pi R(y)c(y)\,dy.$$

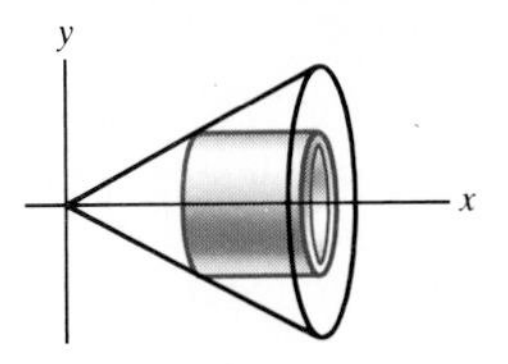

Figure 8.71

To determine $R(y)$ and $c(y)$, look closely at the given triangle, not at the solid of revolution. Inspection of Fig. 8.72 shows that $R(y) = y$. To find $c(y)$, make use of the equation of the line through (0, 0) and (2, 1), namely, $y = x/2$, or $x = 2y$. Inspection of the diagram then shows that

$$c(y) = 2 - 2y.$$

The interval of integration is [0, 1]. Thus

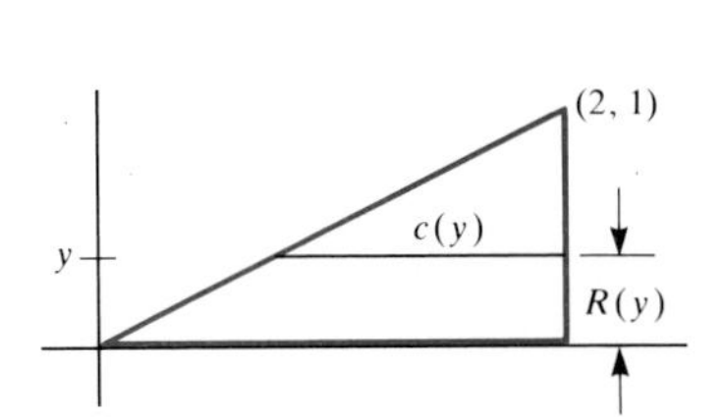

Figure 8.72

$$\begin{aligned}
\text{Volume} &= \int_0^1 2\pi y(2 - 2y)\, dy \\
&= 4\pi \int_0^1 (y - y^2)\, dy \\
&\underset{\text{FTC}}{=} 4\pi\left(\frac{y^2}{2} - \frac{y^3}{3}\right)\Bigg|_0^1 \\
&= 4\pi\left[\left(\frac{1^2}{2} - \frac{1^3}{3}\right) - \left(\frac{0^2}{2} - \frac{0^3}{3}\right)\right] \\
&= 4\pi\left(\frac{1}{2} - \frac{1}{3}\right) = \frac{2\pi}{3}. \quad \blacksquare
\end{aligned}$$

EXAMPLE 3 The region below $y = 1 + \sin x$, above the x axis, and situated between $x = 0$ and $x = 2\pi$ is revolved around the line $x = -2$. Find the volume of the resulting solid of revolution, shown in Fig. 8.73.

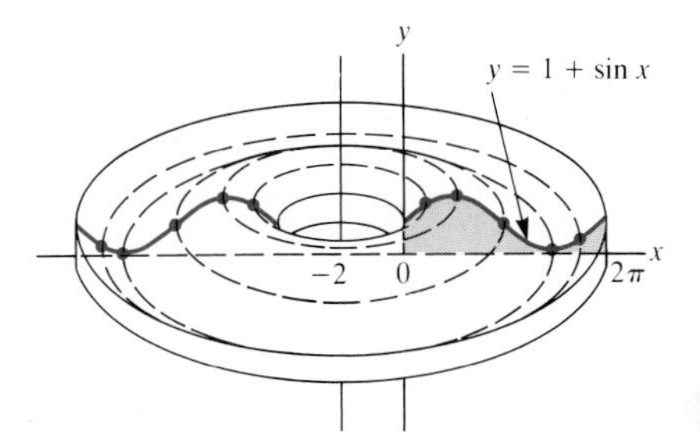

Figure 8.73

SOLUTION (Note that cross sections by planes perpendicular to the y axis would be quite messy. For y between 1 and 2, the cross section is one ring whose radii would be expressed in terms of the arcsine function. For y between 0 and 1, the cross section consists of two pieces.)

The method of concentric shells is much easier:

$$\text{Volume} = \int_a^b 2\pi R(x)c(x)\, dx = \int_0^{2\pi} 2\pi(x + 2)(1 + \sin x)\, dx.$$

Since the derivative of $x + 2$ is simpler than $x + 2$, let us use integration by parts:

$$\begin{aligned}
\int_0^{2\pi} \underbrace{2\pi(x + 2)}_{u}\, \underbrace{(1 + \sin x)\, dx}_{dv} &= \underbrace{2\pi(x + 2)}_{u}\, \underbrace{(x - \cos x)}_{v}\Bigg|_0^{2\pi} - \int_0^{2\pi} \underbrace{(x - \cos x)}_{v}\, \underbrace{2\pi\, dx}_{du} \\
&= 2\pi\,[(2\pi + 2)(2\pi - 1) - 2(-1)] - 2\pi\left(\frac{x^2}{2} - \sin x\right)\Bigg|_0^{2\pi} \\
&= 2\pi\,(4\pi^2 + 2\pi) - 2\pi\left[\left(\frac{(2\pi)^2}{2} - 0\right) - \left(\frac{0^2}{2} - 0\right)\right] \\
&= 2\pi\,(4\pi^2 + 2\pi) - 2\pi \cdot 2\pi^2 = 4\pi^3 + 4\pi^2
\end{aligned}$$

Thus the volume is $4\pi^3 + 4\pi^2$. ■

EXERCISES FOR SEC. 8.4: THE SHELL TECHNIQUE

In Exercises 1 to 6 a region R in the plane is revolved around a line to produce a solid of revolution. In each case, (*a*) draw the region, (*b*) draw the solid of revolution, (*c*) draw the typical approximating shell and label its three dimensions $[c(x), R(x), dx]$ or $[c(y), R(y), dy]$, (*d*) set up a definite integral for the volume, (*e*) evaluate the integral.

1 R, bounded by $y = x^2$, $y = 0$, and $x = 1$, is revolved around the y axis.

2 R, given in Exercise 1, is revolved around the x axis.

3 R, the finite region bounded by $y = \sqrt{x}$ and $y = \sqrt[3]{x}$, is revolved around the x axis.

4 R, given in Exercise 3, is revolved around the y axis.

5 R, bounded by $y = \sin x$ and the x axis between $x = 0$ and $x = \pi/2$, is revolved around the y axis.

6 R, given in Exercise 5, is revolved around the x axis.

7 Find the volume of a sphere of radius a by the shell method.

8 Find the volume of a right circular cone of radius a and height h by the shell method.

In Exercises 9 and 10 use the shell technique to compute the volumes of the given solids of revolution.

9 The region bordered by $y = \sqrt{x}$, $x = 1$, $x = 2$, and the x axis is revolved around the line $x = -1$.

10 The region in Exercise 9 is revolved around the line $y = -2$.

11 Find the volume of the solid of revolution formed by revolving the region bounded by $y = 2 + \cos x$, $x = \pi$, $x = 10\pi$, and the x axis around (*a*) the y axis, (*b*) the x axis.

12 The disk bounded by the circle $(x - b)^2 + y^2 = a^2$, $0 < a < b$, is revolved around the y axis. Find the volume of the doughnut (torus) produced.

13 The region below $y = \cos x$, above the x axis, and between $x = 0$ and $x = \pi/2$ is revolved around the x axis. Find the volume of the resulting solid of revolution by (*a*) parallel cross sections, (*b*) concentric shells.

In Exercises 14 to 16 find the volumes of the solids of revolution given.

14 The region between $y = e^{x^2}$, the x axis, $x = 0$, and $x = 1$ is revolved around the y axis. (It is interesting to note that the fundamental theorem of calculus is of no use in evaluating the area of this region.)

15 The region between $y = \sqrt{1 + x^2}$, $y = 1$, and $x = 1$ is revolved around the line $y = 1$.

16 The region between $y = \ln x$, the x axis, and $x = e$ is revolved around (*a*) the y axis, (*b*) the line $y = 1$, (*c*) the line $y = -1$.

■

17 Let a and b be positive numbers and $y = f(x)$ be a decreasing differentiable function of x such that $f(0) = b$ and $f(a) = 0$. Prove that $\int_0^a 2xy\, dx = \int_0^b x^2\, dy$, (*a*) by considering the volume of a certain solid, (*b*) by integration by parts.

18 Let f in Exercise 17 be elementary. (*a*) Show that, if $x^2 f'$ has an elementary integral, so does xf, and conversely. (*b*) Consider the solid obtained by rotating the region bounded by the curve $y = f(x)$, the x axis, and the y axis around the y axis. Show that its volume expressed by the shell technique involves an elementary integral only when its volume by the cross-section technique involves an elementary integral.

19 Let R be a region in the first quadrant. When it is revolved around the x axis, a solid of revolution is produced. When it is revolved around the y axis, another solid of revolution is produced. Give an example of such a region R with the property that the volume of the first solid *cannot* be evaluated by the fundamental theorem of calculus, but the volume of the second solid can be.

20 When a region R of area A situated to the right of the y axis is revolved around the y axis, the resulting solid of revolution has volume V. When R is revolved around the line $x = -k$, the volume of the resulting solid is V^*. Express V^* in terms of k, A, and V.

21 The region R below $y = (e^x \sin^2 x)/x$ and above $[\pi, 10\pi]$ is rotated around the y axis to produce a solid of revolution. (*a*) Try to find the volume of this solid by parallel cross sections and by the shell technique. (*b*) Which method is easier? (*c*) Why?

8.5 THE CENTROID OF A PLANE REGION

This section introduces the centroid of a plane region and shows its relation to the volume of a solid of revolution. In Sec. 8.8 it will also be of use in the study of work.

The Center of Mass of n Point Masses A small boy on one side of a seesaw (which we regard as weightless) can balance a bigger boy on the other

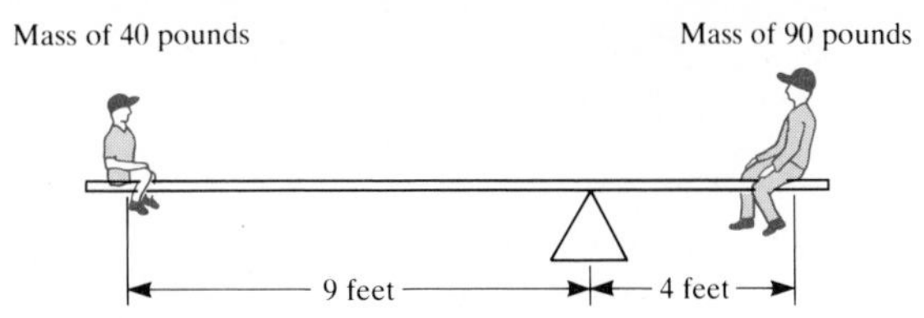

Figure 8.74

side. For example, the two boys in Fig. 8.74 balance. (According to physical laws, each boy exerts a force, due to gravitational attraction, on the seesaw proportional to his mass.)

The small mass with the long lever arm balances the large mass with the small lever arm. Each contributes the same tendency to turn—but in opposite directions. To be more precise, introduce on the seesaw an x axis with its origin 0 at the fulcrum, the point on which the seesaw rests. Define the moment about 0 of a mass m located at the point x on the x axis to be the product mx. Then the bigger boy has a moment (90)(4), while the smaller boy has a moment (40)(-9). The total moment of the lever-mass system is 0, and the masses balance. (See Fig. 8.75.)

Moment is (mass) · (lever arm), where lever arm can be positive or negative.

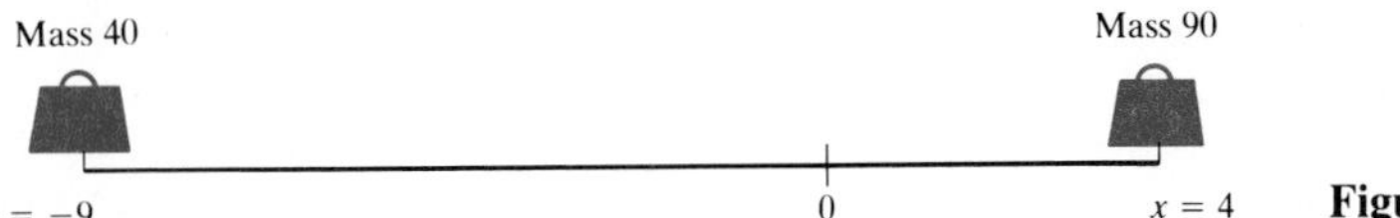

Figure 8.75

If a mass m is located on a line at coordinate x, define its moment about the point having coordinate k as the product $m(x - k)$. (See Fig. 8.76.)

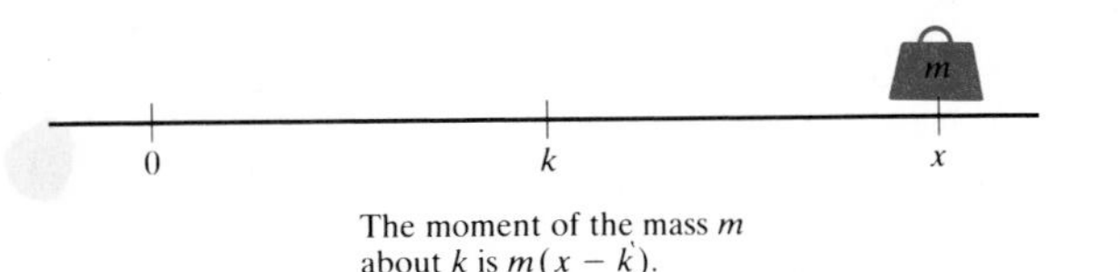

Figure 8.76

Now consider several masses $m_1, m_2, \ldots, m_n$. If mass m_i is located at x_i, with $i = 1, 2, \ldots, n$, then $\Sigma_{i=1}^{n} m_i (x_i - k)$ is the total moment of all the masses about the point k. If a fulcrum is placed at k, then the seesaw rotates clockwise if the total moment is greater than 0, rotates counterclockwise if it is less than 0, and is in equilibrium if the total moment is 0.

EXAMPLE 1 Where should the fulcrum be placed so that the three masses in Fig. 8.77 will be in equilibrium?

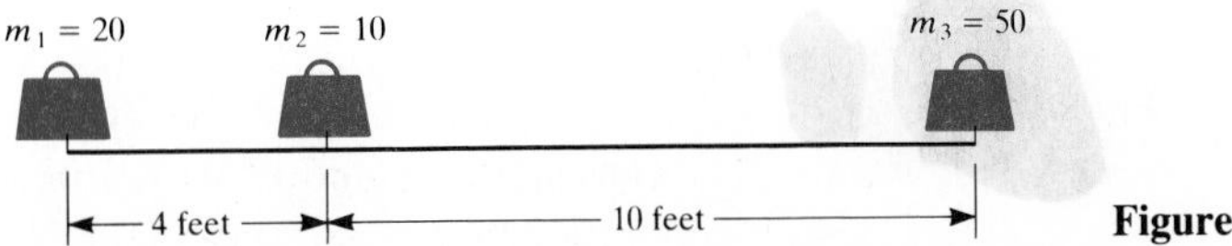

Figure 8.77

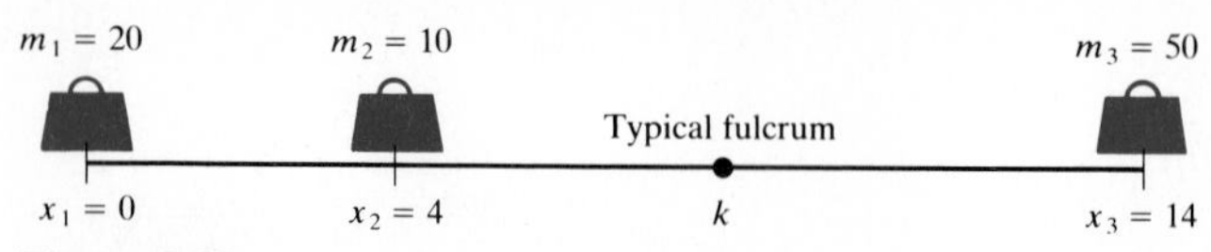

Figure 8.78

SOLUTION Introduce an x axis with origin at mass m_1 and compute the moments about a typical fulcrum having coordinate k; then select k to make the total moment 0. (See Fig. 8.78.)

The total moment about k is

$$20(0 - k) + 10(4 - k) + 50(14 - k).$$

We seek k such that this expression is equal to 0, or equivalently,

$$20 \cdot 0 + 10 \cdot 4 + 50 \cdot 14 = k(20 + 10 + 50).$$

Hence
$$k = \frac{20 \cdot 0 + 10 \cdot 4 + 50 \cdot 14}{80} = 9.25.$$

This means the fulcrum is to the right of the midpoint, which was to be expected. ■

The balancing point of masses $m_1, m_2, \ldots, m_n$, located respectively at $x_1, x_2, \ldots, x_n$ on an x axis, is found by solving the equation

$$\sum_{i=1}^{n} m_i(x_i - k) = 0 \tag{1}$$

for the number k. Expanding Eq. (1), we obtain

$$\sum_{i=1}^{n} m_i x_i - \sum_{i=1}^{n} m_i k = 0$$

or
$$\sum_{i=1}^{n} m_i x_i = \sum_{i=1}^{n} m_i k.$$

Thus
$$k \sum_{i=1}^{n} m_i = \sum_{i=1}^{n} m_i x_i,$$

and finally,
$$k = \frac{\sum_{i=1}^{n} m_i x_i}{\sum_{i=1}^{n} m_i}. \tag{2}$$

This idea is the key to the section.

The number k given by Eq. (2) is called the **center of mass** of the system of masses. *The center of mass is found by dividing the total moment about 0 by the total mass.*

Finding the center of mass of a finite number of "point masses" involves only arithmetic, no calculus. Now let us turn our attention to finding the center of mass of a continuous distribution of matter in the plane. For this purpose, definite integrals will be needed.

The Centroid of a Plane Region

The case where σ is not constant is treated in Chap. 13.

Let R be a region in the plane. Imagine that R is occupied by a thin piece of metal that has a density of σ (sigma) grams per square centimeter. Throughout this section σ will be assumed to be constant (that is, the

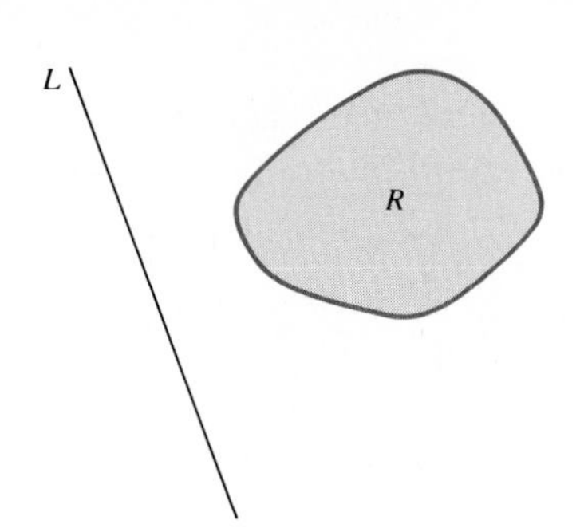

Figure 8.79

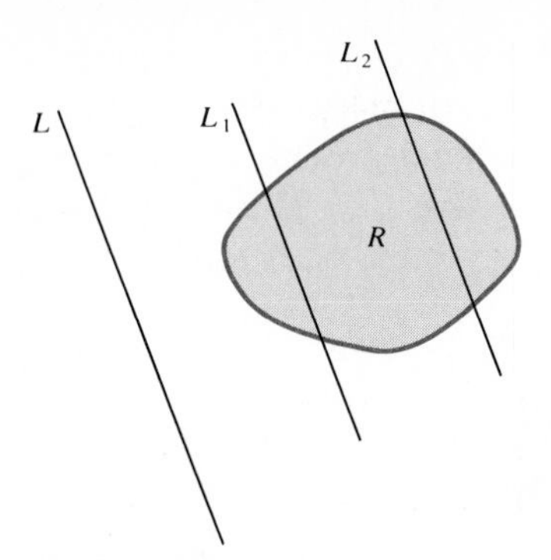

Figure 8.80

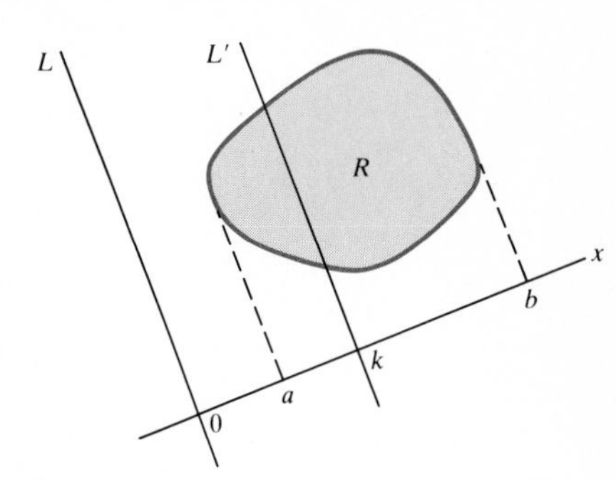

Figure 8.81

metal is "homogeneous"). Let L be a line in the plane. (See Fig. 8.79.) Is there a line parallel to L on which R balances?

The search for a balancing line begins.

Consider the lines L_1 and L_2 in Fig. 8.80. If R were placed on L_1, it would turn one way; if it were placed on L_2, it would turn the other way. It seems reasonable to expect there to be a "balancing" line parallel to L and somewhere between L_1 and L_2. To find that line it is necessary to (*a*) compute the moment of R about any line L' parallel to L and then (*b*) find the line L' for which the moment is 0.

Let L' be any line parallel to L. To compute the moment of R about L, introduce an x axis perpendicular to L with its origin at its intersection with L. Assume that L' passes through the x axis at the point $x = k$, as in Fig. 8.81. In addition, assume that each line parallel to L meets R either in a line segment or at a point on the boundary of R. The lever arm of the mass distributed throughout R varies from point to point. However, the length of the lever arm is almost constant for the mass located between two lines parallel to L and close to each other.

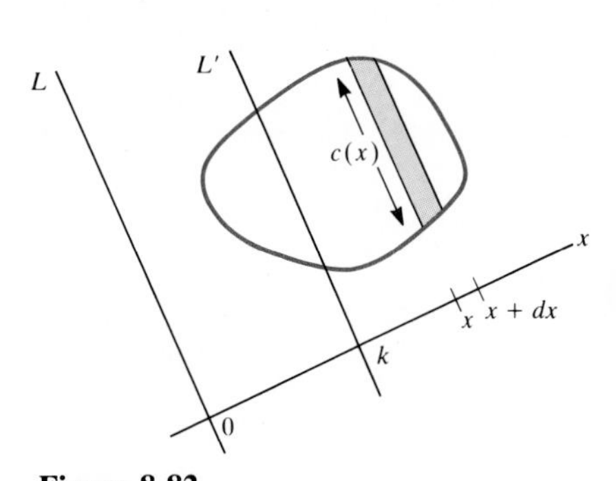

Figure 8.82

Consider the moment about L' of the mass in R located between the lines parallel to L and passing through the points on the x axis with coordinates x and $x + dx$, where dx is a small positive number. (See Fig. 8.82.)

Let $c(x)$ be the length of the cross section of R at x. Then the area of that portion of R between the lines passing through x and $x + dx$ is approximately $c(x)\,dx$. The mass of that portion is consequently about $\sigma c(x)\,dx$. The lever arm around L' of this mass is about $x - k$, which may be positive or negative. The local approximation of the moment of the mass in R about L' is therefore

The all-important local approximation

$$\underbrace{(x - k)}_{\text{lever arm}}\ \underbrace{\sigma \underbrace{c(x)\,dx}_{\text{area}}}_{\text{mass}}.$$

Following the informal approach, we conclude that

The moment of a plane homogeneous mass about a line

$$\boxed{\text{Moment of the mass in } R \text{ about } L' = \int_a^b (x - k)\sigma c(x)\,dx.} \tag{3}$$

We will use Eq. (3) as the formal definition of the moment of a plane distribution of matter about a line.

The moment around L' may or may not be 0. Let us determine k, hence the line L', such that the moment given by formula (3) is 0. To do this, solve the equation

$$0 = \int_a^b (x - k)\sigma c(x)\,dx$$

for k.

We then have

$$0 = \int_a^b x\sigma c(x)\,dx - \int_a^b k\sigma c(x)\,dx,$$

k and σ are constants.

$$0 = \sigma \int_a^b xc(x)\,dx - k\sigma \int_a^b c(x)\,dx,$$

$$k\sigma \int_a^b c(x)\,dx = \sigma \int_a^b xc(x)\,dx,$$

σ cancels.

$$k \int_a^b c(x)\,dx = \int_a^b xc(x)\,dx.$$

Formula for balancing line $x = k$ (σ constant)

$$\boxed{k = \frac{\int_a^b xc(x)\,dx}{\int_a^b c(x)\,dx}.} \tag{4}$$

The numerator is called **the moment of** R **about** L. The density σ does not appear in this moment, since it is assumed to be constant. It is convenient to think of σ as being 1. Then mass and area have the same numerical value. *Henceforth, assume that* $\sigma = 1$. The denominator $\int_a^b c(x)\,dx$ is the area of R. Formula (4) shows that there is a unique balancing line $x = k$ parallel to L. Its coordinate is given by

$$k = \frac{\text{Moment of } R \text{ about } L}{\text{Area of } R}.$$

Assume now that the plane is furnished with an xy coordinate system. There is a unique balancing line parallel to the y axis. Its x coordinate equals

$$\bar{x} = \frac{\int_a^b xc(x)\,dx}{\text{Area of } R}. \tag{5}$$

Similarly, there is a unique balancing line parallel to the x axis. Its y coordinate is given by

$$\bar{y} = \frac{\int_c^d yc(y)\,dy}{\text{Area of } R}, \tag{6}$$

where $c(y)$ is the cross-section function of R for lines parallel to the x axis and $[c, d]$ is the interval of integration.

Definition *Centroid of R.* The **centroid of** R is defined as the point $(\bar{x}, \bar{y})$, where

$$\bar{x} = \frac{\text{Moment of } R \text{ about } y \text{ axis}}{\text{Area of } R} \quad \text{and} \quad \bar{y} = \frac{\text{Moment of } R \text{ about } x \text{ axis}}{\text{Area of } R},$$

Formula for the centroid

$$\bar{x} = \frac{\int_a^b xc(x)\,dx}{\text{Area of } R} \quad \text{and} \quad \bar{y} = \frac{\int_c^d yc(y)\,dy}{\text{Area of } R}. \tag{7}$$

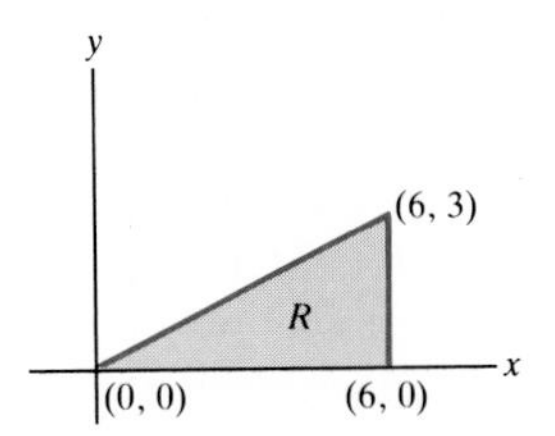

Figure 8.83

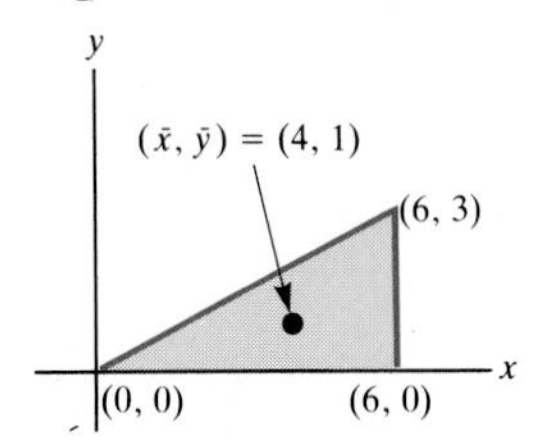

Figure 8.84

It can be shown that the region R *balances on any line through its centroid.* (See Exercises 34 and 35 in Sec. 13.3.) Moreover, if R is suspended motionless from a string attached at its centroid, it will remain in equilibrium.

EXAMPLE 2 Let R be the triangle in the xy plane with vertices at (0, 0), (6, 0), and (6, 3), as shown in Fig. 8.83. Find (*a*) $\bar{x}$, and (*b*) $\bar{y}$.

SOLUTION

(*a*) To find $\bar{x}$, we first compute the moment of R around the y axis,

$$\int_0^6 xc(x)\,dx,$$

where $c(x)$ is the cross-section function for R. To find $c(x)$, use the equation of the line through (0, 0) and (6, 3), namely, $y = x/2$. Thus $c(x) = x/2$. Consequently, the moment of R around the y axis is

$$\int_0^6 x \cdot \frac{x}{2}\,dx = \frac{1}{2}\int_0^6 x^2\,dx \underset{\text{FTC}}{=} \frac{1}{2}\frac{x^3}{3}\Big|_0^6 = \frac{1}{2}\frac{216}{3} = 36.$$

To use formula (7), we divide 36 by the area of the triangle, which is $\frac{1}{2} \cdot 6 \cdot 3 = 9$. Thus

$$\bar{x} = \tfrac{36}{9} = 4.$$

The balancing line parallel to the y axis has the equation $x = 4$.

(*b*) To find $\bar{y}$, first compute the moment $\int_0^3 yc(y)\,dy$. Since the line through (0, 0) and (6, 3) has the equation $y = x/2$, on that line $x = 2y$. Inspection of Fig. 8.83 shows that $c(y) = 6 - 2y$. Thus

$$\int_0^3 yc(y)\,dy = \int_0^3 y(6 - 2y)\,dy = \int_0^3 (6y - 2y^2)\,dy$$

$$\underset{\text{FTC}}{=} \left(3y^2 - \frac{2y^3}{3}\right)\Big|_0^3 = (27 - 18) - 0 = 9.$$

Hence
$$\bar{y} = \frac{9}{\text{Area of } R} = \frac{9}{9} = 1.$$

Thus
$$(\bar{x}, \bar{y}) = (4, 1),$$

which is shown in Fig. 8.84. ■

The centroid of a triangle is always two-thirds the way from each vertex to the midpoint of the opposite side.

Using formula (7), we can show that the centroid of a rectangle is its center. This fact is plausible, since, by symmetry, a rectangle would balance on a line through its center and parallel to an edge, as shown in Fig. 8.85.

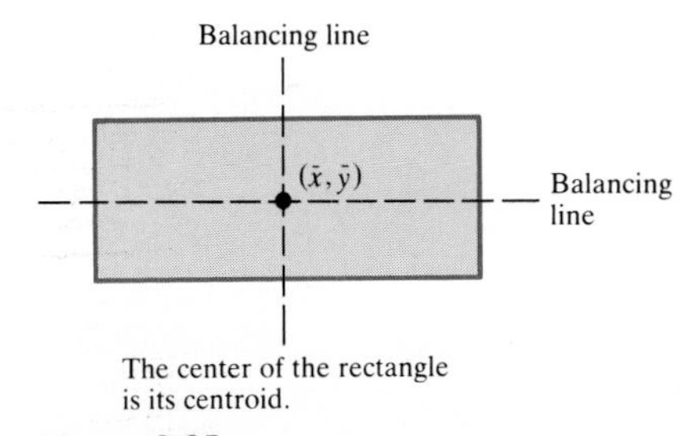

Figure 8.85

Consider a rectangle resting on the x axis, as in Fig. 8.86. Let $\bar{y}$ be the y coordinate of its center of mass. Then

$$\bar{y} = \frac{\text{Moment of rectangle about } x \text{ axis}}{\text{Area of rectangle}},$$

or
$$\begin{pmatrix}\text{Moment of rectangle}\\ \text{about } x \text{ axis}\end{pmatrix} = \begin{pmatrix}y \text{ coordinate of}\\ \text{centroid of rectangle}\end{pmatrix} \cdot \begin{pmatrix}\text{Area of}\\ \text{rectangle}\end{pmatrix} \tag{8}$$

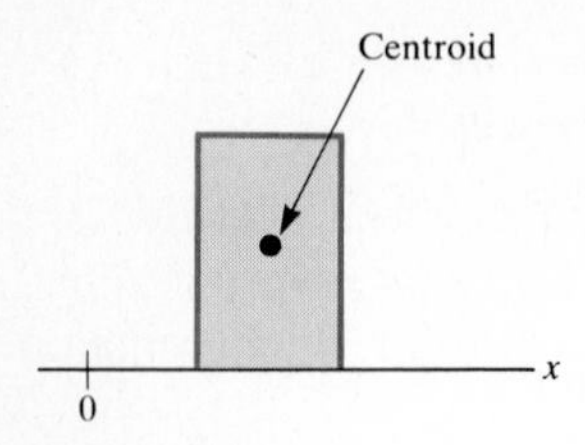

Figure 8.86

Formula (8) is the basis of a shortcut for computing $\bar{y}$ of a region under a curve and above the x axis. We now develop this shortcut.

Let $y = f(x)$ be a continuous function such that $f(x) \geq 0$ for x in $[a, b]$. Let R be the region below the curve $y = f(x)$ and above $[a, b]$. The moment around the x axis of the portion of R between two lines parallel to the y axis, one with x coordinate x and one with x coordinate $x + dx$, where dx is a small positive number, can be estimated easily. (See Fig. 8.87, where this narrow band is shaded.) The narrow band has area approximately $f(x)\,dx$, and the y coordinate of its center is approximately $f(x)/2$. By formula (8), its moment around the x axis is approximately

$$\underbrace{\frac{f(x)}{2}}_{\substack{y \text{ coordinate} \\ \text{of centroid}}} \cdot \underbrace{f(x)\,dx}_{\text{area}}. \tag{9}$$

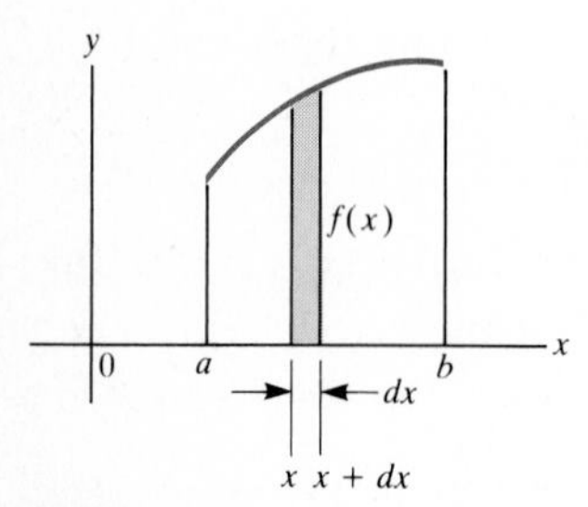

Figure 8.87

From the local approximation (9), we conclude that

$$\begin{pmatrix} \text{Moment of } R \\ \text{about } x \text{ axis} \end{pmatrix} = \int_a^b \frac{[f(x)]^2}{2}\,dx. \tag{10}$$

Consequently, $\bar{y}$, the y coordinate of the centroid of R, equals

$\bar{y}$ for the region under $y = f(x)$

$$\boxed{\bar{y} = \frac{\displaystyle\int_a^b \frac{[f(x)]^2}{2}\,dx}{\text{Area of } R}.} \tag{11}$$

This formula is preferable to formula (7) if R is given as a region under a curve, $y = f(x)$.

EXAMPLE 3 Find the centroid of the semicircular region of radius a shown in Fig. 8.88.

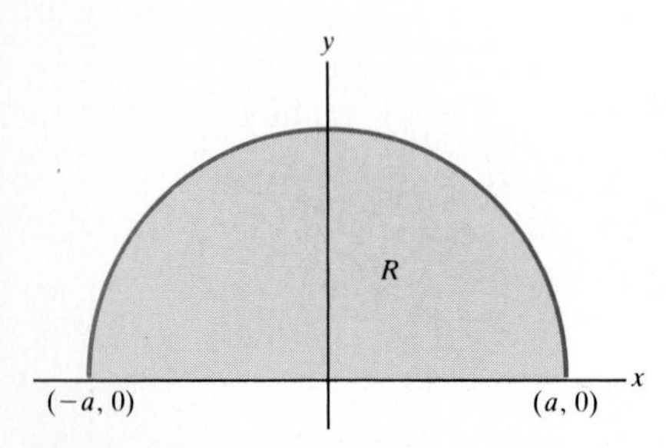

Figure 8.88

SOLUTION By symmetry, $\bar{x} = 0$.

To find $\bar{y}$, use formulas (10) and (11). The function f in this case is given by the formula $f(x) = \sqrt{a^2 - x^2}$. Thus the moment of R about the x axis is

For an even function $f(x)$, $\int_{-a}^{a} f(x) = 2 \int_0^a f(x)$.

$$\int_{-a}^{a} \frac{(\sqrt{a^2 - x^2})^2}{2}\,dx = \int_{-a}^{a} \frac{a^2 - x^2}{2}\,dx = 2\int_0^a \frac{a^2 - x^2}{2}\,dx$$

$$= \int_0^a (a^2 - x^2)\,dx \underset{\text{FTC}}{=} \left(a^2x - \frac{x^3}{3}\right)\Bigg|_0^a$$

$$= \left(a^3 - \frac{a^3}{3}\right) - 0 = \tfrac{2}{3}a^3.$$

Thus
$$\bar{y} = \frac{\frac{2}{3}a^3}{\text{Area of } R} = \frac{\frac{2}{3}a^3}{\frac{1}{2}\pi a^2} = \frac{4a}{3\pi}.$$

(Since $4/(3\pi) \approx 0.42$, the center of gravity of R is at a height of about $0.42a$.) ■

The Centroid and the Volume of a Solid of Revolution The relation between the centroid of a plane region R and the solid of revolution obtained by revolving R about a line is expressed in the following theorem, due to the fourth-century Greek mathematician Pappus.

Theorem *Pappus's theorem.* Let R be a region in the plane and L a line in the plane that either does not meet R or else just meets R at its border. Then the volume of the solid formed by revolving R about L is equal to the product:

$$(\text{Distance centroid of } R \text{ is revolved}) \cdot (\text{Area of } R).$$

Proof Introduce an xy coordinate system in such a way that L is the y axis and R lies to the right of it. The volume of the solid of revolution is

$$V = 2\pi \int_a^b xc(x)\,dx \qquad \text{(shell technique)}.$$

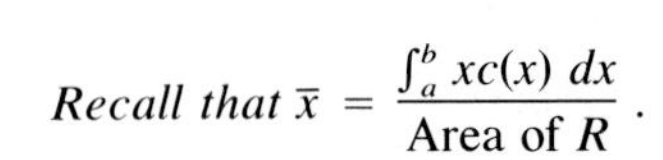

By formula (7),
$$\int_a^b xc(x)\,dx = \bar{x} \cdot \text{Area of } R.$$

Hence
$$V = 2\pi\bar{x} \cdot \text{Area of } R.$$

Since $2\pi\bar{x}$ is the distance the centroid of R is revolved, this proves the theorem. ■

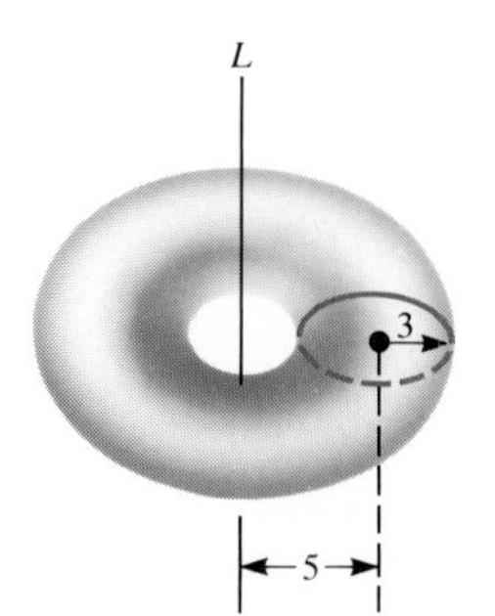

Figure 8.89

EXAMPLE 4 Use Pappus's theorem to find the volume of the "doughnut" formed by revolving a circle of radius 3 inches about a line 5 inches from the center, as shown in Fig. 8.89.

SOLUTION In this case, the area of R is $\pi \cdot 3^2 = 9\pi$. The centroid of a circle is its center. Hence the distance it is revolved is $2\pi \cdot 5 = 10\pi$. The volume of the doughnut is

$$10\pi \cdot 9\pi = 90\pi^2 \text{ cubic inches.} \quad ■$$

Using Pappus's theorem in reverse

Pappus's theorem can be used to find the centroid of a region R if the volume of the solid of revolution obtained by revolving R is known. The next example should be contrasted with Example 3.

EXAMPLE 5 Find the centroid of the half disk R of radius a shaded in Fig. 8.90.

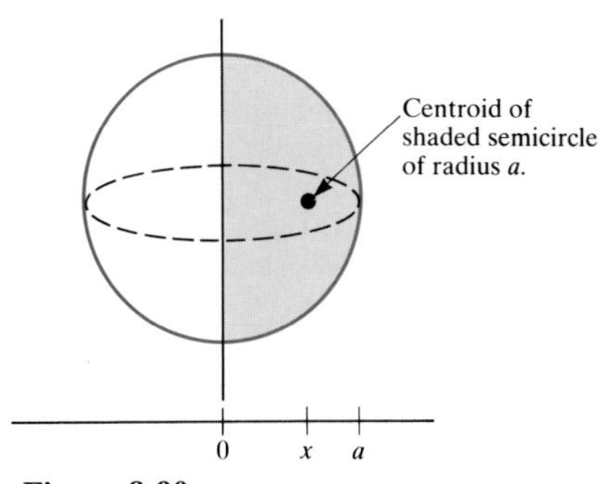

Figure 8.90

SOLUTION By symmetry, the centroid lies somewhere on the radius that is perpendicular to the diameter of the semicircle. Let $\bar{x}$ be its distance from the diameter. When R is revolved about its diameter, it produces a sphere of radius a. The volume of a sphere of radius a is $4\pi a^3/3$.

By Pappus's theorem,

$$2\pi\bar{x} \cdot \text{Area of } R = \text{Volume of sphere of radius } a,$$

or

$$2\pi\bar{x}\,\frac{\pi a^2}{2} = \frac{4}{3}\,\pi a^3.$$

Consequently, $$\pi^2 a^2 \bar{x} = \frac{4}{3}\pi a^3,$$

Compare with Example 3.

and thus $$\bar{x} = \frac{4\pi a^3}{3\pi^2 a^2} = \frac{4a}{3\pi}. \quad ■$$

Two Key Formulas in This Section

Let R be the region below the curve $y = f(x)$ and above $[a, b]$. Then

Don't memorize these formulas; be able to derive them.

$$\bar{x} = \frac{\int_a^b xf(x)\,dx}{\text{Area of } R} \quad \text{and} \quad \bar{y} = \frac{\int_a^b \{[f(x)]^2/2\}\,dx}{\text{Area of } R}.$$

EXERCISES FOR SEC. 8.5: THE CENTROID OF A PLANE REGION

In each of Exercises 1 to 6 find the moment of the given region R about the given line L.

1 R is bounded by $y = \sin x$ and the x axis and lies between $x = 0$ and $x = \pi$. L is the x axis.

2 R is bounded by $y = \cos x$ and the x axis and lies between $x = 0$ and $x = \pi/2$. L is the x axis.

3 R is the same as in Exercise 1. L is the y axis.

4 R is the same as in Exercise 2. L is the y axis.

5 R is the rectangle with vertices $(0, 0)$, $(a, 0)$, (a, b), $(0, b)$, where $a > 0$, $b > 0$. L is the x axis.

6 R is the same as in Exercise 5. L is the y axis.

In Exercises 7 to 18 find the centroids of the indicated regions.

7 The triangle bounded by $y = x$, $x = 1$, and the x axis.

8 The triangle bounded by the two axes and the line $x/4 + y/3 = 1$.

9 The triangle bounded by $y = x$, $y = 2x$, and $x = 2$.

10 The region bounded by $y = x^2$, the x axis, and $x = 1$.

11 The region bounded by $y = x^2$ and the line $y = 4$. (R lies in quadrants I and II.)

12 The region bounded by $y = e^x$ and the x axis, between the lines $x = 1$ and $x = 2$.

13 The region bounded by $y = \sin 2x$ and the x axis, between the lines $x = 0$ and $x = \pi/2$.

14 The region bounded by $y = \sqrt{1 + x^2}$ and the x axis, between the lines $x = 1$ and $x = 2$.

15 The region bounded by $y = \ln x$ and the x axis, between the lines $x = 1$ and $x = e$.

16 The region between $y = 1/\sqrt{x^2 - 1}$, between the lines $x = \sqrt{2}$ and $x = 2$.

17 The triangle whose vertices are $(0, 0)$, $(a, 0)$, $(0, b)$, where a and b are positive numbers.

18 The top half of the region within the ellipse $x^2/a^2 + y^2/b^2 = 1$.

■

Exercise 19 is used in Exercises 20 to 22.

19 Let f and g be continuous functions such that $f(x) \geq g(x) \geq 0$ for x in $[a, b]$. Let R be the region above $[a, b]$ which is bounded by the curves $y = f(x)$ and $y = g(x)$.

(*a*) Set up a definite integral (in terms of f and g) for the moment of R about the y axis.

(*b*) Set up a definite integral with respect to x (in terms of f and g) for the moment of R about the x axis.

In Exercises 20 to 22 find (*a*) the moment of the given region R about the y axis, (*b*) the moment of R about the x axis, (*c*) the area of R, (*d*) $\bar{x}$, (*e*) $\bar{y}$.

20 R is bounded by the curves $y = x^2$ and $y = x^3$, between $x = 0$ and $x = 1$.

21 R is bounded by the curves $y = 3^x$ and $y = 2^x$, between $x = 1$ and $x = 2$.

22 R is bounded by the curves $y = x - 1$ and $y = \ln x$, between $x = 1$ and $x = e$.

23 In a letter of 1680 Leibniz wrote:

> Huygens, as soon as he had published his book on the pendulum, gave me a copy of it; and at that time I was quite ignorant of Cartesian algebra and also of the method of indivisibles, indeed I did not know the correct definition of the center of gravity. For, when by chance I spoke of it to Huygens, I let him know that I thought that a straight line drawn through the center of gravity always cut a figure into two equal parts; since that clearly happened in the case of a square, or a circle, an ellipse, and other figures that have a center of magnitude, I imagined that it was the same for all other figures. Huygens laughed when he heard this, and told me that nothing was further from the truth. (Quoted in C. H. Edwards, *The Historical Development of the Calculus*, p. 239, Springer-Verlag, New York, 1979.)

Give an example showing that "nothing is further from the truth."

24 Let a be a constant ≥ 1. Let R be the region below $y = x^a$, above the x axis, between the lines $x = 0$ and $x = 1$.

(*a*) Sketch R for large a.

(*b*) Compute the centroid $(\bar{x}, \bar{y})$ of R.
(*c*) Find $\lim_{a\to\infty} \bar{x}$ and $\lim_{a\to\infty} \bar{y}$.
(*d*) Show that for large a the centroid of R lies in R.

25 (Contributed by Jeff Lichtman) Let f and g be two continuous functions such that $f(x) \geq g(x) \geq 0$ for x in $[0, 1]$. Let R be the region under $y = f(x)$ and above $[0, 1]$; let R^* be the region under $y = g(x)$ and above $[0, 1]$.
(*a*) Do you think the center of mass of R is at least as high as the center of mass of R^*? (An opinion only.)
(*b*) Let $g(x) = x$. Define $f(x)$ to be $\frac{1}{3}$ for $0 \leq x \leq \frac{1}{3}$ and $f(x)$ to be x if $\frac{1}{3} \leq x \leq 1$. (Note that f is continuous.) Find $\bar{y}$ for R and also for R^*. (Which is larger?)
(*c*) Let a be a constant, $0 \leq a \leq 1$. Let $f(x) = a$ for $0 \leq x \leq a$, and let $f(x) = x$ for $a \leq x \leq 1$. Find $\bar{y}$ for R.
(*d*) Show that the number a for which $\bar{y}$ defined in part (*c*) is a minimum is a root of the equation $x^3 + 3x - 1 = 0$.
(*e*) Show that the equation in (*d*) has only one real root q. (It is approximately 0.32219.)
(*f*) For the functions described in (*c*), how small can the corresponding $\bar{y}$ be? Lichtman showed that if $f(x) \geq x$ for x in $[0, 1]$, then $\bar{y}$ of the corresponding R is at least q.

26 (Using the centroid to compute the force of water against a surface) Let R be a submerged portion of the vertical wall of a pool.
(*a*) Express the force of the water against R as a definite integral. *Suggestion:* Review Sec. 8.3 and assume that each horizontal line in the plane of R meets R in a line segment or a point.
(*b*) Using (*a*), show that the force of the water against R is equal to the pressure at the centroid of R times the area of R.

27 Cut an irregular shape out of cardboard and find three balancing lines for it experimentally. Are they concurrent; that is, do they pass through a common point?

28 This exercise shows that the three medians of a triangle meet at the centroid of the triangle. (A **median** of a triangle is a line that passes through a vertex and the midpoint of the opposite edge.)

Let R be a triangle with vertices A, B, and C. It suffices to show that the centroid of R lies on the median through C and the midpoint M of the edge AB. Introduce an xy coordinate system such that the origin is at A and B lies on the x axis, as in Fig. 8.91.
(*a*) Compute $(\bar{x}, \bar{y})$.
(*b*) Find the equation of the median through C and M.
(*c*) Verify that the centroid lies on the median computed in (*b*).
(*d*) Why would you expect the centroid to lie on each median? It's not because a median divides a triangle into two regions of equal area. (See Exercise 23.)

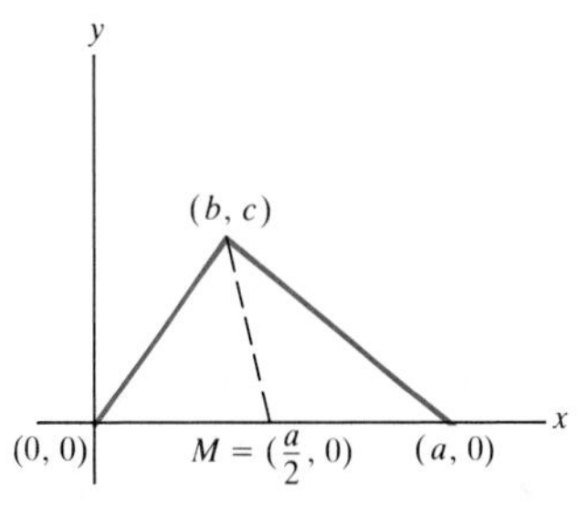

Figure 8.91

Skip

8.6 INEQUALITIES INVOLVING DEFINITE INTEGRALS

In Sec. 5.5 it was stated that if $f(x) \geq g(x)$ for all x in $[a, b]$, then

$$\int_a^b f(x)\,dx \geq \int_a^b g(x)\,dx.$$

In this section we will develop other useful inequalities for integrals.

The first inequality compares $\left|\int_a^b f(x)\,dx\right|$ with $\int_a^b |f(x)|\,dx$. If $f(x) \geq 0$ for all x in $[a, b]$, these two quantities are equal, since both equal $f(x)\,dx$. But if $f(x)$ may be negative for some x and positive for some x, then $\left|\int_a^b f(x)\,dx\right|$ may be quite small, even 0, while $\int_a^b |f(x)|\,dx$ may be large.

Theorem 1 Assume that $\int_a^b f(x)\,dx$ and $\int_a^b |f(x)|\,dx$ both exist. Then

$$\left|\int_a^b f(x)\,dx\right| \leq \int_a^b |f(x)|\,dx.$$

Proof For all x in $[a, b]$, we have

$$-|f(x)| \leq f(x) \leq |f(x)|.$$

Thus
$$-\int_a^b |f(x)|\,dx \leq \int_a^b f(x)\,dx \leq \int_a^b |f(x)|\,dx,$$

from which it follows that

$$\left|\int_a^b f(x)\,dx\right| \leq \int_a^b |f(x)|\,dx. \quad \blacksquare$$

EXAMPLE 1 Discuss the size of $\displaystyle\int_\pi^{10\pi} \frac{\cos x}{x^2}\,dx$.

SOLUTION By Theorem 1,

$$\left|\int_\pi^{10\pi} \frac{\cos x}{x^2}\,dx\right| \leq \int_\pi^{10\pi} \frac{|\cos x|}{x^2}\,dx.$$

Since $|\cos x| \leq 1$,

$$\int_\pi^{10\pi} \frac{|\cos x|}{x^2}\,dx \leq \int_\pi^{10\pi} \frac{dx}{x^2},$$

which is easily shown to be $9/(10\pi)$. ■

The next inequality concerns $\int_a^b f(x)g(x)\,dx$, where the integrand is the product of two functions.

Theorem 2 Assume that $f(x)$ and $g(x)$ are defined for all x in $[a, b]$ and that $g(x) \geq 0$ for all x in $[a, b]$. Assume also that $\int_a^b f(x)g(x)\,dx$ and $\int_a^b g(x)\,dx$ exist and that there are constants m and M such that

$$m \leq f(x) \leq M$$

for all x in $[a, b]$. Then

$$m\int_a^b g(x)\,dx \leq \int_a^b f(x)g(x)\,dx \leq M\int_a^b g(x)\,dx. \tag{1}$$

Proof We multiply the inequality

$$m \leq f(x) \leq M$$

by the *nonnegative number* $g(x)$ and obtain the inequality

$$mg(x) \leq f(x)g(x) \leq Mg(x).$$

Integration of this inequality yields (1). ■

Corollary 1 *The second mean-value theorem for integrals.* Assume that $f(x)$ and $g(x)$ are defined for all x in $[a, b]$ and that $g(x) \geq 0$ for all x in $[a, b]$. Assume also that f is continuous and that $\int_a^b f(x)g(x)\,dx$ and $\int_a^b g(x)\,dx$ exist. Then there is a number c in $[a, b]$ such that

$$\int_a^b f(x)g(x)\,dx = f(c)\int_a^b g(x)\,dx.$$

Proof By the maximum- (and minimum-) value theorem of Sec. 2.6, $f(x)$ assumes a maximum value M and a minimum value m for x in $[a, b]$. By Theorem 2,

$$m \int_a^b g(x)\, dx \le \int_a^b f(x)g(x)\, dx \le M \int_a^b g(x)\, dx.$$

For t in $[a, b]$, let $h(t) = f(t) \int_a^b g(x)\, dx$. Since h is continuous, it assumes all values between its minimum, $m\int_a^b g(x)\, dx$, and its maximum, $M\int_a^b g(x)\, dx$. (Recall the intermediate-value theorem of Sec. 2.6.) Thus there is a number c in $[a, b]$ such that

$$f(c) \int_a^b g(x)\, dx = \int_a^b f(x)g(x)\, dx. \quad \blacksquare$$

The next example will be used in Chap. 10.

EXAMPLE 2 Show that for any continuous function $f(x)$ on $[a, b]$,

$$\int_a^b f(x)(b - x)^n\, dx = f(c)\frac{(b - a)^{n+1}}{n + 1}$$

for some c in $[a, b]$.

SOLUTION The function $(b - x)^n$ is nonnegative for x in $[a, b]$. Corollary 1 shows that there is a number c in $[a, b]$ such that

$$\int_a^b f(x)(b - x)^n\, dx = f(c) \int_a^b (b - x)^n\, dx.$$

However,

$$\int_a^b (b - x)^n\, dx = -\frac{(b - x)^{n+1}}{n + 1}\Bigg|_a^b = -0 - \left[-\frac{(b - a)^{n+1}}{n + 1}\right] = \frac{(b - a)^{n+1}}{n + 1}.$$

This completes the solution. $\blacksquare$

The next inequality relates $\int_a^b f(x)g(x)dx$ to $\int_a^b [f(x)]^2 dx$ and $\int_a^b [g(x)]^2 dx$.

Theorem 3 *The Schwarz inequality.* Let f and g be continuous functions on $[a, b]$. Then

$$\left|\int_a^b f(x)g(x)\, dx\right| \le \left\{\int_a^b [f(x)]^2\, dx\right\}^{1/2} \left\{\int_a^b [g(x)]^2\, dx\right\}^{1/2}. \quad \blacksquare$$

The proof of this is outlined in Exercise 13.

The next example applies the Schwarz inequality to a problem about centroids.

EXAMPLE 3 Let $f(x)$ be continuous for x in $[a, b]$, and assume that $f(x) \ge 0$. Let R be the region below $y = f(x)$ and above $[a, b]$. Show that the y coordinate of the centroid is at least half the average value of $f(x)$ for x in $[a, b]$.

SOLUTION By definition, the average value of $f(x)$ in $[a, b]$ is

$$\frac{\int_a^b f(x)\, dx}{b - a}.$$

By Sec. 8.5, the height of the centroid of R is

$$\bar{y} = \frac{1}{2}\frac{\int_a^b [f(x)]^2\,dx}{\int_a^b f(x)\,dx}.$$

We wish to show, therefore, that

$$\frac{1}{2}\frac{\int_a^b [f(x)]^2\,dx}{\int_a^b f(x)\,dx} \geq \frac{1}{2}\frac{\int_a^b f(x)\,dx}{b-a},$$

or
$$\left[\int_a^b f(x)\,dx\right]^2 \leq \left\{\int_a^b [f(x)]^2\,dx\right\}(b-a),$$

or
$$\left|\int_a^b f(x)\,dx\right| \leq \left\{\int_a^b [f(x)]^2\,dx\right\}^{1/2}(b-a)^{1/2}.$$

But this last inequality is a special case of the Schwarz inequality with $g(x) = 1$ for all x in $[a, b]$. ■

In Exercise 14 the Schwarz inequality is applied to a question about speed.

EXERCISES FOR SEC. 8.6: INEQUALITIES INVOLVING DEFINITE INTEGRALS

1 Use Theorem 1 to find

$$\lim_{a\to\infty}\int_a^{2a}\frac{\cos x}{x^2}\,dx.$$

2 Use Theorem 1 to find

$$\lim_{a\to\infty}\int_a^{2a}\frac{\sin^2 x}{x^3}\,dx.$$

3 Show that inequality (1) with $\leq$ replaced by $\geq$ holds if the condition $g(x) \geq 0$ is replaced by $g(x) \leq 0$.

4 If in Theorem 2 the assumption $g(x) \geq 0$ is removed, could it happen that $\int_a^b f(x)g(x)dx$ is less than both $m\int_a^b g(x)dx$ and $M\int_a^b g(x)\,dx$?

5 Let f and g be continuous on $[a, b]$. By considering $\int_a^b [f(x) - g(x)]^2\,dx$, show that

$$\int_a^b f(x)g(x)\,dx \leq \frac{\int_a^b [f(x)]^2\,dx + \int_a^b [g(x)]^2\,dx}{2}.$$

6 (See Example 3.) Give an example of a region R for which $\bar{y}$ is half the average value of $f(x)$.

7 (See Example 3.) Compare $\bar{y}$ and the average of $f(x)$ in the case $f(x) = x$, $[a, b] = [0, 1]$.

8 Assume that you drive 30 miles per hour for 1 hour and then 50 miles per hour for 1 hour.
(*a*) Graph your speed as a function of time.
(*b*) What is your average speed as a function of time?
(*c*) During this trip you went 80 miles. Graph your speed as a function of distance.
(*d*) What is your average speed as a function of distance?
(*e*) Compare (*b*) and (*d*). Is this result to be expected? (Exercise 14 generalizes this exercise.)

By the formula for the sum of a geometric progression, $1/(1+t) = 1 - t + t^2 - t^3 + t^4/(1+t)$ for $t \neq 1$. (This can be checked by multiplying both sides by $1 + t$.) This formula is needed in Exercises 9 and 10.

9 (*a*) Using the identity given above, show that for $x > -1$,

$$\ln(1+x) = x - \frac{x^2}{2} + \frac{x^3}{3} - \frac{x^4}{4} + \int_0^x \frac{t^4}{1+t}\,dt.$$

(*b*) Show that for $0 \leq x < 1$,

$$0 \leq \int_0^x \frac{t^4}{1+t}\,dt \leq \frac{x^5}{5}.$$

(*c*) Show that if $x - \frac{x^2}{2} + \frac{x^3}{3} - \frac{x^4}{4}$ is used as an estimate of $\ln(1+x)$ for $0 \leq x \leq \frac{1}{2}$, the error is less than $\frac{1}{160}$.
(*d*) Use the formula in (*c*) to estimate ln 1.2. What does (*b*) tell you about the error?

10 (*a*) Using the identity preceding Exercise 9, with t replaced throughout by t^2, show that for $x \geq 0$,

$$\tan^{-1} x = x - \frac{x^3}{3} + \frac{x^5}{5} - \frac{x^7}{7} + \int_0^x \frac{t^8}{1+t^2}\,dt.$$

(*b*) Show that for $0 \leq x \leq 1$,

$$0 \le \int_0^x \frac{t^8}{1+t^2}\,dt \le \frac{x^9}{9}.$$

(c) Show that if $x - \frac{x^3}{3} + \frac{x^5}{5} - \frac{x^7}{7}$ is used as an estimate of $\tan^{-1} x$ for $0 \le x \le \frac{1}{2}$, the error is less than $\frac{1}{4608}$.

(d) Use the formula in (c) to estimate $\tan^{-1} 0.2$. What does (b) tell you about the error?

In Exercises 11 and 12 find a value of c described in Corollary 1.

11 $f(x) = x^2$, $g(x) = \cos x$, $[a, b] = [0, \pi/2]$.

12 $f(x) = e^{2x}$, $g(x) = \sin^2 x$, $[a, b] = [0, \pi/2]$.

■

13 This exercise obtains the Schwarz inequality:

$$\left|\int_A^B f(x)g(x)\,dx\right| \le \left\{\int_A^B [f(x)]^2\,dx\right\}^{1/2}\left\{\int_A^B [g(x)]^2\,dx\right\}^{1/2}.$$

(a) Prove that if the equation $at^2 + bt + c = 0$ has at most one (real) root, then $b^2 - 4ac \le 0$. (When does it have exactly one root?) Assume that $a \ne 0$.

(b) Let f and g be continuous functions, neither of which is the constant function 0 on $[A, B]$. Define a third function h as follows:

$$h(t) = \int_A^B [tf(x) - g(x)]^2\,dx.$$

Show that h is a quadratic polynomial in t, $h(t) = at^2 + bt + c$ for suitable constants, $a, b, c, a \ne 0$.

(c) Express the coefficients a, b and c of h in terms of f and g.

(d) Combining (a), (b), and (c), derive the Schwarz inequality.

(e) When does equality occur in the Schwarz inequality?

14 Imagine computing the average speed of a race car during a race. There are two ways to do this. In one approach, you record its speed at equally spaced instants of time, say, every 5 seconds. In the other approach, you record its speed at equally spaced observation posts, say, every 100 feet. In either approach, you then average the observed speeds. The first case produces the average with respect to time. The second case produces the average with respect to distance. Would you expect the two averages to be equal? If not, which would be larger? In this exercise it is shown that the average of velocity with respect to time is less than or equal to the average of velocity with respect to distance. (See Exercise 8.)

Assume for convenience that the velocity $v = dx/dt$ is positive, hence that speed equals velocity. Assume that from time a to time b the object moves from x_1 to x_2 on the x axis. Let v as a function of time be given by $v = f(t)$; let v as a function of distance be given by $v = g(x)$.

(a) Show that the average of velocity with respect to distance equals

$$\frac{\int_{x_1}^{x_2} g(x)\,dx}{\int_a^b f(t)\,dt}.$$

(b) Show that $\int_{x_1}^{x_2} g(x)\,dx = \int_a^b [f(t)]^2\,dt$.

(c) Using the Schwarz inequality, show that the average of velocity with respect to time is less than or equal to the average of velocity with respect to distance.

15 Let f be a continuous function such that $f(x)$ is always positive. Prove that

$$\left[\int_a^b f(x)\,dx\right]\left[\int_a^b \frac{1}{f(x)}\,dx\right] \ge (b - a)^2.$$

Hint: Use the Schwarz inequality.

16 Let $a_1, \ldots, a_n$ and $b_1, \ldots, b_n$ be $2n$ positive real numbers. Prove that

$$\sum_{i=1}^n a_i b_i \le \left(\sum_{i=1}^n a_i^2\right)^{1/2}\left(\sum_{i=1}^n b_i^2\right)^{1/2}.$$

Hint: Consider $h(t) = \sum_{i=1}^n (ta_i - b_i)^2$ and proceed as in Exercise 13, steps (b) to (e).

17 In the first t seconds a falling body drops $16t^2$ feet. Let its position relative to the y axis at time t be $16t^2$. (We aim the positive part of the axis downward.)

(a) Compute velocity as a function of time. (Note that it is positive.)

(b) Compute velocity as a function of y.

(c) Find the average of velocity with respect to time during the first t seconds.

(d) Find the average of velocity with respect to distance during the first t seconds.

18 Find the average length of the vertical cross section AB of the quadrant of a circle of radius r pictured in Fig. 8.92

(a) As a function of x.

(b) As a function of θ.

(c) Why would you expect the answer to (b) to be less than the answer to (a)?

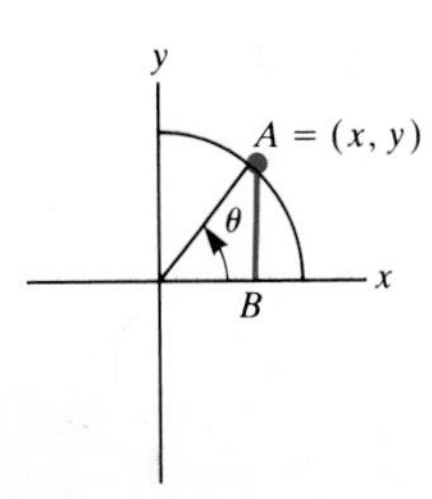

Figure 8.92

19 A bus dispatcher dispatches $n + 1$ buses at time intervals of $a_1, a_2, a_3, \ldots, a_n$ minutes. So the average time interval between buses *from the dispatcher's point of view* is $\Sigma_{i=1}^n a_i/n$. A person who arrives at random to wait at a bus stop has a different view of the average time between buses. (Assume that the distance between buses remains fixed.) The likelihood that this person arrives during the ith time interval is $a_i/\Sigma_{i=1}^n a_i$, so he notices a long interval more than a short interval between buses. The average time between buses *from the rider's point of view* is $(\Sigma_{i=1}^n a_i^2)/(\Sigma_{i=1}^n a_i)$. (His average waiting time would be half this quantity.)

(*a*) Compute the two averages in the case $a_1 = 30$ minutes, $a_2 = 20$ minutes, $a_3 = 10$ minutes. (See Fig. 8.93.)

(*b*) Using Exercise 16, show that the rider's average is at least as large as the dispatcher's average.

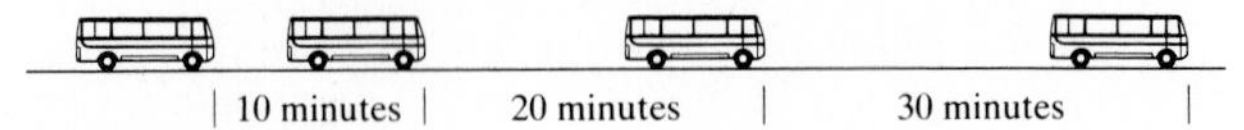

Figure 8.93

8.7 IMPROPER INTEGRALS

Consider the volume of the solid obtained by revolving about the x axis the region bordered by $y = 1/x$ and the x axis, to the right of $x = 1$, as shown in Fig. 8.94. The typical cross section made by a plane perpendicular to the x axis is a circle of radius $1/x$. We might therefore be tempted to say that the volume is $\int_1^\infty \pi(1/x)^2\,dx$. Unfortunately, the symbol $\int_a^\infty f(x)\,dx$ has not been given any meaning so far in this book. The definition of the definite integral involves sums of the form $\Sigma_{i=1}^n f(c_i)(x_i - x_{i-1})$. If a section in the partition has infinite length, such a sum is meaningless.

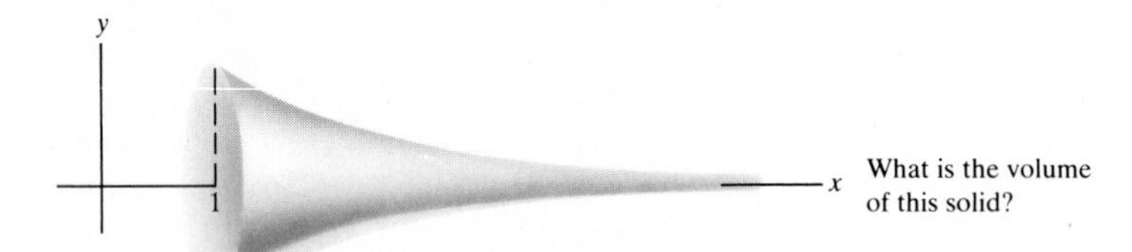

Figure 8.94

It does make sense, however, to examine the volume of that part of the solid from $x = 1$ to $x = b$, where b is some number greater than 1, and then to determine what happens to this volume as $b \to \infty$. In other words, consider $\lim_{b\to\infty} \int_1^b \pi(1/x)^2\,dx$. Now,

$$\int_1^b \pi\left(\frac{1}{x}\right)^2 dx \underset{\text{FTC}}{=} -\frac{\pi}{x}\bigg|_1^b = -\frac{\pi}{b} - \left(-\frac{\pi}{1}\right) = \pi - \frac{\pi}{b}.$$

Thus $\lim_{b\to\infty} \int_1^b \pi(1/x)^2\,dx = \pi - 0 = \pi$. The volume of the endless solid is finite.

This approach suggests a way to give meaning to the symbol $\int_a^\infty f(x)\,dx$.

Improper Integrals: Interval of Integration Unbounded

Definition *Convergent improper integral* $\int_a^\infty f(x)\,dx$. Let f be continuous for $x \geq a$. If $\lim_{b\to\infty} \int_a^b f(x)\,dx$ exists, the function f is said to have a

convergent improper integral from a to ∞. The value of the limit is denoted by $\int_a^\infty f(x)\,dx$.

It was shown above that $\int_1^\infty \pi(1/x)^2\,dx$ is a convergent improper integral, with value π.

Definition *Divergent improper integral* $\int_a^\infty f(x)\,dx$. Let f be a continuous function. If $\lim_{b\to\infty} \int_a^b f(x)\,dx$ does not exist, the function f is said to have a **divergent improper integral** from a to ∞.

EXAMPLE 1 Determine the area of the region below $y = 1/x$, above the x axis, and to the right of $x = 1$.

SOLUTION The area in question is given by

$$\int_1^\infty \frac{1}{x}\,dx = \lim_{b\to\infty} \int_1^b \frac{1}{x}\,dx = \lim_{b\to\infty} (\ln b - \ln 1) = \lim_{b\to\infty} \ln b.$$

Divergence due to integral approaching ∞

Since $\lim_{b\to\infty} \ln b = \infty$, the area is infinite: $\int_1^\infty 1/x\,dx = \infty$; $\int_1^\infty 1/x\,dx$ is a divergent improper integral. ■

The improper integral $\int_1^\infty dx/x$ is divergent because $\int_1^b dx/x \to \infty$ as $b \to \infty$. But an improper integral $\int_a^\infty f(x)\,dx$ can be divergent without being infinite. Consider, for instance, $\int_0^\infty \cos x\,dx$. We have

$$\int_0^b \cos x\,dx = \sin x \Big|_0^b = \sin b.$$

Divergence due to integral oscillating

As $b \to \infty$, $\sin b$ does not approach a limit, nor does it become arbitrarily large. As $b \to \infty$, $\sin b$ just keeps going up and down in the range -1 to 1 infinitely often.

The improper integral $\int_{-\infty}^b f(x)\,dx$ is defined similarly by considering

$$\int_a^b f(x)\,dx$$

for negative values of a of large absolute value. If

The improper integral $\int_{-\infty}^b f(x)\,dx$

$$\lim_{a\to-\infty} \int_a^b f(x)\,dx$$

exists, it is denoted

$$\int_{-\infty}^b f(x)\,dx.$$

In such a case, the improper integral $\int_{-\infty}^b f(x)\,dx$ is said to be convergent. If

$$\lim_{a\to-\infty} \int_a^b f(x)\,dx$$

does not exist, then the improper integral

$$\int_{-\infty}^b f(x)\,dx$$

is said to be divergent.

The improper integral $\int_{-\infty}^{\infty} f(x)\,dx$

To deal with improper integrals over the entire x axis, define

$$\int_{-\infty}^{\infty} f(x)\,dx$$

to be the sum
$$\int_{-\infty}^{0} f(x)\,dx + \int_{0}^{\infty} f(x)\,dx,$$

which will be called **convergent** if both

$$\int_{-\infty}^{0} f(x)\,dx \quad \text{and} \quad \int_{0}^{\infty} f(x)\,dx$$

are convergent. [If at least one of the two is divergent, $\int_{-\infty}^{\infty} f(x)\,dx$ will be called **divergent**.]

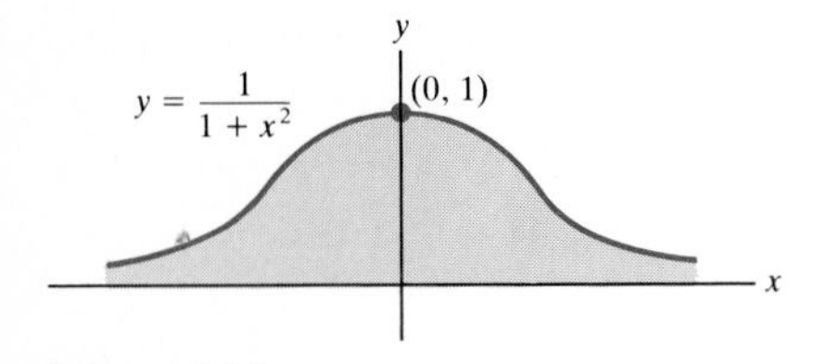

Figure 8.95

EXAMPLE 2 Determine the area of the region bounded by the curve $y = 1/(1 + x^2)$ and the x axis, as indicated in Fig. 8.95.

SOLUTION The area in question equals $\int_{-\infty}^{\infty} \frac{dx}{1 + x^2}$.

Now,
$$\int_{0}^{\infty} \frac{dx}{1 + x^2} = \lim_{b\to\infty} \int_{0}^{b} \frac{dx}{1 + x^2}$$

$$\underset{\text{FTC}}{=} \lim_{b\to\infty} (\tan^{-1} b - \tan^{-1} 0) = \frac{\pi}{2}.$$

By symmetry,
$$\int_{-\infty}^{0} \frac{dx}{1 + x^2} = \frac{\pi}{2}.$$

Hence
$$\int_{-\infty}^{\infty} \frac{dx}{1 + x^2} = \frac{\pi}{2} + \frac{\pi}{2},$$

and the area in question is π. ■

Sometimes $\int_{a}^{\infty} f(x)\,dx$ can be shown to be convergent by comparing it to another improper integral $\int_{a}^{\infty} g(x)\,dx$. This fact rests on the following principle:

Let $h(x)$ be defined for all $x \geq a$ and have the property that $x_1 > x_2$ implies $h(x_1) \geq h(x_2)$. Assume that there is a number B such that $h(x) \leq B$ for all $x \geq a$. Then as $x \to \infty$, $h(x)$ approaches a limit L, and $L \leq B$. [We will assume this result, which is a fundamental property of the real number system. Figure 8.96 suggests that it is plausible. The graph of the function $y = h(x)$ stays below the line $y = B$ and must have a horizontal asymptote.]

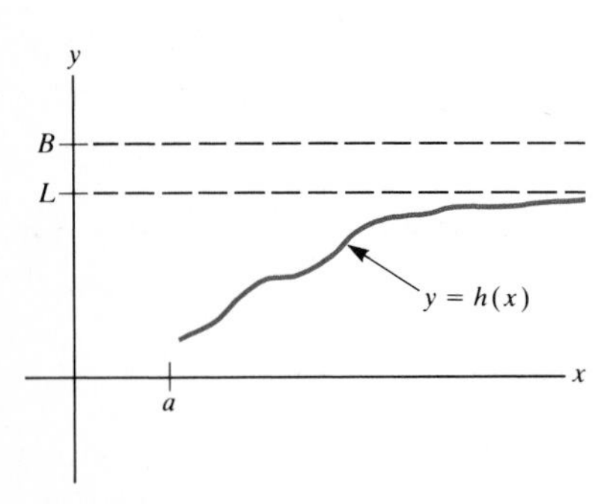

Figure 8.96

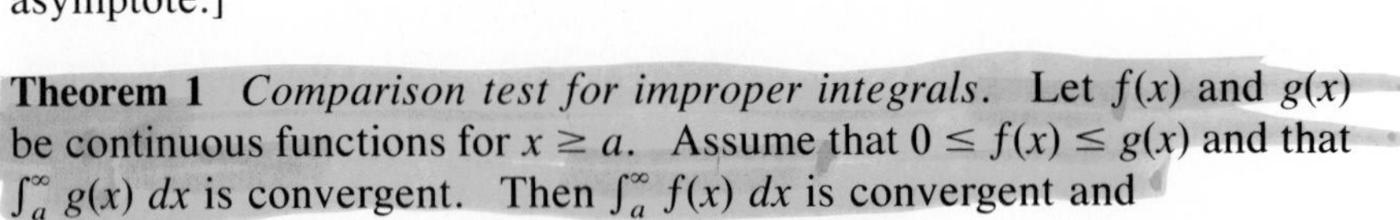

Theorem 1 *Comparison test for improper integrals.* Let $f(x)$ and $g(x)$ be continuous functions for $x \geq a$. Assume that $0 \leq f(x) \leq g(x)$ and that $\int_{a}^{\infty} g(x)\,dx$ is convergent. Then $\int_{a}^{\infty} f(x)\,dx$ is convergent and

$$\int_{a}^{\infty} f(x)\,dx \leq \int_{a}^{\infty} g(x)\,dx.$$

Proof Let $h(b) = \int_{a}^{b} f(x)\,dx$ for $b \geq a$. Since $f(x) \geq 0$, it follows that $x_1 > x_2$ implies $h(x_1) \geq h(x_2)$. Moreover,

$$h(b) = \int_{a}^{b} f(x)\,dx \leq \int_{a}^{b} g(x)\,dx \leq \int_{a}^{\infty} g(x)\,dx.$$

Since $h(b)$ never exceeds $B = \int_a^\infty g(x)\,dx$, it follows that $\lim_{b\to\infty} h(b)$ exists and is not larger than $\int_a^\infty g(x)\,dx$. Thus $\int_a^\infty f(x)\,dx$ exists, and

$$\int_a^\infty f(x)\,dx \le \int_a^\infty g(x)\,dx. \quad ■$$

EXAMPLE 3 The improper integral $\int_0^\infty e^{-x^2}\,dx$ arises in statistics. Show that it is convergent and put a bound on it.

SOLUTION Since e^{-x^2} does not have an elementary antiderivative, we cannot evaluate $\int_0^b e^{-x^2}\,dx$ and use the result to determine the behavior of $\int_0^b e^{-x^2}\,dx$ as $b \to \infty$.

However, we can compare $\int_0^\infty e^{-x^2}\,dx$ to an improper integral that we know converges.

For $x \ge 1$, $x^2 \ge x$; hence $e^{-x^2} \le e^{-x}$. Now,

$$\int_1^b e^{-x}\,dx = -e^{-x}\Big|_1^b = e^{-1} - e^{-b}.$$

Thus
$$\lim_{b\to\infty} \int_1^b e^{-x}\,dx = \frac{1}{e},$$

and $\int_1^\infty e^{-x}\,dx$ is convergent.

Since $0 < e^{-x^2} \le e^{-x}$ for $x \ge 1$, the comparison test tells us that $\int_1^\infty e^{-x^2}\,dx$ is convergent. Furthermore,

$$\int_1^\infty e^{-x^2}\,dx \le \int_1^\infty e^{-x}\,dx = \frac{1}{e}.$$

Thus
$$\int_0^\infty e^{-x^2}\,dx \le \int_0^1 e^{-x^2}\,dx + \frac{1}{e}.$$

Since $e^{-x^2} \le 1$ for $0 < x \le 1$, we conclude that

$$\int_0^\infty e^{-x^2}\,dx \le 1 + \frac{1}{e}. \quad ■$$

Theorem 1 does not cover integrands that assume both negative and positive values, such as $e^{-x^2} \sin x$. As we will see in Example 4, $\int_0^\infty e^{-x^2} \sin x\,dx$ is convergent. To show that it is, we will need the following theorem.

Theorem 2 Assume that $f(x)$ is continuous for $x \ge a$, and assume that $\int_a^\infty |f(x)|\,dx$ is convergent. Then $\int_a^\infty f(x)\,dx$ is convergent.

Proof Write $f(x) = [f(x) + |f(x)|] - |f(x)|$. Note that

$$0 \le f(x) + |f(x)| \le 2|f(x)|.$$

By Theorem 1,
$$\int_a^\infty [f(x) + |f(x)|]\,dx$$

is convergent. Thus

$$\lim_{b\to\infty} \int_a^b f(x)\,dx = \lim_{b\to\infty} \left\{ \int_a^b [f(x) + |f(x)|]\,dx - \int_a^b |f(x)|\,dx \right\}$$

$$= \int_a^\infty [f(x) + |f(x)|]\,dx - \int_a^\infty |f(x)|\,dx.$$

Hence $\int_a^\infty f(x)\,dx$ is convergent. ■

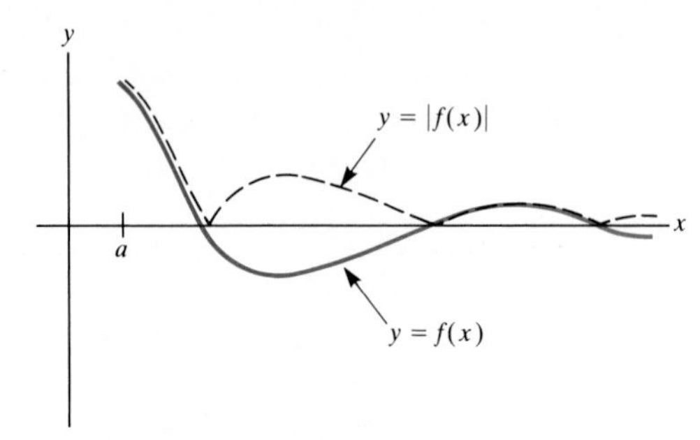

Figure 8.97

A glance at Fig. 8.97 suggests that Theorem 2 is to be expected. If the area under $y = |f(x)|$ is finite, then the area between $y = f(x)$ and the x axis should be finite as well, since areas dipping below the x axis are recorded as negative and cancel out some of the area above the x axis. (See discussion in Sec. 5.4.)

EXAMPLE 4 Show that $\int_0^\infty e^{-x^2} \sin x\,dx$ is convergent.

SOLUTION Since $|\sin x| \le 1$, $|e^{-x^2} \sin x| \le e^{-x^2}$. Since $\int_0^\infty e^{-x^2}\,dx$ converges (by Example 3), the comparison test tells us that $\int_0^\infty |e^{-x^2} \sin x|\,dx$ converges. Theorem 2 then implies that $\int_0^\infty e^{-x^2} \sin x\,dx$ converges. ■

Improper Integrals: Integrand Unbounded

There is a second type of improper integral, in which the function is unbounded in the interval $[a, b]$. If $f(x)$ becomes arbitrarily large in the interval $[a, b]$, then it is possible to have arbitrarily large approximating sums $\Sigma_{i=1}^n f(c_i)(x_i - x_{i-1})$ no matter how fine the partition may be by choosing a c_i that makes $f(c_i)$ large. The next example shows how to get around this difficulty.

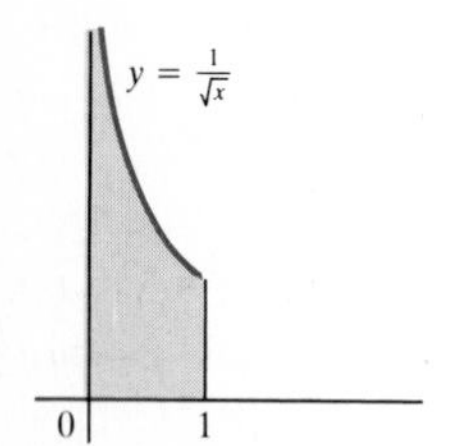

Figure 8.98

EXAMPLE 5 Determine the area of the region bounded by $y = 1/\sqrt{x}$, $x = 1$, and the coordinate axes shown in Fig. 8.98.

SOLUTION Resist for the moment the temptation to write "Area = $\int_0^1 1/\sqrt{x}\,dx$," for $\int_0^1 1/\sqrt{x}\,dx$ does not exist according to the definition of the definite integral given in Chap. 5, since its integrand is unbounded in $[0, 1]$. (Note also that the integrand is not defined at 0.) Instead, consider the behavior of $\int_t^1 1/\sqrt{x}\,dx$ as t approaches 0 from the right. Since

$$\int_t^1 \frac{1}{\sqrt{x}}\,dx = 2\sqrt{x}\,\Big|_t^1 = 2\sqrt{1} - 2\sqrt{t} = 2(1 - \sqrt{t}),$$

it follows that

$$\lim_{t \to 0^+} \int_t^1 \frac{dx}{\sqrt{x}} = 2.$$

The area in question is 2.

The reader should check and see that this is the same value for the area that can be obtained by taking horizontal cross sections and evaluating an improper integral from 0 to ∞. ■

The reasoning in Example 5 motivates the definition of the second type of improper integral, in which the function rather than the interval is unbounded.

Definition *Convergent and divergent improper integrals* $\int_a^b f(x)\,dx$. Let f be continuous at every number in $[a, b]$ except a. If $\lim_{t\to a^+} \int_t^b f(x)\,dx$ exists, the function f is said to have a **convergent improper integral** from a to b. The value of the limit is denoted $\int_a^b f(x)\,dx$. If $\lim_{t\to a^+} \int_t^b f(x)\,dx$ does not exist, the function f is said to have a **divergent improper integral** from a to b; in brief, $\int_a^b f(x)\,dx$ does not exist.

In a similar manner, if f is not defined at b, define $\int_a^b f(x)\,dx$ as $\lim_{t\to b^-} \int_a^t f(x)\,dx$, if this limit exists.

Example 5 is summarized in the statement, "The improper integral $\int_0^1 1/\sqrt{x}\,dx$ is convergent and has the value 2."

More generally, if a function $f(x)$ is not defined at certain isolated numbers, break the domain of $f(x)$ into intervals $[a, b]$ for which $\int_a^b f(x)\,dx$ is either improper or "proper"—that is, an ordinary definite integral.

For instance, the improper integral $\int_{-\infty}^{\infty} 1/x^2\,dx$ is troublesome for four reasons: $\lim_{x\to 0^-} 1/x^2 = \infty$, $\lim_{x\to 0^+} 1/x^2 = \infty$, and the range extends infinitely to the left and also to the right. To treat the integral, write it as the sum of four improper integrals of the two basic types:

$$\int_{-\infty}^{\infty} \frac{1}{x^2}\,dx = \int_{-\infty}^{-1} \frac{1}{x^2}\,dx + \int_{-1}^{0} \frac{1}{x^2}\,dx + \int_{0}^{1} \frac{1}{x^2}\,dx + \int_{1}^{\infty} \frac{1}{x^2}\,dx.$$

All four of the integrals on the right have to be convergent for $\int_{-\infty}^{\infty} 1/x^2\,dx$ to be convergent. As a matter of fact, only the first and last are, so $\int_{-\infty}^{\infty} 1/x^2\,dx$ is divergent.

Just as substitution in a definite integral is valid as long as the same substitution is applied to the limits of integration, substitution in improper integrals is also permissible, as illustrated in Example 6.

EXAMPLE 6 Evaluate $\displaystyle\int_0^{\infty} \frac{dx}{x^2 + 9}$.

SOLUTION Make the substitution $x = 3u$; hence $dx = 3\,du$. As x goes from 0 to ∞, $u = x/3$ also goes from 0 to ∞. Thus

$$\int_0^{\infty} \frac{dx}{x^2 + 9} = \int_0^{\infty} \frac{3\,du}{9u^2 + 9} = \frac{1}{3}\int_0^{\infty} \frac{du}{u^2 + 1}.$$

By Example 2,

$$\int_0^{\infty} \frac{du}{u^2 + 1} = \frac{\pi}{2}.$$

Thus

$$\int_0^{\infty} \frac{dx}{x^2 + 9} = \frac{1}{3}\left(\frac{\pi}{2}\right) = \frac{\pi}{6}. \quad ■$$

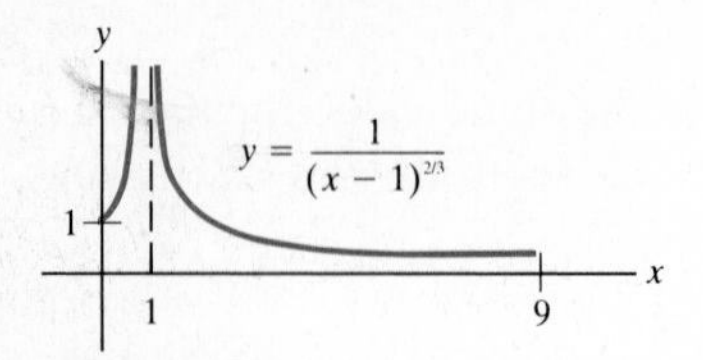

Figure 8.99

EXAMPLE 7 Examine the improper integral $\int_0^9 \frac{dx}{(x-1)^{2/3}}$. (See Fig. 8.99.)

SOLUTION At $x = 1$ the integrand is undefined, and near $x = 1$ it is unbounded. It is necessary to examine the two improper integrals

$$\int_0^1 \frac{dx}{(x-1)^{2/3}} \quad \text{and} \quad \int_1^9 \frac{dx}{(x-1)^{2/3}}.$$

To treat $\int_0^1 \frac{dx}{(x-1)^{2/3}}$, consider $\lim_{t\to 1^-} \int_0^t \frac{dx}{(x-1)^{2/3}}$:

$$\int_0^t \frac{dx}{(x-1)^{2/3}} = 3(x-1)^{1/3}\Big|_0^t$$

$$= 3(t-1)^{1/3} - 3(0-1)^{1/3}.$$

Thus
$$\lim_{t\to 1^-} \int_0^t \frac{dx}{(x-1)^{2/3}} = 0 + 3 = 3.$$

Hence
$$\int_0^1 \frac{dx}{(x-1)^{2/3}} = 3.$$

Next consider $\lim_{t\to 1^+} \int_t^9 \frac{dx}{(x-1)^{2/3}}$:

$$\int_t^9 \frac{dx}{(x-1)^{2/3}} = 3(x-1)^{1/3}\Big|_t^9$$

$$= 3(9-1)^{1/3} - 3(t-1)^{1/3}.$$

Thus
$$\lim_{t\to 1^+} \int_t^9 \frac{dx}{(x-1)^{2/3}} = 3 \cdot 8^{1/3} = 6.$$

Hence $\int_0^9 \frac{dx}{(x-1)^{2/3}}$ is convergent and equals $3 + 6 = 9$. ■

An Improper Integral in Economics The final example illustrates the use of improper integrals in economics.

EXAMPLE 8 (*Present value of future income.*) Both business and government frequently face the question: "What is 1 dollar t years in the future worth today?" Implicit in this question are such considerations as the present value of a business being dependent on its future profit and the cost of a dam being weighed against its future revenue. Determine the present value of a business whose rate of profit t years in the future is $f(t)$ dollars per year.

SOLUTION To begin the analysis, assume that the annual interest rate r remains constant and that 1 dollar deposited today is worth e^{rt} dollars t years from now. This assumption corresponds to continuously compounded interest or to natural growth. Thus A dollars today will be worth Ae^{rt} dollars t years from now. What is the present value of the promise of 1 dollar t years from now? In other words, what amount A invested today will be worth 1 dollar t years from now? To find out, solve the equation $Ae^{rt} = 1$ for A.

The solution is $$A = e^{-rt}. \tag{1}$$

The present value of \$1 t years from now is \$$e^{-rt}$.

Now consider the present value of the future profit of a business (or future revenue of a dam). Assume that the profit flow t years from now is $f(t)$ dollars per year. This rate may vary within the year; consider f to be a continuous function of time. The profit in the small interval of time dt, from time t to time $t + dt$, would be approximately $f(t)\,dt$. The total future profit $F(T)$ from now, when $t = 0$, to some time T in the future is therefore

$$F(T) = \int_0^T f(t)\,dt. \tag{2}$$

But the **present value** of the future profit is *not* given by Eq. (2). It is necessary to consider the present value of the profit earned in a typical short interval of time from t to $t + dt$. According to Eq. (1), its present value is approximately

$$e^{-rt}f(t)\,dt.$$

Hence the present value of future profit from $t = 0$ to $t = T$ is given by

$$\int_0^T e^{-rt}f(t)\,dt. \tag{3}$$

The present value of all future profit is therefore the improper integral $\int_0^\infty e^{-rt}f(t)\,dt$.

To see what influence the interest rate r has, denote by $P(r)$ the present value of all future revenue when the interest rate is r; that is,

$$P(r) = \int_0^\infty e^{-rt}f(t)\,dt. \tag{4}$$

If the interest rate r is raised, then according to Eq. (4) the present value of a business declines. An investor choosing between investing in a business or placing her money in a bank account finds the bank account more attractive when r is raised.

Laplace transform

Equation (4) assigns to a profit function f (which is a function of time t) a present-value function P, which is a function of r, the interest rate. In the theory of differential equations, P is called the **Laplace transform of f** (See Exercises 61 to 67.) ■

EXERCISES FOR SEC. 8.7: IMPROPER INTEGRALS

Exercises 1 to 18 concern improper integrals in which the range of integration is unbounded. In each case determine whether the improper integral is convergent or divergent. Evaluate the convergent ones if possible.

1 $\int_1^\infty \frac{dx}{x^3}$ 2 $\int_1^\infty \frac{dx}{\sqrt[3]{x}}$

3 $\int_1^\infty \frac{\ln x\,dx}{x}$ 4 $\int_0^\infty e^{-x}\,dx$

5 $\int_0^\infty \frac{dx}{x^2 + 4}$ 6 $\int_0^\infty \frac{dx}{x + 100}$

7 $\int_0^\infty \frac{x^3\,dx}{x^4 + 1}$ 8 $\int_{-\infty}^\infty \frac{x\,dx}{x^4 + 1}$

9 $\int_1^\infty x^{-1.01}\,dx$

10 $\int_1^\infty x^{-0.99}\,dx$

11 $\int_0^\infty \frac{dx}{(x+2)^3}$

12 $\int_2^\infty \frac{dx}{x^2-1}$

13 $\int_0^\infty \sin 2x\,dx$

14 $\int_0^\infty \sin^2 x\,dx$

15 $\int_0^\infty \frac{dx}{\sqrt{1+x^3}}$

16 $\int_1^\infty xe^{-x}\,dx$

17 $\int_0^\infty \frac{\sin x}{x^2}\,dx$

18 $\int_0^\infty e^{-x}\sin x^2\,dx$

In Exercises 19 to 24 decide whether the integrals are improper or proper.

19 $\int_0^2 \frac{dx}{x^2-1}$

20 $\int_2^3 \frac{dx}{x^2+1}$

21 $\int_0^1 \frac{\cos x}{x^2}\,dx$

22 $\int_0^\pi x\tan x\,dx$

23 $\int_0^2 \frac{dx}{\sqrt{4-x^2}}$

24 $\int_0^2 \frac{dx}{e-e^x}$

Exercises 25 to 32 concern improper integrals in which the integrand is unbounded. In each case determine whether the improper integral is convergent or divergent. Evaluate the convergent ones if possible.

25 $\int_0^1 x^{-1.01}\,dx$

26 $\int_0^1 x^{-0.99}\,dx$

27 $\int_0^1 \frac{dx}{\sqrt{1-x}}$

28 $\int_{-1}^1 \frac{dx}{\sqrt[3]{x}}$

29 $\int_0^1 \ln x\,dx$

30 $\int_0^{\pi/2} \cot x\,dx$

31 $\int_1^2 \frac{dx}{x\sqrt{x^2-1}}$

32 $\int_7^{16} \frac{dx}{\sqrt[3]{x-8}}$

■

33 (*a*) Show that $\int_1^\infty \frac{\cos x}{x^2}\,dx$ is convergent.

(*b*) Show that $\int_1^\infty \frac{\sin x}{x}\,dx$ is convergent. (*Hint:* Start with integration by parts.)

(*c*) Show that $\int_0^\infty \frac{\sin x}{x}\,dx$ is convergent.

(*d*) Show that $\int_0^\infty \sin e^x\,dx$ is convergent.

34 In R. P. Feynman, *Lectures on Physics*, Addison-Wesley, Reading, Mass., 1963, appears this remark: ". . . the expression becomes

$$\frac{U}{V} = \frac{(kT)^4}{\hbar^3\pi^2c^3}\int_0^\infty \frac{x^3\,dx}{e^x-1}.$$

This integral is just some number that we can get, approximately, by drawing a curve and taking the area by counting squares. It is roughly 6.5. The mathematicians among us can show that the integral is exactly $\pi^4/15$."

Show at least that the integral is convergent.

35 The function $f(x) = (\sin x)/x$ for $x \neq 0$ and $f(0) = 1$ occurs in communication theory. Show that the energy E of the signal represented by f is finite, where

$$E = \int_{-\infty}^\infty [f(x)]^2\,dx.$$

36 Plankton are small football-shaped organisms. The resistance they meet when falling through water is proportional to the integral

$$\int_0^\infty \frac{dx}{\sqrt{(a^2+x)(b^2+x)(c^2+x)}},$$

where a, b, and c describe the dimensions of the plankton. Is this improper integral convergent or divergent?

37 The following is an excerpt from an article on corporate investment:

> It follows, therefore, that the present value of the incremental stream of benefits resulting from the marginal investment can be described as follows if the maintenance of the stockholder's optimum share value is not to be disturbed at the margin:

$$1 = \int_0^\infty (1-b)r'e^{brt}e^{-kt}\,dt, \qquad (i)$$

or

$$1 = \frac{r'(1-b)}{k-rb}. \qquad (ii)$$

(The constants b, k, r, and r' are positive, and $br < k$.) Derive (*ii*) from (*i*).

38 Find the error in the following computations: The substitution $x = y^2$, $dx = 2y\,dy$, yields

$$\int_0^1 \frac{1}{x}\,dx = \int_0^1 \frac{2y}{y^2}\,dy = \int_0^1 \frac{2}{y}\,dy$$
$$= 2\int_0^1 \frac{1}{y}\,dy = 2\int_0^1 \frac{1}{x}\,dx.$$

Hence

$$\int_0^1 \frac{1}{x}\,dx = 2\int_0^1 \frac{1}{x}\,dx,$$

from which it follows that $\int_0^1 dx/x = 0$.

39 Find the error in the following computations: Using the substitution $u = 1/x$, $du = -1/x^2\,dx$, we have

$$\int_{-1}^1 \frac{1}{1+x^2}\,dx = \int_{-1}^1 \frac{1}{1+1/u^2}\left(-\frac{1}{u^2}\,du\right)$$
$$= -\int_{-1}^1 \frac{1}{1+u^2}\,du.$$

Thus $\int_{-1}^{1} 1/(1 + x^2)\, dx$, being equal to its negative, is 0.

40 Let $f(x)$ and $g(x)$ be continuous functions defined on $[a, \infty)$. Show that if $\int_a^\infty [f(x)]^2\, dx$ and $\int_a^\infty [g(x)]^2\, dx$ are convergent, then so is $\int_a^\infty f(x)g(x)\, dx$.

41 The following quote is taken from an article on the energy stored in solar ponds:

> The effect of the free surface of the pond on the temperature at the bottom of the pond is given by an expression involving the integral
>
> $$\int_a^\infty \left(1 - \frac{k}{v^2}\right) e^{-v^2}\, dv.$$

Show that the integral is convergent. (The constants a and k are positive.)

42 (*The gamma function*) For a real number $n > 0$, define $\Gamma(n)$ to be $\int_0^\infty e^{-x}x^{n-1}\, dx$.
(*a*) Evaluate $\Gamma(1)$.
(*b*) Show that $\Gamma(n + 1) = n\Gamma(n)$.
(*c*) Using (*a*) and (*b*), evaluate $\Gamma(2)$, $\Gamma(3)$, $\Gamma(4)$, and $\Gamma(5)$.
(*d*) What is the relationship between $n!$ and $\Gamma(n)$?
The gamma function generalizes the factorial, which is defined only at integers, to all positive real numbers. Handbooks of mathematical tables usually include values of the gamma function.

In Exercises 43 to 53 determine whether the improper integrals are convergent or divergent. Evaluate the convergent ones.

43 $\int_0^\infty e^{-x} \sin 3x\, dx$ **44** $\int_0^\infty x^{-3}\, dx$

45 $\displaystyle\int_0^\infty \frac{dx}{\sqrt[3]{x - 2}}$ **46** $\displaystyle\int_1^\infty \frac{dx}{\sqrt{x - 1}}$

47 $\displaystyle\int_0^1 \frac{e^x\, dx}{e^x - 1}$ **48** $\displaystyle\int_0^\infty x \ln x\, dx$

49 $\displaystyle\int_0^2 \frac{dx}{x^2 - 1}$ **50** $\displaystyle\int_0^\infty \tan^{-1} x\, dx$

51 $\displaystyle\int_0^2 \frac{dx}{(x - 1)^2}$ **52** $\displaystyle\int_{-\infty}^\infty \frac{dx}{x^2 + 2x + 1}$

53 $\displaystyle\int_{-\infty}^\infty \frac{dx}{x^2 + 2x - 2}$

54 Let R be the region between the curves $y = 1/x$ and $y = 1/(x + 1)$ to the right of the line $x = 1$. Is the area of R finite or infinite? If it is finite, evaluate it.

55 Let R be the region between the curves $y = 1/x$ and $y = 1/x^2$ to the right of $x = 1$. Is the area of R finite or infinite? If it is finite, evaluate it.

56 Find the area of the region bounded by the curve $y = 1/(x^2 + 6x + 10)$ and the x axis, to the right of the y axis.

57 Show that $\int_1^\infty x/\sqrt{1 + x^5}\, dx$ is convergent.

58 Show that $\int_0^\infty e^{-t}t^{-1/2}\, dt$ is convergent.

59 If the profit flow in Example 8 remains constant, say, $f(t) = k > 0$, the total future profit is obviously infinite. Show that the present value is k/r, which is finite.

60 Show that if $\int_{-\infty}^\infty f(x)\, dx$ is convergent, it equals $\lim_{L\to\infty} \int_{-L}^{L} f(x)\, dx$, but that $\lim_{L\to\infty} \int_{-L}^{L} f(x)\, dx$ may be finite while $\int_{-\infty}^\infty f(x)\, dx$ is a divergent improper integral.

Let $f(t)$ be a continuous function defined for $t \geq 0$. Assume that, for certain fixed positive numbers r, $\int_0^\infty e^{-rt}f(t)\, dt$ converges and that $e^{-rt}f(t) \to 0$ as $t \to \infty$. Define $P(r)$ to be $\int_0^\infty e^{-rt}f(t)\, dt$. The function P is called the **Laplace transform of the function** f. (For its economic interpretation, see Example 8.) It is an important tool for solving differential equations. In Exercises 61 to 65 find the Laplace transforms of the given functions.

61 $f(t) = t$ **62** $f(t) = t^2$

63 $f(t) = e^t$ (assume $r > 1$)

64 $f(t) = \sin t$ **65** $f(t) = \cos t$

66 Let f and its derivative f' both have Laplace transforms. Let P be the Laplace transform of f, and let Q be the Laplace transform of f'. Show that

$$Q(r) = -f(0) + rP(r).$$

67 Let P be the Laplace transform of f. Let a be a positive constant, and let $g(t) = f(at)$. Let P be the Laplace transform of f, and let Q be the Laplace transform of g. Show that $Q(r) = (1/a)P(r/a)$.

68 When computing the internal energy of a crystal, Claude Garrod, in *Twentieth Century Physics*, Faculty Publishing, Davis, Calif., 1984, p. 326, states that the integral

$$\int_0^{\pi/2} \frac{\sin x}{e^{0.26 \sin x} - 1}\, dx$$

"cannot be evaluated analytically. However, it can easily be computed numerically using Simpson's rule. The result is 5.56."
(*a*) Is the integral proper or improper?
(*b*) How does the integrand behave when x is near 0?
(*c*) What does "cannot be evaluated analytically" mean?
(*d*) Use Simpson's rule with $n = 6$ to estimate the integral.

69 Describe how you would go about estimating $\int_0^\infty e^{-x^2}\, dx$ with an error less than 0.01. Do not go through the arithmetic.

70 Describe how you would go about estimating $\int_0^\infty dx/\sqrt{1 + x^4}$ with an error less than 0.01. Do not go through the arithmetic.

8.8 WORK

The work required to raise a weight of W pounds a distance D feet is defined to be $W \times D$ foot-pounds. An elevator that lifts a 150-pound person 100 feet thus accomplishes

$$150 \times 100 \text{ foot-pounds,}$$

or

$$15{,}000 \text{ foot-pounds of work.}$$

When all parts of an object are lifted the same distance, the work is simply the product of two numbers. We now pose a problem in which different parts of an object are lifted different distances.

A tank is full of water, which weighs 62.4 pounds per cubic foot. The water is pumped out an outlet which is above the level of the water. How much work is accomplished in emptying the tank? (See Fig. 8.100.) Water at the bottom of the tank must be pumped farther than the water at the top: The lower the water is in the tank, the farther it has to be raised.

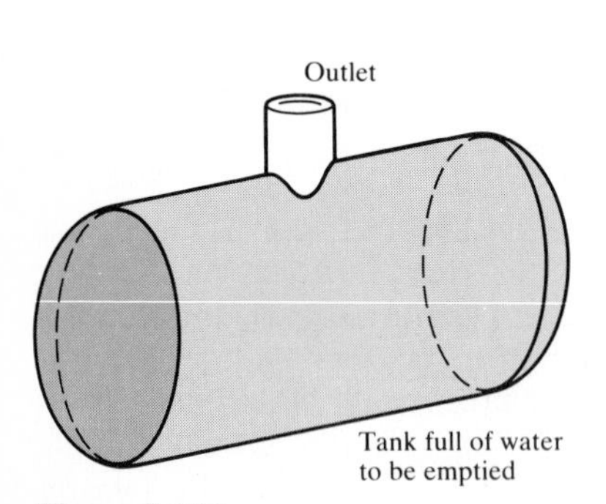

Figure 8.100

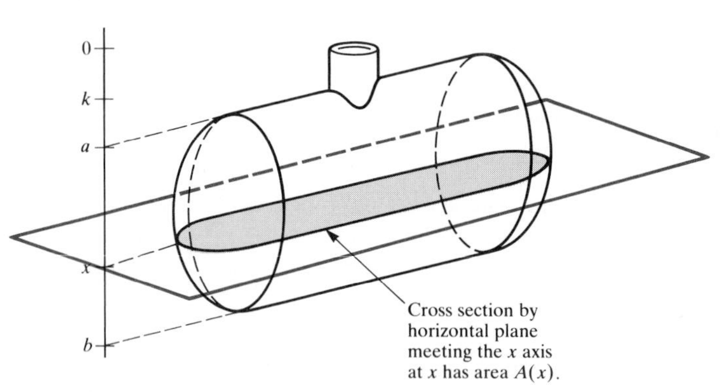

Figure 8.101

To treat this problem mathematically, introduce a vertical x axis with positive part below the origin, as in Fig. 8.101. The plane perpendicular to the x axis at level x has area $A(x)$. [The formula for $A(x)$ depends on the shape of the tank and where the origin of the x axis is placed.] Assume that the tank extends from level $x = a$ to level $x = b$, $b > a$, and that the level of the outlet is k. Observe that water at level x is lifted a distance $x - k$ feet, which we will denote $D(x)$.

To find out how much work is accomplished in emptying the tank, consider a thin horizontal slab of water (all the water in the layer is raised about the same distance). The layer consists of all water with x coordinate between x and $x + dx$, where dx is a small positive number. (See Fig. 8.102.) Approximate this thin layer by a cylinder whose height is dx and whose base is determined by the cross section at level x and thus has area $A(x)$ square feet.

The volume of the cylindrical slab in Fig. 8.103 is $A(x)\,dx$ cubic feet.

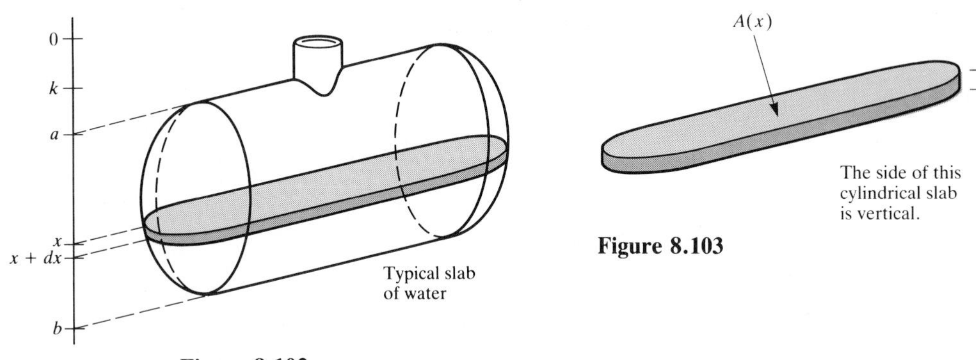

Figure 8.102

Figure 8.103

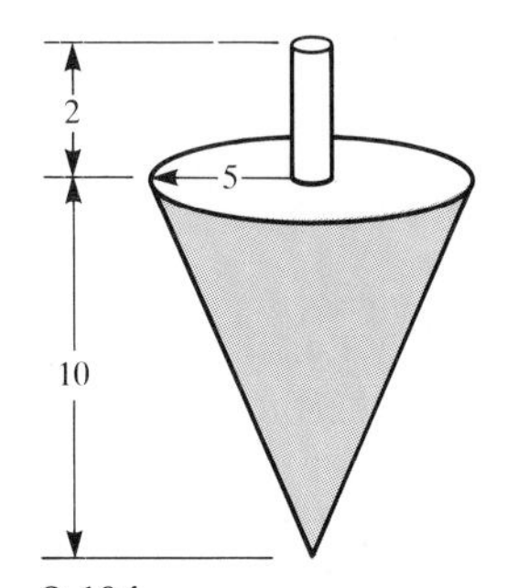

Figure 8.104

The work needed to raise the slab is approximately

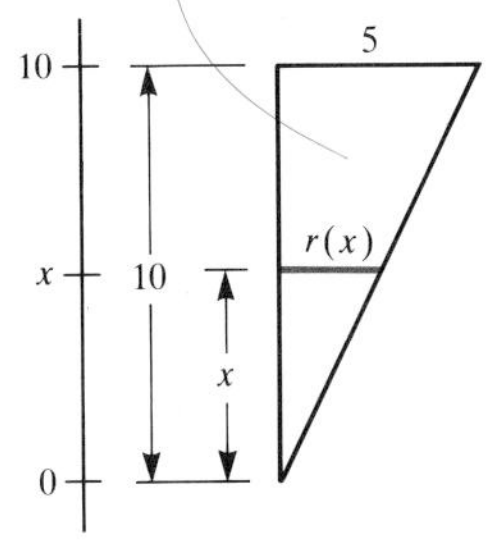

$$\underbrace{\underbrace{A(x)\,dx}_{\text{volume}}\ \underbrace{62.4}_{\text{density}}}_{\text{weight}}\ \underbrace{D(x)}_{\substack{\text{distance}\\ \text{raised}}} \qquad \text{foot-pounds.}$$

Thus the local approximation to the work accomplished is

$$62.4D(x)A(x)\,dx \qquad \text{foot-pounds.}$$

Using the informal approach to setting up definite integrals, we conclude that the total work in emptying the tank is

$$\boxed{\text{Total work} = 62.4 \int_a^b D(x)A(x)\,dx \qquad \text{foot-pounds.}} \tag{1}$$

To apply this formula it is necessary to introduce a vertical x axis. Where its origin should be placed is a choice made to achieve the simplest integrand in Eq. (1). It is also sometimes convenient to have the positive part of the axis point upward.

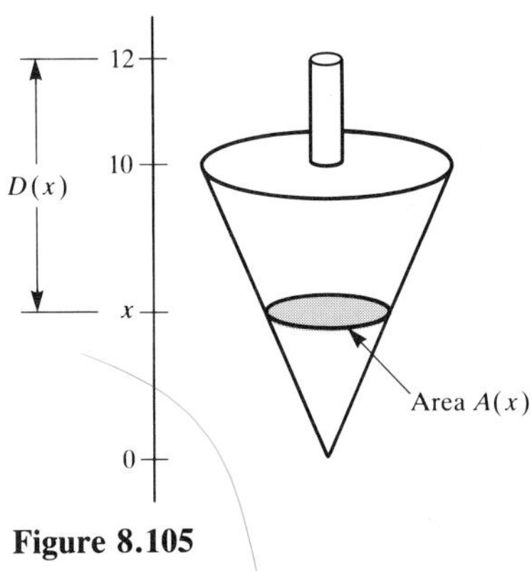

Figure 8.105

EXAMPLE 1 A conical tank of radius 5 feet and height 10 feet is full of water. (See Fig. 8.104.) How much work is required to empty it through an outlet 2 feet above the tank?

Figure 8.106

SOLUTION Place the origin of the vertical x axis level with the point of the cone in order to obtain a convenient description of the cone. For the same reason, put the positive part of the x axis above the origin. (See Fig. 8.105.) Then

$$\text{Work} = 62.4 \int_0^{10} D(x)A(x)\,dx \qquad \text{foot-pounds,}$$

where $D(x)$ is the distance the section at level x is lifted and $A(x)$ is the area of that section. Inspection of Fig. 8.105 shows that $D(x) = 12 - x$. To find $A(x)$, draw the similar triangles shown in Fig. 8.106.

Let $r(x)$ be the radius of the cross-sectional disk corresponding to x. Then

$$\frac{r(x)}{5} = \frac{x}{10}.$$

Hence $r(x) = x/2$. Thus $A(x) = \pi(x/2)^2$, and we have

$$\text{Work} = 62.4 \int_0^{10} D(x)A(x)\,dx$$

$$= 62.4 \int_0^{10} (12 - x)\pi\left(\frac{x}{2}\right)^2 dx \qquad \text{foot-pounds.}$$

Evaluation of the integral shows that

$$\text{Work} = 62.4(375\pi) \qquad \text{foot-pounds.} \quad ■$$

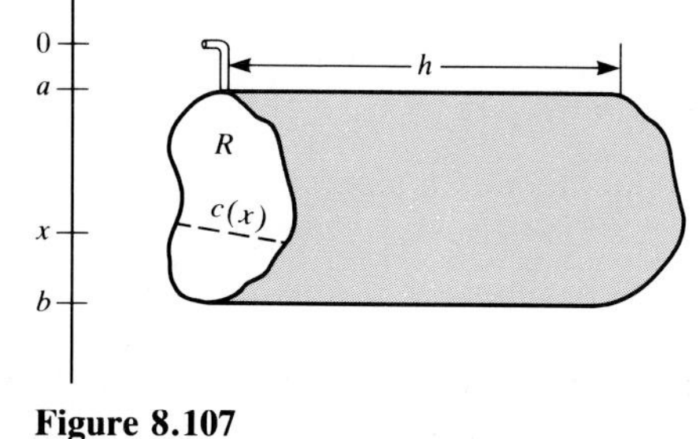

Figure 8.107

In case the tank has the shape of a cylinder, there is a close relation between the work done in emptying it and a centroid.

Consider a cylindrical tank whose base is the plane region R (not necessarily a disk). The base is vertical. The length of the cylinder is h, as shown in Fig. 8.107. Assume that the tank is full of water. Place the x axis so the outlet corresponds to $x = 0$.

R could be a disk, a triangle, a square, etc.

Let the cross section of R at level x have length $c(x)$ feet. By Eq. (1), the work accomplished in emptying the tank out the outlet is

In this case, $A(x) = c(x)h$, the area of a rectangle.

$$62.4 \int_a^b xc(x)h\,dx \qquad \text{foot-pounds.} \tag{2}$$

If the centroid of R is known, it is not necessary to use (2). It turns out that the work required is the same as if all the water were at the depth of the centroid of R. This fact is the substance of the following theorem.

Theorem The work required to pump the water out of a full cylindrical tank of the type described is the product

$$W \cdot D \text{ foot-pounds,}$$

where W is the total weight of water in the tank and D is the distance the water at the centroid of R is lifted.

Proof Place the x axis with its positive part directed downward and with its origin at the outlet, as in Fig. 8.107. The total work is then

$$62.4h \int_a^b xc(x)\,dx \qquad \text{foot-pounds.}$$

Recall the formula for $\bar{x}$.

Since
$$\bar{x} = \frac{\int_a^b xc(x)\,dx}{\text{Area of } R},$$

this equals
$$62.4h\bar{x} \cdot \text{Area of } R \qquad \text{foot-pounds.}$$

But
$$62.4h \cdot \text{Area of } R$$

is the weight W of water in the tank. Hence

$$\text{Work} = W \cdot \bar{x}.$$

Since $\bar{x}$ is the distance the water at the centroid is lifted, this proves the theorem. ■

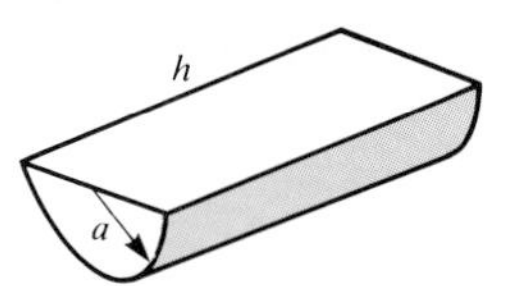

Figure 8.108

EXAMPLE 2 A tank has a semicircular base of radius a feet and length h feet, as shown in Fig. 8.108. If the tank is full of water, how much work is required to empty it over its rim?

SOLUTION By Example 3 in Sec. 8.5, the depth of the centroid of the disk is $4a/3\pi$ feet. The weight of the water in the tank is $62.4h\pi a^2/2$ pounds. By the theorem, the total work required to empty the tank is

$$\underbrace{\frac{4a}{3\pi}}_{\substack{\text{depth}\\ \text{of centroid}}} \cdot \underbrace{62.4\pi\frac{a^2h}{2}}_{\substack{\text{weight of}\\ \text{water}}} \quad \text{foot-pounds,}$$

or

$$62.4\,\frac{2a^3h}{3} \text{ foot-pounds.} \quad ■$$

Work If Force of Gravity Varies

In calculating the work done in emptying a tank we assumed that the force of gravity does not depend on the height of the object. In fact, the farther an object is from the earth, the less it weighs. (Its mass remains constant, but its weight diminishes.) Thus the work required to raise an object 1 foot at sea level is greater than the work required to raise the same object 1 foot at the top of a mountain. However, the difference in altitudes is so small in comparison to the radius of the earth that the difference in work is negligible. On the other hand, when an object is rocketed into space, the fact that the force of gravity diminishes with distance from the center of the earth is critical. Example 3 shows why.

EXAMPLE 3 How much work is required to lift a 1-pound payload from the surface of the earth to "infinity"? (If this work should turn out to be infinite, then it would require an infinite amount of fuel to send rockets off on unlimited orbits. Fortunately, it is finite, as will now be shown.)

SOLUTION The work W necessary to lift an object a distance x against a constant vertical force F is the product of force times distance:

$$W = F \cdot x.$$

Since the gravitational pull of the earth on the payload *changes* with distance from the earth, an (improper) integral will be needed to express the total work required to lift the load to "infinity."

The payload weighs 1 pound at the surface of the earth. The farther it is from the center of the earth, the less it weighs, for the force of the earth on the mass is inversely proportional to the square of the distance of the mass from the center of the earth. Thus the force on the payload is given by k/r^2 pounds, where k is a constant, which will be determined in a moment, and r is the distance in miles from the payload to the center of the earth. When $r = 4{,}000$ (miles), the force is 1 pound; thus

$$1 = \frac{k}{4{,}000^2}.$$

From this it follows that $k = 4{,}000^2$, and therefore the gravitational force

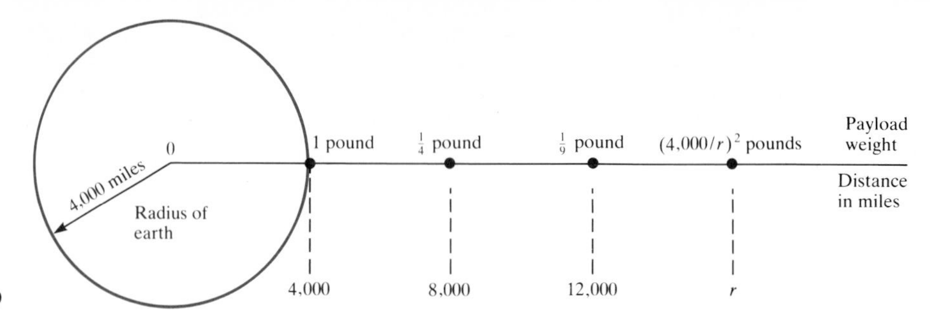

Figure 8.109

on a 1-pound mass is, in general, $(4{,}000/r)^2$ pounds. As the payload recedes from the earth, it loses weight (but not mass), as recorded in Fig. 8.109. The work done in lifting the payload from point r to point $r + dr$ is approximately

$$\underbrace{\left(\frac{4{,}000}{r}\right)^2}_{\text{force}} \quad \underbrace{(dr)}_{\text{distance}} \quad \text{mile-pounds.}$$

Hence the work required to move the 1-pound mass from the surface of the earth to infinity is given by the improper integral $\int_{4{,}000}^{\infty} (4{,}000/r)^2\, dr$. Now,

$$\int_{4{,}000}^{\infty} \left(\frac{4{,}000}{r}\right)^2 dr = \lim_{b\to\infty} \int_{4{,}000}^{b} \left(\frac{4{,}000}{r}\right)^2 dr = \lim_{b\to\infty} \left.\frac{-4{,}000^2}{r}\right|_{4{,}000}^{b}$$

$$= \lim_{b\to\infty} \left(\frac{-4{,}000^2}{b} + \frac{4{,}000^2}{4{,}000}\right) = 4{,}000 \text{ mile-pounds.}$$

The total work is finite because the improper integral is convergent. It is just as if the payload were lifted 4,000 miles against a constant gravitational force equal to that at the surface of the earth. ■

EXERCISES FOR SEC. 8.8: WORK

1 Find the work required to empty the tank shown in Fig. 8.110. The tank consists of a right-circular cylinder surmounted by a hemisphere with an outlet at its top. The tank is full of water.

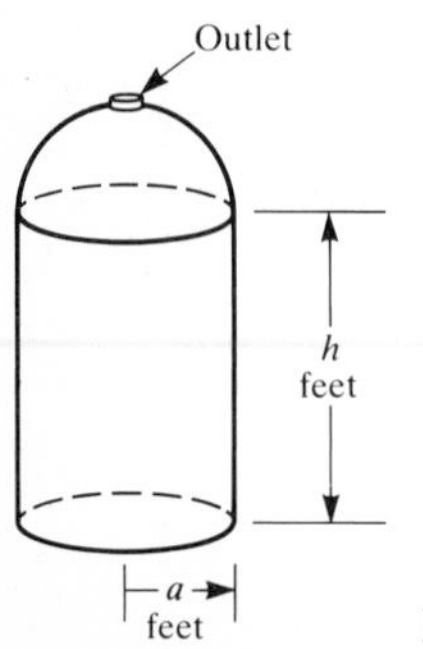

Figure 8.110

2 Find the work required to empty the pyramidal tank shown in Fig. 8.111. The base of the pyramid is square. The outlet is 2 feet above this base.

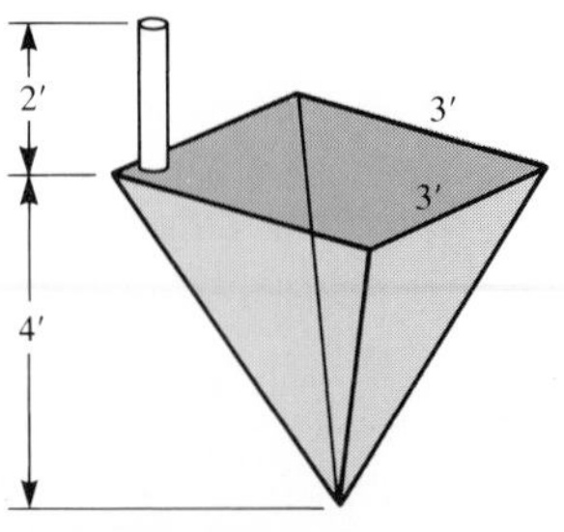

Figure 8.111

In Exercises 3 to 8 find the work required to pump the water out of the tanks shown in Figs. 8.112 to 8.117, either out the spigot, if there is one, or over the rim of the tank. Use the theorem in this section concerning the centroid.

3

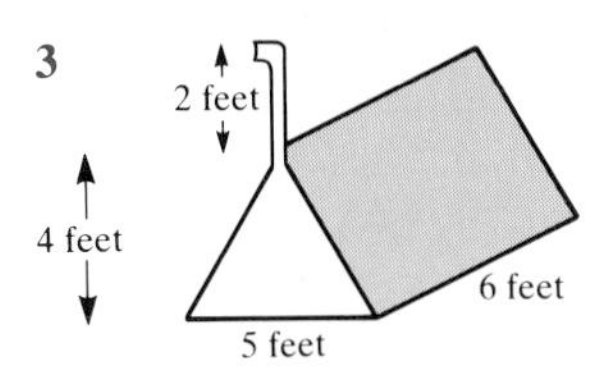

Tank with triangular base is full of water.

Figure 8.112

4

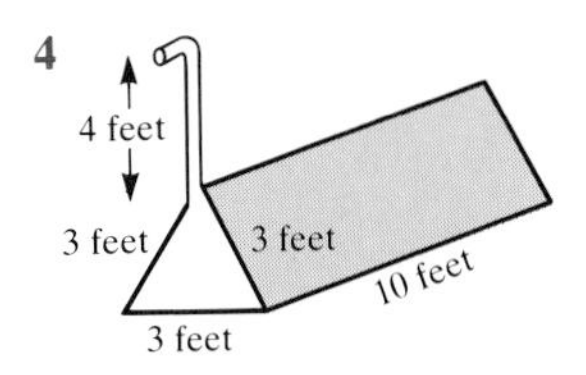

Tank is full of water.

Figure 8.113

5

10 feet

6 feet

Tank with semicircular base is full of water.

Figure 8.114

6

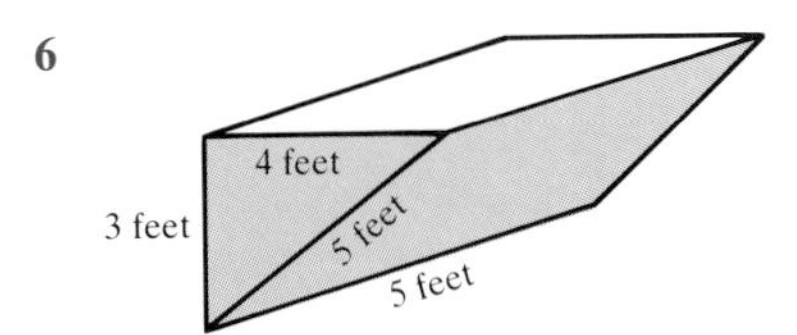

Tank is full of water.

Figure 8.115

7

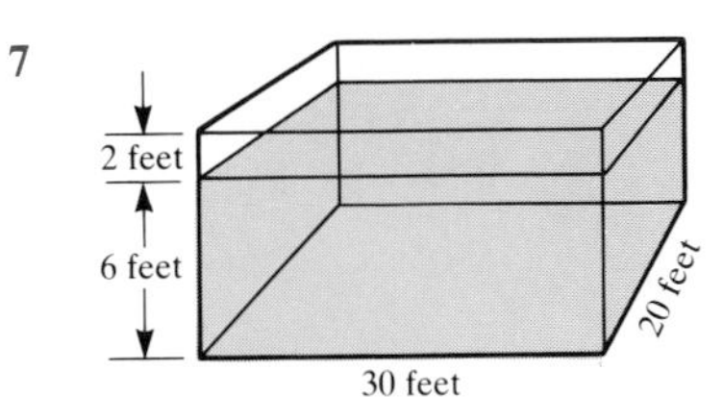

Surface of water is 2 feet below top of tank.

Figure 8.116

8

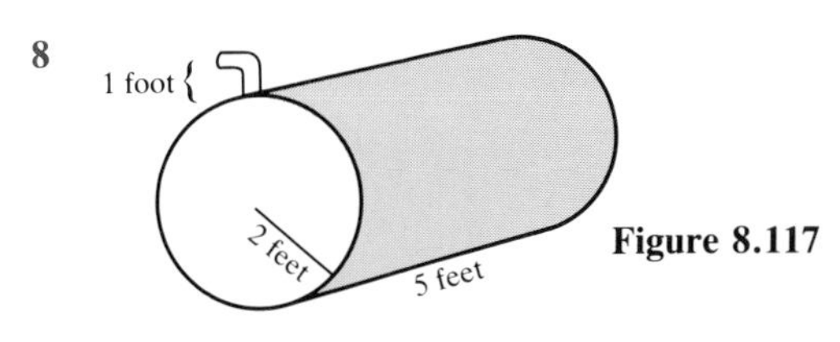

Tank with circular base is full of water.

Figure 8.117

Exercises 9 to 12 refer to Example 3.

9 How much work is done in lifting the 1-pound payload the first 4,000 miles of its journey to infinity?

10 (*a*) If the force of gravity were of the form $k/r^{1.01}$, could a payload be sent to infinity? (*b*) If the force of gravity were of the form k/r, could a payload be sent to infinity?

11 Assume that the force of gravity obeys an inverse cube law, so that the force on a 1-pound payload a distance r miles from the center of the earth ($r \geq 4{,}000$) is $(4{,}000/r)^3$ pounds. How much work would be required to lift a 1-pound payload from the surface of the earth to "infinity"?

12 If a mass which weighs 1 pound at the surface of the earth were launched from a position 20,000 miles from the center of the earth, how much work would be required to send it to "infinity"?

■

13 The force required to hold one end of a spring stretched x feet from its equilibrium position is proportional to x. The force is of the form kx pounds for some positive constant k.
(*a*) How much work is required to stretch the string the first foot from its equilibrium position?
(*b*) How much work is required to stretch it the next foot?

14 Geologists, when considering the origin of mountain ranges, estimate the energy required to lift a mountain up from sea level. Assume that two mountains are composed of the same type of matter, which weighs k pounds per cubic foot. Both are right circular cones in which the height is equal to the radius. One mountain is twice as high as the other. The base of each is at sea level. If the work required to lift the matter in the smaller mountain above sea level is W, what is the corresponding work for the larger mountain?

15 (See Exercise 14.) Assume that Mt. Everest has the shape of a right circular cone of height 30,000 feet and radius 150,000 feet. Assume its density throughout is 200 pounds per cubic foot.
(*a*) How much work was required to lift the material in Mt. Everest if it was initially all at sea level?
(*b*) How does this work compare with the energy of a 1-megaton H bomb? (One megaton is the energy in a million tons of TNT. This is about 3×10^{14} foot-pounds.)

16 Throughout this section the density of the water was assumed to be constant. Assume instead that the liquid being lifted has a density that depends on depth.
(*a*) How should formula (1) be modified to cover this more general case?
(*b*) Does the theorem in this section concerning the centroid still hold?

17 A particle is drawn toward a fixed particle by the force of gravitational attraction. Show that if the force is proportional to r^{-2}, where r is the distance between the particles, then the total work accomplished is infinite. (This result, which violates common sense, suggests that at atomic distances the force of gravity must vary according to some other formula. This paradox is resolved in quantum mechanics.)

8.S SUMMARY

The following table summarizes most of the applications of the definite integral treated in this chapter:

Section	*Concept*	*Memory Aid*
8.1	Area $= \int_a^b c(x)\,dx$	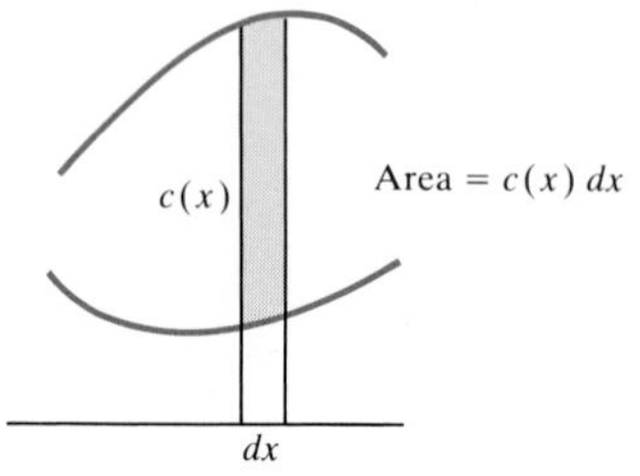 Area $= c(x)\,dx$
8.2	Volume $= \int_a^b A(x)\,dx$ (parallel cross sections) Special case: Solid of revolution where region under $y = f(x)$ and above x axis is revolved about x axis: $$A(x) = \pi[f(x)]^2$$	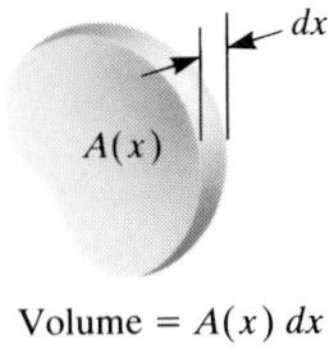 Volume $= A(x)\,dx$
8.3	Force of water $= 62.4 \int_a^b (x - k)c(x)\,dx$ pounds Force $= \underbrace{62.4(x - k)}_{\text{pressure}}\ \underbrace{c(x)\,dx}_{\text{area}}$ pounds	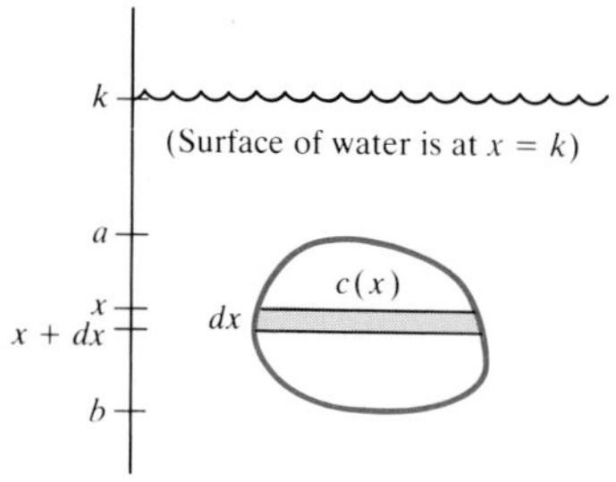 (Surface of water is at $x = k$)
8.4	Volume of solid of revolution $= \int_a^b 2\pi R(x)c(x)\,dx$ (by shells)	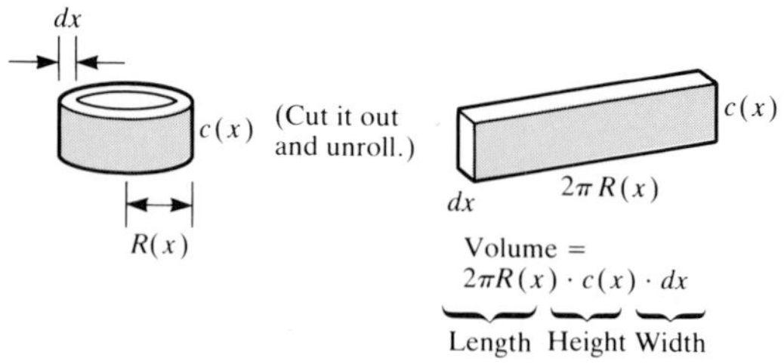 (Cut it out and unroll.) Volume $= \underbrace{2\pi R(x)}_{\text{Length}} \cdot \underbrace{c(x)}_{\text{Height}} \cdot \underbrace{dx}_{\text{Width}}$
8.5	Moment $= \int_a^b xc(x)\,dx$	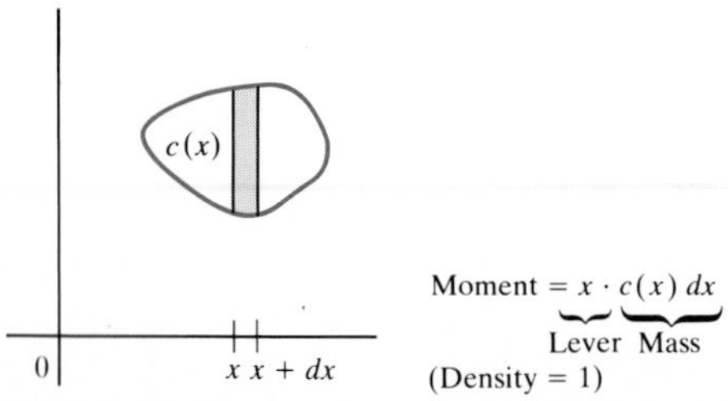 Moment $= \underbrace{x}_{\text{Lever}} \cdot \underbrace{c(x)\,dx}_{\text{Mass}}$ (Density = 1)

Section	Concept	Memory Aid
8.8	Work emptying tank $= \int_a^b 62.4\, D(x)A(x)\, dx$ foot-pounds	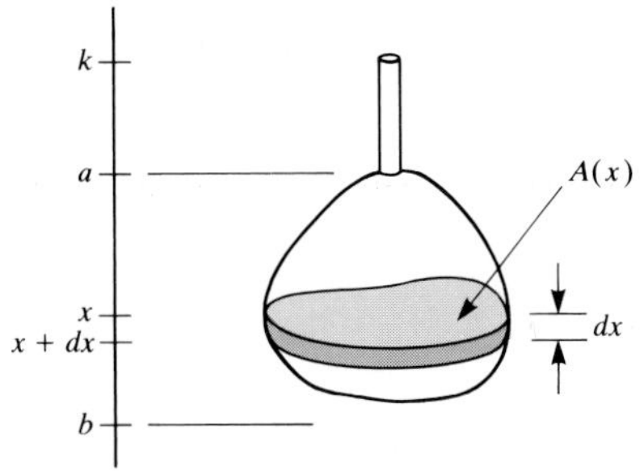 Work $= \underbrace{(x-k)}_{\text{Distance}} \cdot \underbrace{\underbrace{62.4}_{\text{Density}} \cdot \underbrace{A(x)\,dx}_{\text{Volume}}}_{\text{Weight}}$ foot-pounds

Vocabulary and Symbols

cross-sectional length $c(x)$, $c(y)$
cross-sectional area $A(x)$
solid of revolution
shell technique
lever arm
moment about a line
centroid, center of mass, $(\bar{x}, \bar{y})$
Pappus's theorem
force of water
Schwarz inequality
improper integral (convergent and divergent)
work

Key Facts

The centroid $(\bar{x}, \bar{y})$ of a plane region R is given by

$$\bar{x} = \frac{\int_a^b xc(x)\, dx}{\text{Area of } R} \quad \text{and} \quad \bar{y} = \frac{\int_c^d yc(y)\, dy}{\text{Area of } R}.$$

If R is the region below $y = f(x)$ and above $[a, b]$, then

$$\bar{x} = \frac{\int_a^b xf(x)\, dx}{\text{Area of } R} \quad \text{and} \quad \bar{y} = \frac{\frac{1}{2}\int_a^b [f(x)]^2\, dx}{\text{Area of } R}.$$

If both integrals exist,

$$\left| \int_a^b f(x)\, dx \right| \leq \int_a^b |f(x)|\, dx.$$

Extended mean-value theorem for integrals: If $g(x) \geq 0$ for x in $[a, b]$ and f is continuous, then there is a number c in $[a, b]$ such that

$$\int_a^b f(x)g(x)\, dx = f(c) \int_a^b g(x)\, dx.$$

(It is also assumed that both integrals exist.) The same conclusion holds if $g(x) \leq 0$ for x in $[a, b]$.

The Schwarz inequality: If f and g are continuous on $[a, b]$, then

$$\left| \int_a^b f(x)g(x)\, dx \right| \leq \left\{ \int_a^b [f(x)]^2\, dx \right\}^{1/2} \left\{ \int_a^b [g(x)]^2\, dx \right\}^{1/2}.$$

If $0 \leq f(x) \leq g(x)$ and $\int_a^\infty g(x)\, dx$ is convergent, so is $\int_a^\infty f(x)\, dx$ (comparison test).

If $\int_a^\infty |f(x)|\, dx$ is convergent, so is $\int_a^\infty f(x)\, dx$.

THREE USES OF THE CENTROID

The volume of a solid of revolution formed by revolving R equals

Distance centroid of R moves · Area of R (Pappus's theorem).

The force of water against a submerged vertical flat surface R equals

Pressure at centroid of R · Area of R.

The work required to empty a cylindrical tank of water is the same as if all the water were at the depth of the centroid.

GUIDE QUIZ ON CHAP. 8: APPLICATIONS OF THE DEFINITE INTEGRAL

In Exercises 1 to 4 R is the region bounded by $y = e^x$ and the x axis, between $x = 1$ and $x = 2$.

1 Find the area of R, using (*a*) vertical cross sections, (*b*) horizontal cross sections.

2 (*a*) Find the moment of R about the x axis.
(*b*) Find the moment of R about the y axis.
(*c*) Find $(\bar{x}, \bar{y})$, the centroid of R.

3 A solid of revolution is made by revolving R around the line $y = -1$. Find its volume by (*a*) parallel cross sections, (*b*) shells, (*c*) Pappus's theorem.

4 A solid of revolution is made by revolving R around the y axis. Find its volume by the three methods given in Exercise 3.

5 Let R be the region below the curve $y = f(x)$ and above $[a, b]$. Give an intuitive argument that the moment of R around the x axis is $\frac{1}{2}\int_a^b [f(x)]^2\, dx$.

6 State the Schwarz inequality.

7 For which exponents a (if any) is
(*a*) $\int_1^\infty x^a\, dx$ convergent?
(*b*) $\int_0^1 x^a\, dx$ convergent?
(*c*) $\int_0^\infty x^a\, dx$ convergent?

8 Is $\displaystyle\int_0^\infty \frac{dx}{x + e^x}$ convergent or divergent?

9 (*a*) Explain in detail why the centroid is related to the volume of a solid of revolution.
(*b*) Explain in detail why the centroid is related to the work in emptying a cylindrical tank that is full of water.

REVIEW EXERCISES FOR CHAP. 8: APPLICATIONS OF THE DEFINITE INTEGRAL

In Exercises 1 to 4 set up integrals for the given quantities; do not evaluate them.

1 The area of the region above the parabola $y = x^2$ and below the line $y = 2x$, using (*a*) vertical cross sections, (*b*) horizontal cross sections.

2 The volume of the wedge cut from a right circular cylinder of height 5 inches and radius 3 inches by a plane that bisects one base and touches the other base at one point.

3 The volume of the solid obtained by revolving the triangle whose vertices are (2, 0), (2, 1), and (3, 2) about the x axis. (Use the shell technique.)

4 The moment of the region in the first quadrant bounded by $y = x^2$ and $y = x^3$, about the line $y = -2$.

In Exercises 5 to 9
(*a*) Find the area of R.
(*b*) Find the volume of the solid of revolution formed by revolving R about the x axis.
(*c*) The same as (*b*), but around the y axis.
(*d*) The same as (*b*), but around the line $y = -1$.

5 R is the region below the curve $y = x/(1 + x)$ and above [1, 2].

6 R is the region below $y = 1/(1 + x)^2$ and above [0, 1].

7 R is the region below $y = \sin 2x$ and above $[0, \pi/2]$.

8 R is the region below $y = \sqrt{x^2 - 9}$ and above [3, 4].

9 R is the region below $y = 1/(2x + 1)$ and above [0, 1].

In Exercises 10 to 13 find the moments of the given regions R about the given lines L.

10 R: below $y = \sec x$, above $[\pi/6, \pi/4]$; L: the x axis.

11 R: below $y = (\sin x)/x$, above $[\pi/2, \pi]$; L: the y axis.

12 R: below $y = 1/\sqrt{x^2 + 1}$, above [0, 1]; L: the x axis.

13 R: below $y = 1/\sqrt{x^2 + 1}$, above [0, 1]; L: the y axis.

In Exercises 14 to 18 find the areas of the given regions R.

14 R is below $y = 1/(x^2 + 3x + 2)$ and above $y = 1/(x^2 + 3x + 4)$, between $x = 0$ and $x = 1$.

15 R is below $y = x$ and above $y = \tan^{-1} x$, between $x = 0$ and $x = 1$.

16 R is below $y = x\sqrt{2x + 1}$ and above [0, 4].

17 R is below $y = \cos^3 x$ and above $y = \sin^3 x$, between $x = 0$ and $x = \pi/4$.

18 R is below $y = 1/\sqrt{4 - x^2}$ and above $y = x/\sqrt{4 - x^2}$, between $x = 0$ and $x = 1$.

19 Find the area of the region between the curves $y = 1/(x^2 - x)$ and $y = 1/(x^3 - x)$, (*a*) between $x = 2$ and $x = 3$, (*b*) to the right of $x = 3$.

20 Let R be the region below $y = \tan x$ and above $[0, \pi/4]$.
(*a*) Find the area of R.
(*b*) Find the moment of R about the x axis.
(*c*) Find $\bar{y}$.
(*d*) Set up an integral for the moment of R about the y axis. (Don't try to evaluate it; the fundamental theorem of calculus is useless here.)
(*e*) Find the volume of the solid of revolution formed by revolving R around the line $y = -1$.

21 Let R be the region below $y = \sin^2 x$ and above $y = \sin^3 x$, between $x = 0$ and $x = \pi/2$.
(*a*) Find the area of R.
(*b*) Find the moment of R about the x axis. (Recall the formula for $\int_0^{\pi/2} \sin^n x\, dx$.)
(*c*) Find $\bar{y}$.

22 Is $\int_0^\infty (dx/\sqrt{x}\sqrt{x+1}\sqrt{x+2})$ convergent or divergent?

23 A drill of radius a inches bores a hole through the center of a sphere of radius b inches, leaving a ring whose height is 2 inches. Find the volume of the ring.

24 A barrel is made by rotating an ellipse around one of its axes and then cutting off equal caps, top and bottom. It is 3 feet high and 3 feet wide at its midsection. Its top and bottom have a diameter of 2 feet. What is its volume?

25 Let R be the region to the right of the y axis, below $y = e^{-x}$ and above the x axis.
(*a*) Find the area of R.
(*b*) Find the volume of the region obtained by revolving R about the x axis.
(*c*) Find the volume of the region obtained by revolving R about the y axis.

26 Let l be a line which intersects the triangle ABC and is parallel to BC. Suppose that l is twice as far from the point A as from the line BC. (See Fig. 8.118.) Show that the centroid of ABC is on l.

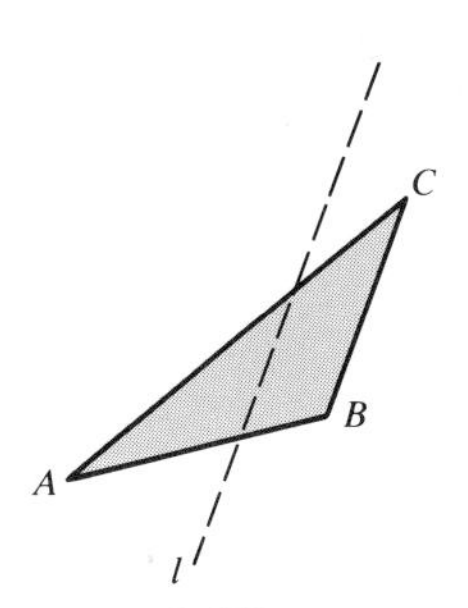

Figure 8.118

27 (*a*) Develop the shell formula for the volume of a solid of revolution.
(*b*) What is the device for remembering the formula?

28 Show that $\int_0^\infty e^{-rx} \sin ax\, dx$ equals $a/(a^2 + r^2)$, where $r > 0$ and a are constant.

29 Show that $\int_0^\infty e^{-rx} \cos ax\, dx$ equals $r/(a^2 + r^2)$, where $r > 0$ and a are constant.

30 Let f be a continuous function such that $f(0) = 2$ and $f(x) \to 3$ as $x \to \infty$. Find the limit of $(1/b) \int_0^b f(x)\, dx$ as (*a*) $b \to 0$, (*b*) $b \to \infty$.

31 By interpreting these improper integrals as expressions for the area of a certain region, show that

$$\int_0^\infty \frac{dx}{1 + x^2} = \int_0^1 \sqrt{\frac{1 - y}{y}}\, dy.$$

32 Define $G(a) = \int_0^\infty a/(1 + a^2x^2)\, dx$.
(*a*) Compute $G(0)$.
(*b*) Compute $G(a)$ if a is negative.
(*c*) Compute $G(a)$ if a is positive.
(*d*) Graph G.

33 Prove that $\int_0^\infty (\sin x^2)/x\, dx = \frac{1}{2} \int_0^\infty (\sin x)/x\, dx$.

34 Is $\displaystyle\int_0^\infty \frac{dx}{(x - 1)^2}$ convergent or divergent?

35 Evaluate $\int_0^\infty e^{-x} \sin (2x + 3)\, dx$.

36 (*a*) Sketch $y = e^{-x}(1 + \sin x)$ for $x \geq 0$.
(*b*) The region beneath the curve in (*a*) and above the positive x axis is revolved around the y axis. Find the volume of the resulting solid.

37 From the fact that $\int (e^x/x)\, dx$ is not elementary, deduce that $\int e^x \ln x\, dx$ is not elementary.

38 Is this computation correct?

$$\int_{-2}^1 \frac{dx}{2x + 1} = \frac{1}{2} \ln |2x + 1| \Big|_{-2}^1 = \tfrac{1}{2} \ln 3 - \tfrac{1}{2} \ln 3 = 0.$$

39 Find the error in the following computations:

$$\int_{-1}^1 \frac{1}{x^2}\, dx = \frac{-1}{x}\Big|_{-1}^1 = \frac{-1}{1} - \frac{-1}{-1} = -2.$$

(The integrand is positive, yet the integral is negative.)

40 It can be proved that $\int_0^\infty x^{n-1}/(1 + x)\, dx = \pi \csc n\pi$ for $0 < n < 1$. Verify that this equation is correct for $n = \frac{1}{2}$.

41 Compute $\int_0^1 x^4 \ln x\, dx$.

42 Show that $\displaystyle\int_0^\infty \frac{dx}{1 + x^4} = \int_0^\infty \frac{x^2\, dx}{1 + x^4}$.

Hint: Let $x = 1/y$.

43 Show that $\displaystyle\int_0^1 (-\ln x)^3\, dx = \int_0^\infty x^3 e^{-x}\, dx$.

44 Assume that the density of the earth x miles from its center is $g(x)$ tons per cubic mile. Set up (informally) an integral for the total mass of the earth.

45 Find the centroid of the region bounded by the parabola $y = x^2$ and the line $y = 3x - 2$.

46 Find the centroid of the finite region bounded by $y = 2^x$ and $y = x^2$, to the right of the y axis.

47 Is $\int_0^\infty \frac{x^2 - 5x^3}{x^6 + 1} \sin 3x \, dx$ convergent or divergent?

48 Is $\int_0^1 \frac{\ln x}{1 - x^2} dx$ convergent or divergent?

49 It follows from Exercise 42 of Sec. 8.7 that $\int_0^\infty x^n e^{-x} dx = n!$. Use this to find $\int_0^\infty x^n e^{-ax} dx$, where a is a positive constant.

50 At time t, $0 \le t \le \pi/2$, a particle is at the point $x = A \sin t$ on the x axis.
(*a*) Find the average of the square of its speed with respect to distance.
(*b*) Find the average of the square of its speed with respect to time.

51 Water flows out of a hole in the bottom of a cylindrical tank of radius r and height h at the rate of $\sqrt{y}$ cubic feet per second when the depth of the water is y feet. Initially the tank is full. (See Fig. 8.119.)
(*a*) How long will it take to become half full?
(*b*) How long will it take to empty?

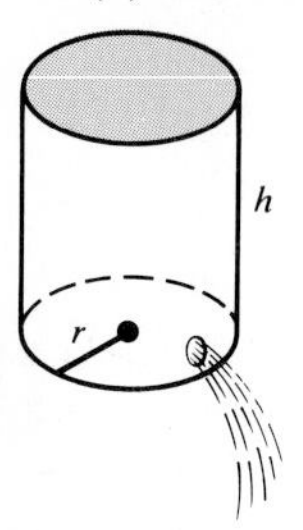

Figure 8.119

52 Is the area under the curve $y = (\ln x)/x^2$, above the x axis and to the right of the line $x = 1$, finite or infinite?

53 (*a*) Let $G(a) = \int_0^\infty 1/[(1 + x^a)(1 + x^2)]\, dx$. Evaluate $G(0)$, $G(1)$, $G(2)$.
(*b*) Show, using the substitution $x = 1/y$, that

$$G(a) = \int_0^\infty \frac{x^a \, dx}{(1 + x^a)(1 + x^2)}.$$

(*c*) From (*b*), show that $G(a) = \pi/4$, independent of a.

54 There are two values of a for which $\int \sqrt{1 + a \sin^2 \theta}\, d\theta$ is elementary. What are they?

55 From Exercise 54 deduce that there are two values of a for which

$$\int \frac{\sqrt{1 + ax^2}}{\sqrt{1 - x^2}} dx$$

is elementary.

56 There are three values of b for which $\int \sqrt{1 + b \cos \theta}\, d\theta$ is elementary. What are they?

57 From Exercise 56 deduce that there are three values of b for which

$$\int \frac{\sqrt{1 + bx}}{\sqrt{1 - x^2}} dx$$

is elementary.

58 Let f be a function such that $f(x) > 0$. Assume that f has derivatives of all orders and that $\ln f(x) = f(x)\int_0^x f(t)dt$. Find (*a*) $f(0)$, (*b*) $f^{(1)}(0)$, (*c*) $f^{(2)}(0)$.

59 Find the number a, $0 \le a \le 2\pi$ that maximizes the function

$$f(a) = \int_0^{2\pi} \sin x \sin (x + a)\, dx.$$

A nonnegative function $f(x)$, such that $\int_{-\infty}^\infty f(x)\, dx = 1$, is called a **probability distribution** [The probability that a certain variable observed in an experiment is between x and $x + \Delta x$ is approximately $f(x)\, \Delta x$.] The improper integral $\int_{-\infty}^\infty xf(x)\, dx$, if it exists, is called the **mean** of the distribution and is denoted μ. The improper integral $\int_{-\infty}^\infty (x - \mu)^2 f(x)\, dx$, if it exists, is called the **variance** of the distribution and is denoted μ_2. The square root of μ_2 is called the **standard deviation** of the distribution and is denoted σ.

60 Let k be a positive constant. Define $f(x)$ to be ke^{-kx} if $x > 0$ and 0 if $x \le 0$.
(*a*) Show that $\int_{-\infty}^\infty f(x)\, dx = 1$.
(*b*) Find μ. (*c*) Find μ_2. (*d*) Find σ.

61 Let k be a positive constant. Define $f(x)$ to be $e^{-x^2/2k^2}/(\sqrt{2\pi}\, k)$. This is a **normal distribution**
(*a*) Show that $\int_{-\infty}^\infty f(x)\, dx = 1$. *Suggestion:* Assume that $\int_0^\infty e^{-x^2} dx = \sqrt{\pi}/2$, a result established in Exercise 28 of Sec. 13.4.
(*b*) Find μ. (*c*) Find μ_2. (*d*) Find σ.

9 Plane Curves and Polar Coordinates

This chapter uses the derivative and the definite integral in the study of the length of a curve, the area within a curve, the area of a curved surface, and motion along a curve.

9.1 POLAR COORDINATES

Rectangular coordinates are only one of the ways to describe points in the plane by pairs of numbers. In this section another system is described, called **polar coordinates**.

The rectangular coordinates x and y describe a point P in the plane as the intersection of a vertical line and a horizontal line. Polar coordinates describe a point P as the intersection of a circle and a ray from the center of that circle. They are defined as follows.

Select a point in the plane and a ray emanating from this point. The point is called the **pole**, and the ray the **polar axis**. (See Fig. 9.1.) Measure positive angles θ counterclockwise from the polar axis and negative angles clockwise. Now let r be a number. To plot the point P that corresponds to the pair of numbers r and θ, proceed as follows:

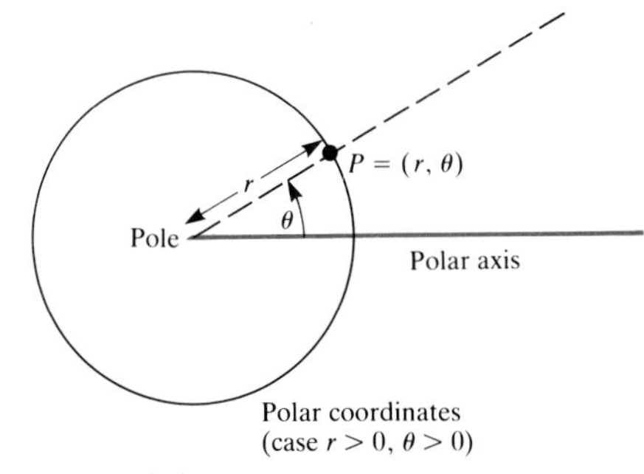

Figure 9.1

If r is positive, P is the intersection of the circle of radius r whose center is at the pole and the ray of angle θ emanating from the pole.

If r is 0, P is the pole, no matter what θ is.

If r is negative, P is at a distance $|r|$ from the pole on the ray directly opposite the ray of angle θ.

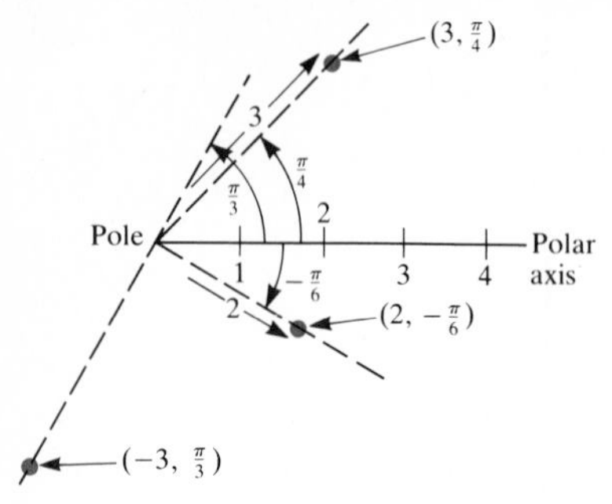

Figure 9.2

In each case P is denoted (r, θ), and the pair r and θ are called **polar coordinates** of P. Note that the point (r, θ) is on the circle of radius $|r|$ whose center is the pole. Also observe that the pole is the midpoint of the points (r, θ) and $(-r, \theta)$. Notice that the point $(-r, \theta + \pi)$ is the same as the point (r, θ). Moreover, changing the angle by 2π does not change the point; that is, $(r, \theta) = (r, \theta + 2\pi) = (r, \theta + 4\pi) = \cdots = (r, \theta + 2k\pi)$ for any integer k.

EXAMPLE 1 Plot the points $(3, \pi/4)$, $(2, -\pi/6)$, $(-3, \pi/3)$ in polar coordinates.

SOLUTION To plot $(3, \pi/4)$, go out a distance 3 on the ray of angle $\pi/4$ (shown in Fig. 9.2). To plot $(2, -\pi/6)$, go out a distance 2 on the ray of angle $-\pi/6$. To plot $(-3, \pi/3)$, draw the ray of angle $\pi/3$, and then go a distance 3 in the *opposite* direction from the pole. (See Fig. 9.2.) ■

The relation between polar and rectangular coordinates

It is customary to have the polar axis coincide with the positive x axis as in Fig. 9.3. In that case, inspection of the diagram shows the following relation between the rectangular coordinates (x, y) and the polar coordinates of the point P:

$$x = r\cos\theta, \qquad y = r\sin\theta,$$

and

$$r^2 = x^2 + y^2, \qquad \tan\theta = \frac{y}{x}.$$

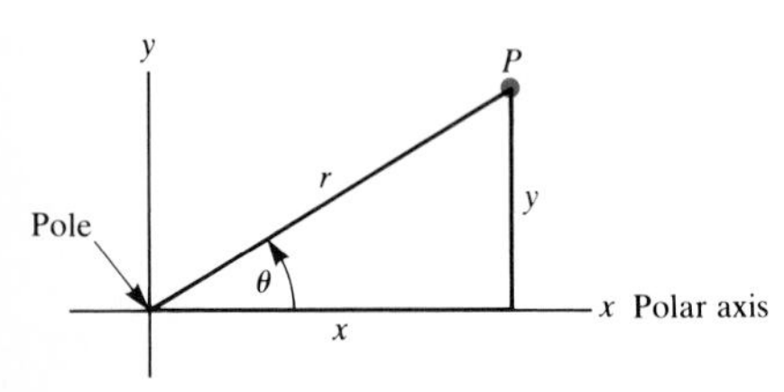

Figure 9.3

(These equations hold even if r is negative.)

Just as we may graph the set of points (x, y), where x and y satisfy a certain equation, so may we graph the set of points (r, θ), where r and θ satisfy a certain equation. It is important, however, to keep in mind that although each point in the plane is specified by a unique ordered pair (x, y) in rectangular coordinates, there are *many ordered pairs (r, θ) in polar coordinates which specify each point.* For instance, the point whose rectangular coordinates are $(1, 1)$ has polar coordinates $(\sqrt{2}, \pi/4)$ or $(\sqrt{2}, \pi/4 + 2\pi)$ or $(\sqrt{2}, \pi/4 + 4\pi)$ or $(-\sqrt{2}, \pi/4 + \pi)$ and so on.

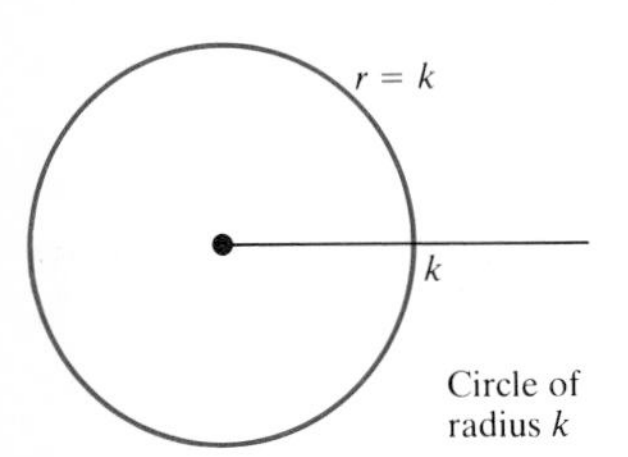

Figure 9.4

The simplest equation in polar coordinates has the form $r = k$, where k is a positive constant. Its graph is the circle of radius k, centered at the pole. (See Fig. 9.4.) The graph of $\theta = \alpha$, where α is a constant, is the line of inclination α. If we restrict r to be nonnegative, then $\theta = \alpha$ describes the ray ("half-line") of angle α. (See Fig. 9.5.)

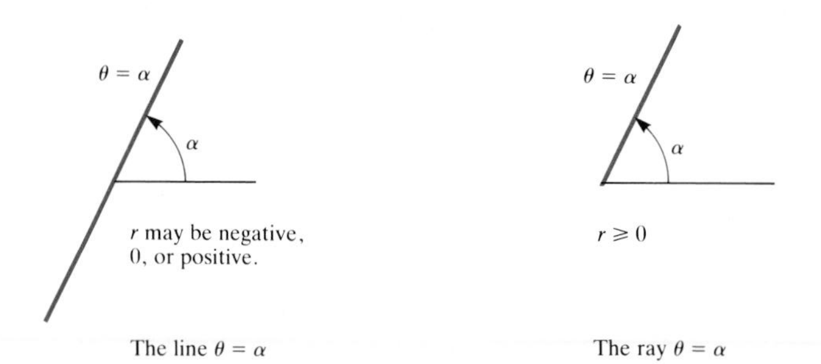

EXAMPLE 2 Graph $r = 1 + \cos\theta$.

SOLUTION Begin by making a table:

θ	0	$\frac{\pi}{4}$	$\frac{\pi}{2}$	$\frac{3\pi}{4}$	π	$\frac{5\pi}{4}$	$\frac{3\pi}{2}$	$\frac{7\pi}{4}$	2π
r	2	$1 + \frac{\sqrt{2}}{2} \approx 1.7$	1	$1 - \frac{\sqrt{2}}{2} \approx 0.3$	0	$1 - \frac{\sqrt{2}}{2} \approx 0.3$	1	$1 + \frac{\sqrt{2}}{2} \approx 1.7$	2

As θ goes from 0 to π, r decreases; as θ goes from π to 2π, r increases. The last point is the same as the first. The graph begins to repeat itself. This heart-shaped curve, shown in Fig. 9.6, is called a **cardioid** ■

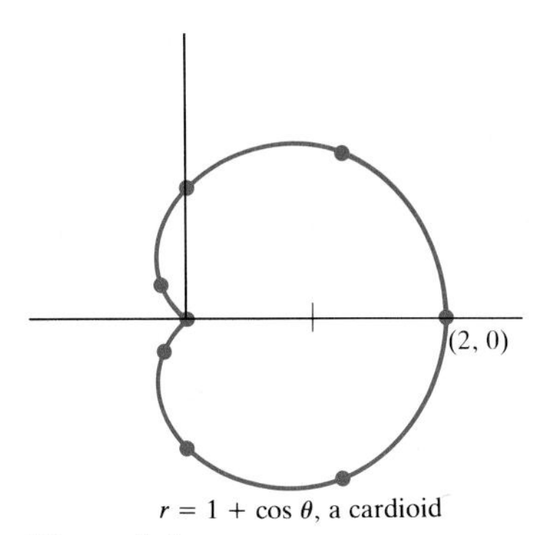

$r = 1 + \cos\theta$, a cardioid

Figure 9.6

The cardioid in Example 2 is tangent to the x axis at the origin. In general, a polar graph that passes through the origin when $\theta = \theta_0$ is tangent to the ray $\theta = \theta_0$ there.

Spirals turn out to be quite easy to describe in polar coordinates. This will be illustrated by the graph of $r = 2\theta$ in the next example.

EXAMPLE 3 Graph $r = 2\theta$ for $\theta \geq 0$.

SOLUTION First make a table:

θ	0	$\frac{\pi}{2}$	π	$\frac{3\pi}{2}$	2π	$\frac{5\pi}{2}$	...
r	0	π	2π	3π	4π	5π	...

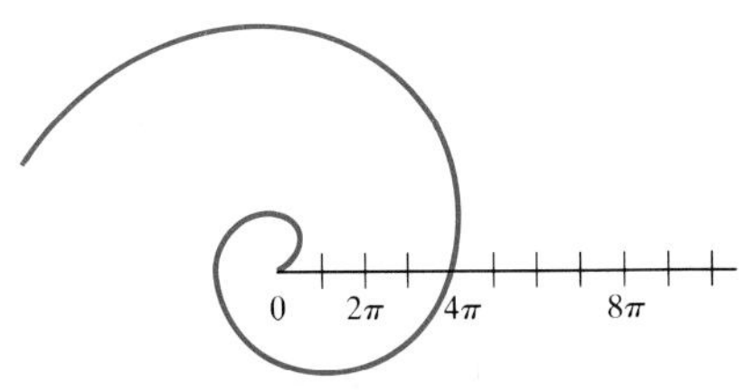

$r = 2\theta$, an archimedean spiral

Figure 9.7

Increasing θ by 2π does *not* produce the same value of r. As θ increases, r increases. The graph for $\theta \geq 0$ is an endless spiral, going infinitely often around the pole. It is indicated in Fig. 9.7. ■

If a is a nonzero constant, the graph of $r = a\theta$ is called an **archimedean spiral**. Example 3 illustrates the case $a = 2$.

Polar coordinates are also convenient for describing loops arranged like the petals of a flower, as Example 4 shows.

EXAMPLE 4 Graph $r = \sin 3\theta$.

SOLUTION As θ increases from 0 up to $\pi/3$, 3θ increases from 0 up to π. Thus r, which is $\sin 3\theta$, goes from 0 up to 1 and then back to 0, for θ in $[0, \pi/3]$. This gives one of the three loops that make up the graph of $r = \sin 3\theta$. For θ in $[\pi/3, 2\pi/3]$, $r = \sin 3\theta$ is negative (or 0). This yields the lower loop in Fig. 9.8. For θ in $[2\pi/3, \pi]$, r is again positive, and we obtain the upper left loop. Further choices of θ lead only to repetition of the loops already shown. ■

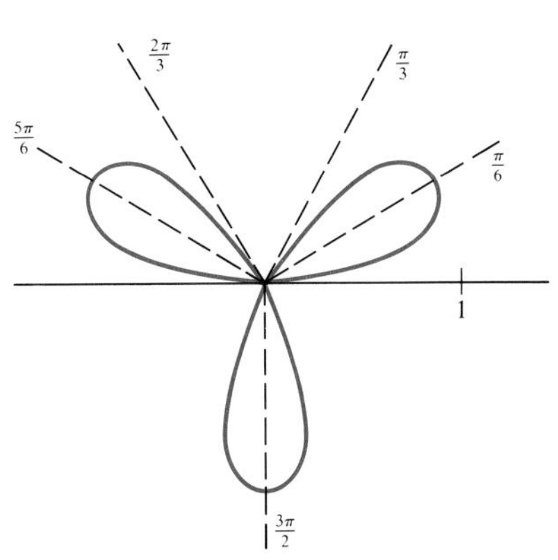

$r = \sin 3\theta$, a three-leaved rose

Figure 9.8

The graph of $r = \sin n\theta$ or $r = \cos n\theta$ has n loops when n is an odd positive integer and $2n$ loops when n is an even positive integer. The next example illustrates the case when n is even.

EXAMPLE 5 Graph the four-leaved rose, $r = \cos 2\theta$.

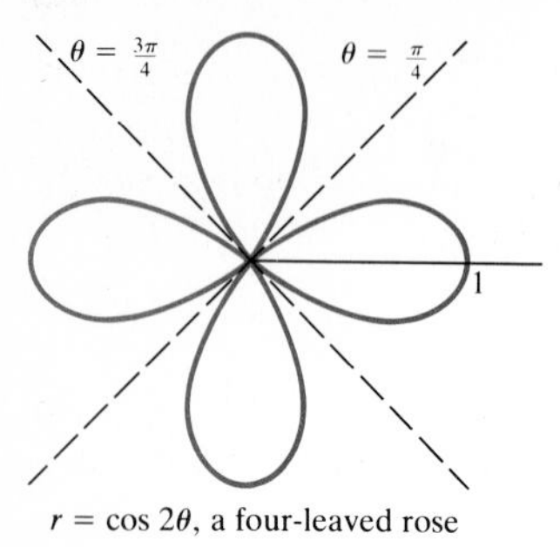

$r = \cos 2\theta$, a four-leaved rose

Figure 9.9

SOLUTION To isolate one loop, find the two smallest nonnegative values of θ for which $\cos 2\theta = 0$. These values are the θ that satisfy $2\theta = \pi/2$ and $2\theta = 3\pi/2$; thus $\theta = \pi/4$ and $\theta = 3\pi/4$. One leaf is described by letting θ go from $\pi/4$ to $3\pi/4$. For θ in $[\pi/4, 3\pi/4]$, 2θ is in $[\pi/2, 3\pi/2]$. Since 2θ is then a second- or third-quadrant angle, $r = \cos 2\theta$ is *negative* or 0. In particular, when $\theta = \pi/2$, $\cos 2\theta$ reaches its smallest value, -1. This loop is the bottom one in Fig. 9.9. The other loops are obtained similarly. ■

EXAMPLE 6 Transform the equation $y = 2$, which describes a horizontal straight line, into polar coordinates.

SOLUTION Since $y = r \sin \theta$, $r \sin \theta = 2$,

or
$$r = \frac{2}{\sin \theta} = 2 \csc \theta.$$

This is more complicated than the original equation, but it is still sometimes useful. ■

EXAMPLE 7 Transform the equation $r = 2 \cos \theta$ into rectangular coordinates and graph it.

SOLUTION Since $r^2 = x^2 + y^2$ and $r \cos \theta = x$, first multiply the equation $r = 2 \cos \theta$ by r, obtaining

$$r^2 = 2r \cos \theta.$$

Hence
$$x^2 + y^2 = 2x.$$

To graph this curve, rewrite the equation as

$$x^2 - 2x + y^2 = 0$$

and complete the square, obtaining

$$(x - 1)^2 + y^2 = 1.$$

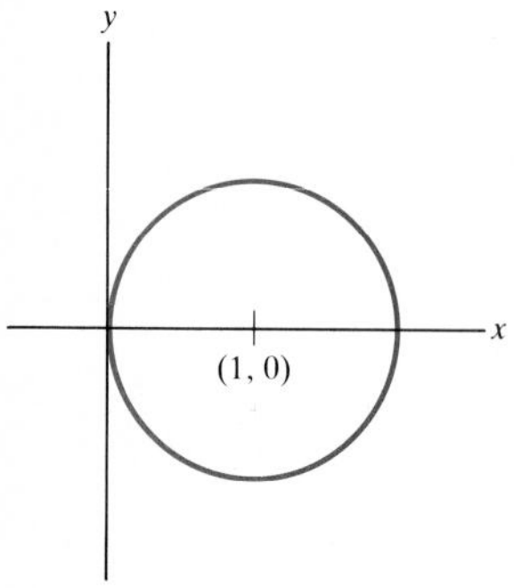

The graph is a circle of radius 1 and center at (1, 0) in rectangular coordinates. It is graphed in Fig. 9.10. ■

$r = 2 \cos \theta$, a circle

Figure 9.10

Remark: In multiplying $r = 2 \cos \theta$ by r, we obtain $r^2 = 2r \cos \theta$. Clearly, the pole lies on the graph of the new equation. Fortunately, it was already present in the graph of $r = 2 \cos \theta$, being obtained when $\theta = \pi/2$; thus we did not change the graph by multiplying by r.

Finding the intersection of two curves in polar coordinates is complicated by the fact that a given point has many descriptions in polar coordinates. Example 8 illustrates how to find the intersection.

EXAMPLE 8 Find the intersection of the curve $r = 1 - \cos \theta$ and the circle $r = \cos \theta$.

SOLUTION First graph the curves. The curve $r = \cos \theta$ is a circle half the size of the one in Example 7. The curve $r = 1 - \cos \theta$ is shown in

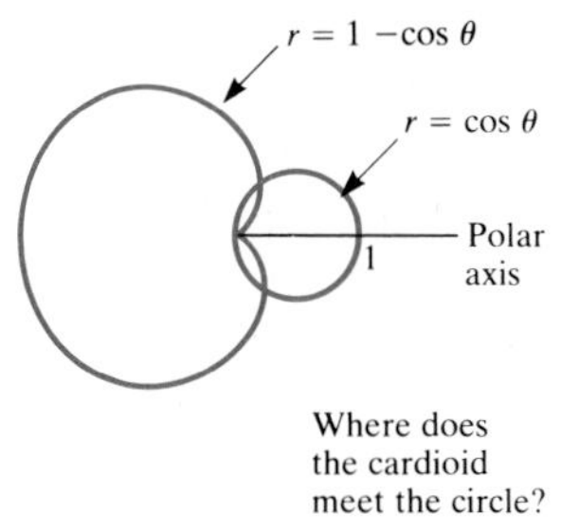

Figure 9.11

Fig. 9.11. (It is a cardioid, being congruent to $r = 1 + \cos\theta$.) It appears that there are three points of intersection.

If a point of intersection is produced because the same value of θ yields the same value of r in both equations, we would have

$$1 - \cos\theta = \cos\theta.$$

Hence $\cos\theta = \frac{1}{2}$. Thus $\theta = \pi/3$ or $\theta = -\pi/3$ (or any angle differing from these by $2n\pi$, n an integer). This gives two of the three points, but it fails to give the origin. Why?

How does the origin get to be on the circle $r = \cos\theta$? Because, when $\theta = \pi/2$, $r = 0$. How does it get to be on the cardioid $r = 1 - \cos\theta$? Because, when $\theta = 0$, $r = 0$. The origin lies on both curves, but we would not learn this by simply equating $1 - \cos\theta$ and $\cos\theta$. ■

When checking for the intersection of two curves, $r = f(\theta)$ and $r = g(\theta)$ in polar coordinates, examine the origin separately. The curves may also intersect at other points not obtainable by setting $f(\theta) = g(\theta)$. This possibility is due to the fact that a given point P has an infinite number of descriptions in polar coordinates: (r, θ) is the same as the points $(r, \theta + 2n\pi)$ and $(-r, \theta + (2n + 1)\pi)$ for any integer n. The safest procedure is to graph the two curves first and then see why they intersect at the points suggested by the graphs.

EXERCISES FOR SEC. 9.1: POLAR COORDINATES

1 Plot the points whose polar coordinates are
(*a*) $(1, \pi/6)$ (*b*) $(2, \pi/3)$
(*c*) $(2, -\pi/3)$ (*d*) $(-2, \pi/3)$
(*e*) $(2, 7\pi/3)$ (*f*) $(0, \pi/4)$

2 Find the rectangular coordinates of the points in Exercise 1.

3 Give at least three pairs of polar coordinates (r, θ) for the point $(3, \pi/4)$, (*a*) with $r > 0$, (*b*) with $r < 0$.

4 Find polar coordinates (r, θ), with $0 \le \theta < 2\pi$ and r positive, for the point whose rectangular coordinates are
(*a*) $(\sqrt{2}, \sqrt{2})$ (*b*) $(-1, \sqrt{3})$
(*c*) $(-5, 0)$ (*d*) $(-\sqrt{2}, -\sqrt{2})$
(*e*) $(0, -3)$ (*f*) $(1, 1)$

In Exercises 5 to 10 transform the equation into one in rectangular coordinates.

5 $r = \sin\theta$ 6 $r = \csc\theta$

7 $r = 3/(4\cos\theta + 5\sin\theta)$

8 $r = 4\cos\theta + 5\sin\theta$

9 $r = \sin 2\theta$ *Hint:* Use the identity for $\sin 2\theta$.

10 $r = 6$

In Exercises 11 to 16 transform the equation into one in polar coordinates.

11 $x + 2y = 3$ 12 $x^2 + y^2 = 5$

13 $xy = 1$ 14 $x^2 + y^2 = 4x$

15 $x = -2$ 16 $y = x^2$

In Exercises 17 to 22 graph the given equations.

17 $r = 1 + \sin\theta$ 18 $r = 3 + 2\cos\theta$

19 $r = 1/\theta,\ \theta > 0$ 20 $r = e^\theta,\ \theta \ge 0$

21 $r = \cos 3\theta$ 22 $r = \sin 2\theta$

23 Where does the cardioid $r = 1 + \cos\theta$ intersect the cardioid $r = \cos\theta - 1$?

24 Where does the four-leaved rose $r = \sin 2\theta$ intersect the circle $r = 1$?

25 Where does the three-leaved rose $r = \sin 3\theta$ intersect the three-leaved rose $r = \cos 3\theta$?

26 Where does the four-leaved rose $r = 2\sin 2\theta$ intersect the circle $r = 1$?

In Exercises 27 and 28 find the intersections of the given curves.

27 $r = \sin\theta,\ r = \cos 2\theta$ 28 $r = \cos\theta,\ r = \cos 2\theta$

29 Explain in detail why the graph of $r = \sin 2\theta$ has four leaves while the graph of $r = \sin 3\theta$ has three leaves.

30 Find an equation in rectangular coordinates for the cardioid $r = 1 + \sin\theta$.

■

31 (See Example 7.) Use elementary geometry and trigonometry to show that any point on the circle of Fig. 9.10 satisfies the equation $r = 2 \cos \theta$.

32 Show that all chords through the pole of the cardioid $r = 1 + \cos \theta$ have the same length.

The curve $r = 1 + a \cos \theta$ (or $r = 1 + a \sin \theta$) is called a **limaçon**. Its shape depends on the choice of the constant a. For $a = 1$, we have the cardioid of Example 2. Exercises 33 and 34 concern other choices of a.

33 Graph $r = 1 + 2 \cos \theta$. (If $|a| > 1$, the graph of $r = 1 + a \cos \theta$ crosses itself and forms a loop.)

34 Graph $r = 1 + \frac{1}{2} \cos \theta$.

35 Consider the curve $r = 1 + a \cos \theta$, where a is fixed, $0 \le a \le 1$.
(*a*) Relative to the same polar axis, graph the curves corresponding to $a = 0, \frac{1}{4}, \frac{1}{2}, \frac{3}{4}, 1$.
(*b*) For $a = \frac{1}{4}$ the graph in (*a*) is convex, but not for $a = 1$. Show that for $\frac{1}{2} < a \le 1$ the curve is not convex. ("Convex" is defined on p. 59.) *Hint:* Find the points on the curve farthest to the left and compare them to the point on the curve corresponding to $\theta = \pi$.

36 (*a*) Obtain the rectangular form of the equation $r^2 = \cos 2\theta$.
(*b*) Graph the curve in part (*a*) using polar coordinates. Note that, if $\cos 2\theta$ is negative, r is not defined and that, if $\cos 2\theta$ is positive, there are two values of r, $\sqrt{\cos 2\theta}$ and $-\sqrt{\cos 2\theta}$.

In Appendix F.4 it is shown that the graph of $r = 1/(1 + e \cos \theta)$ is a parabola if $e = 1$, an ellipse if $0 \le e < 1$, and a hyperbola if $e > 1$. ("e" here is not related to $e \approx 2.718$.) Exercises 37 to 39 concern such graphs.

37 Find an equation in rectangular coordinates for the curve $r = 1/(1 + \cos \theta)$.

38 (*a*) Graph $$r = \frac{1}{1 - \frac{1}{2} \cos \theta}.$$
(*b*) Find an equation in rectangular coordinates for the curve in (*a*).

39 (*a*) Graph $$r = \frac{1}{1 + 2 \cos \theta}.$$
(*b*) What angles do the asymptotes to the graph in (*a*) make with the positive x axis?
(*c*) Find an equation in rectangular coordinates for the curve in (*a*).

40 (*a*) Graph $r = 3 + \cos \theta$.
(*b*) Find the point on the graph in (*a*) that has the maximum y coordinate.

41 Find the y coordinate of the highest point on the right-hand leaf of the four-leaved rose $r = \cos 2\theta$.

42 Sketch (*a*) $r = \sin 4\theta$, (*b*) $r = \sin 5\theta$.

43 Where do the spirals $r = \theta$ and $r = 2\theta$, for $\theta \ge 0$, intersect?

9.2 AREA IN POLAR COORDINATES

Section 5.3 showed how to compute the area of a region if the lengths of parallel cross sections are known. Sums based on estimating rectangles led to the formula

$$\text{Area} = \int_a^b c(x)\, dx,$$

where $c(x)$ denotes the cross-sectional length. Now consider quite a different situation, in which sectors of a circle, not rectangles, provide an estimate of the area.

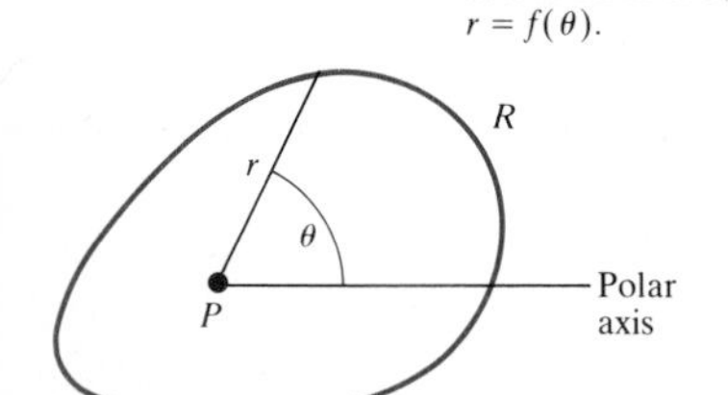

The distance r for any ray from P is known as a function of θ, $r = f(\theta)$.

Figure 9.12

Let R be a region in the plane and P a point inside it. Assume that the distance r from P to any point on the boundary of R is known as a function $r = f(\theta)$. (For convenience, assume that any ray from P meets the boundary of R just once, as in Fig. 9.12.)

The cross sections made by the rays from P are *not* parallel. Instead, like spokes in a wheel, they all meet at the point P. It would be unnatural to use rectangles to estimate the area, but it is reasonable to use sectors of circles that have P as a common vertex.

Begin by recalling that in a circle of radius r a sector of central angle θ has area $(\theta/2)r^2$. (See Fig. 9.13.) This formula plays the same role now as the formula for the area of a rectangle did in Sec. 5.3.

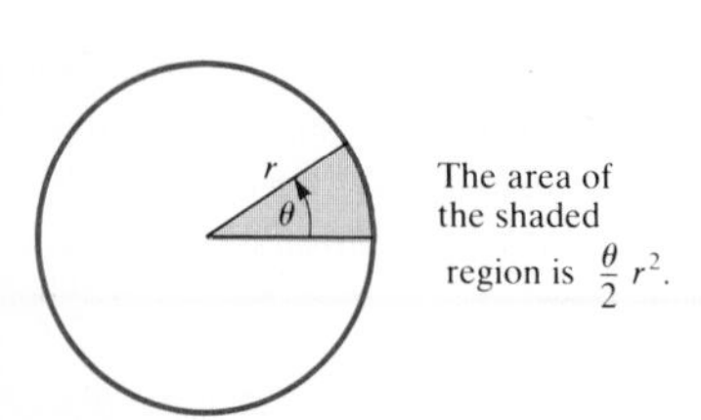

The area of the shaded region is $\frac{\theta}{2} r^2$.

Figure 9.13

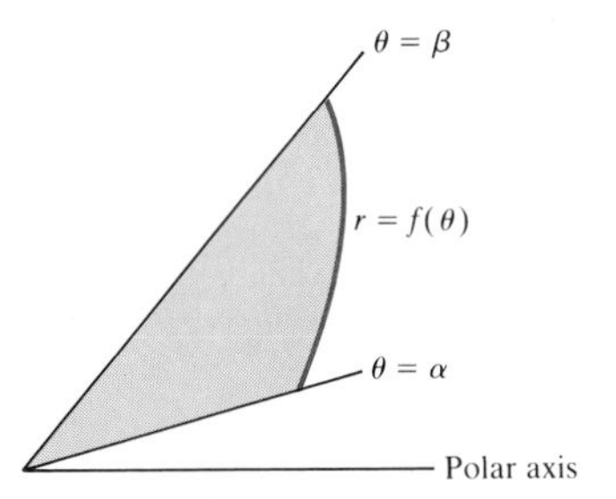

Figure 9.14

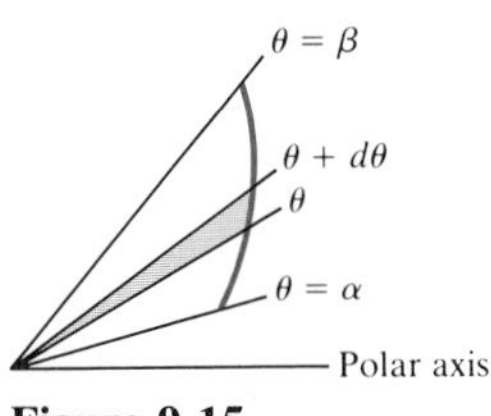

Figure 9.15

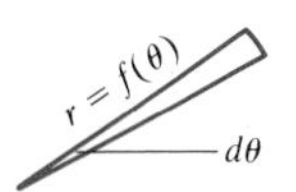

This narrow sector of a circle approximates the narrow wedge in Fig. 9.15.

Figure 9.16

Area in Polar Coordinates (informal approach)

Assume $f(\theta) \geq 0$.

This local estimate is the key to computing area in polar coordinates.

Let R be the region bounded by the rays $\theta = \alpha$ and $\theta = \beta$ and by the curve $r = f(\theta)$, as shown in Fig. 9.14. To obtain a **local estimate** for the area of R, consider the portion of R between the rays corresponding to the angles θ and $\theta + d\theta$, where $d\theta$ is a small positive number. (See Fig. 9.15.) The area of the narrow wedge which is shaded in Fig. 9.15 is approximately that of a sector of a circle of radius $r = f(\theta)$ and angle $d\theta$, shown in Fig. 9.16. The area of the sector in Fig. 9.16 is

$$\frac{[f(\theta)]^2\,d\theta}{2}. \tag{1}$$

Having found the local estimate of area (1), we conclude that the area of R is

$$\int_\alpha^\beta \frac{[f(\theta)]^2\,d\theta}{2}.$$

How to Find Area in Polar Coordinates In summary, the area of the region bounded by the rays $\theta = \alpha$ and $\theta = \beta$ and by the curve $r = f(\theta)$ is

$$\boxed{\int_\alpha^\beta \frac{[f(\theta)]^2}{2}\,d\theta \quad \text{or} \quad \int_\alpha^\beta \frac{r^2\,d\theta}{2}.} \tag{2}$$

[Assume $f(\theta) \geq 0$.] It must be emphasized that no ray from the origin between α and β can cross the curve twice.

Formula (2) is applied in Sec. 14.8 to the motion of satellites and planets.

It may seem surprising to find $[f(\theta)]^2$, not just $f(\theta)$, in the integrand. But remember that area has the dimension "length times length." Since θ, given in radians, is dimensionless, being defined as "length of circular arc divided by length of radius," $d\theta$ is also dimensionless. Hence $f(\theta)\,d\theta$, having the dimension of length, not of area, could not be correct. But $\frac{1}{2}[f(\theta)]^2\,d\theta$, having the dimension of area (length times length), is plausible. For rectangular coordinates, in the expression $f(x)\,dx$, both $f(x)$ and dx have the dimension of length, one along the y axis, the other along the x axis; thus $f(x)\,dx$ has the dimension of area.

Memory device

As an aid in remembering the area of the narrow sector in Fig. 9.16, note that it resembles a triangle of height r and base $r\,d\theta$. Its area is in fact

$$\frac{1}{2}\cdot r \cdot r\,d\theta = \frac{r^2\,d\theta}{2}.$$

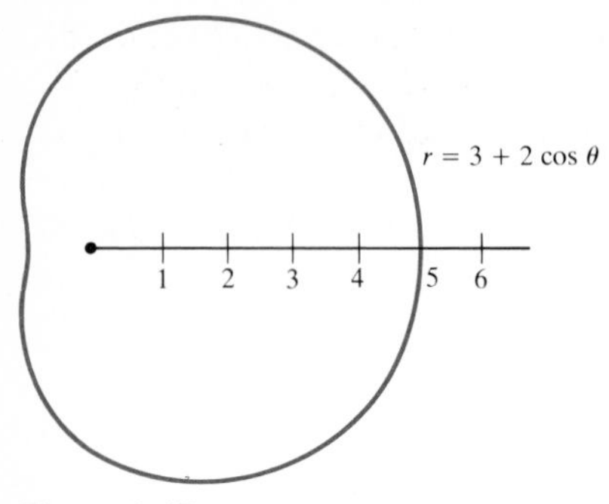

Figure 9.17

EXAMPLE 1 Find the area of the region bounded by the curve $r = 3 + 2\cos\theta$, shown in Fig. 9.17.

SOLUTION By the formula just obtained, this area is

$$\int_0^{2\pi} \tfrac{1}{2}(3 + 2\cos\theta)^2\,d\theta = \frac{1}{2}\int_0^{2\pi}(9 + 12\cos\theta + 4\cos^2\theta)\,d\theta$$

$$= \tfrac{1}{2}(9\theta + 12\sin\theta + 2\theta + \sin 2\theta)\Big|_0^{2\pi} = 11\pi. \quad \blacksquare$$

Observe that any line through the origin intersects the region of Example 1 in a segment of length 6, since $(3 + 2\cos\theta) + [3 + 2\cos(\theta + \pi)] = 6$ for any θ. Also, any line through the center of a circle of radius 3 intersects the circle in a segment of length 6. Thus two sets in the plane can have equal corresponding cross-sectional lengths through a fixed point and yet have different areas: the set in Example 1 has area 11π, while the circle of radius 3 has area 9π. *Knowing the lengths of all the cross sections of a region through a given point is not enough to determine the area of the region!*

EXAMPLE 2 Find the area of the region inside one of the eight loops of the eight-leaved rose $r = \cos 4\theta$.

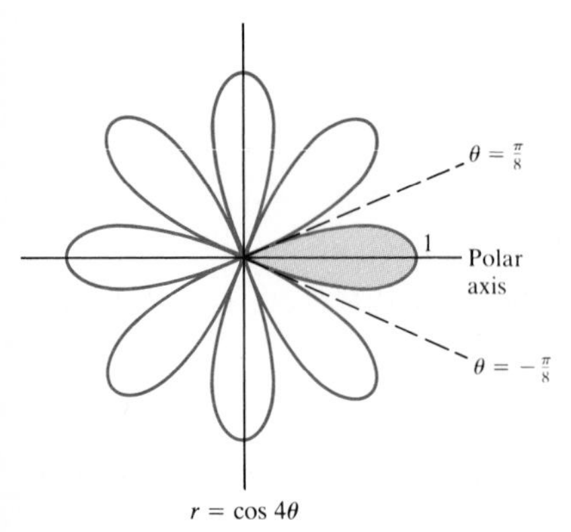

Figure 9.18

SOLUTION To graph one of the loops, start with $\theta = 0$. For that angle, $r = \cos(4 \cdot 0) = \cos 0 = 1$. The point $(r, \theta) = (1, 0)$ is the outer tip of a loop. As θ increases from 0 to $\pi/8$, $\cos 4\theta$ decreases from $\cos 0 = 1$ to $\cos(\pi/2) = 0$. One of the eight loops is therefore bounded by the rays $\theta = \pi/8$ and $\theta = -\pi/8$. It is shown in Fig. 9.18, which, for good measure, displays all eight loops.

The area of the loop which is bisected by the polar axis is

$$\int_{-\pi/8}^{\pi/8} \frac{r^2}{2}\,d\theta = \int_{-\pi/8}^{\pi/8} \frac{\cos^2 4\theta}{2}\,d\theta$$

$$= \int_{-\pi/8}^{\pi/8} \frac{1 + \cos 8\theta}{4}\,d\theta$$

$$= \left(\frac{\theta}{4} + \frac{\sin 8\theta}{32}\right)\Bigg|_{-\pi/8}^{\pi/8}$$

$$= \left(\frac{\pi}{32} + \frac{\sin\pi}{32}\right) - \left[\frac{-\pi}{32} + \frac{\sin(-\pi)}{32}\right]$$

$$= \frac{\pi}{16}.$$

Incidentally, since the part of the loop above the polar axis has the same area as the part below,

$$\int_{-\pi/8}^{\pi/8} \frac{r^2}{2}\,d\theta = 2\int_0^{\pi/8} \frac{r^2}{2}\,d\theta.$$

This shortcut simplifies the arithmetic a little, reducing the chance for error when evaluating the integral. $\blacksquare$

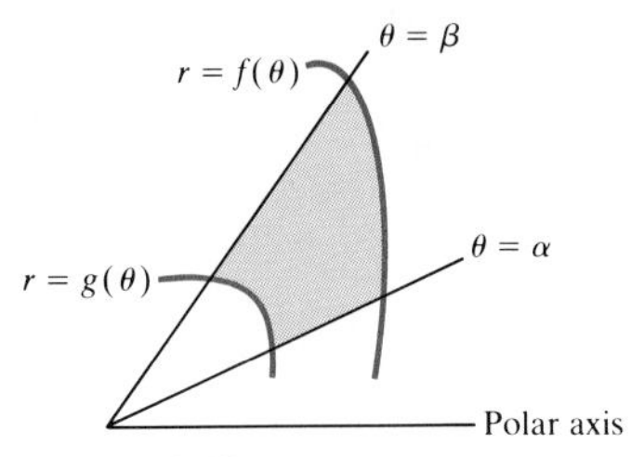

Figure 9.19

The Area between Two Curves Assume that $r = f(\theta)$ and $r = g(\theta)$ describe two curves in polar coordinates and that $f(\theta) \geq g(\theta) \geq 0$ for θ in $[\alpha, \beta]$. Let R be the region between these two curves and the rays $\theta = \alpha$ and $\theta = \beta$, as shown in Fig. 9.19.

The area of R is obtained by subtracting the area within the inner curve, $r = g(\theta)$, from the area within the outer curve, $r = f(\theta)$. That is,

$$\text{Area between two curves} = \int_\alpha^\beta \tfrac{1}{2}[f(\theta)]^2\,d\theta - \int_\alpha^\beta \tfrac{1}{2}[g(\theta)]^2\,d\theta = \frac{1}{2}\int_\alpha^\beta \{[f(\theta)]^2 - [g(\theta)]^2\}\,d\theta. \tag{3}$$

EXAMPLE 3 Find the area of the region between the curves $r = 3 + \sin\theta$ and $r = 2 + \sin\theta$.

SOLUTION Each curve is swept out once as θ goes from 0 to 2π. Moreover, $3 + \sin\theta \geq 2 + \sin\theta \geq 0$. Thus the area between them is

$$\frac{1}{2}\int_0^{2\pi} [(3 + \sin\theta)^2 - (2 + \sin\theta)^2]\,d\theta$$

$$= \frac{1}{2}\int_0^{2\pi} [(9 + 6\sin\theta + \sin^2\theta) - (4 + 4\sin\theta + \sin^2\theta)]\,d\theta$$

$$= \frac{1}{2}\int_0^{2\pi} (5 + 2\sin\theta)\,d\theta$$

$$= \frac{1}{2}(5\theta - 2\cos\theta)\Big|_0^{2\pi}$$

$$= 5\pi. \quad \blacksquare$$

The next example shows that care must be taken when finding the area between two curves given in polar coordinates.

EXAMPLE 4 Find the area of the top half of the region inside the cardioid $r = 1 + \cos\theta$ and outside the circle $r = \cos\theta$.

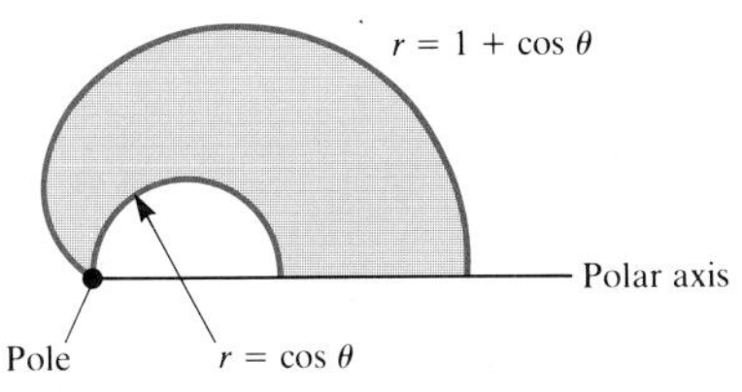

Figure 9.20

SOLUTION The region is shown in Fig. 9.20.

The top half of the circle $r = \cos\theta$ is swept out as θ goes from 0 to $\pi/2$. The top half of the cardioid is swept out as θ goes from 0 to π. Formula (3) for the area between two curves does not cover this situation. The safest thing is to compute the area of the top half of the cardioid and the top half of the circle separately and then subtract the second from the first.

The area of the top half of the cardioid is

$$\frac{1}{2}\int_0^\pi (1 + \cos\theta)^2\,d\theta = \frac{1}{2}\int_0^\pi (1 + 2\cos\theta + \cos^2\theta)\,d\theta.$$

Now,
$$\int_0^\pi \cos\theta\,d\theta = \sin\theta\Big|_0^\pi = 0.$$

Why is $\int_{\pi/2}^{\pi} \cos^2\theta \, d\theta = \int_0^{\pi/2} \cos^2\theta \, d\theta$*?*

Also,

$$\int_0^{\pi} \cos^2 \theta \, d\theta = 2 \int_0^{\pi/2} \cos^2 \theta \, d\theta = 2\left(\frac{\pi}{4}\right) = \frac{\pi}{2},$$

by the trick in Sec. 7.5. Thus the area of the top half of the cardioid is

$$\frac{1}{2}\left(\pi + 0 + \frac{\pi}{2}\right) = \frac{3\pi}{4}.$$

The area of the top half of the circle $r = \cos \theta$ is

Or note that it's half the area of a circle of radius $\frac{1}{2}$.

$$\frac{1}{2}\int_0^{\pi/2} \cos^2 \theta \, d\theta = \frac{\pi}{8}.$$

Thus the area in question is

$$\frac{3\pi}{4} - \frac{\pi}{8} = \frac{5\pi}{8}. \quad \blacksquare$$

Area in Polar Coordinates (formal approach)

We wish to estimate the area of the shaded region in Fig. 9.21. To do so, introduce a partition of $[\alpha, \beta]$:

$$\alpha = \theta_0 < \theta_1 < \cdots < \theta_n = \beta,$$

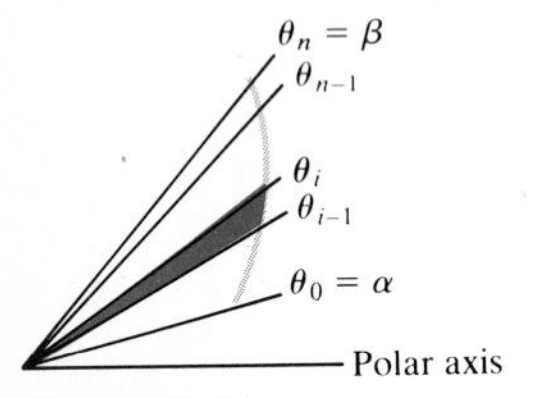

Figure 9.21

and estimate the area of R bounded between the rays whose angles are

$$\theta_{i-1} \quad \text{and} \quad \theta_i,$$

as shown in Fig. 9.22.

Pick any angle θ_i^* in the interval $[\theta_{i-1}, \theta_i]$ and use a sector of a circle whose radius is $f(\theta_i^*)$ to approximate the area of the shaded region. The approximating sector has radius $f(\theta_i^*)$ and angle $\theta_i - \theta_{i-1}$; hence its area is

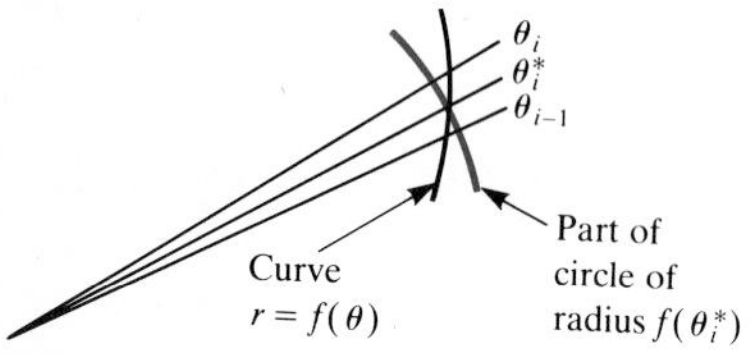

$$\frac{[f(\theta_i^*)]^2}{2}(\theta_i - \theta_{i-1}).$$

θ_i
θ_i^*
θ_{i-1}
Curve $r = f(\theta)$
Part of circle of radius $f(\theta_i^*)$

Figure 9.22

The sum

$$\sum_{i=1}^{n} \frac{[f(\theta_i^*)]^2}{2}(\theta_i - \theta_{i-1})$$

is an estimate of the area of R.

As the mesh of the partition of $[\alpha, \beta]$ approaches 0, these sums become better and better approximations to the area of R and also approach the definite integral

$$\int_\alpha^\beta \frac{[f(\theta)]^2}{2} \, d\theta,$$

which, therefore, is the area of R. This establishes formula (2).

EXERCISES FOR SEC. 9.2: AREA IN POLAR COORDINATES

In each of Exercises 1 to 6 find the area of the region bounded by the indicated curve and rays.

1 $r = 2\theta, \alpha = 0, \beta = \dfrac{\pi}{2}$.

2 $r = \sqrt{\theta}, \alpha = 0, \beta = \pi$.

3 $r = \dfrac{1}{1 + \theta}, \alpha = \dfrac{\pi}{4}, \beta = \dfrac{\pi}{2}$.

4 $r = \sqrt{\sin\theta}$, $\alpha = 0$, $\beta = \dfrac{\pi}{2}$.

5 $r = \tan\theta$, $\alpha = 0$, $\beta = \dfrac{\pi}{4}$.

6 $r = \sec\theta$, $\alpha = \dfrac{\pi}{6}$, $\beta = \dfrac{\pi}{4}$.

7 (*a*) Graph the curve $r = 2\sin\theta$.
(*b*) Compute the area inside it.

8 Find the area within the first turn of the spiral $r = e^\theta$, that is, for $0 \le \theta \le 2\pi$.

9 Find the area of the region inside the cardioid $r = 3 + 3\sin\theta$ and outside the circle $r = 3$.

10 (*a*) Graph the curve $r = \sqrt{\cos 2\theta}$. (Note that r is not defined for all θ.)
(*b*) Find the area inside one of its two loops.

In Exercises 11 to 18 find the areas of the regions described.

11 Inside one loop of the curve $r = \sin 3\theta$.

12 Inside one loop of the curve $r = \cos 2\theta$.

13 Inside one loop of the curve $r = \sin 4\theta$.

14 Inside one loop of the curve $r = \cos 5\theta$.

15 Inside one loop of the curve $r = 2\cos 2\theta$, but outside the circle $r = 1$.

16 Inside the cardioid $r = 1 + \cos\theta$, but outside the circle $r = \sin\theta$. (*Suggestion:* Graph both curves first.)

17 Inside the circle $r = \sin\theta$, but outside the circle $r = \cos\theta$.

18 Inside the curve $r = 4 + \sin\theta$, but outside the curve $r = 3 + \sin\theta$.

■

19 (*a*) Show that the area of the triangle in Fig. 9.23 is $\int_0^\beta \frac{1}{2}\sec^2\theta\, d\theta$.
(*b*) From (*a*) and the fact that the area of a triangle is $\frac{1}{2}$(base)(height), show that $\tan\beta = \int_0^\beta \sec^2\theta\, d\theta$.
(*c*) With the aid of this equation, obtain another proof that $(\tan x)' = \sec^2 x$.

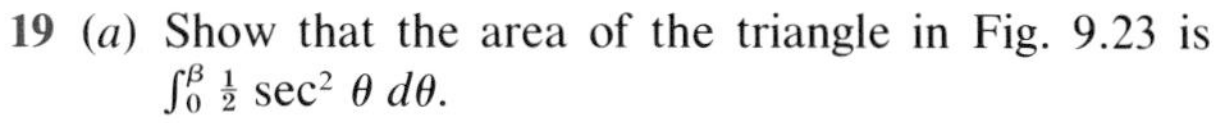

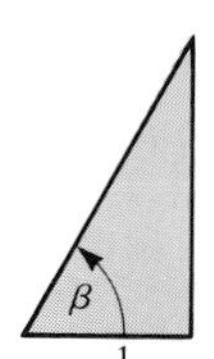

Figure 9.23

20 Show that the area of the shaded crescent between the two circular arcs is equal to the area of square $ABCD$. (See Fig. 9.24.) This type of result encouraged mathematicians from the time of the Greeks to try to find a method using only straightedge and compass for constructing a square whose area equals that of a given circle. This was proved impossible at the end of the nineteenth century.

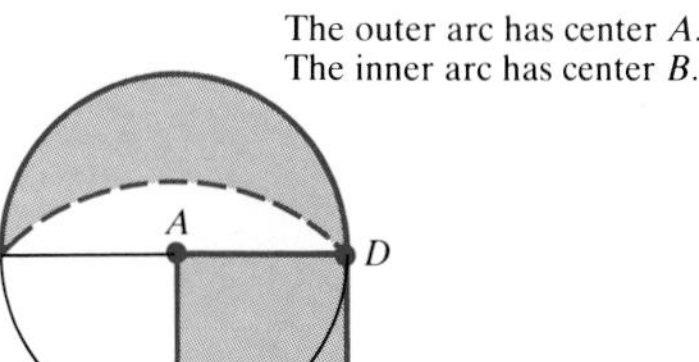

Figure 9.24

21 (*a*) Graph $r = 1/\theta$ for $0 < \theta \le \pi/2$.
(*b*) Is the area of the region bounded by the curve drawn in (*a*) and the rays $\theta = 0$ and $\theta = \pi/2$ finite or infinite?

22 (*a*) Sketch the curve $r = 1/(1 + \cos\theta)$.
(*b*) What is the equation of the curve in (*a*) in rectangular coordinates?
(*c*) Find the area of the region bounded by the curve in (*a*) and the rays $\theta = 0$ and $\theta = 3\pi/4$, using (2).
(*d*) Solve (*c*) using rectangular coordinates and the equation in (*b*).

23 A point P in a region R bounded by a closed curve has the property that each chord through P cuts R into two regions of equal area. Must P bisect each chord through P? Explain. (Assume that each line through P meets the curve at two points.)

24 Let R be a region in the plane bounded by a loop whose equation is $r = f(\theta)$ in polar coordinates. Assume that every chord of R that passes through the pole has length at least 1.
(*a*) Draw at least two examples of such an R.
(*b*) Make a general conjecture about the area of R.
(*c*) Prove it.

25 Like Exercise 24, except each chord through the pole has length at most 1.

26 The curve $r = 3 + 2\cos\theta$ was sketched in Example 1.
(*a*) Find the x coordinate of the points furthest to the left.
(*b*) Find the y coordinate of the highest point.

27 Prove that if a region in the plane has the property that any two points in it are within a distance d of each other, then its area is at most $\pi d^2/4$. *Hint:* Use polar coordinates with the pole on the border of the region, and consider $r(\theta)$ and $r(\theta + \pi/2)$.

28 Sketch the graph of $r = 4 + \cos\theta$. Is it a circle?

9.3 PARAMETRIC EQUATIONS

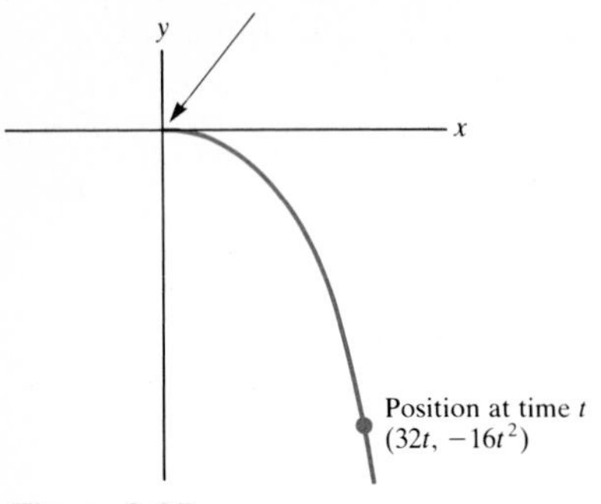

Figure 9.25

If a ball is thrown horizontally with a speed of 32 feet per second, it falls in a curved path. Air resistance disregarded, its position t seconds later is given by $x = 32t$, $y = -16t^2$ relative to the coordinate system in Fig. 9.25. Here the curve is completely described, not by expressing y as a function of x, but by expressing both x and y as functions of a third variable t. The third variable is called a **parameter** (*para* meaning "together," *meter* meaning "measure"). The equations $x = 32t$, $y = -16t^2$ are called **parametric equations** for the curve.

In this example it is easy to eliminate t and so find a direct relation between x and y:

$$t = \frac{x}{32}.$$

Hence $$y = -16\left(\frac{x}{32}\right)^2 = -\frac{16}{(32)^2}x^2 = -\frac{1}{64}x^2.$$

The path of the falling ball is part of the parabola $y = -\frac{1}{64}x^2$.

EXAMPLE 1 Let $\begin{cases} x = \cos 2t \\ y = \sin t \end{cases}$ describe a curve parametrically. Graph this curve and find an equation linking x and y.

SOLUTION Since x and y are the sine and cosine of angles, $|x|$ and $|y|$ never exceed 1. Moreover when t increases by 2π, we obtain the same point again. Here is a table showing a few points on the graph.

t	0	$\frac{\pi}{4}$	$\frac{\pi}{2}$	$\frac{3\pi}{4}$	π	$\frac{5\pi}{4}$	$\frac{3\pi}{2}$	$\frac{7\pi}{4}$	2π
x	1	0	-1	0	1	0	-1	0	1
y	0	$\frac{\sqrt{2}}{2}$	1	$\frac{\sqrt{2}}{2}$	0	$\frac{-\sqrt{2}}{2}$	-1	$\frac{-\sqrt{2}}{2}$	0

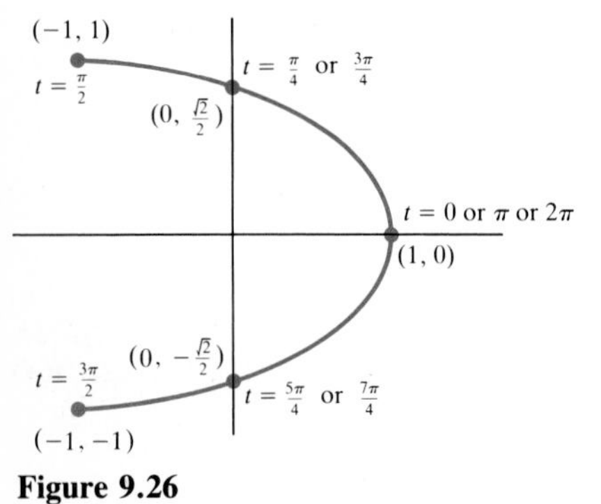

Figure 9.26

The last point duplicates the first. The first eight points suggest the shape of the graph, which is shown in Fig. 9.26. For t in the interval $[0, 2\pi]$, the path runs over the graph twice.

The curve looks like part of a parabola. Fortunately, the equations for x and y are sufficiently simple that t can be eliminated and a relation between x and y found:

$$\begin{aligned} x = \cos 2t &= \cos^2 t - \sin^2 t \\ &= 1 - 2\sin^2 t \\ &= 1 - 2y^2. \end{aligned}$$

Thus the path lies on the parabola

$$x = 1 - 2y^2.$$

Appendix F shows that this is a parabola.

But note that it is only a small part of the parabola and sweeps out this part infinitely often, like a pendulum. ■

In Example 1 it was easy to eliminate the parameter t, thus obtaining a simple equation involving only x and y. In the next example, elimination of the parameter would lead to a complicated equation involving x and y. One advantage of parametric equations is that they can provide a simple description of a curve, although it may be impossible to find an equation in x and y which describes the curve.

EXAMPLE 2 As a bicycle wheel of radius a rolls along, a tack stuck in its circumference traces out a curve called a **cycloid**, which consists of a sequence of arches, one arch for each revolution of the wheel. (See Fig. 9.27.)

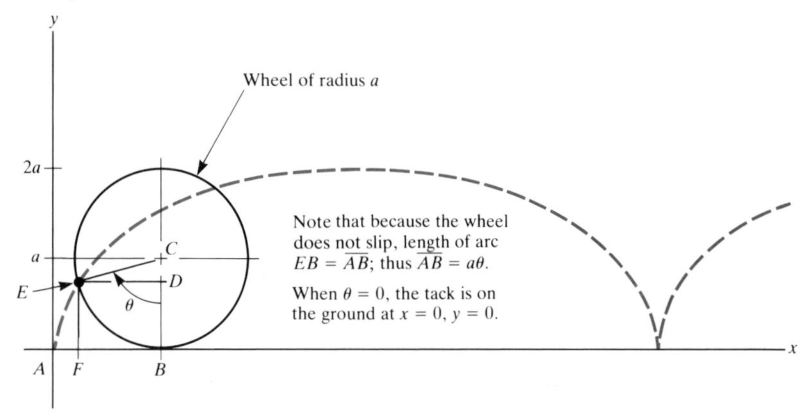

Figure 9.27

Find the position of the tack as a function of the angle θ through which the wheel turns.

SOLUTION The x coordinate of the tack, corresponding to θ, is

$$\overline{AF} = \overline{AB} - \overline{ED} = a\theta - a \sin \theta,$$

and the y coordinate is

$$\overline{EF} = \overline{BC} - \overline{CD} = a - a \cos \theta.$$

Then the position of the tack, as a function of the parameter θ, is

$$\begin{cases} x = a\theta - a \sin \theta \\ y = a - a \cos \theta. \end{cases}$$

In this case, eliminating θ would lead to a complicated relation between x and y. ■

Any curve $y = f(x)$ can be given parametrically: $x = t$, $y = f(t)$.

Any curve can be described parametrically. For instance, consider the curve $y = e^x + x$. It is perfectly legal to introduce a parameter t equal to x and write

$$\begin{cases} x = t \\ y = e^t + t. \end{cases}$$

This device may seem a bit artificial, but it will be useful in the next section in order to apply results for curves expressed by means of parametric equations to curves given in the form $y = f(x)$.

How can we find the *slope of a curve which is described parametrically* by the equations

$$x = g(t), \qquad y = h(t)?$$

An often difficult, perhaps impossible, approach is to solve the equation $x = g(t)$ for t as a function of x and plug the result into the equation $y = h(t)$, thus expressing y explicitly in terms of x; then differentiate the result to find dy/dx. Fortunately, there is a very easy way, which we will now describe.

Assume that y is a differentiable function of x. Then, by the chain rule,

$$\frac{dy}{dt} = \frac{dy}{dx}\frac{dx}{dt},$$

from which it follows that

Formula for the slope of a parameterized curve

$$\boxed{\frac{dy}{dx} = \frac{dy/dt}{dx/dt}.} \tag{1}$$

It is assumed that in formula (1) dx/dt is not 0.

A similar approach gives d^2y/dx^2. By the chain rule,

$$\frac{d}{dt}\left[\frac{dy}{dx}\right] = \frac{d}{dx}\left[\frac{dy}{dx}\right]\frac{dx}{dt} = \frac{d^2y}{dx^2}\frac{dx}{dt}.$$

Solving this equation for d^2y/dx^2 gives

$$\frac{d^2y}{dx^2} = \frac{\dfrac{d}{dt}\left[\dfrac{dy}{dx}\right]}{\dfrac{dx}{dt}}.$$

EXAMPLE 3 At what angle does the arch of the cycloid shown in Example 2 meet the x axis at the origin?

SOLUTION The parametric equations of the cycloid are

$$x = a\theta - a\sin\theta, \qquad y = a - a\cos\theta.$$

Here θ is the parameter. Then

$$\frac{dx}{d\theta} = a - a\cos\theta \quad \text{and} \quad \frac{dy}{d\theta} = a\sin\theta.$$

Consequently,

$$\frac{dy}{dx} = \frac{dy/d\theta}{dx/d\theta} = \frac{a\sin\theta}{a - a\cos\theta} = \frac{\sin\theta}{1 - \cos\theta}.$$

When θ is near 0, (x, y) is near the origin. How does the slope, $\sin\theta/(1 - \cos\theta)$, behave as $\theta \to 0^+$? L'Hôpital's rule applies, and we have

$$\lim_{\theta \to 0^+} \frac{\sin \theta}{1 - \cos \theta} = \lim_{\theta \to 0^+} \frac{\cos \theta}{\sin \theta} = \infty.$$

Thus the cycloid comes in vertically at the origin. ■

Geometry of the Rotary Engine (Optional)

The next two examples use parametric equations to describe the geometric principles of the rotary engine recognized by Felix Wankel in 1954. He found that it is possible for an equilateral triangle to revolve in a certain curve in such a way that its corners maintain contact with the curve and its centroid sweeps out a circle.

EXAMPLE 4 Let b and R be fixed positive numbers and consider the curve given parametrically by

$$x = b \cos 3\theta + R \cos \theta \quad \text{and} \quad y = b \sin 3\theta + R \sin \theta.$$

Show that an equilateral triangle can revolve in this curve while its centroid describes a circle of radius b.

SOLUTION Figure 9.28 shows the typical point $P = (x, y)$ that corresponds to the parameter value θ. As θ increases by $2\pi/3$ from any given angle, the point Q goes once around the circle of radius b and returns to its initial position. During this revolution of Q the point P moves to a point P_1 whose angle, instead of being θ, is $\theta + 2\pi/3$. Thus, if P is on the curve, so are the points P_1 and P_2 shown in Fig. 9.29; these form an equilateral triangle.

Consequently, each vertex of the equilateral triangle sweeps out the curve once, while the centroid Q goes three times around the circle of radius b. ■

What does the curve described in Example 4 look like? Wankel graphed it without knowing that mathematicians had met it long before in a different setting, described in Example 5, which provides a way of graphing the curve.

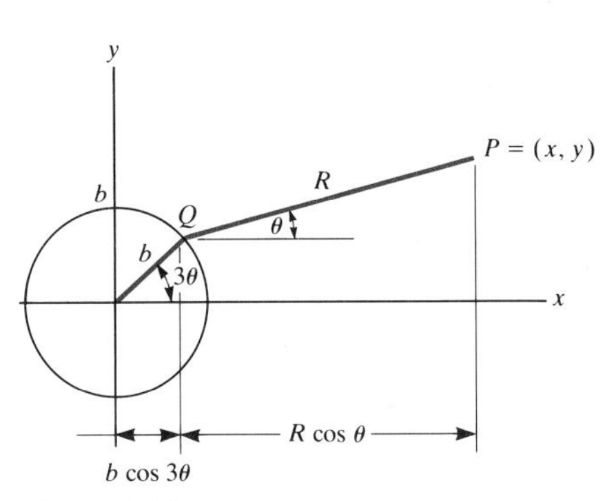

Figure 9.28

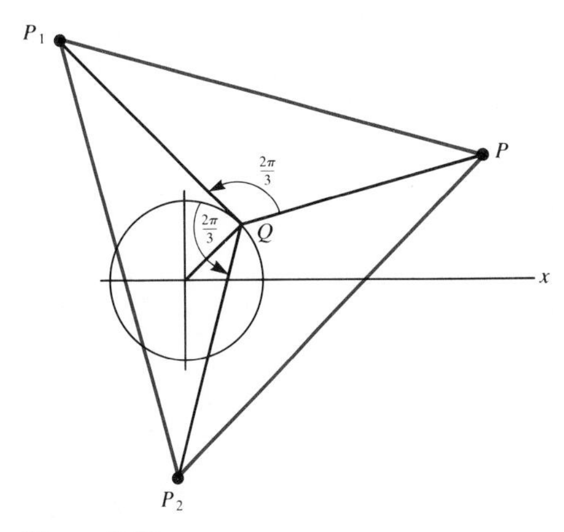

Figure 9.29

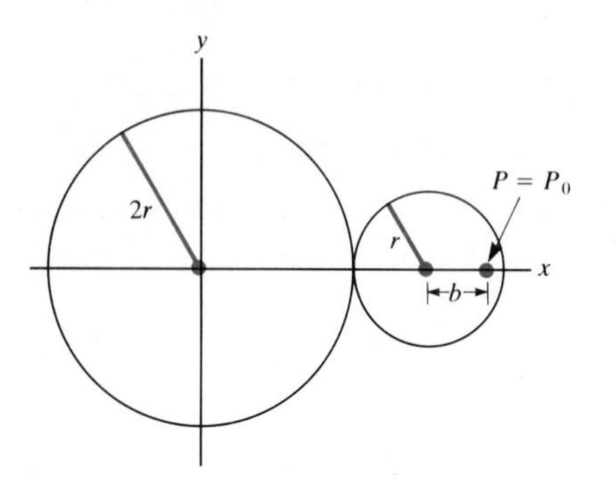

Figure 9.30

Figure 9.31

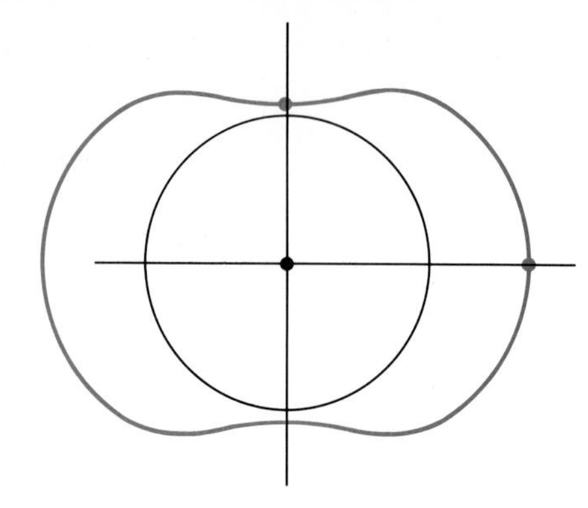

Figure 9.32

EXAMPLE 5 A circle of radius r rolls without slipping around a fixed circle of radius $2r$. Describe the path swept out by a point P located at a distance b from the center of the moving circle, $0 \le b \le r$.

Why is angle ACB equal to 2θ?

SOLUTION Place the rolling circle as shown in Fig. 9.30. Note that the center C of the rolling circle traces out a circle of radius $3r$. Let $R = 3r$.

As the little circle rolls counterclockwise around the fixed circle without slipping, the point P traces out a path whose initial point P_0 is shown in Fig. 9.30. The typical point P on the path as the circle rolls around the larger circle is shown in Fig. 9.31. Since the radius of the rolling circle is half that of the fixed circle (and there is no slipping), angle ACB is 2θ. Thus the angle that CP makes with the x axis is the sum of θ and 2θ, which is 3θ. Consequently, $P = (x, y)$ has coordinates given parametrically as

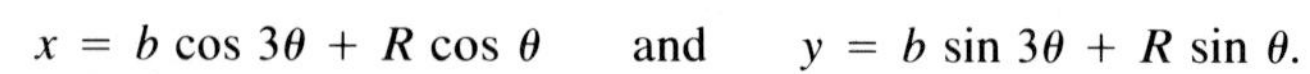

$$x = b \cos 3\theta + R \cos \theta \qquad \text{and} \qquad y = b \sin 3\theta + R \sin \theta.$$

Thus the curve swept out by P is precisely the curve Wankel studied.

Long known to mathematicians as an **epitrochoid**, it appears typically as shown in Fig. 9.32. ■

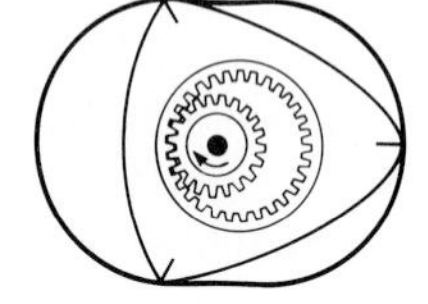

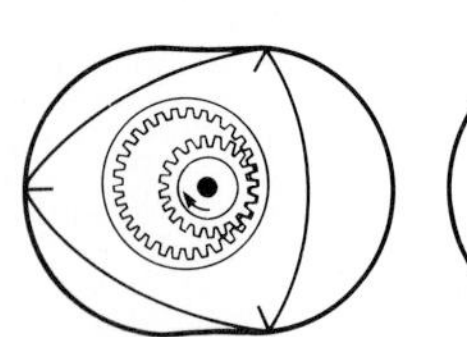

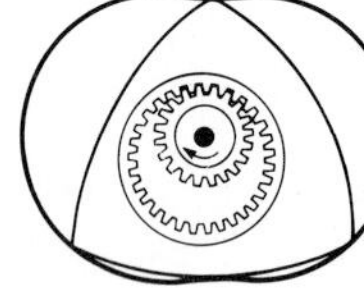

Figure 9.33

In order that the moving rotor in the rotary engine can turn the drive shaft, teeth are placed in it along a circle of radius $2b$ which engage teeth in the drive shaft, which has radius b. (See Fig. 9.33.) For each complete rotation of the rotor the drive shaft completes three rotations.

It was a Stuttgart professor, Othmar Baier, who showed that Wankel's curve was an epitrochoid. This insight was of aid in simplifying the machining of the working surface of the motor.

EXERCISES FOR SEC. 9.3: PARAMETRIC EQUATIONS

1 Consider the parametric equations $x = 2t + 1$, $y = t - 1$.

(*a*) Fill in this table:

t	-2	-1	0	1	2
x					
y					

(*b*) Plot the five points (x, y) obtained in (*a*).
(*c*) Graph the curve given by the parametric equations $x = 2t + 1$, $y = t - 1$.
(*d*) Eliminate t to find an equation for the graph involving only x and y.

2 Consider the parametric equations $x = t + 1$, $y = t^2$.
(*a*) Fill in this table:

t	-2	-1	0	1	2
x					
y					

(*b*) Plot the five points (x, y) obtained in (*a*).
(*c*) Graph the curve.
(*d*) Find an equation in x and y that describes the curve.

3 Consider the parametric equations $x = t^2$, $y = t^2 + t$.
(*a*) Fill in this table:

t	-3	-2	-1	0	1	2	3
x							
y							

(*b*) Plot the seven points (x, y) obtained in (*a*).
(*c*) Graph the curve given by $x = t^2$, $y = t^2 + t$.
(*d*) Eliminate t and find an equation for the graph in terms of x and y.

4 Consider the parametric equations $x = 2 \cos t$, $y = 3 \sin t$.
(*a*) Fill in this table, expressing the entries decimally.

t	0	$\frac{\pi}{4}$	$\frac{\pi}{2}$	$\frac{3\pi}{4}$	π	$\frac{5\pi}{4}$	$\frac{3\pi}{2}$	$\frac{7\pi}{4}$	2π
x									
y									

(*b*) Plot the eight distinct points in (*a*).
(*c*) Graph the curve given by $x = 2\cos t$, $y = 3 \sin t$.
(*d*) Using the identity $\cos^2 t + \sin^2 t = 1$, eliminate t.

In Exercises 5 and 6 express the curves parametrically with parameter x.

5 $y = \sqrt{1 + x^3}$ **6** $y = \tan^{-1} 3x$

In Exercises 7 and 8 express the curves parametrically with parameter θ.

7 $r = \cos 2\theta$ **8** $r = 3 + \cos \theta$

In Exercises 9 to 12 find dy/dx and d^2y/dx^2 for the given curves. (*Remember:* d^2y/dx^2 is *not* the quotient of d^2y/dt^2 and d^2x/dt^2.)

9 $x = t^3 + t$, $y = t^7 + t + 1$.

10 $x = \sin 3t$, $y = \cos 4t$.

11 $r = \cos 3\theta$.

12 $r = 2 + 3 \sin \theta$.

13 A curve is given parametrically by the equations

$$x = t^5 + \sin 2\pi t, \qquad y = t + e^t.$$

Find the slope of the curve at the point corresponding to $t = 1$. (This is a case where you cannot eliminate the parameter.)

14 (*a*) Letting $t = -1$, 0, and 1, find three points on the curve

$$\begin{cases} x = t^7 + t^2 + 1 \\ y = 2t^6 + 3t + 1. \end{cases}$$

(*b*) When t is large, what happens to y/x?
(*c*) What is the slope of the curve at the point corresponding to $t = 1$?

15 A ball is thrown at an angle α and initial velocity v_0, as sketched in Fig. 9.34. It can be shown that if time is in seconds and distance in feet, then t seconds later the ball is at the point

$$\begin{cases} x = (v_0 \cos \alpha)t \\ y = (v_0 \sin \alpha)t - 16t^2. \end{cases}$$

Eliminate t. The resulting equation shows that the path is a parabola.

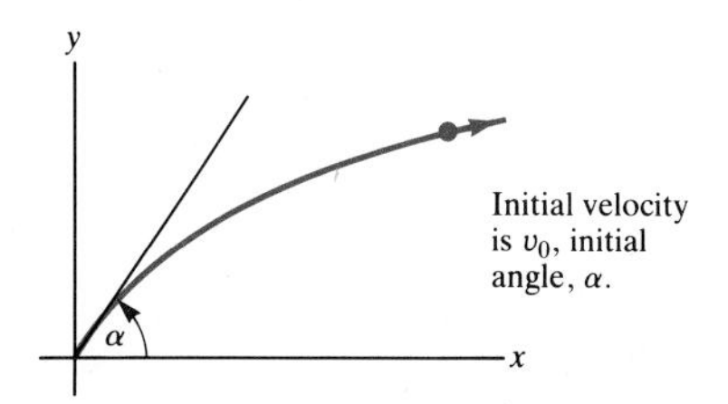

Figure 9.34

16 Eliminate t and plot

$$\begin{cases} x = e^t \\ y = e^{2t}. \end{cases}$$

17 Let a and b be positive numbers. Consider the curve given parametrically by the equations

$$x = a \cos t, \qquad y = b \sin t.$$

(*a*) Show that the curve is the ellipse

$$\frac{x^2}{a^2} + \frac{y^2}{b^2} = 1.$$

(*b*) Find the area of the region bounded by the ellipse in (*a*) by making a substitution that expresses $4 \int_0^a y\,dx$ in terms of an integral in which the variable is t and the range of integration is $[0, \pi/2]$.

18 Consider the curve given parametrically by

$$x = t^2 + e^t, \qquad y = t + e^t \qquad \text{for } t \text{ in } [0, 1].$$

(*a*) Plot the points corresponding to $t = 0, \frac{1}{2}$, and 1.
(*b*) Find the slope of the curve at the point (1, 1).
(*c*) Find the area of the region under the curve and above the interval $[1, e + 1]$.

19 In Example 2, what is the value of θ when the point E is (*a*) at the top of the first arch? (*b*) at the right end of the first arch? (*c*) at the top of the second arch? (*d*) at the right end of the second arch?

■

A curve given in polar coordinates as $r = f(\theta)$ is given in rectangular coordinates parametrically as $x = f(\theta) \cos \theta$, $y = f(\theta) \sin \theta$. This observation is useful in Exercises 20 to 23.

20 Find the slope of the cardioid $r = 1 + \cos \theta$ at the point where it crosses the positive y axis.

21 (*a*) Find the slope of the cardioid $r = 1 + \cos \theta$ at the point (r, θ).
(*b*) What happens to the slope in (*a*) as $\theta \to \pi^-$?

22 Find the slope of the three-leaved rose, $r = \sin 3\theta$, at the point $(r, \theta) = (\sqrt{2}/2, \pi/12)$.

23 The region under the arch of the cycloid

$$\begin{cases} x = a\theta - a \sin \theta \\ y = a - a \cos \theta \end{cases} \quad (0 \le \theta \le 2\pi)$$

and above the x axis is revolved around the x axis. Find the volume of the solid of revolution produced.

24 The same as the preceding exercise, except the region is revolved around the y axis instead of the x axis. Find the volume (*a*) by integration, (*b*) by Pappus's theorem.

25 L'Hôpital's rule in Sec. 6.9 asserts that if $\lim_{t\to 0} f(t) = 0$, $\lim_{t\to 0} g(t) = 0$, and $\lim_{t\to 0} [f'(t)/g'(t)]$ exists, then $\lim_{t\to 0} [f(t)/g(t)] = \lim_{t\to 0} [f'(t)/g'(t)]$. Interpret that rule in terms of the parameterized curve $x = g(t)$, $y = f(t)$. [Make a sketch of the curve near (0, 0) and show on it the geometric meaning of the quotients $f(t)/g(t)$ and $f'(t)/g'(t)$.]

26 Let a be a positive constant. Consider the curve given parametrically by the equations $x = a \cos^3 t$, $y = a \sin^3 t$.
(*a*) Sketch the curve.
(*b*) Find the slope of the curve at the point corresponding to the parameter value t.

27 Consider a tangent line to the curve in Exercise 26 at a point P in the first quadrant. Show that the length of the segment of that line intercepted by the coordinate axes is a.

This exercise is related to Examples 4 and 5, which concern the rotary engine.

28 In Example 5 a circle of radius r rolled around a circle of radius $2r$. Instead, consider the curve produced by a point P at a distance b from the center of the rolling circle, $0 \le b \le r$, if the radius of the fixed circle is $3r$.
(*a*) Find parametric equations of the curve produced.
(*b*) Sketch the curve.
(*c*) The curve in (*b*) is called a three-lobed epitrochoid. Show that a square rotor can revolve in it, as the triangular rotor did in Example 4. Engines with this design have been tried but are not as efficient as the standard rotary engine.

29 A circle of radius a is situated above the xy plane in a plane that is inclined at an angle to the xy plane. The shadow of the circle cast on the xy plane by light perpendicular to the xy plane is an oval curve. Find parametric equations for this curve and use them to show that the shadow is an ellipse.

30 A **trochoid**, a generalization of a cycloid, is defined as follows. A reflector is attached to a bicycle wheel of radius a at a distance b from the center of the wheel. As the wheel rolls without slipping, the reflector sweeps out a curve, called the **trochoid**. Find parametric equations for the trochoid. *Suggestion:* Use as parameter the angle that the wheel turns and assume that initially the reflector is directly below the center of the wheel.

31 A circle of radius b rolls without slipping on the inside of a circle of radius a, $a > b$. A point P on the circumference of the moving circle traces out a curve called a **hypocycloid**. Assume that initially the point P is at the point of contact of the two circles and that the coordinates of the center of the moving circle are initially $(a - b, 0)$. Using Fig. 9.35, which shows the angles θ and ϕ, obtain parametric equations for the hypocycloid, as follows:
(*a*) Show that $b\phi = a\theta$.
(*b*) Show that $x = (a - b) \cos \theta + b \cos (\phi - \theta)$ and $y = (a - b) \sin \theta - b \sin (\phi - \theta)$.

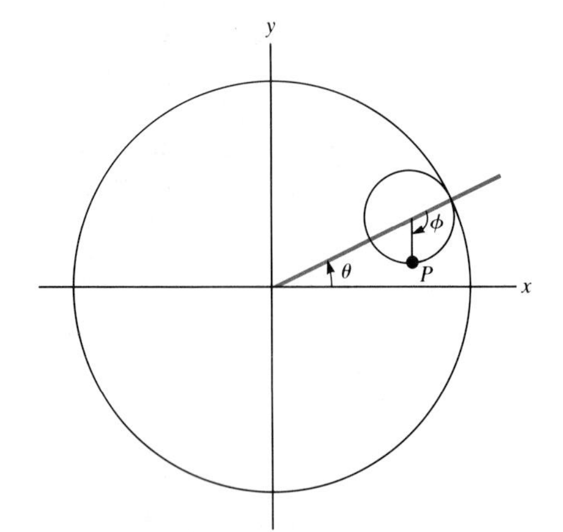

Figure 9.35

32 (See Exercise 31.) When $b = a/4$, the equations for the hypocycloid take a very simple form, obtained as follows:
(*a*) Show that $\sin 3\theta = 3 \sin \theta - 4 \sin^3 \theta$ and $\cos 3\theta = 4 \cos^3 \theta - 3 \cos \theta$.

(*b*) Using part (*a*) and also Exercise 31, show that $x = a\cos^3\theta$ and $y = a\sin^3\theta$.

(*c*) From (*b*) deduce that the hypocycloid in the special case $b = a/4$ has the equation $x^{2/3} + y^{2/3} = a^{2/3}$.

(*d*) Graph the equation $x^{2/3} + y^{2/3} = 1$.

33 A circle of radius b rolls without slipping on the outside of a circle of radius a. A point P on the circumference of the moving circle traces out a curve called an **epicycloid**. Find parametric equations for an epicycloid. For convenience, assume that the center of the fixed circle is $(0, 0)$, the moving circle initially has center $(a + b, 0)$, and initially P is the point of contact of the two circles, $(a, 0)$.

34 (*a*) Sketch the curve

$$\begin{cases} x = t - \tanh t \\ y = \operatorname{sech} t \end{cases}$$

for $t \geq 0$. (This curve is called the **tractrix**.)

(*b*) Let P be any point on the curve and let Q be the point where the tangent at P intersects the x axis. Show that the length of PQ is independent of P.

Because of (*b*), the tractrix is the curve traced out by a weight at one end of a string when the other end is pulled along a straight line that does not pass through the weight.

9.4 ARC LENGTH AND SPEED ON A CURVE

The path of some particle is given parametrically:

$$\begin{cases} x = g(t) \\ y = h(t). \end{cases}$$

(Think of t as time.) A physicist might ask the following questions: How far does the particle travel from time $t = a$ to time $t = b$? What is the speed of the particle at time t?

Consider the "distance traveled" question first. The second will then be easy to answer, making use of the derivative. (In our reasoning we shall call the parameter t and think of it as time, but the results apply to any parameter.)

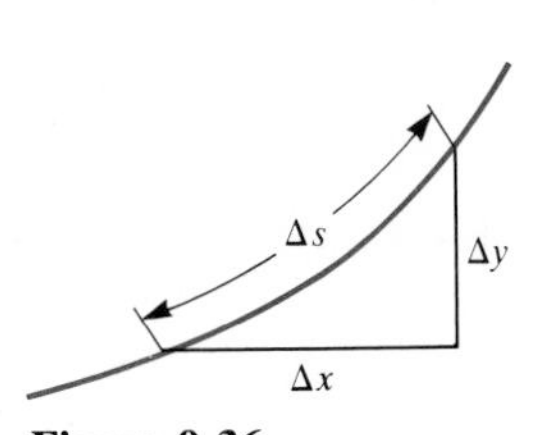

Figure 9.36

Assume that $x = g(t)$ and $y = h(t)$ have continuous derivatives. Let us make a local estimate of the arc length swept out on the path during the short interval of time from t to $t + dt$.

During the time dt the change in the x coordinate, Δx, is approximately $dx = g'(t)\,dt$ and in the y coordinate, Δy, it is approximately $dy = h'(t)\,dt$. The corresponding change in arc length is approximately ds. (The symbol s will stand for arc length along the path.) Now, a very small piece of the curve resembles a straight line. A relation between dx, dy, and ds is therefore suggested by Fig. 9.36, which is a "right triangle" whose longest side is almost straight.

s denotes arc length.

In view of the pythagorean theorem, it is reasonable to suspect that

$$(ds)^2 = (dx)^2 + (dy)^2$$

or

$$ds = \sqrt{(dx)^2 + (dy)^2} \tag{1}$$

This formula is the key memory device for dealing with arc length.

Rewriting Eq. (1) in the form

$$ds = \sqrt{\left(\frac{dx}{dt}\right)^2 + \left(\frac{dy}{dt}\right)^2}\,dt \tag{2}$$

gives us the local estimate of arc length. From this we conclude that the arc length of the curve corresponding to t in $[a, b]$ is given by the integral

Formula for arc length of parameterized curve

$$\text{Arc length} = \int_a^b \sqrt{\left(\frac{dx}{dt}\right)^2 + \left(\frac{dy}{dt}\right)^2}\, dt. \tag{3}$$

This formula holds for a curve given parametrically. If the curve is given in the form $y = f(x)$, it may be put in parametric form, with x as the parameter:

$$y = f(x), \qquad x = x.$$

Since $dx/dx = 1$, formula (3) for the arc length of the curve $y = f(x)$ for x in $[a, b]$ takes the following form:

Formula for arc length of curve $y = f(x)$

$$\text{Arc length} = \int_a^b \sqrt{1 + \left(\frac{dy}{dx}\right)^2}\, dx. \tag{4}$$

(It is assumed that the derivative dy/dx is continuous.)

Three examples will show how these formulas are applied. The first goes back to the year 1657, when the 20-year-old Englishman, William Neil, became the first person to find the length of a curve that was neither a line, a polygon, nor a circle.

EXAMPLE 1 Find the arc length of the curve $y = x^{3/2}$ for x in $[0, 1]$.

SOLUTION By formula (4),

$$\text{Arc length} = \int_0^1 \sqrt{1 + \left(\frac{dy}{dx}\right)^2}\, dx.$$

Since $y = x^{3/2}$, $dy/dx = \frac{3}{2}x^{1/2}$. Thus

$$\begin{aligned}
\text{Arc length} &= \int_0^1 \sqrt{1 + (\tfrac{3}{2}x^{1/2})^2}\, dx \\
&= \int_0^1 \sqrt{1 + \frac{9x}{4}}\, dx \\
&= \int_1^{13/4} \sqrt{u} \cdot \frac{4}{9}\, du \qquad \text{where } u = 1 + (9x/4),\ du = \frac{9}{4}\, dx \\
&\underset{\text{FTC}}{=} \frac{4}{9} \cdot \frac{2}{3} u^{3/2} \Big|_1^{13/4} = \frac{8}{27}\left[\left(\frac{13}{4}\right)^{3/2} - 1^{3/2}\right] \\
&= \frac{8}{27}\left(\frac{13^{3/2}}{8} - 1\right) = \frac{13^{3/2} - 8}{27}. \quad \blacksquare
\end{aligned}$$

Incidentally, the length of the curve $y = x^a$, where a is a rational number, usually *cannot* be computed with the aid of the fundamental theorem. The only cases in which it can be computed by the fundamental theorem are $a = 1$ (the graph is the line $y = x$) and $a = 1 + 1/n$, where n is an integer. Exercise 26 treats this question.

EXAMPLE 2 Find the distance s which the ball described at the beginning of Sec. 9.3 travels during the first b seconds.

SOLUTION Here

$$x = 32t \quad \text{and} \quad y = -16t^2.$$

Thus

$$\frac{dx}{dt} = 32 \quad \text{and} \quad \frac{dy}{dt} = -32t.$$

By formula (3),

$$s = \int_0^b \sqrt{(32)^2 + (-32t)^2}\, dt = 32 \int_0^b \sqrt{1 + t^2}\, dt,$$

a definite integral that can be evaluated with the aid of a table or the substitution $t = \tan\theta$; its value is

$$16b\sqrt{1 + b^2} + 16 \ln (b + \sqrt{1 + b^2}). \quad ■$$

EXAMPLE 3 Find the length of one arch of the cycloid in Example 2 of Sec. 9.3.

SOLUTION Here the parameter is θ, and we compute $dx/d\theta$ and $dy/d\theta$:

$$\frac{dx}{d\theta} = \frac{d}{d\theta}(a\theta - a \sin \theta) = a - a \cos \theta,$$

and

$$\frac{dy}{d\theta} = \frac{d}{d\theta}(a - a \cos \theta) = a \sin \theta.$$

To complete one arch, θ varies from 0 to 2π. By formula (3), the length of one arch is $\int_0^{2\pi} \sqrt{(a - a \cos \theta)^2 + (a \sin \theta)^2}\, d\theta$. Thus

$$\begin{aligned}
\text{Length of arch} &= a \int_0^{2\pi} \sqrt{(1 - \cos \theta)^2 + (\sin \theta)^2}\, d\theta \\
&= a \int_0^{2\pi} \sqrt{1 - 2 \cos \theta + (\cos^2 \theta + \sin^2 \theta)}\, d\theta \\
&= a \int_0^{2\pi} \sqrt{2 - 2 \cos \theta}\, d\theta \\
&= a\sqrt{2} \int_0^{2\pi} \sqrt{1 - \cos \theta}\, d\theta \\
&= a\sqrt{2} \int_0^{2\pi} \sqrt{2} \sin \frac{\theta}{2}\, d\theta \qquad \text{trigonometry} \\
&= 2a \int_0^{2\pi} \sin \frac{\theta}{2}\, d\theta \underset{\text{FTC}}{=} 2a \left(-2 \cos \frac{\theta}{2}\right)\Bigg|_0^{2\pi} \\
&= 2a[-2(-1) - (-2)(1)] = 8a.
\end{aligned}$$

This step works since $\sin(\theta/2) \geq 0$ *for* $0 \leq \theta \leq 2\pi$.

This means that while θ varies from 0 to 2π, a bicycle travels a distance $2\pi a \approx 6.28a$, and the tack in the tire travels a distance $8a$. ■

Speed of a Particle Moving on a Curve

In practice we are not interested as much in the length of the path as in the speed of the particle as it moves along the path. The work done so far in this section helps us find this speed easily.

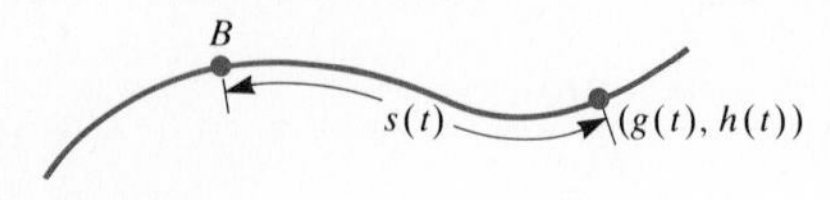

Figure 9.37

Consider a particle which at time t is at the point $(x, y) = (g(t), h(t))$. Choose a point B on the curve from which to measure distance along the curve, as shown in Fig. 9.37. Let $s(t)$ denote the distance from B to $(g(t), h(t))$. We shall always assume that B has been chosen in such a way that $s(t)$ is an *increasing* function of t.

Definition *Speed on a curved path.* If ds/dt exists, it is called the **speed of the particle**

Since $s(t)$ is assumed to be an increasing function, speed is not negative.

As early as Sec. 3.1 we were able to treat the speed of a particle moving in a straight path. Now it is possible to compute the speed of a particle moving on a curved path.

Speed of a particle on a curved path

> If a particle at time t is at the point $(x, y) = (g(t), h(t))$, where g and h are functions having continuous derivatives, then its speed at time t is equal to
>
> $$\sqrt{[g'(t)]^2 + [h'(t)]^2}.$$

The argument is short. Let $s(t)$ denote the arc length along the curve from some base point B to the particle at time t.

Now,
$$s(t) = \int_a^t \sqrt{[g'(T)]^2 + [h'(T)]^2}\, dT.$$

(The letter T is introduced, since t is already used to describe the interval of integration.) Differentiation of this relation with respect to t (using the second fundamental theorem of calculus) yields

$$\frac{ds}{dt} = \sqrt{[g'(t)]^2 + [h'(t)]^2}.$$

EXAMPLE 4 Find the speed at time t of the ball described at the beginning of Sec. 9.3.

SOLUTION At time t the ball is at the point

$$(x, y) = (32t, -16t^2).$$

Newton's dot notation:
$$\dot{x} = \frac{dx}{dt},\ \dot{y} = \frac{dy}{dt}$$

Thus dx/dt, usually written $\dot{x}$, is 32, and $\dot{y}$ is $-32t$. The speed of the ball is

$$\frac{ds}{dt} = \sqrt{\dot{x}^2 + \dot{y}^2} = \sqrt{32^2 + (-32t)^2}$$
$$= 32\sqrt{1 + t^2} \quad \text{feet per second.} \quad \blacksquare$$

So far in this section curves have been described in rectangular coordinates. Next consider a curve given in polar coordinates by the equation $r = f(\theta)$.

How to Find the Arc Length of $r = f(\theta)$

The length of the curve $r = f(\theta)$ for θ in $[\alpha, \beta]$ is equal to

$$\int_\alpha^\beta \sqrt{[f(\theta)]^2 + [f'(\theta)]^2}\, d\theta$$

or

$$\int_\alpha^\beta \sqrt{r^2 + (r')^2}\, d\theta.$$

(Assume that f has a continuous derivative.)

This formula can be derived from that for the arc length of a parameterized curve in rectangular coordinates, as follows. Find the rectangular coordinates of the point whose polar coordinates are

$$(r, \theta) = (f(\theta), \theta).$$

They are

$$\begin{cases} x = f(\theta) \cos \theta \\ y = f(\theta) \sin \theta. \end{cases}$$

The curve is now given in rectangular form with parameter θ. Thus its length is

$$\int_\alpha^\beta \sqrt{\left(\frac{dx}{d\theta}\right)^2 + \left(\frac{dy}{d\theta}\right)^2}\, d\theta.$$

Now,

$$\frac{dx}{d\theta} = f(\theta)(-\sin \theta) + f'(\theta) \cos \theta,$$

and

$$\frac{dy}{d\theta} = f(\theta) \cos \theta + f'(\theta) \sin \theta.$$

Hence

$$\left(\frac{dx}{d\theta}\right)^2 + \left(\frac{dy}{d\theta}\right)^2 = [f(\theta)]^2 \sin^2 \theta - 2f(\theta)f'(\theta) \sin \theta \cos \theta + [f'(\theta)]^2 \cos^2 \theta$$
$$+ [f(\theta)]^2 \cos^2 \theta + 2f(\theta)f'(\theta) \sin \theta \cos \theta + [f'(\theta)]^2 \sin^2 \theta,$$

which, by the identity $\sin^2 \theta + \cos^2 \theta = 1$, simplifies to $[f(\theta)]^2 + [f'(\theta)]^2$. This justifies the formula.

EXAMPLE 5 Find the length of the spiral $r = e^{-3\theta}$ for θ in $[0, 2\pi]$.

SOLUTION First compute

$$r' = \frac{dr}{d\theta} = -3e^{-3\theta},$$

and then use the formula

$$\text{Arc length} = \int_\alpha^\beta \sqrt{r^2 + (r')^2}\, d\theta = \int_0^{2\pi} \sqrt{(e^{-3\theta})^2 + (-3e^{-3\theta})^2}\, d\theta$$

$$= \int_0^{2\pi} \sqrt{e^{-6\theta} + 9e^{-6\theta}}\, d\theta = \sqrt{10} \int_0^{2\pi} \sqrt{e^{-6\theta}}\, d\theta$$

$$= \sqrt{10} \int_0^{2\pi} e^{-3\theta}\, d\theta \underset{\text{FTC}}{=} \sqrt{10}\, \frac{e^{-3\theta}}{-3} \Bigg|_0^{2\pi}$$

$$= \sqrt{10} \left(\frac{e^{-3 \cdot 2\pi}}{-3} - \frac{e^{-3 \cdot 0}}{-3} \right) = \sqrt{10} \left(\frac{e^{-6\pi}}{-3} + \frac{1}{3} \right)$$

$$= \frac{\sqrt{10}}{3}(1 - e^{-6\pi}). \quad \blacksquare$$

Memory aid for arc length in polar coordinates

Think of the shaded region in Fig. 9.38 as almost a right triangle. [The "hypotenuse," which is part of the curve $r = f(\theta)$, is not straight. The "leg" of length $r\, d\theta$ is part of a circle of radius r.] It suggests the equation

$$(ds)^2 = (r\, d\theta)^2 + (dr)^2. \tag{5}$$

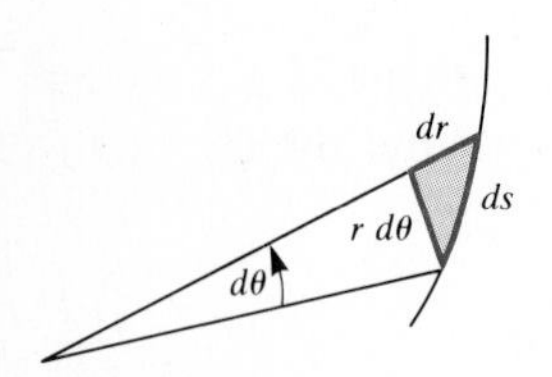

Figure 9.38

Division of Eq. (5) by $(d\theta)^2$ gives

$$\left(\frac{ds}{d\theta}\right)^2 = r^2 + \left(\frac{dr}{d\theta}\right)^2.$$

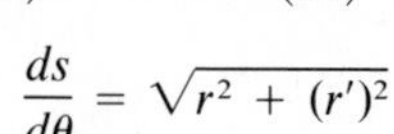

Hence

$$\frac{ds}{d\theta} = \sqrt{r^2 + (r')^2}$$

is the integrand needed in computing arc length in polar coordinates.

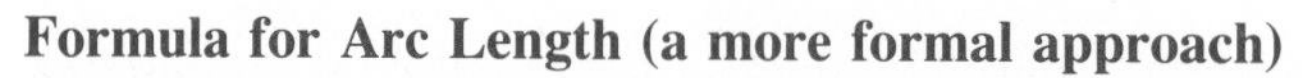

Formula for Arc Length (a more formal approach)

Partition the time interval $[a, b]$ and use this partition to inscribe a polygon in the curve of the moving particle, as shown in Fig. 9.39.

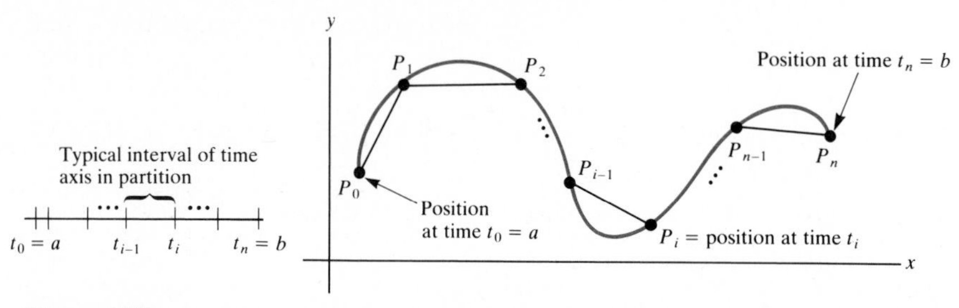

Figure 9.39

We are assuming $x = g(t)$ and $y = h(t)$ have continuous derivatives.

The length of such a polygon should approach the arc length as the mesh of the partition of $[a, b]$ shrinks toward 0 since the points P_i along the curve will get closer and closer together. The length of the typical straight-line segment $P_{i-1}P_i$, where $P_{i-1} = (g(t_{i-1}), h(t_{i-1}))$ and $P_i = (g(t_i), h(t_i))$, is (by the distance formula)

$$\sqrt{[g(t_i) - g(t_{i-1})]^2 + [h(t_i) - h(t_{i-1})]^2},$$

and so the length of the polygon is the sum

$$\sum_{i=1}^{n} \sqrt{[g(t_i) - g(t_{i-1})]^2 + [h(t_i) - h(t_{i-1})]^2}. \tag{6}$$

We shall relate this sum to sums of the type appearing in the definition of a definite integral over $[a, b]$.

By the mean-value theorem there exist numbers T_i^* and T_i^{**}, both in the interval $[t_{i-1}, t_i]$, such that $g(t_i) - g(t_{i-1}) = g'(T_i^*)(t_i - t_{i-1})$ and $h(t_i) - h(t_{i-1}) = h'(T_i^{**})(t_i - t_{i-1})$. Thus the sum (6) can be rewritten

$$\sum_{i=1}^{n} \sqrt{[g'(T_i^*)]^2 + [h'(T_i^{**})]^2}\,(t_i - t_{i-1}). \tag{7}$$

If T_i^{**} were equal to T_i^*, then this sum (7) would be an approximating sum used in defining

$$\int_a^b \sqrt{[g'(t)]^2 + [h'(t)]^2}\,dt.$$

To get around this difficulty, notice that since h' is continuous, $h'(T_i^*)$ is near $h'(T_i^{**})$ when the mesh of the partition of $[a, b]$ is small. If the sum (7) is a good approximation to the arc length, then presumably so is the sum

This step is justified in advanced calculus by Duhamel's principle.

$$\sum_{i=1}^{n} \sqrt{[g'(T_i^*)]^2 + [h'(T_i^*)]^2}\,(t_i - t_{i-1}). \tag{8}$$

In other words, it is reasonable to expect that

$$\text{Arc length} = \lim_{\text{mesh}\to 0} \sum_{i=1}^{n} \sqrt{[g'(T_i^*)]^2 + [h'(T_i^*)]^2}\,(t_i - t_{i-1}).$$

But that limit is precisely the definition of the definite integral

$$\int_a^b \sqrt{[g'(t)]^2 + [h'(t)]^2}\,dt.$$

This shows why this definite integral should yield the arc length.

EXERCISES FOR SEC. 9.4: ARC LENGTH AND SPEED ON A CURVE

In Exercises 1 to 14 find the arc lengths of the given curves over the given intervals.

1 $y = x^{3/2}$, x in $[1, 2]$

2 $y = x^{2/3}$, x in $[0, 1]$

3 $y = x^3/3 + 1/(4x)$, x in $[1, 2]$

4 $y = 1/(2x^2) + x^4/16$, x in $[2, 3]$

5 $y = x^{4/3}$, x in $[0, 1]$

6 $y = x^{5/4}$, x in $[0, 1]$

7 $y = (e^x + e^{-x})/2$, x in $[0, b]$

8 $y = x^2/2 - (\ln x)/4$, x in $[2, 3]$

9 $x = \cos^3 t$, $y = \sin^3 t$, t in $[0, \pi/2]$

10 $x = \cos t + t \sin t$, $y = \sin t - t \cos t$, t in $[\pi/6, \pi/4]$

11 $r = e^\theta$, θ in $[0, 2\pi]$

12 $r = 1 + \cos \theta$, θ in $[0, \pi]$

13 $r = \cos^2 (\theta/2)$, θ in $[0, \pi]$

14 $r = \sin^2 (\theta/2)$, θ in $[0, \pi]$

In each of Exercises 15 to 18 find the speed of the particle at time t, given the parametric description of its path.

15 $x = 50t$, $y = -16t^2$

16 $x = \sec 3t$, $y = \sin^{-1} 4t$

17 $x = t + \cos t$, $y = 2t - \sin t$

18 $x = \csc 3t$, $y = \tan^{-1} \sqrt{t}$

■

19 (*a*) How far does a bug travel from time $t = 1$ to time $t = 2$ if at time t it is at the point (t^2, t^3)?
(*b*) How fast is it moving at time t?
(*c*) Graph its path relative to an xy coordinate system. Where is it at $t = 1$? At $t = 2$?
(*d*) Eliminate t to find y as a function of x for $t \geq 0$.

20 (Calculator)
(*a*) Graph $y = x^3/3$.
(*b*) Estimate its arc length from (0, 0) to (3, 9) by an inscribed polygon whose vertices have x coordinates 0, 1, 2, 3.
(*c*) Set up a definite integral for the arc length in question.
(*d*) Estimate the definite integral in (*c*) by using a par-

tition of [0, 3] into three sections, each of length 1, and the trapezoidal approximation.

21 Assume that a curve is described in rectangular coordinates in the form $x = f(y)$. Show that

$$\text{Arc length} = \int_c^d \sqrt{1 + \left(\frac{dx}{dy}\right)^2}\, dy,$$

where y ranges in the interval $[c, d]$.

22 (See Exercise 21.) Consider the arc length of the curve $y = x^{2/3}$ for x in the interval [1, 8].
(*a*) Set up a definite integral for this arc length using x as the parameter.
(*b*) Set up a definite integral for this arc length using y as the parameter.
(*c*) Evaluate the easier of the integrals in (*a*) and (*b*).

23 (*a*) At time t a particle has polar coordinates $r = g(t)$, $\theta = h(t)$. How fast is it moving?
(*b*) Use the formula in (*a*) to find the speed of a particle which at time t is at the point $(r, \theta) = (e^t, 5t)$.

24 Let $P = (x, y)$ depend on θ as shown in Fig. 9.40.
(*a*) Sketch the curve that P sweeps out.
(*b*) Show that $P = (2 \cos \theta, \sin \theta)$.
(*c*) Set up a definite integral for the length of the curve described by P. (Do not evaluate it.)
(*d*) Eliminate θ and show that P is on the ellipse

$$\frac{x^2}{4} + \frac{y^2}{1} = 1.$$

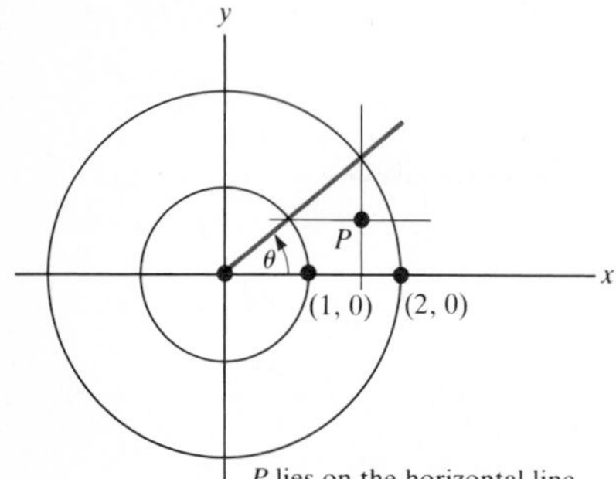

P lies on the horizontal line through $(\cos \theta, \sin \theta)$ and the vertical line through $(2 \cos \theta, 2 \sin \theta)$.

Figure 9.40

25 Show that if

$$y = \frac{x^{m+1}}{m+1} + \frac{x^{1-m}}{4(m-1)},$$

where m is any number other than 1 or -1, then the definite integral for the arc length of this curve can be computed with the aid of the fundamental theorem of calculus. Consider only arcs corresponding to x in $[a, b]$, $0 < a < b$.

26 Consider the length of the curve $y = x^m$, where m is a rational number. Show that the fundamental theorem of calculus is of aid in computing this length only if $m = 1$ or if m is of the form $1 + 1/n$ for some integer n. *Hint:* The analog of Chebyshev's theorem in Exercise 204 in Chap. 7 Review Exercises holds for $\int x^p(1 + x)^q\, dx$.

27 Consider the cardioid $r = 1 + \cos \theta$ for θ in $[0, \pi]$. We may consider r as a function of θ or as a function of s, arc length along the curve, measured, say, from (2, 0).
(*a*) Find the average of r with respect to θ.
(*b*) Find the average of r with respect to s. *Hint:* Express all quantities appearing in this average in terms of θ.

28 The function $r = f(\theta)$ describes, for θ in $[0, 2\pi]$, a curve in polar coordinates. Assume r' is continuous and $f(\theta) > 0$. Prove that the average of r as a function of arc length is at least as large as the quotient $2A/s$, where A is the area swept out by the radius and s is the arc length of the curve. When is the average equal to $2A/s$?

29 Find the arc length of the archimedean spiral $r = a\theta$ for θ in $[0, 2\pi]$.

30 Let $y = f(x)$ for x in [0, 1] describe a curve that starts at (0, 0), ends at (1, 1), and lies in the square with vertices (0, 0), (1, 0), (1, 1), and (0, 1). Assume f has a continuous derivative.
(*a*) What can be said about the arc length of the curve? How small and how large can it be?
(*b*) Answer (*a*) if it is assumed also that $f'(x) \geq 0$ for x in [0, 1].

9.5 AREA OF A SURFACE OF REVOLUTION

This section develops a technique for computing the surface area of a solid of revolution, such as a sphere. The approach will be intuitive and will only justify the plausibility of defining the area of a surface of revolution as a certain definite integral.

Begin by considering the area of a rather simple surface of revolution, the curved part of a cone whose base has radius r and whose slant height

Take a cone first.

is l. (See Fig. 9.41.) If this cone is cut along a line through its point and laid flat, it becomes a sector of a circle of radius l, as shown in Fig. 9.42.

Now, the area of a sector of radius l and angle θ (in radians) is $\frac{1}{2}l^2\theta$. Since $\theta = 2\pi r/l$, the area of this sector is $\frac{1}{2}l^2(2\pi r/l)$, which equals πrl. Thus

$$\text{Area of curved surface of cone} = \pi rl. \tag{1}$$

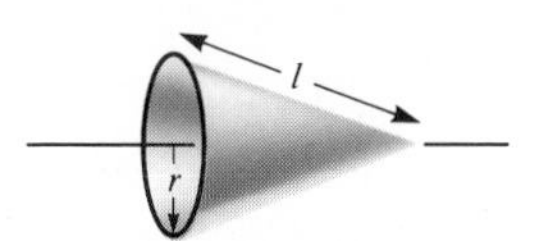

Figure 9.41

From formula (1) we will obtain the key fact needed in this section, namely, the formula for the area swept out by revolving a line segment around a line.

Consider a line segment of length L in the plane which does not meet a certain line in the plane, called the axis. (See Fig. 9.43.) When the line segment is revolved around the axis, it sweeps out a curved surface. We will show that the area of this surface equals

$$2\pi rL, \tag{2}$$

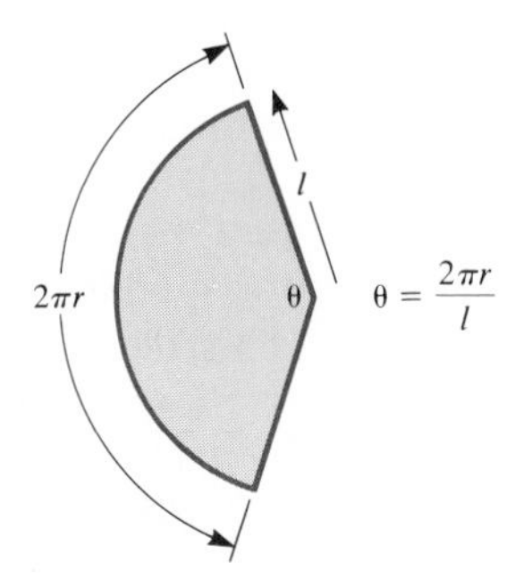

The cone laid flat is a sector of a circle.

Figure 9.42

where r is the distance from the midpoint of the line segment to the axis. (Note the resemblance to Pappus's theorem.) The surface in Fig. 9.43 is called a **frustum of a cone**

The argument for (2) depends on the fact that the frustum in Fig. 9.43 is the difference of two cones, one of slant height l_1 and radius r_1 and one of slant height l_2 and radius r_2, as shown in Fig. 9.44. The area of the frustum in Fig. 9.43 is therefore

$$\pi r_1 l_1 - \pi r_2 l_2. \tag{3}$$

It will now be shown that (3) reduces to $2\pi rL$, where $r = (r_1 + r_2)/2$.

By similar triangles in Fig. 9.44,

$$\frac{r_1}{r_2} = \frac{l_1}{l_2}.$$

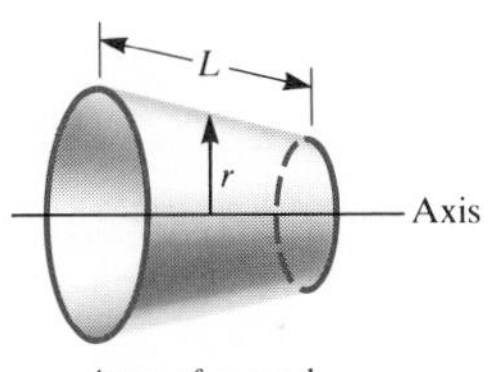

Area of curved surface equals $2\pi rL$.

Figure 9.43

Hence $r_1 l_2 = r_2 l_1$. This fact will be used below to replace $r_1 l_2$ by $r_2 l_1$.

Now,

$$\begin{aligned} \pi r_1 l_1 - \pi r_2 l_2 &= \pi r_1(l_2 + L) - \pi r_2 l_2 = \pi r_1 l_2 + \pi r_1 L - \pi r_2 l_2 \\ &= \pi r_2 l_1 + \pi r_1 L - \pi r_2 l_2 = \pi r_2(l_1 - l_2) + \pi r_1 L \\ &= \pi r_2 L + \pi r_1 L = \pi(r_1 + r_2)L \\ &= 2\pi rL. \end{aligned}$$

This establishes (2).

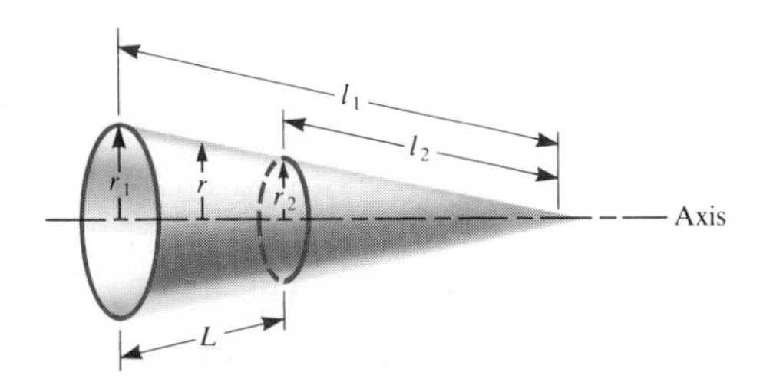

Figure 9.44

Formula (2) is the basis for the following definition for the area of the surface swept out by revolving a curve around an axis. The definition will be justified informally.

Definition *Area of a surface of revolution.* Consider a curve given by the parametric equations $x = g(t)$, $y = h(t)$, where g and h have continuous derivatives and $h(t) \geq 0$. Let C be that portion of the curve corresponding to t in $[a, b]$. Then the **area of the surface of revolution** formed by revolving C about the x axis is

$$\int_a^b 2\pi h(t)\sqrt{[g'(t)]^2 + [h'(t)]^2}\,dt, \tag{4}$$

or equivalently, $$\int_a^b 2\pi y \sqrt{\left(\frac{dx}{dt}\right)^2 + \left(\frac{dy}{dt}\right)^2}\,dt.$$

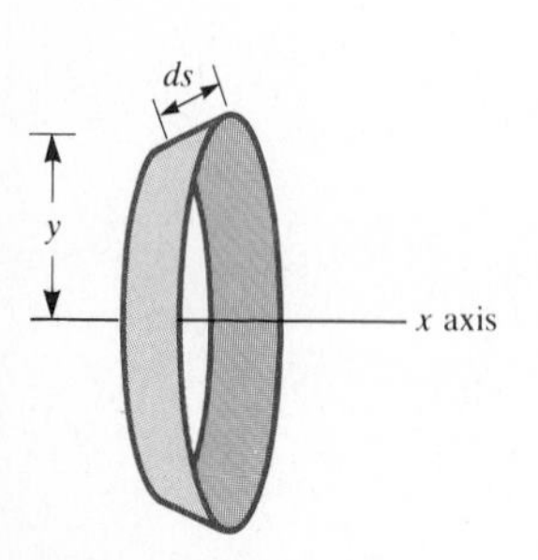

Figure 9.45

Justification (informal) As the parameter goes from the value t to the value $t + dt$, where dt is a small positive number, the corresponding point on the curve sweeps out a short section of the curve. This short section is almost straight. Call its length ds. Moreover, this section is approximately a fixed distance y from the x axis. The area of the narrow circular band swept out by revolving this section of the curve around the x axis can be estimated with the aid of formula (2). (See Fig. 9.45.) By formula (2), the area of the band is about

$$2\pi y\,ds. \tag{5}$$

But $ds = \sqrt{(dx/dt)^2 + (dy/dt)^2}\,dt$. So the area of the narrow band in Fig. 9.45 is approximately

$$2\pi y \sqrt{\left(\frac{dx}{dt}\right)^2 + \left(\frac{dy}{dt}\right)^2}\,dt. \tag{6}$$

With the aid of this local approximation, we immediately conclude that

$$\text{Surface area} = \int_a^b 2\pi y \sqrt{\left(\frac{dx}{dt}\right)^2 + \left(\frac{dy}{dt}\right)^2}\,dt.$$

This completes the motivation of formula (4).

If a curve is given by $y = f(x)$, where f has a continuous derivative and $f(x) \geq 0$, it may be parameterized by the equations $x = t$ and $y = f(t)$. Then $dx/dt = dx/dx = 1$ and $dy/dt = dy/dx$. Hence the surface area swept out by revolving about the x axis that part of the curve above $[a, b]$ is

A formula for surface area if $y = f(x)$ is revolved around x axis

$$\text{Surface area} = \int_a^b 2\pi y \sqrt{1 + \left(\frac{dy}{dx}\right)^2}\,dx.$$

As the formulas are stated, they seem to refer only to surfaces obtained by revolving a curve about the x axis. In fact, they refer to revolution about any line. The factor y in the integrand,

$$2\pi y \sqrt{\left(\frac{dx}{dt}\right)^2 + \left(\frac{dy}{dt}\right)^2},$$

is the distance from the typical point on the curve to the axis of revolution. Replace y by R (for *radius*) to free ourselves from coordinate systems. (Use capital R to avoid confusion with polar coordinates.) The expression

$$\sqrt{\left(\frac{dx}{dt}\right)^2 + \left(\frac{dy}{dt}\right)^2}\, dt$$

is simply ds, since $$\frac{ds}{dt} = \sqrt{\left(\frac{dx}{dt}\right)^2 + \left(\frac{dy}{dt}\right)^2}.$$

The simplest way to write the formula for surface area of revolution is then

General formula for surface area

$$\boxed{\text{Surface area} = \int_c^d 2\pi R\, ds,}$$

where the interval $[c, d]$ refers to the parameter s. However, in practice s is seldom used as the parameter. Instead, x, y, t, or θ is used and the interval of integration describes the interval through which the parameter varies.

To remember this formula, think of a narrow circular band of width ds and radius R as analogous to the rectangle shown in Fig. 9.46.

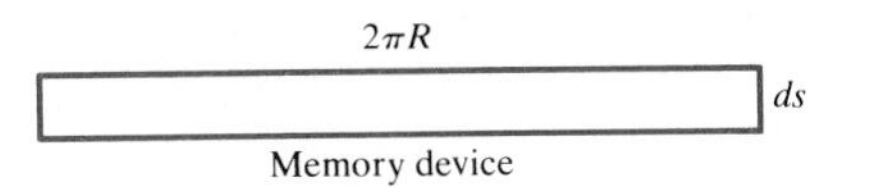

Figure 9.46

EXAMPLE 1 Find the area of the surface obtained by revolving around the y axis the part of the parabola $y = x^2$ that lies between $x = 1$ and $x = 2$. (See Fig. 9.47.)

R is found by inspection of a diagram.

SOLUTION The surface area is $\int_a^b 2\pi R\, ds$. Since the curve is described as a function of x, choose x as the parameter. By inspection of Fig. 9.47, $R = x$. Next, note that

$$ds = \frac{ds}{dx}\, dx = \sqrt{1 + \left(\frac{dy}{dx}\right)^2}\, dx.$$

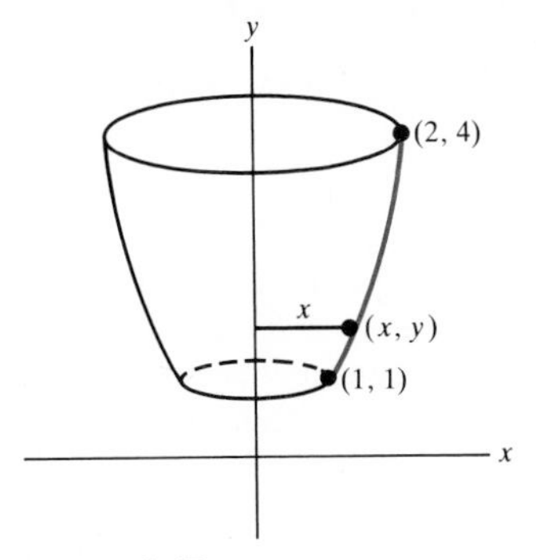

Figure 9.47

The surface area is therefore

$$\int_1^2 2\pi x\sqrt{1 + 4x^2}\, dx.$$

To evaluate the integral, use the substitution

$$u = 1 + 4x^2, \qquad du = 8x\, dx.$$

Hence $x\, dx = du/8$. The new limits of integration are $u = 5$ and $u = 17$. Thus

$$\text{Surface area} = \int_5^{17} 2\pi\sqrt{u}\,\frac{du}{8} = \frac{\pi}{4}\int_5^{17} \sqrt{u}\, du$$

$$= \frac{\pi}{4}\cdot\frac{2}{3}u^{3/2}\Big|_5^{17} = \frac{\pi}{6}(17^{3/2} - 5^{3/2}). \quad ■$$

EXAMPLE 2 Find the surface area of a sphere of radius a.

SOLUTION The circle of radius a has the equation $x^2 + y^2 = a^2$. The top half has the equation $y = \sqrt{a^2 - x^2}$. The sphere of radius a is formed

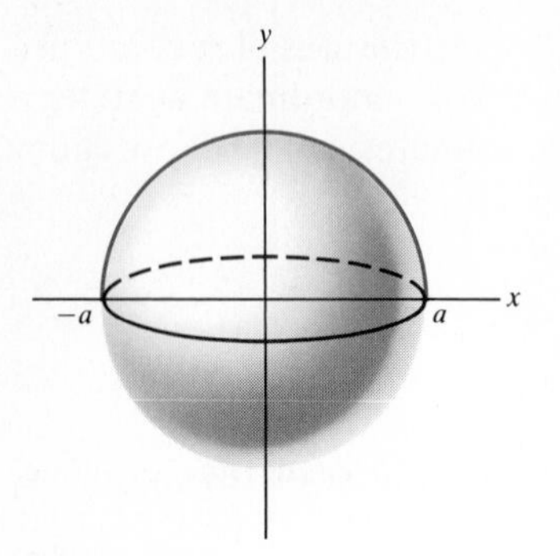

Figure 9.48

by revolving this top half around the x axis. (See Fig. 9.48.) Using the memory device "$2\pi R\ ds$," we have

$$\text{Surface area of sphere} = \int_{-a}^{a} 2\pi R \frac{ds}{dx}\,dx.$$

Now, $R = y$ and $ds/dx = \sqrt{1 + (dy/dx)^2}$, where we have $dy/dx = -x/\sqrt{a^2 - x^2}$. Thus

$$\text{Surface area of sphere} = \int_{-a}^{a} 2\pi y \sqrt{1 + \left(\frac{-x}{\sqrt{a^2 - x^2}}\right)^2}\,dx$$

$$= \int_{-a}^{a} 2\pi\sqrt{a^2 - x^2}\sqrt{1 + \frac{x^2}{a^2 - x^2}}\,dx$$

$$= \int_{-a}^{a} 2\pi\sqrt{a^2 - x^2}\sqrt{\frac{a^2}{a^2 - x^2}}\,dx$$

All this work for a constant integrand.

$$= \int_{-a}^{a} 2\pi a\,dx = 2\pi a x \Big|_{-a}^{a}$$

$$= 4\pi a^2.$$

Exercise 10 has another approach.

The surface area of a sphere is 4 times the area of its equatorial cross section. ■

EXERCISES FOR SEC. 9.5: AREA OF A SURFACE OF REVOLUTION

In each of Exercises 1 to 4 set up a definite integral for the area of the indicated surface using the suggested parameter. Show the radius R on a diagram; do *not* evaluate the definite integrals.

1 The curve $y = x^3$; x in $[1, 2]$; revolved about the x axis; parameter x.

2 The curve $y = x^3$; x in $[1, 2]$; revolved about the line $y = -1$; parameter x.

3 The curve $y = x^3$; x in $[1, 2]$, revolved about the y axis; parameter y.

4 The curve $y = x^3$; x in $[1, 2]$; revolved about the y axis; parameter x.

5 Find the area of the surface obtained by rotating about the x axis that part of the curve $y = e^x$ that lies above $[0, 1]$.

6 Find the area of the surface formed by rotating one arch of the curve $y = \sin x$ about the x axis.

7 One arch of the cycloid given parametrically by the formula $x = \theta - \sin\theta$, $y = 1 - \cos\theta$ is revolved around the x axis. Find the area of the surface produced.

8 The curve given parametrically by $x = e^t \cos t$, $y = e^t \sin t$, t in $[0, \pi/2]$, is revolved around the x axis. Find the area of the surface produced.

■

The formula in Exercise 9 is used in Exercises 10 to 12.

9 Show that the area of the surface obtained by revolving the curve $r = f(\theta)$, $\alpha \le \theta \le \beta$, around the polar axis is

$$\int_{\alpha}^{\beta} 2\pi r \sin\theta\,\sqrt{r^2 + (r')^2}\,d\theta.$$

10 Use the formula in Exercise 9 to find the surface area of a sphere of radius a.

11 Find the area of the surface formed by revolving the portion of the curve $r = 1 + \cos\theta$ in the first quadrant about (*a*) the x axis, (*b*) the y axis. [In (*b*) the identity $1 + \cos\theta = 2\cos^2(\theta/2)$ may help.]

12 The curve $r = \sin 2\theta$, θ in $[0, \pi/2]$, is revolved around the polar axis. Set up an integral for the surface area.

13 Consider the smallest tin can that contains a given sphere. (The height and diameter of the tin can equal the diameter of the sphere.)

(*a*) Compare the volume of the sphere with the volume of the tin can. Archimedes, who obtained the solution about 2200 years ago, considered it his greatest accomplishment. Cicero wrote, about two centuries after Archimedes' death:

> I shall call up from the dust [the ancient equivalent of a blackboard] and his measuring-rod an obscure, insignificant person belonging to the same city [Syracuse], who lived many years after, Archimedes. When I was quaestor I tracked out his grave, which was unknown to the Syracusans (as they totally denied its existence), and found it enclosed all round and covered with brambles and thickets; for I re-

membered certain doggerel lines inscribed, as I had heard, upon his tomb, which stated that a sphere along with a cylinder had been set up on the top of his grave. Accordingly, after taking a good look all around (for there are a great quantity of graves at the Agrigentine Gate), I noticed a small column rising a little above the bushes, on which there was the figure of a sphere and a cylinder. And so I at once said to the Syracusans (I had their leading men with me) that I believed it was the very thing of which I was in search. Slaves were sent in with sickles who cleared the ground of obstacles, and when a passage to the place was opened we approached the pedestal fronting us; the epigram was traceable with about half the lines legible, as the latter portion was worn away.

(Cicero, *Tusculan Disputations,* vol. 23, translated by J. E. King, Loeb Classical Library, Harvard University, Cambridge, 1950.) Archimedes was killed by a Roman soldier in 212 B.C. Cicero was quaestor in 75 B.C.

(*b*) Compare the surface area of the sphere with the area of the curved side of the can.

14 (*a*) Compute the area of the portion of a sphere of radius a that lies between two parallel planes at distances c and $c + h$ from the center of the sphere $(0 \le c \le c + h \le a)$.
(*b*) The result in (*a*) depends only on h, not on c. What does this mean geometrically?

15 The portion of the curve $x^{2/3} + y^{2/3} = 1$ situated in the first quadrant is revolved around the x axis. Find the area of the surface produced.

In each of Exercises 16 to 23 find the area of the surface formed by revolving the indicated curve about the indicated axis. Leave the answer as a definite integral, but indicate how it could be evaluated by the fundamental theorem of calculus.

16 $y = 2x^3$ for x in $[0, 1]$; about the x axis.
17 $y = 1/x$ for x in $[1, 2]$; about the x axis.
18 $y = x^2$ for x in $[1, 2]$; about the x axis.
19 $y = x^{4/3}$ for x in $[1, 8]$; about the y axis.
20 $y = x^{2/3}$ for x in $[1, 8]$; about the line $y = 1$.
21 $y = x^3/6 + 1/(2x)$ for x in $[1, 3]$; about the y axis.
22 $y = x^3/3 + 1/(4x)$ for x in $[1, 2]$; about the line $y = -1$.
23 The arc in Example 1; about the line $y = -1$.

24 Although the fundamental theorem of calculus is of no use in computing the perimeter of the ellipse $x^2/a^2 + y^2/b^2 = 1$, it is useful in computing the surface area of the "football" formed when the ellipse is rotated about one of its axes. Assuming that $a > b$ and that the ellipse is revolved around the x axis, find that area. Does your answer give the correct formula for the surface area of a sphere of radius a, $4\pi a^2$? (Let $b \to a^-$.)

25 The region bounded by $y = 1/x$ and the x axis and situated to the right of $x = 1$ is revolved around the x axis.
(*a*) Show that its volume is finite but its surface area is infinite.
(*b*) Does this mean that an infinite area can be painted by pouring a finite amount of paint into this solid?

26 Check that the formula for surface area agrees with our formula for the area of a cone.

27 If the band formed by revolving a line segment is cut along the rotated segment and laid out in the plane, what shape will it be? *Warning:* It is generally *not* a rectangle. With this approach, compute its area.

28 Consider a solid of revolution. Its volume is approximated closely by the sum of the volumes of thin parallel circular slabs. Is the area of the surface of the solid approximated closely by the sum of the areas of the curved surfaces of the slabs?

29 Define the moment of a curve around the x axis to be $\int_a^b y\,ds$, where a and b refer to the range of the arc length, s. The moment of the curve around the y axis is defined as $\int_a^b x\,ds$. The centroid of the curve, $(\bar{x}, \bar{y})$, is defined by setting

$$\bar{x} = \frac{\int_a^b x\,ds}{\text{Length of curve}}, \qquad \bar{y} = \frac{\int_a^b y\,ds}{\text{Length of curve}}.$$

Find the centroid of the top half of the circle $x^2 + y^2 = a^2$.

30 (See Exercise 29.) Show that the area of the surface obtained by revolving a curve around the x axis is equal to the length of the curve times the distance that the centroid of the curve moves. (This is a variant of Pappus's theorem.)

31 Use Exercise 30 to find the surface area of the doughnut formed by revolving a circle of radius a around a line a distance b from its center, $b \ge a$.

32 Use Exercise 30 to find the area of the curved part of a cone of radius a and height h.

33 A disk of radius a is covered by a finite number of strips (perhaps overlapping). Prove that the sum of their widths is at least $2a$. (If the strips are parallel, the assertion is clearly true; do not assume that the strips are parallel.) A strip consists of the points between two parallel lines. *Hint:* Think of the sphere of which the disk is an equatorial cross section and use Exercise 14.

34 Assume that f is a continuous function, $f(x) \ge 0$ for all x, and $\int_{-\infty}^{\infty} f(x)\,dx$ is convergent. The area under the curve and above the x axis is finite. Does it follow that the area of the surface formed by revolving the curve around the x axis is finite also?

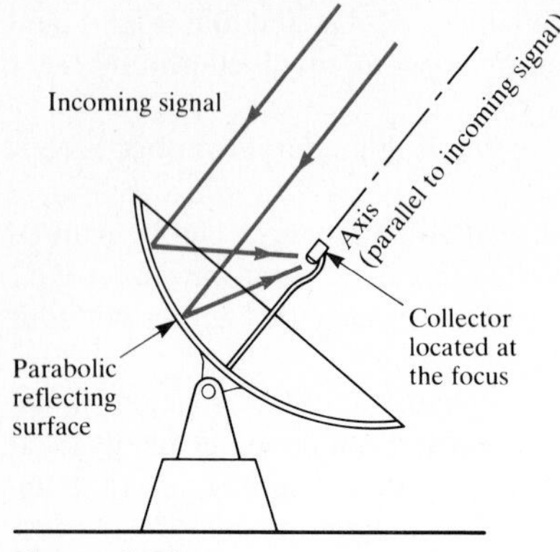

Figure 9.49

9.6 THE REFLECTION PROPERTY OF THE PARABOLA AND ELLIPSE

A satellite dish receiver and a microwave reflector are parabolic in shape. The reason is that all rays of light parallel to the axis of the parabola, after bouncing off the parabola, pass through a common point. This point is called the **focus** of the parabola. (See Fig. 9.49.) Similarly, the reflector behind a flashlight bulb is parabolic.

This section shows that the parabola has this reflection property. It also describes a similar property for an ellipse and some of its applications. The main tool will be a formula that determines the angle between two lines in terms of their slopes.

Consider a line L in the xy plane. It forms an **angle of inclination** α, $0 \le \alpha < \pi$, with the positive x axis. The slope m of L is $\tan \alpha$. (If $\alpha = \pi/2$, the slope is not defined.) See Fig. 9.50.

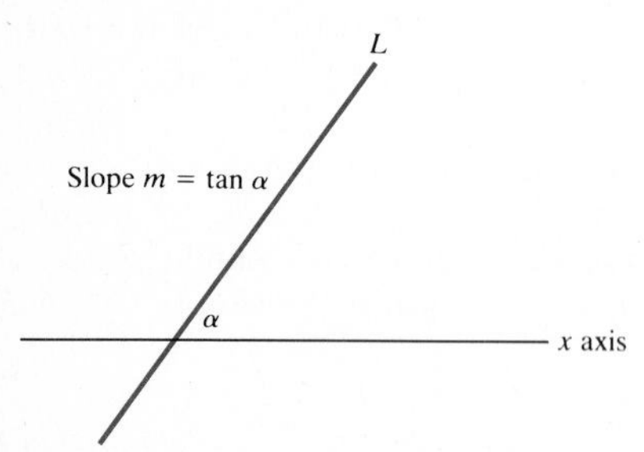

Figure 9.50

Consider two lines L and L' with angles of inclination α and α' and slopes m and m', respectively, as in Fig. 9.51. There are two (supplementary) angles between the two lines. The following definition serves to distinguish one of these two angles as *the* angle between L and L'.

Definition *Angle between two lines.* Let L and L' be two lines in the xy plane so named that L has the *larger* angle of inclination, $\alpha > \alpha'$. The angle θ between L and L' is defined to be

$$\theta = \alpha - \alpha'.$$

If L and L' are parallel, define θ to be 0.

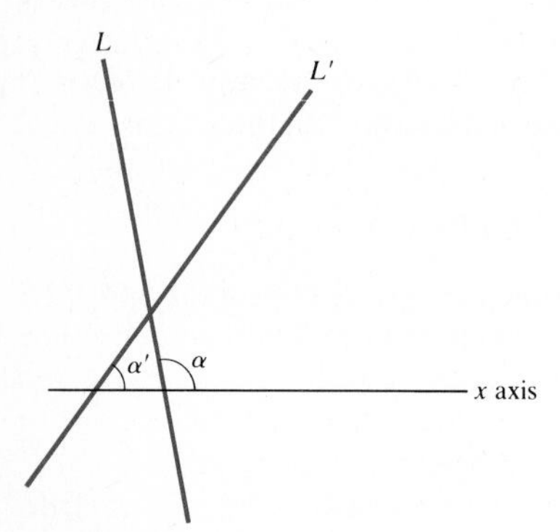

Figure 9.51

Figure 9.52 shows θ for some typical L and L'. In each case, θ is the counterclockwise angle from L' to L. Note that θ depends on the choice of the x axis and that $0 \le \theta < \pi$.

The tangent of θ is easily expressed in terms of the slopes m and m'. We have

$$\begin{aligned} \tan \theta &= \tan (\alpha - \alpha') \\ &= \frac{\tan \alpha - \tan \alpha'}{1 + \tan \alpha \tan \alpha'} \qquad \text{by the identity for } \tan (A - B) \\ &= \frac{m - m'}{1 + mm'}. \end{aligned}$$

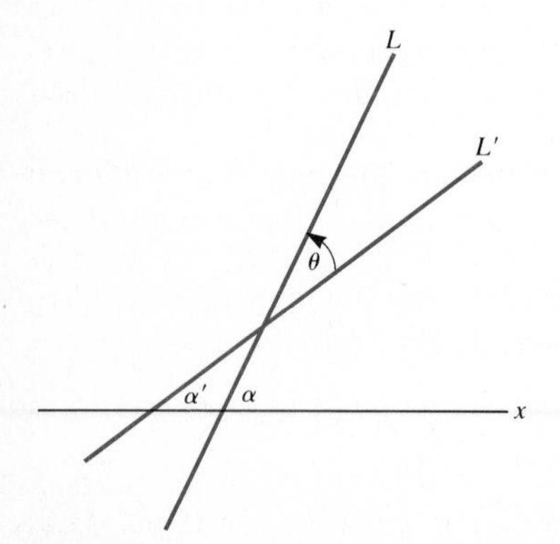
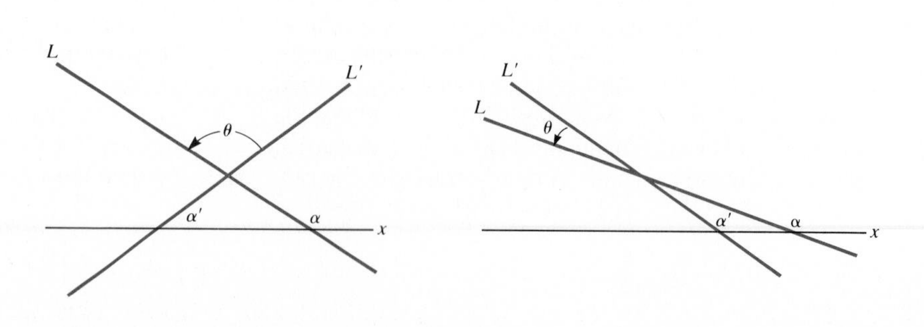

Figure 9.52

Formula for tangent of angle between two lines in terms of their slopes

Thus
$$\tan\theta = \frac{m - m'}{1 + mm'}, \tag{1}$$
where m is the slope of the line with larger angle of inclination. If $mm' = -1$, then $\theta = \pi/2$; this corresponds to the fact that as mm' approaches -1, $|\tan\theta| \to \infty$.

Now we are ready to show why parabolas make ideal collectors. Our argument uses the fact that the angle of reflection of a light ray equals its angle of incidence, as shown in Fig. 9.53.

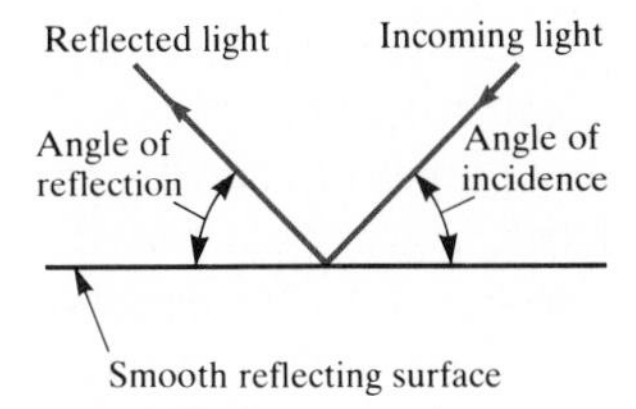

Figure 9.53

Consider a parabola with focus $F = (c, 0)$, $c > 0$, and directrix $x = -c$. Its equation is $y^2 = 4cx$. The top half has the equation $y = 2\sqrt{cx}$. (See Fig. 9.54.)

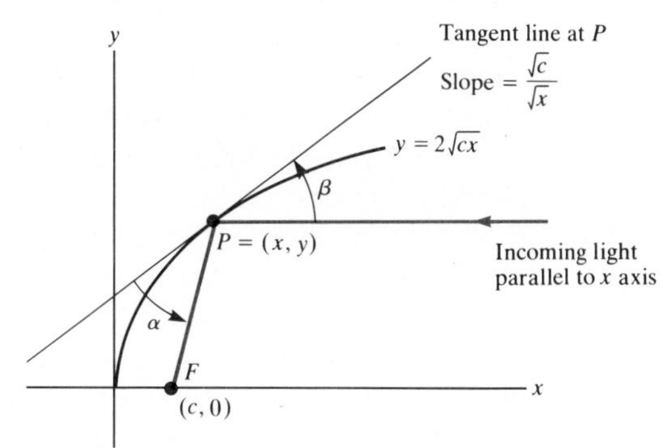

Figure 9.54

Let P be an arbitrary point on the curve $y = 2\sqrt{cx}$. We must show that the angle α between the line FP and the tangent line to the parabola at P equals the angle β between the tangent line and the line through P parallel to the x axis. To do this, we will show that $\tan\alpha = \tan\beta$.

The slope of the tangent line at a point on the parabola $y = 2\sqrt{cx}$ is $\sqrt{c}/\sqrt{x}$. The slope of the line FP is, by the two-point formula,
$$\frac{y - 0}{x - c} = \frac{y}{x - c}.$$

By formula (1),
$$\tan\alpha = \frac{\dfrac{y}{x - c} - \dfrac{\sqrt{c}}{\sqrt{x}}}{1 + \left(\dfrac{y}{x - c}\right)\left(\dfrac{\sqrt{c}}{\sqrt{x}}\right)}.$$

Also,
$$\tan\beta = \frac{\sqrt{c}}{\sqrt{x}}, \tag{2}$$
since it is the slope of the tangent line at P.

Some algebra, together with the equation $y = 2\sqrt{c}\sqrt{x}$, simplifies the formula for $\tan\alpha$, as follows:
$$\tan\alpha = \frac{y\sqrt{x} - \sqrt{c}\,(x - c)}{(x - c)\sqrt{x} + y\sqrt{c}} = \frac{2\sqrt{c}\sqrt{x}\sqrt{x} - \sqrt{c}\,x + \sqrt{c}\,c}{x\sqrt{x} - c\sqrt{x} + 2\sqrt{c}\sqrt{x}\sqrt{c}}$$
$$= \frac{\sqrt{c}\,x + \sqrt{c}\,c}{x\sqrt{x} + c\sqrt{x}} = \frac{\sqrt{c}(x + c)}{\sqrt{x}(x + c)} = \frac{\sqrt{c}}{\sqrt{x}}.$$

By Eq. (2), $\tan\alpha = \tan\beta$. Thus the reflection property of the parabola is established.

Diocles, in his book *On Burning Mirrors,* written around the year 190 B.C., studied both spherical and parabolic reflectors, both of which had been considered by earlier scientists. Some had thought that a spherical reflector focuses incoming light at a single point. This is false, and Diocles showed that a spherical reflector subtending an angle of 60° reflects light that is parallel to its axis of symmetry to points on this axis that occupy about one-thirteenth of the radius. He proposed an experiment, "Perhaps you would like to make two examples of a burning-mirror, one spherical, one parabolic, so that you can measure the burning power of each." *On Burning Mirrors* contains the first known proof that a parabola has the reflecting property described in this section. (See Diocles, *On Burning Mirrors,* edited by G. J. Toomer, Springer, New York, 1976.) [It is a simple exercise in trigonometry to show that a spherical mirror of radius r and subtending an angle of 60° causes light parallel to its axis of symmetry to reflect and meet the axis in an interval of length $r\,(1/\sqrt{3} - \frac{1}{2}) \approx r/12.9$. So even a spherical reflector makes an effective solar oven. After all, a potato or hamburger is not a point.]

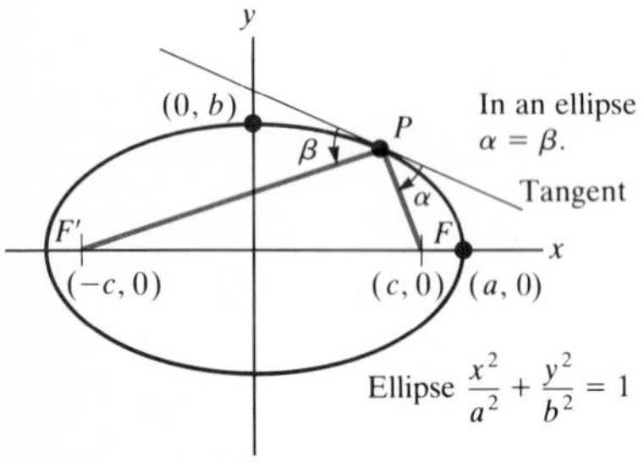

Figure 9.55

Now let us turn to the reflecting property of an ellipse.

An ellipse with foci F and F' has the property that sound (or light) starting at F passes through F' after bouncing off the ellipse. This follows from the following geometric property. For any point P on the ellipse, the lines PF and PF' make equal angles α and β with the tangent line to the ellipse at P. (See Fig. 9.55.)

The argument follows the same general lines. First, write the equation of the ellipse in the form

$$\frac{x^2}{a^2} + \frac{y^2}{b^2} = 1.$$

The foci are at $(c, 0)$ and $(-c, 0)$, where $c^2 = a^2 - b^2$. Then compute the slopes of PF and PF' by the two-point formula and the slope of the tangent line at P by differentiation (either explicit or implicit). Finally, use formula (1) and check that $\tan \alpha = \tan \beta$. The details are left as Exercise 11.

This reflecting property of the ellipse has diverse applications. For instance, in the construction of chips for computers it is necessary to bake a photo mask onto the surface of a silicon wafer, a process that requires focusing heat at the mask. This is accomplished by placing a heat source at one focus of an ellipse and the wafer at the other focus, as in Fig. 9.56.

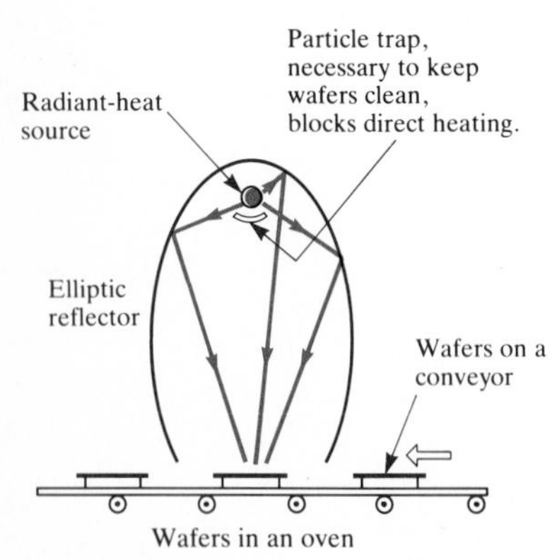

Figure 9.56

The reflection property is used in wind tunnel tests of aircraft noise. The test is run in an elliptical chamber, with the aircraft model at one focus and a microphone at the other.

Whispering rooms, such as the rotunda in the Capitol in Washington, D.C., are based on the same principle. A person talking quietly at one focus can be heard easily at the other focus. (Fortunately, since all the paths of the sound from F to F' have the same length, a whisper at F arrives at F' at one time.)

The reflecting property of the parabola was known to Greek mathematicians before Diocles. Legend has it that Archimedes (287–212 B.C.) arranged mirrors in a parabolic arc to burn the ships of an invader. Appolonius, a contemporary of Diocles, proved that the lines from a point

P on an ellipse to the foci make equal angles with the tangent at P. However, there is no evidence that this property was put to any use in ancient times.

For a geometric (noncalculus) proof of the reflection property of the ellipse, see the high school text *Geometry: A Guided Inquiry,* by G. D. Chakerian, C. Crabill, and S. Stein, Sunburst, Pleasantville, N.Y., 1986. (The argument is closely connected with finding the shortest path joining two given points and touching a given line, where both points lie on the same side of the line.)

EXERCISES FOR SEC. 9.6: THE REFLECTION PROPERTY OF THE PARABOLA AND ELLIPSE

In each of Exercises 1 and 2 find the angle between the two given lines.

1 A line of inclination $\pi/4$ and a line of inclination $3\pi/4$.

2 A line of inclination $5\pi/6$ and a line of inclination $\pi/6$.

In each of Exercises 7 to 10 find the tangent of the angle between the two curves at the indicated point of intersection.

3 Slopes 2 and 3.

4 Slopes 2 and $-\frac{1}{2}$.

5 Slopes -2 and -3.

6 Slopes $\sqrt{3}$ and $-\sqrt{3}$.

In each of Exercises 7 to 10 find the tangent of the angle between the two curves at the indicated point of intersection.

7 $y = \sin x$ and $y = \cos x$ at $(\pi/4, \sqrt{2}/2)$.

8 $y = x^2$ and $y = x^3$ at $(1, 1)$.

9 $y = e^x$ and $y = e^{-x}$ at $(0, 1)$.

10 $y = \sec x$ and $y = \sqrt{2} \tan x$ at $(\pi/4, \sqrt{2})$.

■

11 Establish the reflection property of the ellipse.

Exercises 12 to 18 concern the angle γ shown in Fig. 9.57. The curve is the graph of $r = f(\theta)$ in polar coordinates. P is a point on the curve. The angle between the ray from the pole O through P and the tangent line at P is called γ.

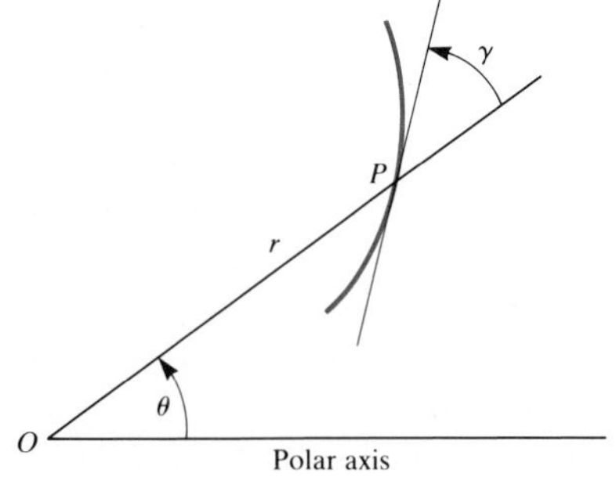

Figure 9.57

12 Find $\tan \gamma$ as follows:

(*a*) Express the slope of the ray OP in terms of θ.

(*b*) Obtain the slope of the tangent line by expressing the curve parametrically as

$$\begin{cases} x = f(\theta) \cos \theta \\ y = f(\theta) \sin \theta. \end{cases}$$

[This slope will be expressed in terms of $f(\theta)$, $f'(\theta)$, $\cos \theta$, and $\sin \theta$.]

(*c*) Using (*a*), (*b*), and formula (1), show that

$$\tan \gamma = \frac{f(\theta)}{f'(\theta)},$$

if $f'(\theta) \neq 0$.

(*d*) The answer in (*c*) is quite simple. Can you draw a picture showing why it is to be expected?

13 Find γ for the spiral $r = e^\theta$.

14 Show that for the cardioid $r = 1 - \cos \theta$, $\gamma = \theta/2$.

15 If for the curve $r = f(\theta)$, γ always equals θ, what are all the possibilities for f?

16 (*a*) For the cardioid $r = 1 + \cos \theta$ find $\lim_{\theta \to \pi^-} \gamma$.

(*b*) Sketch $r = 1 + \cos \theta$ using the information obtained in (*a*).

17 If for the curve $r = f(\theta)$, γ is independent of θ, what are all the possibilities for f?

18 Four dogs are chasing each other counterclockwise at the same speed. Initially they are at the four vertices of a square of side a. As they chase each other, each running directly toward the dog in front, they approach the center of the square in spiral paths. How far does each dog travel? (See Fig. 9.58.)

(*a*) First find the equation of the dog's path in polar coordinates; then find its arc length.

(*b*) Answer the question without calculus.

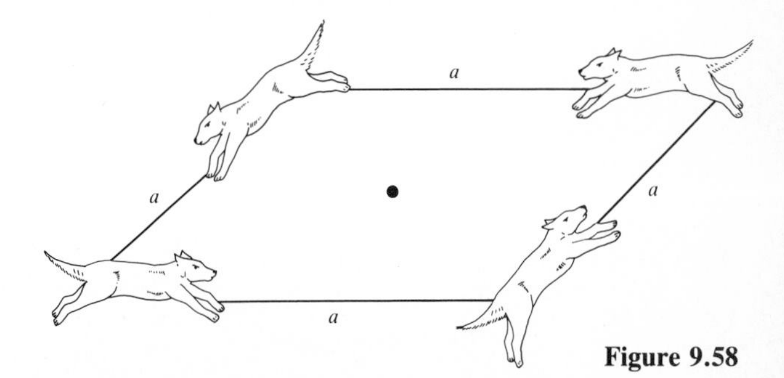

Figure 9.58

9.7 CURVATURE

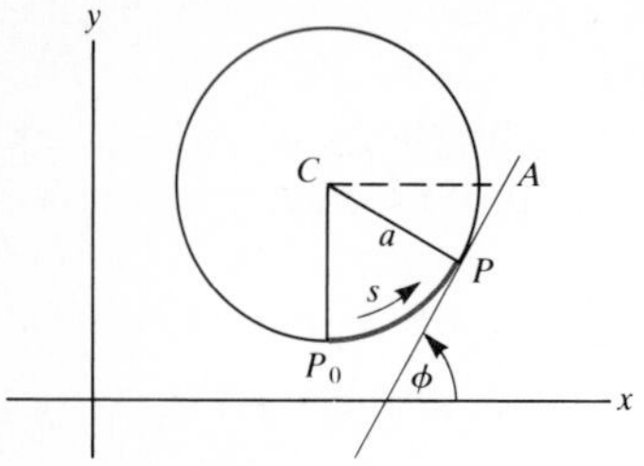

The tangent line turns as P moves counterclockwise and s increases.

Figure 9.59

The rate of change in y with respect to x measures the steepness of a curve. What is a reasonable measure of its "curviness" or curvature? A line, being perfectly straight, has no curvature. It seems reasonable that a line, therefore, should have curvature 0. Moreover, a large circle should have less curvature than a small circle.

When you walk around a small circle, your direction changes much more rapidly than when you walk around a large circle. To make this idea more precise, and to obtain a measure of the curvature of a circle, consider the diagram in Fig. 9.59 of a circle of radius a and a line tangent to it. Start at the bottom of the circle at P_0 and walk counterclockwise. At a distance s along the curve from P_0 the direction is given by the angle ϕ from the positive x axis to the tangent line at P; the angle ϕ depends on s. Define **the curvature of a circle** to be the absolute value of the rate at which ϕ changes with respect to s, that is, $|d\phi/ds|$.

The next theorem shows that the curvature is small for a large circle and large for a small circle; in fact, it is simply the reciprocal of the radius.

Theorem 1 For a circle of radius a, the curvature $|d\phi/ds|$ is constant and equals $1/a$, the reciprocal of the radius.

Proof It is necessary to express ϕ as a function of s. To do this, introduce the line CA parallel to the x axis, as in Fig. 9.59. Then ϕ equals angle PAC (alternate interior angles). Thus angle ϕ is the complement of angle PCA (by right triangle CPA). But angle P_0CP is also the complement of angle PCA. Thus ϕ = angle P_0CP, which, by the definition of radian measure, has measure s/a. Thus

$$\phi = \frac{s}{a} \quad \text{and} \quad \frac{d\phi}{ds} = \frac{1}{a},$$

and the theorem is proved. ■

If we measured s in the clockwise direction, $d\phi/ds$ would be $-1/a$.

Since $|d\phi/ds|$, the absolute value of the rate of change of direction with respect to arc length, gives a reasonable measure of curvature for a circle—the larger the circle, the less its curvature—it is common to use it as a measure of curvature for other curves.

Before defining curvature in general, we should discuss the arc length s and the angle ϕ in a little more detail. First of all, let us agree to measure arc length in such a way that it increases as we move along the curve away from the base point.

Second, consider the angle ϕ. There is ambiguity in the choice of ϕ, for if ϕ describes the line, so does $\phi + n\pi$ for any integer n. The particular choice is not of any importance; what does matter is that ϕ should vary continuously as we traverse the curve. Look back at Fig. 9.59, which showed the typical tangent line to a circle of radius a. If we choose $\phi = 0$ for the (horizontal) tangent line at P_0, then our choice of ϕ for all other tangent lines to the circle, as we traverse the circle counterclockwise, is determined. Figure 9.60 shows that as P goes once around the circle, ϕ increases by 2π.

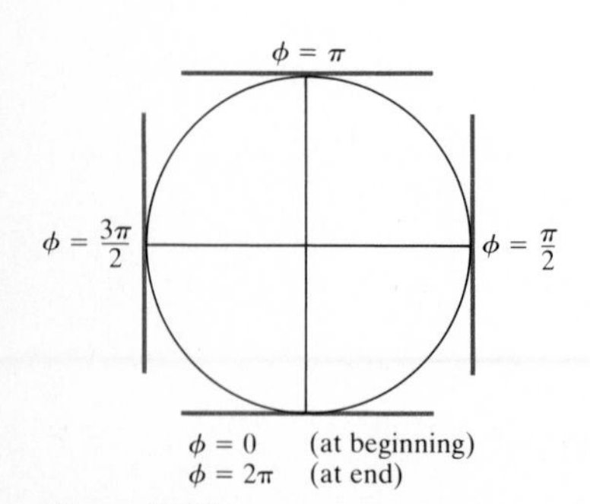

Figure 9.60

If we had chosen ϕ for the tangent line at P_0 initially to be, say, π, then, as we traverse the circle once, ϕ would increase to 3π. Since it is the *rate* at which ϕ changes that concerns us, not its actual value, this ambiguity, which may occur with any curve, does not affect the following definition.

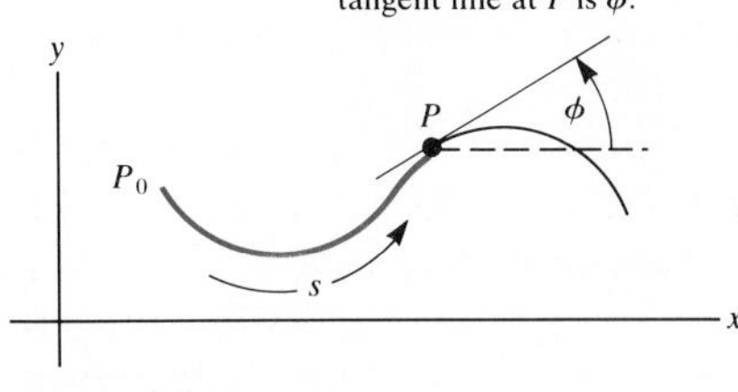

Figure 9.61

Definition *Curvature.* Assume that a curve is given parametrically, with the parameter of the typical point P being s, the distance along the curve from a fixed point P_0 to P. Let ϕ be the angle between the tangent line at P and the positive part of the x axis. The **curvature** κ at P is the absolute value of the derivative, $d\phi/ds$:

$$\kappa = \left|\frac{d\phi}{ds}\right|$$

(if this derivative exists). (See Fig. 9.61.)

κ is the Greek letter "kappa."

Observe that a straight line has zero curvature everywhere, since ϕ is constant.

When the curve is given in the form $y = f(x)$, the curvature depends on dy/dx and d^2y/dx^2, as the next theorem shows.

Theorem 2 Let arc length s be measured along the curve $y = f(x)$ from a fixed point P_0. Assume that x increases as s increases. Then

Formula for curvature of $y = f(x)$

$$\kappa = \text{curvature} = \frac{|d^2y/dx^2|}{[1 + (dy/dx)^2]^{3/2}}.$$

Proof By the chain rule,

$$\frac{d\phi}{ds} = \frac{d\phi/dx}{ds/dx}.$$

As was shown in Sec. 9.4,

$$\frac{ds}{dx} = \left[1 + \left(\frac{dy}{dx}\right)^2\right]^{1/2}.$$

All that remains is to express $d\phi/dx$ in terms of dy/dx and d^2y/dx^2. Note that, in Fig. 9.62,

$$\frac{dy}{dx} = \tan\phi \qquad \text{(slope of tangent line).}$$

Thus
$$\frac{d^2y}{dx^2} = \sec^2\phi\,\frac{d\phi}{dx} = (1 + \tan^2\phi)\frac{d\phi}{dx} = \left[1 + \left(\frac{dy}{dx}\right)^2\right]\frac{d\phi}{dx},$$

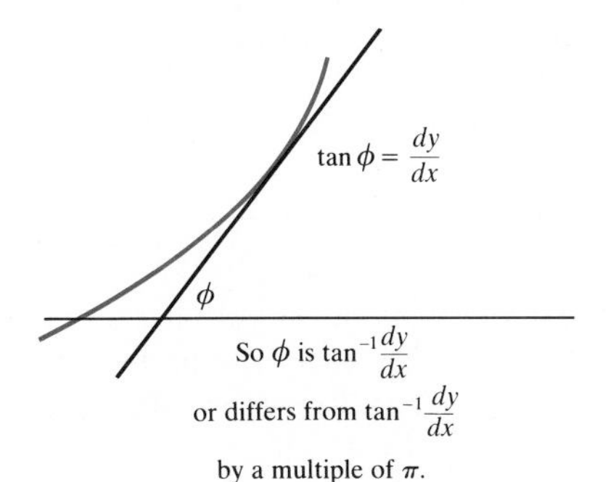

Figure 9.62

and we have

$$\frac{d\phi}{dx} = \frac{d^2y/dx^2}{1 + (dy/dx)^2}.$$

Consequently,
$$\frac{d\phi}{ds} = \frac{d\phi/dx}{ds/dx} = \frac{d^2y/dx^2}{[1 + (dy/dx)^2]\sqrt{1 + (dy/dx)^2}},$$

and the theorem is proved. ■

EXAMPLE 1 Find the curvature at a typical point (x, y) on the curve $y = x^2$.

SOLUTION In this case, $dy/dx = 2x$ and $d^2y/dx^2 = 2$. Thus the curvature at (x, y) is

$$\kappa = \frac{|d^2y/dx^2|}{[1 + (dy/dx)^2]^{3/2}} = \frac{2}{[1 + (2x)^2]^{3/2}}.$$

The maximum curvature occurs when $x = 0$. As $|x|$ increases, the curvature approaches 0, and the curve gets straighter. ■

Theorem 2 tells how to find the curvature if y is given as a function of x. If a curve is described parametrically, its curvature can be found with the aid of the next theorem. The proof, which rests on Theorem 2, is outlined in Exercise 27.

Theorem 3 If, as we move along the parameterized curve $x = g(t)$, $y = f(t)$ to a point P, both x and the arc length s from a point P_0 increase as t increases, then

See Exercise 22 for the formula in polar coordinates.

$$\text{Curvature} = \frac{|\dot{x}\ddot{y} - \dot{y}\ddot{x}|}{[(\dot{x})^2 + (\dot{y})^2]^{3/2}}.$$

(The dot notation for derivatives shortens the formula; $\dot{x} = dx/dt$, $\ddot{x} = d^2x/dt^2$, etc.) ■

EXAMPLE 2 The cycloid determined by a wheel of radius 1 has the parametric equations

$$x = \theta - \sin\theta \quad \text{and} \quad y = 1 - \cos\theta,$$

as shown in Sec. 9.3. (See Fig. 9.63.) Find the curvature at a typical point on this curve.

SOLUTION To do so, use Theorem 3, noting that θ plays the role of the parameter t. First,

$$\frac{dx}{d\theta} = 1 - \cos\theta, \qquad \frac{dy}{d\theta} = \sin\theta,$$

$$\frac{d^2x}{d\theta^2} = \sin\theta, \qquad \frac{d^2y}{d\theta^2} = \cos\theta.$$

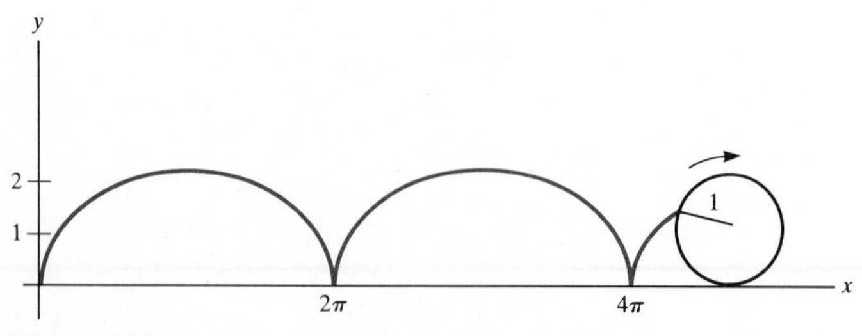

Figure 9.63

Then, by Theorem 3,

$$\text{Curvature} = \frac{|(1 - \cos\theta)\cos\theta - (\sin\theta)\sin\theta|}{[(1 - \cos\theta)^2 + (\sin\theta)^2]^{3/2}}$$

$$= \frac{|\cos\theta - (\cos^2\theta + \sin^2\theta)|}{[1 - 2\cos\theta + (\cos^2\theta + \sin^2\theta)]^{3/2}}$$

$$= \frac{|\cos\theta - 1|}{(2 - 2\cos\theta)^{3/2}} = \frac{1 - \cos\theta}{2^{3/2}(1 - \cos\theta)^{3/2}}$$

$$= \frac{1}{\sqrt{8}} \frac{1}{(1 - \cos\theta)^{1/2}}.$$

Since $y = 1 - \cos\theta$, the curvature is simply $1/\sqrt{8y}$. ■

As Theorem 1 shows, a circle with curvature κ has radius $1/\kappa$. This suggests the following definition.

Definition *Radius of curvature.* The **radius of curvature** of a curve at a point is the reciprocal of the curvature:

A large radius of curvature implies a small curvature.

$$\text{Radius of curvature} = \frac{1}{\text{Curvature}} = \frac{1}{\kappa}.$$

As can easily be checked, the radius of curvature of a circle of radius a is, fortunately, a.

The cycloid in Example 2 has radius of curvature at the point (x, y) equal to

$$\frac{1}{1/\sqrt{8y}} = \sqrt{8y}.$$

In particular, at the top of an arch, the cycloid's radius of curvature is $\sqrt{8 \cdot 2} = 4$.

The osculating circle

At a given point P on a curve, the osculating circle at P is defined to be that circle which (a) passes through P, (b) has the same slope at P as the curve does, and (c) has the same second derivative there. By Theorem 2, the osculating circle and the curve have the same curvature at P. Hence they have the same radius of curvature.

For instance, consider the parabola $y = x^2$ of Example 1. When $x = 1$, the curvature is $2/5^{3/2}$ and the radius of curvature is $5^{3/2}/2 \approx 5.6$. The osculating circle at (1, 1) is shown in Fig. 9.64.

Observe that the osculating circle in Fig. 9.64 *crosses the parabola* as it passes through the point (1, 1). Although this may be surprising, a little reflection will show why it is to be expected.

Think of driving along the parabola $y = x^2$. If you start at (1, 1) and drive up along the parabola, the curvature diminishes. It is smaller than that of the circle of curvature at (1, 1). Hence you would be turning your steering wheel to the left and would be traveling *outside* the circle of curvature at (1, 1). On the other hand, if you start at (1, 1) and move to the left on the parabola, the curvature increases and is greater than that of the osculating circle at (1, 1), so you would be driving *inside* the

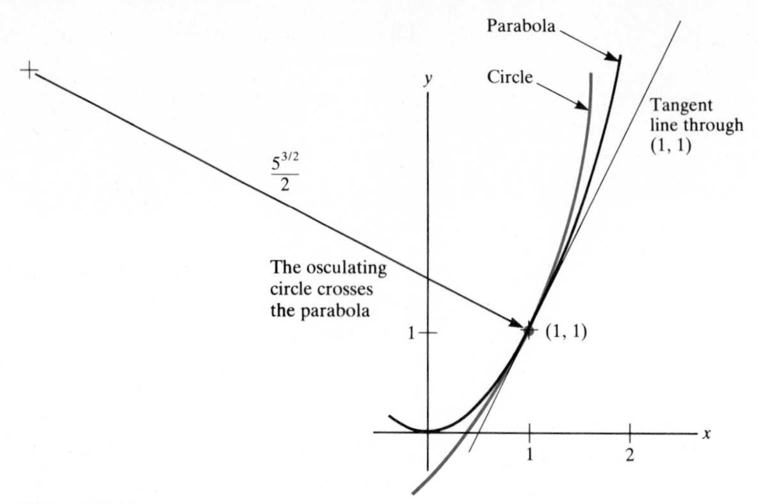

Figure 9.64

osculating circle at (1, 1). This informal argument shows why the osculating circle crosses the curve in general. At a point where the curvature is neither a local maximum nor a local minimum, the osculating circle crosses the curve. In the case of the curve $y = x^2$, the only osculating circle that does *not* cross the curve at its point of tangency is the one that is tangent at (0, 0), where the curvature is a maximum.

EXERCISES FOR SEC. 9.7: CURVATURE

In each of Exercises 1 to 6 find the curvature and radius of curvature of the given curve at the given point.

1 $y = x^2$ at (1, 1) **2** $y = \cos x$ at (0, 1)

3 $y = e^{-x}$ at $(1, 1/e)$ **4** $y = \ln x$ at $(e, 1)$

5 $y = \tan x$ at $(\pi/4, 1)$ **6** $y = \sec 2x$ at $(\pi/6, 2)$

In Exercises 7 to 10 find the curvatures of the given curves for the given values of the parameter.

7 $\begin{cases} x = 2\cos 3t \\ y = 2\sin 3t \end{cases}$ at $t = 0$

8 $\begin{cases} x = 1 + t^2 \\ y = t^3 + t^4 \end{cases}$ at $t = 2$

9 $\begin{cases} x = e^{-t}\cos t \\ y = e^{-t}\sin t \end{cases}$ at $t = \pi/6$

10 $\begin{cases} x = \cos^3\theta \\ y = \sin^3\theta \end{cases}$ at $\theta = \pi/3$

11 (*a*) Compute the curvature and radius of curvature for the curve $y = (e^x + e^{-x})/2$.
(*b*) Show that the radius of curvature at (x, y) is y^2.

12 Find the radius of curvature along the curve $y = \sqrt{a^2 - x^2}$, where a is a constant. (Since the curve is part of a circle of radius a, the answer should be a).

■

13 For what value of x is the radius of curvature of $y = e^x$ smallest? *Hint:* How does one find the minimum of a function?

14 For what value of x is the radius of curvature of $y = x^2$ smallest?

15 (*a*) Show that where a curve has its tangent parallel to the x axis its curvature is simply the absolute value of second derivative d^2y/dx^2.
(*b*) Show that the curvature is never larger than the absolute value of d^2y/dx^2.

16 An engineer lays out a railroad track as indicated in Fig. 9.65. BC is part of a circle. AB and CD are straight and tangent to the circle. After the first train runs over this track, the engineer is fired because the curvature is not a continuous function. Why should it be?

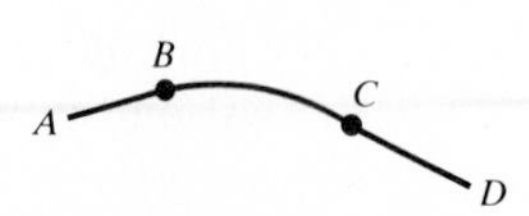

Figure 9.65

17 Railroad curves are banked to reduce wear on the rails and flanges. The greater the radius of curvature, the less the curve must be banked. The best bank angle A satisfies the equation $\tan A = v^2/(32R)$, where v is speed in feet per second and R is radius of curvature in feet. A train travels in the elliptical track

$$\frac{x^2}{1{,}000^2} + \frac{y^2}{500^2} = 1$$

(where x and y are measured in feet) at 60 miles per hour (equals 88 feet per second). Find the best angle A at the points (1,000, 0) and (0, 500).

18 The flexure formula in the theory of beams asserts that the bending moment M required to bend a beam is proportional to the desired curvature, $M = c/R$, where c is a constant depending on the beam and R is the radius of curvature. A beam is bent to form the parabola $y = x^2$. What is the ratio between the moments required at (0, 0) and at (2, 4)? (See Fig. 9.66.)

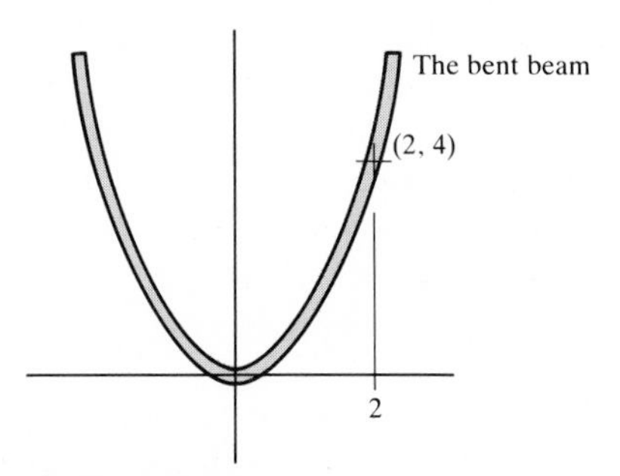

Figure 9.66

Exercises 19 to 21 are related.

19 Find the radius of curvature at a typical point on the ellipse

$$\begin{cases} x = a \cos \theta \\ y = b \sin \theta. \end{cases}$$

20 (*a*) Show, by eliminating θ, that the curve in Exercise 19 is the ellipse

$$\frac{x^2}{a^2} + \frac{y^2}{b^2} = 1.$$

(*b*) What is the radius of curvature of this ellipse at $(a, 0)$? at $(0, b)$?

21 An ellipse has a major diameter of length 6 and a minor diameter of length 4. Draw the circles that most closely approximate this ellipse at the four points which lie at the extremities of its diameters. (See Exercises 19 and 20.)

22 Theorem 2 gives a formula for curvature if the curve is given in rectangular form, $y = f(x)$. If the curve is given in polar form, $r = f(\theta)$, show that curvature equals the absolute value of $[r^2 + 2(r')^2 - rr'']/[r^2 + (r')^2]^{3/2}$. *Hint:* Consider the parametric representation of the curve as $x = r \cos \theta$, $y = r \sin \theta$, where $r = f(\theta)$.

23 Use the formula in Exercise 22 to show that the cardioid $r = 1 + \cos \theta$ has curvature $3\sqrt{2}/(4\sqrt{r})$ at (r, θ).

24 Use the formula in Exercise 22 to find the curvature of $r = a \cos \theta$.

25 Use the formula in Exercise 22 to find the curvature of $r = \cos 2\theta$.

26 If, on a curve, $dy/dx = y^3$, express the curvature in terms of y.

27 (*a*) Show that if x and y are given parametrically as functions of t, then

$$\frac{d^2y}{dx^2} = \frac{\dot{x}\ddot{y} - \dot{y}\ddot{x}}{(\dot{x})^3}.$$

(*b*) Prove Theorem 3.

28 At the top of the cycloid in Example 2 the radius of curvature is twice the diameter of the rolling circle. What would you have guessed the radius of curvature to be at this point? Why is it not simply the diameter of the wheel, since the wheel at each moment is rotating about its point of contact with the ground?

29 As is shown in physics, the larger the radius of curvature of a turn, the faster a given car can travel around that turn. The radius of curvature required is proportional to the square of the maximum speed. Or conversely, the maximum speed around a turn is proportional to the square root of the radius of curvature. If a car moving on the path $y = x^3$ (x and y measured in miles) can go 30 miles per hour at (1, 1) without sliding off, how fast can it go at (2, 8)?

9.S SUMMARY

The following table provides memory aids for most of the formulas in the chapter:

Section	*Concept*	*Memory Aid*
9.2	Area $= \int_\alpha^\beta \frac{r^2}{2}\,d\theta$	Area $= \frac{1}{2}r \cdot r\,d\theta$
9.4	Arc length $= \int_a^b \sqrt{\left(\frac{dx}{dt}\right)^2 + \left(\frac{dy}{dt}\right)^2}\,dt$ Arc length $= \int_a^b \sqrt{1 + \left(\frac{dy}{dx}\right)^2}\,dx$	$(ds)^2 = (dx)^2 + (dy)^2$
	Arc length $= \int_\alpha^\beta \sqrt{r^2 + (r')^2}\,d\theta$ Speed $= \sqrt{\left(\frac{dx}{dt}\right)^2 + \left(\frac{dy}{dt}\right)^2}$	$(ds)^2 = (r\,d\theta)^2 + (dr)^2$
9.5	Area of surface of revolution $= \int_c^d 2\pi R\,ds$	Area $= 2\pi R \cdot ds$

Vocabulary and Symbols

polar coordinates (r, θ)
pole, polar axis
parametric representation
arc length s
surface of revolution
curvature κ
radius of curvature
angle between two lines

Key Facts

Polar and rectangular coordinates are related through the equations $x = r\cos\theta$, $y = r\sin\theta$, $r^2 = x^2 + y^2$, $\tan\theta = y/x$.

The slope of a curve given parametrically as $x = g(t)$, $y = h(t)$ is

$$\frac{dy}{dx} = \frac{dy/dt}{dx/dt} = \frac{\dot{y}}{\dot{x}}.$$

The area $\int_a^b y\,dx$ under a curve given parametrically is obtained by a substitution in which y, dx, and the limits of integration are all expressed in terms of the parameter t. [See Exercise 17(*b*) of Sec. 9.3.] The parameter may be denoted by a different letter, such as θ, for instance.

The angle between two lines in the plane equals the larger inclination minus the smaller inclination. The tangent of this angle is

$$\frac{m - m'}{1 + mm'},$$

where m is the slope of the line of larger inclination and m' is the slope of the other line.

The curvature of a curve is the absolute value of the rate at which the angle of inclination of the tangent line changes with respect to arc length: $|d\phi/ds|$. The curvature of $y = f(x)$ is

$$\frac{|d^2y/dx^2|}{[1 + (dy/dx)^2]^{3/2}}.$$

Curvature can also be computed for curves given parametrically, and thus in polar coordinates.

The radius of curvature is the reciprocal of the curvature.

GUIDE QUIZ ON CHAP. 9: PLANE CURVES AND POLAR COORDINATES

In Exercises 1 to 7 set up definite integrals for the given quantities, but do not evaluate them.

1 The area of the region within one loop of the curve $r = \cos 5\theta$.

2 The length of the curve $y = x^4$ from $x = 1$ to $x = 2$.

3 The area of the surface obtained by revolving the curve in Exercise 2 around the x axis.

4 The area of the surface obtained by revolving the curve in Exercise 2 around the line $x = 1$.

5 The length of the curve $r = 2 + \sin \theta$, $0 \le \theta \le 2\pi$.

6 The area of the surface obtained by revolving the top half of the curve in Exercise 5 around the polar axis.

7 The length of the curve $x = 5t^2$, $y = \sqrt{t}$, $0 \le t \le 1$.

8 At time t a moving particle is at the point $x = 12t$, $y = -16t^2 + 5t$. What is its speed when $t = 1$?

9 A curve is given parametrically as $x = \tan t$, $y = \sec t$. Eliminate t to find an equation in x and y for the curve.

10 (*a*) Find the inclination of the tangent line to the curve $r = \cos 2\theta$ at the point for which $\theta = \pi/8$.
(*b*) Graph the curve and check that the answer in (*a*) is reasonable.

11 What is the radius of the circle that best approximates the curve $y = 1/x$ at (1, 1) in the sense that it has the same radius of curvature as the curve does at (1, 1)?

REVIEW EXERCISES FOR CHAP. 9: PLANE CURVES AND POLAR COORDINATES

1 (*a*) Develop the formula for area in polar coordinates.
(*b*) What is the device for remembering the formula?

2 (*a*) Develop the formula for arc length with parameter t.
(*b*) What is the device for remembering the formula?

3 (*a*) Develop the formula for the area of a surface of revolution with parameter x.
(*b*) What is the device for remembering the formula?

4 See Exercise 12 in Sec. 9.6.
(*a*) Develop the formula for $\tan \gamma$, where γ is the angle between the ray from the pole to the point P on the curve $r = f(\theta)$ and the tangent line to the curve at P.
(*b*) Draw the memory device for the formula in (*a*).

5 (*a*) Define "curvature."
(*b*) From (*a*), obtain the formula for the curvature of the curve $y = f(x)$.

6 Consider the curve $y = e^x$ for x in [0, 1].
(*a*) Set up integrals for its arc length and for the areas of the surfaces obtained by rotating the curve around the x axis and also about the y axis.
(*b*) Two of the three integrals are elementary. Evaluate them.

7 Consider the curve $y = \sin x$ for x in $[0, \pi]$. Proceed as in Exercise 6. This time, however, only one of the three integrals is elementary. Evaluate it.

8 Graph $r = 3/(\cos \theta + 2 \sin \theta)$ after first finding the rectangular form of the equation.

9 Find the maximum y coordinate of the curve $r = 1 + \cos\theta$.

10 Find the minimum x coordinate of the curve $r = 1 + \cos\theta$.

11 Is the total length of the curve $r = 1/(1 + \theta)$, $\theta \geq 0$, finite or infinite?

12 Is the total length of the curve $r = 1/(1 + \theta^2)$, $\theta \geq 0$, finite or infinite?

13 What is the length of the curve $r = e^{-\theta}$, $\theta \geq 0$?

14 Assume that x and y are functions of t. Obtain the formula for d^2y/dx^2 in terms of the derivatives $\dot{x}$, $\ddot{x}$, $\dot{y}$, $\ddot{y}$.

15 If

$$\begin{cases} x = \cos 2t \\ y = \sin 3t, \end{cases}$$

express dy/dx and d^2y/dx^2 in terms of t.

16 Find the radius of curvature of the curve $y = \ln x$ at $(e, 1)$.

17 Let $r = e^{\theta}$, $0 \leq \theta \leq \pi/4$, describe a curve in polar coordinates. In parts (b) to (f) set up definite integrals for the quantities, and show that they could be evaluated by the fundamental theorem of calculus, but do not evaluate them.

(a) Sketch the curve.

(b) The area of the region R below the curve and above the interval $[1, e^{\pi/4}\sqrt{2}/2]$ on the x axis.

(c) The volume of the solid obtained by revolving R, defined in (b), about the x axis.

(d) The volume of the solid obtained by revolving R about the y axis.

(e) The area of the surface obtained by revolving the curve in (a) around the x axis.

(f) The area of the surface obtained by revolving the curve in (a) around the y axis.

Rare is the curve for which the corresponding five integrals are all elementary.

18 Find the length of the curve $y = \ln x$ from $x = 1$ to $x = \sqrt{3}$. An integral table will save a lot of work.

■

19 "If $dy/dx = \dot{y}/\dot{x}$, why is not d^2y/dx^2 equal to $\ddot{y}/\ddot{x}$?" How would you answer this question?

20 A set R in the plane bounded by a curve is **convex** if, whenever P and Q are points in R, the line segment PQ also lies in R. A curve is convex if it is the boundary of a convex set. Consider a curve with no line segments.

(a) Show why the average radius of curvature with respect to angle ϕ as you traverse a convex curve equals (Length of curve)/2π.

(b) Deduce from (a) that a convex curve of length L has a radius of curvature equal to $L/2\pi$ somewhere on the curve.

21 Prove that the average value of the curvature as a function of arc length s as you sweep out a convex curve in the counterclockwise direction is 2π/Length of curve. (See Exercise 20 for the definition of convex curve.)

22 A curve has the property that any of its chords makes equal angles with the tangent lines to the curve at the ends of the chord, as shown in Fig. 9.67. Show that the curve is part of a circle. *Hint:* Keep A fixed and let B vary. Take A as the pole of a polar coordinate system and the tangent line at that point as the polar axis.

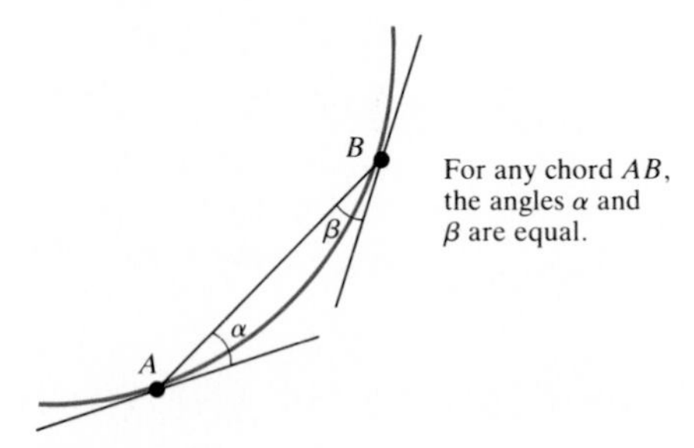

Figure 9.67

23 When calculating the surface-to-volume ratio of the rotary engine, engineers had to determine the arc length of a portion of the epitrochoid.

(a) Show that the length of a general arc of the epitrochoid given parametrically as in Example 5 of Sec. 9.3 is

$$\int_{\alpha}^{\beta} \sqrt{9b^2 + R^2 + 6bR\cos 2\theta}\, d\theta.$$

(b) Show that the integral in (a) equals

$$\int_{\alpha}^{\beta} (3b + R)\sqrt{1 - k^2\sin^2\theta}\, d\theta,$$

where $k^2 = 12bR/(3b + R)^2$.

(c) Show that $k^2 \leq 1$. Thus the integral in (b) is an elliptic integral, which is tabulated in many mathematical handbooks.

24 Let a be a rational number. Consider the curve $y = x^a$ for x in the interval $[1, 2]$. Show that the area of the surface obtained by rotating this curve around the x axis can be evaluated by the fundamental theorem of calculus in the cases $a = 1$ or $1 + 2/n$, where n is any nonzero integer. (These are the only rational a for which the pertinent integral is elementary.) Assume Chebyshev's theorem, which asserts that if p and q are rational numbers, then $\int x^p(1 + x)^q\, dx$ is elementary only when p is an integer, q is an integer, or $p + q$ is an integer.

25 Read the generalized mean-value theorem stated in Exercise 177 of Sec. 6.S.

(*a*) What does it say about the curve given parametrically by the equations $x = g(t)$, $y = f(t)$? (Here t plays the role of the x in Exercise 177.)

(*b*) Let $h(t)$, for t in $[a, b]$, be defined as the vertical distance from $(g(t), f(t))$ to the line that passes through $(g(a), f(a))$ and $(g(b), h(b))$, as in the proof of the mean-value theorem in Sec. 4.1. Use the function h to prove the generalized mean-value theorem.

26 Let a be a number, and let $x = g(t)$ and $y = h(t)$, $t \geq a$, describe a curve. Let $P = (g(a), h(a))$, and let $Q(t) = (g(t), h(t))$, $t > a$.

(*a*) Sketch the chord $PQ(t)$ and the tangent line $T(t)$ at $Q(t)$.

(*b*) What is the slope of $PQ(t)$? What is the slope of $T(t)$?

(*c*) Under what assumptions will the two slopes in (*b*) have the same limit as $t \to a^+$?

10 Series and Complex Numbers

How is it possible to compute $\sin x$ to eight decimal places? Certainly not by drawing a large right triangle with angle x, measuring the side opposite the angle and dividing that number by the length of the hypotenuse. We would be lucky to get two decimal places correct.

Functions such as $\sin x$, $\cos x$, and e^x can be computed by approximating them by polynomials. Since a polynomial can be evaluated by additions and multiplications, we then have a practical way of calculating $\sin x$, $\cos x$, and e^x.

Two questions must be considered: How can we find polynomials with which to approximate a given function? How large is the possible error in the approximation?

These questions provide the guiding theme of this chapter.

10.1 SEQUENCES

A **sequence** of real numbers,

$$a_1, a_2, a_3, \ldots, a_n, \ldots,$$

is a function that assigns to each positive integer n a number a_n. The number a_n is called the **nth term** of the sequence. For example, the sequence

$$\left(1+\frac{1}{1}\right)^1, \left(1+\frac{1}{2}\right)^2, \left(1+\frac{1}{3}\right)^3, \ldots, \left(1+\frac{1}{n}\right)^n, \ldots$$

was considered in Sec. 6.2 in the study of the number e. In this case,

$$a_n = \left(1+\frac{1}{n}\right)^n.$$

As another example, Newton's method of estimating a root of $f(x) = 0$ involves the construction of a sequence $x_1, x_2, \ldots$, where

$$x_{n+1} = x_n - \frac{f(x_n)}{f'(x_n)}.$$

Sometimes the notation $\{a_n\}$ is used as an abbreviation of the sequence $a_1, a_2, \ldots, a_n, \ldots$. For instance, e involves the sequence $\{(1 + 1/n)^n\}$.

The limit of a sequence

If, as n gets larger, a_n approaches a number L, then L is called the **limit** of the sequence. If the sequence $a_1, a_2, \ldots$ has a limit L, we write

$$\lim_{n\to\infty} a_n = L.$$

For instance, we write

$$\lim_{n\to\infty} \left(1 + \frac{1}{n}\right)^n = e.$$

If a_n becomes and remains arbitrarily large as n gets larger, we write $\lim_{n\to\infty} a_n = \infty$. (The limit does not exist in this case.) For instance, $\lim_{n\to\infty} 2^n = \infty$.

To show that $\lim_{n\to\infty} a_n = 0$, show that $\lim_{n\to\infty} |a_n| = 0$.

The assertion that "a_n approaches the number L" is equivalent to the assertion that $|a_n - L| \to 0$ as $n \to \infty$. In particular, if $L = 0$, the assertion that $a_n \to 0$ as $n \to \infty$ is equivalent to the assertion that $|a_n| \to 0$ as $n \to \infty$. This means that if the absolute value of a_n approaches 0 as $n \to \infty$, then a_n approaches 0 as $n \to \infty$. This observation will be of use in this section and later.

A sequence need not begin with the term a_1. Later in the chapter sequences of the form $a_0, a_1, a_2, \ldots$ will be considered. In such a case, a_0 is called **the zeroth term**. Or we may consider a "tail end" of a sequence, a sequence that begins with the term a_k: $a_k, a_{k+1}, a_{k+2}, \ldots$.

The next example introduces a simple but important sequence.

EXAMPLE 1 *The sequence $\{(\frac{1}{2})^n\}$.* A certain radioactive substance decays, losing half its mass in an hour. In the long run how much is left?

SOLUTION If the initial mass is 1 gram, after an hour only $\frac{1}{2}$ gram remains. During the next hour, half this amount is lost and half remains. Thus after 2 hours

$$(\tfrac{1}{2})^2 = 0.25 \text{ gram}$$

remains. After 3 hours

$$(\tfrac{1}{2})^3 = 0.125 \text{ gram}$$

remains. In general, after n hours $(\frac{1}{2})^n$ gram remains.

When n is large, the amount remaining is very small. In other words, the sequence

Since $2^n \to \infty$, it follows that $\dfrac{1}{2^n} \to 0$.

$$\{(\tfrac{1}{2})^n\}$$

approaches 0 when n gets large. This is summarized by the equation

$$\lim_{n\to\infty} (\tfrac{1}{2})^n = 0. \quad \blacksquare$$

An important fact about sequences to keep in mind at all times is that the terms of the sequence $\{a_n\}$ may perhaps never equal the value of their limit L but merely approach it arbitrarily closely.

Furthermore, not every sequence has a limit, as we saw in the case $\{2^n\}$. However, a sequence may fail to have a limit without approaching infinity, as the next example illustrates.

EXAMPLE 2 Let $a_n = (-1)^n$ for $n = 1, 2, 3, \ldots$. What happens to a_n when n is large? Does the sequence have a limit?

SOLUTION The first four terms of the sequence are

$$a_1 = (-1)^1 = -1,$$

$$a_2 = (-1)^2 = 1,$$

$$a_3 = (-1)^3 = -1,$$

$$a_4 = (-1)^4 = 1.$$

The numbers of this sequence continue to alternate $-1, 1, -1, 1, \ldots$. This sequence does not approach a single number. Therefore, it does not have a limit. ■

Definition *Convergent and divergent sequences.* A sequence that has a limit is said to **converge** or to be **convergent**. A sequence that does not have a limit is said to **diverge** or to be **divergent**.

The sequence $\{(\frac{1}{2})^n\}$ of Example 1 converges to 0. The sequence $\{(-1)^n\}$ of Example 2 is divergent. There is no general procedure for deciding whether a sequence is convergent or divergent. However, there are methods for dealing with many sequences that arise in practice.

EXAMPLE 3 A certain appliance depreciates in value over the years. In fact, at the end of any year it has only 80 percent of the value it had at the beginning of the year. What happens to its value in the long run if its value when new is \$1?

SOLUTION Let a_n be the value of the appliance at the end of the nth year. Thus $a_1 = 0.8$ and $a_2 = (0.8)(0.8) = 0.8^2 = 0.64$. Similarly, $a_3 = 0.8^3$. The question concerns the sequence $\{0.8^n\}$.

This table lists a few values of 0.8^n, rounded off to four decimal places:

n	1	2	3	4	5	10	20
0.8^n	0.8	0.64	0.512	0.4096	0.3277	0.1074	0.0115

The entries in the table suggest that

$$\lim_{n \to \infty} 0.8^n = 0.$$

To verify this assertion, it is necessary to show that the decreasing sequence $\{0.8^n\}$ gets arbitrarily small. To indicate why it does, let us estimate how large n must be so that

$$0.8^n < 0.0001.$$

Taking logarithms to base 10 translates this inequality into the inequality

$$n \log_{10} 0.8 < -4,$$

Division of an inequality by a negative number reverses the direction of the inequality.

or

$$n > \frac{-4}{\log_{10} 0.8}.$$

Note that

$$\frac{-4}{\log_{10} 0.8} \approx \frac{-4}{-0.0969} \approx 41.3.$$

So for $n \geq 42$, the number 0.8^n is less than 0.0001. A similar argument shows that 0.8^n approaches 0 as closely as we please when n is large. Consequently,

$$\lim_{n \to \infty} 0.8^n = 0,$$

just as, in Example 1,

$$\lim_{n \to \infty} (\tfrac{1}{2})^n = 0.$$

[The only difference is that $(0.8)^n$ approaches 0 more slowly than $(\frac{1}{2})^n$ does.]

In the long run the appliance will be worth less than a nickel, then less than a penny, etc. ■

The result in Example 3 generalizes to the following theorem.

Theorem 1 If r is a number in the open interval $(-1, 1)$, then

$$\lim_{n \to \infty} r^n = 0.$$

Proof If $0 < r < 1$, an argument like that in Example 3 can be given. (See Exercise 31.) If $r = 0$, the sequence $\{r^n\}$ is just the constant sequence $0, 0, 0, \ldots$, which has the limit 0. If $-1 < r < 0$, consider the sequence $\{|r^n|\}$. Since $|r^n| = |r|^n$, we know that $|r^n| \to 0$. Thus $r^n \to 0$ as $n \to \infty$. ■

In Example 4 a type of sequence is introduced that occurs later in the chapter in the study of $\sin x$, $\cos x$, and e^x.

$n!$ is defined in Appendix C.

EXAMPLE 4 Does the sequence defined by $a_n = 3^n/n!$ converge or diverge?

SOLUTION The first terms of this sequence are computed (to two decimal places) with the aid of this table:

n	1	2	3	4	5	6	7	8
3^n	3	9	27	81	243	729	2,187	6,561
$n!$	1	2	6	24	120	720	5,040	40,320
$a_n = \dfrac{3^n}{n!}$	3.00	4.50	4.50	3.38	2.03	1.01	0.43	0.16

Although a_2 is larger than a_1 and a_3 is equal to a_2, from a_4 through a_8, as the table shows, the terms decrease.

The numerator 3^n becomes large as $n \to \infty$, influencing a_n to grow large. But the denominator $n!$ also becomes large as $n \to \infty$, influencing the quotient a_n to shrink toward 0. For $n = 1$ and 2 the first influence dominates, but then, as the table shows, the denominator $n!$ seems to grow faster than the numerator 3^n, forcing a_n toward 0.

To see why
$$\frac{3^n}{n!} \to 0$$
as $n \to \infty$, consider, for instance, a_{10}. Express a_{10} as the product of 10 fractions:

$$a_{10} = \tfrac{3}{1}\tfrac{3}{2}\tfrac{3}{3}\tfrac{3}{4}\tfrac{3}{5}\tfrac{3}{6}\tfrac{3}{7}\tfrac{3}{8}\tfrac{3}{9}\tfrac{3}{10}.$$

The first three fractions are ≥ 1. But all the seven remaining fractions are $\leq \frac{3}{4}$. Thus

$$a_{10} < \tfrac{3}{1}\tfrac{3}{2}\tfrac{3}{3}\left(\tfrac{3}{4}\right)^7.$$

Similarly,
$$a_{100} < \tfrac{3}{1}\tfrac{3}{2}\tfrac{3}{3}\left(\tfrac{3}{4}\right)^{97}.$$

By Theorem 1,
$$\lim_{n\to\infty} \left(\tfrac{3}{4}\right)^n = 0.$$

Thus
$$\lim_{n\to\infty} a_n = 0.$$

Similar reasoning shows that for any fixed number k,

This limit will be used often.

$$\lim_{n\to\infty} \frac{k^n}{n!} = 0.$$

This means that the factorial function grows faster than any exponential k^n. ■

A definite integral can be expressed as a limit of a sequence, as Example 5 illustrates.

EXAMPLE 5 Express $\int_0^2 x^3\,dx$ as a limit of a sequence.

SOLUTION For a positive integer n, partition [0, 2] into n sections of equal length, $2/n$, using the numbers

$$x_0 = 0,\ x_1 = \frac{2}{n},\ x_2 = 2\cdot\frac{2}{n},\ \ldots,\ x_i = i\cdot\frac{2}{n},\ \ldots,\ x_n = n\cdot\frac{2}{n} = 2.$$

As the sampling number c_i in the ith section use x_i. Then

$$\sum_{i=1}^{n} (x_i)^3(x_i - x_{i-1}) = \sum_{i=1}^{n}\left(i\cdot\frac{2}{n}\right)^3\frac{2}{n} = \frac{16}{n^4}\sum_{i=1}^{n} i^3$$

is an approximation of $\int_0^2 x^3\,dx$, and we have

$$\lim_{n\to\infty}\frac{16}{n^4}\sum_{i=1}^{n} i^3 = \int_0^2 x^3\,dx.$$

Thus the integral is the limit of the sequence $\{a_n\}$, where

$$a_n = \frac{16}{n^4}\sum_{i=1}^{n} i^3. \quad ■$$

The following theorem will be used several times to show that a sequence converges. It says that if a sequence is nondecreasing ($a_1 \le a_2 \le a_3 \le \cdots$) but does not get arbitrarily large, then it must be convergent.

Theorem 2 Let $\{a_n\}$ be a nondecreasing sequence with the property that there is a number B such that $a_n \le B$ for all n. That is, $a_1 \le a_2 \le a_3 \le a_4 \le \cdots \le a_n \le \cdots$ and $a_n \le B$ for all n. Then the sequence $\{a_n\}$ is convergent and a_n approaches a number L less than or equal to B.

Similarly, if $\{a_n\}$ is a nonincreasing sequence and there is a number B such that $a_n \ge B$ for all n, then the sequence $\{a_n\}$ is convergent and its limit is greater than or equal to B. ■

The proof, which depends on a fundamental property of the real numbers, called "completeness," is omitted. (The **completeness** of the real numbers amounts to the fact that there are no holes in the x axis.) Figure 10.1 shows that Theorem 2 is at least plausible. (The a_n's increase but, being less than B, approach some number L, and that number is not larger than B.)

Figure 10.1

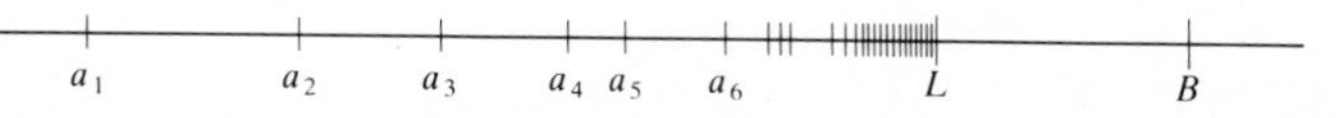

The limits of sequences $\{a_n\}$ behave like the limits of functions $f(x)$, as discussed in Sec. 2.2. The following theorem will be given without proof.

Theorem 3 If $\lim_{n\to\infty} a_n = A$ and $\lim_{n\to\infty} b_n = B$, then

Remember that A and B are real numbers (not "infinity").

(*a*) $\lim_{n\to\infty} (a_n + b_n) = A + B.$

(*b*) $\lim_{n\to\infty} (a_n - b_n) = A - B.$

(*c*) $\lim_{n\to\infty} a_n b_n = AB.$

(*d*) $\lim_{n\to\infty} \frac{a_n}{b_n} = \frac{A}{B} \qquad (B \ne 0).$

(*e*) If k is a constant, $\lim_{n\to\infty} ka_n = kA.$ ■

For instance,

$$\lim_{n\to\infty}\left[\frac{3}{n} + \left(\frac{1}{2}\right)^n\right] = 3\lim_{n\to\infty}\frac{1}{n} + \lim_{n\to\infty}\left(\frac{1}{2}\right)^n$$
$$= 3 \cdot 0 + 0 = 0.$$

Techniques for dealing with $\lim_{x\to\infty} f(x)$ can often be applied to determining $\lim_{n\to\infty} a_n$. The essential point is

$$\text{If } \lim_{x\to\infty} f(x) = L, \qquad \text{then} \qquad \lim_{n\to\infty} f(n) = L.$$

Example 6 illustrates this observation.

EXAMPLE 6 Find $\lim_{n\to\infty} (n/2^n)$.

SOLUTION Consider the function $f(x) = x/2^x$. By l'Hôpital's rule (infinity-over-infinity case),

$$\lim_{x\to\infty} \frac{x}{2^x} = \lim_{x\to\infty} \frac{1}{2^x \ln 2} = 0.$$

Thus $$\lim_{n\to\infty} \frac{n}{2^n} = 0. \quad \blacksquare$$

The converse of the statement "if $\lim_{n\to\infty} f(x) = L$, then $\lim_{n\to\infty} f(n) = L$" is not true. For example, take $f(x) = \sin \pi x$. Then $\lim_{n\to\infty} f(n) = 0$, but $\lim_{x\to\infty} f(x)$ does not exist.

In Secs. 2.7 and 2.8 various limit concepts were given precise, as contrasted to informal, definitions. The following definition is in the same spirit.

Precise definition of limit of a sequence

Definition *Limit of a sequence.* The number L is the **limit of the sequence** $\{a_n\}$ if for each $\varepsilon > 0$ there is an integer N such that

$$|a_n - L| < \varepsilon$$

for all integers $n > N$.

EXERCISES FOR SEC. 10.1: SEQUENCES

In each of Exercises 1 to 18 determine whether the sequence with the given value of a_n converges or diverges. If the sequence converges, give its limit.

1 $\{(0.3)^n\}$
2 $\{(0.99)^n\}$
3 $\{1^n\}$
4 $\{(1.01)^n\}$
5 $\{(-0.5)^n\}$
6 $\{(-\frac{9}{10})^n\}$
7 $\{(\frac{4}{3})^n\}$
8 $\{1/n!\}$
9 $\{2^n\}$
10 $\{n!\}$
11 $\left\{3\left(\frac{n+1}{n}\right) + \frac{n}{2^n}\right\}$
12 $\left\{\frac{5n+1}{3n-1} \cdot \frac{6}{n!}\right\}$
13 $\{\cos n\pi\}$
14 $\{\sin 2n\pi\}$
15 $\left\{\left(1 + \frac{2}{n}\right)^n\right\}$
16 $\left\{\left(\frac{n-1}{n}\right)^n\right\}$
17 $\left\{\frac{10^n}{n!}\right\}$
18 $\left\{\frac{(-100)^n}{n!}\right\}$

■

19 Let $a_n = 100^n/n!$.
(*a*) Show that $a_1 < a_2 < \cdots < a_{99}$.
(*b*) Show that $a_{99} = a_{100}$.
(*c*) Show that $a_{100} > a_{101} > a_{102} > \cdots$.

20 Let $a_n = 200^n/n!$. What is the smallest n such that a_n is larger than the succeeding term a_{n+1}?

21 Examine these sequences for convergence or divergence:
(*a*) $\left\{n \sin \frac{1}{n}\right\}$, (*b*) $\{n \sin \pi n\}$, (*c*) $\left\{\frac{\sin n}{n}\right\}$.

22 How large must n be so that $(0.99)^n$ is less than 0.001?

23 Assume that each year inflation eats away 10 percent of the value of a dollar. Let a_n be the value of a dollar after n years of such inflation.
(*a*) Find a_4. (*b*) Find $\lim_{n\to\infty} a_n$.

24 (Calculator)
(*a*) Compute $4^n/n!$ for $n = 1, 2, 3, 4, 5$, and 15 to three decimal places.
(*b*) What is the largest term in the sequence $\{4^n/n!\}$?
(*c*) Find $\lim_{n\to\infty} (4^n/n!)$.

25 If $\lim_{n\to\infty} x^{2n}/(1 + x^{2n})$ exists, call it $f(x)$. (x is fixed, n varies.)
(*a*) Compute $f(\frac{1}{2})$, $f(2)$, $f(1)$.
(*b*) For which x is $f(x)$ defined? Graph $y = f(x)$.
(*c*) Where is f continuous?

In Exercises 26 to 28 determine the given limits by first showing that each limit is a definite integral $\int_a^b f(x)\,dx$ for a suitable interval $[a, b]$ and function f.

26 $\lim_{n\to\infty} \sum_{i=1}^{n} \left(\frac{i}{n}\right)^2 \frac{1}{n}$

27 $\lim_{n\to\infty}\left(\frac{1}{n+1}+\frac{1}{n+2}+\cdots+\frac{1}{2n}\right)$

28 $\lim_{n\to\infty}\sum_{i=1}^{n}\frac{n}{n^2+i^2}$

29 Let $a_n = \frac{1}{1\cdot 2}+\frac{1}{2\cdot 3}+\cdots+\frac{1}{n(n+1)}$.

(*a*) Compute a_n for $n = 1, 2, 3, 4$, in each case expressing a_n as a single fraction.

(*b*) Find a short formula for a_n and explain your answer.

(*c*) Show that $\lim_{n\to\infty} a_n = 1$.

30 Let $a_n = \frac{1}{2^2}+\frac{1}{3^2}+\cdots+\frac{1}{n^2}$ for $n \geq 2$.

Show that $\lim_{n\to\infty} a_n$ exists and is ≤ 1. *Hint:* See Exercise 29 and use Theorem 2.

31 This exercise completes the proof of Theorem 1 in the style of Example 3. Let r be a positive number less than 1. Let p be a positive number. Show that there is an integer n such that $r^n < p$. *Hint:* Use logarithms.

32 This exercise provides a neat proof of Theorem 1 in the case $0 < r < 1$. Let r be a positive number less than 1.

(*a*) Using Theorem 2, show that $\lim_{n\to\infty} r^n$ exists.

(*b*) Call the limit in (*a*) L. Why is $rL = L$?

(*c*) From (*b*) deduce that $L = 0$.

33 Let $P(x)$ and $Q(x)$ be polynomials. Assume that $Q(n)$ is not 0 for any positive integer n. What relation must there be between the degree of $P(x)$ and the degree of $Q(x)$ if the sequence $\{P(n)/Q(n)\}$ is convergent?

34 (Calculator or computer) The Fibonacci sequence was first considered in the *Liber Abaci,* by Fibonacci, around the year 1200. (This was the book that introduced decimal notation to Europe.) The nth term of this sequence is usually denoted F_n, with $F_1 = 1$, $F_2 = 1$, and $F_{n+2} = F_{n+1} + F_n$, for $n \geq 1$. For instance, $F_3 = 1 + 1 = 2$, $F_4 = F_3 + F_2 = 2 + 1 = 3$, $F_5 = 5$, $F_6 = 8$. (Each term from the third on is the sum of the two preceding terms.)

(*a*) Compute F_7, F_8, F_9, F_{10}.

(*b*) Compute F_{n+1}/F_n, for $n = 1, 2, \ldots, 10$.

(*c*) What do you think happens to the sequence $a_n = F_{n+1}/F_n$ as $n \to \infty$?

35 (Calculator) Define a sequence $\{a_n\}$ as follows: $a_1 = 1$, $a_2 = 3$, and $a_{n+2} = (1 + a_{n+1})/a_n$, for $n \geq 1$. Examine the behavior of a_n as $n \to \infty$.

10.2 SERIES

Frequently a new sequence is formed by summing terms of a given sequence. Example 1 illustrates this way of constructing a sequence.

EXAMPLE 1 Given the sequence $1, 0.8, 0.8^2, 0.8^3, \ldots$, form a new sequence $\{S_n\}$ as follows:

S is short for "sum."

$$S_1 = 1,$$

$$S_2 = 1 + 0.8,$$

$$S_3 = 1 + 0.8 + 0.8^2,$$

and, in general, $S_n = 1 + 0.8 + 0.8^2 + \cdots + 0.8^{n-1}$.

Each S_n is the sum of the first n terms of the sequence $1, 0.8, 0.8^2, 0.8^3, \ldots$. Examine the behavior of S_n as $n \to \infty$.

SOLUTION

$$S_1 = 1,$$

$$S_2 = 1 + 0.8 = 1.8,$$

$$S_3 = 1 + 0.8 + 0.8^2 = 1 + 0.8 + 0.64 = 2.44,$$

$$S_4 = 1 + 0.8 + 0.8^2 + 0.8^3 = 1 + 0.8 + 0.64 + 0.512 = 2.952,$$

$$S_5 = 1 + 0.8 + 0.8^2 + 0.8^3 + 0.8^4$$

$$= 1 + 0.8 + 0.64 + 0.512 + 0.4096 = 3.3616.$$

Similar computations show that $S_{50} \approx 4.99993$.

Two influences on $\{S_n\}$

Two influences affect the growth of S_n as n increases. On the one hand, the number of summands increases, causing S_n to get larger. On the other hand, the summands approach 0, so that S_n grows more and more slowly as n increases. How these two forces balance is answered by Theorem 1 of this section, which shows that the sequence $\{S_n\}$ converges and that its limit is 5:

$$\lim_{n\to\infty} S_n = 5. \quad \blacksquare$$

The rest of this section extends the ideas introduced in Example 1.

Let $a_1, a_2, a_3, \ldots, a_n, \ldots$ be a sequence. From this sequence a new sequence $S_1, S_2, S_3, \ldots, S_n, \ldots$ can be formed:

$$S_1 = a_1,$$

$$S_2 = a_1 + a_2,$$

$$S_3 = a_1 + a_2 + a_3,$$

$$\ldots\ldots\ldots\ldots\ldots,$$

Summation notation was introduced in Sec. 5.2.

$$S_n = a_1 + a_2 + a_3 + \cdots + a_n = \sum_{i=1}^{n} a_i.$$

The sequence of sums $S_1, S_2 \ldots$ is called the **series** obtained from the sequence $a_1, a_2, \ldots$. Traditionally, it is referred to as **the series whose nth term is a_n**. Common notations for the sequence $\{S_n\}$ are $\Sigma_{n=1}^{\infty} a_n$ and $a_1 + a_2 + a_3 + \cdots + a_n + \cdots$. The sum

$$S_n = a_1 + a_2 + \cdots + a_n = \sum_{i=1}^{n} a_i$$

is called a **partial sum** or the **nth partial sum**. If the sequence of partial sums of a series converges to L, then L is called the **sum** of the series and the series is said to be **convergent**. Frequently one writes $L = a_1 + a_2 + \cdots + a_n + \cdots$. Remember, however, that we do not add an infinite number of numbers; we take the limit of finite sums. A series that is not convergent is called **divergent**.

Only finitely many summands are ever added up.

Example 1 concerns the series whose nth term is 0.8^{n-1}:

$$S_n = 1 + 0.8^1 + 0.8^2 + \cdots + 0.8^{n-1}.$$

It is a special case of a geometric series, which will now be defined.

Appendix C treats geometric series with a finite number of terms.

Definition *Geometric series.* Let a and r be real numbers. The series

$$a + ar + ar^2 + \cdots + ar^{n-1} + \cdots$$

is called the **geometric series with initial terms a and ratio r**.

The series in Example 1 is a geometric series with initial term 1 and ratio 0.8.

Theorem 1 If $-1 < r < 1$, the geometric series

$$a + ar + \cdots + ar^{n-1} + \cdots$$

converges to $a/(1 - r)$.

Proof Let S_n be the sum of the first n terms:

$$S_n = a + ar + \cdots + ar^{n-1}.$$

By the formula in Appendix C for the sum of a finite geometric series,

$$S_n = \frac{a(1 - r^n)}{1 - r}.$$

By Theorem 1 in Sec. 10.1, $\lim_{n\to\infty} r^n = 0.$

Thus $$\lim_{n\to\infty} S_n = \frac{a}{1 - r},$$

proving the theorem. ■

In particular, if $a = 1$ and $r = 0.8$, as in Example 1, the geometric series has the sum

$$\frac{1}{1 - 0.8} = \frac{1}{0.2} = 5.$$

Theorem 1 says nothing about geometric series in which the ratio r is ≥ 1 or ≤ -1. The next theorem, which concerns series in general, not just geometric series, will be of use in settling this case.

Theorem 2 *The nth-term test for divergence.* If $\lim_{n\to\infty} a_n \neq 0$, then the series $a_1 + a_2 + \cdots + a_n + \cdots$ diverges. (The same conclusion holds if $\{a_n\}$ has no limit.)

Proof Assume that the series $a_1 + a_2 + \cdots$ converges. Since S_n is the sum $a_1 + a_2 + \cdots + a_n$, while S_{n-1} is the sum of the first $n - 1$ terms, it follows that $S_n = S_{n-1} + a_n$, or

$$a_n = S_n - S_{n-1}.$$

Let $$S = \lim_{n\to\infty} S_n.$$

Then we also have $$S = \lim_{n\to\infty} S_{n-1},$$

since S_{n-1} runs through the same numbers as S_n. Thus

$$\begin{aligned}\lim_{n\to\infty} a_n &= \lim_{n\to\infty} (S_n - S_{n-1})\\ &= \lim_{n\to\infty} S_n - \lim_{n\to\infty} S_{n-1}\\ &= S - S\\ &= 0.\end{aligned}$$

So if a series converges, its nth term must approach 0.

This proves the theorem. ■

Theorem 2 implies that if $a \neq 0$ and $r \geq 1$, the geometric series

$$a + ar + \cdots + ar^{n-1} + \cdots$$

diverges. For instance, if $r = 1$,

$$\lim_{n\to\infty} ar^n = \lim_{n\to\infty} a1^n = a,$$

which is not 0. If $r > 1$, then r^n gets arbitrarily large as n increases; hence $\lim_{n\to\infty} ar^n$ does not exist. Similarly, if $r \leq -1$, $\lim_{n\to\infty} ar^n$ does not exist. The above results and Theorem 1 can be summarized by this statement: The geometric series

$$\sum_{n=1}^{\infty} ar^{n-1} = a + ar + ar^2 + \cdots + ar^{n-1} + \cdots,$$

for $a \neq 0$, converges if and only if $|r| < 1$.

Warning: Even if the nth term approaches 0, the series can diverge!

This is a good time to read Appendix K.

Theorem 2 tells us that if the series $a_1 + a_2 + a_3 + \cdots$ converges, then a_n approaches 0 as $n \to \infty$. The converse of this statement is *not true*. If a_n approaches 0 as $n \to \infty$, it does *not* follow that the series $a_1 + a_2 + a_3 + \cdots$ converges. Be careful to make this distinction. To emphasize this point, the next example presents *a series that diverges even though its nth term approaches 0 as $n \to \infty$.*

EXAMPLE 2 Show that the series

$$\frac{1}{\sqrt{1}} + \frac{1}{\sqrt{2}} + \frac{1}{\sqrt{3}} + \cdots + \frac{1}{\sqrt{n}} + \cdots$$

diverges.

SOLUTION Consider

$$S_n = \frac{1}{\sqrt{1}} + \frac{1}{\sqrt{2}} + \cdots + \frac{1}{\sqrt{n}}.$$

Each of the n summands in S_n is $\geq 1/\sqrt{n}$. Hence

$$S_n \geq \underbrace{\frac{1}{\sqrt{n}} + \frac{1}{\sqrt{n}} + \cdots + \frac{1}{\sqrt{n}}}_{n \text{ summands}} = \frac{n}{\sqrt{n}} = \sqrt{n}.$$

$\frac{1}{\sqrt{n}} \to 0$ so slowly that $\frac{1}{\sqrt{1}} + \frac{1}{\sqrt{2}} + \cdots + \frac{1}{\sqrt{n}}$ gets arbitrarily large.

As n increases, $\sqrt{n}$ increases without bound. Since $S_n \geq \sqrt{n}$, we conclude that $\lim_{n\to\infty} S_n$ does not exist. In fact, we write $\lim_{n\to\infty} S_n = \infty$.

In short, the series

$$\frac{1}{\sqrt{1}} + \frac{1}{\sqrt{2}} + \cdots + \frac{1}{\sqrt{n}} + \cdots$$

diverges even though its nth term, $1/\sqrt{n}$, approaches 0. ■

The harmonic series was so named by the Greeks because of the role of $1/n$ in musical harmony.

In the next example, the nth term approaches 0 much faster than $1/\sqrt{n}$ does. Still the series diverges. The series in this example is called the **harmonic series**. The argument that it diverges is due to the French mathematician Nicolas of Oresme, who presented it about the year 1360.

EXAMPLE 3 Show that the harmonic series $\frac{1}{1} + \frac{1}{2} + \cdots + \frac{1}{n} + \cdots$ diverges.

SOLUTION Collect the summands in longer and longer groups in the manner indicated below (each group from the third on has twice the number of summands as the previous group):

$$\underbrace{\tfrac{1}{1}} + \underbrace{\tfrac{1}{2}} + \underbrace{\tfrac{1}{3} + \tfrac{1}{4}} + \underbrace{\tfrac{1}{5} + \tfrac{1}{6} + \tfrac{1}{7} + \tfrac{1}{8}} + \underbrace{\tfrac{1}{9} + \tfrac{1}{10} + \cdots + \tfrac{1}{16}} + \underbrace{\tfrac{1}{17} + \cdots}.$$

The sum of the terms in each group is at least $\frac{1}{2}$. For instance,

$$\tfrac{1}{5} + \tfrac{1}{6} + \tfrac{1}{7} + \tfrac{1}{8} > \tfrac{1}{8} + \tfrac{1}{8} + \tfrac{1}{8} + \tfrac{1}{8} = \tfrac{4}{8} = \tfrac{1}{2},$$

and
$$\tfrac{1}{9} + \tfrac{1}{10} + \cdots + \tfrac{1}{16} > \tfrac{1}{16} + \tfrac{1}{16} + \cdots + \tfrac{1}{16} = \tfrac{8}{16} = \tfrac{1}{2}.$$

Since the repeated addition of $\frac{1}{2}$'s produces sums as large as we please, the series diverges. ■

An important moral: The nth-term test is only a test for divergence.

If the series $a_1 + a_2 + \cdots + a_n + \cdots$ converges, it follows that $a_n \to 0$. However, if $a_n \to 0$, it *does not necessarily follow* that $a_1 + a_2 + \cdots + a_n + \cdots$ converges. Indeed, there is no general, practical rule for determining whether a series converges or diverges. Fortunately, a few rules suffice to decide on the convergence or divergence of the most common series; they will be presented in this chapter.

Two basic properties of series, similar to those of definite integrals, are recorded in the next theorem. Exercise 30 asks for the proof.

Theorem 3

(*a*) If $\Sigma_{n=1}^{\infty} a_n$ is a convergent series with sum L, and if c is a number, then $\Sigma_{n=1}^{\infty} ca_n$ is convergent and has the sum cL.

(*b*) If, furthermore, $\Sigma_{n=1}^{\infty} b_n$ is a convergent series with sum M, then $\Sigma_{n=1}^{\infty} (a_n + b_n)$ is a convergent series with sum $L + M$. ■

Front ends don't affect convergence.

Keep in mind that you can disregard any finite number of terms when deciding whether a series is convergent or divergent. If you delete a finite number of terms from a series and what is left converges, then the series you started with converges. Another way to look at this is to note that a "front end," $a_1 + a_2 + \cdots + a_n$, does not influence convergence or divergence. It is rather a "tail end," $a_{n+1} + a_{n+2} + \cdots$, that matters. The sum of the series is the sum of any tail end plus the sum of the corresponding front end; that is, for any positive integer m,

$$\sum_{n=1}^{\infty} a_n = \underbrace{\sum_{n=1}^{m} a_n}_{\text{front end}} + \underbrace{\sum_{n=m+1}^{\infty} a_n}_{\text{tail end}}.$$

EXERCISES FOR SEC. 10.2: SERIES

In each of Exercises 1 to 8 determine whether the given geometric series converges. If it does, find its sum.

1 $1 + \frac{1}{2} + \frac{1}{4} + \frac{1}{8} + \cdots + (\frac{1}{2})^{n-1} + \cdots$

2 $1 - \frac{1}{3} + \frac{1}{9} - \frac{1}{27} + \cdots + (-\frac{1}{3})^{n-1} + \cdots$

3 $\displaystyle\sum_{n=1}^{\infty} 10^{-n}$

4 $\displaystyle\sum_{n=1}^{\infty} 10^{n}$

5 $\displaystyle\sum_{n=1}^{\infty} 5(0.99)^{n}$

6 $\displaystyle\sum_{n=1}^{\infty} 7(-1.01)^{n}$

7 $\displaystyle\sum_{n=1}^{\infty} 4\left(\frac{2}{3}\right)^{n}$

8 $\displaystyle -\frac{3}{2} + \frac{3}{4} - \frac{3}{8} + \cdots + \frac{3}{(-2)^n} + \cdots$

In Exercises 9 to 16 determine whether the given series converge or diverge. Find the sums of the convergent series.

9 $-5 + 5 - 5 + 5 - \cdots + (-1)^n 5 + \cdots$

10 $\sum_{n=1}^{\infty} \frac{1}{[1 + (1/n)]^n}$

11 $\sum_{n=1}^{\infty} \frac{2}{n}$

12 $\sum_{n=1}^{\infty} \frac{n}{2n + 1}$

13 $\sum_{n=1}^{\infty} 6\left(\frac{4}{5}\right)^n$

14 $\sum_{n=1}^{\infty} 100\left(\frac{-8}{9}\right)^n$

15 $\sum_{n=1}^{\infty} (2^{-n} + 3^{-n})$

16 $\sum_{n=1}^{\infty} (4^{-n} + n^{-1})$

■

17 This is a quote from an economics text: "The present value of the land, if a new crop is planted at time t, $2t$, $3t$, etc., is

$$P = g(t)e^{-rt} + g(t)e^{-2rt} + g(t)e^{-3rt} + \cdots.$$

Note that each term is the previous term multiplied by e^{-rt}. By the formula for the sum of a geometric series,

$$P = \frac{g(t)e^{-rt}}{1 - e^{-rt}}."$$

Check that the missing step, which simplified the formula for P, was correct.

18 (Calculator) Let $S_n = \sum_{i=1}^{n} \frac{1}{i(i + 1)}$.

(*a*) Compute S_5.
(*b*) Compute S_{10}.
(*c*) What do you think happens to S_n as $n \to \infty$?
(*d*) Find the sum of the series, using partial fractions to represent

$$\frac{1}{i(i + 1)}.$$

19 A certain rubber ball, when dropped on concrete, always rebounds 90 percent of the distance it falls.
(*a*) If the ball is dropped from a height of 6 feet, how far does it travel during the first three descents and ascents?
(*b*) How far does it travel before coming to rest?

20 A patient takes A grams of a certain medicine every 6 hours. The amount of each dose active in the body t hours later is Ae^{-kt} grams, where k is a positive constant and time is measured in hours. *Hint:* See Review Exercise 71 in Sec. 10.S.
(*a*) Show that immediately after taking the medicine for the nth time, the amount active in the body is

$$S_n = A + Ae^{-6k} + Ae^{-12k} + \cdots + Ae^{-6(n-1)k}.$$

(*b*) If, as $n \to \infty$, $S_n \to \infty$, the patient would be in danger. Does $S_n \to \infty$? If not, what is $\lim_{n\to\infty} S_n$?

21 The decimal 0.3333 $\cdots$ stands for the sum of the geometric series $\sum_{n=1}^{\infty} 3(10^{-n})$. Use this fact to show that 0.3333 $\cdots$ is equal to $\frac{1}{3}$.

22 Write the decimal

$$0.525252 \cdots \qquad \text{(52's continuing)}$$

as a geometric series and use Theorem 1 to find its value.

23 A gambler tosses a penny until a head appears. On the average, how many times does she toss a penny to get a head? Parts (*a*) and (*b*) concern this question.
(*a*) Experiment with a penny on 10 runs. Each run consists of tossing a penny until a head appears. Average the lengths of the 10 runs.
(*b*) The probability of a run of length one is $\frac{1}{2}$, since a head must appear on the first toss. The probability of a run of length two is $(\frac{1}{2})^2$. The probability of having a head appear for the first time on the nth toss is $(\frac{1}{2})^n$. It is shown in probability theory that the average number of tosses to get a head is $\sum_{n=1}^{\infty} n/2^n$. (This is a theoretical average approached as the experiment is repeated many times.) Compute $\sum_{n=1}^{8} n/2^n$. (The next exercise sums the infinite series.)

24 This exercise concerns the sum of the series

$$1 + 2x + 3x^2 + 4x^3 + \cdots + nx^{n-1} + \cdots.$$

(*a*) Use l'Hôpital's rule to show that if $0 < x < 1$, then $\lim_{t\to\infty} tx^{t-1} = 0$. ($x$ is fixed and t varies through real numbers.)
(*b*) Use (*a*) to show that if $|x| < 1$, $nx^{n-1} \to 0$ as $n \to \infty$.
(*c*) By differentiating the equation

$$1 + x + x^2 + \cdots + x^n = \frac{1 - x^{n+1}}{1 - x},$$

show that

$$1 + 2x + 3x^2 + \cdots + nx^{n-1} = \frac{1 - x^{n+1} + (n + 1)x^{n+1} - (n + 1)x^n}{(1 - x)^2}.$$

(*d*) With the aid of (*c*), show that, for $|x| < 1$,

$$\sum_{n=1}^{\infty} nx^{n-1} = \frac{1}{(1 - x)^2}.$$

(*e*) Use (*d*) to evaluate $\sum_{n=1}^{\infty} n/2^{n-1}$ and then $\sum_{n=1}^{\infty} n/2^n$, the series needed in Exercise 23.

25 Oresme, around the year 1360, summed the series $\sum_{n=1}^{\infty} n/2^n$ by drawing the endless staircase shown in Fig. 10.2, in which each stair has width 1 and is twice as high as the stair immediately to its right.

(*a*) By looking at the staircase in two ways, show that

$$1 + \tfrac{1}{2} + \tfrac{1}{4} + \tfrac{1}{8} + \cdots = \tfrac{1}{2} + \tfrac{2}{4} + \tfrac{3}{8} + \cdots.$$

(*b*) Use (*a*) to sum $\Sigma_{n=1}^{\infty} n/2^n$.

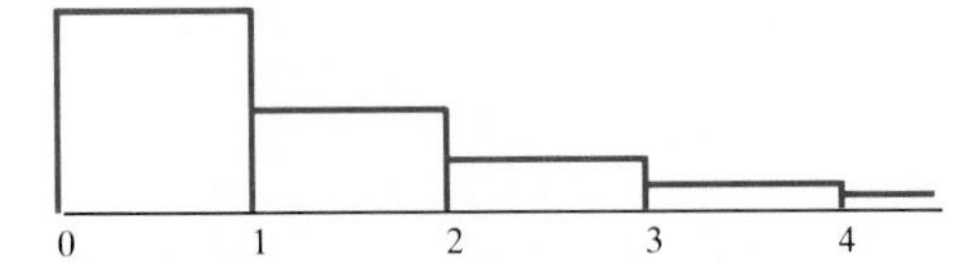

Figure 10.2

26 A plodding snail is at the point $(1/t, (\sin t)/t)$ at time $t \geq 1$. (*a*) Show that, as $t \to \infty$, the snail approaches the origin. (*b*) Show that the length of the snail's journey is infinite. *Suggestion:* Sketch the path; don't try to find its arc length by integration.

27 Using the formula for the sum of a geometric series, express the repeating decimal $3.45212121 \cdots$ as the quotient of two integers.

28 Deficit spending by the federal government inflates the nation's money supply. However, much of the money paid out by the government is spent in turn by those who receive it, thereby producing additional spending. This produces a chain reaction, called by economists the **multiplier effect**. It results in much greater total spending than the government's original expenditure. To be specific, suppose the government spends 1 billion dollars and that the recipients of that expenditure in turn spend 80 percent while retaining 20 percent. Let S_n be the *total* spending generated after n transactions in the chain, 80 percent of receipts being expended at each step.

(*a*) Show that $S_n = 1 + 0.8 + 0.8^2 + \cdots + 0.8^{n-1}$ billion dollars.

(*b*) Show that as n increases, the total spending approaches 5 billion dollars. (The number 5 is called the **multiplier**.)

(*c*) What would be the total spending if 90 percent of receipts is spent at each step instead of 80 percent?

29 (*How banks, with the assistance of the public, create money*) If a deposit of A dollars is made at a bank, the bank can lend out most of this amount. However, it cannot lend out all the amount, for it must keep a reserve to meet the demands of depositors who may withdraw money from their accounts. The government stipulates what this reserve is, usually between 16 and 20 percent of the amount deposited. Assume that a bank is allowed to lend 80 percent of the amount deposited. If a person deposits \$1,000, then the bank can lend another person \$800. Assume that this borrower deposits all the amount; then the bank can lend a third person \$640 of that deposit. This process can go on indefinitely, through a fourth person, a fifth, and so on. If this process continues indefinitely, how large will all the deposits total in the long run? Note that this total is much larger than the initial deposit. The banks have created money.

30 (This exercise contrasts Theorem 2 for convergent series with the behavior of convergent improper integrals.)

(*a*) If $\int_0^{\infty} f(x)\,dx$ is convergent, it does *not* follow that $\lim_{x\to\infty} f(x) = 0$, even for $f(x) \geq 0$. Show this by considering the function whose graph is shown in Fig. 10.3. The width of the triangle in the interval $[n, n+1]$ is 2^{-n} and its height is 1.

(*b*) However, show that if $\int_0^{\infty} g(x)\,dx$ is convergent, then $\int_n^{n+1} g(x)\,dx$ approaches 0 as $n \to \infty$.

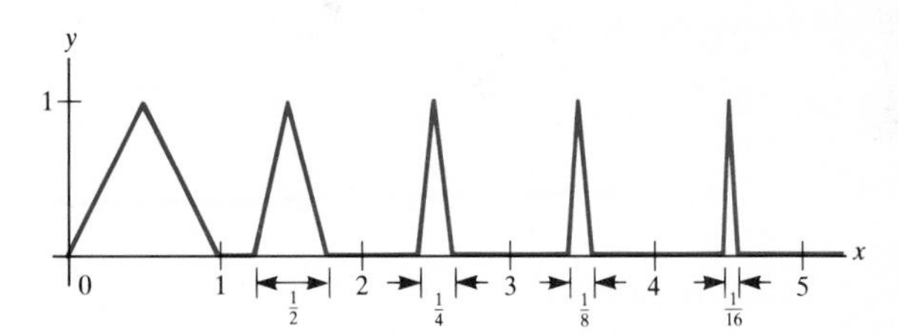

Figure 10.3

10.3 THE INTEGRAL TEST

In this section and the next only series with positive terms will be considered. The following test for convergence or divergence applies to many series with positive terms. In particular, it will show that the series

$$\frac{1}{1} + \frac{1}{2} + \frac{1}{3} + \frac{1}{4} + \cdots + \frac{1}{n} + \cdots$$

diverges, but that the series

$$\frac{1}{1^{1.01}} + \frac{1}{2^{1.01}} + \frac{1}{3^{1.01}} + \frac{1}{4^{1.01}} + \cdots + \frac{1}{n^{1.01}} + \cdots$$

converges.

A function is "positive" if its values are positive for the domain of interest.

Theorem 1 *Integral test.* Let $f(x)$ be a continuous decreasing positive function for $x \geq 1$. Then
(*a*) if $\int_1^\infty f(x)\,dx$ is convergent, $\Sigma_{n=1}^\infty f(n)$ is convergent;
(*b*) if $\int_1^\infty f(x)\,dx$ is divergent, $\Sigma_{n=1}^\infty f(n)$ is divergent.

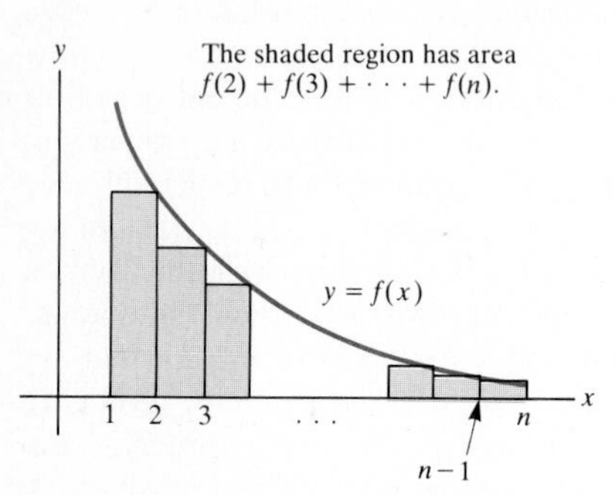

Figure 10.4

Proof

(*a*) Assume $\int_1^\infty f(x)\,dx$ is convergent.
Consider the total area of the $n - 1$ rectangles shown in Fig. 10.4. Each rectangle has width 1. The height of the rectangle above [1, 2] is $f(2)$. Hence the rectangle above [1, 2] has area $f(2) \cdot 1 = f(2)$. Similarly, the rectangle above [2, 3] has area $f(3)$. The total area of the $n - 1$ shaded rectangles in Fig. 10.4 is then

$$f(2) + f(3) + \cdots + f(n).$$

Since these rectangles lie below the curve $y = f(x)$,

$$f(2) + f(3) + \cdots + f(n) < \int_1^n f(x)\,dx.$$

Adding $f(1)$ to both sides of this inequality yields the inequality

$$f(1) + f(2) + \cdots + f(n) < f(1) + \int_1^n f(x)\,dx.$$

Since $\int_1^\infty f(x)\,dx$ is a convergent integral, the sums $f(1) + f(2) + \cdots + f(n)$ do not get arbitrarily large. Hence, by Theorem 2 in Sec. 10.1, $\Sigma_{n=1}^\infty f(n)$ is a convergent series.

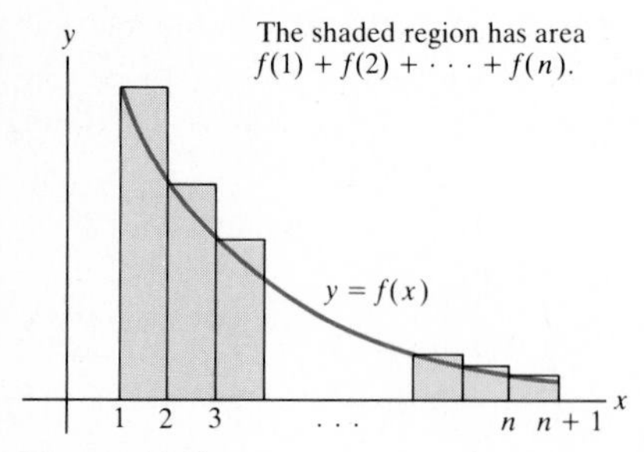

Figure 10.5

(*b*) Figure 10.5 is the key to showing that, if $\int_1^\infty f(x)\,dx$ is divergent, so is $\Sigma_{n=1}^\infty f(n)$.
In contrast to Fig. 10.4, each rectangle in Fig. 10.5 has a height equal to the value of f at its *left* abscissa. That is, the rectangle above the interval $[i, i + 1]$, $i = 1, 2, \ldots, n$, has height $f(i)$ and thus has area $f(i)$. Since the n shaded rectangles in Fig. 10.5 contain the region under the curve $y = f(x)$ and above the interval $[1, n + 1]$, we have

$$f(1) + f(2) + \cdots + f(n) > \int_1^{n+1} f(x)\,dx.$$

Since $\int_1^\infty f(x)\,dx = \infty$, it follows that $\Sigma_{n=1}^\infty f(n)$ is divergent. ■

Note that the integral test does *not* give the value of the convergent series, $\Sigma_{n=1}^\infty f(n)$.

The integral test will show that the series

$$\frac{1}{1^{1.01}} + \frac{1}{2^{1.01}} + \frac{1}{3^{1.01}} + \cdots + \frac{1}{n^{1.01}} + \cdots \tag{1}$$

is convergent. That this series is convergent may be surprising since it resembles the harmonic series

$$\frac{1}{1} + \frac{1}{2} + \frac{1}{3} + \cdots + \frac{1}{n} + \cdots$$

which was shown in Sec. 10.2 to be divergent.

Series (1) is an example of a p series, which will now be defined in general.

Definition *p series.* For a fixed real number p, the series

$$\sum_{n=1}^{\infty} \frac{1}{n^p} = \frac{1}{1^p} + \frac{1}{2^p} + \frac{1}{3^p} + \cdots$$

is called the p **series** (with exponent p).

When $p = 1$, the p series is the harmonic series. When $p = 1.01$, the p series is the series (1).

Theorem 2 (*The p test*) The p series $\Sigma_{n=1}^{\infty} 1/n^p$ converges if $p > 1$. It diverges if $p \leq 1$.

Proof Analyze the cases $p < 0$, $p = 0$, and $p > 0$ separately.

If p is negative, then $1/n^p = n^{-p} \to \infty$ as $n \to \infty$. By the nth-term test for divergence, $\Sigma_{n=1}^{\infty} 1/n^p$ is divergent.

If $p = 0$, then $1/n^p = 1/n^0 = 1$. All the terms of the p series in this case are 1. Since the nth term does not approach 0 as $n \to \infty$, the nth-term test for divergence shows that the series diverges.

If p is positive, we make use of the integral test. The function $f(x) = 1/x^p = x^{-p}$ is decreasing for $x \geq 1$. Thus the integral test may be applied. Now if $p \neq 1$,

$$\int_1^{\infty} x^{-p}\, dx = \lim_{b\to\infty} \int_1^b x^{-p}\, dx$$

$$= \lim_{b\to\infty} \left. \frac{x^{1-p}}{1-p} \right|_1^b \qquad \text{(since } p \neq 1\text{)}$$

$$= \lim_{b\to\infty} \left(\frac{b^{1-p}}{1-p} - \frac{1}{1-p} \right).$$

For $p > 1$, this limit is $-1/(1-p) = 1/(p-1)$, since

$$\lim_{b\to\infty} b^{1-p} = 0.$$

Thus $\int_1^{\infty} x^{-p}\, dx$ is convergent. By the integral test the p series converges for $p > 1$.

For $p < 1$, b^{1-p} gets arbitrarily large as $b \to \infty$; hence $\int_1^{\infty} x^{-p}\, dx$ is divergent and the p series diverges for $p < 1$.

If $p = 1$,

$$\int_1^{\infty} x^{-p}\, dx = \int_1^{\infty} \frac{1}{x}\, dx$$

$$= \lim_{b\to\infty} \int_1^b \frac{1}{x}\, dx$$

$$= \lim_{b\to\infty} (\ln b - \ln 1).$$

The divergence of the harmonic series was also shown in Sec. 10.2.

Since $\ln b$ gets arbitrarily large as $b \to \infty$, $\int_1^{\infty} 1/x\, dx$ is divergent. Thus the harmonic series diverges. ■

In particular, series (1) converges, since it is the p series with $p = 1.01$, which is greater than 1. However, the series $\Sigma_{n=1}^{\infty} 1/\sqrt{n}$ diverges, since it is the p series with $p = \frac{1}{2}$. (A different argument for the divergence of this series was given in Sec. 10.2.)

When we use a front end of a series to estimate the sum of the whole series, there will be an error, namely, the sum of the corresponding tail end. For the sum of a front end to be a good estimate of the sum of the whole series, we must be sure that the sum of the corresponding tail end is small. Otherwise, we would be like the carpenter who measures a board as "5 feet long with an error of perhaps as much as 5 feet." That is why we wish to be sure that the sum of the tail end is small.

Let S_n be the sum of the first n terms of a convergent series $\Sigma_{n=1}^{\infty} a_n$ whose sum is S. The difference

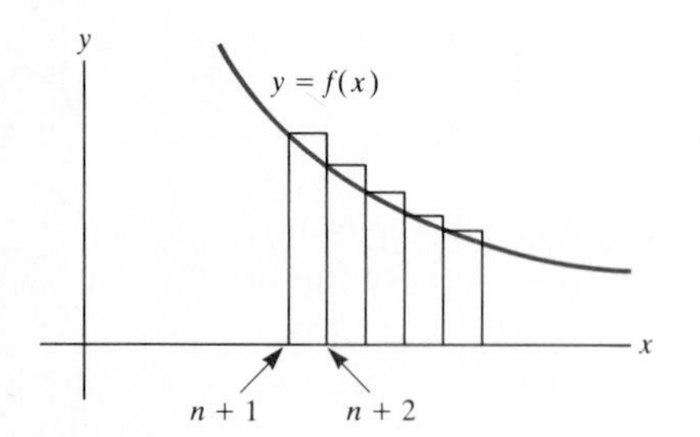

Figure 10.6

$$R_n = S - S_n = a_{n+1} + a_{n+2} + a_{n+3} + \cdots$$

is called the **remainder** or **error** in using the sum of the first n terms to approximate the sum of the series. For the series of the special type considered in this section, it is possible to use an improper integral to estimate the error. The reasoning depends once again on comparing a staircase of rectangles with the area under a curve.

Recall that $f(x)$ is a continuous decreasing positive function. The error in using $S_n = f(1) + f(2) + \cdots + f(n) = \Sigma_{i=1}^{n} f(i)$ to approximate $\Sigma_{i=1}^{\infty} f(i)$ is the sum $\Sigma_{i=n+1}^{\infty} f(i)$. This sum is the area of the endless staircase of rectangles shown in Fig. 10.6. Comparing the rectangles with the region under the curve $y = f(x)$, we conclude that

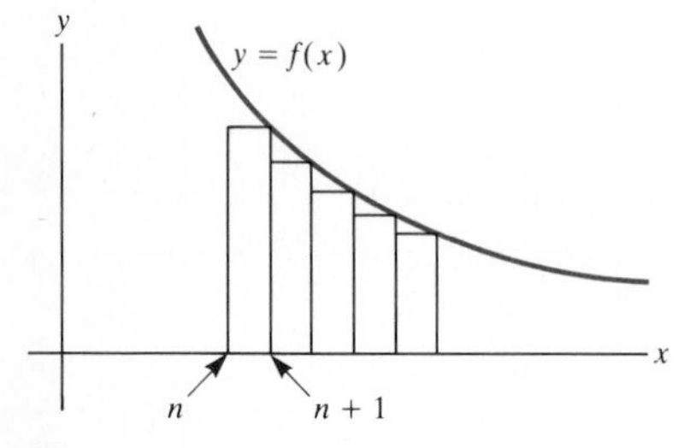

Figure 10.7

$$R_n = f(n + 1) + f(n + 2) + \cdots > \int_{n+1}^{\infty} f(x)\, dx. \tag{2}$$

Inequality (2) gives a *lower* estimate of the error.

The staircase in Fig. 10.7, which lies below the curve, gives an *upper* estimate of the error. Inspection of Fig. 10.7 shows that

$$R_n = f(n + 1) + f(n + 2) + \cdots < \int_{n}^{\infty} f(x)\, dx.$$

Putting these observations together yields the following estimate of the error.

Theorem 3 Let $f(x)$ be a continuous decreasing positive function such that $\int_1^{\infty} f(x)\, dx$ is convergent. Then the error R_n in using $f(1) + f(2) + \cdots + f(n)$ to estimate $\Sigma_{i=1}^{\infty} f(i)$ satisfies the inequality

Estimating the error

$$\int_{n+1}^{\infty} f(x)\, dx < R_n < \int_{n}^{\infty} f(x)\, dx. \quad \blacksquare \tag{3}$$

EXAMPLE The first five terms of the p series $\dfrac{1}{1^2} + \dfrac{1}{2^2} + \cdots + \dfrac{1}{n^2} + \cdots$ are used to estimate the sum of the series.

(*a*) Estimate the error in using just the first five terms.

(*b*) Estimate $\Sigma_{n=1}^{\infty} 1/n^2$.

SOLUTION

(*a*) By inequality (3), the error R_5 satisfies the inequality

$$\int_6^\infty \frac{dx}{x^2} < R_5 < \int_5^\infty \frac{dx}{x^2}.$$

Now,
$$\int_5^\infty \frac{dx}{x^2} = -\frac{1}{x}\bigg|_5^\infty = 0 - \left(-\frac{1}{5}\right) = \frac{1}{5}.$$

Similarly,
$$\int_6^\infty \frac{dx}{x^2} = \frac{1}{6}.$$

Thus
$$\tfrac{1}{6} < R_5 < \tfrac{1}{5}.$$

(*b*) The sum of the first five terms of the series is

$$\frac{1}{1^2} + \frac{1}{2^2} + \frac{1}{3^2} + \frac{1}{4^2} + \frac{1}{5^2} \approx 1.4636.$$

Since the sum of the remaining terms (the "tail end") is between $\frac{1}{6}$ and $\frac{1}{5}$, the sum of the series is between $1.463 + \frac{1}{6} > 1.629$ and $1.464 + \frac{1}{5} = 1.664$. ■

The integral test works if the function $f(x)$ decreases from some point on.

So far it has been assumed that the function $f(x)$ is decreasing for $x \geq 1$. If, instead, $f(x)$ decreases for $x \geq a$, where a is some constant, similar results hold. For instance, $f(x) = x^2/e^x$ increases in the interval $[1, 2]$ and decreases in the interval $[2, \infty)$. The series

$$\frac{1^2}{e^1} + \frac{2^2}{e^2} + \frac{3^2}{e^3} + \cdots + \frac{n^2}{e^n} + \cdots \tag{4}$$

can be examined with a slight modification of the integral test. Just chop off the first term, that is, $1^2/e^1$. Since $f(x) = x^2/e^x$ is decreasing for $x \geq 2$, the integral test may be applied to the tail end,

$$\frac{2^2}{e^2} + \frac{3^2}{e^3} + \cdots + \frac{n^2}{e^n} + \cdots \qquad (n \geq 2). \tag{5}$$

The improper integral $\int_2^\infty x^2 e^{-x}\,dx$ can be evaluated and shown to be convergent. Thus (5) is convergent and consequently so is (4).

EXERCISES FOR SEC. 10.3: THE INTEGRAL TEST

In Exercises 1 to 8 use the integral test to determine whether the series diverge or converge.

1 $\sum_{n=1}^{\infty} \frac{1}{n^{1.1}}$ **2** $\sum_{n=1}^{\infty} \frac{1}{n^{0.9}}$

3 $\sum_{n=1}^{\infty} \frac{n}{n^2 + 1}$ **4** $\sum_{n=1}^{\infty} \frac{1}{n^2 + 1}$

5 $\sum_{n=2}^{\infty} \frac{1}{n \ln n}$ **6** $\sum_{n=1}^{\infty} \frac{1}{n + 1{,}000}$

7 $\sum_{n=1}^{\infty} \frac{\ln n}{n}$ **8** $\sum_{n=1}^{\infty} \frac{n^3}{e^n}$

In Exercises 9 to 12 use the p test to determine whether the series diverge or converge.

9 $\sum_{n=1}^{\infty} \frac{1}{\sqrt[3]{n}}$ **10** $\sum_{n=1}^{\infty} \frac{1}{n^3}$

11 $\sum_{n=1}^{\infty} \frac{1}{n^{1.001}}$ **12** $\sum_{n=1}^{\infty} \frac{1}{n^{0.999}}$

In each of Exercises 13 to 16 (*a*) compute the sum of the first four terms of the series to four decimal places, (*b*) give upper and lower bounds on the error R_4, (*c*) combine (*a*) and (*b*) to estimate the sum of the series.

13 $\sum_{n=1}^{\infty} \frac{1}{n^3}$ **14** $\sum_{n=1}^{\infty} \frac{1}{n^4}$

15 $\sum_{n=1}^{\infty} \frac{1}{n^2+1}$ **16** $\sum_{n=1}^{\infty} \frac{1}{n^2+n}$

■

The ideas used in establishing the integral test can also be used to find estimates for the sum of the first n terms of the series $\Sigma_{n=1}^{\infty} f(n)$ discussed in this section. Exercises 17 to 21 give some examples.

17 Let $f(x)$ be a decreasing continuous positive function for $x \geq 1$. Show, by diagrams, that for $n \geq 2$,

$$\int_1^{n+1} f(x)\,dx < \sum_{i=1}^{n} f(i) < f(1) + \int_1^n f(x)\,dx.$$

18 Use the inequality in Exercise 17 to show that for $n \geq 2$,

$$2\sqrt{n+1} - 2 < \sum_{i=1}^{n} \frac{1}{\sqrt{i}} < 2\sqrt{n} - 1.$$

19 Show that for $n \geq 2$,

$$\ln(n+1) < \frac{1}{1} + \frac{1}{2} + \frac{1}{3} + \cdots + \frac{1}{n} < 1 + \ln n.$$

20 (Calculator) Show that the sum of the first million terms of the harmonic series is between 13.8 and 14.9.

21 (*a*) By comparing the sum with integrals, show that

$$\ln \tfrac{201}{100} < \tfrac{1}{100} + \tfrac{1}{101} + \tfrac{1}{102} + \cdots + \tfrac{1}{200} < \ln \tfrac{200}{99}.$$

(*b*) Show that $\lim_{n\to\infty} \sum_{i=n}^{2n} \frac{1}{i} = \ln 2.$

22 According to Theorem 3, how many terms of the series $\Sigma_{n=1}^{\infty} 1/n^2$ could you use to estimate the sum of the series with an error (*a*) at most 0.01? (*b*) at most 0.001?

23 According to Theorem 3, how many terms of the series $\Sigma_{n=1}^{\infty} 1/n^3$ could you use to estimate the sum of the series with an error (*a*) at most 0.01? (*b*) at most 0.001?

24 (*a*) Let $f(x)$ be a decreasing continuous positive function for $x \geq 1$ such that $\int_1^{\infty} f(x)\,dx$ is convergent. Show that

$$\int_1^{\infty} f(x)\,dx < \sum_{n=1}^{\infty} f(n) < f(1) + \int_1^{\infty} f(x)\,dx.$$

(*b*) Use (*a*) to estimate $\Sigma_{n=1}^{\infty} 1/n^2$.

Exercises 25 and 26 concern products, rather than sums, of numbers.

25 Let $\{a_n\}$ be a sequence of positive numbers. Denote the product $(1 + a_1)(1 + a_2) \ldots (1 + a_n)$ by $\Pi_{i=1}^{n} (1 + a_i)$.
(*a*) Show that $\Sigma_{i=1}^{n} a_i \leq \Pi_{i=1}^{n} (1 + a_i)$.
(*b*) Show that if $\lim_{n\to\infty} \Pi_{i=1}^{n} (1 + a_i)$ exists, then $\Sigma_{n=1}^{\infty} a_n$ is convergent.

26 (See Exercise 25.)
(*a*) Show that $1 + a_i \leq e^{a_i}$. *Hint:* Show that $1 + x \leq e^x$ for $x \geq 0$.
(*b*) Show that if $\Sigma_{n=1}^{\infty} a_n$ is convergent, then $\lim_{n\to\infty} \Pi_{i=1}^{n} (1 + a_i)$ exists.

27 Here is an argument that there is an infinite number of primes. Assume that there is only a finite number of primes, $p_1, p_2, \ldots, p_m$.
(*a*) Show that

$$\frac{1}{1 - (1/p_i)} = 1 + \frac{1}{p_i} + \frac{1}{p_i^2} + \frac{1}{p_i^3} + \cdots.$$

(*b*) Show then that

$$\frac{1}{1 - 1/p_1} \frac{1}{1 - 1/p_2} \cdots \frac{1}{1 - 1/p_m} = \sum_{n=1}^{\infty} \frac{1}{n}.$$

(Assume that the series can be multiplied term by term.)
(*c*) From (*b*) obtain a contradiction.

28 Cards of unit length from a playing deck are piled on top of each other, as shown in Fig. 10.8.
(*a*) If there are n cards, show that it is possible to arrange them so that the total distance that the right edge of the top card extends beyond the table is $\Sigma_{i=1}^{n} (1/2i)$. (The length of a card is 1.)
(*b*) Show that it is possible to place three cards in such a pile that the right edge of the top card extends $\frac{11}{12}$ of a card beyond the table.
(*c*) If you have 52 cards, estimate how far beyond the table the top card can extend.
(*d*) If you have an unlimited supply of cards, how far beyond the table can you arrange to have the top card extend?

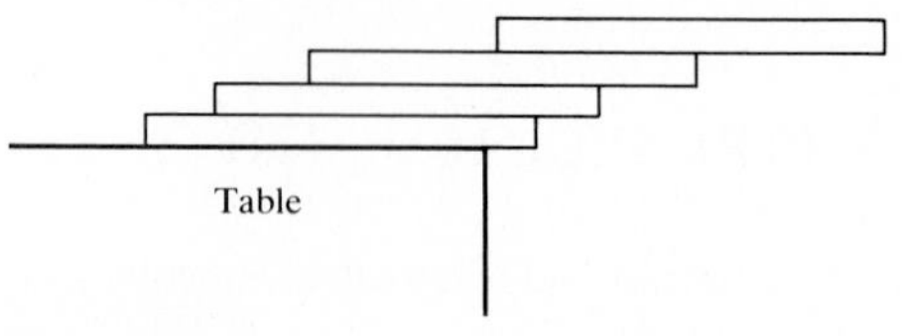

Figure 10.8

29 (Compare with Exercise 22.) Let $S_n = \Sigma_{i=1}^{n} 1/i^2$, $R_n = \Sigma_{i=n+1}^{\infty} 1/i^2$ and $S = \Sigma_{i=1}^{\infty} 1/i^2$.
(*a*) Show that

$$\frac{1}{n+1} < R_n < \frac{1}{n}.$$

(*b*) Show that

$$\left| S - \left[S_n + \frac{1}{2}\left(\frac{1}{n} + \frac{1}{n+1}\right)\right]\right| < \frac{1}{2}\left(\frac{1}{n} - \frac{1}{n+1}\right).$$

(*c*) If you use $S_n + \frac{1}{2}[1/n + 1/(n + 1)]$ as an estimate of S, how large should n be in order to make

$$S - \left[S_n + \frac{1}{2}\left(\frac{1}{n} + \frac{1}{n+1}\right)\right]$$

less than 0.01? Less than 0.001?

30 (See Exercise 29.) Using the method in Exercise 29, develop a formula for estimating $S = \sum_{i=1}^{\infty} 1/i^3$ that is more efficient than just using the partial sum $S_n = \sum_{i=1}^{n} 1/i^3$.

10.4 THE COMPARISON TEST AND THE RATIO TEST

So far in this chapter three tests for convergence (or divergence) of a series have been presented. The first concerned a special type of series, a geometric series. The second, the nth-term test for divergence, asserts that if the nth term of a series does *not* approach 0, the series diverges. The third, the integral test, applies to certain series of positive terms. In this section four further tests are developed: the comparison, limit comparison, ratio, and root tests. They are the ones most frequently applied. This section concerns only series with all terms positive.

The first test is similar to the comparison test for improper integrals given in Sec. 8.7.

Theorem 1 *The comparison test.*
(*a*) If the series

$$c_1 + c_2 + \cdots + c_n + \cdots$$

Comparing two series term by term

converges, where each c_n is positive, and if

$$0 \leq p_n \leq c_n$$

for each n, then the series

$$p_1 + p_2 + \cdots + p_n + \cdots$$

converges.
(*b*) If the series

$$d_1 + d_2 + \cdots + d_n + \cdots$$

diverges, where each d_n is positive, and if

$$p_n \geq d_n$$

for each n, then the series

$$p_1 + p_2 + \cdots + p_n + \cdots$$

diverges.

Proof
(*a*) Let the sum of the series $c_1 + c_2 + \cdots$ be C. Let S_n denote the partial sum $p_1 + p_2 + \cdots + p_n$. Then, for each n,

$$S_n = p_1 + p_2 + \cdots + p_n \leq c_1 + c_2 + \cdots + c_n < C.$$

Since the p_n's are nonnegative,

$$S_1 \leq S_2 \leq \cdots \leq S_n \leq \cdots.$$

$S_1 \leq S_2 \leq \cdots < C.$

Since each S_n is less than C, Theorem 2 of Sec. 10.1 assures us that the sequence

$$S_1, S_2, \ldots, S_n, \ldots$$

converges to a number L (less than or equal to C). In other words, the series $p_1 + p_2 + \cdots$ converges (and its sum is less than or equal to the sum $c_1 + c_2 + \cdots$).

(*b*) The divergence test follows immediately from the convergence test. If the series $p_1 + p_2 + \cdots$ converged, so would the series $d_1 + d_2 + \cdots$, which is assumed to diverge. ■

To apply the comparison tests, compare a series to one known to converge (or diverge).

EXAMPLE 1 Show that the series $\frac{2}{3}\frac{1}{1^2} + \frac{3}{4}\frac{1}{2^2} + \cdots + \frac{n+1}{n+2}\frac{1}{n^2} + \cdots$ converges.

SOLUTION The series resembles the series

$$\frac{1}{1^2} + \frac{1}{2^2} + \cdots + \frac{1}{n^2} + \cdots,$$

which was shown by the integral test to be convergent. Since the fraction $(n + 1)/(n + 2)$ is less than 1,

$$\frac{n+1}{n+2}\frac{1}{n^2} < \frac{1}{n^2}.$$

Thus, by the comparison test for convergence, the series

$$\frac{2}{3}\frac{1}{1^2} + \frac{3}{4}\frac{1}{2^2} + \cdots + \frac{n+1}{n+2}\frac{1}{n^2} + \cdots$$

also converges. ■

The next test is similar to the comparison test. It is useful when two series of positive terms resemble each other a great deal, even though the terms of one series are not necessarily less than the terms of the other.

Theorem 2 *The limit-comparison test.* Let $p_1 + p_2 + \cdots + p_n + \cdots$ be a series of positive terms to be tested for convergence or divergence.

(*a*) Let $c_1 + c_2 + \cdots + c_n + \cdots$ be a convergent series of positive terms. If $\lim_{n\to\infty} p_n/c_n$ exists, then $p_1 + p_2 + \cdots + p_n + \cdots$ also converges.

(*b*) Let $d_1 + d_2 + \cdots + d_n + \cdots$ be a divergent series of positive terms. If $\lim_{n\to\infty} p_n/d_n$ exists and is not 0 or if it is infinite, then $p_1 + p_2 + \cdots + p_n + \cdots$ also diverges.

Proof We shall prove part (*a*). Let $\lim_{n\to\infty} p_n/c_n = a$. Since as $n \to \infty$,

$p_n/c_n \to a$, there must be an integer N such that, for all $n \geq N$, p_n/c_n remains less than, say, $a + 1$. Thus

$$p_n < (a + 1)c_n \qquad n \geq N.$$

Now the series

$$(a + 1)c_N + (a + 1)c_{N+1} + \cdots + (a + 1)c_n + \cdots,$$

being $a + 1$ times the tail end of a convergent series, is itself convergent. By the comparison test,

$$p_N + p_{N+1} + \cdots + p_n + \cdots$$

is convergent. Hence $p_1 + p_2 + \cdots + p_n + \cdots$ is convergent.

Part (b) can be proved similarly. ■

Note that in Theorem 2(b) nothing is said about $\lim_{n\to\infty} p_n/d_n = 0$. The reason is that in this circumstance $\Sigma_{n=1}^{\infty} p_n$ can either converge or diverge. For instance, take $\Sigma_{n=1}^{\infty} d_n$ to be the divergent series $\Sigma_{n=1}^{\infty} 1/\sqrt{n}$. The series $\Sigma_{n=1}^{\infty} 1/n^2$ is convergent and $\lim_{n\to\infty} (1/n^2)/(1/\sqrt{n}) = 0$. On the other hand, the harmonic series $\Sigma_{n=1}^{\infty} 1/n$ is divergent and again $\lim_{n\to\infty} (1/n)/(1/\sqrt{n}) = 0$.

EXAMPLE 2 Show that $\displaystyle\sum_{n=1}^{\infty} \frac{(1 + 1/n)^n[1 + (-\frac{1}{2})^n]}{2^n}$ converges.

SOLUTION Note that as $n \to \infty$, $(1 + 1/n)^n \to e$ and $1 + (-\frac{1}{2})^n \to 1$. The major influence is the 2^n in the denominator. So use the limit-comparison test, with the convergent series $c_1 + c_2 + \cdots + c_n + \cdots = 1 + \frac{1}{2} + \frac{1}{4} + \cdots + 1/2^n + \cdots$, which the given series resembles. Then

$$\lim_{n\to\infty} \frac{\dfrac{\left(1 + \dfrac{1}{n}\right)^n \left[1 + \left(-\dfrac{1}{2}\right)^n\right]}{2^n}}{\dfrac{1}{2^n}} = \lim_{n\to\infty} \left(1 + \frac{1}{n}\right)^n \left[1 + \left(-\frac{1}{2}\right)^n\right] = e \cdot 1 = e.$$

Since $\Sigma_{n=1}^{\infty} 2^{-n}$ is convergent, so is the given series. ■

The next test is suggested by the test for the convergence of a geometric series. In a geometric series the ratio between consecutive terms is constant. The next test concerns series where this ratio is "almost constant."

Theorem 3 *The ratio test.* Let $p_1 + p_2 + \cdots + p_n + \cdots$ be a series of positive terms.

(a) If $\displaystyle\lim_{n\to\infty} \frac{p_{n+1}}{p_n}$ exists and is less than 1, the series converges.

(b) If $\displaystyle\lim_{n\to\infty} \frac{p_{n+1}}{p_n}$ exists and is greater than 1 or infinite, the series diverges.

Proof

(a) Let $$\lim_{n\to\infty} \frac{p_{n+1}}{p_n} = s < 1.$$

Select a number r such that $s < r < 1$. Then there is an integer N such that for all $n \geq N$,

$$\frac{p_{n+1}}{p_n} < r,$$

and, therefore, $$p_{n+1} < rp_n.$$

Thus $$p_{N+1} < rp_N$$

$$p_{N+2} < rp_{N+1} < r(rp_N) = r^2 p_N$$

$$p_{N+3} < rp_{N+2} < r(r^2 p_N) = r^3 p_N,$$

and so on.

Thus the terms of the series

$$p_N + p_{N+1} + p_{N+2} + \cdots$$

are less than the corresponding terms of the geometric series

$$p_N + rp_N + r^2 p_N + \cdots$$

(except for the first term p_N, which equals the first term of the geometric series). Since $r < 1$, the latter series converges. By the comparion test, $p_N + p_{N+1} + p_{N+2} + \cdots$ converges. Adding in the front end,

$$p_1 + p_2 + \cdots + p_{N-1},$$

still results in a convergent series.

(*b*) If $\lim_{n\to\infty} p_{n+1}/p_n$ is greater than 1 or is infinite, then for all n from some point on, p_{n+1} is larger than p_n. Thus the nth term of the series $p_1 + p_2 + \cdots$ cannot approach 0. By the nth-term test for divergence the series diverges. ■

No information if ratio approaches 1

No mention is made in Theorem 3 of the case $\lim_{n\to\infty} p_{n+1}/p_n = 1$. The reason for this omission is that anything can happen; the series may diverge or it may converge. (Exercise 27 illustrates these possibilities.) Also, $\lim_{n\to\infty} p_{n+1}/p_n$ may not exist. In these cases, one must look to other tests to determine whether the series diverges or converges.

The ratio test is a natural one to try if the nth term of a series involves powers of a fixed number, as the next example shows.

EXAMPLE 3 Show that the series $p + 2p^2 + 3p^3 + \cdots + np^n + \cdots$ converges for any fixed number p for which $0 < p < 1$.

SOLUTION Let a_n denote the nth term of the series. Then

$$a_n = np^n \quad \text{and} \quad a_{n+1} = (n+1)p^{n+1}.$$

The ratio between consecutive terms is

$$\frac{a_{n+1}}{a_n} = \frac{(n+1)p^{n+1}}{np^n} = \frac{n+1}{n}p.$$

To see what its sum is, look at Exercise 24 in Sec. 10.2.

Thus
$$\lim_{n\to\infty} \frac{a_{n+1}}{a_n} = p,$$
and the series converges. ■

EXAMPLE 4 Find for which positive values of x the series

$$\frac{1}{0!} + \frac{x}{1!} + \frac{x^2}{2!} + \frac{x^3}{3!} + \cdots + \frac{x^n}{n!} + \cdots$$

converges and for which it diverges. (Each choice of x determines a specific series with constant terms.)

SOLUTION If we start the series at $n = 0$, then the nth term, a_n, is $x^n/n!$ Thus

$$a_{n+1} = \frac{x^{n+1}}{(n+1)!},$$

and therefore
$$\frac{a_{n+1}}{a_n} = \frac{\dfrac{x^{n+1}}{(n+1)!}}{\dfrac{x^n}{n!}} = x\,\frac{n!}{(n+1)!} = \frac{x}{n+1}.$$

Since x is fixed,
$$\lim_{n\to\infty} \frac{x}{n+1} = 0.$$

In the next section, it will be shown to converge for negative x too.

By the ratio test, the series converges for all positive x. ■

The next example uses the ratio test to establish divergence.

EXAMPLE 5 Show that the series $\frac{2}{1} + \frac{2^2}{2} + \cdots + \frac{2^n}{n} + \cdots$ diverges.

SOLUTION In this case, $a_n = 2^n/n$ and

$$\frac{a_{n+1}}{a_n} = \frac{\dfrac{2^{n+1}}{n+1}}{\dfrac{2^n}{n}}$$
$$= \frac{2^{n+1}}{n+1}\,\frac{n}{2^n} = 2\,\frac{n}{n+1}.$$

So the series is like a geometric series with ratio 2.

Thus
$$\lim_{n\to\infty} \frac{a_{n+1}}{a_n} = 2,$$
which is larger than 1. By the ratio test, the series diverges. ■

It is not really necessary to call on the powerful ratio test to establish the divergence of the series in Example 5. Since $\lim_{x\to\infty} 2^x/x = \infty$, its nth term gets arbitrarily large; by the nth-term test, the series diverges. Comparison with the harmonic series also demonstrates divergence.

The next test, closely related to the ratio test, is of use when the nth term contains only powers, such as n^n or 3^n. It is not useful if factorials, such as $n!$, are present.

Theorem 4 *The root test.* Let $\Sigma_{n=1}^{\infty} p_n$ be a series of positive terms. Then

(*a*) if $\lim_{n\to\infty} \sqrt[n]{p_n}$ exists and is less than 1, $\Sigma_{n=1}^{\infty} p_n$ converges.
(*b*) if $\lim_{n\to\infty} \sqrt[n]{p_n}$ exists and is greater than 1 or is infinite, $\Sigma_{n=1}^{\infty} p_n$ diverges.
(*c*) if $\lim_{n\to\infty} \sqrt[n]{p_n} = 1$, no conclusion can be drawn ($\Sigma_{n=1}^{\infty} p_n$ may converge or may diverge). ■

The proof of the root test is outlined in Exercises 25 and 30.

EXAMPLE 6 Use the root test to determine whether $\sum_{n=1}^{\infty} \frac{3^n}{n^n}$ converges or diverges.

SOLUTION

$$\lim_{n\to\infty} \sqrt[n]{\frac{3^n}{n^n}} = \lim_{n\to\infty} \frac{3}{n} = 0.$$

By the root test, the series converges. ■

EXERCISES FOR SEC. 10.4: THE COMPARISON TEST AND THE RATIO TEST

In Exercises 1 to 4 use the comparison test to determine whether the series converge or diverge.

1 $\sum_{n=1}^{\infty} \frac{1}{n^2+3}$

2 $\sum_{n=1}^{\infty} \frac{n+2}{(n+1)\sqrt{n}}$

3 $\sum_{n=1}^{\infty} \frac{\sin^2 n}{n^2}$

4 $\sum_{n=1}^{\infty} \frac{1}{n2^n}$

In Exercises 5 to 8 use the limit-comparison test to determine whether the series converge or diverge.

5 $\sum_{n=1}^{\infty} \frac{5n+1}{(n+2)n^2}$

6 $\sum_{n=1}^{\infty} \frac{2^n+n}{3^n}$

7 $\sum_{n=1}^{\infty} \frac{n+1}{(5n+2)\sqrt{n}}$

8 $\sum_{n=1}^{\infty} \frac{(1+1/n)^n}{n^2}$

In Exercises 9 to 18 use any test to determine convergence or divergence.

9 $\sum_{n=1}^{\infty} \frac{n^3}{2^n}$

10 $\sum_{n=1}^{\infty} \frac{(n+1)^2}{n!}$

11 $\sum_{n=1}^{\infty} \frac{n!}{n^n}$

12 $\sum_{n=1}^{\infty} \frac{4n+1}{(2n+3)n^2}$

13 $\sum_{n=1}^{\infty} \frac{1}{n^n}$

14 $\sum_{n=1}^{\infty} \frac{(2n+1)(2^n+1)}{3^n+1}$

15 $\sum_{n=1}^{\infty} \frac{1+\cos n}{n^2}$

16 $\sum_{n=1}^{\infty} \frac{\ln n}{n^2}$

17 $\sum_{n=1}^{\infty} \frac{2^n}{(n+1)^n}$

18 $\sum_{n=1}^{\infty} \frac{5^n}{n(n^n)}$

In Exercises 19 to 22 determine for which positive numbers x the series (*a*) converge, (*b*) diverge.

19 $\sum_{n=1}^{\infty} \frac{x^n}{n} = \frac{x}{1} + \frac{x^2}{2} + \frac{x^3}{3} + \cdots + \frac{x^n}{n} + \cdots$

20 $\sum_{n=0}^{\infty} \frac{x^n}{2^n} = 1 + \frac{x}{2} + \frac{x^2}{4} + \frac{x^3}{8} + \cdots + \frac{x^n}{2^n} + \cdots$

21 $\sum_{n=0}^{\infty} \frac{2^n x^n}{n!} = 1 + \frac{2x}{1} + \frac{4x^2}{2} + \frac{8x^3}{6} + \cdots + \frac{2^n x^n}{n!} + \cdots$

22 $\sum_{n=1}^{\infty} n^5 x^n = x + 2^5x^2 + 3^5x^3 + \cdots + n^5x^n + \cdots$

■

23 Use the result of Example 4 to show that, for $x > 0$, $\lim_{n\to\infty} x^n/n! = 0$. (This fact was established directly in Sec. 10.1.)

24 Use Exercise 24 of Sec. 10.2 to find the sum of the series in Example 3.

25 This exercise shows that the root test gives no infor-

mation if $\lim_{n\to\infty} \sqrt[n]{p_n} = 1$.

(a) Show that for $p_n = 1/n$, $\sum_{n=1}^{\infty} p_n$ diverges and $\lim_{n\to\infty} \sqrt[n]{p_n} = 1$.

(b) Show that for $p_n = 1/n^2$, $\sum_{n=1}^{\infty} p_n$ converges and $\lim_{n\to\infty} \sqrt[n]{p_n} = 1$.

26 Solve Example 4 using the root test.

27 This exercise shows that the ratio test is useless when $\lim_{n\to\infty} p_{n+1}/p_n = 1$.

(a) Show that if $p_n = 1/n$, then $\sum_{n=1}^{\infty} p_n$ diverges and $\lim_{n\to\infty} p_{n+1}/p_n = 1$.

(b) Show that if $p_n = 1/n^2$, then $\sum_{n=1}^{\infty} p_n$ converges and $\lim_{n\to\infty} p_{n+1}/p_n = 1$.

28 (a) Show that $\sum_{n=1}^{\infty} 1/(1 + 2^n)$ converges.

(b) Show that the sum of the series in (a) is between 0.64 and 0.77. (Use the first three terms and control the sum of the rest of the series by comparing it to the sum of a geometric series.)

29 (a) Show that $\sum_{n=1}^{\infty} n/[(n + 1)2^n]$ converges.

(b) Show that the sum of the series in (a) is between

$$\tfrac{1}{2}\cdot\tfrac{1}{2} + \tfrac{2}{3}\cdot\tfrac{1}{4} + \tfrac{3}{4}\cdot\tfrac{1}{8}$$

and

$$\tfrac{1}{2}\cdot\tfrac{1}{2} + \tfrac{2}{3}\cdot\tfrac{1}{4} + \tfrac{3}{4}\cdot\tfrac{1}{8} + \tfrac{1}{8}.$$

30 (Proof of the root test, Theorem 4)

(a) Assume that $\lim_{n\to\infty} \sqrt[n]{p_n} = L < 1$. Pick any r, $L < r < 1$, and then N such that $\sqrt[n]{p_n} < r$ for $n > N$. Show that $p_n < r^n$ for $n > N$ and compare a tail end of $\sum_{n=1}^{\infty} p_n$ to a geometric series.

(b) Assume that $\sqrt[n]{p_n} = L > 1$. Pick any number r, $1 < r < L$, and then N such that $\sqrt[n]{p_n} > r$ for $n > N$. Show that $p_n > r^n$ for $n > N$. From this conclude that $\sum_{n=1}^{\infty} p_n$ diverges.

31 Determine whether $\sum_{n=2}^{\infty}(1/\ln n)$ converges or diverges.

32 Prove the following result, which is used in the statistical theory of stochastic processes: Let $\{a_n\}$ and $\{c_n\}$ be two sequences of nonnegative numbers such that $\sum_{n=1}^{\infty} a_n c_n$ converges and $\lim_{n\to\infty} c_n = 0$. Then $\sum_{n=1}^{\infty} a_n c_n^2$ converges.

33 Prove part (b) of Theorem 2.

10.5 THE ALTERNATING-SERIES AND ABSOLUTE-CONVERGENCE TESTS

The tests for convergence or divergence in Secs. 10.3 and 10.4 concerned series whose terms are positive. This section examines series which may have both positive and negative terms. Two tests for the convergence of such a series are presented. The alternating-series test applies to series whose terms alternate in sign, +, −, +, −, . . . and decrease in absolute value. In the absolute-convergence test the signs may vary in any way.

Definition *Alternating series.* If $p_1, p_2, \ldots, p_n, \ldots$ is a sequence of positive numbers, then the series

$$\sum_{n=1}^{\infty} (-1)^{n+1} p_n = p_1 - p_2 + p_3 - p_4 + \cdots + (-1)^{n+1}p_n + \cdots$$

and the series

$$\sum_{n=1}^{\infty} (-1)^n p_n = -p_1 + p_2 - p_3 + p_4 - \cdots + (-1)^n p_n + \cdots$$

are called **alternating series**.

For instance,

$$1 - \frac{1}{3} + \frac{1}{5} - \frac{1}{7} + \cdots + (-1)^{n+1}\frac{1}{2n-1} + \cdots$$

and

$$-1 + 1 - 1 + 1 - \cdots + (-1)^n + \cdots$$

are alternating series.

By the nth-term test, the second series diverges. The following theorem implies that the first series converges.

Theorem 1 *The alternating-series test.* If $p_1, p_2, \ldots, p_n, \ldots$ is a decreasing sequence of positive numbers such that $\lim_{n\to\infty} p_n = 0$, then the series whose nth term is $(-1)^{n+1} p_n$,

$$\sum_{n=1}^{\infty} (-1)^{n+1}p_n = p_1 - p_2 + p_3 - \cdots + (-1)^{n+1}p_n + \cdots,$$

converges.

Proof The idea of the proof is easily conveyed by a specific case. For the sake of concreteness and simplicity, consider the series in which $p_n = 1/n$, that is, the series

$$1 - \frac{1}{2} + \frac{1}{3} - \frac{1}{4} + \cdots + (-1)^{n+1}\frac{1}{n} + \cdots.$$

Consider first the partial sums of an *even* number of terms, S_2, S_4, $S_6, \ldots$. For clarity, group the summands in pairs:

$$S_2 = (1 - \tfrac{1}{2})$$

$$S_4 = (1 - \tfrac{1}{2}) + (\tfrac{1}{3} - \tfrac{1}{4}) = S_2 + (\tfrac{1}{3} - \tfrac{1}{4})$$

$$S_6 = (1 - \tfrac{1}{2}) + (\tfrac{1}{3} - \tfrac{1}{4}) + (\tfrac{1}{5} - \tfrac{1}{6}) = S_4 + (\tfrac{1}{5} - \tfrac{1}{6})$$

$$\cdots\cdots\cdots\cdots\cdots\cdots\cdots\cdots\cdots\cdots$$

Since $\frac{1}{3}$ is larger than $\frac{1}{4}$, $\frac{1}{3} - \frac{1}{4}$ is positive. Therefore S_4, which equals $S_2 + (\frac{1}{3} - \frac{1}{4})$, is larger than S_2. Similarly,

$$S_6 > S_4.$$

More generally, then, $S_2 < S_4 < S_6 < S_8 < \cdots$.

The sequence $\{S_{2n}\}$ is increasing.

Next, it will be shown that S_{2n} is less than 1, the first term of the given sequence. First of all,

$$S_2 = 1 - \tfrac{1}{2} < 1.$$

Next, consider S_4:

$$\begin{aligned} S_4 &= 1 - \tfrac{1}{2} + \tfrac{1}{3} - \tfrac{1}{4} \\ &= 1 - (\tfrac{1}{2} - \tfrac{1}{3}) - \tfrac{1}{4} \\ &< 1 - (\tfrac{1}{2} - \tfrac{1}{3}). \end{aligned}$$

Since $\frac{1}{2} - \frac{1}{3}$ is positive, this shows that

$$S_4 < 1.$$

Similarly,

$$\begin{aligned} S_6 &= 1 - (\tfrac{1}{2} - \tfrac{1}{3}) - (\tfrac{1}{4} - \tfrac{1}{5}) - \tfrac{1}{6} \\ &< 1 - (\tfrac{1}{2} - \tfrac{1}{3}) - (\tfrac{1}{4} - \tfrac{1}{5}) \\ &< 1. \end{aligned}$$

In general then,

$$S_{2n} < 1$$

for all n.

The sequence

$$S_2, S_4, S_6, \ldots$$

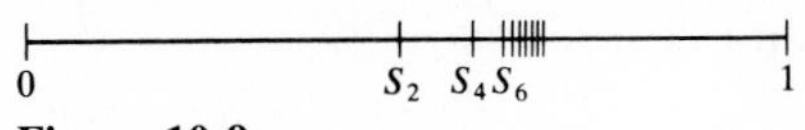

Figure 10.9

is therefore increasing and yet bounded by the number 1, as indicated in Fig. 10.9. By Theorem 2 of Sec. 10.1, $\lim_{n\to\infty} S_{2n}$ exists. Call this limit S.

Next look at $S_1, S_3, S_5, \cdots$.

All that remains to be shown is that the numbers

$$S_1, S_3, S_5, \ldots$$

also converge to S.

Note that
$$S_3 = 1 - \tfrac{1}{2} + \tfrac{1}{3} = S_2 + \tfrac{1}{3}$$
$$S_5 = 1 - \tfrac{1}{2} + \tfrac{1}{3} - \tfrac{1}{4} + \tfrac{1}{5} = S_4 + \tfrac{1}{5}.$$

The term $1/(2n + 1)$ will be p_{2n+1} in the general case.

In general,
$$S_{2n+1} = S_{2n} + \frac{1}{2n+1}.$$

Thus
$$\lim_{n\to\infty} S_{2n+1} = \lim_{n\to\infty}\left(S_{2n} + \frac{1}{2n+1}\right)$$
$$= \lim_{n\to\infty} S_{2n} + \lim_{n\to\infty}\frac{1}{2n+1}$$
$$= S + 0$$
$$= S.$$

Since the partial sums

$$S_2, S_4, S_6, \ldots$$

and the partial sums

$$S_1, S_3, S_5, \ldots$$

both have the same limit S, it follows that

$$\lim_{n\to\infty} S_n = S.$$

Thus the series

$$1 - \tfrac{1}{2} + \tfrac{1}{3} - \tfrac{1}{4} + \tfrac{1}{5} - \cdots$$

converges. ■

A similar argument applies to any alternating series whose nth term approaches 0 and whose terms decrease in absolute value.

Decreasing alternating series

An alternating series whose terms decrease in absolute value as n increases will be called a **decreasing alternating series**. Theorem 1 shows that a decreasing alternating series whose nth term approaches 0 as $n \to \infty$ converges.

EXAMPLE 1 Estimate the sum S of the series $1 - \frac{1}{2} + \frac{1}{3} - \frac{1}{4} + \cdots$.

SOLUTION These are the first five partial sums:

$$S_1 = 1 = 1.00$$
$$S_2 = 1 - \tfrac{1}{2} = 0.500$$
$$S_3 = 1 - \tfrac{1}{2} + \tfrac{1}{3} \approx 0.500 + 0.3333 = 0.8333$$

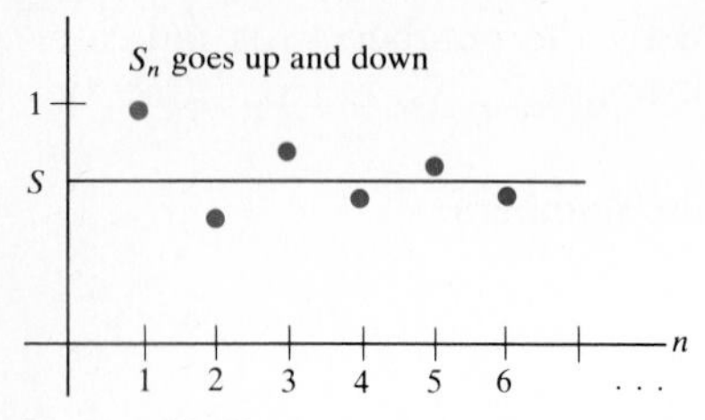

Figure 10.10

$$S_4 = S_3 - \tfrac{1}{4} \approx 0.8333 - 0.250 = 0.5833$$

$$S_5 = S_4 + \tfrac{1}{5} \approx 0.5833 + 0.200 = 0.7833.$$

Figure 10.10 is a graph of S_n as a function of n. The sums $S_1, S_3, \ldots$ approach S from above. The sums $S_2, S_4, \ldots$ approach S from below. For instance,

$$S_4 < S < S_5$$

gives the information that $0.583 < S < 0.784$. (See Fig. 10.11.) ■

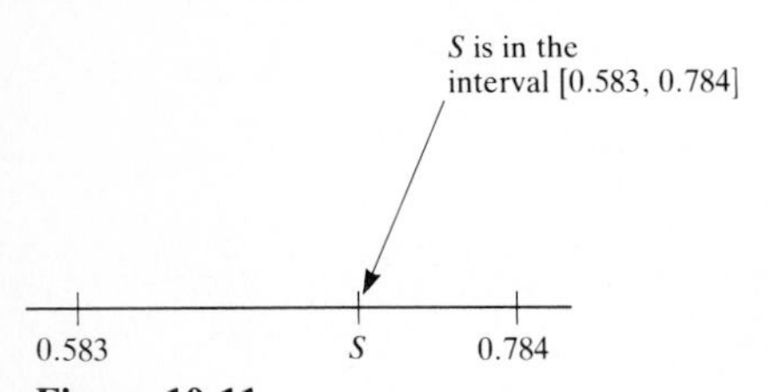

Figure 10.11

The error in estimating the sum of a decreasing alternating series

As Fig. 10.10 suggests, any partial sum of a series satisfying the hypothesis of the alternating-series test differs from the sum of the series by less than the absolute value of the first omitted term. That is, if S_n is the sum of the first n terms of the series and S is the sum of the series, then the error.

$$R_n = S - S_n,$$

has absolute value at most p_{n+1}, which is the absolute value of the first omitted term. Moreover, S is between S_n and S_{n+1} for every n.

EXAMPLE 2 Does the series

$$\frac{3}{1!} - \frac{3^2}{2!} + \frac{3^3}{3!} - \frac{3^4}{4!} + \frac{3^5}{5!} - \cdots + (-1)^{n+1}\frac{3^n}{n!} + \cdots$$

converge or diverge?

SOLUTION This is an alternating series. By Example 4 of Sec. 10.1, its nth term approaches 0. Let us see whether the absolute values of the terms decrease in size, term by term. The first few absolute values are

At first the terms increase.

$$\frac{3}{1!} = 3$$

$$\frac{3^2}{2!} = \frac{9}{2} = 4.5$$

$$\frac{3^3}{3!} = \frac{27}{6} = 4.5$$

$$\frac{3^4}{4!} = \frac{81}{24} = 3.375.$$

At first they increase. However, the fourth term is less than the third. Let us show that the rest of the terms decrease in size. For instance,

But then they decrease.

$$\frac{3^5}{5!} = \frac{3}{5}\frac{3^4}{4!} < \frac{3^4}{4!},$$

and, for $n \geq 3$,

$$\frac{3^{n+1}}{(n+1)!} = \frac{3}{n+1}\frac{3^n}{n!} < \frac{3^n}{n!}.$$

By the alternating-series test, the series that begins

$$\frac{3^3}{3!} - \frac{3^4}{4!} + \frac{3^5}{5!} - \frac{3^6}{6!} + \cdots$$

converges. Call its sum S. If the two terms

$$\frac{3}{1!} - \frac{3^2}{2!}$$

are added on, we obtain the original series, which therefore converges and has the sum

$$\frac{3}{1!} - \frac{3^2}{2!} + S. \quad ■$$

In the alternating-series test the absolute values of the terms must eventually be decreasing.

As Example 2 illustrates, the alternating-series test works as long as the nth term approaches 0 and the terms decrease in size from some point on in the series.

It may seem that any alternating series whose nth term approaches 0 converges. *This is not the case,* as is shown by this series:

$$\frac{2}{1} - \frac{1}{1} + \frac{2}{2} - \frac{1}{2} + \frac{2}{3} - \frac{1}{3} + \frac{2}{4} - \frac{1}{4} + \cdots, \tag{1}$$

whose terms alternate $2/n$ and $-1/n$.

Let S_n be the sum of the first n terms of (1). Then

$$S_2 = \frac{2}{1} - \frac{1}{1} = \frac{1}{1},$$

$$S_4 = \left(\frac{2}{1} - \frac{1}{1}\right) + \left(\frac{2}{2} - \frac{1}{2}\right) = \frac{1}{1} + \frac{1}{2},$$

$$S_6 = \left(\frac{2}{1} - \frac{1}{1}\right) + \left(\frac{2}{2} - \frac{1}{2}\right) + \left(\frac{2}{3} - \frac{1}{3}\right) = \frac{1}{1} + \frac{1}{2} + \frac{1}{3},$$

and more generally,

$$S_{2n} = \frac{1}{1} + \frac{1}{2} + \frac{1}{3} + \cdots + \frac{1}{n}.$$

Since S_{2n} gets arbitrarily large as $n \to \infty$ (the harmonic series diverges), the series (1) diverges.

Also, an alternating series whose terms decrease in size from some point on need not converge. Consider, for instance, the series

$$\frac{2}{1} - \frac{3}{2} + \frac{4}{3} - \frac{5}{4} + \cdots + (-1)^{n+1}\left(\frac{n+1}{n}\right) + \cdots.$$

Since the absolute value of the nth term approaches 1, the nth term does not approach 0. By the nth-term test for divergence, the series diverges.

Absolute Convergence

Consider a series $\qquad a_1 + a_2 + \cdots + a_n + \cdots,$

whose terms may be positive, negative, or zero. It is reasonable to expect it to behave at least as "nicely" as the series

$$|a_1| + |a_2| + \cdots + |a_n| + \cdots,$$

Glance back at Sec. 8.7 before reading Theorem 2.

since by making all the terms positive we give the series more chance to diverge. This is similar to the case with improper integrals in Sec. 8.7, where it was shown that if $\int_a^\infty |f(x)|\,dx$ converges, then so does $\int_a^\infty f(x)\,dx$. The next theorem (and its proof) is similar to Theorem 2 in Sec. 8.7.

Theorem 2 *Absolute-convergence test.* If the series

$$|a_1| + |a_2| + \cdots + |a_n| + \cdots$$

converges, then so does the series

$$a_1 + a_2 + \cdots + a_n + \cdots.$$

Proof Since the series $|a_1| + |a_2| + \cdots$ converges, so does the series $2|a_1| + 2|a_2| + \cdots$. (Its sum is $2\sum_{n=1}^\infty |a_n|$.)

Next, introduce the series whose nth term is

$$a_n + |a_n|.$$

Note that if a_n is negative, $a_n + |a_n| = 0$, while if a_n is nonnegative, $a_n + |a_n| = 2|a_n|$. Hence, for all n,

$$0 \le a_n + |a_n| \le 2|a_n|.$$

Comparing what to what?

By the comparison test, $\sum_{n=1}^\infty (a_n + |a_n|)$ converges. Let

$$\sum_{n=1}^\infty (a_n + |a_n|) = A \qquad \text{and} \qquad \sum_{n=1}^\infty |a_n| = B.$$

Now

$$\sum_{n=1}^k a_n = \sum_{n=1}^k (a_n + |a_n|) - \sum_{n=1}^k |a_n|.$$

Thus, as $k \to \infty$, $\sum_{n=1}^k a_n \to A - B$, and the theorem is proved. ■

The next example is a typical illustration of how Theorem 2 is applied.

EXAMPLE 3 Show that $\frac{1}{1^2} + \frac{1}{2^2} - \frac{1}{3^2} + \frac{1}{4^2} + \frac{1}{5^2} - \frac{1}{6^2} + \cdots$ (two positive terms alternating with one negative term) converges.

SOLUTION The series whose nth term is the absolute value of the nth term of the given series is

The p series with p = 2

$$\frac{1}{1^2} + \frac{1}{2^2} + \cdots + \frac{1}{n^2} + \cdots.$$

In Sec. 10.3 this series was shown to converge (by the integral test). By the absolute-convergence test, the original series, with +'s and −'s, converges. ■

The alternating series

$$1 - \tfrac{1}{2} + \tfrac{1}{3} - \tfrac{1}{4} + \cdots$$

converges, as shown by Theorem 1. However, when all the terms are replaced by their absolute values, the resulting series, the harmonic series, does not converge; that is,

$$1 + \tfrac{1}{2} + \tfrac{1}{3} + \tfrac{1}{4} + \cdots$$

diverges. Thus the converse of Theorem 2 is false.

The following definitions are frequently used in describing these various cases of convergence or divergence.

Definition *Absolute convergence.* A series $a_1 + a_2 + \cdots$ is said to **converge absolutely** if the series $|a_1| + |a_2| + \cdots$ converges.

Theorem 2 can be stated simply: "If a series converges absolutely, then it converges."

Definition *Conditional convergence.* A series $a_1 + a_2 + \cdots$ is said to **converge conditionally** if it converges but does *not* converge absolutely.

$1 - \frac{1}{2} + \frac{1}{3} - \frac{1}{4} + \cdots$ *converges conditionally.*

For instance, the **alternating harmonic series** $1 - \frac{1}{2} + \frac{1}{3} - \frac{1}{4} + \cdots$ is conditionally convergent.

The next example shows how the absolute convergence test can be combined with other tests to establish convergence of a series.

EXAMPLE 4 Show that

$$\frac{3}{1}\left(\frac{1}{2}\right) - \frac{5}{2}\left(\frac{1}{2}\right)^2 + \frac{7}{3}\left(\frac{1}{2}\right)^3 + \cdots + (-1)^{n+1}\frac{2n+1}{n}\left(\frac{1}{2}\right)^n + \cdots \tag{2}$$

converges.

SOLUTION Consider the series of positive terms

$$\frac{3}{1}\left(\frac{1}{2}\right) + \frac{5}{2}\left(\frac{1}{2}\right)^2 + \frac{7}{3}\left(\frac{1}{2}\right)^3 + \cdots + \frac{2n+1}{n}\left(\frac{1}{2}\right)^n + \cdots.$$

The fact that $(2n + 1)/n \to 2$ as $n \to \infty$ suggests use of the limit-comparison test, comparing the series to the convergent geometric series $\sum_{n=1}^{\infty} (\frac{1}{2})^n$. We have

$$\lim_{n\to\infty} \frac{\dfrac{2n+1}{n}\left(\dfrac{1}{2}\right)^n}{(\frac{1}{2})^n} = 2.$$

Thus $\sum_{n=1}^{\infty} [(2n + 1)/n](\frac{1}{2})^n$ converges. Consequently, the given series (2), with positive and negative terms, converges absolutely. Thus it converges. ■

Example 4 suggests the following useful generalization of the limit-comparison test.

Theorem 3 *The absolute-limit-comparison test.* Let $\Sigma_{n=1}^{\infty} a_n$ be a series whose terms may be negative or positive.

(*a*) Let $c_1 + c_2 + \cdots + c_n + \cdots$ be a convergent series of positive terms. If $\lim_{n\to\infty} |a_n/c_n|$ exists, then $\Sigma_{n=1}^{\infty} a_n$ is absolutely convergent, hence convergent.

(*b*) Let $d_1 + d_2 + \cdots + d_n + \cdots$ be a divergent series of positive terms. If $\lim_{n\to\infty} |a_n/d_n|$ exists and is not 0, then $\Sigma_{n=1}^{\infty} |a_n|$ diverges. $\Sigma_{n=1}^{\infty} a_n$ may or may not diverge. ■

The ratio test of Sec. 10.4 also has an analog when absolute values are introduced.

Theorem 4 *The absolute-ratio test.* Let $\Sigma_{n=1}^{\infty} a_n$ be a series such that

$$\lim_{n\to\infty} \left|\frac{a_{n+1}}{a_n}\right| = L < 1.$$

The absolute-ratio test

Then $\Sigma_{n=1}^{\infty} a_n$ converges. If $L > 1$ or if $\lim_{n\to\infty} |a_{n+1}/a_n| = \infty$, then $\Sigma_{n=1}^{\infty} a_n$ diverges.

Proof Take the case $L < 1$. By the ratio test, $\Sigma_{n=1}^{\infty} |a_n|$ converges. Since $\Sigma_{n=1}^{\infty} |a_n|$ converges, it follows that $\Sigma_{n=1}^{\infty} a_n$ converges also. The case $L > 1$ is treated in Exercise 36. The case $L = \infty$ can be treated as follows. If $\lim_{n\to\infty} |a_{n+1}/a_n| = \infty$, the ratio $|a_{n+1}|/|a_n|$ gets arbitrarily large as $n \to \infty$. So from some point on the numbers $|a_n|$ increase. By the nth-term test for divergence, $\Sigma_{n=1}^{\infty} a_n$ is divergent. ■

The absolute-ratio test simplifies work with minus signs.

Theorem 4 would establish the convergence of the series in Example 4 as follows. Let $a_n = (-1)^{n+1}[(2n+1)/n](\frac{1}{2})^n$. Then

$$\left|\frac{a_{n+1}}{a_n}\right| = \left|\frac{(-1)^{n+2}\dfrac{2n+3}{n+1}\left(\dfrac{1}{2}\right)^{n+1}}{(-1)^{n+1}\dfrac{2n+1}{n}\left(\dfrac{1}{2}\right)^{n}}\right| = \frac{2n+3}{2n+1}\cdot\frac{n}{n+1}\cdot\frac{1}{2},$$

which approaches $\frac{1}{2}$ as $n \to \infty$. Thus $\Sigma_{n=1}^{\infty} a_n$ converges (in fact, absolutely).

EXAMPLE 5 Examine the series

$$\frac{\cos x}{1^2} + \frac{\cos 2x}{2^2} + \frac{\cos 3x}{3^2} + \cdots + \frac{\cos nx}{n^2} + \cdots \tag{3}$$

for convergence or divergence.

SOLUTION The number x is fixed. The numbers $\cos nx$ may be positive, negative, or zero, in an irregular manner. However, for all n, $|\cos nx| \le 1$.

Recall that the series

$$\frac{1}{1^2} + \frac{1}{2^2} + \frac{1}{3^2} + \cdots + \frac{1}{n^2} + \cdots$$

converges, as shown in Sec. 10.3. Since $|\cos nx|/n^2 \le 1/n^2$, the series

Advanced calculus shows that for $0 \le x \le 2\pi$, series (3) sums to $\frac{3x^2 - 6\pi x + 2\pi^2}{12}$.

$$\frac{|\cos x|}{1^2} + \frac{|\cos 2x|}{2^2} + \frac{|\cos 3x|}{3^2} + \cdots + \frac{|\cos nx|}{n^2} + \cdots$$

converges by the comparison test. Series (3) thus converges absolutely for all x. Hence it converges. ■

A series that converges absolutely has the property that no matter how the terms are rearranged the new series converges and has the same sum as the original series. It might be expected that all convergent series have this property, but this is not the case. For instance, the alternating harmonic series

$$\tfrac{1}{1} - \tfrac{1}{2} + \tfrac{1}{3} - \tfrac{1}{4} + \tfrac{1}{5} - \cdots \tag{4}$$

does not. To show this, rearrange the summands so that two positive summands alternate with one negative summand, as follows:

$$\tfrac{1}{1} + \tfrac{1}{3} - \tfrac{1}{2} + \tfrac{1}{5} + \tfrac{1}{7} - \tfrac{1}{4} + \cdots. \tag{5}$$

Rearranging a conditionally convergent series is dangerous.

The positive summands in (5) have much more influence than the negative summands. In the battle between the positives and the negatives, the positives will win by a bigger margin in (5) than in (4). In fact, as shown in Exercise 35, the sum of (5) is $\frac{3}{2} \ln 2$. [But the sum of (4) is just $\ln 2$, as shown in Exercise 28(*e*) of Sec. 10.7.]

In advanced calculus it is demonstrated that a conditionally convergent series can be rearranged to converge to any preassigned sum or even to diverge to ∞ or $-\infty$.

EXERCISES FOR SEC. 10.5: THE ALTERNATING-SERIES AND ABSOLUTE-CONVERGENCE TESTS

In Exercises 1 to 8, which concern alternating series, determine which series converge and which diverge. Explain your answers.

1 $\frac{1}{2} - \frac{2}{3} + \frac{3}{4} - \frac{4}{5} + \cdots + (-1)^{n+1}\frac{n}{n+1} + \cdots$

2 $-\frac{1}{1+\frac{1}{2}} + \frac{1}{1+\frac{1}{4}} - \frac{1}{1+\frac{1}{8}} + \cdots + (-1)^n\frac{1}{1+2^{-n}} + \cdots$

3 $\frac{1}{\sqrt{1}} - \frac{1}{\sqrt{2}} + \frac{1}{\sqrt{3}} - \frac{1}{\sqrt{4}} + \cdots + (-1)^{n+1}\frac{1}{\sqrt{n}} + \cdots$

4 $\frac{5}{1!} - \frac{5^2}{2!} + \frac{5^3}{3!} - \frac{5^4}{4!} + \cdots + (-1)^{n+1}\frac{5^n}{n!} + \cdots$

5 $\frac{3}{\sqrt{1}} - \frac{2}{\sqrt{1}} + \frac{3}{\sqrt{2}} - \frac{2}{\sqrt{2}} + \frac{3}{\sqrt{3}} - \frac{2}{\sqrt{3}} + \cdots$

6 $\sqrt{1} - \sqrt{2} + \sqrt{3} - \sqrt{4} + \cdots + (-1)^{n+1}\sqrt{n} + \cdots$

7 $\frac{1}{3} - \frac{2}{5} + \frac{3}{7} - \frac{4}{9} + \frac{5}{11} - \cdots + (-1)^{n+1}\frac{n}{2n+1} + \cdots$

8 $\frac{1}{1^2} - \frac{1}{2^2} + \frac{1}{3^2} - \frac{1}{4^2} + \cdots + (-1)^{n+1}\frac{1}{n^2} + \cdots$

9 Consider the alternating harmonic series

$$\sum_{n=1}^{\infty} \frac{(-1)^{n+1}}{n}.$$

(*a*) Compute S_5 and S_6 to five decimal places.
(*b*) Is the estimate S_5 smaller or larger than the sum of the series?
(*c*) Use (*a*) and (*b*) to find two numbers between which the sum of the series must lie.

10 Consider the series $\sum_{n=1}^{\infty} (-1)^{n+1}2^{-n}/n$.
(*a*) Estimate the sum of the series using S_6.
(*b*) Estimate the error R_6.

In Exercises 11 to 26 determine which series diverge, converge absolutely, or converge conditionally. Explain your answers.

11 $\sum_{n=1}^{\infty} \frac{(-1)^n}{\sqrt[3]{n^2}}$

12 $\sum_{n=1}^{\infty} (-1)^n \ln \frac{1}{n}$

13 $\sum_{n=2}^{\infty} \frac{(-1)^n}{n \ln n}$

14 $\sum_{n=1}^{\infty} \frac{\sin n}{n^{1.01}}$

15 $\sum_{n=1}^{\infty} \left(1 - \cos \frac{\pi}{n}\right)$

16 $\sum_{n=1}^{\infty} (-1)^n \cos\left(\frac{\pi}{n^2}\right)$

17 $\sum_{n=1}^{\infty} \sin\left(\frac{\pi}{n^2}\right)$

18 $\sum_{n=1}^{\infty} \frac{(-2)^n}{n!}$

19 $\frac{1}{1^2} + \frac{1}{2^2} - \frac{1}{3^2} - \frac{1}{4^2} + \frac{1}{5^2} + \frac{1}{6^2} - \cdots$

(Two +'s alternating with two −'s.)

20 $\sum_{n=1}^{\infty} \frac{(-3)^n(1 + n^2)}{n!}$

21 $\sum_{n=1}^{\infty} \frac{\cos n\pi}{2n + 1}$

22 $\sum_{n=1}^{\infty} \frac{(-1)^n(n + 5)}{n^2}$

23 $\sum_{n=1}^{\infty} \frac{(-9)^n}{10^n + n}$

24 $\sum_{n=1}^{\infty} \frac{(-1)^n}{\sqrt[3]{n}}$

25 $\sum_{n=1}^{\infty} \frac{(-1.01)^n}{n!}$

26 $\sum_{n=1}^{\infty} \frac{(-\pi)^{2n+1}}{(2n + 1)!}$

■

27 The series $\Sigma_{n=1}^{\infty} (-1)^{n+1}2^{-n}$ is both a geometric series and a decreasing alternating series whose nth term approaches 0.
(*a*) Compute S_6 to three decimal places.
(*b*) Using the fact that the series is a decreasing alternating series, put a bound on R_6.
(*c*) Using the formula for the sum of a geometric series, compute R_6 exactly.

28 Show that the sum of the series $\Sigma_{n=1}^{\infty} (-1)^{n+1}/n$ is between 0.63 and 0.76. *Suggestion:* Compute S_7 and S_8.

29 Show that the sum of the series $\Sigma_{n=1}^{\infty} (-1)^{n+1}/(2n - 1)$ is between 0.72 and 0.84.

30 Show that the sum of the series $\Sigma_{n=1}^{\infty} (-1)^{n+1}/n!$ is between 0.625 and 0.634.

31 Show that the sum of the series $\Sigma_{n=1}^{\infty} (-1)^{n+1}/(2n - 1)!$ is between 0.833 and 0.842.

32 Let $P(x)$ and $Q(x)$ be two polynomials of degree at least one. Assume that for $n \geq 1$, $Q(n) \neq 0$. What relation must there be between the degrees of $P(x)$ and $Q(x)$ if
(*a*) $P(n)/Q(n) \to 0$ as $n \to \infty$?
(*b*) $\Sigma_{n=1}^{\infty} P(n)/Q(n)$ converges absolutely?
(*c*) $\Sigma_{n=1}^{\infty} (-1)^n P(n)/Q(n)$ converges conditionally?

33 Prove Theorem 1 as follows:
(*a*) Show that $S_2 < S_4 < S_6 < \cdots < p_1$.
(*b*) Show that $\lim_{n\to\infty} S_{2n}$ exists.
(*c*) Call the limit in (*b*) "S." Show that

$$\lim_{n\to\infty} S_{2n+1} = S.$$

Exercises 34 and 35 are related.

34 Use elementary geometry to show that the area of that part of the endless staircase shown in Fig. 10.12 above the curve $y = 1/x$, is between $\frac{1}{2}$ and 1.

The area of the shaded region above the curve is known as **Euler's constant** γ, whose decimal representation begins 0.577. Thus, if a_n is defined by $a_n = \Sigma_{i=1}^{n} 1/i - \ln n$, then $a_n \to \gamma$ as $n \to \infty$. It is not known whether γ is rational or irrational. (See Exercise 34 in Sec. 6.3.)

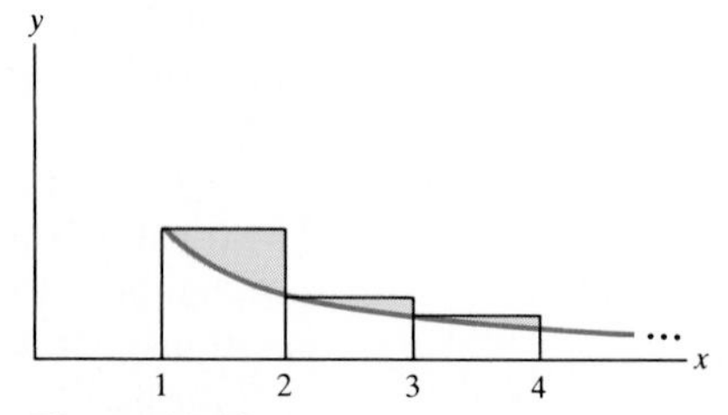

Figure 10.12

35 This exercise shows that the sum of the series (5) equals $\frac{3}{2}$ ln 2. The argument depends on the formula $\Sigma_{i=1}^{n} 1/i = \ln n + a_n$, where a_n approaches γ as $n \to \infty$.
(*a*) Show that $\Sigma_{i=1}^{n} (1/2i) = (\ln n)/2 + (a_n)/2$.
(*b*) Show that

$$\frac{1}{1} + \frac{1}{3} + \frac{1}{5} + \cdots + \frac{1}{4n - 1} = \sum_{i=1}^{2n} \frac{1}{2i - 1}$$

$$= \sum_{i=1}^{4n} \frac{1}{i} - \frac{1}{2}\sum_{i=1}^{2n} \frac{1}{i}.$$

(*c*) From (*a*) and (*b*) deduce that

$$\sum_{i=1}^{2n} \frac{1}{2i - 1} = \ln 4 - \frac{\ln 2}{2} + \frac{\ln n}{2} + a_{4n} - \frac{a_{2n}}{2}.$$

(*d*) Show that

$$1 + \frac{1}{3} - \frac{1}{2} + \frac{1}{5} + \frac{1}{7} - \frac{1}{4} + \frac{1}{9} + \frac{1}{11} - \frac{1}{6}$$

$$+ \cdots + \frac{1}{4n - 3} + \frac{1}{4n - 1} - \frac{1}{2n}$$

equals

$$\ln 4 - \frac{\ln 2}{2} + a_{4n} - \frac{a_{2n}}{2} - \frac{a_n}{2}.$$

(*e*) From (*d*) deduce that the sum of (5) is $\frac{3}{2}$ ln 2. [In Exercise 28(*e*) of Sec. 10.7 it is shown that $1 - \frac{1}{2} + \frac{1}{3} - \frac{1}{4} + \cdots = \ln 2$. However, you may use the method of this exercise to obtain this fact as well.]

36 This exercise treats the second half of the absolute-ratio test.

(a) Show that if

$$\lim_{n\to\infty} \left| \frac{a_{n+1}}{a_n} \right| = L > 1.$$

then $|a_n| \to \infty$ as $n \to \infty$. *Suggestion:* First show that there is a number r, $r > 1$, such that for some integer N, $|a_{n+1}| > r|a_n|$ for all $n \geq N$.

(b) From (a) deduce that a_n does not approach 0 as $n \to \infty$.

37 (a) Give an example of a series that converges, but not absolutely.

(b) Give an example of a function $f(x)$ such that $\int_a^\infty f(x)\,dx$ converges but $\int_a^\infty |f(x)|\,dx$ diverges.

10.6 POWER SERIES

This section is concerned with series whose terms depend on x. Such series arise in approximating a function $f(x)$ by a polynomial $P(x)$.

Let a and x be real numbers and $\{a_n\}$ be a sequence. The series

$$\sum_{n=0}^{\infty} a_n(x-a)^n = a_0 + a_1(x-a) + a_2(x-a)^2 + \cdots + a_n(x-a)^n + \cdots \qquad (1)$$

Power series

is called a **power series in $x - a$** If $a = 0$, we obtain a **power series in x**,

$$\sum_{n=0}^{\infty} a_n x^n = a_0 + a_1 x + a_2 x^2 + \cdots + a_n x^n + \cdots. \qquad (2)$$

For instance, the geometric series

$$1 + x + x^2 + \cdots + x^n + \cdots \qquad (3)$$

The power series $1 + x + x^2 + x^3 + \cdots$ represents $1/(1-x)$ for $|x| < 1$.

is a power series. If the ratio x has absolute value less than 1, this series converges and has the sum $1/(1 - x)$:

$$\frac{1}{1-x} = \sum_{n=0}^{\infty} x^n \qquad |x| < 1. \qquad (4)$$

If $|x| \geq 1$, the geometric series (3) diverges by the nth-term test. As (4) illustrates, a power series may converge for some values of x and diverge for other values of x. In addition, Eq. (4) shows that some functions that are not polynomials may, however, be represented as power series, which we may think of as "polynomials of infinite degree."

As will be shown in Sec. 10.8, e^x, $\sin x$, and $\cos x$ can be represented by power series in x that converge for all values of x. Power series in x are frequently called **Maclaurin series**.

For each fixed choice of x, a power series becomes a series with constant terms.

The power series $a_0 + a_1x + a_2x^2 + \cdots$ certainly converges when $x = 0$. It may or may not converge for other choices of x. However, as Theorem 1 will show, if the series converges at a certain value c, it converges at any number x whose absolute value is less than $|c|$.

Theorem 1 Let c be a nonzero number. Assume that

$$a_0 + a_1 c + a_2 c^2 + \cdots$$

converges. Then, if $|x| < |c|$,

$$a_0 + a_1 x + a_2 x^2 + \cdots$$

converges. In fact, it converges absolutely.

Proof Since $\sum_{n=0}^{\infty} a_n c^n$ converges, the nth term $a_n c^n$ approaches 0 as $n \to \infty$. Thus there is an integer N such that for $n \geq N$, $|a_n c^n| \leq 1$. From now on in the proof, consider only $n \geq N$. Now,

$$a_n x^n = a_n c^n \left(\frac{x}{c}\right)^n.$$

Since

$$|a_n x^n| = |a_n c^n| \left|\frac{x}{c}\right|^n,$$

it follows that for $n \geq N$,

$$|a_n x^n| \leq \left|\frac{x}{c}\right|^n \qquad (\text{since } |a_n c^n| \leq 1 \text{ for } n \geq N).$$

The series

$$\sum_{n=N}^{\infty} \left|\frac{x}{c}\right|^n$$

is a geometric series with ratio $|x/c| < 1$. Hence it converges.

Since

$$|a_n x^n| \leq \left|\frac{x}{c}\right|^n$$

for $n \geq N$, the series

$$\sum_{n=N}^{\infty} |a_n x^n|$$

converges (by the comparison test). Thus $\sum_{n=N}^{\infty} a_n x^n$ converges (in fact, absolutely). Putting in the front end $\sum_{n=0}^{N-1} a_n x^n$, we conclude that the series $\sum_{n=0}^{\infty} a_n x^n$ converges absolutely if $|x| < |c|$. ■

The x's for which the series converges form an unbroken set.

By Theorem 1, the set of numbers x such that $\sum_{n=0}^{\infty} a_n x^n$ converges has no holes. In other words, the set of such x consists of one unbroken piece, which includes the number 0. Moreover if c is in that set, so is the entire open interval $(-|c|, |c|)$.

There are two possibilities. In the first case, there are arbitrarily large c such that the series converges in $(-c, c)$. This means that the series converges for all x. In the second case, there is a bound on the numbers c such that the series converges for x in $(-c, c)$. In this case, there is a smallest bound on the c's; call it R. Consequently, either

1 $a_0 + a_1 x + a_2 x^2 + \cdots$ converges for all x, or
2 there is a number R such that $a_0 + a_1 x + a_2 x^2 + \cdots$ converges for all x such that $|x| < R$ but diverges when $|x| > R$.

The radius of convergence

In the second case, R is called the **radius of convergence** of the series. In case 1, the radius of convergence is said to be infinite, $R = \infty$. For the geometric series $1 + x + x^2 + \cdots + x^n + \cdots$, $R = 1$, since the series converges when $|x| < 1$ and diverges when $|x| > 1$. (It also diverges when $x = 1$ and $x = -1$.) A power series with radius of convergence R may or may not converge at $x = R$ and at $x = -R$. For convenient reference, these observations are stated as Theorem 2.

Theorem 2 Associated with the power series $\sum_{n=0}^{\infty} a_n x^n$ is a radius of convergence R. If $R = 0$, the series converges only for $x = 0$. If R is a positive real number, the series converges for $|x| < R$ and diverges for $|x| > R$. If R is ∞, the series converges for all x. ■

EXAMPLE 1 Find all values of x for which $x - \dfrac{x^2}{2} + \dfrac{x^3}{3} - \dfrac{x^4}{4} + \cdots + \dfrac{(-1)^{n+1}x^n}{n} + \cdots$ converges.

SOLUTION Because of the presence of x^n, use the absolute-ratio test. The absolute value of the ratio of successive terms is

$$\left| \frac{\dfrac{(-1)^{n+2}x^{n+1}}{n+1}}{\dfrac{(-1)^{n+1}x^n}{n}} \right| = |x| \frac{n}{n+1}.$$

As $n \to \infty$, $n/(n + 1) \to 1$. Thus,

$$\text{if } |x| < 1, \qquad \lim_{n\to\infty} |x| \frac{n}{n+1} = |x| < 1;$$

$$\text{if } |x| > 1, \qquad \lim_{n\to\infty} |x| \frac{n}{n+1} = |x| > 1.$$

The absolute-ratio test takes care of $|x| < 1$ and $|x| > 1$.

By the absolute-ratio test, the series converges for $|x| < 1$ and diverges for $|x| > 1$. It remains to see what happens when $x = 1$ or $x = -1$.

Checking the behavior at $x = 1$

For $x = 1$, the alternating harmonic series

$$1 - \tfrac{1}{2} + \tfrac{1}{3} - \tfrac{1}{4} + \cdots$$

is obtained. This series converges, by the alternating-series test. Thus $x - x^2/2 + x^3/3 - x^4/4 + \cdots$ converges when $x = 1$. Hence the series $x - x^2/2 + x^3/3 - x^4/4 + \cdots$ converges for $-1 < x \le 1$.

Checking the behavior at $x = -1$

What about $x = -1$? The series then becomes

$$(-1) - \frac{(-1)^2}{2} + \frac{(-1)^3}{3} - \frac{(-1)^4}{4} + \cdots + \frac{(-1)^{n+1}(-1)^n}{n} + \cdots$$

or

$$-1 - \frac{1}{2} - \frac{1}{3} - \frac{1}{4} - \cdots - \frac{1}{n} - \cdots,$$

which, being the negative of the harmonic series, diverges.

The radius of convergence is $R = 1$. Figure 10.13 records the information obtained about the series.

The series $\sum_{n=1}^{\infty} (-1)^{n+1} \dfrac{x^n}{n}$

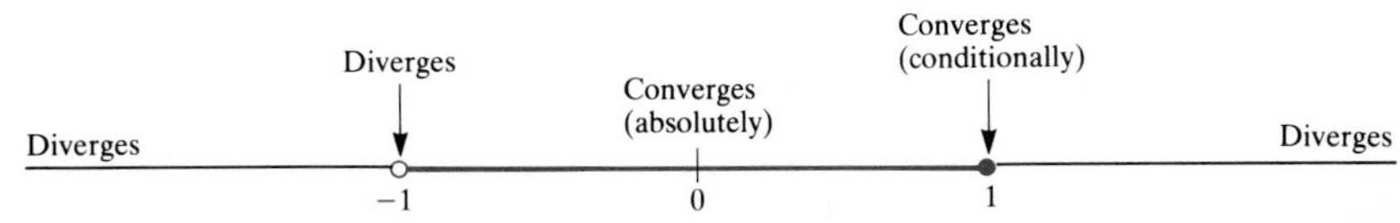

Figure 10.13 ■

EXAMPLE 2 Find the radius of convergence of

$$\sum_{n=0}^{\infty} \frac{x^n}{n!} = 1 + x + \frac{x^2}{2!} + \frac{x^3}{3!} + \cdots + \frac{x^n}{n!} + \cdots.$$

SOLUTION Because of the presence of the power x^n and the factorial $n!$ and the fact that x may be negative, the absolute-ratio test is the logical test to use. The absolute value of the ratio between successive terms is

$$\left| \frac{\dfrac{x^{n+1}}{(n+1)!}}{\dfrac{x^n}{n!}} \right| = |x| \frac{n!}{(n+1)!} = \frac{|x|}{n+1}.$$

Since

$$\lim_{n \to \infty} \frac{|x|}{n+1} = 0,$$

the limit of the ratio between successive terms is 0, which is less than 1. Consequently, the series converges for all x. That is, the radius of convergence R is infinite. ■

A case where $R = \infty$

The next example represents the opposite extreme, $R = 0$.

EXAMPLE 3 Find the radius of convergence of the series $\sum_{n=1}^{\infty} n^n x^n = 1x + 2^2x^2 + 3^3x^3 + \cdots + n^n x^n + \cdots$.

SOLUTION The series converges for $x = 0$.

If $x \neq 0$, consider the nth term $n^n x^n$, which can be written as $(nx)^n$. As $n \to \infty$, $|nx| \to \infty$. Thus the nth term does not approach 0 as $n \to \infty$. By the nth-term test, the series diverges. In short, the series converges only when $x = 0$. The radius of convergence in this case is $R = 0$. ■

A case where $R = 0$

Just as a power series in x has an associated radius of convergence, so does a power series in $x - a$. To see this, consider any such power series,

$$\sum_{n=0}^{\infty} a_n(x-a)^n = a_0 + a_1(x-a) + a_2(x-a)^2 + \cdots. \tag{5}$$

Let $u = x - a$. Then series (5) becomes

$$\sum_{n=0}^{\infty} a_n u^n = a_0 + a_1 u + a_2 u^2 + \cdots. \tag{6}$$

Series (6) has a certain radius of convergence R. That is, (6) converges for $|u| < R$ and diverges for $|u| > R$. Consequently (5) converges for $|x - a| < R$ and diverges for $|x - a| > R$. The number R is called the radius of convergence of the series (5). (R may be infinite.) As Fig. 10.14 suggests, the series $\sum_{n=0}^{\infty} a_n(x-a)^n$ converges in an interval whose midpoint is a. The question marks in Fig. 10.14 indicate that the series may converge or may diverge at the numbers $a - R$ and $a + R$. These cases must be looked at separately.

If $\sum_{n=a}^{\infty} a_n (x - a)^n$ *has radius of convergence R*

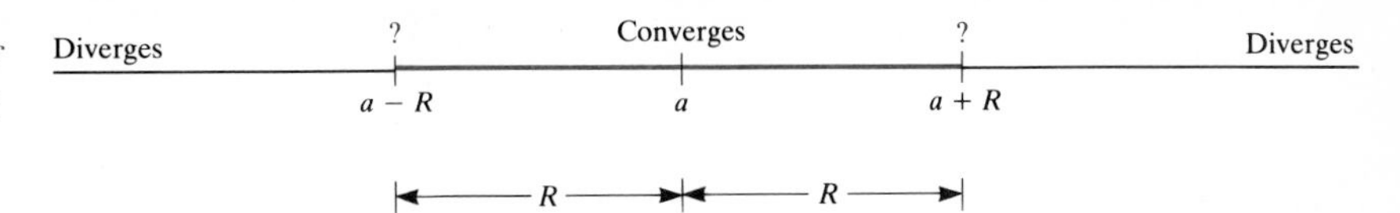

Figure 10.14

These observations are summarized in the following theorem.

Theorem 3 Associated with the power series $\Sigma_{n=0}^{\infty} a_n(x - a)^n$ is a radius of convergence R. If $R = 0$, the series converges only for $x = a$. If R is a positive real number, the series converges for $|x - a| < R$ and diverges for $|x - a| > R$. If $R = \infty$, the series converges for all x. ■

EXAMPLE 4 Find all values of x for which

$$(x - 1) - \frac{(x - 1)^2}{2} + \frac{(x - 1)^3}{3} - \frac{(x - 1)^4}{4} + \cdots \tag{7}$$

converges.

SOLUTION Note that this is Example 1 with x replaced by $x - 1$. Thus $x - 1$ plays the role that x played in Example 1. Consequently, series (7) converges for $-1 < x - 1 \leq 1$, that is, for $0 < x \leq 2$ and diverges for all other values of x. Its radius of convergence is $R = 1$. The set of values where the series converges is an interval whose midpoint is $x = 1$. The convergence of (7) is recorded in Fig. 10.15.

The series $\sum_{n=1}^{\infty} {}^{n+1} \frac{(x - 1)^n}{n}$

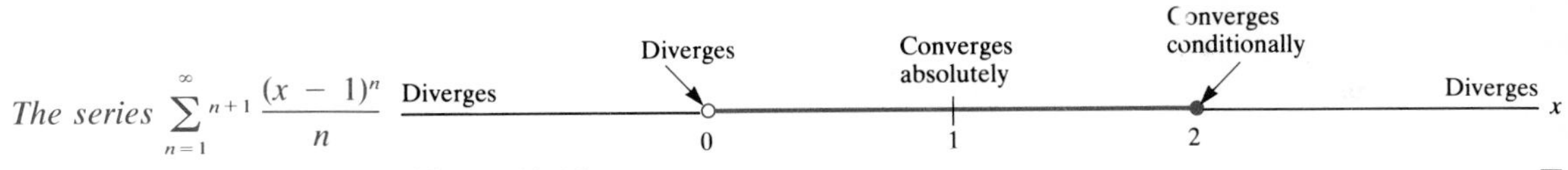

Figure 10.15 ■

EXAMPLE 5 Find the radius of convergence of

$$\sum_{n=0}^{\infty} n \left(\frac{x - 3}{2}\right)^n. \tag{8}$$

SOLUTION The absolute value of the ratio of successive terms is

$$\left| \frac{(n + 1)\left(\frac{x - 3}{2}\right)^{n+1}}{n\left(\frac{x - 3}{2}\right)^n} \right| = \left(1 + \frac{1}{n}\right) \left|\frac{x - 3}{2}\right|.$$

Since $1 + 1/n$ approaches 1 as $n \to \infty$, we see by the absolute-ratio test that series (8) converges (absolutely) if

$$\left|\frac{x - 3}{2}\right| < 1.$$

and diverges if

$$\left|\frac{x-3}{2}\right| > 1.$$

The inequality $|x - 3|/2 < 1$ is equivalent to $|x - 3| < 2$, which means that $R = 2$. Thus the series converges for

$$-2 < x - 3 < 2,$$

or, upon addition of 3 to both sides,

$$1 < x < 5.$$

What happens at $x = 1$ and $x = 5$?

When $x = 1$, series (8) becomes $\Sigma_{n=1}^{\infty} n(-1)^n$, which diverges, since its nth term does not approach 0 as $n \to \infty$. When $x = 5$, we obtain $\Sigma_{n=1}^{\infty} n$, which diverges for the same reason. Figure 10.16 records this information.

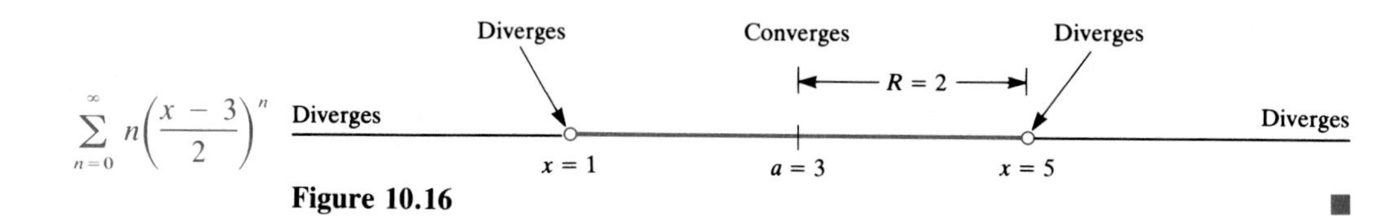

Figure 10.16 ■

EXERCISES FOR SEC. 10.6: POWER SERIES

In Exercises 1 to 12 draw the appropriate diagrams (like Fig. 10.15 or 10.16) showing where the series converge or diverge. Explain your work.

1 $\sum_{n=1}^{\infty} \frac{x^n}{n^2}$ 2 $\sum_{n=1}^{\infty} \frac{x^n}{\sqrt{n}}$

3 $\sum_{n=0}^{\infty} \frac{x^n}{3^n}$ 4 $\sum_{n=1}^{\infty} nx^n$

5 $\sum_{n=0}^{\infty} \frac{2n^2+1}{n^2-5} x^n$ 6 $\sum_{n=1}^{\infty} \frac{x^n}{n}$

7 $\sum_{n=0}^{\infty} \frac{x^n}{(2n)!}$ 8 $\sum_{n=0}^{\infty} \frac{2^n x^n}{n!}$

9 $\sum_{n=0}^{\infty} \frac{x^n}{(2n+1)!}$ 10 $\sum_{n=0}^{\infty} n!x^n$

11 $\sum_{n=1}^{\infty} \frac{(-1)^{n+1}x^n}{n}$ 12 $\sum_{n=1}^{\infty} \frac{2^n x^n}{n}$

13 Assume that $\Sigma_{n=0}^{\infty} a_n x^n$ converges when $x = 9$ and diverges when $x = -12$. What, if anything, can be said about
(*a*) convergence when $x = 7$?
(*b*) absolute convergence when $x = -7$?
(*c*) absolute convergence when $x = 9$?
(*d*) convergence when $x = -9$?
(*e*) divergence when $x = 10$?
(*f*) divergence when $x = -15$?
(*g*) divergence when $x = 15$?

14 Assume that $\Sigma_{n=0}^{\infty} a_n x^n$ converges when $x = -5$ and diverges when $x = 8$. What, if anything, can be said about
(*a*) convergence at $x = 4$?
(*b*) absolute convergence at $x = 4$?
(*c*) convergence at $x = 7$?
(*d*) absolute convergence at $x = -5$?
(*e*) convergence at $x = -9$?
(*f*) convergence at $x = -8$?

15 If the series $\Sigma_{n=0}^{\infty} a_n x^n$ converges whenever x is positive, must it converge whenever x is negative?

16 If $\Sigma_{n=0}^{\infty} a_n 6^n$ converges, what can be said about the convergence of

(*a*) $\sum_{n=0}^{\infty} a_n(-6)^n$? (*b*) $\sum_{n=0}^{\infty} a_n 5^n$? (*c*) $\sum_{n=0}^{\infty} a_n(-5)^n$?

In Exercises 17 to 24 draw the appropriate diagrams showing where the series converge and diverge.

17 $\sum_{n=0}^{\infty} \frac{(x-2)^n}{n!}$

18 $\sum_{n=1}^{\infty} \frac{(x-1)^n}{n3^n}$

19 $\sum_{n=0}^{\infty} \frac{(x-1)^n}{n+3}$

20 $\sum_{n=0}^{\infty} \frac{(x-4)^n}{2n+1}$

21 $\sum_{n=1}^{\infty} \frac{n(x-2)^n}{2n+3}$

22 $\sum_{n=2}^{\infty} \frac{(x-5)^n}{n \ln n}$

23 $\sum_{n=0}^{\infty} \frac{(x+3)^n}{5^n}$

24 $\sum_{n=1}^{\infty} n(x+1)^n$

■

25 (*a*) If the power series $\Sigma_{n=0}^{\infty} a_n x^n$ diverges when $x = 3$, at which x must it diverge?
(*b*) If the power series $\Sigma_{n=0}^{\infty} a_n(x+5)^n$ diverges when $x = -2$, at which x must it diverge?

26 If $\Sigma_{n=0}^{\infty} a_n(x-3)^n$ converges for $x = 7$, at what other values of x must the series necessarily converge?

27 Find the radius of convergence of $\Sigma_{n=0}^{\infty} x^{2n+1}/(2n+1)!$.

28 (*a*) Letting $a_0 = 1$, $a_1 = -x^2/2!, \ldots$, find the nth term of the series

$$1 - \frac{x^2}{2!} + \frac{x^4}{4!} - \frac{x^6}{6!} + \cdots.$$

(*b*) Show that the series in (*a*) converges for all x.

29 (The binomial series) The binomial theorem in Appendix C asserts that if n is a positive integer,

$$(1+x)^n = 1 + nx + \frac{n(n-1)}{2}x^2 + \cdots + x^n.$$

Newton generalized the binomial theorem to exponents other than positive integers, as follows:

Let a be a real number that is not 0 or a positive integer and form the series

$$1 + \frac{a}{1}x + \frac{a(a-1)}{1 \cdot 2}x^2 + \frac{a(a-1)(a-2)}{1 \cdot 2 \cdot 3}x^3 + \cdots.$$

(*a*) Show that, for $x \neq 0$, no term is 0.
(*b*) Show that, for $a = -1$ and $|x| < 1$, the sum of the series is $(1+x)^a$.
(*c*) Show that the series converges absolutely (hence converges) whenever $|x| < 1$.

In Sec. 10.12 it is shown that the series converges to $(1+x)^a$ whenever $|x| < 1$.

30 If $\Sigma_{n=0}^{\infty} a_n x^n$ has a radius of convergence 3 and $\Sigma_{n=0}^{\infty} b_n x^n$ has a radius convergence 5, what can be said about the radius of convergence of $\Sigma_{n=0}^{\infty} (a_n + b_n)x^n$?

31 Using the ideas of Exercise 9 in Sec. 8.6, show that for $0 \le x < 1$,

$$x - \frac{x^2}{2} + \frac{x^3}{3} - \frac{x^4}{4} + \cdots = \ln(1+x).$$

32 Using the ideas of Exercise 10 in Sec. 8.6, show that for $0 \le x < 1$,

$$x - \frac{x^3}{3} + \frac{x^5}{5} - \frac{x^7}{7} + \cdots = \tan^{-1} x.$$

10.7 MANIPULATING POWER SERIES

In advanced calculus it is proved that within its interval of convergence a power series behaves in many ways like a polynomial. In particular, it can be differentiated term by term,

$$(a_0 + a_1x + a_2x^2 + a_3x^3 + \cdots)' = a_1 + 2a_2x + 3a_3x^2 + \cdots.$$

In Sec. 3.4 it was shown that the sum of a *finite* number of functions can be differentiated term by term. The proof of this for power series is far more involved. In Theorem 1 the differentiation rule for power series in x is stated precisely. (A similar theorem holds for power series in $x - a$.)

Theorem 1 *Differentiating a power series.* Assume that $R > 0$ and that $\Sigma_{n=0}^{\infty} a_n x^n$ converges to $f(x)$ for $|x| < R$. Then for $|x| < R$, f is differentiable, $\Sigma_{n=1}^{\infty} na_n x^{n-1}$ converges, and

$$f'(x) = a_1 + 2a_2x + 3a_3x^2 + \cdots. \quad ■$$

The proof of Theorem 1 is left to advanced calculus.

This theorem is *not* covered by the fact that the derivative of the sum of a *finite* number of functions is the sum of their derivatives.

EXAMPLE 1 Obtain a power series for the function $1/(1 - x)^2$ from that for $1/(1 - x)$.

SOLUTION From the formula for the sum of a geometric series, we know that

$$\frac{1}{1 - x} = 1 + x + x^2 + x^3 + \cdots \qquad \text{for } |x| < 1.$$

According to Theorem 1, if we differentiate both sides of this equation, we obtain a true equation, namely,

$$\frac{1}{(1 - x)^2} = 0 + 1 + 2x + 3x^2 + \cdots \qquad \text{for } |x| < 1.$$

Thus

$$\frac{1}{(1 - x)^2} = 1 + 2x + 3x^2 + \cdots = \sum_{n=0}^{\infty} (n + 1)x^n \qquad \text{for } |x| < 1. \quad ■$$

Suppose that $f(x)$ has a power-series representation $a_0 + a_1x + a_2x^2 + \cdots$; Theorem 1 enables us to find what the coefficients a_0, a_1, a_2, . . . must be. The formula for a_n appears in Theorem 2. In this theorem, $f^{(n)}$ denotes the nth derivative of f; $f^{(1)} = f'$, $f^{(0)}$ stands for f itself, and $0! = 1$.

Theorem 2 *Formula for a_n.* Let R be a positive number and suppose that $f(x)$ is represented by the power series $\Sigma_{n=0}^{\infty} a_nx^n$ for $|x| < R$, that is,

$$f(x) = a_0 + a_1x + \cdots + a_nx^n + \cdots \qquad \text{for } |x| < R.$$

Then

$$a_n = \frac{f^{(n)}(0)}{n!}. \tag{1}$$

Proof When $x = 0$ we obtain $f(0) = a_0 + a_10 + a_20^2 + \cdots$. Hence

$$f(0) = a_0,$$

Compare this to the argument in Example 3 of Sec. 4.3.

which is (1) for $n = 0$. To obtain a_1, differentiate $f(x)$ and get

$$f^{(1)}(x) = a_1 + 2a_2x + 3a_3x^2 + \cdots + na_nx^{n-1} + \cdots. \tag{2}$$

Set $x = 0$ in (2) and obtain $\quad f^{(1)}(0) = a_1.$

This establishes (1) for $n = 1$.

Getting a_2

To obtain a_2, differentiate (2) and get

$$f^{(2)}(x) = 2a_2 + 3 \cdot 2a_3x + \cdots + n(n - 1)a_nx^{n-2} + \cdots. \tag{3}$$

Letting $x = 0$ gives $\quad f^{(2)}(0) = 2a_2.$

Hence

$$a_2 = \frac{f^{(2)}(0)}{2},$$

and (1) is established for $n = 2$.

Getting a_3

To obtain a_3, differentiate (3) as follows:

$$f^{(3)}(x) = 3 \cdot 2a_3 + 4 \cdot 3 \cdot 2a_4x + \cdots + n(n - 1)(n - 2)a_nx^{n-3} + \cdots. \tag{4}$$

Set $x = 0$, obtaining $$f^{(3)}(0) = 3 \cdot 2a_3,$$

or $$a_3 = \frac{f^{(3)}(0)}{3!}.$$

This establishes (1) for $n = 3$ and also shows why the factorial appears in the denominator of (1). The reader should differentiate (4) and verify (1) for $n = 4$. The argument applies for all n and can be completed by induction. ■

Just keep on differentiating.

Theorem 2 also tells us that there can be only one series of the form $\Sigma_{n=0}^{\infty} a_n x^n$ that represents $f(x)$, for the coefficients a_n are completely determined by the function $f(x)$ and its derivatives.

The next example applies Theorem 2 to a particular function, e^x. In this example it is assumed that e^x can be represented by a power series. This assumption will be justified in Sec. 10.8.

EXAMPLE 2 If e^x can be represented as a power series in x, what must that series be?

SOLUTION Let $e^x = a_0 + a_1 x + a_2 x^2 + \cdots + a_n x^n + \cdots$ for all x. By Theorem 2,

$$a_n = \frac{n\text{th derivative of } e^x \text{ at } 0}{n!}.$$

Now $(e^x)' = e^x$, $(e^x)'' = e^x$, and so on. All the higher derivatives of e^x are e^x again. At $x = 0$ they all have the value 1. Thus

$$a_n = \frac{1}{n!}$$

If e^x can be represented as a power series, we must have

Power series for e^x

$$e^x = \frac{1}{0!} + \frac{x}{1!} + \frac{x^2}{2!} + \frac{x^3}{3!} + \frac{x^4}{4!} + \cdots + \frac{x^n}{n!} + \cdots = \sum_{n=0}^{\infty} \frac{x^n}{n!}.$$

(In Sec. 10.8 it will be shown that e^x is indeed represented by this series.) ■

Calculations similar to those in Example 2 show that *if* $\sin x$ is representable as a power series in x, then

Power series for $\sin x$

$$\sin x = x - \frac{x^3}{3!} + \frac{x^5}{5!} - \frac{x^7}{7!} + \cdots = \sum_{n=0}^{\infty} \frac{(-1)^n x^{2n+1}}{(2n+1)!}. \tag{5}$$

Differentiation of (5) then yields

Power series for $\cos x$

$$\cos x = 1 - \frac{x^2}{2!} + \frac{x^4}{4!} - \frac{x^6}{6!} + \cdots = \sum_{n=0}^{\infty} \frac{(-1)^n x^{2n}}{(2n)!}. \tag{6}$$

The next theorem justifies the term-by-term integration of power series.

Theorem 3 *Integrating a power series.* Assume that $R > 0$ and

$$f(x) = a_0 + a_1x + a_2x^2 + \cdots + a_nx^n + \cdots \qquad \text{for } |x| < R.$$

Then
$$a_0x + \frac{a_1x^2}{2} + \frac{a_2x^3}{3} + \cdots + \frac{a_nx^{n+1}}{n+1} + \cdots$$

converges for $|x| < R$, and

$$\int_0^x f(t)\,dt = a_0x + \frac{a_1x^2}{2} + \frac{a_2x^3}{3} + \cdots. \quad ■$$

Note that the t is used to avoid writing $\int_0^x f(x)\,dx$, an expression in which x describes both the interval of integration $[0, x]$ and the independent variable of the function. To avoid use of x with two different meanings, a different letter should be used in the integrand. The next example demonstrates the power of Theorem 3.

EXAMPLE 3 Integrate the power series for $1/(1 + x)$ to obtain a power series for $\ln(1 + x)$.

SOLUTION Start with the series $1/(1- x) = 1 + x + x^2 + \cdots$ for $|x| < 1$.
Replace x by $-x$ and obtain

$$\frac{1}{1+x} = 1 - x + x^2 - x^3 + x^4 - \cdots \qquad \text{for } |x| < 1.$$

By Theorem 3,
$$\int_0^x \frac{dt}{1+t} = x - \frac{x^2}{2} + \frac{x^3}{3} - \frac{x^4}{4} + \cdots \qquad \text{for } |x| < 1.$$

Now,
$$\int_0^x \frac{dt}{1+t} = \ln(1+t)\Big|_0^x$$
$$= \ln(1+x) - \ln(1+0)$$
$$= \ln(1+x).$$

The power series for $\ln(1 + x)$

Thus
$$\ln(1+x) = x - \frac{x^2}{2} + \frac{x^3}{3} - \frac{x^4}{4} + \cdots \qquad \text{for } |x| < 1. \quad ■$$

One reason for representing a function as a power series is to be able to estimate its values to any desired degree of accuracy. The next example illustrates this by computing ln 1.2 to several decimal places.

EXAMPLE 4 Find ln 1.2 to three decimal places.

SOLUTION By the preceding example, with $x = 0.2$,

$$\ln 1.2 = 0.2 - \frac{(0.2)^2}{2} + \frac{(0.2)^3}{3} - \frac{(0.2)^4}{4} + \cdots \tag{7}$$

Making the error so small should give us "three-decimal accuracy."

How many terms of Eq. (7) should we use to obtain an estimate with an error less than 0.0001? Since Eq. (7) is an alternating decreasing series, the error is less than the absolute value of the first omitted term. So let us find the first term whose absolute value is less than 0.0001.

We are seeking a term $(0.2)^m/m$ less than 0.0001. For simplicity, let us make $(0.2)^m$ less than 0.0001 [it would follow that $(0.2)^m/m$ would also be less than 0.0001]. For which m is $(0.2)^m < 0.0001$? Testing $m = 1$, 2, 3, 4, 5, and 6 with a little arithmetic shows that $m = 6$ meets our demand, since $(0.2)^6 \approx 0.00006$. Thus

Or check $(0.2)^m/m$ for $m = 1, 2, \cdots$ on a calculator

$$0.2 - \frac{(0.2)^2}{2} + \frac{(0.2)^3}{3} - \frac{(0.2)^4}{4} + \frac{(0.2)^5}{5} \tag{8}$$

is an estimate of ln 1.2 with an error less than 0.0001. Since the next term is negative, (8) overestimates ln 1.2. Thus

$$0.2 - \frac{(0.2)^2}{2} + \frac{(0.2)^3}{3} - \frac{(0.2)^4}{4} + \frac{(0.2)^5}{5} - \frac{(0.2)^6}{6} < \ln 1.2$$
$$< 0.2 - \frac{(0.2)^2}{2} + \frac{(0.2)^3}{3} - \frac{(0.2)^4}{4} + \frac{(0.2)^5}{5}.$$

A little arithmetic gives us, therefore,

$$0.18232 < \ln 1.2 < 0.18234.$$

So, to four places, ln 1.2 is 0.1823, and to three places it is 0.182. ■

In addition to differentiating and integrating power series, we may also add, subtract, multiply, and divide them just like polynomials. The following theorem states the rules for these operations. The first two are easy to establish; proofs of the latter two are reserved for an advanced calculus course.

Theorem 4 *The algebra of power series.* Assume that

$$f(x) = a_0 + a_1x + a_2x^2 + \cdots \quad \text{and} \quad g(x) = b_0 + b_1x + b_2x^2 + \cdots$$

for $|x| < R$. Then for $|x| < R$,
(*a*) $f(x) + g(x) = \sum_{n=0}^{\infty} (a_n + b_n)x^n$.
(*b*) $f(x) - g(x) = \sum_{n=0}^{\infty} (a_n - b_n)x^n$.
(*c*) $f(x)g(x) = a_0b_0 + (a_0b_1 + a_1b_0)x + (a_0b_2 + a_1b_1 + a_2b_0)x^2 + \cdots$
(*d*) $f(x)/g(x)$ is obtainable by long division, if $g(x) \neq 0$ for $|x| < R$. ■

Two examples will illustrate the usefulness of Theorem 4.

EXAMPLE 5 Find the first few terms of the power series in x for $e^x/(1 - x)$.

SOLUTION By Theorem 4, we just multiply the series as we would polynomials:

$$e^x \frac{1}{1-x} = \left(1 + x + \frac{x^2}{2!} + \frac{x^3}{3!} + \cdots\right)(1 + x + x^2 + x^3 + \cdots)$$
$$= 1 \cdot 1 + (1 \cdot 1 + 1 \cdot 1)x + \left(1 \cdot 1 + 1 \cdot 1 + \frac{1}{2!} \cdot 1\right)x^2$$
$$+ \left(1 \cdot 1 + 1 \cdot 1 + \frac{1}{2!} \cdot 1 + \frac{1}{3!} \cdot 1\right)x^3 + \cdots$$
$$= 1 + 2x + \tfrac{5}{2}x^2 + \tfrac{8}{3}x^3 + \cdots \qquad |x| < 1.$$ ■

EXAMPLE 6 Find the first four terms of the power series in x for $e^x/\cos x$.

SOLUTION Write down the power series in x for e^x and $\cos x$ up through the term of degree 3 and arrange the long division as follows:

$$\begin{array}{r|l}
 & 1 + x + x^2 + \frac{2x^3}{3} + \cdots \\
\hline
1 + 0x - \frac{x^2}{2} + 0x^3 + \cdots & 1 + x + \frac{x^2}{2} + \frac{x^3}{6} + \cdots \\
 & 1 + 0x - \frac{x^2}{2} + 0x^3 + \cdots \\
\cline{2-2}
 & \quad x + x^2 + \frac{x^3}{6} + \cdots \\
 & \quad x + 0x^2 - \frac{x^3}{2} + \cdots \\
\cline{2-2}
 & \qquad x^2 + \frac{2x^3}{3} + \cdots \\
 & \qquad x^2 + 0x^3 - \cdots \\
\cline{2-2}
 & \qquad\quad \frac{2x^3}{3} + \cdots \\
 & \qquad\quad \frac{2x^3}{3} + \cdots \\
\cline{2-2}
 & \qquad\qquad \cdots
\end{array}$$

Thus the power series in x for $e^x/\cos x$ begins

$$1 + x + x^2 + \frac{2x^3}{3}. \quad ■$$

Section 6.9 presented l'Hôpital's rule as a method for dealing with $\lim_{x\to a} f(x)/g(x)$ when both numerator and denominator approach 0 as $x \to a$. The next example shows that to calculate such limits, power series may on occasion be far more efficient.

EXAMPLE 7 Find $\lim_{x\to 0} \dfrac{\sin^2 x - x^2}{(e^{x^2} - 1)^2}$.

SOLUTION Since

$$\sin x = x - \frac{x^3}{6} + \frac{x^5}{120} - \cdots,$$

the power series in x for the numerator begins

$$\begin{aligned}
\sin^2 x - x^2 &= \left(x - \frac{x^3}{6} + \frac{x^5}{120} - \cdots\right)^2 - x^2 \\
&= \left(x^2 - \frac{x^4}{3} + \cdots\right) - x^2 \\
&= -\frac{x^4}{3} + \text{terms of degree more than 4.}
\end{aligned}$$

$\frac{-x^4}{3}$ is the lead term.

To develop the power series for the denominator, first replace x in the relation

$$e^x = 1 + x + \frac{x^2}{2} + \frac{x^3}{6} + \cdots$$

by x^2 and obtain $$e^{x^2} = 1 + x^2 + \frac{x^4}{2} + \frac{x^6}{6} + \cdots.$$

Thus $$(e^{x^2} - 1)^2 = \left(x^2 + \frac{x^4}{2} + \frac{x^6}{6} + \cdots\right)^2$$

x^4 is the lead term.

$$= x^4 + \text{terms of higher degree.}$$

Hence $$\frac{\sin^2 x - x^2}{(e^{x^2} - 1)^2} = \frac{-x^4/3 + \text{terms of degree more than 4}}{x^4 + \text{terms of degree more than 4}}$$

$$= \frac{x^4(-\frac{1}{3} + \text{terms of degree at least 1})}{x^4(1 + \text{terms of degree at least 1})}.$$

Cancel the x^4 in numerator and denominator and obtain

$$\frac{\sin^2 x - x^2}{(e^{x^2} - 1)^2} = \frac{-\frac{1}{3} + \text{terms of degree at least 1}}{1 + \text{terms of degree at least 1}}.$$

Thus $$\lim_{x \to 0} \frac{\sin^2 x - x^2}{(e^{x^2} - 1)^2} = \frac{-\frac{1}{3}}{1} = -\frac{1}{3}.$$

The reader may find the limit by l'Hôpital's rule, but the computations are messier. ■

Power series in $x - a$

The various theorems and methods of this section were stated for power series in x. But analogous theorems hold for power series in $x - a$. Such series may be differentiated and integrated inside the interval in which they converge. For instance, Theorem 2 generalizes to the following assertion.

Theorem 5 *Formula for a_n.* Let R be a positive number and suppose that $f(x)$ is represented by $\Sigma_{n=0}^{\infty} a_n(x - a)^n$ for $|x - a| < R$. Then

$$a_n = \frac{f^{(n)}(a)}{n!}. \quad ■ \tag{9}$$

The proof is similar to that of Theorem 2: Differentiate n times and replace x by a.

The next example illustrates the use of formula (9).

EXAMPLE 8 If $\sin x$ can be represented as a series in powers of $x - \pi/4$ for all x, find what the series must be.

SOLUTION In this case, $f(x) = \sin x$ and $a = \pi/4$. In order to use formula (9), it is necessary to evaluate all the derivatives of $\sin x$ at $\pi/4$. This table records the computations:

n	$f^{(n)}(x)$	$f^{(n)}(\pi/4)$	$a_n = \frac{f^{(n)}(\pi/4)}{n!}$
0	$\sin x$	$\sqrt{2}/2$	$(\sqrt{2}/2)/0!$
1	$\cos x$	$\sqrt{2}/2$	$(\sqrt{2}/2)/1!$
2	$-\sin x$	$-\sqrt{2}/2$	$-(\sqrt{2}/2)/2!$
3	$-\cos x$	$-\sqrt{2}/2$	$-(\sqrt{2}/2)/3!$
4	$\sin x$	$\sqrt{2}/2$	$(\sqrt{2}/2)/4!$
5	$\cos x$	$\sqrt{2}/2$	$(\sqrt{2}/2)/5!$
..	......	....	....

The higher derivatives of $\sin x$ repeat in blocks of four:

$$\sin x, \qquad \cos x, \qquad -\sin x, \qquad -\cos x.$$

The series for $\sin x$ in powers of $x - \pi/4$ begins, therefore,

$$\sin x = \frac{\sqrt{2}}{2} + \frac{\sqrt{2}}{2}\left(x - \frac{\pi}{4}\right) - \frac{\sqrt{2}/2}{2!}\left(x - \frac{\pi}{4}\right)^2 - \frac{\sqrt{2}/2}{3!}\left(x - \frac{\pi}{4}\right)^3 + \frac{\sqrt{2}/2}{4!}\left(x - \frac{\pi}{4}\right)^4 + \cdots.$$

Two +'s continue to alternate with two −'s. ■

HOW SOME CALCULATORS FIND e^x

The power series in x for e^x is

$$1 + x + \frac{x^2}{2} + \frac{x^3}{3} + \cdots + \frac{x^n}{n} + \cdots. \qquad (10)$$

For $x = 10$, this would give

$$e^{10} = 1 + 10 + \frac{10^2}{2!} + \frac{10^3}{3!} + \frac{10^4}{4!} + \cdots + \frac{10^n}{n!} + \cdots.$$

Although the terms eventually become very small, the first few terms are quite large. (For instance, the fifth term, $10^4/4!$, is about 417.) So when x is large, series (10) provides a time-consuming procedure for calculating e^x.

Some calculators use the following method instead.

The values of e^x at certain inputs are built into the memory:

$$\begin{aligned} e^1 &\approx 2.718281828459, \\ e^{10} &\approx 22{,}026.46579, \\ e^{100} &\approx 2.6881171 \times 10^{43}, \\ e^{0.1} &\approx 1.1051709181, \\ e^{0.01} &\approx 1.0100501671, \\ e^{0.001} &\approx 1.0010005002. \end{aligned}$$

[This may be done with the aid of series (10).] Then, to find $e^{315.425}$, say, the calculator computes

$$(e^{100})^3(e^{10})^1(e^1)^5(e^{0.1})^4(e^{0.01})^2(e^{0.001})^5.$$

Here it makes use of the identities $e^{x+y} = e^x e^y$ and $(e^x)^y = e^{xy}$. [Similarly, $\sin x$ and $\cos x$ can be found by using the identities for $\sin(x + y)$ and $\cos(x + y)$.]

EXERCISES FOR SEC. 10.7: MANIPULATING POWER SERIES

1 This exercise concerns the power series in x for $\sin x$.

(*a*) Copy and fill in this table for $f(x) = \sin x$.

n	$f^{(n)}(x)$	$f^{(n)}(0)$	$f^{(n)}(0)/n!$
0			
1			
2			
3			
4			
5			

(*b*) Use (*a*) to write out the first six terms of the power series in x for $\sin x$ (including terms that are 0).

(*c*) Show that the nth nonzero term in the power series in (*b*) is

$$(-1)^{n+1}\frac{x^{2n-1}}{(2n-1)!}.$$

2 Carry out the analog of Exercise 1 for $f(x) = \cos x$.

3 Find the power series in x for $\cos x$ by differentiating that for $\sin x$. [See Equations (5) and (6).]

4 Differentiate the series $1 + x + x^2/2! + \cdots + x^n/n! + \cdots$ and show that the derivative equals the given series.

In Exercises 5 to 8 determine with the aid of Theorem 2 the first three nonzero terms of the power series in x for the given functions.

5 $\tan x$ **6** $\tan^{-1} x$

7 $\sin^{-1} x$ **8** $\sqrt{1 + x}$

9 (*a*) Show that for $|t| < 1$,

$$\frac{1}{1+t^2} = 1 - t^2 + t^4 - t^6 + \cdots.$$

(*b*) Use Theorem 3 to show that for $|x| < 1$,

$$\tan^{-1} x = x - \frac{x^3}{3} + \frac{x^5}{5} - \frac{x^7}{7} + \cdots.$$

(*c*) Give the formula for the nth term of the series in (*b*).

10 Using Theorem 3, show that for $|x| < 1$,

(*a*) $$\int_0^x \frac{dt}{1+t^3} = x - \frac{x^4}{4} + \frac{x^7}{7} - \frac{x^{10}}{10} + \cdots.$$

(*b*) Give a formula for the nth term of the series in (*a*).

In Exercises 11 and 12 obtain the first three nonzero terms in the power series in x for the indicated functions by algebraic operations with known series.

11 $e^x \sin x$ **12** $\dfrac{x}{\cos x}$

In Exercises 13 to 16 use power series to determine the limits.

13 $\lim_{x\to 0} \dfrac{\sin^2 x^3}{(1 - \cos x^2)^3}$

14 $\lim_{x\to 0} \left[\dfrac{1}{\sin x} - \dfrac{1}{\ln(1+x)}\right]$

15 $\lim_{x\to 0} \dfrac{(e^{x^2} - 1)^3}{\sin^3 x^2}$ **16** $\lim_{x\to 0} \dfrac{\sin x\,(1 - \cos x)}{e^{x^3} - 1}$

In Exercises 17 to 20 use the formula $a_n = f^{(n)}(a)/n!$ to obtain the indicated series. Write out the first three nonzero terms.

17 Series for $\sin x$ in powers of $x - \pi/6$.

18 Series for $\cos x$ in powers of $x + \pi/4$.

19 Series for e^x in powers of $x - 1$.

20 Series for $\tan x$ in powers of $x - \pi/4$.

■

21 Let $f(x) = a_0 + a_1 x + a_2 x^2 + \cdots$ for $|x| < R$.

(*a*) If only even powers appear, that is, $a_n = 0$ for all odd n, show that $f(-x) = f(x)$, that is, f is an even function.

(*b*) If only odd powers appear, that is, $a_n = 0$ for all even n, show that $f(-x) = -f(x)$, that is, f is an odd function.

22 (*a*) Use the first five terms of the power series in x for e^x to estimate $e = e^1$.

(*b*) Show that the error in (*a*) is less than the sum of the geometric progression

$$\frac{1}{5!} + \frac{1}{6 \cdot 5!} + \frac{1}{6^2 \cdot 5!} + \cdots.$$

(*c*) Deduce from (*a*) and (*b*) that

$$2.708 < e < 2.719.$$

23 (*a*) Noting that $\sqrt{e} = e^{1/2}$, use the first five terms of the power series in x for e^x to estimate $\sqrt{e}$.

(*b*) Estimate the error in (*a*) by comparing it to the sum of a geometric progression.

24 Obtain formula (9) in Theorem 5 for $n = 0, 1, 2, 3$.

In Exercises 25 to 27 use a calculator.

25 (*a*) Use the first 10 terms of the series $e^x = \Sigma_{n=0}^{\infty} x^n/n!$ to estimate $e = e^1$.

(*b*) Show that the error in the estimate in (*a*) is less than $11/(10 \cdot 10!) \approx 0.0000003$.

26 (*a*) Use the first three nonzero terms of the power series in x for $\sin x$ to estimate $\sin \pi/5$ ($= \sin 36°$).

(*b*) Show that the error in the estimate in (*a*) is less than $(\pi/5)^7/7! \approx 0.000008$.

27 The integral $\int_0^1 e^{-x^2}\,dx$ cannot be evaluated by the fundamental theorem of calculus.

(*a*) Replacing x in the power series $e^x = \Sigma_{n=0}^{\infty} x^n/n!$ by $-x^2$, obtain the power series for e^{-x^2}.

(*b*) Show that

$$\int_0^1 e^{-x^2}\,dx = 1 - \frac{1}{3} + \frac{1}{5 \cdot 2!} - \frac{1}{7 \cdot 3!} + \cdots.$$

(*c*) Use (*b*) to estimate $\int_0^1 e^{-x^2}\,dx$ to three-decimal-place accuracy.

28 In this exercise the power series for $\ln(1 + x)$ will be obtained without borrowing from advanced calculus the result on integration of power series. It has the further advantage that it takes care of $x = 1$.

(*a*) Show that for $t \neq -1$,

$$\frac{1}{1+t} = 1 - t + \cdots + (-1)^{n-1}t^{n-1} + (-1)^n \frac{t^n}{1+t}.$$

(*b*) Use the identity in (*a*) to show that for $x > -1$,

$$\ln(1+x) = x - \frac{x^2}{2} + \frac{x^3}{3} - \cdots + (-1)^{n-1}\frac{x^n}{n} + (-1)^n \int_0^x \frac{t^n}{1+t}\,dt.$$

(*c*) Show that if x is in $[0, 1]$, then $\int_0^x t^n/(1+t)\,dt$ approaches 0 as $n \to \infty$. *Hint:* $1 + t \geq 1$.

(*d*) Show that if $-1 < x \leq 0$, then $\int_0^x t^n/(1+t)\,dt$ approaches 0 as $n \to \infty$. *Hint:* $1 + t \geq 1 + x$.

(*e*) Conclude that if $-1 < x \leq 1$, then

$$\ln(1+x) = x - \frac{x^2}{2} + \frac{x^3}{3} - \cdots + (-1)^{n-1}\frac{x^n}{n} + \cdots.$$

Note that $x = 1$ yields the alternating harmonic series and the equation

$$\ln 2 = 1 - \tfrac{1}{2} + \tfrac{1}{3} - \tfrac{1}{4} + \cdots + (-1)^{n+1}\frac{1}{n} + \cdots$$

29 In this exercise the power series for $\tan^{-1} x$ will be obtained without using the result from advanced calculus concerning the integration of power series. Moreover, it shows that $\tan^{-1} x = x - x^3/3 + x^5/5 - \cdots$, even when $|x| = 1$.

(*a*) Using the identity in Exercise 28(*a*), show that

$$\frac{1}{1+t^2} = 1 - t^2 + \cdots + (-1)^{n-1}t^{2n-2} + (-1)^n \frac{t^{2n}}{1+t^2}.$$

(*b*) From (*a*) deduce that

$$\tan^{-1} x = x - \frac{x^3}{3} + \frac{x^5}{5} - \cdots$$
$$+ (-1)^{n-1}\frac{x^{2n-1}}{2n-1} + (-1)^n \int_0^x \frac{t^{2n}}{1+t^2}\,dt.$$

(*c*) Show that if $0 \le x \le 1$, $\int_0^x t^{2n}/(1+t^2)\,dt \to 0$ as $n \to \infty$.

(*d*) From (*c*) deduce that for $|x| \le 1$,

$$\tan^{-1} x = x - \frac{x^3}{3} + \frac{x^5}{5} - \frac{x^7}{7} + \cdots.$$

(*e*) Show that

$$\sum_{n=0}^{\infty} \frac{(-1)^n}{2n+1} = 1 - \frac{1}{3} + \frac{1}{5} - \frac{1}{7} + \cdots = \frac{\pi}{4}.$$

30 This exercise presents a function so "flat" at the origin that all its derivatives are 0 there, yet it is not the constant function $f(x) = 0$.

Let $f(x) = e^{-1/x^2}$ if $x \neq 0$, and let $f(0) = 0$.

(*a*) Show that f is continuous.

(*b*) Show that f is differentiable.

(*c*) Show that $f^{(1)}(0) = 0$ and $f^{(2)}(0) = 0$.

(*d*) Explain why $f^{(n)}(0) = 0$ for all $n \ge 0$.

(*e*) Show that $f(x)$ is not representable by a power series in x with a nonzero radius of convergence.

31 What theorems in the text justify the assertion that

$$\lim_{x\to 0} (a_0 + a_1x + a_2x^2 + a_3x^3 + \cdots) = a_0?$$

Assume the series has a nonzero radius of convergence.

32 Show that $\sqrt{x}$ cannot be represented by a power series in x with a nonzero radius of convergence.

33 Show that $\sqrt[3]{x}$ cannot be represented by a power series in x with a nonzero radius of convergence.

10.8 TAYLOR'S FORMULA

In Sec. 10.7 it was pointed out that if $f(x)$ can be represented as a power series in $x - a$ over some interval around a, then that series must be

$$f(a) + f^{(1)}(a)(x-a) + \frac{f^{(2)}(a)}{2!}(x-a)^2 + \cdots + \frac{f^{(n)}(a)}{n!}(x-a)^n + \cdots. \tag{1}$$

Letting $f^{(0)}(x)$ denote $f(x)$ and recalling that $0! = 1$, we can rewrite series (1) in the form

$$\sum_{n=0}^{\infty} \frac{f^{(n)}(a)}{n!}(x-a)^n. \tag{2}$$

Series (2) may or may not sum to $f(x)$.

The series (2) is called the **Taylor series at $x = a$ associated with the function $f(x)$**. When $a = 0$, the series (2) is called the **Maclaurin series associated with $f(x)$**. To show that (2) equals $f(x)$, we must show that

$$\lim_{n\to\infty}\left[f(a) + f^{(1)}(a)(x-a) + \cdots + \frac{f^{(n)}(a)}{n!}(x-a)^n\right] = f(x).$$

In other words, it is necessary to show that

$$\lim_{n\to\infty}\left\{f(x) - \left[f(a) + f^{(1)}(a)(x-a) + \cdots + \frac{f^{(n)}(a)}{n!}(x-a)^n\right]\right\} = 0. \tag{3}$$

This section obtains short formulas for the expression in the braces in (3). With the aid of these formulas we then show, for instance, that $e^x = \sum_{n=0}^{\infty} x^n/n!$. (In Sec. 10.7 this equation was obtained with two assumptions: first, that e^x has a power series; second, that it is legal to differentiate a power series term by term.)

Definition *Taylor polynomial of degree n, $P_n(x; a)$.* If the function f has derivatives through order n at a, then the **Taylor polynomial of degree n of f at a** is defined as

$$f(a) + f'(a)(x - a) + \frac{f^{(2)}(a)}{2!}(x - a)^2 + \cdots + \frac{f^{(n)}(a)}{n!}(x - a)^n.$$

This polynomial of degree n is denoted $P_n(x; a)$. It is just the first $n + 1$ terms, a front end, of the Taylor series.

In Sec. 10.7 the power series $\sum_{n=0}^{\infty} a_n(x - a)^n$ was constructed to have the same derivatives at a as $f(x)$ has. (Recall that the coefficients a_n were found by repeated differentiation.) So the polynomial $P_n(x; a)$ can be described as that polynomial of degree at most n whose derivatives at a up through order n coincide with those of $f(x)$ at a. This means that P_n and f have the same value at a, the same slope at a, the same second derivative at a, . . . , the same nth derivative at a. (The approximating Taylor polynomials P_0, P_1, P_2, P_3, P_4 for e^x are shown in Fig. 10.17.)

EXAMPLE 1 Find the Taylor polynomial of degree 4 associated with e^x at $a = 0$.

SOLUTION In this case, $f(x) = e^x$. Repeated differentiation yields

$$f^{(1)}(x) = e^x, \qquad f^{(2)}(x) = e^x, \qquad f^{(3)}(x) = e^x, \qquad \text{and} \qquad f^{(4)}(x) = e^x.$$

At $x = 0$, all these derivatives have the value 1. The Taylor polynomial of degree 4 at 0 is therefore

$$P_4(x; 0) = 1 + \frac{1}{1!}(x - 0) + \frac{1}{2!}(x - 0)^2 + \frac{1}{3!}(x - 0)^3 + \frac{1}{4!}(x - 0)^4,$$

or simply,

$$1 + x + \frac{x^2}{2!} + \frac{x^3}{3!} + \frac{x^4}{4!}. \quad ■$$

Taking front ends of $P_4(x; 0)$ gives us the Taylor polynomials of lower degrees:

$$P_0(x; 0) = 1,$$

$$P_1(x; 0) = 1 + x,$$

$$P_2(x; 0) = 1 + x + \frac{x^2}{2!},$$

$$P_3(x; 0) = 1 + x + \frac{x^2}{2!} + \frac{x^3}{3!}.$$

In Fig. 10.17 the function $f(x) = e^x$ and the Taylor polynomials up to degree 4 are graphed. Note that the higher the degree, the closer the Taylor polynomial approximates the function e^x near $x = 0$. However, since e^x grows much faster than any polynomial as $x \to \infty$, no fixed polynomial can approximate e^x closely throughout the x axis. Furthermore, as $x \to -\infty$, $e^x \to 0$, but a polynomial of degree at least 1 approaches ∞ or $-\infty$.

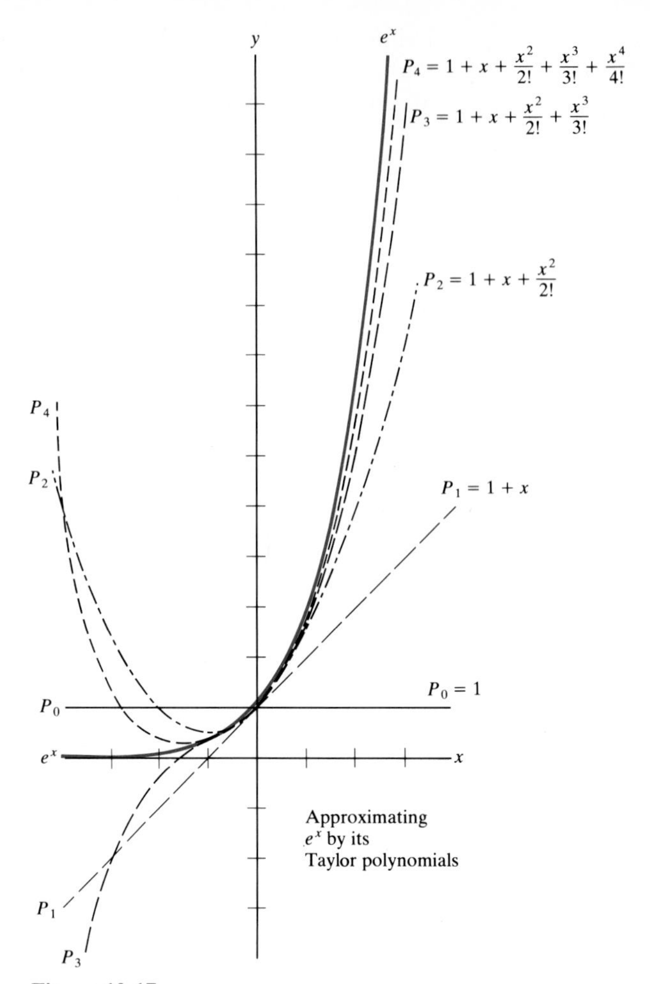

Figure 10.17

Definition *The remainder (error)* $R_n(x; a)$. Let f be a function and let $P_n(x; a)$ be the associated Taylor polynomial of degree n at a. The number $R_n(x; a)$ defined by the equation

$$f(x) = P_n(x; a) + R_n(x; a)$$

is called the **remainder** (or **error**) in using the Taylor polynomial $P_n(x; a)$ to approximate $f(x)$.

Our hope is that for well-behaved functions f,

In $P_n(x; a)$, think of a and x as fixed and $n \to \infty$.

$$P_n(x; a) \to f(x)$$

as $n \to \infty$. But $R_n(x; a) = f(x) - P_n(x; a)$, so we hope that $R_n(x; a) \to 0$ as $n \to \infty$. If the remainder $R_n(x; a)$ does approach 0 as $n \to \infty$, then the function $f(x)$ is represented by its Taylor series. Now, Theorem 1 expresses $R_n(x; a)$ as an integral; Theorem 2 expresses $R_n(x; a)$ in terms of a derivative. These formulas enable us to show that for $f(x) = e^x$, $\sin x$, or $\cos x$, $R_n(x; a)$ indeed approaches 0 as $n \to \infty$. They also can be used to establish general properties of functions.

The fundamental theorem of calculus asserts that if $f'(x)$ is continuous, then

This equation is the basis of our approach.

$$f(b) - f(a) = \int_a^b f'(x)\,dx,$$

or

$$f(b) = f(a) + \int_a^b f'(x)\,dx, \tag{4}$$

which says, "the error in using $f(a)$ to approximate $f(b)$ is expressed as an integral involving $f'(x)$." Now, $f(a)$ is the Taylor polynomial of degree 0 (which is constant). So the error in using $P_0(b; a)$ to approximate $f(b)$ is expressible as a definite integral involving $f^{(1)}(x)$. [We use the notation $P_0(b; a)$ instead of $P_0(x; a)$ because we are approximating $f(b)$.] Theorem 1 asserts that the error in using $P_n(b; a)$ to approximate $f(b)$ is expressible in terms of an integral involving $f^{(n+1)}(x)$. The proof of Theorem 1 depends on the following lemma, whose proof is a direct application of integration by parts.

Lemma Assume that the function $f(x)$ has continuous derivatives up through order $n + 1$ in the interval $[a, b]$. Then, for $n \geq 1$,

$$\int_a^b f^{(n)}(x)(b - x)^{n-1}\,dx = f^{(n)}(a)\frac{(b - a)^n}{n} + \frac{1}{n}\int_a^b f^{(n+1)}(x)(b - x)^n\,dx.$$

Proof Use integration by parts, with $u = f^{(n)}(x)$ and $dv = (b - x)^{n-1}\,dx$, $du = f^{(n+1)}(x)\,dx$, and $v = -(b - x)^n/n$. Then

$$\int_a^b \underbrace{f^{(n)}(x)}_{u}\underbrace{(b - x)^{n-1}\,dx}_{dv} = \underbrace{f^{(n)}(x)}_{u}\left[\underbrace{\frac{-(b - x)^n}{n}}_{v}\right]\Bigg|_a^b - \int_a^b \underbrace{-\frac{(b - x)^n}{n}}_{v}\underbrace{f^{(n+1)}(x)\,dx}_{du}$$

$$= f^{(n)}(a)\frac{(b - a)^n}{n} + \frac{1}{n}\int_a^b f^{(n+1)}(x)\,(b - x)^n\,dx,$$

and the lemma is established. ■

The lemma expresses $\int_a^b f^{(n)}(x)(b - x)^{n-1}\,dx$ as the sum of two terms: One term involves an integral with a similar integrand, in which n is replaced by $n + 1$, and the other involves $f^{(n)}(a)$. Every time the lemma is applied in the following proof it replaces an integral by another term in the Taylor series and an integral.

Theorem 1 *Integral form of the remainder.* Assume that a function f has continuous derivatives up through order $n + 1$ in the interval $[a, b]$. Let $P_n(b; a)$ be the Taylor polynomial of degree n associated with f in powers of $b - a$, and let $R_n(b; a)$ be the difference between $f(b)$ and the estimate $P_n(b; a)$:

$$f(b) = P_n(b; a) + R_n(b; a).$$

Then

$$R_n(b; a) = \frac{1}{n!}\int_a^b (b - x)^n f^{(n+1)}(x)\,dx. \tag{5}$$

Proof (We prove the theorem for $n = 0, 1, 2, 3$; mathematical induction would establish it for all n.)

For $n = 0$, the theorem asserts that

$$f(b) = P_0(b; a) + \frac{1}{0!}\int_a^b (b - x)^0 f^{(1)}(x)\, dx. \tag{6}$$

Since $P_0(b; a)$ is $f(a)$, $0! = 1$ and $(b - x)^0 = 1$ for $x \neq b$, Eq. (6) reduces to Eq. (4), which is true. The theorem is verified for $n = 0$.

To verify it for $n = 1$, apply the lemma with $n = 1$ to the relation

$$f(b) = f(a) + \int_a^b (b - x)^0 f^{(1)}(x)\, dx,$$

obtaining $$f(b) = f(a) + \frac{f^{(1)}(a)}{1}(b - a) + \frac{1}{1}\int_a^b f^{(2)}(x)(b - x)^1\, dx. \tag{7}$$

Since $P_1(b; a) = f(a) + f^{(1)}(a)(b - a)$, this proves the theorem for $n = 1$.

To establish the theorem for $n = 2$, apply the lemma with $n = 2$ to Eq. (7), obtaining

$$f(b) = f(a) + f^{(1)}(a)(b - a) + \frac{f^{(2)}(a)}{2}(b - a)^2 + \frac{1}{2}\int_a^b f^{(3)}(x)(b - x)^2\, dx. \tag{8}$$

This establishes the theorem for $n = 2$.

Next, apply the lemma with $n = 3$ to Eq. (8), obtaining

$$f(b) = f(a) + f^{(1)}(a)(b - a) + \frac{f^{(2)}(a)}{2}(b - a)^2 + \frac{1}{2}\left[\frac{f^{(3)}(a)(b - a)^3}{3} + \frac{1}{3}\int_a^b f^{(4)}(x)(b - x)^3\, dx\right],$$

or $$f(b) = f(a) + f^{(1)}(a)(b - a) + \frac{f^{(2)}(a)}{2}(b - a)^2 + \frac{1}{3!}f^{(3)}(a)(b - a)^3 + \frac{1}{3!}\int_a^b f^{(4)}(x)(b - x)^3\, dx. \tag{9}$$

Inspection of Eq. (9) shows that it confirms the theorem for $n = 3$. Equation (9) also shows why the factorial $n!$ appears. (The reader may check the case $n = 4$ and see why 4! appears.) ■

Theorem 1 also holds when $b < a$ and the interval is $[b, a]$.

In the usual notation for the remainder, $R_n(x; a)$, we have

$$R_n(x; a) = \frac{1}{n!}\int_a^x (x - t)^n f^{(n+1)}(t)\, dt. \tag{10}$$

Theorem 1 expresses the error as an integral. It is possible to express the error also as a derivative, which is much easier to work with. The derivative form is good enough to show that $R_n(x; a) \to 0$ as $n \to \infty$ for e^x, $\sin x$, and $\cos x$, but it is not strong enough to deal with $\sqrt{1 + x}$. (The integral form is required in Sec. 10.12 to treat this case, which is often applied in engineering and physics.).

The derivative form is too weak to show that for $|x| < 1$,
$$\frac{1}{1 - x} = 1 + x + \cdots + x^n + \cdots.$$

The derivative form for $R_n(x; a)$ follows from the integral form. It is often called **Lagrange's formula for the remainder**.

Theorem 2 *Derivative form of the remainder.* Assume that a function f has continuous derivatives of orders through $n + 1$ in an interval that includes the numbers a and b. Let $P_n(b; a)$ be the Taylor polynomial of degree n associated with f in powers of $b - a$. Then there is a number c_n in the closed interval with endpoints a and b such that

$$R_n(b; a) = f(b) - P_n(b; a) = \frac{f^{(n+1)}(c_n)}{(n+1)!}(b - a)^{n+1}.$$

Proof By Theorem 1, $R_n(b; a) = \frac{1}{n!}\int_a^b f^{(n+1)}(x)(b - x)^n \, dx.$

Since $(b - x)^n \geq 0$ for $a \leq x \leq b$, the second mean-value theorem for integrals (see Sec. 8.6) applies, and we know that there is a number c_n in $[a, b]$ such that

$$\frac{1}{n!}\int_a^b f^{(n+1)}(x)(b - x)^n \, dx = \frac{f^{(n+1)}(c_n)}{n!}\int_a^b (b - x)^n \, dx.$$

But
$$\int_a^b (b - x)^n \, dx = -\frac{(b - x)^{n+1}}{n + 1}\bigg|_a^b = \frac{(b - a)^{n+1}}{n + 1}.$$

Hence
$$R_n(b; a) = \frac{f^{(n+1)}(c_n)}{n!}\frac{(b - a)^{n+1}}{n + 1},$$

from which the theorem follows.

A similar argument takes care of the case $b < a$. ■

As a special case of Theorem 2, we have the useful formula

$$R_n(x; 0) = \frac{f^{(n+1)}(c_n)x^{n+1}}{(n + 1)!} \tag{11}$$

for some number c_n between 0 and x.

Note that c_n depends on x and n.

EXAMPLE 2 Show that the power series in x associated with $f(x) = e^x$ represents $f(x)$ for all x.

SOLUTION Take the case $x > 0$. We must show that $R_n(x; 0) \to 0$ as $n \to \infty$. By Theorem 2, there is for each positive integer n a number c_n between 0 and x such that

$$R_n(x; 0) = \frac{f^{(n+1)}(c_n)}{(n + 1)!}x^{n+1}.$$

As observed in Example 1, all the derivatives of the function e^x are simply e^x itself. Hence $f^{(n+1)}(c_n) = e^{c_n}$. Since $0 \leq c_n \leq x$ and $f(x)$ is increasing, we have

$$1 = e^0 \leq e^{c_n} \leq e^x.$$

Thus

$$\frac{1 \cdot x^{n+1}}{(n + 1)!} \leq R_n(x; 0) \leq \frac{e^x x^{n+1}}{(n + 1)!}.$$

Once we know that $R_n(x; 0) \to 0$ as $n \to \infty$, we know $f(x) = \Sigma_{n=0}^{\infty} f^{(n)}(0)x^n/n!$.

Since x is fixed and $n \to \infty$, $R_n(x; 0) \to 0$ as $n \to \infty$.

For $x < 0$, the reasoning is similar to the case $x > 0$, the main difference being that we now have $x \leq c_n \leq 0$. The reader may carry out the details.

Since the series for e^x was already found in Example 2 of Sec. 10.7, we can therefore write

$$e^x = 1 + x + \frac{x^2}{2!} + \frac{x^3}{3!} + \cdots + \frac{x^n}{n!} + \cdots = \sum_{n=0}^{\infty} \frac{x^n}{n!}.$$

This equation clearly holds when $x = 0$, since it then reduces to the equation $e^0 = 1$. ■

EXAMPLE 3 Show that the Maclaurin series associated with $f(x) = \sin x$ represents $f(x)$ for all x.

SOLUTION All that is needed is to show that $R_n(x; 0) \to 0$ as $n \to \infty$. Now, by the derivative form of the remainder,

$$R_n(x; 0) = \frac{f^{(n+1)}(c_n)x^{n+1}}{(n+1)!},$$

where c_n is between 0 and x.

If $f(x) = \sin x$, then $f^{(1)}(x) = \cos x$, $f^{(2)}(x) = -\sin x$, $f^{(3)}(x) = -\cos x$, $f^{(4)}(x) = \sin x$, and so on. The higher derivatives are either $\pm\sin x$ or $\pm\cos x$. Thus, for any nonnegative integer n and real number c,

$$|f^{(n+1)}(c)| \leq 1.$$

Consequently, $$|R_n(x; 0)| = \frac{|f^{(n+1)}(c_n)x^{n+1}|}{(n+1)!} \leq \frac{|x|^{n+1}}{(n+1)!},$$

which approaches 0 as $n \to \infty$.

Hence the Maclaurin series associated with $\sin x$ represents $\sin x$ for all x. Since that series is $\sum_{n=0}^{\infty} (-1)^n x^{2n+1}/(2n+1)!$, we have

$$\sin x = x - \frac{x^3}{3!} + \frac{x^5}{5!} - \cdots + (-1)^n \frac{x^{2n+1}}{(2n+1)!} + \cdots.$$

(Terms in the Maclaurin series with value 0 are not shown.) ■

Using a front end of a power series to estimate $\sin x$

Example 3 provides a remarkably efficient way to estimate $\sin x$ for x in the range, say, 0° to 45°: Use the polynomial $x - x^3/3! + x^5/5!$ or $x - x^3/3!$ (x is the radian measure of the angle). The error in using the first estimate is less than $(\pi/4)^7/7! < 0.00004$. (Why?) The error in using the second estimate, $x - x^3/6$, is less than $(\pi/4)^5/5! < 0.003$.

The front end of the power-series representation of a function $f(x)$ is often taken as an approximation of the function. The next example illustrates this use of power series.

EXAMPLE 4 Estimate $\sqrt{e} = e^{1/2}$ using the first four terms of the Maclaurin series for e^x. Discuss the error.

SOLUTION We have $f(x) = e^x = 1 + x + \dfrac{x^2}{2!} + \dfrac{x^3}{3!} + \cdots$.

Thus $e^{1/2}$ is approximated by

$$1 + \frac{1}{2} + \frac{(\frac{1}{2})^2}{2!} + \frac{(\frac{1}{2})^3}{3!} = 1 + \frac{1}{2} + \frac{1}{8} + \frac{1}{48} \approx 1.64583.$$

By the derivative form of the remainder, the error is of the form

$$\frac{f^{(4)}(c)(\frac{1}{2})^4}{4!}$$

for some number c in $[0, \frac{1}{2}]$. Since $f(x) = e^x$, $f^{(4)}(x) = e^x$. Thus the error is of the form

$$\frac{e^c(\frac{1}{16})}{24} = \frac{e^c}{384} \quad \text{for } c \text{ in } [0, \tfrac{1}{2}].$$

Since $e^c \le e^{1/2} < 2$ (because $e < 4$), we see that the error is positive but less than $\frac{2}{384} = \frac{1}{192} \approx 0.0052$. Thus, to two decimal places, $e^{1/2} \approx 1.65$. ■

Using a front end of a power series to estimate $\int_a^b f(x)$

In statistics, the integral $\int_{-\infty}^{b} (1/\sqrt{2\pi})\, e^{-x^2/2}\, dx$ is of major importance. Since $e^{-x^2/2}$ does not have an elementary antiderivative, the integral must be estimated by other means and its values tabulated. For instance, Burington's *Handbook of Mathematical Tables and Formulas,* lists it for b in the range $[0, 4]$ at intervals of 0.01.

The next example shows how to evaluate $\int_a^b f(x)\, dx$ when $f(x)$ is representable by a power series.

EXAMPLE 5 Use the Maclaurin series for e^x to estimate $\displaystyle\int_0^1 e^{-x^2}\, dx$.

SOLUTION

$$e^x = 1 + x + \frac{x^2}{2!} + \frac{x^3}{3!} + \cdots.$$

Replacing x by $-x^2$ yields

$$e^{-x^2} = 1 - x^2 + \frac{x^4}{2!} - \frac{x^6}{3!} + \cdots. \tag{12}$$

For $0 < |x| \le 1$, series (12) is a decreasing alternating series. Thus

$$1 - x^2 + \frac{x^4}{2!} - \frac{x^6}{3!} < e^{-x^2} < 1 - x^2 + \frac{x^4}{2!} - \frac{x^6}{3!} + \frac{x^8}{4!}.$$

Hence

$$\int_0^1 \left(1 - x^2 + \frac{x^4}{2!} - \frac{x^6}{3!}\right) dx < \int_0^1 e^{-x^2}\, dx < \int_0^1 \left(1 - x^2 + \frac{x^4}{2!} - \frac{x^6}{3!} + \frac{x^8}{4!}\right) dx,$$

or

$$1 - \frac{1}{3} + \frac{1}{5 \cdot 2!} - \frac{1}{7 \cdot 3!} < \int_0^1 e^{-x^2}\, dx < 1 - \frac{1}{3} + \frac{1}{5 \cdot 2!} - \frac{1}{7 \cdot 3!} + \frac{1}{9 \cdot 4!}.$$

From this it follows that

Supply the omitted arithmetic

$$0.742 < \int_0^1 e^{-x^2}\, dx < 0.748. \quad ■$$

In Sec. 10.12 the integral form of $R_n(x; a)$ is used to show that the Maclaurin series for $(1 + x)^a$, where a is any real number and $|x| < 1$, converges to $(1 + x)^a$. The Lagrange form of the remainder would fail in this case, and it is interesting to see why it is inadequate.

Take the case $a = -1$. Then $(1 + x)^a = 1/(1 + x)$. We know that

$$f(x) = \frac{1}{1 + x} = 1 - x + x^2 - x^3 + \cdots \qquad \text{for } |x| < 1,$$

since the right-hand side of the equation is a geometric series with first term 1 and ratio $-x$. Moreover, it is easy to check that the right side of the equation is the Maclaurin series for $1/(1 + x)$, so we know that $R_n(x; 0) \to 0$ as $n \to \infty$.

Let us see what happens when we use Lagrange's formula in the case $x = -\frac{1}{2}$. Theorem 2 says that there is a number c_n in $[-\frac{1}{2}, 0]$ such that

$$R_n(-\tfrac{1}{2}; 0) = \frac{f^{(n+1)}(c_n)}{(n + 1)!}\left(-\frac{1}{2}\right)^{n+1}.$$

Carry out the computation

A straightforward computation shows that

$$f^{(n+1)}(c_n) = (-1)^{n+1}(n + 1)!(1 + c_n)^{-n-2}.$$

Hence
$$\left|R_n\left(-\frac{1}{2}; 0\right)\right| = \frac{1}{(1 + c_n)^{n+2}}\left(\frac{1}{2}\right)^{n+1} = \frac{1}{(2 + 2c_n)^{n+1}}\,\frac{1}{1 + c_n}. \tag{13}$$

All that we know about c_n is that it is in $[-\frac{1}{2}, 0]$. If it is close to $-\frac{1}{2}$, then (13) could be very near 2. So the Lagrange formula for the remainder fails to show that $R_n(-\frac{1}{2}; 0)$ approaches 0 as $n \to \infty$. However, when $-\frac{1}{2} < x < 1$, Theorem 2 is strong enough to show that $R_n(x; 0) \to 0$ as $n \to \infty$. (See Exercise 19.) As shown in Sec. 10.12, the integral form of $R_n(x; 0)$ takes care of all x in $(-1, 1)$.

Remark: The basic idea of Taylor's series is to approximate a function by a polynomial whose derivatives up through order n at some point coincide with those of the function.

Another way of approximating $f(x)$ by a polynomial; alas, it doesn't work.

We might expect that the following procedure would produce a good polynomial approximation to a given function $f(x)$ throughout an interval $[a, b]$. Divide the interval into n sections of equal length by $n + 1$ points, of which the leftmost is a and the rightmost is b. There is then a unique polynomial $P(x)$ of degree at most n that coincides with the given function at these $n + 1$ values. We would expect that when n is large, $|P(x) - f(x)|$ would be small for all x in $[a, b]$.

This is not the case, even for such a pleasant function as $f(x) = 1/(1 + x^2)$ and the interval $[-5, 5]$. Say you pick $5m + 1$ equally spaced numbers (where m is odd) in $[-5, 5]$ and let $P(x)$ be the polynomial of degree at most $5m$ that coincides with the function at those numbers. Then it can be shown by the methods of numerical analysis that $|f(4.875) - P(4.875)| > 1.8^m/451$, which approaches ∞ as $m \to \infty$. This means that $P(x)$ departs wildly from $f(x)$.

EXERCISES FOR SEC. 10.8: TAYLOR'S FORMULA

In each of Exercises 1 to 6 find the integral form of the remainder $R_3(x; 0)$ for the given function.

1 $\dfrac{1}{1 + x}$

2 $\dfrac{1}{1 - x}$

3 $\ln(1 + x)$

4 $\ln(1 - x)$

5 $\cos x$

6 $\sin x$

In Exercises 7 to 10 find the Taylor polynomials $P_3(x; 0)$ for the given functions.

7 $\sin x$

8 $\cos x$

9 $\tan x$ **10** $\sqrt{1+x}$

In Exercises 11 to 14 graph the given functions and the Taylor polynomials $P_0(x; 0)$, $P_1(x; 0)$, $P_2(x; 0)$, and $P_3(x; 0)$. In each exercise graph the functions relative to the same axes.

11 $1/(1+x)$ **12** $\ln(1+x)$

13 $\sin x$ **14** $\cos x$

In Exercises 15 to 20 show for the given functions that $R_n(x; 0) \to 0$ as $n \to \infty$, using the derivative form for $R_n(x; 0)$.

15 e^x **16** e^{-x}

17 $\sin 2x$ **18** $\cos x$

19 $\dfrac{1}{1+x}$, $-\frac{1}{2} < x < 1$

20 $\ln(1+x)$, $-\frac{1}{2} < x < 1$

21 Show that for $f(x) = \sin x$, $R_n(x; a) \to 0$ as $n \to \infty$ for any x and a.

22 Show that for $f(x) = e^x$, $R_n(x; a) \to 0$ as $n \to \infty$ for any x and a.

In each of Exercises 23 to 28 use the first three nonzero terms of a Maclaurin series to estimate the given number. Also use the derivative form of the remainder to put an upper bound on the error.

23 $\sqrt[3]{e}$ **24** e^2

25 $\sin 1$

26 $\sin 20°$ *Hint:* First convert to radians.

27 $\cos 10°$ **28** $\cos \frac{1}{3}$

■

29 In Example 4 it was shown that the error is less than $\frac{1}{192}$. Using the derivative form of the remainder, show that the error is greater than $\frac{1}{384}$.

30 Show that $\sin x \approx x - x^3/3!$ with an error at most $\frac{1}{120}$ for $|x| \le 1$.

31 Show that $\cos x \approx 1 - x^2/2 + x^4/24$ with an error at most $\frac{1}{720}$ for $|x| \le 1$.

32 From the Maclaurin series for e^x obtain the Maclaurin series for (*a*) e^{x^3}, (*b*) e^{-x}, (*c*) $(e^x - 1)/x$.

33 (Calculator) Graph, relative to the same axes, $y = \sin x$ and its Taylor polynomial $x - x^3/6 + x^5/120$.

34 By differentiating, show that $f(x)$ and $P_3(x; a)$ have the same derivatives through order 3 at a.

35 (*a*) From the Maclaurin series for $\cos x$ obtain that for $\cos x^2$.

(*b*) Use (*a*) to estimate $\int_0^{1/2} \cos x^2\, dx$ to three-decimal-place accuracy.

36 Use the Maclaurin series for $\ln(1+x)$ to estimate $\ln(1.1)$ to three-decimal-place accuracy.

37 Estimate $\int_0^1 \sqrt{x} \sin x\, dx$ to two-decimal-place accuracy.

38 Justify this statement, found in a biological monograph: "Expanding the equation

$$a \cdot \ln(x+p) + b \cdot \ln(y+q) = M,$$

we obtain

$$a\left(\ln p + \frac{x}{p} - \frac{x^2}{2p^2} + \frac{x^3}{3p^3} - \cdots\right) + b\left(\ln q + \frac{y}{q} - \frac{y^2}{2q^2} + \frac{y^3}{3q^3} - \cdots\right) = M."$$

39 Justify the second sentence in this statement, quoted from a biological monograph: "Hence the probability of extinction $1 - y$ will be given by $1 - y = e^{-(1+k)y}$. If k is small, y is approximately equal to $2k$."

40 Let $f(x)$ be a function such that for n sufficiently large, $|f^{(n)}(x)| \le n^5$. Deduce that for all x and a,

$$f(x) = f(a) + f^{(1)}(a)(x-a) + \frac{f^{(2)}(a)}{2!}(x-a)^2 + \cdots.$$

41 In R. P. Feynman, *Lectures on Physics,* Addison-Wesley, Reading, Mass., 1963, appears this remark:

> Thus the average energy is
>
> $$\langle E \rangle = \frac{\hbar\omega(0 + x + 2x^2 + 3x^3 + \cdots)}{1 + x + x^2 + \cdots}.$$
>
> Now the two sums which appear here we shall leave for the reader to play with and have some fun with. When we are all finished summing and substituting for x in the sum, we should get—if we make no mistakes in the sum—
>
> $$\langle E \rangle = \frac{\hbar\omega}{e^{\hbar\omega/kT} - 1}.$$
>
> This, then, was the first quantum-mechanical formula ever known, or ever discussed, and it was the beautiful culmination of decades of puzzlement.

Have the aforementioned fun, given that $x = e^{-\hbar\omega/kT}$.

42 Assume that $f(x)$ has a continuous fourth derivative. Let M_4 be the maximum of $|f^{(4)}(x)|$ for x in $[-1, 1]$. Show that

$$\left|\int_{-1}^{1} f(x)\, dx - f\left(\frac{1}{\sqrt{3}}\right) - f\left(-\frac{1}{\sqrt{3}}\right)\right| \le \frac{7M_4}{270}.$$

Suggestion: Use the representation $f(x) = f(0) + f^{(1)}(0)x + f^{(2)}(0)x^2/2 + f^{(3)}(0)x^3/6 + f^{(4)}(c)\, x^4/24$, where c depends on x.

43 In *Optimal Replacement Policy,* a text on the maintenance of equipment, D. W. Jorgenson, J. J. McCall, and R. Radner presented the following argument: "Suppose that the distribution of time to failure of the equipment is exponential. Then we have

$$\frac{e^{\lambda N}}{\lambda} - N = \frac{1}{\lambda} + K,$$

an equation for N, the replacement time period. A quadratic approximation to $e^{\lambda N}$ results in the formula $N = \sqrt{2K/\lambda}$." (*Continued on page 548.*)

In this argument λ and K are constants: λ determines the failure distribution and K is the time to effect a repair.

Explain how the formula for N is obtained. (*Hint:* Use the approximation $e^x = 1 + x + x^2/2$.)

44 What is the power series in x for $f(x) = 1 + 3x + 5x^2$?

10.9 COMPLEX NUMBERS

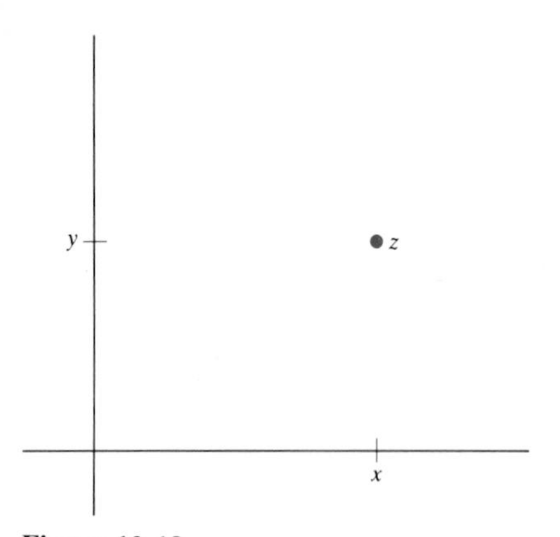

Figure 10.18

Complex numbers, together with power series, in Sec. 10.10 will yield a surprising relation between e^x, sin x, and cos x.

Let us think of the number line of real numbers as coinciding with the x axis of an xy coordinate system. This number line, with its addition, subtraction, multiplication, and division, is but a small part of a number system that occupies the plane and which obeys the usual rules of arithmetic. This section describes that system, known as the **complex numbers**. One of the important properties of the complex numbers is that any nonconstant polynomial has a root; in particular, the equation $x^2 = -1$ has two solutions.

By a complex number z we shall mean an expression of the form $x + iy$, where x and y are real numbers and i is a symbol such that $i^2 = -1$. This expression will be identified with the point (x, y) in the xy plane, as in Fig. 10.18. Every point in the plane may therefore be thought of as a complex number.

To add or multiply two complex numbers, follow the usual rules of arithmetic of real numbers, with one new proviso:

Whenever you see i^2, replace it by -1.

For instance, to add the complex numbers $3 + 2i$ and $-4 + 5i$, just collect like terms:

$$(3 + 2i) + (-4 + 5i) = [3 + (-4)] + (2i + 5i) = -1 + 7i.$$

Addition does not make use of the fact that $i^2 = -1$. However, multiplication does, as Example 1 shows.

EXAMPLE 1 Compute the product $(2 + i)(3 + i)$.

SOLUTION By the distributive law, $a(b + c) = ab + ac$,

$$(2 + i)(3 + i) = (2 + i)3 + (2 + i)i.$$

By the distributive law, $(b + c)a = ba + ca$,

$$(2 + i)3 = 2 \cdot 3 + i3 = 6 + 3i \qquad \text{and} \qquad (2 + i)i = 2i + i^2 = 2i - 1.$$

Thus
$$(2 + i)(3 + i) = (6 + 3i) + (2i - 1) = 5 + 5i.$$

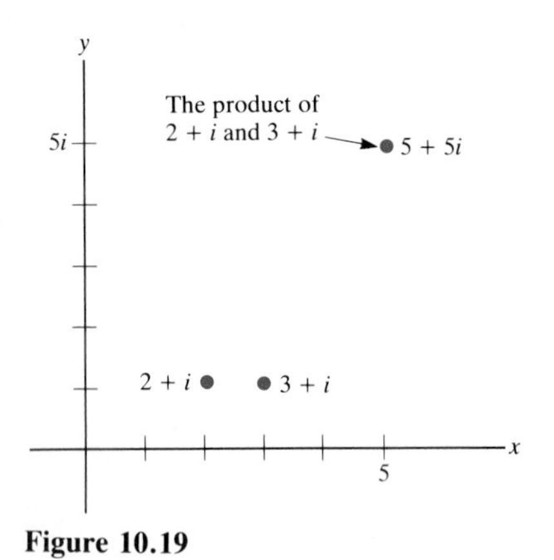

Figure 10.19

Figure 10.19 shows the complex numbers $2 + i$, $3 + i$, and their product, $5 + 5i$. ■

-1 has two square roots.

Note that $(-i)(-i) = i^2 = -1$. Both i and $-i$ are square roots of -1. The symbol $\sqrt{-1}$ traditionally denotes i.

Real numbers are on the x axis, imaginary on the y axis.

A complex number that lies on the y axis is called **imaginary**. Every complex number z is the sum of a real number and an imaginary number,

$z = x + iy$. The number x is called the **real part of** z, and y is called the **imaginary part**. One writes "Re $z = x$" and "Im $z = y$."

We have seen how to add and multiply complex numbers. Subtraction is straightforward. For instance,

$$(3 + 2i) - (4 - i) = (3 - 4) + [2i - (-i)]$$
$$= -1 + 3i.$$

Division of complex numbers requires rationalizing the denominator, as Example 2 illustrates.

EXAMPLE 2 Compute $\frac{1 + 5i}{3 + 2i}$.

SOLUTION To divide, "conjugate the denominator":

$$\frac{1 + 5i}{3 + 2i} \cdot \frac{3 - 2i}{3 - 2i} = \frac{3 + 15i - 2i + 10}{9 - 4i^2} = \frac{13 + 13i}{13} = 1 + i \quad ■$$

Conjugate of z

The solution of Example 2 used the **conjugate** of a complex number. The conjugate of the complex number $z = x + yi$ is the complex number $x - yi$, which is denoted $\bar{z}$. Note that

$$z\bar{z} = (x + yi)(x - yi) = x^2 + y^2$$

and

$$z + \bar{z} = (x + yi) + (x - yi) = 2x.$$

Thus both $z\bar{z}$ and $z + \bar{z}$ are real.

Figure 10.20 shows the relation between z and $\bar{z}$, which is the mirror image of z in the x axis.

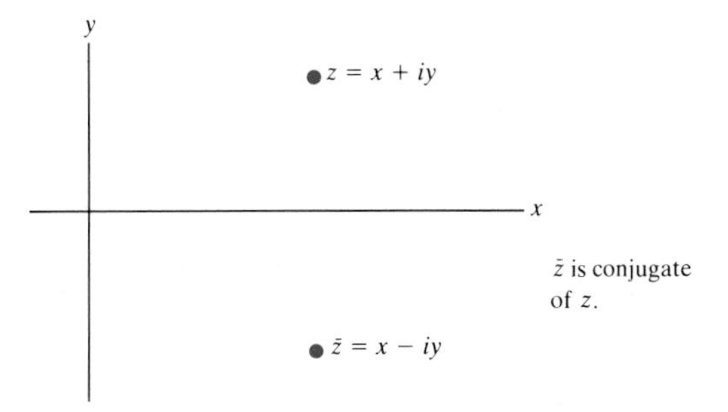

Figure 10.20

Every polynomial has a root in the complex numbers.

The complex numbers provide the equation $x^2 + 1 = 0$ with two solutions, i and $-i$. This illustrates an important property of the complex numbers: If $f(x) = a_n x^n + a_{n-1}x^{n-1} + \cdots + a_0$ is any polynomial of degree $n \geq 1$, with real or complex coefficients, then there is a complex number z such that $f(z) = 0$. This fact is known as the **fundamental theorem of algebra**. Its proof requires advanced mathematics. Example 3 illustrates this theorem.

EXAMPLE 3 Solve the quadratic equation $z^2 - 4z + 5 = 0$.

SOLUTION By the quadratic formula, the solutions are

$$z = \frac{-(-4) \pm \sqrt{(-4)^2 - 4 \cdot 1 \cdot 5}}{2}$$

$$= \frac{4 \pm \sqrt{-4}}{2} = \frac{4 \pm 2i}{2} = 2 \pm i.$$

The solutions are $2 + i$ and $2 - i$.

These solutions can be checked by substitution in the original equation. For instance,

$$(2 + i)^2 - 4(2 + i) + 5 = (4 + 4i + i^2) - 8 - 4i + 5$$
$$= 4 + 4i - 1 - 8 - 4i + 5 = 0 + 0i = 0.$$

Yes, it checks. The solution $2 - i$ can be checked similarly. ■

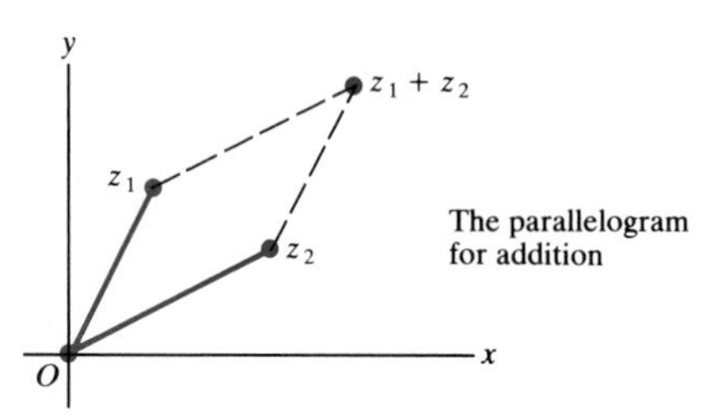

Figure 10.21

The sum of the complex numbers z_1 and z_2 is the fourth vertex (opposite O) in a parallelogram determined by the origin O and the points z_1 and z_2, as shown in Fig. 10.21.

The geometric relation between z_1, z_2, and their product, z_1z_2, is easily described in terms of the magnitude and argument of a complex number. Each complex number z other than the origin is at some distance r from the origin and has a polar angle θ relative to the x axis as polar axis. The distance r is called the **magnitude of** z, and θ is called an **argument of** z. A complex number has an infinity of arguments differing from each other by an integer multiple of 2π. The complex number 0, which lies at the origin, has magnitude 0 and any angle as argument. In short, we may think of magnitude and argument as polar coordinates r and θ of z, with the restriction that r is nonnegative. The magnitude of z is denoted $|z|$. The symbol arg z denotes any of the arguments of z, it being understood that if θ is an argument of z, then so is $\theta + 2n\pi$ for any integer n.

The symbols $|z|$ and arg z

EXAMPLE 4

(a) Draw all complex numbers of magnitude 3.

(b) Draw the complex number z of magnitude 3 and argument $\pi/6$.

SOLUTION

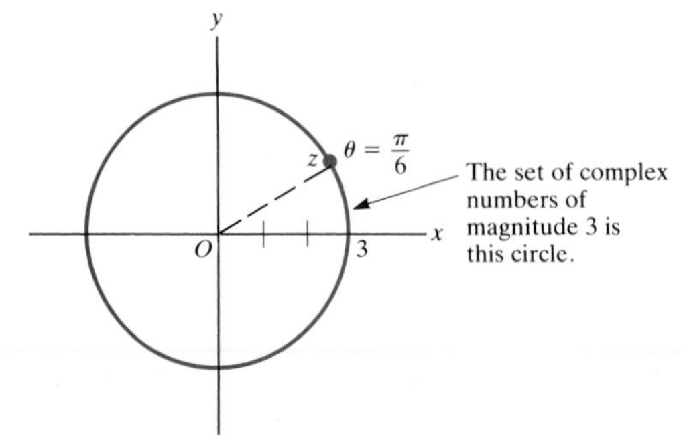

Figure 10.22

(*a*) The complex numbers of magnitude 3 form a circle of radius 3 with center at O. (See Fig. 10.22.)
(*b*) The complex number of magnitude 3 and argument $\pi/6$ is shown in Fig. 10.22. ■

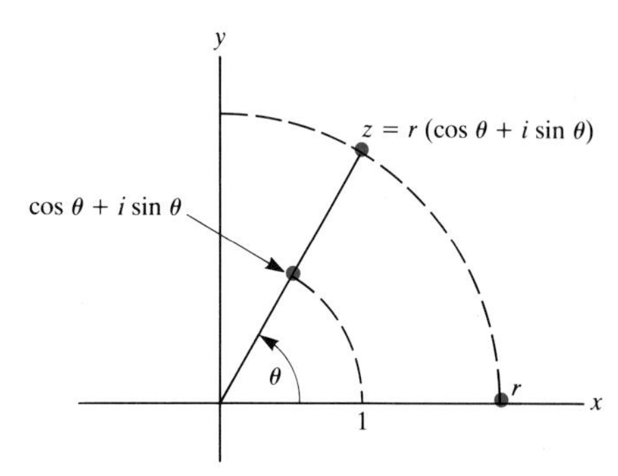

Figure 10.23

Observe that $|x + iy| = \sqrt{x^2 + y^2}$, by the pythagorean theorem. Each complex number $z = x + iy$ other than 0 can be written as the product of a positive real number and a complex number of magnitude 1. To show this, let $z = x + iy$ have magnitude r and argument θ. Recalling the relation between polar and rectangular coordinates, we conclude that

$$z = r\cos\theta + i\,r\sin\theta$$
$$= r(\cos\theta + i\sin\theta).$$

The number r is a positive real number; the number $\cos\theta + i\sin\theta$ has magnitude 1 since $\sqrt{\cos^2\theta + \sin^2\theta} = 1$. Figure 10.23 shows the numbers r and $\cos\theta + i\sin\theta$, whose product is z.

The following theorem describes *how to multiply two complex numbers if they are given in polar form,* that is, in terms of their magnitudes and arguments.

Theorem Assume that z_1 has magnitude r_1 and argument θ_1 and that z_2 has magnitude r_2 and argument θ_2. Then the product z_1z_2 has magnitude r_1r_2 and argument $\theta_1 + \theta_2$.

Proof

$$z_1z_2 = r_1(\cos\theta_1 + i\sin\theta_1)r_2(\cos\theta_2 + i\sin\theta_2)$$
$$= r_1r_2(\cos\theta_1 + i\sin\theta_1)(\cos\theta_2 + i\sin\theta_2)$$
$$= r_1r_2[\cos\theta_1\cos\theta_2 - \sin\theta_1\sin\theta_2 + i(\sin\theta_1\cos\theta_2 + \cos\theta_1\sin\theta_2)]$$
$$= r_1r_2[\cos(\theta_1 + \theta_2) + i\sin(\theta_1 + \theta_2)] \quad \text{(by trigonometric identities).}$$

Recall the identities for $\cos(A + B)$ *and* $\sin(A + B)$.

How to multiply in terms of magnitude and argument

Thus the argument of z_1z_2 is $\theta_1 + \theta_2$ and the magnitude of z_1z_2 is r_1r_2. This proves the theorem. ■

In practical terms, the theorem says, "*to multiply two complex numbers just add their arguments and multiply their magnitudes.*"

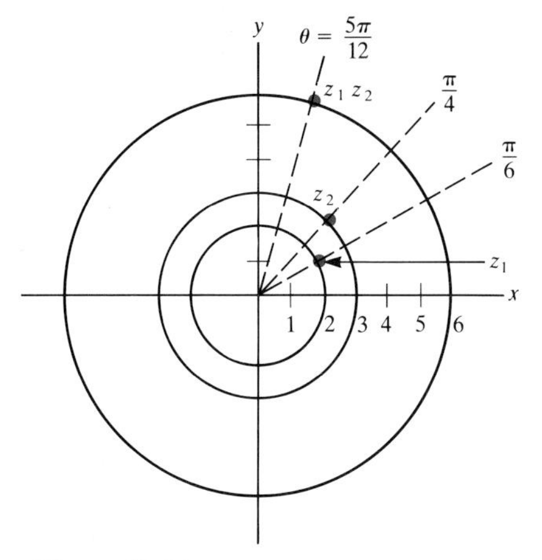

Figure 10.24

EXAMPLE 5 Find z_1z_2 for z_1 and z_2 in Fig. 10.24.

SOLUTION z_1 has magnitude 2 and argument $\pi/6$; z_2 has magnitude 3 and argument $\pi/4$. Thus z_1z_2 has magnitude $2 \times 3 = 6$ and argument $\pi/6 + \pi/4 = 5\pi/12$. It is shown in Fig. 10.24. ■

EXAMPLE 6 Using the geometric description of multiplication, find the product of the real numbers -2 and -3.

SOLUTION The number -2 has magnitude 2 and argument π. The number -3 has magnitude 3 and argument π. Thus $(-2) \times (-3)$ has magnitude $2 \times 3 = 6$ and argument $\pi + \pi = 2\pi$. The complex number with magnitude 6 and argument 2π is just our old friend, the real number 6. Thus $(-2) \times (-3) = 6$ in agreement with the customary product of two negative real numbers. (See Fig. 10.25.)

As expected, "negative times negative is positive."

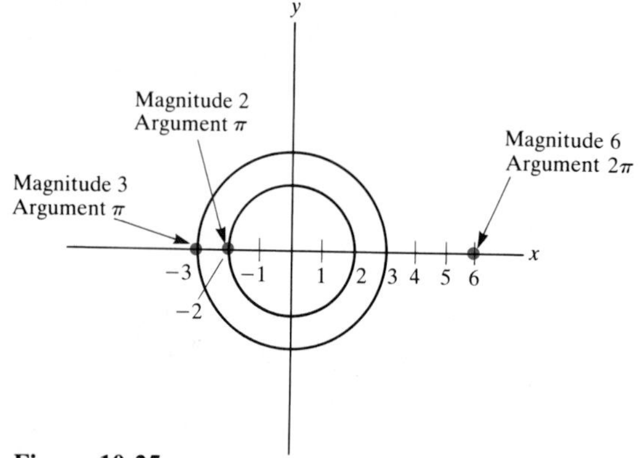

Figure 10.25 ■

See Exercise 17.

Division of complex numbers given in polar form is similar, except that the magnitudes are divided and the arguments subtracted:

$$\frac{r_1(\cos \theta_1 + i \sin \theta_1)}{r_2(\cos \theta_2 + i \sin \theta_1)} = \left(\frac{r_1}{r_2}\right) [\cos (\theta_1 - \theta_2) + i \sin (\theta_1 - \theta_2)].$$

EXAMPLE 7 Let $z_1 = 6\left(\cos \frac{\pi}{2} + i \sin \frac{\pi}{2}\right)$ and $z_2 = 3\left(\cos \frac{\pi}{6} + i \sin \frac{\pi}{6}\right)$. Find (*a*) $z_1 z_2$ and (*b*) z_1/z_2.

SOLUTION

(*a*)

$$\begin{aligned} z_1 z_2 &= 6 \cdot 3 \left[\cos\left(\frac{\pi}{2} + \frac{\pi}{6}\right) + i \sin\left(\frac{\pi}{2} + \frac{\pi}{6}\right)\right] \\ &= 18\left(\cos \frac{2\pi}{3} + i \sin \frac{2\pi}{3}\right) \\ &= 18\left(-\frac{1}{2} + i\frac{\sqrt{3}}{2}\right) = -9 + 9\sqrt{3}i. \end{aligned}$$

(*b*)

$$\begin{aligned} \frac{z_1}{z_2} &= \frac{6\left(\cos \frac{\pi}{2} + i \sin \frac{\pi}{2}\right)}{3\left(\cos \frac{\pi}{6} + i \sin \frac{\pi}{6}\right)} \\ &= 2\left(\cos \frac{\pi}{3} + i \sin \frac{\pi}{3}\right) \\ &= 2\left(\frac{1}{2} + \frac{\sqrt{3}}{2} i\right) = 1 + \sqrt{3}i. \end{aligned}$$ ■

EXAMPLE 8 Compute $(1 + i)(3 + 2i)$ and check the answer in terms of magnitudes and arguments.

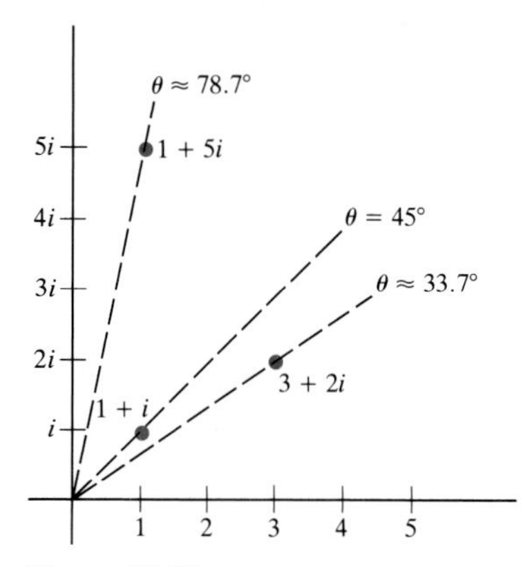

Figure 10.26

SOLUTION

$$\begin{aligned}(1 + i)(3 + 2i) &= (1 + i)(3) + (1 + i)(2i) \\ &= 3 + 3i + 2i + 2i^2 \\ &= 3 + 3i + 2i - 2 \\ &= 1 + 5i.\end{aligned}$$

As a check, let us see if $|1 + 5i| = |1 + i|\ |3 + 2i|$. We have

$$\begin{aligned}|1 + 5i| &= \sqrt{1^2 + 5^2} = \sqrt{26}, \\ |1 + i| &= \sqrt{1^2 + 1^2} = \sqrt{2}, \\ |3 + 2i| &= \sqrt{3^2 + 2^2} = \sqrt{13}.\end{aligned}$$

Since $\sqrt{26} = \sqrt{2}\sqrt{13}$, the magnitude of $1 + 5i$ is the product of the magnitudes of $1 + i$ and $3 + 2i$.

$\arg(x + iy) = \tan^{-1}(y/x)$ for $x + iy$ in first or fourth quadrants

Next consider arguments. First of all, $\arg(1 + 5i) = \tan^{-1} 5 \approx 1.3734$. Similarly, $\arg(1 + i) = \tan^{-1} 1 \approx 0.7854$ and $\arg(3 + 2i) = \tan^{-1} \frac{2}{3} \approx 0.5880$. Note that $0.7854 + 0.5880 = 1.3734$. (See Fig. 10.26.) ■

When the polar coordinates of z are known, it is easy to compute the powers $z^2, z^3, z^4, \ldots$. Let z have magnitude r and argument θ. Then $z^2 = z \cdot z$ has magnitude $r \cdot r = r^2$ and argument $\theta + \theta = 2\theta$. So, to square a complex number, just square its magnitude and double its angle.

How to compute z^n

More generally, for any positive integer n, to compute z^n, find $|z|^n$ and multiply the argument of z by n. In short,

This equation is known as DeMoivre's law.

$$[r(\cos\theta + i\sin\theta)]^n = r^n(\cos n\theta + i\sin n\theta).$$

Example 9 illustrates this geometric view of computing powers.

EXAMPLE 9 Let z have magnitude 1 and argument $2\pi/5$. Compute and sketch z, z^2, z^3, z^4, z^5, and z^6.

SOLUTION Since $|z| = 1$, it follows that $|z^2| = |z|^2 = 1^2 = 1$. Similarly, for all positive integers n, $|z^n| = 1$; that is, z^n is a point on the unit circle with center 0. All that remains is to examine the arguments of z^2, z^3, etc.

The argument of z^2 is $2(2\pi/5) = 4\pi/5$. Similarly, $\arg z^3 = 6\pi/5$, $\arg z^4 = 8\pi/5$, $\arg z^5 = 10\pi/5 = 2\pi$, and $\arg z^6 = 12\pi/5$. Observe that $z^5 = 1$, since it has magnitude 1 and argument 2π. Similarly, $z^6 = z$, since both z and z^6 have magnitude 1 and arguments that differ by an integer multiple of 2π. (Or, algebraically, $z^6 = z^5 \cdot z = 1z = z$.) The powers of z form the vertices of a regular pentagon, as shown in Fig. 10.27. ■

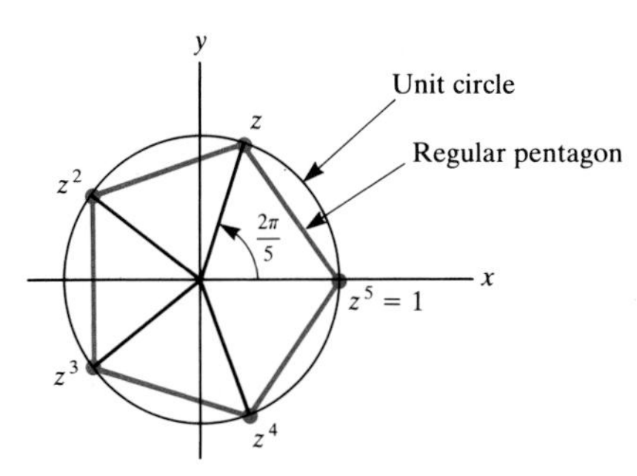

Figure 10.27

The equation $x^5 = 1$ has only one real root, namely, 1. However, it has five complex roots. For instance, the number z shown in Fig. 10.27 is a solution of $x^5 = 1$, since $z^5 = 1$. Another root is z^2, since $(z^2)^5 = z^{10} = (z^5)^2 = 1^2 = 1$. Similarly, z^3 and z^4 are roots of $x^5 = 1$. The roots are $1, z, z^2, z^3$, and z^4.

The powers of i

The powers of i will be needed in the next section. They are $i^2 = -1$, $i^3 = i^2 \cdot i = (-1)i = -i$, $i^4 = i^3 \cdot i = (-i)i = -i^2 = 1$, $i^5 = i^4 \cdot i = i$, and so on. They repeat in blocks of four: for any integer n, $i^{n+4} = i^n$.

ALTERNATING CURRENT AND COMPLEX NUMBERS

As early as their sophomore year many students will take a course in electric circuits, where they will find the complex numbers used in the analysis of alternating currents. They will use a different notation, which is described in this table:

Standard	*Electrical Engineering*
i	j (i is used for current)
$\bar{z}$	z^*
$r(\cos\theta + i\sin\theta)$	$r\angle\theta$

The symbol j is used in the following discussion of alternating current.

The complex numbers, introduced by mathematicians in the course of their pure research, were accepted by them as a legitimate structure early in the nineteenth centu y. At that time, electricity was only an object of laboratory interest; it had little practical significance. Yet before the century was over, the discoveries concerning electricity and magnetism were to transform our world, and complex numbers were to serve as a tool in that transformation by simplifying the algebra of alternating currents. The details are worth sketching, not only to show the importance of complex numbers, but also to demonstrate how the "pure" knowledge of one generation can become the "practical" technique of a later generation.

In the case of a **direct current**, such as that provided by a battery, the voltage E is constant. This constant voltage, working against a resistance R, produces a constant current I. The three real numbers E, I, and R are related by the equation

$$E = IR,$$

which says, "current is proportional to voltage drop."

For the long-distance transmission of electric power, **alternating currents** are far more efficient than direct currents. An alternating current is produced by rotating a coiled wire in a fixed magnetic field. Charles Proteus Steinmetz, an engineer at General Electric when the United States was starting to bring electricity to the cities, found the algebra of alternating currents unwieldly. As he wrote in 1893,

> The current rises from zero to a maximum; then decreases again to nothing, reverses and rises to a maximum in the opposite direction; decreases to zero, again reverses and rises to a maximum in the first direction—and so on.
>
> Thus in all calculations with alternating current, instead of a simple mechanical value of direct current theory, the investigator had to use a complicated function of time to represent the alternating current. The theory of alternating current apparatus thereby became so complicated that the investigator never got very far. . . .
>
> The idea suggested itself at length of representing the alternating current by a single complex number. . . . This proved the solution of the alternating current calculation.
>
> It gave to the alternating current a single numerical value, just as to the direct current, instead of the complicated function of time of the previous theory; and thereby it made alternating current calculations as simple as direct current calculations.
>
> The introduction of the complex number has eliminated the function of time from the alternating current theory, and has made the alternating current theory the simple algebra of the complex number, just as the direct current theory leads to the simple algebra of the real number.

Let us see how Steinmetz, as has been said, "generated electricity out of the square root of -1."

To describe the motor which produces the varying voltage, he used a single complex number $\mathbf{E}$. The magnitude of $\mathbf{E}$ is the maximum voltage that the motor produces. The argument of $\mathbf{E}$ is determined by the initial position of the rotating coil.

Corresponding to the alternating current is the complex number $\mathbf{I}$. The magnitude of $\mathbf{I}$ is the maximum current flowing through the circuit. The argument of $\mathbf{I}$ records the lag in the current. (It does not necessarily reach its maximum when the voltage does.)

Within the typical circuit are, in addition, a resistance, a capacitor (for storing electric charge), and a fixed coil (which produces a magnetic field when current passes through it). The resistance is described by a real number R; the capacitor, by the real number X_C; the fixed coil, by the real number X_L. Steinmetz introduced the single complex number

$$\mathbf{R} = R + (X_L - X_C)j,$$

(where j stands for $\sqrt{-1}$) called the **complex impedance** and then described the basic behavior of alternating currents in the single equation

$$\mathbf{E} = \mathbf{IR}.$$

[For instance, this equation records the fact that the maximum voltage is equal to the product of the maximum current and the number $\sqrt{R^2 + (X_L - X_C)^2}$. Why?]

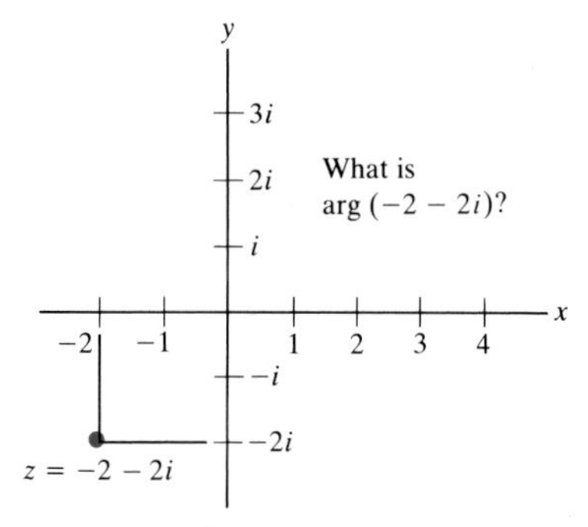

Figure 10.28

It is often useful to express a complex number $z = x + iy$ in polar form. Note first that $|z| = \sqrt{x^2 + y^2}$. To find θ, it is best to sketch z in order to see in which quadrant it lies. Although $\tan\theta = y/x$, we cannot say that $\theta = \tan^{-1}(y/x)$, since $\tan^{-1} u$ lies between $\pi/2$ and $-\pi/2$ for any real number u. However, θ may be a second- or third-quadrant angle. For instance, to put $z = -2 - 2i$ in polar form, first sketch z, as in Fig. 10.28. We have $|z| = \sqrt{(-2)^2 + (-2)^2} = \sqrt{8}$ and $\arg z = 5\pi/4$. Thus

$$z = \sqrt{8}\left(\cos\frac{5\pi}{4} + i\sin\frac{5\pi}{4}\right).$$

Note that $\tan^{-1}[-2/(-2)]$, which is $\pi/4$, is *not* an argument of z.

Roots of a complex number

Each complex number z, other than 0, has exactly n nth roots for each positive integer n. These roots can be found by expressing z in polar coordinates. If $z = r(\cos\theta + i\sin\theta)$, that is, has magnitude r and argument θ, then one nth root is

$$r^{1/n}\left(\cos\frac{\theta}{n} + i\sin\frac{\theta}{n}\right).$$

To check that this is an nth root of z, just raise it to the nth power.

To find the other nth roots of z, change the argument of z from θ to $\theta + 2k\pi$, $k = 1, 2, \ldots, n - 1$. Then

$$r^{1/n}\left(\cos\frac{\theta + 2k\pi}{n} + i\sin\frac{\theta + 2k\pi}{n}\right)$$

is also an nth root of z. (Why?)

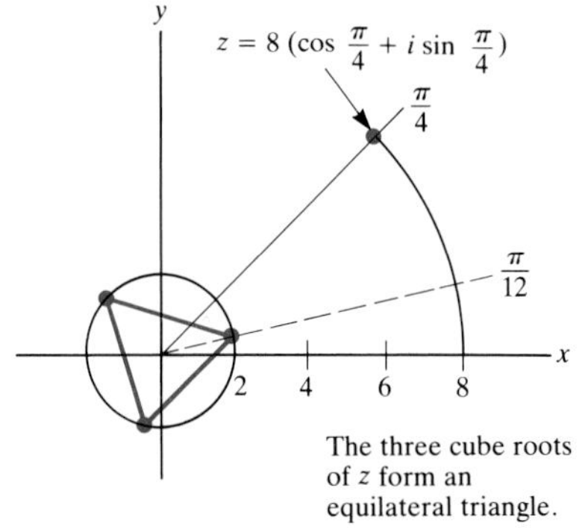

Figure 10.29

For instance, let $z = 8[\cos(\pi/4) + i\sin(\pi/4)]$. Then the three cube roots of z all have magnitude $8^{1/3} = 2$. Their arguments are

$$\frac{\pi/4}{3} = \frac{\pi}{12}, \qquad \frac{\pi/4 + 2\pi}{3} = \frac{\pi}{12} + \frac{2\pi}{3}, \qquad \frac{\pi/4 + 4\pi}{3} = \frac{\pi}{12} + \frac{4\pi}{3}.$$

These three roots are shown in Fig. 10.29, along with z.

EXERCISES FOR SEC. 10.9: COMPLEX NUMBERS

In Exercises 1 and 2 compute the given quantities:

1 (*a*) $(2 + 3i) + (\frac{1}{2} - 2i)$ (*b*) $(2 + 3i)(2 - 3i)$

(*c*) $\dfrac{1}{2 - i}$ (*d*) $\dfrac{3 + 2i}{4 - i}$

2 (*a*) $(2 + 3i)^2$ (*b*) $\dfrac{4}{3 - i}$

(*c*) $(1 + i)(3 - i)$ (*d*) $\dfrac{1 + 5i}{2 - 3i}$

3 Let z_1 have magnitude 2 and argument $\pi/6$, and let z_2 have magnitude 3 and argument $\pi/3$.
(*a*) Plot z_1 and z_2.
(*b*) Find z_1z_2 using the polar form.
(*c*) Write z_1 and z_2 in the form $x + iy$.
(*d*) With the aid of (*c*) compute z_1z_2.

4 Let z_1 have magnitude 2 and argument $\pi/4$, and let z_2 have magnitude 1 and argument $3\pi/4$.
(*a*) Plot z_1 and z_2.
(*b*) Find their product using the polar form.
(*c*) Write z_1 and z_2 in rectangular form, $x + iy$.
(*d*) With the aid of (*c*), compute z_1z_2.

5 The complex number z has argument $\pi/3$ and magnitude 1. Find and plot (*a*) z^2, (*b*) z^3, (*c*) z^4.

6 Find (*a*) i^3, (*b*) i^4, (*c*) i^5, (*d*) i^{73}.

7 If z has magnitude 2 and argument $\pi/6$, what are the magnitude and argument of (*a*) z^2? (*b*) z^3? (*c*) z^4? (*d*) z^n? (*e*) Sketch z, z^2, z^3, z^4.

8 Let z have magnitude 0.9 and argument $\pi/4$. (*a*) Find and plot z^2, z^3, z^4, z^5, z^6. (*b*) What happens to z^n as $n \to \infty$?

9 Find and plot all solutions of the equation $z^5 = 32[\cos(\pi/4) + i\sin(\pi/4)]$.

10 Find and plot all solutions of $z^4 = 8 + 8\sqrt{3}\,i$. (*Hint:* First draw $8 + 8\sqrt{3}i$.)

11 Let P have magnitude r and argument θ. Let Q have magnitude $1/r$ and argument $-\theta$. Show that $P \times Q = 1$. (Q is called the **reciprocal of** P, denoted P^{-1} or $1/P$.)

12 Find P^{-1} if $P = 4 + 4i$. (Use the definition in Exercise 11.)

13 (*a*) By substitution, verify that $2 + 3i$ is a solution of the equation $x^2 - 4x + 13 = 0$.
(*b*) Use the quadratic formula to find all solutions of the equation $x^2 - 4x + 13 = 0$.

14 (*a*) Use the quadratic formula to find the solutions of the equation $x^2 + x + 1 = 0$.
(*b*) Plot the solutions in (*a*).

15 Write in polar form: (*a*) $5 + 5i$, (*b*) $-\frac{1}{2} - \frac{\sqrt{3}}{2}i$, (*c*) $-\frac{\sqrt{2}}{2} + \frac{\sqrt{2}}{2}i$, (*d*) $3 + 4i$.

16 Write in rectangular form:

(*a*) $3\left(\cos\frac{3\pi}{4} + i\sin\frac{3\pi}{4}\right)$

(*b*) $2\left(\cos\frac{\pi}{6} + i\sin\frac{\pi}{6}\right)$

(*c*) $10(\cos\pi + i\sin\pi)$

(*d*) $\frac{1}{5}(\cos 22^\circ + i\sin 22^\circ)$ (Express the answer to three places.)

17 Let z_1 have magnitude r_1 and argument θ_1, and let z_2 have magnitude r_2 and argument θ_2.
(*a*) Explain why the magnitude of z_1/z_2 is r_1/r_2.
(*b*) Explain why the argument of z_1/z_2 is $\theta_1 - \theta_2$.

18 Compute

$$\frac{\cos\frac{5\pi}{4} + i\sin\frac{5\pi}{4}}{\cos\frac{3\pi}{4} + i\sin\frac{3\pi}{4}}$$

two ways: (*a*) by the result in Exercise 17, (*b*) by conjugating the denominator.

19 Compute

(*a*) $(2 + 3i)(1 + i)$ (*b*) $\dfrac{2 + 3i}{1 + i}$

(*c*) $(7 - 3i)(\overline{7 - 3i})$

(*d*) $3(\cos 42^\circ + i\sin 42^\circ) \cdot 5(\cos 168^\circ + i\sin 168^\circ)$

(*e*) $\dfrac{\sqrt{8}(\cos 147^\circ + i\sin 147^\circ)}{\sqrt{2}(\cos 57^\circ + i\sin 57^\circ)}$

(*f*) $1/(3 - i)$

(*g*) $[3(\cos 52^\circ + i\sin 52^\circ)]^{-1}$

(*h*) $\left(\cos\frac{\pi}{6} + i\sin\frac{\pi}{6}\right)^{12}$

20 Compute

(*a*) $(3 + 4i)(3 - 4i)$ (*b*) $\dfrac{3 + 5i}{-2 + i}$

(*c*) $\dfrac{1}{2 + i}$ (*d*) $\left(\cos\frac{\pi}{12} + i\sin\frac{\pi}{12}\right)^{20}$

(*e*) $[r(\cos\theta + i\sin\theta)]^{-1}$

(*f*) $\text{Re}\,((r(\cos\theta + i\sin\theta))^{10})$

(*g*) $\dfrac{3\left(\cos\frac{\pi}{6} + i\sin\frac{\pi}{6}\right)}{5 - 12i}$

21 Find and plot all solutions of $z^3 = i$.

22 Sketch all complex numbers z such that (*a*) $z^6 = 1$, (*b*) $z^6 = 64$.

■

23 Using the fact that

$$(\cos\theta + i\sin\theta)^n = \cos n\theta + i\sin n\theta$$

find formulas for $\cos 3\theta$ and $\sin 3\theta$ in terms of $\cos\theta$ and $\sin\theta$.

24 (*a*) If $|z_1| = 1$ and $|z_2| = 1$, how large can $|z_1 + z_2|$ be? (*Hint:* Draw some pictures.)
(*b*) If $|z_1| = 1$ and $|z_2| = 1$, what can be said about $|z_1 z_2|$?

25 Show that (*a*) $\overline{z_1 z_2} = \bar{z}_1\bar{z}_2$, (*b*) $\overline{z_1 + z_2} = \bar{z}_1 + \bar{z}_2$.

26 If arg z is θ, what is the argument of (*a*) $\bar{z}$, (*b*) $1/z$?

27 For which complex numbers z is $\bar{z} = 1/z$?

28 Let $z = \dfrac{1}{\sqrt{2}} + \dfrac{i}{\sqrt{2}}$.

(*a*) Compute z^2 algebraically.
(*b*) Compute z^2 by putting z into polar form.

29 Let $z = \dfrac{1}{2} + \dfrac{i}{2}$.

(*a*) Sketch the numbers z^n for $n = 1, 2, 3, 4$, and 5.
(*b*) What happens to z^n as $n \to \infty$?

30 Let $z = 1 + i$.
(*a*) Sketch the numbers $z^n/n!$ for $n = 1, 2, 3, 4$, and 5.
(*b*) What happens to $z^n/n!$ as $n \to \infty$?

31 Let a, b, and c be real numbers such that $b^2 - 4ac < 0$.
(*a*) Show that the equation $az^2 + bz + c = 0$ has two complex roots, r_1 and r_2.
(*b*) Show that $az^2 + bz + c = a(z - r_1)(z - r_2)$.
(*c*) Show that r_1 and r_2 are conjugates, $r_2 = \bar{r}_1$.

32 Let a, b, and c be complex numbers such that $a \neq 0$ and $b^2 - 4ac \neq 0$. Show that $az^2 + bz + c = 0$ has two roots.

33 The real number system occupies a line and the complex number system occupies a plane. It might be conjectured that there is a number system containing the complex numbers that occupies space and obeys the usual rules of arithmetic. *This exercise shows that no such system exists.*

Assume that each point in space corresponds to a unique "number" $a + bi + cj$, where $i^2 = -1$ and a, b, and c are real. Denote the product of i and j, ij, by $a + bi + cj$.

(*a*) Assuming the usual rules of arithmetic, show that $-j = ai - b + c(ij)$.
(*b*) From (*a*) deduce that j is of the form $x + iy$ for some real numbers x and y.
(*c*) From (*b*) obtain a contradiction.

34 Compute the roots of the following equations and plot them relative to the same axes.

(*a*) $x^2 - 3x + 2 = 0$ (*b*) $x^2 - 3x + 2.25 = 0$
(*c*) $x^2 - 3x + 2.5 = 0$

35 Let z_1 and z_2 be the roots of $ax^2 + bx + c = 0$, $a \neq 0$.

(*a*) Using the quadratic formula (or by other means), show that $z_1 + z_2 = -b/a$ and $z_1 z_2 = c/a$.
(*b*) From (*a*) deduce that

$$ax^2 + bx + c = a(x - z_1)(x - z_2).$$

(*c*) With the aid of (*b*) show that

$$\frac{1}{ax^2 + bx + c} = \frac{1}{a(z_1 - z_2)}\left(\frac{1}{x - z_1} - \frac{1}{x - z_2}\right).$$

Part (*c*) shows that the theory of partial fractions, described in Sec. 7.4, becomes much simpler when complex numbers are allowed as the coefficients of the polynomials. Only partial fractions of the type $k/(ax + b)^n$ are needed.

36 Find and plot the roots of $x^2 + ix + 3 - i = 0$.

37 Let $f(x) = a_0 + a_1x + a_2x^2 + a_3x^3 + a_4x^4$, where each coefficient a_i is real.

(*a*) Show that if c is a root of $f(x) = 0$, then so is $\bar{c}$.
(*b*) Show that if c is a root of $f(x) = 0$ and is not real, then $(x - c)(x - \bar{c})$ divides $f(x)$.
(*c*) Using the fundamental theorem of algebra, show that any fourth-degree polynomial with real coefficients can be expressed as the product of polynomials of degree at most 2 with real coefficients.

10.10 THE RELATION BETWEEN THE EXPONENTIAL AND THE TRIGONOMETRIC FUNCTIONS

With the aid of complex numbers, Euler in 1743 discovered that the trigonometric functions can be expressed in terms of the exponential function e^z, where z is complex. This section will retrace his discovery. In particular, it will show that

Surprise: Pulling trigonometry out of e^x

$$e^{i\theta} = \cos\theta + i\sin\theta, \qquad \cos\theta = \frac{e^{i\theta} + e^{-i\theta}}{2}, \qquad \text{and} \qquad \sin\theta = \frac{e^{i\theta} - e^{-i\theta}}{2i}.$$

The Maclaurin series for e^x when x is real suggests the following definition.

Definition *e^z for complex z.* Let z be a complex number. The sequence of complex numbers

$$1,\ 1 + z,\ 1 + z + \frac{z^2}{2!},\ \ldots,\ 1 + z + \frac{z^2}{2!} + \cdots + \frac{z^n}{n!},\ \ldots$$

approaches a complex number which will be called e^z. In short,

$$e^z = 1 + z + \frac{z^2}{2!} + \cdots + \frac{z^n}{n!} + \cdots = \sum_{n=0}^{\infty} \frac{z^n}{n!}.$$

We shall leave it to a more advanced course to show that the series converges, that is, that there is a complex number L such that

$$\lim_{k\to\infty}\left|\sum_{n=0}^{k}\frac{z^n}{n!} - L\right| = 0.$$

It can be shown that $e^{z_1+z_2} = e^{z_1}e^{z_2}$ in accordance with the basic law of exponents.

The following theorem, due to Euler, provides the key link between the exponential function e^z and the trigonometric functions $\cos\theta$ and $\sin\theta$.

Theorem 1 Let θ be a real number. Then

$$e^{i\theta} = \cos\theta + i\sin\theta.$$

Proof By definition of e^z for any complex number,

$$e^{i\theta} = 1 + i\theta + \frac{(i\theta)^2}{2!} + \frac{(i\theta)^3}{3!} + \frac{(i\theta)^4}{4!} + \cdots.$$

Thus

$$e^{i\theta} = 1 + i\theta + \frac{i^2\theta^2}{2!} + \frac{i^3\theta^3}{3!} + \frac{i^4\theta^4}{4!} + \cdots$$

Recall that $i^2 = -1$, $i^3 = -i$, $i^4 = 1$, $i^5 = i, \ldots$.

$$= 1 + i\theta - \frac{\theta^2}{2!} - \frac{i\theta^3}{3!} + \frac{\theta^4}{4!} + \cdots$$

$$= \left(1 - \frac{\theta^2}{2!} + \frac{\theta^4}{4!} - \cdots\right) + i\left(\theta - \frac{\theta^3}{3!} + \cdots\right)$$

$$= \cos\theta + i\sin\theta.$$

Figure 10.30 shows $e^{i\theta}$, which lies on the standard unit circle.

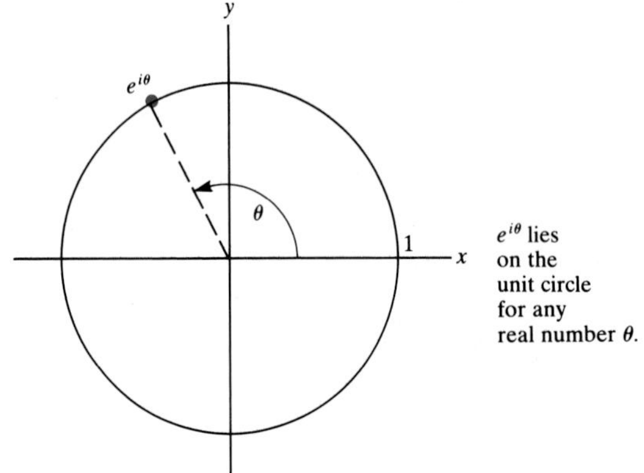

Figure 10.30 ■

Theorem 1 asserts, for instance, that

$$e^{\pi i} = \cos\pi + i\sin\pi = -1 + i\cdot 0 = -1,$$

or

$$e^{\pi i} = -1,$$

an equation that links e (the fundamental number in calculus), π (the fundamental number in trigonometry), i (the fundamental complex number), and the negative number -1. The history of that short equation would recall the struggles of hundreds of mathematicians to create the number system that we now take for granted.

With the aid of Theorem 1, both $\cos\theta$ and $\sin\theta$ may be expressed in terms of the exponential function.

Theorem 2 Let θ be a real number. Then

$$\cos\theta = \frac{e^{i\theta} + e^{-i\theta}}{2} \quad \text{and} \quad \sin\theta = \frac{e^{i\theta} - e^{-i\theta}}{2i}.$$

Proof By Theorem 1,

$$e^{i\theta} = \cos\theta + i\sin\theta. \tag{1}$$

Replacing θ by $-\theta$ in Eq. (1), we obtain

$$e^{-i\theta} = \cos\theta - i\sin\theta. \tag{2}$$

Addition of Eqs. (1) and (2) yields

$$e^{i\theta} + e^{-i\theta} = 2\cos\theta.$$

Hence

$$\cos\theta = \frac{e^{i\theta} + e^{-i\theta}}{2}.$$

Subtraction of Eq. (2) from Eq. (1) yields

$$e^{i\theta} - e^{-i\theta} = 2i\sin\theta.$$

Hence

$$\sin\theta = \frac{e^{i\theta} - e^{-i\theta}}{2i}.$$

This establishes the theorem. ■

Recall Sec. 6.11, where $\cosh x$ *and* $\sinh x$ *were defined.*

The hyperbolic functions $\cosh x$ and $\sinh x$ were defined in terms of the exponential function by

$$\cosh x = \frac{e^{x} + e^{-x}}{2} \quad \text{and} \quad \sinh x = \frac{e^{x} - e^{-x}}{2}.$$

Theorem 2 shows that the trigonometric functions could be similarly defined in terms of the exponential function—if complex numbers were available.

Old saying: "God created the complex numbers; anything less is the work of man."

Indeed, from the complex numbers and e^z we could even obtain the derivative formulas for $\sin\theta$ and $\cos\theta$. For instance,

$$(\sin\theta)' = \left(\frac{e^{i\theta} - e^{-i\theta}}{2i}\right)' = \frac{i\,e^{i\theta} + i\,e^{-i\theta}}{2i} = \cos\theta.$$

(That the familiar rules for differentiation extend to complex functions is justified in a course in complex variables.)

The magnitude and argument of e^{x+iy}

If $z = x + iy$, the evaluation of e^z can be carried out as follows:

$$e^z = e^{x+iy} = e^x e^{iy} = e^x(\cos y + i\sin y).$$

Thus the magnitude of e^{x+iy} is e^x and the argument of e^{x+iy} is y.

EXAMPLE 1 Compute and sketch (*a*) $e^{2+(\pi/6)i}$, (*b*) $e^{2+\pi i}$, and (*c*) $e^{2+3\pi i}$.

SOLUTION (*a*) $e^{2+(\pi/6)i}$ has magnitude e^2 and argument $\pi/6$. (*b*) $e^{2+\pi i}$ has magnitude e^2 and argument π; it equals $-e^2$. (*c*) $e^{2+3\pi i}$ has magnitude

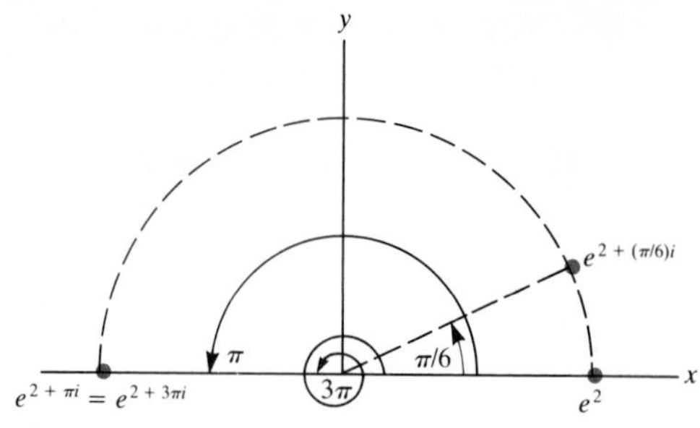

Figure 10.31

e^2 and argument 3π, so is the same number as the number in (*b*). The results are sketched in Fig. 10.31. ■

The next example illustrates a typical computation in alternating currents.

EXAMPLE 2 Find the real part of $100\, e^{j(\pi/6)}e^{j\omega t}$. Here t refers to time, ω is a real constant related to frequency, and j is the electrical engineers' symbol for $\sqrt{-1}$.

SOLUTION

$$\begin{aligned} 100\, e^{j(\pi/6)}e^{j\omega t} &= 100\, e^{j(\pi/6)+j\omega t} \\ &= 100\, e^{j[(\pi/6)+\omega t]} \\ &= 100\left[\cos\left(\frac{\pi}{6}+\omega t\right) + j\sin\left(\frac{\pi}{6}+\omega t\right)\right]. \end{aligned}$$

Thus

$$\text{Re}\,(100\, e^{j(\pi/6)}e^{j\omega t}) = 100\cos\left(\frac{\pi}{6}+\omega t\right). \quad ■$$

It is often convenient to think of $\cos\theta$ as $\text{Re}\,(e^{i\theta})$. The next example exploits this point of view.

EXAMPLE 3 Evaluate $\sum_{n=0}^{\infty} \frac{\cos n\theta}{n!}$.

SOLUTION Recall that $e^{in\theta} = \cos n\theta + i\sin n\theta$. Hence $\cos n\theta = \text{Re}\,(e^{in\theta})$, and we have

Recall the definition of e^z.

$$\begin{aligned} \sum_{n=0}^{\infty}\frac{\cos n\theta}{n!} &= \text{Re}\sum_{n=0}^{\infty}\frac{e^{in\theta}}{n!} = \text{Re}\sum_{n=0}^{\infty}\frac{(e^{i\theta})^n}{n!} \\ &= \text{Re}\,(e^{(e^{i\theta})}) \\ &= \text{Re}\,(e^{\cos\theta + i\sin\theta}) \\ &= \text{Re}\,(e^{\cos\theta}e^{i\sin\theta}) \\ &= \text{Re}\,\{e^{\cos\theta}[\cos(\sin\theta) + i\sin(\sin\theta)]\}. \end{aligned}$$

Hence

$$\sum_{n=0}^{\infty}\frac{\cos n\theta}{n!} = e^{\cos\theta}\cos(\sin\theta). \quad ■$$

EXERCISES FOR SEC. 10.10: THE RELATION BETWEEN THE EXPONENTIAL AND THE TRIGONOMETRIC FUNCTIONS

In Exercises 1 to 6 sketch the numbers given and state their real and imaginary parts.

1 $e^{(5\pi i/4)}$ **2** $5e^{(\pi i/4)}$

3 $2e^{(\pi i/4)} + 3e^{(\pi i/6)}$ **4** e^{2+3i}

5 $e^{(\pi i/6)}e^{(3\pi i/4)}$ **6** $2e^{\pi i} \cdot 3e^{-(\pi i/3)}$

In Exercises 7 to 10 express the given numbers in the form $re^{i\theta}$, for a positive real number r and argument θ, $-\pi < \theta \leq \pi$.

7 $\dfrac{e^2}{\sqrt{2}} - \dfrac{e^2}{\sqrt{2}}i$ **8** $3\left(\cos\dfrac{\pi}{4} + i\sin\dfrac{\pi}{4}\right)$

9 $5\left(\cos\dfrac{\pi}{6} + i\sin\dfrac{\pi}{6}\right) \cdot 3\left(\cos\dfrac{\pi}{2} + i\sin\dfrac{\pi}{2}\right)$

10 $7\left(\cos\dfrac{7\pi}{3} + i\sin\dfrac{7\pi}{3}\right)$

In Exercises 11 to 14 plot the given numbers.

11 $e^{(\pi i/4 + 3\pi i)}$ **12** $e^{1+(9\pi/4)i}$

13 $e^{2-(\pi/3)i}$ **14** $e^{[-1+(17\pi/6)i]}$

15 Let $z = e^{a+bi}$. Find (*a*) $|z|$, (*b*) $\bar{z}$, (*c*) z^{-1}, (*d*) Re z, (*e*) Im z, (*f*) arg z. [In (*f*) assume a and b are positive.]

16 For which values of a and b is $\lim_{n\to\infty} (e^{a+ib})^n = 0$?

17 Find all z such that $e^z = 1$.

18 Find all z such that $e^z = -1$.

■

19 In Claude Garrod's *Twentieth Century Physics,* Faculty Publishing, Davis, Calif., p. 107, there is the remark: "Using the fact that

$$(e^{-i\omega_0 t})^*(e^{-i\omega_0 t}) = 1,$$

we can easily evaluate the probability density for these standing waves." In this text, z^* denotes the conjugate of z and ω_0 is real. Justify the equation.

20 Use the fact that $1 + \cos\theta + \cos 2\theta + \cdots + \cos(n-1)\theta$ is the real part of $1 + e^{\theta i} + e^{2\theta i} + \cdots + e^{(n-1)\theta i}$ to find a short formula for that trigonometric sum.

21 Find all z such that $e^z = 3 + 4i$. (*Hint:* Write $z = x + iy$.)

22 Assuming that $e^{z_1+z_2} = e^{z_1}e^{z_2}$ for complex numbers z_1 and z_2, obtain the trigonometric identities for $\cos(A + B)$ and $\sin(A + B)$.

23 Evaluate $\displaystyle\sum_{n=0}^{\infty} \frac{\cos n\theta}{2^n}$.

24 Find $\displaystyle\sum_{n=0}^{\infty} \frac{\sin n\theta}{n!}$.

Exercises 25 and 26 treat the complex logarithms of a complex number. They show that $z = \ln w$ is not single valued.

25 Let w be a nonzero complex number. Show that there are an infinite number of z such that $e^z = w$.

26 (See Exercise 25.) If $e^z = w$, write $z = \ln w$, although $\ln w$ is not a uniquely defined number. If b is a nonzero complex number and q is a complex number, define b^q to be $e^{q \ln b}$. Since $\ln b$ is not unique, b^q is usually not unique. List all possible values of (*a*) $(-1)^i$, (*b*) $10^{1/2}$, (*c*) 10^3.

27 (*a*) How would you define $\sinh z$ and $\cosh z$ for a complex number z?
(*b*) Show that $\sinh(ix) = i\sin x$ and $\cosh(ix) = \cos x$. [Note the close relation that (*b*) establishes between the hyperbolic and the circular functions.]

28 (*a*) How would you define $\sin z$ and $\cos z$ for complex z?
(*b*) When z is real, $|\sin z| \leq 1$ and $|\cos z| \leq 1$. Do these inequalities hold for all complex z?

29 Let $z = \dfrac{1+i}{\sqrt{2}}$.
(*a*) Plot z, $z^2/2!$, $z^3/3!$, $z^4/4!$.
(*b*) Plot $1 + z + z^2/2! + z^3/3! + z^4/4!$, which is an estimate of $\exp[(1 + i)/\sqrt{2}]$.
(*c*) Plot $\exp[(1 + i)/\sqrt{2}]$ relative to the same axes.

30 An integral table lists $\int xe^{ax}\,dx = e^{ax}(ax - 1)/a^2$. At first glance, finding the integral of $xe^{ax}\cos bx$ may appear to be a much harder problem. However, by noticing that $\cos bx = \text{Re}(e^{ibx})$, we can reduce it to a simpler problem. Following this approach, find $\int xe^{ax}\cos bx\,dx$. *Suggestion:* The formula for $\int xe^{ax}\,dx$ holds when a is complex.

31 Let a and b be real numbers. Explain why $\int \sin^m ax \cos^n bx\,dx$ is an elementary function for any positive integers m and n. (Do not evaluate the integral.)

10.11 LINEAR DIFFERENTIAL EQUATIONS WITH CONSTANT COEFFICIENTS (Optional)

This section treats a type of differential equation that many engineering and physics students may meet even before they take a D.E. course. It is intended to serve as a reference. (In Theorem 4 it makes use of the complex numbers.)

The differential equation $dy/dx = ay$, or equivalently,

$$\frac{dy}{dx} - ay = 0 \tag{1}$$

was solved in Sec. 6.7. Any solution has to be of the form $y = Ae^{ax}$ for some constant A. This section is concerned with generalizations of Eq. (1).

First, we consider differential equations of the form

$$\frac{dy}{dx} + ay = f(x), \tag{2}$$

where a is a real constant and $f(x)$ is some function of x. [Equation (1) is the special case where $f(x) = 0$.] Equation (2) is called a **first-order linear differential equation with constant coefficients**.

Second, we consider the second-order equation

$$\frac{d^2y}{dx^2} + b\frac{dy}{dx} + cy = f(x), \tag{3}$$

where b and c are real constants. For some b and c, solving Eq. (3) may use complex numbers even though the solution will be a real function.

An engineer or physicist will meet Eq. (3) in the form

$$L\frac{d^2q}{dt^2} + R\frac{dq}{dt} + \frac{q}{C} = V \sin \omega t$$

in the study of electric currents. Here q is a charge that varies with time, dq/dt is current, $V \sin \omega t$ describes an applied voltage, R is resistance, L is inductance, and C is a constant describing the capacitor. They also meet Eq. (3) in the study of motion in the form

$$m\frac{d^2x}{dt^2} + b\frac{dx}{dt} + kx = F_0 \sin \omega t.$$

Here x describes the location of a particle moving on a line, $F_0 \sin \omega t$ is an applied force, $b(dx/dt)$ describes a damping effect, kx describes the force of a spring, and m is the mass. (See A. Hudson and R. Nelson, *University Physics*, p. 685, Harcourt Brace Jovanovich, New York, 1982.)

Imagine for the moment that you have found a particular solution y_p of Eq. (2) and a solution y_1 of the associated *homogeneous* equation obtained from Eq. (2) by replacing $f(x)$ by 0,

The homogeneous case

$$\frac{dy}{dx} + ay = 0. \tag{4}$$

A straightforward computation then shows that $y_p + y_1$ is a solution of Eq. (2), as follows:

If you know one solution of Eq. (2) and all solutions of Eq. (4), you know all solutions of Eq. (2).

$$\begin{aligned}\frac{d}{dx}(y_p + y_1) + a(y_p + y_1) &= \frac{dy_p}{dx} + \frac{dy_1}{dx} + ay_p + ay_1 \\ &= \left(\frac{dy_p}{dx} + ay_p\right) + \left(\frac{dy_1}{dx} + ay_1\right) \\ &= f(x) + 0 = f(x).\end{aligned}$$

Now, the function $y_1 = Ce^{-ax}$, for any constant C, is a solution of Eq. (4). Thus, if y_p is a solution of Eq. (2), then so is $y_p + Ce^{-ax}$. In fact, each solution of Eq. (2) must be of the form $y_p + Ce^{-ax}$. To see why, assume that y_p and y both satisfy Eq. (2). Then

$$\begin{aligned}\frac{d}{dx}(y - y_p) + a(y - y_p) &= \left(\frac{dy}{dx} + ay\right) - \left(\frac{dy_p}{dx} + ay_p\right) \\ &= f(x) - f(x) = 0.\end{aligned}$$

Thus $y - y_p$, being a solution of Eq. (4), must be of the form Ce^{-ax} for some constant C. Thus $y = y_p + Ce^{-ax}$.

These observations are summarized in the following theorem.

Theorem 1 Let y_p be a particular solution of the differential equation

$$\frac{dy}{dx} + ay = f(x).$$

Then the most general solution is

$$y_p + Ce^{-ax}. \quad \blacksquare$$

EXAMPLE 1 Solve the differential equation

$$\frac{dy}{dx} + 3y = 12.$$

SOLUTION One solution is the constant function $y_p = 4$. The most general solution is, therefore, $4 + Ce^{-3x}$ for any constant C. ■

Once a particular solution y_p has been found, Theorem 1 provides the general solution. Example 2 illustrates one technique for finding y_p.

EXAMPLE 2 Find all solutions of the differential equation

$$\frac{dy}{dx} - y = \sin x. \tag{5}$$

Start by guessing what a solution might look like.

SOLUTION First find one solution. Since $f(x) = \sin x$, let us see if there is a solution of the form $y_p = A \cos x + B \sin x$, for some constants A and B. Substitution in Eq. (5) yields

$$\frac{d}{dx}(A \cos x + B \sin x) - (A \cos x + B \sin x) = \sin x.$$

So we want

$$-A \sin x + B \cos x - A \cos x - B \sin x = \sin x,$$

or simply, $$(-A - B) \sin x + (B - A) \cos x = \sin x.$$

We are just equating coefficients of $\sin x$ *and* $\cos x$ *on both sides.*

Choose A and B such that $-A - B = 1$ and $B - A = 0$. It follows that $B = A$ and that $-A - (A) = 1$ or $A = -\frac{1}{2}$. Consequently,

$$y_p = -\tfrac{1}{2}\cos x - \tfrac{1}{2}\sin x$$

is a solution of Eq. (5), as may be checked by substitution in Eq. (5).

The general solution of the homogeneous equation $dy/dx - y = 0$ is Ce^x, so the general solution of Eq. (5) is

$$y = -\tfrac{1}{2}\cos x - \tfrac{1}{2}\sin x + Ce^x. \quad \blacksquare$$

Example 2 uses the method of **undetermined coefficients**: Guess a general form of the solution and see if the unknown constants can be chosen properly to yield a solution of the differential equation.

Before turning to solutions of Eq. (3), consider the special case when $f(x)$ is identically 0, the so-called **homogeneous** case.

Let us find all solutions of the homogeneous equation

Homogeneous linear differential equation of second order

$$\frac{d^2y}{dx^2} + b\frac{dy}{dx} + cy = 0. \tag{6}$$

If y_1 and y_2 are both solutions of Eq. (6), a straightforward computation shows that $C_1y_1 + C_2y_2$ is also a solution of Eq. (6) for any choice of constants C_1 and C_2. [Since Eq. (6) involves the second derivative of y, we expect the general solution for y to contain two arbitrary constants.]

EXAMPLE 3 Solve

$$\frac{d^2y}{dx^2} - 3\frac{dy}{dx} + 2y = 0. \tag{7}$$

SOLUTION Recalling our experience with Eq. (1), we are tempted to look for a solution of the form e^{kx} for some constant k.

Substitution of e^{kx} into Eq. (7) yields

$$\frac{d^2}{dx^2}(e^{kx}) - 3\frac{d}{dx}(e^{kx}) + 2e^{kx} = 0,$$

or

$$k^2e^{kx} - 3ke^{kx} + 2e^{kx} = 0,$$

which is equivalent to

$$k^2 - 3k + 2 = 0. \tag{8}$$

By the quadratic formula, $k = 1$ or $k = 2$. Thus $y_1 = e^x$ and $y_2 = e^{2x}$ are solutions of Eq. (7). Consequently,

$$C_1e^x + C_2e^{2x} \tag{9}$$

is a solution of Eq. (7) for any choice of constants C_1 and C_2. (It can be proved that there are no other solutions.) ■

The most general solution of the differential equation

$$\frac{d^2y}{dx^2} + 6\frac{dy}{dx} + 9y = 0 \tag{10}$$

is of a different form. If we try $y = e^{kx}$, we obtain

$$k^2e^{kx} + 6ke^{kx} + 9e^{kx} = 0$$
$$e^{kx}(k^2 + 6k + 9) = 0$$
$$(k + 3)^2 = 0$$
$$k = -3.$$

A repeated root produces only one solution of the form e^{kx}.

This gives only the solutions of the form $y = Ce^{-3x}$. However, a second-order equation should possess a solution containing *two* arbitrary constants. Let us seek all solutions of the form

$$y = v(x)e^{-3x},$$

hoping to find some not of the form Ce^{-3x}.

Straightforward computations give

$$\frac{dy}{dx} = v(x)(-3e^{-3x}) + v'(x)e^{-3x} = -3v(x)e^{-3x} + v'(x)e^{-3x}$$

and
$$\frac{d^2y}{dx^2} = 9v(x)e^{-3x} - 6v'(x)e^{-3x} + v''(x)e^{-3x}.$$

Substituting into Eq. (10) yields

$$9v(x)e^{-3x} - 6v'(x)e^{-3x} + v''(x)e^{-3x} - 18v(x)e^{-3x} + 6v'(x)e^{-3x} + 9v(x)e^{-3x} = 0,$$

Check the algebra

which simplifies to
$$v''(x)e^{-3x} = 0;$$

hence to
$$v''(x) = 0.$$

Therefore, $v(x) = C_1 + C_2x$, and our general solution is

$$y = C_1e^{-3x} + C_2xe^{-3x},$$

for arbitrary constants C_1 and C_2.

The key to the nature of the solutions of Eq. (6) lies in the **associated quadratic equation**

$$t^2 + bt + c = 0. \tag{11}$$

The type of solution to Eq. (6) depends on the nature of the roots of Eq. (11). There are three cases: two distinct real roots, a repeated root (necessarily real), and two distinct complex roots. Each case will be described by a corresponding theorem.

Distinct real roots

Theorem 2 If $b^2 - 4c$ is positive, Eq. (11) has two distinct real roots, r_1 and r_2. In this case, the general solution of Eq. (6) is

$$C_1e^{r_1x} + C_2e^{r_2x}. \quad ■$$

The proof that $C_1e^{r_1x} + C_2e^{r_2x}$ is a solution is left to the reader. Theorem 2 covers the differential equation (7).

EXAMPLE 4 Solve $\dfrac{d^2y}{dx^2} + 5\dfrac{dy}{dx} + y = 0.$

SOLUTION In this case, $b^2 - 4c = 21$, which is positive. The roots of the associated quadratic equation are

$$r_1 = \frac{-5 + \sqrt{21}}{2} \quad \text{and} \quad r_2 = \frac{-5 - \sqrt{21}}{2}.$$

The general solution of the differential equation is

$$C_1e^{[(-5+\sqrt{21})/2]x} + C_2e^{[(-5-\sqrt{21})/2]x}. \quad ■$$

The next theorem concerns the special case when the associated quadratic equation $t^2 + bt + c = 0$ has a repeated root, r.

Repeated root

Theorem 3 If $b^2 - 4c = 0$, Eq. (11) has a repeated root r. In this case, the general solution of Eq. (6) is

$$C_1e^{rx} + C_2xe^{rx} = (C_1 + C_2x)e^{rx}. \quad ■$$

That $(C_1 + C_2x)e^{rx}$ is a solution is left to the reader to check by substitution. Theorem 3 is illustrated by the solution of Eq. (10).

Distinct complex (nonreal) roots

Theorem 4 If $b^2 - 4c$ is negative, Eq. (11) has two distinct complex roots $r_1 = p + iq$ and $r_2 = p - iq$. In this case, the general solution of Eq. (6) is

$$(C_1 \cos qx + C_2 \sin qx)\, e^{px}.$$

Proof Just as in Theorem 2,

Here the complex numbers enter.

$$y = A_1e^{r_1x} + A_2e^{r_2x} \tag{12}$$

is a solution of Eq. (6) for any choice of constants A_1 and A_2, even complex. Unfortunately, (12) will usually be complex. In order to find a *real* function that satisfies Eq. (6), expand (12):

$$\begin{aligned} A_1e^{r_1x} + A_2e^{r_2x} &= A_1e^{(p+iq)x} + A_2e^{(p-iq)x} \\ &= A_1e^{px}e^{iqx} + A_2e^{px}e^{-iqx} \\ &= e^{px}[A_1(\cos qx + i \sin qx) + A_2(\cos(-qx) + i \sin(-qx))] \\ &= e^{px}[(A_1 + A_2)\cos qx + i(A_1 - A_2)\sin qx]. \end{aligned}$$

Getting some real solutions

Appropriate choices of A_1 and A_2 will generate the desired solution. Choosing $A_1 = A_2 = \frac{1}{2}$ produces the real solution $e^{px} \cos qx$. Next, choose A_1 and A_2 so that $A_1 + A_2 = 0$ and $i(A_1 - A_2) = 1$. This will produce the real solution $e^{px} \sin qx$. [To find A_1 and A_2, solve the simultaneous equations $A_1 + A_2 = 0$ and $i(A_1 - A_2) = 1$. The solutions are $A_1 = -i/2$, and $A_2 = i/2$, which may be found by algebra.]

Thus

$$C_1e^{px} \cos qx + C_2e^{px} \sin qx \tag{13}$$

is a real-valued solution of Eq. (6) for any choice of real constants C_1 and C_2. It can be proved that there are no other real solutions. ■

EXAMPLE 5 Find the general solution of the differential equation of harmonic motion,

$$\frac{d^2y}{dx^2} = -k^2y, \tag{14}$$

where k is a constant.

SOLUTION Rewrite Eq. (14) in the form

$$\frac{d^2y}{dx^2} + k^2y = 0,$$

which has the associated quadratic equation $t^2 + k^2 = 0$. The roots of this equation are $0 + ki$ and $0 - ki$. By Theorem 4, the general solution of Eq. (14) is

$$C_1 e^{0x} \cos kx + C_2 e^{0x} \sin kx,$$

or simply,

$$C_1 \cos kx + C_2 \sin kx. \quad \blacksquare$$

Equation (14) describes the motion of a mass bobbing at the end of a spring. The height of the mass at time x is y. Since the motion is oscillatory, it is plausible that it is described by a combination of $\cos kx$ and $\sin kx$. (See Sec. 4.5 for further discussion, where the motion is described as "harmonic.")

The equation $\frac{d^2y}{dx^2} + b\frac{dy}{dx} + cy = f(x)$

If y_p is any particular solution of

$$\frac{d^2y}{dx^2} + b\frac{dy}{dx} + cy = f(x), \tag{15}$$

and y^* is a solution of the associated homogeneous equation (6), then $y_p + y^*$ is a solution of Eq. (15), as may be checked by a straightforward calculation. Since we know how to find the general solution of Eq. (6), all that remains is to find a particular solution of Eq. (15). This can often be accomplished by a shrewd guess and the use of undetermined coefficients, as illustrated by the following example.

EXAMPLE 6 Solve the differential equation

$$\frac{d^2y}{dx^2} + \frac{dy}{dx} + 2y = 2x^2 + 5. \tag{16}$$

If y has degree n why does the left side of (16) have degree n?

SOLUTION Since $2x^2 + 5$ is a polynomial, let us seek a polynomial solution. If there is such a solution, it cannot have degree greater than 2, since the right-hand side of Eq. (16) has degree 2. So try $y = Ax^2 + Bx + C$; hence $y' = 2Ax + B$ and $y'' = 2A$. Substitution in Eq. (16) gives

$$2A + (2Ax + B) + 2(Ax^2 + Bx + C) = 2x^2 + 5,$$

or

$$2Ax^2 + (2A + 2B)x + (2A + B + 2C) = 2x^2 + 5.$$

Comparing coefficients gives $2A = 2$, $2A + 2B = 0$, and $2A + B + 2C = 5$. Thus $A = 1$, $B = -1$, and $C = 2$.

Consequently, $y_p = x^2 - x + 2$ is a particular solution of Eq. (16).

Next, turn to solving the associated homogeneous equation

$$\frac{d^2y}{dx^2} + \frac{dy}{dx} + 2y = 0. \tag{17}$$

Here $b = 1$ and $c = 2$, so $b^2 - 4c = -7$. The roots of the associated quadratic equation $t^2 + t + 2 = 0$ are

$$\frac{-1 \pm \sqrt{-7}}{2} = \frac{-1}{2} \pm \frac{\sqrt{7}}{2} i.$$

By Theorem 4, the general solution of Eq. (17) is

$$y^* = C_1 e^{-x/2} \cos \frac{\sqrt{7}}{2} x + C_2 e^{-x/2} \sin \frac{\sqrt{7}}{2} x.$$

Putting everything together, we obtain the general solution of Eq. (16), namely,

$$y = x^2 - x + 2 + C_1e^{-x/2}\cos\frac{\sqrt{7}}{2}x + C_2e^{-x/2}\sin\frac{\sqrt{7}}{2}x. \quad \blacksquare$$

Guessing a particular solution of Eq. (15) depends on the form of $f(x)$. This table describes the most common cases:

Form of $f(x)$	*Guess for* y_p
A polynomial	Another polynomial
e^{kx} (k not a root of associated quadratic equation)	Ae^{kx}
xe^{kx} (k not a root of the associated quadratic equation)	$(A + Bx)e^{kx}$
$e^{kx}\sin qx$ or $e^{kx}\cos qx$ ($k + qi$ not a root of the associated quadratic equation)	$Ae^{kx}\cos qx + Be^{kx}\sin qx$

A complete handbook of mathematical tables includes several pages of specific solutions for a much wider variety of functions $f(x)$ that appear on the right side of Eq. (15).

EXERCISES FOR SEC. 10.11: LINEAR DIFFERENTIAL EQUATIONS WITH CONSTANT COEFFICIENTS

In Exercises 1 to 22 find all solutions of the given differential equations.

1 $y' + 2y = 0.$
2 $y' + 2y = \cos x$
3 $3y' + 12y = x$
4 $y' - \frac{1}{3}y = e^{2x}$
5 $y' - y = x^2$
6 $y'/2 + y = xe^{2x}$
7 $y'' - 2y' - 3y = 0$
8 $y'' + 5y' + 6y = 0$
9 $2y'' - y' - 3y = 0$
10 $2y'' - y' + 3y = 0$
11 $4y'' - 12y' + 9y = 0$
12 $4y'' + 9y = 0$
13 $y'' - 3y' + y = 0$
14 $3y'' - 2y' + 3y = 0$
15 $y'' - 6y' + 9y = 0$
16 $y'' + y' + y = 0$
17 $y'' - 2\sqrt{11}y' + 11y = 0$
18 $y'' - 3y' + 4y = 0$
19 $y'' - 2y' - 3y = e^{2x}$
20 $y'' + y' + y = x^2$
21 $y'' - 4y' + y = \cos 3x$
22 $y'' + 3y' + 2y = e^{-2x}\sin x + \cos 3x$

■

23 (*a*) Show that $y = e^{-ax}\int e^{ax}f(x)\,dx$ is a solution of $y' + ay = f(x)$.
(*b*) Use (*a*) to find a solution of $y' + y = 1/(1 + e^x)$.
(*c*) Find all solutions of the equation in (*b*).

24 Check that $C_1e^{-3x} + C_2xe^{-3x}$ is a solution of Eq. (10).

25 (*a*) Show that if $b^2 - 4c = 0$, then $t^2 + bt + c = 0$ has only one root, r, and $(t - r)^2 = t^2 + bt + c$.
(*b*) Check that $C_1e^{rx} + C_2xe^{rx}$ is a solution of $y'' + by' + c = 0$.

26 Let k be a nonzero constant. Find all solutions of the equation $y'' = k^2y$.

In some tables of solutions to differential equations, y'' is written D^2y and y' is written Dy. The equation $y'' + by' + c = f(x)$ is then expressed as $(D^2 + bD + c)y = f(x)$. Similarly, $y'' - 2ay' + a^2y = (D^2 - 2aD + a^2)y = (D - a)^2y$.

27 Verify that $x^2e^{ax}/2$ is a solution of $(D - a)^2y = e^{ax}$.

28 Verify that $e^{rx}/(r - a)^2$ is a solution of $(D - a)^2y = e^{rx}$, if $r \neq a$.

29 Verify that $(-x\cos bx)/(2b)$ is a solution of $(D^2 + b^2)y = \sin bx$, $b \neq 0$.

30 Verify that $(\sin sx)/(b^2 - s^2)$ is a solution of $(D^2 + b^2)y = \sin sx$, $s^2 \neq b^2$.

10.12 THE BINOMIAL THEOREM FOR ANY EXPONENT (Optional)

If n is a positive integer and x is any number, then

$$(1 + x)^n = 1 + nx + \frac{n(n-1)}{1 \cdot 2}x^2 + \frac{n(n-1)(n-2)}{1 \cdot 2 \cdot 3}x^3 + \cdots + \frac{n(n-1)\cdots 1}{1 \cdot 2 \cdots n}x^n, \quad (1)$$

as stated in Sec. 4.3. In fact, the right side of Eq. (1) is the Maclaurin series for $(1 + x)^n$; all powers from x^{n+1} on have coefficient 0. In this section we examine the Maclaurin series for $(1 + x)^r$, where r is not a positive integer or 0. It turns out that for $|x| < 1$, the function $(1 + x)^r$ is indeed represented by its associated Maclaurin series. However, as mentioned in Sec. 10.8, the formula

$$R_n(x; 0) = \frac{f^{(n+1)}(c_n)x^{n+1}}{(n+1)!}$$

is inadequate for showing that $\lim_{n\to\infty} R_n(x; 0) = 0$. The integral form of the remainder is needed.

Consider the Maclaurin series for $f(x) = (1 + x)^r$, where r is *not* a positive integer or 0. The following table will help in computing $f^{(n)}(0)$:

n	$f^{(n)}(x)$	$f^{(n)}(0)$
0	$(1 + x)^r$	1
1	$r(1 + x)^{r-1}$	r
2	$r(r-1)(1 + x)^{r-2}$	$r(r-1)$
3	$r(r-1)(r-2)(1 + x)^{r-3}$	$r(r-1)(r-2)$
⋯	⋯	⋯
n	$r(r-1)\cdots(r-n+1)(1 + x)^{r-n}$	$r(r-1)(r-2)\cdots(r-n+1)$

Consequently, the Maclaurin series associated with $(1 + x)^r$ is

$$1 + rx + \frac{r(r-1)}{1 \cdot 2}x^2 + \frac{r(r-1)(r-2)}{1 \cdot 2 \cdot 3}x^3 + \cdots. \quad (2)$$

The two key questions in this section

Note that the series does not stop, for r is not a positive integer or 0. For which x does series (2) converge? If it does converge, does it represent $(1 + x)^r$?

To get a feeling for series (2), consider the case $r = -1$. When $r = -1$, series (2) becomes

$$1 + (-1)x + \frac{(-1)(-2)}{1 \cdot 2}x^2 + \frac{(-1)(-2)(-3)}{1 \cdot 2 \cdot 3}x^3 + \cdots,$$

So this geometric series is a special case of the binomial series.

or

$$1 - x + x^2 - x^3 + \cdots.$$

This series converges for $|x| < 1$. Moreover, it does represent the function $(1 + x)^r = (1 + x)^{-1}$, for it is a geometric series with first term 1 and ratio $-x$.

EXAMPLE 1 Show that series (2) converges when $|x| < 1$ and diverges when $|x| > 1$.

SOLUTION For $x = 0$, the series clearly converges. So consider $x \neq 0$. Let a_n be the term containing the power x^n. Then

$$a_n = \frac{r(r-1)(r-2)\cdots(r-n+1)}{1\cdot 2\cdot 3\cdots n}x^n,$$

and

$$a_{n+1} = \frac{r(r-1)(r-2)\cdots(r-n)}{1\cdot 2\cdot 3\cdots(n+1)}x^{n+1}.$$

Thus

$$\left|\frac{a_{n+1}}{a_n}\right| = \left|\frac{\dfrac{r(r-1)(r-2)\cdots(r-n)}{1\cdot 2\cdot 3\cdots(n+1)}x^{n+1}}{\dfrac{r(r-1)(r-2)\cdots(r-n+1)}{1\cdot 2\cdot 3\cdots n}x^n}\right|$$

$$= \left|\frac{r-n}{n+1}x\right|.$$

Since r is fixed, $\lim_{n\to\infty}\left|\frac{a_{n+1}}{a_n}\right| = |x|.$

By the absolute-ratio test, series (2) converges when $|x| < 1$ and diverges when $|x| > 1$. ■

The series (2) converges when $|x| < 1$, *but does it converge to* $f(x) = (1 + x)^r$? This will be answered with the aid of the integral form of the remainder and the following lemma.

Lemma Let x be a fixed number in $(-1, 1)$. Then if t is in the closed interval whose endpoints are 0 and x,

$$\left|\frac{x-t}{1+t}\right| \le |x|. \quad ■$$

The proof is outlined in Exercise 16.

To show that series (2) converges to $(1 + x)^r$, we must show that $R_n(x; 0)$ approaches 0 as $n \to \infty$.

By Sec. 10.8,

$$R_n(x; 0) = \frac{1}{n!}\int_0^x f^{(n+1)}(t)(x-t)^n\,dt.$$

Now, $f(x) = (1 + x)^r$. Repeated differentiation shows that

$$f^{(n+1)}(t) = r(r-1)(r-2)\cdots(r-n)(1+t)^{r-n-1}.$$

Thus

$$R_n(x; 0) = \frac{r(r-1)(r-2)\cdots(r-n)}{n!}\int_0^x \left(\frac{x-t}{1+t}\right)^n (1+t)^{r-1}\,dt. \quad (3)$$

Consider x in $(-1, 1)$. By the lemma and Sec. 8.6,

$$\left|\int_0^x \left(\frac{x-t}{1+t}\right)^n (1+t)^{r-1}\,dt\right| \le \left|\int_0^x |x^n(1+t)^{r-1}|\,dt\right|$$

$$= |x|^n \left|\int_0^x (1+t)^{r-1}\,dt\right|.$$

Thus

$$|R_n(x; 0)| \leq \left|\frac{r(r-1)(r-2)\cdots(r-n)}{n!}\right| |x|^n \left|\int_0^x (1+t)^{r-1}\,dt\right|. \tag{4}$$

The integral in series (4) does not depend on n. The remaining part of series (4) is simply the absolute value of a typical term in series (2); hence it approaches 0 as $n \to \infty$. [Example 1 showed that series (2) converges; therefore its nth term approaches 0 as $n \to \infty$.] Consequently, $|R_n(x; 0)| \to 0$ as $n \to \infty$. This establishes the binomial theorem for arbitrary exponents and $|x| < 1$.

BINOMIAL THEOREM

If $|x| < 1$ and r is any real number, then

$$(1+x)^r = 1 + rx + \frac{r(r-1)}{2!}x^2 + \cdots + \frac{r(r-1)\cdots(r-n+1)}{n!}x^n + \cdots.$$

No mention has been made of the cases $x = 1$ or -1. When $x = 1$, the series diverges for $r \leq -1$, but for $r > -1$ it converges to $(1 + x)^r = 2^r$. When $x = -1$, the series converges to $(1 + x)^r = 0$ when $r \geq 0$, but it diverges when $r < 0$. These assertions are usually justified in advanced calculus.

To put the binomial theorem in summation notation, introduce the generalized binomial coefficient

$$\binom{r}{n} = \frac{r(r-1)\cdots(r-n+1)}{n!}$$

for any real number r and nonnegative integer n. Then

$$(1+x)^r = \sum_{n=0}^{\infty} \binom{r}{n} x^n \qquad \text{for } |x| < 1.$$

EXAMPLE 2 Write out the first four terms of the binomial expansion of $1/\sqrt{1+x} = (1+x)^{-1/2}$, for $|x| < 1$.

SOLUTION In this case, $r = -\frac{1}{2}$. Thus series (2) becomes

$$(1+x)^{-1/2} = 1 + (-\tfrac{1}{2})x + \frac{(-\frac{1}{2})(-\frac{3}{2})}{2!}x^2 + \frac{(-\frac{1}{2})(-\frac{3}{2})(-\frac{5}{2})}{3!}x^3 + \cdots$$

$$= 1 - \frac{1}{2}x + \frac{1 \cdot 3}{2^2 \cdot 2!}x^2 - \frac{1 \cdot 3 \cdot 5}{2^3 \cdot 3!}x^3 + \cdots. \quad \blacksquare$$

The binomial series for $1/\sqrt{1-x^2}$ is often used (see Exercises 9 and 18 to 23).

EXAMPLE 3 Find the Maclaurin series for $1/\sqrt{1-x^2}$, if $|x| < 1$.

SOLUTION Replace x in $1/\sqrt{1+x}$ by $-x^2$. Then, by Example 2,

$$\frac{1}{\sqrt{1-x^2}} = (1-x^2)^{-1/2}$$
$$= 1 - \frac{1}{2}(-x^2) + \frac{1\cdot 3}{2^2\cdot 2!}(-x^2)^2 - \frac{1\cdot 3\cdot 5}{2^3\cdot 3!}(-x^2)^3 + \cdots$$
$$= 1 + \frac{1}{2}x^2 + \frac{1\cdot 3}{2^2\cdot 2!}x^4 + \frac{1\cdot 3\cdot 5}{2^3\cdot 3!}x^6 + \cdots. \quad ■$$

EXAMPLE 4 Write out the first four terms of the binomial series for $1/(1+x)^2$.

SOLUTION In this case, $r = -2$. Thus

$$\frac{1}{(1+x)^2} = (1+x)^{-2}$$
$$= 1 + (-2)x + \frac{(-2)(-3)}{2!}x^2 + \frac{(-2)(-3)(-4)}{3!}x^3 + \cdots$$
$$= 1 - 2x + 3x^2 - 4x^3 + \cdots. \quad ■$$

EXERCISES FOR SEC. 10.12: THE BINOMIAL THEOREM FOR ANY EXPONENT

In Exercises 1 to 4 write out the first five terms of the binomial expansions of the given functions.

1 $(1+x)^{1/2}$
2 $(1+x)^{1/3}$
3 $(1+x)^{-3}$
4 $(1+x)^{-4}$

In Exercises 5 to 7 write the first four nonzero terms in the Maclaurin series for the given functions.

5 $(1+x^3)^{1/2}$
6 $(1+x^2)^{1/3}$
7 $(1-x^2)^{1/3}$

8 Show that if $|x| > 1$ and r is not a nonnegative integer, the binomial series (2) does not converge.

9 In R. P. Feynman, *Lectures on Physics,* Addison-Wesley, Reading, Mass., 1963, this statement appears in sec. 15.8 of vol. 1:

> An approximate formula to express the increase of mass, for the case when the velocity is small, can be found by expanding $m_0/\sqrt{1-v^2/c^2} = m_0(1-v^2/c^2)^{-1/2}$ in a power series, using the binomial theorem. We get
>
> $$m_0\left(1-\frac{v^2}{c^2}\right)^{-1/2} = m_0\left(1+\frac{1}{2}\frac{v^2}{c^2}+\frac{3}{8}\frac{v^4}{c^4}+\cdots\right).$$
>
> We see clearly from the formula that the series converges rapidly when v is small and the terms after the first two or three are negligible.

Check the expansion and justify the equation.

10 In *Introduction to Fluid Mechanics,* by Stephen Whitaker, Krieger, New York, 1981, the following argument appears in the discussion of flow through a nozzle:

> The pressure p equals
>
> $$\left(1+\frac{\gamma-1}{2}M^2\right)^{\gamma/(1-\gamma)}.$$
>
> By the binomial theorem and the fact that $v^2 = M^2\gamma RT$,
>
> $$p = 1 - \frac{1}{2}\frac{v^2}{RT} + \frac{\gamma(2\gamma-1)}{8}M^4 + \cdots.$$

Fill in the steps. (γ is specific heat, which is about 1.4, and M is a Mach number, which is in the range 1 to 2.)

■

11 Using the first four nonzero terms of the Maclaurin series for $\sqrt{1+x^3}$, estimate $\int_0^1 \sqrt{1+x^3}\,dx$, an integral that cannot be evaluated by the fundamental theorem of calculus.

Exercises 12 to 15 concern the Lagrange form of $R_n(x;0)$ for $f(x) = (1+x)^r$.

12 Let $f(x) = (1+x)^r$. Show that the Lagrange form of $R_n(x;0)$ is

$$R_n(x;0) = \frac{r(r-1)\cdots(r-n)}{(n+1)!}(1+c)^r\left(\frac{x}{1+c}\right)^{n+1},$$

where c is between 0 and x. Note that c depends both on x and n.

13 Let $0 < x < 1$. Using the formula in Exercise 12, show that $R_n(x;0) \to 0$ as $n \to \infty$.

14 Let $-\frac{1}{2} < x < 0$. Using the formula in Exercise 12, show that $R_n(x; 0) \to 0$, following these steps:
(*a*) Show that $1 + c \geq 1 + x$.
(*b*) Show that $\left|\dfrac{x}{1+c}\right| \leq \dfrac{|x|}{1+x}$.
(*c*) Show that $|x|/(1 + x) < \frac{1}{2}$.
(*d*) Show that $R_n(x; 0) \to 0$ as $n \to \infty$.

15 Let $x = -\frac{1}{2}$. Explain why the Lagrange form of $R_n(x; 0)$ is not of use in showing that $R_n(x; 0)$ in Exercise 12 approaches 0 as $n \to \infty$. (The same trouble occurs for any x, $-1 < x \leq -\frac{1}{2}$.)

16 (Proof of Lemma) Consider a fixed number x in $(-1, 1)$.
(*a*) For $0 \leq x < 1$ and $0 \leq t \leq x$ show that

$$-x \leq \frac{x-t}{1+t} \leq x$$

by clearing the denominator.
(*b*) For $-1 < x \leq 0$ and $x \leq t \leq 0$ show that

$$x \leq \frac{x-t}{1+t} \leq -x.$$

(*c*) Combining (*a*) and (*b*), show that if $|x| < 1$ and t is between 0 and x, then

$$\left|\frac{x-t}{1+t}\right| \leq |x|.$$

Exercises 17 to 23 outline an argument due to Euler that

$$\frac{1}{1^2} + \frac{1}{2^2} + \frac{1}{3^2} + \cdots = \frac{\pi^2}{6}.$$

17 Show that if

$$\frac{1}{1^2} + \frac{1}{3^2} + \frac{1}{5^2} + \frac{1}{7^2} + \cdots + \frac{1}{(2n-1)^2} + \cdots = \frac{\pi^2}{8},$$

then

$$\frac{1}{1^2} + \frac{1}{2^2} + \frac{1}{3^2} + \cdots + \frac{1}{n^2} + \cdots = \frac{\pi^2}{6}.$$

Hint: Break the second series into its odd and even terms.

18 Show that

$$\int_0^1 \frac{\sin^{-1} x}{\sqrt{1-x^2}}\,dx = \frac{\pi^2}{8}.$$

19 Use Example 3 to show that if $|t| < 1$ then

$$\frac{1}{\sqrt{1-t^2}} = 1 + \frac{1}{2}t^2 + \frac{1 \cdot 3}{2 \cdot 4}t^4 + \frac{1 \cdot 3 \cdot 5}{2 \cdot 4 \cdot 6}t^6 + \cdots.$$

20 (See Exercise 19.) Show that

$$\sin^{-1} x = x + \frac{1}{2}\frac{x^3}{3} + \frac{1 \cdot 3}{2 \cdot 4}\frac{x^5}{5} + \frac{1 \cdot 3 \cdot 5}{2 \cdot 4 \cdot 6}\frac{x^7}{7} + \cdots$$

for $|x| < 1$. This equation is also valid when $x = 1$ or -1.

21 Show that

$$\int_0^1 \frac{x^{2n+1}}{\sqrt{1-x^2}}\,dx = \int_0^{\pi/2} \sin^{2n+1}\theta\,d\theta.$$

22 Assuming that it is safe to integrate term by term, even in the case of an improper integral, show that

$$\int_0^1 \frac{\sin^{-1} x}{\sqrt{1-x^2}}\,dx = \frac{1}{1^2} + \frac{1}{3^2} + \cdots + \frac{1}{(2n-1)^2} + \cdots.$$

23 Deduce that

$$\sum_{n=1}^{\infty} n^{-2} = \frac{\pi^2}{6}.$$

24 (*a*) The ellipse $x^2/a^2 + y^2/b^2 = 1$ for $a \leq b$, has the parameterization

$$x = a \cos t, \qquad y = b \sin t.$$

Show that the arc length of one quadrant of an ellipse is

$$\int_0^{\pi/2} b\sqrt{1 - [1 - (a/b)^2]\sin^2 t}\,dt.$$

The integrand does not have an elementary antiderivative.
(*b*) Assume that in (*a*) $a < b$. Then the arc length integral has the form $\int_0^{\pi/2} b\sqrt{1 - k^2 \sin^2 t}\,dt$, where $0 < k < 1$. The "elliptic integral"

$$E = \int_0^{\pi/2} \sqrt{1 - k^2 \sin^2 \theta}\,d\theta$$

is tabulated in mathematical handbooks for many values of k in $[0, 1]$. Using the binomial theorem and the formula for $\int_0^{\pi/2} \sin^n \theta\,d\theta$ (Exercise 88 in Sec. 10.S), obtain the first six terms for E as a series in powers of k^2.

10.S SUMMARY

This chapter opened with sequences and then examined series with constant terms, first with only positive terms and then with arbitrary terms. It then considered power series, where each term involved a power of x (or of $x - a$).

With the aid of complex numbers and series it was shown that the trigonometric and exponential functions are intimately connected. The function e^z played a critical role in the solution of a general type of second-order differential equation.

The chapter closed by using the integral form of the remainder to obtain the expansion of $(1 + x)^r$ as a power series for $|x| < 1$ and any exponent.

Vocabulary and Symbols

sequence a_n or $\{a_n\}$
convergent, divergent sequence
limit of a sequence
series
partial sum S_n
convergent, divergent series
nth term of series
geometric series
nth-term test for divergence
integral test
p series
harmonic series
comparison test
limit-comparison test
ratio test
root test
alternating series
decreasing alternating series
alternating-series test
absolute convergence
conditional convergence
absolute-convergence test
absolute-ratio test
power series
Maclaurin series
radius of convergence
Taylor series $\Sigma_{n=0}^{\infty} f^{(n)}(a)(x - a)^n/n!$
Taylor polynomial $P_n(x; a)$
remainder (error) in Taylor series $R_n(x; a) = f(x) - P_n(x; a)$
complex number
magnitude and argument of a complex number
particular solution
homogeneous equation
general solution

Key Facts

If $|r| < 1$, $r^n \to 0$ as $n \to \infty$.
$x^n/n! \to 0$ as $n \to \infty$.
A geometric series $\Sigma_{n=0}^{\infty} ax^n$ converges to $a/(1 - x)$ if $|x| < 1$.
A p series $\Sigma_{n=1}^{\infty} 1/n^p$ converges if $p > 1$ and diverges if $p \leq 1$.

Associated with a power series $\Sigma_{n=0}^{\infty} a_n(x - a)^n$ is its radius of convergence, R. (R may be 0, a positive number, or infinite.) The series converges for $|x - a| < R$ and diverges for $|x - a| > R$. It may or may not converge when $|x - a| = R$, that is, when $x = a + R$ or $x = a - R$.

Within its radius of convergence a power series may be differentiated or integrated term by term.

The binomial series

$$1 + rx + \frac{r(r - 1)}{2!}x^2 + \frac{r(r - 1)(r - 2)}{3!}x^3 + \cdots$$

converges to $(1 + x)^r$ for $|x| < 1$ and any exponent r. The series has only a finite number of nonzero terms only if r is a nonnegative integer.

Tests for Convergence (or Divergence) of $\Sigma_{n=1}^{\infty} a_n$

Name	*Brief Formulation*	*When Used*	*Example*
1. nth-term test for divergence	If a_n does *not* approach 0 as $n \to \infty$, $\Sigma_{n=1}^{\infty} a_n$ diverges.	When you suspect a_n does not approach 0.	$\sum_{n=1}^{\infty} \frac{n}{2n+1}$ diverges.
2. Integral test	If $f(x) > 0$ decreases and $\int_1^{\infty} f(x)\,dx$ converges, $\Sigma_{n=1}^{\infty} a_n$ converges. If $\int_1^{\infty} f(x)\,dx$ diverges, $\Sigma_{n=1}^{\infty} a_n$ diverges.	When you have a *positive decreasing* series $a_n = f(n)$ and $\int f(x)\,dx$ is easy to calculate.	$\sum_{n=1}^{\infty} \frac{1}{n^2}$ converges, since $\int_1^{\infty} x^{-2}\,dx$ is convergent.
3. Comparison test	If $0 \le a_n \le c_n$ and $\Sigma_{n=1}^{\infty} c_n$ converges, so must $\Sigma_{n=1}^{\infty} a_n$. If $a_n \ge d_n \ge 0$ and $\Sigma_{n=1}^{\infty} d_n$ diverges, so must $\Sigma_{n=1}^{\infty} a_n$.	When you have a *positive* series that can be compared to a series known to converge or diverge.	$\sum_{n=1}^{\infty} \frac{n+1}{n+2}\frac{1}{2^n}$ converges, since $\sum_{n=1}^{\infty} \frac{1}{2^n}$ does.
4. Limit-comparison test	If $a_n/c_n \to$ nonzero limit, then $\Sigma_{n=1}^{\infty} a_n$ converges if $\Sigma_{n=1}^{\infty} c_n$ does and diverges if $\Sigma_{n=1}^{\infty} c_n$ does. (For a more general statement, see Theorem 2, Sec. 10.4.)	When you have a positive series very much like a series known to converge or diverge.	$\sum_{n=1}^{\infty} \frac{1+(-\frac{1}{2})^n}{2^n}$ converges, since $\sum_{n=1}^{\infty} \frac{1}{2^n}$ converges.
5. Ratio test		See "absolute-ratio" test, which is more useful.	$\sum_{n=1}^{\infty} \frac{n}{2^n}$ converges.
6. Decreasing-alternating-series test	A decreasing alternating series whose nth term $\to 0$ converges.	When you have an *alternating series* whose terms diminish in absolute value from some point on and approach 0.	$\sum_{n=1}^{\infty} \frac{(-1)^n}{n}$ converges.
7. Absolute-convergence test	If $\Sigma_{n=1}^{\infty} \lvert a_n \rvert$ converges, so does $\Sigma_{n=1}^{\infty} a_n$.	When you feel that the series would converge even if its terms were all made positive.	$\sum_{n=1}^{\infty} \frac{\cos n}{n^2}$ converges.
8. Absolute-ratio test	If $\lvert a_{n+1}/a_n \rvert \to L < 1$, $\Sigma_{n=1}^{\infty} a_n$ converges (absolutely). If $\lvert a_{n+1}/a_n \rvert \to L > 1$, $\Sigma_{n=1}^{\infty} a_n$ diverges. If $L = 1$, no information.	Especially suitable for power series.	$\sum_{n=0}^{\infty} \frac{x^n}{n!}$ converges absolutely for all x.
9. Root test	If $\sqrt[n]{\lvert a_n \rvert} \to L < 1$, $\Sigma_{n=1}^{\infty} a_n$ converges.	If something like n^n appears, try root test.	$\sum_{n=1}^{\infty} \frac{3^n}{n^n}$ converges.

Estimating the Error

Assume that $\Sigma_{n=1}^{\infty} a_n = S$. Let $\Sigma_{i=1}^{n} a_i = S_n$. The error, or remainder, R_n is defined to be the difference $S - S_n$.

If $\Sigma_{n=1}^{\infty} a_n$ is a decreasing alternating series whose nth term approaches 0, then $\lvert R_n \rvert < \lvert a_{n+1} \rvert$. If $\Sigma_{n=1}^{\infty} a_n$ is a positive series to which the integral test applies, then $\int_{n+1}^{\infty} f(x)\,dx < R_n < \int_n^{\infty} f(x)\,dx$. If $\Sigma_{n=1}^{\infty} a_n$ is a convergent geometric series, $a_n = ar^{n-1}$, then $R_n = ar^n/(1-r)$.

In the case of the Maclaurin series associated with the function f, the error $R_n(x; 0)$ is the difference between $f(x)$ and the sum of the first $n + 1$ terms:

$$R_n(x; 0) = f(x) - \sum_{k=0}^{n} \frac{f^{(k)}(0)x^k}{k!}.$$

There are two formulas for $R_n(x; 0)$:

$$R_n(x; 0) = \frac{1}{n!}\int_0^x f^{(n+1)}(t)(x - t)^n\,dt \qquad \text{(integral form)}$$

and

$$R_n(x; 0) = \frac{f^{(n+1)}(c_n)x^{n+1}}{(n+1)!}, \qquad c_n \text{ between } 0 \text{ and } x \qquad \text{(derivative or Lagrange form)}$$

The second formula is easier to use and is strong enough to deal with e^x, $\sin x$, and $\cos x$. It is not strong enough for $(1 + x)^r$, however. Similar formulas hold for Taylor series in powers of $x - a$.

Function	*Maclaurin Series*	*Reference*
e^x	$1 + x + \frac{x^2}{2!} + \frac{x^3}{3!} + \cdots = \sum_{n=0}^{\infty} \frac{x^n}{n!}$	Example 2, Sec. 10.8
$\sin x$	$x - \frac{x^3}{3!} + \frac{x^5}{5!} - \frac{x^7}{7!} + \cdots = \sum_{n=0}^{\infty} (-1)^n \frac{x^{2n+1}}{(2n+1)!}$	Secs. 10.7 and 10.8
$\cos x$	$1 - \frac{x^2}{2!} + \frac{x^4}{4!} - \frac{x^6}{6!} + \cdots = \sum_{n=0}^{\infty} (-1)^n \frac{x^{2n}}{(2n)!}$	Secs. 10.7 and 10.8
$\frac{1}{1 - x}$	$1 + x + x^2 + \cdots = \sum_{n=0}^{\infty} x^n,\ \lvert x\rvert < 1$	Geometric series, Sec. 10.2
$\ln(1 + x)$	$x - \frac{x^2}{2} + \frac{x^3}{3} - \cdots = \sum_{n=1}^{\infty} (-1)^{n+1} \frac{x^n}{n},\ -1 < x \le 1$	Example 3 and Exercise 28, Sec. 10.7
$\tan^{-1} x$	$x - \frac{x^3}{3} + \frac{x^5}{5} - \cdots = \sum_{n=0}^{\infty} (-1)^n \frac{x^{2n+1}}{2n+1},\ \lvert x\rvert \le 1$	Exercise 29, Sec. 10.7
$\sin^{-1} x$	$x + \frac{1}{2}\frac{x^3}{3} + \frac{1 \cdot 3}{2 \cdot 4}\frac{x^5}{5} + \frac{1 \cdot 3 \cdot 5}{2 \cdot 4 \cdot 6}\frac{x^7}{7} + \cdots,\ \lvert x\rvert \le 1$	Exercise 20, Sec. 10.12
$(1 + x)^r$	$1 + rx + \frac{r(r-1)}{2!}x^2 + \frac{r(r-1)(r-2)}{3!}x^3 + \cdots,\ \lvert x\rvert < 1$	General binomial theorem, Sec. 10.12

Uses of Power Series

Computational

A partial sum of a power series provides a numerical approximation of its sum.

Theoretical

The formula for the remainder is of use in analyzing the efficiency of numerical procedures. The formula for $R_1(x; 0)$ was used to show when Newton's method works. (Exercise 22 in Sec. 6.10.) The formula for $R_3(x; 0)$ was used to estimate the size of

$$\int_{-1}^{1} f(x)\,dx - f\left(\frac{1}{\sqrt{3}}\right) - f\left(-\frac{1}{\sqrt{3}}\right). \qquad \text{(Exercise 42 in Sec. 10.8)}$$

The power series for e^x, $\cos x$, and $\sin x$ were used in Sec. 10.10 to obtain the fundamental equation $e^{i\theta} = \cos\theta + i \sin\theta$.

GUIDE QUIZ ON CHAP. 10: SERIES AND COMPLEX NUMBERS

1 A Maclaurin series converges for $x = 2$. For which other values of x must it converge?

2 The series $\sum_{n=0}^{\infty} a_n(x-2)^n$ diverges for $x = 5$. For which other values of x must it diverge?

3 Give an example of a Maclaurin series that converges for $|x| < 1$ and no other values of x.

4 Estimate $\int_0^1 \sin x^2\, dx$ to two decimal places.

5 State the integral and derivative forms of the remainder in a Taylor series, (*a*) $R_n(x; 0)$, (*b*) $R_n(x; a)$.

6 Give the Maclaurin series and their radii of convergence for the following functions:

(*a*) $\dfrac{1}{1+x^2}$ (*b*) e^{-x} (*c*) $\cos x$

(*d*) $\ln(1-x)$ (*e*) $\dfrac{1}{1-2x}$ (*f*) $1 + 3x + 5x^2$

7 Assume that $f(x)$ is represented by a Maclaurin series for some open interval $(-R, R)$, $f(x) = \sum_{n=0}^{\infty} a_n x^n$. Explain why $a_n = f^{(n)}(0)/n!$.

8 Determine for which x the following series converge or diverge:

(*a*) $\displaystyle\sum_{n=1}^{\infty} \frac{n}{1+n^2}(x-2)^n$ (*b*) $\displaystyle\sum_{n=0}^{\infty} \frac{2^n}{n!}(x+3)^n$

9 Solve these differential equations:

(*a*) $2\dfrac{d^2y}{dx^2} + 3\dfrac{dy}{dx} - 2y = 0$

(*b*) $4\dfrac{d^2y}{dx^2} - 12\dfrac{dy}{dx} + 9y = \cos x$

(*c*) $\dfrac{d^2y}{dx^2} + \dfrac{dy}{dx} + y = x$

(*d*) $3\dfrac{dy}{dx} + 2y = \cos x$

REVIEW EXERCISES FOR CHAP. 10: SERIES AND COMPLEX NUMBERS

1 Give the formula for the nth nonzero term of the Maclaurin series for

(*a*) $\tan^{-1} 3x$ (*b*) $\ln(1+x^2)$ (*c*) e^{-x}

(*d*) $\sin x^2$ (*e*) $\cos x^2$

In some cases it may be simpler to start at $n = 0$.

2 (*a*) Show that

$$\sum_{n=0}^{\infty} \frac{\cos(2n+1)t}{(2n+1)^2}$$

converges for all t.

(*b*) In the theory of Fourier series it is shown that for $0 \le t < \pi$, the sum of the series in (*a*) is $(\pi^2 - 2\pi t)/8$. Deduce that

$$\frac{\pi^2}{8} = \frac{1}{1^2} + \frac{1}{3^2} + \frac{1}{5^2} + \cdots.$$

3 A text on the economics of equipment maintenance contains this argument:

> Adding the discounted net benefits over all the cycles yields
>
> $B(N) =$
>
> $$\sum_{n=1}^{\infty} e^{-(n-1)(N+K)\alpha}\left[\int_0^N e^{-\alpha t}b(t)\, dt - e^{-\alpha N}c(N)\right].$$
>
> Summing the geometric series gives
>
> $$B(N) = \frac{\int_0^N e^{-\alpha t}b(t)\, dt - e^{-\alpha N}c(N)}{1 - e^{-\alpha(N+K)}}.$$

Sum the geometric series to check that the missing algebra is correct.

4 Obtain the first three nonzero terms of the Maclaurin series for $\sin 2x$:

(*a*) by replacing x by $2x$ in the Maclaurin series for $\sin x$,

(*b*) by using the formula $a_n = f^{(n)}(0)/n!$,

(*c*) by using the identity $\sin 2x = 2 \sin x \cos x$ and the Maclaurin series for $\sin x$ and $\cos x$.

5 Obtain the first three nonzero terms of the Maclaurin series for $\sin^2 x$:

(*a*) by using the formula $a_n = f^{(n)}(0)/n!$,

(*b*) by using the identity $\sin^2 x = (1 - \cos 2x)/2$ and the series for $\cos 2x$.

6 An engineer wishes to use a partial sum of the Maclaurin series for e^x to approximate e^x. How many terms of the series should be used to be sure that the error is less than

(*a*) 0.01 for $|x| \le 1$? (*b*) 0.001 for $|x| \le 1$?

(*c*) 0.01 for $|x| \le 2$? (*d*) 0.001 for $|x| \le 2$?

In each of Exercises 7 to 31 determine whether the series converges or diverges. If you can give the sum of the series, do so.

7 $\displaystyle\sum_{n=0}^{\infty} e^{-n}$

8 $\displaystyle\sum_{n=1}^{\infty} \sin\frac{1}{n}$

9 $\displaystyle\sum_{n=1}^{\infty} \frac{1+(-1)^n}{n^2}$

10 $\displaystyle\sum_{n=1}^{\infty} \frac{n^3}{2^n}$

11 $\sum_{n=1}^{\infty} \frac{\cos^3 n}{n^2}$

12 $\sum_{n=0}^{\infty} (-1)^n \frac{\pi^{2n+1}}{2^{2n+1}(2n+1)!}$

13 $\sum_{n=1}^{\infty} \frac{1}{n\sqrt{n}}$ **14** $\sum_{n=0}^{\infty} (-\frac{3}{4})^n$

15 $\sum_{n=1}^{\infty} (-1)^n \ln\left(\frac{n+1}{n}\right)$ **16** $\sum_{n=1}^{\infty} \frac{(-2)^n}{n}$

17 $\sum_{n=1}^{\infty} n \sin \frac{1}{n}$ **18** $\sum_{n=0}^{\infty} \frac{5n^2 - 3n + 1}{2n^3 + n^2 - 1}$

19 $\sum_{n=0}^{\infty} \frac{5n^3 + 6n + 1}{n^5 + n^3 + 2}$ **20** $\sum_{n=1}^{\infty} \ln\left(\frac{n+1}{n}\right)$

21 $\sum_{n=1}^{\infty} \frac{2^{-n}}{n}$ **22** $\sum_{n=0}^{\infty} (-1)^n \frac{\pi^{2n}}{(2n)!}$

23 $\sum_{n=0}^{\infty} \frac{10^n}{n!}$ **24** $\sum_{n=1}^{\infty} \frac{\ln n}{n}$

25 $\sum_{n=0}^{\infty} \frac{(-1)^n(\frac{1}{2})^n}{n!}$ **26** $\sum_{n=1}^{\infty} \frac{\sqrt{n+1} - \sqrt{n}}{n}$

27 $\sum_{n=0}^{\infty} \frac{n+2}{n+1}\left(\frac{2}{3}\right)^n$ **28** $\sum_{n=1}^{\infty} \frac{n^2}{1+n^3}$

29 $\sum_{n=1}^{\infty} \frac{n \cos n}{1+n^4}$ **30** $\sum_{n=1}^{\infty} \frac{(-1)^n n^2}{(2n)!}$

31 $\sum_{n=1}^{\infty} \frac{n-3}{n\sqrt{n}}$

In each of Exercises 32 to 35 use the first three nonzero terms of a Maclaurin series to estimate the quantity. Also put an upper bound on the error.

32 $\sqrt[10]{e}$ **33** $\cos \frac{1}{3}$

34 $\sin 28°$ **35** $1/\sqrt{e}$

In each of Exercises 36 to 43 determine for which x the series diverges, converges absolutely, and converges conditionally. Give the radius of convergence in each case and the sum of the series if it is easily determined.

36 $\sum_{n=1}^{\infty} \frac{2^n x^n}{n}$ **37** $\sum_{n=1}^{\infty} nx^{n-1}$

38 $\sum_{n=0}^{\infty} \frac{(x-3)^n}{n!}$ **39** $\sum_{n=0}^{\infty} \frac{x^{2n}}{n!}$

40 $\sum_{n=1}^{\infty} (-n)^n x^n$ **41** $\sum_{n=1}^{\infty} \frac{x^n}{n}$

42 $\sum_{n=0}^{\infty} \frac{3^n(x-\frac{2}{3})^n}{4^n}$ **43** $\sum_{n=0}^{\infty} \frac{n^5+2}{n^3+1}(x+1)^n$

In each of Exercises 44 to 53 determine the radius of convergence and for which x the series converges or diverges.

44 $\sum_{n=1}^{\infty} \frac{(n!)^2 x^n}{n^{2n}}$ **45** $\sum_{n=1}^{\infty} \frac{2^n(x-1)^n}{[1+(1/n)]^n}$

46 $\sum_{n=1}^{\infty} (n^2 x)^n$ **47** $\sum_{n=0}^{\infty} \frac{(2n+1)x^n}{n!}$

48 $\sum_{n=1}^{\infty} \frac{(-1)^n n^2 (x-3)^n}{n^3+1}$ **49** $\sum_{n=2}^{\infty} \frac{(x-1)^n}{\ln n}$

50 $\sum_{n=2}^{\infty} (\ln n)(x+2)^n$ **51** $\sum_{n=1}^{\infty} \frac{(x+3)^n}{\sqrt[3]{n}}$

52 $\sum_{n=1}^{\infty} \frac{(-1)^n x^{2n}}{\sqrt{n}}$ **53** $\sum_{n=1}^{\infty} \left(\frac{n+1}{n+3}\right)^{n^2} x^n$

54 Suppose $\sum_{n=0}^{\infty} a_n(x-2)^n$ converges for $x = 7$ and diverges for $x = -3$. What can be said about its radius of convergence?

In Exercises 55 to 57 determine the limits using power series. It might be instructive to solve the problems by l'Hôpital's rule also.

55 $\lim_{x\to 0} \frac{\ln(1+x^2) - \sin^2 x}{\tan x^2}$

56 $\lim_{x\to 0} \frac{(e^{x^2}-1)^2}{1 - x^2/2 - \cos x}$ **57** $\lim_{x\to 0} \frac{(1-\cos x^2)^5}{(x - \sin x)^{20}}$

58 Put an upper bound on the error in using the sum of the first 1000 terms to estimate the sum of the series:

(*a*) $\sum_{n=1}^{\infty} n^{-2}$ (*b*) $\sum_{n=1}^{\infty} (-1)^n n^{-2}$ (*c*) $\sum_{n=1}^{\infty} (n!)^{-1}$

59 Estimate or compute exactly

$$\sum_{n=1}^{\infty} (-2)^{-n} - \sum_{n=1}^{10} (-2)^{-n}$$

(*a*) by noticing that the series is alternating decreasing, (*b*) by noticing that the absolute value of the error is less than $\sum_{n=11}^{\infty} 2^{-n}$, (*c*) by noticing that $\sum_{n=1}^{\infty} (-2)^{-n}$ is a geometric series, (*d*) by considering $f(x) = (1+x)^{-1}$ for $x = \frac{1}{2}$ and using Lagrange's form of the remainder.

60 Estimate the positive root of the equation $e^x = 2x + 1$
(*a*) using a Taylor polynomial of degree 3 to approximate e^x,
(*b*) using Newton's method for the function $e^x - 2x - 1$.

61 Express $x^2 + x + 2$ as a polynomial in powers of $x - 5$.

62 In each of the following integrals, use the first three nonzero terms of the Maclaurin series for the integrand to estimate the integral. Put an upper bound on the size of the error.

(*a*) $\int_0^{1/2} \cos x^3 \, dx$ (*b*) $\int_0^1 \sin x^2 \, dx$

(c) $\int_0^{1/2} \sqrt[3]{1 + x^3}\, dx$ (d) $\int_1^2 e^{-x^3}\, dx$

In Exercises 63 to 65 find all solutions of the differential equations.

63 (a) $\dfrac{dy}{dx} + 4y = 12$ (b) $\dfrac{dy}{dx} + y = \cos x$

64 (a) $\dfrac{d^2y}{dx^2} + 2\dfrac{dy}{dx} - 3y = 0$

(b) $\dfrac{d^2y}{dx^2} + 2\dfrac{dy}{dx} + 3y = 0$

65 (a) $2\dfrac{d^2y}{dx^2} + 5\dfrac{dy}{dx} + 3y = 3x^2$

(b) $2\dfrac{d^2y}{dx^2} + 5\dfrac{dy}{dx} + 3y = 3x^2 + 2\sin x$

■

66 For $|x| \le 1$, $\tan^{-1} x = \sum_{n=0}^{\infty} (-1)^n x^{2n+1}/(2n+1)$. Thus $\pi/4 = 1 - \frac{1}{3} + \frac{1}{5} - \frac{1}{7} + \cdots$, which converges slowly. The identity

$$\frac{\pi}{4} = 4\tan^{-1}\frac{1}{5} - \tan^{-1}\frac{1}{70} + \tan^{-1}\frac{1}{99} \qquad (1)$$

provides a faster way to estimate π.

(a) Using the first three nonzero terms of the Maclaurin series for $\tan^{-1} x$ and identity (1), estimate $\pi/4$ and hence π.

(b) Discuss the size of the error in (a) in the estimate of $\pi/4$.

(c) How would you show that identity (1) is true? (Describe a method, but don't carry out the computations.)

67 Let $f(x) = \sum_{n=0}^{\infty} 2^n x^n$. Find $f^{(33)}(0)$.

68 Give an example of a Maclaurin series whose radius of convergence is 1 and which

(a) converges at 1 and -1,

(b) diverges at 1 and -1,

(c) converges at 1 and diverges at -1.

69 Show that if $\sum_{n=0}^{\infty} a_n 3^n$ converges, then so does $\sum_{n=0}^{\infty} n a_n 2^n$.

70 Find $f^{(99)}(0)$, $f^{(100)}(0)$, and $f^{(101)}(0)$ if $f(x)$ is

(a) $\tan^{-1} x$ (b) e^{x^2}

71 Disregard air resistance and assume that an ideal tennis ball, when dropped on a concrete surface, always rebounds 80 percent of the distance that it falls. If it is dropped from a height of 6 feet, it then bounces infinitely often. Is the total time for these bounces finite or infinite? (An object falling from rest drops $16t^2$ feet in t seconds. The time of each rise equals the time of the following descent.)

72 A certain function f has $f(0) = 3$, $f^{(1)}(0) = 2$, $f^{(2)}(0) = 5$, $f^{(3)}(0) = \frac{1}{2}$, and $f^{(j)}(x) = 0$ if $j > 3$. Give an explicit formula for $f(x)$.

73 (a) Show that if $\{a_n\}$ is a sequence of positive terms and $\sum_{n=1}^{\infty} a_n$ converges, so does $\sum_{n=1}^{\infty} a_n^2$.

(b) Give an example of a sequence $\{a_n\}$ such that $\sum_{n=1}^{\infty} a_n$ converges but $\sum_{n=1}^{\infty} a_n^2$ does not.

74 Define $f(x)$ to be $(1 + x)^{1/x}$ if $x > -1$ but $x \ne 0$; define $f(0) = e$. Find $f'(0)$.

75 Give the formula for the nth nonzero term of the Maclaurin series for (a) $\tan^{-1} x$, (b) $\ln(1 + x^2)$, (c) e^{-x^2}, (d) $\sin x^2$, (e) $\cos x$.

76 (a) Show that if $\sum_{n=1}^{\infty} a_n^2$ and $\sum_{n=1}^{\infty} b_n^2$ converge, so does $\sum_{n=1}^{\infty} a_n b_n$.

(b) Show that if $\sum_{n=1}^{\infty} a_n^2$ converges, so does $\sum_{n=1}^{\infty} a_n/n$.

77 Prove that $\sin x$ cannot be written as a polynomial $P(x)$, even if we demand that $\sin x = P(x)$ only throughout a small interval $[a, b]$.

78 The zeta function $\zeta(p)$ is defined for $p > 1$ as $\zeta(p) = \sum_{n=1}^{\infty} n^{-p}$.

(a) Examine $\lim_{p\to 1^+} \zeta(p)$.

(b) Show that $(p - 1)^{-1} < \zeta(p) < p(p - 1)^{-1}$.

(c) Show that $\lim_{p\to 1^+} \zeta(p)(p - 1) = 1$.

79 Let f be a function such that $f^{(3)}$ is continuous for all x. Assume that $\lim_{x\to\infty} f(x) = 1$ and $\lim_{x\to\infty} f^{(3)}(x) = 0$. Prove that $\lim_{x\to\infty} f^{(1)}(x) = 0 = \lim_{x\to\infty} f^{(2)}(x)$. *Hint:* Express $f(a + 1)$ and $f(a - 1)$ in terms of derivatives of f at a.

80 Let f be a function such that $f^{(2)}$ is continuous for all x. Prove that if $|f(x)| \le 1$ and $|f^{(2)}(x)| \le 1$ for all x in $[0, 2]$, then $|f^{(1)}(x)| \le 2$ for all x in $[0, 2]$. *Hint:* Express both $f(0)$ and $f(2)$ in terms of derivatives of f at x.

81 Assume that $f^{(2)}$ is continuous and $f^{(2)}(a) \ne 0$. By the mean-value theorem, $f(a + h) = f(a) + hf'(a + \theta h)$ for some θ in $[0, 1]$.

(a) When h is small, why is θ unique?

(b) Prove that $\theta \to \frac{1}{2}$ as $h \to 0$.

82 Although $f'(a)$ is the limit of $[f(a + \Delta x) - f(a)]/\Delta x$, there is a better way to estimate $f'(a)$ than by that quotient. Assume that $f^{(3)}$ is continuous. Show that

(a) $$\frac{f(a + \Delta x) - f(a)}{\Delta x} = f'(a) + \frac{f^{(2)}(c_1)}{2}\Delta x$$

for some c_1 between a and $a + \Delta x$, and

(b) $$\frac{f(a + \Delta x) - f(a - \Delta x)}{2\Delta x} = f'(a) + \frac{f^{(3)}(c_2)}{6}(\Delta x)^2,$$

where c_2 is in $[a - \Delta x, a + \Delta x]$. [Since the error in using the quotient in (b) involves $(\Delta x)^2$, while the error in using the standard quotient involves Δx, the quotient in (b) is more accurate when Δx is small.] Test this observation on the function $y = x^3$ at $a = 2$.

83 In advanced mathematics a certain function $E(x)$ is *defined* as the sum $\sum_{n=0}^{\infty} x^n/n!$. Pretending that you have never heard of e or e^x, solve the following problems:

(a) Show that $E(0) = 1$.

(*b*) Show that $E'(x) = E(x)$.
(*c*) Show that $E(x)E(-x) = 1$. *Hint:* Differentiate $E(x)E(-x)$ and use (*a*) and (*b*).
(*d*) Deduce that $E(x + y)/E(x)$ is independent of x.
(*e*) Deduce that $E(x + y) = E(x)E(y)$.

84 Consider $\int_0^b xe^{-x}\,dx$, when b is a small positive number. Since e^{-x} is then close to $1 - x$, the definite integral behaves like $\int_0^b (x - x^2)\,dx = b^2/2 - b^3/3$, hence approximately like $b^2/2$. On the other hand, $\int_0^b xe^{-x}\,dx = 1 - e^{-b}(1 + b)$, and, since e^{-b} is approximately $1 - b$, we have $1 - e^{-b}(1 + b)$ approximately equal to $1 - (1 - b)(1 + b) = b^2$. Hence $\int_0^b xe^{-x}\,dx$ behaves like b^2. Which is correct, $b^2/2$ or b^2? Find the error.

In Exercises 85 to 90 a short formula for estimating $n!$ is obtained.

85 Let f have the properties that for $x \geq 1$, $f(x) \geq 0$, $f'(x) > 0$, and $f''(x) < 0$. Let a_n be the area of the region below the graph of $y = f(x)$ and above the line segment that joins $(n, f(n))$ with $(n + 1, f(n + 1))$.
(*a*) Draw a large-scale version of Fig. 10.32. The individual regions of areas a_1, a_2, a_3, and a_4 should be clear and not too narrow.
(*b*) Using geometry, show that the series $a_1 + a_2 + a_3 + \cdots$ converges and has a sum no larger than the area of the triangle with vertices $(1, f(1))$, $(2, f(2))$, $(1, f(2))$.

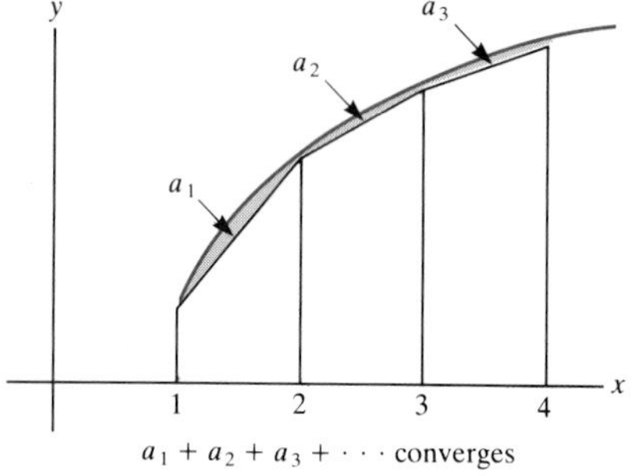

Figure 10.32

86 Let $y = \ln x$.
(*a*) Using Exercise 85, show that as $n \to \infty$,

$$\int_1^n \ln x\,dx - \left[\frac{\ln 1 + \ln 2}{2} + \frac{\ln 2 + \ln 3}{2} + \cdots + \frac{\ln (n - 1) + \ln n}{2}\right]$$

has a limit; denote this limit by C.
(*b*) Show that (*a*) is equivalent to the assertion

$$\lim_{n\to\infty} (n \ln n - n + 1 - \ln n! + \ln \sqrt{n}) = C.$$

87 From Exercise 86(*b*), deduce that there is a constant k such that

$$\lim_{n\to\infty} \frac{n!}{k(n/e)^n\sqrt{n}} = 1.$$

Exercises 88 and 89 are related. Review Exercise 41 of Sec. 7.2 first.

88 Let $I_n = \int_0^{\pi/2} \sin^n x\,dx$.
(*a*) Evaluate I_0 and I_1.
(*b*) Show that

$$I_{2n} = \frac{2n - 1}{2n}\,\frac{2n - 3}{2n - 2}\cdots\frac{3}{4}\,\frac{1}{2}\,\frac{\pi}{2}$$

and $$I_{2n+1} = \frac{2n}{2n + 1}\,\frac{2n - 2}{2n - 1}\cdots\frac{4}{5}\,\frac{2}{3}.$$

(*c*) Show that

$$\frac{I_7}{I_6} = \frac{6}{7}\,\frac{6}{5}\,\frac{4}{5}\,\frac{4}{3}\,\frac{2}{3}\,\frac{2}{1}\,\frac{2}{\pi}.$$

(*d*) Show that

$$\frac{I_{2n+1}}{I_{2n}} = \frac{2n}{2n + 1}\,\frac{2n}{2n - 1}\,\frac{2n - 2}{2n - 1}\cdots\frac{2}{3}\,\frac{2}{1}\,\frac{2}{\pi}.$$

(*e*) Show that

$$\frac{2n}{2n + 1} I_{2n} < \frac{2n}{2n + 1} I_{2n+1} < I_{2n},$$

and thus $$\lim_{n\to\infty} \frac{I_{2n+1}}{I_{2n}} = 1.$$

(*f*) From (*d*) and (*e*), deduce that

$$\lim_{n\to\infty} \frac{2 \cdot 2}{1 \cdot 3}\,\frac{4 \cdot 4}{3 \cdot 5}\,\frac{6 \cdot 6}{5 \cdot 7}\cdots\frac{(2n)(2n)}{(2n - 1)(2n + 1)} = \frac{\pi}{2}.$$

This is **Wallis's formula**, usually written in shorthand as

$$\frac{2 \cdot 2}{1 \cdot 3}\,\frac{4 \cdot 4}{3 \cdot 5}\,\frac{6 \cdot 6}{5 \cdot 7}\cdots = \frac{\pi}{2}.$$

89 (*a*) Show that $2 \cdot 4 \cdot 6 \cdot 8 \cdots 2n = 2^n n!$.
(*b*) Show that $1 \cdot 3 \cdot 5 \cdots (2n - 1) = (2n)!/(2^n n!)$.
(*c*) From Exercise 88, deduce that

$$\lim_{n\to\infty} \frac{(n!)^2 4^n}{(2n)!\sqrt{2n + 1}} = \sqrt{\frac{\pi}{2}}.$$

90 (*a*) Using Exercise 89(*c*), show that k in Exercise 87 equals $\sqrt{2\pi}$. Thus a good estimate of $n!$ is provided by the formula

$$n! \approx \sqrt{2\pi n}\left(\frac{n}{e}\right)^n.$$

This is known as **Stirling's formula**.
(*b*) Using the factorial key on a calculator, compute 20!. Then compute the ratio $\sqrt{2\pi n}(n/e)^n/n!$ for $n = 20$.

91 Determine for which x the series $\sum_{n=1}^{\infty} n^n x^n/n!$ converges or diverges.

The next exercise is used in Exercise 93.

92 Let Δf denote $f(x + 1) - f(x)$ and Df denote, as usual, the derivative of f. Just as we speak of $D^n f$, we can define $\Delta^n f$ by repeated applications of Δ to f. For instance,

$$(\Delta^2 f)(x) = (\Delta f)(x + 1) - (\Delta f)(x)$$
$$= [f(x + 2) - f(x + 1)]$$
$$- [f(x + 1) - f(x)].$$

(*a*) Show that $D\Delta f = \Delta D f$.

(*b*) Show that $(\Delta f)(x) = (Df)(c_1)$ for some number c_1, $x < c_1 < x + 1$.

(*c*) Show that for any positive integer k there is a number c_k, $x < c_k < x + k$, such that

$$(\Delta^k f)(x) = (D^k f)(c_k).$$

(Assume that f has derivatives of all orders.)

(*d*) Let r be a positive number that is not an integer. Show that

$$\Delta^k(x^r) = r(r - 1) \cdots (r - k + 1)c_k^{r-k}$$

for some c_k, $x < c_k < x + k$.

93 Let r be a positive number such that n^r is an integer for all positive integers n. Show that r is a positive integer. The preceding exercise may be of aid.

Incidentally, it is known that if p^r is an integer for three distinct primes p, then r is an integer. It is not known whether the assumption that p^r is an integer for two distinct primes forces r to be an integer.

94 (*a*) From the Maclaurin series for $\cos x$ in powers of x, obtain the Maclaurin series for $\cos 2x$.

(*b*) Exploiting the identity $\sin^2 x = (1 - \cos 2x)/2$, obtain the Maclaurin series for $(\sin^2 x)/x^2$.

(*c*) Estimate $\int_0^1 [(\sin x)/x]^2\, dx$ using the first three nonzero terms of the series in (*b*).

(*d*) Find a bound on the error in the estimate in (*c*).

Notes For Student Use

Appendix A
Real Numbers

This appendix discusses rational and irrational numbers, division by 0, inequalities, intervals, and absolute value.

The **positive integers** are the counting numbers, 1, 2, 3, . . . ; the **negative integers** are $-1, -2, -3, \ldots$. The set of positive and negative integers, together with 0, is called the **set of integers**. The integers appear on the number line as regularly spaced points, as shown in Fig. A.1.

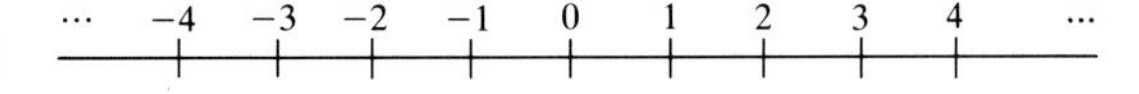

Figure A.1

Every point on the number line corresponds to a **real number**. For instance, $\frac{11}{3}$, $-\sqrt{2}$, π, and 1.13 are real numbers and their positions are shown in Fig. A.2.

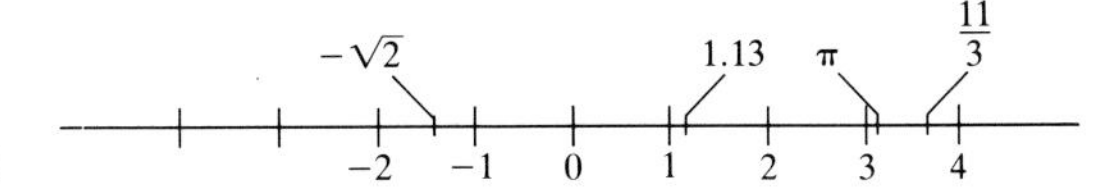

Figure A.2

A real number that can be written as a fraction or ratio p/q, where p is an integer and q is a positive integer, is called a **rational number**. For instance, $\frac{11}{3}$ is rational, as are $8 = \frac{8}{1}$ and $7\sqrt{2}/(3\sqrt{2}) = \frac{7}{3}$. A real number that is not rational is called **irrational**. Greek mathematicians some 2,400 years ago showed that $\sqrt{2}$ is irrational; Johann Lambert in 1761 showed that π is irrational. This means that neither $\sqrt{2}$ nor π can be written as the quotient of two integers, p/q. (See Exercise 37.)

If a and b are real numbers, we may always form their sum $a + b$, their difference $a - b$, and their product ab. If b is not 0, then there is a unique number x such that $bx = a$. This number x, the **quotient** of a

by b, is denoted a/b. For instance, since the equation $2x = 6$ has the unique solution 3, we write $6/2 = 3$.

Why division by zero is meaningless

However, when $b = 0$, solving the equation $bx = a$ runs into some problems. For instance, the equation

$$0x = 6$$

has no solution whatsoever, since the product of 0 and any real number is 0. Thus the symbol $\frac{6}{0}$ is totally meaningless. When both a and b are 0, the equation $bx = a$ runs into a different trouble: there are too many solutions. The equation is now

$$0x = 0.$$

Every real number is a solution; for instance, $0 \cdot 5 = 0$, $0 \cdot \pi = 0$, and $0(-3) = 0$. The symbol $\frac{0}{0}$ is meaningless because it does not describe a single number.

WRONG WAY $\frac{a}{0}$

In short, "division by zero" makes no sense. If you find yourself dividing by zero while working a problem, turn back, as though you had met a sign warning "WRONG WAY." In particular, resist the temptation to say that $\frac{0}{0}$ is equal to either 1 or 0. This temptation will be placed before you often in the study of limits and derivatives. Although $\frac{3}{3} = 1$ and $\frac{0}{3} = 0$, the expression $\frac{0}{0}$ is utterly devoid of meaning.

Inequalities If the point that represents the number a lies to the left of the point that represents the number b on the number line, we write $a < b$ ("a is less than b") or $b > a$ ("b is greater than a"). For instance, $5 < 7$, $7 > 5$, $-8 < 2$, and $-8 < -3$.

The expression $a \leq b$ means that a is either less than b or equal to b. Thus $3 \leq 4$ and $4 \leq 4$. Read "$a \leq b$" as "a is less than or equal to b." If $a \leq b$, we also write $b \geq a$.

Inequalities behave nicely with respect to addition and subtraction but demand great care in the case of multiplication and division. Inspection of the following list of properties of inequalities shows the difference.

The properties of inequalities.

1 If $a < b$ and $c < d$, then $a + c < b + d$. (You can add two inequalities that are in the same direction.)
2 If $a < b$ and c is any number, then $a + c < b + c$. (You can add the same number to both sides of an inequality.)
3 If $a < b$ and c is any number, then $a - c < b - c$. (You can subtract the same number from both sides of an inequality.)

Multiplication or division by a positive number preserves an inequality but multiplication or division by a negative number reverses *an inequality.*

4 If $a < b$ and c is a *positive* number, then $ac < bc$. (Multiplication by a positive number preserves an inequality.)
5 If $a < b$ and c is a *negative* number, then $ac > bc$. (Multiplication by a negative number reverses an inequality.)
6 If $a < b$ and $c < d$ and a, b, c, and d are *positive,* then $ac < bd$. (You can multiply two inequalities, if they are in the same direction and involve only positive numbers.)
7 If $a < b$ and c is a *positive* number, then $a/c < b/c$. (Division by a positive number preserves an inequality.)

8 If $a < b$ and c is *negative*, then $a/c > b/c$. (Division by a negative number reverses the inequality.)

$2 < 3$ but $\frac{1}{2} > \frac{1}{3}$.

9 If a and b are positive numbers and $a < b$, then

$$\frac{1}{a} > \frac{1}{b}.$$

(Taking reciprocals of an inequality between positive numbers reverses the inequality.)

You cannot subtract one inequality from another; for instance, $5 < 8$ and $1 < 6$, but it is not true that $5 - 1$ is less than $8 - 6$.

The notation $a < x < b$ in the following definition is short for the two inequalities $a < x$ and $x < b$. Thus, $a < x < b$ is short for "x is larger than a and less than b."

Definition *Open interval.* Let a and b be real numbers, with $a < b$. The **open interval** (a, b) consists of all numbers x such that $a < x < b$.

Definition *Closed interval.* Let a and b be real numbers, with $a < b$. The **closed interval** $[a, b]$ consists of all numbers x such that $a \leq x \leq b$.

The meaning of ● and ○ in diagrams

The open interval (a, b) is obtained from the closed interval $[a, b]$ by the deletion of the endpoints a and b. Some intervals are shown in Fig. A.3. The solid dot (●) indicates the presence of a point (included point); the hollow dot (○) indicates the absence of a point (excluded point). This notation is used in diagrams throughout the text.

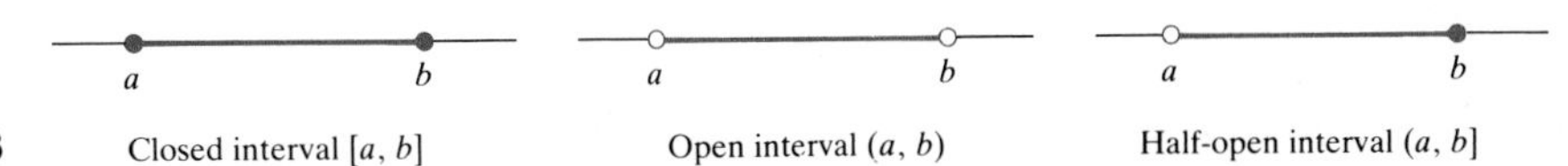

Figure A.3 Closed interval $[a, b]$ Open interval (a, b) Half-open interval $(a, b]$

Infinite intervals

Infinite intervals are also of use. For convenience they are listed in the following table:

Symbol	*Description*	*Picture*	*In Words*
$[a, \infty)$	$x \geq a$	a $[a, \infty)$	Closed interval to the right of a
(a, ∞)	$x > a$	a (a, ∞)	Open interval to the right of a
$(-\infty, a]$	$x \leq a$	$(-\infty, a]$ a	Closed interval to the left of a
$(-\infty, a)$	$x < a$	$(-\infty, a)$ a	Open interval to the left of a
$(-\infty, \infty)$	all x		Entire x axis

The symbols ∞ and $-\infty$ do not refer to numbers but provide a convenient shorthand.

The following three examples apply the properties of inequalities and illustrate the interval notation.

EXAMPLE 1 Find all numbers x such that $3x + 1 < 5x + 2$.

SOLUTION Gradually transform the given inequality until one side of the inequality consists of just x and the other side does not involve x.

$3x + 1 < 5x + 2$	given
$3x < 5x + 1$	(subtracting 1 from both sides, rule 3)
$-2x < 1$	(subtracting $5x$ from both sides, rule 3)
$x > -\frac{1}{2}$	(dividing by the negative number -2, rule 8)

Each of these steps can be reversed. So the given inequality is equivalent to the inequality $x > -\frac{1}{2}$ and so has as solutions all numbers in the interval $(-\frac{1}{2}, \infty)$. ■

EXAMPLE 2 Describe the set of numbers x such that

$$(x + 2)(x - 1)(x - 3) > 0.$$

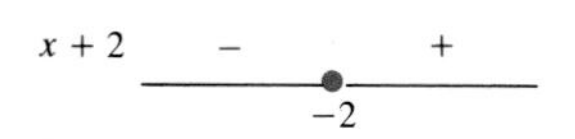

Figure A.4

SOLUTION First check where each of the three factors is positive or negative. For example, $x + 2$ is 0 if $x = -2$, positive if $x > -2$, and negative if $x < -2$. This information is recorded in Fig. A.4. Similarly, $x - 1$ changes sign at $x = 1$ and $x - 3$ changes sign at $x = 3$. This information is collected in the following table:

x		-2	1	3
$x + 2$	−	+	+	+
$x - 1$	−	−	+	+
$x - 3$	−	−	−	+
$(x + 2)(x - 1)(x - 3)$	−	+	−	+

The product of the three factors is positive when x is in $(-2, 1)$ or $(3, \infty)$. The solutions of the inequality $(x + 2)(x - 1)(x - 3) > 0$ thus fill out the two intervals shown in Fig. A.5.

Figure A.5

■

The absolute value of a real number tells "how far it is from 0." The following definition makes this precise.

Definition *Absolute value.* The **absolute value** of a positive number x is x itself. The absolute value of a negative number x is its negative $-x$. The absolute value of 0 is 0.

The absolute value of x is denoted $|x|$. By definition, $|3| = 3$ and $|-3| = -(-3) = 3$. For all x, $|-x| = |x|$; x and $-x$ have the same absolute value. On some programmable calculators, the "absolute value" key is labeled "abs." On occasion, we use abs x to denote $|x|$.

The absolute value behaves better with respect to multiplication than with respect to addition. For any real numbers x and y,

$$|xy| = |x|\,|y| \qquad \text{(The absolute value of the product is the product of the absolute values.)}$$

and

$$|x + y| \leq |x| + |y| \qquad \text{(The absolute value of the sum is less than or equal to the sum of the absolute values.)}$$

The second assertion is known as the **triangle inequality**. For instance, $|(-3) + 7| \leq |-3| + |7|$, as a little arithmetic verifies.

EXAMPLE 3 Sketch on the number line the set of numbers x such that $|x| < 2$.

SOLUTION A positive number has an absolute value less than 2 if the number itself is less than 2. A negative number has an absolute value less than 2 if the number itself is greater than -2. Also, the absolute value of 0 is less than 2. All told, the set of numbers x such that $|x| < 2$ is the open interval $(-2, 2)$, sketched in Fig. A.6. The interval $(-2, 2)$ consists of all numbers within a distance 2 of the origin. ■

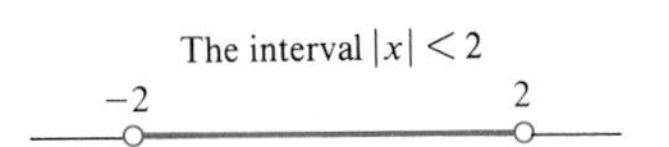

Figure A.6

Observe that for any number $a > 0$, $|x| < a$ describes the open interval $(-a, a)$.

EXAMPLE 4 Sketch on the number line the set of numbers x such that $|5(x - 1)| < 3$.

SOLUTION

$\lvert 5(x-1)\rvert < 3$	given
$5\lvert x-1\rvert < 3$	absolute value of product
$\lvert x-1\rvert < \frac{3}{5}$	dividing by 5
$-\frac{3}{5} < x - 1 < \frac{3}{5}$	$x - 1$ has absolute value less than $\frac{3}{5}$
$1 - \frac{3}{5} < x < 1 + \frac{3}{5}$	adding 1 to an inequality
$\frac{2}{5} < x < \frac{8}{5}$	arithmetic

In short, x is in the open interval $(\frac{2}{5}, \frac{8}{5})$, sketched in Fig. A.7. ■

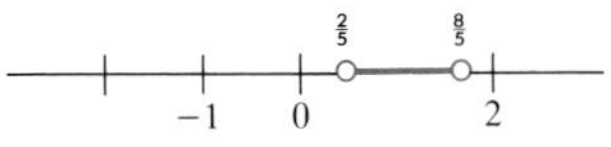

Figure A.7

EXERCISES FOR APPENDIX A: REAL NUMBERS

In Exercises 1 to 6 assume that a and b are positive numbers and $a < b$. Place the correct symbol, $<$ or $>$, in each blank.

1 $a + 3$ __ $b + 3$

2 $a - 2$ __ $b - 2$

3 $5a$ __ $5b$

4 $(-2)a$ __ $(-2)b$

5 $1/a$ __ $1/b$

6 $3 - a$ __ $3 - b$

In Exercises 7 to 20 describe where the inequalities hold. (Use interval notation.)

7 $2x + 7 < 4x + 9$

8 $3x - 5 < 7x + 11$

9 $-3x + 2 > 5x + 18$

10 $2x > 3x + 7$

11 $(x - 1)(x - 3) > 0$

12 $(x + 1)(x - 3) < 0$

13 $(x + 2)(x + 3) < 0$

14 $(2x + 6)(x + 3) < 0$

15 $x(x - 1)(x + 1) > 0$

16 $(x - 2)(x - 3)(x - 4) > 0$

17 $x(x + 3)(x + 5) > 0$

18 $(x - 1)^2(x - 2) > 0$

19 $(3x - 1)(2x - 1) > 0$

20 $x(2x + 1)(3x - 1) > 0$

In each of Exercises 21 to 24 sketch the intervals for which the inequality holds.

21 $|x - 3| < 2$

22 $|2x - 4| < 1$

23 $|3(x - 1)| < 6$

24 $|4(x + 2)| < 2$

■

25 For which x is $x^2(x - 3) > 0$?

26 For which x is $(x - 1)^2(x - 2)^2 < 0$?

27 For which positive numbers x is $x < x^2$?

28 For which positive numbers x is $x^2 < x^3$?

In Exercises 29 and 30 give an example of numbers a and b, $a < b$, neither 0, for which the stated inequality is *false*.

29 $a^2 < b^2$

30 $1/a > 1/b$

31 Give an example of numbers a, b, c, and d such that $a < b$, $c < d$, and $ac > bd$.

32 Express 3.1416 in the form p/q, where p and q are integers. This shows that 3.1416 is a rational number. It is commonly used as a rational approximation of the irrational number π.

33 Let $a = 2.3474747\ldots$, an endless decimal in which the block "47" continues to repeat without end. This exercise shows that a is rational.
(*a*) Compute $100a$.
(*b*) Compute $100a - a$.
(*c*) From (*b*) deduce that $a = 232.4/99$.
(*d*) From (*c*) deduce that a can be written as the quotient of two integers and is therefore rational. It can be shown that any number whose decimal representation from some point on consists of a block repeated endlessly is rational.

34 Here is a proposed proof that $0 = 1$. Let $x = 0$. Then $x = x^2$. Cancellation of x from both sides of the equation show that $1 = x$. Thus we have $x = 0$ and $x = 1$ and conclude that $0 = 1$. Where is the error?

35 Here is another proof that $0 = 1$. Let $x = 1$. Then $x^2 = x^3$ or $x^2(x - 1) = 0$. Thus $x^2 = 0$, from which it follows that $x = 0$. Since $x = 1$ and $x = 0$, we conclude that $0 = 1$. Where is the error?

36 (*a*) Show that, if x and y are rational numbers, so is their product. (*Hint:* $x = m/n$ and $y = p/q$, where m, n, p, and q are integers.)
(*b*) Give an example of irrational numbers x and y whose product is rational.

37 This is how the Greeks proved that $\sqrt{2}$ is irrational. Assume that $\sqrt{2}$ is rational. Then $\sqrt{2} = m/n$, where m and n are positive integers. Moreover, we may assume that m and n have no common factor other than 1 (for if there were a common factor, we could cancel it in numerator and denominator). They then showed that both m and n must be even, which is a contradiction, since two even numbers have the common factor 2. Their argument depends on the fact that if the square of an integer is even, then the integer itself must be even.
(*a*) Show that $2n^2 = m^2$.
(*b*) From (*a*) deduce that m is even; hence $m = 2p$, for some integer p.
(*c*) Deduce that $n^2 = 2p^2$.
Thus m and n are both even, a contradiction.

Exercises 38 and 39 concern an inequality that arose in the study of bias in admissions at a university.

38 Let a_1, b_1, a_2, b_2, A_1, B_1, A_2, and B_2 be positive numbers such that

$$\frac{a_1}{b_1} < \frac{A_1}{B_1} \quad \text{and} \quad \frac{a_2}{b_2} < \frac{A_2}{B_2}.$$

Does it follow that

$$\frac{a_1 + a_2}{b_1 + b_2} < \frac{A_1 + A_2}{B_1 + B_2}?$$

One department admits a_1 out of b_1 women applicants and A_1 out of B_1 men applicants. Since $a_1/b_1 < A_1/B_1$, we may say that the department seems biased against women. Another department admits a_2 out of b_2 women and A_2 out of B_2 men; it also appears to be biased against women. The two departments, considered together, admit $a_1 + a_2$ out of $b_1 + b_2$ women and $A_1 + A_2$ out of $B_1 + B_2$ men. Can we be sure that the two, taken as a whole, are biased against women, that is, that

$$\frac{a_1 + a_2}{b_1 + b_2} < \frac{A_1 + A_2}{B_1 + B_2}?$$

39 (See Exercise 38.) Jane has a higher batting average than Jim in May. In June she again has a higher batting average. Does this imply that for the 2-month period taken as a whole, Jane has a higher batting average than Jim? (Batting average = hits/at bats.)

Appendix B
Graphs and Lines

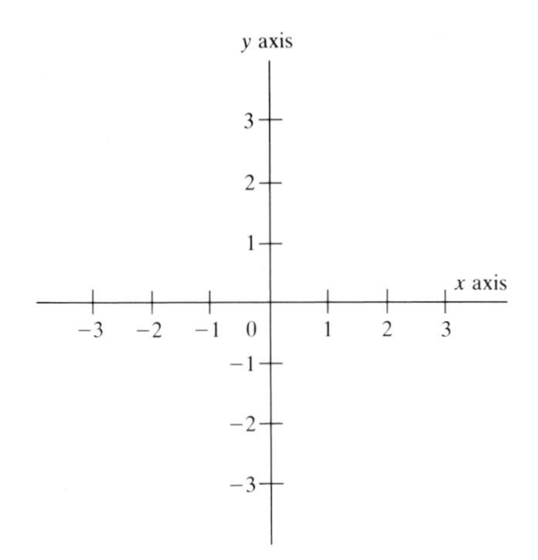

Figure B.1

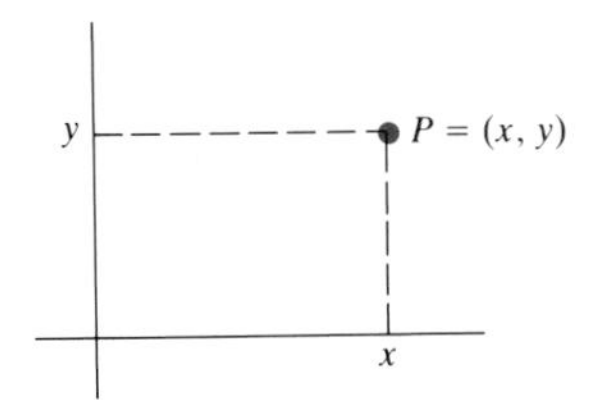

Figure B.2

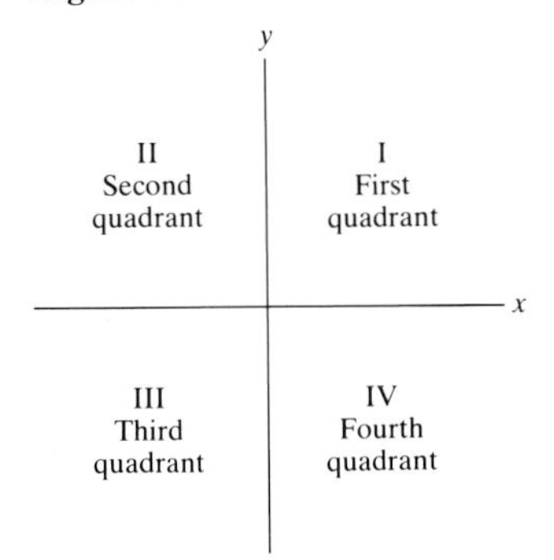

Figure B.3

This appendix reviews coordinate systems, graphs, lines, and their slopes—concepts that play a key role in calculus.

B.1 COORDINATE SYSTEMS AND GRAPHS

Just as each point on a line can be described by a number, each point in the plane can be described by a pair of numbers. To do this, choose two perpendicular lines furnished with identical scales, as in Fig. B.1. One is called the x **axis** and the other, the y **axis**. Usually the x **axis** is horizontal, as in Fig. B.1. Any point P in the plane can then be described by a pair of numbers. The line through P parallel to the y axis meets the x axis at a number x, called the x **coordinate** or **abscissa** of P. The line through P parallel to the x axis meets the y axis at a number y, called the y **coordinate** or **ordinate** of P. P is then denoted (x, y), as in Fig. B.2. The point $(0, 0)$, where the two axes cross, is called the **origin**.

The two axes cut the plane into four parts, called **quadrants**, numbered as in Fig. B.3. In the first quadrant, both the x and y coordinates are positive; in the second, x is negative and y is positive; in the third, both x and y are negative; in the fourth, x is positive and y is negative.

EXAMPLE 1 Plot the points $(1, 2)$, $(-3, 1)$, $(-\frac{1}{2}, -2)$, and $(3, -1)$ and specify the quadrant in which each lies.

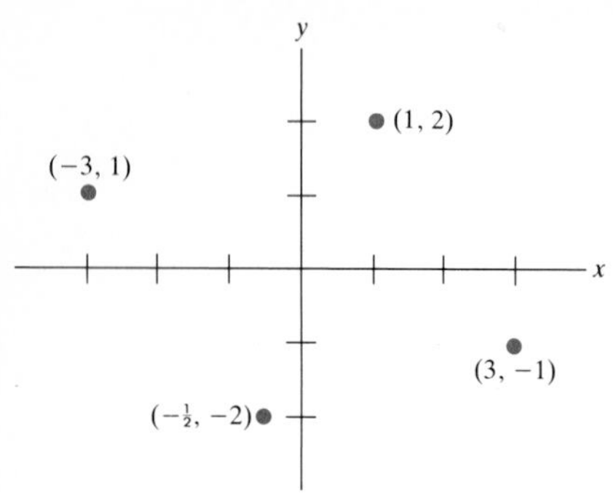

Figure B.4

SOLUTION As shown in Fig. B.4, the point (1, 2) is in the first quadrant, (−3, 1) is in the second, $(-\frac{1}{2}, -2)$ is in the third, and (3, −1) is in the fourth. ■

The *distance d between two points* $P_1 = (x_1, y_1)$ and $P_2 = (x_2, y_2)$ can be found with the aid of the pythagorean theorem. Form a right triangle whose hypotenuse is the line segment joining P_1 to P_2 and whose legs are parallel to the axes, as in Fig. B.5.

The length of the horizontal leg is $x_2 - x_1$ if $x_2 \geq x_1$, as in Fig. B.5. However, if $x_1 > x_2$, the length is $-(x_2 - x_1)$. Similarly, the length of the vertical leg is either $y_2 - y_1$ or its negative. In either case, since the negative of a number has the same square as the number, we have, by the pythagorean theorem,

Recall that $(-a)^2 = a^2$.

$$d^2 = (x_2 - x_1)^2 + (y_2 - y_1)^2, \tag{1}$$

The distance formula

or

$$d = \sqrt{(x_2 - x_1)^2 + (y_2 - y_1)^2}. \tag{2}$$

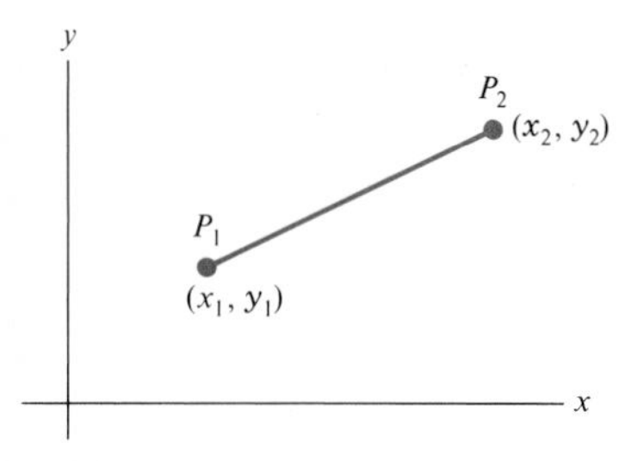

Figure B.5

Keep both Eqs. (1) and (2) in mind. While Eq. (2) gives d explicitly, Eq. (1) is often preferable in computations since it does not involve square roots.

EXAMPLE 2 Find the distance between the points (−1, 3) and (2, −5).

SOLUTION Let $(x_1, y_1) = (-1, 3)$ and $(x_2, y_2) = (2, -5)$. Then

$$d^2 = [2 - (-1)]^2 + [(-5) - 3]^2 = 3^2 + (-8)^2 = 73.$$

Thus $\quad d = \sqrt{73}.$

We could just as well have labeled the points in the opposite order, $(x_1, y_1) = (2, -5)$ and $(x_2, y_2) = (-1, 3)$. The arithmetic is slightly different, but the result is the same:

$$d^2 = [(-1) - 2]^2 + [3 - (-5)]^2 = (-3)^2 + 8^2 = 73$$

and $\quad d = \sqrt{73}.$ ■

The distance formula gives us a way of dealing algebraically with the geometric notion of a *circle of radius r and center* (0, 0). A point (x, y) lies on this circle if its distance from (0, 0) is r, that is, if

$$\sqrt{(x - 0)^2 + (y - 0)^2} = r$$

or, more simply, if

$$x^2 + y^2 = r^2.$$

The converse is true also: If $x^2 + y^2 = r^2$, then $\sqrt{(x - 0)^2 + (y - 0)^2} = r$, so the point (x, y) lies on the circle of radius r and center (0, 0). For the sake of brevity, we may speak of "the circle $x^2 + y^2 = r^2$," which is the circle of radius r centered at the origin.

EXAMPLE 3 Determine which of these points lie on the circle of radius 13 and center at the origin: (5, 12), (−5, 12), (10, 7).

SOLUTION To test whether the point (x, y) lies on the circle of radius 13 and center at the origin, check whether

$$x^2 + y^2 = 13^2,$$

that is, whether $x^2 + y^2 = 169$.

Since
$$5^2 + 12^2 = 25 + 144 = 169,$$

the point (5, 12) lies on the circle. So does the point (−5, 12), since $(-5)^2 + 12^2 = 25 + 144 = 169$.

Does (10, 7) also lie on this circle? We find that $10^2 + 7^2 = 100 + 49 = 149$. Thus (10, 7) does not lie on the circle. (See Fig. B.6.) ■

The graph of an equation consists of the points for which the equation holds.

The circle shown in Fig. B.6 is called the **graph** of the equation $x^2 + y^2 = 169$. The graph of any equation involving one or both of the letters x and y consists of those points (x, y) whose coordinates satisfy the equation.

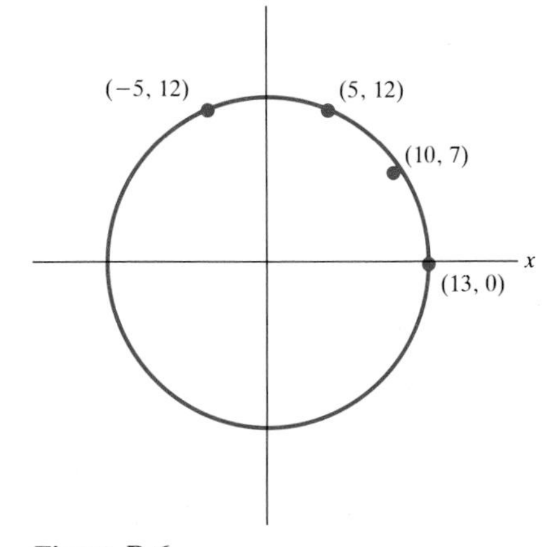

Figure B.6

EXAMPLE 4 Graph the curve $y = x^2$.

SOLUTION To begin, find a few points on the graph by choosing some specific values of x and calculating the corresponding values of y. Let us use $x = 0, 1, 2, 3, -1, -2$, and -3 and fill in this table:

x	0	1	2	3	−1	−2	−3
$y = x^2$	0	1	4	9	1	4	9

The table provides seven points on the graph of $y = x^2$, as shown in Fig. B.7. Note that, as $|x|$ increases, so does $y = x^2$. The graph goes arbitrarily high. ■

The graph of $y = x^3$ can be sketched in a similar manner, by starting with a table of values. (Calculus provides additional techniques for graphing.) It is sketched in Fig. B.8.

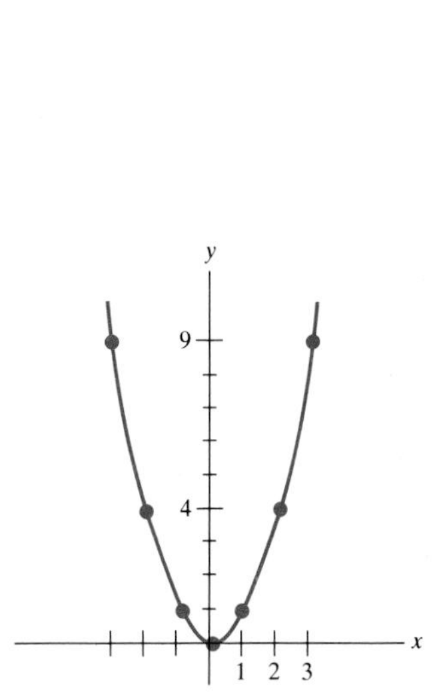

Figure B.7

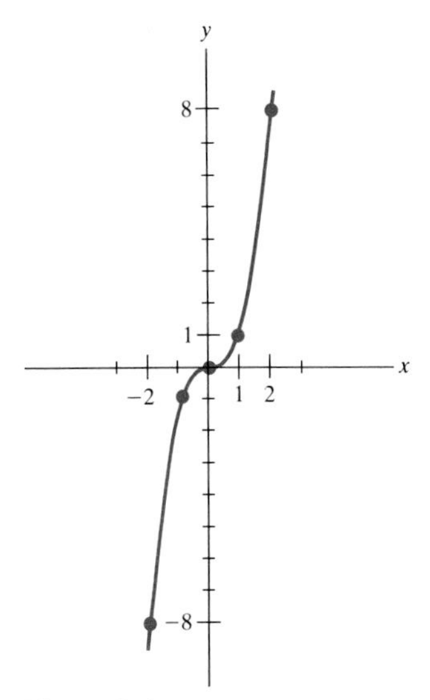

Figure B.8

Near the point (0, 0) the graph of $y = x^3$ looks almost like the x axis. You can see why if you plot the points $(\frac{1}{2}, \frac{1}{8})$ and $(\frac{1}{4}, \frac{1}{64})$ on a large scale. Both points lie on $y = x^3$.

The graphs of $x^2 + y^2 = r^2$, $y = x^2$, and $y = x^3$ possess some **symmetry**. For instance, the part of the parabola $y = x^2$ left of the y axis is a mirror image of the part right of the y axis. Had we known this in advance, we could have sketched the graph in the first quadrant and then copied this graph in the second quadrant. Since many graphs have some symmetry, the following tests for symmetry can be quite useful.

Definition The graph of an equation is **symmetric with respect to the y axis** if the equation is unchanged when x is replaced by $-x$.

Symmetry with respect to the y axis

For example, when x is replaced by $-x$ in the equation $y = x^2$, we obtain the equation $y = (-x)^2$, which simplifies to the original equation $y = x^2$. If x appears only to *even* powers in an equation, the graph of that equation will be symmetric with respect to the y axis.

Definition The graph of an equation is **symmetric with respect to the x axis** if the equation is unchanged when y is replaced by $-y$.

Symmetry with respect to the x axis

For example, the graph of $x^2 + y^2 = 169$ is symmetric with respect to the x axis, since the equation $x^2 + (-y)^2 = 169$ reduces to $x^2 + y^2 = 169$, the original equation. The graph is also symmetric with respect to the y axis, since $(-x)^2 + y^2 = 169$ also reduces to $x^2 + y^2 = 169$.

The graph of $y = x^3$ is not symmetric with respect to the x axis or the y axis. But it does have the symmetry described in the next definition.

Definition The graph of an equation is **symmetric with respect to the origin** if we obtain an equivalent equation when we replace y by $-y$ and x by $-x$ at the same time.

Symmetry with respect to the origin

If in the equation $y = x^3$ we replace y by $-y$ and x by $-x$, we obtain the equation $-y = (-x)^3$, which a little algebra reduces to $-y = -x^3$, and thus, $y = x^3$, the original equation. Thus the cubic $y = x^3$ is symmetric with respect to the origin.

Figure B.9 is a memory device showing these three types of symmetry.

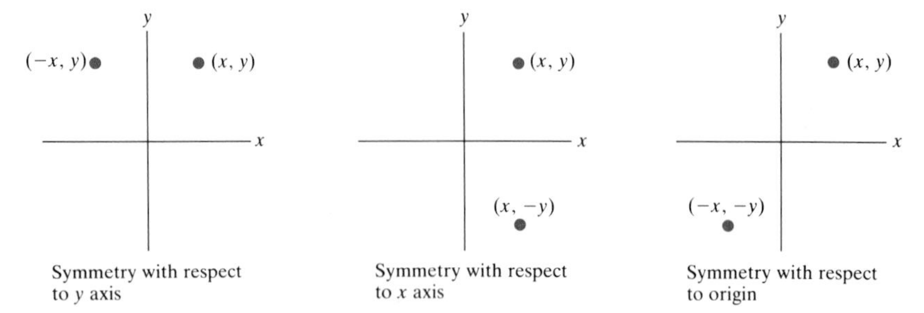

Figure B.9

The x coordinates of the points where a graph meets the x axis are the **x intercepts** of the graph. The y coordinates of the points where a graph meets the y axis are the **y intercepts** of the graph.

How to find intercepts

To find the x intercepts of a graph, set $y = 0$ in its equation. To find the y intercepts, set $x = 0$.

For instance, to find the x intercepts of the circle $x^2 + y^2 = 169$, set $y = 0$, obtaining the equation

$$x^2 = 169.$$

The x intercepts are therefore 13 and -13, which can be checked by inspecting Fig. B.6.

Graph of $y = x^2 - 4x - 5$

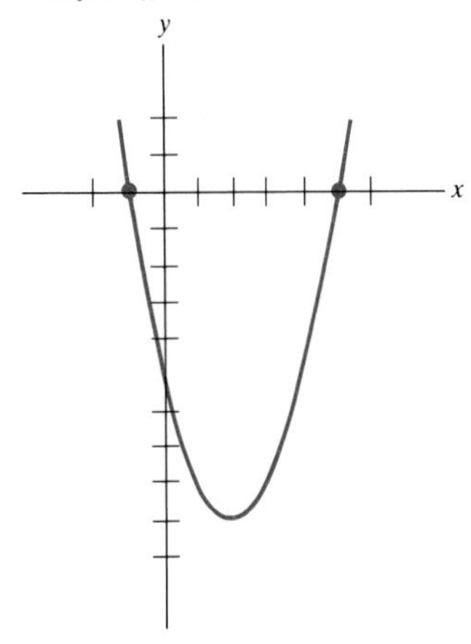

Figure B.10

EXAMPLE 5 Find the intercepts of the graph of $y = x^2 - 4x - 5$.

SOLUTION To find the x intercepts, set $y = 0$, obtaining

$$0 = x^2 - 4x - 5.$$

By the quadratic formula, $x = \dfrac{4 \pm \sqrt{16 + 20}}{2} = \dfrac{4 \pm \sqrt{36}}{2}$

$$= \frac{4 \pm 6}{2}$$

$$= 5 \text{ and } -1.$$

To find y intercepts, set $x = 0$, obtaining

$$y = 0^2 - 4 \cdot 0 - 5 = -5.$$

There is only one y intercept, namely, -5.

This is a parabola. The graph and intercepts are shown in Fig. B.10. ■

The following table lists some of the most common graphs:

Equation	*Graph*	*Comment*
(1) $y = ax^2 + bx + c$ (A parabola)	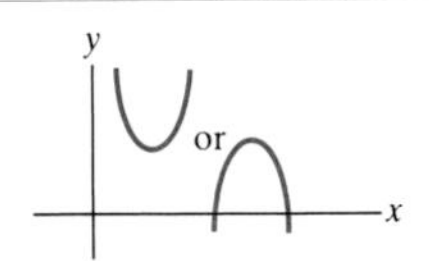	If $a > 0$, the graph opens upward. If $a < 0$, the graph opens downward.
(2) $y = ax^3 + bx^2 + cx + d$ (A cubic)	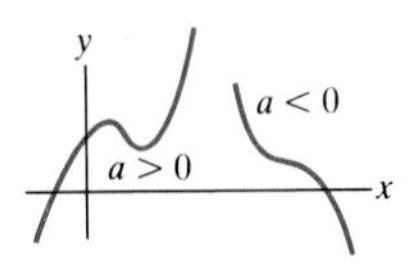	S-shaped
(3) $\dfrac{x^2}{a^2} + \dfrac{y^2}{b^2} = 1$ (An ellipse)	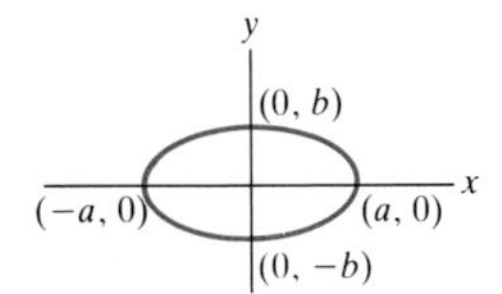	When $b = a$, the ellipse is a circle.
(4) $\dfrac{x^2}{a^2} - \dfrac{y^2}{b^2} = 1$ (A hyperbola)	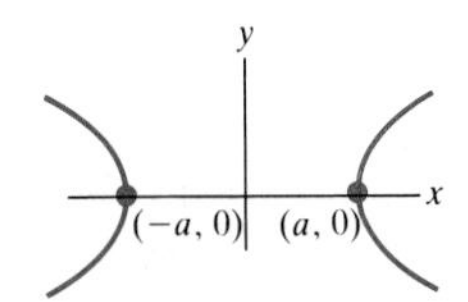	The two pieces, called **branches**, are unbounded.

More general equations whose graphs are also parabolas, ellipses, and hyperbolas are discussed in Appendix F, where the conic sections are defined.

Note that an ellipse is symmetric with respect to both axes. Moreover, its x intercepts are a and $-a$; its y intercepts are b and $-b$. For $a > b$, the ellipse is wider than high; for $b > a$, it is higher than wide.

EXAMPLE 6 Sketch the curve $\dfrac{x^2}{8} + \dfrac{y^2}{4} = 1$.

SOLUTION In this case, $a^2 = 8$ and $b^2 = 4$, so $a = \sqrt{8} \approx 2.8$ and $b = 2$. First plot the four points where the ellipse crosses the axes, then fill in the curve smoothly freehand, as shown in Fig. B.11.

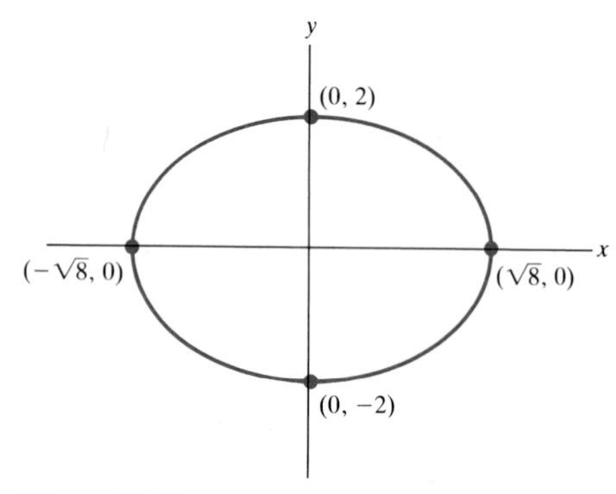

Figure B.11 ■

Although Eq. (4) looks a lot like Eq. (3) in the table, its graph is quite different. First of all, the hyperbola given in the table has no y intercept. To see why, consider the equation

$$\frac{0^2}{a^2} - \frac{y^2}{b^2} = 1,$$

which reduces to

$$y^2 = -b^2.$$

Since b^2 is positive, $-b^2$ is negative. But there is no real number y whose square is negative. Thus the hyperbola (4) has no y intercept.

Second, the hyperbola extends arbitrarily far from the origin. To show this, solve for y:

$$\frac{y^2}{b^2} = \frac{x^2}{a^2} - 1$$

$$y^2 = b^2\left(\frac{x^2}{a^2} - 1\right)$$

$$y^2 = \frac{b^2}{a^2}(x^2 - a^2)$$

$$y = \pm\frac{b}{a}\sqrt{x^2 - a^2}.$$

For $|x| \geq a$, the expression in the radical is not negative. So, for $|x| \geq a$, hence for large x, there will be a corresponding value of y [namely, $y = \pm(b/a)\sqrt{x^2 - a^2}$].

When x is large, $\sqrt{x^2 - a^2}$ is very close to x. (Check this for $x = 20$ and $a = 3$.) So for large x, the graph is close to the line $y = (b/a)x$.

To graph the hyperbola $x^2/a^2 - y^2/b^2 = 1$, follow these steps:

1. Plot the points $(a, 0)$ and $(-a, 0)$, where the hyperbola meets the x axis.
2. Plot the point (a, b) and the line through it and the origin. Do the same for the point $(a, -b)$.
3. Sketch the hyperbola freehand, using the two lines in step 2 as guides. (As shown in Exercise 52, the hyperbola approaches these lines arbitrarily closely.) The two lines are called **asymptotes** of the hyperbola.

EXAMPLE 7 Graph the hyperbola $\dfrac{x^2}{9} - \dfrac{y^2}{4} = 1$.

SOLUTION In this case, $a^2 = 9$ and $b^2 = 4$, so $a = 3$ and $b = 2$. The x intercepts are 3 and -3. One asymptote passes through (0, 0) and (3, 2) and the other through (0, 0) and (3, -2). With the aid of the intercepts and the asymptotes, the graph is easy to sketch freehand. (See Fig. B.12.) Note that since x and y appear only to even powers, the graph is symmetric with respect to both axes.

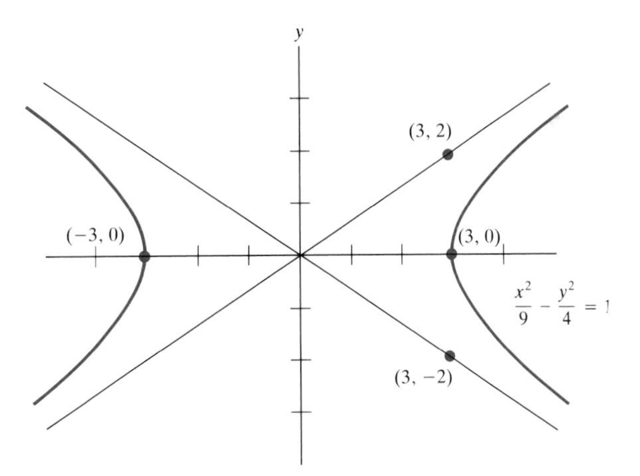

Figure B.12 ■

A more general equation whose graph is a hyperbola

As mentioned in Appendix F, when $B^2 - 4AC$ is positive, the graph of $Ax^2 + Bxy + Cy^2 + Dx + Ey + F = 0$ is a hyperbola (or, in special cases, a pair of intersecting lines). However, in general the x and y axes are not the lines of symmetry of the hyperbola. In particular, the graph of $xy = a$, where a is a nonzero constant, is a hyperbola.

EXAMPLE 8 Graph the hyperbola $xy = 1$.

SOLUTION Since the product of x and y is 1, neither x nor y can be 0. Thus the hyperbola $xy = 1$ has no intercepts. Moreover, since $x \neq 0$, we can solve for y in terms of x,

$$y = \frac{1}{x},$$

and make a table of values:

x	$\frac{1}{10}$	10	1	2	$-\frac{1}{10}$	-10	-1	-2
$y = \frac{1}{x}$	10	$\frac{1}{10}$	1	$\frac{1}{2}$	-10	$-\frac{1}{10}$	-1	$-\frac{1}{2}$

With the aid of these eight points, sketch the graph, which consists of two identical pieces. Far to the right and to the left the graph approaches the x axis. Also, as x approaches 0, the graph approaches the y axis. In this case, the x and y axes are the asymptotes of the hyperbola. (See Fig. B.13.)

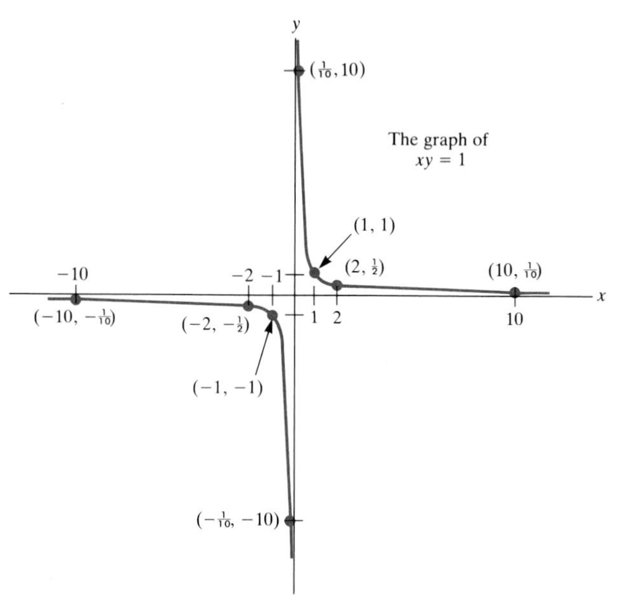

Figure B.13

■

EXERCISES FOR SEC. B.1: COORDINATE SYSTEMS AND GRAPHS

In Exercises 1 and 2 plot the points and state in which quadrant each lies.

1 (*a*) $(-3, 5)$ (*b*) $(4, -3)$ (*c*) $(2, 7)$ (*d*) $(-2, -7)$

2 (*a*) $(1, 3)$ (*b*) $(-3, -\frac{1}{2})$ (*c*) $(5, -3)$ (*d*) $(-3, 4)$

In Exercises 3 and 4 find the distances between the given points.

3 (*a*) $(4, 1)$ and $(-2, 9)$ (*b*) $(6, 3)$ and $(1, 15)$ (*c*) $(4, 0)$ and $(7, 0)$

4 (*a*) $(3, 4)$ and $(5, 6)$ (*b*) $(-3, -2)$ and $(4, -2)$ (*c*) $(-3, -4)$ and $(3, 4)$

In Exercises 5 to 16 graph the equations.

5 $x^2 + y^2 = 49$
6 $x^2 + y^2 = 1$
7 $y = 2x^2$
8 $y = -2x^2$
9 $y = -x^2$
10 $y = -x^2/2$
11 $y = -x^3$
12 $y = 2x^3$
13 $y = x^4$ (Include the points for which $x = 0, \frac{1}{2}$, and 1.)
14 $y = x^5$ (Include the points for which $x = 0, \frac{1}{2}$, and 1.)
15 $y = x - 1$
16 $y = -2x + 3$

In Exercises 17 to 22 find the x and y intercepts, if there are any.

17 $y = 2x + 6$ **18** $y = 3x - 6$

19 $xy = 6$ **20** $x^2 - y^2 = 1$

21 $y = 2x^2 + 5x - 3$ **22** $y = 4 - x^2$

In Exercises 23 to 32 graph the equations.

23 $\frac{x^2}{25} + \frac{y^2}{16} = 1$ **24** $\frac{x^2}{16} + \frac{y^2}{25} = 1$

25 $\frac{x^2}{16} - \frac{y^2}{25} = 1$ **26** $\frac{x^2}{25} - \frac{y^2}{16} = 1$

27 $xy = 8$ **28** $xy = -1$

29 $y = x^2 - 2x + 3$ **30** $y = -x^2 + 3x + 4$

31 $\frac{y^2}{4} - \frac{x^2}{9} = 1$ (Note that the negative sign is with x^2, not y^2.)

32 $y^2 - x^2 = 1$

In Exercises 33 to 36 graph the cubics.

33 $y = x^3 - 3x^2 + 3x$ **34** $y = 2x^3 - 3x^2$

35 $y = x^3 + x$ **36** $y = -x^3 + x$

In Exercises 37 to 44 graph the equations.

37 $y = x - x^2$ **38** $y = 1 + x^2$

39 $y = (x - 1)^2$ **40** $y = \sqrt{x}$

41 $y = |x|$ **42** $y = \sqrt[3]{x}$

43 $y = (x - 1)(x - 2)$ **44** $y = \sqrt[3]{x - 1}$

In Exercises 45 and 46 find an equation of the circle of given radius and center.

45 Radius 7, center (2, 1) **46** Radius $\frac{1}{2}$, center $(-2, 3)$

47 (*a*) What symmetry does the graph of $x = y^2$ have?
(*b*) Graph $x = y^2$.

48 (*a*) What symmetry does the graph of $y^2 = x^3$ have?
(*b*) Graph $y^2 = x^3$.

■

49 (*a*) Graph $y = x^2 + 3x + 3$.
(*b*) What is the lowest point on the graph? [*Hint:* $x^2 + 3x + 3 = (x + \frac{3}{2})^2 + 3 - \frac{9}{4}$.]

50 (*a*) Graph $y = x^2 + 4x + 9$.
(*b*) What point on the graph in (*a*) has the smallest y coordinate? [*Hint:* See the hint for part (*b*) of Exercise 49.]

51 Let $y = ax^2 + bx + c$, where a is positive and the discriminant, $b^2 - 4ac$, is positive.
(*a*) Find the value of x for which y is as small as possible. (*Hint:* First rewrite $ax^2 + bx + c$ by completing the square, which is reviewed in Appendix C.)
(*b*) Show that the number x in (*a*) is the average of the two x intercepts of the graph.

52 This exercise shows why the part of the hyperbola $x^2/a^2 - y^2/b^2 = 1$ in the first quadrant approaches the line $y = bx/a$ when x is large.
Consider the point $P = (x, y)$ on the hyperbola, $x \geq a$, $y \geq 0$.
(*a*) Show that $y = (b/a)\sqrt{x^2 - a^2}$.
(*b*) Let $Q = (x, bx/a)$ be the point on the line $y = bx/a$ with the same x coordinate as P. Show that Q is near P for large x, by showing that

$$\frac{bx}{a} - \frac{b}{a}\sqrt{x^2 - a^2}$$

approaches 0 when x is large. [*Hint:* Rationalize by multiplying by $(x + \sqrt{x^2 - a^2})/(x + \sqrt{x^2 - a^2})$.]

B.2 LINES AND THEIR SLOPES

The vertical change, $y_2 - y_1$, is the "rise"; the horizontal change, $x_2 - x_1$, is the "run"; slope is "rise over run."

Consider a line that is not parallel to the y axis. Select distinct points $P_1 = (x_1, y_1)$ and $P_2 = (x_2, y_2)$ on it as in Fig. B.14. As we move from P_1 to P_2, the vertical change is $y_2 - y_1$ and the horizontal change is $x_2 - x_1$, as shown in Fig. B.15. The quotient $(y_2 - y_1)/(x_2 - x_1)$ is a measure of the steepness of the line and is called its **slope**.

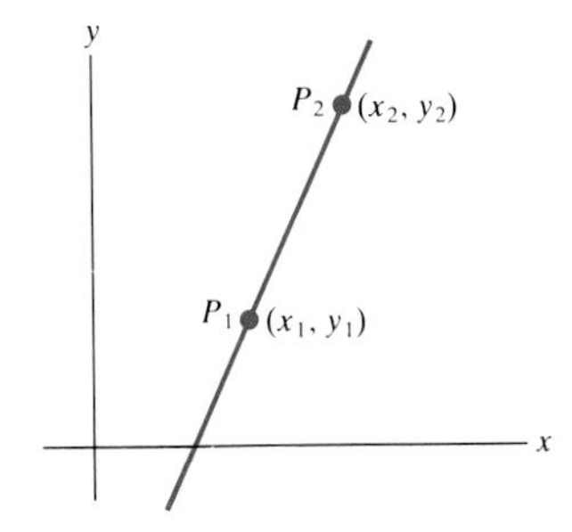

Figure B.14

Definition *Slope of a line.* Consider a line that is not parallel to the y axis. Let $P_1 = (x_1, y_1)$ and $P_2 = (x_2, y_2)$ be two distinct points on the line. The **slope** of the line is

$$m = \frac{y_2 - y_1}{x_2 - x_1}.$$

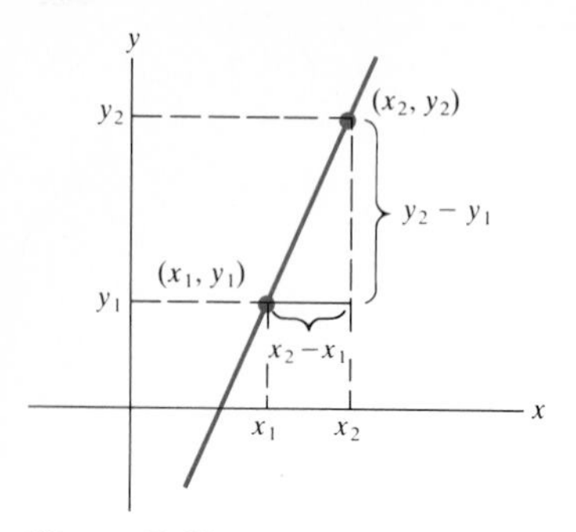

Figure B.15

The slope of a line parallel to the y axis is not defined. Accordingly, a vertical line is said to have "no slope."

EXAMPLE 1 Find the slope of the line through the points (1, 4) and (3, 5).

SOLUTION Let P_1 be (1, 4) and P_2 be (3, 5). Then

$$m = \frac{5 - 4}{3 - 1} = \frac{1}{2}. \quad ■$$

In finding the slope in Example 1 we could have chosen P_1 to be (3, 5) and P_2 to be (1, 4). The arithmetic would be slightly different, but the result would be the same, since in this case

$$m = \frac{4 - 5}{1 - 3} = \frac{-1}{-2} = \frac{1}{2}.$$

Order of points does not affect slope.

The order of the two points does not affect the slope, since

$$\frac{y_2 - y_1}{x_2 - x_1} = \frac{y_1 - y_2}{x_1 - x_2}.$$

Moreover, the slope of a line does not depend on the particular pair of points selected on it. If one pair is $P_1 = (x_1, y_1)$ and $P_2 = (x_2, y_2)$ and the other pair is $P_1' = (x_1', y_1')$ and $P_2' = (x_2', y_2')$, then

$$\frac{y_2 - y_1}{x_2 - x_1} = \frac{y_2' - y_1'}{x_2' - x_1'}.$$

This equation follows from the fact that the ratios between the lengths of corresponding sides of similar triangles are equal. (See Fig. B.16, which shows the case where the slope is positive; a similar argument works for the case where the slope is negative.)

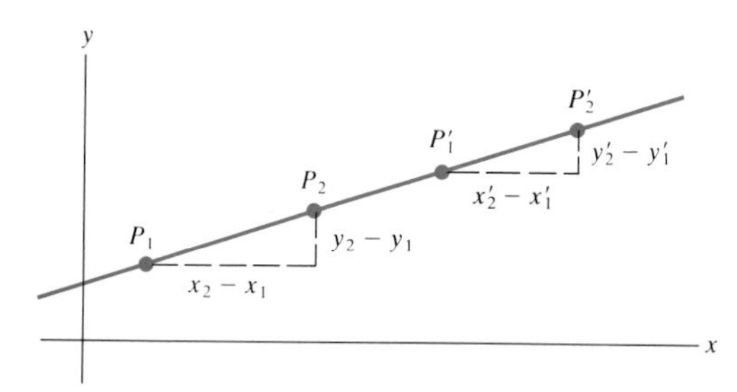

Figure B.16

EXAMPLE 2 Find the slopes of the lines through the given points. Then draw the lines.

(*a*) (0, 2) and (1, 8) (*b*) (1, 1) and (5, 2)
(*c*) (2, 6) and (4, 3) (*d*) (5, 1) and (−2, 1)

SOLUTION

(*a*) $m = \dfrac{8 - 2}{1 - 0} = 6.$ (*b*) $m = \dfrac{2 - 1}{5 - 1} = \dfrac{1}{4}.$

(*c*) $m = \dfrac{3 - 6}{4 - 2} = -\dfrac{3}{2}.$ (*d*) $m = \dfrac{1 - 1}{(-2) - 5} = 0.$

The lines are graphed in Fig. B.17.

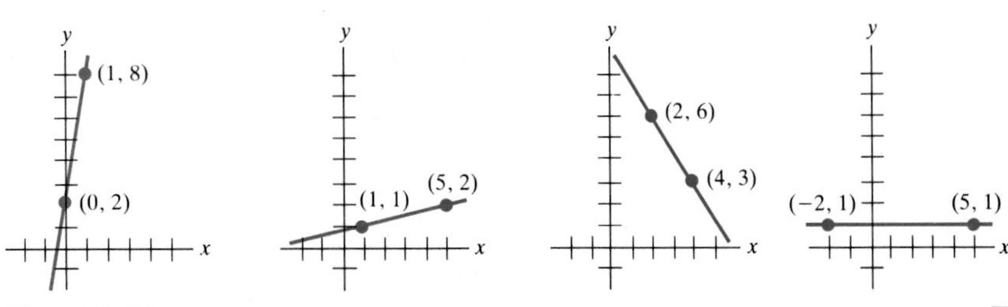

Figure B.17 ■

As suggested by the lines in Example 2, a *positive slope tells us that as we move from left to right on the line, we go up.* (As the x coordinate increases, so does the y coordinate.) *A negative slope tells us that as we move from left to right on the line, we go down.* (As the x coordinate increases, the y coordinate decreases.) A slope of 0 corresponds to a *horizontal* line, that is, a line parallel to the x axis.

Meaning of positive and negative slopes

Given the slopes of two lines, we can determine whether they are parallel or perpendicular.

Test for parallel lines

A line of slope m_1 is *parallel* to a line of slope m_2 if and only if $m_1 = m_2$.

The test for perpendicularity is a little fancier.

Test for perpendicular lines

A line of slope m_1 is *perpendicular* to a line of slope m_2 if and only if the *product of the two slopes* is -1, $m_1 m_2 = -1$.

(In other words,

$$m_2 = -\frac{1}{m_1}.$$

The slope m_2 is the negative of the reciprocal of m_1.) It is assumed that neither line is parallel to the y axis.

Proofs of these two tests are outlined in Exercises 35 and 36.

EXAMPLE 3 Determine whether the line through $(-1, -2)$ and $(2, 2)$ is parallel to the line through $(6, 5)$ and $(1, 1)$.

SOLUTION The slope of the first line is

$$\frac{2 - (-2)}{2 - (-1)} = \frac{4}{3}.$$

The slope of the second line is

$$\frac{1 - 5}{1 - 6} = \frac{-4}{-5} = \frac{4}{5}.$$

Since $\frac{4}{3} \neq \frac{4}{5}$, the lines are not parallel. ■

EXAMPLE 4 Determine whether the line through (1, 5) and (4, 4) is perpendicular to the line through (2, 1) and (3, 4).

SOLUTION The first line has slope

$$\frac{4-5}{4-1} = \frac{-1}{3}.$$

The second line has slope

$$\frac{4-1}{3-2} = 3.$$

The product of the two slopes is $3(-\frac{1}{3}) = -1$, so the two lines are perpendicular. ■

Having discussed the geometry and slopes of lines, let us find equations whose graphs are lines.

A line parallel to the y axis has the equation $x = c$. If a line is not parallel to the y axis, it has a slope m and a y intercept b. The point $(0, b)$ is on the line. Now take any other point (x, y) on the line. Since both $(0, b)$ and (x, y) are on the line,

$$\frac{y-b}{x-0} = m,$$

or simply,

$$y = mx + b. \tag{1}$$

The steps are reversible. That is, if (x, y) satisfies Eq. (1), then (x, y) lies on the line. Equation (1) is called the **slope-intercept equation** of a line. Each line not parallel to the y axis has a unique description of the form (1). If $m = 0$, the line is parallel to the x axis and has the equation $y = 0x + b$, which is just $y = b$. In this case, the symbol x does not appear.

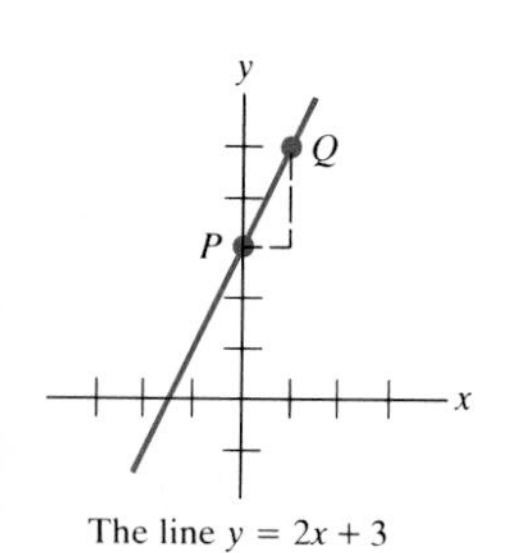

The line $y = 2x + 3$

Figure B.18

EXAMPLE 5 Graph the lines (*a*) $y = 2x + 3$ and (*b*) $y = -4x/3 + 1$.

SOLUTION

(*a*) $y = 2x + 3$. Here the y intercept is 3, so the point $P = (0, 3)$ is on the line. Plot this point. To obtain a second point on the line, write the slope 2 in the form $\frac{2}{1}$. Plot a second point, Q, 1 unit to the right of P and 2 units above it, as shown in Fig. B.18. The line through P and Q has slope 2 and y intercept 3.

(*b*) $y = -4x/3 + 1$. The y intercept is 1 and the slope is $-\frac{4}{3}$. The point $P = (0, 1)$ is on the line. Next draw the point Q that is 3 units to the right of P and 4 units lower, as shown in Fig. B.19. The line through P and Q has slope $-\frac{4}{3}$ and y intercept 1. ■

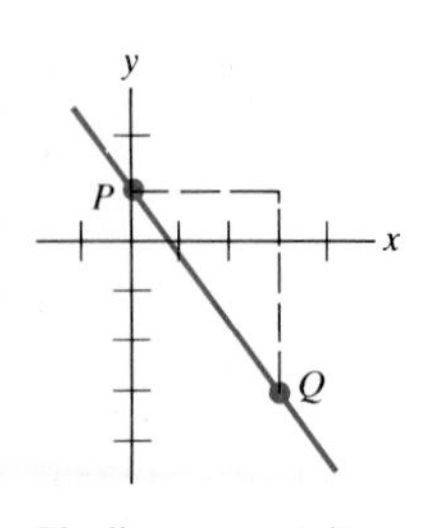

The line $y = -4x/3 + 1$

Figure B.19

Sometimes we may wish to find an equation of the line that passes through a certain point $P_1 = (x_1, y_1)$ and has a certain slope m. To find such an equation, consider the typical point $P = (x, y)$ on the line, $x \neq x_1$. The slope determined by P and P_1 must be the same as the prescribed slope m, that is,

$$\frac{y-y_1}{x-x_1} = m.$$

Clearing the denominator provides an equation in x and y, which is called the **point-slope equation** of the line. It is useful if you know the slope and a point on the line. The line through (x_1, y_1) of slope m is described by the equation

$$y - y_1 = m(x - x_1).$$

EXAMPLE 6 Find the point-slope equation of the line through (1, 3) of slope -2.

SOLUTION The point-slope formula gives

$$y - 3 = -2(x - 1).$$

This equation could be rewritten in several ways. For instance, $y - 3 = -2x + 2$ or $y = -2x + 5$ or $2x + y = 5$. ■

The point-slope formula gives a quick way to find an equation of the line through two given points $P_1 = (x_1, y_1)$ and $P_2 = (x_2, y_2)$. (Assume that $x_1 \neq x_2$. If $x_1 = x_2$, the line is parallel to the y axis and has the equation $x = x_1$.)

If $x_1 \neq x_2$, the slope of the line through P_1 and P_2 is

$$\frac{y_2 - y_1}{x_2 - x_1}.$$

How to find an equation of the line through two given points

Since the line passes through (x_1, y_1) and has slope $(y_2 - y_1)/(x_2 - x_1)$, its point-slope equation is

$$y - y_1 = \frac{y_2 - y_1}{x_2 - x_1}(x - x_1).$$

EXAMPLE 7 Find an equation of the line through $(-1, 4)$ and $(2, 3)$.

SOLUTION The slope of the line is

$$\frac{3 - 4}{2 - (-1)} = -\frac{1}{3}.$$

Since the point $(-1, 4)$ is on the line, the line has the point-slope equation

$$y - 4 = -\tfrac{1}{3}[x - (-1)].$$

Thus

$$y - 4 = -\frac{x}{3} - \frac{1}{3},$$

or

$$y = -\frac{x}{3} + \frac{11}{3},$$

which is the slope-intercept equation of the line. ■

Let A, B, and C be fixed numbers such that at least one of A and B is not zero. With the aid of the slope-intercept formula it will be shown that *the graph of the equation* $Ax + By + C = 0$ *is a line*. For this reason, the equation $Ax + By + C = 0$ is said to be "a linear equation in x and y."

To show that the graph of $Ax + By + C = 0$ is a line, consider two cases: (1) $B = 0$ and (2) $B \neq 0$.

Case 1: $B = 0$. In this instance, the equation is just $Ax + C = 0$, which is equivalent to $x = -C/A$. So the graph is a line parallel to the y axis.

Case 2: $B \neq 0$. In this case rewrite the equation $Ax + By + C = 0$ as follows:

$$By = -Ax - C,$$

or

$$y = -\frac{A}{B}x - \frac{C}{B}.$$

The graph is therefore the line of slope $-A/B$ and y intercept $-C/B$, by the slope-intercept formula.

EXAMPLE 8 Find the intercepts and slope of the line $2x + 3y - 6 = 0$.

SOLUTION To find the x intercept, set $y = 0$, obtaining

$$2x + 3 \cdot 0 - 6 = 0$$
$$2x = 6$$
$$x = 3.$$

The x intercept is 3.

To find the y intercept, set $x = 0$, obtaining

$$2 \cdot 0 + 3y - 6 = 0$$
$$3y = 6$$
$$y = 2.$$

The y intercept is 2.

See Exercise 31 for a way to get slope directly from intercepts.

To find the slope, rewrite the equation $2x + 3y - 6 = 0$ in the slope-intercept form by solving for y:

$$2x + 3y - 6 = 0$$
$$3y = -2x + 6$$
$$y = (-\tfrac{2}{3})x + 2.$$

The slope is $-\frac{2}{3}$. ■

EXERCISES FOR SEC. B.2: LINES AND THEIR SLOPES

In Exercises 1 to 6 plot the points and find the slopes of the lines through them.

1 $(-1, 1)$ and $(4, 2)$
2 $(3, 1)$ and $(-2, -1)$
3 $(-1, 4)$ and $(3, -1)$
4 $(1, 5)$ and $(5, 1)$
5 $(1, 7)$ and $(11, 7)$
6 $(-2, -5)$ and $(0, -5)$

7 Describe the slope of each line in Fig. B.20 as positive, negative, or 0.

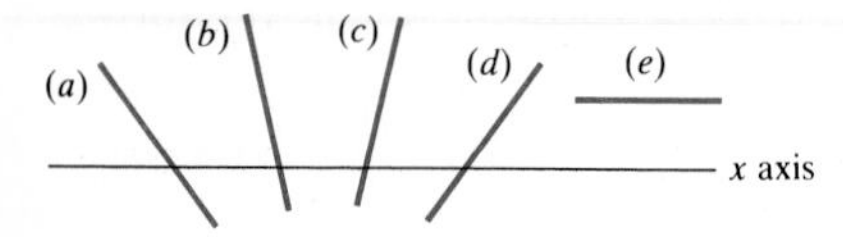

Figure B.20

8 Describe the slope of each line in Fig. B.21 as positive, negative, or 0.

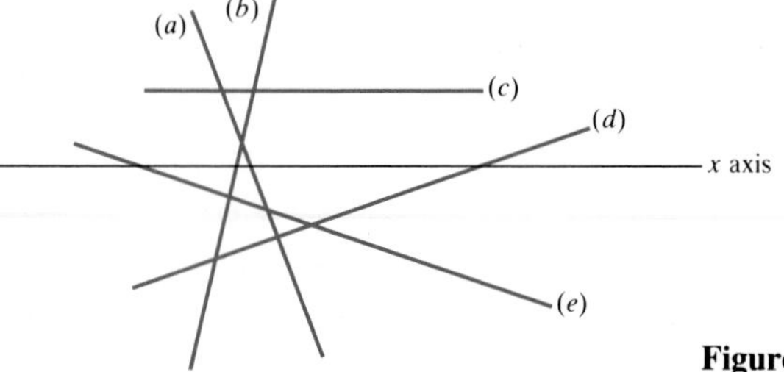

Figure B.21

9 A line L has slope 4. (*a*) What is the slope of a line parallel to L? (*b*) What is the slope of a line perpendicular to L?

10 A line L has slope $-\frac{1}{2}$. (*a*) What is the slope of a line parallel to L? (*b*) What is the slope of a line perpendicular to L?

11 Is the line through (2, 4) and (3, 7) parallel to the line through (1, 7) and (−1, 1)?

12 Is the line through (1, 2) and (−3, 5) parallel to the line through (3, −1) and (−1, 4)?

13 Is the line through (0, 0) and (4, 6) perpendicular to the line through (5, −1) and (3, 2)?

14 Is the line through (−1, −2) and (6, 2) perpendicular to the line through (2, 3) and (8, −7)?

In Exercises 15 and 16 find the slope-intercept equations for the lines:

15 (*a*) with y intercept 2 and slope 3, (*b*) with y intercept -2 and slope $\frac{2}{3}$, (*c*) with y intercept 0 and slope -3.

16 (*a*) with y intercept $\frac{3}{4}$ and slope $-\frac{1}{2}$, (*b*) with y intercept -2 and slope 2, (*c*) with y intercept 3 and slope $\frac{3}{5}$.

In Exercises 17 and 18 find the slopes and y intercepts of the given lines and graph them.

17 (*a*) $y = 3x - 1$
(*b*) $y = -2x + 1$
(*c*) $y = 3x/5$

18 (*a*) $y = -x/2 + 3$
(*b*) $y = -3x + 4$
(*c*) $y = -2x/3 + 1$

In Exercises 19 and 20 find point-slope equations of the given lines:

19 (*a*) Slope 3, passing through (1, 2)
(*b*) Slope -2, passing through (3, −1)

20 (*a*) Slope $\frac{4}{3}$, passing through the origin
(*b*) Slope 0, passing through (2, 5)

In Exercises 21 and 22 find point-slope equations of the lines through the given points.

21 (*a*) (1, 2) and (5, 3)
(*b*) (−1, 2) and (3, 1)
(*c*) (4, 5) and (2, 3)

22 (*a*) (4, 4) and (6, 6)
(*b*) (−2, −5) and (3, 4)
(*c*) (0, 0) and (3, 5)

In Exercises 23 and 24 find the slope-intercept equations of the given lines and graph the lines.

23 (*a*) $x + y + 1 = 0$ (*b*) $-2x + y = 0$
(*c*) $2x + 3y - 12 = 0$

24 (*a*) $x - y = 0$ (*b*) $-2x - 5y + 10 = 0$
(*c*) $x - 2y = 4$

■

25 Does the line through (4, 1) and (7, 2) pass through the point (*a*) (10, 3)? (*b*) (14, 5)?

26 Find an equation of the line with y intercept 2 and perpendicular to the line $y = (-2x/3) + 7$.

27 Find an equation of the line through (1, 3) and perpendicular to the line through (4, 1) and (2, 5).

28 Find an equation of the line parallel to the line $y = x + 6$ and passing through the origin.

29 Find an equation of the line parallel to $y = 3x + 2$ with y intercept 5.

Exercises 30 to 33 concern another form of an equation of a line, the "intercept form," which is convenient for working with the intercepts.

30 Let a and b be nonzero numbers. Show that the line $x/a + y/b = 1$ has x intercept a and y intercept b. (See Fig. B.22.) The equation $x/a + y/b = 1$ is called the **intercept equation** of the line through $(a, 0)$ and $(0, b)$.

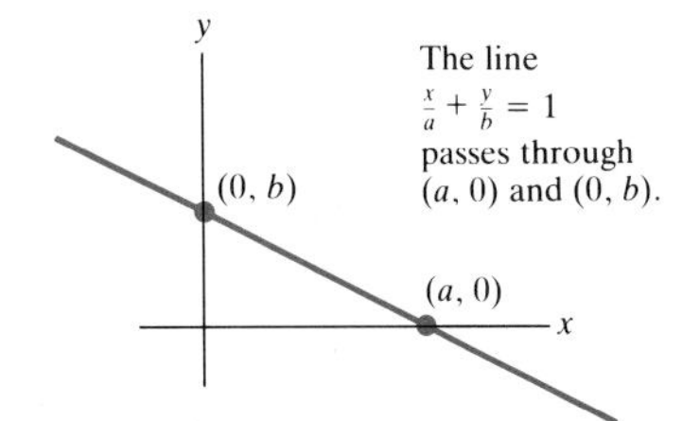

Figure B.22

31 Find the slope of the line whose intercept equation is $x/a + y/b = 1$.

32 Find the intercept equation of the line with (*a*) x intercept 3 and y intercept 5, (*b*) x intercept -1 and y intercept 2, (*c*) x intercept $-\frac{1}{2}$ and y intercept -3.

33 Find the x intercept of the line (*a*) with slope 4 and y intercept -7, (*b*) through (1, 3) with slope $-\frac{2}{3}$, (*c*) $x/3 - y/4 = 1$, (*d*) $-2x + 3y = 12$.

34 Find the intersection of the line through (0, 1) and (4, 7) with the line through (3, 3) and (5, 1).

35 This exercise outlines a proof that parallel lines have equal slopes. Let L and L' be parallel lines. Let the fixed distance between them, measured parallel to the y axis, be d. That is, if (x, y) is on L, then $(x, y + d)$ is on L'. Let $P_1 = (x_1, y_1)$ and $P_2 = (x_2, y_2)$ be points on L. Then $P_1' = (x_1, y_1 + d)$ and $P_2' = (x_2, y_2 + d)$ are points on L'.
(*a*) Find the slope of L using P_1 and P_2.
(*b*) Find the slope of L' using P_1' and P_2'.
Putting (*a*) and (*b*) together shows that the slopes of parallel lines are equal.

36 This exercise outlines a proof that the product of the slopes of perpendicular lines in the xy plane (neither parallel to the y axis) is -1. Assume that L_1 and L_2 are perpendicular lines. Let L_1 have slope m_1 and L_2 have slope m_2. For convenience, assume that both lines pass through the origin. A sketch shows that one line has positive slope and one line has negative slope. Say that m_1 is positive and m_2 is negative.
(*a*) Show that the point $(1, m_1)$ lies on L_1.
(*b*) Show that the point $(-m_1, 1)$ lies on L_2. (Recall that L_2 is perpendicular to L_1.)
(*c*) Deduce that $m_2 = -1/m_1$.
(*d*) If the two perpendicular lines do not pass through the origin, show that the product of their slopes is -1. [*Hint:* Use (*c*) and Exercise 35.]

37 (*a*) Sketch the lines $\frac{x}{3} + \frac{y}{4} = 1$ and $\frac{x}{4} - \frac{y}{3} = 1$.

(*b*) Find their slopes.

(*c*) Are they perpendicular?

38 Is the line through $(-2, 1)$ and $(3, 8)$ parallel to the line through $(3, -2)$ and $(10, 8)$?

39 A line has the equation $y = mx + b$. What can be said about m if the line (*a*) is nearly vertical (almost parallel to the y axis)? (*b*) is nearly horizontal (almost parallel to the x axis)? (*c*) slopes downward as you move to the right? (*d*) slopes upward as you move to the right?

40 Write in the form $y = mx + b$ the equation of the line through $(1, 2)$ and $(0, 5)$.

41 Find a point P on the x axis such that the line through P and $(1, 1)$ is perpendicular to the line through P and $(-3, 4)$.

Appendix C Topics in Algebra

This appendix presents several topics from algebra: rationalizing an expression, completing the square, the quadratic formula, the binomial theorem, geometric series, and roots of polynomials.

The identities $\sqrt{a}\sqrt{a} = a$ and $(a + b)(a - b) = a^2 - b^2$ enable us *to remove a square root* from certain algebraic expressions. Three examples illustrate the technique.

Rationalizing

EXAMPLE 1 Remove the square root from the denominator of the expression $4/\sqrt{2}$. ("Rationalize" the denominator.)

SOLUTION

$$\frac{4}{\sqrt{2}} = \frac{4}{\sqrt{2}} \cdot \frac{\sqrt{2}}{\sqrt{2}} = \frac{4\sqrt{2}}{2} = 2\sqrt{2} \quad \blacksquare$$

EXAMPLE 2 Remove the square root from the denominator of the fraction $2/(1 + \sqrt{5})$. (Rationalize the denominator.)

SOLUTION

$$\frac{2}{1+\sqrt{5}} = \frac{2}{1+\sqrt{5}} \cdot \frac{1-\sqrt{5}}{1-\sqrt{5}} = \frac{2-2\sqrt{5}}{1^2-(\sqrt{5})^2} = \frac{2-2\sqrt{5}}{1-5} = \frac{2-2\sqrt{5}}{-4}$$

$$= \frac{\sqrt{5}-1}{2}. \quad \blacksquare$$

In the next example the numerator is rationalized.

EXAMPLE 3 Remove the square roots from the numerator of the fraction $(\sqrt{x} - \sqrt{2})/(x - 2)$, $x \neq 2$. (Rationalize the numerator.)

SOLUTION

$$\frac{\sqrt{x}-\sqrt{2}}{x-2} = \frac{\sqrt{x}-\sqrt{2}}{x-2}\cdot\frac{\sqrt{x}+\sqrt{2}}{\sqrt{x}+\sqrt{2}} = \frac{(\sqrt{x})^2-(\sqrt{2})^2}{(x-2)(\sqrt{x}+\sqrt{2})}$$

$$= \frac{x-2}{(x-2)(\sqrt{x}+\sqrt{2})} = \frac{1}{\sqrt{x}+\sqrt{2}}. \quad ■$$

The expression $x^2 + bx + c$ can be written in the form $(x + k)^2 + d$ for a suitable choice of constants k and d. Finding the numbers k and d is called **completing the square**.

Completing the square

Recall that $$(x + k)^2 = x^2 + 2kx + k^2.$$

Therefore, if $$(x + k)^2 + d = x^2 + bx + c,$$

then we have $$x^2 + 2kx + k^2 + d = x^2 + bx + c.$$

Consequently, $2k$ must be b. In other words, k must be $b/2$. The next two examples show why this observation is the key to completing the square.

EXAMPLE 4 Complete the square in $x^2 + 6x + 11$.

SOLUTION In this case, $b = 6$ and $b/2 = 3$. Thus $k = 3$. To find d, proceed as follows:

Leaving space to complete the square

$$\begin{aligned} x^2 + 6x + 11 &= (x^2 + 6x \qquad) + 11 \\ &= (x^2 + 6x + 3^2) + 11 - 3^2 \\ &= (x + 3)^2 + 11 - 9 \\ &= (x + 3)^2 + 2. \quad ■ \end{aligned}$$

EXAMPLE 5 Complete the square in $x^2 - 5x + 2$.

SOLUTION First of all, $k = -\frac{5}{2}$. Then

$$\begin{aligned} x^2 - 5x + 2 &= [x^2 - 5x + (-\tfrac{5}{2})^2] + 2 - (-\tfrac{5}{2})^2 \\ &= (x - \tfrac{5}{2})^2 + 2 - \tfrac{25}{4} \\ &= (x - \tfrac{5}{2})^2 - \tfrac{17}{4}. \quad ■ \end{aligned}$$

A slight variation in the technique illustrated in Examples 4 and 5 can be used to complete the square in the expression $ax^2 + bx + c$, that is, to write it in the form $e(x + k)^2 + d$, for suitable constants k, d, and e. The next example shows how to do this.

EXAMPLE 6 Complete the square in $2x^2 + 5x + 3$.

SOLUTION First, factor the coefficient of x^2 out of the first two terms, obtaining

$$2x^2 + 5x + 3 = 2\,(x^2 + \tfrac{5}{2}x) + 3.$$

Then complete the square in the expression $x^2 + 5x/2$ as follows:

$$x^2 + \tfrac{5}{2}x = x^2 + \tfrac{5}{2}x + (\tfrac{5}{4})^2 - (\tfrac{5}{4})^2$$
$$= (x + \tfrac{5}{4})^2 - (\tfrac{5}{4})^2.$$

Thus
$$2x^2 + 5x + 3 = 2[(x + \tfrac{5}{4})^2 - (\tfrac{5}{4})^2] + 3$$
$$= 2[(x + \tfrac{5}{4})^2 - \tfrac{25}{16}] + 3$$
$$= 2(x + \tfrac{5}{4})^2 - \tfrac{25}{8} + 3$$
$$= 2(x + \tfrac{5}{4})^2 - \tfrac{1}{8}. \quad ■$$

Completing the square permits us to analyze the real solutions, if there are any, of the quadratic equation

$$ax^2 + bx + c = 0.$$

The **quadratic formula**,

$$x = \frac{-b \pm \sqrt{b^2 - 4ac}}{2a}, \tag{1}$$

which describes the solutions, is obtained as follows:

$$ax^2 + bx + c = 0 \qquad \text{given}$$
$$a\left(x^2 + \frac{b}{a}x\right) + c = 0$$
$$a\left[\left(x + \frac{b}{2a}\right)^2 - \left(\frac{b}{2a}\right)^2\right] + c = 0 \qquad \text{completing the square}$$
$$a\left(x + \frac{b}{2a}\right)^2 - a\left(\frac{b}{2a}\right)^2 + c = 0$$
$$a\left(x + \frac{b}{2a}\right)^2 = a\left(\frac{b}{2a}\right)^2 - c$$
$$\left(x + \frac{b}{2a}\right)^2 = \left(\frac{b}{2a}\right)^2 - \frac{c}{a} = \frac{b^2}{4a^2} - \frac{c}{a} = \frac{b^2 - 4ac}{4a^2}$$
$$x + \frac{b}{2a} = \pm\sqrt{\frac{b^2 - 4ac}{4a^2}} \qquad \text{taking square roots}$$
$$= \pm\frac{\sqrt{b^2 - 4ac}}{2a}$$
$$x = \frac{-b}{2a} \pm \frac{\sqrt{b^2 - 4ac}}{2a}$$

and finally
$$x = \frac{-b \pm \sqrt{b^2 - 4ac}}{2a}.$$

The number $b^2 - 4ac$ which appears under the square root in Eq. (1) is called the **discriminant** of the quadratic expression $ax^2 + bx + c$. Note that if the discriminant is negative, the equation $ax^2 + bx + c = 0$ has no real solutions; complex solutions are treated in Sec. 10.9.

INFORMATION PROVIDED BY THE DISCRIMINANT $b^2 - 4ac$

If $b^2 - 4ac < 0$, $ax^2 + bx + c = 0$ has no real solutions.

If $b^2 - 4ac > 0$, $ax^2 + bx + c = 0$ has two distinct real solutions.

If $b^2 - 4ac = 0$, $ax^2 + bx + c = 0$ has one real solution.

EXAMPLE 7 Discuss the solutions of the equations
(*a*) $2x^2 + x + 3 = 0$
(*b*) $2x^2 - 7x + 4 = 0$
(*c*) $x^2 - 6x + 9 = 0$

SOLUTION

(*a*) Here $a = 2$, $b = 1$, and $c = 3$. Thus the discriminant $b^2 - 4ac$ equals $1^2 - 4 \cdot 2 \cdot 3 = -23$. Since the discriminant is negative there are no real solutions.

(*b*) Here $a = 2$, $b = -7$, and $c = 4$. The discriminant is therefore $(-7)^2 - 4 \cdot 2 \cdot 4 = 17$, which is positive. Thus there are two real solutions, $(7 + \sqrt{17})/4$ and $(7 - \sqrt{17})/4$.

(*c*) Here $a = 1$, $b = -6$, and $c = 9$. Thus

$$x = \frac{-(-6) \pm \sqrt{(-6)^2 - 4 \cdot 1 \cdot 9}}{2 \cdot 1} = \frac{6 \pm \sqrt{36 - 36}}{2} = \frac{6}{2}.$$

There is only one solution, namely, 3. This reflects the fact that when $x^2 - 6x + 9$ is factored, the factor $x - 3$ is repeated, that is, $x^2 - 6x + 9 = (x - 3)^2$. ■

Let n be a positive integer and let x be a real number. When $(1 + x)^n$ is multiplied out, it leads to a sum of terms, each of which is of the form "a coefficient times a power of x." For instance,

$$(1 + x)^2 = 1 + 2x + x^2,$$

$$(1 + x)^3 = 1 + 3x + 3x^2 + x^3,$$

and

$$(1 + x)^4 = 1 + 4x + 6x^2 + 4x^3 + x^4.$$

The **binomial theorem** provides a formula for the expansion of $(1 + x)^n$ for any positive integer n. It asserts that

$$(1 + x)^n = 1 + nx + \frac{n(n - 1)}{1 \cdot 2}x^2 + \frac{n(n - 1)(n - 2)}{1 \cdot 2 \cdot 3}x^3 + \cdots + x^n. \quad (2)$$

The coefficient of x^k in the expansion of $(1 + x)^n$ is

$$\frac{n(n - 1)(n - 2) \cdots (n - k + 1)}{1 \cdot 2 \cdot 3 \cdots k}. \quad (3)$$

This coefficient is denoted $\binom{n}{k}$ or C_k^n and is called **a binomial coefficient**. To help remember formula (3) for $\binom{n}{k}$, note that the denominator is the product of the integers from 1 through k. The numerator has the same number of factors as the denominator, starting with n and going down to its smallest factor, which turns out to be $n - k + 1$. Simply write each factor in the numerator directly above a factor in the denominator, and you will stop at the right number. For instance,

$$\binom{7}{3} = C_3^7 = \frac{7 \cdot 6 \cdot 5}{1 \cdot 2 \cdot 3}.$$

In the $\binom{n}{k}$ notation the binomial theorem reads

$$(1 + x)^n = 1 + \binom{n}{1} x + \binom{n}{2} x^2 + \binom{n}{3} x^3 + \cdots + x^n,$$

where n is a positive integer. In Sec. 4.3, the binomial theorem is obtained by calculus.

EXAMPLE 8 Find the coefficient of x^4 in the expansion of $(1 + x)^9$.

SOLUTION In this case, $n = 9$ and $k = 4$. Thus the coefficient is

$$\frac{9 \cdot 8 \cdot 7 \cdot 6}{1 \cdot 2 \cdot 3 \cdot 4},$$

which equals 126. ■

The product of all the integers from 1 through k, $1 \cdot 2 \cdot \cdots \cdot k$, is called k **factorial** and is denoted "$k!$". Thus $5! = 1 \cdot 2 \cdot 3 \cdot 4 \cdot 5 = 120$. Formula (3) for the binomial coefficient can also be written

$$\frac{n(n-1)(n-2)\cdots(n-k+1)}{k!}. \tag{4}$$

In fact, formula (3) can be replaced by a formula that uses only factorials. To see this, note that

$$\begin{aligned} n(n-1)(n-2)\cdots(n-k+1) &= \frac{n(n-1)(n-2)\cdots(n-k+1)\,(n-k)\cdot\cdots\cdot 3\cdot 2\cdot 1}{(n-k)\cdot\cdots\cdot 3\cdot 2\cdot 1} \\ &= \frac{n!}{(n-k)!}. \end{aligned}$$

Combining this information with formula (3) shows that the binomial coefficient $\binom{n}{k}$ can be written completely in terms of factorials:

$$\binom{n}{k} = \frac{n!}{k!\,(n-k)!}. \tag{5}$$

When using a hand-held calculator you may find this the most useful of the three formulas, since many calculators have a factorial key that gives $k!$ for k up to around 69.

It is necessary to define $0!$ to be 1 in order that formula (5) remain valid even in the cases $k = 0$ and $k = n$. For instance, if $k = n$, formula (5) gives

$$\frac{n!}{n!\,0!} = 1,$$

which is what we want: the coefficient of x^n in the expansion of $(1 + x)^n$ is 1.

EXAMPLE 9 Find the coefficient of x^5 in the expansion of $(1 + x)^{11}$.

SOLUTION By formula (3),

$$\binom{11}{5} = \frac{11 \cdot 10 \cdot 9 \cdot 8 \cdot 7}{1 \cdot 2 \cdot 3 \cdot 4 \cdot 5} = 462.$$

If instead you use formula (5) with a calculator, the computations would be

$$\binom{11}{5} = \frac{11!}{5!6!} = \frac{39{,}916{,}800}{120 \cdot 720} = 462. \quad \blacksquare$$

The binomial theorem is frequently stated in terms of the expansion of $(a + b)^n$ instead of that of $(1 + x)^n$, which is the special case in which $a = 1$ and $b = x$. It asserts that

$$(a + b)^n = a^n + na^{n-1}b + \frac{n(n-1)}{1 \cdot 2} a^{n-2}b^2 + \cdots + b^n. \tag{6}$$

The coefficient of $a^{n-k}b^k$ is

$$\binom{n}{k} = \frac{n(n-1)(n-2)\cdots(n-k+1)}{1 \cdot 2 \cdot 3 \cdot \cdots \cdot k} = \frac{n!}{k!\,(n-k)!}.$$

For instance, $(a + b)^4 = a^4 + 4a^3b + 6a^2b^2 + 4ab^3 + b^4$.

Another algebraic identity provides a short formula for the sum

$$a + ax + ax^2 + \cdots + ax^{n-1}, \tag{7}$$

So named by the Greeks because corresponding lengths in similar geometric figures are in the same ratio.

where a and x are real numbers and n is a positive integer. Note that there are n summands in (7). Such an expression is called a **finite geometric series** with **first term** a and **ratio** x. Each term in (7) is obtained from the one before by multiplying by the ratio x. We present a shortcut for evaluating the sum $a + ax + ax^2 + \cdots + ax^{n-1}$.

A short formula for the sum of a finite geometric series

Let a and x be real numbers, with $x \neq 1$. Let n be a positive integer. Then

$$a + ax + ax^2 + \cdots + ax^{n-1} = \frac{a(1 - x^n)}{1 - x}. \tag{8}$$

To show that formula (8) is true, multiply both sides by $1 - x$. The multiplication of $a + ax + ax^2 + \cdots + ax^{n-1}$ by $1 - x$ leads, happily, to many cancellations:

$$\begin{aligned}(1 - x)&(a + ax + ax^2 + \cdots + ax^{n-1})\\ &= (a + ax + ax^2 + \cdots + ax^{n-1}) - (ax + ax^2 + \cdots + ax^n)\\ &= a - ax^n\\ &= a(1 - x^n).\end{aligned}$$

Division by $1 - x$ then establishes formula (8).

EXAMPLE 10 Use formula (8) to evaluate $1 + \frac{1}{2} + (\frac{1}{2})^2 + (\frac{1}{2})^3 + (\frac{1}{2})^4$.

SOLUTION In this case, $a = 1$, $x = \frac{1}{2}$, and $n = 5$. Thus

$$1 + \tfrac{1}{2} + (\tfrac{1}{2})^2 + (\tfrac{1}{2})^3 + (\tfrac{1}{2})^4 = \frac{1[1 - (\frac{1}{2})^5]}{1 - \frac{1}{2}}$$

$$= \frac{1 - (\frac{1}{2})^5}{\frac{1}{2}}$$

$$= 2(1 - \tfrac{1}{32}) = 2 - \tfrac{1}{16} = \tfrac{31}{16}. \quad \blacksquare$$

Next we examine the relation between the roots of a polynomial and its first-degree factors.

If $x - a$ is a factor, then a is a root.

Let a be a real number and $P(x)$ a polynomial that has $x - a$ as a factor. This means that there is a polynomial $Q(x)$ such that $P(x) = Q(x)(x - a)$. Replacing x by a gives us $P(a) = Q(a)(a - a)$. Since $a - a = 0$, it follows that $P(a) = 0$; that is, "a is a root of the polynomial $P(x)$."

How about the reverse direction? If a is a root of $P(x)$, that is, $P(a) = 0$, does it follow that $x - a$ must be a factor of $P(x)$?

In any case, whether $x - a$ is or is not a factor of $P(x)$, we may divide $x - a$ into $P(x)$ by long division to get a quotient $Q(x)$ and a remainder, which is 0 or a polynomial of lower degree than $x - a$. This means the remainder is just a constant k. So we may write

$$P(x) = Q(x)(x - a) + k. \tag{9}$$

Replacing x by a gives us

$$P(a) = Q(a)(a - a) + k.$$

If a is a root, then $x - a$ is a factor.

Since $P(a) = 0$ and $a - a = 0$, we see that $k = 0$. This means, according to Eq. (9), that $x - a$ divides $P(x)$.

Shortcut for finding first-degree factors

Thus we have this neat test for deciding whether $P(x)$ has a first-degree factor: Look for a root of $P(x)$. If you can find one, say, a, then $x - a$ is a factor. If there are no roots, then $P(x)$ has no first-degree factors. This shortcut is known as the **factor theorem**.

The factor theorem does *not* test for factors of degree higher than 1. For instance, $x^4 + 4$ has no real roots, yet it has a second-degree factor, since

$$x^4 + 4 = (x^2 + 2x + 2)(x^2 - 2x + 2),$$

as can be checked by multiplying out the right side.

There is also a shortcut for searching for rational roots of a polynomial that has integer coefficients. Example 11 illustrates the technique.

EXAMPLE 11 Does the equation $4x^3 + 3x + 2 = 0$ have any rational roots?

SOLUTION Let us see whether the rational number a/b is a root. (Here a and b are integers with no common divisor larger than 1. We may assume that b is positive.) If a/b is a root of the equation, then

Plug in $x = a/b$ and clear denominator.

$$4a^3 + 3ab^2 + 2b^3 = 0. \tag{10}$$

Now, a divides $4a^3 + 3ab^2$; hence a divides $2b^3 = -(4a^3 + 3ab^2)$. But a and b have no common divisors other than 1 and -1, so a divides 2.

Similarly, b divides $4a^3$; hence b divides 4. Since $b > 0$, we have $b = 1, 2,$ or 4. All that remains is to check the various combinations of a and b, namely, the candidates

$$\frac{1}{1}, \frac{1}{2}, \frac{1}{4}, \frac{-1}{1}, \frac{-1}{2}, \frac{-1}{4}, \frac{2}{1}, \frac{2}{2}, \frac{2}{4}, \frac{-2}{1}, \frac{-2}{2}, \frac{-2}{4}.$$

(There are duplications, for instance, $\frac{1}{1} = \frac{2}{2}$.)

Test each of these possibilities by plugging them into Eq. (10). If none produces a root, then the equation $4x^3 + 3x + 2 = 0$ has no rational roots. As you may check, only $-\frac{1}{2}$ is a root. ■

EXERCISES FOR APPENDIX C: TOPICS IN ALGEBRA

In Exercises 1 to 8 rationalize the denominators.

1 $10/\sqrt{5}$ **2** $3/\sqrt{x}$

3 $(2 + 4\sqrt{2})/\sqrt{2}$ **4** $x^3/\sqrt{x}$

5 $4/(3 - \sqrt{3})$ **6** $2/(3 + \sqrt{2})$

7 $x/(1 - \sqrt{x})$ **8** $1/(\sqrt{2} - \sqrt{3})$

In Exercises 9 to 12 rationalize the numerators.

9 $(3 + \sqrt{2})/5$ **10** $(\sqrt{x+1} - \sqrt{x})/x$

11 $(\sqrt{x} - \sqrt{5})/(x - 5)$ **12** $(\sqrt{u} - \sqrt{v})/5$

In Exercises 13 to 18 complete the squares.

13 (*a*) $x^2 + 8x + 13$ (*b*) $x^2 - 8x + 23$ (*c*) $x^2 - x + 2$

14 (*a*) $x^2 + 6x + 3$ (*b*) $x^2 - 6x + 3$ (*c*) $x^2 + 3x + 5$

15 (*a*) $x^2 + 3x - 2$ (*b*) $x^2 + 3x + 7$ (*c*) $x^2 + 5x/2 + 4$

16 (*a*) $x^2 + x/3 + 1$ (*b*) $x^2 + 2x/3 - 7$ (*c*) $x^2 + 10x + 25$

17 (*a*) $2x^2 - 5x + 3$ (*b*) $2x^2 + 6x + 7$ (*c*) $3x^2 + 5x + 1$

18 (*a*) $4x^2 - 8x + 5$ (*b*) $3x^2 + 2x + 1$ (*c*) $3x^2 - 2x + 1$

In Exercises 19 and 20 use the quadratic formula to find all real solutions of the equations.

19 (*a*) $x^2 + x + 1 = 0$ (*b*) $x^2 + x - 1 = 0$ (*c*) $x^2 + 2x + 1 = 0$

20 (*a*) $2x^2 + 5x - 7 = 0$ (*b*) $3x^2 + x + 5 = 0$ (*c*) $4x^2 - 4x + 1 = 0$

Using the discriminant, determine how many distinct real solutions there are for each of the equations in Exercises and 22.

21 (*a*) $2x^2 + 5x + 6 = 0$ (*b*) $2x^2 + 5x + 2 = 0$ (*c*) $4x^2 - 12x + 9 = 0$

22 (*a*) $9x^2 - 30x + 25 = 0$ (*b*) $x^2 + x + 1 = 0$ (*c*) $x^2 - x - 1 = 0$

23 Find the coefficient of x^2 in the expansion of (*a*) $(1 + x)^5$, (*b*) $(1 + x)^6$, (*c*) $(1 + x)^{10}$.

24 Find the coefficient of x^3 in the expansion of (*a*) $(1 + x)^5$, (*b*) $(1 + x)^6$, (*c*) $(1 + x)^{10}$.

25 Use the binomial theorem to expand $(1 + x)^7$.

26 Use the binomial theorem to find the first four terms (starting with 1) in the expansion of $(1 + x)^{12}$.

27 What is the coefficient of a^8b^2 in the expansion of $(a + b)^{10}$?

28 What is the coefficient of a^8b^3 in the expansion of $(a + b)^{11}$?

In Exercises 29 and 30 use the formula for the sum of a geometric series to obtain short expressions for the given sums.

29 (*a*) $1 + \frac{1}{3} + \frac{1}{3^2} + \frac{1}{3^3} + \frac{1}{3^4} + \frac{1}{3^5} + \frac{1}{3^6}$

(*b*) $5 + 5 \cdot 2 + 5 \cdot 2^2 + 5 \cdot 2^3 + 5 \cdot 2^4 + 5 \cdot 2^5 + 5 \cdot 2^6 + 5 \cdot 2^7$

(*c*) $6 - 6 \cdot \frac{1}{2} + 6 \cdot (\frac{1}{2})^2 - 6 \cdot (\frac{1}{2})^3 + 6 \cdot (\frac{1}{2})^4$

30 (*a*) $1 - \frac{1}{2} + (\frac{1}{2})^2 - (\frac{1}{2})^3 + (\frac{1}{2})^4 - (\frac{1}{2})^5$

(*b*) $8 + \frac{8}{10} + \frac{8}{10^2} + \frac{8}{10^3} + \frac{8}{10^4} + \frac{8}{10^5} + \frac{8}{10^6} + \frac{8}{10^7} + \frac{8}{10^8} + \frac{8}{10^9}$

(*c*) $\frac{2}{3} + (\frac{2}{3})^2 + (\frac{2}{3})^3 + (\frac{2}{3})^4 + (\frac{2}{3})^5 + (\frac{2}{3})^6 + (\frac{2}{3})^7 + (\frac{2}{3})^8$

In Exercises 31 to 36 find all factors of the form $x - a$ where a is a rational root of the given polynomial. (There may be none.)

31 $x^3 + 2x^2 - x - 2$ **32** $2x^3 - x^2 - 2x + 1$

33 $x^3 - x^2 - 2x + 2$ **34** $4x^2 + 9x + 3$

35 $6x^3 - 7x^2 - 7x + 6$

36 $x^4 + x^3 - 3x^2 - 4x - 4$

In Exercises 37 and 38 express the polynomial as a product of first-degree factors.

37 $x^2 + 5x + 2$ (*Hint:* First find the roots.)

38 $3x^2 - 2x - 2$

■

39 The decimal 0.333333333 is actually a finite geometric progression in disguise, namely,

$$\frac{3}{10} + \frac{3}{10^2} + \frac{3}{10^3} + \frac{3}{10^4} + \frac{3}{10^5} + \frac{3}{10^6} + \frac{3}{10^7} + \frac{3}{10^8} + \frac{3}{10^9}.$$

Using this fact, find a simpler expression for 0.333333333.

40 A flea, to amuse itself, jumps $\frac{1}{2}$ meter to the right, then $\frac{1}{4}$ meter to the left, then $\frac{1}{8}$ meter to the right, then $\frac{1}{16}$ meter to the left, and so on, as shown in Fig. C.1. On the nth jump the flea travels $1/2^n$ meter, but continues alternating right and left. The flea starts at the number 0 on the number line.

(*a*) Show that after n jumps, where n is odd, the flea is at the number $\frac{1}{3}(1 + 1/2^n)$.

(*b*) Show that after n jumps, where n is even, the flea is at the number $\frac{1}{3}(1 - 1/2^n)$.

(*c*) As the flea keeps on jumping, what single number does it keep jumping over?

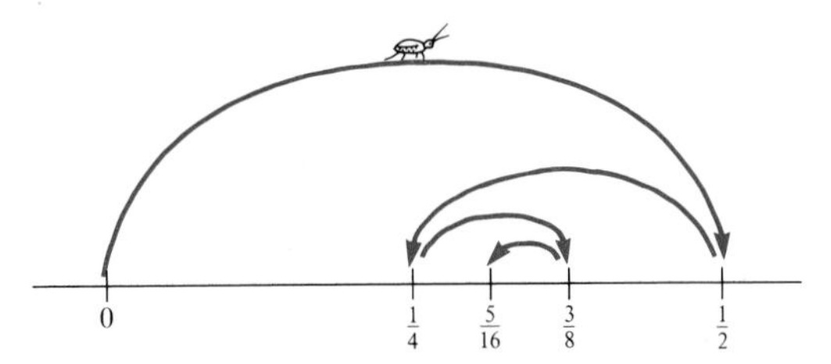

Figure C.1

41 From the fact that

$$(1 + x)^4 = 1 + 4x + 6x^2 + 4x^3 + x^4,$$

deduce that

$$(a + b)^4 = a^4 + 4a^3b + 6a^2b^2 + 4ab^3 + b^4.$$

Hint: First show that

$$(a + b)^4 = a^4\left[1 + \left(\frac{b}{a}\right)\right]^4.$$

(Assume $a \neq 0$.)

42 Use the binomial theorem to show that when $|x|$ is very small, $(1 + x)^{10}$ is approximately $1 + 10x$. If you have a calculator, compare the two quantities when $x = 0.01$.

43 Using a formula for the binomial coefficients, show that in the expansion of $(1 + x)^n$ the powers x^k and x^{n-k} have equal coefficients.

44 Show that for all values of x, $x^2 + x + 1 \geq \frac{3}{4}$. *Hint:* Complete the square.

Appendix D Exponents

For each fixed positive number b there is an exponential function $f(x) = b^x$. Every scientific calculator has a y^x-key, and before the advent of the calculator, mathematical tables listing the values of exponential functions to many decimal places were common. This appendix first reviews the definition of b^x for positive b and then examines the situation when b is replaced by a negative number.

In the expression b^x, where b is positive and x is any real number, b is called the **base** and x is called the **exponent**. We will review the definition of b^x in a sequence of steps:

1 x is a positive integer.
2 x is 0.
3 x is a negative integer.
4 x is the reciprocal of a positive integer, $x = 1/n$.
5 x is rational, $x = m/n$, where m is an integer and n is a positive integer.
6 x is irrational.

For the sake of simplicity, let us take the special case $b = 2$. All the ideas show up here.

Definition If *x is a positive integer,* define 2^x to be the product of x 2's.

2^x when $x = 1, 2, 3, \ldots$

For instance, $$2^5 = 2 \cdot 2 \cdot 2 \cdot 2 \cdot 2 = 32,$$

$$2^3 = 2 \cdot 2 \cdot 2 = 8, \qquad 2^2 = 2 \cdot 2 = 4, \qquad \text{and} \qquad 2^1 = 2.$$

Note that $$2^2 \cdot 2^3 = (2 \cdot 2)(2 \cdot 2 \cdot 2) = 2^{2+3}.$$

More generally, for any positive integers x and y,

$$2^{x+y} = 2^x \cdot 2^y.$$

This **basic law of exponents** (which holds for any positive base) will serve as a guide to the definition of 2^x when x is not a positive integer. The demand that $2^{x+y} = 2^x \cdot 2^y$ for *all real numbers* x and y will tell us how the exponential function 2^x must be defined. Note that $2^{x+y+z} = 2^{(x+y)+z} = 2^{x+y} \cdot 2^z = (2^x \cdot 2^y)2^z = 2^x \cdot 2^y \cdot 2^z$. Thus the basic law of exponents extends to the case when the exponent is the sum of three numbers: $2^{x+y+z} = 2^x \cdot 2^y \cdot 2^z$. In a similar manner, it can be extended to the case where the exponent is the sum of any finite number of numbers.

For instance, *what should the numerical value of* 2^0 *be?* If the basic law of exponents is to be true for all exponents, then, in particular, this equation must hold:

$$2^{0+1} = 2^0 \cdot 2^1;$$

hence

$$2^1 = 2^0 \cdot 2^1.$$

But $2^1 = 2$, so 2^0 must satisfy the equation

$$2 = 2^0 \cdot 2.$$

2^0 must be 1.

There is no choice; 2^0 must be 1.

What should $2^{-1}, 2^{-2}, 2^{-3}, \ldots$ *be?* To preserve the basic law of exponents, we must have, for instance,

$$2^3 \cdot 2^{-3} = 2^{3+(-3)}.$$

But $2^{3+(-3)} = 2^0 = 1$. Thus $2^3 \cdot 2^{-3} = 1$. This shows that if 2^{-3} is to have meaning, it must be the reciprocal of 2^3, that is,

$$2^{-3} = \frac{1}{2^3} = \frac{1}{8}.$$

For this reason, *we define*

2^{-n} must be $1/(2^n)$.

$$2^{-n} \quad \textit{to be} \quad \frac{1}{2^n}$$

for any positive integer n.

The exponential 2^x has now been defined for any *integer* x. Let us graph $y = 2^x$ for integer values of x, as in Fig. D.1.

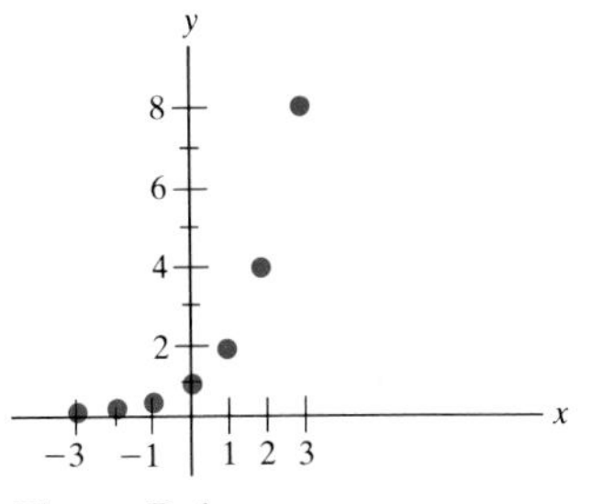

Figure D.1

It is tempting to draw a smooth curve through these points, but 2^x has not yet been defined if x is not an integer.

How should $2^{1/2}$ *be defined?* To preserve the basic law of exponents, it is necessary that

$$2^{1/2} \cdot 2^{1/2} = 2^{1/2+1/2};$$

thus

$$2^{1/2} \cdot 2^{1/2} = 2^1 = 2.$$

Hence, $2^{1/2}$ is a solution of the equation

$$x^2 = 2.$$

Why $2^{1/2}$ should be the square root of 2

Should it be the positive or the negative solution? The graph just sketched suggests that $2^{1/2}$ should be positive. Thus define $2^{1/2}$ to be $\sqrt{2}$, the square root of 2. Note that $2^{1/2} \approx 1.4$, which fits nicely into the preceding graph, as is shown in Fig. D.2.

Similarly, $2^{1/3}$ can be determined by the basic law of exponents:

$$2^{1/3} \cdot 2^{1/3} \cdot 2^{1/3} = 2^{1/3+1/3+1/3} = 2^1 = 2.$$

Thus $2^{1/3}$ must be a solution of the equation

$$x^3 = 2.$$

Why $2^{1/3}$ should be the cube root of 2

There is only one solution, and it is called the **cube root of** 2, denoted also $\sqrt[3]{2}$. Thus define

$$2^{1/3} = \sqrt[3]{2},$$

which is about 1.26.

Similarly, *define* $2^{1/n}$ *for any positive integer n to be the positive solution of the equation* $x^n = 2$, that is,

$$2^{1/n} = \sqrt[n]{2}.$$

What should $2^{3/4}$ be?

How should $2^{3/4}$ be defined? To preserve the basic law of exponents, we must have

$$2^{1/4} \cdot 2^{1/4} \cdot 2^{1/4} = 2^{3/4}.$$

In short, we must have $\quad 2^{3/4} = (2^{1/4})^3.$

Defining $2^{m/n}$

This suggests that for *any integer m and positive integer n we should define*

$$2^{m/n} \quad \textit{to be} \quad (2^{1/n})^m.$$

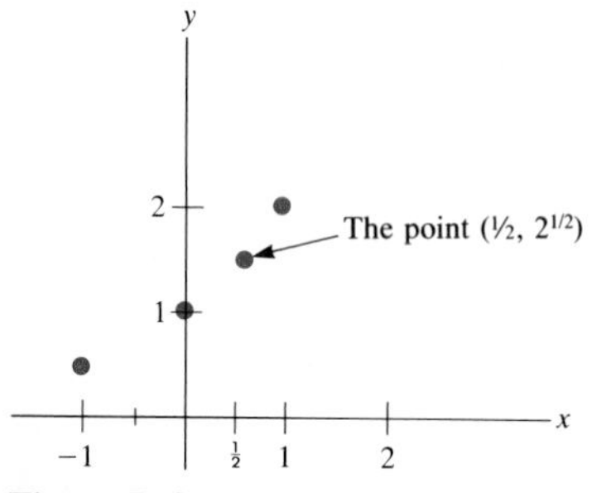

Figure D.2

With this step we have defined 2^x for every rational number x.

How should 2^x be defined when x is not rational? For instance, how should $2^{\sqrt{2}}$ be defined? To begin, consider the decimal representation of $\sqrt{2}$, namely,

$$\sqrt{2} = 1.41421356 \ldots.$$

The successive decimals 1.4, 1.41, 1.414, . . . are rational (for instance, $1.41 = \frac{141}{100}$). Thus $2^{1.4}, 2^{1.41}, 2^{1.414}, \ldots$ are already defined. This table indicates some of their values:

x	1.4	1.41	1.414	1.4142	1.41421
2^x	2.63901 . . .	2.65737 . . .	2.66474 . . .	2.66511 . . .	2.66513 . . .

By going further in this table we can get better approximations of $2^{\sqrt{2}}$. This procedure, though cumbersome, would enable us to find $2^{\sqrt{2}}$ to any number of decimal places. To five decimal places,

$$2^{\sqrt{2}} \approx 2.66514.$$

The graph of $y = 2^x$ is shown in Fig. D.3.

For any positive number b, the definition of b^x follows the same outline as the definition of 2^x. For $b > 1$, the graph of $f(x) = b^x$ resembles that

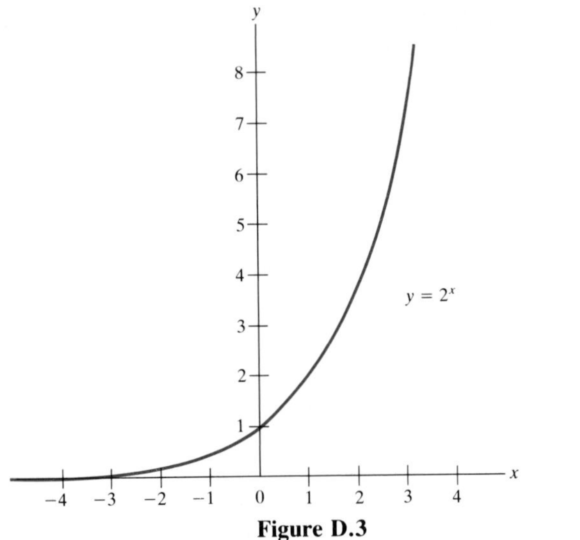

Figure D.3

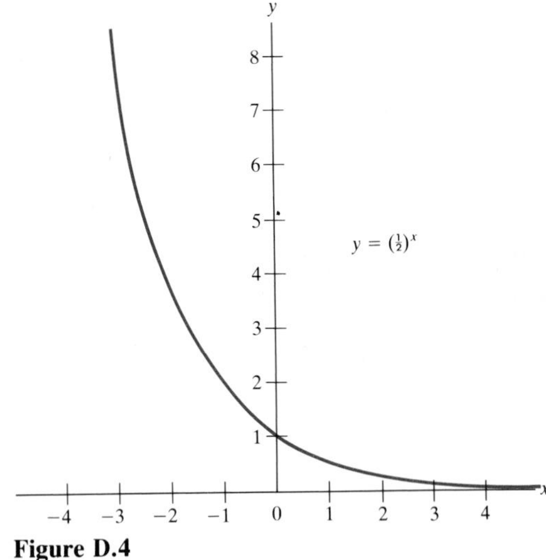

Figure D.4

of 2^x. Far to the left it gets near the x axis; far to the right it gets arbitrarily high. Moreover, as x increases, so does b^x, if $b > 1$.

For $0 < b < 1$, the graph of $f(x) = b^x$ resembles the graph of $(\frac{1}{2})^x$, which is shown in Fig. D.4.

EXAMPLE 1 Evaluate $64^{1/3}$, $64^{2/3}$, $64^{1/2}$, and $64^{-2/3}$.

SOLUTION $64^{1/3}$ is the cube root of 64, $\sqrt[3]{64}$, which is 4.

$$64^{2/3} = (64^{1/3})^2 = 4^2 = 16;$$

$$64^{1/2} = \sqrt{64} = 8;$$

$$64^{-2/3} = (64^{1/3})^{-2} = 4^{-2} = \tfrac{1}{16} = 0.0625. \quad \blacksquare$$

EXAMPLE 2 Find a decimal approximation of $2^{1/3}$, the cube root of 2.

SOLUTION The quickest way is to use a calculator. Either the y^x-key, with $y = 2$ and $x = \frac{1}{3}$, or the $\sqrt[x]{y}$-key, with $y = 2$ and $x = 3$, will do the trick and show that $2^{1/3} \approx 1.25992105$.

If a calculator is not available, brute-force arithmetic could be used. For instance, by straightforward multiplication, $1.2^3 = 1.728$ and $1.3^3 = 2.197$. Thus

$$1.2 < 2^{1/3} < 1.3.$$

To find the next decimal place, compute $1.21^3, 1.22^3, \ldots, 1.29^3$. Since $1.25^3 = 1.953125$ and $1.26^3 = 2.000376$,

$$1.25 < 2^{1/3} < 1.26.$$

With enough time and pencils, the curious could find as many decimal places of $2^{1/3}$ as needed. $\blacksquare$

In addition to the basic law of exponents, there are other laws that hold for the exponential functions for any positive base. The laws are displayed here:

LAWS OF EXPONENTS

The bases are positive, the exponents any real numbers.

$b^{x+y} = b^x b^y$	basic law of exponents
$b^{x-y} = \dfrac{b^x}{b^y}$	difference of exponents
$(b^x)^y = b^{xy}$	power of a power
$(ab)^x = a^x b^x$	power of a product
$\left(\dfrac{a}{b}\right)^x = \dfrac{a^x}{b^x}$	power of a quotient
$b^0 = 1$	definition
$b^1 = b$	definition

The laws of exponents help in simplifying certain algebraic expressions, as illustrated by Example 3.

EXAMPLE 3 Let b be a positive number. Write each of the following in the form b^x for a suitable exponent x: (*a*) $(\sqrt{b})^3$, (*b*) b^7/b^3, (*c*) $\sqrt{b}\sqrt[3]{b}$, (*d*) $1/\sqrt{b}$.

SOLUTION

(*a*) $(\sqrt{b})^3 = (b^{1/2})^3 = b^{(1/2)(3)} = b^{3/2}$.

(*b*) $\dfrac{b^7}{b^3} = b^{7-3} = b^4$.

(*c*) $\sqrt{b}\sqrt[3]{b} = b^{1/2}b^{1/3} = b^{1/2+1/3} = b^{5/6}$.

Parts (a) and (d) used the "power of a power" law.

(*d*) $\dfrac{1}{\sqrt{b}} = (\sqrt{b})^{-1} = (b^{1/2})^{-1} = b^{-1/2}$. ■

For certain values of the exponent x it is possible to define b^x even if b is not positive.

First of all, 0^x is defined for *positive* x to be 0. However, 0^0 is not defined, nor is 0^x, when x is negative.

If the base b is negative and n is an *odd* integer, $b^{1/n}$ is defined to be $\sqrt[n]{b}$, just as in the case when b is positive. For instance,

$$(-8)^{1/3} = -2.$$

More generally, b^x can be defined for any negative base and any exponent x of the form m/n, where m is an integer and n is a positive *odd* integer, just as in the case b positive.

Definition Let b be negative, m an integer, and n a positive odd integer. Then $b^{m/n}$ is

$$(\sqrt[n]{b})^m.$$

The equation $(b^x)^y = b^{xy}$ does not always hold when b is negative. For instance, $[(-1)^2]^{1/2} = 1^{1/2} = 1$, but $(-1)^{2 \cdot 1/2} = (-1)^1 = -1$.

EXAMPLE 4 Graph $f(x) = x^{1/3}$, the cube root function.

SOLUTION This table records a few values of the function:

x	-8	-1	0	1	8
$f(x) = x^{1/3}$	-2	-1	0	1	2

The graph is shown in Fig. D.5. The curve passes through the origin vertically.

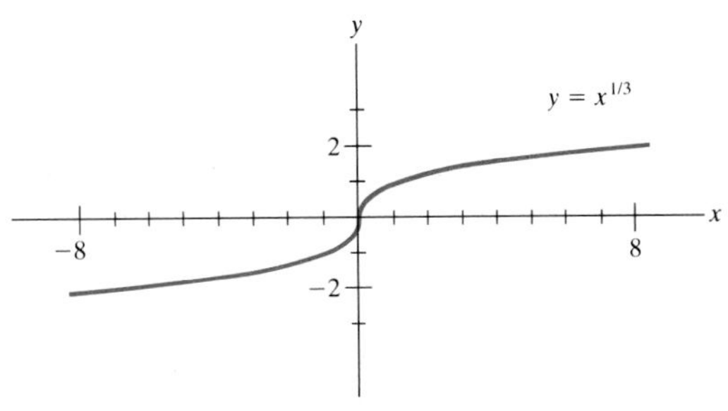

Figure D.5

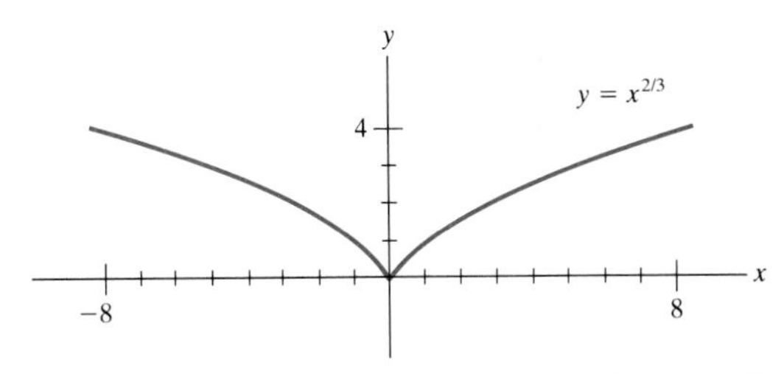

Figure D.6 ■

EXAMPLE 5 Graph $f(x) = x^{2/3}$.

SOLUTION Again, begin with a few convenient inputs x. For instance,

$$(-8)^{2/3} = (\sqrt[3]{-8})^2 = (-2)^2 = 4.$$

x	-8	-1	0	1	8
$x^{2/3}$	4	1	0	1	4

The graph is symmetric with respect to the y axis and comes to a point or "cusp" at the origin, as shown in Fig. D.6. ■

Note that when b is negative and n is an *even* positive integer, b has no real nth root. For instance, $(-9)^{1/2}$ makes no sense, since -9 does not have a real square root.

EXERCISES FOR APPENDIX D: EXPONENTS

1 Evaluate 32^x for x equal to (*a*) 0, (*b*) 1, (*c*) $\frac{1}{5}$, (*d*) $-\frac{1}{5}$, (*e*) $\frac{3}{5}$.

2 Evaluate 64^x for x equal to (*a*) $\frac{1}{2}$, (*b*) $-\frac{1}{2}$, (*c*) $\frac{2}{3}$, (*d*) 0, (*e*) $\frac{5}{6}$.

In Exercises 3 to 6 graph the given functions.

3 $f(x) = 4^x$
4 $f(x) = 5^x$
5 $f(x) = (\frac{1}{4})^x$
6 $f(x) = (\frac{1}{5})^x$

7 Express in the form 2^x for a suitable exponent x:
(*a*) 16 (*b*) $\frac{1}{8}$ (*c*) $\sqrt{2}$ (*d*) 1 (*e*) 0.25

8 Express in the form 4^x for a suitable exponent x:
(*a*) 16 (*b*) 8 (*c*) $1/\sqrt{2}$ (*d*) $4\sqrt{2}$ (*e*) 16^{35}

9 Let b be a positive number. Express in the form b^x:
(*a*) $b\sqrt{b}$ (*b*) $(\sqrt{b})^3$ (*c*) $1/\sqrt[3]{b}$
(*d*) $\sqrt[3]{b^2}$ (*e*) $\sqrt{\sqrt{b}}$

10 Let a be a positive number. Express in the form a^x:
(*a*) $(\sqrt{a})^2$ (*b*) $1/a^2$ (*c*) 1
(*d*) $(a^3/a^5)^{10}$ (*e*) $a/\sqrt[3]{a}$

In Exercises 11 and 12 give the domains and ranges of the functions. (See Sec. 1.1 for definitions of "domain" and "range.")

11 (*a*) 2^x (*b*) $x^{1/3}$ (*c*) $x^{1/4}$ (*d*) $x^{2/3}$

12 (*a*) $(\frac{1}{2})^x$ (*b*) x^3 (*c*) $x^{4/5}$ (*d*) $x^{3/4}$

13 (*a*) Compute $x^{3/5}$ for $x = -32, -1, 0, 1, 32$.
(*b*) Using (*a*), graph $y = x^{3/5}$.

14 (*a*) Compute $x^{4/5}$ for $x = -32, -1, 0, 1, 32$.
(*b*) Using (*a*), graph $y = x^{4/5}$. (Use a large graph.)

15 A square root table lists 2.236 for $\sqrt{5}$. Which is larger, 2.236 or $\sqrt{5}$?

16 A square root table lists 2.646 for $\sqrt{7}$. Which is larger, 2.646 or $\sqrt{7}$?

■

17 For which positive numbers x is (*a*) $x < x^2$? (*b*) $x = x^2$? (*c*) $x > x^2$?

18 There is only one pair of positive integers x and y, $x \neq y$, such that $x^y = y^x$. What is that pair? (It is not necessary to prove that it is unique.)

19 Graph $y = 2^{-x}$.

20 Graph $y = 2^x + 2^{-x}$.

21 Express each of these numbers in the form 10^x:
(*a*) 1,000 (*b*) 0.0001
(*c*) 1,000,000 (*d*) 0.0000001

22 Each positive number can be written in the form $A \cdot 10^x$, where $1 \le A < 10$ and x is an integer. This is the scientific notation for real numbers. Express each of the following in scientific notation:
(*a*) 900 (*b*) 957 (*c*) 0.095
(*d*) 15,000 (*e*) 0.0015

23 There are two ways to calculate $64^{2/3}$ by hand, either as $(64^2)^{1/3}$ or as $(64^{1/3})^2$. Which is easier?

24 If a calculator has a square root key and a reciprocal key, how could you use them to calculate (*a*) $5^{1/4}$, (*b*) $5^{1/8}$, (*c*) $5^{3/8}$, (*d*) $5^{-1/4}$?

25 (Calculator)
(*a*) Using the y^x-key, evaluate $(2^h - 1)/h$ for $h = 0.1$, 0.01, and 0.001.
(*b*) Continue part (*a*) with smaller h. The calculator may give erroneous values when h is chosen too small. Experiment with your calculator with h as small as 0.0000001 or 0.00000001 and smaller. [It is shown in Chap. 6 that as h is chosen smaller and smaller, the quotient in (*a*) approaches a fixed number whose decimal expansion begins 0.693.]

26 This exercise shows why b^x is not defined when b is negative and x is irrational. For instance, consider $(-1)^{\sqrt{2}}$.
(*a*) Show that $\frac{15}{11} < \sqrt{2} < \frac{16}{11}$.
(*b*) Evaluate $(-1)^{15/11}$ and $(-1)^{16/11}$.
(*c*) What trouble arises when you use rational numbers, m/n, near $\sqrt{2}$ and the values of $(-1)^{m/n}$ as a way to find out what $(-1)^{\sqrt{2}}$ "should be"?

27 The ability of a nuclear weapon to destroy hard missile targets is called its **lethality**, denoted K. K is a function of the explosive power of the warhead, denoted Y, measured in megatons, and the accuracy of the missile. A megaton is equivalent to 1 million tons of TNT. The accuracy is measured by the radius of the circle around the target such that the warhead has a 50 percent chance of landing within that circle. This radius, measured in nautical miles, is denoted CEP (circle of error probable). The lethality is given by the formula

$$K = Y^{2/3}(\text{CEP})^{-2}.$$

A nautical mile is about 6,076 feet.
(*a*) Find the total lethality of one Trident 2 missile with fourteen 150,000-ton-yield warheads and CEP = 200 feet. (Multiply the lethality of one warhead by 14.)
(*b*) Find the total lethality of one SS-NX-18 missile with three 500,000-ton-yield warheads and CEP = 0.25 nautical mile.
(*c*) Which increases K more: to make the warhead 8 times as large (in megatonnage) or to make CEP 3 times as small?

Appendix E Trigonometry

The part of trigonometry most useful in calculus concerns primarily three functions: sine, cosine, and tangent, which are defined in terms of a circle.

In daily life angles are measured in degrees, with 360° measuring the full circular angle. The number 360 was chosen by the Babylonian astronomers, perhaps because there are close to 360 days in a year and 360 has many divisors. This somewhat arbitrary measure would complicate the calculus of trigonometric functions. In calculus a much more natural and geometric measure is used, called **radian measure**, which is defined in terms of how much arc an angle cuts off on a circle.

To measure the size of an angle, such as the angle ABC shown in Fig. E.1, draw a circle with center at B. If the circle has radius r and the angle intercepts an arc of length s, then the quotient s/r shall be the measure of the angle, and we say that the angle has a measure of s/r

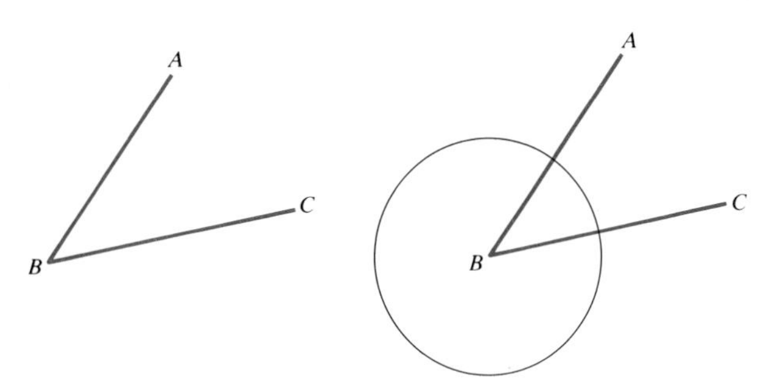

Figure E.1

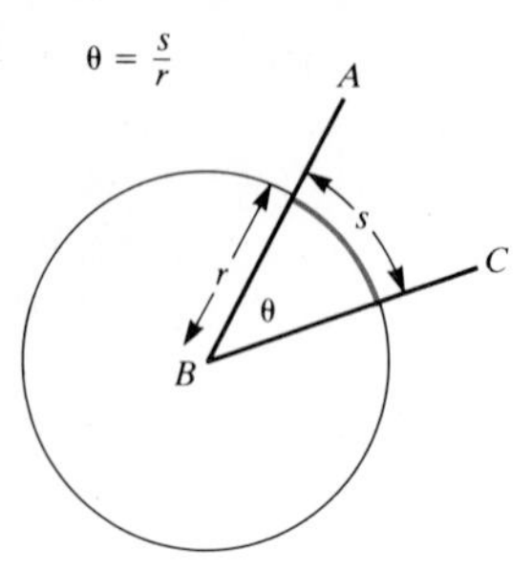

Figure E.2

radians. It is frequently convenient to denote the measure of an angle by θ, and then write $\theta = s/r$. (See Fig. E.2.) Since both s and r measure lengths, their ratio is dimensionless. Moreover, the radian measure of an angle does not depend on the size of the circle.

EXAMPLE 1 Find the radian measure of the right angle ABC in Fig. E.3.

SOLUTION Draw a circle of radius r centered at B and compute the quotient s/r. The circumference of a circle of radius r is $2\pi r$. The right angle intercepts a quarter of the circumference. Thus

$$s = \tfrac{1}{4}(2\pi r) = \frac{\pi r}{2} \quad \text{and} \quad \theta = \frac{s}{r} = \frac{\pi r/2}{r} = \frac{\pi}{2}.$$

So a right angle has the measure $\pi/2$ radians (about 1.57 radians). ■

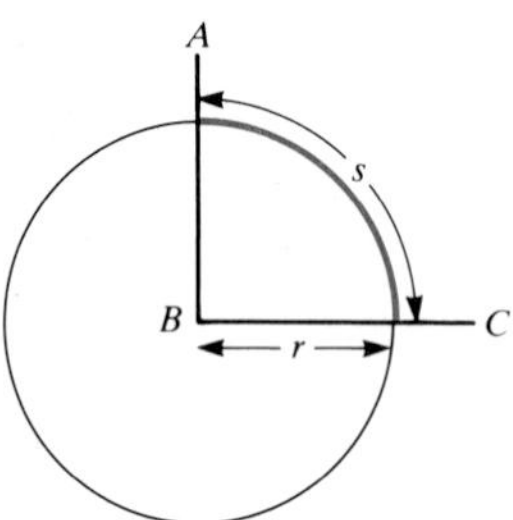
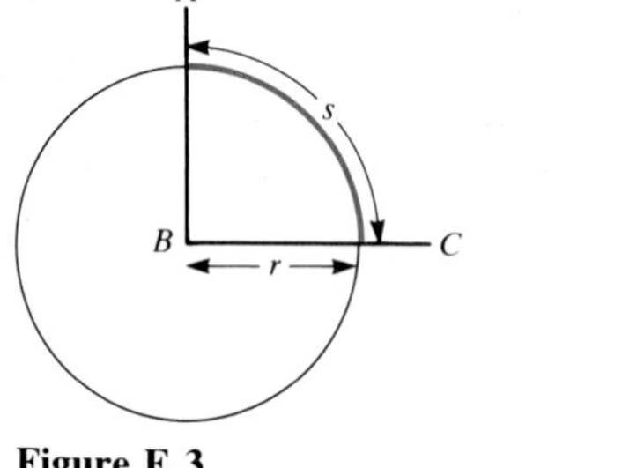

Figure E.3

The straight angle, which has 180°, consists of two right angles and therefore has a measure of π radians. This fact is helpful in translating from degrees to radians and from radians to degrees: Simply use the proportion

$$\frac{\text{Number of degrees}}{180} = \frac{\text{Number of radians}}{\pi}.$$

EXAMPLE 2 What is the measure in radians of the 30° angle?

SOLUTION The proportion in this case becomes

$$\frac{30}{180} = \frac{\text{Number of radians}}{\pi},$$

from which it follows that

$$\text{Number of radians} = \pi\frac{30}{180} = \frac{\pi}{6}. \quad ■$$

EXAMPLE 3 How many degrees are there in an angle of 1 radian?

SOLUTION In this case, the proportion is

$$\frac{\text{Number of degrees}}{180} = \frac{1}{\pi};$$

hence

$$\text{Number of degrees} = \frac{180}{\pi} \approx \frac{180}{3.14} \approx 57.3°.$$

So an angle of 1 radian is about 57.3°, a little less than 60°. ■

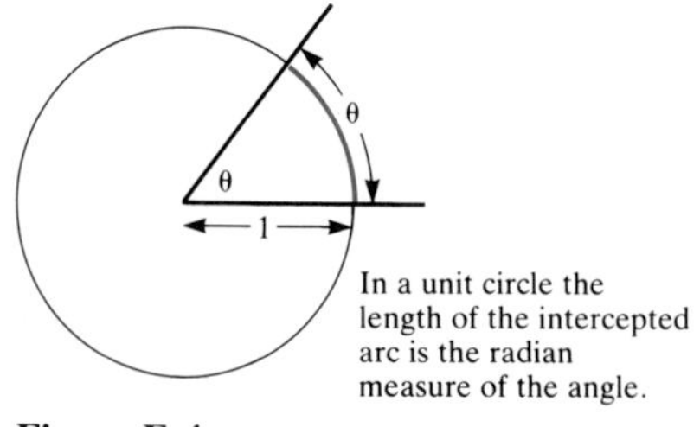

In a unit circle the length of the intercepted arc is the radian measure of the angle.

Figure E.4

In the case of the **unit circle**, the circle whose radius is 1, the formula $\theta = s/r$ becomes $\theta = s/1 = s$. In that case, the length of arc intercepted equals the measure of the angle in radians, as shown in Fig. E.4.

Angles θ larger than 2π

So far, an angle has been associated with each number θ in the interval $[0, 2\pi]$. Next, associate an angle with any *positive* number θ. For convenience, a unit circle will be used, and one arm of the angle will be placed along the positive x axis. To draw the second arm of the angle of θ radians, go around the unit circle in a counterclockwise direction a distance θ. The point P reached determines the second arm. For instance, if $\theta = 5\pi/2$ radians, it is necessary to travel clear around the circle once and then reach the point P above the center of the circle. In this case we obtain the right angle, also described by $\pi/2$ radians and shown in Figure E.5.

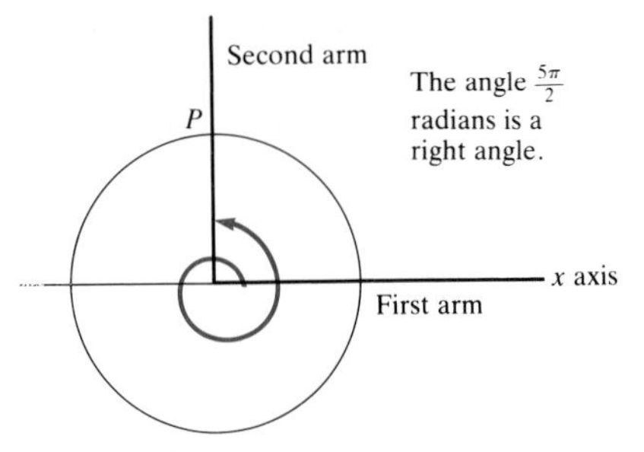

Figure E.5

Every time we travel around the unit circle, we increase the measure of an angle by 2π radians. Thus the right angle of $\pi/2$ radians has an endless supply of descriptions:

$$\frac{\pi}{2}, \quad \frac{\pi}{2} + 2\pi = \frac{5\pi}{2}, \quad \frac{\pi}{2} + 4\pi = \frac{9\pi}{2}, \quad \ldots .$$

Negative θ

To associate angles with the negative number θ, go *clockwise* around the unit circle through an angle $|\theta|$. For instance, to draw the angle $-\pi/2$, start at the point (1, 0) and move along the unit circle clockwise through a right angle until reaching the point P directly below the center of the circle. Note in Fig. E.6 that an angle of $-\pi/2$ radians is coterminal with an angle of $3\pi/2$ radians.

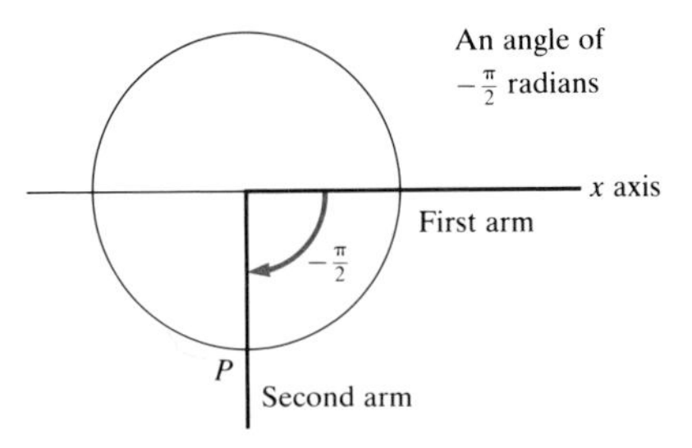

Figure E.6

The two fundamental functions of trigonometry, sine and cosine, can now be defined.

Definition *The sine and cosine functions.* For each number θ, the sine and cosine of θ are defined as follows. Draw the angle of θ radians whose first arm is the positive x axis and whose vertex is at (0, 0). The second arm meets the unit circle whose center is at (0, 0) in a point P. The x coordinate of P is called the cosine of θ and is denoted **cos** θ. The y coordinate of P is called the sine of θ and is denoted **sin** θ. (See Fig. E.7.)

This diagram is the most important in this section. It is the basis of trigonometry.

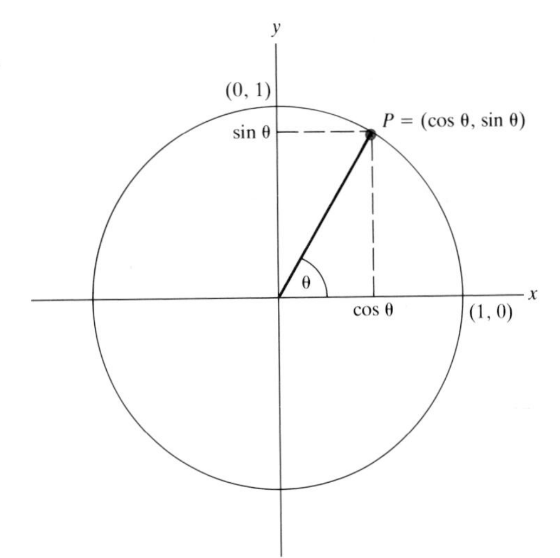

Figure E.7

Find cos $(\pi/2)$ and sin $(\pi/2)$.

If $\theta = \pi/2$, then the angle is a right angle, and $P = (0, 1)$. Hence

$$\cos\frac{\pi}{2} = 0 \quad \text{and} \quad \sin\frac{\pi}{2} = 1. \quad ■$$

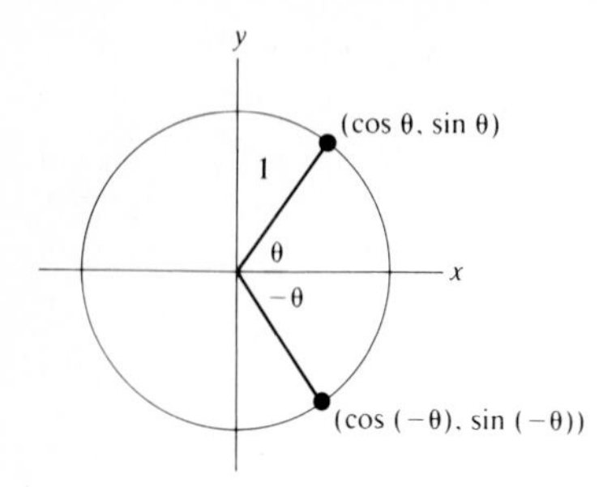

Diagram showing that $\cos(-\theta) = \cos\theta$ and $\sin(-\theta) = -\sin\theta$.

Figure E.8

EXAMPLE 5 Find $\cos(-\pi)$ and $\sin(-\pi)$.

SOLUTION If $\theta = -\pi$, P is the point $(-1, 0)$. Hence

$$\cos(-\pi) = -1 \quad \text{and} \quad \sin(-\pi) = 0. \quad \blacksquare$$

The trigonometric functions satisfy various identities. First of all, since a change of 2π in θ leads to the same point P on the circle,

$$\cos(\theta + 2\pi) = \cos\theta$$

and

$$\sin(\theta + 2\pi) = \sin\theta.$$

One says that the cosine and sine functions have **period** 2π. Second, inspection of the unit circle in Fig. E.8 shows that

$$\cos(-\theta) = \cos\theta \quad \text{and} \quad \sin(-\theta) = -\sin\theta.$$

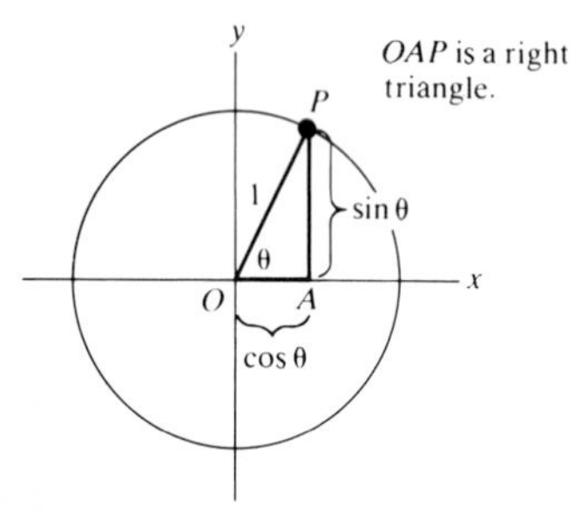

OAP is a right triangle.

Figure E.9

The numbers $\cos\theta$ and $\sin\theta$ are related by the equation

$$\cos^2\theta + \sin^2\theta = 1.$$

[$\cos^2\theta$ is short for $(\cos\theta)^2$.] To establish this, apply the pythagorean theorem to the right triangle *OAP* shown in Fig. E.9.

With the aid of this relation between $\cos\theta$ and $\sin\theta$ the next two examples determine $\cos(\pi/4)$, $\sin(\pi/4)$, $\cos(\pi/3)$, and $\sin(\pi/3)$.

EXAMPLE 6 Find $\cos(\pi/4)$ and $\sin(\pi/4)$.

SOLUTION When the angle is $\pi/4$ (45°), a quick sketch shows that the cosine equals the sine:

$$\cos\frac{\pi}{4} = \sin\frac{\pi}{4}.$$

Thus

$$\cos^2\frac{\pi}{4} + \cos^2\frac{\pi}{4} = 1,$$

from which it follows that

$$\cos^2\frac{\pi}{4} = \frac{1}{2}.$$

Since $\cos(\pi/4)$ is positive,

$$\cos\frac{\pi}{4} = \sqrt{\frac{1}{2}} = \frac{\sqrt{2}}{2} \approx 0.707,$$

and

$$\sin\frac{\pi}{4} = \frac{\sqrt{2}}{2}. \quad \blacksquare$$

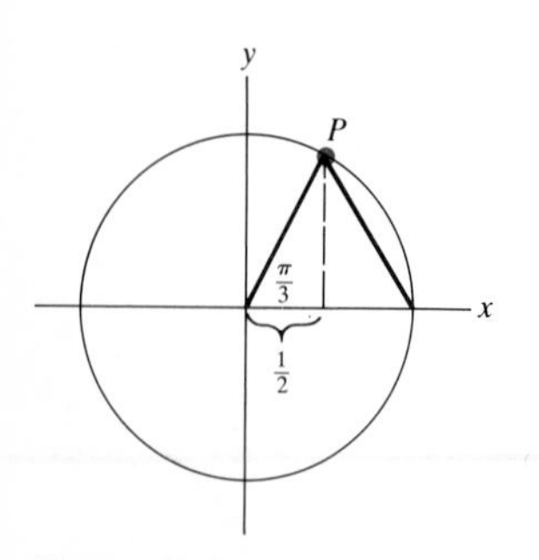

Figure E.10

EXAMPLE 7 Find $\cos(\pi/3)$ and $\sin(\pi/3)$.

SOLUTION The angle $\pi/3$ (60°) is the angle in an equilateral triangle. Place such a triangle in the unit circle as shown in Fig. E.10. Inspection of the figure shows that

$$\cos\frac{\pi}{3} = \frac{1}{2}.$$

Then,
$$\left(\frac{1}{2}\right)^2 + \sin^2\frac{\pi}{3} = 1$$
$$\sin^2\frac{\pi}{3} = \frac{3}{4}$$
$$\sin\frac{\pi}{3} = \frac{\sqrt{3}}{2} \approx 0.866. \quad ■$$

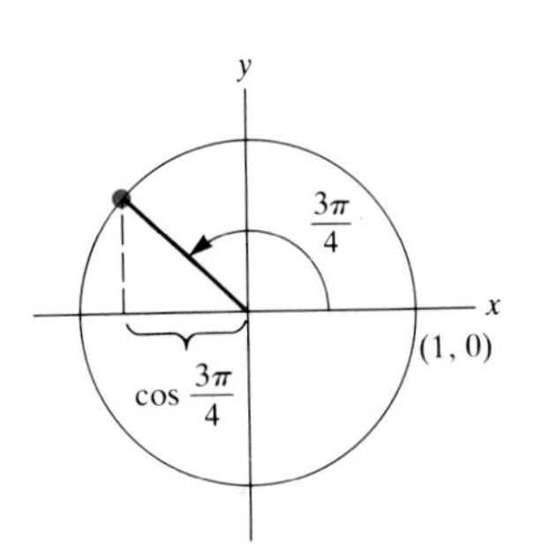

Figure E.11

Once cos ($\pi/4$) is known, the cosines of multiples of $\pi/4$ can be found by sketching the unit circle. For instance, to find cos ($3\pi/4$), draw an angle of $3\pi/4$ radians, as in Fig. E.11. It is clear that cos ($3\pi/4$) is negative and that $|\cos(3\pi/4)| = \sqrt{2}/2$. Hence

$$\cos\frac{3\pi}{4} = -\frac{\sqrt{2}}{2}.$$

A similar method can be used to compute the cosines of multiples of $\pi/6$. With the aid of such computations, this table for the cosine function can be obtained:

θ	0	$\frac{\pi}{6}$	$\frac{\pi}{4}$	$\frac{\pi}{3}$	$\frac{\pi}{2}$	$\frac{2\pi}{3}$	$\frac{3\pi}{4}$	$\frac{5\pi}{6}$	π	$\frac{7\pi}{6}$	$\frac{4\pi}{3}$	$\frac{3\pi}{2}$	2π
$\cos\theta$	1	$\frac{\sqrt{3}}{2}$	$\frac{\sqrt{2}}{2}$	$\frac{1}{2}$	0	$\frac{-1}{2}$	$\frac{-\sqrt{2}}{2}$	$\frac{-\sqrt{3}}{2}$	-1	$\frac{-\sqrt{3}}{2}$	$\frac{-1}{2}$	0	1

(It is easier to draw the unit circle and angle each time you need cos θ than to memorize this table.) This table provides enough information to graph the cosine function. Since $\cos(\theta + 2\pi) = \cos\theta$, the graph consists of the portion from 0 to 2π endlessly repeated. It is sketched in Fig. E.12. The graph of the sine function can be sketched in a similar manner. It is shown in Fig. E.13.

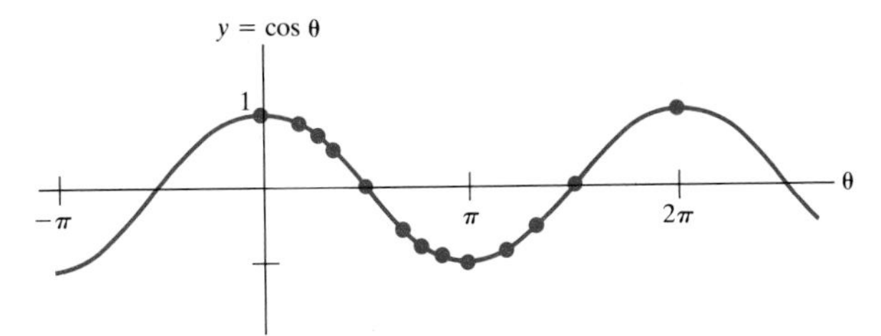

Figure E.12

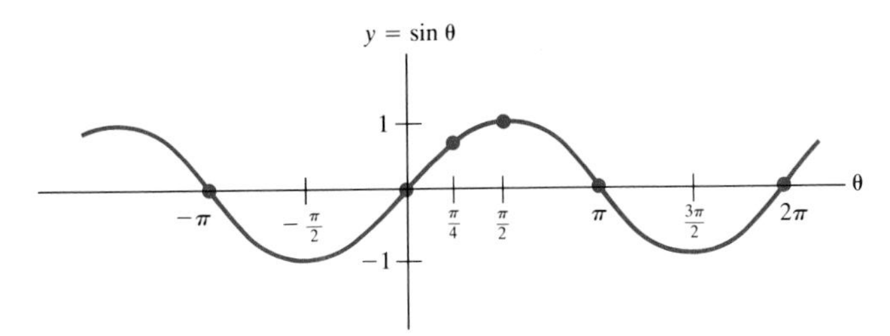

Figure E.13

Four important identities relate the cosine and sine of the sum and difference of two angles to the values of the cosine and sine of the angles:

$$\cos(A + B) = \cos A \cos B - \sin A \sin B,$$

$$\sin(A + B) = \sin A \cos B + \cos A \sin B,$$

$$\cos(A - B) = \cos A \cos B + \sin A \sin B,$$

$$\sin(A - B) = \sin A \cos B - \cos A \sin B.$$

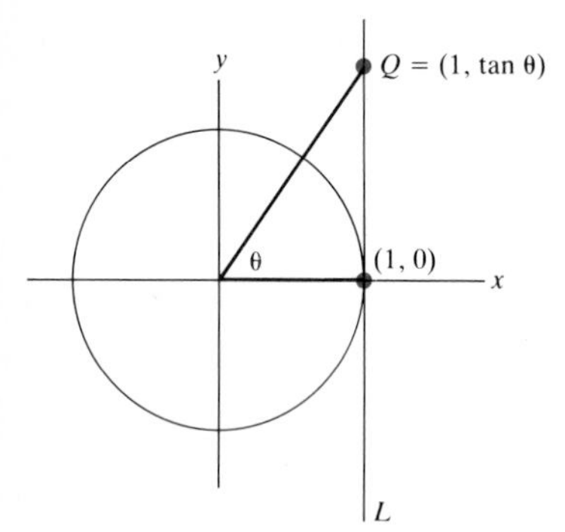

Figure E.14

(These identities are obtained in Exercises 43 to 48.) From these follow the "double angle" and "half angle" identities:

$$\cos 2\theta = \cos^2 \theta - \sin^2 \theta,$$

$$\cos 2\theta = 2\cos^2 \theta - 1,$$

$$\cos 2\theta = 1 - 2\sin^2 \theta,$$

$$\sin 2\theta = 2 \sin \theta \cos \theta,$$

$$\sin^2 \theta = \frac{1 - \cos 2\theta}{2},$$

$$\cos^2 \theta = \frac{1 + \cos 2\theta}{2}.$$

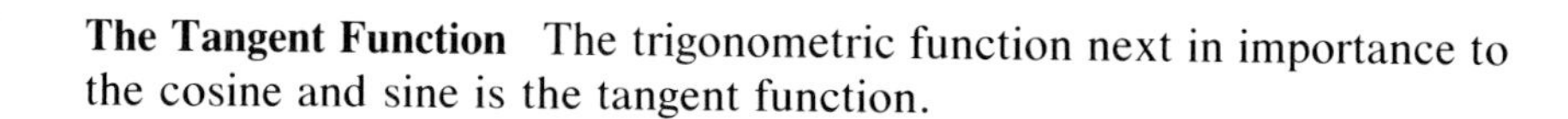

The Tangent Function The trigonometric function next in importance to the cosine and sine is the tangent function.

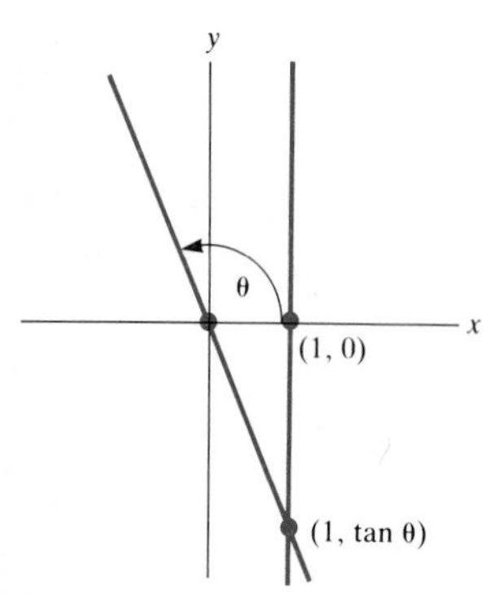

Figure E.15

Definition *The tangent function.* For each number θ that does not differ from $\pi/2$ by a multiple of π, the tangent of θ, denoted $\tan \theta$, is defined as follows: Draw the angle of θ radians whose first arm is the positive x axis and whose vertex is (0, 0). Let L be the line through (1, 0) parallel to the y axis. The line on the second arm of the angle meets the line L at a point Q. The y coordinate of Q is called **tan θ**. (See Fig. E.14.)

Note from Fig. E.14 that for θ near $\pi/2$ but less than $\pi/2$, $\tan \theta$ becomes very large. While $\cos \theta$ and $\sin \theta$ never exceed 1, $\tan \theta$ takes arbitrarily large values. Note that for $\pi/2 < \theta < \pi$ (a second-quadrant angle), $\tan \theta$ is negative, as shown in Fig. E.15. The behavior of θ for θ near $\pi/2$ is described by these two limits:

$$\lim_{\theta \to (\pi/2)^-} \tan \theta = \infty \quad \text{and} \quad \lim_{\theta \to (\pi/2)^+} \tan \theta = -\infty.$$

It follows from the definition of the tangent function that

$$\tan(\theta + \pi) = \tan \theta.$$

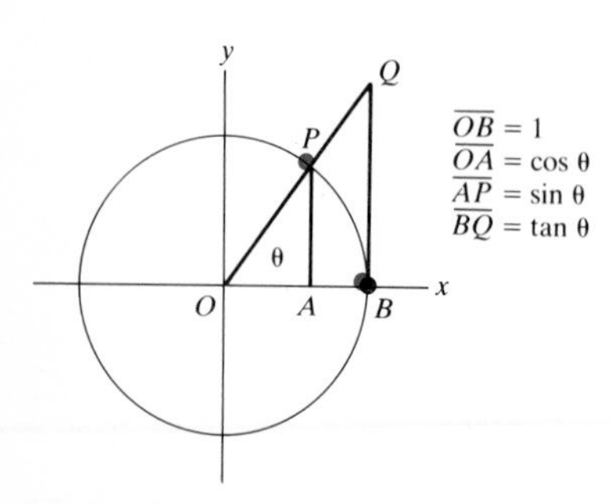

Figure E.16

While cosine and sine have period 2π, the tangent function has period π.

The functions $\cos \theta$, $\sin \theta$, and $\tan \theta$ are available on many calculators. These functions are related by the equation

$$\tan \theta = \frac{\sin \theta}{\cos \theta}.$$

This can easily be deduced from inspection of Fig. E.16. By the similarity of triangles OBQ and OAP,

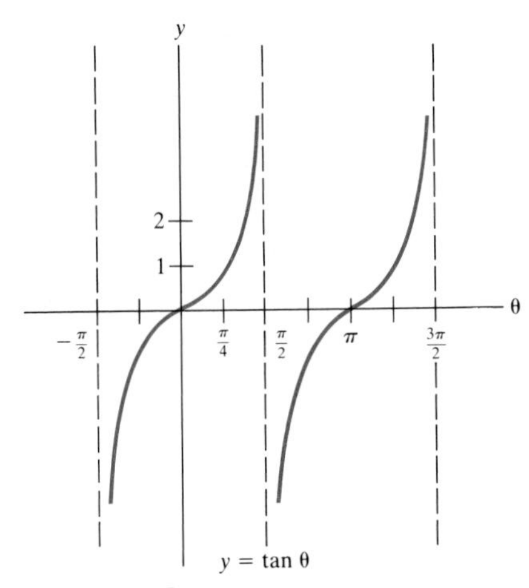

Figure E.17

$$\frac{\overline{BQ}}{\overline{OB}} = \frac{\overline{AP}}{\overline{OA}},$$

or

$$\frac{\tan\theta}{1} = \frac{\sin\theta}{\cos\theta}.$$

Thus

$$\tan\theta = \frac{\sin\theta}{\cos\theta}.$$

The following identities involving the tangent function are of use:

$$\tan(A + B) = \frac{\tan A + \tan B}{1 - \tan A \tan B},$$

$$\tan(A - B) = \frac{\tan A - \tan B}{1 + \tan A \tan B},$$

$$\tan 2\theta = \frac{2\tan\theta}{1 - \tan^2\theta}.$$

It is easy to see that $\tan 0 = 0$ and $\tan \pi/4 = 1$. Figure E.17 presents the graph of the tangent function.

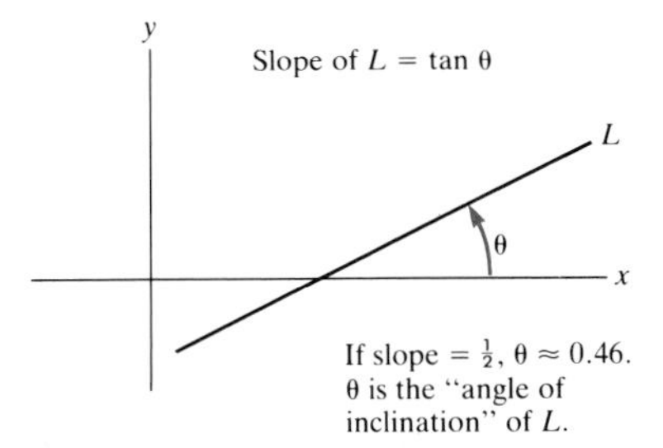

Figure E.18

The tangent function can be used to describe the slope of a line. Let a line L make an angle θ with the positive x axis. (This angle is taken counterclockwise from the x axis upward to the line; thus $0 \le \theta < \pi$.) The angle θ is called the **angle of inclination** of the line L. Then the slope of L is equal to $\tan\theta$. If the slope is known, then the angle of inclination can be estimated with the aid of a calculator. For instance, when the slope is $\frac{1}{2}$, as in the case of the line L in Fig. E.18, the angle is about 0.46 radian or 26.6°.

Three other trigonometric functions will be needed in calculus. They are the reciprocals of the cosine, sine, and tangent functions, and are called, respectively, **secant**, **cosecant**, and **cotangent**:

$$\sec\theta = \frac{1}{\cos\theta}, \qquad \csc\theta = \frac{1}{\sin\theta}, \qquad \cot\theta = \frac{\cos\theta}{\sin\theta}.$$

Consider the function $\sec\theta$. First of all, it is not defined when $\cos\theta = 0$. For instance, $\pi/2$ is not in the domain of $\sec\theta$. However, for θ near $\pi/2$, $|\sec\theta|$ is large. Second, since $|\cos\theta| \le 1$ for all θ, $|\sec\theta| \ge 1$ for all θ for which $\sec\theta$ is defined. Third, $\sec\theta$ has period 2π since $\cos\theta$ does. Fourth, $\sec\theta$ is positive when $\cos\theta$ is positive and negative when $\cos\theta$ is negative. Figure E.19 shows the graph of $y = \sec\theta$ and, for comparison, the graph of $\cos\theta$, which is shown dashed.

The graph of $y = \csc\theta$ bears a similar relation to the graph of $\sin\theta$. It is simply the graph of $\sec\theta$ moved $\pi/2$ units to the right. (See Fig. E.20.)

The graph of $y = \cot\theta$, along with that of $\tan\theta$, is shown in Fig. E.21. Note that when $\tan\theta$ is large, $\cot\theta$ is near 0; when $\tan\theta$ is near 0, $|\cot\theta|$ is large.

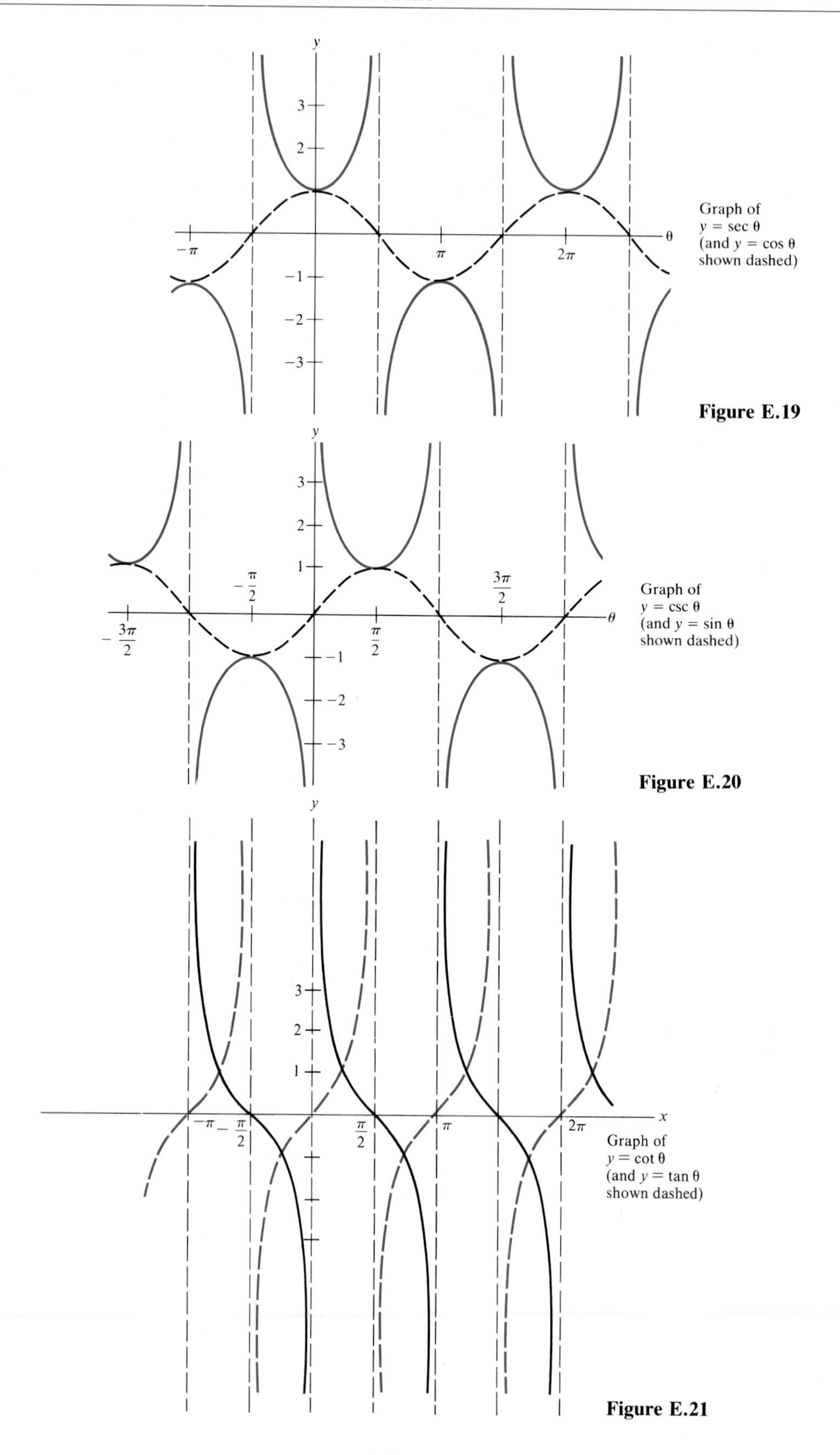

Figure E.19

Figure E.20

Figure E.21

The Law of Cosines If the lengths of two sides of a triangle, a and b, and the angle θ between these sides are known, the length of the third side c is determined. The formula for finding c, the **law of cosines**, is

$$c^2 = a^2 + b^2 - 2ab \cos \theta.$$

(When $\theta = \pi/2$, the law of cosines is the pythagorean theorem.) This formula is used several times in the text. A proof is outlined in Exercise 53. Further trigonometric identities are developed in the exercises.

EXERCISES FOR APPENDIX E: TRIGONOMETRY

1 What are the radian measures of the following angles?
(*a*) 90° (*b*) 30°
(*c*) 120° (*d*) 360°

2 What are the radian measures of the following angles?
(*a*) 180° (*b*) 60°
(*c*) 45° (*d*) 0°

3 How many degrees are there in an angle whose radian measure is (*a*) $3\pi/4$, (*b*) $\pi/3$, (*c*) $2\pi/3$, (*d*) 4π?

4 How many degrees are there in an angle whose radian measure is (*a*) $\pi/4$, (*b*) $\pi/6$, (*c*) 2π, (*d*) π?

5 An angle intercepts an arc of 5 inches in a circle of radius 3 inches.
(*a*) What is the measure of the angle in radians?
(*b*) What is the measure of the angle in degrees?

6 How long an arc of a circle of radius 3 inches is intercepted by an angle of 0.5 radian?

7 (*a*) Express an angle of 50° in radians.
(*b*) Express an angle of 2 radians in degrees.

8 (*a*) Express an angle of 3 radians in degrees.
(*b*) Express an angle of 1 degree in radians.

9 How long an arc does an angle of 1.5 radians intercept in a circle of radius (*a*) 3 inches, (*b*) 4 inches, (*c*) 5 inches?

10 How would you draw an angle of 2 radians (*a*) with a protractor that shows angles in degrees? (*b*) with a string?

11 Draw an angle of (*a*) $13\pi/6$ radians, (*b*) -3π radians.

12 Draw an angle of (*a*) 6π radians, (*b*) $-\pi/3$ radians.

13 In Example 7 it was shown that $\cos (\pi/3) = \frac{1}{2}$. Using the identity $\sin \theta = \cos (\pi/2 - \theta)$, deduce that $\sin (\pi/6) = \frac{1}{2}$.

14 In Example 7 it was shown that $\sin (\pi/3) = \sqrt{3}/2$. Using the identity $\cos \theta = \sin (\pi/2 - \theta)$, deduce that $\cos (\pi/6) = \sqrt{3}/2$.

15 Fill in this table by making a quick sketch of the unit circle:

θ	0	$\frac{\pi}{6}$	$\frac{\pi}{4}$	$\frac{\pi}{3}$	$\frac{\pi}{2}$	π	$\frac{3\pi}{2}$	2π
$\sin \theta$								

16 By sketching the angles in a unit circle and using the information that $\cos (\pi/6) = \sqrt{3}/2$ and $\sin (\pi/6) = \frac{1}{2}$, find
(*a*) $\cos (-\pi/6)$ (*b*) $\sin (-\pi/6)$
(*c*) $\cos (5\pi/6)$ (*d*) $\sin (5\pi/6)$
(*e*) $\cos (13\pi/6)$ (*f*) $\sin (13\pi/6)$

17 Check the identity for $\cos (A + B)$ when $A = \pi/6$ and $B = \pi/3$.

18 Check the identity for $\sin (A + B)$ when $A = \pi/4$ and $B = \pi/4$.

19 Find $\sin \frac{5\pi}{12}$. *Hint:* $\frac{5\pi}{12} = \frac{\pi}{4} + \frac{\pi}{6}$.

20 Find $\sin \frac{\pi}{12}$. *Hint:* $\frac{\pi}{12} = \frac{\pi}{4} - \frac{\pi}{6}$.

21 Deduce the identity $\cos 2\theta = \cos^2 \theta - \sin^2 \theta$ from the identity for $\cos (A + B)$.

22 Deduce the identity $\sin 2\theta = 2 \sin \theta \cos \theta$ from the identity for $\sin (A + B)$.

23 Deduce the identity $\cos 2\theta = 2 \cos^2 \theta - 1$ from Exercise 21.

24 Deduce the identity $\cos 2\theta = 1 - 2 \sin^2 \theta$ from Exercise 21.

25 Deduce from Exercise 23 that $\cos^2 \theta = (1 + \cos 2\theta)/2$.

26 Deduce from Exercise 24 that $\sin^2 \theta = (1 - \cos 2\theta)/2$.

27 Use a sketch of the angle in a unit circle to determine the sign (+ or −) of the function in each case:
(*a*) $\sin \theta$, $\pi < \theta < 2\pi$ (*b*) $\tan \theta$, $\pi < \theta < 3\pi/2$
(*c*) $\cos \theta$, $-\pi/2 < \theta < \pi/2$
(*d*) $\tan \theta$, $\pi/2 < \theta < \pi$

28 Give the sign (+ or −) of $\cos \theta$, $\sin \theta$, and $\tan \theta$ for (*a*) $0 < \theta < \pi/2$, (*b*) $-\pi/2 < \theta < 0$.

29 (*a*) Using the definition of $\tan \theta$ and a sketch, show that $\tan (\pi/4) = 1$.
(*b*) Using the identity $\tan \theta = (\sin \theta)/\cos \theta$, show that $\tan (\pi/4) = 1$.

30 Find (*a*) $\tan (\pi/6)$, (*b*) $\tan (\pi/3)$.

31 Using a calculator, find the slope of a line whose angle of inclination is
(*a*) 10° (*b*) 70° (*c*) 110°
(*d*) 135° (*e*) 0°

32 Using a calculator, find the angle of inclination, both in degrees and in radians, of a line whose slope is
(*a*) -1 (*b*) 1 (*c*) 2 (*d*) 3 (*e*) -3

33 Find the angle of inclination in degrees and in radians of a line whose slope is
(*a*) $\sqrt{3}$ (*b*) $1/\sqrt{3}$

34 (Calculator) Find the angle of inclination in degrees and in radians of a line whose slope is
(*a*) $\frac{1}{10}$ (*b*) 10

35 Consider an acute angle θ in any right triangle, as in Fig. E.22. With the aid of the similar triangles in Fig. E.23, show that (*a*) $\cos\theta = a/c$, (*b*) $\sin\theta = b/c$, and (*c*) $\tan\theta = b/a$. These formulas for cosine, sine, and tangent are sometimes taken as their definitions for θ in the first quadrant.

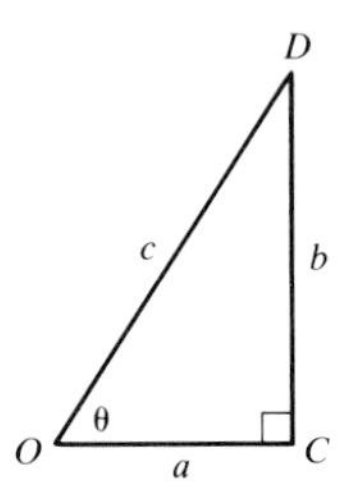

Figure E.22

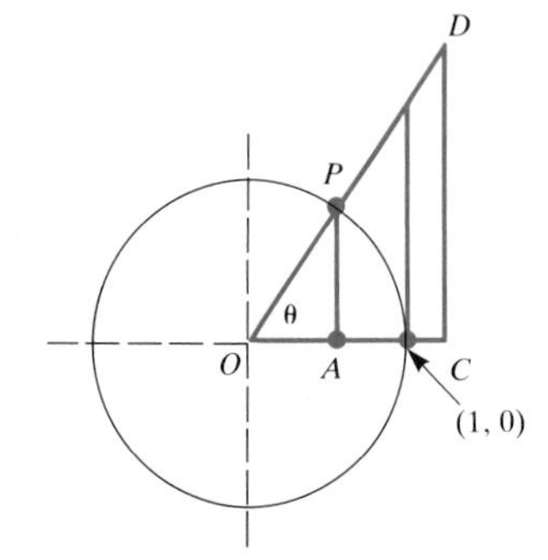

Figure E.23

36 (See Exercise 35, Fig. E.22.)
(*a*) Express b in terms of θ and c.
(*b*) Express a in terms of θ and c.
(*c*) Express b in terms of θ and a.

37 In Fig. E.24 the two acute angles in a right triangle are labeled α and β.
(*a*) Express $\cos\alpha$ in terms of the lengths of the sides.
(*b*) Express $\sin\beta$ in terms of the lengths of the sides.
(*c*) Express $\tan\alpha$ in terms of the lengths of the sides.

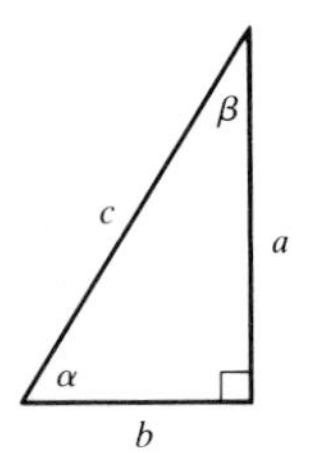

Figure E.24

38 (See Exercise 35.) Use the triangle OCD in Fig. E.25 to obtain $\cos(\pi/3)$, $\sin(\pi/3)$ and $\tan(\pi/3)$. (ODE is equilateral.) A quick sketch of OCD will remind you of the values of $\cos(\pi/3)$, $\sin(\pi/3)$, and $\tan(\pi/3)$ when you need them.

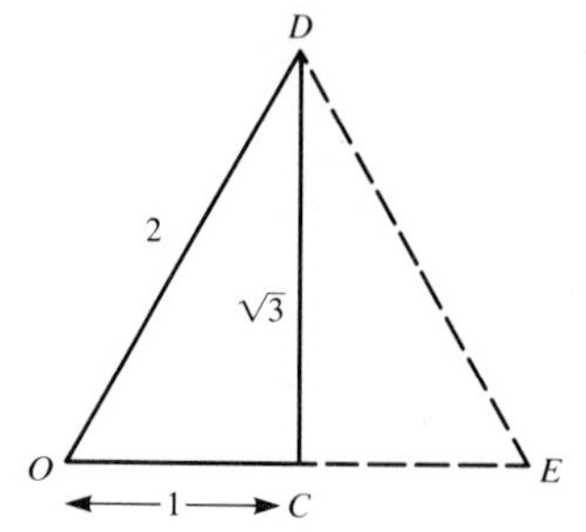

Figure E.25

39 Use OCD in Exercise 38 to compute $\cos(\pi/6)$, $\sin(\pi/6)$, and $\tan(\pi/6)$. *Hint:* See Exercise 35.

40 (See Exercise 35.) Solve for the length x in each triangle in Fig. E.26.

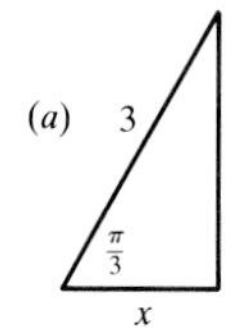

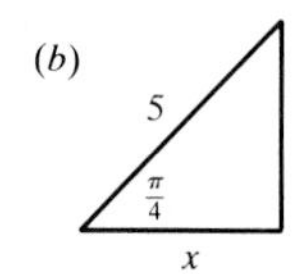

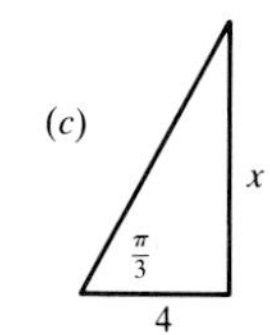

Figure E.26

41 (*a*) Evaluate $\sec\theta$ for $\theta = 0, \pi/6, \pi/4, \pi/3$.
(*b*) Plot the four points given by (*a*) and graph $\sec\theta$.

42 (*a*) Evaluate $\csc\theta$ for $\theta = \pi/6, \pi/4, \pi/3, \pi/2$.
(*b*) Plot the four points given by (*a*) and graph $\csc\theta$.

■

43 This exercise outlines a proof that $\cos(A + B) = \cos A\cos B - \sin A\sin B$. Figure E.27 shows two diagrams involving the unit circle.
(*a*) Show that the line segments PQ and RS in Fig. E.27 have the same length.

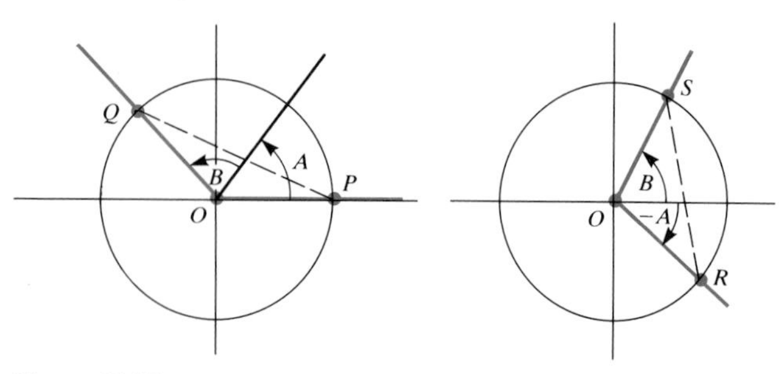

Figure E.27

(*b*) Show that

$$Q = (\cos(A + B), \sin(A + B)),$$
$$S = (\cos B, \sin B),$$
$$R = (\cos A, -\sin A).$$

(*c*) Using the coordinates in (*b*), show that

$$\overline{PQ}^2 = 2 - 2 \cos (A + B).$$

(*d*) Using the coordinates in (*b*), show that

$$\overline{RS}^2 = 2 - 2 \cos A \cos B + 2 \sin A \sin B.$$

(*e*) Deduce the identity for $\cos (A + B)$.

44 Replacing B by $-B$ in the identity for $\cos (A + B)$, obtain the identity

$$\cos (A - B) = \cos A \cos B + \sin A \sin B.$$

45 Using the identity for $\cos (A - B)$, show that

$$\sin \theta = \cos \left(\frac{\pi}{2} - \theta\right).$$

46 Using the identity $\sin \theta = \cos (\pi/2 - \theta)$ from Exercise 45, show that $\cos \theta = \sin (\pi/2 - \theta)$.

47 With the aid of the identities in Exercises 44 to 46, show that $\sin (A + B) = \sin A \cos B + \cos A \sin B$, as follows:

(*a*) Show that $\sin (A + B) = \cos [\pi/2 - (A + B)] = \cos [(\pi/2 - A) - B]$.

(*b*) Apply the identity in Exercise 44 to $\cos (x - y)$ with $x = \pi/2 - A$ and $y = B$.

(*c*) Use the identities in Exercises 45 and 46 to remove $\pi/2$ from the identity you obtain in (*b*).

48 Use the identity for $\sin (A + B)$ obtained in Exercise 47 to show that $\sin (A - B) = \sin A \cos B - \cos A \sin B$.

49 (*a*) From the identity $\cos 2\theta = 2 \cos^2 \theta - 1$ deduce that $\cos \theta = \pm\sqrt{(1 + \cos 2\theta)/2}$.

(*b*) Use the identity in (*a*) to find $\cos (\pi/4)$.

(*c*) Use the identity in (*a*) to find $\cos (3\pi/4)$.

50 (*a*) From the identity $\cos 2\theta = 1 - 2 \sin^2 \theta$ deduce that $\sin \theta = \pm\sqrt{(1 - \cos 2\theta)/2}$.

(*b*) Use the identity in (*a*) to find $\sin (\pi/4)$.

(*c*) Use the identity in (*a*) to find $\sin (-\pi/4)$.

51 Using the identities for $\cos (A - B)$ and $\sin (A - B)$, prove that

$$\tan (A - B) = \frac{\tan A - \tan B}{1 + \tan A \tan B}.$$

52 Show that

(*a*) $\sec^2 \theta = 1 + \tan^2 \theta$,

(*b*) $\csc^2 \theta = 1 + \cot^2 \theta$.

53 This exercise outlines a proof of the law of cosines in the case $0 < \theta < \pi/2$. Consider Fig. E.28.

(*a*) Show that $\overline{CD} = a \cos \theta$ and $\overline{AD} = b - a \cos \theta$.

(*b*) Show that

$$a^2 - a^2 \cos^2 \theta = h^2 = c^2 - (b - a \cos \theta)^2.$$

(*c*) From (*b*) deduce the law of cosines.

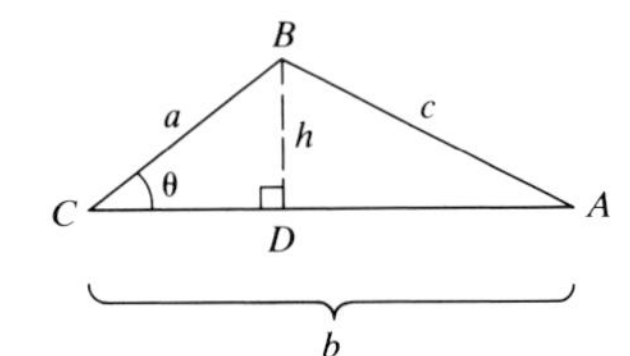

Figure E.28

Appendix F Conic Sections

This appendix defines the ellipse, hyperbola, and parabola and develops their equations in rectangular coordinates (Secs. F.1, F.2, and F.3) and in polar coordinates (Sec. F.4).

F.1 CONIC SECTIONS

The intersection of a plane and the surface of a double cone is called a **conic section** If the plane cuts off a bounded curve, that curve is called an **ellipse** (In particular, a circle is an ellipse.) See Fig. F.1.

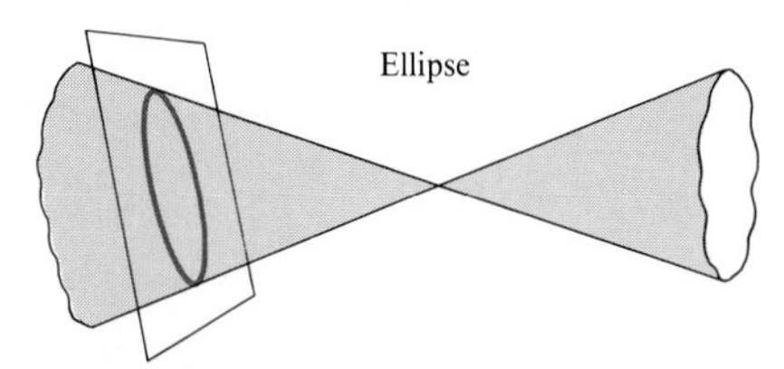

A plane may cut a double cone in an ellipse

Figure F.1

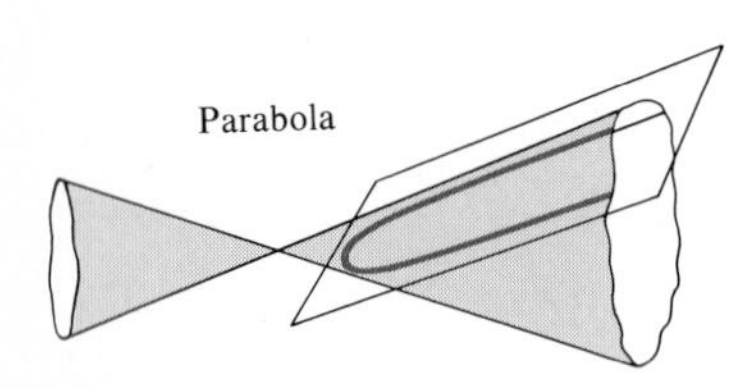

Figure F.2

If the plane is parallel to the edge of the double cone, as in Fig. F.2, the intersection is called a **parabola** In the cases of the ellipse and the parabola, the plane generally meets just one of the two cones.

If the plane meets both parts of the cone and is not parallel to an edge, the intersection is called a **hyperbola** The hyperbola consists of two separate pieces. It can be proved that these two pieces are congruent and that they are *not* congruent to parabolas. (See Fig. F.3.)

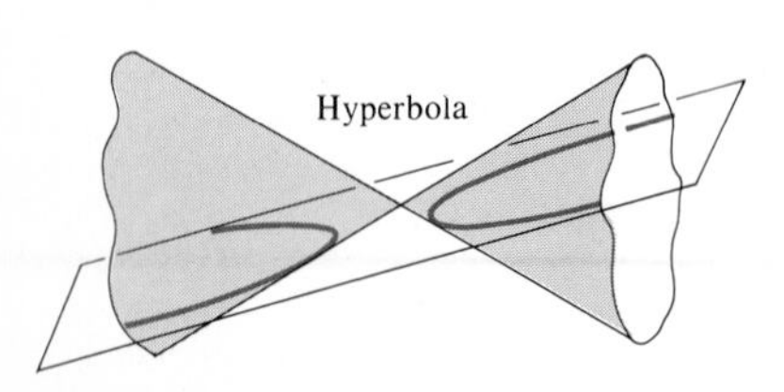

Figure F.3

For the sake of simplicity, we shall use a definition of the conic sections that depends only on the geometry of the plane. It is shown in geometry courses that the two approaches yield the same curves.

Definition *Ellipse.* Let F and F' be points in the plane and let a be a fixed positive number such that $2a$ is greater than the distance between F and F'. A point P in the plane is on the **ellipse** determined by F, F', and $2a$ if and only if the sum of the distances from P to F and from P to F' equals $2a$. Points F and F' are the **foci** of the ellipse.

"Foci" (pronounced "foe-sigh") is the plural of "focus."

To construct an ellipse, place two tacks in a piece of paper, tie a string of length $2a$ to them, and trace out a curve with a pencil held against the string, keeping the string taut by means of the pencil point. (See Fig. F.4.) The foci are at the tacks. (Note that when $F = F'$, the ellipse is a circle of radius a.) The four points on the ellipse that are furthest from or closest to the center are called **vertices** A circle does not have vertices.

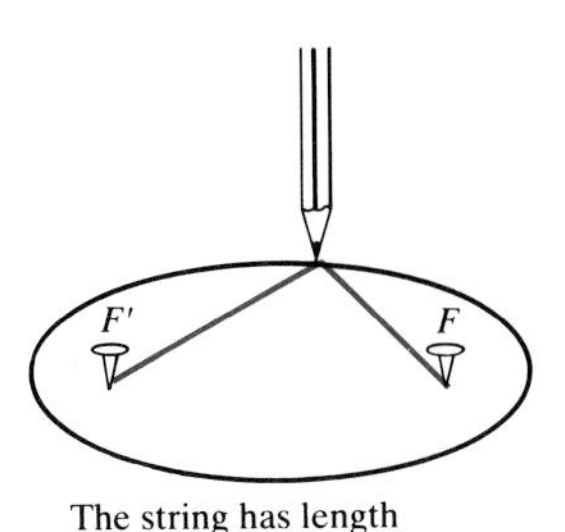

The string has length $2a$, greater than the distance between the tacks.

Figure F.4

The equation of a circle whose center is at (0, 0) and whose radius is a is

$$x^2 + y^2 = a^2,$$

or

$$\frac{x^2}{a^2} + \frac{y^2}{a^2} = 1. \tag{1}$$

Let us generalize this result by determining the equation of an ellipse. To make the equation as simple as possible, introduce the x and y axes in such a way that the x axis contains the foci and the origin is midway between them, as in Fig. F.5. Thus $F = (c, 0)$ and $F' = (-c, 0)$, where $c \geq 0$ and $2c < 2a$; hence $c < a$.

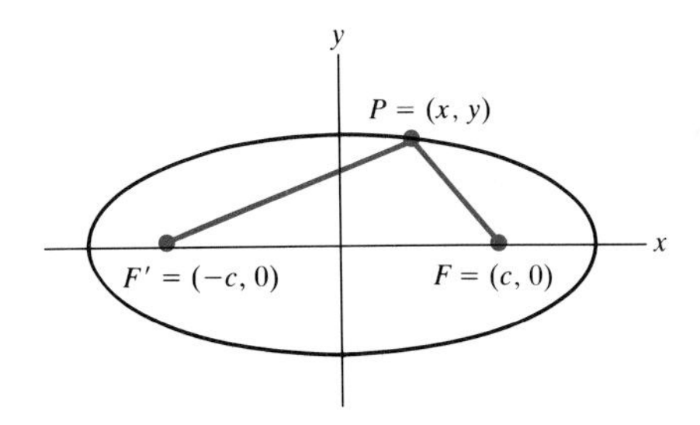

Figure F.5

Now translate into algebra this assertion: The sum of the distances from $P = (x, y)$ to F and from P to F' equals $2a$. By the distance formula, the distance from P to F is

$$\sqrt{(x - c)^2 + (y - 0)^2}.$$

Similarly, the distance from P to F' is

$$\sqrt{(x + c)^2 + (y - 0)^2}.$$

Thus the point (x, y) is on the ellipse if and only if

$$\sqrt{(x - c)^2 + y^2} + \sqrt{(x + c)^2 + y^2} = 2a.$$

A few algebraic steps will transform this equation into an equation without square roots.

First, write the equation as

$$\sqrt{(x + c)^2 + y^2} = 2a - \sqrt{(x - c)^2 + y^2}.$$

Then square both sides, obtaining

$$(x + c)^2 + y^2 = 4a^2 - 4a\sqrt{(x - c)^2 + y^2} + (x - c)^2 + y^2.$$

Expanding yields

$$x^2 + 2cx + c^2 + y^2 = 4a^2 - 4a\sqrt{(x - c)^2 + y^2} + x^2 - 2cx + c^2 + y^2,$$

which a few cancellations reduce to

$$2cx = 4a^2 - 4a\sqrt{(x - c)^2 + y^2} - 2cx,$$

or $$4cx - 4a^2 = -4a\sqrt{(x-c)^2 + y^2}.$$

Dividing by -4 yields

$$a^2 - cx = a\sqrt{(x-c)^2 + y^2}.$$

Squaring gets rid of the square root:

$$a^4 - 2a^2cx + c^2x^2 = a^2(x^2 - 2cx + c^2 + y^2),$$
$$= a^2x^2 - 2a^2cx + a^2c^2 + a^2y^2,$$

or $$a^4 + c^2x^2 = a^2x^2 + a^2c^2 + a^2y^2.$$

This equation can be transformed to

$$(a^2 - c^2)x^2 + a^2y^2 = a^2(a^2 - c^2).$$

Dividing both sides by $a^2(a^2 - c^2)$ results in the equation

$$\frac{x^2}{a^2} + \frac{y^2}{a^2 - c^2} = 1. \qquad (2)$$

Since $a^2 - c^2 > 0$, there is a number b such that

$$b^2 = a^2 - c^2 \qquad b > 0,$$

and thus Eq. (2) takes the shorter form

Equation of ellipse in standard position, foci on x axis

$$\boxed{\frac{x^2}{a^2} + \frac{y^2}{b^2} = 1.} \qquad (3)$$

[Note that Eq. (3) generalizes Eq. (1), the equation for a circle.]

We now find the four vertices of the ellipse by checking where the curve intersects the x and y axes. Setting $y = 0$ in Eq. (3), we obtain $x = a$ or $x = -a$; if we set $x = 0$ in Eq. (3), we obtain $y = b$ or $-b$. Thus the four "extreme" points of the ellipse have coordinates $(a, 0)$, $(-a, 0)$, $(0, b)$, and $(0, -b)$, as shown in Fig. F.6. Observe that the distance from F or F' to $(0, b)$ is a, which is half the length of string. The right triangle in the diagram, with vertices F, $(0, b)$, and the origin, is a reminder of the fact that $b^2 = a^2 - c^2$. Keep in mind that in the above ellipse a is larger than b. The **semimajor axis** is said to have length a; the **semiminor axis** has length b. Observe that we could interchange the roles of x and y and produce an ellipse with foci on the y axis. In this case, y would have the larger denominator (the square of the semimajor axis) in the equation in standard form, which becomes $y^2/a^2 + x^2/b^2 = 1$, $a > b$.

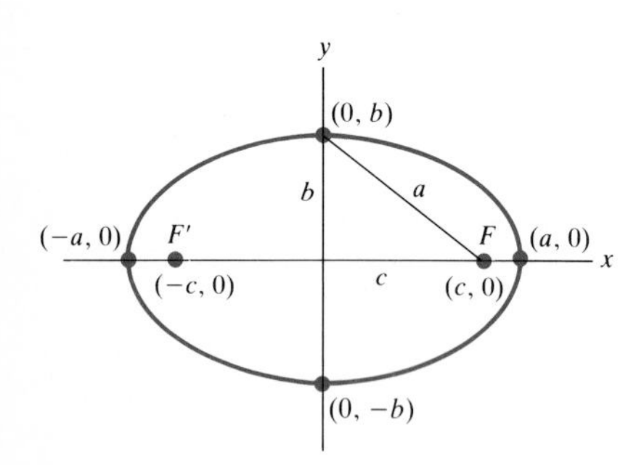

Figure F.6

EXAMPLE 1 Discuss the foci and "length of string" of the ellipse whose equation is $\frac{x^2}{25} + \frac{y^2}{9} = 1$.

SOLUTION Since the larger denominator is with the x^2, the foci lie on the x axis. In this case, $a = 5$ and $b = 3$. The length of string is $2a = 10$. The foci are at a distance

$$c = \sqrt{a^2 - b^2} = \sqrt{25 - 9} = 4$$

from the origin, as shown in Fig. F.7.

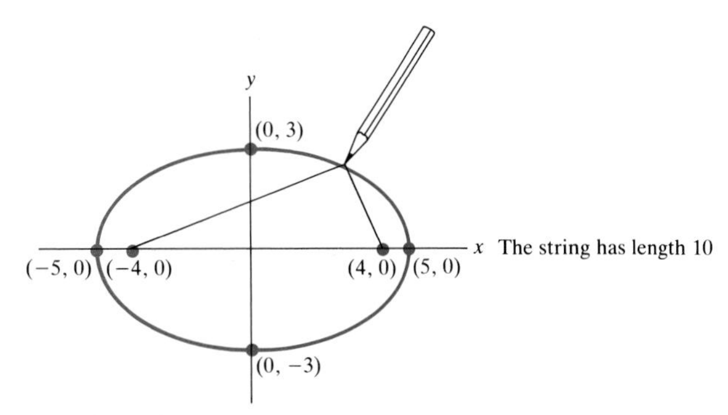

Figure F.7 ■

EXAMPLE 2 Discuss the foci and "length of string" of the ellipse whose equation is $\dfrac{x^2}{9} + \dfrac{y^2}{25} = 1.$

SOLUTION This is similar to Example 1. The only difference is that the roles of x and y are interchanged. The foci are at $(0, 4)$ and $(0, -4)$, and the ellipse is longer in the y direction than in the x direction. ■

The definition of the hyperbola is similar to that of the ellipse.

Definition *Hyperbola.* Let F and F' be points in the plane and let a be a fixed positive number such that $2a$ is less than the distance between F and F'. A point P in the plane is on the **hyperbola** determined by F, F', and $2a$ if and only if the difference between the distances from P to F and from P to F' equals $2a$ (or $-2a$). Points F and F' are called the **foci** of the hyperbola. The points V_1 and V_2 are called the **vertices** of the hyperbola. (See Fig. F.8.)

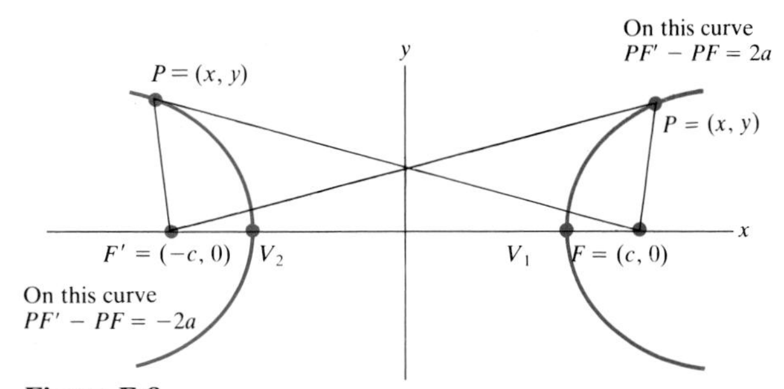

Figure F.8

A hyperbola consists of two separate curves. On one of the curves $\overline{PF'} - \overline{PF} = 2a$; on the other, $\overline{PF'} - \overline{PF} = -2a$. If the distance $\overline{FF'}$ is $2c$, then $2a < 2c$; hence $a < c$. Again place the axes in such a way that $F = (c, 0)$ and $F' = (-c, 0)$. Let $P = (x, y)$ be a typical point on the hyperbola. Then x and y satisfy the equation

$$\sqrt{(x - c)^2 + y^2} - \sqrt{(x + c)^2 + y^2} = \pm 2a. \tag{4}$$

Some algebra similar to that used in simplifying the equation of the ellipse transforms Eq. (4) into

$$\frac{x^2}{a^2} + \frac{y^2}{a^2 - c^2} = 1. \tag{5}$$

But now $a^2 - c^2$ is *negative* and can be expressed as $-b^2$ for some number $b > 0$. Hence the hyperbola has the equation

Equation of hyperbola in standard position

$$\boxed{\frac{x^2}{a^2} - \frac{y^2}{b^2} = 1.} \tag{6}$$

If the foci are on the y axis, the equation is

$$\boxed{\frac{y^2}{a^2} - \frac{x^2}{b^2} = 1.}$$

In both cases, $c^2 = a^2 + b^2$.

EXAMPLE 3 Sketch the hyperbola $\dfrac{x^2}{9} - \dfrac{y^2}{16} = 1$.

SOLUTION Since the minus sign is with the y^2, the foci are on the x axis. In this case, $a^2 = 9$ and $b^2 = 16$. Observe that the hyperbola meets the x axis at $(3, 0)$ and $(-3, 0)$. The hyperbola does not meet the y axis. The distance c from the origin to a focus is determined by the equation

$$c^2 = 9 + 16 = 25.$$

Hence

$$c = 5.$$

The hyperbola

$$\frac{x^2}{9} - \frac{y^2}{16} = 1$$

is shown in Fig. F.9. ■

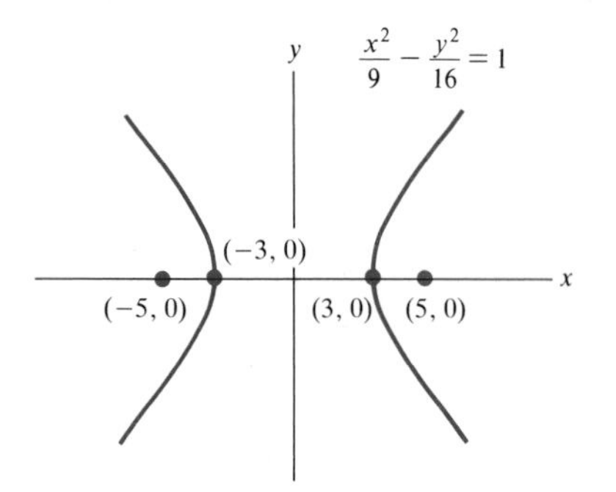

Figure F.9

The definition of a parabola involves the distance to a point and the distance to a line.

Definition *Parabola.* Let L be a line in the plane and let F be a point in the plane which is not on the line. A point P in the plane is on the **parabola** determined by F and L if and only if the distance from P to F equals the distance from P to the line L. Point F is the **focus** of the parabola; line L is its **directrix**. The point V in Fig. F.10 is called the **vertex** of the parabola. It is the point on the parabola nearest the directrix.

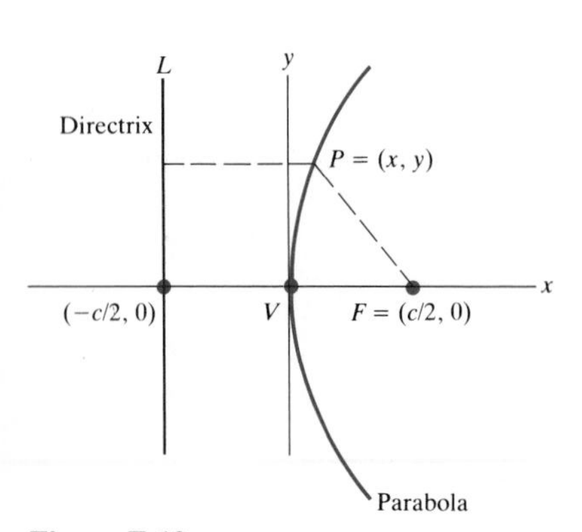

Figure F.10

To obtain an algebraic equation for a parabola, denote the distance from point F to line L by c, and introduce axes in such a way that $F = (c/2, 0)$ and L has the equation $x = -c/2$ with $c > 0$. The distance from $P = (x, y)$ to F is $\sqrt{(x - c/2)^2 + (y - 0)^2}$. The distance from P to the line L is $x + c/2$. Thus the equation of the parabola is

$$\sqrt{\left(x - \frac{c}{2}\right)^2 + y^2} = x + \frac{c}{2}. \tag{7}$$

Squaring and simplifying reduces Eq. (7) to

$$\boxed{y^2 = 2cx \qquad c > 0,} \tag{8}$$

which is the equation of a parabola in "standard position."

If the focus is at $(0, c/2)$ and the directrix is the line $y = -c/2$, the parabola has the equation

$$\boxed{x^2 = 2cy.}$$

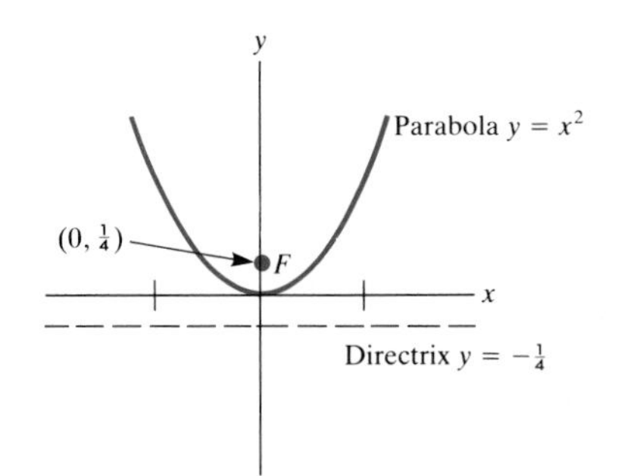

Figure F.11

EXAMPLE 4 Sketch the parabola $y = x^2$, showing its focus and directrix.

SOLUTION The equation $y = x^2$ is equivalent to $x^2 = 2cy$, where $c = \frac{1}{2}$. The focus is on the y axis at $(0, \frac{1}{4})$. The directrix is the line $y = -\frac{1}{4}$. The parabola is sketched in Fig. F.11. ■

This table lists the equations of conics in standard position:

Conic Section	*Equation in Standard Position*	*Location of Foci*
Ellipse	$\frac{x^2}{a^2} + \frac{y^2}{b^2} = 1$	$a > b$, foci on x axis
	$\frac{y^2}{a^2} + \frac{x^2}{b^2} = 1$	$a > b$, foci on y axis
Circle	$\frac{x^2}{a^2} + \frac{y^2}{a^2} = 1$	Both foci at $(0, 0)$
Hyperbola	$\frac{x^2}{a^2} - \frac{y^2}{b^2} = 1$	Foci on x axis
	$\frac{y^2}{a^2} - \frac{x^2}{b^2} = 1$	Foci on y axis
Parabola	$y^2 = 2cx$	Focus at $(c/2, 0)$
	$x^2 = 2cy$	Focus at $(0, c/2)$

EXERCISES FOR SEC. F.1: CONIC SECTIONS

1 Find the equation of the ellipse with foci at $(2, 0)$ and $(-2, 0)$ such that the sum of the distances from a point on the ellipse to the two foci is 10.

2 Find the equation of the ellipse with foci at $(0, 3)$ and $(0, -3)$ such that the sum of the distances from a point on the ellipse to the two foci is 14.

3 Sketch the hyperbola $x^2 - y^2 = 1$ and its foci.

4 Sketch the hyperbola $y^2 - x^2 = 1$ and its foci.

5 Sketch the parabola $y = 6x^2$, its focus, and its directrix.

6 Sketch the parabola $x = -6y^2$, its focus, and its directrix.

7 What is the equation of the parabola whose focus is at $(3, 0)$ and whose directrix is the line $x = -3$?

8 What is the equation of the parabola whose focus is at $(0, -5)$ and whose directrix is the line $y = 5$?

9 Obtain Eq. (5) from Eq. (4).

10 Obtain Eq. (8) from Eq. (7).

In Exercises 11 to 18 sketch the graphs of the given equations, the foci in the case of an ellipse or a hyperbola, and the focus and directrix in the case of a parabola.

11 $\frac{x^2}{49} + \frac{y^2}{25} = 1$ **12** $\frac{x^2}{4} + \frac{y^2}{36} = 1$

13 $\frac{x^2}{49} - \frac{y^2}{25} = 1$ **14** $\frac{y^2}{49} - \frac{x^2}{25} = 1$

15 $y^2 = 5x$ **16** $x^2 = 3y$

17 $y^2 = -5x$ **18** $x^2 = -3y$

■

19 (*a*) Using the definition of the hyperbola, show that the hyperbola that has its foci at $(\sqrt{2}, \sqrt{2})$ and $(-\sqrt{2}, -\sqrt{2})$ and for which $2a = 2\sqrt{2}$ has the equation $xy = 1$.
(*b*) Graph $xy = 1$ and show the foci.

20 How would you inscribe an elliptical garden in a rectangle whose dimensions are 8 by 10 feet? (The line through the foci is to be parallel to an edge of the rectangle.)

21 In the definition of the hyperbola it was assumed that $2a$ is less than the distance between the foci. Show that if $2a$ were greater than the distance between the foci, the hyperbola would have no points.

22 Find the equation of the parabola whose focus is (2, 4) and whose directrix is the line $y = -3$.

Exercises 23 and 24 concern the determination of the position of an object with the aid of conic sections.

23 The location of a submarine can be found as follows. A small explosion is set off at point D. The sound from this explosion bounces off the submarine and is picked up by hydrophones at points A, B, and C. The time of the explosion and the times of reception of the sound at A, B, and C are known. Show how to locate the submarine as the point where three ellipses meet.

24 The sound of the shooting of a cannon arrives 1 second later at point B than at point A.
(*a*) Show that the cannon is somewhere on a certain hyperbola whose foci are A and B.
(*b*) On which of the two pieces (branches) of the hyperbola is the cannon located?
With the aid of a third listening post the location of the cannon can be more precisely determined. LORAN, a system for long-range navigation, is based on a similar use of hyperbolas.

25 A plane intersects the surface of a right circular cylinder in a curve. Prove that this curve is an ellipse, as defined in terms of foci and sum of distances. *Hint:* Consider the two spheres inscribed in the cylinder and tangent to the plane, the spheres being on opposite sides of the plane. Let $2a$ denote the distance between the equators of the spheres perpendicular to the axis of the cylinder, and let F and F' be the points at which the spheres touch the plane.

F.2 TRANSLATION OF AXES AND THE GRAPH OF $Ax^2 + Cy^2 + Dx + Ey + F = 0$

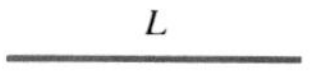

L

Figure F.12

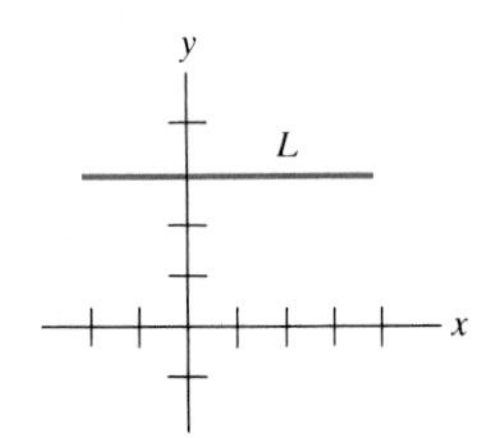

Figure F.13

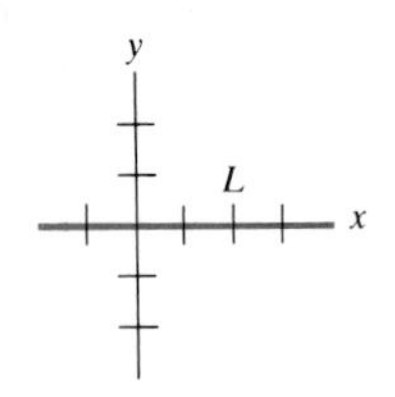

Figure F.14

The equation of a particular geometric object such as a line, a circle, or a conic section depends on where we choose to place the axes. Consider, for instance, the line L in Fig. F.12. Relative to the axes in Fig. F.13 it has the equation $y = 3$. Relative to the axes in Fig. F.14 it has the equation $y = 0$.

Clearly, a wise choice of axes may yield a simpler equation for a given line or curve. This section shows one way to choose convenient axes and uses the method to analyze equations of the form $Ax^2 + Cy^2 + Dx + Ey + F = 0$.

A point P has coordinates (x, y) relative to a given choice of axes. Another pair of axes is chosen parallel to the first pair with its origin at the point (h, k). Call the second pair of axes the $x'y'$ **axes**. (See Fig. F.15.) Inspection of Fig. F.15 shows that

$$x' = x - h \quad \text{and} \quad y' = y - k, \tag{1}$$

or equivalently,
$$x = x' + h \quad \text{and} \quad y = y' + k. \tag{2}$$

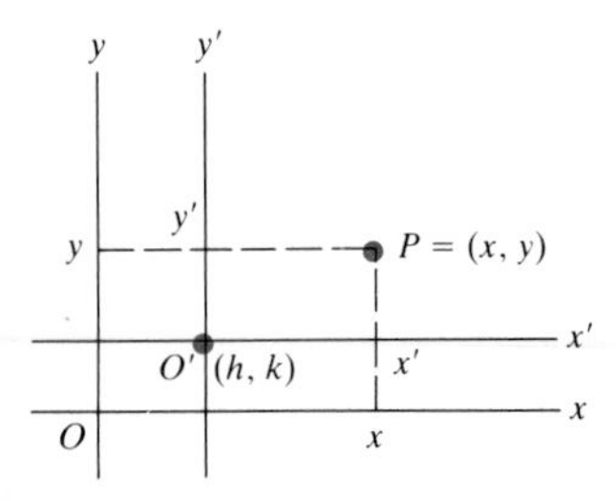

Figure F.15

The coordinates change by fixed amounts, h and k. This observation is applied in the following three examples.

EXAMPLE 1 Find an equation of the parabola whose focus is at (3, 7) and whose directrix is the line $y = 1$, as shown in Fig. F.16.

SOLUTION Introduce an $x'y'$-coordinate system whose origin is at the

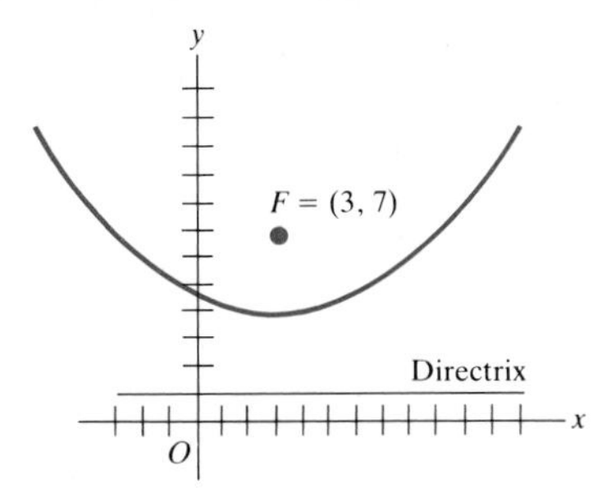

Figure F.16

point (3, 4), which is midway between the focus and the directrix, as shown in Fig. F.17.

Relative to the $x'y'$ axes the parabola is in standard position. Since the distance from the focus to the directrix is $c = 6$, the parabola has an equation $(x')^2 = 2 \cdot 6y'$, that is, $(x')^2 = 12y'$ relative to the $x'y'$ axes. By Eq. (1),

$$(x - 3)^2 = 12(y - 4). \tag{3}$$

Equation (3) describes the parabola relative to the xy axes. It could be rewritten as

$$y = \frac{x^2 - 6x + 57}{12}. \quad \blacksquare$$

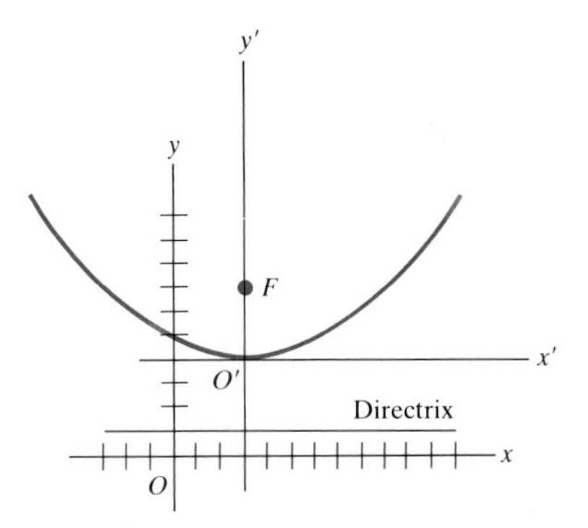

Figure F.17

EXAMPLE 2 Graph the equation $x^2 - 2x + y^2 - 6y - 15 = 0$.

SOLUTION Complete the square in order to find axes relative to which the equation of the curve is simpler. We are looking for h and k, the coordinates of the "new" origin. The details run like this:

$$x^2 - 2x + y^2 - 6y - 15 = 0, \tag{4}$$

$$(x^2 - 2x \qquad) + (y^2 - 6y \qquad) = 15,$$

$$(x^2 - 2x + 1) + (y^2 - 6y + 9) = 15 + 1 + 9,$$

$$(x - 1)^2 + (y - 3)^2 = 25. \tag{5}$$

Equation (5) suggests that we introduce $x'y'$ axes with origin at the point (1, 3). In this case, $h = 1$, $k = 3$, and $x' = x - 1$ and $y' = y - 3$. For (x', y') on the graph, x' and y' satisfy the equation

$$(x')^2 + (y')^2 = 25. \tag{6}$$

The graph of Eq. (6) is a circle of radius 5 centered at the origin of the $x'y'$-coordinate system. This is also the graph of Eq. (4). It is shown in Fig. F.18.

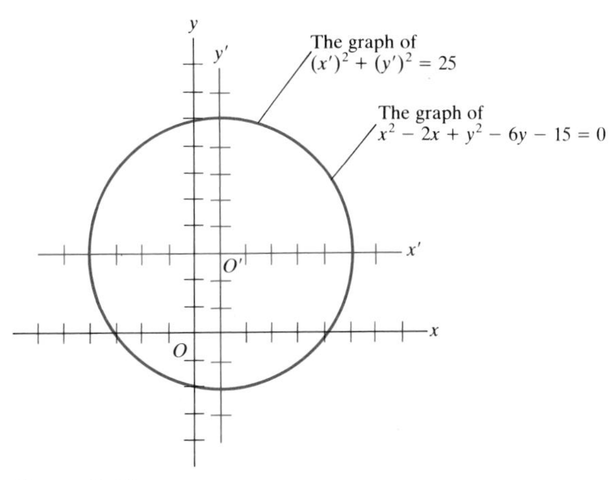

Figure F.18

■

The method of completing the square illustrated by Example 2 works equally well for any equation which has the general form $Ax^2 + Cy^2 + Dx + Ey + F = 0$. It can be shown that the graph is a conic section (perhaps empty or one or two intersecting lines) or a pair of parallel lines.

EXAMPLE 3 Show that the graph of $x^2 + 4y^2 - 6x + 8y + 9 = 0$ is a conic section and graph it.

SOLUTION Begin by completing the square, much as in Example 2:

$$x^2 + 4y^2 - 6x + 8y + 9 = 0, \tag{7}$$

$$(x^2 - 6x \quad) + (4y^2 + 8y \quad) = -9,$$

$$(x^2 - 6x \quad) + 4(y^2 + 2y \quad) = -9,$$

$$(x^2 - 6x + 9) + 4(y^2 + 2y + 1) = -9 + 9 + 4,$$

$$(x - 3)^2 + 4(y + 1)^2 = 4,$$

$$\frac{(x-3)^2}{4} + \frac{(y+1)^2}{1} = 1. \tag{8}$$

Equation (8) suggests that we choose $x'y'$ axes with origin at $(3, -1)$. The graph then has the equation

$$\frac{(x')^2}{4} + \frac{(y')^2}{1} = 1 \tag{9}$$

relative to the $x'y'$ axes. Equation (9) describes an ellipse in standard position with $a^2 = 4$ and $b^2 = 1$. It is graphed in Fig. F.19.

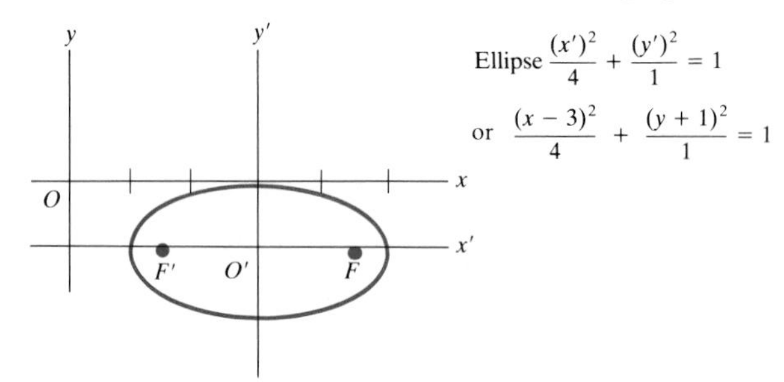

Figure F.19

Since $4 > 1$, the foci are on the x' axis. To find them, solve for c in the equation $c^2 = a^2 - b^2$, that is, $c^2 = 4 - 1 = 3$, so $c = \sqrt{3}$. The foci F and F' are at $(\sqrt{3}, 0)$ and $(-\sqrt{3}, 0)$ relative to the $x'y'$ axes. Since $x = x' + 3$ and $y = y' - 1$, the foci are at $(3 - \sqrt{3}, -1)$ and $(3 + \sqrt{3}, -1)$ relative to the xy axes. ■

EXERCISES FOR SEC. F.2: TRANSLATION OF AXES AND THE GRAPH OF $Ax^2 + Cy^2 + Dx + Ey + F = 0$

Using a suitable translation of axes, graph the equations in Exercises 1 to 14 relative to the xy axes. Identify the type of conic and give the foci, vertices, and asymptotes, as appropriate.

1 $y = (x - 2)^2$

2 $y - 1 = 4(x - 1)^2$

3 $y + 1 = 2(x - 3)^2$

4 $y = x^2 + 6x + 9$

5 $y = 3x^2 + 12x + 13$

6 $y = 2x^2 - 12x + 20$

7 $x^2 + y^2 - 2x - 4y + 4 = 0$

8 $x^2 + y^2 + 6x - 7 = 0$

9 $x^2 - y^2 - 4x + 4y - 1 = 0$

10 $9x^2 - 4y^2 - 18x - 27 = 0$

11 $-4x^2 + 9y^2 + 24x + 36y - 36 = 0$

12 $4x^2 + y^2 - 16x + 12 = 0$

13 $25x^2 + 4y^2 + 100x + 24y + 36 = 0$

14 $x^2 + 4y^2 - 2x - 16y + 21 = 0$

In Exercises 15 to 22 a conic is described relative to xy axes. Choose $x'y'$ axes such that the equation of the conic is as simple as possible and give the equation in the x' and y' coordinates.

15 The ellipse that has vertices (1, 0), (4, 2), (1, 4), and (−2, 2)

16 The ellipse that has one focus at (1, 0) and vertices at (−1, 3) and (3, 3)

17 The hyperbola that has asymptotes $y = \frac{2}{3}x + \frac{1}{3}$ and $y = -\frac{2}{3}x + \frac{5}{3}$ and one focus at (3, 1)

18 The hyperbola that has vertices (1, 3) and (1, 1) and one focus at (1, 4)

19 The ellipse whose axes are parallel to the xy axes and which is inscribed in the rectangle defined by $x = 7$, $x = 1$, $y = -2$, and $y = 2$

20 The ellipse with foci (2, 1) and (8, 1) with constant sum of distances (string length) 10

21 The parabola with focus (7, 3) and directrix $x = 1$

22 The parabola with focus (5, 1) and directrix $x = 3$

■

23 This exercise concerns the graph of $Ax^2 + Cy^2 + F = 0$ for nonzero constants A, C, and F.
(*a*) Show that if A and C are positive and F is negative, then the graph is an ellipse.
(*b*) Show that if A, C, and F are positive, then the graph is empty.
(*c*) Show that if A and C have opposite signs, then the graph is a hyperbola.
(*d*) When is the graph a circle?

24 This exercise concerns the graph of the equation $Ax^2 + Cy^2 + Dx + Ey + F = 0$, where A, C, and F are nonzero constants and D and E are constants which may be 0.
(*a*) Show that if A and C have the same sign, then the graph is an ellipse or else is empty.
(*b*) Show that if A and C have opposite signs, then the graph is a hyperbola.
(*c*) When is the graph a circle?

25 Show that if A is 0 and C and D are not 0, then the graph of $Ax^2 + Cy^2 + Dx + Ey + F = 0$ is a parabola.

26 Show that the graph of $y = ax^2 + bx + c$, $a \neq 0$, is a parabola by finding the equation of the graph relative to $x'y'$ axes whose origin is at the point $(-b/2a, c - b^2/4a)$.

F.3 ROTATION OF AXES AND THE GRAPH OF $Ax^2 + Bxy + Cy^2 + Dx + Ey + F = 0$

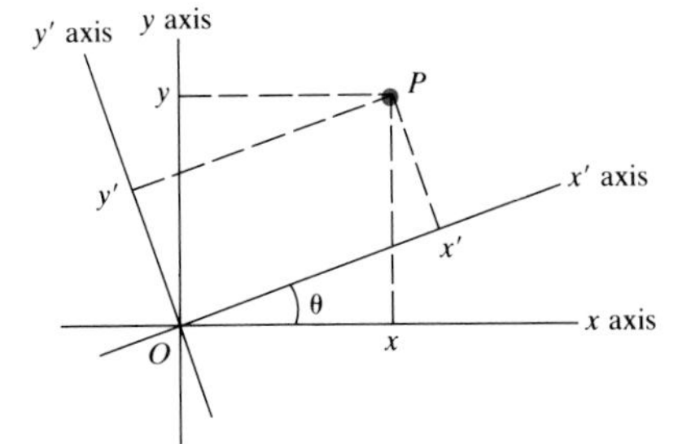

Figure F.20

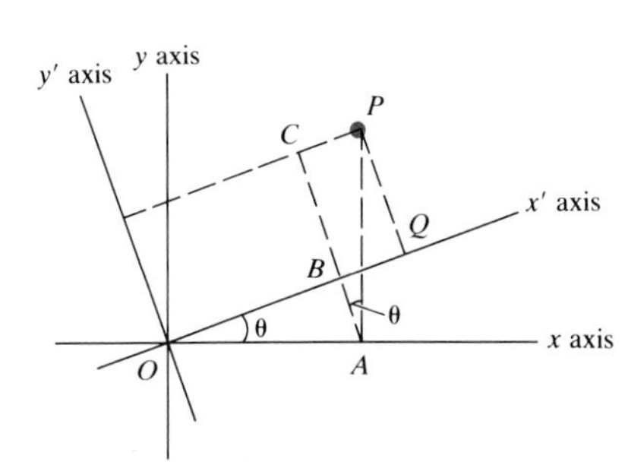

Figure F.21

This section examines the graph of $Ax^2 + Bxy + Cy^2 + Dx + Ey + F = 0$, which is a conic (including degenerate cases when it is a line, a pair of lines, a point, or empty). The key idea is to remove the term Bxy by choosing $x'y'$ axes that are tilted with respect to the xy axes.

A point P has coordinates (x, y) relative to a given choice of axes. Another pair of axes is chosen with the same origin but rotated by an angle θ. Call these tilted axes the $x'y'$ axes. How are x' and y', the coordinates of P in the $x'y'$ axes, related to x and y? (See Fig. F.20.)

To determine the relation, introduce the line shown in Fig. F.21. By Fig. F.21,

$$x' = \overline{OB} + \overline{BQ} = \overline{OB} + \overline{CP} = \overline{OA}\cos\theta + \overline{AP}\sin\theta = x\cos\theta + y\sin\theta$$

and $$y' = \overline{AC} - \overline{AB} = \overline{AP}\cos\theta - \overline{OA}\sin\theta = y\cos\theta - x\sin\theta.$$

Thus $$\begin{cases} x' = x\cos\theta + y\sin\theta \\ y' = -x\sin\theta + y\cos\theta. \end{cases} \qquad (1)$$

Just as the $x'y'$ axes are obtained from the xy axes through a rotation by the angle θ, the xy axes are obtained from the $x'y'$ axes by the rotation by the angle $-\theta$. In view of (1), then,

$$\begin{cases} x = x' \cos(-\theta) + y' \sin(-\theta) \\ y = -x' \sin(-\theta) + y' \cos(-\theta), \end{cases}$$

which reduces to

$$\begin{cases} x = x' \cos\theta - y' \sin\theta \\ y = x' \sin\theta + y' \cos\theta. \end{cases} \tag{2}$$

(These are the key equations of this section.)

EXAMPLE 1 Find an equation of the graph of $xy = 1$ relative to the $x'y'$ axes obtained by rotating the xy axes $\pi/4$ radians ($= 45°$). (The equation $xy = 1$ is graphed in Example 8 of Sec. B.1.)

SOLUTION Use Eq. (2) with $\theta = \pi/4$ radians. We have

$$\begin{cases} x = x' \cos\dfrac{\pi}{4} - y' \sin\dfrac{\pi}{4} \\ y = x' \sin\dfrac{\pi}{4} + y' \cos\dfrac{\pi}{4}, \end{cases}$$

or

$$\begin{cases} x = \dfrac{\sqrt{2}}{2}(x' - y') \\ y = \dfrac{\sqrt{2}}{2}(x' + y'). \end{cases} \tag{3}$$

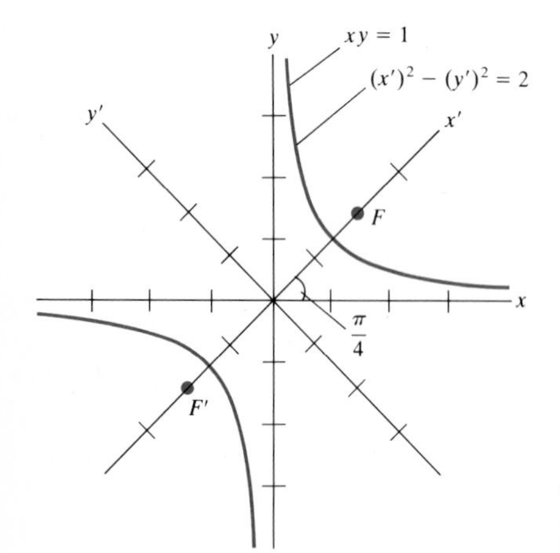

Figure F.22

Substitution of Eqs. (3) into the equation $xy = 1$ yields

$$\frac{\sqrt{2}}{2}(x' - y')\frac{\sqrt{2}}{2}(x' + y') = 1,$$

$$\tfrac{1}{2}[(x')^2 - (y')^2] = 1, \qquad \text{multiplying out}$$

$$(x')^2 - (y')^2 = 2, \qquad \text{clearing denominator} \tag{4}$$

which is the equation of a hyperbola with foci on the x' axis.

To graph Eq. (4), rewrite it as

$$\frac{(x')^2}{(\sqrt{2})^2} - \frac{(y')^2}{(\sqrt{2})^2} = 1. \tag{5}$$

This is the hyperbola $(x')^2/a^2 - (y')^2/b^2 = 1$ with $a = \sqrt{2}$ and $b = \sqrt{2}$. Its foci are at $(x', y') = (c, 0)$ and $(-c, 0)$, where $c = \sqrt{a^2 + b^2} = \sqrt{2 + 2} = 2$. The hyperbola is sketched in Fig. F.22. In the xy system, the foci are at $(\sqrt{2}, \sqrt{2})$ and $(-\sqrt{2}, -\sqrt{2})$. ■

In Example 1 rotation of the axes by $\pi/4$ radians resulted in an equation with no $x'y'$ term. [In other words, the coefficient of the product $x'y'$ is 0 in (4).] Less precisely, we "got rid of the xy term." It turns out that for any equation of the form $Ax^2 + Bxy + Cy^2 + Dx + Ey + F = 0$, it is always possible to rotate the axes in such a way that the $x'y'$ term disappears.

We sketch the algebra, leaving the details to be filled in by the reader. Start with the equation

$$Ax^2 + Bxy + Cy^2 + Dx + Ey + F = 0, \tag{6}$$

and make the substitution (2). After some multiplications and collections, we obtain an equation in x' and y' of the form

$$A'(x')^2 + B'x'y' + C'(y')^2 + D'x' + E'y' + F' = 0 \tag{7}$$

for certain constants A', B', C', D', E', and F'.

At this point only the coefficient of $x'y'$, namely, B', draws our interest, since we want to find θ that makes B' equal to 0. The formula for B' is

$$B' = 2(C - A) \sin \theta \cos \theta + B(\cos^2 \theta - \sin^2 \theta), \tag{8}$$

which simplifies to

$$B' = (C - A) \sin 2\theta + B \cos 2\theta.$$

To make $B' = 0$, find θ such that

$$(C - A) \sin 2\theta + B \cos 2\theta = 0.$$

Thus

$$\frac{\sin 2\theta}{\cos 2\theta} = \frac{B}{A - C}$$

or

$$\tan 2\theta = \frac{B}{A - C}. \tag{9}$$

For θ in $(0, \pi/4)$, $\tan 2\theta$ sweeps through all positive numbers. For θ in $(\pi/4, \pi/2)$, $\tan 2\theta$ sweeps through all negative numbers. If $C = A$, use $\theta = \pi/4$; by Eq. (8), B' then is 0. Thus it is always possible to find a first-quadrant angle θ such that Eq. (9) holds. For that θ, B' is 0 and Eq. (7) takes the form

$$A'(x')^2 + C'(y')^2 + D'x' + E'y' + F' = 0. \tag{10}$$

This type of equation was discussed in the preceding section. Its graph is a conic section except in certain degenerate cases. Thus the same holds for the graph of Eq. (6).

Getting Rid of the xy Term. We have seen that there is an angle θ such that, relative to $x'y'$ axes inclined at an angle θ to the xy axes, Eq. (7) takes the form of Eq. (10). First, find θ such that $\tan 2\theta = B/(A - C)$. Then, find $\cos \theta$ and $\sin \theta$ and use the substitution (2). The steps are

1 Find $\tan 2\theta$.
2 Find $\cos 2\theta$.
3 Find $\sin \theta = \sqrt{\dfrac{1 - \cos 2\theta}{2}}$ and $\cos \theta = \sqrt{\dfrac{1 + \cos 2\theta}{2}}$.

Since θ is in the first quadrant, the positive square roots are used.

Example 2 illustrates the technique.

EXAMPLE 2 Choose $x'y'$ axes so that the equation for the graph of

$$-7x^2 + 48xy + 7y^2 - 25 = 0 \tag{11}$$

has no $x'y'$ term. Then graph the curve.

SOLUTION In this case, $A = -7$, $B = 48$, $C = 7$. Hence θ must be

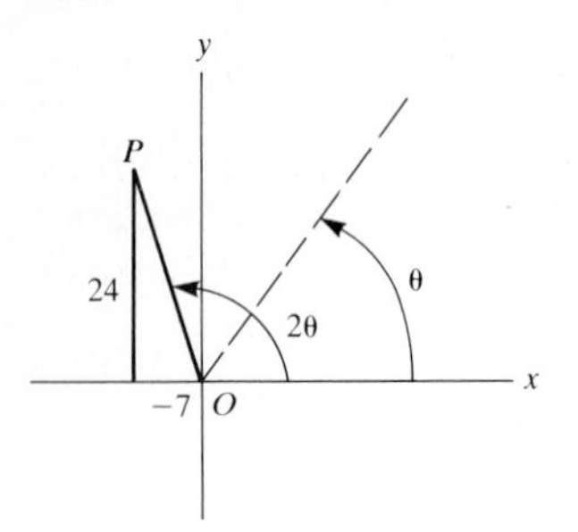

Figure F.23

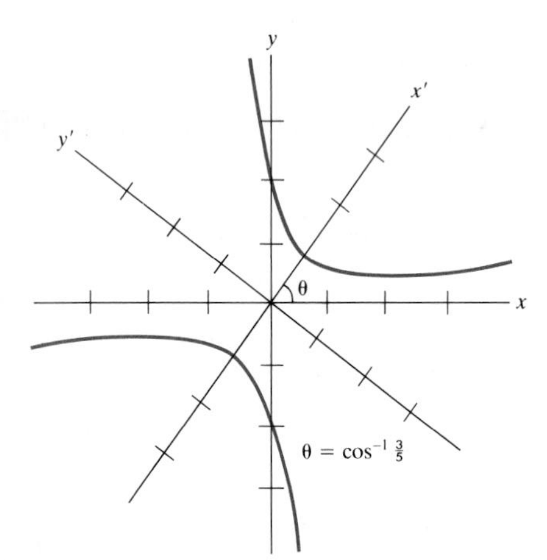

Figure F.24

chosen so that

$$\tan 2\theta = \frac{48}{(-7) - 7} = \frac{48}{-14} = \frac{24}{-7}. \tag{12}$$

Figure F.23 shows 2θ and θ. By the pythagorean theorem, $\overline{OP}$ in Fig. F.23 is $\sqrt{(24)^2 + (-7)^2} = 25$. Thus $\cos 2\theta = (-7)/25$. Consequently,

$$\sin\theta = \sqrt{\frac{1 - \left(\frac{-7}{25}\right)}{2}} = \frac{4}{5},$$

and

$$\cos\theta = \sqrt{\frac{1 + \left(\frac{-7}{25}\right)}{2}} = \frac{3}{5}.$$

Thus (2) becomes

$$\begin{cases} x = \frac{3}{5}x' - \frac{4}{5}y' \\ y = \frac{4}{5}x' + \frac{3}{5}y'. \end{cases} \tag{13}$$

Substitution of (13) into Eq. (11) gives

$$-7(\tfrac{3}{5}x' - \tfrac{4}{5}y')^2 + 48(\tfrac{3}{5}x' - \tfrac{4}{5}y')(\tfrac{4}{5}x' + \tfrac{3}{5}y') + 7(\tfrac{4}{5}x' + \tfrac{3}{5}y')^2 - 25 = 0,$$

which simplifies to

$$625(x')^2 - 625(y')^2 = 625,$$

or

$$(x')^2 - (y')^2 = 1,$$

which describes a hyperbola in standard position relative to the $x'y'$ axes. Its foci are on the x' axis. It is sketched in Fig. F.24. ■

The reasoning in this section and the one preceding shows that if the graph of (6) is not degenerate, it is a conic. *To determine what type of conic it is, compute the discriminant,* $\mathcal{D} = B^2 - 4AC$. If it is negative, the conic is an ellipse; if it is positive, the conic is a hyperbola; if it is zero, the conic is a parabola. Remember this with the nonsense word "neph" (*n*egative, *e*llipse, *p*ositive, *h*yperbola). Exercises 18 to 20 explain why this test works. For emphasis, we state this test formally: To determine what type of conic the equation

Quick way to identify curve: neph

$$Ax^2 + Bxy + Cy^2 + Dx + Ey + F = 0$$

describes, compute $\mathcal{D} = B^2 - 4AC$. When $\mathcal{D} < 0$, the conic is an ellipse; when $\mathcal{D} > 0$, the conic is a hyperbola. If $\mathcal{D} = 0$, the conic is a parabola.

There are also degenerate cases.

EXERCISES FOR SEC. F.3: ROTATION OF AXES AND THE GRAPH OF $Ax^2 + Bxy + Cy^2 + Dx + Ey + F = 0$

Graph each of the curves in Exercises 1 to 8 using the technique described in this section. In the case of hyperbolas, draw the asymptotes too.

1 $x^2 - 4xy - 2y^2 - 6 = 0$

2 $41x^2 + 24xy + 34y^2 - 25 = 0$

3 $x^2 + xy + y^2 - 12 = 0$

4 $5x^2 + 6xy + 5y^2 - 8 = 0$

5 $23x^2 + 26\sqrt{3}\,xy - 3y^2 - 144 = 0$

6 $3x^2 + 2\sqrt{3}xy + y^2 + 2x + 2\sqrt{3}y = 0$
7 $6x^2 - 12xy + 6y^2 - \sqrt{2}x + \sqrt{2}y = 0$
8 $7x^2 - 48xy - 7y^2 - 25 = 0$

In Exercises 9 to 12 graph the equations.

9 $y = \dfrac{x^2 + 3}{x - 1}$

10 $y = \dfrac{x + 1}{x + 3}$

11 $xy + x + y + 1 = 0$

12 $y = \dfrac{x^2 + 2x + 1}{x - 1}$

In each of Exercises 13 to 16 use the discriminant $\mathcal{D} = B^2 - 4AC$ to determine what type of conic the equation describes.

13 $-x^2 + 24xy + 6y^2 - 10x + 13y + 5 = 0$
14 $x^2 + xy + y^2 + 3x + 2y = 0$
15 $x^2 - 2xy + y^2 + x + 3y + 5 = 0$
16 $3x^2 - xy - y^2 = 1$
17 For each of the following equations the graph is degenerate; it is either one or two lines, a point, or empty. Graph each of them.
(*a*) $x^2 - y^2 = 0$ (*b*) $x^2 + 2xy + y^2 = 0$
(*c*) $3x^2 + 4y^2 = 0$
(*d*) $3x^2 + 2xy + 3y^2 + 1 = 0$

Exercises 18 to 20 are related.

18 Show that A', B', and C' in (7) are given by

$$A' = A\cos^2\theta + B\cos\theta\sin\theta + C\sin^2\theta,$$
$$B' = 2(C - A)\sin\theta\cos\theta + B(\cos^2\theta - \sin^2\theta),$$
$$C' = A\sin^2\theta - B\cos\theta\sin\theta + C\cos^2\theta.$$

19 Use Exercise 18 to show that $(B')^2 - 4A'C' = B^2 - 4AC$.
20 Assume that $x'y'$ axes have been chosen to make $B' = 0$. Thus the graph of (6) is given by the equation $A'(x')^2 + C'(y')^2 + D'x' + E'y' + F' = 0$. Assume this graph is a conic (that is, is not degenerate).
(*a*) Show that if $(B')^2 - 4A'C'$ is negative, the graph is an ellipse. (Keep in mind that $B' = 0$.)
(*b*) Show that if $(B')^2 - 4A'C'$ is positive, the graph is a hyperbola.
(*c*) Show that if $(B')^2 - 4A'C'$ is 0, the graph is a parabola.
(*d*) Combining these facts with Exercise 19, explain the "neph" memory device.
21 Show that if the graph of $Ax^2 + Bxy + Cy^2 + Dx + Ey + F = 0$ is a circle, then $B = 0$ and $A = C$.
22 Sketch the graphs of
(*a*) $(3x - y + 1)^2 = 0$
(*b*) $(x + y + 1)(x - y - 2) = 0$
(*c*) $x^2 - y^2 = 0$
[Note that (*a*) and (*b*), when multiplied out, produce equations of the form $Ax^2 + Bxy + Cy^2 + Dx + Ey + F = 0$. *Hint:* When is the product of two numbers equal to 0?]

In Exercises 23 to 25 give the coordinates of the foci relative to the $x'y'$ axes and their relation to the xy axes for the given conic.

23 The conic of Example 2
24 The conic of Exercise 1
25 The conic of Exercise 3
26 The conic of Exercise 6

F.4 CONIC SECTIONS IN POLAR COORDINATES

This section depends on Sec. 9.1 and is used in Sec. 14.8.

For the study of the conic sections in terms of polar coordinates, it is convenient to use definitions that depend on the ratios of distances rather than on their sums or differences. (Note that the definition of the parabola involves essentially the ratio of two distances being equal to 1.)

Consider the ellipse whose foci are at $F = (c, 0)$ and $F' = (-c, 0)$ and whose "length of string" is $2a$. In the algebraic treatment of the ellipse in Sec. F.1, the equation

$$a^2 - cx = a\sqrt{(x - c)^2 + y^2}$$

appears.

Some algebraic manipulations of this equation will show that this ellipse can be defined in terms of the focus F, a line, and a fixed ratio of the distances from P to F and from P to the line.

First, observe that this equation asserts that, for $P = (x, y)$,

$$a^2 - cx = a\overline{PF},$$

or, equivalently,

$$\overline{PF} = a - \frac{c}{a}x.$$

Denote the quotient c/a, which is less than 1, by e. The number e is called the **eccentricity** of the ellipse. (When $e = 0$, the ellipse is a circle.) Thus

$$\overline{PF} = a - ex = e\left(\frac{a}{e} - x\right),$$

an equation which is meaningful if the ellipse is not a circle. Now, $(a/e) - x$ is the distance from P to the vertical line through $(a/e, 0)$, a line which will be denoted L. Letting $Q = (a/e, y)$, the point on L and on the horizontal line through P, we have

$$\overline{PF} = e\overline{PQ}.$$

(See Fig. F.25.)

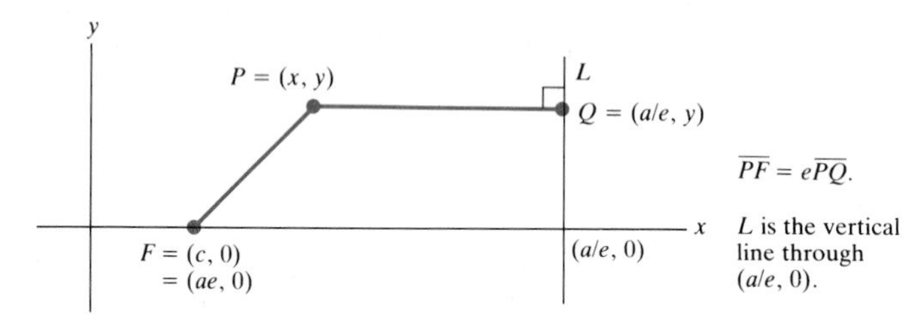

$\overline{PF} = e\overline{PQ}$. L is the vertical line through $(a/e, 0)$.

Figure F.25

In other words, *the ratio $\overline{PF}/\overline{PQ}$ has a constant value, e,* less than 1. Thus the ellipse, like the parabola, can be defined in terms of a point F and a line L.

EXAMPLE 1 Find the eccentricity and draw the line L for the ellipse $\frac{x^2}{25} + \frac{y^2}{9} = 1$.

SOLUTION In this ellipse, $a = 5$ and $b = 3$. Thus $c = \sqrt{a^2 - b^2} = \sqrt{25 - 9} = 4$. Consequently, $e = c/a = \frac{4}{5}$. The line L has the equation $x = a/e$, or $x = 5/\frac{4}{5}$; hence

$$x = \tfrac{25}{4} = 6.25.$$

The ellipse is sketched in Fig. F.26. Note that for each point P on the ellipse,

$$\frac{\overline{PF}}{\overline{PQ}} = \frac{4}{5}. \quad ■$$

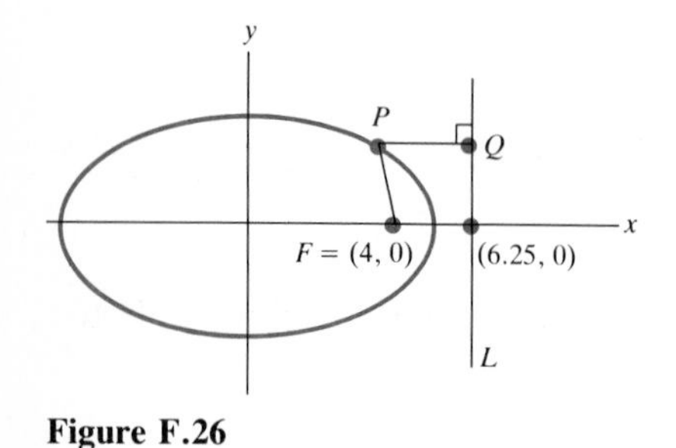

Figure F.26

The hyperbola can be treated in a similar manner. The main difference is that the eccentricity of a hyperbola, again defined as c/a, is greater than 1. With this background, we now describe the approach to the conic sections in terms of the ratios of certain distances.

Definition *Conic section.* Let L be a line in the plane and let F be a point in the plane but not on the line. Let e be a positive number. A point P in the plane is on the **conic section** determined by F, L, and e if and only if

$$\frac{\text{Distance from } P \text{ to } F}{\text{Distance from } P \text{ to } L} = e.$$

When $e = 1$, the conic section is a parabola (this is the definition of the parabola used in the preceding section). When $e < 1$, it is an ellipse. When $e > 1$, it is a hyperbola. The point F is called a **focus**; the line L is called the **directrix**.

To obtain the simplest description of the conic sections in polar coordinates, place the pole at the focus F. Let the polar axis make an angle B with a line perpendicular to the directrix. Figure F.27 shows a typical point $P = (r, \theta)$ on the conic section, as well as the point Q on the directrix nearest P. Let the distance from F to the directrix be p. Then $\overline{PF} = r$, $\overline{PQ} = p - r\cos(\theta - B)$, and $\overline{PF}/\overline{PQ} = e$. Thus

$$\frac{r}{p - r\cos(\theta - B)} = e. \tag{1}$$

Solving Eq. (1) for r yields *the equation of a conic section in polar coordinates,*

$$r = \frac{ep}{1 + e\cos(\theta - B)}. \tag{2}$$

Usually B is chosen to be 0 or π.

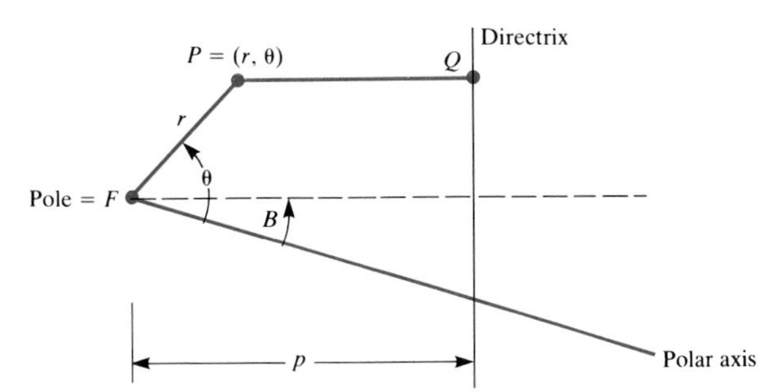

Figure F.27

EXAMPLE 2 Show that the graph of the equation $r = \dfrac{8}{5 + 6\cos\theta}$ is a conic section.

SOLUTION This can be put in the form of Eq. (2) by dividing both numerator and denominator by 5:

$$r = \frac{\frac{8}{5}}{1 + \frac{6}{5}\cos\theta} = \frac{(\frac{6}{5})(\frac{8}{6})}{1 + \frac{6}{5}\cos\theta}.$$

Hence the graph is a conic section for which $p = \frac{8}{6}$ and $e = \frac{6}{5}$. It is a hyperbola, since $e > 1$. ■

EXERCISES FOR SEC. F.4: CONIC SECTIONS IN POLAR COORDINATES

1 (*a*) On the graph of $r = 10/(3 + 2\cos\theta)$ sketch the four points corresponding to $\theta = 0, \pi/2, \pi, 3\pi/2$.
(*b*) Using Eq. (2), show that the curve in (*a*) is an ellipse.

2 Obtain Eq. (2) from Eq. (1).

3 Find the eccentricities of these conics:
(*a*) $r = 5/(3 + 4\cos\theta)$ (*b*) $r = 5/(4 + 3\cos\theta)$
(*c*) $r = 5/(3 + 3\cos\theta)$ (*d*) $r = 5/(3 - 4\cos\theta)$

4 (*a*) Show that $r = 8/(1 - \frac{1}{2}\cos\theta)$ is the equation of an ellipse. *Hint:* Set $B = \pi$ in Eq. (2).
(*b*) Graph the ellipse and its foci.
(*c*) Find a, where $2a$ is the fixed sum of the distances from points on the ellipse to the foci.

5 In rectangular coordinates the focus of a certain parabola is at $(-1, 0)$ and its directrix is the line $x = 1$.
(*a*) Show that its equation is $y^2 = -4x$.
(*b*) Find the equation of the parabola relative to a polar coordinate system whose pole is at F and whose polar axis contains the positive x axis.
(*c*) Find the equation of the parabola relative to a polar coordinate system whose pole is at the origin of the rectangular coordinate system and whose polar axis coincides with the positive x axis.

6 Assume that an ellipse has the equation $x^2/a^2 + y^2/b^2 = 1$ relative to a rectangular coordinate system. Place a polar coordinate system so that the polar axis coincides with the positive x axis (the pole thus being at the center of the ellipse, *not at a focus*). Show that the polar equation of the ellipse is (the relatively complicated)

$$r^2 = \frac{a^2b^2}{b^2\cos^2\theta + a^2\sin^2\theta}.$$

Appendix G
Theory of Limits

In Sec. 2.8 a precise definition of $\lim_{x\to a} f(x)$ was given. This definition is the basis of this appendix, where the fundamental properties of limits will be obtained. In particular, it will be proved that any polynomial is continuous.

G.1 PROOFS OF SOME THEOREMS ABOUT LIMITS

With the aid of the ϵ,δ definition of limit in Sec. 2.8 it can be established that if

$$\lim_{x\to a} f(x) = A \quad \text{and} \quad \lim_{x\to a} g(x) = B,$$

then $$\lim_{x\to a} [f(x) + g(x)] = A + B, \qquad \lim_{x\to a} [f(x)g(x)] = AB,$$

and $$\lim_{x\to a} \frac{f(x)}{g(x)} = \frac{A}{B} \qquad \text{(if } B \text{ is not 0).}$$

The first two assertions will be proved in this section.

Theorem 1 If $\lim_{x\to a} f(x) = A$ and $\lim_{x\to a} g(x) = B$, then

$$\lim_{x\to a} [f(x) + g(x)] = A + B.$$

Proof It must be shown that given any $\epsilon > 0$, no matter how small, there exists a number $\delta > 0$ depending on ϵ such that

$$|[f(x) + g(x)] - (A + B)| < \epsilon \tag{1}$$

whenever $|x - a| < \delta$ and $x \neq a$.

Rewrite $[f(x) + g(x)] - (A + B)$ as $[f(x) - A] + [g(x) - B]$, which is a sum of two quantities known to be small when x is near a. Since the absolute value of the sum of two numbers is not larger than the sum of their absolute values,

$$|[f(x) - A] + [g(x) - B]| \le |f(x) - A| + |g(x) - B|. \tag{2}$$

Since $\lim_{x\to a} f(x) = A$, there is a positive number δ_1 such that

$$|f(x) - A| < \frac{\epsilon}{2}$$

when $|x - a| < \delta_1$ and $x \ne a$. (Why we pick $\epsilon/2$ rather than ϵ will be clear in a moment.) Similarly, there is a positive number δ_2 such that

$$|g(x) - B| < \frac{\epsilon}{2}$$

when $|x - a| < \delta_2$ and $x \ne a$.

Now let δ be the smaller of δ_1 and δ_2. For any x (not equal to a) such that $|x - a| < \delta$, we have both

$$|x - a| < \delta_1 \qquad \text{and} \qquad |x - a| < \delta_2$$

and, therefore, simultaneously,

$$|f(x) - A| < \frac{\epsilon}{2} \qquad \text{and} \qquad |g(x) - B| < \frac{\epsilon}{2}. \tag{3}$$

Combining (2) and (3), we conclude that when $|x - a| < \delta$ but $x \ne a$,

$$|[f(x) + g(x)] - (A + B)| < \frac{\epsilon}{2} + \frac{\epsilon}{2} = \epsilon.$$

Thus for each $\epsilon > 0$ there exists a suitable $\delta > 0$, depending, of course, on ϵ, f, and g. This ends the proof. ■

Theorem 1 is the basis of the proof of the next theorem.

Theorem 2 The sum of two functions that are continuous at a is itself continuous at a.

Proof Let f and g be continuous at a. Let h be their sum; that is, let $h(x) = f(x) + g(x)$. We wish to show that h is continuous at a.

In view of the definition of continuity given in Sec. 2.5, it must be shown that $h(a)$ is defined and that $\lim_{x\to a} h(x) = h(a)$.

Since f and g are defined at a, so is h, and $h(a) = f(a) + g(a)$. All that remains is to show that $\lim_{x\to a} [f(x) + g(x)] = f(a) + g(a)$. By Theorem 1, $\lim_{x\to a} [f(x) + g(x)] = \lim_{x\to a} f(x) + \lim_{x\to a} g(x)$. Since f and g are continuous at a, $\lim_{x\to a} f(x) = f(a)$ and $\lim_{x\to a} g(x) = g(a)$. This concludes the proof. ■

Theorem 3 If $\lim_{x\to a} f(x) = A$ and $\lim_{x\to a} g(x) = B$, then

$$\lim_{x\to a} f(x)g(x) = AB.$$

Plan of proof: We know that $|f(x) - A|$ and $|g(x) - B|$ are small when x is near a. We wish to conclude that $|f(x)g(x) - AB|$ is small when x is near a. The algebraic identity

$$f(x)g(x) - AB = f(x)[g(x) - B] + B[f(x) - A] \tag{4}$$

will be of use. From Eq. (4) and properties of the absolute value, it follows that

$$|f(x)g(x) - AB| \le |f(x)||g(x) - B| + |B||f(x) - A|. \tag{5}$$

Now $|B|$ is fixed, and $|f(x) - A|$ and $|g(x) - B|$ are small when x is near a. The real problem is to control $|f(x)|$. Watch carefully the way in which $|f(x)|$ is treated in the proof.

Proof Consider the case $B \neq 0$. Let $\epsilon > 0$ be given. We wish to show that there is a number $\delta > 0$ such that $|f(x)g(x) - AB| < \epsilon$ when $|x - a| < \delta$ but $x \neq a$. Observe that

$$\begin{aligned} |f(x)g(x) - AB| &= |f(x)[g(x) - B] + B[f(x) - A]| \\ &\le |f(x)||g(x) - B| + |B||f(x) - A|. \end{aligned} \tag{6}$$

Since $\lim_{x \to a} f(x) = A$, there is a number $\delta_1 > 0$ such that

$$|f(x) - A| < \frac{\epsilon}{2|B|}$$

when $|x - a| < \delta_1$ but $x \neq a$. [It will be clear in a moment why we want $|f(x) - A|$ to be less than $\epsilon/(2|B|)$.] Thus the second summand on the right side of (6) is less than

$$\frac{|B|\epsilon}{2|B|} = \frac{\epsilon}{2}.$$

For x such that $|x - a| < \delta_1$, $|f(x)|$ does not become arbitrarily large, since $|f(x) - A| < \epsilon/(2|B|)$. Indeed, for such values of x

$$|f(x)| = |A + [f(x) - A]| \le |A| + |f(x) - A| < |A| + \frac{\epsilon}{2|B|}.$$

Letting $C = |A| + \epsilon/(2|B|)$, we have $|f(x)| < C$ when $|x - a| < \delta_1$ but $x \neq a$. [This controls the size of $|f(x)|$.]

Since $\lim_{x \to a} g(x) = B$, there is a $\delta_2 > 0$ such that $|g(x) - B| < \epsilon/(2C)$ when $|x - a| < \delta_2$ but $x \neq a$.

Now let δ be the smaller of δ_1 and δ_2. When $|x - a| < \delta$ and $x \neq a$, both $|x - a| < \delta_1$ and $|x - a| < \delta_2$ hold; hence $|f(x) - A| < \epsilon/(2|B|)$, $|f(x)| < C$, and $|g(x) - B| < \epsilon/(2C)$. Inspection of (6) then shows that for such x,

$$|f(x)g(x) - AB| < C\frac{\epsilon}{2C} + |B|\frac{\epsilon}{2|B|} = \frac{\epsilon}{2} + \frac{\epsilon}{2} = \epsilon.$$

The proof is completed. The case $B = 0$ is left to the reader as Exercise 7. ■

Theorem 4 The product of two functions that are continuous at a is itself continuous at a. ■

The proof is similar to that of Theorem 2, but depends on Theorem 3 instead of Theorem 1.

The fact that the sum and product of continuous functions are continuous is the basis of the proof that *any polynomial is continuous.* It is necessary to prove first that the function $f(x) = x$ is continuous and then that any constant function is continuous.

Theorem 5 The function f, such that $f(x) = x$, is continuous everywhere. So are the functions $x^2, x^3, x^4, \ldots$.

Proof Since $f(a) = a$, it must be shown that $|f(x) - a|$ is small whenever $|x - a|$ is sufficiently small. More precisely, for $\epsilon > 0$ we wish to exhibit $\delta > 0$ such that $|f(x) - a| < \epsilon$ whenever $|x - a| < \delta$. But $f(x) = x$; hence $|f(x) - a|$ is simply $|x - a|$. Let $\delta = \epsilon$. Thus when $|x - a| < \delta$, it follows that $|f(x) - a| < \epsilon$. This shows that the function x is continuous.

Since the function x^2 is the product of the function x and the function x, Theorem 4 implies that x^2 is continuous. Similarly, x^3 is continuous, for $x^3 = x^2x$. Mathematical induction establishes the continuity of x^n for all positive integers n. ■

Theorem 6 Any constant function is continuous everywhere.

Proof Let $f(x) = c$ for all x. For any $\epsilon > 0$, choose $\delta = 722$, a perfectly fine positive number. Now $|f(x) - f(a)| = |c - c| = 0 < \epsilon$ for any x, and hence for x such that $|x - a| < 722$. (Any $\delta > 0$ will do.) Thus f is continuous at any number a. ■

Theorem 7 Any polynomial is continuous everywhere.

Proof We illustrate the idea of the proof by showing that $6x^2 - 5x + 1$ is continuous everywhere.

By Theorem 6, the constant functions $f(x) = 6$, $g(x) = -5$, and $h(x) = 1$ are continuous everywhere. By Theorem 5, the functions x and x^2 are continuous everywhere. By Theorem 4, the functions $6x^2$ and $(-5)x$ are continuous. By Theorem 2, the function $[6x^2 + (-5)x]$ is continuous. Again by Theorem 2, the function $[6x^2 + (-5)x] + 1$ is continuous. Thus $6x^2 - 5x + 1$ is continuous. The same argument applies to any polynomial. ■

EXERCISES FOR SEC. G.1: PROOFS OF SOME THEOREMS ABOUT LIMITS

1 Prove that $f(x) = 1/x$ is continuous at any number $a \neq 0$.

2 Prove that if g is continuous at a and f is continuous at $g(a)$, then the composite function $h = f \circ g$ is continuous at a.

3 Using Exercises 1 and 2, show that the function $1/(x^2 + 1)$ is continuous.

4 Using Exercises 1 and 2, show that if $g(x)$ is continuous at a and $g(a) \neq 0$, then $1/g(x)$ is continuous at a.

5 Let $f(x)$ and $g(x)$ be continuous at a and $g(a) \neq 0$. Combining Exercise 4 with a result in this section, show that $f(x)/g(x)$ is continuous at a.

6 Let $\lim_{x \to a} g(x) = B \neq 0$. Show that $\lim_{x \to a} [1/g(x)] = 1/B$.

7 Prove Theorem 3 in the case $B = 0$.

8 What theorems assure us that $(x^3 - 1)/(x^2 - 5)$ is continuous throughout its domain?

9 Assume that f is continuous at a and $f(a)$ is positive. Prove that there is an open interval (b, c) including a such that $f(x)$ is positive for all x in that interval.

In the remaining exercises use the definitions from Sec. 2.7.

10 Prove that if $\lim_{x\to\infty} f(x) = \infty$ and $\lim_{x\to\infty} g(x) = \infty$, then $\lim_{x\to\infty} [f(x) + g(x)] = \infty$.

11 Prove that if $\lim_{x\to\infty} f(x) = \infty$ and $\lim_{x\to\infty} g(x) = A$, then $\lim_{x\to\infty} [f(x) + g(x)] = \infty$.

12 Prove that if $\lim_{x\to\infty} f(x) = A$ and $\lim_{x\to\infty} g(x) = B$, then $\lim_{x\to\infty} [f(x) + g(x)] = A + B$.

13 Prove that if $\lim_{x\to\infty} f(x) = \infty$ and $\lim_{x\to\infty} g(x) = \infty$, then $\lim_{x\to\infty} f(x)g(x) = \infty$.

14 Prove that if $\lim_{x\to\infty} f(x) = \infty$ and $\lim_{x\to\infty} g(x) = A > 0$, then $\lim_{x\to\infty} f(x)g(x) = \infty$.

15 Prove that if $\lim_{x\to\infty} f(x) = A$ and $\lim_{x\to\infty} g(x) = B$, then $\lim_{x\to\infty} f(x)g(x) = AB$.

G.2 DEFINITIONS OF OTHER LIMITS

Other formal definitions in the spirit of the ϵ, δ definition of Sec. 2.8 are given in the text. A formal definition of the limit of a sequence appears at the end of Sec. 10.1. However, the formal definition of a limit of $f(x, y)$, a function defined in the plane, was not given. The definition, similar to that for the limit of $f(x)$, will be given now.

Let $P_0 = (a, b)$ be a point in the xy plane. Let $f(P) = f(x, y)$ be defined for all points $P = (x, y)$ in the plane sufficiently near P_0, except perhaps at P_0 itself. To be precise, let $d(P_0, P)$ denote the distance between P_0 and P. Assume that there is a positive number r such that f is defined for all points P satisfying the inequality

$$0 < d(P_0, P) < r.$$

(See Fig. G.1.)

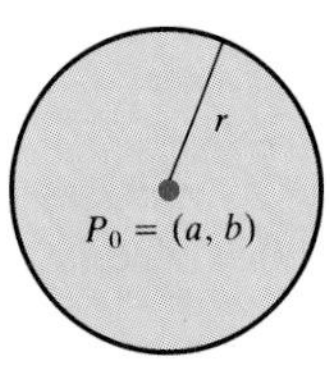

Figure G.1

This assumption on f assures us that the domain of f includes points arbitrarily near P_0.

The definition of "$\lim_{P\to P_0} f(P) = L$" can now be given. It will say that if P is close enough to P_0, then $f(P)$ is as close as we please to $f(P_0)$.

Definition $\lim_{P\to P_0} f(P) = L$. Let $P_0 = (a, b)$ be a point in the xy plane and let L be a real number. Assume that $f(P)$ is defined for all P such that $0 < d(P_0, P) < r$ for some positive real number r. Assume that for each positive number ϵ there is a positive number δ such that, for all points P in the domain of f that satisfy the inequality

$$0 < d(P_0, P) < \delta,$$

it is true that
$$|f(P) - L| < \epsilon.$$

Then we say that "the limit of $f(P)$ as P approaches P_0 is L" and write

$$\lim_{P\to P_0} f(P) = L,$$

or
$$\lim_{(x, y)\to(a, b)} f(x, y) = L.$$

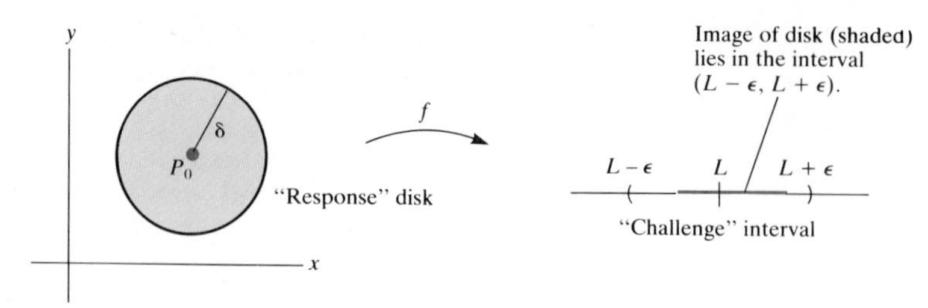

Figure G.2

Figure G.2 shows the meaning of the challenge ϵ and the response δ. No matter how small the challenge interval around L is, a disk around P_0 can be found such that, for any point P in the disk (except perhaps P_0), $f(P)$ lies within that interval.

Continuity of $f(x, y)$ at $P_0 = (a, b)$ is then defined in the same way as continuity of $f(x)$ at a. First of all, f must be defined at P_0 and throughout some disk centered at P_0. Then,

1 $\lim_{P \to P_0} f(P)$ exists.
2 That limit is $f(P_0)$.

In Sec. 13.8 mappings F from the uv plane to the xy plane were discussed. Limits and continuity for such functions are defined very much like the corresponding concepts for $f(x)$ (a function from a line to a line) and $f(x, y)$ (a function from a plane to a line).

Rather than define the limit of such a function and then define continuity, let us go directly to continuity.

Let F be a mapping from the uv plane to the xy plane. Assume that F is defined at P_0 and throughout some disk centered at P_0. We shall say that the mapping F is **continuous** at P_0 if, for every $\epsilon > 0$, there is a $\delta > 0$ such that if

$$d(P, P_0) < \delta,$$

then
$$d(F(P), F(P_0)) < \epsilon.$$

Geometrically, the definition asserts that no matter how small a disk is drawn around $F(P_0)$, it is possible to find a (response) disk around P_0 such that every point in the second disk is mapped by F into the first (challenge) disk. See Fig. G.3.

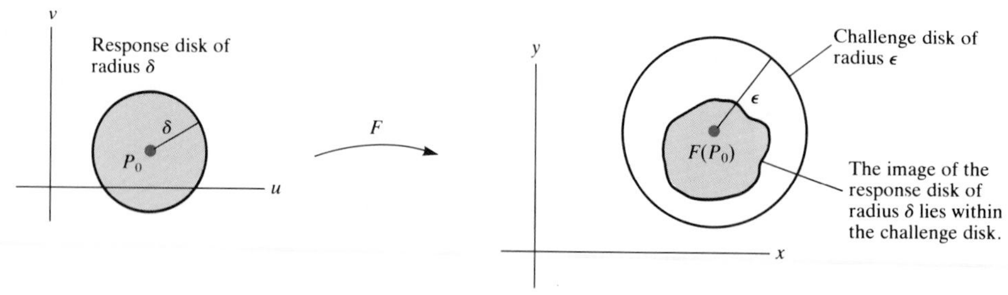

Figure G.3

It can be shown that if $F(u, v) = (f(u, v), g(u, v))$, then the continuity of F is equivalent to the continuity of f and g.

The definition of the definite integral $\int_a^b f(x)\,dx$ in Sec. 5.3 is informal in the sense that it uses such terms as "approaches a certain number" and "shrinks toward 0." The following definition, in the spirit of Sec. 2.8, is more precise.

Definition The number L is the **definite integral of f over $[a, b]$** if the following condition is met: For any positive number ϵ (however small) there must exist a positive number δ, depending on ϵ, such that any sum

$$\sum_{i=1}^{n} f(c_i)(x_i - x_{i-1})$$

formed with any partition of $[a, b]$ of mesh less than δ (no matter how c_i is chosen in $[x_{i-1}, x_i]$) differs from L by less than ϵ, that is, satisfies the inequality

$$\left| \sum_{i=1}^{n} f(c_i)(x_i - x_{i-1}) - L \right| < \epsilon.$$

EXERCISES FOR SEC. G.2: DEFINITIONS OF OTHER LIMITS

1 Prove that if $f(x, y)$ is positive and continuous at $P_0 = (a, b)$, then there is a disk with center P_0 where f remains positive.

2 Prove that the mapping $F = (f, g)$ is continuous at P_0 if and only if both f and g are continuous.

Appendix H Logarithms and Exponentials Defined through Calculus

This appendix provides a definition of logarithms and exponentials completely different from the definitions met in high school algebra. The main assumption is that $\int_1^x dt/t$ exists. In contrast to the algebraic approach, logarithms are defined first and exponentials after.

H.1 THE NATURAL LOGARITHM DEFINED AS A DEFINITE INTEGRAL

Recall how the logarithm and exponential functions were defined. First, the exponential function b^x $(b > 0)$ was built up in stages:

$$b^n = b \cdot b \cdot \cdots \cdot b \ (n \text{ times})$$

for $n = 1, 2, 3, \ldots$; $b^0 = 1$; $b^{-n} = 1/b^n$ for $n = 1, 2, 3, \ldots$; $b^{1/n} = \sqrt[n]{b}$, the positive nth root of b for $n = 1, 2, 3, \ldots$; $b^{m/n} = (b^{1/n})^m$ for m an integer and n a positive integer; and finally, b^x equals the limit of $b^{m/n}$ as $m/n \to x$ for irrational x.

A thorough treatment of the exponential functions based on this approach encounters many difficulties, such as: How do we know that b has an nth root? If $b > 1$, is b^x increasing? Does $\lim_{m/n \to x} b^{m/n}$ exist? After answering these questions, we would still be left with showing that b^x is continuous and that $b^x b^y = b^{x+y}$.

Assuming that b^x has these desired properties, we then defined $\log_b x$. Then, to obtain the derivative of $\log_b x$ we had to assume that

$$\lim_{h \to 0} (1 + h)^{1/h}$$

exists and that $\log_b x$ is a continuous function.

The present section takes a completely different approach. It first defines a function $L(x)$ with the aid of the definite integral. [It will turn out that $L(x) = \ln x$.] In Sec. H.2 the exponential function will be defined as the inverse of the function $L(x)$ and it will be shown that $\lim_{h \to 0} (1 + h)^{1/h}$ exists and equals e.

Basic to the following argument is the theorem (see Sec. 5.3) that a continuous function has a definite integral over each interval $[a, b]$ in its domain.

Definition *The function* $L(x)$.

$$= \int_1^x \frac{1}{t}\,dt, \qquad x > 0.$$

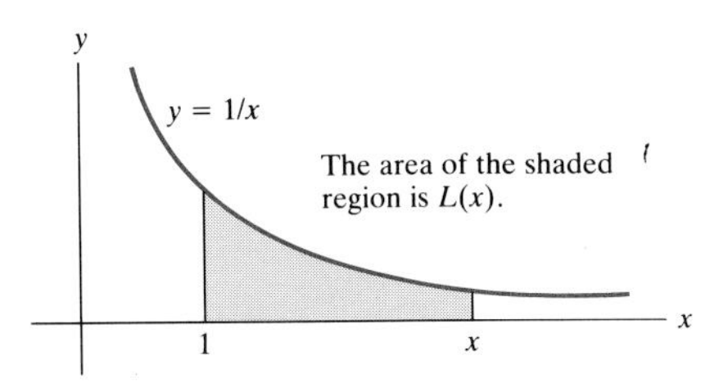

Figure H.1

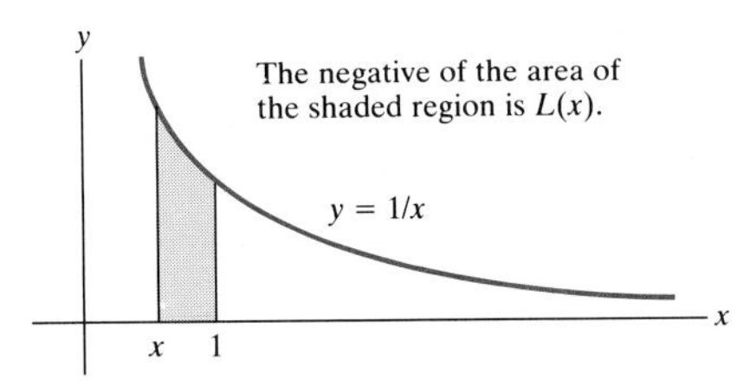

Figure H.2

Observe that $L(1) = \int_1^1 (1/t)\,dt = 0$; if $x > 1$, $L(x) > 0$ and is the area of the shaded region in Fig. H.1. If $0 < x < 1$, then $L(x) = \int_1^x (1/t)\,dt = -\int_x^1 (1/t)\,dt$, the negative of the area of the shaded region in Fig. H.2. Thus, if $0 < x < 1$, we have $L(x) < 0$. $L(x)$ resembles a logarithm function also in that $L(x)$ is defined only for $x > 0$.

By the second fundamental theorem of calculus,

$$L'(x) = \frac{1}{x}. \tag{1}$$

The information that $L(1) = 0$ and that $L'(x) = 1/x$ already shows that $L(x) = \ln x$. After all, since $L(x)$ and $\ln x$ have the same derivative, they differ by a constant,

$$L(x) = \ln x + C. \tag{2}$$

To find the constant C, set $x = 1$ in Eq. (2), obtaining

$$L(1) = \ln (1) + C.$$

Since both $L(1)$ and $\ln (1)$ equal 0, it follows that $C = 0$. Thus $L(x) = \ln x$.

It is reassuring to know that $L(x)$, though defined so differently, is $\ln x$, defined in Sec. 6.3. However, in this section, *no use will be made of this knowledge,* since the present goal is to build up the theory of logarithms from scratch, with minimal assumptions.

Incidentally, the fact that the area under the curve $y = 1/x$ behaves like a logarithm was first noted by Gregory St. Vincent and his friend A. A. de Sarasa in 1647, some twenty years before the work of Newton and Leibniz.

Property 1 $L(xy) = L(x) + L(y)$, for $x, y > 0$.

Proof Hold y fixed at the value k. It will be shown that

$$L(xk) = L(x) + L(k), \qquad x > 0.$$

Introduce two functions of x,

$$f(x) = L(xk) \quad \text{and} \quad g(x) = L(x) + L(k).$$

We wish to show that $f(x) = g(x)$.

To begin, compute their derivatives:

$$f'(x) = [L(xk)]'$$

$$= \frac{1}{xk}(xk)' \qquad \text{by Eq. (1) and the chain rule}$$

$$= \frac{k}{xk} = \frac{1}{x},$$

and

$$g'(x) = [L(x) + L(k)]'$$

$$= L'(x) = \frac{1}{x}. \qquad k \text{ is a constant}$$

Since $f(x)$ and $g(x)$ have the same derivative, they differ by a constant, that is,

$$L(xk) = L(x) + L(k) + C \qquad C \text{ constant.} \tag{3}$$

To find C, set $x = 1$ in Eq. (3), obtaining

$$L(k) = L(1) + L(k) + C.$$

Since $L(1) = 0$, this equation shows that $C = 0$, and Property 1 is established. ■

Property 2 For any integer n, $L(x^n) = nL(x)$, $x > 0$.

Proof Differentiate both $L(x^n)$ and $nL(x)$:

$$[L(x^n)]' = \frac{1}{x^n} nx^{n-1} \qquad \text{by Eq. (1) and the chain rule}$$

$$= \frac{n}{x}$$

and

$$[nL(x)]' = \frac{n}{x} \qquad \text{by Eq. (1).}$$

Thus there is a constant C such that

$$L(x^n) = nL(x) + C. \tag{4}$$

Setting $x = 1$ in Eq. (4) shows that

$$L(1) = nL(1) + C,$$

or

$$0 = 0 + C.$$

Hence $C = 0$ and Property 2 is established. ■

Property 2, with $n = -1$, implies that

$$L(x^{-1}) = (-1)L(x),$$

or simply,

$$L\left(\frac{1}{x}\right) = -L(x),$$

an equation that should not come as a surprise.

Property 3 $\lim_{x \to \infty} L(x) = \infty$ and $\lim_{x \to 0^+} L(x) = -\infty$.

Proof Since $L'(x)$ is positive, $L(x)$ is an increasing function. Moreover, $L(2) > 0$ and $L(2^n) = nL(2)$. Thus $L(2^n)$ gets arbitrarily large as $n \to \infty$, so

$$\lim_{x\to\infty} L(x) = \infty.$$

To show that

$$\lim_{x\to 0^+} L(x) = -\infty,$$

replace x by $1/t$, where $t \to \infty$. Then we have

$$\begin{aligned}\lim_{x\to 0^+} L(x) &= \lim_{t\to\infty} L\left(\frac{1}{t}\right)\\ &= \lim_{t\to\infty} [-L(t)] \qquad \text{by Property 2}\\ &= -\lim_{t\to\infty} L(t) = -\infty. \qquad \blacksquare\end{aligned}$$

Thus the range of L consists of all real numbers.

Since $L(1) = 0$ and $L(x)$ is an increasing continuous function that takes on arbitrarily large values, there must exist a unique number x such that $L(x) = 1$. (The intermediate-value theorem guarantees that there is at least one x. Why is there only one?)

Definition e is the unique solution to the equation $L(x) = 1$, that is, $L(e) = 1$.

In Sec. H.2, after the exponential functions are defined, it will be shown that $e = \lim_{h\to 0} (1 + h)^{1/h}$.

EXERCISES FOR SEC. H.1: THE NATURAL LOGARITHM DEFINED AS A DEFINITE INTEGRAL

1 Show that L is continuous.

2 Show that the graph of $L(x)$ is concave downward.

3 (*a*) Compute the area of the eight rectangles of the same width in Fig. H.3.

(*b*) From (*a*) deduce that $L(3) > 1$.

(*c*) From (*b*) deduce that $e < 3$.

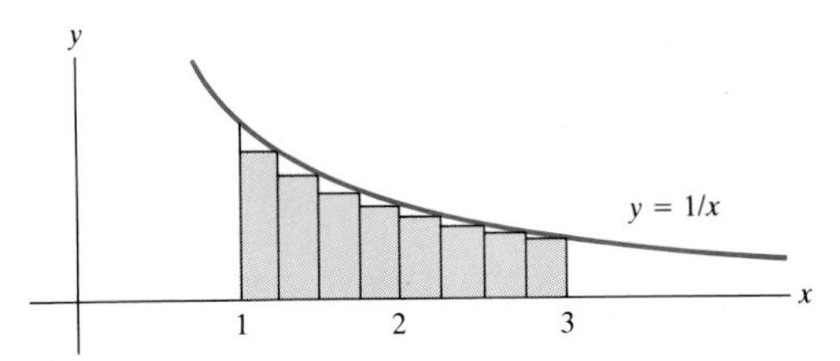

Figure H.3

4 Using an argument like that in Exercise 3, show that $e > 2$.

5 Let n be an integer larger than 1. By considering appropriate rectangles of width 1, show that

$$L(n) > \frac{1}{2} + \frac{1}{3} + \frac{1}{4} + \cdots + \frac{1}{n}.$$

6 (*a*) Let n be an integer larger than 1. Show that

$$L(n + 1) < \frac{1}{1} + \frac{1}{2} + \frac{1}{3} + \cdots + \frac{1}{n}.$$

(*b*) From (*a*) deduce that $\lim_{n\to\infty} \sum_{i=1}^{n} 1/i = \infty$. (This was first proved by Oresme around the year 1360.)

Exercises 7 and 8 outline a way to compute $L(2)$ with any degree of accuracy desired.

7 (*a*) By using a partition of $[1, 2]$ of n sections of equal length and left endpoints as sampling points, show that

$$L(2) < \frac{1}{n} + \frac{1}{n + 1} + \cdots + \frac{1}{2n - 1}.$$

(*b*) Deduce that $L(2) < 0.74$.

8 Show that for any positive integer n,

$$L(2) > \frac{1}{n + 1} + \frac{1}{n + 2} + \cdots + \frac{1}{2n}.$$

Suggestion: Form a suitable partition of [1, 2] and use right endpoints as sampling points.

■

9 This exercise shows that $L(ab) = L(a) + L(b)$ without using the fundamental theorem of calculus.

(*a*) Prove that for $b > 1$, we have $\int_a^{ab} (1/x)\, dx = \int_1^b (1/x)\, dx$ by observing that if the sum

$$\sum_{i=1}^n (1/c_i)(x_i - x_{i-1})$$

is an approximation of the integral $\int_1^b (1/x)\, dx$, then $\sum_{i=1}^n (1/ac_i)(ax_i - ax_{i-1})$ is an approximation of $\int_a^{ab} (1/x)\, dx$.

(*b*) If $a, b > 1$, deduce that $\int_1^{ab} (1/x)\, dx - \int_1^a (1/x)\, dx = \int_1^b (1/x)\, dx$.

(*c*) From (*b*), show that if $a, b > 1$, then $L(ab) = L(a) + L(b)$.

H.2 EXPONENTIAL FUNCTIONS DEFINED IN TERMS OF LOGARITHMS

In Sec. H.1 the function $L(x) = \int_1^x dt/t$, $x > 0$, was introduced and shown to have the familiar properties of logarithms. It was pointed out that if all the gaps in the development of $\ln x$ were filled in, then $L(x)$ could be shown to coincide with $\ln x$. The number e was defined in Sec. H.1 by the condition $L(e) = 1$, that is, $\int_1^e dx/x = 1$.

The present section introduces a function E, which will turn out to be the exponential function to the base e. However, it is important in this section to make no use at all of b^x, as defined in Appendix D, since the object of this section is to construct these functions with the fewest assumptions.

Recall that the function L is increasing for $x > 0$. It thus has an inverse.

Definition *The function E.* The inverse of the function L will be denoted E.

Since L has domain $(0, \infty)$ and range $(-\infty, \infty)$, E has domain $(-\infty, \infty)$ and range $(0, \infty)$. Note that E is a one-to-one function.

Since $L(1) = 0$ and E is the inverse of L, $E(0) = 1$, just like an exponential function. Moreover, since $L(e) = 1$, it follows that $E(1) = e$. Keep in mind that $E(L(x)) = x$ for $x > 0$ and $L(E(x)) = x$ for all x on the x axis. Theorem 1 shows that the function E has the basic property of an exponential function.

Theorem 1 $E(x + y) = E(x)E(y)$.

Proof Write x as $L(u)$ and y as $L(v)$. This is possible because the range of L is $(-\infty, \infty)$. Then

$$\begin{aligned} E(x + y) &= E(L(u) + L(v)) \\ &= E(L(uv)) \qquad L(uv) = L(u) + L(v) \\ &= uv \qquad E \text{ and } L \text{ are inverses} \\ &= E(x)E(y). \quad ■ \end{aligned}$$

Next, the exponential function b^x will be defined for $b > 0$. In the traditional approach, there is the identity

$$b^x = e^{x \ln b}.$$

This suggests how to define the function b^x in the present approach.

Definition *The exponential function* b^x. If $b > 0$, define b^x to be

$$E(xL(b)).$$

Theorem 2 If b is positive, then

$$b^0 = 1, \tag{1}$$

$$b^1 = b, \tag{2}$$

$$b^{x+y} = b^x b^y. \tag{3}$$

Proof

(1) $b^0 = E(0 \cdot L(b)) = E(0) = 1.$

(2) $b^1 = E(1 \cdot L(b)) = E(L(b)) = b.$

(3)

$b^{x+y} = E((x + y)L(b))$	definition of b^{x+y}
$= E(xL(b) + yL(b))$	algebra
$= E(xL(b))E(yL(b))$	Theorem 1
$= b^x b^y$	definition of b^x and b^y. ■

According to the definition of b^x just given, when the base b is chosen to be e, we have

$$e^x = E(xL(e)) = E(x \cdot 1) = E(x).$$

So e^x is another name for $E(x)$. But do not think of it as the function e^x that was built up laboriously in Appendix D first for integer x, then for rational x, and finally, by limits, for irrational x.

Theorem 3 $L(b^x) = xL(b)$.

Proof

$b^x = E(xL(b))$	definition of b^x
$L(b^x) = L(E(xL(b)))$	
$= xL(b)$	L and E are inverses of each other. ■

Theorem 4 $(b^x)^y = b^{xy}$.

Proof Since L is one-to-one, it suffices to show that $L(b^{xy}) = L((b^x)^y)$. This will be established by three applications of Theorem 3:

	$L(b^{xy}) = xyL(b)$	Theorem 3,
while	$L((b^x)^y) = yL(b^x)$	Theorem 3
	$= y(xL(b))$	Theorem 3
	$= xyL(b)$	algebra.

Thus $L(b^{xy}) = L((b^x)^y)$, and the proof is complete. ■

If $b \neq 1$, the function b^x is one-to-one. To show this, assume that

$$b^{x_1} = b^{x_2}, \tag{4}$$

and show that $x_1 = x_2$. By the definition of b^x, Eq. (4) can be written

$$E(x_1 L(b)) = E(x_2 L(b)).$$

Because E is one-to-one, $x_1 L(b) = x_2 L(b)$. Since $b \neq 1$, $L(b) \neq 0$, and cancellation of $L(b)$ yields $x_1 = x_2$.

Definition *The function* $\log_b x$. If $b > 0$, define $\log_b x$ to be

$$\frac{L(x)}{L(b)}.$$

From the properties of $L(x)$, it can be shown that $\log_b x$ is the inverse of the function b^x. Theorem 5 shows that $\log_b x$ has the traditional properties of a logarithm.

Theorem 5 If $b > 0$, $b \neq 1$, then

$$\log_b 1 = 0, \tag{5}$$

$$\log_b b = 1, \tag{6}$$

$$\log_b xy = \log_b x + \log_b y \qquad x, y > 0, \tag{7}$$

$$\log_b x^y = y \log_b x \qquad x > 0. \quad \blacksquare \tag{8}$$

The proof is left to the reader.

Since E is the inverse of the increasing differentiable function L, it is also differentiable. To find its derivative, let

$$y = E(x). \tag{9}$$

Hence

$$x = L(y). \tag{10}$$

Implicit differentiation of Eq. (10) with respect to x yields

$$1 = \frac{1}{y}\frac{dy}{dx}.$$

Thus

$$\frac{dy}{dx} = y.$$

In short,

$$\frac{d}{dx}(E(x)) = E(x). \tag{11}$$

This proves the next theorem.

Theorem 6 The derivative of E is E:

$$\frac{d}{dx}(E(x)) = E(x). \quad \blacksquare$$

With the help of Theorem 6 it is not hard to prove the following generalization.

Theorem 7 $\dfrac{d}{dx}(b^x) = b^x L(b)$. $\blacksquare$

The proofs of Theorem 7 and of the following theorem are left to the reader.

Theorem 8 $\dfrac{d}{dx}(\log_b x) = \dfrac{\log_b e}{x}$. ■

It is now possible to show that

$$\lim_{x \to 0} (1 + x)^{1/x} = e.$$

[Keep in mind that e is defined in App. H.1 by the demand that $L(e) = 1$, that is, $\int_1^e dt/t = 1$.]

Theorem 9 $\lim_{x \to 0} (1 + x)^{1/x} = e.$

Proof The derivative of $L(x)$ at $x = 1$ is $\frac{1}{1}$, or 1. On the other hand, by the definition of the derivative,

$$\begin{aligned} L'(1) &= \lim_{x \to 0} \frac{L(1 + x) - L(1)}{x} \\ &= \lim_{x \to 0} \frac{L(1 + x)}{x} \qquad \text{since } L(1) = 0 \\ &= \lim_{x \to 0} \frac{1}{x} L(1 + x) \\ &= \lim_{x \to 0} L[(1 + x)^{1/x}] \qquad \text{Theorem 3.} \end{aligned}$$

Hence

$$1 = \lim_{x \to 0} L[(1 + x)^{1/x}],$$

so

$$E(1) = E\left\{\lim_{x \to 0} L[(1 + x)^{1/x}]\right\}.$$

Since E is continuous, E and lim can be switched; thus

$$E(1) = \lim_{x \to 0} E[L(1 + x)^{1/x}]. \tag{12}$$

(See Example 4 of Sec. 2.5.) Since $E(1) = e$, and E and L are inverse functions, Eq. (12) implies that

$$e = \lim_{x \to 0} (1 + x)^{1/x}. \quad ■$$

Theorem 9 shows that the definition of e in Sec. 6.2 as a limit is consistent with the definition in Sec. H.1.

EXERCISES FOR SEC. H.2: EXPONENTIAL FUNCTIONS DEFINED IN TERMS OF LOGARITHMS

In the exercises prove the given statements using the definitions in this appendix.

1 Theorem 5

2 Theorem 7

3 Theorem 8

4 $L(x)/L(b)$ is the inverse of the function b^x.

Appendix I The Taylor Series for $f(x, y)$

The higher partial derivatives of $z = f(x, y)$ were introduced in Sec. 12.3. Recall that

$$\frac{\partial}{\partial y}\left(\frac{\partial z}{\partial x}\right) \quad \text{is denoted} \quad \frac{\partial^2 z}{\partial y\, \partial x}, \quad \frac{\partial^2 f}{\partial y\, \partial x} \quad \text{or} \quad z_{xy}.$$

There are four partial derivatives of the second order:

$$z_{xx}, \quad z_{xy}, \quad z_{yx}, \quad \text{and} \quad z_{yy}.$$

If they are continuous, z_{xy} equals z_{yx}. Each of these four partial derivatives may be differentiated with respect to x or with respect to y. Thus there are eight possible partial derivatives of order 3. For instance, two of them are

$$\frac{\partial(z_{xx})}{\partial x} \quad \text{and} \quad \frac{\partial(z_{xx})}{\partial y},$$

which will be denoted z_{xxx} and z_{xxy}.

The eight are

$$z_{xxx}, \quad z_{xxy}, \quad z_{xyx}, \quad z_{xyy}, \quad z_{yxx}, \quad z_{yxy}, \quad z_{yyx}, \quad \text{and} \quad z_{yyy}.$$

If they are continuous, however, many of them are equal. For instance,

$$z_{xxy} = z_{xyx},$$

for

$$z_{xxy} = (z_x)_{xy} \quad \text{and} \quad z_{xyx} = (z_x)_{yx}.$$

In general, the order of differentiation does not affect the result; all differentiations with respect to x may be done first, and afterward the differentiations with respect to y. Thus

$$z_{xyy} = z_{yxy} = z_{yyx},$$

or, in the ∂ notation, $$\frac{\partial^3 z}{\partial y^2\, \partial x} = \frac{\partial^3 z}{\partial y\, \partial x\, \partial y} = \frac{\partial^3 z}{\partial x\, \partial y^2}.$$

Similar statements and notations hold for partial derivatives of higher orders. Thus

$$z_{xyxyy} = z_{xxyyy} = \frac{\partial^5 z}{\partial y^3\, \partial x^2}.$$

EXAMPLE 1 Compute the partial derivatives of x^4y^7 up through order 3.

SOLUTION To begin:

$$z_x = 4x^3y^7 \quad \text{and} \quad z_y = 7x^4y^6.$$

Then, $z_{xx} = 12x^2y^7$, $z_{xy} = z_{yx} = 28x^3y^6$, and $z_{yy} = 42x^4y^5$. The third-order partial derivatives are

$$z_{xxx} = 24xy^7,$$

$$z_{xxy} = z_{xyx} = z_{yxx} = 84x^2y^6,$$

$$z_{xyy} = z_{yxy} = z_{yyx} = 168x^3y^5,$$

and

$$z_{yyy} = 210x^4y^4.$$

Note that on account of duplication there are in practice only four partial derivatives of order 3. Similarly, there are in practice only five different partial derivatives of order 4, and $n + 1$ different partial derivatives of order n. ■

Just as many functions $f(x)$ of a single variable can be expressed as power series in $x - a$, so can many functions $f(x, y)$ of two variables be expressed as series in powers of $x - a$ and $y - b$. Such a series begins

$$c_1 + c_2(x - a) + c_3(y - b) + c_4(x - a)^2 + c_5(x - a)(y - b) + c_6(y - b)^2 + \cdots,$$

where the c's are constants. Frequently, in applications, only this much of the series is used to approximate the function near the point (a, b). The typical term has the form $c(x - a)^r(y - b)^s$. The **degree** of the term is $r + s$. The terms are written from left to right in increasing degree; within terms of a given degree, the terms are written in increasing degree of $y - b$.

We shall obtain a formula for this series in terms of the partial derivatives of f evaluated at (a, b). The argument begins by relating $f(x, y)$ to a function of one variable.

Let f be a function of x and y that possesses partial derivatives of all orders. Let a, b, h, and k be fixed numbers. Define a function g as follows:

$$g(t) = f(a + th, b + tk).$$

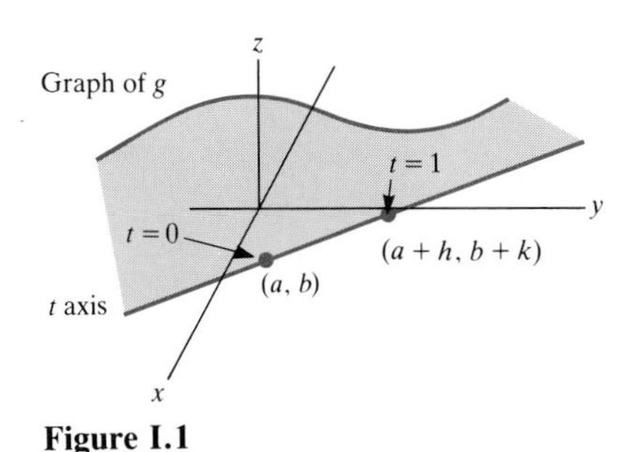

Figure I.1

The Taylor series for f can be obtained from that for g by expressing the derivatives $g'(0)$, $g^{(2)}(0)$, $g^{(3)}(0)$, . . . in terms of partial derivatives of f.

To compute $g'(t)$, observe that g is a composite function:

$$g(t) = f(x, y), \quad \text{where } x = a + th \text{ and } y = b + tk. \tag{1}$$

(See Fig. I.1.) By the chain rule (Theorem 1 of Sec. 12.5),

$$g'(t) = \frac{\partial f}{\partial x}\frac{dx}{dt} + \frac{\partial f}{\partial y}\frac{dy}{dt}.$$

Now by (1), $$\frac{dx}{dt} = h \quad \text{and} \quad \frac{dy}{dt} = k.$$

Thus $$g'(t) = f_x \cdot h + f_y \cdot k, \tag{2}$$

where f_x and f_y are evaluated at $(a + th, b + tk)$. In particular,

$$g'(0) = f_x(a, b)h + f_y(a, b)k. \tag{3}$$

Next, express $g^{(2)}(t)$ in terms of partial derivatives of the function f. To do so, differentiate Eq. (2) with respect to t:

$$\begin{aligned} g^{(2)}(t) &= \frac{d}{dt}[g'(t)] = \frac{d}{dt}(f_x h + f_y k) \\ &= \frac{\partial}{\partial x}(f_x h + f_y k)\frac{dx}{dt} + \frac{\partial}{\partial y}(f_x h + f_y k)\frac{dy}{dt} \\ &= \frac{\partial}{\partial x}(f_x h + f_y k)h + \frac{\partial}{\partial y}(f_x h + f_y k)k \\ &= (f_{xx} h + f_{yx} k)h + (f_{xy} h + f_{yy} k)k. \end{aligned}$$

Hence $$g^{(2)}(t) = f_{xx} h^2 + 2f_{xy} hk + f_{yy} k^2, \tag{4}$$

where all the partial derivatives are evaluated at $(a + th, b + tk)$. Thus

$$g^{(2)}(0) = f_{xx}(a, b)h^2 + 2f_{xy}(a, b)hk + f_{yy}(a, b)k^2. \tag{5}$$

Notice the similarity of the right side of Eq. (5) to the binomial expansion

$$(c + d)^2 = c^2 + 2cd + d^2,$$

in the coefficients, the powers of h and k, and the subscripts. To make use of this similarity, introduce the expression

$$(h\,\partial_x + k\,\partial_y)^2 f,$$

where $(h\,\partial_x + k\,\partial_y)^2$ is treated formally like an algebraic product: for instance, $(\partial_x\,\partial_x)f$ is interpreted as f_{xx}.

Thus Eq. (5) may be written in this shorthand as

$$g^{(2)}(0) = (h\,\partial_x + k\,\partial_y)^2 f \Big|_{(a,\,b)}. \tag{6}$$

[The symbol $|_{(a,\,b)}$ means "evaluated at (a, b)."]

Differentiating Eq. (4) with respect to t sufficiently often we can show similarly that

$$g^{(n)}(t) = (h\,\partial_x + k\,\partial_y)^n f \Big|_{(a+th,\,b+tk)} \tag{7}$$

for $n = 1, 2, 3, \ldots$.

Theorem *Taylor series for a function of two variables.* Let f have continuous partial derivatives of all orders up to and including $n + 1$ at and near the point (a, b). If $(x, y) = (a + h, b + k)$ is sufficiently near (a, b), then

$$f(x, y) = f(a, b) + (h\,\partial_x + k\,\partial_y)f \Big|_{(a,b)} + \frac{(h\,\partial_x + k\,\partial_y)^2 f}{2!}\Big|_{(a,b)}$$

$$+ \cdots + \frac{(h\,\partial_x + k\,\partial_y)^n f}{n!}\Big|_{(a,b)} + \frac{(h\,\partial_x + k\,\partial_y)^{n+1} f}{(n+1)!}\Big|_{(X,Y)},$$

where (X, Y) is some point on the line segment joining (a, b) and (x, y).
In sigma notation this reads

$$f(x, y) = \sum_{i=0}^{n} \frac{(h\,\partial_x + k\,\partial_y)^i f}{i!}\Big|_{(a,b)} + \frac{(h\,\partial_x + k\,\partial_y)^{n+1} f}{(n+1)!}\Big|_{(X,Y)}.$$

Proof This theorem follows from Theorem 2 of Sec. 10.8. Once again we make use of the function g, defined as follows:

$$g(t) = f(a + th, b + tk).$$

(See Fig. I.1.) Observe that

$$g(0) = f(a, b) \tag{8}$$

and

$$g(1) = f(a + h, b + k) = f(x, y). \tag{9}$$

By Lagrange's formula for the remainder $R_n(1; 0)$,

$$g(1) = g(0) + g'(0) \cdot 1 + \frac{g^{(2)}(0)}{2!} 1^2 + \cdots + \frac{g^{(n)}(0)}{n!} 1^n + \frac{g^{(n+1)}(T)}{(n+1)!} 1^{n+1} \tag{10}$$

for a suitable number T, $0 \le T \le 1$. Combining Eqs. (7) and (10) completes the proof. ■

As a consequence of the theorem, the coefficient of $h^r k^s$ in the Taylor series for f is

$$\frac{1}{(r+s)!}\binom{r+s}{r}\frac{\partial^{r+s} f}{\partial x^r\,\partial y^s},$$

where the partial derivative is evaluated at (a, b), and $\binom{r+s}{r}$ denotes the binomial coefficient

$$\frac{(r+s)!}{r!\,s!}.$$

Observe that $(r + s)!$ can be canceled in the numerator and denominator. Thus the coefficient of $h^r k^s$ is

$$\frac{1}{r!\,s!}\frac{\partial^{r+s} f}{\partial x^r\,\partial y^s}.$$

EXAMPLE 2 Use the theorem to express $f(x, y) = x^2 y$ in powers of $x - 1$ and $y - 2$.

SOLUTION In this case, $a = 1$, $b = 2$, $h = x - 1$, and $k = y - 2$. To begin, compute the partial derivatives of f at (1, 2). We have

$$f_x = 2xy, \qquad f_{xx} = 2y, \qquad f_{xy} = 2x, \qquad f_{xxy} = 2, \qquad \text{and} \qquad f_y = x^2.$$

All higher partial derivatives of f are identically 0. Thus $f(1, 2) = 2$, $f_x(1, 2) = 2 \cdot 1 \cdot 2 = 4$, and so on. Therefore,

$$\begin{aligned} f(x, y) &= f(1 + h, 2 + k) \\ &= f(1, 2) + [hf_x(1, 2) + kf_y(1, 2)] \\ &\quad + \left[\frac{h^2 f_{xx}(1, 2) + 2hkf_{xy}(1, 2) + k^2 f_{yy}(1, 2)}{2!}\right] \\ &\quad + \left[\frac{h^3 f_{xxx}(1, 2) + 3h^2 k f_{xxy}(1, 2) + 3hk^2 f_{xyy}(1, 2) + k^3 f_{yyy}(1, 2)}{3!}\right] \\ &= 2 + 4h + k + \frac{4h^2 + 4hk + 0k^2}{2!} + \frac{6h^2 k}{3!}; \end{aligned}$$

that is,

$$x^2 y = 2 + 4(x - 1) + (y - 2) + 2(x - 1)^2 + 2(x - 1)(y - 2) + (x - 1)^2(y - 2). \qquad (11)$$

This can be checked by expanding the right side of the equation. ■

According to the theorem, the Taylor series associated with $f(x, y)$ in powers of $x - a$ and $y - b$ begins

$$f(a, b) + f_x(a, b)(x - a) + f_y(a, b)(y - b) + \frac{f_{xx}(a, b)}{2!}(x - a)^2 + \frac{2f_{xy}(a, b)}{2!}(x - a)(y - b) + \frac{f_{yy}(a, b)}{2!}(y - b)^2 + \cdots.$$

Whether the series converges and whether it converges to $f(x, y)$ depend on the behavior of the higher-order partial derivatives of f.

Taylor series for functions of three or more variables can be developed in a similar way.

Sometimes only the terms up through degree 1 are used to approximate the function. For instance, Harold A. Thomas, in an essay entitled "Population Dynamics of Primitive Societies" (included in *Is There an Optimum Level of Population?* edited by S. Fred Singer), wrote,

> For a small three-dimensional region about the equilibrium point $(\overline{P}, \overline{x}, \overline{w})$ the utility function may be approximated by the first terms of a Taylor's series:
>
> $$U(P, x, w) = U(\overline{P}, \overline{x}, \overline{w}) + b_1(P - \overline{P}) + b_2(x - \overline{x}) + b_3(w - \overline{w}) + \cdots,$$
>
> where b_1, b_2, and b_3 represent the partial derivatives $\partial U/\partial P$, $\partial U/\partial x$, and $\partial U/\partial w$ evaluated at the equilibrium point.

The notation is different from that of this section, but the idea is the same: To approximate the behavior of a function near a point by a polynomial in several variables.

EXERCISES FOR APPENDIX I: THE TAYLOR SERIES FOR $f(x, y)$

In each of Exercises 1 to 4 compute all eight partial derivatives of the third order of the given function.

1 $x^5 y^7$
2 x/y
3 e^{2x+3y}
4 $\sin(x^2 + y^3)$

5 Verify Eq. (11) by expanding the right side of the equation.

6 Using Eq. (11), compute the difference in the volumes of these two boxes: One has a square base of side 1 foot and height 2 feet; the other has a square base of side 1.1 feet and height 2.1 feet.

7 (*a*) Using partial derivatives, obtain the first four nonzero terms in the Taylor series for e^{x+y^2} in powers of x and y.

(*b*) Noticing that $e^{x+y^2} = e^x e^{y^2}$ and using a few terms of the Maclaurin series for e^x and e^{y^2}, solve (*a*) again.

8 (*a*) Using partial derivatives, obtain the first four nonzero terms in the Taylor series for $\cos(x + y)$ in powers of x and y.

(*b*) Using the Maclaurin series for $\cos t$ and replacing t by $x + y$, obtain the terms described in (*a*).

9 Assume that f has continuous partial derivatives of all orders, $f(0, 0) = 2$, $f_x(0, 0) = 3$, $f_y(0, 0) = -5$, $f_{xx}(0, 0) = 6$, $f_{xy}(0, 0) = 7$, and $f_{yy}(0, 0) = 1$. Write out the Taylor series in powers of x and y associated with f up through terms of degree two.

10 The Taylor series for a certain function $f(x, y)$ begins $5 + 6x + 11y - 2x^2 - 3xy + 7y^2 + \cdots$. Use this information to determine $f(0, 0)$ and the first- and second-order partial derivatives of f at $(0, 0)$.

11 Verify that the expansion of $\sqrt{1 + x + y}$ begins with $1 + \frac{1}{2}x + \frac{1}{2}y - \frac{1}{8}x^2 - \frac{1}{4}xy - \frac{1}{8}y^2 + \cdots$ by using the theorem of this section.

12 (*a*) Prove that if all the partial derivatives of f are continuous, then

$$f_{xyxy} = f_{xyyx}.$$

(*b*) Prove that each of the 16 partial derivatives of f of the fourth order is equal to one of these five:

$$\frac{\partial^4 f}{\partial x^4}, \quad \frac{\partial^4 f}{\partial y\, \partial x^3}, \quad \frac{\partial^4 f}{\partial y^2\, \partial x^2}, \quad \frac{\partial^4 f}{\partial y^3\, \partial x}, \quad \frac{\partial^4 f}{\partial y^4}.$$

13 What is the coefficient of x^5y^7 in the Taylor series expansion of the function f around the point $(a, b) = (0, 0)$?

Appendix J The Interchange of Limits

This appendix provides proofs of two theorems whose validity was assumed in the text. In Sec. J.1 the equality of the mixed partial derivatives is examined. It can be read after Sec. 12.3. Section J.2 concerns the differentiation of $\int_a^b f(x, y)\,dx$ with respect to y. Section J.3, besides presenting an example of a function $f(x, y)$ whose two mixed partials are not equal, gives a gentle introduction to advanced calculus.

J.1 THE EQUALITY OF f_{xy} AND f_{yx}

For most common functions $f(x, y)$,

$$\frac{\partial}{\partial y}\left(\frac{\partial f}{\partial x}\right) \quad \text{equals} \quad \frac{\partial}{\partial x}\left(\frac{\partial f}{\partial y}\right);$$

that is, the order in which we compute partial derivatives does not affect the result. This assertion is justified in the following theorem.

Theorem Let f be a function defined on the xy plane. If f_{xy} and f_{yx} exist and if at least one of them is continuous at all points, then they are equal.

Proof We assume f_{yx} is continuous. To keep the proof uncluttered, it will be shown that $f_{xy}(0, 0)$ equals $f_{yx}(0, 0)$. The identical argument holds for any point (a, b).

To begin, consider the definition of $f_{xy}(0, 0)$:

$$f_{xy}(0, 0) = \left.\frac{\partial(f_x)}{\partial y}\right|_{(0,\,0)} = \lim_{k\to 0} \frac{f_x(0, k) - f_x(0, 0)}{k}.$$

But, by definition of the partial derivative f_x,

$$f_x(0, k) = \lim_{h\to 0} \frac{f(h, k) - f(0, k)}{h} \quad \text{and} \quad f_x(0, 0) = \lim_{h\to 0} \frac{f(h, 0) - f(0, 0)}{h}.$$

Thus

$$\begin{aligned} f_{xy}(0, 0) &= \lim_{k\to 0} \frac{f_x(0, k) - f_x(0, 0)}{k} \\ &= \lim_{k\to 0} \frac{\lim_{h\to 0} \dfrac{f(h, k) - f(0, k)}{h} - \lim_{h\to 0} \dfrac{f(h, 0) - f(0, 0)}{h}}{k} \\ &= \lim_{k\to 0} \left\{ \lim_{h\to 0} \frac{[f(h, k) - f(0, k)] - [f(h, 0) - f(0, 0)]}{hk} \right\}. \end{aligned} \tag{1}$$

Let us focus our attention on the numerator in (1):

$$\text{Numerator} = [f(h, k) - f(0, k)] - [f(h, 0) - f(0, 0)]. \tag{2}$$

Note that the second bracketed expression is obtained from the first bracketed expression by replacing k by 0. Define, for fixed h, a function

$$u(y) = f(h, y) - f(0, y). \tag{3}$$

Then (2) takes the simple form

$$u(k) - u(0). \tag{4}$$

By the mean-value theorem (see Sec. 4.1),

$$u(k) - u(0) = u'(K)k \tag{5}$$

for some K between 0 and k. But by the definition of the function u, given in Eq. (3),

$$u'(K) = f_y(h, K) - f_y(0, K). \tag{6}$$

Thus, by the mean-value theorem applied to the function $f_y(x, K)$, for fixed K,

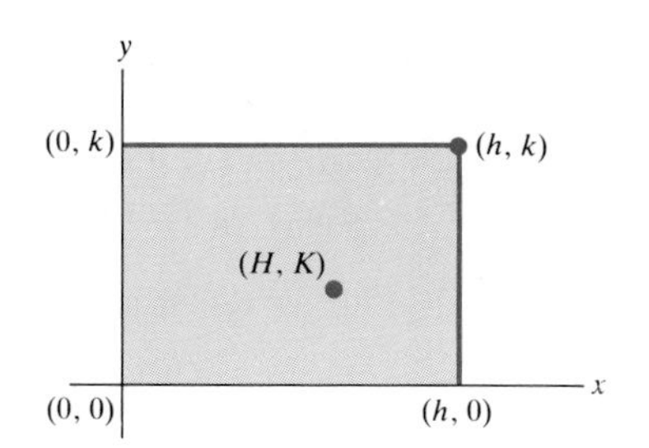

Figure J.1

$$u'(K) = f_{yx}(H, K)h \tag{7}$$

for some H between 0 and h.

Thus Eq. (2) becomes

$$\text{Numerator} = f_{yx}(H, K)hk \tag{8}$$

for some point (H, K) in the rectangle with vertices (0, 0), $(h, 0)$, (h, k), and $(0, k)$. (See Fig. J.1.) Substituting Eq. (8) in Eq. (1), we obtain

$$f_{xy}(0, 0) = \lim_{k\to 0} [\lim_{h\to 0} f_{yx}(H, K)].$$

Since H is between 0 and h and K is between 0 and k, it follows that H and K both approach 0 as $h \to 0$ and $k \to 0$. By the continuity of f_{yx}, $f_{yx}(H, K)$ approaches $f_{yx}(0, 0)$. Hence,

$$f_{xy}(0, 0) = f_{yx}(0, 0),$$

as asserted. ■

Example 3 in Sec. J.3 presents a function whose two mixed partial derivatives f_{xy} and f_{yx} are *not* equal at (0, 0).

EXERCISES FOR SEC. J.1: THE EQUALITY OF f_{xy} AND f_{yx}

1 Prove the theorem without looking at the text.
2 Prove the theorem at an arbitrary point (a, b).
3 Read Example 3 in Sec. J.3.

J.2 THE DERIVATIVE OF $\int_a^b f(x, y)\,dx$ WITH RESPECT TO y

The integral

$$\int_a^b f(x, y)\,dx$$

depends on y. Let

$$F(y) = \int_a^b f(x, y)\,dx.$$

It makes sense to speak of the derivative of F with respect to y. It turns out that for most common functions f,

$$\frac{dF}{dy} = \int_a^b \frac{\partial f}{\partial y}\,dx.$$

That is,

$$\frac{d}{dy}\left[\int_a^b f(x, y)\,dx\right] = \int_a^b \frac{\partial f}{\partial y}\,dx.$$

Generally, it is safe to differentiate the integral by differentiating the integrand.

The reader may check this assertion for $f(x, y) = x^3 + xy^2$ before going through the proof of the theorem.

Theorem Let f be defined on the xy plane, and assume that f and f_y are continuous. Assume also that f_{yy} is defined on the xy plane and that it is bounded on each rectangle. [That is, if R is a rectangle, then there is a number M, depending on R, such that $|f_{yy}(x, y)| \le M$ for all (x, y) in R.] Let F be defined by

$$F(y) = \int_a^b f(x, y)\,dx.$$

Then F is differentiable, and

$$\frac{dF}{dy} = \int_a^b \frac{\partial f}{\partial y}\,dx.$$

Proof To show that

$$\lim_{h\to 0} \frac{F(y + h) - F(y)}{h} = \int_a^b \frac{\partial f}{\partial y}\,dx,$$

consider, for a fixed y, the difference

$$\frac{F(y + h) - F(y)}{h} - \int_a^b f_y(x, y)\,dx, \tag{1}$$

which, by the definition of F, equals

$$\frac{\int_a^b f(x, y + h)\,dx - \int_a^b f(x, y)\,dx}{h} - \int_a^b f_y(x, y)\,dx. \tag{2}$$

Now, (2) equals

$$\int_a^b \left[\frac{f(x, y + h) - f(x, y)}{h} - f_y(x, y) \right] dx. \tag{3}$$

To show that (1) approaches 0 as $h \to 0$, it suffices to show that the integrand in (3) is small when h is small. It may be assumed now that $|h| \leq 1$.

First of all, the expression

$$\frac{f(x, y + h) - f(x, y)}{h}$$

in the integrand in (3) equals

$$\frac{hf_y(x, y + H)}{h} = f_y(x, y + H)$$

for some number H between 0 and h, by the mean-value theorem. (H depends on x, y, and h.)

Thus the integrand in (3) equals

$$f_y(x, y + H) - f_y(x, y). \tag{4}$$

By the mean-value theorem, (4) equals

$$Hf_{yy}(x, y + H^*) \tag{5}$$

for some number H^* between 0 and H.

Since $|H^*| \leq |H| \leq |h| \leq 1$, the point $(x, y + H^*)$ lies somewhere in the rectangle whose vertices are

$$(a, y - 1), \qquad (a, y + 1), \qquad (b, y - 1), \qquad (b, y + 1).$$

By assumption, $|f_{yy}| \leq M$ on this rectangle. By (4) and (5), the integrand in (3) has absolute value at most

$$|H|\ M \leq |h|\ M.$$

Thus the absolute value of (3) is at most

$$|h|\ M(b - a),$$

which approaches 0 as $h \to 0$, since M and $b - a$ are fixed numbers. This proves the theorem. ■

The assumption made in the theorem that f_{yy} is bounded in each rectangle is satisfied if f_{yy} is continuous. So the theorem does cover the cases commonly encountered. In advanced calculus the theorem is proved without any assumption on f_{yy}. (See, for instance, R. C. Buck, *Advanced Calculus,* 3d ed., p. 118, McGraw-Hill, New York, 1978.)

EXERCISES FOR SEC. J.2: THE DERIVATIVE OF $\int_a^b f(x, y)\,dx$ WITH RESPECT TO y

1 Verify the theorem for $f(x, y) = x^3y^4$.

2 Verify the theorem for $f(x, y) = \cos xy$.

3 For what value of y does the function

$$F(y) = \int_0^{\pi/2} (y - \cos x)^2\,dx$$

have a minimum? (Use the theorem of this section.)

4 Let $G(u, v, w) = \int_u^v f(w, x)\,dx$. Find (*a*) $\partial G/\partial v$, (*b*) $\partial G/\partial u$, (*c*) $\partial G/\partial w$.

5 Let $G(u) = \int_0^u f(u, x)\,dx$. Find dG/du.

6 Let $G(u, v) = \int_0^u e^{-vx^2}\,dx$. Find (*a*) $\partial G/\partial u$, (*b*) $\partial G/\partial v$.

7 Let $F(y) = \int_0^1 [(x^y - 1)/\ln x]\,dx$ for $y \geq 0$.

(*a*) Assuming that one may differentiate F by differentiating under the integral sign, show that $dF/dy = 1/(1 + y)$.

(*b*) From (*a*) deduce that $F(y) = \ln(1 + y) + C$.

(*c*) Show that the constant C in part (*b*) is 0 by examining the case $y = 0$.

J.3 THE INTERCHANGE OF LIMITS

Although the two theorems in Secs. J.1 and J.2 are independent, they both illustrate a certain type of problem which students who go on to advanced calculus will study, namely, the interchange of limits. To see what this means, let us take a new look at the theorem in Sec. J.1, which concerns the equality of the mixed partials. By (1) in Sec. J.1,

$$f_{xy}(0, 0) = \lim_{k\to 0}\left[\lim_{h\to 0}\frac{f(h, k) - f(0, k) - f(h, 0) + f(0, 0)}{hk}\right].$$

Similarly, from the definition of f_{yx}, it can be shown that

$$f_{yx}(0, 0) = \lim_{h\to 0}\left[\lim_{k\to 0}\frac{f(h, k) - f(0, k) - f(h, 0) + f(0, 0)}{hk}\right].$$

Note that the two quotients are identical, but that f_{xy} involves

$$\lim_{k\to 0}\left(\lim_{h\to 0}\right),$$

while f_{yx} involves

$$\lim_{h\to 0}\left(\lim_{k\to 0}\right).$$

It is tempting to claim that the order of taking limits should not matter. But it does. (In Example 3 it is shown that f_{xy} does not always equal f_{yx}.) This instance raises the general question, "When can one interchange limits?" Example 1 presents a simple case in which the order of taking the limits *does* matter.

EXAMPLE 1 Let $f(x, y) = x^y$ for $x > 0$ and $y > 0$. Evaluate the two "repeated limits"

$$\lim_{y\to 0^+}\left(\lim_{x\to 0^+} x^y\right) \quad \text{and} \quad \lim_{x\to 0^+}\left(\lim_{y\to 0^+} x^y\right).$$

SOLUTION

$$\lim_{y\to 0^+}\left(\lim_{x\to 0^+} x^y\right) = \lim_{y\to 0^+} 0 = 0.$$

On the other hand,

$$\lim_{x\to 0^+}\left(\lim_{y\to 0^+} x^y\right) = \lim_{x\to 0^+} 1 = 1.$$

This shows that *the order of taking limits may affect the result*. Moreover, it suggests why the symbol 0^0 is not given any meaning. ■

The next example also illustrates the effect of switching the order of taking limits. In Example 3 it becomes the basis of an illustration that shows the mixed partial derivatives f_{xy} and f_{yx} are not always equal.

EXAMPLE 2 Let

$$g(x, y) = \begin{cases} \dfrac{x^2 - y^2}{x^2 + y^2} & \text{if } (x, y) \neq (0, 0), \\ 0 & \text{if } (x, y) = (0, 0). \end{cases}$$

Show that $$\lim_{x \to 0} \left[\lim_{y \to 0} g(x, y) \right] \neq \lim_{y \to 0} \left[\lim_{x \to 0} g(x, y) \right].$$

SOLUTION

$$\lim_{y \to 0} g(x, y) = \lim_{y \to 0} \frac{x^2 - y^2}{x^2 + y^2} = \frac{x^2}{x^2} = 1.$$

Thus $$\lim_{x \to 0} \left[\lim_{y \to 0} g(x, y) \right] = \lim_{x \to 0} 1 = 1.$$

On the other hand, $$\lim_{x \to 0} g(x, y) = \lim_{x \to 0} \frac{x^2 - y^2}{x^2 + y^2} = \frac{-y^2}{y^2} = -1.$$

Thus $$\lim_{y \to 0} \left[\lim_{x \to 0} g(x, y) \right] = \lim_{y \to 0} (-1) = -1. \quad \blacksquare$$

EXAMPLE 3 Let $f(x, y) = xyg(x, y)$, where g is given in Example 2. Show that $f_{xy}(0, 0) \neq f_{yx}(0, 0)$.

SOLUTION That f_x, f_y, f_{xy}, f_{yx} exist at all points is left for the reader to demonstrate. (See Exercise 8.) Note that $f(x, y) = 0$ whenever x or y is 0.

By (1) in Sec. J.1,

$$f_{xy}(0, 0) = \lim_{k \to 0} \left\{ \lim_{h \to 0} \frac{[f(h, k) - f(0, k)] - [f(h, 0) - f(0, 0)]}{hk} \right\}$$

$$= \lim_{k \to 0} \left[\lim_{h \to 0} \frac{f(h, k)}{hk} \right] = \lim_{k \to 0} \left[\lim_{h \to 0} g(h, k) \right] = -1$$

from Example 2. Similarly,

$$f_{yx}(0, 0) = \lim_{h \to 0} \left[\lim_{k \to 0} g(h, k) \right] = 1.$$

Therefore, $f_{xy}(0, 0) \neq f_{yx}(0, 0)$. $\blacksquare$

The theorem in Sec. J.2, which asserts that in general

$$\frac{d}{dy} \left[\int_a^b f(x, y)\, dx \right] = \int_a^b f_y(x, y)\, dx,$$

also concerns the validity of switching limits. After all, both the derivative and the definite integral are defined as limits.

The exercises present more examples of the interchange of limits. The main point of the examples and the exercises is that the interchange of limits is a risky business. Fortunately, there are theorems that imply that sometimes the order of taking limits has no effect on the outcome.

EXERCISES FOR SEC. J.3: THE INTERCHANGE OF LIMITS

1 Let $f(x, y) = 1$ if $y \geq x$ and let $f(x, y) = 0$ if $y < x$.
(*a*) Shade in the part of the plane where $f(x, y) = 1$.
(*b*) Show that $\lim_{x\to\infty} [\lim_{y\to\infty} f(x, y)] = 1$.
(*c*) Show that $\lim_{y\to\infty} [\lim_{x\to\infty} f(x, y)] = 0$.

2 Show that $\lim_{x\to 0^+} [\lim_{n\to\infty} nx/(1 + nx)] = 1$, while $\lim_{n\to\infty} [\lim_{x\to 0^+} nx/(1 + nx)] = 0$.

3 Let $f_n(x) = nx/(1 + n^2x^4)$. Show that

$$\int_0^\infty \lim_{n\to\infty} f_n(x)\,dx = 0,$$

but

$$\lim_{n\to\infty} \int_0^\infty f_n(x)\,dx = \frac{\pi}{4}.$$

4 Show that $\lim_{x\to 0} [\lim_{y\to 0} x^2/(x^2 + y^2)]$ is not equal to $\lim_{y\to 0} [\lim_{x\to 0} x^2/(x^2 + y^2)]$.

5 Let $f_n(x) = n\pi \sin(n\pi x)$ if $0 \leq x \leq 1/n$ and 0 otherwise.
(*a*) Graph f_1, f_2, and f_3.
(*b*) Show that $\lim_{n\to\infty} \int_0^1 f_n(x)\,dx = 2$, but that $\int_0^1 \lim_{n\to\infty} f_n(x)\,dx = 0$.

6 Compare

$$\lim_{x\to\infty}\left(\lim_{y\to\infty} \frac{x^2}{x^2 + y^2 + 1}\right)$$

and

$$\lim_{y\to\infty}\left(\lim_{x\to\infty} \frac{x^2}{x^2 + y^2 + 1}\right).$$

7 Let $f_n(x) = (1/n) \sin nx$ for all x and all positive integers n. Show that

$$\lim_{n\to\infty}\left[\lim_{h\to 0} \frac{f_n(h) - f_n(0)}{h}\right] = 1,$$

while

$$\lim_{h\to 0}\left[\lim_{n\to\infty} \frac{f_n(h) - f_n(0)}{h}\right] = 0.$$

8 Show that f_{xy} and f_{yx} exist at all points in the plane, where f is the pathological function in Example 3.

9 Show that

$$\int_0^\infty \left[\int_0^1 (2xy - x^2y^2)e^{-xy}\,dx\right] dy = 1,$$

but

$$\int_0^1 \left[\int_0^\infty (2xy - x^2y^2)e^{-xy}\,dy\right] dx = 0.$$

10 Show that l'Hôpital's rule in the zero-over-zero case concerns the equality of these two limits:

$$\lim_{\Delta t\to 0}\left[\lim_{t\to a} \frac{f(t + \Delta t) - f(t)}{g(t + \Delta t) - g(t)}\right]$$

and

$$\lim_{t\to a}\left[\lim_{\Delta t\to 0} \frac{f(t + \Delta t) - f(t)}{g(t + \Delta t) - g(t)}\right].$$

11 Let $f_n(x)$ be defined for each x in $[0, 1]$ and each positive integer n. Assume that f_n is continuous for each n and that $\lim_{n\to\infty} f_n(x)$ exists for each x in $[0, 1]$. Call this limit $f(x)$,

$$\lim_{n\to\infty} f_n(x) = f(x).$$

(*a*) Show that the statement "f is continuous at a" is equivalent to the equation

$$\lim_{n\to\infty}\left[\lim_{x\to a} f_n(x)\right] = \lim_{x\to a}\left[\lim_{n\to\infty} f_n(x)\right].$$

(*b*) In particular, let $f_n(x) = x^n$. Show that in this case the two repeated limits in (*a*) are not necessarily equal.

Appendix K The Converse of a Statement

Many mathematical assertions can be put into the form "if A, then C," where A is an assumption and C is a conclusion. For instance, the assertion "the product of a multiple of 2 and a multiple of 3 is a multiple of 6" can be restated as "if a is a multiple of 2 and b is a multiple of 3, then ab is a multiple of 6." (Here a and b refer to integers.) As another example, the assertion "the square of an even integer is always even" can be rephrased as "if a is even, then a^2 is even." (a refers to integers.)

Starting with a statement of the form "if A, then C," we can construct a new statement, "if C, then A," which is called the **converse** of "if A, then C." Even if the original statement, "if A, then C," is true, its converse may be false. Example 1 illustrates this case. In Example 2 the converse happens also to be true.

EXAMPLE 1 Is the converse of "if a is a multiple of 2 and b is a multiple of 3, then ab is a multiple of 6" true or false?

SOLUTION The converse is "if ab is a multiple of 6, then a is a multiple of 2 and b is a multiple of 3." (The symbols a and b refer to integers.) The counterexample $a = 6$ and $b = 1$ shows that it is not true for all possible pairs a and b under consideration. Hence the converse is false. ■

EXAMPLE 2 Is the converse of "if a is even, then a^2 is even" true or false?

SOLUTION The converse is "if a^2 is even, then a is even." (a refers to integers.) This assertion is true. If a were odd, a^2 would be odd, since the product of two odd numbers is always odd. ■

There are two important reasons to distinguish between a statement "if A, then C" and its converse "if C, then A." First of all, they make completely different claims. Second, although the first statement may be true, the converse may be true or it may be false. As it happens, calculus contains many statements whose converses are false. Therefore, failure to distinguish a statement from its converse can lead to confusion. The following exercises should help you avoid potential disaster.

EXERCISES FOR APPENDIX K: THE CONVERSE OF A STATEMENT

In each of Exercises 1 to 6 two true statements are given. However, the converse of one is true and the converse of the other is false. In each case write out the converse. If it is true, explain why it holds. It it is false, give a counterexample.

1 (*a*) If $a = b$, then $a^2 = b^2$. (a and b refer to real numbers.)
(*b*) If $a = b$, then $a^3 = b^3$. (a and b refer to real numbers.)

2 (*a*) If a is odd, then a^2 is odd. (a refers to integers.)
(*b*) If a is odd, then $a + a$ is even. (a refers to integers.)

3 (*a*) If $b = c$, then $ab = ac$. (a, b, and c refer to real numbers.)
(*b*) If $b = c$, then $a + b = a + c$. (a, b, and c refer to real numbers.)

4 (*a*) If a is rational, then a^2 is rational. (a refers to real numbers.)
(*b*) If a is rational, then $2a$ is rational. (a refers to real numbers.)

5 (*a*) If a and b are odd, then ab is odd. (a and b refer to integers.)
(*b*) If a and b are even, then ab is even. (a and b refer to integers.)

6 (*a*) If a is a multiple of 6, then a^2 is a multiple of 6. (a refers to integers.)
(*b*) If a is a multiple of 4, then a^2 is a multiple of 4. (a refers to integers.)

In each of Exercises 7 to 12 decide whether the converse of the given statement is true or false.

7 If the three sides of triangle T are equal, then the three angles of triangle T are equal.

8 If quadrilateral Q is a square, then all four sides of quadrilateral Q are equal.

9 If quadrilateral Q is a parallelogram, then opposite sides of quadrilateral Q are equal.

10 If hexagon H is regular, then opposite sides of hexagon H are equal.

11 [$P(x)$ denotes a polynomial in x with real coefficients, and a denotes a real number.] If $P(a) = 0$, then $x - a$ divides $P(x)$. (See "factor theorem" in Appendix C, page S-29.)

12 If m is even, then the polynomial $x^2 - 1$ divides the polynomial $x^m - 1$. (m refers to nonnegative integers.)

Exercises 13 to 20 refer to concepts developed in the text. In each case, decide whether the converse of the given statement is true or false. The functions f and g are assumed to be defined on the entire x axis.

13 If f and g are one-to-one, then $f \circ g$ is one-to-one (Sec. 1.3).

14 If f is a one-to-one function, then $f \circ f$ is a one-to-one function (Sec. 1.3).

15 If $f'(x) = 0$ for all x, then $f(x)$ is constant (Sec. 4.1).

16 If $f'(x)$ is positive for all x, then f is increasing (Sec. 4.1).

17 If f is a polynomial of degree 2, then $f^{(3)}(x) = 0$ for all x (Sec. 4.1).

18 If f is differentiable at a, then f is continuous at a (Sec. 3.3).

19 If f has a relative maximum or minimum at a, then $f'(a) = 0$ (Sec. 4.1).

20 If f is elementary, then f' is elementary (Sec. 5.4).

21 If $\Sigma_{n=1}^{\infty} a_n$ converges, then $\lim_{n\to\infty} a_n = 0$ (Sec. 10.2).

22 If $\Sigma_{n=1}^{\infty} |a_n|$ converges, then $\Sigma_{n=1}^{\infty} a_n$ converges (Sec. 10.5).

23 If **F** is a conservative vector field, then **F** is a gradient field (Sec. 15.3).

24 If **F** is a conservative vector field, then **curl F** = **0** (Secs. 15.3 and 15.7).

■

25 Is either of these two statements true? If so, which? "If a is irrational, then a^2 is irrational." "If a^2 is irrational, then a is irrational." (a refers to real numbers.)

26 Taking the converse of a statement can sometimes raise interesting questions. For instance, consider the statement "if a is a nonnegative integer, then 2^a, 3^a, and 5^a are integers." (a refers to real numbers.) Its converse is "if 2^a, 3^a, and 5^a are integers, then a is a nonnegative integer." This converse has been shown to be true by some very difficult, advanced mathematics. However, it is not known whether the converse of "if a is a nonnegative integer, then 2^a and 3^a are integers" is true. Do you think it is? Is the converse of the statement "if a is a nonnegative integer, then 2^a is an integer" true or false?

Answers to Selected Odd-Numbered Exercises and to Guide Quizzes

CHAPTER 1. FUNCTIONS

Sec. 1.1. Functions

1.

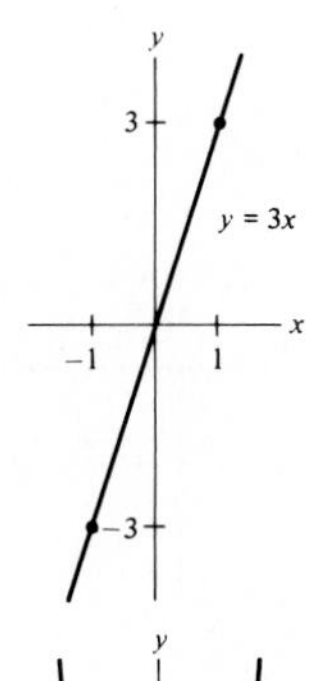

3.

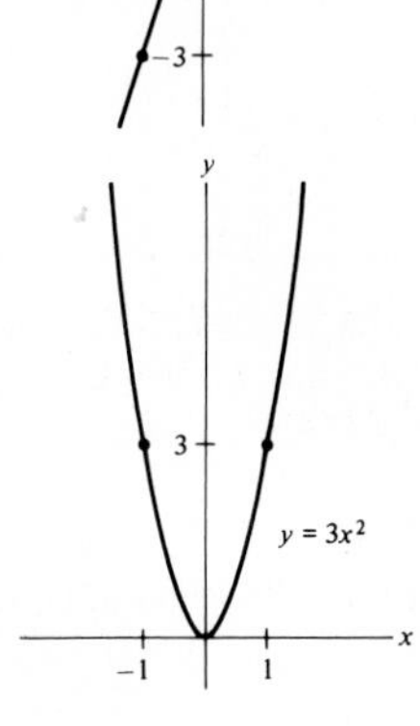

5.

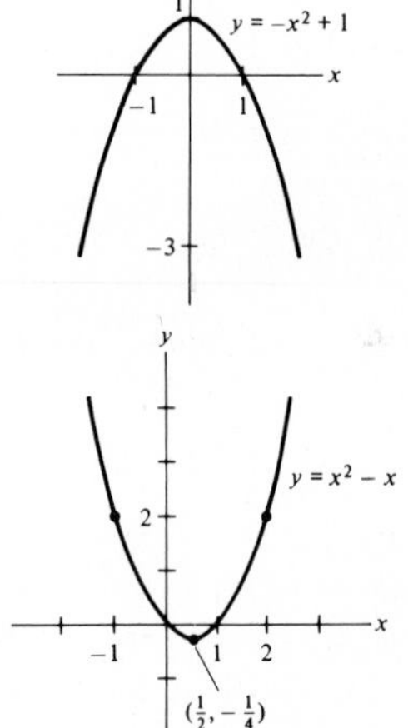

7.

y = x² − x

2

−1

1 2

$(\frac{1}{2}, -\frac{1}{4})$

9.

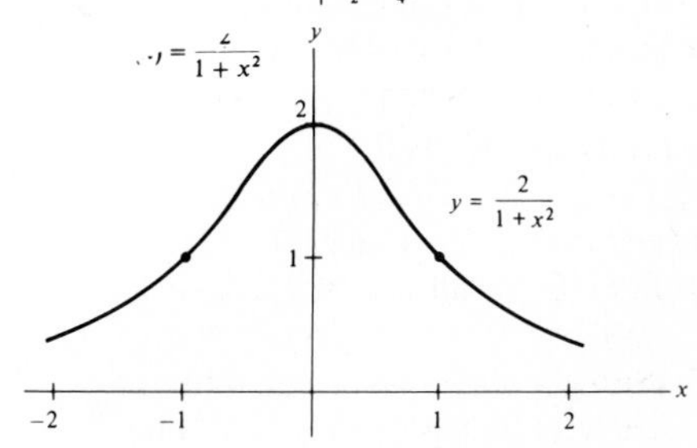

11. Domain: $[0, \infty)$; range: $[0, \infty)$
13. Domain: $[-2, -2]$; range: $[0, 2]$
15. Domain: all nonzero x; range: $(0, \infty)$
17. Domain: all nonzero x; range: all nonzero y
19. Domain: $(0, \infty)$; range: $(0, \infty)$
21. (*b*) and (*c*) are; (*a*) is not.
23. (*a*) 0 (*b*) 4 (*c*) 2.25 (*d*) 1
25. (*a*) 27 (*b*) 27
27. $3a^2 + 3a + 1$
29. $-(c + d)/(c^2d^2)$
31. $1 - (1/uv)$
33. $\sqrt{x^2 + 16} + \sqrt{x^2 - 8x + 25}$
35. $2\pi x(x + y)$
37.

x	-2	-1	0	$\frac{1}{2}$	1	2	3
$x^2 + x$	2	0	0	$\frac{3}{4}$	2	6	12

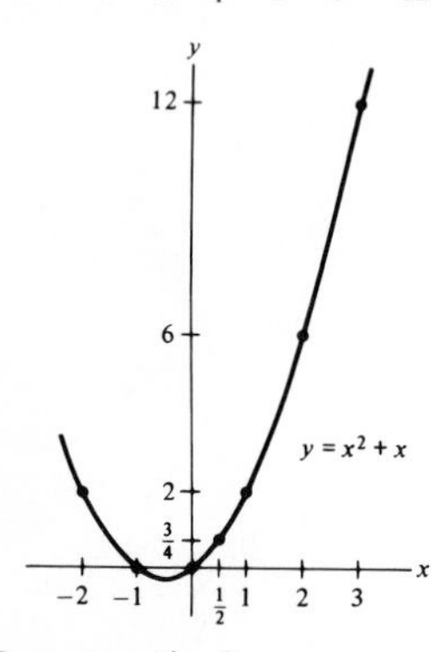

39. (*a*) 0, 3, 6, 9, respectively
(*b*)

x	0	1	2	3
$f(x)$	0	3	6	9

(*c*)

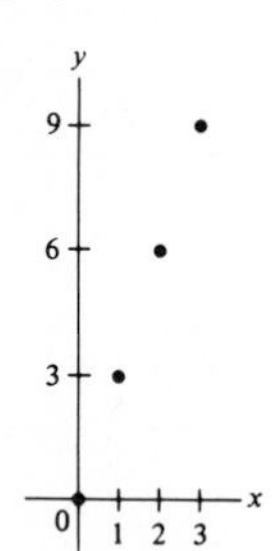

(*d*) Neither
41. (*a*) 7.0000 (*b*) 6.0100 (*c*) 5.9900 (*d*) 6.0001
(*e*) Approaches 6
43. (*b*) and (*c*)
45. (*c*) Equals -3. (*d*) Equals 9, equals -9.
(*e*) $f(x) = 3x$ for all x
47. $f(x) = 137 \cdot 2^x$; $f(x) = 0$; $f(x) = 2^{\lfloor x \rfloor}$, where $\lfloor x \rfloor$ is the largest integer less than or equal to x.
49. $f(x) = 0$, $f(x) = \log_3 x$, $f(x) = -4 \log_6 x$

Sec. 1.2. Composite Functions

1. $y = x^6$
3. $y = \sqrt{1 + 2x^3}$
5. $y = u^2$, $u = 1 + x^3$ (for example)
7. $y = \sqrt[3]{u}$, $u = 1/v$, $v = 1 + x^2$ (for example)
9. $w = 27t^3$
11. $w = \sqrt[5]{1 + x^{-2}}$
13. $b = a + 1$
15. (*b*) No
17. $f(0) = 0$
19. Yes; one
21. Infinitely many
23. (*a*) Odd (*b*) Even (*c*) Even (*d*) Neither
(*e*) Even (*f*) Neither (*g*) Neither (*h*) Odd

Sec. 1.3. One-to-One Functions and Their Inverse Functions

1. (*a*) No (*b*) Yes; $x = \sqrt[4]{y}$
3. (*a*) Yes; $x = \sqrt[5]{y - 1}$ (*b*) Yes; $x = \sqrt[5]{y - 1}$
5. Yes

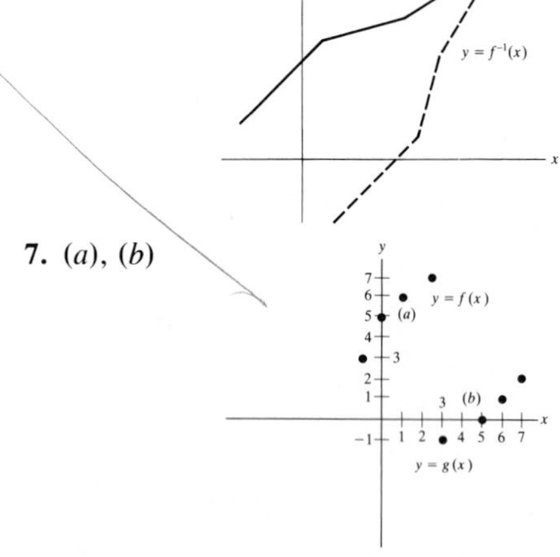

7. (*a*), (*b*)

9. Yes
11. (*a*)

x	1	2	3
$f(x)$	a	b	c

x	1	2	3
$f(x)$	a	c	b

x	1	2	3
$f(x)$	b	c	a

x	1	2	3
$f(x)$	c	a	b

x	1	2	3
$f(x)$	b	a	c

x	1	2	3
$f(x)$	c	b	a

(*b*) For example, the inverse of the third function in the first column is

y	a	b	c
$f^{-1}(y)$	2	1	3

13. True
15. False; $f(x) = x^2$, $g(x) = 2^x$
17. Increasing
19. Decreasing
21. Yes
23. Yes; $x = y/(1 - y)$
25. (*a*) No (*b*) No (*c*) No (*d*) Yes

Sec. 1.S. Guide Quiz

1. (*a*) No (*b*) Yes
2. (*a*), (*c*)

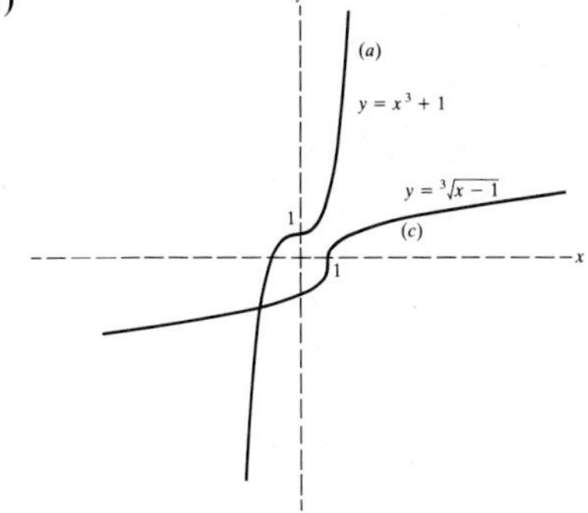

(*b*) It is increasing. (*d*) $x = \sqrt[3]{y - 1}$
3. (*a*) $3x^2 + 3xh + h^2$ (*b*) $-1/[x(x + h)]$
4. Approaches 0
5. $y = f(g(h(x)))$, where $f(u) = u^3 + u$, $g(v) = \sqrt{v}$, $h(x) = 1 + 2x$ (for example)

Sec. 1.S. Review Exercises

1. $\pi x^2\sqrt{a^2 - x^2}$
3. $ad \neq bc$
5. Domain: all x; range: all y
7. Domain: all x; range: $[0, \infty)$
9. Domain: $(-1, \infty)$; range: $(0, \infty)$
11. 0.41
13. $-1/[(a + h + 1)(a + 1)]$
15. $u^2 + uv + v^2 - 3$
17.

19.

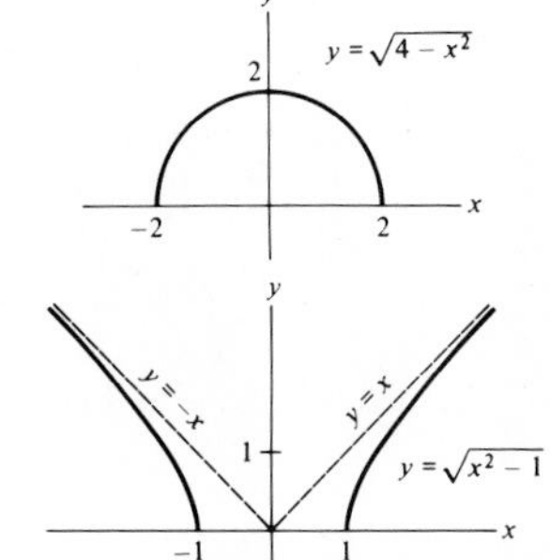

21.

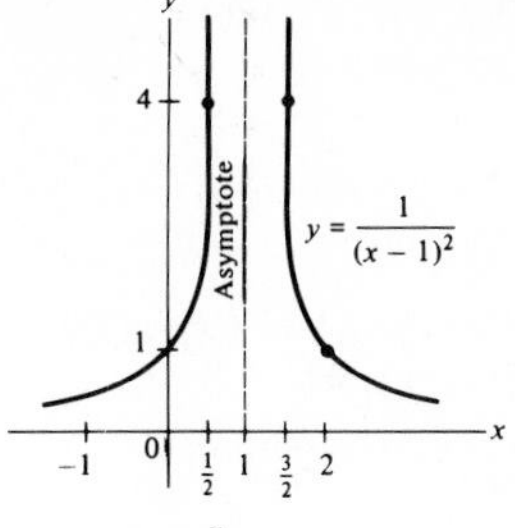

23.

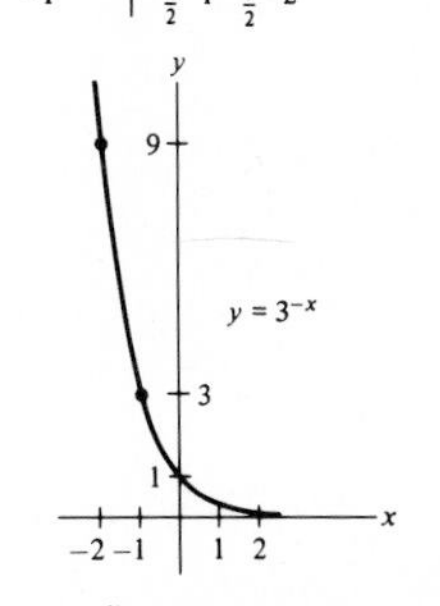

25.

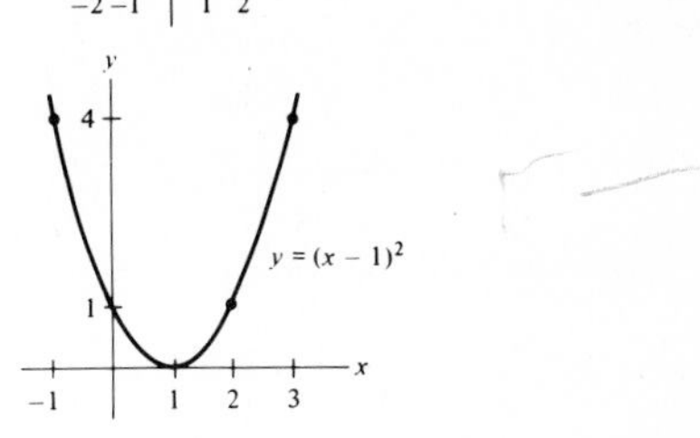

27. (*b*) and (*c*)
29. (*a*), (*b*)

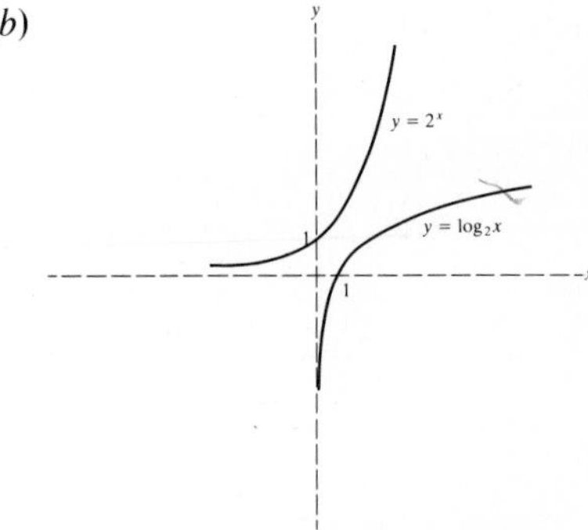

(*c*) $\log_2 x$

31. (*a*)

x	0.001	0.01	0.1	1	2	10	100
$f(x)$	2.717	2.705	2.594	2	1.732	1.271	1.047

(*b*)

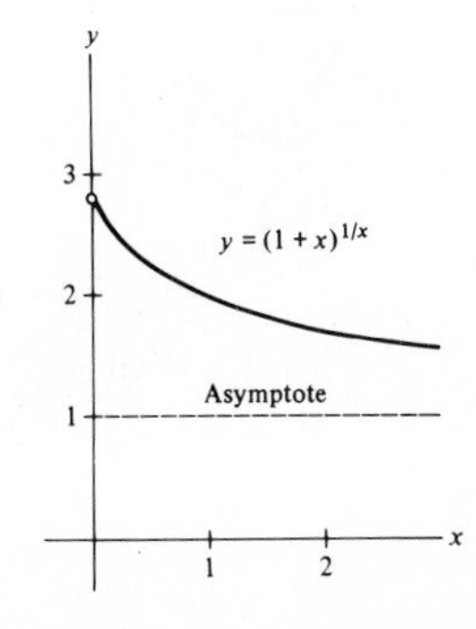

33. (*a*) $x(50 - x)$ (*b*) (0, 50)
35. (*b*) Infinitely many
37. No
39. (*a*) One (*b*) No

CHAPTER 2. LIMITS AND CONTINUOUS FUNCTIONS

Sec. 2.1. The Limit of a Function

1. 12
3. 4
5. $\frac{4}{3}$
7. $\frac{1}{5}$
9. 25
11. 0
13. 1
15. 3
17. $-\frac{1}{4}$
19. $\frac{1}{4}$
21. 2
23. 1
25. (*a*) 2 (*b*) 1 (*c*) 1 (*d*) 2
27. (*a*) 0 (*b*) 0 (*c*) 4
29. (*a*)

x	1	0.1	0.01	0.001	−1	−0.01	−0.001
$f(x)$	2	1.618	1.588	1.585	1.333	1.582	1.585

(*b*) Yes; ≈ 1.585
31. 1
33. (*a*)

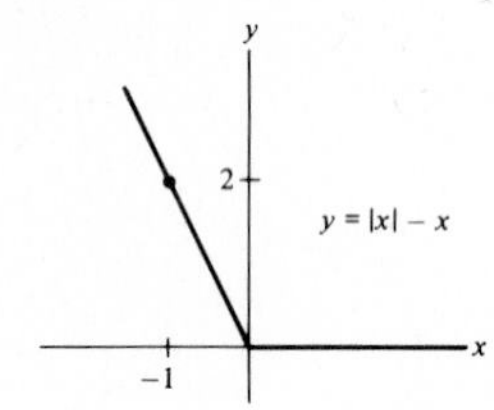

(*b*) For all *a*
35. (*a*)

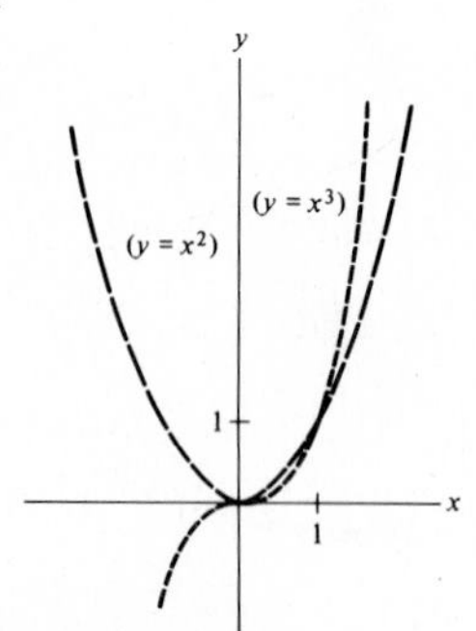

(*b*) No (*c*) $\lim_{x\to 1} f(x) = 1$
(*d*) $\lim_{x\to 0} f(x) = 0$ (*e*) $a = 0$ or 1

Sec. 2.2. Computations of Limits

1. ∞
3. $-\infty$
5. ∞
7. 0
9. ∞
11. 0
13. ∞
15. ∞
17. 0
19. 50
21. $\frac{2}{3}$
23. $\frac{2}{3}$
25. (*a*) ∞ (*b*) $-\infty$ (*c*) does not exist
27. Both are wrong; the limit is −13.
29. (*a*) ∞ (*b*) Indeterminate (*c*) ∞
(*d*) Indeterminate
31. (*a*) a/b (*b*) ∞ (*c*) 0
33. No limit; oscillates
35. (*a*) 2 (*b*) Does not exist

Sec. 2.3. Asymptotes and Their Use in Graphing

1.

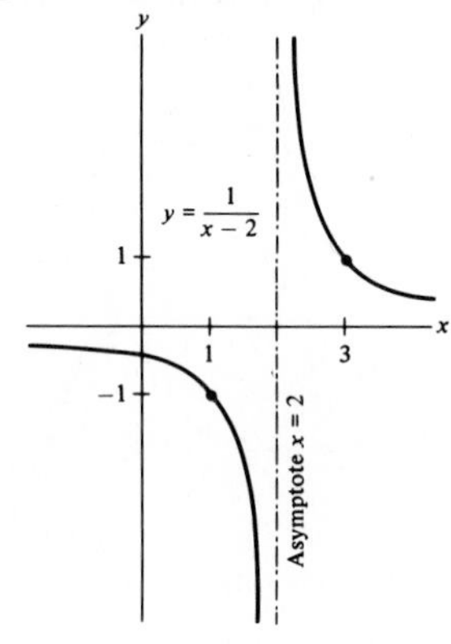

3.

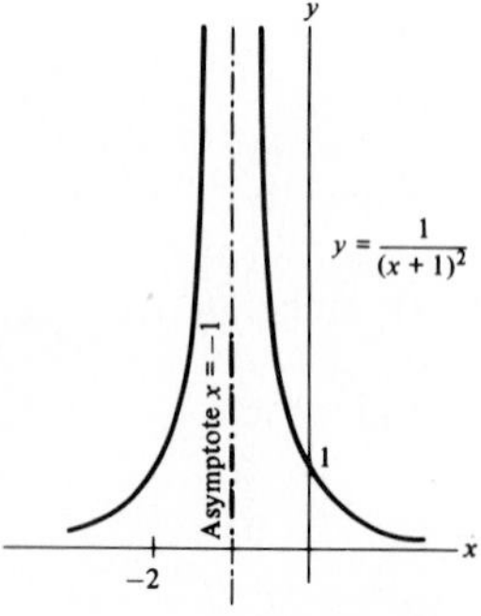

5.

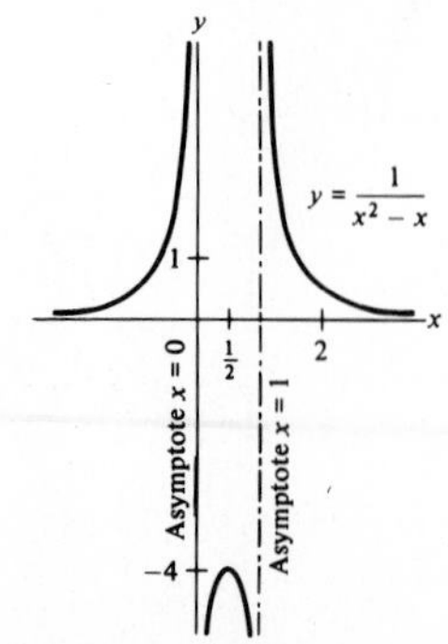

7.

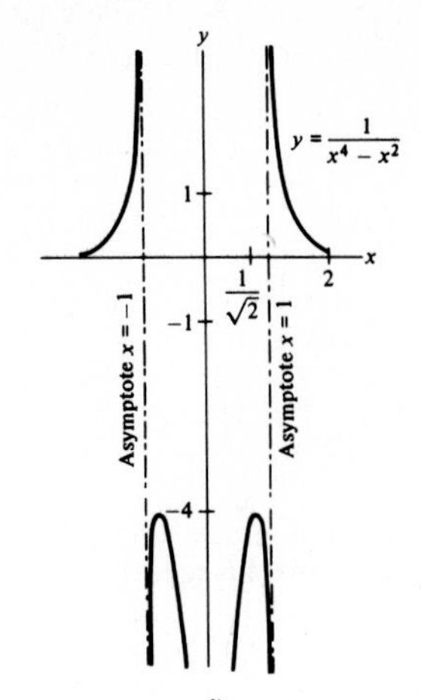

9.

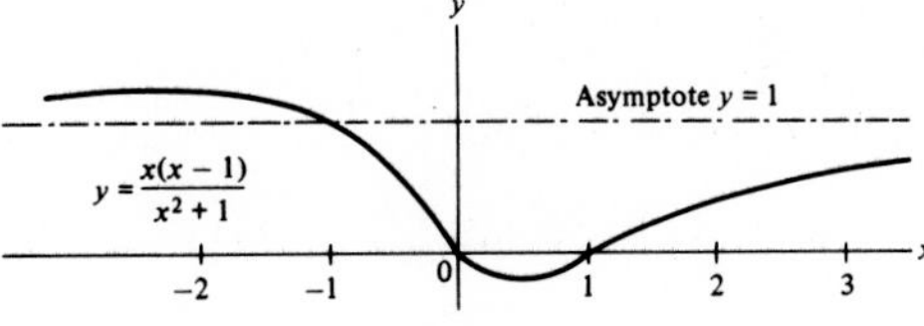

11.

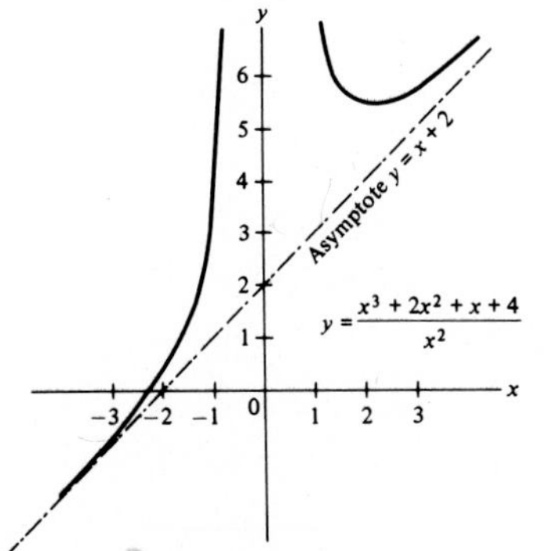

13.

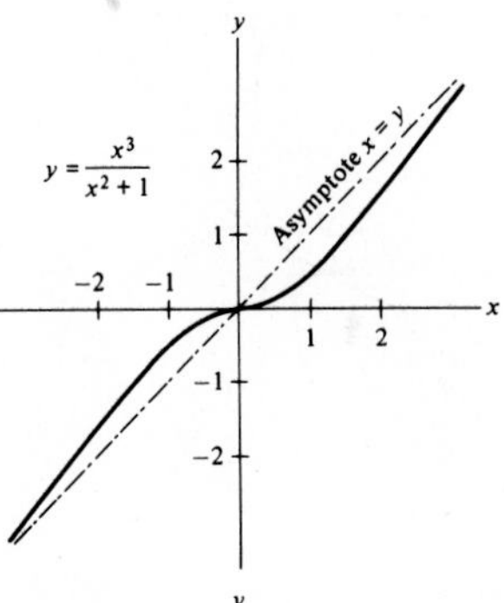

15.

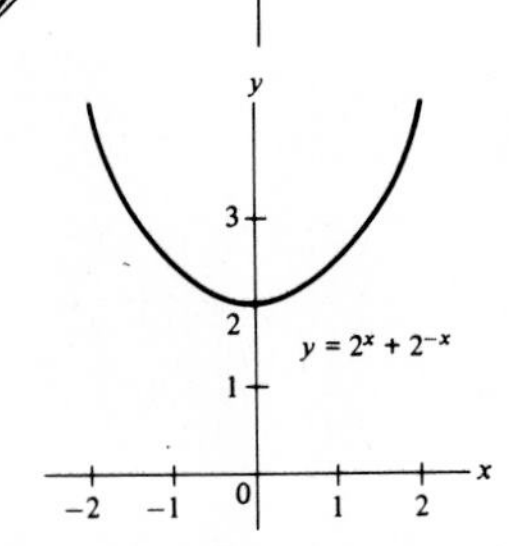

17.

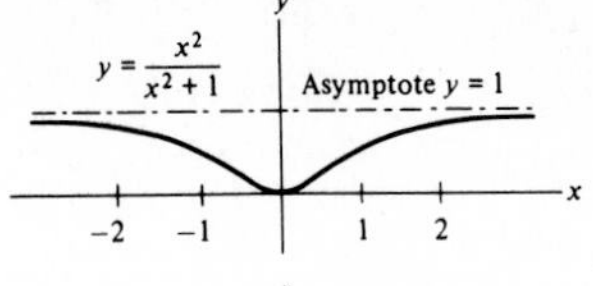

19.

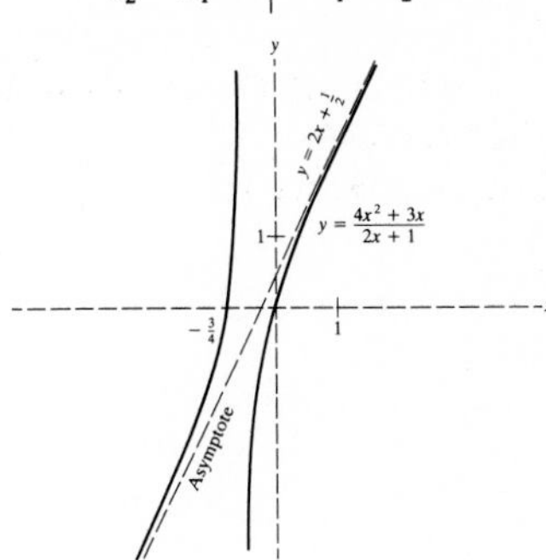

21.

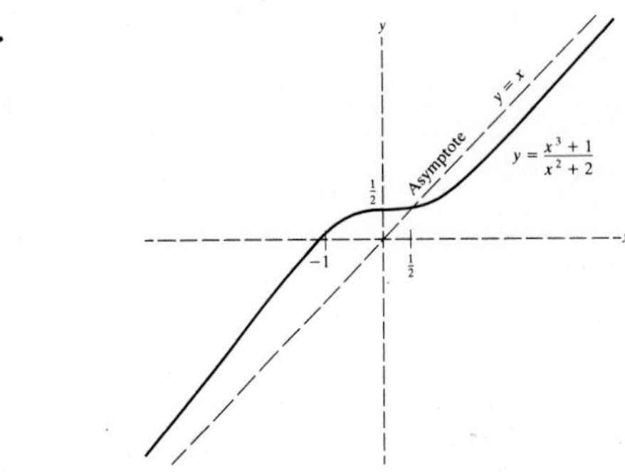

23. $y = ax + (b - a)$

25. (*b*) $-\frac{1}{4}$ (*c*) -4 (*d*) $(\frac{1}{2}, -4)$

Sec. 2.4. The Limit of $(\sin \theta)/\theta$ as θ Approaches 0

1. (*a*) $9\pi/4$ (*b*) $\theta/2$ (*c*) 2θ

3. $\frac{1}{2}$

5. $\frac{3}{5}$

7. 0

9. 0

11. $\frac{1}{2}$

13. ∞

15. ∞

17. (*a*) All nonzero x (*c*) 0
(*d*) $x = n\pi$, n a nonzero integer
(*e*)

x	0.1	$\pi/2$	$3\pi/2$	2π	$5\pi/2$	3π	$7\pi/2$
$\sin x$	0.10	1	−1	0	1	0	−1
$\dfrac{\sin x}{x}$	1.00	0.64	−0.21	0	0.13	0	−0.09

(*f*), (*g*)

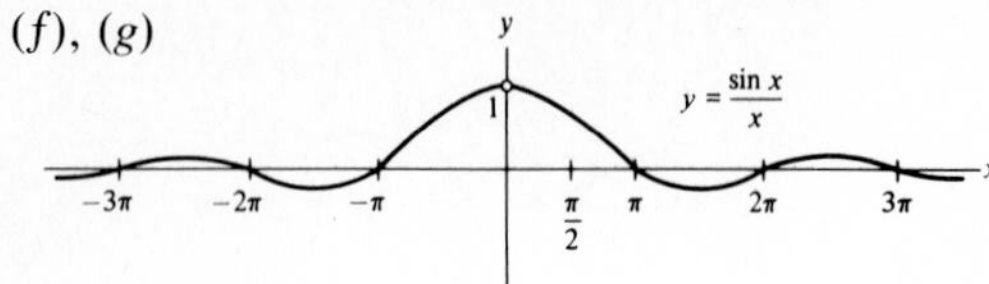

(*h*) 1

19. 0

21. (*b*)

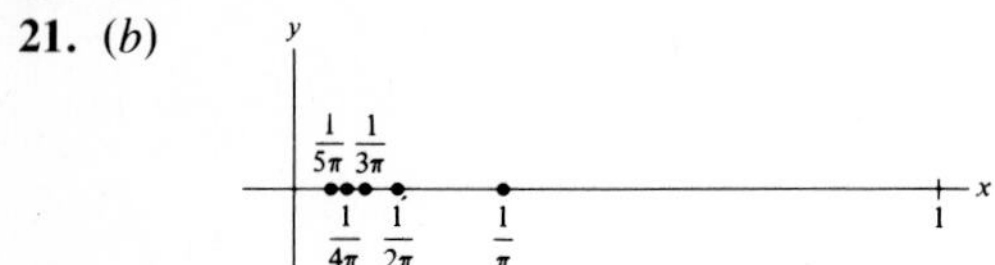

23. (*b*)

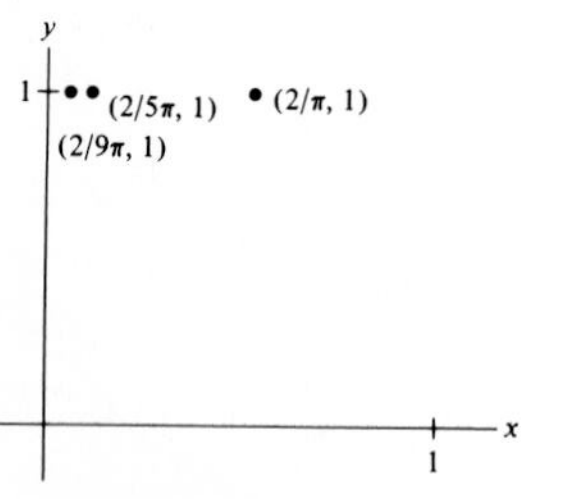

25. (*a*)

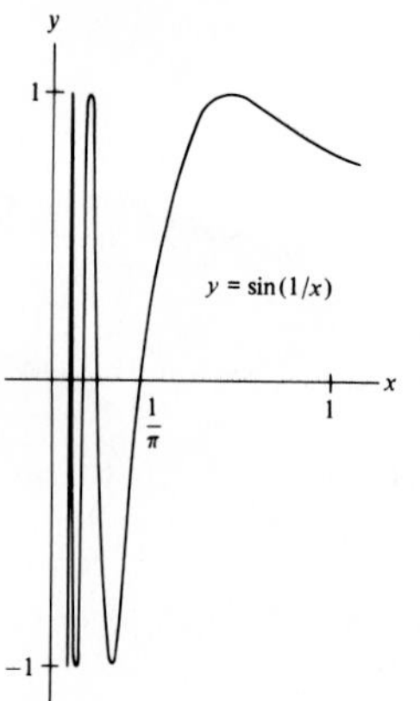

(*b*) No

27. Near 0.1667

29. (*b*) $\cos 0.1 \approx 0.995$, $\cos 0.01 \approx 0.99995$
(*c*) Calculator agrees to 5 places in first value, 9 places in second.

Sec. 2.5. Continuous Functions

1. (*a*) Yes; 1 (*b*) Yes

3. (*a*) No (*b*) No

5. (*a*) No (*b*) No

7. (*a*) Yes (*b*) No (*c*) No (*d*) No (*e*) Yes
(*f*) No

9. (*a*)

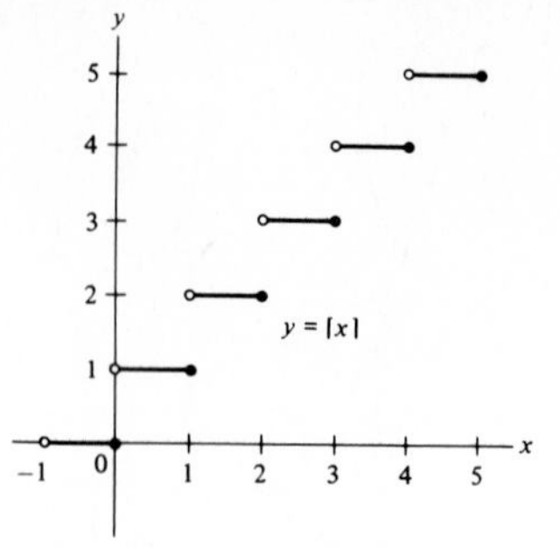

(*b*) Yes; 4 (*c*) Yes; 5 (*d*) No (*e*) No
(*f*) Nonintegers (*g*) Integers

11. Yes

13. (*a*)

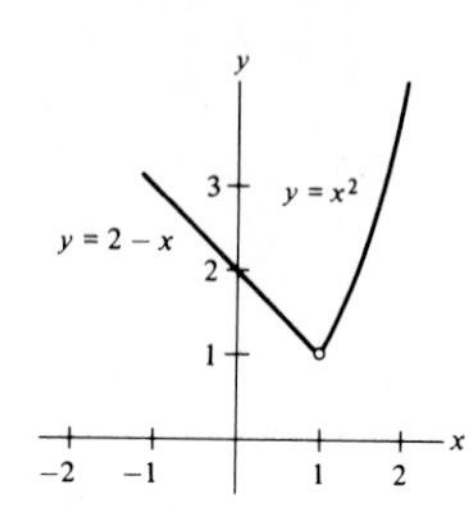

(*b*) Yes

15. (*a*)

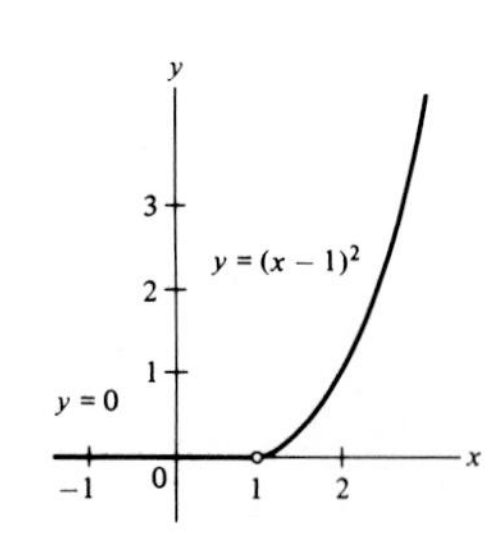

(*b*) Yes

17. (*a*) 1 (*b*) 1 (*c*) No (*d*) Yes
(*e*)

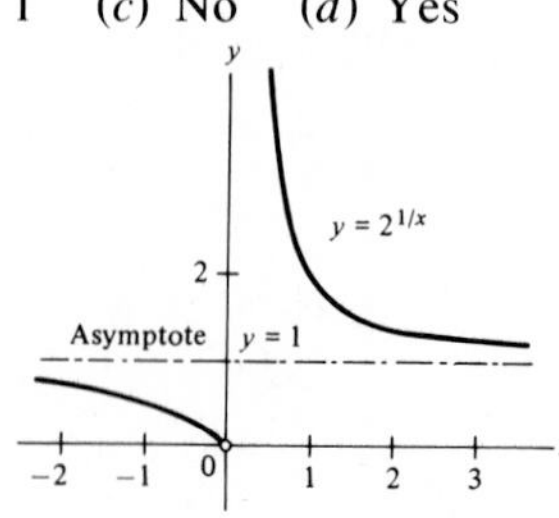

(*f*) No

19. (*a*)

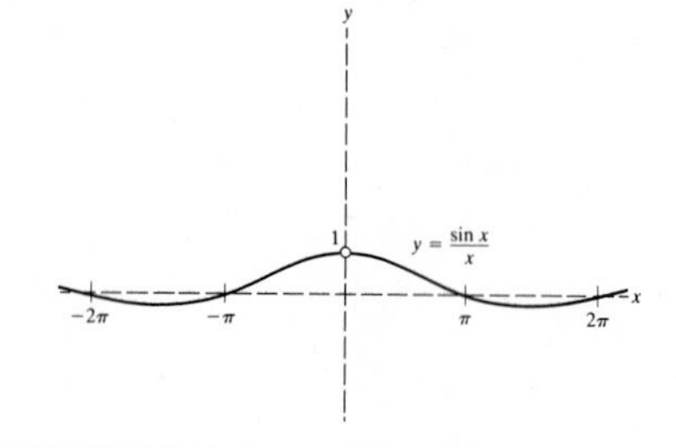

(*b*) Yes

21. (*a*)

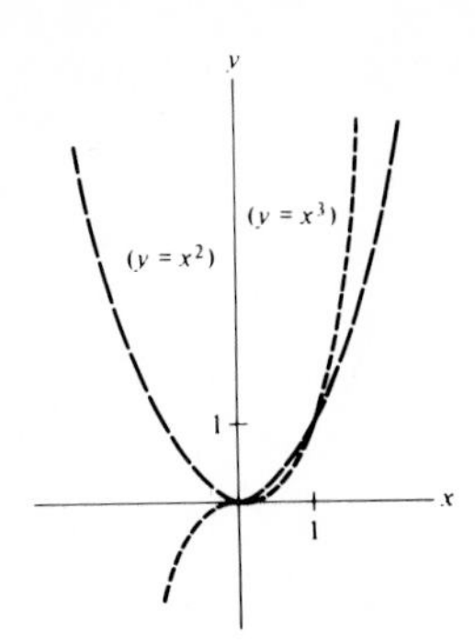

(*b*) 0, 1

Sec. 2.6. The Maximum-Value Theorem and the Intermediate-Value Theorem

1. (*a*) Yes (*b*) Yes
3. (*a*) Yes (*b*) Yes (*c*) No
5. (*a*) Yes (*b*) No
9. $c = \frac{5}{3}$
11. $c = \pi, 2\pi, 3\pi, 4\pi, 5\pi$
13. $c = \pi/3, 5\pi/3, 7\pi/3, 11\pi/3, 13\pi/3$
15. $c = -1, 0, 1$
17. Yes
25. 2 and 2
29. Yes

Sec. 2.7. Precise Definitions of "$\lim_{x\to\infty} f(x) = \infty$" and "$\lim_{x\to\infty} f(x) = L$"

1. (*a*), (*b*) Any two numbers ≥ 200 (*c*) 200
3. (*a*) 400 (*b*) 2,000
5. Let $D = E/3$.
7. Let $D = E - 5$.
9. Let $D = \frac{1}{2}(E - 4)$.
11. Let $D = \frac{1}{4}(E + 100)$.
13. (*a*) 10 (*b*) $\sqrt{E}$ (*c*) Any value will work. (*d*) Let $D = \sqrt{|E|}$.
15. (*a*), (*b*) Any two numbers ≥ 10 (*c*) 10 (*d*) Let $D = 1/\epsilon$.
17. Let $D = 1/\epsilon$.
19. Let $D = 2/\sqrt{\epsilon}$.
21. Let $D = 100 + (1/\epsilon)$.
27. For each number E there is a number D such that $f(x) < E$ for all $x > D$.

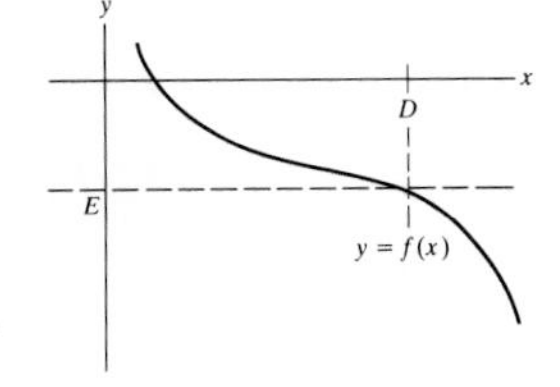

29. For each number E there is a number D such that $f(x) < E$ for all $x < D$.

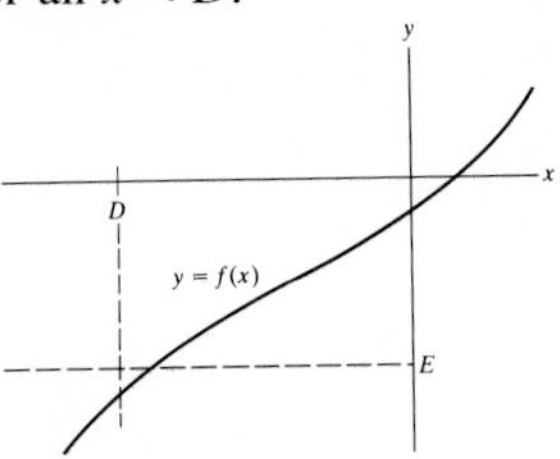

Sec. 2.8. Precise Definition of "$\lim_{x\to a} f(x) = L$"

5. $\frac{1}{5}$ (or any positive $\delta \leq \sqrt{5} - 2$)
7. (*b*) Let $\delta = \min(1, \epsilon/7)$. ("min" means the lesser of the two numbers.)
9. (*b*) Let $\delta = \min(1, \epsilon/12)$.
11. For each positive number ϵ there is a positive number δ such that $|f(x) - L| < \epsilon$ for all x between a and $a + \delta$.
13. For each number E there is a positive number δ such that $f(x) > E$ for all x satisfying $0 < |x - a| < \delta$.
15. For any number E there is a positive number δ such that $f(x) > E$ for all x satisfying $a < x < a + \delta$.
17. (*a*) $\frac{1}{30}$ (*b*) $\frac{1}{3}\sqrt{\epsilon}$
19. Let $\epsilon = \frac{1}{2}$.

Sec. 2.S. Guide Quiz

3. (*a*) Yes (*b*) No (*c*) No (*d*) No (*e*) No (*f*) Yes
4. (*a*) 6 (*b*) $\frac{3}{5}$ (*c*) -3 (*d*) 0 (*e*) $\frac{1}{2}$ (*f*) ∞ (*g*) 0 (*h*) 4 (*i*) 12 (*j*) Does not exist (*k*) Does not exist (*l*) 0 (*m*) 1 (*n*) $\frac{1}{8}$
5. (*a*) 12 (*b*) $\frac{3}{4}$ (*c*) 7 (*d*) Indeterminate (*e*) Indeterminate
6. (*a*) ∞ (*b*) 0 (*c*) Indeterminate (*d*) ∞ (*e*) Indeterminate

Sec. 2.S. Review Exercises

1. 1
3. $\frac{8}{3}$
5. $\frac{1}{2}$
7. $-\infty$
9. $\frac{1}{4}$
11. 2
13. ∞
15. 5
17. ∞
19. 1
21. 1
23. 0
25. $\frac{1}{3}$
27. 0
29. 0
31. 0

33. $\pi^2/16\sqrt{2}$
35. 1
37. $f(x) = 5x$, $g(x) = x$
39. $f(x) = 1/x$, $g(x) = x^2$
41. $f(x) = x^2$, $g(x) = x$
43. (*a*) Yes (*b*) No
45. (*a*) Yes (*b*) No
49. (*a*) Yes, 2 (*b*) Yes, 0
51. $c = 2$
53. $c = \pi/4$
55. 0
57. Yes; let $f(0) = 0$.
61. (*c*) No (*d*) Yes
63. (*a*) No (*b*) Yes

CHAPTER 3. THE DERIVATIVE

Sec. 3.1. Four Problems with One Theme

1. 6
3. -4
5. 12
7. (*a*) 0
(*b*)

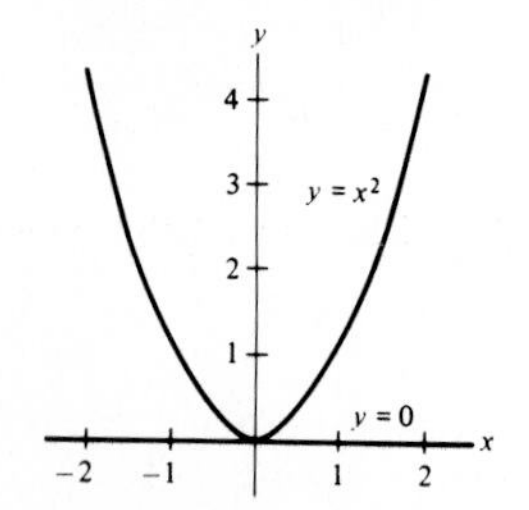

9. 96 ft/sec
11. 32 ft/sec
13. (*a*) 1.261 ft (*b*) 12.61 ft/sec
(*c*) $12 + 6h + h^2$ ft/sec (*d*) 12 ft/sec
15. (*a*) 3.9 (*b*) 3.99 (*c*) $4 + h$
17. (*a*), (*b*)

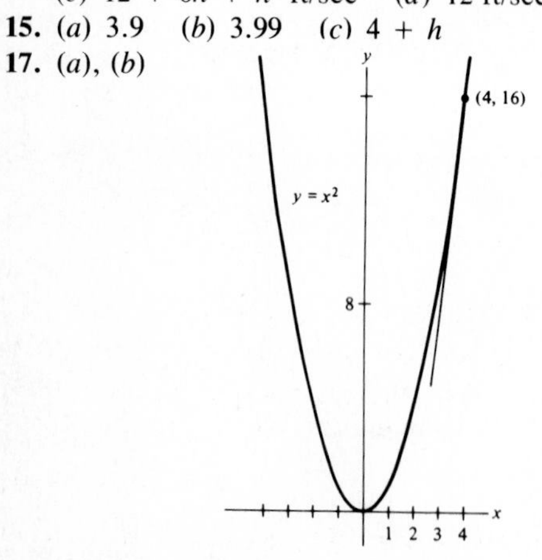

(*d*) 8.01 (*e*) 7.99
19. (*a*) 2.1 (*b*) 2.01 (*c*) 2.001 (*d*) 2
21. (*a*) 0.99 (*b*) 0.999 (*c*) 1
23. (*a*) 0.0601 (*b*) 6.01 (*c*) 5.99 (*d*) 6 (*e*) 6
25. $y = -2x - 1$
27. (*a*) 0.0401 (*b*) 4.01 (*c*) 4
29. (*a*), (*b*), (*c*)

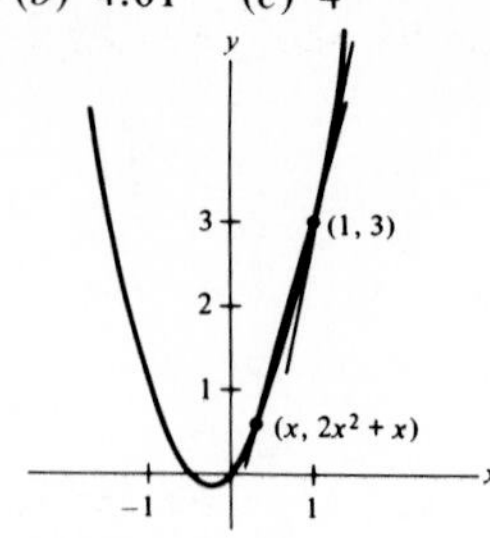

(*d*) $2x + 3$ (*e*) 5
31. $2x$
33. $2x$
35. (*a*)

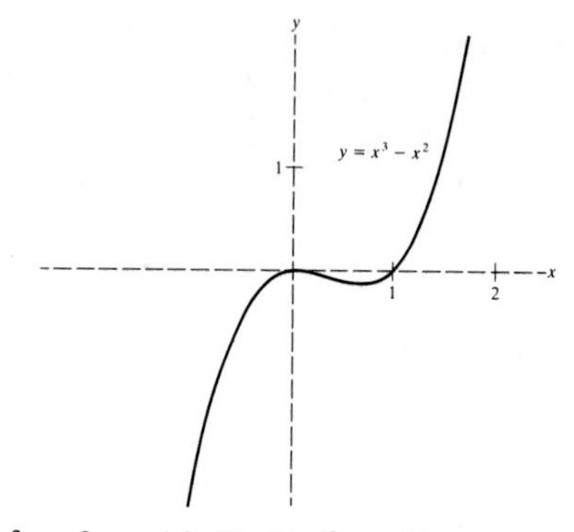

(*b*) $3x^2 - 2x$ (*c*) $(0, 0)$, $(\frac{2}{3}, -\frac{4}{27})$
(*d*) $(-\frac{1}{3}, -\frac{4}{27})$, $(1, 0)$
37. No
39. $(1, 0)$ and $(1 - \sqrt{3}, 9 - 5\sqrt{3})$

Sec. 3.2. The Derivative

1. $4x^3$
3. 2
5. $2x$
7. $-10x + 4$
9. $7/(2\sqrt{x})$
11. $-1/x^2$
13. $-2/x^3$
15. $6/x^2$
17. -4
19. $5a^4$
21. $\frac{1}{12}$
23. $\frac{1}{32}$
25. (*a*) 4.641 (*b*) 4
27. (*a*) 4.060401 (*b*) 4
29. (*b*)

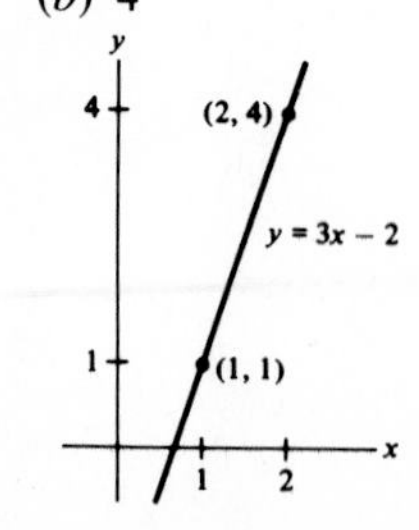

31. (*a*) $a = 2$ (*b*) $a = 1, 2$ (*c*) $a = 1, 2, 3$

Sec. 3.3. The Derivative and Continuity; Antiderivatives

1. $3x^2$
3. $1/(2\sqrt{x})$
5. $10x$
7. $-(3/x^2)$
9. $-(3/x^2) - 4$
11. (*a*) 0.21 (*b*) -0.59
13. (*a*) 6 (*b*) 6 (*c*) 6 (*d*) 6
(*e*) $6x + C$; choose five different values of C.
15. (*a*) $4x^3$ (*b*) $68x^3$ (*c*) $4kx^3$ (*d*) $x^4/4$
17. (*a*) $6kx^5$ (*b*) $(x^6/6) + C$; choose three different values of C.
19. (*a*) $a = 0$ (*b*) $a = 1, 3$
21. (*a*), (*b*)

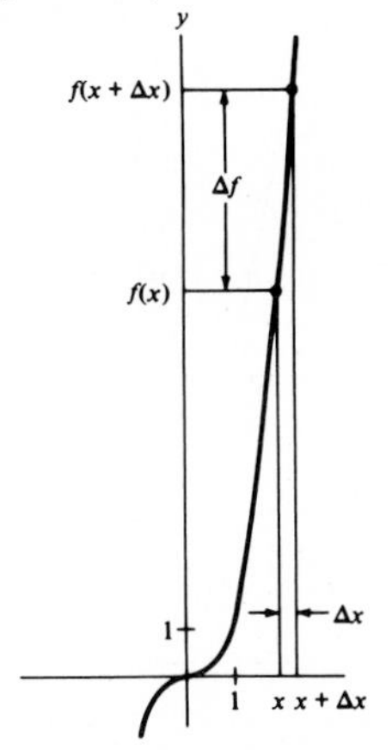

23. $-3/(3x + 5)^2$
25. $-\sin x$
27. (*a*) $x^2 + C$, for any constant C
(*b*) $\frac{1}{2}x^2 + C$, for any constant C
(*c*) $7x + C$, for any constant C
(*d*) $\frac{1}{3}x^3 + C$, for any constant C

Sec. 3.4. The Derivatives of the Sum, Difference, Product, and Quotient

1. $5x^4 - 4x$
3. $8x^3 - 12x + 5$
5. $5x^4 + 12x^3 - 3x^2 - 12x - 2$
7. $\frac{1}{7}(12x^3 - 2x + 5)$
9. $5/(2\sqrt{x})$
11. $-12/x^2$
13. $\dfrac{3 - 6x - x^2}{(3 + x^2)^2}$
15. $\dfrac{-t^4 + 6t^3 - 3t^2 + 2t - 3}{(t^3 + 1)^2}$
17. $3x^2 + \frac{7}{2}x^{5/2}$
19. $24x^2$
21. $\dfrac{1}{x^2} - \dfrac{2}{x^3}$
23. $\dfrac{-3x^2 - 2}{(x^3 + 2x + 1)^2}$
25. $\dfrac{15x^6 + 15x^4 + 5x^2 + 16x - 3}{(5x^2 + 3)^2}$
27. $8x + 4$
29. $\dfrac{-2}{(x - 1)^2}$
31. $\dfrac{-1}{2x^{3/2}}$
33. $\dfrac{-x^5 - 9x^4 - 20x^3 - 5x^2 + 23x + 2}{2\sqrt{x}(x^3 - 5x + 2)^2}$
35. (*a*) $x^4/4 + C$, for any two values of C
(*b*) $x^2/2 + C$, for any two values of C
(*c*) $-1/x + C$, for any two values of C
(*d*) $-1/(2x^2) + C$, for any two values of C
37. 3
39. $\frac{41}{2}$
41. $\frac{5}{6}, \frac{5}{6}$
43. $\frac{1}{12}$
45. $\frac{8}{3}$
53. Yes; $y = (1 + \sqrt{3})x - (2 + \sqrt{3})/2$ and $y = (1 - \sqrt{3})x - (2 - \sqrt{3})/2$
57. $1, -3, \frac{1}{9}(-13 \pm \sqrt{610})$

Sec. 3.5. The Derivatives of the Trigonometric Functions

1. $x \sin x$
3. $x^2 \sin x$
5. $\tan^2 x$
7. $x^2 \cos x$
9. $\dfrac{1 + \sin x}{\cos^2 x}$
11. $\dfrac{-1 - 3\cos x}{\sin^2 x}$
13. $\dfrac{-\csc x\,(3x \cot x + 1)}{3x^{4/3}}$
15. $\sin x\,(\sec^2 x + 1)$
17. $\dfrac{-(1 + x^2)\csc^2 x - 2x \cot x}{(1 + x^2)^2}$
21. (*a*) 0 (*b*) $-\frac{1}{2}$ (*c*) $-1/\sqrt{2}$ (*d*) $-\frac{1}{2}\sqrt{3}$ (*e*) -1
23. (*a*) 3 (*b*) -3 (*c*) $3, -3$ (*d*) $3, 3$
25. $\pi/4$
27. (*a*) $\sin x + C$, for any two values of C
(*b*) $5 \sin x + C$, for any two values of C
(*c*) $-\cos x + C$, for any two values of C
(*d*) $3 \cos x + C$, for any two values of C
31. $-\csc \theta \cot \theta$

Sec. 3.6. Composite Functions and the Chain Rule

1. $y = u^{50}$, $u = x^3 + x^2 - 2$
3. $y = \sqrt{u}$, $u = x + 3$
5. $y = \sin u$, $u = 2x$
7. $y = u^3$, $u = \cos v$, $v = 2x$
9. $40(2x^3 - x)^{39}(6x^2 - 1)$

11. $-\dfrac{12}{x^2}\left(1+\dfrac{3}{x}\right)^3$

13. $\dfrac{3x^2+1}{2\sqrt{x^3+x+2}}$

15. $3\cos 3x$

17. $\dfrac{-(x+x^3)\sec^2(1/x)+(3x^2+x^4)\tan(1/x)}{(1+x^2)^2}$

19. $(2x+1)^4(3x+1)^6(72x+31)$

21. $x\sin^4 3x\,(15x\cos 3x+2\sin 3x)$

23. $\dfrac{-10}{(2x+3)^6}$

25. $\dfrac{-6(2x+1)^2(x+1)}{(3x+1)^5}$

27. $\dfrac{(x^3-1)^5\cot^3 5x}{x}\left(\dfrac{15x^2}{x^3-1}-\dfrac{15\csc^2 5x}{\cot 5x}-\dfrac{1}{x}\right)$

29. $\dfrac{-4(1+2x)^3}{(1+3x)^5}$

31. $\dfrac{1}{\sqrt{x}}\tan\sqrt{x}\sec^2\sqrt{x}$

33. $\dfrac{x}{(1-x^2)^{3/2}}$

35. $\dfrac{30(3x-2)^3}{\sqrt{5(3x-2)^4+1}}$

37. $4\cos 3x\cos 4x-3\sin 3x\sin 4x$

39. $\frac{7}{5}x^{2/5}$

41. $\dfrac{7x^2}{3}(x^3+2)^{-2/9}$

43. $\frac{10}{3}x^{7/3}$

45. $5y^4\dfrac{dy}{dx}$

47. $(\cos y)\dfrac{dy}{dx}$

49. (*a*) $3(\cos 3x)\cos(\sin 3x)$
(*b*) $-10(\sec^2 5x)\sin(\tan 5x)\cos(\tan 5x)$

51. $\frac{1}{3}\sin 3x$

53. $-\frac{1}{2}\cos 2x$

55. $\sin^3 3x$

57. $\dfrac{10\tan 2x\sec^2 2x}{\sqrt{4+\tan^2 2x}}$

59. (*a*) $y=\dfrac{x}{\sqrt{2}}+\dfrac{4-\pi}{4\sqrt{2}}$ (*b*) $y=8x+\left(\sqrt{3}-\dfrac{4\pi}{3}\right)$

Sec. 3.S. Guide Quiz

2. (*a*) $15x^2-2$ (*b*) $\dfrac{-15}{(3x+2)^2}+6$ (*c*) $6\cos 2x$
(*d*) $-2x^{-3}$

3. (*a*) $\dfrac{5}{2\sqrt{x}}$ (*b*) $\dfrac{6x(1-x^2)}{\sqrt{3-2x^2}}$ (*c*) $-5\sin 5x$
(*d*) $\dfrac{3x}{2}(1+x^2)^{-1/4}$ (*e*) $\dfrac{2\sec^2 6x}{(\tan 6x)^{2/3}}$
(*f*) $5x^3\cos 5x+3x^2\sin 5x$ (*g*) $\dfrac{-1}{(2x+1)^{3/2}}$
(*h*) $\dfrac{-4(10x^4-3x^2)}{(2x^5-3x^3)^5}$ (*i*) $\dfrac{x^2}{(x^3-3)^{2/3}}$
(*j*) $\dfrac{6(2x^3+x^2-1)}{(3x+1)^2}$ (*k*) $\dfrac{-10x}{(5x^2+1)^2}$ (*l*) $\dfrac{-30}{(3x+2)^{11}}$
(*m*) $x^2(1+2x)^4\sec 3x\,[16x+3+3x(1+2x)\tan 3x]$
(*n*) $\dfrac{-\csc\sqrt{x}\cot\sqrt{x}}{2\sqrt{x}}$ (*o*) $\dfrac{24\csc^2 4x}{(1+3\cot 4x)^3}$

4. (*a*), (*b*)

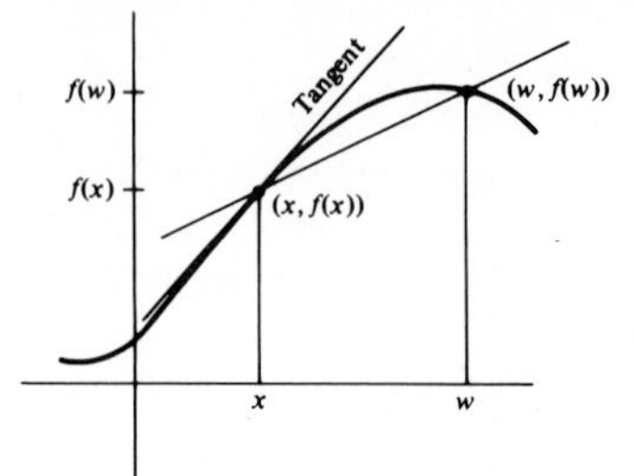

5. (*a*)

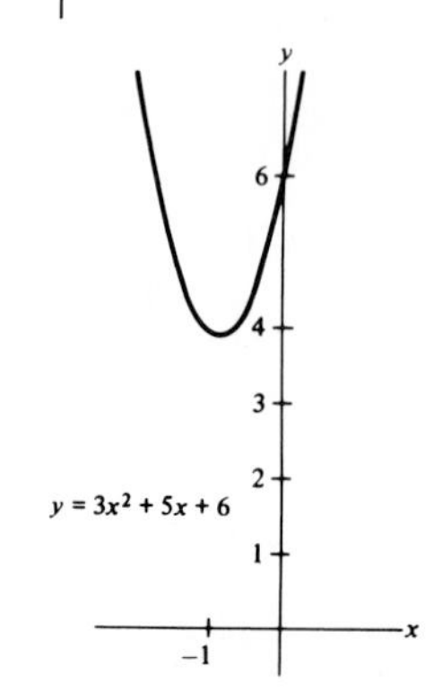

(*c*) $x=-\frac{5}{6}$

6. (*a*)

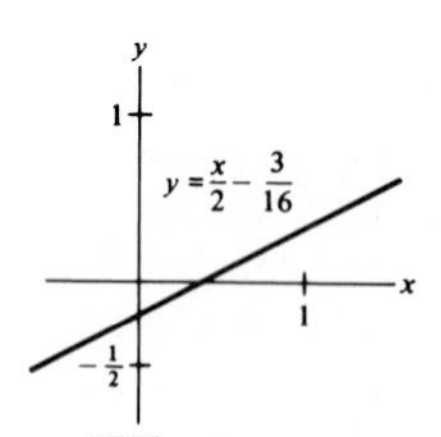

(*b*) $y=\dfrac{x}{2}-\dfrac{3}{16}$

7. (*a*)

x	−2	−1	0	1	2	3
x^3-12x	16	11	0	−11	−16	−9

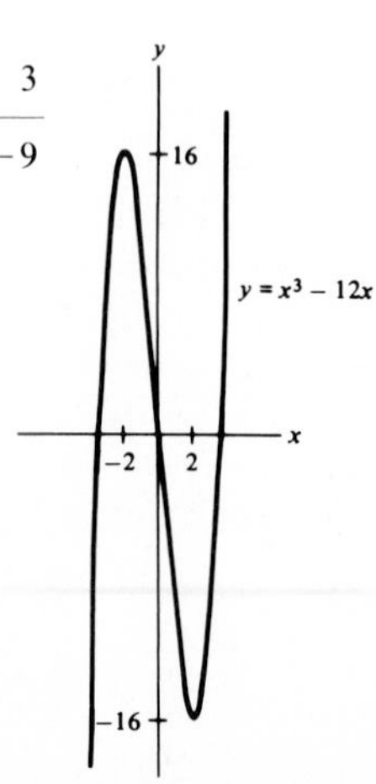

(*b*) $3x^2 - 12$ (*c*) $x = \pm 2$ (*d*) $(-2, 16), (2, -16)$

8. (*a*) $f(x + h) - f(x)$ is the change in height during the time interval of duration h. The quotient is the average velocity. (*b*) $f(x + h) - f(x)$ is the growth of the culture during the time interval of duration h. The quotient is the average rate of growth.
(*c*) See Problem 4 in Section 3.1.
(*d*) See Problem 3 in Section 3.1.

9. (*a*) $2t - 2$ m/sec (*b*) $-\frac{3}{2}$ m/sec
(*c*) $\frac{3}{2}$ m/sec (*d*) Left

10. (*a*) $\frac{1}{3}x^3$ (*b*) $-1/x$ (*c*) $-\frac{1}{3}\cos 3x$
(*d*) $x^3 + 2x^2 + 5x$ (*e*) $\tan x$ (*f*) $\frac{1}{2}\sin^2 x$

Sec. 3.S. Review Exercises

1. $15x^2$

3. $\dfrac{-1}{(x + 3)^2}$

5. $-3 \sin 3x$

7. $10x^4 + 3x^2 - 1$

9. $\dfrac{4x^2 + 2x}{(4x + 1)^2}$

11. $\dfrac{3x + 1}{\sqrt{3x^2 + 2x + 4}}$

13. $\dfrac{4}{3\sqrt[3]{2t - 1}}$

15. $10 \sin 5x \cos 5x$

17. $\frac{20}{7}(5x + 1)^3$

19. $3x \cos 3x + \sin 3x$

21. $\dfrac{4 \tan \sqrt[3]{1 + 2x} \sec^2 (\sqrt[3]{1 + 2x})}{3(1 + 2x)^{2/3}}$

23. $\dfrac{(x^4 + 3x^2) \cos 2x - (2x^3 + 2x^5) \sin 2x}{(1 + x^2)^2}$

25. $\dfrac{x}{3\sqrt{x^2 + 3}\,(1 + \sqrt{x^2 + 3})^{2/3}}$

27. $-\frac{35}{3} \csc^2 5x (\cot 5x)^{4/3}$

29. $\dfrac{4}{\sqrt{8x + 3}}$

31. $\dfrac{2x - x^4}{(x^3 + 1)^2}$

33. $\dfrac{-5[4(x^2 + 3x)^3(2x + 3) + 1]}{7[(x^2 + 3x)^4 + x]^{12/7}}$

35. $\dfrac{3x + 1}{5\sqrt{2x + 1}} \cos^2 6x - \dfrac{12x}{5} \sqrt{2x + 1} \cos 6x \sin 6x$

37. $x \sin ax$

39. $\sin^2 ax$

41. (*b*) 64, 32, 0, -32 ft/sec (*c*) 64, 32, 0, 32 ft/sec
(*d*) Rising: $0 < t < 2$; falling: $t > 2$

43. $y = -x$

45. (*a*) 12 g/cm

47. (*a*) $\frac{3}{2}$ (*b*) $\frac{1}{2}$ (*c*) $\frac{1}{4}$ (*d*) $\frac{1}{6}$

49. (*a*)

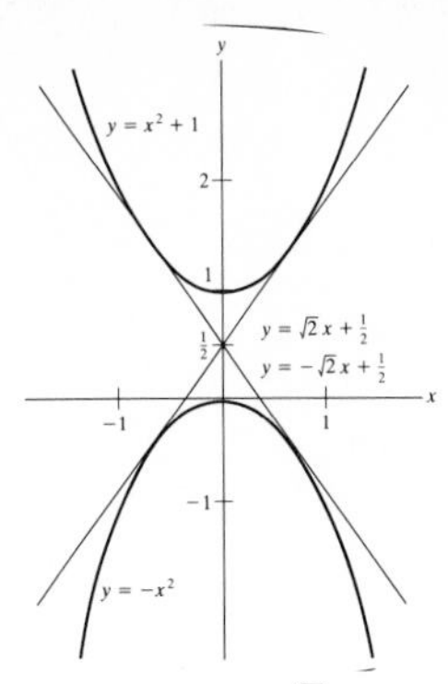

(*b*) $y = \sqrt{2}x + \frac{1}{2}$, $y = -\sqrt{2}x + \frac{1}{2}$

51. (*a*) 12.61 (*b*) 11.41 (*c*) 12

53. (*a*) 1,000 (*b*) $5 + \dfrac{x}{100}$ (*c*) 5.1 (*d*) 5.105

55. (*a*) 5 (*b*) 2.25 (*c*) 4.5 (*d*) 4

57. (*a*) $3y^2 \dfrac{dy}{dx}$ (*b*) $(-\sin y) \dfrac{dy}{dx}$ (*c*) $-\dfrac{1}{y^2} \dfrac{dy}{dx}$

59. In each part, pick any two values of C. (*a*) $x^4 + C$
(*b*) $\dfrac{x^4}{4} + C$ (*c*) $\dfrac{x^5}{5} + \dfrac{x^4}{4} + \sin x + C$
(*d*) $\dfrac{x^4}{4} - \cos x + C$ (*e*) $\dfrac{x^5}{5} + \dfrac{2x^3}{3} + x + C$

61. (*a*) All values of a (*b*) $a = 1$ (*c*) $a = 0, 2, 3, 4$

67. $\dfrac{\cos \sqrt{3}}{2\sqrt{3}}$

69. (*a*), (*b*)

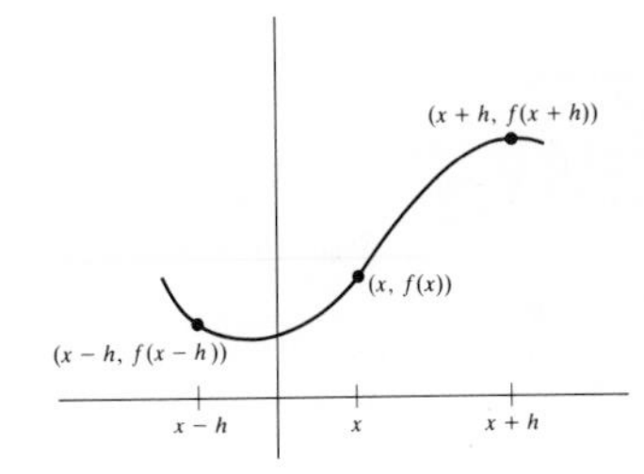

(*d*) $f'(x)$ (*e*) $3x^2$

CHAPTER 4. APPLICATIONS OF THE DERIVATIVE

Sec. 4.1. Rolle's Theorem and the Mean-Value Theorem

1. (*a*)

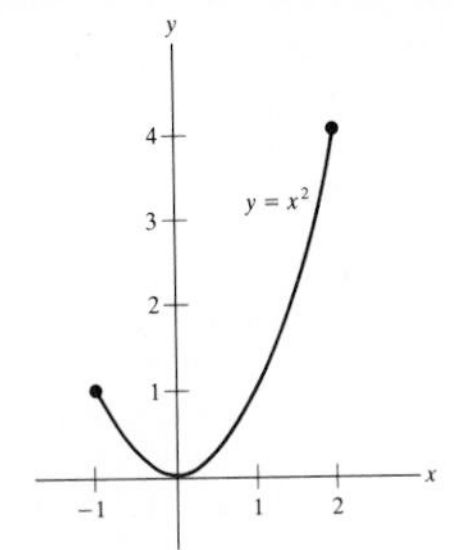

(*b*) 4 (*c*) Yes (*d*) No (*e*) Yes

3. (*a*)

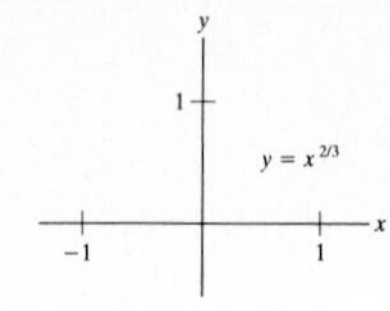

(*c*) No

5. $-1, 0, 1$

7. $\pi, 2\pi, 3\pi, 4\pi, 5\pi$

9. $\frac{1}{3}(-3 + \sqrt{57})$

11. (*a*) $2 \sec^2 x \tan x$ (*b*) 1

13. 0 or 1

15. 0, 1, 2, or 3

17. All intervals

19. (*a*) $2x^4 + C$ (*b*) $-\frac{1}{2} \cos 2x + C$ (*c*) $-1/x + C$ (*d*) $-\frac{1}{2}(x + 1)^{-2} + C$

21. (*a*) Corollary 2 (*b*) Corollary 1

23. (*a*) $-1, 1$ (*c*) 110

27. (*a*) $3\sqrt{1 - x^2} \cos 3x - \dfrac{x}{\sqrt{1 - x^2}} \sin 3x$

(*b*) $\dfrac{1 - 5x^2}{3x^{2/3}(x^2 + 1)^2}$ (*c*) $\dfrac{-4}{(2x + 1)^3} \sec^2 \left[\dfrac{1}{(2x + 1)^2}\right]$

29. No

33. The mean-value theorem

35. (*b*) If $r > 1$ and $x > 0$, then $(1 + x)^r > 1 + rx$.

39. (*b*) $13, -3$

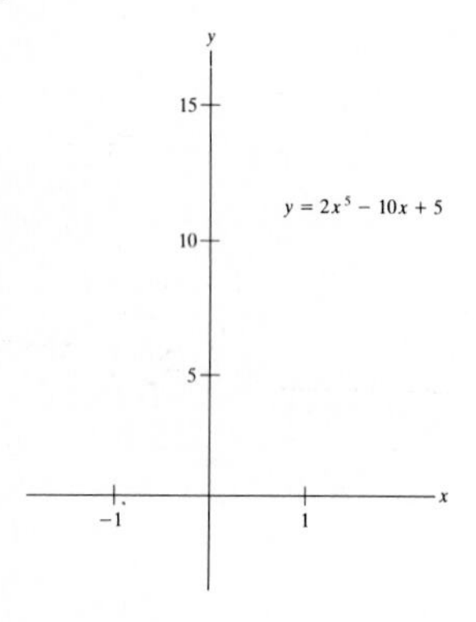

41. No

Sec. 4.2. Using the Derivative and Limits when Graphing a Function

9. Critical point at (1, 1)

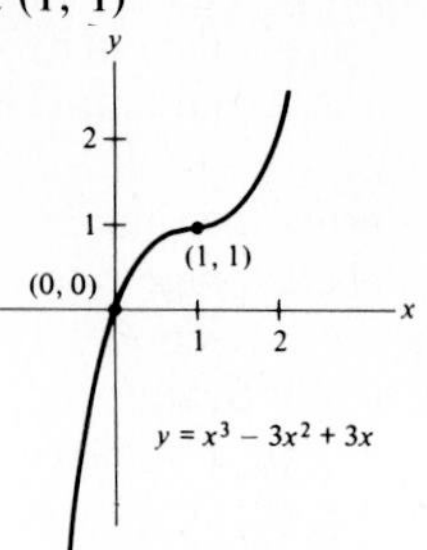

11. Critical point and global minimum at (1, 0)

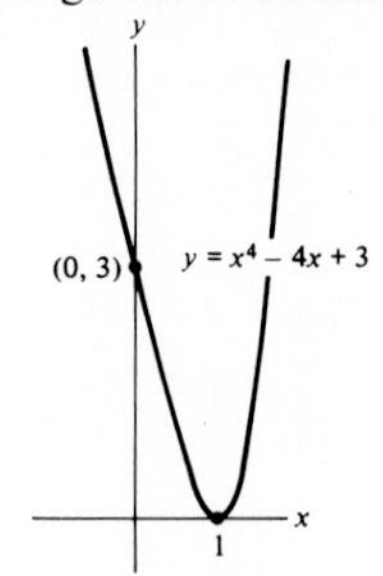

13. Critical point and global minimum at $(3, -4)$

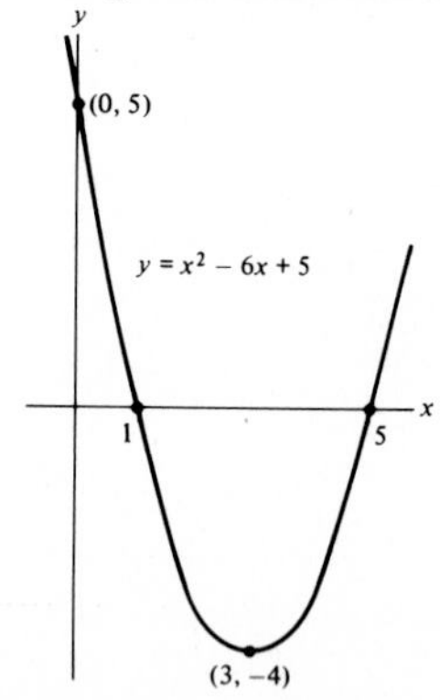

15. Critical point and local maximum at (0, 0); critical point and local minimum at $(b, f(b))$ for $b = \frac{1}{4}(-3 + \sqrt{33})$; critical point and global minimum at $(a, f(a))$ for $a = \frac{1}{4}(-3 - \sqrt{33})$

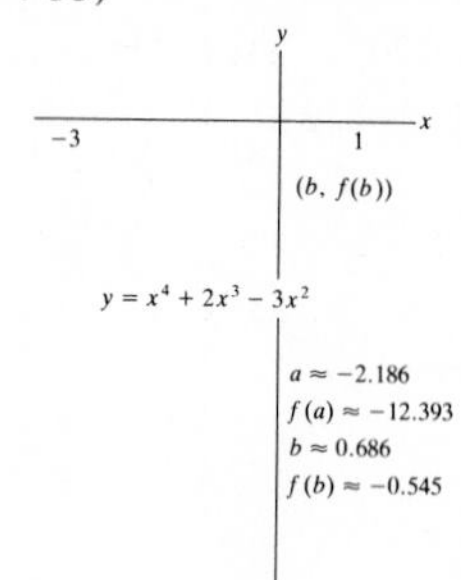

17. Asymptotes $x = \frac{1}{3}$ and $y = 1$

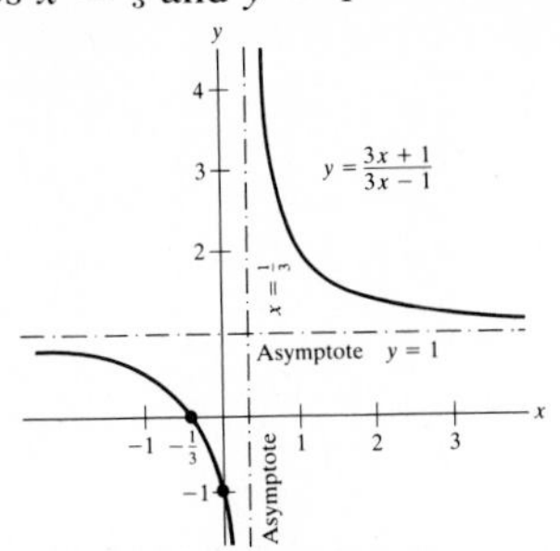

19. Critical point and global maximum at $(1, \frac{1}{2})$; critical point and global minimum at $(-1, -\frac{1}{2})$; asymptote $y = 0$

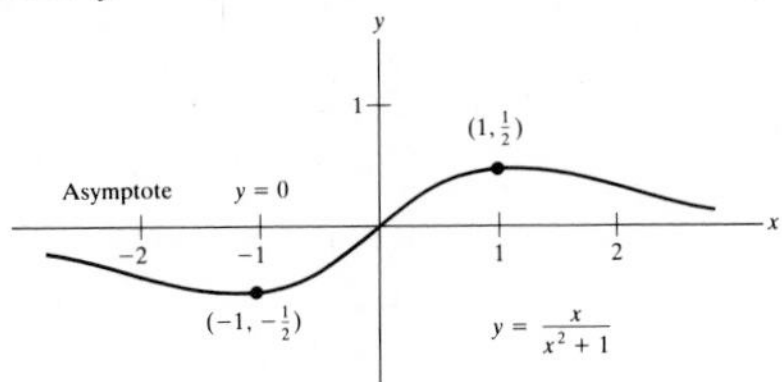

21. Critical point and local maximum at $(\frac{1}{4}, -8)$; asymptotes $x = 0$, $x = \frac{1}{2}$, and $y = 0$

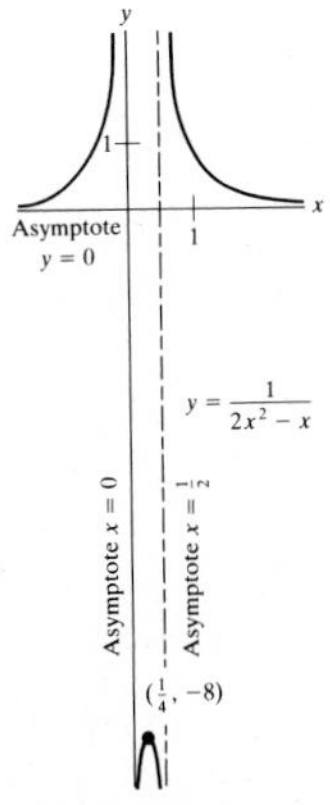

23. Critical point and local maximum at $(0, -\frac{3}{4})$; asymptotes $x = -2$, $x = 2$, $y = 1$

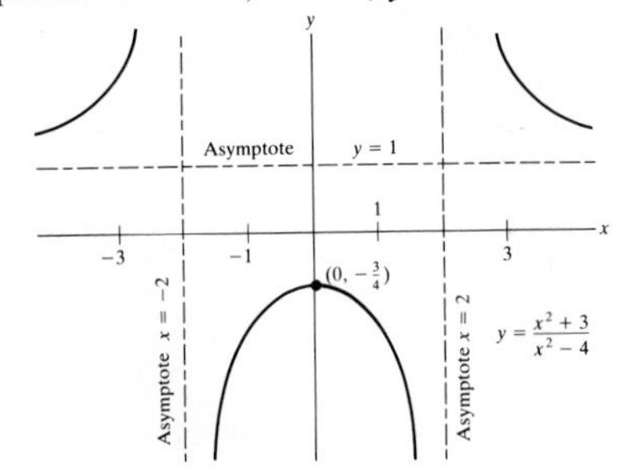

25. $\frac{1}{4}$, 0

27. 3, 0

29. 24, -8

31. 2, 0

33.

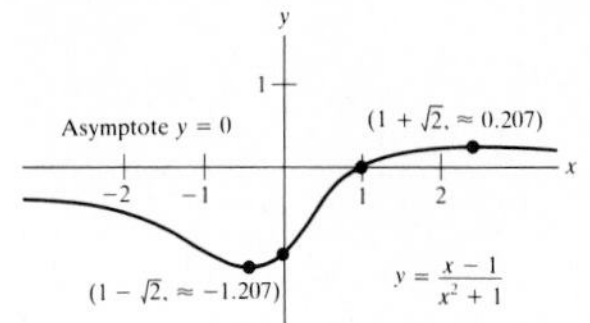

35.

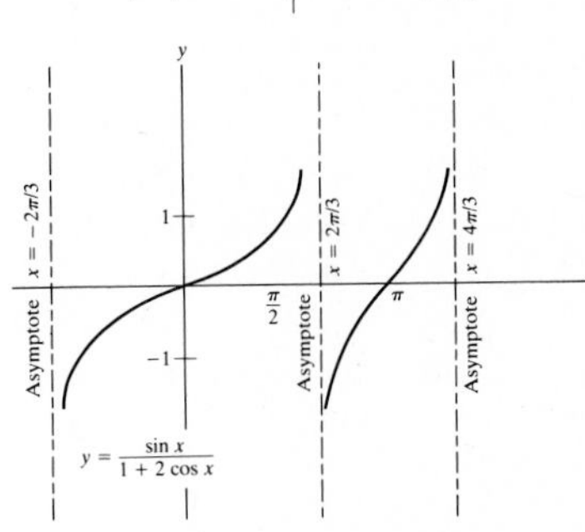

37.

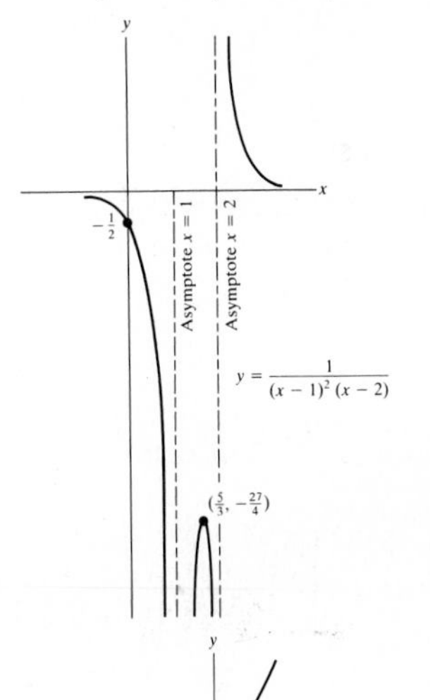

39.

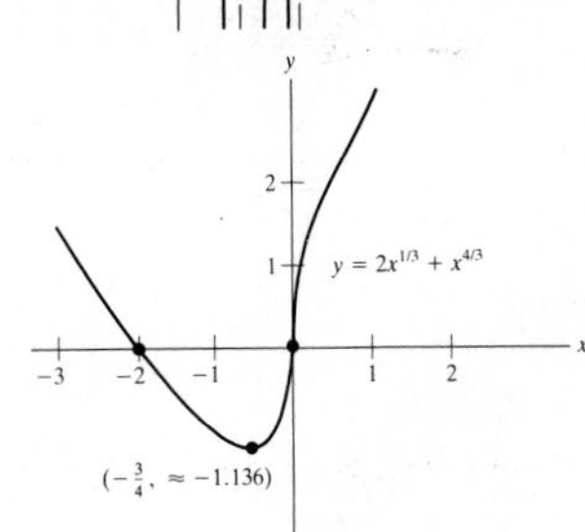

41.

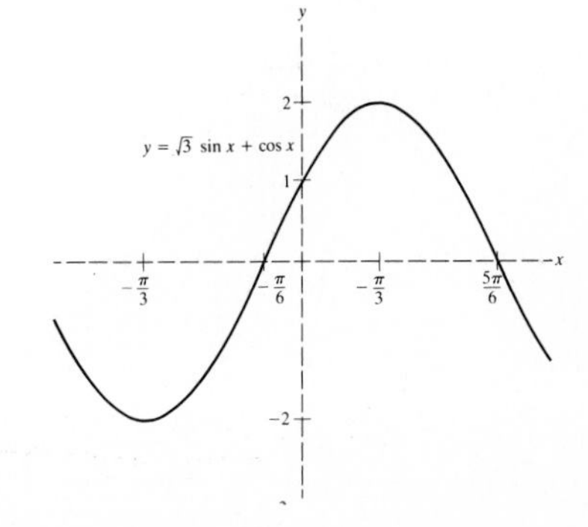

Sec. 4.3. Higher Derivatives and the Binomial Theorem

1. $2x - 1, 2, 0, 0$
3. $3x^2, 6x, 6, 0$
5. $2\cos 2x, -4\sin 2x, -8\cos 2x, 16\sin 2x$
7. $-\frac{2}{x^3}, \frac{6}{x^4}, -\frac{24}{x^4}, \frac{120}{x^6}$
9. (a) $8\sec^2 2x \tan 2x$ (b) $-(1+2x)^{-3/2}$
   (c) $\frac{\csc^2\sqrt{x}}{4x^{3/2}} + \frac{\csc^2\sqrt{x}\cot\sqrt{x}}{2x}$
11. Yes, one
13. $f(x) = cx + d - \frac{2}{9}\sin 3x$, for constants c and d
15. (a) 1, 3, 6, 6 (b) $a_0, a_1, 2a_2, 6a_3$
17. $a^4 + 4a^3b + 6a^2b^2 + 4ab^3 + b^4$

Sec. 4.4. Concavity and the Second Derivative

1. Maximum at $x = 0$; minimum at $x = \frac{2}{3}$; inflection point at $x = \frac{1}{3}$

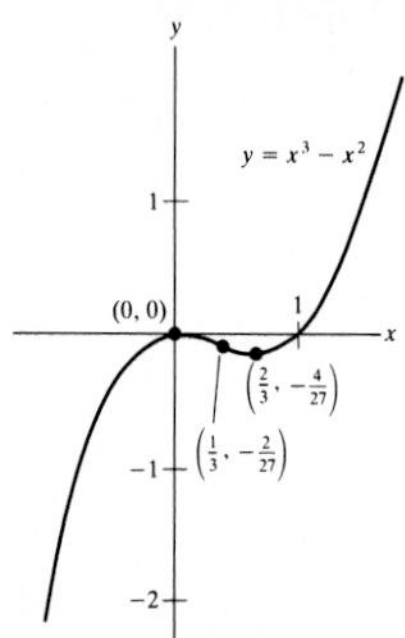

3. Minimum at $x = -\frac{3}{2}$; inflection points at $x = 0$ and $x = -1$

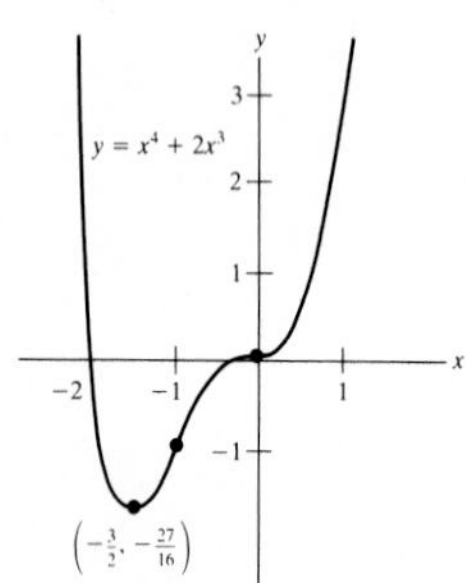

5. Minimum at $x = 3$; inflection points at $x = 0$ and $x = 2$

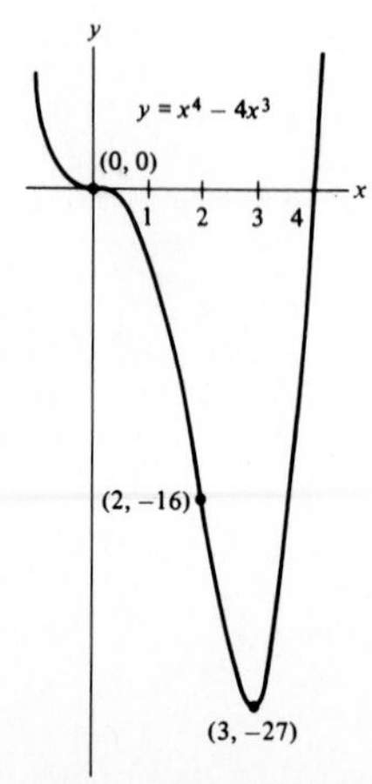

7. Maximum at $x = 0$; inflection points at $x = \pm 1/\sqrt{3}$

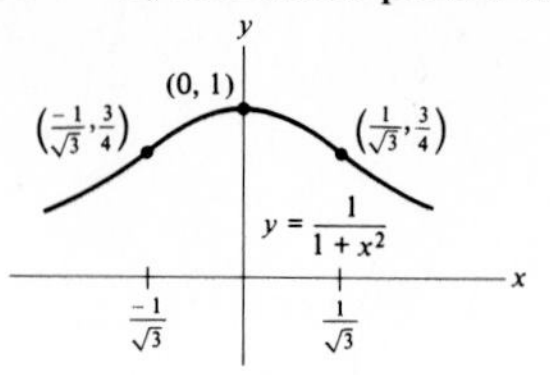

9. Maximum at $x = -1$; minimum at $x = 5$; inflection point at $x = 2$

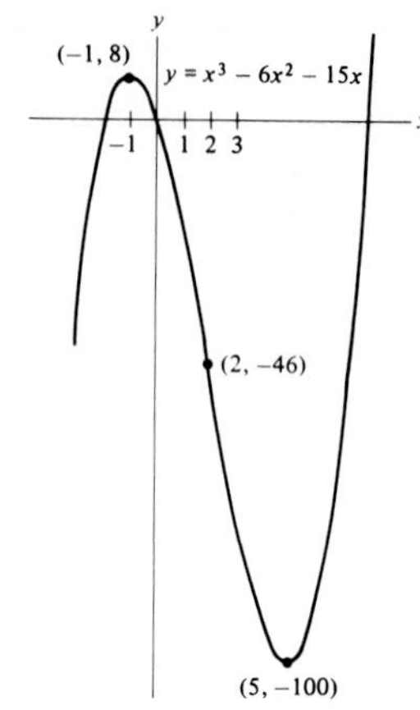

11. Inflection points at $x = 0$ and $x = 1/\sqrt[3]{2}$

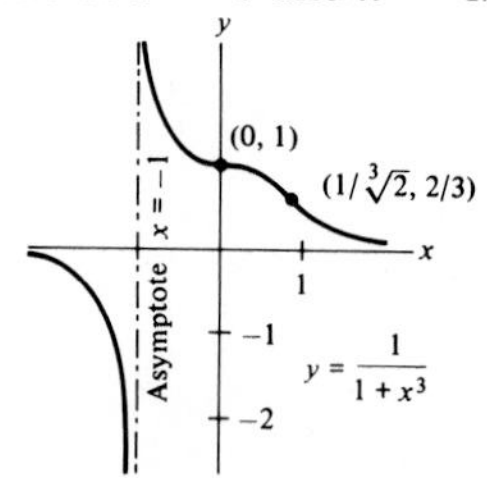

13. Inflection points at $x = n\pi$, $n = 0, \pm 1, \pm 2, \pm 3, \ldots$

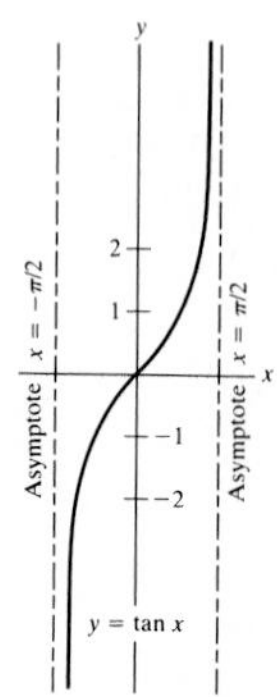

15. Maximum at $x = 0$; inflection points at $x = \pm\frac{1}{3}$

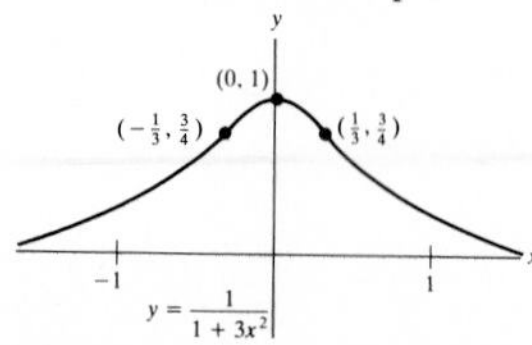

17. Minimum at $x = 1$

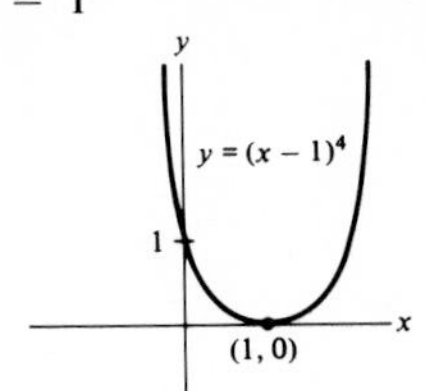

19.

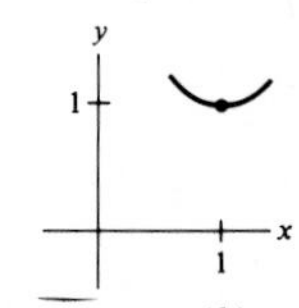

21. (*a*) (*b*) (*c*) (*d*)

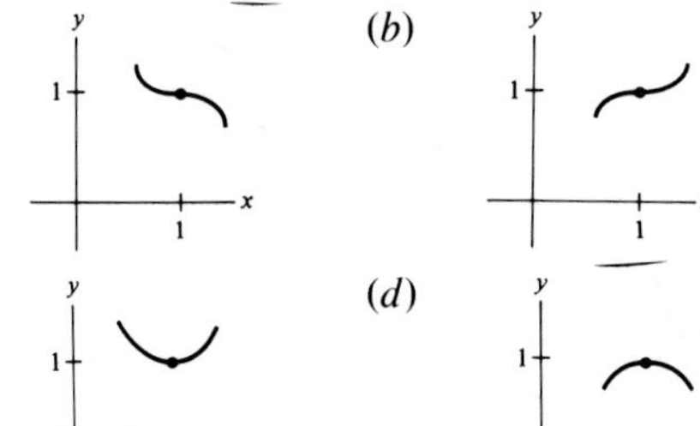

23.

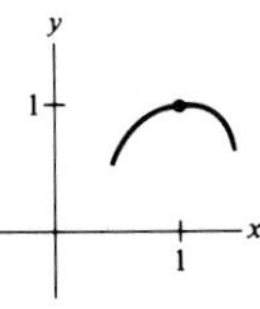

25.

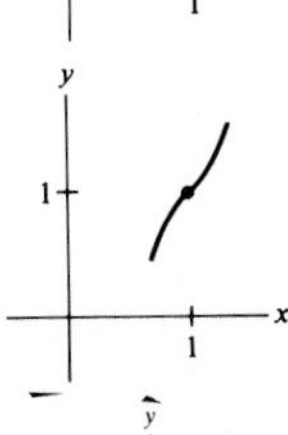

33.

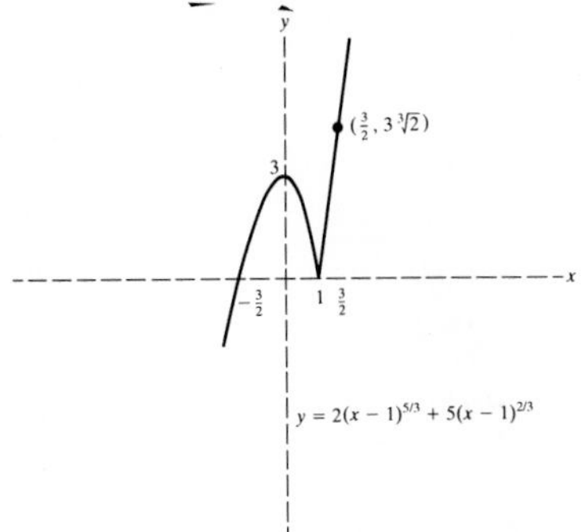

35. (*a*) (*b*)

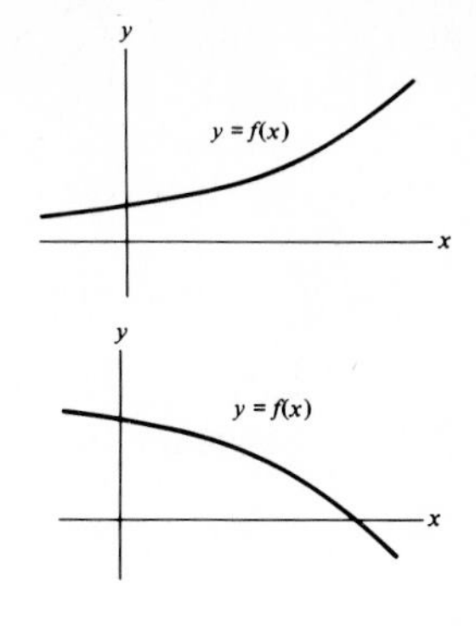

(*c*) No

41. No; no

Sec. 4.5. Motion and the Second Derivative

1. $f'' > 0$

3. (*a*) 4 sec (*b*) Velocity $= -64$ ft/sec; speed $= 64$ ft/sec

7. $3t^2 - 3t + 3$ ft

9. $-16t^2$

11. (*a*) $|A|$ (*b*) $|A|\sqrt{c}$, at 0 (*c*) 0, at $\pm A$

13. (*a*)

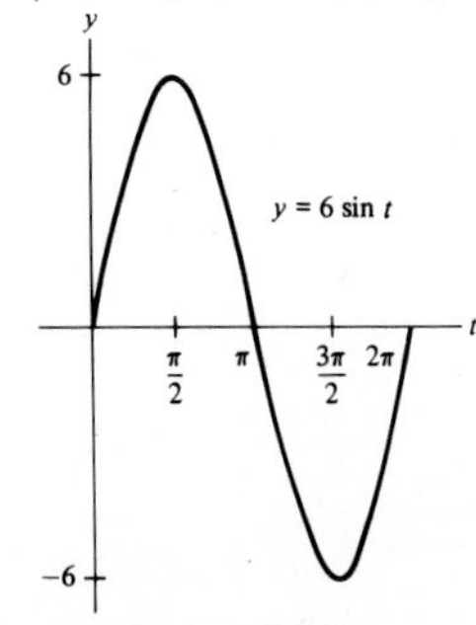

(*b*) 6 in (*d*) $y = 0$

(*e*) $y = \pm 6$ inches

17. 16,000 mi

19. $\lim_{r\to\infty} v = 0$

21. No

Sec. 4.6. Applied Maximum and Minimum Problems

1. $\frac{5}{6}$ in

3. $\frac{1}{3}(6 - 2\sqrt{3})$ in

5. $x = 25$, $y = 50$ ft

7. $h = \sqrt[3]{\dfrac{400}{\pi}}$ in

9. 10 in × 10 in × 10 in

11. $\theta = \pi/4$; $w = h = \sqrt{2}a$

15. 40 ft × 60 ft

17. (*a*) $x = 1$, $y = 0$ (*b*) $x = \frac{1}{2}$, $y = \frac{1}{2}$

19. $x = 8$ ft

21. $r = 36/\pi$ in, $h = 36$ in

23. 18 in × 18 in × 36 in

25. (*c*) $\sqrt[3]{25/\pi}$ in (*d*) $4\sqrt[3]{25/\pi}$ in
27. Length $= 2\sqrt[3]{75/7}$ in, height $= \frac{7}{3}\sqrt[3]{75/7}$ in
29. 414 on the first road and 586 on the second
33. (*a*) 10 ft × 20 ft (*b*) 25 feet square
(*c*) 20 feet square
35. Walk from F to the midpoint of an edge containing neither F nor S, and then to S.
37. (*a*), (*b*) Walk across the grass to B. (*c*) Walk across the grass to the point on the sidewalk $\frac{1}{4}$ mile from B.
39. $r = (\sqrt{2/3})a$, $h = (2/\sqrt{3})a$
41. Length $= \sqrt[3]{2V/3}$, height $= \frac{3}{2}\sqrt[3]{2V/3}$
43. (*a*) SD (*b*) 63
47. $13\sqrt{13}$
49. $\sqrt{a(a+b)}$ ft
51. \$130
53. (*b*) $\dfrac{ALc}{2v}x^3 + \dfrac{RLv}{2A}x - K$

Sec. 4.7. Implicit Differentiation

1. -4
3. $-7/15$
5. $-\pi/2$
7. $-\frac{8}{9}$
9. $r = \sqrt[3]{50/\pi}$ in, $h = 2\sqrt[3]{50/\pi}$ in
11. $x = 25$ ft, $y = 50$ ft
15. 18 in × 18 in × 36 in
17. $-\dfrac{y^3 + \sec^2(x+y)}{3xy^2 + \sec^2(x+y)}$
19. $\dfrac{7x - 24y}{24x + 7y}$
21. (*b*) $-\frac{3}{5}$ (*d*) $-\frac{168}{125}$
23. $-2, -6$
25. $(-2, 4), (2, -4)$
29. (*b*) 16 at $(\pm 4, 0)$
(*c*)

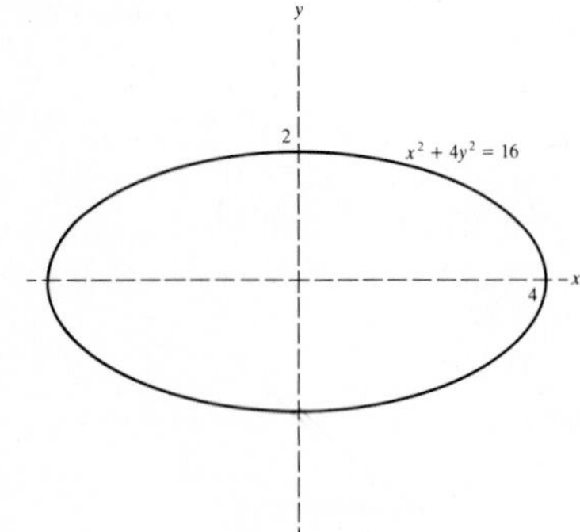

Sec. 4.8. The Differential

1. $dy = 0.6$, $\Delta y = 0.69$

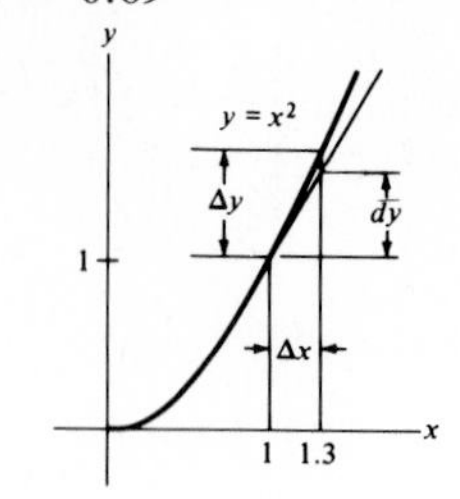

3. $dy = -\frac{1}{3}$, $\Delta y = \sqrt{7} - 3 \approx -0.354$

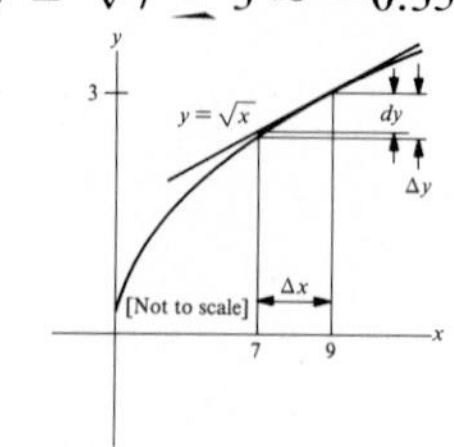

5. $dy = \pi/9 \approx 0.349$, $\Delta y = 1 - 1/\sqrt{3} \approx 0.423$

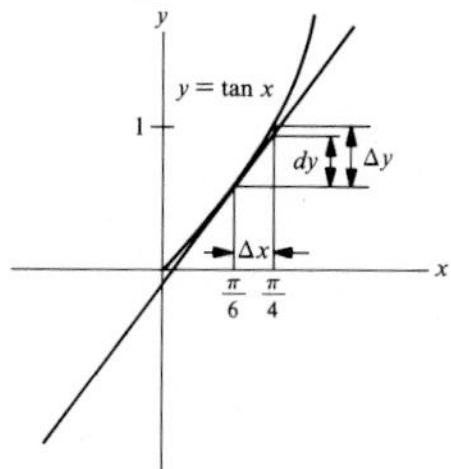

11. $-(3/x^4)\,dx$
13. $2\cos 2x\,dx$
15. $-\csc x \cot x\,dx$
17. $-(1/x^2)(5x \csc^2 5x + \cot 5x)\,dx$
19. 9.9
21. 2.9259
23. 0.98
25. 0.8560
27. 0.13
29. 0.248125
31. 0.5302
33. 0.8835
35. About $p/2$ percent
37. (*a*) $df = 2x\,\Delta x$, $\Delta f = 2x(\Delta x) + (\Delta x)^2$
(*b*)

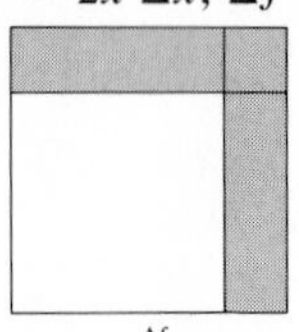

(*c*)

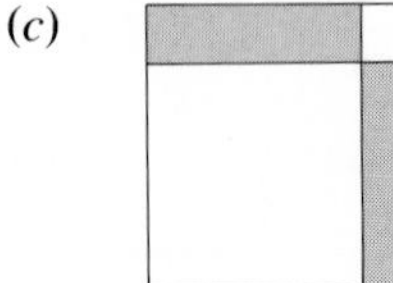

41. (*a*) $\sin\theta \approx \theta$, $\cos\theta \approx 1$, $\tan\theta \approx \theta$

Sec. 4.S. Guide Quiz

3. (*a*) The graph crosses the x axis at $x = a$.
(*b*) There is a maximum at $x = a$.

(*c*) The concavity changes from upward to downward; $x = a$ is an inflection point.

6. (*a*) $-\frac{1}{3}\cos 3x + C$ (*b*) Corollary 2 in Sec. 4.1

7. $\frac{1}{3}(4 - \sqrt{7})$ is squared, $\frac{1}{3}(-1 + \sqrt{7})$ is cubed.

8. Maximum at $x = 0$; minimum at $x = L/2$ $(r = 0)$

9. (*c*) 8.1

10.

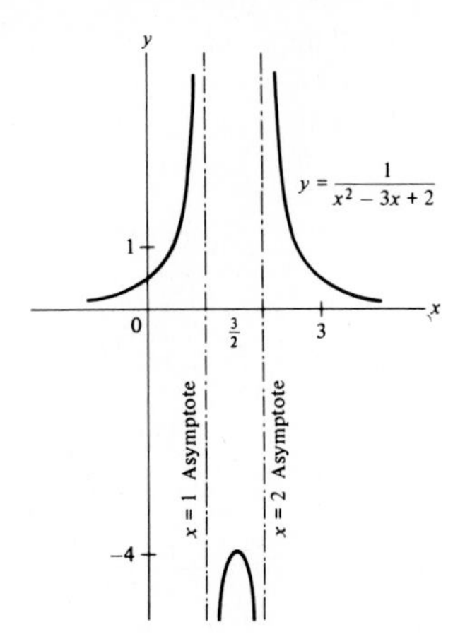

11. $0, -\frac{1}{3}$

12. (*a*) $\frac{x^4}{4} + \frac{2x^3}{3} + C$ (*b*) $-\frac{1}{2x^2} + C$

(*c*) $-\frac{5}{2}\cos 2x + C$ (*d*) $\sqrt{1 + x^2} + C$

14.

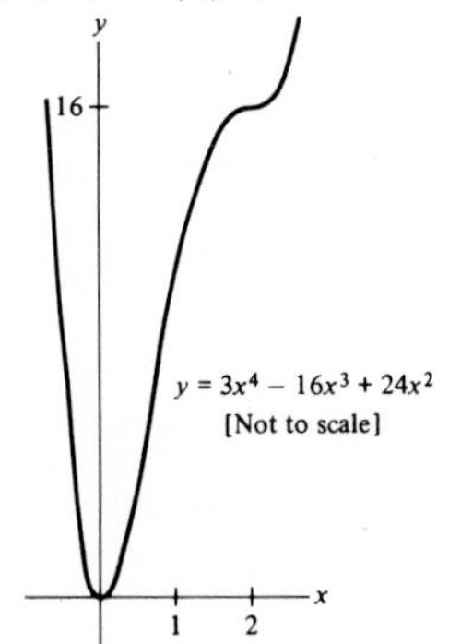

15. (*a*) $240x - \frac{24}{x^5}$ (*b*) $16\cos 2x$ (*c*) 0

(*d*) $-\frac{15}{16}x^{-7/2}$

18.

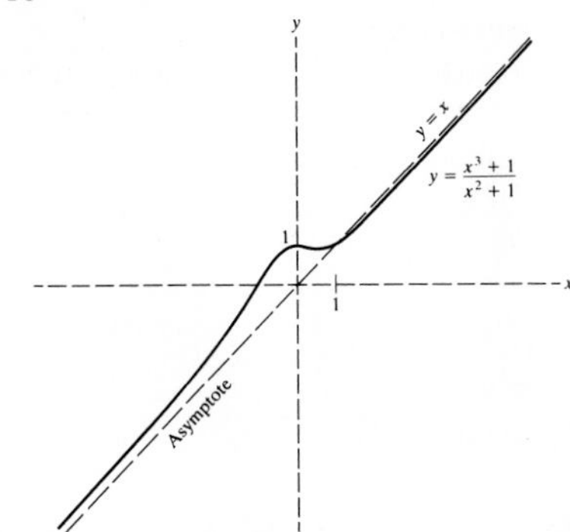

Sec. 4.S. Review Exercises

3. (*b*) No

5. $y = \frac{1}{x^2} + \frac{1}{x - 1}$

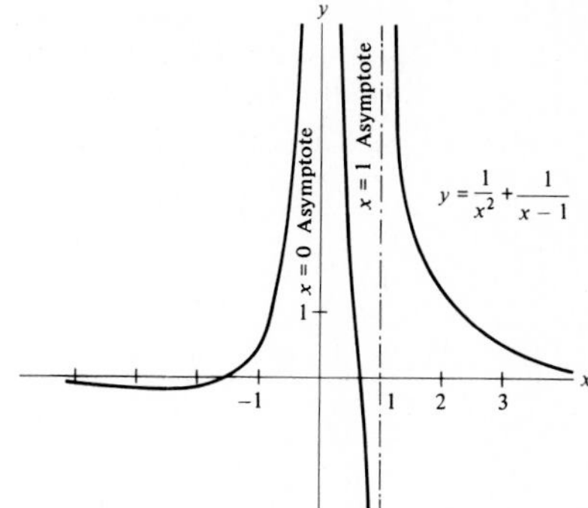

7. (*a*), (*d*)

(*b*) $\frac{1}{4}$ (*c*) $\sqrt{5} - 2 \approx 0.236$

11. (*a*) $9L/(4\sqrt{3} + 9)$ for the triangle, $4\sqrt{3}L/(4\sqrt{3} + 9)$ for the square (*b*) All for the square

13. $f(3) \geq 3$

15. -1

17. $y = \frac{\sqrt{x}}{1 + x}$

21. (*a*) $\frac{4x^3 + 12x^2 - 2}{(x + 2)^2}$

(*b*) $\frac{3x^5}{2\sqrt{1 + 3x}} + 5x^4\sqrt{1 + 3x}$ (*c*) $\frac{10}{7}(2x - 1)^4$

(*d*) $\frac{2}{\sqrt{x}}\sin^3\sqrt{x}\cos\sqrt{x}$ (*e*) $\frac{3}{x^4}\sin\frac{1}{x^3}$

(*f*) $\frac{-x}{\sqrt{1 - x^2}}\sec^2\sqrt{1 - x^2}$

23. -12

25. (*a*) 1.01 (*b*) 0.988

27. $y = x + 1$, $y = -x - 1$

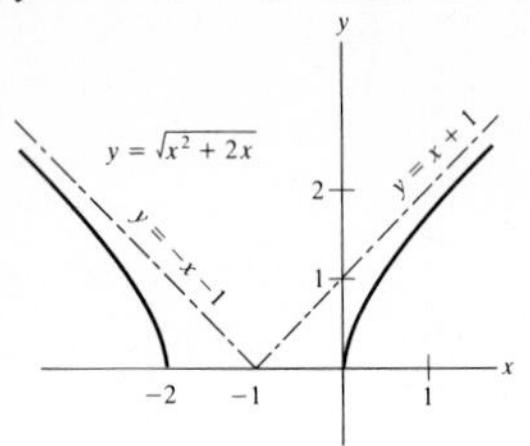

35. $\sqrt{2}a \times \sqrt{2}b$
37. $a/\sqrt{2}$
41. (*a*) Yes

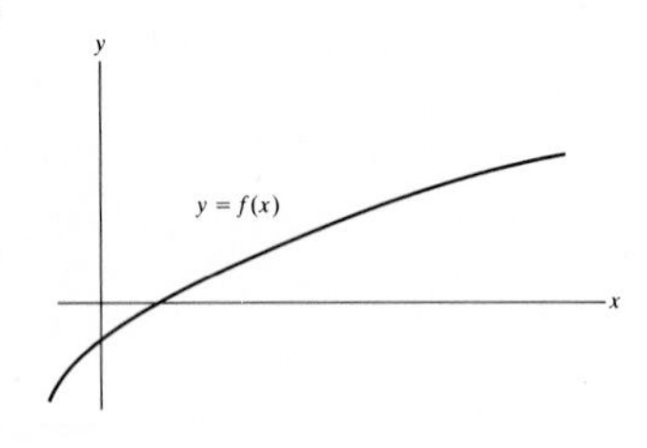

(*b*) No
(*c*) Yes

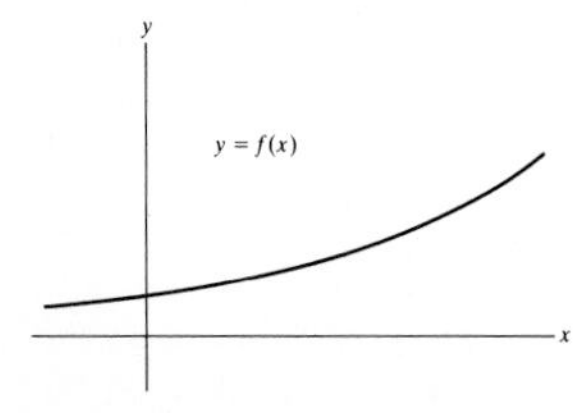

43. (*a*) 2, $\frac{11}{2}$ (*b*) $-1, 4, 6$ (*c*) $-1, 1, 4, \frac{11}{2}, 6$
(*d*) None (*e*) -1
45. (*a*) $36x - 3x^2$ (*b*) Midpoint
47. $h = \sqrt{8r}$
49. (*a*) 5 (*b*) $\sqrt{A^2 + B^2}$
55. Yes

CHAPTER 5. THE DEFINITE INTEGRAL

Sec. 5.1. Estimates in Four Problems

1. (*a*) 8.75 (*b*) 14
(*c*) More

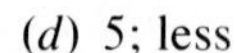
(*d*) 5; less

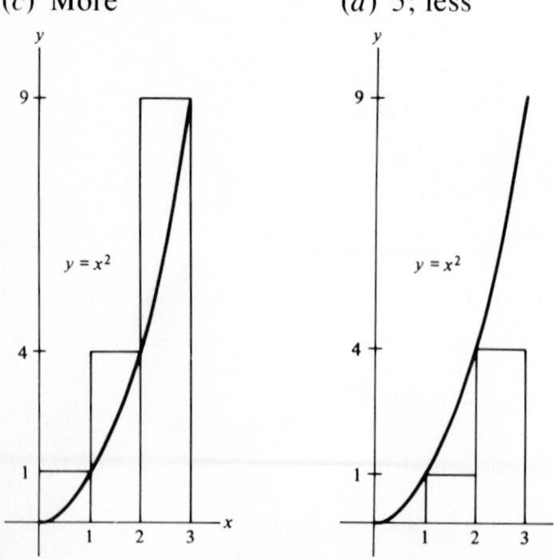

(*e*) 14, 5
3. (*a*) 44.55 g (*b*) 59.4 g (*c*) 32.4 g
(*d*) 59.4 g, 32.4 g
5. Problem 1: 2,413/300 ≈ 8.043
Problem 2: 2,413/60 ≈ 40.217 g
Problem 3: 4,826/75 ≈ 64.347 mi
Problem 4: 2,413/300 ≈ 8.043 ft^3
7. 7.695 ft^3 (*b*) 10.395 ft^3
9. 143/16 = 8.9375 (millions of dollars)
11. (*a*)

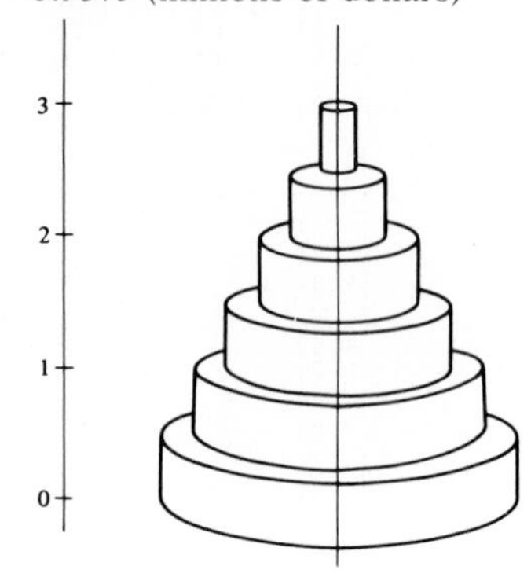

(*b*) $143\pi/16 \approx 28.078$ ft^3
13. (*a*) 1,879/2,520 ≈ 0.7456 (*b*) 1,627/2,520 ≈ 0.6456
15. (*a*) $\frac{1}{4}(x + 2)^4$ (*b*) $\frac{1}{5}x^4 + \frac{2}{3}x^3 + x$ (*c*) $-\frac{1}{2}\cos x^2$
(*d*) $x^4/4 - 1/(2x^2)$ (*e*) $2\sqrt{x}$
17. Underestimates (for a perfect graph)

Sec. 5.2. Summation Notation and Approximating Sums

1. (*a*) 6 (*b*) 20 (*c*) 14
3. (*a*) 4 (*b*) 1 (*c*) 450
5. (*a*) $\Sigma_{i=0}^{100} 2^i$ (*b*) $\Sigma_{j=3}^{7} x^j$ (*c*) $\Sigma_{k=3}^{102} \frac{1}{k}$
7. (*a*) $\Sigma_{i=1}^{3} x_{i-1}^2(x_i - x_{i-1})$ (*b*) $\Sigma_{i=1}^{3} x_i^2(x_i - x_{i-1})$
9. (*a*) $2^{100} - 1$ (*b*) $-99/100$ (*c*) $-100/101$
13. 13.50
15. 10.00
17. 1.07
19. (*a*) 5,050 (*b*) 1,501,500
23. $11(b - a)$
25. 6
27. 117/160 = 0.73125
29. About 32,000 years
31. (*a*) 1.5, 1.083, 0.95, and 0.885, respectively

Sec. 5.3. The Definite Integral

1. (*a*) 1 (*b*) 2 (*c*) 3
3. (*a*) 125/3 (*b*) 64/3 (*c*) 61/3
5. (*a*) a (*b*) $a + \dfrac{b - a}{n}$

(*c*) $a + 2\dfrac{b - a}{n}$, $a + i\dfrac{b - a}{n}$

(*d*) $\frac{b-a}{6} \times \left[2(a^2+ab+b^2) + \frac{3(b^2-a^2)}{n} + \frac{(b-a)^2}{n^2}\right]$

(*e*) $\frac{b^3}{3} - \frac{a^3}{3}$

7. (*a*) Duration of the *i*th time interval
(*b*) Speed at some time in the *i*th time interval
(*c*) Estimate of distance traveled during the *i*th time interval (*d*) Estimate of total distance traveled
(*e*) Actual total distance traveled

9. (*a*) Length of the *i*th time interval
(*b*) Rate of water flow at an instant in the *i*th time interval
(*c*) Approximate amount of water flowing into the lake during the *i*th time interval
(*d*) Estimated total volume of water flow into the lake
(*e*) Actual volume of water flow

11. $\frac{26}{3}$ g
13. (*a*) $\frac{77}{60} \approx 1.2833$ (*b*) $\frac{19}{20} = 0.95$
15. (*a*) $\frac{4}{25} = 0.16$ (*b*) $\frac{9}{25} = 0.36$
17. (*a*) 0.77 (*b*) 1.20
19. (*a*) $b^4/4$ (*b*) $\frac{15}{4}$
21. $1{,}669/1{,}800 \approx 0.9272$
27. (*a*)

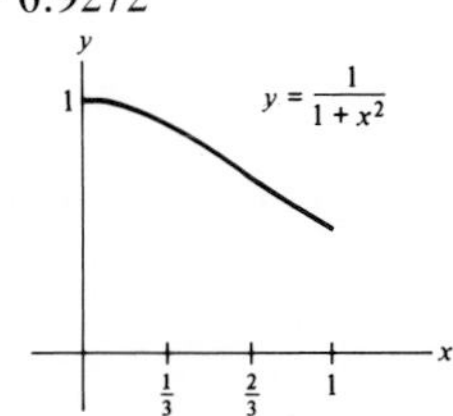

(*d*) $0.7337 < A < 0.8338$
29. (*c*) Logarithmic functions
31. $n \geq 10{,}000$

Sec. 5.4. The Fundamental Theorems of Calculus

3. 63
5. $\frac{12}{5}$
7. 10.5 ft
9. 18.75 g
11. 936 cm^3
13. 3.75
15. 27
17. 190/3
19. 18.6
21. x^4
23. x^5
25. $(\sqrt{1+\sin^3 x})(\cos x)$
27. (*c*) $\frac{4}{3}\pi r^3$
29. (*a*) $\pi[f(x)]^2$ (*b*) $\int_a^b \pi[f(x)]^2\,dx$
31. $\pi/4$
33. $37\pi/3$
35. (*b*) 1
37. (*a*) Cannot apply FTC (*b*) $\frac{3}{4}(2^{4/3} - 1)$
(*c*) $\frac{3}{2}(2^{2/3} - 1)$

Sec. 5.5. Properties of the Antiderivative and the Definite Integral

1. $x^2 - \frac{1}{4}x^4 + \frac{1}{6}x^6 + C$
3. $\frac{3}{2}x^2 - 2\cos 2x + 1/(4x) + C$
5. $\frac{1}{3}(2x+5)^{3/2} + C$
7. $\frac{1}{2}\sec 2x + C$
9. $x - \frac{1}{2}\sin 2x + C$
11. (*a*) $\frac{7}{3}$ (*b*) $\frac{1}{6}$ (*c*) $-\frac{1}{3}(\cos 6 - \cos 3)$
13. (*a*) 3 (*b*) 0 (*c*) 0 (*d*) -2
15. (*a*) Function (*b*) Number (*c*) Number
17. (*a*) True (*b*) False
19. (*a*) $\sin(x^2)$ (*b*) 3 (*c*) $\sin(x^2)$
23. 0
25. -42
27. $-\sin^3 x$
29. $9x\tan 3x - 4x\tan 2x$
31. $2x\cos x^2$
33. $-\frac{x^2}{2}(1+3x)^{-3/2}(6\sin 2x + 15x\sin 2x + 4x\cos 2x + 12x^2\cos 2x)$
35. $c = 0, \pi, 2\pi, 3\pi$, or 4π
37. $\sqrt[3]{9/2}$
39. $1, 9, \frac{13}{3}$
41. $\frac{1}{9}, 1, \frac{1}{3}$
43. $-\sin^3 x$
45. Distance traveled
47. It makes no difference.
49. (*a*) Displacement from time *a* to time *b*
(*b*) Acceleration

Sec. 5.6. Proofs of the Two Fundamental Theorems of Calculus

5. (*a*) sin 25
(*b*) $\frac{8x^2}{1-5x^6} + \frac{40x^6+4}{(1-5x^6)^{3/2}}\int_0^{x^2}\frac{dt}{\sqrt{1-5t^3}}$

Sec. 5.S. Guide Quiz

2. (*a*) $\frac{29}{18} \approx 1.6111$ (*b*) $\frac{163}{252} \approx 0.6468$
3. (*a*) $-2/(2x+3)^2$ (*b*) $\frac{1}{15}$
4. $3\pi/4$
5. $\frac{4}{3}(3 - \sqrt{3})$
6. $6x\cos(3x^3) - 27x^4\sin(3x^3) - 2\cos(3x^2) + 12x^2\sin(3x^2)$

Sec. 5.S. Review Exercises

1. $\frac{15}{2}$
3. $\frac{1}{3}$

5. $\frac{5}{72}$
7. $\tan x + C$
9. $\sec x + C$
11. $-4 \csc x + C$
13. $\frac{1}{7}x^7 + \frac{1}{2}x^4 + x + C$
15. $5x^{20} + C$
17. (*a*) $18x^2(x^3 + 1)^5$ (*b*) $\frac{1}{18}(x^3 + 1)^6 + C$
19. (*a*) $d + d^2 + d^3$ (*b*) $x + x^2 + x^3 + x^4$
(*c*) $0 + \frac{1}{2} + \frac{1}{2} + \frac{3}{8} = \frac{11}{8}$ (*d*) $\frac{3}{2} + \frac{4}{3} + \frac{5}{4} + \frac{6}{5} = \frac{317}{60}$
(*e*) $(\frac{1}{2} - \frac{1}{3}) + (\frac{1}{3} - \frac{1}{4}) + (\frac{1}{4} - \frac{1}{5}) = \frac{3}{10}$
(*f*) $\sin(\pi/4) + \sin(\pi/2) + \sin(3\pi/4) + \sin \pi = 1 + \sqrt{2}$
21. (*b*) $1/n$ (*c*) $1/n$ (*d*) Right endpoint
23. (*a*) $\int_0^2 x^3\,dx$ (*b*) $\int_0^1 x^4\,dx$ (*c*) $\int_1^3 x^5\,dx$
25. $\sin \sqrt{3}$

CHAPTER 6. TOPICS IN DIFFERENTIAL CALCULUS

Sec. 6.1. Review of Logarithms

1. (*a*) $5 = \log_2 32$ (*b*) $4 = \log_3 81$
(*c*) $-3 = \log_{10}(0.001)$ (*d*) $0 = \log_5 1$
(*e*) $\frac{1}{3} = \log_{1,000}(10)$ (*f*) $\frac{1}{2} = \log_{49} 7$
3. (*a*)

x	$\frac{1}{9}$	$\frac{1}{3}$	1	3	9
$\log_3 x$	−2	−1	0	1	2

(*b*)

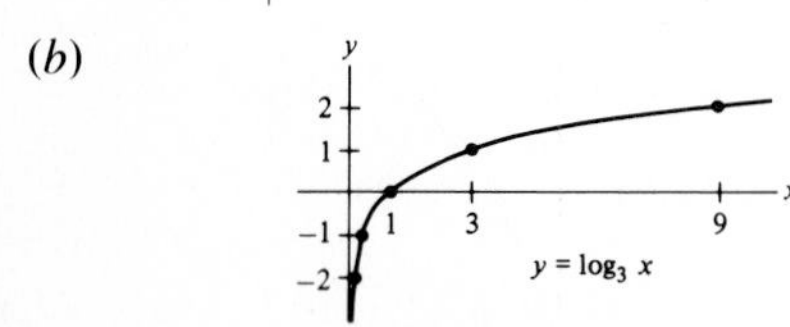

5. (*a*) $2^x = 7$ (*b*) $5^s = 2$ (*c*) $3^{-1} = \frac{1}{3}$
(*d*) $7^2 = 49$
7. (*a*) 16 (*b*) $\frac{1}{2}$ (*c*) 7 (*d*) g
9. (*a*) $\frac{1}{2}$ (*b*) 5 (*c*) −3
11. $x = \log_3 \frac{7}{2} \approx 1.1403$
13. $x = 0$
15. (*a*) 3, $\frac{1}{3}$, 1
17. (*a*) 0.60 (*b*) 0.70 (*c*) 0.78 (*d*) 0.90 (*e*) 0.96
(*f*) 0.18 (*g*) 0.08 (*h*) 0.12 (*i*) 1.3 (*j*) 2.3
(*k*) −2.22
19. (*a*) 1.46 (*b*) 1.58
21. (*a*) $\log_{1/2} x = -\log_2 x$
(*b*)

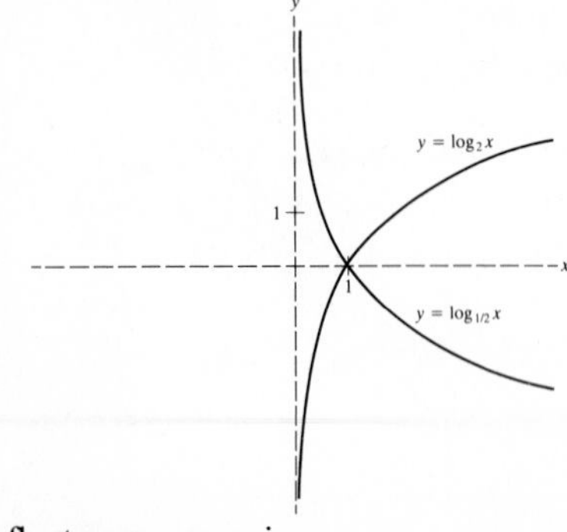

(*c*) Reflect across x axis.

25. $7 \log_{10}(\cos x) + \frac{3}{2}\log_{10}(x^2 + 5) - \log_{10}(4 + \tan^2 x)$
27. $x^2[\log_{10} x + \frac{1}{2}\log_{10}(2 + \cos x)]$
29. $\log_2 10$
35. (*a*) 65,050 (*b*) 10.1640625

Sec. 6.2. The Number e

1. 1
3. $e^{3/4}$
5. e
7. Approaches ∞
9. $A\left(1 + \dfrac{rt}{100n}\right)^n$
11. (*a*) \$1,500 (*b*) \$1,562.50 (*c*) \$1,632.09
(*d*) \$1,648.16 (*e*) \$1,648.72
13. (*a*) ≈0.71773, ≈0.69556, ≈0.69075 (*b*) ≈0.693
(*c*) ≈1.10 (*d*) ≈1

15.

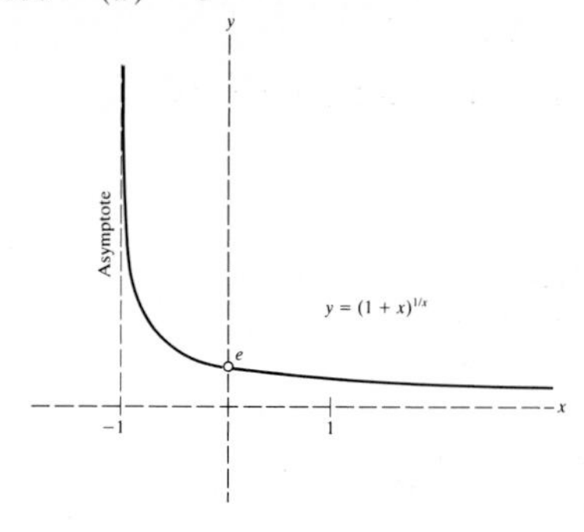

Sec. 6.3. The Derivatives of the Logarithmic Functions

1. $\dfrac{2x}{1 + x^2}$
3. $\dfrac{1 - \ln x}{x^2}$
5. $\dfrac{1}{x}\sin 4x + 4 \cos 4x \ln 3x$
7. $2 \cot x \ln(\sin x)$
9. $\dfrac{2}{2x + 3}$
11. $\dfrac{x}{(5x + 2)^2}$
13. $\dfrac{1}{\sqrt{x^2 - 5}}$
15. $\dfrac{1}{x(3x + 5)}$
17. $\dfrac{1}{25 - x^2}$
19. $\dfrac{6x}{x^2 + 1} + \dfrac{20x^4}{x^5 + 1}$
21. $(1 + 3x)^5(\sin 3x)^6\left(\dfrac{15}{1 + 3x} + 18 \cot 3x\right)$
23. $x^{-1/2}(\sec 4x)^{5/3}\sin^3 2x\left(\dfrac{20}{3}\tan 4x + 6 \cot 2x - \dfrac{1}{2x}\right)$
25. (*a*) $\ln|5x + 1| + C$ (*b*) $\frac{1}{2}\ln(x^2 + 5) + C$
(*c*) $\ln|\sin x| + C$ (*d*) $\ln|\ln x| + C$

27.

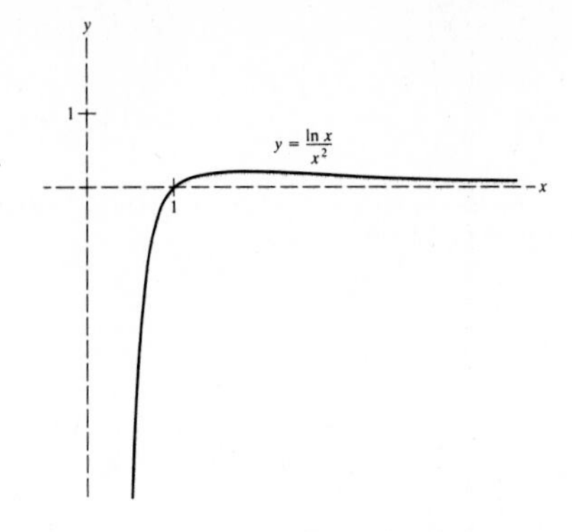

29. $\frac{3}{2}$

31. $\dfrac{1}{(3 \ln 10)x}$

35. (*a*) $\ln b$ (*c*) $\pi\left(1 - \dfrac{1}{b}\right)$
(*e*) See the Student Solutions Manual.

Sec. 6.4. The Derivative of b^x

1. $2xe^{x^2}$
3. $2(x^2 + x)e^{2x}$
5. $-x(\ln 2)2^{-x^2+1}$
7. $x^{(x^2)}(x + 2x \ln x)$
9. 3
11. $x^{\tan 3x}\left(3 \sec^2 3x \ln x + \dfrac{1}{x} \tan 3x\right)$
13. $\dfrac{-(1 + e^x)(5 \sin 5x) - (\cos 5x)(4 + 5e^x)}{e^{4x}(1 + e^x)^2}$
15. $x^{\sqrt{3}-1} e^{x^2}(2x^2 \sin 3x + 3x \cos 3x + \sqrt{3} \sin 3x)$
17. $\dfrac{1}{x + \sqrt{1 + e^{3x}}}\left(1 + \dfrac{3e^{3x}}{2\sqrt{1 + e^{3x}}}\right)$
19. xe^{ax}
21. $e^{ax} \sin bx$
23. (*a*) $\frac{1}{3}(e^{15} - e^3)$ (*b*) $(\pi/6)(e^{30} - e^6)$
25. (*a*) $999/(\ln 10)$ (*b*) $999{,}999\pi/(2 \ln 10)$
27. $1 + x$
29. $1 + 2.30x$
31. x
33. $0.43x$
35. (*a*) (0, 0) (*b*) $(1, 1/e)$
(*c*) $(1, 1/e)$, global maximum (*d*) $(2, 2/e^2)$
(*e*) $y = 0$
(*f*)

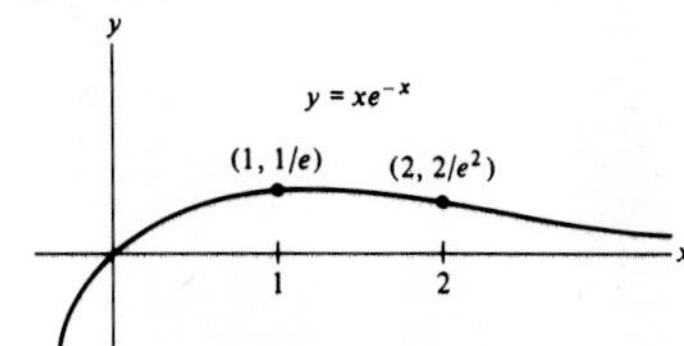

37. (*a*) (0, 0) (*b*) (0, 0), $(3, 27/e^3)$
(*c*) $(3, 27/e^3)$, global maximum
(*d*) (0, 0), $(3 - \sqrt{3}, \approx 0.574)$, $(3 + \sqrt{3}, \approx 0.933)$
(*e*) $y = 0$
(*f*)

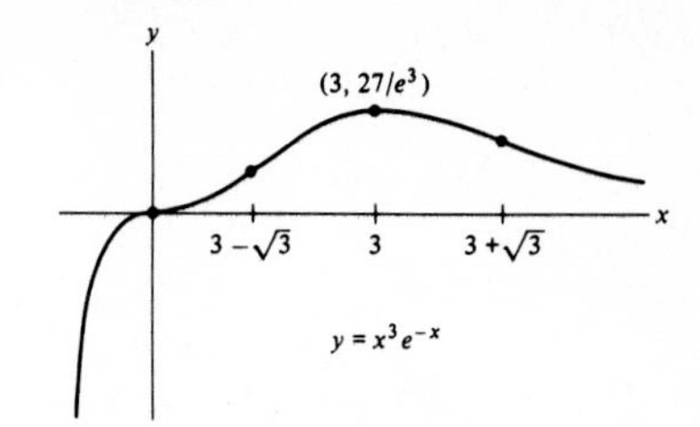

39. (*a*) (0, 0), (1, 0) (*b*) $\left(\dfrac{3 - \sqrt{5}}{2}, \approx 0.161\right)$, $\left(\dfrac{3 + \sqrt{5}}{2}, \approx -0.309\right)$ (*c*) $\left(\dfrac{3 - \sqrt{5}}{2}, \approx 0.161\right)$, global maximum; $\left(\dfrac{3 + \sqrt{5}}{2}, \approx -0.309\right)$, local minimum
(*d*) (1, 0), $(4, -12/e^4)$ (*e*) $y = 0$
(*f*)

y
1
$y = (x - x^2)e^{-x}$
$(a, f(a))$
x
1
$(b, f(b))$ $(4, -12/e^4)$
$a = (3 - \sqrt{5})/2$
$b = (3 + \sqrt{5})/2$

41. (*a*) 1 (*b*) $2 \ln 2$ (*c*) $\ln 10$
45. (*a*)

t	0	$\pi/2$	π	$3\pi/2$	2π	$5\pi/2$	3π	$7\pi/2$	4π
$e^{-t} \sin t$	0	0.2079	0	−0.0090	0	0.0004	0	−0.000017	0

(*b*)

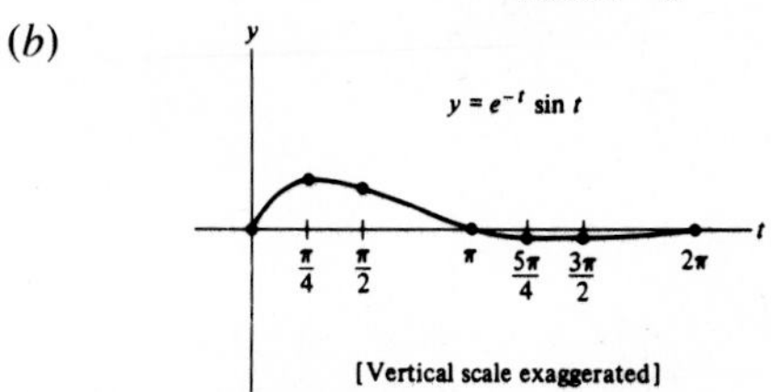

(*c*) $(\pi/4, \approx 0.3224)$, $(5\pi/4, \approx -0.0139)$
47. (*a*) 60 m (*b*) 40 m
51. (*b*) Yes (*c*) No

Sec. 6.5. The Derivatives of the Inverse Trigonometric Functions

3. (*a*) 1.2490 (*b*) −1.1071 (*c*) 0.4115 (*d*) 1.2310
5. (*a*) and (*d*)
7.

x	−1	−0.8	−0.6	−0.4	−0.2	0	0.2	0.4	0.6	0.8	1
$\sin^{-1} x$	−1.57	−0.93	−0.64	−0.41	−0.20	0	0.20	0.41	0.64	0.93	1.57

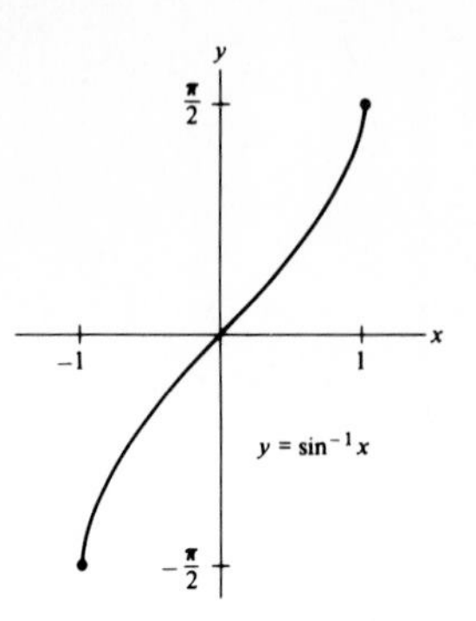

11. $1/\sqrt{2}$

13. -1

15. 0.3

17. $5/\sqrt{1-25x^2}$

19. $1/(|x|\sqrt{9x^2-1})$

21. $1/(3x^{2/3}(1+x^{2/3}))$

23. $x/(2\sqrt{x-1}) + 2x\sec^{-1}\sqrt{x}$

25. $\dfrac{3\sin 3x}{\sqrt{1-9x^2}} + 3\cos 3x \sin^{-1} 3x$

27. $e^{-2x}\left[\dfrac{x}{|x|\sqrt{9x^2-1}} + (1-2x)\sec^{-1} 3x\right]$

29. $\dfrac{1}{2\sqrt{x}(1+x)}$

31. $\dfrac{1}{2x\sqrt{x-1}\,\sec^{-1}\sqrt{x}}$

33. $\dfrac{1}{\tan^{-1} 10^x} - \dfrac{x\,10^x \ln 10}{(\tan^{-1} 10^x)^2(1+10^{2x})}$

35. $\sqrt{\dfrac{1+x}{1-x}}$

37. $\dfrac{6(\tan^{-1} 2x)^2}{1+4x^2}$

39. $\sqrt{2-x^2}$

41. $\dfrac{5}{3x\sqrt{3x^5-1}}$

43. $\sqrt{\dfrac{2-x}{1+x}}$

45. $(\sin^{-1} 2x)^2$

47. (*a*) $\dfrac{1}{\sqrt{x^2-9}}$ (*b*) $\dfrac{1}{\sqrt{9-x^2}}$

49. (*a*) $\dfrac{1}{x\sqrt{2x^2+1}}$ (*b*) $\dfrac{1}{|x|\sqrt{2x^2-1}}$

51. (*b*) $\sin^{-1}\left(\dfrac{x}{5}\right) + C$ (*c*) $\sin^{-1}\left(\dfrac{x}{\sqrt{5}}\right) + C$

53. (*a*)

(*b*) $\ln(1+b)$ (*c*) Infinite

Sec. 6.6. Related Rates

1. $2\sqrt{901}$ ft/sec ≈ 60.03 ft/sec

3. (*a*) $\frac{3}{4}$ ft/sec (*b*) $\frac{4}{3}$ ft/sec (*c*) $9/\sqrt{19}$ ft/sec

5. (*b*) $x = 3\tan\theta$

7. (*a*) -1 ft²/sec (*b*) 8 ft/sec (*c*) 69/13 ft/sec

9. (*a*) $5/(2\pi)$ yd/hr (*b*) $1/(10\pi)$ yd/hr

11. $27/(2\sqrt{2})$ ft²/sec, increasing

13. (*a*) -0.00014 ft/sec² (*b*) -7.56 ft/sec²

15. (*a*) $6x$ (*b*) 18

17. $\frac{19}{4}$ ft/sec

19. (*a*) $-25/\sqrt{29}$ ft/sec (*b*) Decreasing

21. (*a*) $5\pi/2$ ft/sec (*b*) $5\sqrt{3}\pi/2$ ft/sec

23. $211{,}210/\sqrt{42{,}241} \approx 1{,}027.7$ ft/sec

Sec. 6.7. Separable Differential Equations

7. $\frac{1}{5}y^5 = \frac{1}{4}x^4 + C$

9. $y = kx$

11. $\sin^{-1} y = \tan^{-1} x + C$, or

$$y = \frac{kx}{\sqrt{1+x^2}} \pm \frac{\sqrt{1-k^2}}{\sqrt{1+x^2}}, \text{ where } k = \cos C$$

13. $\frac{1}{3}y^3 = -\frac{1}{2}x^2 + C$

15. $\frac{1}{2}\ln(1+y^2) = \frac{1}{3}x^3 + C$

17. $\pm\sec^{-1} y = \frac{1}{3}x^3 + C$

19. $\sin^{-1} y = -2(3-x)^{1/2} + C$

21. (*a*) $\dfrac{di}{dt} = \dfrac{E - Ri}{L}$ (*b*) $i = \dfrac{1}{R}\left[E - (E - Ri_0)e^{-Rt/L}\right]$

23. (*b*) $f(t) = f(0)e^{kt}$

25. (*c*) $\dfrac{k}{k^2+a^2}\sin ax - \dfrac{a}{k^2+a^2}\cos ax + ce^{-kx}$

Sec. 6.8. Natural and Inhibited Growth

3. (*a*) 0.98 (*b*) -0.0202

5. (*a*) 0.10 (*b*) 1.10 (*c*) 0.0953 (*d*) 7.2725 hours

7. (*a*) 10 g (*b*) 1.0986 (*c*) 200 percent

9. (*a*) 38.85 years (*b*) 77.71 years (*c*) 169.06 years

11. The year 2040 (in 53.5 years)

13. (*a*) 6:18 P.M. (*b*) 0.2618 (*c*) 29.93 percent

15. (*a*) $0.25A$ (*b*) $0.785A$

17. 282.8 g

19. (*a*) $Ae^{-0.005t}$ (*b*) 138.6 days

23. (*a*) $Y = \frac{3}{2}x + 1$ (*b*) $y = 2^{(3/2)x+1}$ (*c*) A line

25. (*a*)

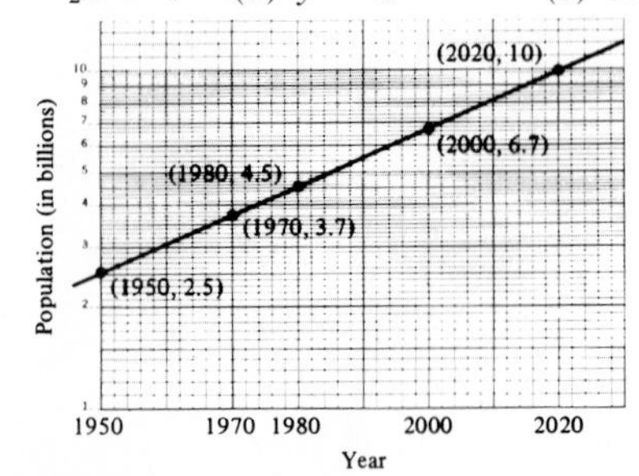

(*b*) 6.7 billion (*c*) 2020

29. (*b*) $P(t) = \frac{1}{k}\{[kP(0) - h]e^{kt} + h\}$

31. (Let p be the interest rate) (*a*) $\frac{\ln 2}{\ln(1 + p/100)}$

(*b*) $\frac{100 \ln 2}{p}$

39. No, unless the relative growth rates are equal.

Sec. 6.9. L'Hôpital's Rule

1. 3
3. $\frac{3}{2}$
5. 0
7. $\frac{1}{2}$
9. 3
11. 0
13. e^{-2}
15. 1
17. 1
19. 0
21. ln 3/ln 2
23. Does not exist
25. 1
27. -1
29. Does not exist
31. 0
33. 1
35. Does not exist
37. 1
39. 0
41. $\ln \frac{5}{3}$
43. $\frac{16}{9}$
45. $\frac{1}{6}$
47. Approaches (3, 0)
49. (*b*) $\frac{3}{4}$
53. $\sqrt{2}$
57. (*a*) 0 (*b*) 0 (*c*) Cannot be determined (*d*) Does not exist (*e*) 1 (*f*) 1 (*g*) 1
59. (*a*) Cannot be determined (*b*) ∞ (*c*) 0 (*d*) 0 (*e*) 0 (*f*) Cannot be determined (*g*) Cannot be determined
61. (*b*) $\sqrt{e}$
63. 0
67. $e^{-1/6}$

Sec. 6.10. Newton's Method for Solving an Equation

3. $x_2 = 4.375$, $x_3 \approx 4.359$
5. (*b*) $x_2 = 3$, $x_3 \approx 2.259$ (*c*) $x_2 \approx 1.917$, $x_3 \approx 1.913$
7. (*b*) $x_2 \approx 0.857$
9. (*a*)

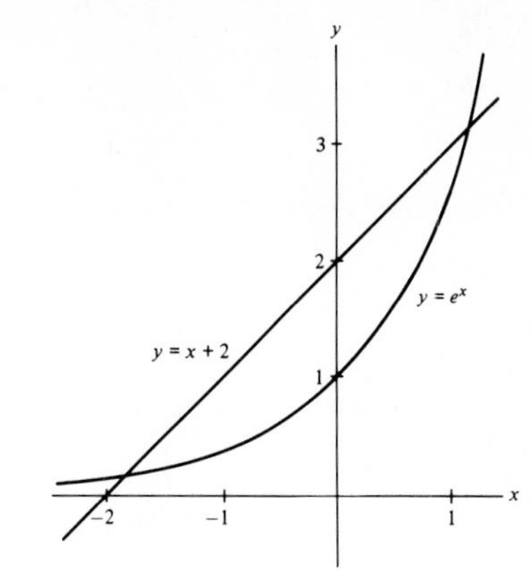

(*c*) ≈ 1.15

11. (*a*)

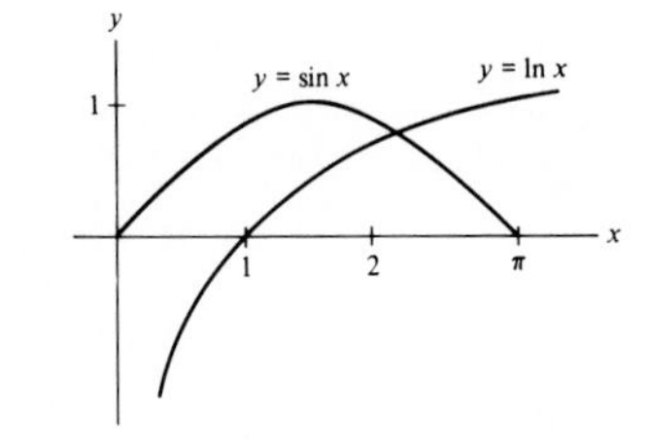

(*c*) If $x_1 = 2$, then $x_2 \approx 2.236$.

13. (*a*)

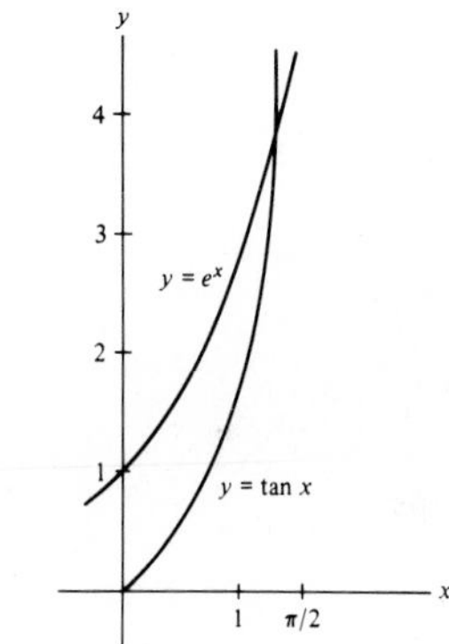

(*c*) If $x_1 = 1.3$, then $x_2 \approx 1.307$.

15. (*b*) $x_2 = 1.5$, $x_3 \approx 1.316$

(*c*)

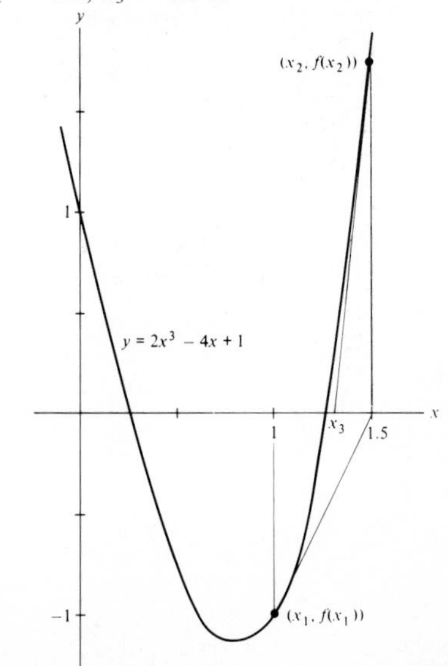

17. $x_2 = -1/\sqrt{5}$, $x_3 = 1/\sqrt{5}$
19. 0.95
21. (*a*) 1 (*b*) 0.86

Sec. 6.11. The Hyperbolic Functions and Their Inverses

3. $\cosh t$
5. $-\text{sech}\, t \tanh t$
7. $3 \sinh 3x$
9. $(\text{sech}^2 \sqrt{x})/(2\sqrt{x})$
11. $e^{3x} (\cosh x + 3 \sinh x)$
13. $(4 \sinh 4x)(\coth 5x)(\text{csch}\, x^2) + (\cosh 4x)(-5 \,\text{csch}^2\, 5x)(\text{csch}\, x^2) + (\cosh 4x)(\coth 5x)(-2x \,\text{csch}\, x^2 \coth x^2)$
29. $\pi/4$
31. (*a*)

t	−3	−2	−1	0	1	2	3
$\cosh t$	10.068	3.762	1.543	1	1.543	3.762	10.068
$\sinh t$	−10.018	−3.627	−1.175	0	1.175	3.627	10.018

(*b*)

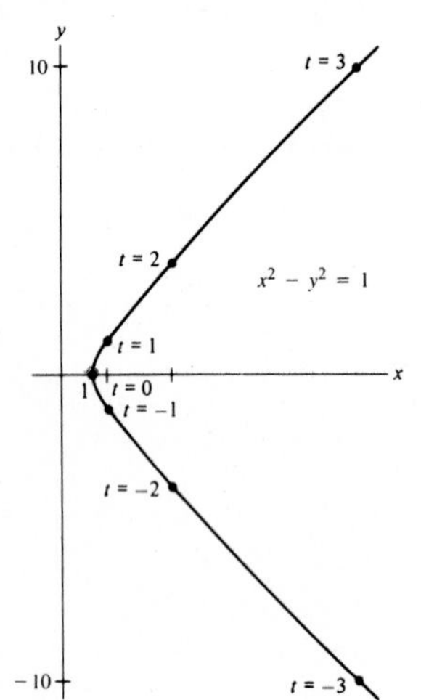

33. (*a*)

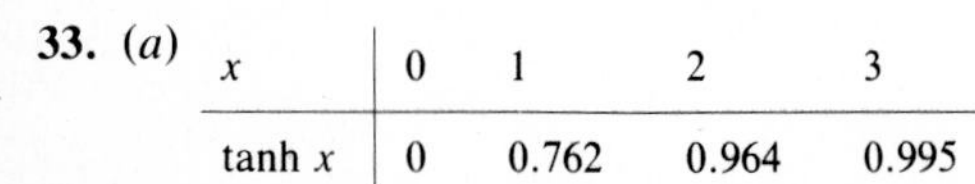

x	0	1	2	3
$\tanh x$	0	0.762	0.964	0.995

(*b*)

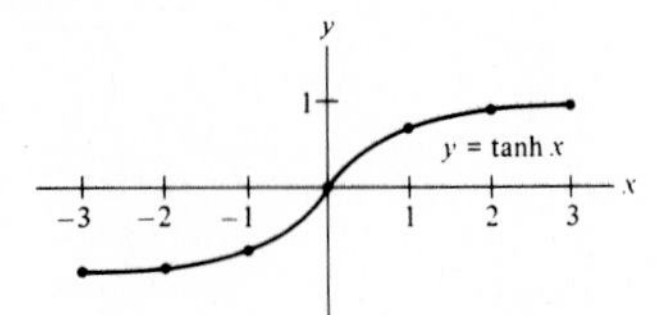

35. (*a*) $v(t) = V \tanh (gt/V)$ (*d*) $dv/dt = g \,\text{sech}^2 (gt/V)$
37. (0, 0)

Sec. 6.S. Guide Quiz

1. (*a*) $\dfrac{3e^{\sin^{-1} 3x}}{\sqrt{1 - 9x^2}}$ (*b*) $\dfrac{8 \cos [(\tan^{-1} 4x)^2] \tan^{-1} 4x}{1 + 16x^2}$

(*c*) $\left(\dfrac{3x^2 + 5}{x^3 + 5x} + \dfrac{2 + \ln x}{2\sqrt{x}}\right)(x^3 + 5x)x^{\sqrt{x}}$

(*d*) $(\sec^{-1} 2x)^{-2} \times \left[(\ln 5)\, 5^x \sec^{-1} 2x \sec^2 5^x - \dfrac{\tan 5^x}{|x|\sqrt{4x^2 - 1}}\right]$

(*e*) $5\left[\dfrac{16x}{(x^2 + 1)\ln(x^2 + 1)} - \dfrac{3}{x} + \dfrac{4}{3} \tan 2x\right] \times \left\{x^{-3} \dfrac{[\ln(x^2 + 1)]^8}{\sqrt[3]{\cos^2 2x}}\right\}^5$

2. (*a*) $e^{ax} \cos bx$ (*b*) $\sin^{-1} ax$ (*c*) $\tan^{-1} ax$ (*d*) $\sec^{-1} ax$

3. $\dfrac{-2}{e + 1}$, $\dfrac{-(3e^2 + 2e - 5)}{(e + 1)^3}$

4. (*a*) $\dfrac{5\sqrt{5}}{3}$ ft/sec (*b*) $\dfrac{3\sqrt{11}}{2}$ ft/sec

5. (*a*) $\frac{1}{2}e^{2y} = \frac{1}{4}x^4 + C$ (*b*) $\frac{1}{8} \ln(4y^2 + 1) = x + C$

6. 12.706 days

7. (*a*) $\frac{2}{3}$ (*b*) Does not exist (*c*) 0 (*d*) 1 (*e*) 1

8. $\log_{10} x$, $\ln x$, x^3, $(1.001)^x$, 2^x

9. (*a*) $x > 0$ (*b*) $x = 1$ (*c*) $x = e$ (*d*) $(e, 1/e)$ is a maximum (*e*) $0 < x < e$, $x > e$ (*f*) No; yes (*g*) $-\infty$ (*h*) 0
(*i*)

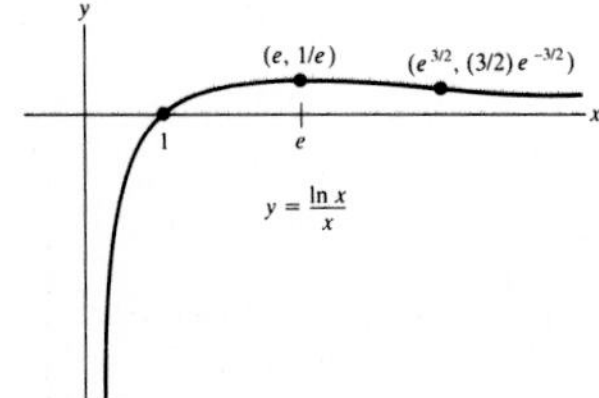

10. $e = \lim_{x\to 0} (1 + x)^{1/x}$ (*b*) $e \approx 2.718$

11. Divide $\log_{10} 7$ by $\log_{10} e$.

Sec. 6.S. Review Exercises

1. $\dfrac{3x^2}{2(1 + x^3)^{1/2}}$

3. $\dfrac{1}{2\sqrt{x}}$

5. $-6 \cos 3x \sin 3x$

7. $\frac{3}{2}\sqrt{x}$

9. $\dfrac{\cos x}{2\sqrt{\sin x}}$

11. $-2x \csc^2 x^2$

13. $e^{x^2}\left(\dfrac{2x^2 - 1}{2x^2}\right)$

15. $\dfrac{x^{5/6}}{\sqrt{1 - x^2}} + \dfrac{5}{6} x^{-1/6} \sin^{-1} x$

17. $e^{3x}(3x^2 + 2x)$

19. $3 \sec 3x$

21. $\dfrac{-\sin\sqrt{x}}{2\sqrt{x}}$

23. $\sec x$

25. $\dfrac{-3x}{(6+3x^2)^{3/2}}$

27. $\sqrt{\frac{15}{2}(5x+7)}$

29. $\dfrac{x}{3x+4}$

31. $(1+x^2)^4[(1+x^2)\,3\cos 3x + 10x\sin 3x]$

33. $(24x^2-8)(2x^3-2x+5)^3$

35. $(1+2x)^4[10\cos 3x - 3(1+2x)\sin 3x]$

37. $4\cos 3x\cos 4x - 3\sin 3x\sin 4x$

39. $-6x\csc 3x^2\cot 3x^2$

41. $\left(\dfrac{x}{1+x}\right)^4(7x^2+4x^3)$

43. $(\sec 3x)(3x\tan 3x + 1)$

45. $\dfrac{(x^3-1)^3(x^{10}+1)^4}{(2x+1)^5}\left(\dfrac{9x^2}{x^3-1}+\dfrac{40x^9}{x^{10}+1}-\dfrac{10}{2x+1}\right)$

47. $\dfrac{1}{\sqrt{x^2+1}}$

49. $e^{-x}\left(\dfrac{2x}{1+x^4}-\tan^{-1}x^2\right)$

51. $e^{\sqrt{x}}$

53. $\dfrac{1-2\ln x}{x^3}$

55. $\dfrac{(\sin x)(\cos^2 x+1)}{\cos^2 x}$

57. $\dfrac{x^2\ln(1+x^2)}{|x|\sqrt{9x^2-1}} + 2x\sec^{-1}3x\ln(1+x^2) + \dfrac{2x^3}{1+x^2}\sec^{-1}3x$

59. $1-\dfrac{2}{x-1}-\dfrac{1}{(x+1)^2}$

61. $\dfrac{-12x-3}{6x^2+3x+1}$

63. $\dfrac{15}{5x+1}+\dfrac{12}{6x+1}-\dfrac{8}{2x+1}$

65. $\sqrt{9-x^2}$

67. $3\tan^2 3x$

69. $\dfrac{1}{\sqrt{x^2+25}}$

71. $\sin^{-1}x$

73. $\sec 3x$

75. $\sin^3 3x$

77. $e^{3x}\sin 2x\,(3\sin 2x\tan x + 4\cos 2x\tan x + \sin 2x\sec^2 x)$

79. $\sqrt{4x^2+3}$

81. $\dfrac{x^2}{(x+1)^3}(5x^5+7x^4-2x^2-x+9)$

83. $(1+3x)^{x^2}\left[\dfrac{3x^2}{1+3x}+2x\ln(1+3x)\right]$

85. $\dfrac{1+x^2+\ln x-9x^2\ln x}{(1+x^2)^6}$

87. (*a*)

x	$\frac{1}{8}$	$\frac{1}{4}$	$\frac{1}{2}$	1	2	4	8
$\log_2 x$	-3	-2	-1	0	1	2	3

(*b*)

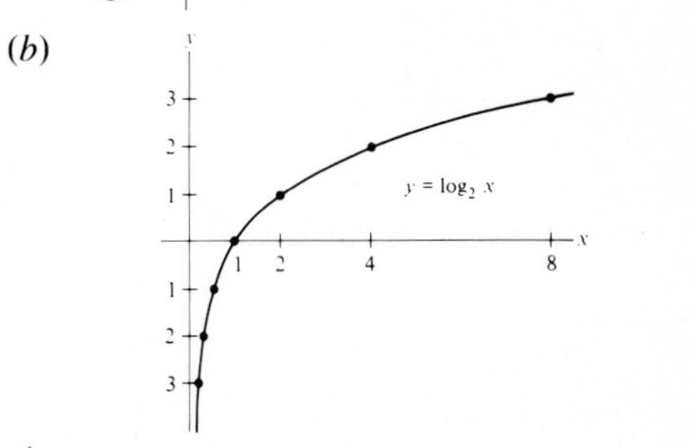

(*c*) ∞ (*d*) $-\infty$

89. (*b*) 0.0217

101. (*a*) $2/(1-x^2)$ (*b*) $\frac{1}{2}\ln 3$

103. $\dfrac{ab}{\sqrt{ax+b}\,(ax+b-b^2)}$

105. (*a*) $\ln|x^3+x-6|+C$ (*b*) $\frac{1}{2}\ln|\sin 2x|+C$ (*c*) $\frac{1}{5}\ln|5x+3|+C$ (*d*) $\dfrac{-1}{5(5x+3)}+C$

107. (*a*) $\dfrac{4}{3}\sec 4x\csc 4x - \dfrac{10}{1-2x}-\dfrac{2}{1+3x}$

(*b*) $\dfrac{(1+x^2)^3\sqrt{1+x}}{\sin 3x}\times\left[\dfrac{6x}{1+x^2}+\dfrac{1}{2(1+x)}-3\cot 3x\right]$

109. (*a*) Does not exist (*b*) 0

113. $\frac{1}{2}$

115.

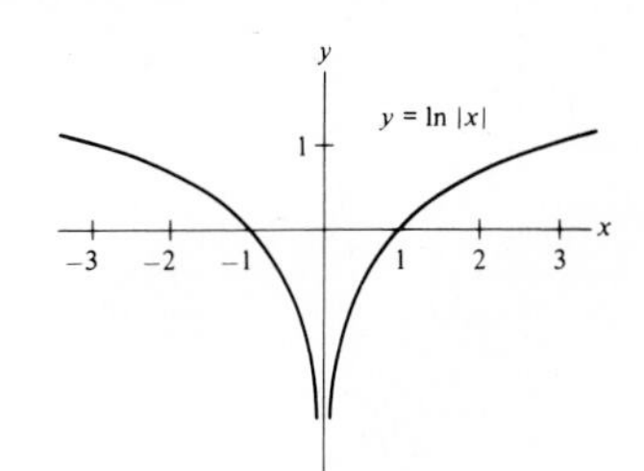

117. $(e, 1)$

119. $y'(0) = -2$

121. 2

123. e^3

125. $-\frac{1}{2}$

127. e^2

129. -1

131. 0

133. $\frac{2}{3}$

135. 3
137. 3
139. 1
141. $\frac{8}{3}$
143. $-\infty$
145. 0
147. $\ln \frac{5}{3}$
149. e^6
151. 1
153. 0
155. 2
157. 1
159. 0
163. (*a*) 0 (*b*) 1 (*c*) Cannot tell (*d*) Cannot tell (*e*) Cannot tell (*f*) 0
167. (*a*) e^x (*b*) $\ln x$ (*c*) $\sqrt[3]{x}$ (*d*) $x/3$ (*e*) x^3 (*f*) $\sin x$
169. 1
173. 1 if $0 < a < 1$; a if $a > 1$
175. 29 days
179. (*a*) Yes (*b*) No (*c*) Yes (*d*) No (*e*) Yes (*f*) Yes
181. 1
183. $\frac{2}{3}$
185. 2.10
187. (*a*) 10 m, about 6 m, about 10 m
(*b*)

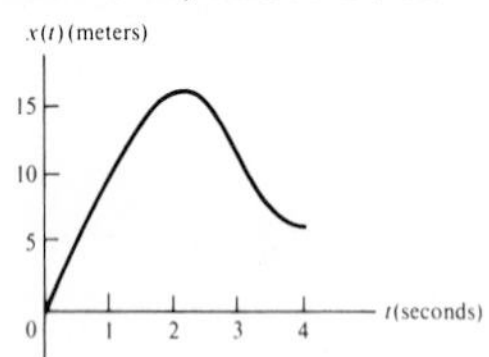

(*c*)

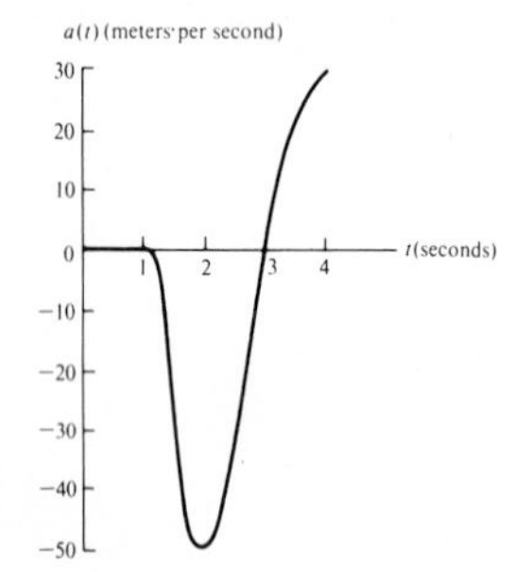

CHAPTER 7. COMPUTING ANTIDERIVATIVES

Sec. 7.1. The Substitution Technique

1. $\frac{5}{4}x^4 + C$
3. $\frac{3}{4}x^{4/3} + C$
5. $6\sqrt{x} + C$
7. $-\frac{5}{2}e^{-2x} + C$
9. $6 \sec^{-1} x + C$
11. $\frac{1}{4} \ln (1 + x^4) + C$
13. $-\ln (1 + \cos x) + C$
15. $\ln |x + x^2| + C$
17. $\dfrac{x^5}{5} + 2x^3 + 9x + C$
19. $\dfrac{x^3}{3} + \dfrac{3}{4}x^4 + C$
21. $\frac{2}{7}x^{7/2} + C$
23. $\ln |x| + 2\sqrt{x} + C$
25. $\frac{1}{6}(1 + 3x)^6 + C$
27. $\sqrt{2} - 1$
29. $-\frac{1}{2} \cos 2x + C$
31. $\frac{1}{3}(e^6 - e^{-3})$
33. $\frac{1}{3} \sin^{-1} 3x + C$
35. $\frac{1}{3}$
37. $\frac{1}{5}(\ln x)^5 + C$
39. $-\frac{1}{12}(1 - x^2)^6 + C$
41. $\frac{3}{8}(1 + x^2)^{4/3} + C$
43. $2e^{\sqrt{t}} + C$
45. $-\frac{1}{3} \cos 3\theta + C$
47. $\frac{2}{7}(x - 3)^{7/2} + C$
49. $\ln |x^2 + 3x + 2| + C$
51. $\frac{1}{2}e^{2x} + C$
53. $-\frac{1}{5} \cos x^5 + C$
55. $\frac{1}{2} \tan^{-1} (x^2) + C$
57. $x - \ln |1 + x| + C$
59. $\frac{1}{2}[\ln(3x)]^2 + C$
61. $\frac{1}{3} \int_1^8 e^u \, du$
63. $\int_1^2 (u^{-2} - 2u^{-3} - 2u^{-4}) \, du$
65. $\int_0^1 u^3 \, du$
67. $\dfrac{1}{a^3}\left(\dfrac{1}{2} a^2x^2 - abx + b^2 \ln |ax + b|\right) + C$
69. $\dfrac{1}{a^3}\left(ax - 2b \ln |ax + b| - \dfrac{b^2}{ax + b}\right) + C$
71. They are both right.
73. Jill is right.

Sec. 7.2. Integration by Parts

1. $\frac{1}{4}e^{2x}(2x - 1) + C$
3. $-\dfrac{x}{2} \cos 2x + \dfrac{1}{4} \sin 2x + C$
5. $\frac{1}{2}x^2 \ln 3x - \frac{1}{4}x^2 + C$
7. $\dfrac{5e - 10}{e^2}$
9. $\sin^{-1} 1 - 1$

11. $\frac{1}{3}x^3 \ln x - \frac{1}{9}x^3 + C$
13. $\frac{1}{1188}(3x + 5)^{11}(33x - 5) + C$
15. $3(\ln 3)^2 - 6 \ln 3 - 2(\ln 2)^2 + 4 \ln 2 + 2$
17. $(e - 2)/e$
19. $\frac{1}{2}e^x(\sin x - \cos x) + C$
21. $-\frac{\ln(1 + x^2)}{x} + 2 \tan^{-1} x + C$
25. (a), (b) $(x + 1) \ln(x + 1) - x + C$
(c) (b) is easier.
27. $\int (\ln x)^n\, dx = x(\ln x)^n - n \int (\ln x)^{n-1}\, dx$
29. $2(\sin \sqrt{x} - \sqrt{x} \cos \sqrt{x}) + C$
31. $2(\exp \sqrt{x})(\sqrt{x} - 1) + C$
33. (a) $\frac{1}{2}(x - \sin x \cos x) + C$
(b) $-\frac{1}{4} \sin^3 x \cos x + \frac{3}{8}(x - \sin x \cos x) + C$
(c) $-\frac{1}{6} \sin^5 x \cos x - \frac{5}{24} \sin^3 x \cos x + \frac{5}{16}(x - \sin x \cos x) + C$
37. $\frac{2}{9}(3x + 7)^{3/2}\left(\frac{3x}{5} - \frac{14}{15}\right) + C$
39. $\frac{1}{20a^2}(ax + b)^4(4ax - b) + C$
41. (a) $I_0 = \frac{\pi}{2}$, $I_1 = 1$ (c) $I_2 = \frac{\pi}{4}$, $I_3 = \frac{2}{3}$
(d) $I_4 = \frac{3\pi}{16}$, $I_5 = \frac{8}{15}$
(e) $I_n = \frac{2 \cdot 4 \cdot \cdots \cdot (n - 1)}{3 \cdot 5 \cdot \cdots \cdot n}$,
when n is odd, $n \geq 3$
(f) $I_n = \left(\frac{1 \cdot 3 \cdot \cdots \cdot (n - 1)}{2 \cdot 4 \cdot \cdots \cdot n}\right)\frac{\pi}{2}$,
when n is even, $n \geq 2$
45. (e) $f(b) = f(0) + f^{(1)}(0)b + \frac{f^{(2)}(0)b^2}{2} + \frac{f^{(3)}(0)b^3}{6} + \frac{1}{6}\int_0^b f^{(4)}(x)(b - x)^3\, dx$
47. 0
49. $\frac{1}{81}(30 \sin 3 - 53 \cos 3 + 8)$

Sec. 7.3. How to Integrate Certain Rational Functions

1. $\frac{1}{3} \ln |3x - 4| + C$
3. $-\frac{5}{2(2x + 7)} + C$
5. $\frac{1}{3} \tan^{-1}\left(\frac{x}{3}\right) + C$
7. $\frac{1}{2} \ln (x^2 + 9) + C$
9. $\ln (x^2 + 9) + \tan^{-1}\left(\frac{x}{3}\right) + C$
11. $\frac{1}{20} \tan^{-1}\left(\frac{4x}{5}\right) + C$
13. $\frac{1}{32} \ln (16x^2 + 25) + C$
15. $\frac{1}{18} \ln (9x^2 + 4) + \frac{1}{3} \tan^{-1}\left(\frac{3x}{2}\right) + C$
17. $\frac{1}{\sqrt{6}} \tan^{-1}\left(\frac{\sqrt{6}x}{3}\right) + C$
19. $\frac{1}{\sqrt{2}} \tan^{-1}\left(\frac{x + 1}{\sqrt{2}}\right) + C$
21. $\frac{1}{\sqrt{2}} \tan^{-1}\left(\frac{x - 1}{\sqrt{2}}\right) + C$
23. $\frac{2}{\sqrt{23}} \tan^{-1}\left(\frac{4x + 1}{\sqrt{23}}\right) + C$
25. $\frac{1}{\sqrt{3}} \tan^{-1}\left(\frac{x + 2}{\sqrt{3}}\right) + C$
27. $\frac{1}{\sqrt{10}} \tan^{-1}\left(\frac{2x + 2}{\sqrt{10}}\right) + C$
29. $\ln (x^2 + 2x + 3) - \sqrt{2} \tan^{-1}\left(\frac{x + 1}{\sqrt{2}}\right) + C$
31. $\frac{3}{10} \ln (5x^2 + 3x + 2) - \frac{9}{5\sqrt{31}} \tan^{-1}\left(\frac{10x + 3}{\sqrt{31}}\right) + C$
33. $\frac{1}{2} \ln (x^2 + x + 1) + \frac{1}{\sqrt{3}} \tan^{-1}\left(\frac{2x + 1}{\sqrt{3}}\right) + C$
35. $\frac{1}{2} \ln (3x^2 + 2x + 1) + 2\sqrt{2} \tan^{-1}\left(\frac{3x + 1}{\sqrt{2}}\right) + C$
37. (a) $(x + 3)(x - 3)$ (b) $(x + \sqrt{5})(x - \sqrt{5})$
(c) Irreducible (d) $(x + 1)(x + 2)$ (e) $(x + 3)^2$
(f) Irreducible
(g) $[x + \frac{1}{2}(5 - \sqrt{17})][x + \frac{1}{2}(5 + \sqrt{17})]$
39. $\frac{1}{\sqrt{ac}} \tan^{-1}\left(\frac{\sqrt{ac}x}{c}\right) + K$
47. $\frac{-(13 + 4x)}{72(4x^2 + 8x + 13)} - \frac{1}{108} \tan^{-1}\left[\frac{2}{3}(x + 1)\right] + C$

Sec. 7.4. Integration of Rational Functions by Partial Fractions

1. $\frac{k_1}{x + 1} + \frac{k_2}{x + 2}$
3. $\frac{k_1}{x - 2} + \frac{k_2}{x + 2}$
5. $\frac{k_1}{x + 1} + \frac{k_2}{(x + 1)^2} + \frac{k_3}{(x + 1)^3}$
7. $1 + \frac{k_1}{x - 1} + \frac{k_2}{x + 1}$
9. $x - 1 + \frac{k_1}{x - 1} + \frac{k_2}{x + 2}$
11. $\frac{c_1x + d_1}{x^2 + x + 1} + \frac{c_2x + d_2}{(x^2 + x + 1)^2} + \frac{c_3x + d_3}{(x^2 + x + 1)^3}$
13. $x + \frac{3}{x + 1}$
15. $x + \frac{1}{2} + \frac{5/2}{2x + 1}$

17. $\frac{1}{x+1} - \frac{2}{x+2}$

19. $-\frac{1}{x} + \frac{2}{x+1} - \frac{3}{(x+1)^2}$

21. $\frac{2}{x+1} - \frac{1}{(x+1)^2} + \frac{2}{(x+1)^3}$

23. $\frac{1}{x+1} + \frac{3}{(x+1)^2}$

25. $x - 3 + \frac{4x+15}{x^2+3x+5}$

27. $x + \frac{1}{x} + \frac{2x-1}{x^2+x+1}$

29. $\ln\left|\frac{x+1}{(x+2)^2}\right| + C$

31. $\ln\left|\frac{(x+1)^2}{x}\right| + \frac{3}{x+1} + C$

33. $\frac{x^2}{2} - 3x + 2\ln(x^2+3x+5) + \frac{18}{\sqrt{11}}\tan^{-1}\left(\frac{2x+3}{\sqrt{11}}\right) + C$

35. $\frac{x^2}{2} + \ln|x(x^2+x+1)| - \frac{4}{\sqrt{3}}\tan^{-1}\left(\frac{2x+1}{\sqrt{3}}\right) + C$

37. $\frac{3}{2}x^2 + \ln|x^2-1| + C$

39. $\frac{1}{2}\ln\frac{32}{27}$

41. $\frac{1}{2}\ln\left|\frac{y-1}{y+1}\right| = \ln\left|\frac{x+1}{x}\right| - \frac{1}{x} + C$

43. (*a*) $\frac{1}{4}\ln(x^4+1) + C$ (*b*) $\frac{1}{2}\tan^{-1}x^2 + C$
(*c*) $\frac{\sqrt{2}}{8}\ln\left(\frac{x^2+\sqrt{2}x+1}{x^2-\sqrt{2}x+1}\right) + \frac{\sqrt{2}}{4}[\tan^{-1}(\sqrt{2}x+1) + \tan^{-1}(\sqrt{2}x-1)] + C$

45. (*a*) One choice is $\frac{5}{6} = \frac{1}{2} + \frac{1}{3}$.
(*b*) One choice is $\frac{4}{15} = \frac{2}{3} - \frac{2}{5}$.
(*c*) One choice is $\frac{19}{15} = 1 + \frac{2}{3} - \frac{2}{5}$.
(*d*) One choice is $\frac{7}{27} = \frac{2}{9} + \frac{1}{27}$.

47. (*b*) k_2: cover $x - q_2$ and plug in q_2. k_3: cover $x - q_3$ and plug in q_3. (*c*) $\frac{3/2}{x} - \frac{5}{x-1} + \frac{9/2}{x-2}$
(*d*) $\frac{9/5}{2x+1} - \frac{2/5}{x+3}$

Sec. 7.5. How to Integrate Powers of Trigonometric Functions

1. $\frac{\theta}{2} - \frac{\sin 2\theta \cos 2\theta}{4} + C$

3. $\frac{\pi}{8}$

5. $-\frac{1}{2}(\cos 2\theta - \frac{2}{3}\cos^3 2\theta + \frac{1}{5}\cos^5 2\theta) + C$

7. $\frac{2}{3}$

9. $\frac{\cos^7\theta}{7} - \frac{\cos^5\theta}{5} + C$

11. $\frac{\sin^4\theta}{4} - \frac{\sin^6\theta}{6} + C$

13. $\frac{\pi}{32} + \frac{1}{4} - \frac{9\sqrt{3}}{64} \approx 0.1046$

15. $\frac{\tan^5\theta}{5} + C$

17. $\frac{\sec^3\theta}{3} - \sec\theta + C$

19. $\frac{\tan^2\theta}{2} + \ln|\cos\theta| + C$

21. $\frac{4}{3}$

23. $\frac{\sin^3\theta}{3} - \frac{\sin^5\theta}{5} + C$

25. $\ln|\sin\theta| + C$

27. $\frac{1}{3}\ln|\sec 3\theta + \tan 3\theta| + C$

29. $\frac{\cos^8\theta}{8} - \frac{\cos^6\theta}{6} + C$

31. $\frac{\csc^4\theta}{4} - \frac{\csc^6\theta}{6} + C$ or $-\left(\frac{\cot^6\theta}{6} + \frac{\cot^4\theta}{4}\right) + C$

33. $-\left(\frac{\cot^5\theta}{5} + \frac{\cot^7\theta}{7}\right) + C$

35. $\sec\theta - \frac{\cos^3\theta}{3} + 2\cos\theta + C$

37. $\frac{8}{15}$

39. $5\pi/8$

41. $\frac{1}{8}\left(\theta - \frac{\sin 4\theta}{4}\right) + C$

45. $-\ln|\csc\theta + \cot\theta| + C = \ln|\csc\theta - \cot\theta| + C$

Sec. 7.6. How to Integrate Rational Functions of sin θ and cos θ

1. $2\int \frac{u^2-2u-1}{(u^2-3)(u^2+1)}\,du$

3. $\int \frac{3u^2-2u-3}{(u^2-u-1)(1+u^2)}\,du$

5. $\frac{1}{5}\ln\left|\frac{1+2\tan\theta/2}{2-\tan\theta/2}\right| + C$

7. $\frac{1}{2}\ln\left(\frac{3+2\sqrt{2}}{3}\right)$

9. $\frac{1}{7}\ln\left|\frac{2\tan(\theta/2)+1}{\tan(\theta/2)-3}\right| + C$

11. $2\theta - 2\tan\frac{\theta}{2} + 2\ln\left|\sec\frac{\theta}{2}\right| + C$

13. $\frac{2}{125}\left[6\ln\left|\frac{\tan(\theta/2)+2}{2\tan(\theta/2)-1}\right| - \frac{20+15\tan(\theta/2)}{\sec^2(\theta/2)}\right] + C$

15. $\frac{1}{3}\ln\left|\frac{\tan(\theta/2)+3}{\tan(\theta/2)-3}\right| + C$

17. $\frac{1}{50}\ln[25\tan^2(\theta/2) + 4\tan(\theta/2) + 13] -$

$$\frac{2}{25\sqrt{321}} \tan^{-1}\left[\frac{25 \tan(\theta/2) + 2}{\sqrt{321}}\right] + C$$

Sec. 7.7. How to Integrate Rational Functions of x and $\sqrt{a^2 - x^2}$, $\sqrt{a^2 + x^2}$, $\sqrt{x^2 - a^2}$, or $\sqrt[n]{ax + b}$

1. $x = 2 \tan \theta$
3. $x = \sqrt{5} \sin \theta$
5. $2\sqrt{2} \int_\alpha^\beta \frac{\sin^3 \theta}{\cos^2 \theta\,(2 + \cos^2 \theta)}\, d\theta$; $x = \sqrt{2} \tan \theta$, $\alpha = \tan^{-1}\left(\frac{1}{\sqrt{2}}\right)$, $\beta = \tan^{-1}\left(\frac{3}{\sqrt{2}}\right)$
7. $\frac{\sqrt{3}}{2} \int_\alpha^\beta (\sin \theta + \sqrt{3} \sin^2 \theta)\, d\theta$; $x = \frac{2}{\sqrt{3}} \sec \theta$, $\alpha = \cos^{-1}\left(\frac{1}{\sqrt{3}}\right)$, $\beta = \cos^{-1}\left(\frac{2}{3\sqrt{3}}\right)$
9. $\frac{2\pi}{3} - \frac{\sqrt{3}}{2}$
11. $10 - 2\sqrt{7} + \frac{9}{2} \ln\left(\frac{9}{4 + \sqrt{7}}\right)$
13. $\ln|\sqrt{9 + x^2} + x| + C$
15. $2[\sqrt{x} - 3 \ln(\sqrt{x} + 3)] + C$
17. $2 + 4 \ln \frac{3}{2}$
19. $2\sqrt{2x + 1} + \ln\left|\frac{\sqrt{2x + 1} - 1}{\sqrt{2x + 1} + 1}\right| + C$
21. $\frac{(3x + 2)^{7/3}}{21} - \frac{(3x + 2)^{4/3}}{6} + C$
23. $x + 10\sqrt{x} + 20 \ln|\sqrt{x} - 2| + C$
25. $\frac{1}{2}(a^2 \sin^{-1}(x/a) + x\sqrt{a^2 - x^2}) + C$
27. $\frac{1}{2}[x\sqrt{a^2 + x^2} + a^2 \ln(\sqrt{a^2 + x^2} + x)] + C$
29. $\frac{1}{5} \ln|5x + \sqrt{25x^2 - 16}| + C$
31. $(64 - 33\sqrt{3})/480$

Sec. 7.8. What to Do in the Face of an Integral

1. Division, power rule
3. Partial fractions
5. Integration by parts
7. Repeated integration by parts
9. Substitution
11. Substitution, power rule
13. Substitution, division, power rule
15. Trigonometric identity
17. Substitution, power rule
19. Substitution, integration by parts
21. Recursion formula, completing the square, substitution
23. Partial fractions
25. Substitution, power rule
27. Substitution, division, methods of Sec. 7.3
29. Trigonometric substitution
31. Substitution, partial fractions
33. Logarithm rules, power rule
35. Repeated integration by parts, trigonometric identity
37. Division, substitution, power rule
39. Substitution, power rule
41. Substitution, power rule
43. Substitution, partial fractions
45. Substitution, division, power rule
47. Partial fractions
49. Substitution, power rule
51. Rules of logarithms and exponents
53. Break in two, substitution on both parts
55. Substitution, power rule
57. Substitution, power rule
59. Substitution, power rule
61. Trigonometric identity, recursion formula of Example 7, Sec. 7.2
63. Recursion formula of Example 7, Sec. 7.2
65. Trigonometric substitution, substitution
67. Substitution, power rule
69. Substitution
71. Substitution, power rule
73. Completing the square, substitution
75. Substitution, power rule
77. Substitution, power rule
79. $n = 1$: $\frac{2}{3}(1 + x)^{3/2} + C$; $n = 2$: $\frac{x\sqrt{1 + x^2}}{2} + \frac{1}{2} \ln(\sqrt{1 + x^2} + x) + C$
81. $n = 1$: $-\frac{4}{3}(1 - x)^{3/2} + \frac{2}{5}(1 - x)^{5/2} + C$; $n = 2$: $\frac{x}{2}\sqrt{1 - x^2} + \frac{1}{2} \sin^{-1} x + C$
83. $\pm 2\sqrt{2} \cos \frac{x}{2} + C$
85. $\frac{1}{20} \ln\left(\frac{x^{20}}{x^{20} + 1}\right) + C$

Sec. 7.9. Estimates of Definite Integrals

1. (*a*) $\frac{19}{54}$ (*b*) $\frac{1}{3}$; error $= \frac{1}{54}$
3. (*a*) $\frac{28}{15}$ (*b*) ln 5; error ≈ 0.25723
5. $54{,}493/65{,}520 \approx 0.83170$
7. $76/45 \approx 1.68889$
9. $15/7 \approx 2.14286$
11. $82{,}141/98{,}280 \approx 0.83579$
13. (*a*) 3.06339 (*b*) 3.05914
19. (*a*) $\frac{1}{3} \ln 2 + \frac{\pi}{3\sqrt{3}} \approx 0.835649$ (*b*) $\frac{3{,}231{,}532}{3{,}866{,}940} \approx 0.835682$ (*c*) ≈ 0.000033
27. $\frac{3}{2}$
29. (*a*) $192{,}199/179{,}129 \approx 1.072964$ (*b*) $567{,}896/537{,}387 \approx 1.056773$ (*c*) $\frac{1}{4\sqrt{2}}\left[\ln\left(\frac{17 + 4\sqrt{2}}{17 - 4\sqrt{2}}\right) + 2 \tan^{-1}(4\sqrt{2} + 1) + 2 \tan^{-1}(4\sqrt{2} - 1)\right] \approx 1.105521$

31. (*a*) 1,369/3 (*b*) 1,369/18

Sec. 7.S. Guide Quiz

1. $\frac{1}{4}\ln(17/2)$

2. $\frac{2}{3}\sin^{-1}\left(\frac{3x}{2}\right) + \frac{x\sqrt{4-9x^2}}{2} + C$

3. $\frac{1}{4}\ln\left|\frac{x-1}{x+1}\right| - \frac{1}{2}\tan^{-1}x + C$

4. $\frac{\tan^6 2x}{12} + C$

5. $x + \frac{1}{4}\ln\left|\frac{x-1}{x+1}\right| - \frac{1}{2}\tan^{-1}x + C$

6. $\frac{1}{3}\ln(1+\sqrt{2})$

7. $\sqrt{x} + \sqrt[4]{x} + \frac{1}{2}\ln|2\sqrt[4]{x} - 1| + C$

8. $\frac{1}{\sqrt{6}}\ln\left|\frac{\sqrt{x+3}-\sqrt{6}}{\sqrt{x+3}+\sqrt{6}}\right| + C$

9. $\frac{1}{2\sqrt{3}}\ln\left|\frac{x+\sqrt{3}}{x-\sqrt{3}}\right| + C$

10. $-\frac{1}{2}(\cos 2x - \frac{2}{3}\cos^3 2x + \frac{1}{5}\cos^5 2x) + C$

11. $\frac{1}{5}e^x(\cos 2x + 2\sin 2x) + C$

12. $-\frac{1}{12}\csc^3 3x\cot 3x - \frac{1}{8}\csc 3x\cot 3x + \frac{1}{8}\ln|\csc 3x - \cot 3x| + C$

13. $\frac{1}{4}\tan^{-1}x^4 + C$

14. $\frac{1}{2}\ln\left|\frac{\sqrt{4+x^2}-2}{x}\right| + C$

15. $\frac{x\sqrt{x^2-9}}{2} + \frac{9}{2}\ln|x+\sqrt{x^2-9}| + C$

16. $\frac{x}{8}(-2x^2+45)\sqrt{9-x^2} + \frac{243}{8}\sin^{-1}\left(\frac{x}{3}\right) + C$

17. $\frac{1}{\sqrt{2}}\tan^{-1}\left[\frac{\tan(x/2)}{\sqrt{2}}\right] + C$

18. $-\frac{\sqrt{x^2+25}}{25x} + C$

19. $\frac{x^2}{2} + \frac{2}{3}x^{-3/2} + C$

20. $\frac{1}{4}\cos 2x - \frac{1}{16}\cos 8x + C$

21. (*a*) See Sec. 7.9 (*b*) Linear functions (*c*) See Sec. 7.9. (*d*) Decrease by a factor of 100

22. (*a*) See Sec. 7.9. (*b*) Polynomials of degree ≤ 3 (*c*) See Sec. 7.9. (*d*) Decrease by a factor of 10,000

Sec. 7.S. Review Exercises

1. (*a*) $\int_1^2 u^{3/2}\,du$ (*b*) $\frac{2}{5}(4\sqrt{2}-1)$

3. (*a*) $\frac{373}{14}$ (*b*) $\frac{721}{9}$

5. (*a*) $-\frac{1}{2x^2} + C$ (*b*) $2\sqrt{x+1} + C$ (*c*) $\frac{1}{5}\ln(1+5e^x) + C$

7. $-\frac{1}{4}(1+x^2)^{-2} + \frac{1}{6}(1+x^2)^{-3} + C$

9. $\frac{x^3+1}{3}\ln(1+x) - \frac{x^3}{9} + \frac{x^2}{6} - \frac{x}{3} + C$

11. (*a*) $\frac{1}{2\sqrt{2}}\ln\left|\frac{(2-\sqrt{2})+(1-\sqrt{2})\cos\theta}{(2+\sqrt{2})+(1+\sqrt{2})\cos\theta}\right| + C$
(*b*) $\frac{1}{2\sqrt{2}}\ln\left|\frac{\cos\theta-\sqrt{2}}{\cos\theta+\sqrt{2}}\right| + C$

13. (*a*) Not elementary (*b*) $\pm 2\sin\theta + C$ (*c*) $\pm 2\sqrt{1-\cos\theta} + C$

15. (*a*) $\frac{6}{11}$

17. $\frac{2}{x-1} + \frac{1}{x^2+x+1}$

19. $\frac{-1}{x+1} + \frac{x}{x^2-x+1}$

21. $\frac{x}{x^2+\sqrt{2}x+1} - \frac{1}{x^2-\sqrt{2}x+1}$

23. $5x + 6 + \frac{1}{x} - \frac{1}{x+1}$

25. (*a*) m odd (*b*) n even (*c*) Either m or n odd (*d*) m even or n odd (*e*) m odd or n even

27. $\int \frac{8\cos^3\theta + 1}{64\cos^6\theta + 5}\,d\theta$

29. $5^7\sqrt{5}\int \frac{\sec\theta\tan^{15}\theta\,d\theta}{5\sec^2\theta + 3 + \sqrt{5}\tan\theta}$

31. $\int \frac{3[2u/(1-u^2)]^2 + [(1+u^2)/(1-u^2)] + 1}{2 + [2u/(1-u^2)] + [(1-u^2)/(1+u^2)]} \times \frac{2\,du}{1+u^2}$

33. $\frac{1}{12}\ln|\sin x - 2| - \frac{1}{24}\ln(\sin^2 x + 2\sin x + 4) - \frac{1}{4\sqrt{3}}\tan^{-1}\left(\frac{\sin x + 1}{\sqrt{3}}\right) + C$

35. $-\frac{(x^2+1)^{3/2}}{3x^3} + C$

37. $-\ln(3+\cos x) + C$

39. $\frac{2}{9}(x^3-1)^{3/2} + C$

41. $\frac{1}{16}\left(\tan^{-1}\frac{x}{2} + \frac{2x}{4+x^2}\right) + C$

43. $\frac{x}{8} - \frac{\sin 12x}{96} + C$

45. $\frac{1}{9}\tan^3 3\theta - \frac{1}{3}\tan 3\theta + \theta + C$

47. $\frac{x^2}{2} - \frac{1}{2x^2} + \ln|x| + C$

49. $\frac{10^x}{\ln 10} + C$

51. $\frac{x^2}{4(x^4+1)} + \frac{1}{4}\tan^{-1}x^2 + C$

53. $\frac{1}{4}(2x + \sin 2x) + C$

55. $\frac{1}{3}(x^2+4)^{3/2} + C$

57. $\frac{1}{3}\tan^{-1}x^3 + C$

59. $-\frac{1}{3}\cos x^3 + C$

61. $\frac{x^5}{5}\ln x - \frac{1}{25}x^5 + C$

63. $2e^{\sqrt{x}} + C$

65. $(x-1)\ln(x^3-1) - 3x + \dfrac{3}{2}\ln(x^2+x+1) + \sqrt{3}\tan^{-1}\left(\dfrac{2x+1}{\sqrt{3}}\right) + C$

67. $-\dfrac{1}{\sqrt{x^2+1}} + C$

69. $-\dfrac{2}{\sqrt{x+1}} + C$

71. $3\tan^{-1}(x+2) + C$

73. $\frac{3}{5}x^{5/3} - \frac{3}{4}x^{4/3} + x - \frac{3}{2}x^{2/3} + 3x^{1/3} - 3\ln|1+x^{1/3}| + C$

75. $\frac{1}{7}(x^2+1)^{7/2} - \frac{3}{5}(x^2+1)^{5/2} + (x^2+1)^{3/2} - (x^2+1)^{1/2} + C$

77. $-\dfrac{\tan^{-1}x}{x} + \ln|x| - \dfrac{1}{2}\ln(x^2+1) + C$

79. $\frac{1}{10}e^x(\sin 3x - 3\cos 3x) + C$

81. $\dfrac{x}{4\sqrt{4-x^2}} + C$

83. $\dfrac{1}{8}\ln\left|\dfrac{x^2-3}{x^2+1}\right| + C$

85. $\frac{3}{8}(x-1)^{8/3} + \frac{6}{5}(x-1)^{5/3} + \frac{3}{2}(x-1)^{2/3} + C$

87. $\sqrt{x^2+4} + 2\ln\left|\dfrac{\sqrt{x^2+4}-2}{x}\right| + C$

89. $\dfrac{\sec^5\theta}{5} + C$

91. $-\dfrac{1}{3}\ln\left|\dfrac{3+\sqrt{x^2+9}}{x}\right| + C$

93. $\frac{3}{2}x^{2/3} - \frac{6}{5}x^{5/3} + \frac{3}{8}x^{8/3} + C$

95. $\frac{4}{3}\sin^3 x - \frac{4}{5}\sin^5 x + C$

97. $\frac{1}{2}e^{2x} - 2x - \frac{1}{2}e^{-2x} + C$

99. $\frac{1}{2}(x^2\sin^{-1}x^2 + \sqrt{1-x^4}) + C$

101. $-\dfrac{x}{25} - \dfrac{1}{5}e^{-x} + \dfrac{1}{25}\ln(e^x+5) + C$

103. $\frac{1}{135}(36x+14)(3x+2)^{3/2} + C$

105. $\ln|x-1| - \dfrac{2}{x-1} - \dfrac{1}{2(x-1)^2} + C$

107. $-x + 2\ln|e^x-1| + C$

109. $x + 2x^3 + \frac{9}{5}x^5 + C$

111. $x - \dfrac{1}{3}\ln|x+1| + \dfrac{1}{6}\ln(x^2-x+1) - \dfrac{1}{\sqrt{3}}\tan^{-1}\left(\dfrac{2x-1}{\sqrt{3}}\right) + C$

113. $\sqrt{2x+1} + C$

115. $\dfrac{1}{2\sqrt{17}}\ln\left|\dfrac{2x^2-3-\sqrt{17}}{2x^2-3+\sqrt{17}}\right| + C$

117. $\tan^{-1}e^x + C$

119. $\ln\left|\dfrac{x+2}{x+3}\right| + C$

121. $2\ln|x^2+5x+6| + C$

123. $\dfrac{2}{\sqrt{23}}\tan^{-1}\left(\dfrac{4x+5}{\sqrt{23}}\right) + C$

125. $\dfrac{1}{\sqrt{73}}\ln\left|\dfrac{4x+5-\sqrt{73}}{4x+5+\sqrt{73}}\right| + C$

127. $-\cot x + C$

129. $-\dfrac{\cot^3 x}{3} - \cot x + C$

131. $\sin^{-1}(2x-5) + C$

133. $\frac{1}{4}(\sin^{-1}2x + 2x\sqrt{1-4x^2}) + C$

135. $2\sqrt{x^2+1} + C$

137. $x + 3\ln|x-1| + \ln|x+1| + 4\tan^{-1}x + C$

139. $-\dfrac{1}{x+2} - 3\ln|x+1| + C$

141. $3\ln|x| + \tan^{-1}2x + C$

143. $x^2 + 4\ln|\sqrt{3}x-1| + \ln|\sqrt{3}x+1| + C$

145. $\frac{1}{3}\sin^{-1}3x + C$

147. $\dfrac{x}{2\sqrt{3x^2+2}} + C$

149. $\frac{1}{4}\ln|\sec 4x + \tan 4x| + C$

151. $\frac{1}{2}e^x - \frac{1}{10}e^x\cos 2x - \frac{1}{5}e^x\sin 2x + C$

153. $\frac{8}{315}(1+\sqrt{1+\sqrt{x}})^{5/2}(61 + 35\sqrt{x} - 65\sqrt{1+\sqrt{x}}) + C$

155. $(4x^{3/4} - 12x^{1/2} + 24x^{1/4} - 24)e^{\sqrt[4]{x}} + C$

157. $\frac{1}{2}(\sec x\tan x + \ln|\sec x + \tan x|) + C$

159. $\frac{1}{3}[(x^3+1)\ln(x^3+1) - x^3] + C$

161. $9\ln|x| + 3x^2 + (x^4/4) + C$

163. $10\ln|\csc x - \cot x| + 6\ln|\sin x| + 9\cos x + C$

165. $e^x - e^{-x} + C$

167. $\ln|x| + \dfrac{1}{2}\ln(x^2+2x+3) + \dfrac{1}{\sqrt{2}}\tan^{-1}\left(\dfrac{x+1}{\sqrt{2}}\right) + C$

169. $\ln|\tan\theta| + C$

171. $\frac{1}{4}(x^2 - 2x\sin 2x - \cos 2x) + C$

173. $x\tan x - (x^2/2) + \ln|\cos x| + C$

175. (*a*) $\dfrac{1}{2}\ln\left|\dfrac{x+1}{x+3}\right| + C$ (*b*) $-\dfrac{1}{x+2} + C$
(*c*) $\tan^{-1}(x+2) + C$
(*d*) $\dfrac{1}{2\sqrt{6}}\ln\left|\dfrac{x+2-\sqrt{6}}{x+2+\sqrt{6}}\right| + C$

177. (*a*) $\frac{1}{8}(2\sec^3 x\tan x + 3\sec x\tan x + 3\ln|\sec x + \tan x|) + C$
(*b*) $(\sec^5 x)/5 + C$
(*c*) $\frac{1}{2}\sec^2 x + C$

179. $-\frac{1}{7}(1-x^2)^{7/2} + C$

181. (*a*) $2\int(u^2-1)^2\,du$ (*b*) $\displaystyle\int\frac{(u-1)^2}{u^{1/2}}\,du$
(*c*) $2\int\tan^5\theta\sec\theta\,d\theta$
(*d*) $\frac{2}{5}(x+1)^{5/2} - \frac{4}{3}(x+1)^{3/2} + 2(x+1)^{1/2} + C$

183. $\frac{21}{8}$

185. $2(\sqrt{2}-1)$

187. (*a*) $x^2\sqrt{1+x^2} - \int 2x\sqrt{1+x^2}\,dx$
(*b*) $\int\tan^3\theta\sec\theta\,d\theta$ (*c*) $\int(u^2-1)\,du$

189. $\frac{2}{5}(1+x)^{5/2} - \frac{2}{3}(1+x)^{3/2} + C$

191. (*a*) e^{x^2} does not have an elementary antiderivative.
(*b*) 1.4537

193. 0

195. (*a*) $\frac{1}{3}\left(\frac{1}{x-1} - \frac{x-1}{x^2+x+1}\right)$

(*b*) $\frac{1}{9}\left(-\frac{1}{x-1} + \frac{3}{(x-1)^2} + \frac{1}{x+2}\right)$

(*c*) $x^2 - \frac{1}{3}\left(\frac{1}{x+1} + \frac{2x-1}{x^2-x+1}\right)$

(*d*) $\frac{1}{12}\left(\frac{1}{x+2} - \frac{x-4}{x^2-2x+4}\right)$

197. No, there is no error.

211. (*b*) $n = 1$: $\frac{2}{3}(1+x)^{3/2} - 2(1+x)^{1/2} + C$;
$n = 2$: $\sqrt{1+x^2} + C$;
$n = 4$: $\frac{1}{2}\ln(x^2 + \sqrt{1+x^4}) + C$

213. (*a*) n is odd. (*b*) $n = 3$: $\frac{1}{2}\sqrt{1+x^4} + C$;
$n = 5$: $\frac{x^2}{4}\sqrt{1+x^4} - \frac{1}{4}\ln(x^2 + \sqrt{1+x^4}) + C$

215. $\frac{2}{15}(8\sqrt{2} - 4\sqrt{3} - \sqrt{6} + 1)$
$+ \frac{16}{15\sqrt{30}}\left[\tan^{-1}\left(\frac{3\sqrt{5}}{4}\right) - \tan^{-1}\left(\frac{\sqrt{30}}{8}\right)\right.$
$\left. - \tan^{-1} 2\sqrt{15} + \tan^{-1}\left(\frac{3\sqrt{5}}{\sqrt{2}}\right)\right]$

219. $h \le \frac{1}{4}$

CHAPTER 8. APPLICATIONS OF THE DEFINITE INTEGRAL

Sec. 8.1. Computing Area by Parallel Cross Sections

1. $\frac{1}{12}$

3. $\frac{1}{3}$

5. $\frac{21}{4}$

7. $\sqrt{2} - 1$

9. $\frac{1}{6}$

11. $\frac{8}{3}$

13. $\frac{2}{3}$

15. 1

17. 25π

19. $\frac{1}{\sqrt{3}}\tan^{-1}\left(\frac{2}{\sqrt{3}}\right) - \frac{\pi}{8} - \frac{1}{4}\ln\frac{7}{6}$

21. $\frac{9}{2} - \sqrt{2}\ln(3 + \sqrt{8})$

23. $\frac{10^{10} + 91}{\ln 10} - 10$

25. $\ln 2$

27. $\frac{1}{4}\ln\frac{45}{8} + \frac{\pi}{8} - \frac{1}{2}\tan^{-1} 2$

29. $c(x) = \begin{cases} x + \sqrt{9 - x^2} & -3/\sqrt{2} \le x \le 3/\sqrt{2} \\ 2\sqrt{9 - x^2} & 3/\sqrt{2} \le x \le 3 \end{cases}$

31. (*b*) $c(x) = \sin x,\ 0 \le x \le \pi$
(*c*) $c(y) = \pi - 2\sin^{-1} y,\ 0 \le y \le 1$

33. (*b*) $c(x) = \begin{cases} x/e & 0 \le x \le 1 \\ x/e - \ln x & 1 \le x \le e \end{cases}$
(*c*) $c(y) = e^y - ey,\ 0 \le y \le 1$

35. $\frac{\pi}{4} + \frac{1}{2}$

37. $\frac{2\pi}{3} - \frac{\sqrt{3}}{2}$

39. (*a*) 1 (*c*) 0.36

43. $\frac{1}{2}$

Sec. 8.2. Computing Volume by Parallel Cross Sections

1. (*b*) $\int_0^1 x^4\,dx$ (*c*) $\frac{1}{5}$

3. (*b*) $\int_0^1 (y^{1/3} - y^{1/2})^2\,dy$ (*c*) $\frac{1}{110}$

5. $\int_0^2 \frac{3}{2}x^2\,dx = 4$

7. $\int_0^4 \frac{3}{16}x^2\,dx = 4$

9. (*d*) $\int_1^2 \pi x\,dx$ (*e*) $3\pi/2$

11. (*d*) $\int_0^1 \pi(x^4 - x^6)\,dx$ (*e*) $2\pi/35$

13. (*d*) $\pi\int_0^{\pi/4}(\tan^2 x - \sin^2 x)\,dx$ (*e*) $\frac{\pi}{8}(10 - 3\pi)$

15. (*b*) $A(x) = \frac{\pi a^2 x^2}{h^2},\ 0 \le x \le h$ (*c*) $\int_0^h \frac{\pi a^2 x^2}{h^2}\,dx$
(*d*) $\frac{\pi a^2 h}{3}$

17. (*b*) $A(x) = \frac{1}{2}(a^2 - x^2)\tan\theta,\ -a \le x \le a$
(*c*) $\frac{2}{3}a^3 \tan\theta$

19. (*b*) $A(x) = \begin{cases} 9\pi & -4 \le x \le 4 \\ (25 - x^2)\pi & -5 \le x \le -4,\ 4 \le x \le 5 \end{cases}$
(*c*) $244\pi/3$

21. 144

23. $\pi/120$

25. $\frac{13}{27}$

27. (*a*) Perpendicular to axis
(*b*)

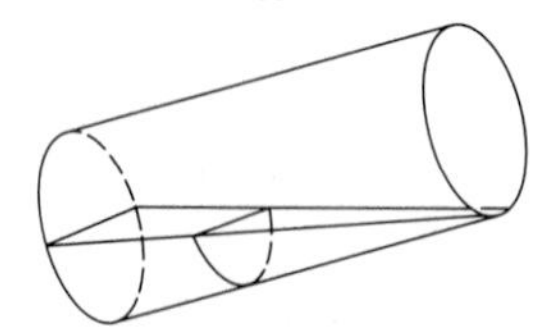

29. (*a*) Parallel to axis and to intersection of base with surface of water
(*b*)

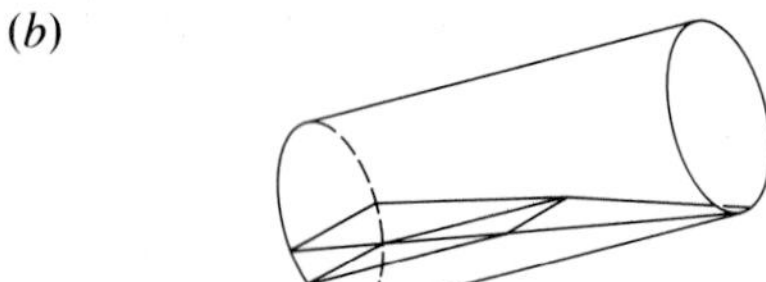

31. (*a*) (*b*)

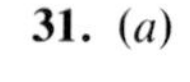

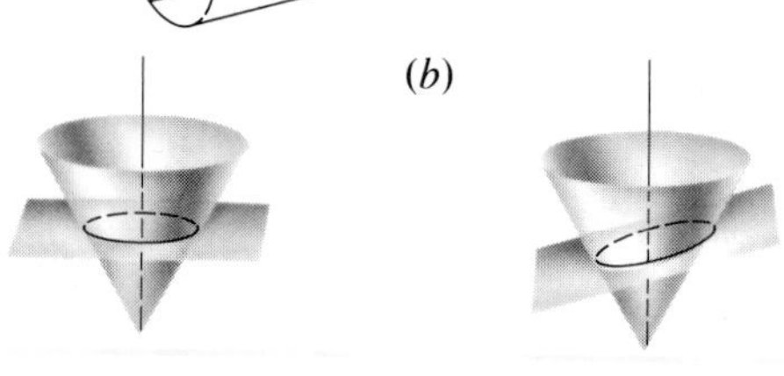

(c) 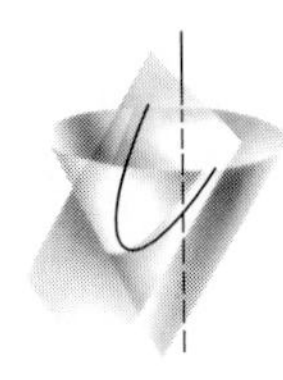 (d) 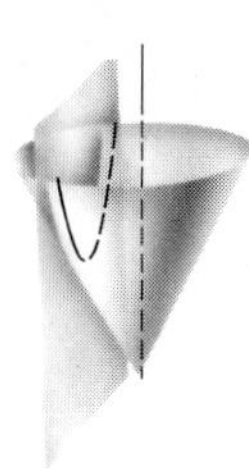

33. $a = 8\sqrt[3]{3}[2\pi(1452 - 845 \sin^{-1} \frac{12}{13})]^{-1/3}$
35. $\frac{2}{3}a^2h$
37. $(\pi/2)a^2h$

Sec. 8.3. How to Set Up a Definite Integral

3. (a) $g(r)(2\pi r\, dr)$ ft^3 (b) $\int_{1,000}^{2,000} 2\pi r g(r)\, dr$
5. (a) $2{,}000\pi^2x^2\, dx$ ergs (b) $\int_{-3}^{3} 2{,}000\pi^2x^2\, dx$ ergs (c) $36{,}000\pi^2$ ergs
7. 6(62.4) lb = 374.4 lb
9. 144(62.4) lb = 8,985.6 lb
11. $64\pi(62.4)$ lb = $3{,}993.6\pi$ lb
13. $\frac{15}{2}(62.4)$ lb = 468 lb
15. (a) $2\pi r g(r)\, dr$ cm^3/sec (b) $\int_0^b 2\pi r g(r)\, dr$
19. (a) \$1.00 (b) Effects of inflation (c) $qg(t)$ dollars (d) $g(t)f(t)\, dt$ dollars (e) $\int_0^b f(t)g(t)\, dt$ dollars
21. $(62.4)\dfrac{175\sqrt{3}}{4}$ lb
23. $dI = \dfrac{3M}{a^3} x^3\sqrt{a^2 - x^2}\, dx$, $I = \frac{2}{5}Ma^2$
25. (a) Approaches 1 (c) 1 (d) $\int_0^{1,000} tF'(t)\, dt$

Sec. 8.4. The Shell Technique

1. (d) $\int_0^1 2\pi x^3\, dx$ (e) $\pi/2$
3. (d) $\int_0^1 2\pi(y^3 - y^4)\, dy$ (e) $\pi/10$
5. (d) $\int_0^{\pi/2} 2\pi x \sin x\, dx$ (e) 2π
7. $\frac{4}{3}\pi a^3$
9. $(4\pi/15)(22\sqrt{2} - 8)$
11. (a) $2\pi(99\pi^2 + 2)$ (b) $81\pi^2/2$
13. $\pi^2/4$
15. $(\pi/3)[7 - 3\sqrt{2} - 3\ln(1 + \sqrt{2})]$
19. Let R be bounded by $y = e^{x^2}$, $y = 0$, $x = 0$, and $x = 1$.
21. (a) $(4\pi/5)(e^{10\pi} - e^{\pi})$ (b) Shell technique

Sec. 8.5. The Centroid of a Plane Region

1. $\pi/4$
3. π
5. $ab^2/2$
7. $(\frac{2}{3}, \frac{1}{3})$
9. $(\frac{4}{3}, 2)$
11. $(0, \frac{12}{5})$
13. $(\pi/4, \pi/8)$
15. $\left(\dfrac{e^2 + 1}{4}, \dfrac{e - 2}{2}\right)$
17. $\left(\dfrac{a}{3}, \dfrac{b}{3}\right)$
19. (a) $\int_a^b x[f(x) - g(x)]\, dx$
(b) $\frac{1}{2}\int_a^b \{[f(x)]^2 - [g(x)]^2\}\, dx$
21. (a) $\dfrac{15}{\ln 3} - \dfrac{6}{(\ln 3)^2} - \dfrac{6}{\ln 2} + \dfrac{2}{(\ln 2)^2}$
(b) $3\left(\dfrac{6}{\ln 3} - \dfrac{1}{\ln 2}\right)$ (c) $2\left(\dfrac{3}{\ln 3} - \dfrac{1}{\ln 2}\right)$
(d) $\dfrac{15/(\ln 3) - 6/(\ln 3)^2 - 6/(\ln 2) + 2/(\ln 2)^2}{2[3/(\ln 3) - 1/(\ln 2)]}$
(e) $\dfrac{3(6\ln 2 - \ln 3)}{2(3\ln 2 - \ln 3)}$
23. Let R be the triangle bounded by the positive axes and the line $x + y = 1$.
25. (b) $\bar{y}_R = \frac{29}{90}$, $\bar{y}_{R^*} = \frac{1}{3}$ (c) $\dfrac{2a^3 + 1}{3(a^2 + 1)}$
(f) ≈ 0.322185

Sec. 8.6. Inequalities Involving Definite Integrals

1. 0
7. $\bar{y} = \frac{1}{3}$; average $= \frac{1}{2}$
9. (d) $1{,}367/7{,}500 = 0.18226666\cdots$; $0 \le \text{error} \le 0.000064$
11. $\frac{1}{2}\sqrt{\pi^2 - 8}$
13. (c) $a = \int_A^B [f(x)]^2\, dx$, $b = -2\int_A^B f(x)g(x)\, dx$, $c = \int_A^B [g(x)]^2\, dx$ (e) $f = 0$ or $g = Kf$ for some constant K.
17. (a) $v(t) = 32t$ ft/sec (b) $v(y) = 8\sqrt{y}$ ft/sec (c) $16t$ ft/sec (d) $64t/3$ ft/sec
19. (a) Dispatcher's average = 20 minutes; rider's average = 70/3 minutes.

Sec. 8.7. Improper Integrals

1. Convergent, $\frac{1}{2}$
3. Divergent
5. Convergent, $\pi/4$
7. Divergent
9. Convergent, 100
11. Convergent, $\frac{1}{8}$
13. Divergent
15. Convergent
17. Divergent
19. Improper
21. Improper
23. Improper
25. Divergent
27. Convergent, 2
29. Convergent, -1
31. Convergent, $\pi/3$
39. $u = 1/x$ is not defined at $x = 0$.
43. Convergent, $\frac{3}{10}$
45. Divergent
47. Divergent
49. Divergent
51. Divergent
53. Divergent

55. Infinite

61. $P(r) = \dfrac{1}{r^2}$

63. $P(r) = \dfrac{1}{r - 1}$

65. $P(r) = \dfrac{r}{r^2 + 1}$

69. Estimate $\int_0^{1.6} e^{-x^2}\,dx$ by Simpson's method with $n = 8$ and add 0.05. (Other methods are possible.)

Sec. 8.8. Work

1. $62.4\pi(\frac{5}{12}a^4 + a^3h + \frac{1}{2}a^2h^2)$ ft · lb
3. $280 \cdot 62.4$ ft · lb = 17,472 ft · lb
5. $180 \cdot 62.4$ ft · lb = 11,232 ft · lb
7. $18{,}000 \cdot 62.4$ ft · lb = 1,123,200 ft · lb
9. 2,000 mi · lb
11. 2,000 mi · lb
13. (*a*) $k/2$ ft · lb (*b*) $3k/2$ ft · lb
15. (*a*) $3.375 \times 10^{20}\pi$ ft · lb
(*b*) About 3.5×10^6 times as much.

Sec. 8.S. Guide Quiz

1. $e(e - 1)$
2. (*a*) $\frac{1}{4}e^2(e - 1)(e + 1)$ (*b*) e^2
(*c*) $\left(\dfrac{e}{e - 1}, \dfrac{e(e + 1)}{4}\right)$
3. $\dfrac{\pi e}{2}(e^3 + 3e - 4)$
4. $2\pi e^2$
6. See Theorem 3 in Sec. 8.6.
7. (*a*) $a < -1$ (*b*) $a > -1$ (*c*) Never convergent
8. Convergent
9. (*a*) See Pappus's theorem in Sec. 8.5.
(*b*) See the theorem in Sec. 8.8.

Sec. 8.S. Review Exercises

1. (*a*) $\int_0^2 (2x - x^2)\,dx$ (*b*) $\int_0^4 [\sqrt{y} - (y/2)]\,dy$
3. $\int_0^1 \pi y^2\,dy + \int_1^2 \pi y(2 - y)\,dy$
5. (*a*) $1 + \ln\frac{2}{3}$ (*b*) $\pi(\frac{7}{6} + 2\ln\frac{2}{3})$
(*c*) $2\pi(\frac{1}{2} + \ln\frac{2}{3})$ (*d*) $\pi(\frac{19}{6} + 4\ln\frac{2}{3})$
7. (*a*) 1 (*b*) $\pi^2/4$ (*c*) $\pi^2/2$ (*d*) $\pi^2/4 + 2\pi$
9. (*a*) $\frac{1}{2}\ln 3$ (*b*) $\pi/3$ (*c*) $(\pi/2)(2 - \ln 3)$
(*d*) $(\pi/3)(1 + 3\ln 3)$
11. 1
13. $\sqrt{2} - 1$
15. $\frac{1}{2}(2 + 2\ln 2 - \pi)$
17. $\frac{5}{6}\sqrt{2} - \frac{2}{3}$
19. (*a*) $\frac{1}{2}\ln\frac{3}{2}$ (*b*) $\frac{1}{2}\ln 2$
21. (*a*) $\dfrac{3\pi - 8}{12}$ (*b*) $\pi/64$ (*c*) $\dfrac{3\pi}{16(3\pi - 8)}$
23. $4\pi/3$
25. (*a*) 1 (*b*) $\pi/2$ (*c*) 2π
27. (*b*) See Fig. 8.68.
35. $\frac{1}{5}(\sin 3 + 2\cos 3)$
39. The integral is improper; the integrand is undefined at 0.
41. $-\frac{1}{25}$
45. $(\frac{3}{2}, \frac{12}{5})$
47. Convergent
49. $n!/a^{n+1}$
51. (*a*) $\pi r^2\sqrt{h}(2 - \sqrt{2})$ seconds (*b*) $2\pi r^2\sqrt{h}$ seconds
53. (*a*) $G(0) = G(1) = G(2) = \pi/4$
55. $a = 0, -1$
57. $b = -1, 0, 1$
59. $a = 0$
61. (*b*) $\mu = 0$ (*c*) $\mu_2 = k^2$ (*d*) $\sigma = k$

CHAPTER 9. PLANE CURVES AND POLAR COORDINATES

Sec. 9.1. Polar Coordinates

1.

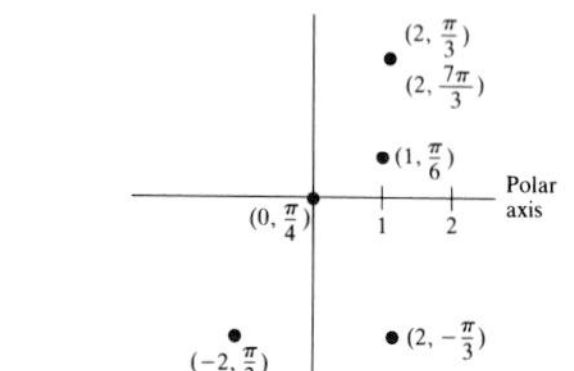

3. (*a*) $(3, \pi/4 + 2\pi k)$ for any integer k
(*b*) $(-3, 5\pi/4 + 2\pi k)$ for any integer k
5. $x^2 + y^2 = y$
7. $4x + 5y = 3$
9. $(x^2 + y^2)^3 = 4x^2y^2$
11. $r = \dfrac{3}{\cos\theta + 2\sin\theta}$
13. $r^2 = 2\csc 2\theta$
15. $r = -2\sec\theta$
17.

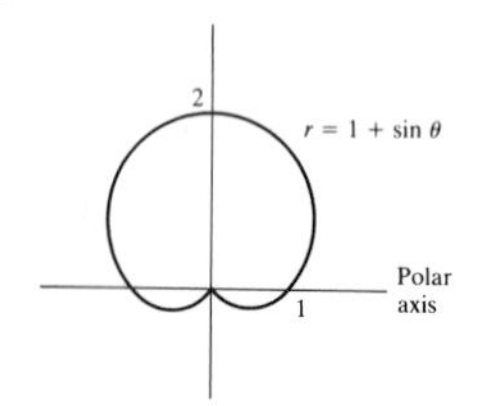

19.

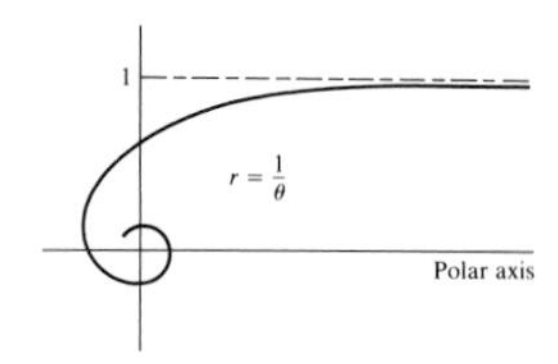

21.

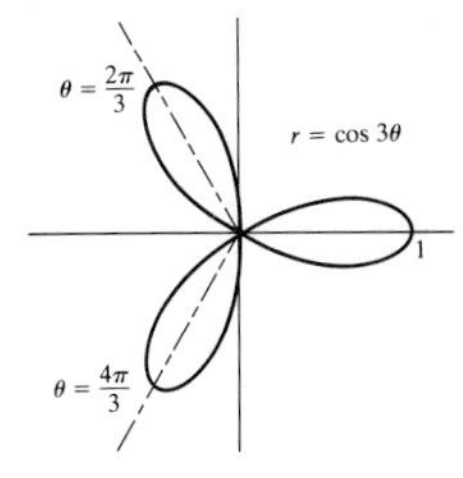

23. The curves are identical.

25. $(0, 0)$, $(1/\sqrt{2}, \pi/12)$, $(1/\sqrt{2}, 3\pi/4)$, $(1/\sqrt{2}, 17\pi/12)$

27. $(0, 0)$, $(\frac{1}{2}, \pi/6)$, $(1, \pi/2)$, $(\frac{1}{2}, 5\pi/6)$

33.

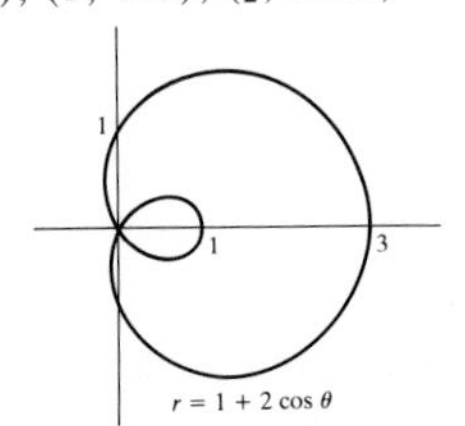

35. (*a*)

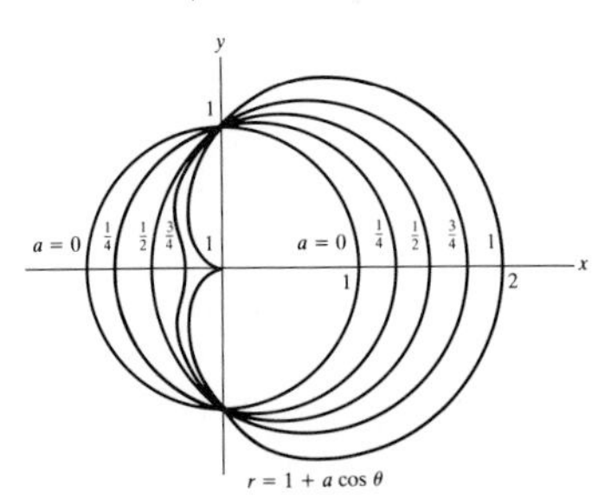

37. $y^2 = 1 - 2x$

39. (*a*)

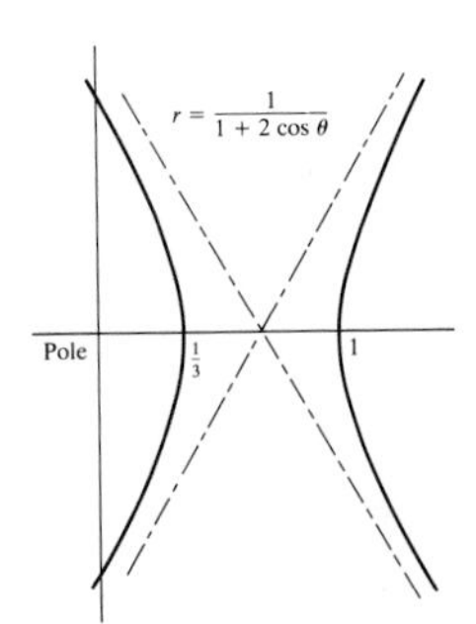

(*b*) $\pm 2\pi/3$ (*c*) $y^2 = 3x^2 - 4x + 1$

41. $\sqrt{6}/9$

43. $(4n\pi, 0)$, n an integer, $n \geq 0$

Sec. 9.2. Area in Polar Coordinates

1. $\pi^3/12$

3. $\dfrac{\pi}{(4 + \pi)(2 + \pi)}$

5. $\dfrac{4 - \pi}{8}$

7. (*a*)

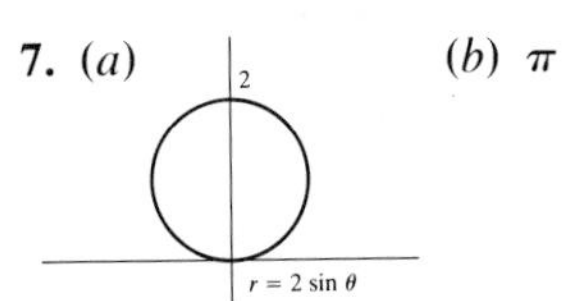

(*b*) π

9. $18 + 9\pi/4$

11. $\pi/12$

13. $\pi/16$

15. $\pi/6 + \sqrt{3}/4$

17. $(\pi + 2)/8$

21. (*a*)

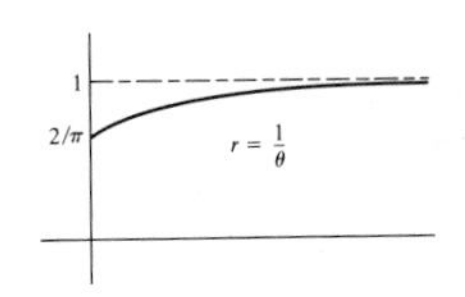

(*b*) Infinite

23. Yes

Sec. 9.3. Parametric Equations

1. (*a*)

t	−2	−1	0	1	2
x	−3	−1	1	3	5
y	−3	−2	−1	0	1

(*b*), (*c*)

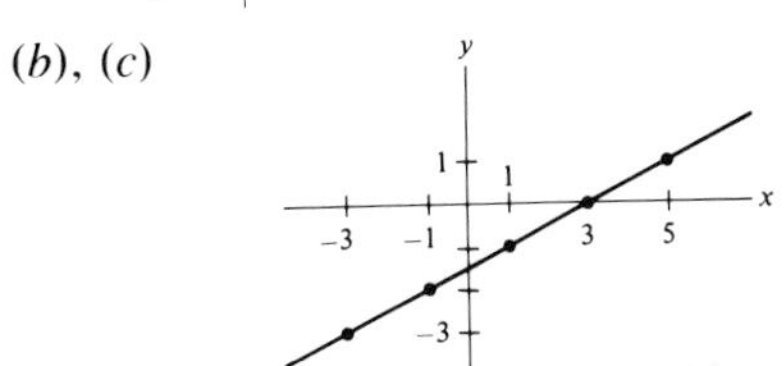

(*d*) $x - 2y = 3$

3. (*a*)

t	−3	−2	−1	0	1	2	3
x	9	4	1	0	1	4	9
y	6	2	0	0	2	6	12

(*b*), (*c*)

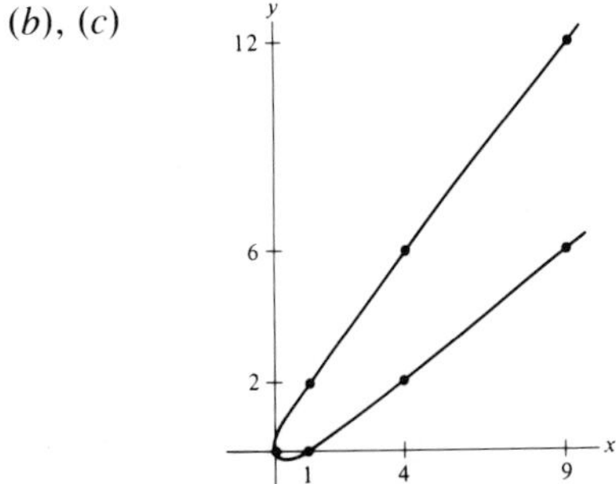

(*d*) $x^2 + y^2 - 2xy - x = 0$

5. $x = t$, $y = \sqrt{1 + t^3}$

7. $\theta = t$, $r = \cos 2t$

9. $\dfrac{dy}{dx} = \dfrac{7t^6 + 1}{3t^2 + 1}, \dfrac{d^2y}{dx^2} = \dfrac{84t^7 + 42t^5 - 6t}{(3t^2 + 1)^3}$

11. $\dfrac{dy}{dx} = \dfrac{3 \sin 3\theta \sin \theta - \cos 3\theta \cos \theta}{3 \sin 3\theta \cos \theta + \cos 3\theta \sin \theta}$;

$\dfrac{d^2y}{dx^2} = \dfrac{10 + 8 \sin^2 3\theta}{(3 \sin 3\theta \cos \theta + \cos 3\theta \sin \theta)^3}$

13. $\dfrac{1 + e}{5 + 2\pi}$

15. $y = x \tan \alpha - \dfrac{16x^2}{v_0^2 \cos^2 \alpha}$

17. (*b*) πab

19. (*a*) π (*b*) 2π (*c*) 3π (*d*) 4π

21. (*a*) $-\dfrac{\cos \theta + \cos 2\theta}{\sin \theta + \sin 2\theta}$ (*b*) It approaches 0.

(*c*) 0

23. $5\pi^2a^3$

29. $x = a \cos \theta$, $y = a \sin \theta \cos \alpha$, where α is the tilt of the circle.

33. $x = (a + b) \cos \theta - b \cos \left(\dfrac{a + b}{b} \theta\right)$;

$y = (a + b) \sin \theta - b \sin \left(\dfrac{a + b}{b} \theta\right)$ (θ is the polar angle of the center of the circle of radius b.)

Sec. 9.4. Arc Length and Speed on a Curve

1. $\dfrac{22\sqrt{22} - 13\sqrt{13}}{27}$

3. $\frac{59}{24}$

5. $\frac{1}{512}(820 - 81 \ln 3)$

7. $\frac{1}{2}(e^b - e^{-b})$

9. $\frac{3}{2}$

11. $\sqrt{2}\,(e^{2\pi} - 1)$

13. 2

15. $2\sqrt{625 + 256t^2}$

17. $\sqrt{6 - 2 \sin t - 4 \cos t}$

19. (*a*) $\frac{1}{27}(40^{3/2} - 13^{3/2})$ (*b*) $t\sqrt{4 + 9t^2}$

(*c*)

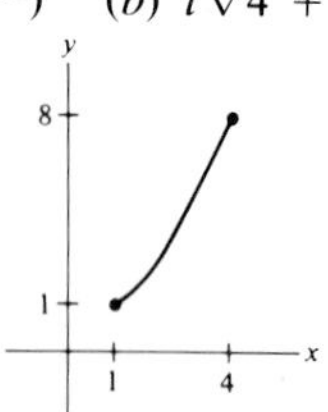

(*d*) $y = x^{3/2}$

23. (*a*) $\sqrt{[g'(t)]^2 + [g(t)h'(t)]^2}$ (*b*) $\sqrt{26}e^t$

27. (*a*) 1 (*b*) $\frac{4}{3}$

29. $\dfrac{a}{2}\,[2\pi\sqrt{4\pi^2 + 1} + \ln (2\pi + \sqrt{4\pi^2 + 1})]$

Sec. 9.5. Area of a Surface of Revolution

1. $\int_1^2 2\pi x^3\sqrt{1 + 9x^4}\, dx$

3. $\int_1^8 2\pi y^{1/3}\sqrt{1 + \frac{1}{9}y^{-4/3}}\, dy$

5. $\pi[e\sqrt{1 + e^2} + \ln(e + \sqrt{1 + e^2}) - \sqrt{2} - \ln(1 + \sqrt{2})]$

7. $64\pi/3$

11. (*a*) $\frac{1}{5}(32 - 4\sqrt{2})\pi$ (*b*) $\frac{24}{5}\sqrt{2}\pi$

13. (*a*) The volume of the sphere is $\frac{2}{3}$ the volume of the cylinder. (*b*) Both surface areas are $4\pi a^2$.

15. $6\pi/5$

17. $2\pi \displaystyle\int_1^2 \frac{\sqrt{1 + x^4}}{x^3}\, dx$; use trigonometric substitution.

19. $\dfrac{2\pi}{3} \displaystyle\int_1^8 x\sqrt{9 + 16x^{2/3}}\, dx$; use trigonometric substitution.

21. $\pi \displaystyle\int_1^3 \left(x^3 + \frac{1}{x}\right) dx$; use the power rule.

23. $2\pi \int_1^2 (x^2 + 1)\sqrt{1 + 4x^2}\, dx$

27. It will be a section between two concentric circles and two rays from their center. The area is $\pi RL - \pi rl$.

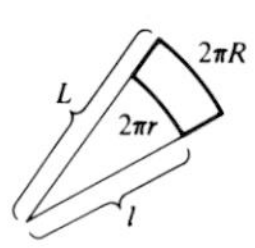

29. $\bar{x} = 0$, $\bar{y} = 2a/\pi$

31. $4\pi^2ab$

Sec. 9.6. The Reflection Property of the Parabola and Ellipse

1. $\pi/2$

3. $\tan^{-1} \frac{1}{7} \approx 0.142$ radians

5. $\tan^{-1} \frac{1}{7} \approx 0.142$ radians

7. $-2\sqrt{2}$

9. The tangent is undefined; the curves are perpendicular.

13. $\pi/4$

15. $f(\theta) = a \sin \theta$, for some constant a

17. $f(\theta) = Ae^{k\theta}$, for constants A and k

Sec. 9.7. Curvature

1. $2/(5\sqrt{5})$, $5\sqrt{5}/2$

3. $e^2/(e^2 + 1)^{3/2}$, $(e^2 + 1)^{3/2}/e^2$

5. $4/(5\sqrt{5})$, $5\sqrt{5}/4$

7. $\frac{1}{2}$

9. $e^{\pi/6}/\sqrt{2}$

11. (*a*) $4/(e^x + e^{-x})^2$, $(e^x + e^{-x})^2/4$

13. $-\frac{1}{2} \ln 2$

17. At (1000, 0), $A = \tan^{-1} 0.968 \approx 0.77$ radians ($\approx 44°$). At (0, 500), $A = \tan^{-1} 0.121 \approx 0.12$ radians ($\approx 7°$).

19. $(a^4y^2 + b^4x^2)^{3/2}/(a^4b^4)$

21.

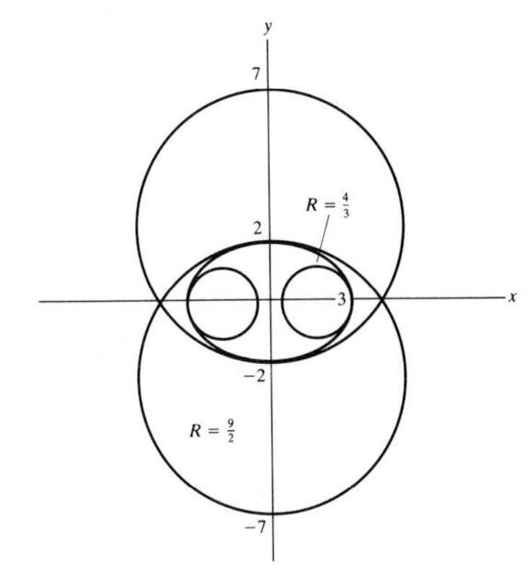

25. $(5 + 3\sin^2 2\theta)/(1 + 3\sin^2 2\theta)^{3/2}$
29. ≈ 157.6 mi/h

Sec. 9.S. Guide Quiz

1. $\int_{-\pi/10}^{\pi/10} \frac{1}{2}\cos^2 5\theta\, d\theta$
2. $\int_1^2 \sqrt{1 + 16x^6}\, dx$
3. $\int_1^2 2\pi x^4\sqrt{1 + 16x^6}\, dx$
4. $\int_1^2 2\pi(x - 1)\sqrt{1 + 16x^6}\, dx$
5. $\int_0^{2\pi} \sqrt{5 + 4\sin\theta}\, d\theta$
6. $\int_0^{\pi} 2\pi(2 + \sin\theta)\sin\theta\sqrt{5 + 4\sin\theta}\, d\theta$
7. $\int_0^1 \sqrt{100t^2 + 1/(4t)}\, dt$
8. $\sqrt{873}$
9. $y^2 = x^2 + 1$
10. (*a*) $\phi = \dfrac{\pi}{8} + \gamma \approx 3.07$ radians ($\approx 176°$)
(*b*) See Fig. 9.9.
11. $\sqrt{2}$

Sec. 9.S. Review Exercises

1. (*a*) See Sec. 9.2. (*b*) See Fig. 9.16.
3. (*a*) See Sec 9.5. (*b*) See Fig. 9.46.
5. (*a*) Curvature = $|d\phi/ds|$
(*b*) See Theorem 2 of Sec. 9.7.
7. (*a*) $\int_0^{\pi} \sqrt{1 + \cos^2 x}\, dx$, $\int_0^{\pi} 2\pi\sin x\sqrt{1 + \cos^2 x}\, dx$,
$\int_0^{\pi} 2\pi x\sqrt{1 + \cos^2 x}\, dx$
(*b*) Surface area (about x axis) =
$2\pi[\sqrt{2} + \ln(1 + \sqrt{2})]$.
9. $3\sqrt{3}/4$
11. Infinite
13. $\sqrt{2}$

15. $\dfrac{dy}{dx} = -\dfrac{3\cos 3t}{2\sin 2t}$,
$\dfrac{d^2y}{dx^2} = \dfrac{18\sin 2t\sin 3t + 12\cos 2t\cos 3t}{(-2\sin 2t)^3}$
17. (*a*)

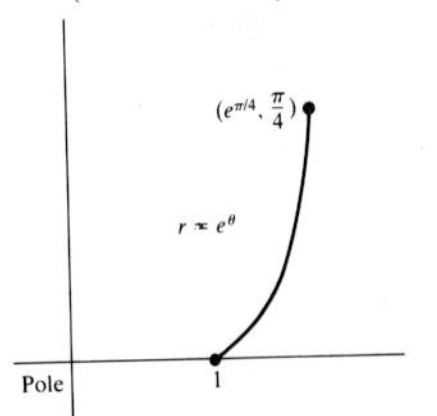

(*b*) $\int_0^{\pi/4} e^{2\theta}\sin\theta(\cos\theta - \sin\theta)\, d\theta$
(*c*) $\int_0^{\pi/4} \pi e^{3\theta}\sin^2\theta(\cos\theta - \sin\theta)\, d\theta$
(*d*) $\int_0^{\pi/4} 2\pi e^{3\theta}\sin\theta\cos\theta(\cos\theta - \sin\theta)\, d\theta$
(*e*) $\int_0^{\pi/4} 2\sqrt{2}\pi e^{2\theta}\sin\theta\, d\theta$
(*f*) $\int_0^{\pi/4} 2\sqrt{2}\pi e^{2\theta}\cos\theta\, d\theta$

CHAPTER 10. SERIES AND COMPLEX NUMBERS

Sec. 10.1. Sequences

1. 0
3. 1
5. 0
7. 0
9. Diverges to ∞
11. 3
13. Diverges, oscillates
15. e^2
17. 0
21. (*a*) 1 (*b*) 0 (*c*) 0
23. (*a*) 0.6561 (*b*) 0
25. (*a*) 0, 1, $\frac{1}{2}$
(*b*) All x

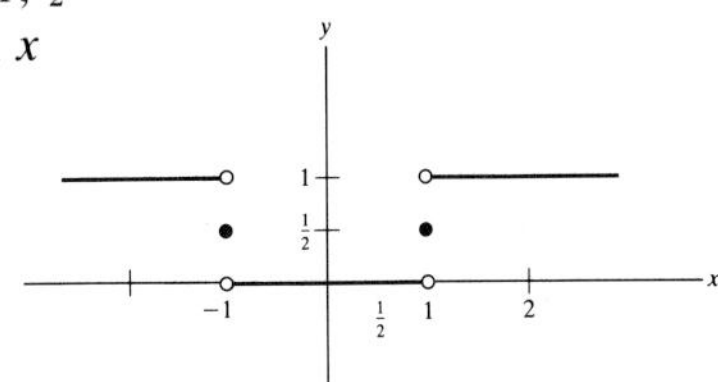

(*c*) $x \neq \pm 1$
27. ln 2
29. (*a*) $\frac{1}{2}, \frac{2}{3}, \frac{3}{4}, \frac{4}{5}$ (*b*) $n/(n+1)$
33. The degree of $Q(x)$ is greater than or equal to the degree of $P(x)$.
35. Oscillates with period 5: 1, 3, 4, $\frac{5}{3}$, $\frac{2}{3}$, 1, 3, 4, ...

Sec. 10.2. Series

1. 2
3. $\frac{1}{9}$
5. 495
7. 8

9. Diverges
11. Diverges to ∞
13. 24
15. $\frac{3}{2}$
19. (*a*) 30.8940 ft (*b*) 114 ft
23. (*b*) 251/128
25. (*b*) 2
27. 2,848/825
29. $5,000

Sec. 10.3. The Integral Test

1. Converges
3. Diverges
5. Diverges
7. Diverges
9. Diverges
11. Converges
13. (*a*) 1.1777 (*b*) $\frac{1}{50} \le R_4 \le \frac{1}{32}$ (*c*) $1.1977 \le \Sigma_{n=1}^{\infty} 1/n^3 \le 1.2089$
15. (*a*) 0.8588 (*b*) $0.1974 \le R_4 \le 0.2450$ (*c*) $1.0562 \le \Sigma_{n=1}^{\infty} 1/(n^2+1) \le 1.1038$
23. (*a*) 8 (*b*) 23
29. (*c*) 7, 22

Sec. 10.4. The Comparison Test and the Ratio Test

1. Converges
3. Converges
5. Converges
7. Diverges
9. Converges
11. Converges
13. Converges
15. Converges
17. Converges
19. (*a*) $0 < x < 1$ (*b*) $x \ge 1$
21. (*a*) $x > 0$ (*b*) Nowhere
31. Diverges

Sec. 10.5. The Alternating-Series and Absolute-Convergence Tests

1. Diverges
3. Converges
5. Diverges
7. Diverges
9. (*a*) $S_5 \approx 0.78333$, $S_6 \approx 0.61667$ (*b*) Larger (*c*) $0.61666 < S < 0.78334$
11. Converges conditionally
13. Converges conditionally
15. Converges absolutely
17. Converges absolutely
19. Converges absolutely
21. Converges conditionally
23. Converges absolutely
25. Converges absolutely
27. (*a*) 0.328 (*b*) $R_6 < 2^{-7} \approx 0.0078$ (*c*) $1/192 \approx 0.0052$
37. (*a*) $a_n = (-1)^{n-1}/n$ (*b*) $f(x) = (-1)^{[2x]}/[x]$, where $[x]$ represents the greatest integer not exceeding x

Sec. 10.6. Power Series

1. 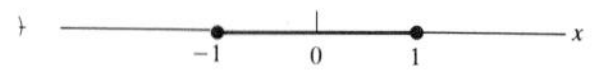
3. 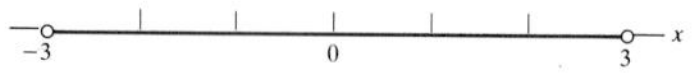
5. 
7. 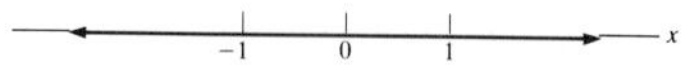
9. 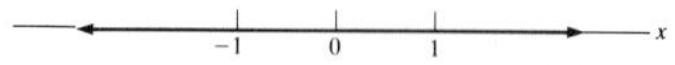
11. 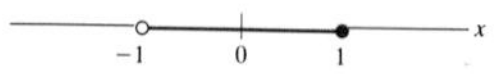
13. (*a*) Converges absolutely (*b*) Converges absolutely (*c*) Cannot be determined (*d*) Cannot be determined (*e*) Cannot be determined (*f*) Diverges (*g*) Diverges
15. Yes
17. 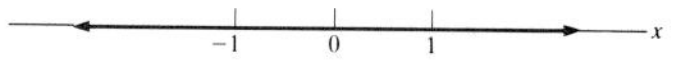
19. 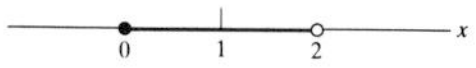
21. 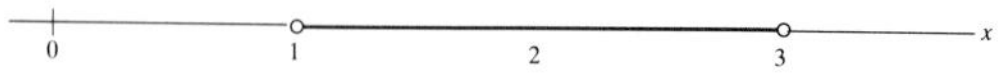
23. 
25. (*a*) $x \ge 3$ and $x < -3$ (*b*) $x < -8$ and $x \ge -2$
27. ∞

Sec. 10.7. Manipulating Power Series

1.

n	$f^{(n)}(x)$	$f^{(n)}(0)$	$f^{(n)}(0)/n!$
0	$\sin x$	0	0
1	$\cos x$	1	1
2	$-\sin x$	0	0
3	$-\cos x$	-1	$-\frac{1}{6}$
4	$\sin x$	0	0
5	$\cos x$	1	$\frac{1}{120}$

(*b*) $0 + x + 0x^2 - (x^3/6) + 0x^4 + (x^5/120)$

5. $x + \dfrac{x^3}{3} + \dfrac{2x^5}{15}$
7. $x + \dfrac{x^3}{6} + \dfrac{3x^5}{40}$
9. (*c*) $a_n = \dfrac{(-1)^n x^{2n+1}}{2n+1}$, $n = 0, 1, 2, \ldots$
11. $x + x^2 + \dfrac{x^3}{3}$
13. ∞

15. 1

17. $\frac{1}{2} + \frac{\sqrt{3}}{2}\left(x - \frac{\pi}{6}\right) - \frac{1}{4}\left(x - \frac{\pi}{6}\right)^2$

19. $e + e(x - 1) + \frac{e}{2}(x - 1)^2$

23. (*a*) 1.6484 (*b*) Error $< 1/3{,}520$

25. $98{,}641/36{,}288 \approx 2.7182815256$

27. (*a*) $1 - x^2 + \frac{x^4}{2!} - \frac{x^6}{3!} + \cdots$ (*c*) 0.747

Sec. 10.8. Taylor's Formula

1. $\int_0^x \frac{4(x - t)^3}{(1 + t)^5}\,dt$

3. $\int_0^x \frac{-(x - t)^3}{(1 + t)^4}\,dt$

5. $\frac{1}{6}\int_0^x (x - t)^3 \cos t\,dt$

7. $x - \frac{x^3}{6}$

9. $x + \frac{x^3}{3}$

11. $P_0(x; 0) = 1$, $P_1(x; 0) = 1 - x$, $P_2(x; 0) = 1 - x + x^2$, $P_3(x; 0) = 1 - x + x^2 - x^3$

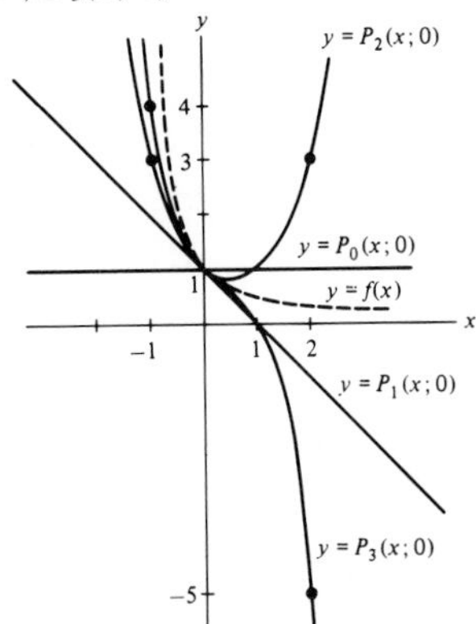

13. $P_0(x; 0) = 0$, $P_1(x; 0) = x$, $P_2(x; 0) = x$, $P_3(x; 0) = x - (x^3/6)$

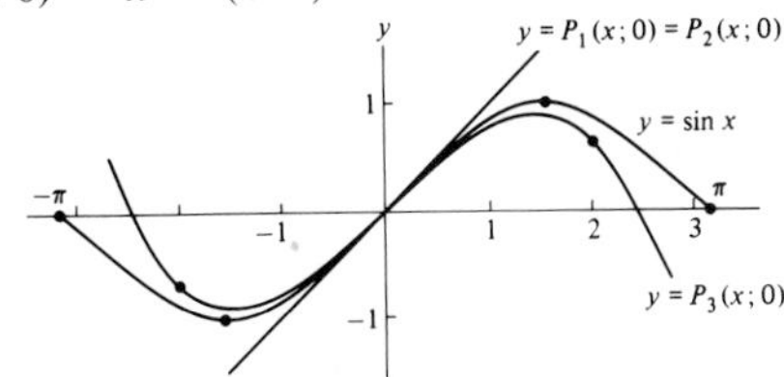

23. $\sqrt[3]{e} \approx 25/18 \approx 1.3889$; error $< 1/81 \approx 0.0123$

25. $\sin 1 \approx 0.841667$; $|\text{error}| < 1/7! \approx 0.000198$

27. $\cos 10° \approx 0.98480779$; $|\text{error}| < 3.9 \times 10^{-8}$

33.

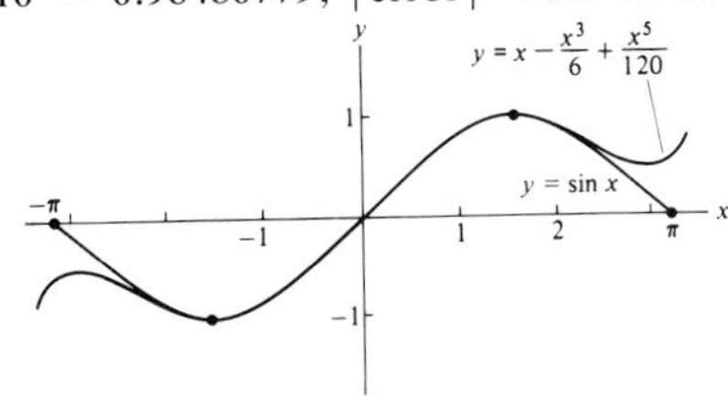

35. (*a*) $1 - \frac{x^4}{2} + \frac{x^8}{4!} - \frac{x^{12}}{6!} + \cdots$ (*b*) 0.497

37. 0.36

Sec. 10.9. Complex Numbers

1. (*a*) $\frac{5}{2} + i$ (*b*) 13 (*c*) $2/5 + i/5$ (*d*) $\frac{10}{17} + \frac{11}{17}i$

3. (*a*)

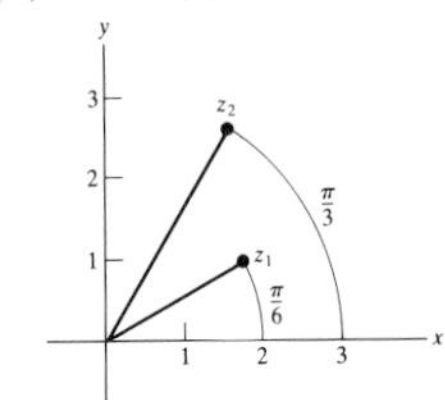

(*b*) $6i$ (*c*) $z_1 = \sqrt{3} + i$, $z_2 = \frac{3}{2} + \frac{3\sqrt{3}}{2}i$ (*d*) $6i$

5.

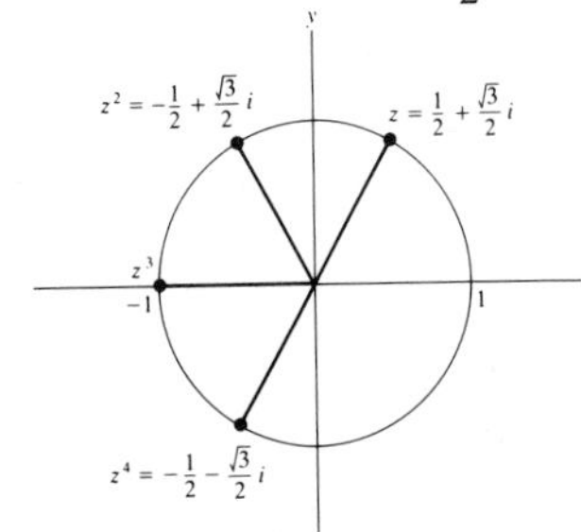

7. (*a*) $|z^2| = 4$, $\arg z^2 = \pi/3$
(*b*) $|z^3| = 8$, $\arg z^3 = \pi/2$
(*c*) $|z^4| = 16$, $\arg z^4 = 2\pi/3$
(*d*) $|z^n| = 2^n$, $\arg z^n = n\pi/6$
(*e*)

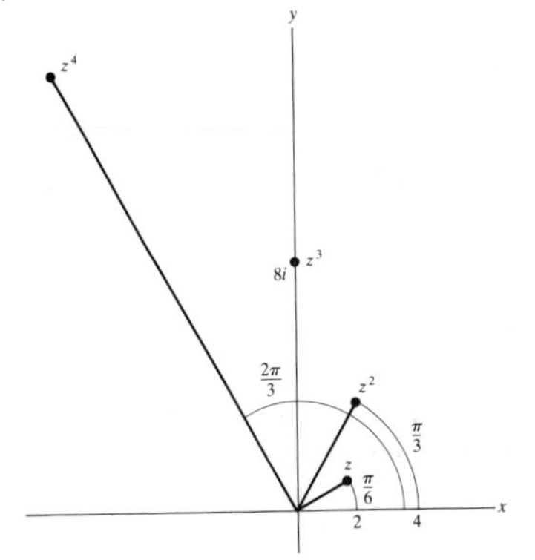

9. All five have magnitude 2. Their arguments are $\pi/20$, $9\pi/20$, $17\pi/20$, $5\pi/4$, $33\pi/20$.

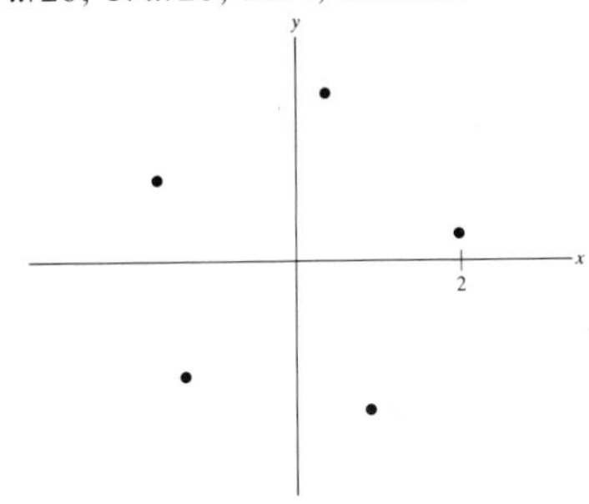

13. (*b*) $2 + 3i,\ 2 - 3i$

15. (*a*) $(5\sqrt{2}, \pi/4)$ (*b*) $(1, 4\pi/3)$
(*c*) $(1, 3\pi/4)$ (*d*) $(5, \tan^{-1}\frac{4}{3})$

19. (*a*) $-1 + 5i$ (*b*) $\dfrac{5}{2} + \dfrac{i}{2}$
(*c*) 58 (*d*) $15\left(-\dfrac{\sqrt{3}}{2} - \dfrac{i}{2}\right)$
(*e*) $2i$ (*f*) $\dfrac{3}{10} + \dfrac{i}{10}$
(*g*) $\frac{1}{3}(\cos 308° + i \sin 308°)$ (*h*) 1

21. $-i,\ \pm\dfrac{\sqrt{3}}{2} + \dfrac{1}{2}i$

23. $\cos 3\theta = 4\cos^3\theta - 3\cos\theta$,
$\sin 3\theta = 3\sin\theta - 4\sin^3\theta$

27. $|z| = 1$

29. (*a*)

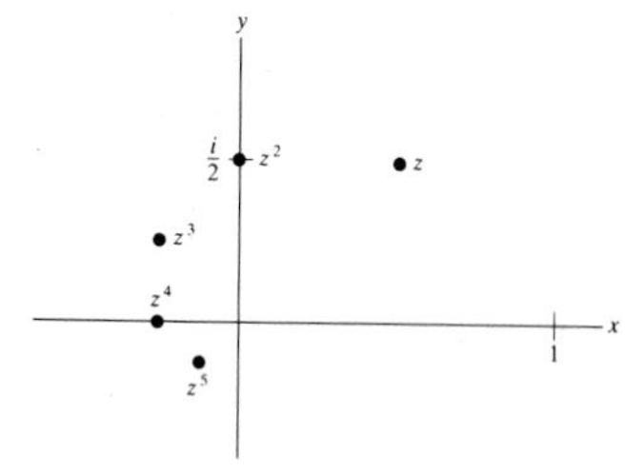

(*b*) $\lim_{n\to\infty} z^n = 0$

Sec. 10.10. The Relation Between the Exponential and the Trigonometric Functions

1. $\text{Re } z = -1/\sqrt{2}$, $\text{Im } z = -1/\sqrt{2}$

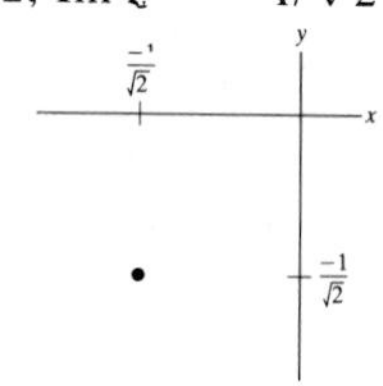

3. $\text{Re } z = \sqrt{2} + 3\sqrt{3}/2$, $\text{Im } z = \sqrt{2} + \frac{3}{2}$

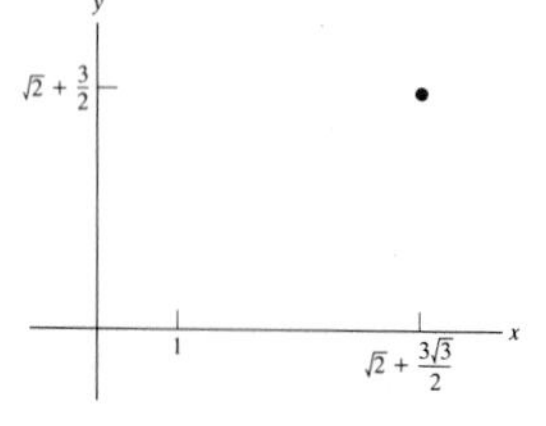

5. $\text{Re } z = \cos(11\pi/12)$, $\text{Im } z = \sin(11\pi/12)$

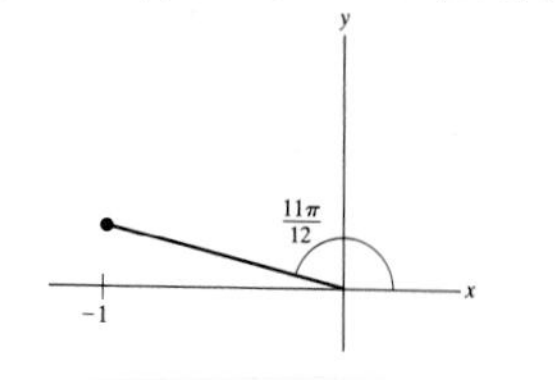

7. $e^2 e^{-\pi i/4}$

9. $15e^{2\pi i/3}$

11.

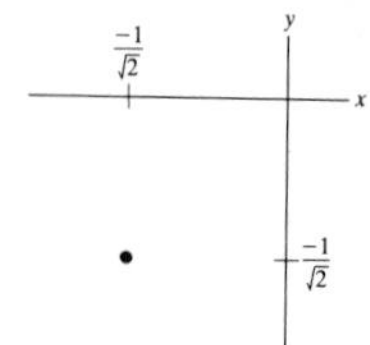

13.

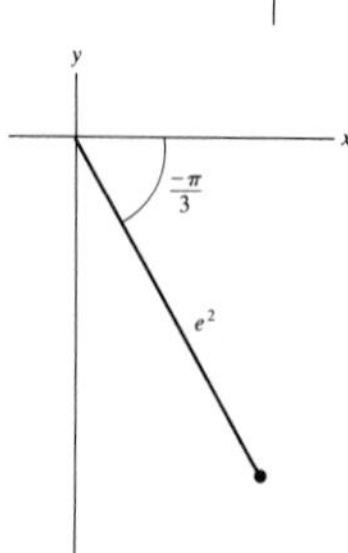

15. (*a*) e^a (*b*) e^{a-bi} (*c*) e^{-a-bi} (*d*) $e^a \cos b$
(*e*) $e^a \sin b$ (*f*) b

17. $z = 2\pi n i$, n an integer

21. $z = \ln 5 + (\alpha + 2\pi n)i$, $\alpha = \tan^{-1}\frac{4}{3}$, n an integer

23. $\dfrac{4 - 2\cos\theta}{5 - 4\cos\theta}$

27. (*a*) $\sinh z = \frac{1}{2}(e^z - e^{-z})$, $\cosh z = \frac{1}{2}(e^z + e^{-z})$

29. (*a*)

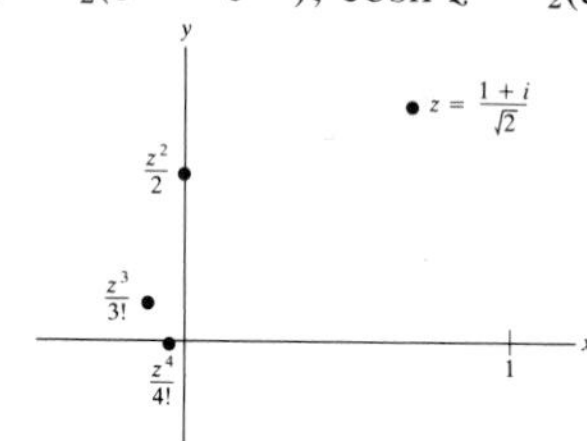

(*b*), (*c*)

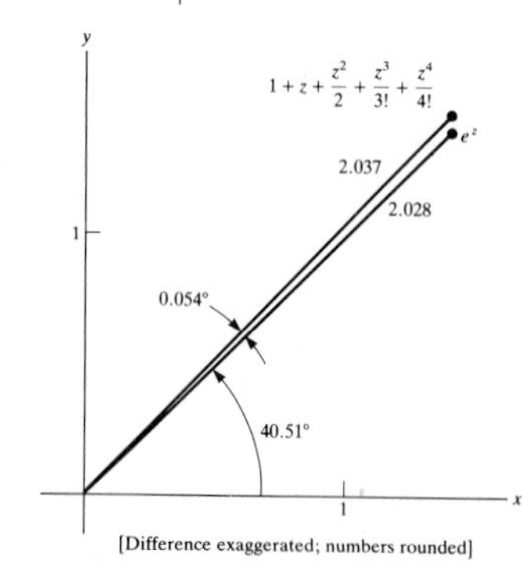

[Difference exaggerated; numbers rounded]

31. Rewrite the trigonometric functions in terms of exponentials using Theorem 2 and expand the product.

Sec. 10.11. Linear Differential Equations with Constant Coefficients

1. $y = Ce^{-2x}$

3. $y = \dfrac{x}{12} - \dfrac{1}{48} + Ce^{-4x}$

5. $y = -x^2 - 2x - 2 + Ce^x$
7. $y = C_1e^{-x} + C_2e^{3x}$
9. $y = C_1e^{3x/2} + C_2e^{-x}$
11. $y = (C_1 + C_2x)e^{3x/2}$
13. $y = e^{3x/2}(C_1e^{\sqrt{5}x/2} + C_2e^{-\sqrt{5}x/2})$
15. $y = C_1e^{3x} + C_2xe^{3x}$
17. $y = C_1e^{\sqrt{11}x} + C_2xe^{\sqrt{11}x}$
19. $y = -\frac{1}{3}e^{2x} + C_1e^{-x} + C_2e^{3x}$
21. $y = -\frac{1}{26}\cos 3x - \frac{3}{52}\sin 3x + e^{2x}(C_1e^{\sqrt{3}x} + C_2e^{-\sqrt{3}x})$
23. (b) $y_p = e^{-x}\ln(1 + e^x)$
(c) $y = e^{-x}\ln(1 + e^x) + Ce^{-x}$

Sec. 10.12. The Binomial Theorem for Any Exponent (Optional)

1. $1 + \frac{1}{2}x - \frac{1}{8}x^2 + \frac{1}{16}x^3 - \frac{5}{128}x^4$
3. $1 - 3x + 6x^2 - 10x^3 + 15x^4$
5. $1 + \frac{1}{2}x^3 - \frac{1}{8}x^6 + \frac{1}{16}x^9$
7. $1 - \frac{1}{3}x^2 - \frac{1}{9}x^4 - \frac{5}{81}x^6$
11. $1{,}247/1{,}120 \approx 1.1134$

Sec. 10.S. Guide Quiz

1. $-2 < x < 2$
2. $x > 5$ and $x < -1$
3. $1/(1 - x) = 1 + x + x^2 + \cdots$
4. 0.31
6. (a) $1 - x^2 + x^4 - x^6 + \cdots, R = 1$
(b) $1 - x + \dfrac{x^2}{2} - \dfrac{x^3}{3!} + \cdots, R = \infty$
(c) $1 - \dfrac{x^2}{2} + \dfrac{x^4}{4!} - \dfrac{x^6}{6!} + \cdots, R = \infty$
(d) $-x - \dfrac{x^2}{2} - \dfrac{x^3}{3} - \dfrac{x^4}{4} - \cdots, R = 1$
(e) $1 + 2x + 4x^2 + 8x^3 + \cdots, R = \frac{1}{2}$
8. (a) $1 \le x < 3$ (b) All x
9. (a) $y = C_1e^{x/2} + C_2e^{-2x}$
(b) $y = \frac{1}{169}(5\cos x - 12\sin x) + (C_1 + C_2x)e^{3x/2}$
(c) $y = x + \left[C_1\cos\left(\dfrac{\sqrt{3}}{2}x\right) + C_2\sin\left(\dfrac{\sqrt{3}}{2}x\right)\right]e^{-x/2}$
(d) $y = \frac{1}{13}(2\cos x + 3\sin x) + Ce^{-2x/3}$

Sec. 10.S. Review Exercises

1. (a) $(-1)^{n+1}\dfrac{(3x)^{2n-1}}{2n - 1}$ (b) $(-1)^{n+1}\dfrac{x^{2n}}{n}$
(c) $(-1)^n\dfrac{x^n}{n!}$ (Sum starts at $n = 0$.)
(d) $(-1)^{n+1}\dfrac{x^{2(2n-1)}}{(2n - 1)!}$
(e) $(-1)^n\dfrac{x^{4n}}{(2n)!}$ (Sum starts at $n = 0$.)
5. (a), (b) $\dfrac{(2x)^2}{2\cdot 2!} - \dfrac{(2x)^4}{2\cdot 4!} + \dfrac{(2x)^6}{2\cdot 6!} - \cdots$
7. Converges, $e/(e - 1)$
9. Converges, $\pi^2/12$
11. Converges
13. Converges
15. Converges, $\ln(2/\pi)$
17. Diverges
19. Converges
21. Converges, $\ln 2$
23. Converges, e^{10}
25. Converges, $e^{-1/2}$
27. Converges, $3 + \frac{3}{2}\ln 3$
29. Converges
31. Diverges
33. $\cos\frac{1}{3} \approx 0.94496$; error $< 3.4 \times 10^{-5}$
35. $1/\sqrt{e} \approx \frac{5}{8}$; error $< \frac{1}{48} \approx 0.0208$
37. Converges absolutely for $|x| < 1$ and diverges elsewhere. $R = 1$ and the sum is $1/(1 - x)^2$.
39. Converges absolutely for all x. $R = \infty$ and the sum is e^{x^2}.
41. Converges absolutely for $-1 < x < 1$ and converges conditionally at $x = -1$. $R = 1$ and the sum is $-\ln(1 - x)$.
43. Converges absolutely for $-2 < x < 0$ and diverges elsewhere; $R = 1$.
45. Converges absolutely for $\frac{1}{2} < x < \frac{3}{2}$ and diverges elsewhere; $R = \frac{1}{2}$.
47. Converges absolutely for all x; $R = \infty$.
49. Converges absolutely for $0 < x < 2$ and converges conditionally at $x = 0$; $R = 1$.
51. Converges absolutely for $-4 < x < -2$ and converges conditionally at $x = -4$; $R = 1$.
53. Converges absolutely for $-e^2 < x < e^2$; $R = e^2$.
55. 0
57. ∞
59. (a) Between $-1/2{,}048$ and $-1/4{,}096$
(b) Absolute value $\le 1/1{,}024$
(c) Equals $-1/3{,}072$
(d) Between $-1/2{,}048$ and $-2/531{,}441$
61. $32 + 11(x - 5) + (x - 5)^2$
63. (a) $y = 3 + Ce^{-4x}$
(b) $y = \frac{1}{2}(\cos x + \sin x) + Ce^{-x}$
65. (a) $y = \frac{1}{9}(9x^2 - 30x + 38) + C_1e^{-x} + C_2e^{-3x/2}$
(b) $y = \frac{1}{9}(9x^2 - 30x + 38) - \frac{1}{13}(5\cos x - \sin x) + C_1e^{-x} + C_2e^{-3x/2}$
67. $2^{33} \cdot 33!$
71. Finite; $(\sqrt{6}/4)(9 + 4\sqrt{5})$ seconds
73. (b) $(-1)^n/\sqrt{n}$
75. (a) $(-1)^{n+1}\dfrac{x^{2n-1}}{2n - 1}$ (b) $(-1)^{n+1}\dfrac{x^{2n}}{n}$
(c) $(-1)^n\dfrac{x^{2n}}{n!}$ (Sum starts at $n = 0$.)

(*d*) $(-1)^{n+1}\dfrac{x^{2(2n-1)}}{(2n-1)!}$

(*e*) $(-1)^n\dfrac{x^{2n}}{(2n)!}$ (Sum starts at $n = 0$.)

91. Converges for $-1/e \le x < 1/e$ and diverges elsewhere.

APPENDIX A. REAL NUMBERS

1. $a + 3 < b + 3$
3. $5a < 5b$
5. $1/a > 1/b$
7. $(-1, \infty)$
9. $(-\infty, -2)$
11. $(-\infty, 1), (3, \infty)$
13. $(-3, -2)$
15. $(-1, 0), (1, \infty)$
17. $(-5, -3), (0, \infty)$
19. $(-\infty, \frac{1}{3}), (\frac{1}{2}, \infty)$
21. 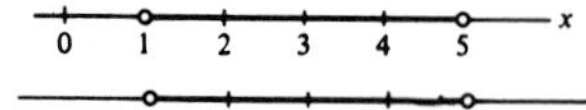

23. 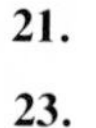

25. $x > 3$
27. $x > 1$
29. $a = -2, b = 1$
31. $a = -1, b = 1, c = -2, d = 1$
33. (*a*) 234.74747 · · · (*b*) 232.4
(*d*) $a = 2{,}324/990$
35. Cancellation of $x - 1$ is division by zero.
39. No

APPENDIX B. GRAPHS AND LINES

Sec. B.1. Coordinate Systems and Graphs

1.

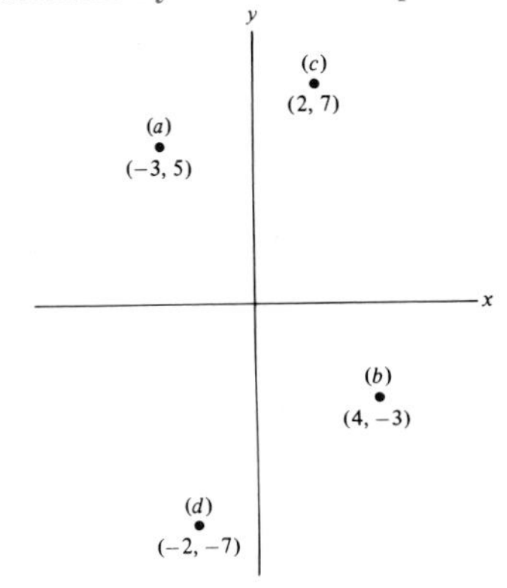

(*a*) Second (*b*) Fourth (*c*) First (*d*) Third

3. (*a*) 10 (*b*) 13 (*c*) 3

5.

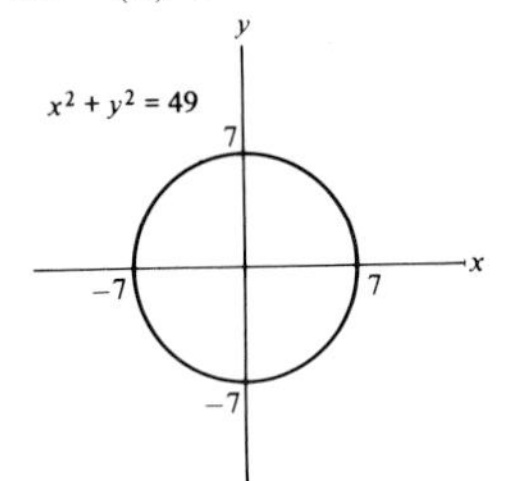

7.

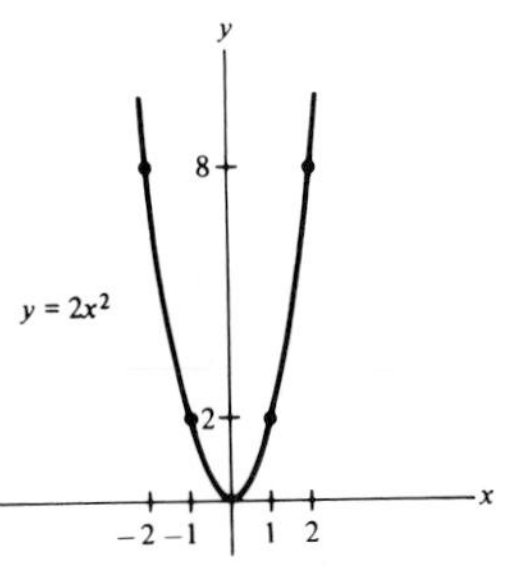

9.

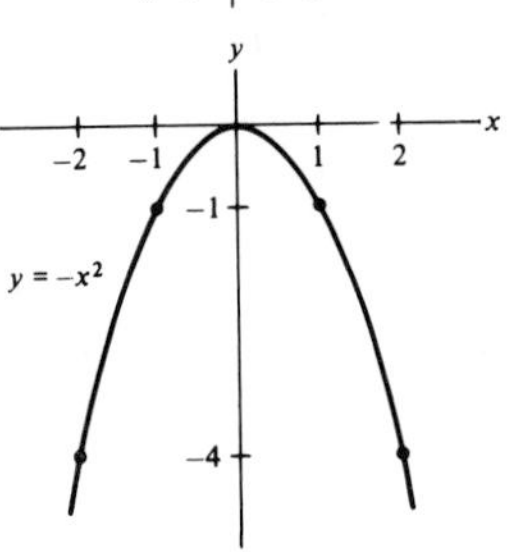

11.

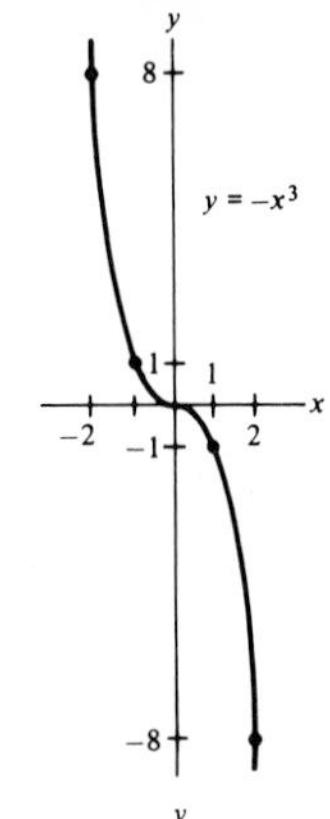

13.

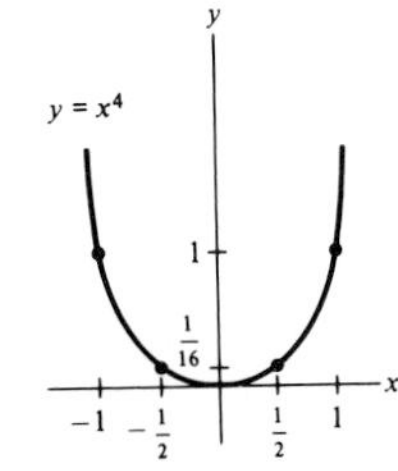

15.

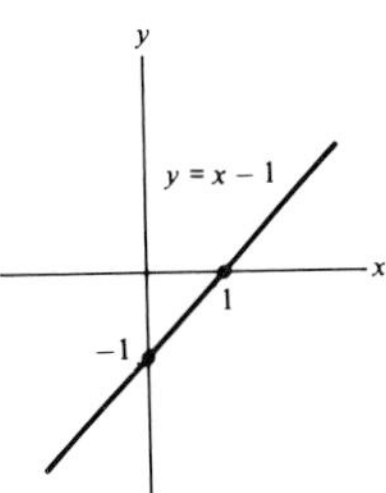

17. x intercept, -3; y intercept, 6

19. No intercepts

21. x intercepts, -3 and $\frac{1}{2}$; y intercept, -3

23.

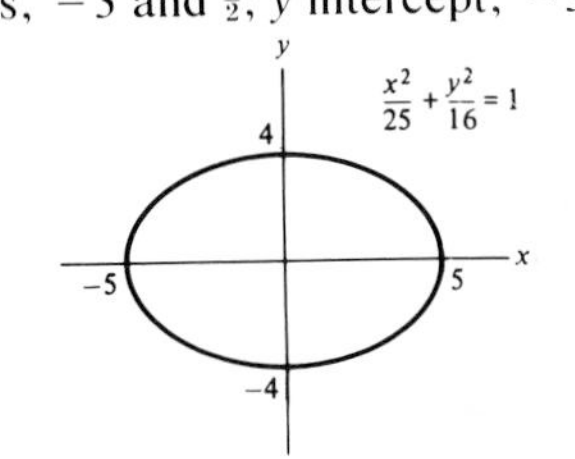

25.

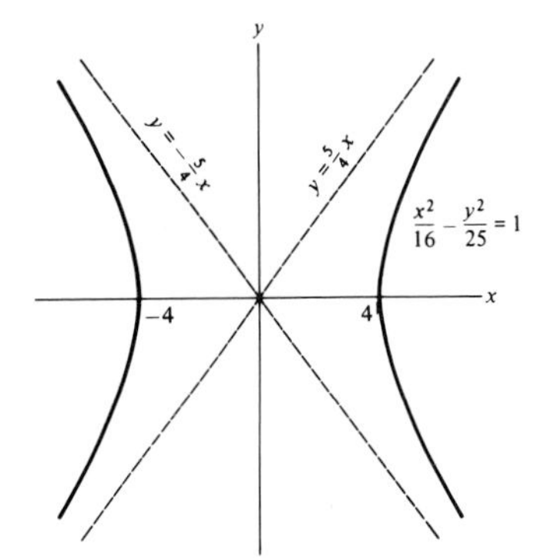

27.

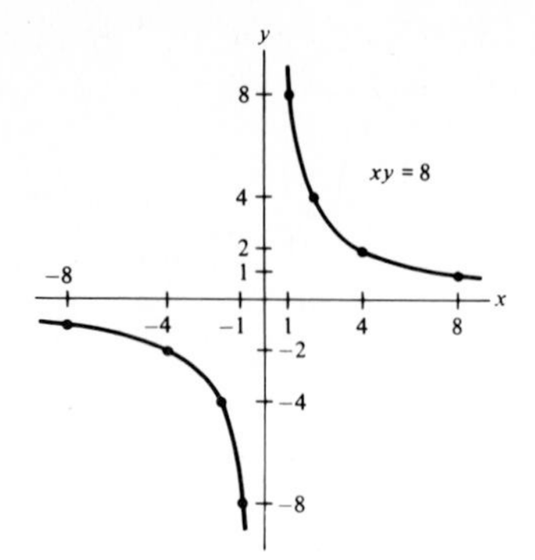

29.

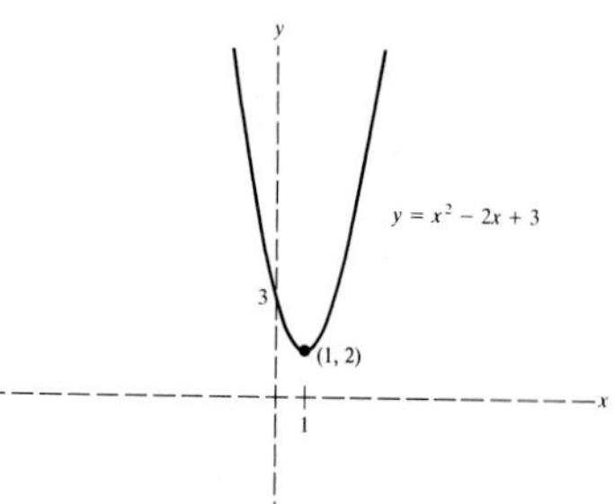

31.

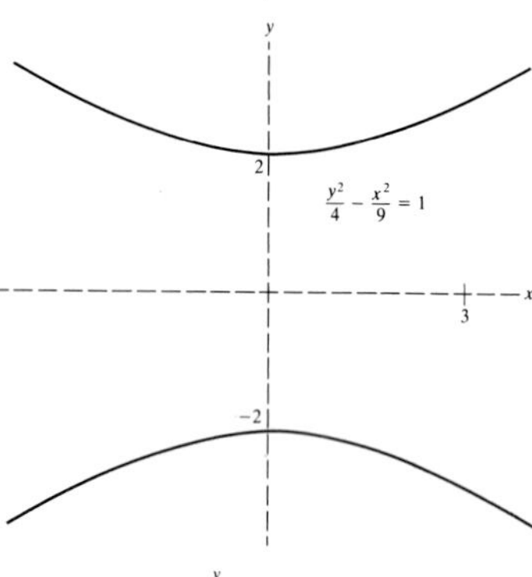

33.

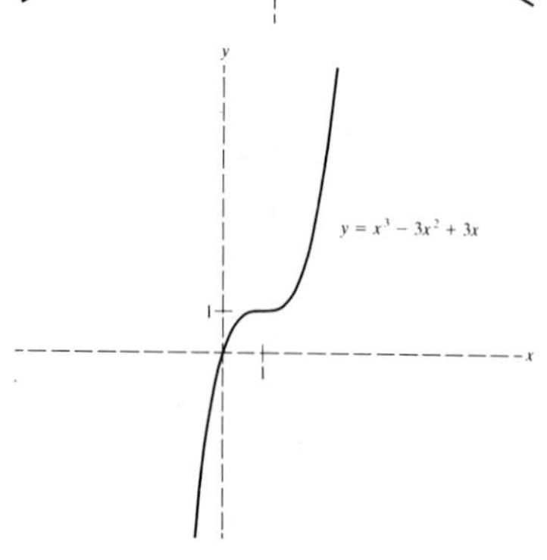

35.

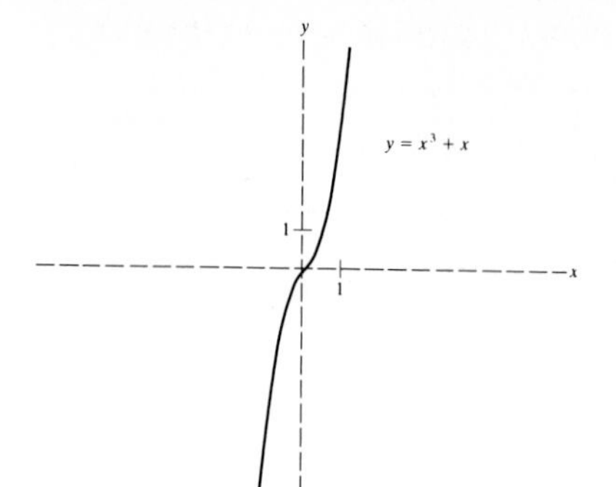

37.

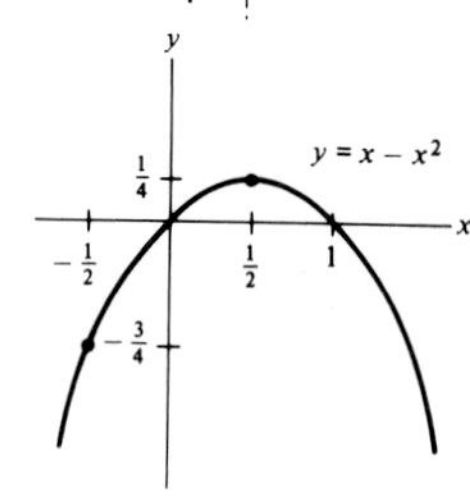

39.

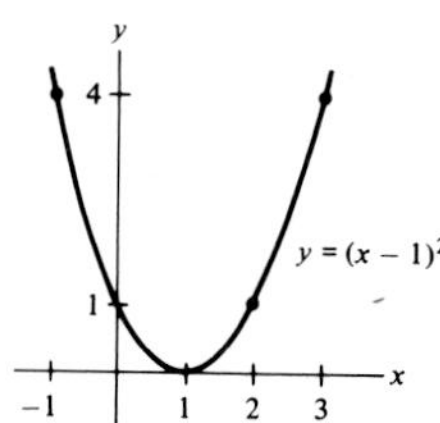

41.

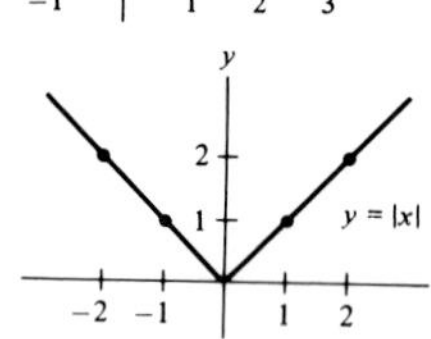

43.

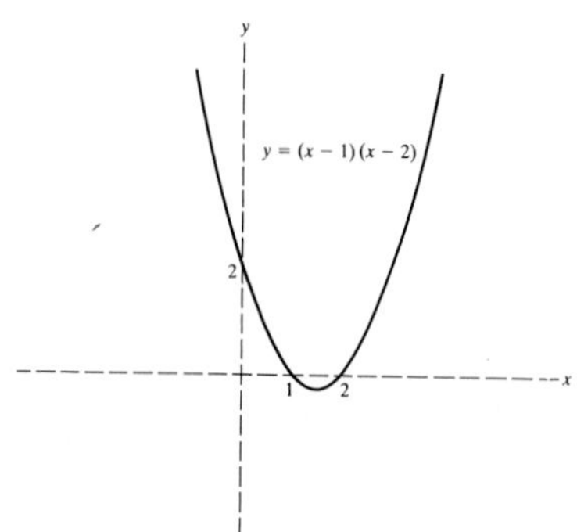

45. $(x - 2)^2 + (y - 1)^2 = 49$

47. (*a*) Symmetric with respect to the x axis
(*b*)

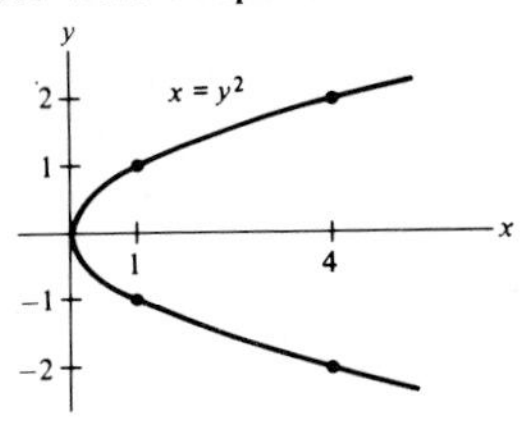

49. (*a*)

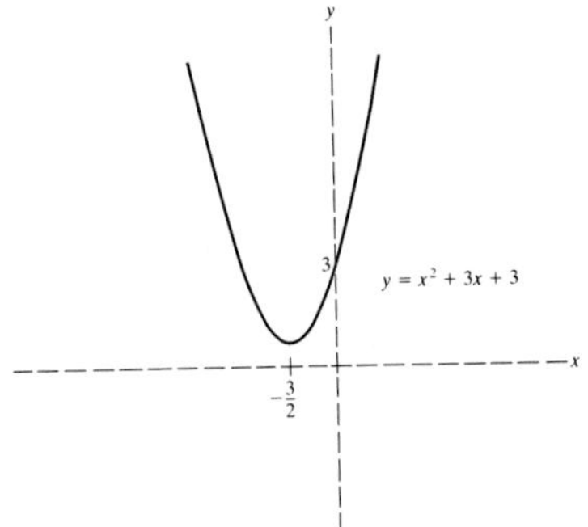

(*b*) $\left(-\frac{3}{2}, \frac{3}{4}\right)$

51. (*a*) $x = -b/(2a)$

Sec. B.2. Lines and Their Slopes

1. $m = \frac{1}{5}$

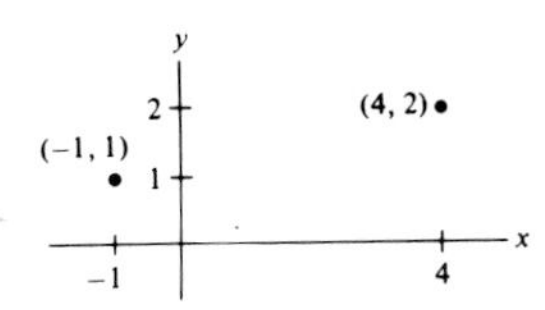

3. $m = -\frac{5}{4}$

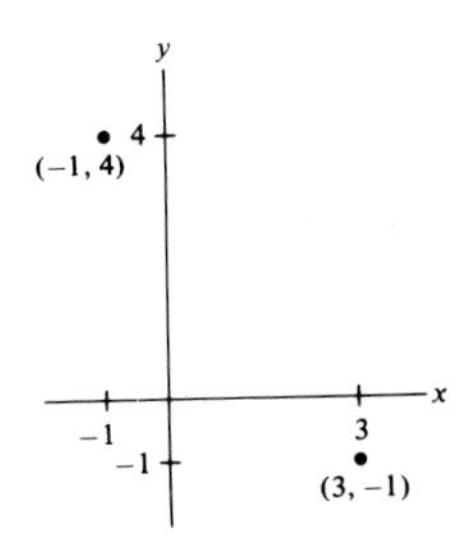

5. $m = 0$

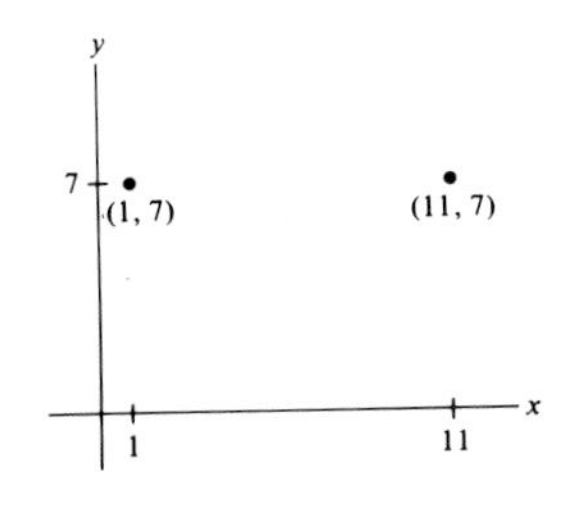

7. (*a*) Negative (*b*) Negative (*c*) Positive
(*d*) Positive (*e*) Zero

9. (*a*) 4 (*b*) $-\frac{1}{4}$

11. Yes

13. No

15. (*a*) $y = 3x + 2$ (*b*) $y = \frac{2}{3}x - 2$ (*c*) $y = -3x$

17. (*a*) $m = 3$, $b = -1$

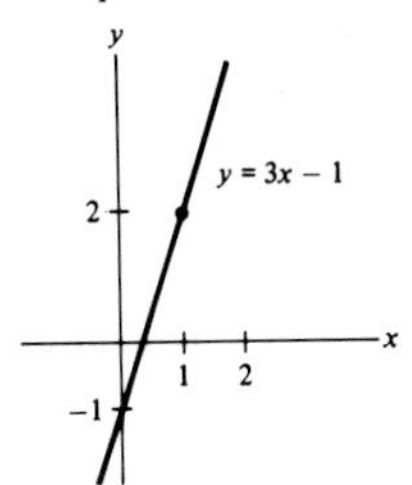

(*b*) $m = -2$, $b = 1$

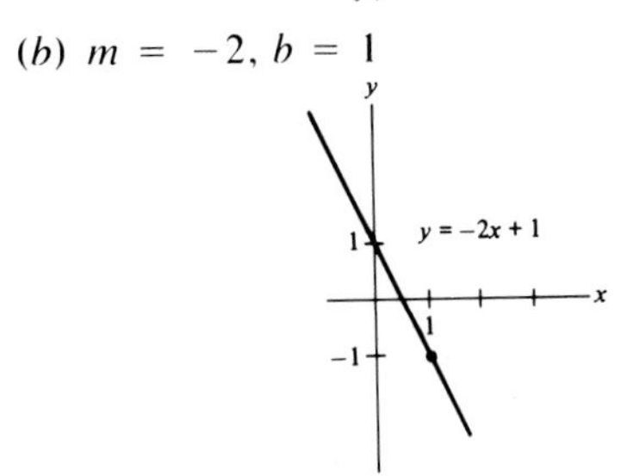

(*c*) $m = \frac{3}{5}$, $b = 0$

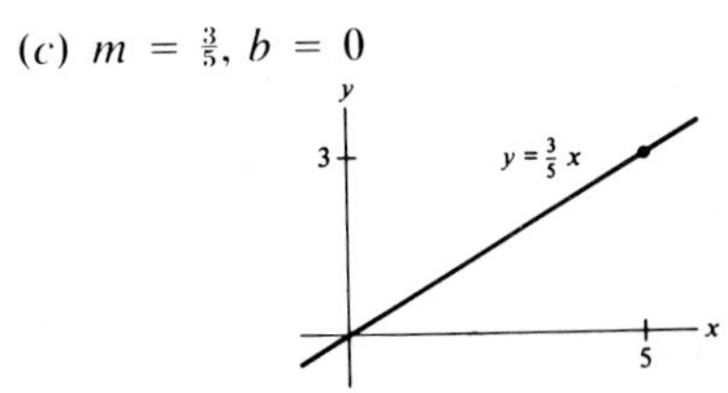

19. (*a*) $y - 2 = 3(x - 1)$ (*b*) $y + 1 = -2(x - 3)$

21. (*a*) $y - 2 = \frac{1}{4}(x - 1)$ (*b*) $y - 2 = -\frac{1}{4}(x + 1)$
(*c*) $y - 5 = x - 4$

23. (*a*)

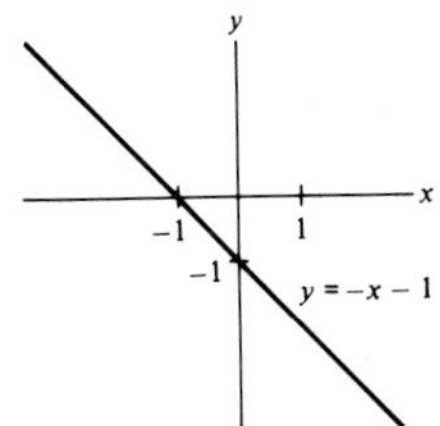

(*b*)

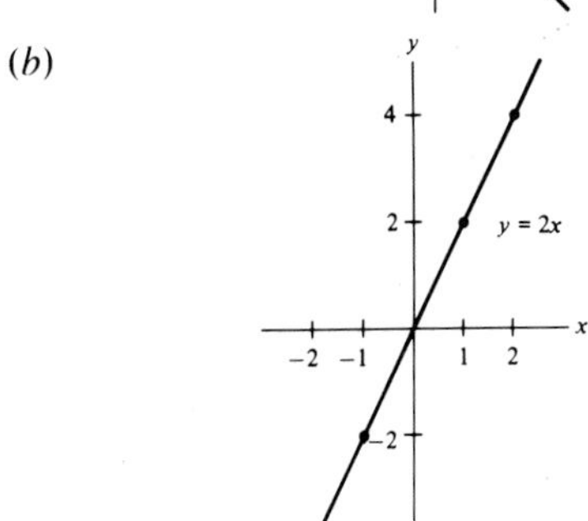

(*c*)

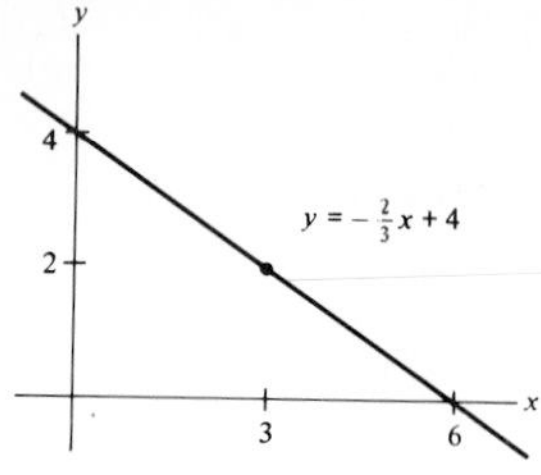

25. (*a*) Yes (*b*) No
27. $y = x/2 + 5/2$
29. $y = 3x + 5$
31. $m = -b/a$
33. (*a*) $\frac{7}{4}$ (*b*) $\frac{11}{2}$ (*c*) 3 (*d*) -6
35. (*a*) $m = (y_2 - y_1)/(x_2 - x_1)$
(*b*) $m' = (y_2 - y_1)/(x_2 - x_1)$
37. (*a*)

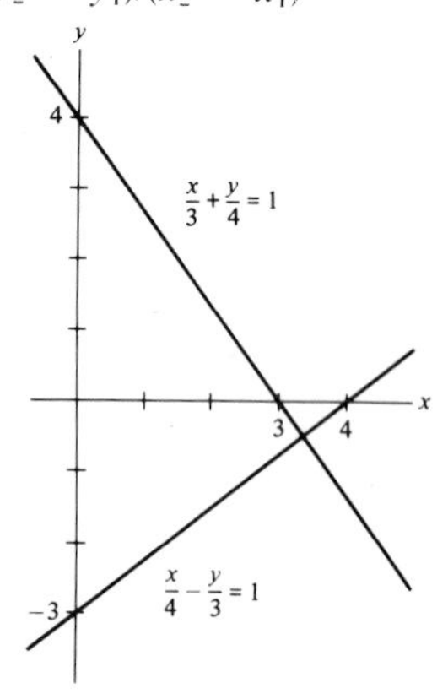

(*b*) $-\frac{4}{3}, \frac{3}{4}$ (*c*) Yes
39. (*a*) $|m|$ is large. (*b*) m is close to 0. (*c*) $m < 0$
(*d*) $m > 0$
41. $P = (-1, 0)$

APPENDIX C. TOPICS IN ALGEBRA

1. $2\sqrt{5}$
3. $\sqrt{2} + 4$
5. $\frac{1}{3}(6 + 2\sqrt{3})$
7. $\dfrac{x(1 + \sqrt{x})}{1 - x}$
9. $\dfrac{7}{5(3 - \sqrt{2})}$
11. $\dfrac{1}{\sqrt{x} + \sqrt{5}}$
13. (*a*) $(x + 4)^2 - 3$ (*b*) $(x - 4)^2 + 7$
(*c*) $(x - \frac{1}{2})^2 + \frac{7}{4}$
15. (*a*) $(x + \frac{3}{2})^2 - \frac{17}{4}$ (*b*) $(x + \frac{3}{2})^2 + \frac{19}{4}$
(*c*) $(x + \frac{5}{4})^2 + \frac{39}{16}$
17. (*a*) $2(x - \frac{5}{4})^2 - \frac{1}{8}$ (*b*) $2(x + \frac{3}{2})^2 + \frac{5}{2}$
(*c*) $3(x + \frac{5}{6})^2 - \frac{13}{12}$
19. (*a*) No real solutions (*b*) $x = \frac{1}{2}(-1 \pm \sqrt{5})$
(*c*) $x = -1$
21. (*a*) None (*b*) Two (*c*) One
23. (*a*) 10 (*b*) 15 (*c*) 45
25. $1 + 7x + 21x^2 + 35x^3 + 35x^4 + 21x^5 + 7x^6 + x^7$
27. 45
29. (*a*) 1,093/729 (*b*) 1,275 (*c*) 33/8
31. $x - (-2), x - (-1), x - 1$
33. $x - 1$
35. $x - (-1), x - \frac{2}{3}, x - \frac{3}{2}$
37. $\left(x + \dfrac{5 + \sqrt{17}}{2}\right)\left(x + \dfrac{5 - \sqrt{17}}{2}\right)$
39. $\dfrac{1}{3}\left(1 - \dfrac{1}{10^9}\right)$

APPENDIX D. EXPONENTS

1. (*a*) 1 (*b*) 32 (*c*) 2 (*d*) $\frac{1}{2}$ (*e*) 8
3.

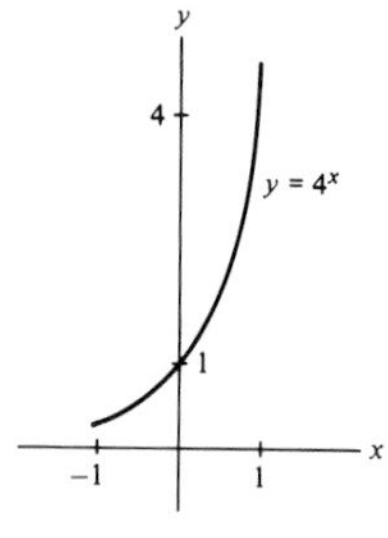

5.

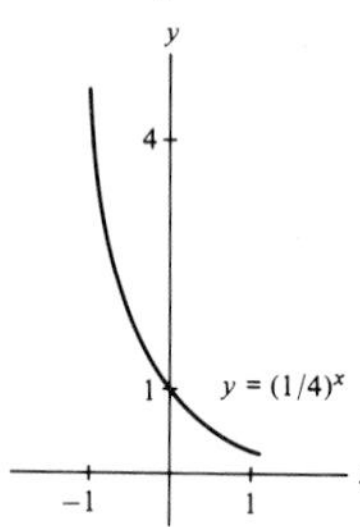

7. (*a*) 2^4 (*b*) 2^{-3} (*c*) $2^{1/2}$ (*d*) 2^0 (*e*) 2^{-2}
9. (*a*) $b^{3/2}$ (*b*) $b^{3/2}$ (*c*) $b^{-1/3}$ (*d*) $b^{2/3}$ (*e*) $b^{1/4}$
11. (*a*) Domain: all x; range: $y > 0$
(*b*) Domain: all x; range: all y
(*c*) Domain: $x \geq 0$; range: $y \geq 0$
(*d*) Domain: all x; range: $y \geq 0$
13. (*a*)

x	−32	−1	0	1	32
$x^{3/5}$	−8	−1	0	1	8

(*b*)

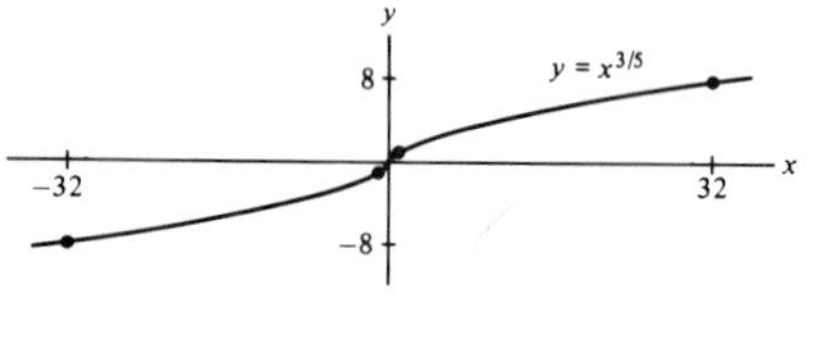

15. $\sqrt{5}$
17. (*a*) $x > 1$ (*b*) $x = 1$ (*c*) $0 < x < 1$

19.

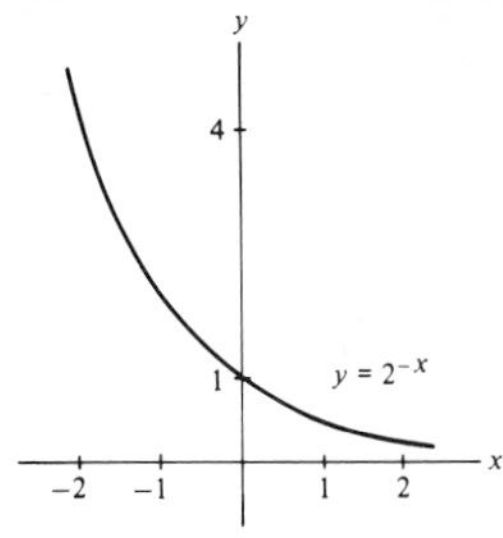

21. (*a*) 10^3 (*b*) 10^{-4} (*c*) 10^6 (*d*) 10^{-7}
23. The second way
25. (*a*)

h	0.1	0.01	0.001
$\frac{2^h - 1}{h}$	0.7177	0.6956	0.6934

27. (*a*) 3,647.8 (*b*) 30.238 (*c*) Reduce CEP.

APPENDIX E. TRIGONOMETRY

1. (*a*) $\pi/2$ (*b*) $\pi/6$ (*c*) $2\pi/3$ (*d*) 2π
3. (*a*) 135 (*b*) 60 (*c*) 120 (*d*) 720
5. (*a*) $\frac{5}{3}$ (*b*) $\approx 95.49°$
7. (*a*) $5\pi/18$ (*b*) $\approx 114.59°$
9. (*a*) 4.5 in (*b*) 6 in (*c*) 7.5 in
11. (*a*)

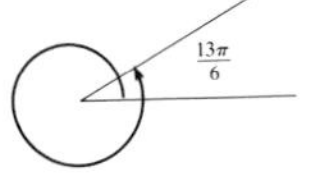

(*b*)

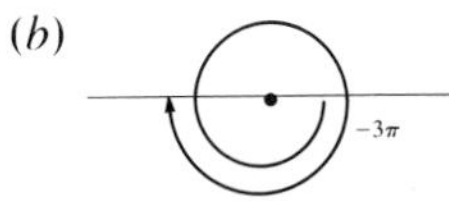

15.

θ	0	$\pi/6$	$\pi/4$	$\pi/3$	$\pi/2$	π	$3\pi/2$	2π
$\sin\theta$	0	$\frac{1}{2}$	$1/\sqrt{2}$	$\sqrt{3}/2$	1	0	-1	0

19. $(1 + \sqrt{3})/(2\sqrt{2})$
27. (*a*) $-$ (*b*) $+$ (*c*) $+$ (*d*) $-$
31. (*a*) ≈ 0.1763 (*b*) ≈ 2.7475 (*c*) ≈ -2.7475 (*d*) -1 (*e*) 0
33. (*a*) $60° = \pi/3$ radians (*b*) $30° = \pi/6$ radians
37. (*a*) $\cos\alpha = b/c$ (*b*) $\sin\beta = b/c$ (*c*) $\tan\alpha = a/b$
39. $\sqrt{3}/2$, $1/2$, $1/\sqrt{3}$
41. (*a*)

θ	0	$\pi/6$	$\pi/4$	$\pi/3$
$\sec\theta$	1	$2/\sqrt{3}$	$\sqrt{2}$	2

(*b*)

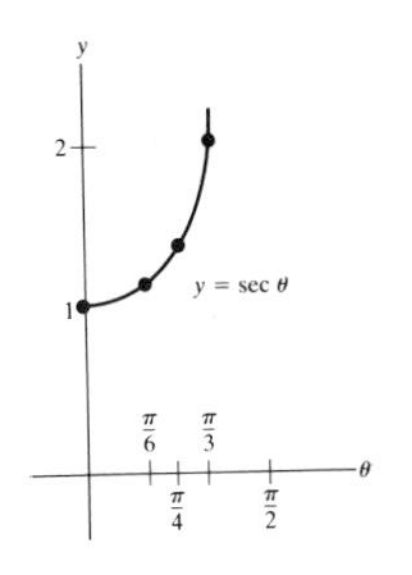

49. (*a*) $1/\sqrt{2}$ (*b*) $-1/\sqrt{2}$

APPENDIX F. CONIC SECTIONS

Sec. F.1. Conic Sections

1. $\frac{x^2}{25} + \frac{y^2}{21} = 1$

3.

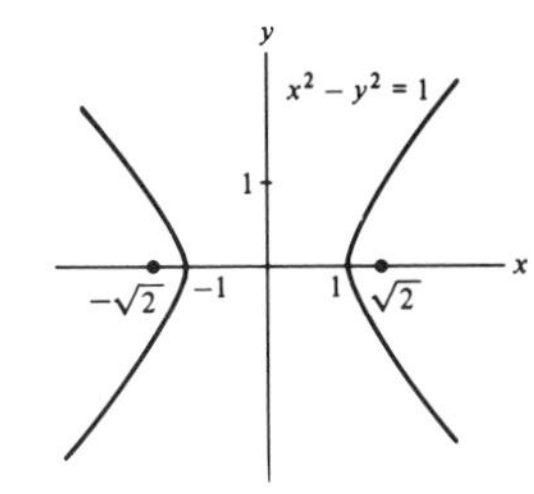

5.

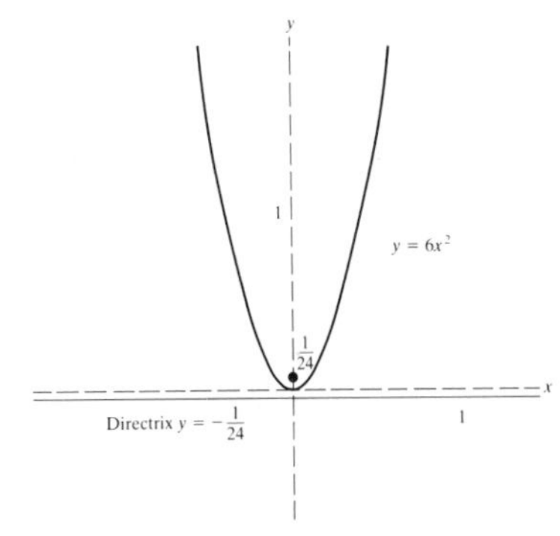

7. $y^2 = 12x$

11.

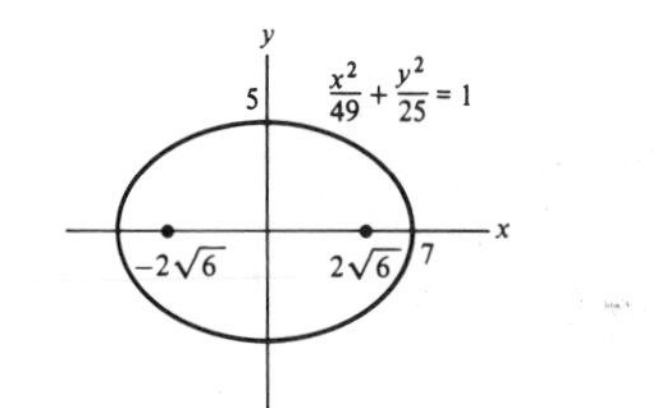

13.

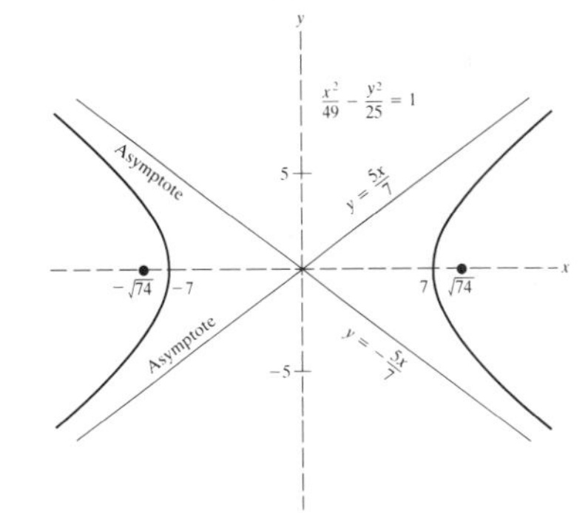

15.

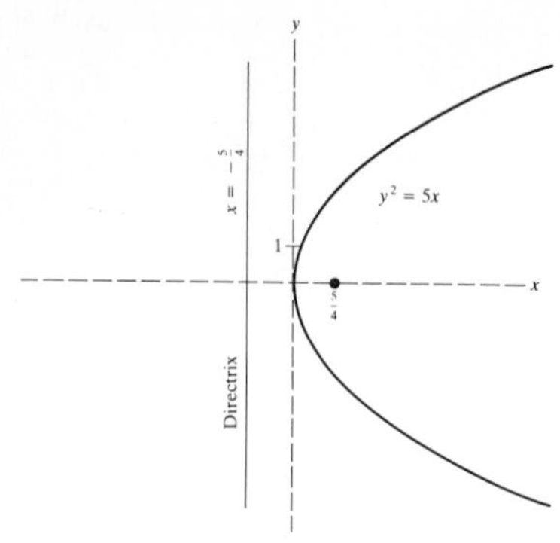

17.

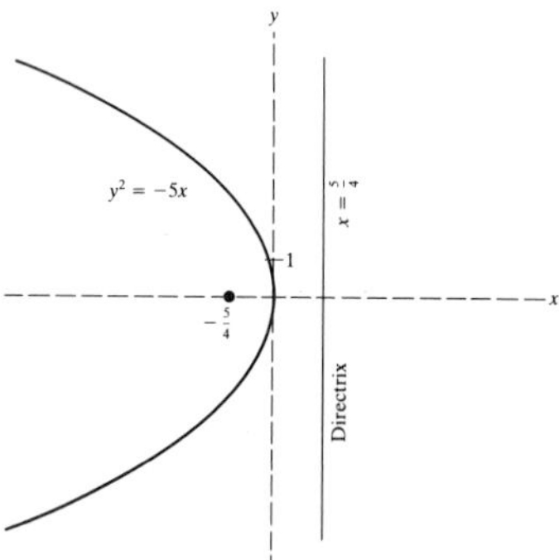

19. (*b*)

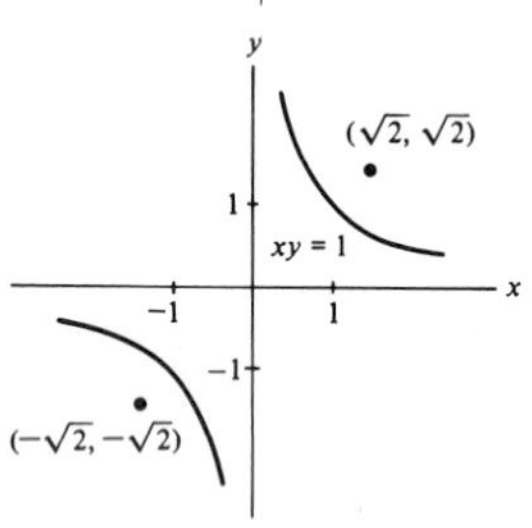

Sec. F.2. Translation of Axes and the Graph of $Ax^2 + Cy^2 + Dx + Ey + F = 0$

1.

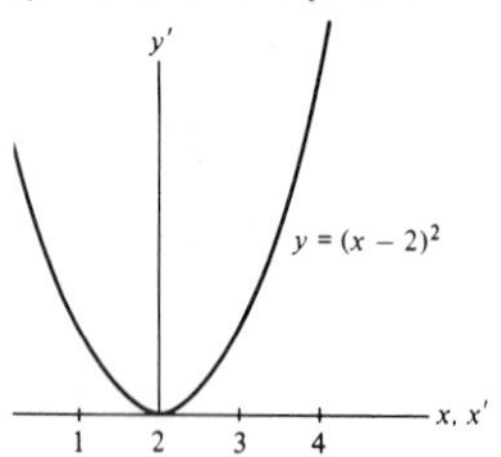

Parabola; focus $= (2, \frac{1}{4})$; vertex $= (2, 0)$

3.

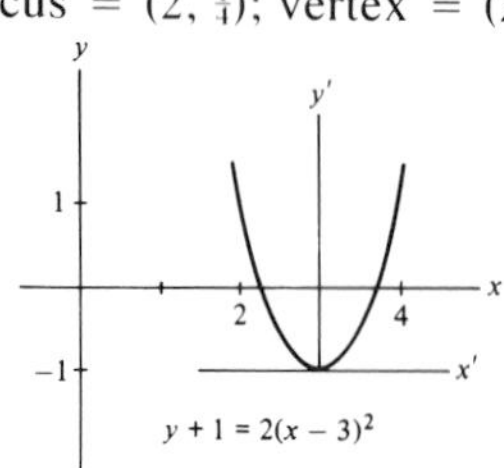

Parabola; focus $= (3, -\frac{7}{8})$; vertex $= (3, -1)$

5.

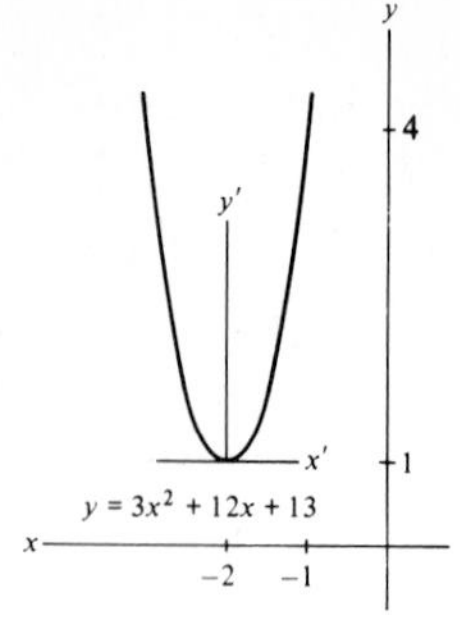

Parabola; focus $= (-2, \frac{13}{12})$; vertex $= (-2, 1)$

7.

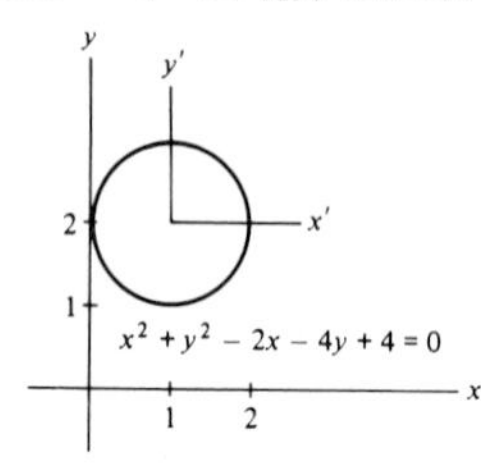

Circle; center (= focus) $= (1, 2)$

9.

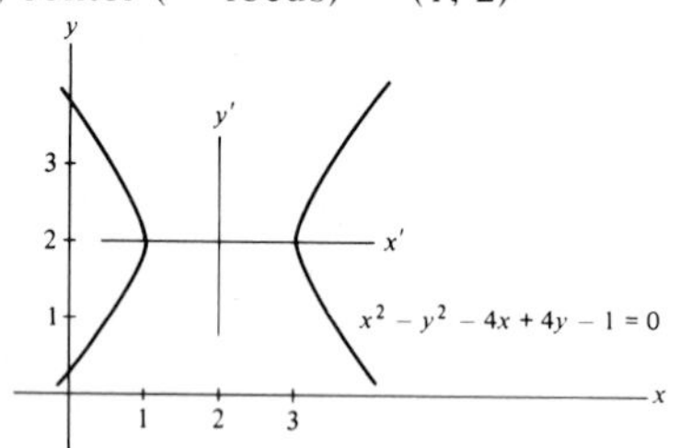

Hyperbola; foci $= (2 \pm \sqrt{2}, 0)$; vertices $= (1, 0)$, $(3, 0)$; asymptotes: $y = x$ and $x + y = 4$

11.

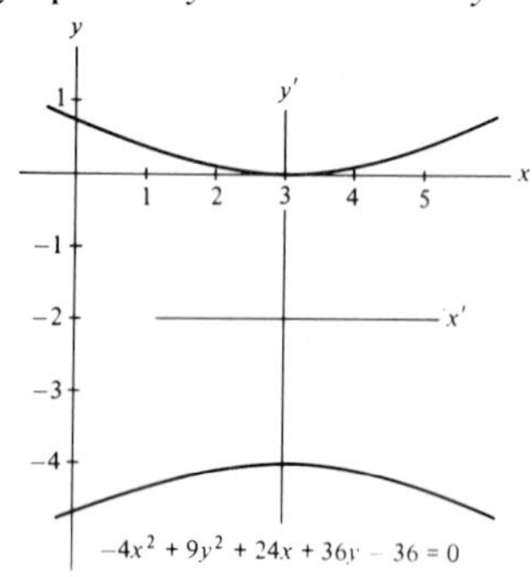

Hyperbola; foci $= (3, -2 \pm \sqrt{13})$; vertices $= (3, 0)$, $(3, -4)$; asymptotes: $y = \frac{2}{3}x - 4$ and $y = -\frac{2}{3}x$

13.

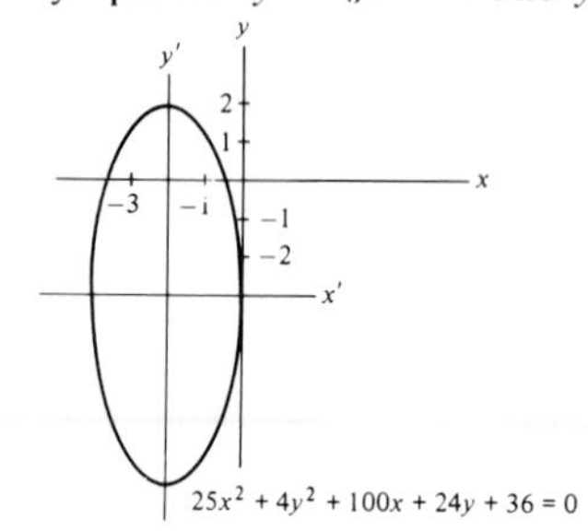

Ellipse; foci $= (-2, -3 \pm \sqrt{21})$; vertices $= (-4, -3), (0, -3), (-2, -8), (-2, 2)$

15. $(x')^2/9 + (y')^2/4 = 1$, where $x' = x - 1$, $y' = y - 2$.

17. $\frac{13}{36}(x')^2 - \frac{13}{16}(y')^2 = 1$, where $x' = x - 1$, $y' = y - 1$.

19. $(x')^2/9 + (y')^2/4 = 1$, where $x' = x - 4$, $y' = y$.

21. $(y')^2 = 12x'$, where $x' = x - 4$, $y' = y - 3$.

23. (*d*) When $A = C$ and their sign is opposite that of F

Sec. F.3. Rotation of Axes and the Graph of $Ax^2 + Bxy + Cy^2 + Dx + Ey + F = 0$

1.

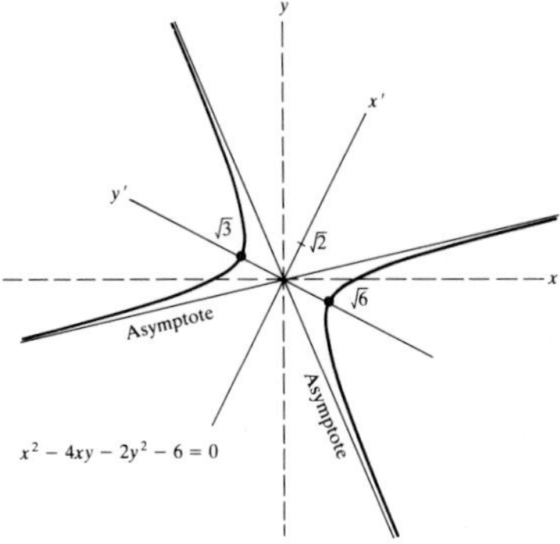

3.

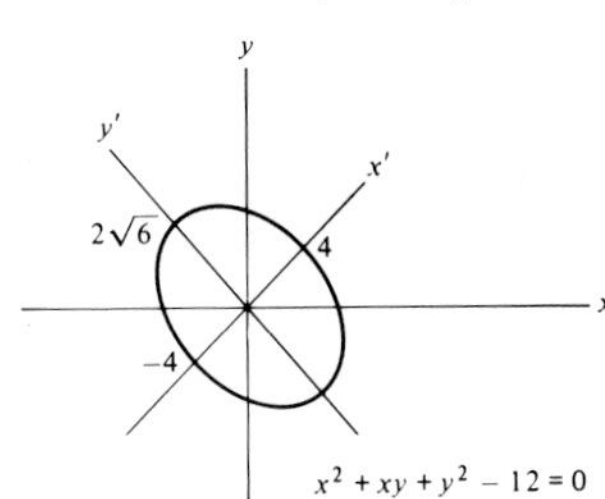

5.

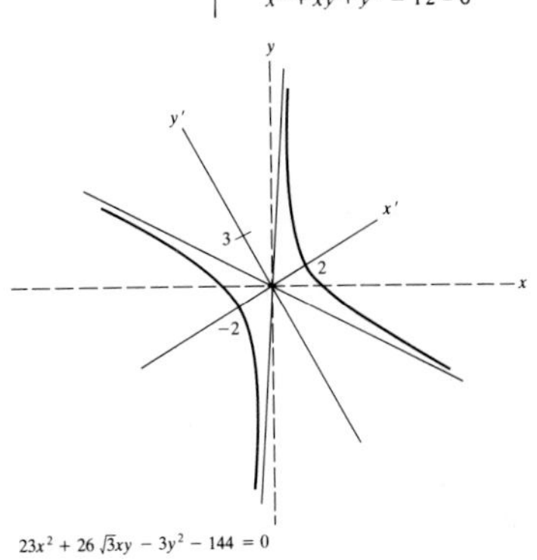

7.

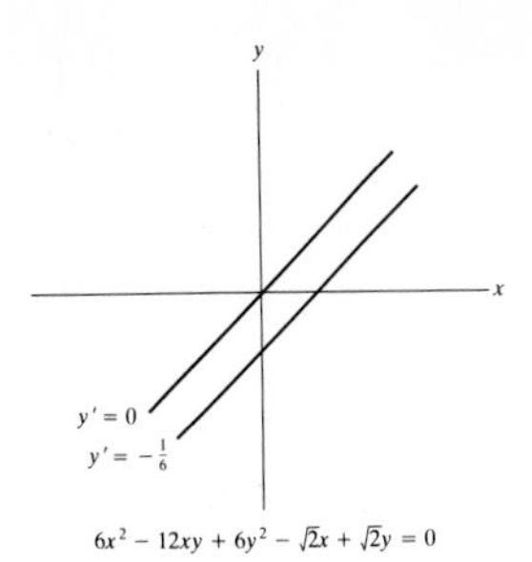

9.

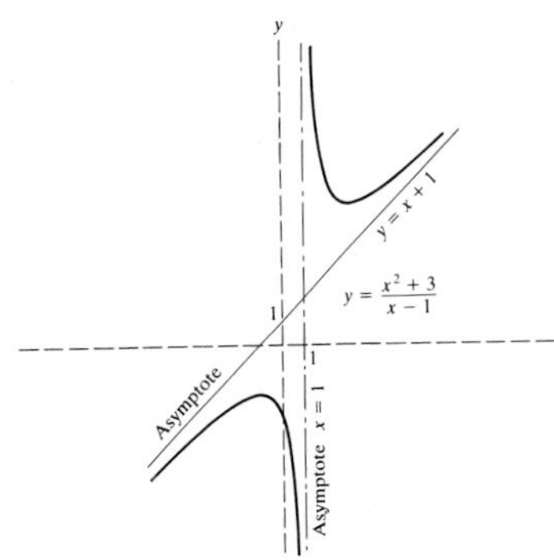

11.

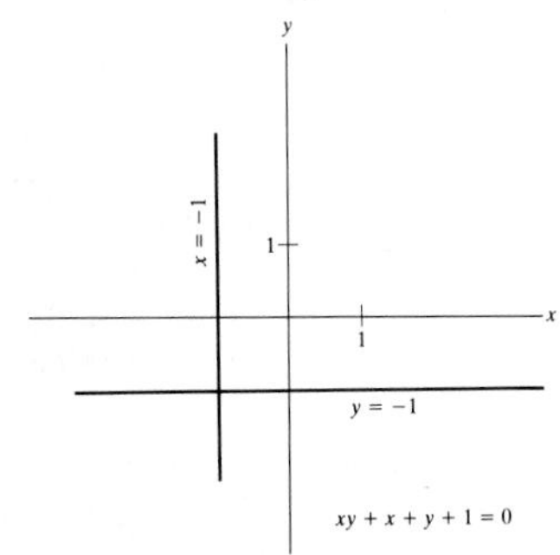

13. Hyperbola ($\mathscr{D} = 600$)

15. Parabola ($\mathscr{D} = 0$)

17. (*a*)

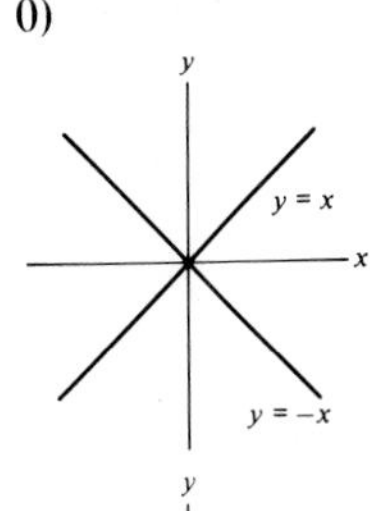

(*b*)

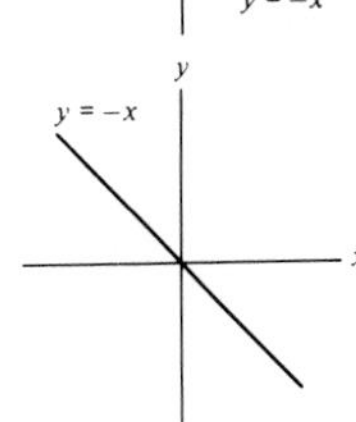

(*c*)

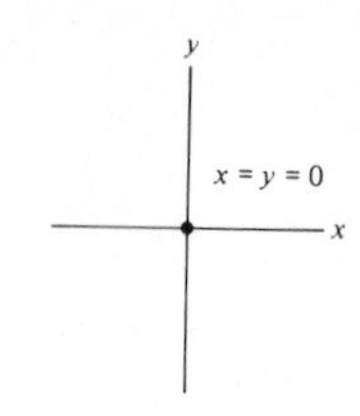

(*d*) Empty

23. $(x', y') = (\pm\sqrt{2}, 0)$; $(x, y) = \pm(\frac{3}{5}\sqrt{2}, \frac{4}{5}\sqrt{2})$

25. $(x', y') = (0, \pm 4)$; $(x, y) = \pm(2\sqrt{2}, -2\sqrt{2})$

Sec. F.4. Conic Sections in Polar Coordinates

1. (*a*)

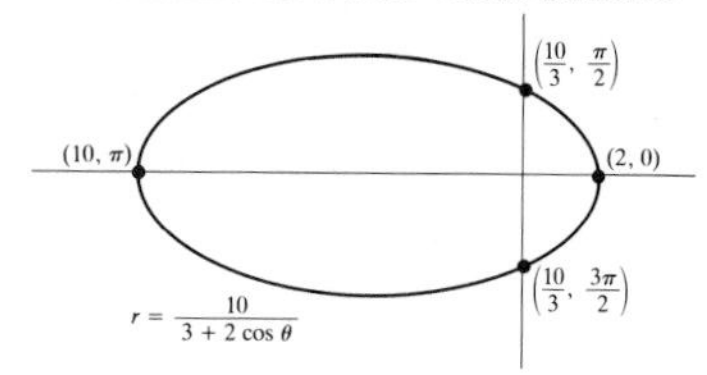

3. (*a*) $e = \frac{1}{3}$ (*b*) $e = \frac{3}{4}$ (*c*) $e = 1$ (*d*) $e = \frac{4}{3}$

5. (*b*) $r = 2/(1 + \cos\theta)$ (*c*) $r = -4 \csc\theta \cot\theta$

APPENDIX H. LOGARITHMS AND EXPONENTIALS DEFINED THROUGH CALCULUS

Sec. H.1. The Natural Logarithm Defined as a Definite Integral

3. (*a*) $28{,}271/27{,}720 \approx 1.0199$

APPENDIX I. THE TAYLOR SERIES FOR $f(x, y)$

1. $f_{xxx} = 60x^2y^7$, $f_{yyy} = 210x^5y^4$, $f_{xxy} = f_{xyx} = f_{yxx} = 140x^3y^6$, $f_{yyx} = f_{xyy} = f_{xyy} = 210x^4y^5$

3. $f_{xxx} = 8e^{2x+3y}$, $f_{yyy} = 27e^{2x+3y}$, $f_{xxy} = f_{xyx} = f_{yxx} = 12e^{2x+3y}$, $f_{yyx} = f_{yxy} = f_{xyy} = 18e^{2x+3y}$

7. (*a*), (*b*) $e^{x+y^2} = 1 + x + (x^2/2) + y^2 + \cdots$

9. $2 + 3x - 5y + 3x^2 + 7xy + \frac{1}{2}y^2$

13. $\frac{1}{5!\,7!}\frac{\partial^{12} f}{\partial x^5 \partial y^7}(0, 0)$

APPENDIX J. THE INTERCHANGE OF LIMITS

Sec. J.2. The Derivative of $\int_a^b f(x, y)\,dx$ with Respect to y

3. $y = 2/\pi$

5. $f(u, u) + \int_0^u f_u(u, x)\,dx$

Sec. J.3. The Interchange of Limits

1. (*a*)

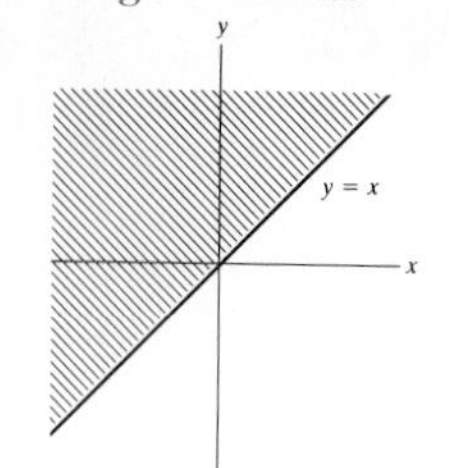

5. (*a*)

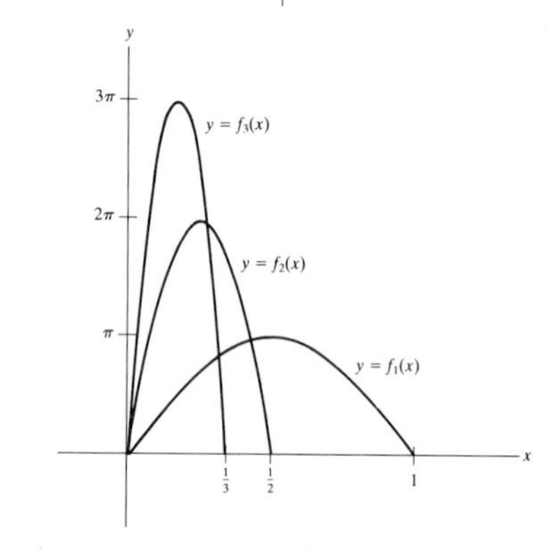

APPENDIX K. THE CONVERSE OF A STATEMENT

1. (*a*) "If $a^2 = b^2$, then $a = b$." is false.
(*b*) "If $a^3 = b^3$, then $a = b$." is true.

3. (*a*) "If $ab = ac$, then $b = c$." is false.
(*b*) "If $a + b = a + c$, then $b = c$." is true.

5. (*a*) "If ab is odd, then a and b are odd." is true.
(*b*) "If ab is even, then a and b are even." is false.

7. True

9. True

11. True

13. False

15. True

17. True

19. False

21. False

23. True

25. Yes, the second.

List of Symbols

Note: Numbers in parentheses refer to exercises on the indicated pages. Numbers prefixed by "S" refer to the pages at the end of the book.

Symbol	*Description*	*Page*
$\lvert x \rvert$, abs x	Absolute value of x	S4, S5
a_N	Normal component of acceleration	788
a_T	Tangential component of acceleration	788
$\overline{AB}$	Length of AB	45
$\widehat{AB}$	Length of arc AB	726
$\rightarrow$	Approaches	23
$x \rightarrow a$	x approaches a	23
$x \rightarrow a^+$	x approaches a from the right	24
$x \rightarrow a^-$	x approaches a from the left	24
$\mathcal{D}$	Discriminant	673
$\bullet$	Included point	25
$\circ$	Excluded point	25
$\approx$	Approximate equality	22
$\stackrel{?}{=}$	Conjectured equality	276
ϵ	Epsilon (challenge)	64
δ	Delta (response)	67
$F(x)\Big\vert_a^b$	$F(b) - F(a)$	193
$\Big\vert_{x=a}^{x=b}$	Evaluation limits	699
$\Big\vert_{(a,b)}$	Evaluation at (a, b)	S84
$[F(x)]_a^b$	$F(b) - F(a)$	
FTC	Fundamental theorem of calculus	212
$\Gamma(x)$	Gamma function	431(42)
∞	Infinity	S3

Symbol	*Description*	*Page*
(a, b)	Open interval	S3
$[a, b]$	Closed interval	S3
$[a, b), (a, b]$	"half-open" intervals	S3
lim	Limit	21
$J, \dfrac{\partial(x, y)}{\partial(u, v)}$	Jacobian	742
$n!$	n factorial	143
(x, y)	Ordered pair	S7
$R(x, y)$	Rational function	353
$R(\cos\theta, \sin\theta)$	Rational function of $\cos\theta$ and $\sin\theta$	353
Σ	Summation notation	193
σ	Density in the plane	410
σ	Standard deviation	442
μ	Mean	442
μ_2	Variance	442
κ	Curvature	479
τ	Torsion	802(27)
ω	Angular velocity	703
$f(x)$	Function of x	2
Δf	Change in f	88, 650
df, dy	Differentials	175, 651
f^{-1}	Inverse of f	15
$f \circ g$	Composite function	9
$A(x)$	Cross-sectional area	206
$c(x), c(y)$	Cross-sectional length	383
e^x	Exponential function	254
$E(x)$	Exponential function	S78
$\exp x$	Exponential function	330
$\ln x$	Natural logarithm	248
$L(x)$	Natural logarithm	S75
$\log_b x$	Logarithm base b	234
$\tan^{-1} x$	Inverse tangent	260
$\arctan x$	Inverse tangent	260
$\sinh x, \cosh x, \tanh x$	Hyperbolic functions	301–302
$\sinh^{-1} x$	Inverse hyperbolic sine	303
$[x], \lfloor x \rfloor$	Greatest integer (floor) function	49
$\lceil x \rceil$	Ceiling function	54(9)
$\dfrac{dy}{dx}, f', Df$	First derivative	82
$\dfrac{d^2y}{dx^2}, f'', D^2f$	Second derivative	141
$\dfrac{d^ny}{dx^n}, f^{(n)}$	nth derivative	142
$\dfrac{\partial y}{\partial x}, f_x, f_1$	Partial derivative	645
$dy \div dx$	Quotient of differentials	177
$\dot{y}, \ddot{y}$	Newton's notation for derivative	90
$\int f(x)\,dx$	Antiderivative	217
$\int_a^b f(x)\,dx$	Definite integral over interval	200
$\int_R f(P)\,dA$	Definite integral over plane region	689
$\int_R f(P)\,dV$	Definite integral over solid region	715

Symbol	*Description*	*Page*
$\int_C f(P)\,ds$	Line integral	812
$\int_C (P\,dx + Q\,dy + R\,dz)$	Line integral	815
$\oint_C f(P)\,ds$	Line integral over closed path	816
$\int_C \mathbf{F} \cdot d\mathbf{r}$	Line integral	815
$\int_{\mathcal{S}} \delta(P)\,dS$	Surface integral	843
$\int_{\mathcal{S}} \mathbf{F} \cdot \mathbf{n}\,dS$	Surface integral	852(17)
i	$\sqrt{-1}$	548
z	Complex number	548
$\lvert z \rvert$	Magnitude of z	550
$\mathbf{Im}\,z$	Imaginary part of z	549
$\mathbf{Re}\,z$	Real part of z	549
z^{-1}	Reciprocal of z	556(11)
$\bar{z}$	Conjugate of z	549
$x + iy$	Complex number	548
$e^{i\theta}$	Complex exponential	557
$\mathbf{A}$	Vector	583
$\lvert \mathbf{A} \rvert$	Magnitude of $\mathbf{A}$	583
PQ	Directed line segment (vector)	583
(x, y)	Vector	584
$\mathbf{A} \cdot \mathbf{B}$	Dot product	595
$\mathbf{A} \times \mathbf{B}$	Cross product	611
$\mathbf{i}, \mathbf{j}, \mathbf{k}$	Basic unit vectors	592
$\mathbf{T}$	Unit tangent vector	784
$\mathbf{N}$	Principal unit normal vector	784
$\mathbf{B}$	Binormal vector	787
∇	Del	667
∇f	Gradient of f	667
$\nabla \cdot \mathbf{F}$	Divergence of $\mathbf{F}$	807
$\text{div}\,\mathbf{F}$	Divergence of $\mathbf{F}$	807
$\nabla \times \mathbf{F}$	Curl of $\mathbf{F}$	809
$\mathbf{curl\,F}$	Curl of $\mathbf{F}$	808
$\nabla^2 f$	Laplacian of f	809
$D_{\mathbf{u}} f$	Directional derivative	667
(x, y)	Rectangular coordinates	S7
(x, y, z)	Rectangular coodinates	588
(r, θ)	Polar coordinates	444
(r, θ, z)	Cylindrical coordinates	722
(ρ, θ, φ)	Spherical coordinates	724
$\bar{x}, \bar{y}, \bar{z}$	Centroid or center of mass	702, 716
M_x, M_y	Moment	702
M_{xy}, M_{yz}, M_{xz}	Moment	716
I_x, I_y, I_z	Moment of inertia	703, 716
$t_{1/2}$	Half-life	282
t_2	Doubling time	281
$\begin{pmatrix} a_1 & a_2 \\ b_1 & b_2 \end{pmatrix}$	Matrix	603
$\begin{vmatrix} a_1 & a_2 \\ b_1 & b_2 \end{vmatrix}$	Determinant	604
$\binom{n}{k}, C_k^n$	Binomial coefficient	S26

Index

Note: Numbers in parentheses refer to exercises on the indicated pages. Page numbers prefixed with 'S' refer to pages at the end of the book.

GEOMETRY

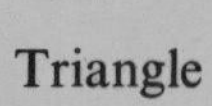
Triangle

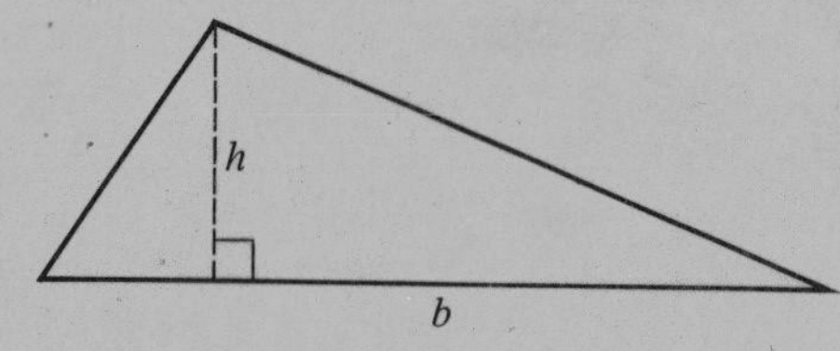

$$\text{Area} = \frac{bh}{2}$$

Triangle

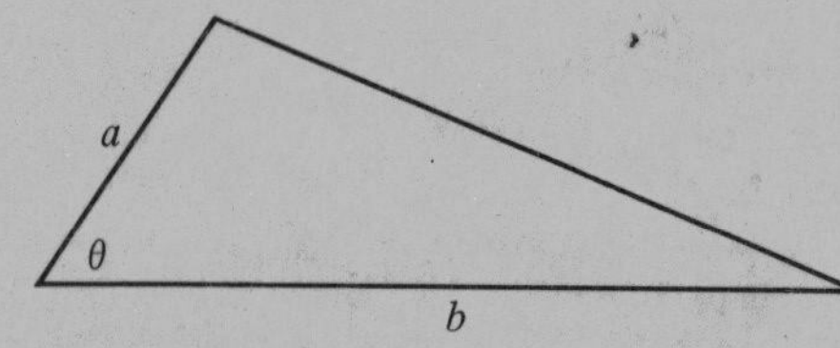

$$\text{Area} = \frac{ab \sin \theta}{2}$$

Equilateral triangle

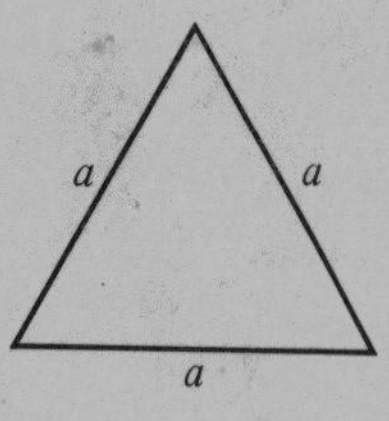

$$\text{Area} = \frac{\sqrt{3}\, a^2}{4}$$

Parallelogram

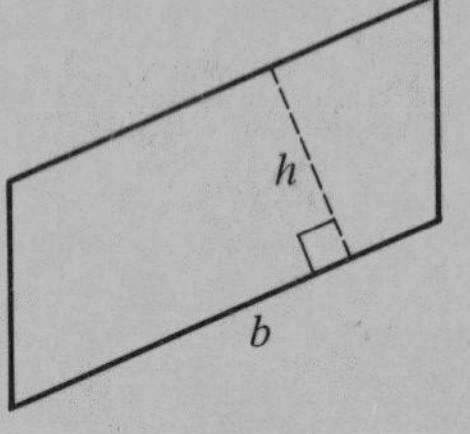

$$\text{Area} = bh$$

Trapezoid

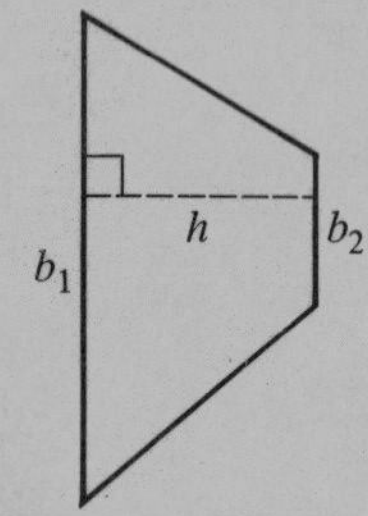

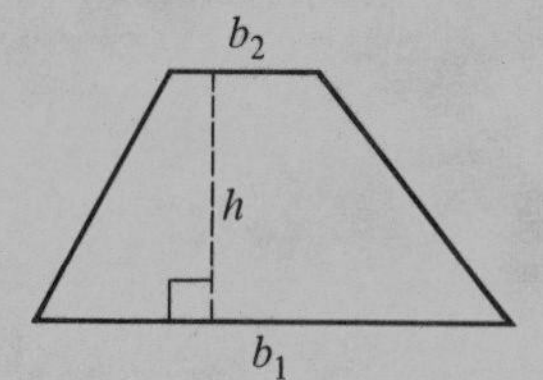

$$\text{Area} = \frac{(b_1 + b_2)h}{2}$$

Disk

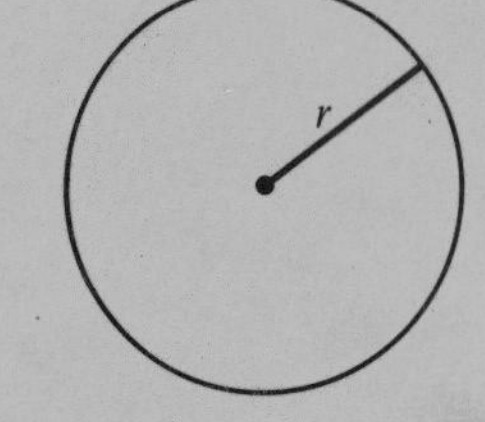

Area $= \pi r^2$
Circumference $= 2\pi r$

Ball

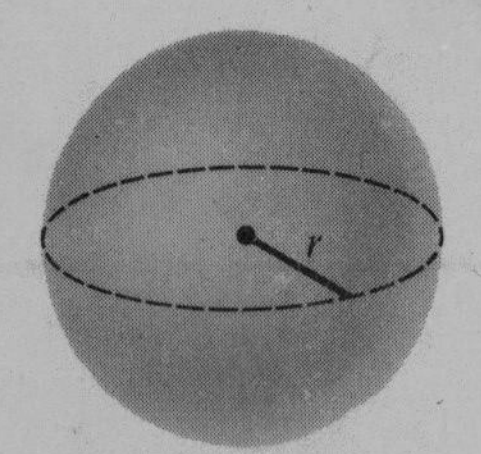

$$\text{Volume} = \frac{4\pi r^3}{3}$$

Area of surface $= 4\pi r^2$